MODERN QUANTUM THEORY

Modern Quantum Theory

From Quantum Mechanics to Entanglement and Quantum Information

Reinhold A. Bertlmann

Faculty of Physics, University of Vienna

Nicolai Friis

Institute of Atomic and Subatomic Physics—Atominstitut, TU Wien

OXFORD
UNIVERSITY PRESS

OXFORD

UNIVERSITY PRESS

Great Clarendon Street, Oxford, OX2 6DP,
United Kingdom

Oxford University Press is a department of the University of Oxford.
It furthers the University's objective of excellence in research, scholarship,
and education by publishing worldwide. Oxford is a registered trade mark of
Oxford University Press in the UK and in certain other countries

Published in the United States of America by Oxford University Press
198 Madison Avenue, New York, NY 10016, United States of America

British Library Cataloguing in Publication Data
Data available

Library of Congress Control Number: 2023935806

ISBN 9780199683338

DOI: 10.1093/oso/9780199683338.001.0001

Printed and bound by
CPI Group (UK) Ltd, Croydon, CR0 4YY

R. A. B. wants to dedicate this book to Renate,
his loving companion through all these years.

N. F. dedicates this book to Alma, Noah, and Verena,
inspired by their endless curiosity and joy to learn.

Acknowledgements

Reinhold Bertlmann

First of all, it was a stroke of luck for me to have met *John Stewart Bell*. His wisdom and advice was invaluable in our collaboration and, through me, has largely flowed into this book. In *Walter Grimus*, a renowned particle physicist, I was also lucky to have found a most interested collaborator, keen on exploring quantum-information phenomena within particle physics. The collaboration with him was indeed very inspiring and a great pleasure. I enjoyed the close collaboration with my former students, in particular with *Beatrix Hiesmayr*, *Philipp Krammer*, *Philipp Köhler*, *Katharina Durstberger*, and *Gabriele Uchida*. I especially want to emphasize the wonderful collaboration and friendship with my former student *Nicolai Friis*, which finally led to writing this book. I am further very grateful to *Heide Narnhofer* and the late *Walter Thirring* for showing me the beauty of mathematical physics in exploring quantum states; the collaboration with them was very enlightening and full of joy. Furthermore, I benefited a lot from the collaboration with the experimentalists in neutron interferometry: *Yuji Hasegawa* and the late *Helmut Rauch*, with *Stefan Filipp* and *Stephan Sponar*. I would also like to thank *Albert Bramon*, *Gianni Garbarino*, and *Dagmar Bruß* for a fruitful collaboration. I am thankful to *Herbert Pietschmann* for his interest and support in the research on quantum physics and to *Brigitte Kromp*, *Alexander Zartl*, and the late *Wolfgang Kerber* of the Central Library of Physics and Chemistry for their continuous help. Special thanks go to *Anton Zeilinger* with whom I had a joyful collaboration over more than two decades. Our joint seminar on the "Foundations of Quantum Mechanics", held over the course of 25 years, and the accompanying quantum workshops were ground-breaking for the education of the students at the University of Vienna, among them my then student-to-be, Nicolai Friis. Furthermore, the two conferences "Quantum [Un]speakables" in 2000 and 2014, organized by Anton Zeilinger and myself in Vienna, became celebrated events. The fruitful outcomes of all of this has flowed into the writing of this book. The discussions, comments, criticisms, and suggestions of my colleagues at the University of Vienna, *Markus Arndt*, *Markus Aspelmeyer*, *Časlav Brukner*, *Philip Walther*, *Marcus Huber*, *Frank Verstrate*, *Helmut Neufeld*, and *Bernhard Baumgartner*, were also highly stimulating and must be duly acknowledged. Finally, and most of all, I am greatly indebted to my wife *Renate Bertlmann*, to whom this book is dedicated, for her never-ending encouragement and support over such a long time.

Nicolai Friis

In March of 2006, during the second year of my undergraduate physics degree at the University of Vienna, I attended the quantum mechanics lectures by a certain *Reinhold Bertlmann*. This was a pivotal moment in my life, at which Reinhold kindled my fascination with quantum physics, setting me on the path that ultimately led to this book. I think many physicists of my generation brought up in the scientific environment of Vienna will agree that his lectures did not just provide technical knowledge, but were full of joy and conveyed a sense of awe of physics that was incredibly inspirational. From this moment on, I followed Reinhold's lectures closely and ended up completing my physics diploma under his supervision, but his inspiration, support, and friendship have been invaluable to me also ever since. Over the following years I had the great luck and pleasure of collaborating with a host of gifted scientists, both at my institutions in Nottingham, Innsbruck, and Vienna, but also internationally, many of whom I consider to be not just colleagues but dear friends, but naming all of them here would go far beyond the scope of these acknowledgements, and so I focus here on thanking those individuals that supported me in various ways (that some of them might perhaps not even have been aware of) in writing this book. Specifically, I would like to thank *Jorma Louko* for sharing his wisdom and attention to detail, especially with regards to complicated mathematical functions and perturbation theory. I am also deeply grateful to *Hans J. Briegel* for giving me the opportunity, and the necessary stability and support, to pursue my academic goals, as well as the freedom to develop my own ideas. I want to thank *Vedran Dunjko* for offering keen advice on many occasions. I am grateful to *Michalis Skotiniotis* for numerous most enjoyable scientific discussions. I would like to thank *Yelena Guryanova*, *Max Lock*, and *Hendrik Poulsen Nautrup* for many stimulating discussions, fruitful collaboration, and for their encouragement to complete this book (and their belief that it would eventually be completed). I am further thankful for the collaboration with my former student *Simon Morelli*, and for his patience with me as I was writing this book. I thank *Florian Kanitschar*, *Markus Miethlinger*, *Ida Mishra*, and *Phila Rembold* for help proofreading the manuscript. I feel indebted to *Marcus Huber*, who indeed I first met in Reinhold Bertlmann's office, for a collaboration and friendship that have accompanied and in many respects shaped my scientific journey for close to fifteen years now. His work and ideas have influenced this book in many ways, far beyond the extent of our joint research, teaching, and other academic adventures that we have embarked upon together during this time. Writing this book would not have been possible without the support of my family and friends. My wonderful partner *Verena Hofstätter* has been a steadfast pillar to me; her counsel, insight, and guidance have been irreplaceable. Our children *Alma* and *Noah*, who this book is also dedicated to, fill our lives with joy every day and continue to fascinate me with their curiosity and ingenuity. I hope this book may one day be a reminder to them of the beauty that can be found in science. I am also especially grateful to all my other wonderful family members, in particular, my parents, *Elisabeth Friis* and *Hans Petter Friis*, and to *Hildegard Hofstätter* and *Wernhard Hofstätter*, whose unwavering support and dedication to their grandchildren has provided the stability to balance academic work and family life that has made writing this book possible in the first place. Finally I would like to thank my dear friend *Michaela Graf* for providing moral support.

Our acknowledgment would not be complete without thanking the tireless staff at Oxford University Press: Mr. Sonke Adlung, Senior Editor, and Mrs. Giulia Lipparini, Project Editor, for their cooperation, advice, and patience.

Preface

In recent decades quantum theory has experienced an extensive revival owing to the rapid development of quantum information and quantum technologies. One of the basic characteristics of quantum states is a form of correlation called entanglement. These quantum correlations give rise to a conceptually fascinating feature of nature dubbed non-locality, which is revealed via the violation of Bell inequalities, but entanglement is also at the heart of quantum communication protocols such as quantum teleportation or (delayed-choice) entanglement swapping, or quantum key distribution, but is also of central importance in quantum computing. A central aim of this book therefore is to provide a detailed introduction to entanglement, its characteristics, its variants and its applications, from a physical perspective.

The inspiration for this book arose from a series of courses and seminars on quantum mechanics and quantum information theory held by the authors: covering material on topics such as the foundations of quantum mechanics, entanglement, and decoherence taught by Reinhold Bertlmann at the University of Vienna over many years for undergraduate and graduate students in physics, and also lectures and seminars by Nicolai Friis at the University of Innsbruck and at the University of Vienna.

We have divided the book into three parts: Part I Quantum Mechanics, Part II Entanglement and Non-Locality, and Part III Advanced Topics in Modern Quantum Physics.

Although there are already many books on quantum mechanics, we felt that there was a certain disconnect between many established books on quantum mechanics on the one hand, which were in many cases written before the advent of quantum information, and, on the other hand, contemporary books on quantum-information processing, which frequently take a view heavily influenced by computer science and classical information processing. Here, we therefore want to offer an introduction to quantum information theory from the perspective of physics. To pave the way, Part I provides a modern view on quantum mechanics, a central topic of theoretical physics that can be considered to be part of the standard repertoire of any physicist.

Part I is therefore intended to be useful for a broad readership, ranging from undergraduate and graduate students to lecturers and researchers especially in physics, but in other natural sciences as well. Quantum mechanics is conceptually rich and technically elaborate, but we aimed to be as self-contained as possible, providing explicit derivations and examples in as much detail as possible, and only relying on certain standard results from mathematics. Our point of view is a practical one. We want to show how to work with quantum mechanics. In this sense, Part I should be accessible to any reader familiar with basic elements of classical mechanics and electrodynamics, but of course we have put emphasis on a pedagogical presentation throughout the book.

Part II is dedicated to the foundations of quantum mechanics and the theory of quantum entanglement. Starting from a presentation of the basic mathematical framework of density operators, we first examine hidden-variable theories and the Einstein–Podolsky–Rosen paradox before we discuss Bell inequalities, both the theoretical and the experimental side, but also touching upon philosophical questions and the issue of the interpretation of quantum mechanics. This discussion then motivates the deeper study of entanglement-based quantum communication protocols, most prominent among them quantum teleportation, before we give a detailed exposition of the theory of entanglement and separability for two or more parties, including techniques for the detection and quantification of entanglement. Part II is thus aimed more directly at readers who wish to familiarize themselves with quantum information theory, and while the initial chapters of Part II are still aimed to be accessible for undergraduate students, the later chapters of this part progressively include more advanced materials that we hope will provide a useful resource also for more experienced readers and researchers.

In Part III we then fully turn to more advanced topics in modern quantum physics. Whereas Parts I and II in a sense represent self-contained story arcs, the third part of the book is intended to be a collection of chapters where each can stand on its own and in this way individually supplement the contents of Parts I and II according to the interest and predilection of the reader. Nevertheless, our selection is to be seen in the light of quantum information theory. Foremost we present an overview of entropies in the domain of classical physics and information processing, as well as to quantify the amount of information inherent in a quantum system. Then we focus on a formal treatment of quantum channels and quantum operations, before turning our view to open quantum systems. There we present the framework of dynamical maps and derive a Markovian master equation to describe possible decoherence of the system, resulting from the system's interaction with its environment. Next we discuss quantum measurements and quantum metrology, before we give a detailed presentation of quantum states of light, with a particular focus on the theory of coherent states and their description in phase space via the Wigner function, before finally studying the more general family of Gaussian states. At the end of the book we turn to a quite different area, namely to particle physics, and show how entanglement and techniques from quantum information play a role in this context. In particular, we discuss the features of "strange" K mesons and "beauty" B mesons. We think that an exploration of concepts from quantum information in this area can provide new tests of quantum mechanics and shows the far reach of ideas from this field.

Vienna, 15 December 2022   Reinhold A. Bertlmann and Nicolai Friis

Contents

Part I

Quantum Mechanics

In Part I we provide a self-contained introduction to quantum mechanics, blending a historical perspective with a modern view on this central topic in theoretical physics. By now, quantum mechanics has become part of the standard knowledge of every physicist, and we have selected the topics covered in Part I to provide the concepts, tools, and methods that form the backbone of a first university-level course in theoretical quantum physics. At the same time, Part I is aimed to be a useful resource for a wider group of physicists, engineers, chemists, biologists, as well as students and researchers in other disciplines in the natural sciences that touch upon quantum physics in one way or another. We have therefore given special attention to the paedagogical aspects. In addition, the presentation in Part I serves as a basis for Parts II and III of this book, which delve deeper into more advanced topics in modern quantum theory, in particular, quantum information theory.

The structure of Part I is intended as a step-by-step guide for understanding the principles governing any quantum-mechanical systems in general, but also more specifically for understanding how quantum mechanics describes the fundamental interactions between matter and (electromagnetic) radiation, i.e., the stability of atoms and their electronic structure.

To this end, we begin by examining some crucial experiments and basic principles of quantum mechanics, which have strongly influenced the development of quantum theory in its modern form, before we turn to its basic equation, the Schrödinger equation and examine properties and implications of the latter. Although we assume that readers are familiar with basic algebra, calculus, and trigonometry, all more advanced mathematics needed to follow the derivations is introduced in the book. In particular, we present the mathematical formalism for states and operators in the Hilbert spaces encountered in quantum theory. We then use this formalism to study the time-dependent and time-independent Schrödinger equation in one spatial dimension by way of examples for typical potentials, including a detailed treatment of the important case of the quantum-mechanical harmonic oscillator.

We then shift our focus to the concept of orbital angular momentum and we elaborate on the three-dimensional Schrödinger equation to examine the spherical potential well, and ultimately the important case of the hydrogen atom. The concept of spin is then introduced and incorporated into the previously established framework, based on which we revise our analysis of the atomic structure. Following on, we consider electromagnetism in quantum mechanics to study the Pauli equation, gauge symmetries, and the Aharonov–Bohm effect. We explore geometric phases like the Berry phase, and we examine theoretical ideas such as the Dirac monopole in terms of differential geometry and topology, briefly introducing the required mathematical formalism of differential geometry on the way. Finally, we discuss perturbative methods in quantum mechanics at the end of Part I. Starting with time-independent perturbation theory to study the fine structure of the hydrogen atom, the Zeeman effect, and the Stark effect, before we complete the first part of the book with a look at time-dependent perturbations.

1
Wave-Particle Duality

1.1 Planck's Law of Black-Body Radiation

1.1.1 Quantization of Energy

The foundation of quantum mechanics was laid in the year 1900 with *Max Planck's* discovery of the quantized nature of energy (Planck, 1900*b*; Planck, 1901). At that time the classical laws of physics were largely undisputed and appeared to be in accordance with experimental observations. Only a few phenomena eluded an explanation in terms of Newtonian mechanics, linear optics, or thermodynamics. Among these was the spectrum of radiation emitted from a glowing object. The classical law of Rayleigh–Jeans did not agree with empirical observations and physicists were at a loss to resolve the discrepancy.

Finally, Max Planck was able to address the problem and derive a formula for the spectrum of black-body radiation that matched the observed phenomena. However, to develop his formula he was forced to assume that the energy that is exchanged between a black body and its thermal (electromagnetic) radiation is quantized. In other words, for a fixed (angular) frequency, energy can be absorbed or emitted by the black body in discrete steps only. Planck's formula is able to explain—as we shall see—all features of the black-body radiation. Let us phrase his finding in the following way:

Proposition 1.1 *Energy is quantized and given in units of $\hbar\omega$,* $E = \hbar\omega$.

Here ω denotes the angular frequency $\omega = 2\pi\nu$, and ν the frequency, but in the following we will drop the prefix "angular" and refer to ω as the frequency unless otherwise mentioned. Let us also keep in mind the relation to the wavelength λ, i.e., $\lambda\nu = c$, where c denotes the speed of light. Furthermore, the oscillation period T is given by $\nu = \frac{1}{T}$.

The fact that the energy is proportional to the frequency, rather than the intensity of the wave—as we would expect from classical physics—appears quite counterintuitive initially. The proportionality constant $\hbar$ is called *Planck's constant* and its numerical value is

$$\hbar = \frac{h}{2\pi} = 1.054 \times 10^{-34} \,\text{Js} = 6.582 \times 10^{-16} \,\text{eVs},$$

$$h = 6.626 \times 10^{-34} \,\text{Js} = 4.136 \times 10^{-15} \,\text{eVs}.$$

1.1.2 Black-Body Radiation

A black body, by definition, is an object that completely absorbs all incident light (radiation). This property makes a black body a perfect source of thermal radiation. Practically, a black body can be thought of as an oven with a small aperture, as illustrated in Fig. 1.1. All radiation that enters through the opening has a negligibly small probability of leaving through it again.

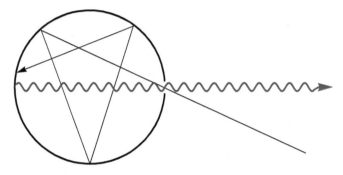

Fig. 1.1 Scheme for realization of a black body. A black body can be modelled as a cavity with a small aperture. The radiation from the aperture is practically thermal since any radiation entering through the hole has a negligible probability of exiting the cavity again.

Consequently, radiation originating from the opening can be attributed to thermal radiation that is emitted by the cavity walls. The energy that is emitted from the aperture at any given frequency can be measured for fixed oven temperatures as shown in Fig. 1.2. Such radiation sources are also referred to as (thermal) cavities.

The classical law that describes such thermal radiation is the law of *Rayleigh–Jeans*. It expresses the *spectral energy density* $u(\omega)$ in terms of the frequency ω and the temperature T. The quantity $u(\omega)$ has units of energy per volume per frequency.

Theorem 1.1 (Rayleigh–Jeans law) $\qquad u(\omega) = \frac{k_B T}{\pi^2 c^3} \, \omega^2.$

Here k_B denotes *Boltzmann's constant*, $k_B = 1.38 \times 10^{-23} \,\text{J K}^{-1}$.

Boltzmann's constant plays a role in classical thermo-statistics where (ideal) gases are analysed whereas here we describe properties of radiation. From the point of view

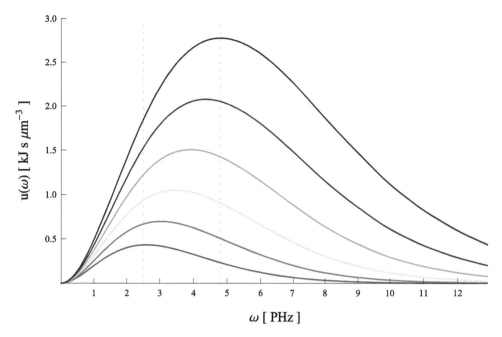

Fig. 1.2 Black-body spectrum. The spectral energy density $u(\omega)$, given by Planck's law (Theorem 1.3), is plotted as a function of the (angular) frequency ω. The different curves correspond to temperatures increasing from $T = 7 \times 10^3$ K (bottom, red) to $T = 1.3 \times 10^4$ K (top, purple) in steps of 1.2×10^3 K. The peaks of the curves are shifted through the visible spectrum, from about 2.5 PHz (1 PHz $= 10^{15}$ Hz) to 4.8 PHz (dotted, vertical lines), according to Wien's displacement law.

of classical physics this is a quite curious fact. The properties of what can classically be considered waves are determined, in part, by a quantity that governs the thermodynamic properties of particles, i.e., gases in the sense of Boltzmann. How does this constant enter into the physical model for black-body radiation?

The quantity $k_{\mathrm{B}}T$ has the dimension of energy. According to the *equipartition theorem*, each degree of freedom (of motion) in a classical system in thermal equilibrium contributes the mean energy of $E = \frac{1}{2}k_{\mathrm{B}}T$ to the total average energy of the system. We shall phrase this more precisely when deriving the law of Rayleigh–Jeans in Section 1.1.3.

However, other problems with the law of Rayleigh–Jeans are more distressing than the enigmatic appearance of Boltzmann's constant in Theorem 1.1. One can see immediately that the integral of the energy density over all possible frequencies is divergent, i.e.,

$$\int_0^\infty d\omega \ u(\omega) \to \infty \,, \tag{1.1}$$

which would imply an infinite amount of energy being radiated from every black body. This paradoxical issue is known as the *ultraviolet catastrophe*. It suggests that the law of Rayleigh–Jeans is valid only for small frequencies.

This obstacle was addressed by *Wilhelm Wien* in 1896, who empirically determined a law, Theorem 1.2, which is valid at higher frequencies, while it fails to match the experimental observations at lower frequencies.

> **Theorem 1.2 (Wien's law)** $\quad u(\omega) \to A\,\omega^3 e^{-B\frac{\omega}{T}} \quad for \quad \omega \to \infty.$

Here A and B are empirical constants.

As already mentioned, the dilemma was resolved by Max Planck. He derived an impressive formula which interpolates between the law of Rayleigh–Jeans and Wien's law, see Fig. 1.3. To succeed with the derivation, Planck had to assume the correctness of Proposition 1.1, i.e., energy can only occur in quanta of $\hbar\omega$. With this assumption he deduced a formula that is valid for all frequency ranges. For this achievement he was awarded the 1918 Nobel Prize in Physics.

> **Theorem 1.3 (Planck's law)** $\quad u(\omega) = \dfrac{\hbar}{\pi^2 c^3} \dfrac{\omega^3}{\exp\left(\hbar\omega/k_{\mathrm{B}}T\right) - 1}.$

From Planck's law we quickly arrive at the already well-known laws of Rayleigh–Jeans and Wien by taking the limits $\omega \to 0$ or $\omega \to \infty$ respectively:

$$\lim_{\omega \to 0} u(\omega) = \frac{k_{\mathrm{B}}T}{\pi^2 c^3} \, \omega^2 \qquad \text{Rayleigh–Jeans,} \tag{1.2a}$$

$$\lim_{\omega \to \infty} u(\omega) = \frac{\hbar}{\pi^2 c^3} \, \omega^3 \, e^{-\hbar\omega/k_{\mathrm{B}}T} \qquad \text{Wien.} \tag{1.2b}$$

Let us point out some of the consequences of Theorem 1.3.

(i) **Wien's displacement law**: The wavelength λ_{max} of maximal spectral energy density is shifted to smaller wavelengths for increasing temperatures T, such that:

$$\boxed{\lambda_{\mathrm{max}}T = const. = 0,29\,\mathrm{cm\,K}.} \tag{1.3}$$

It should be noted here that this law cannot be expressed in terms of the frequency ω by the mere substitution $\lambda \to 2\pi c/\omega$ since the different spectral forms of Planck's law have different units. We have expressed Planck's law for the quantity $u_\omega(\omega, T)$ for the spectral variable ω but we have suppressed the subscript denoting this. The corresponding spectral energy density for the variable λ is $u_\lambda(\lambda, T) = -\frac{d\omega}{d\lambda} u_\omega(\omega(\lambda), T)$.

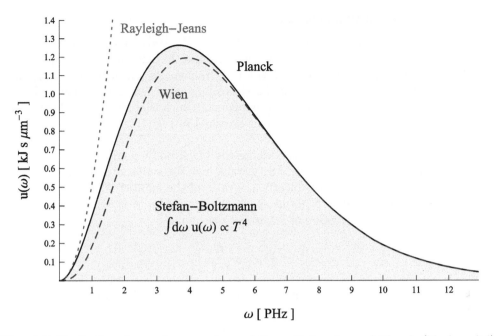

Fig. 1.3 Radiation laws. A comparison of the radiation laws of Planck (black, solid), Rayleigh–Jeans (red, dotted), and Wien (blue, dashed) is shown for the temperature $T = 10^4$ K. Planck's law (Theorem 1.3) interpolates between the laws of Rayleigh–Jeans (Theorem 1.1), valid in the small frequency regime, and Wien (Theorem 1.2), which is a good approximations for large frequencies. The radiative power of the black body is given by the area under the solid curve (Stefan–Boltzmann law), which is proportional to T^4.

(ii) **Stefan–Boltzmann law:** The power P/A radiated from the surface of the black body per unit surface area is proportional to the fourth power of the black-body temperature T. One can quickly see this by noting that the radiative power is proportional to the integral of the spectral energy density

$$\int\limits_0^\infty d\omega\, u(\omega) \;\propto\; T^4 \int\limits_0^\infty d\left(\frac{\hbar\omega}{k_\mathrm{B}T}\right) \frac{\left(\frac{\hbar\omega}{k_\mathrm{B}T}\right)^3}{\exp\left(\hbar\omega/k_\mathrm{B}T\right) - 1}\,. \tag{1.4}$$

Substituting $\frac{\hbar\omega}{kT} = x$ and using the formula

$$\int\limits_0^\infty dx\, \frac{x^3}{e^x - 1} = \frac{\pi^4}{15} \tag{1.5}$$

one obtains the proportionality

$$\boxed{\frac{P}{A} \propto T^4}\,. \tag{1.6}$$

1.1.3 Derivation of Planck's Law

Let us now give a formal derivation of Planck's law. We start by considering a black body realized by a hollow metal cube. We will assume that the metal is composed of atoms with suitable pairs of energy levels for all frequencies permitted by the geometry of the black body (the modes of the electromagnetic field compatible with the boundary conditions, as is explained in the following) such that the material can emit and absorb photons with energy $E = \hbar\omega$ as sketched in Fig. 1.4.

Since we are interested in the case where the black body is emitting thermal radiation we have to assume that the cavity walls and the radiation contained within are in thermal equilibrium. Consequently, the ratio of the number N_e of atoms in the excited state and the number N_g of atoms in the ground state is determined by the (classical) *Boltzmann distribution* of statistical mechanics

$$\frac{N_e}{N_g} = \exp\left(-\frac{E}{k_B T}\right). \tag{1.7}$$

As first pointed out by *Albert Einstein* (1916), three processes play a role in the dynamics of the system at hand: the *absorption* and *spontaneous emission* of photons, along with the *stimulated emission* due to the radiation background, see Fig. 1.4. The probabilities for absorption and stimulated emission are proportional to the (average) number of photons $\bar{n}$. The probability for the spontaneous-emission process, on the other hand, does not depend on the number of photons present, it is just proportional to the transition probability $P_{e\leftrightarrow g}$ between the ground state and the excited state. The absorption rate and the total emission rate are then obtained by multiplication with the numbers N_g and N_e of atoms in the ground state or excited state respectively:

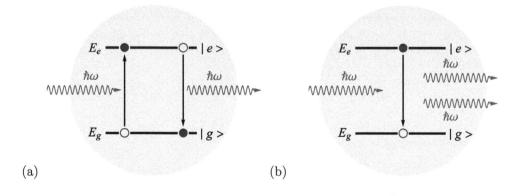

(a)   (b)

Fig. 1.4 Absorption and emission. The atoms of the cavity walls are modelled as two-level systems with ground states $|g\rangle$ of energy E_g and excited states $|e\rangle$ of energy E_e. The cavity walls can absorb or emit photons of energy $E = \hbar\omega = E_e - E_g$. (a) illustrates the *absorption* and *spontaneous emission* of a photon and the corresponding excitation or de-excitation of the atom. (b) depicts the *stimulated emission* due to the presence of other photons.

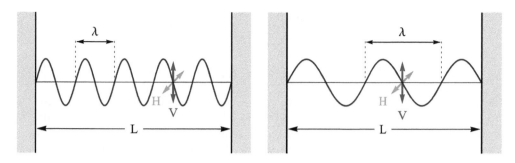

Fig. 1.5 Standing wave. The photons are described as standing waves of wavelength λ, as well as horizontal polarization H and/or vertical polarization V in a cavity of length L.

$$\text{Absorption rate}: \quad N_\text{g}\,\overline{n}\,P_{\text{e}\leftrightarrow\text{g}},$$

$$\text{Total emission rate}: \quad N_\text{e}\,(\overline{n}+1)\,P_{\text{e}\leftrightarrow\text{g}}.$$

In thermal equilibrium the rates for emission and absorption of a photon must be equal, i.e., we have *detailed balance*, $N_\text{g}\,\overline{n}\,P_{\text{e}\leftrightarrow\text{g}} = N_\text{e}\,(\overline{n}+1)\,P_{\text{e}\leftrightarrow\text{g}}$. Combining this with eqn (1.7) we find

$$\frac{\overline{n}}{\overline{n}+1} = \frac{N_\text{e}}{N_\text{g}} = \exp\left(-\frac{E}{k_\text{B}T}\right) = \exp\left(-\frac{\hbar\omega}{k_\text{B}T}\right), \tag{1.8}$$

from which we can express the average photon number in the cavity as

$$\overline{n} = \frac{1}{\exp\left(\hbar\omega/k_\text{B}T\right)-1}, \tag{1.9}$$

and the average photon energy is simply given by $\overline{n}\,\hbar\omega$.

Next, let us turn to the energy density. Here, the boundary conditions that are imposed by the walls of the cavity allow us to consider the photons as standing waves inside the metal cube, as illustrated in Fig. 1.5. We are interested in the number of possible modes of the electromagnetic field[1] inside an infinitesimal element dk of the wave number k, where $k = \frac{2\pi}{\lambda} = \frac{\omega}{c}$. Let us consider a standing wave of fixed wavelength λ that is compatible with the boundary conditions.

The number N of wavelengths λ that fit within a one-dimensional cavity of length L is then given by

$$N = \frac{L}{\lambda} = \frac{L}{2\pi}\frac{2\pi}{\lambda} = \frac{L}{2\pi}k. \tag{1.10}$$

The corresponding infinitesimal element dN can immediately be generalized to three spatial dimensions, i.e.,

---

[1]Here, the term "modes" just refers to the linearly independent solutions of the corresponding equation of motion, Maxwell's equations in the case of the electromagnetic field, subject to the relevant boundary conditions.

$$\text{1-dim:} \quad dN = \frac{L}{2\pi} dk \,, \tag{1.11}$$

$$\text{3-dim:} \quad dN = 2 \frac{L^3}{(2\pi)^3} d^3k \,, \tag{1.12}$$

where we have inserted a factor of 2 in the three-dimensional case to account for the two (polarization) degrees of freedom (see Fig. 1.5) of the photon and we recognize L^3 as the volume of the cube. Subsequently, we switch to spherical polar coordinates in the wave-number space, i.e., we use $d^3k = 4\pi k^2 dk$, before we substitute $k = \frac{\omega}{c}$ for the wave number to obtain

$$dN = 2 \frac{L^3}{(2\pi)^3} \frac{4\pi\omega^2 d\omega}{c^3} \,. \tag{1.13}$$

From this expression we can now calculate the energy density of the photons.

(i) **Classically**: In the classical case we follow the *equipartition theorem* (see, e.g., Landau and Lifshitz, 1980, p. 129), telling us that in thermal equilibrium each degree of freedom whose associated variable enters quadratically into the expression for the total energy contributes an average energy of $E = \frac{1}{2} k_B T$ to the total average energy of the system. Although this does not give the correct predictions for quantum-mechanical systems (especially for low temperatures), as we shall see soon, let us continue with this derivation for the sake of comparison. In this case, one may consider the standing waves as harmonic oscillators with mean energy

$$E = \langle E_{\text{kin}} \rangle + \langle V \rangle = \left\langle \frac{m}{2} v^2 \right\rangle + \left\langle \frac{m\omega}{2} x^2 \right\rangle = \frac{1}{2} k_B T + \frac{1}{2} k_B T = k_B T. \tag{1.14}$$

For the oscillator the kinetic energy and the potential energy contribute equally to the total mean energy, $\langle E_{\text{kin}} \rangle = \langle V \rangle$, and both contributions are proportional to quadratic variables. We can thus write

$$dE = k_B T \, dN \tag{1.15}$$

and, using eqn (1.13), we calculate the (spectral) energy density

$$u(\omega) = \frac{1}{L^3} \frac{dE}{d\omega} = \frac{k_B T}{\pi^2 c^3} \omega^2 \,, \tag{1.16}$$

which we recognize as Theorem 1.1, the law of *Rayleigh–Jeans*.

(ii) **Quantum-mechanically**: In the quantum-mechanical case we use the average photon energy $E = \bar{n}\,\hbar\omega$ with $\bar{n}$ from eqn (1.9), which we calculated by using the quantization hypothesis, Proposition 1.1, to write

$$dE = \bar{n}\,\hbar\omega \, dN \tag{1.17}$$

and again inserting eqn (1.13) we finally arrive at

$$u(\omega) = \frac{1}{L^3}\frac{dE}{d\omega} = \frac{\hbar}{\pi^2 c^3}\frac{\omega^3}{e^{(\hbar\omega/k_{\mathrm{B}}T)} - 1},\qquad (1.18)$$

and we recover Theorem 1.3, *Planck's law* for black-body radiation.

1.2 The Photoelectric Effect

1.2.1 Observation of the Photoelectric Effect

In 1887 *Heinrich Hertz* discovered a phenomenon which became known as the *photoelectric effect*: A metal surface that is illuminated by ultraviolet light emits electrons. Although the effect might appear unremarkable at first, its explanation touches the very foundation of quantum mechanics. The importance of the discovery of the photoelectric effect lies in the inability of classical physics to describe the effect in its full extent. The incompatibility with classical models is based on three simple observations:

(i) The kinetic energy of the emitted electrons is independent of the intensity of the illuminating source of light.

(ii) The kinetic energy of the emitted electrons rises with increasing frequency of the monochromatic light.

(iii) There exists a threshold frequency below which no electrons are emitted.

In terms of classical physics these experimental facts could not be reconciled with the interaction of the electromagnetic field with the metal. From (classical) electrodynamics it was known that the (vacuum) energy density (see, e.g., Ohanian, 1988, p. 249)

$$u_{\mathrm{EM}} = \frac{1}{8\pi}\left(\vec{E}^{\,2} + \vec{B}^{\,2}\right)\qquad (1.19)$$

as well as the (vacuum) energy-flux density given by the Poynting vector

$$\vec{S} = \frac{c}{4\pi}\,\vec{E}\times\vec{B}\qquad (1.20)$$

are both proportional to the intensity. Thus "knocking" electrons out of the metal is possible in the classical model, but there is no threshold that limits this process to certain frequencies. As long as the surface is exposed to the radiation, the electrons are expected to absorb energy until they get detached from the metal.

1.2.2 Einstein's Explanation for the Photoelectric Effect

The explanation for the phenomenon of the photoelectric effect was subsequently given by Albert Einstein. In 1905 he proposed his *photon hypothesis* (*Lichtquantenhypothese*, in German)

Proposition 1.2 (Photon hypothesis)

Light consists of quanta (photons), each carrying an energy of $E = \hbar\omega$.

Using this assumption Einstein was able to provide an explanation of the photoelectric effect, which we will here state in the following way: The incident photons transfer their energy $\hbar\omega$ to the electrons in the metal. Since the electrons are bound to the metal, only photons of sufficiently large frequencies will supply enough energy to overcome the electrostatic barrier of the metal and release electrons from it. The particular threshold energy for this process is called the work function and it depends on the specific metal used. The kinetic energy of the electrons released from the surface is then given by the photoelectric formula:

Proposition 1.3 (Photoelectric formula) $E_{\mathrm{kin},e^-} = \frac{mv^2}{2} = \hbar\omega - W.$

Here, W is the *work function* of the metal, while m, v, and E_{kin,e^-} denote the mass, velocity, and kinetic energy of the emitted electrons, respectively. From Proposition 1.3 we deduce that the kinetic energy of the emitted electrons is bounded by the frequency of the incident light such that $E_{\mathrm{kin},e^-} < \hbar\omega$. One can further observe that a *threshold frequency* ω_0 exists at which the electrons escaping from the metal have vanishing velocity, i.e., $E_{\mathrm{kin},e^-}(\omega_0) = 0$.

$$\omega_0 = W/\hbar. \tag{1.21}$$

For frequencies smaller than the threshold frequency, i.e., if $\omega < \omega_0$, no electrons are emitted and the incident photons are either scattered from the surface or the photon energy is absorbed and contributes to electronic or thermal excitations instead.

Einstein's explanation caused serious debates within the community and was not generally accepted due to conceptual difficulties to unify the quantized (particle) nature of light with Maxwell's equations, which stood on solid ground experimentally and suggested a wave-like behaviour of light. The experiment that was able to convince the majority of physicists was performed by *Robert Andrews Millikan* and we shall discuss it in the following Section 1.2.3. For his achievement of explaining the photoelectric effect using his photon hypothesis, Einstein was awarded the Nobel Prize in Physics in 1921.

1.2.3 The Millikan Experiment

In 1914 Millikan developed an experimental setup, sketched in Fig. 1.6, which allowed him to verify the accuracy of Einstein's photoelectric formula (Proposition 1.3). For his work on the electron charge and the photoelectric effect Millikan was awarded the Nobel Prize in Physics in 1923.

The setting for the experiment used by Millikan is the following: A metal plate is illuminated with ultraviolet light which causes the emission of electrons. This generates an electric current between the plate and an electron collector. The electric current was originally measured by a galvanometer, although, in principle, any type of ammeter can be used. Additionally, a voltage U is applied which impedes the approach of the electrons towards the collector. To reach the collector, the electron energy therefore has to be increased by $q_e U$ with respect to the situation without the additional voltage, where $q_e \approx 1.602 \times 10^{-19}$ C is the elementary charge (the electron charge is $-q_e$).

Consequently, if the kinetic energy of the electrons is below this level, i.e., if $q_e U > E_{\text{kin,e}^-}$, no electrons arrive at the collector. If, however, the kinetic energy is large enough to detect a current, the voltage can be regulated to a value U_0, such that the current measured by the ammeter vanishes. The kinetic energy of the electrons can then be deduced from U_0. This allows one to measure the work function W of the metal at hand.

$$\boxed{q_e U_0 = E_{\text{kin,e}^-} = \hbar\omega - W.}$$
(1.22)

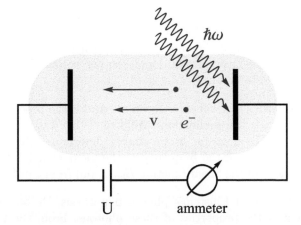

Fig. 1.6 Millikan experiment. A metal surface with work function W is illuminated by ultraviolet light consisting of photons of energy $\hbar\omega$ each. If the frequency ω is larger than $\omega_0 = (W - q_e U)/\hbar$, electrons are released from the surface of the metal and an electric current is registered by the ammeter. By varying the voltage U until the current vanishes the work function of the metal can be determined.

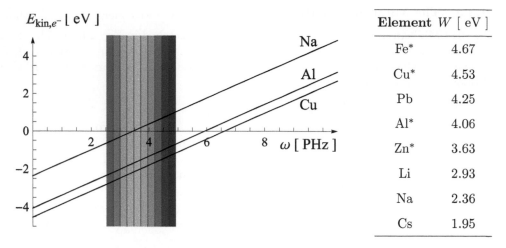

Element	W [eV]
Fe*	4.67
Cu*	4.53
Pb	4.25
Al*	4.06
Zn*	3.63
Li	2.93
Na	2.36
Cs	1.95

Fig. 1.7 Determination of Planck's constant. The kinetic energy E_{kin,e^-} of the electrons that are emitted from the metal surface and the frequency ω of the incident photons have a linear relationship. The slope of the resulting curve is given by Planck's constant $\hbar$. The vertical and horizontal intersections with the axis are determined by the work function W and the threshold frequency ω_0 respectively, see Proposition 1.3. For most metals, e.g., iron (Fe), copper (Cu), lead (Pb), or aluminium (Al), electrons can only be released by ultraviolet radiation. For alkali metals, e.g., lithium (Li), sodium (Na), or caesium (Cs), the photoelectric effect can be observed for light in the visible spectrum. For elements in crystalline form (marked by an asterisk in the table above) the work function depends on the crystal structure (displayed values from the CRC Handbook of Chemistry and Physics, 2005).

The range for work functions of typical metals is approximately between about 2 eV and 6 eV (see, e.g., the CRC Handbook of Chemistry and Physics, 2005). So far, we have implicitly assumed that the anode and the cathode are composed of the same material. If this is not the case an additional contact voltage between the different metals has to be taken into account.

Furthermore, Millikan's experimental setup allows for the precise measurement of Planck's constant. In fact, his publication (Millikan, 1914) was even titled "A Direct Determination of 'h'". To measure $\hbar$ one simply determines the kinetic energy of the electrons in dependence of the frequency of the incident light. Planck's constant is then given as the slope of the linear function $E_{\text{kin},e^-}(\omega)$ from eqn (1.22), see Fig. 1.7.

With the confirmation of Einstein's photon hypothesis, Proposition 1.2, we now want to take a look at the properties of these photons. From the theory of special relativity we know that energy and velocity are related by

$$E = \sqrt{\vec{p}^2 c^2 + m^2 c^4} \quad \text{and} \quad \vec{v} = \frac{\vec{p} c^2}{\sqrt{\vec{p}^2 c^2 + m^2 c^4}}. \tag{1.23}$$

Since the velocity of the photon is $|\vec{v}| = c \cong 2,99.10^8 \text{ms}^{-1}$ we conclude that the photon is massless, i.e.,

$$m_{\text{photon}} = 0. \tag{1.24}$$

Comparing the relativistic and quantum-mechanical expressions for the photon energy, $E = pc$ and $E = \hbar\omega$ respectively, and substituting the wave vector $|\vec{k}| = \omega/c$ one can conclude that the photon carries the momentum

$$\vec{p}_{\text{photon}} = \hbar\vec{k}. \tag{1.25}$$

1.3   The Compton Effect

Let us now turn to another effect that demonstrates the particle aspect of electromagnetic radiation, the *Compton effect*. As we have concluded from eqn (1.25) in Section 1.2, quanta of light carry a momentum $\hbar\vec{k}$. It is thus natural to assume that this momentum can be transferred to other particles in scattering processes. Indeed, such a phenomenon can be observed. In a process called *Compton scattering*, photons are inelastically scattered from electrons, see Fig. 1.8.

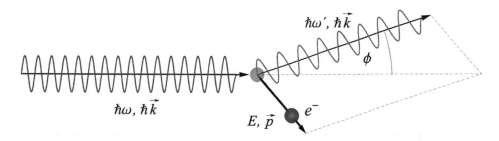

Fig. 1.8 Compton scattering. A photon with momentum $\hbar\vec{k}$ and (kinetic) energy $\hbar\omega$, where $\omega = |\vec{k}|$, is scattered from a resting electron with energy mc^2. The collision between the particles is elastic, i.e., total momentum and energy are conserved, whereas the scattering is inelastic, i.e., the kinetic energy of the photon is (partially) transferred to the electron. The scattered photon and electron have momenta $\vec{k}'$ and $\vec{p}$, as well as energies $\hbar\omega' = \hbar|\vec{k}'|$ and $E = \sqrt{m^2c^4 + \vec{p}^2c^2}$, respectively.

1.3.1   The Compton Shift Formula

To gain a better understanding of this process we can quantify the changes to the frequency and wavelength of the photons. Even though the scattering is *inelastic*, i.e., the kinetic energy of the photon is not conserved, the collision is elastic in the sense that the total energy and momentum is conserved. We can write this as

$$mc^2 + \hbar\omega = E + \hbar\omega', \tag{1.26a}$$

$$\hbar\vec{k} = \vec{p} + \hbar\vec{k}'. \tag{1.26b}$$

We insert eqns (1.26a) and (1.26b) into the relativistic energy-momentum relation $E^2 = (\vec{p}^2 c^2 + m^2 c^4)$, see eqn (1.23), to obtain

$$m^2 c^2 = \frac{1}{c^2} E^2 - \vec{p}^2 = \frac{1}{c^2} \left(mc^2 + \hbar\omega - \hbar\omega' \right)^2 - \hbar^2 \left(\vec{k} - \vec{k}' \right)^2 . \tag{1.27}$$

The last term on the right-hand side can then be rewritten in terms of $k = |\vec{k}|$, $k' = |\vec{k}'|$ and the scattering angle ϕ, see Fig. 1.8, by using the relation $\vec{k}\vec{k}' = kk' \cos\phi$, i.e.,

$$\hbar^2 \left(\vec{k} - \vec{k}' \right)^2 = \hbar^2 \left(k^2 + k'^2 - 2\vec{k}\vec{k}' \right) = \hbar^2 \left(k^2 + k'^2 - 2kk' \cos\phi \right) . \tag{1.28}$$

We can further recall that $\omega = kc$ and $\omega' = k'c$ to write

$$\frac{1}{c^2} \hbar^2 \left(\omega - \omega' \right)^2 = \hbar^2 \left(k - k' \right)^2 = \hbar^2 \left(k^2 + k'^2 - 2kk' \right) .$$

Subsequently, we insert eqns (1.28) and (1.29) into eqn (1.27) and obtain

$$k - k' = \frac{\hbar}{mc} kk' \left(1 - \cos\phi \right) . \tag{1.29}$$

Finally, we express the amplitudes of the wave vectors in terms of their respective wavelengths, i.e., $k = 2\pi/\lambda$ and $k' = 2\pi/\lambda'$, which yields

$$2\pi \left(\frac{1}{\lambda} - \frac{1}{\lambda'} \right) = \frac{\hbar}{mc} \frac{(2\pi)^2}{\lambda\lambda'} 2 \sin^2 \frac{\phi}{2} . \tag{1.30}$$

Multiplying both sides by $\lambda\lambda'/2\pi$ we arrive at

---

Lemma 1.1 (Compton shift formula) $\Delta\lambda = \lambda' - \lambda = 4\pi \frac{\hbar}{mc} \sin^2 \frac{\phi}{2} .$

We observe that the wavelength λ' of the scattered photon has increased with respect to the wavelength λ of the incident photon due to the energy transfer to the electron. The wavelength shift $\Delta\lambda$ is directly related to the *scattering angle* ϕ.

It is customary to define the *Compton wavelength* $\lambda_C = \frac{h}{mc}$ with a value of about 2.43×10^{-10} cm for electrons. The Compton wavelength of a particle corresponds to the wavelength of a photon that has the same energy as the particle in its rest frame.

Since λ_c is very small, high-energy photons, e.g., X-rays or gamma rays, are needed to observe the effect. We can estimate the maximal relative wavelength shift by selecting a typical wavelength of 7×10^{-9} cm for X-rays, for which we get

$$\frac{\Delta\lambda}{\lambda} = \frac{2\lambda_c}{\lambda_{\text{Xray}}} = \frac{2 \times 2.43 \times 10^{-10} \text{ cm}}{7 \times 10^{-9} \text{ cm}} \approx 0.07 = 7\% . \tag{1.31}$$

In the limit of small photon energy the scattering becomes elastic, i.e., the kinetic energy of the radiation can be considered to be unchanged throughout the process. This limiting case is referred to as *Thomson scattering* (see, e.g., Jackson, 1999, Sec. 14.8).

1.3.2 The Experiment of Compton

In 1923 *Arthur Holly Compton* experimentally observed the inelastic scattering of X-rays from nearly free electrons in a graphite block. In 1927 he received the Nobel Prize in Physics for this accomplishment.

The setup used by Compton comprises an X-ray source that illuminates the graphite, an aperture to filter the scattered radiation according to the scattering angle, and a spectrometer to determine the wavelength and intensity of the X-rays, see Fig. 1.9. By rotating the aperture and spectrometer, the dependence on the scattering angle could be determined.

The experiment (Compton, 1923) clearly showed an intensity maximum in the spectrum of the scattered radiation that is shifted with respect to the initial wavelength of the X-rays, see Fig. 1.10. The unshifted peak corresponds to scattering from the nuclei of the carbon atoms in the graphite block. Since the masses of the atoms are much larger than those of the electrons, the corresponding Compton wavelengths are much smaller, such that the scattering becomes effectively elastic.

In conclusion, we find that Compton's results confirm the particle character of light and we therefore assign the energy $E = \hbar\omega$ and the momentum $\vec{p} = \hbar\vec{k}$ to the (indivisible) photons. Furthermore, the appearance of Planck's constant in Compton's formula, Lemma 1.1, suggests that this is a purely quantum-mechanical effect. Classically, no inelastic scattering should occur.

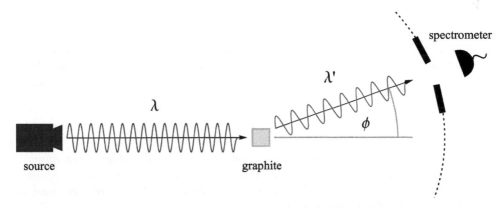

Fig. 1.9 Illustration of Compton's experiment. A graphite target is illuminated by radiation from an X-ray source. The photons of the X-ray beam with wavelength λ are inelastically scattered off the quasi-free electrons in the carbon structure. The photons scattered at the angle ϕ are filtered using an array of slits before their shifted wavelength λ' is determined by a spectrometer.

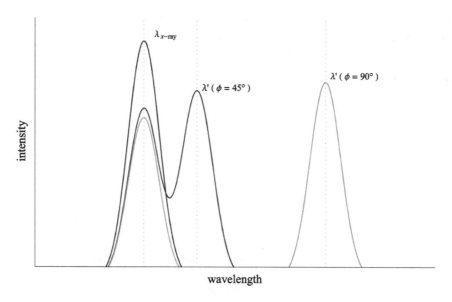

Fig. 1.10 Illustration of the Compton shift. The intensity and wavelength of the scattered radiation is measured at different scattering angles ϕ. As the scattering angle increases from 0 to 180° the initial peak at the wavelength $\lambda_{\text{X-ray}}$ is split in two. The shifted intensity maximum at $\lambda'(\phi)$ corresponds to Compton scattering from the electrons in the graphite block. The curves shown illustrate the measurement outcomes for $\phi = 0$ (purple), $\phi = 45°$ (blue) and $\phi = 90°$ (green).

1.4   Bohr's Theses

In the early years of the twentieth century many elementary questions about our understanding of matter, especially atoms, and radiation were unanswered and created serious problems. In particular, within *Ernest Rutherford's* (1911) atom model of electrons orbiting a central nucleus, it remained unclear why electrons do not tumble into the nucleus. That is, electrons rotating around the nucleus would be in accelerated motion. By the laws of (classical) electrodynamics they would emit radiation, thus losing the energy that is keeping them on their orbit.

Another important open problem concerned the observation of sharp spectral lines of radiation emitted from atoms, e.g., by Rayleigh[2] (Strutt, 1906). In an attempt to overcome these issues, *Niels Bohr* (1913) postulated new rules that are incompatible with classical notions. Here, we will cast these postulates in terms of four statements, Propositions 1.4–1.7, that we will attribute to Bohr even if they are rephrased here with respect to his formulations in 1913.

---

[2]John William Strutt, 3rd Baron Rayleigh, today probably more well known as (Lord) Rayleigh, made invaluable contributions to physics, and, in particular, to the development of early quantum physics. He was awarded the 1904 Nobel Prize in Physics for his studies of the densities of important gases and for the discovery of the noble gas argon.

Proposition 1.4 (Bohr's quantum postulate)

Electrons can only populate certain discrete energy levels in atoms. During the transition from a level with energy E_n to a level with energy E_m a photon with frequency $\omega = \frac{1}{\hbar}|(E_n - E_m)|$ is either emitted $(E_n > E_m)$ or absorbed $(E_n < E_m)$.

Having postulated the correspondence between energy levels and the emitted or absorbed photons, the next step is to postulate how these energy levels come to be.

Proposition 1.5 (Bohr's quantization rule)

The discrete energy levels in atoms correspond to stationary states of the electrons. The orbital angular momenta $\vec{L}$ for these circular orbits are quantized, i.e.,

$$L = n\hbar$$

where $L = |\vec{L}|$ and n is an integer $(n = 1, 2, ...)$.

The energy spectrum for the hydrogen atom, (approximately) given by the *Rydberg formula* and empirically determined by *Johannes Rydberg* (1889), could then be explained in Bohr's theory as the energy of an electron transition between two orbitals labelled by m and $n < m$. Furthermore, assuming that the electrons are held on their circular orbits by the Coulomb attraction, Bohr's model from 1913 allowed the derivation of the Rydberg formula

$$\hbar\omega = E_m - E_n = R\left(\frac{1}{n^2} - \frac{1}{m^2}\right), \qquad (1.32)$$

where $\mathcal{R} = \frac{R}{hc} = \frac{m_e q^4}{8\epsilon_o^2 h^3 c} = 1.097 \times 10^5\,\text{cm}^{-1}$ is the *Rydberg constant*. Alternatively, the Rydberg formula can be expressed for the wavelength λ of the emitted or absorbed photons

$$\frac{1}{\lambda} = \frac{R}{hc}\left(\frac{1}{n^2} - \frac{1}{m^2}\right). \qquad (1.33)$$

Arnold Sommerfeld, a prominent physicists at that time and teacher of Nobel Prize laureates such as *Werner Heisenberg*, *Wolfgang Pauli*, and *Hans Bethe*, later extended Bohr's atom model to explain even the fine structure of hydrogen. We shall discuss the quantum-mechanical model of the hydrogen atom in detail in Section 7.5 of Chapter 7, and elaborate on the fine-structure corrections in Section 10.2 of Chapter 10.

Although Bohr's model allows a very simple and intuitive visualization of the atomic structure, i.e., electrons orbiting on fixed paths around the nucleus in analogy to planetary motion, it turns out that the model is not only too simple but just wrong.

In fact, the perception of particles following distinct paths altogether has no meaning for quantum-mechanical calculations and only certain interpretations of quantum mechanics allow for their existence at all. But generally, as we shall see in Section 1.7, we need to cast aside the concept of trajectories in quantum mechanics.

Nevertheless, Bohr's model represented a great step forward even though it was clear that these *ad hoc* prescriptions could not be considered a definitive theory.

Another ongoing debate concerned the question whether light could be considered to be a wave or if it is composed of particles, concepts which seemed to contradict each other but were both observed under certain conditions. Bohr tried to formulate this problem in the following way:

Proposition 1.6 (Bohr's complementarity principle)

Wave and particle are two aspects of describing physical phenomena that are complementary to each other.

Depending on the measuring instrument used either waves or particles are observed, but never both at the same time, i.e., wave and particle nature are not simultaneously observable.

Similar statements can be made about other *complementary quantities* like position and momentum, energy and time (both of these pairs of quantities will be discussed in this chapter), or spin along orthogonal directions (which we will discuss in Chapter 8).

Some of these questions are still pertinent to ongoing research today and play important roles in experiments regarding "which-way detectors", "quantum erasers", and "continuous complementarity" to name but a few.

In particular, a remaining question that still causes discussion is the connection of quantum mechanics to classical physics, which Bohr phrased in the following way:

Proposition 1.7 (Bohr's correspondence principle)

In the limit of large quantum numbers classical physics is reproduced.

The propositions of this section form the basis for what is usually referred to as the *Copenhagen interpretation of quantum mechanics*, which was developed mainly by Niels Bohr and Werner Heisenberg.

Bohr was *the* leading figure of quantum physics at that time and his institute in Copenhagen became the centre of the avantgarde of physics where physicists from all over the world met, e.g., *Werner Heisenberg, Wolfgang Pauli, George Gamow, Lev D. Landau, Erwin Schrödinger*, and *Hendrik A. Kramers*, to name but a few. In 1922 Bohr was awarded the Nobel Prize in Physics for the investigation of the structure of atoms and their radiation. Bohr continued to influence views about quantum physics well into the second half of the twentieth century. His dispute with Einstein about the foundations of quantum mechanics became legendary and we shall return to it in Chapters 12 and 13.

1.5 Wave Properties of Matter

As we shall discuss in this section, before the formulation of quantum theory, not only electromagnetic radiation but also massive particles required a more sophisticated description than could be provided by classical mechanics. More specifically, the association of microscopical (quantum) objects, e.g., electrons, with ideally localized, point-like particles with sharp momenta, is misleading (see Section 1.6 and Section 2.5) and cannot account for all observed phenomena. A very important step towards a more complete description was *Louis de Broglie's* proposal of wavelike behaviour of matter in 1923, and subsequent work (de Broglie, 1924; 1925; 1927; see also the discussion in Weinberger, 2006), for which he was awarded the Nobel Prize in Physics in 1929.

1.5.1 Louis de Broglie's Hypothesis

In view of particle properties of light (waves)—photons—Louis de Broglie ventured to consider the reverse phenomenon, he proposed to assign wave properties to matter, which we will formulate here in the following proposition:

Proposition 1.8 (Louis de Broglie's hypothesis)

Every particle with mass $m > 0$ and momentum $\vec{p}$ has an associated wavelength

$$\lambda_{\mathrm{dB}} = \frac{h}{|\vec{p}|} = \frac{h}{\sqrt{2\,m\,E_{\mathrm{kin}}}},$$

where $E_{\mathrm{kin}} = \vec{p}^2/2m$ is the particle's kinetic energy.

Proposition 1.5.1 is a simple consequence of assigning frequencies and wavelengths to (kinetic) energies and momenta of particles in (reversed) analogy to photons, i.e.,

$$E_{\mathrm{kin}} = \hbar\omega \quad \text{and} \quad p = \hbar k = h/\lambda_{\mathrm{dB}} \tag{1.34}$$

where $p = |\vec{p}|$ and $k = |\vec{k}|$. The wavelength λ_{dB} can then be expressed in terms of the momentum p. We can then use the expression for the kinetic energy $E_{\mathrm{kin}} = p^2/2m$ to write $p = \sqrt{2\,m\,E_{\mathrm{kin}}}$ to arrive at the expression for the *de Broglie wavelength* λ_{dB} of massive particles as shown in Proposition 1.5.1.

In this context the notion of *matter waves* was introduced. De Broglie developed the idea that *pilot waves* (from French: "l'onde pilote", see, e.g., de Broglie, 1927

p. 240, Sec. V.11) guide the particle's motion on definite trajectories. In his own words[3] (de Broglie, 1924, p. 450) "We are then inclined to admit that any moving body may be accompanied by a wave and that it is impossible to disjoin motion of body and propagation of wave." This point of view—coexistence of wave *and* particle—is in contrast to Bohr's view (see Proposition 1.6) and, moreover, can lead to serious difficulties, as we shall see in Section 1.6.

Note that the wave assignment discussed so far was made for free particles, i.e., particles that are not subject to any external forces. The question whether or not the potential energy corresponding to such forces would influence his hypothesis was raised by de Broglie himself (de Broglie, 1924, p. 450, Sec. IV). We will return to this issue when we consider Schrödinger's theory (see Chapter 2) where we will study the nature of the waves in the light of *Max Born's* probability interpretation (see Section 1.7).

1.5.2 Electron Diffraction from a Crystal

We observe that, following Proposition 1.5.1, electrons with a kinetic energy of several eV and mass $m_e \cong 0.5$ MeV$/c^2$ have a de Broglie wavelength of a few Å. For example, for an energy of 10 eV we obtain $\lambda_{dB} \cong 3.9$ Å, which is the same order of magnitude as the lattice spacing of atoms in crystals. According to de Broglie's hypothesis, it is thus possible to diffract electrons from a crystal lattice analogously to the diffraction of light from a grating.

Corresponding experiments to test de Broglie's hypothesis were independently carried out in 1927 by *Clinton Davisson* and *Lester Germer* (Davisson and Germer, 1928), as well as by *George Paget Thomson* and *Alexander Reid* (Thomson, 1927; Thomson and Reid, 1927). Here, we focus on the description of the experiment by Davisson and Germer, which involved electrons at appropriate velocities that were sent onto a nickel crystal with a lattice spacing of $d \cong 0,92$ Å, see Fig. 1.11.

The intensity of the reflected electron beam was then measured for different angles, reproducing the diffraction pattern postulated by *William Henry Bragg* and his son *William Lawrence Bragg* for X-rays in 1913 (see, e.g., the discussion in Tipler and Llewellyn, 2008, Sec. 3–4). As we can see from Fig. 1.11, the Bragg condition for constructive interference relevant to the Davisson–Germer experiment is

$$\boxed{\sin\theta = \frac{n\lambda}{2d}\,,} \qquad (1.35)$$

where $n \in \mathbb{N}$. Meanwhile, the observation of an intensity maximum of the diffraction (Bragg peak) for a scattering angle $\phi = 50°$, which translates to an angle in the Bragg condition of $\Theta = 65°$, suggests

$$\Rightarrow \quad \lambda = 2 \times 0.92 \text{ Å} \times \sin 65° = 1.67 \text{ Å},$$

which is in perfect accordance with the de Broglie wavelength for an acceleration voltage of $U = 54$ V used in this experiment. The Davisson–Germer experiment thus

---

[3]It is interesting to note though that this quotation is from a paper entitled "A tentative theory of light quanta" and is hence referring to photons, not necessarily massive particles.

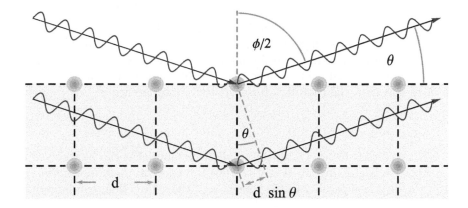

Fig. 1.11 Illustration of the Davisson–Germer experiment. The electron matter waves are scattered from a lattice of nickel atoms with vertical and horizontal spacing d. The rays that are scattered from the second layer of atoms constructively interfere with the rays that are scattered from the surface of the crystal if their path difference $2d\sin(\theta)$ is an integer multiple of the de Broglie wavelength λ_{dB} of the electrons. The corresponding intensity maximum can be experimentally observed at the scattering angle $\phi = (180° - 2\theta)$.

helped to confirm the wavelike nature of matter and to validate the claims of early quantum mechanics. For their experimental discovery of the electron diffraction Davisson and Thomson[4] were awarded the Nobel Prize in Physics in 1937.

1.6   Heisenberg's Uncertainty Principle

We now want to turn our attention to a fundamental quantum-mechanical rule, *Heisenberg's uncertainty principle*, which presents some significant conceptual challenges even though the mathematics behind it is straightforward. In this section, we will illustrate the uncertainty principle using the so-called *Heisenberg microscope* put forward by Werner Heisenberg[5] in 1927. However, the Heisenberg microscope should be seen more as an example for the application of the uncertainty principle, or even just as a heuristic illustration of the motivation behind the principle, rather than a justification of the principle itself. A formal proof of what is called the Heisenberg uncertainty principle was first given by *Earle Hesse Kennard* (1927) and independently by *Hermann Weyl* (see Weyl, 1931, p. 77 and p. 393), and we will present a formal derivation in Section 2.5.

---

[4]Perhaps interestingly from today's point of view, their collaborators, Lester Germer and Alexander Reid, respectively, were not credited with a Nobel Prize for their work. Furthermore, it is quite amusing to note that George Paget Thomson's father, *Joseph John Thomson*, received the Nobel Prize in Physics in 1906 for showing the "opposite", namely that the electron is a particle.

[5]Although Werner Heisenberg is today often associated mostly with the uncertainty principle, and he was the recipient of the 1932 Nobel Prize in Physics, the award was not specifically for the discovery of the uncertainty principle, but rather *"for the creation of quantum mechanics, ..."* (NobelPrize.org).

1.6.1 Heisenberg's Microscope

Let us consider the measurement of an electron's position by illuminating a screen with light that is scattered from the electron. The electron will then appear on the screen as a central dot, i.e., as an intensity maximum, surrounded by concentric dark and bright rings corresponding to higher-order intensity minima and maxima, respectively. Since the electron practically acts as a source of light, we can treat it as an aperture of width d. In this case, the condition for destructive interference is

$$\sin \phi = \frac{n\lambda}{d}, \tag{1.36}$$

where $n \in \mathbb{N}$. According to eqn (1.36) the smallest distance that can be resolved by an optical microscope is given by $d = \lambda / \sin \phi$. Thus the *uncertainty of localization* of the electron in the direction parallel to the screen, e.g., the x direction, can be written as

$$\Delta x = d = \frac{\lambda}{\sin \phi}. \tag{1.37}$$

It seems that, if we were to choose the wavelength λ to be small enough while $\sin \phi \neq 0$ then Δx could become arbitrarily small. But, as we shall see, the precision of the position estimate increases at the expense of the precision of the measurement of the electron's momentum. Why is that? The photons are detected on the screen but the direction of their point of origin is unknown within an angle ϕ. Consequently, the uncertainty of the photon momentum p_{photon} parallel to the screen is given by $p_{\text{photon}} \sin \phi$. The Compton recoil (see Section 1.3) of the electron due to the collision with the photon then has the same uncertainty. In particular, the momentum p_x of the electron along the x direction has an uncertainty proportional to the momentum uncertainty of the photons hitting the screen, i.e.,

$$\Delta p_x = p_{\text{photon}} \sin \phi = \frac{h}{\lambda} \sin \phi, \tag{1.38}$$

where we have inserted the momentum of the photon $p_{\text{photon}} = \hbar k = h/\lambda$ from eqn (1.25) in the last step. Combining the expressions for the position and momentum uncertainties of the electron from eqn (1.37) and eqn (1.38), respectively, we get

$$\boxed{\text{Heisenberg's uncertainty relation:} \quad \Delta x \, \Delta p_x = h.}$$

We will see in Section 2.5 that this is not the highest precision that can be achieved in principle, in fact, the previous relation can be generalized to the statement

$$\boxed{\Delta x \, \Delta p_x \geq \tfrac{\hbar}{2}.} \tag{1.39}$$

This inequality constitutes a fundamental principle that has nothing to do with technical imperfections of the measurement apparatus. We can phrase the uncertainty principle in the following way:

Proposition 1.9 (Heisenberg's uncertainty principle)

The more precise the information is that we have about the position of a particle, the less precise is the information about the momentum of that particle, and vice versa.

We should note here, that the example of the Heisenberg microscope does not serve as a formal proof of the uncertainty principle. It is purely a gedanken experiment that operates on the premise that the electron is a classical particle with fixed, but undetermined, position and momentum. Moreover, the Heisenberg microscope seems to suggest that the uncertainties arise from disturbances of the system due to the measurement process, but, as we shall see in Section 2.5, the formal derivation of eqn (1.39) does not involve any such assumption.

1.6.2   Energy-Time Uncertainty Principle

The concept of intrinsic uncertainty relations of quantum-mechanical measurements is not limited to positions and momenta. To illustrate this, let us construct such a relation for the *energy-time uncertainty*, i.e., between the uncertainties Δt for the duration of a physical process and ΔE for the respective energy. Consider, for example, a particle of mass $m > 0$ travelling along the x axis with velocity $v_x = p_x/m$. Since it takes the particle the time Δt to cover the distance Δx, we can express the time uncertainty, that is, the period of time where the particle is localizable within an interval of length Δx as

$$\Delta t \;=\; \frac{\Delta x}{v_x} \;=\; \frac{m}{p_x}\,\Delta x\,. \tag{1.40}$$

Assuming that the particle is not subject to external forces, the energy uncertainty ΔE can be computed from the variation of the kinetic energy $E = E_{\mathrm{kin}} = \vec{p}^2/2m$ with respect to the momentum p_x in the x direction,

$$\Delta E \;=\; \frac{p_x}{m}\Delta p_x\,. \tag{1.41}$$

We can now combine Δt from eqn (1.40) with ΔE from eqn (1.41) to arrive at

$$\Delta t\,\Delta E \;=\; \frac{m}{p_x}\,\Delta x\,\frac{p_x}{m}\Delta p_x \;=\; \Delta x\,\Delta p_x\,, \tag{1.42}$$

where we can insert eqn (1.39) to obtain

$$\boxed{\Delta E\,\Delta t \;\geq\; \tfrac{\hbar}{2}\,.} \tag{1.43}$$

the energy-time uncertainty relation. We can conclude that there is a fundamental complementarity between energy and time. An important consequence of this observation about energy-time uncertainty is the finite width of the spectral lines in atomic emission and absorption processes. That is, in the long-time limit, transitions between energy levels of atoms due to the interaction with the electromagnetic field occur only with field modes those frequency (multiplied by $\hbar$) matches the energy difference between the respective levels, and we will make this mathematically precise when discussing Fermi's golden rule in Section 10.5.3.

1.7 The Double-Slit Experiment

The conceptual intricacies of the complementary description of quantum objects as particles and waves can be revealed in a particularly simple scenario, the so-called *double-slit experiment*. The interference effect occurring in such a scenario was already well known from classical optics at the time—similar experiments with light had been performed already in 1804 by *Thomas Young*—and could be explained by considering light as (electromagnetic) waves. However, the central issue becomes apparent when light is considered as consisting of individual photons. Alternatively, the diffraction of massive particles, such as electrons, from the double slit can be considered, similarly to the electron diffraction in Section 1.5.2. In this case it turns out that it is not possible to explain the effect in a classical way. *Richard Feynman* described the double slit as "a phenomenon which is impossible, absolutely impossible, to explain in any classical way, and which has in it the heart of quantum mechanics. In reality, it contains the only mystery" (Feynman *et al.*, 1971, Sec. 1–1), emphasizing the great significance of the double slit as *the* fundamental phenomenon of quantum mechanics.

Related experiments have been performed with electrons by *Gottfried Möllenstedt* and *Claus Jönsson* (1959; Jönsson, 1961), with neutrons by *Heinrich Kurz* and *Helmut Rauch* (1969), by *Clifford Glenwood Shull*[6] (1969), as well as by *Anton Zeilinger* [et al.] in the eighties (see, e.g., the review Zeilinger *et al.*, 1988). More recently, in 1999 and subsequent years, *Markus Arndt* and collaborators have performed a series of experiments using large molecules, so-called fullerenes. We shall discuss the fascinating quantum-mechanical behaviour of these large molecules in Section 1.7.3. First, we will give a qualitative description of the double-slit experiment from the point of view of classical and quantum mechanics.

1.7.1 Comparison of Classical and Quantum-Mechanical Results

Let us consider classical particles that pass through an array of two identical slits. Subsequently, the particles are detected on a screen behind the apertures, see Fig. 1.12 (a). When either the first or the second slit is open, the respective distributions P_1 and P_2 of particle detections on the screen are centred directly behind the respective slits. When both of the apertures are open at the same time we would expect the resulting distribution $P_{1,2}^{\text{class}}$ to be just the sum of the previous distributions for individually opened slits, i.e.,

$$P_{1,2}^{\text{class}} = P_1 + P_2. \tag{1.44}$$

For a sufficiently large number of classical objects, such as, e.g., marbles, eqn (1.44) is indeed valid.

---

[6]Shull was awarded the 1994 Nobel Prize in Physics for the development of neutron spectroscopy (NobelPrize.org).

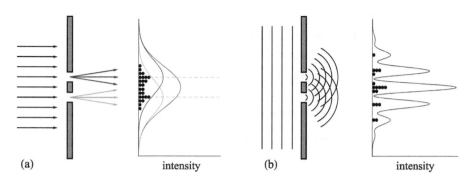

Fig. 1.12 Double slit. (a) The distribution of classical particles on a screen behind a double slit is given by the sum of the respective single-slit distributions. (b) For quantum-mechanical particles, e.g., photons or electrons, this is no longer true. The waves originating from the two slits produce a distinctive interference pattern on the screen.

Now, we consider the same setup as before, but instead of classical particles let us use small quantum objects like electrons or neutrons. Of course this also requires the use of corresponding detection devices that count the number of incident particles at various positions behind the double slit. We can think of such devices as arrays of small detectors that click when a particle is registered. When either one or the other slit is open, the detector response, in other words the "clicking" rate, will be distributed in a similar fashion as in the classical case. In particular, the distributions will be centred directly behind the individually open apertures. For sufficiently narrow slits, the intensity distribution will of course not just have a single peak but for the discussion here the explicit form of P_1 and P_2 does not matter. More important, however, is the case when both slits are open. Then the resulting distribution $P_{1,2}^{\mathrm{QM}}$ is not given by the sum of the single-slit distributions as in the classical case, see Fig. 1.12 (b). Let us formally write

$$P_{1,2}^{\mathrm{QM}} \neq P_1 + P_2 . \tag{1.45}$$

It turns out that the distribution one obtains in an experiment with quantum-mechanical particles resembles an interference pattern of waves originating at the two apertures. The intensity, the clicking rate, is high for constructive interference of these waves while it is low where destructive interference occurs.

1.7.2 Interpretation of Quantum-Mechanical Results

We now want to formulate the ideas of the previous section in a more rigorous way, concentrating on electrons as representatives of quantum-mechanical objects. We have already argued in Section 1.5 that electrons exhibit wave-like behaviour. Mathematically, this behaviour can be described by representing electrons by a so-called *wave function* $\psi(x,t)$, a complex-valued function. However, to establish a connection with the experimental observations discussed in Section 1.7.1 we need to assign an operational interpretation to this mathematical object. Let us phrase this connection in terms of the following proposition:

Proposition 1.10 (Born's probability interpretation)

The probability $P(x,t)$ of finding the electron at a certain position x at the time t is given by the intensity of the wave, that is, the modulus squared of the wave function $\psi(x,t)$, i.e.,

$$P(x,t) = |\psi(x,t)|^2 = \psi^*(x,t)\,\psi(x,t)\,.$$

For this interpretation of the wave function formulated in 1926, *Max Born* was awarded the 1954 Nobel Prize in Physics. Proposition 1.7.2, which we will here also refer to as the *Born rule*, allows us to make the statement of eqn (1.45) more precise. Let us denote the wave functions for the electron when either slit 1 or 2 is open by ψ_1 and ψ_2 respectively. The corresponding probability distributions are given by

$$P_1 = |\psi_1|^2 \quad \text{(slit 1 open)}, \tag{1.46}$$

$$P_2 = |\psi_2|^2 \quad \text{(slit 2 open)}. \tag{1.47}$$

When both apertures are open, the overall wave function ψ is a *superposition* of the individual wave functions, i.e., $\psi = \psi_1 + \psi_2$, and we have the total probability

$$P = |\psi|^2 = |\psi_1 + \psi_2|^2 = |\psi_1|^2 + |\psi_2|^2 + 2\,\text{Re}\,(\psi_1^*\psi_2)\,. \tag{1.48}$$

We see that for two simultaneously open slits the overall distribution is not just the sum of the individual probabilities. The right-hand side of eqn (1.48) features an *interference* term. The sign of the interference term at any point x and time t determines if the waves ψ_1 and ψ_2 interfere constructively or destructively.

Let us briefly recapitulate. A beam of electrons, fundamental quantum objects, is sent through a double slit. The individual electrons then hit a screen or detector where they are registered as a localized lump or cause a click in the detector, thus occurring as definite *particles*. The probability for the detection, on the other hand, is given by the intensity $|\psi|^2$ of the *wave* ψ. In this sense the electrons behave both as particles and waves.

Interestingly, the probability distribution—the interference pattern—is not created, as one might naively assume, by the interference of the wave functions of different electrons that are simultaneously passing through the apertures. The distribution on the screen that is built up by an appropriately large number of electrons remains the same if they are sent through the slits one by one. Furthermore, it is impossible to predict the exact location that a particular electron will hit on the screen. The single detections occur randomly and if only a few electrons have been registered, no specific structure can be observed in the detected spatial distribution. Only when we have gathered plenty of individual detections, say thousands, a noticeable interference pattern emerges, see Fig. 1.14.

However, even if the electrons are sent through the experimental setup one by one it is operationally meaningless to assign specific paths to the individual electrons.

The wave function only provides a probability for the detection at the screen but we cannot obtain information about the particular slits that the electrons have passed through whilst maintaining the interference pattern. If we try to determine whether the electron passes slit 1 or 2, e.g., by placing a light source behind the double slit to illuminate the passing electrons, the interference pattern disappears and we end up with a classical distribution.

Proposition 1.11 (Wave-Particle Duality)

Gaining path information destroys the wave-like behaviour.

Finally, it is of crucial importance to recognize that the electron does not split when it passes through the double slit. Whenever an electron is detected, it is always the whole electron and not part of it. When both slits are open it makes no sense to speak of the electron as following a distinct path since no such information is available[7].

1.7.3 Interferometry with Large Molecules

As we have established so far, small particles such as electrons and neutrons are definitely quantum systems, i.e., they show wave-like behaviour. We know from Louis de Broglie's hypothesis, Proposition 1.5.1, that every particle with momentum $\vec{p}$ can be assigned a wavelength. However, most common objects that surround us in our everyday lives have much larger masses than elementary particles like electrons. Consequently their de Broglie wavelengths are much smaller than their size which makes any attempts of sending them through double slits of comparable dimensions futile. So it is quite natural to ask, loosely speaking, how big or how heavy particles can be in order to maintain their ability for interference. In this sense, one might ask what the boundary for this quantumness is, or even if such a boundary exists at all.

While it might not be possible to give definite answers to these questions, the experimental groups around *Markus Arndt* and *Anton Zeilinger* have been pushing this boundary in a series of experiments with fullerenes (Arndt *et al.*, 1999) and other large molecules (Juffmann *et al.*, 2012). Spherical fullerenes, also called "buckyballs", are C_{60} molecules with high spherical symmetry. Their shape resembles a football with a diameter of approximately $D \approx 1\,\text{nm}$ (given by the outer electron shell of the molecule) and a mass of 720 amu, see Fig. 1.13.

In the experiment reported in (Arndt *et al.*, 1999), fullerenes are heated to 900 K in an oven from which they are emitted with a thermal velocity distribution. With an appropriate mechanism, e.g., a set of slotted disks rotating at constant angular velocity, a narrow range of velocities is selected from the thermal distribution. After

---

[7]Certain interpretations of quantum mechanics assign definite paths to the electrons. However, this requires the introduction of additional parameters which cannot be sufficiently measured (by definition) and thus do not improve the predictive power of the theory. The corresponding paths therefore do not carry any operational significance. For references to the Bohmian interpretation see for example Holland (1993) and for the many worlds interpretation see D'Espagnat (1999).

Fig. 1.13 Fullerenes. Sixty carbon atoms are arranged in pentagonal and hexagonal structures, forming a sphere analogously to a football. Picture used with permission of Markus Arndt.

collimation, the fullerenes pass an SiN lattice with gaps of width $a = 50\,\text{nm}$ and a grating period of $d = 100\,\text{nm}$. Finally, the particles are ionized by a laser before they are detected by an ion counter scanning the target area.

For a velocity $v = 220$ m/s the de Broglie wavelength

$$\lambda_{\text{dB}} = \frac{h}{mv} = 2,5\,\text{pm} \approx \frac{1}{400}\,\text{D} \tag{1.49}$$

of the fullerenes is about 400 times smaller than their size. With these dimensions the slits can be wide enough to allow the particles to pass through, while still being small enough to permit interference for the wavelength λ_{dB} in (1.49). However, the angle Θ between the central and the first-order intensity maxima is very small, i.e.,

$$\Theta \approx \sin\Theta = \frac{\lambda_{\text{dB}}}{d} = 25\,\mu\text{rad} = 5".\tag{1.50}$$

Therefore a good collimation is needed to increase the spatial (transverse) coherence. Furthermore, the whole experiment also has to be performed in a high-vacuum chamber to prevent the loss of coherence due to interactions of the fullerenes with residual gas molecules.

The intensity of the beam is kept very low such that the fullerenes enter the interferometer one by one. Consequently the interference pattern is built up by individual detections. This also ensures that the fullerenes do not interact with each other since the mean distance between them is about 1000 times larger than the range of the intermolecular potentials. As a side effect of the high temperature, the fullerenes emit photons, i.e., they emit black-body radiation. Fortunately, these photons do not influence the interference pattern because their wavelengths $\lambda \approx 5 - 20\,\mu\text{m}$ are much bigger than the grating period. Therefore the photons do not provide any path information and the interference phenomenon can be observed (Arndt *et al.*, 1999).

Since the first attempts to uncover the wave-like behaviour of large molecules, the techniques we have described so far have been further refined. For example, gratings

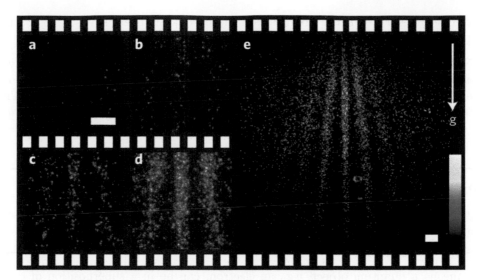

Fig. 1.14 Molecule Interference Pattern. A beam of complex phthalocyanine ($C_{32}H_{18}N_8$) molecules is diffracted from a grating and creates an interference pattern. The molecules are illuminated after the diffraction and their fluorescence is captured by an electron-multiplying charge-coupled device (EMCCD) camera. Selected images show the interference pattern build up. Used with permission of Thomas Juffmann and Markus Arndt (University of Vienna), from Fig. 3 in *Real-time single-molecule imaging of quantum interference*, Juffmann *et al.*, Nat. Nanotechnol. **7**, 297–300 (2012); permission conveyed through Copyright Clearance Center, Inc.

of solid matter can be replaced by standing waves created by reflecting laser beams off a mirror, which results in diffraction gratings made of light (Rasel *et al.*, 1995). This removes the harmful effects of *van der Waals* forces, i.e., the interaction between the molecules in the material of the grating and those in the particle beam. This, along with other experimental advances, allows to observe even the interference of very complicated molecules, see Fig. 1.14.

We can conclude that quantum interference effects can be made visible not only for photons or electrons, but also for complex, large molecules such as fullerenes. The detection counts quite accurately fit the diffraction pattern predicted by quantum theory. The individual particles travel through the interferometer and appear, seemingly random, on the screen or detector. However, when a large number of these particles accumulate, an emerging diffraction pattern reveals their wave-like behaviour.

1.8 Schrödinger's Cat

Thus far we have established that quantum objects, be they massless photons or heavy molecules, behave like waves under certain circumstances, and like particles in other respects. The description as particles and waves is, as phrased in Proposition 1.6, complementary. While this might seem at odds with the classical particle description, there is nothing unclear about the predictions made by quantum mechanics using Born's interpretation—Proposition 1.7.2.

Nevertheless, there is an aspect of the description of quantum systems as waves whose interpretation can create some puzzlement, especially when large quantum objects (such as the fullerenes discussed in the previous section) are concerned: The central feature of the wave nature of quantum systems is the so-called *superposition principle* that we have already encountered in eqn (1.48). Let us phrase it—for now—in terms of the following proposition.

Proposition 1.12 (Superposition principle)

Quantum systems can be in superpositions of states that represent possibilities that are mutually exclusive in classical physics.

This means a quantum object, e.g., an electron, can be in a superposition of completely distinct states, for example, two different positions. As a figure of speech one might say in that case that the electron *is* at two locations at once. However, such a statement is operationally meaningless because any measurement of the position would reveal a single location of the electron. However, the superposition principle does allow the electron to be in a superposition of these distinct states before the measurement. Perhaps prompted by the apparent strangeness of this situation, Erwin Schrödinger (1935a) reflected on the superposition principle by formulating the perhaps somewhat drastic gedanken experiment of a cat in a box, a version of which we will describe now.

Let us assume that a cat is locked into a box together with a quantum system, e.g., an atom and a photon detector. For the sake of the discussion let us further assume that the box is made of some particular material that does not allow any particles or radiation to pass through, i.e., such that there is no way to exchange information with the inside of the box unless we open it. Let one of the electrons in the shell of the entrapped atom be in an excited state, denoted as $|e\rangle$[8]. Then there is a non-zero probability, let us assume a 50% chance for simplicity, for the transition of the electron from the excited state $|e\rangle$ to the ground state $|g\rangle$. In the case such a transition occurs, the atom emits a photon which hits the detector placed in the box. However, since we do not have access to the interior we have to consider the electron to be in a superposition of the states $|e\rangle$ and $|g\rangle$ until we actually open the box. Formally, let us write

$$|e,0\rangle + |g,1\rangle , \qquad (1.51)$$

where 0 and 1 represent the number of photons registered by the detector. From a classical point of view, this situation may already appear be somewhat strange conceptually: We have a superposition of a macroscopic measurement apparatus either

---

[8]We shall introduce the precise mathematical meaning of this notation in Section 3.1.3. For now it suffices to consider $|g\rangle$ and $|e\rangle$ as symbols that represent the wave functions ψ_g and ψ_e respectively.

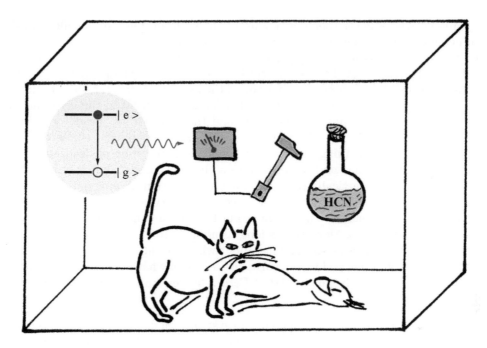

Fig. 1.15 Schrödinger's cat. In the gedanken experiment the photo detector in the box is coupled to a simple mechanism that will break a phial containing hydrogen cyanide (HCN). Once released this highly toxic substance will immediately lead to the demise of the imaginary cat. Quantum theory suggests that, before we open the box, the atom and the cat are in a superposition of being excited and alive, or in the ground state and dead, respectively.

clicking, or not. Schrödinger took this conundrum one step further. Suppose that the box is also equipped with a device, coupled to the photon detector, that will *kill* the cat if a photon is detected, as illustrated in Fig. 1.15.

Suppose then, the cat, consisting of its atoms and molecules, also behaves as a quantum system. By the logic used to write eqn (1.51) we have to conclude that, unless we open the box to check, the total state of the electron and the cat is

$$| e, \text{alive} \rangle + | g, \text{dead} \rangle . \tag{1.52}$$

The extreme fashion in which this issue is formulated has led to Schrödinger's cat becoming a synonym for aspects of quantum mechanics that some might deem to be mysterious. We do not believe in mystifying quantum theory, but Schrödinger's infamous cat nevertheless serves to illustrate several extremely important concepts in quantum mechanics.

First and foremost, Schrödinger's cat can (and should, in our opinion) be understood as a reminder to be careful when using informal language to translate mathematical formalism to statements of fact. In particular, a superposition of the quantum states corresponding to the properties "dead" or "alive" (or $| g \rangle$ and $| e \rangle$ for that matter) is not the same as (being in a state of) being dead *and* alive simultaneously. More generally, when applying the formalism of quantum mechanics one should be aware what the predictions of the formalism actually are and what is perhaps only an intuitive interpretation of the mathematical apparatus. We make predictions for the outcomes of possible measurements. If we open the box, we "measure" the cat, and we can tell whether it is alive or not. Specifically, the measurement in this hypothetical scenario is one with two possible outcomes, and these are either "dead" or "alive", with no other option. As we will learn in Chapter 3, a measurement whose outcome would correspond to anything else, in particular a superposition of dead and alive, is mathematically possible and would require a change of basis[9], but it is not at all clear how one would practically carry out such an operation. A pragmatic stance on this issue would be to simply say that we need not worry about the cat being in a superposition of dead and alive if we cannot put forward any measurement whose outcomes would confirm this. Moreover, as long as no measurement has been performed, here, as long as the box is closed, arguing about what the cat "is" or "is not" remains operationally meaningless and would constitute a case of what is called *counterfactual reasoning*. Or, as famously put by *Asher Peres*: "unperformed experiments have no results" (1978).

Admittedly, describing a cat as a quantum system with two possible states is somewhat of a stretch as far as practical experiments are concerned. Nonetheless, the issue of identifying a practically (at least in principle) realizable measurement to confirm the superposition of two macroscopically distinct states raises a question that we have touched upon already in Section 1.7.3: How large or complex can a system become and still show genuine quantum-mechanical behaviour? On the one hand, as we have argued, this is a problem of designing suitable measurement setups to reveal the wave-like behaviour that we attribute to quantum objects. Approaching this question from a purely theoretical point of view, it can be argued (Skotiniotis *et al.*, 2017) that a measurement apparatus able to confirm superposition states of a macroscopic object the size of the cat would have to be of a size comparable with that of Earth itself. Nevertheless, the exploration of the possibility to test quantum phenomena on the micro-, meso-, and macroscopic scale is still ongoing, see, e.g., the review on macroscopicity by Fröwis *et al.* (2018).

On the other hand, we should note here that, whether a suitable measurement setup is available or not, with increasing size and complexity of the system, it becomes increasingly difficult to isolate it from its environment. That is, there are myriads of interactions between the atoms and molecules of the cat and, e.g., the air surrounding it. These interactions can be viewed as measurements whose outcomes we simply do not know disturbing the system. A whole branch of quantum physics dealing with this phenomenon called *decoherence* has been developed to describe the dynamics of open quantum systems that are not completely isolated from their environment, and we dedicate Chapter 22 of Part III of this book to this question. To briefly comment on this issue already here, let us remark that quantum systems interacting with their

---

[9]This nomenclature will become clear in Chapter 3.

surroundings typically (appear to) equilibrate (Linden *et al.*, 2009), even though the exact mechanisms and conditions that generically lead to (specifically, thermal) equilibrium are a matter of ongoing research, see, for instance, the review by Gogolin and Eisert (2016). Nevertheless, as far as macroscopic quantum states are concerned it can be stated quite plainly that all "macroscopic quantum states are fragile and hard to prepare" (López-Incera *et al.*, 2019).

The perplexing cat and the atom described in eqn (1.52) also already hint at another fascinating concept referred to as *entanglement*. The state of the atom and the cat are clearly correlated: neither has a definite state on its own, while the total system, atom plus cat, is in a definite state. We could argue that it is enough to open the box just enough to make a measurement of the electron to also learn about the cat. In some sense, finding the electron in the ground state would immediately lead us to conclude that the cat has been killed. Part II of this book entirely focuses on the aspects of such correlations in quantum mechanics.

Finally, one should note that Schrödinger's proposal requires very precise control of the atom in the box and the photon that is emitted from it. Exacting such absolute precision over individual quantum systems has long been considered impossible by some, while it seems to have inspired others. In 2012 the Nobel Prize in Physics was awarded jointly to *Serge Haroche* and *David J. Wineland* for their mastery of the control of individual particles, photons and atoms respectively, while preserving their quantum-mechanical nature.

2
The Time-Dependent Schrödinger Equation

2.1 Wave Function and Time-Dependent Schrödinger Equation

In Chapter 1 we have discussed how to understand the wave-particle duality of quantum-mechanical objects. Following Planck and Einstein, we have argued that the energies and (angular) frequencies of light quanta are related by Planck's constant, i.e., $E = \hbar\omega$. By additionally relating the momentum to a wave vector, $\vec{p} = \hbar\vec{k}$, de Broglie extended this dualism to massive particles.

2.1.1 Discovering the Schrödinger Equation

However, it was *Erwin Schrödinger* who reconsidered de Broglie's matter wave concept, elevating it to a dynamical theory. In 1926 he published two papers (Schrödinger, 1926*a*; Schrödinger, 1926*b*) in which he proposed an equation of motion—a wave equation—for the wave functions of particles, that is in accordance with empirical observations. Today, this differential equation is considered to be a fundamental axiom of (non-relativistic) quantum theory that cannot be derived from first principles, much like Newton's laws, and the discovery of this equation led to him being awarded the 1933 Nobel Prize in Physics.

Despite its axiomatic character, we can nonetheless argue for the plausibility of the form of Schrödinger's wave equation. To this end, let us consider representatives for wave functions. In general, any wave function $\psi(t, \vec{x})$ will depend on some spatial coordinate $\vec{x}$ and on time t. A particular example for such a function is a (free) wave packet which we can write in terms of the Fourier integral

$$\psi(t, \vec{x}) = \int \frac{d^3k}{(2\pi)^{3/2}} \, \tilde{\psi}(\vec{k}) \, e^{i(\vec{k}\cdot\vec{x} - \omega t)} , \qquad (2.1)$$

where the function $\tilde{\psi}(\vec{k})$ does not depend on time. We shall discuss such objects in more detail later on in Section 2.5.3. For now, let us focus on the constituent plane waves $e^{i(\vec{k}\cdot\vec{x} - \omega t)}$, which wave packets such as (2.1) are superpositions of. Naively considering plane waves as representatives for wave functions $\psi(t, \vec{x})$, we differentiate with respect to t and with respect to the spatial coordinates x^i $(i = 1, 2, 3)$ with $x^1 = x$, $x^2 = y$, and $x^3 = z$,

$$i\hbar \frac{\partial}{\partial t}\, \psi\,(t, \vec{x}) \,=\, \hbar\omega\,\psi\,(t, \vec{x}) \,=\, E\,\psi\,(t, \vec{x})\,, \tag{2.2a}$$

$$-i\hbar\,\vec{\nabla}\,\psi\,(t, \vec{x}) \,=\, \hbar\vec{k}\,\psi\,(t, \vec{x}) \,=\, \vec{p}\,\psi\,(t, \vec{x})\,, \tag{2.2b}$$

$$-\hbar^2\,\Delta\,\psi\,(t, \vec{x}) \,=\, \hbar^2 |\vec{k}|^2\,\psi\,(t, \vec{x}) \,=\, |\vec{p}|^2\,\psi\,(t, \vec{x})\,, \tag{2.2c}$$

where $(\vec{\nabla})_i = \partial/\partial x^i$, $\Delta = \vec{\nabla} \cdot \vec{\nabla}$ is the so-called *Laplace operator*, and we have made use of the relations between wave and particle properties from eqn (1.34). Inserting eqns (2.2a) and (2.2c) into the non-relativistic energy-momentum relation for a free particle of mass m, i.e., $E = \frac{|\vec{p}|^2}{2m}$, we see that the plane waves satisfy the differential equation

$$i\hbar \frac{\partial}{\partial t}\, \psi\,(t, \vec{x}) \,=\, -\frac{\hbar^2}{2m}\,\Delta\,\psi\,(t, \vec{x})\,. \tag{2.3}$$

Since we have assumed that $\tilde{\psi}(\vec{k})$ does not depend on $\vec{x}$ or t, the (free) wave packet in eqn (2.1) also satisfies this equation. In general, however, there is also a potential-energy contribution and the non-relativistic energy-momentum relation reads

$$E = \frac{|\vec{p}|^2}{2m} + V\,, \tag{2.4}$$

where V is the potential energy. By including this term, we then obtain the following differential equation for the wave function $\psi\,(t, \vec{x})$:

Proposition 2.1 (Time-dependent Schrödinger equation)

$$i\hbar \frac{\partial}{\partial t}\, \psi\,(t, \vec{x}) \,=\, \left(-\frac{\hbar^2}{2m}\,\Delta\, +\, V(\vec{x})\right) \psi\,(t, \vec{x}) \,=\, H\,\psi\,(t, \vec{x})\,.$$

It is customary to collect the operators for the kinetic and potential energy into a single operator, the *Hamiltonian*

$$H := -\frac{\hbar^2}{2m}\,\Delta\, +\, V(\vec{x})\,. \tag{2.5}$$

Schrödinger assumed his equation to hold quite generally[1] for massive particles in an external potential $V(\vec{x})$, where one can assume for simplicity that the potential is in-

---

[1]Historically, Schrödinger, who was informed by Einstein about de Broglie's thesis (de Broglie, 1925), gave a talk at the ETH Zürich in 1925. At this opportunity *Peter J. W. Debye* encouraged him to look for a wave equation, which Schrödinger indeed quickly discovered. However, in his first attempt he used the relativistic energy-momentum relation and found what is nowadays known as the *Klein–Gordon equation*. Since this equation did not reproduce the spectrum of the hydrogen atom correctly he discarded it again. Only later, in 1926, during his Christmas vacation near Arosa (Switzerland), Schrödinger succeeded by assuming a non-relativistic relation between energy and momentum (see, e.g., Kerber *et al.*, 1987 and Moore, 1989). Thus he discovered the famous *"Schrödinger equation"* which describes the spectrum of the hydrogen atom correctly, as we shall see in Chapter 7.

dependent of the time t. The solution $\psi(t, \vec{x})$ of the Schrödinger equation is called the *wave function* and we shall discuss its interpretation in Section 2.1.3.

The Schrödinger equation is a partial differential equation with the following important properties:

(i) It is a *first-order* equation *in time* t: The wave function is therefore determined by initial conditions, which is desirable from a physical point of view.

(ii) It is *linear in* ψ: Consequently, linear combinations of solutions are again solutions, i.e., if ψ_1 and ψ_2 are solutions, then also $\psi = c_1\psi_1 + c_2\psi_2$ with $c_1, c_2 \in \mathbb{C}$ is a solution. In other words, the *superposition principle* of Proposition 1.12 applies.

(iii) It is *homogeneous*: The normalization (see Section 2.1.4) holds for all times t.

2.1.2   First Quantization

As we have seen for plane waves, the physical quantities energy and momentum can be obtained from the wave function by applying differential operators, i.e.,

$$i\hbar \frac{\partial}{\partial t}\psi = E\,\psi \quad \text{and} \quad -i\hbar\,\vec{\nabla}\,\psi = \vec{p}\,\psi\,, \tag{2.6}$$

which allowed us to deduce the Schrödinger equation. This *quantum-mechanical correspondence* between physical quantities and operators, in particular, here the association

$$E \;\to\; i\hbar\,\frac{\partial}{\partial t} \quad \text{and} \quad \vec{p} \;\to\; -i\hbar\,\vec{\nabla}\,, \tag{2.7}$$

is valid quite generally. In fact, as we will discuss in more detail in Section 2.3, *all* physical quantities that can be directly measured can be represented by operators. This association is usually referred to as *"first quantization"*[2] and it immediately enables us to convert known classical relations to quantum-mechanical expressions. For instance, one can quickly find the Schrödinger equation by starting with the classical Hamilton function $H(\vec{x}, \vec{p})$ and "quantizing" it by inserting the operators of eqn (2.7) into the classical equation $H(\vec{x}, \vec{p}) = |\vec{p}|^2/(2m) + V = E$.

2.1.3   The Interpretation of the Wave Function

A closer inspection of the solutions of the Schrödinger equation is now in order. For this purpose let us restrict the discussion to one spatial dimension with coordinate x to keep the notation simple. The operational interpretation of the wave function is due to *Max Born* (see Proposition 1.7.2 and Born 1926) and can be phrased in the following way: The probability for finding a particle described by $\psi(t, x)$ in an infinitesimal interval $[x, x + dx]$ is given by $|\psi(t, x)|^2 dx$. As illustrated in Fig. 2.1, the corresponding probability for a finite interval is then simply obtained by integration.

---

[2]Indeed, in quantum field theory there is also a formal procedure referred to as "second quantization", which we will discuss in Chapter 25 of Part III. However, the distinction between "first" and "second" has historical reasons and can be considered to be purely semantical.

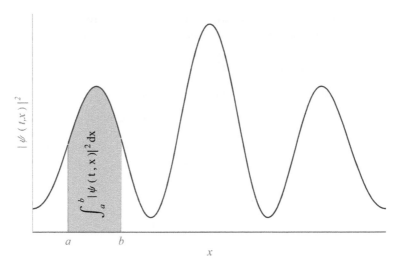

Fig. 2.1 Probability interpretation of the wave function. The probability of finding the particle in an interval $[a, b]$ is given by the area enclosed between $x = a$, $x = b$, the x axis and $|\psi(t, x)|^2$.

Let us examine the statistical interpretation of the wave function a little more closely. For an individual particle we can, at most, identify measurement outcomes that are excluded. For example, we can be certain that we will not find a particle at locations x for which the respective wave function $\psi(t, x)$ vanishes. For all other locations the probability of finding the particle is finite. It thus seems that the wave function provides very little information about the outcome of a single-particle experiment.

However, the interpretation of the modulus squared of the wave function as a probability becomes justified when the experiment is repeated many times for identically prepared particles. In fact, there is no additional information available about the individual particles in the ensemble. The wave function contains *all* the information about the physical system. We should note that several additional, competing interpretations of the wave function exist that differ in philosophical and semantical aspects. Nonetheless, they all agree in their quantitative predictions, based on the rules of quantum mechanics. We shall discuss additional limitations on possible interpretations in Section 13.4 of Part II. For now let us take a pragmatic approach and adopt an interpretation for the wave function in three spatial dimensions that, we believe, should be compatible with any other valid interpretation of quantum mechanics:

Proposition 2.2 (Ensemble interpretation of the wave function)

The probability distribution for the outcomes of any measurement that is carried out many times for an ensemble of identically prepared particles is fully determined by the wave function $\psi(t, \vec{x})$ of these particles.

Interestingly, the statistical interpretation further leads to a principal uncertainty in the localization of a particle. We cannot predict the definite measurement outcome for a specific particle, i.e., its location at a certain time, and it therefore becomes meaningless to assign a path to the particle.

2.1.4   The Normalization of the Wave Function

Since we interpret $|\psi(t, \vec{x})|^2$ as a probability measure for an ensemble of identically prepared particles, it is crucial that this distribution is normalized. In other words, the probability to find a single particle *somewhere* in space should be exactly equal to one if a single particle was prepared. This need not be the case for all wave functions ψ a priori, which requires us to introduce the *normalization condition*

$$\int\limits_{-\infty}^{\infty} d^3x \, |\psi(t, \vec{x})|^2 = 1 \,. \tag{2.8}$$

The normalization thus imposes a restriction on the asymptotic behaviour of the wave function, i.e., $|\psi(t, \vec{x})|^2$ needs to be appropriately bounded as $|\vec{x}| \to \infty$. The wave function must be *square integrable* on the interval $x, y, z \in [-\infty, \infty]$ such that the integral on the left-hand side of eqn (2.8) is finite. To formulate this in more mathematically abstract terms, the wave function must be an element of the space of square-integrable functions in $\mathbb{R}^3$, $\psi(t, \vec{x}) \in L_2(\mathbb{R}^3, d\mu(\vec{x}))$, with respect to a suitable measure of integration $d\mu(\vec{x})$. The notion of such a vector space is a special case of so-called *Lebesgue spaces* fL_p. The theory of such spaces provides a rich and elegant mathematical framework. However, we shall not focus our discussion on these structures, and we are therefore referring the interested reader to available literature on this subject, e.g., Rudin (1991, p. 36) and Adams and Fournier (2003, Chap. 2).

Irrespective of the mathematical intricacies of this problem we will consider formal solutions of the Schrödinger equation that are not normalizable as mathematical tools only, while we require physical states to be normalizable. This interpretation applies, for instance, to plane waves $e^{i(\vec{k}\cdot\vec{x} - \omega t)}$. Solutions of the Schrödinger equation which are of this form can only be normalized within a region of finite volume, for instance a cube of edge length L, i.e.,

$$\psi(t, \vec{x}) = \begin{cases} L^{-3/2} \, e^{i(\vec{k}\cdot\vec{x} - \omega t)} & \text{for } 0 \leq x, y, z \leq L \\ 0 & \text{otherwise} \end{cases}, \tag{2.9}$$

such that the normalization condition of eqn (2.8) is satisfied:

$$\int\limits_{-\infty}^{\infty} d^3x \, \psi^*(t, \vec{x})\psi(t, \vec{x}) = \frac{1}{L^3} \int\limits_{0}^{L} d^3x \, e^{-i(\vec{k}\cdot\vec{x} - \omega t)} \, e^{i(\vec{k}\cdot\vec{x} - \omega t)} = \frac{L^3}{L^3} = 1 \,. \tag{2.10}$$

Similar restrictions apply for any additional spatial dimensions. We conclude that plane waves can only represent physical particles in a box and generally wave packets have to be considered for the description of particles. Nonetheless, plane waves can be useful tools for computations as long as the requirement for normalization is kept in mind for the physical interpretation of the results. In addition to the normalization, the wave functions of particles in a box are subject to boundary conditions that restrict the plane waves in eqn (2.9), as we will discuss in Chapter 4.

2.2   The Continuity Equation

Having established the normalization of the wave function and Born's interpretation let us switch back to a notation that is suitable for any number of spatial dimensions. Specifically, we identify the quantity $|\psi(t, \vec{x})|^2$, where $\vec{x} \in \mathbb{R}^3$, with a probability density in three dimensions.

Definition 2.1   *The **probability density** ρ is given by the modulus squared of the wave function ψ*

$$\rho(t, \vec{x}) := |\psi(t, \vec{x})|^2 = \psi^*(t, \vec{x})\,\psi(t, \vec{x})\,.$$

Given such a density function at time t, one is naturally interested in its time evolution. To construct the equation of motion for $\rho(t, \vec{x})$ we turn to the three-dimensional Schrödinger equation

$$i\hbar\,\frac{\partial}{\partial t}\,\psi(t, \vec{x}) = H\,\psi(t, \vec{x})\,, \tag{2.11}$$

and we form its complex conjugate

$$-i\hbar\,\frac{\partial}{\partial t}\,\psi^*(t, \vec{x}) = H^*\,\psi^*(t, \vec{x})\,. \tag{2.12}$$

Here we note that the Hamiltonian H is a so-called *Hermitian* operator. In the present context[3] this means that we can write $H = H^*$. The expressions in eqns (2.11) and (2.12) can now be compared with the partial derivative of the probability density with respect to time, i.e.,

$$\frac{\partial}{\partial t}\,\rho(t, \vec{x}) = \frac{\partial}{\partial t}\,\psi^*(t, \vec{x})\,\psi(t, \vec{x}) = \left(\frac{\partial}{\partial t}\psi^*\right)\psi + \psi^*\left(\frac{\partial}{\partial t}\psi\right)\,. \tag{2.13}$$

Inserting the Schrödinger equation and its complex conjugate, eqns (2.11) and (2.12), respectively, into eqn (2.13) one easily arrives at

$$\frac{\partial}{\partial t}\rho(t, \vec{x}) = -\frac{1}{i\hbar}\left[(H\,\psi^*)\,\psi - \psi^*H\psi\right] = \frac{\hbar}{2mi}\,\vec{\nabla}\left[(\vec{\nabla}\psi^*)\psi - \psi^*(\vec{\nabla}\psi)\right]\,. \tag{2.14}$$

Note that the potential V is a scalar that can be freely exchanged with the wave function ψ and its complex conjugate, in other words, these quantities *commute*. Inspection of the right-hand side of eqn (2.14) suggests that we should identify a *current density* $\vec{j}(t, \vec{x})$ associated to $\rho(t, \vec{x})$.

---

[3]A mathematically more precise formulation of this criterion will be given later in Section 2.3.2.

Definition 2.2   *The **probability-current density** $\vec{j}$ is defined as*

$$\vec{j}(t, \vec{x}) := \frac{\hbar}{2mi} \left[\psi^* \vec{\nabla} \psi - \psi \vec{\nabla} \psi^* \right].$$

In full analogy to the electric charge density in electrodynamics, the time evolution of the probability density is governed by a *continuity equation*.

Theorem 2.1 (Continuity equation) $\frac{\partial}{\partial t} \rho(t, \vec{x}) + \vec{\nabla} \cdot \vec{j}(t, \vec{x}) = 0$

An important consequence of Theorem 2.1 is the *conservation of probability*,

$$\int_{\mathbb{R}^3} d^3x \, |\psi(t, \vec{x})|^2 = 1 \qquad \forall \, t \geq 0, \tag{2.15}$$

mirroring the conservation of electric charge in electrodynamics. The continuity equation of Theorem 2.1 is hence interpreted as follows: Any change of the probability density within a region of space is balanced by a probability flux through the boundary of that region.

Let us make this statement mathematically more precise. For simplicity, we assume that the probability density is normalized to one at some initial time t_0 and that it is non-zero only within a bounded region $R \in \mathbb{R}^3$. We then take the derivative of eqn (2.15) with respect to time, and, applying the continuity equation (Theorem 2.1), we arrive at

$$\frac{\partial}{\partial t} \int_R d^3x \, \rho(t, \vec{x}) = -\int_R d^3x \, \vec{\nabla} \cdot \vec{j}(t, \vec{x}) \overset{\text{Gauss}}{=} \oint_{\partial R} d\vec{s} \cdot \vec{j}(t, \vec{x}). \tag{2.16}$$

Here, Gauss' divergence theorem has been used to convert the integral over the three-dimensional region R into an integral over its closed (two-dimensional) boundary ∂R with outward pointing unit normal $\vec{n}$ and surface element ds, such that $d\vec{s} = ds\,\vec{n}$. Since the non-vanishing probability density has been restricted to a bounded region in $\mathbb{R}^3$ we may enlarge this region to a sphere with radius r and rewrite the surface integral on the right-hand side of eqn (2.16) in spherical coordinates. The normalization of the wave function further implies that ψ has to fall off stronger than $1/r$ as $r \to \infty$.

Hence, the spatial derivative $\vec{\nabla}\psi$ and its complex conjugate must decrease at least as $1/r^2$, and, consequently, the probability current falls off stronger than $1/r^3$ in that limit. Combining these insights we find that the integral over the surface of the sphere behaves as

$$\int dr \, r^2 \, \frac{1}{r^3} \to 0 \qquad \text{for} \quad r \to \infty. \tag{2.17}$$

Consequently, we can conclude that the overall probability is conserved in time, i.e.,

$$\int_{\mathbb{R}^3} d^3x \; \rho(t, \vec{x}) = \text{const.} = 1 \,. \tag{2.18}$$

The probability current $\vec{j}$ represents a flow of probability. Interestingly, it may be related to the quantum-mechanical *momentum operator* $\vec{P} = -i\hbar\vec{\nabla}$ from eqn (2.7) by noting that

$$\vec{j}(t, \vec{x}) = \frac{1}{2m}\left[\psi^*(-i\hbar\vec{\nabla})\psi + \left(\psi^*(-i\hbar\vec{\nabla})\psi\right)^* \right] = \frac{1}{m}\,\text{Re}\big(\psi^*\vec{P}\psi\big)\,. \tag{2.19}$$

Using a plane wave as an example for the wave function reveals that the probability can be viewed as "flowing" in the direction of the associated particles' velocities, that is

$$\vec{j}(t, \vec{x}) = \frac{1}{m}\,\text{Re}\big(e^{-i(\vec{k}\cdot\vec{x}-\omega t)}(-i\hbar\vec{\nabla})e^{i(\vec{k}\cdot\vec{x}-\omega t)}\big)$$

$$= \frac{1}{m}\,\text{Re}\left(e^{-i(\vec{k}\cdot\vec{x}-\omega t)}\,\hbar\vec{k}\,e^{i(\vec{k}\cdot\vec{x}-\omega t)}\right) = \frac{\vec{p}}{m} = \vec{v}\,, \tag{2.20}$$

where we have inserted $\hbar\vec{k} = \vec{p}$. Phrasing this more directly, the probability-current density replaces the notion of fixed velocities for particles much in the same way as the probability density takes the place of definite locations of quantum objects.

2.3 States and Observables

Within the Schrödinger equation (Proposition 2.1.1) we have encountered two types of objects that encode the structure of our description of the physical world—*wave functions* and *operators*. The wave function contains information about the location of quantum-mechanical particles, while we have argued that operators are associated to physical quantities like energy or momentum. It is now our aim to place these ideas in a more general theoretical construction. However, so far our nomenclature is suitable only for a narrow range of physical situations. The term "wave function" suggests a scalar distribution in space that evolves in time, but we aim to apply the formalism of quantum mechanics also to other degrees of freedom, for instance, a particle's spin. Let us therefore introduce a more practical terminology: The *state* of a physical system is the collection of all accessible information about that system, encoded in an appropriate mathematical object, for instance, a wave function $\psi(t, \vec{x})$.

Definition 2.3   *The **state** of a quantum-mechanical system is represented by a vector ψ in a complex vector space, called* Hilbert space, *that is equipped with a scalar product.*

The construction of Hilbert spaces will receive a rigorous treatment in Chapter 3. The wave function is an example for a state vector and is therefore included in the very general framework of the Hilbert-space description. To keep the discussion simple, for the remainder of Chapter 2 wave functions will be the focus of our attention. The word "operator", on the other hand, is too broad in its meaning for our purposes. Not every mathematical operator can be expected to correspond to a physical quantity or operation. We therefore define:

Definition 2.4   *Physical quantities that can be measured, for instance, energy or momentum, are called **observables**. They are represented by Hermitian operators acting on the vectors of the Hilbert space. The eigenvalues of these operators correspond to the possible outcomes of individual measurements.*

The significance of Hermitian operators derives from their role in the scalar product of the Hilbert space, which we shall discuss shortly. Definitions 2.3 and 2.4 associate physical quantities with mathematical objects. The theory of quantum mechanics provides the rules to combine these concepts and derive predictions for the outcome of experiments. Formalizing these rules requires the introduction of some additional mathematical background, with which we will proceed now.

2.3.1   The Scalar Product

As mentioned in Definition 2.3, the Hilbert space of quantum states is equipped with a *scalar product* that we shall denote as $\langle \cdot | \cdot \rangle$, where the two "slots" indicated by the dots are filled by quantum states, for instance, wave functions ψ and ϕ.

Definition 2.5   *The **scalar product** of two wave functions ψ and ϕ is defined as*

$$\langle \phi | \psi \rangle := \int\limits_{-\infty}^{\infty} d^3x \, \phi^*(t, \vec{x}) \, \psi(t, \vec{x}) \,.$$

Notation employed for the scalar product may vary: For instance, the vertical line between the two vectors could be replaced by a comma, $\langle \psi, \phi \rangle$. However, the convention to write the scalar product in the form $\langle \cdot | \cdot \rangle$ turns out to be advantageous for the so-called *Dirac notation* that allows the modular combination of "*bra*" $\langle \cdot |$ and "*ket*" $| \cdot \rangle$, as will be discussed in Section 3.1.3.

   In addition to (or instead of) variables like spatial position or momentum, which take values in a continuous set, quantum states may also depend on discrete variables. In such a case the integral in Definition 2.5 needs to be amended or replaced by a sum over the discrete degrees of freedom. In general, it will prove useful to use the superscript symbol "†" which combines the transposition in the space of the discrete variables with the usual complex conjugation of the wave function, i.e.,

$$\langle \phi | \psi \rangle = \sum_n \int\limits_{-\infty}^{\infty} d^3x \, \phi_n^*(t, \vec{x}) \, \psi_n(t, \vec{x}) = \int\limits_{-\infty}^{\infty} d^3x \, \phi^\dagger(t, \vec{x}) \, \psi(t, \vec{x}) \,, \qquad (2.21)$$

where ϕ and ψ are interpreted as vectors whose components ϕ_n and ψ_n depend on t and $\vec{x}$. Regardless of the type of variable, both the scalar products in Definition 2.5 and eqn (2.21) have the following properties for all normalizable quantum states ϕ, ϕ_1, ϕ_2, and ψ, ψ_1, ψ_2, and for all $c_1, c_2 \in \mathbb{C}$:

$$\langle \phi | \psi \rangle = \langle \psi | \phi \rangle^*, \tag{2.22a}$$

$$\text{linear in "ket"} \quad \langle \phi | c_1 \psi_1 + c_2 \psi_2 \rangle = c_1 \langle \phi | \psi_1 \rangle + c_2 \langle \phi | \psi_2 \rangle, \tag{2.22b}$$

$$\text{anti-linear in "bra"} \quad \langle c_1 \phi_1 + c_2 \phi_2 | \psi \rangle = c_1^* \langle \phi_1 | \psi \rangle + c_2^* \langle \phi_2 | \psi \rangle, \tag{2.22c}$$

$$\text{positive definite} \quad \langle \psi | \psi \rangle > 0 \quad \forall \psi \neq 0, \tag{2.22d}$$

$$\text{non-degenerate} \quad \langle \psi | \psi \rangle = 0 \quad \Leftrightarrow \quad \psi \equiv 0. \tag{2.22e}$$

Viewing the last property, eqn (2.22d), in terms of the normalization condition of eqn (2.8), it can be seen that we simply require the scalar product of any physical state with itself to evaluate to 1, $\langle \psi | \psi \rangle = 1$. For any two states, ϕ and ψ, the modulus $|\langle \phi | \psi \rangle|$ of the scalar product is then smaller or equal to 1, and can be taken as a measure of their similarity—their *overlap*.

2.3.2 Operators

With the scalar product at hand we can now incorporate measurable quantities into the picture. In particular, it is reasonable to assume that the physical theory describes our system both before and after any measurement[4]. Loosely speaking, physical interventions should be represented by operations that map wave functions to wave functions, or, more broadly speaking, quantum states to quantum states. In terms of mathematics this means that observables are *linear operators* on the Hilbert space $\mathcal{H}$.

Definition 2.6   *A is a **linear operator** with domain $\mathcal{D}(A)$, if for $A\psi_1 = \phi_1$ and $A\psi_2 = \phi_2$, where $\psi_1, \psi_2, \phi_1, \phi_2 \in \mathcal{D}(A)$, and for all $c_1, c_2 \in \mathbb{C}$ follows that $A(c_1 \psi_1 + c_2 \psi_2) = c_1 \phi_1 + c_2 \phi_2$.*

Note that the domain of the operator has been included directly into the definition of linear operators. In fact, it is generally useful to think of an operator as defined by both its action and its domain, and write operators as $(A, \mathcal{D}(A))$ instead of simply A. However, in an attempt to keep the discussion simple we shall refrain from explicitly stating the domains of operators, unless this is specifically required. The operators that we have already encountered, e.g., $\vec{\nabla}, \Delta, \frac{\partial}{\partial t}, V(\vec{x})$, are all linear operators. Linear operators act distributively in the sense that

$$(A + B)\psi = A\psi + B\psi, \tag{2.23a}$$

$$AB\psi = A(B\psi). \tag{2.23b}$$

---

[4]In fact, the entire measurement process may be formulated as a quantum-mechanical interaction. However, since the construction of such a model requires more advanced tools we shall postpone such considerations until Chapter 23 in Part III.

However, note that linear operators do not *commute* in general, i.e., A and B

$$AB \neq BA. \tag{2.24}$$

While this may appear insignificant at the moment, *non-commutativity* is an important feature of quantum mechanics that we will encounter in many peculiar phenomena, for instance, when re-examining the uncertainty principle in Section 2.5. Nonetheless, linear operators on the Hilbert space map vectors to vectors. For example, consider an operator A that maps the wave function ϕ to the wave function ϕ', i.e., $A\phi = \phi'$. Evaluating the scalar product of ϕ' with another wave function ψ, we may write

$$\langle \psi | \phi' \rangle = \langle \psi | A \phi \rangle = \int d^3x \, \psi^*(\vec{x}) \, A \, \phi(\vec{x}), \tag{2.25}$$

where we have omitted the limits of integration. Unless otherwise stated we will assume that the limits of integration are $\pm\infty$ from now on. In eqn (2.25) we have considered the operator A to be acting on ϕ. Another operator—the *adjoint operator*—can be introduced, which acts on ψ instead, while reproducing the same scalar product. Recall that, after all, the scalar product is simply a complex number.

Definition 2.7 $A^\dagger$ *is called the **adjoint operator** to A, if for all $\phi \in \mathcal{D}(A)$ and for all $\psi \in \mathcal{D}(A^\dagger)$ holds that $\langle A^\dagger \psi | \phi \rangle = \langle \psi | A \phi \rangle$.*

Applying Definition 2.7 to eqn (2.25) we write

$$\langle A^\dagger \psi | \phi \rangle = \int d^3x \, \left(A^\dagger \psi^*(\vec{x}) \right) \phi(\vec{x}) = \int d^3x \, \psi^*(\vec{x}) \, A \, \phi(\vec{x}) = \langle \psi | A \phi \rangle. \tag{2.26}$$

In finite-dimensional Hilbert spaces all linear operators can be written as matrices, for which the adjoint operator is simply obtained by *Hermitian conjugation*, i.e., complex conjugation and transposition, $A^\dagger = (A^\top)^* = (A^*)^\top$. In expectation of the Dirac notation that we will discuss in Section 3.1.3 let us introduce a slight notational tweak that will prove to be practical for many computations in quantum mechanics. We write

$$\langle \psi | A \phi \rangle = \langle \psi | A | \phi \rangle. \tag{2.27}$$

Returning to our discussion, we now define an important property of linear operators.

Definition 2.8   *An operator A is called **Hermitian** if it is formally identical to its adjoint, $A^\dagger = A$, and the domains satisfy $D(A^\dagger) \supset D(A)$. If $A^\dagger = A$ and $D(A^\dagger) = D(A)$, then A is called **self-adjoint***

Strictly speaking, only self-adjoint operators can be considered to correspond to measurable quantities (observables) in the sense that their eigenvalues represent the possible measurement outcomes. Typically, the difference between the Hermitian and truly self-adjoint operators is of importance when the state space of an infinite-dimensional

Hilbert space is restricted. For instance, when the wave functions are confined to a finite region of space as in eqn (2.9). However, in such cases Hermitian (sometimes also called symmetric) operators can often be replaced by their so-called *self-adjoint extensions*, which are equally suitable for representing physical quantities. We shall not be concerned with the exact procedure of obtaining these extensions right now, but we will return to this issue when the analysis requires it in Section 3.3 of the next chapter. Meanwhile, we direct the interested reader to a pedagogical review on this topic by Bonneau *et al.* (2001). Let us now consider some examples of observables.

(i) **Position operator X_i**

Consider a quantum system with wave function $\psi(t, \vec{x})$ confined to a bounded region $R \subset \mathbb{R}^3$. We define the position operator

$$X_i \psi(t, \vec{x}) = x_i \psi(t, \vec{x}), \tag{2.28}$$

where x_i ($i = 1, 2, 3$) is the ith component of $\vec{x}$. It can be easily seen from eqn (2.26) that this multiplicative operator is *Hermitian* (and it is indeed also self-adjoint), independently of the restriction to the region R. However, choosing the square-integrable functions over the region R as the domain guarantees that the integral of the scalar product remains finite, i.e.,

$$\int_R d^3x \, \psi^*(t, \vec{x}) \, X_i \, \phi(t, \vec{x}) = \int_R d^3x \, x_i \, \psi^*(t, \vec{x}) \, \phi(t, \vec{x}) \; < \; \infty. \tag{2.29}$$

If we take R to be all of $\mathbb{R}^3$, then the domain of X_i consists of all states for which

$$\int_{\mathbb{R}^3} d^3x \, x_i^2 \, |\psi(t, \vec{x})|^2 < \infty. \tag{2.30}$$

(ii) **Momentum operator P_i**

We have already encountered the momentum operator

$$P_i \psi(t, \vec{x}) = -i\hbar \nabla_i \psi(t, \vec{x}) \tag{2.31}$$

for square-integrable functions, $\psi \in \mathfrak{L}_2(\mathbb{R}^3)$, where $\nabla_i = \partial/\partial x^i$. Since such wave functions must vanish at infinity one can easily see that P_i is Hermitian from an integration by parts. In addition, the domains of P_i and its adjoint coincide. According to Definition 2.8, the momentum operator in $\mathbb{R}^3$ is self-adjoint.

(iii) **Hamiltonian H**

Finally, the free-particle Hamiltonian in $\mathbb{R}^3$, acting on wave functions according to

$$H\psi(t, x) = -\frac{\hbar^2}{2m} \Delta \psi(t, x), \tag{2.32}$$

is a Hermitian (and self-adjoint) operator, which can again be verified by integration by parts.

2.3.3   The Commutator

Now that we have a viable way to represent physical observables, we are immediately faced with a peculiarity. The order in which two (Hermitian) operators A and B are applied to states makes a difference.

$$A B \neq B A. \tag{2.33}$$

Naively, we might consider further restricting the allowed operators by demanding that only interchangeable operators correspond to physical observables. However, it can be seen straight away that this is not tractable, since this would mean that *either* position *or* momentum are not physically measurable quantities. Consider the position operator X_m from eqn (2.28) and the momentum operator P_n from eqn (2.31) for wave functions ψ in $\mathfrak{L}_2(\mathbb{R}^3)$. Assuming that both $(X_m \psi)$ and $(P_n \psi)$ are square-integrable functions (and therefore are in the domain of P_n and X_m, respectively), we find

$$P_n \, X_m \, \psi(\vec{x}) = -i\hbar \nabla_n \left(x_m \, \psi(\vec{x}) \right) = -i\hbar \delta_{mn} \, \psi(\vec{x}) - i\hbar \, x_m \, \nabla_n \, \psi(\vec{x}), \tag{2.34}$$

where δ_{mn} is the Kronecker delta, $\delta_{mn} = 1$ if $m = n$, and $\delta_{mn} = 0$, otherwise. Formally, we can then write the relation of eqn (2.34) as

$$X_m \, P_n \, - \, P_n \, X_m \, = \, i\hbar \delta_{mn}. \tag{2.35}$$

We have to accept that the order of observables in quantum mechanics cannot be freely exchanged—operators do not generally *commute*. In fact, this situation occurs so frequently that a specific notation is introduced.

Definition 2.9   *The **commutator** $[\cdot,\cdot]$ of two operators A and B is given by*

$$[A,B] = A B - B A.$$

As we shall learn in Chapter 3, measurements *change* the state of the system. Consequently, the results of consecutive measurements are not independent of the order in which they are performed. In particular, the *canonical commutation relations* for positions and momenta from eqn (2.35) are of great significance for the structure of the phase space in quantum mechanics.

Theorem 2.2 (Canonical commutation relations)

The components X_m and P_n of the position and momentum operator, respectively, satisfy the canonical commutation relations

$$[X_m, P_n] = i\hbar \delta_{mn}, \quad [X_m, X_n] = [P_m, P_n] = 0.$$

The commutator of Definition 2.9 has some simple algebraic properties. For all linear operators A, B, C, and for all $c \in \mathbb{C}$ the commutator satisfies

$$[A,A] = 0,\tag{2.36a}$$

antisymmetry $\qquad [A,B] = -[B,A]\,,\tag{2.36b}$

bilinearity $\qquad [cA,B] = [A,cB] = c\,[A,B]\,,\tag{2.36c}$

$$[A,B]^{\dagger} = [B^{\dagger},A^{\dagger}] = -[A^{\dagger},B^{\dagger}]\,,\tag{2.36d}$$

$$[A,BC] = B\,[A,C] + [A,B]\,C,\tag{2.36e}$$

Jacobi identity $\quad [A,[B,C]] + [B,[C,A]] + [C,[A,B]] = 0\,.\tag{2.36f}$

Here, a specialist remark on the mathematical structure is in order. The relations of eqns (2.36a), (2.36b), and (2.36f) suggest that the commutator acts as the *Lie* bracket for the *Lie* algebra formed by the linear operators. The corresponding *Lie*-group elements, which are also linear operators, are obtained by exponentiation,

$$e^{A} = \sum_{n=0}^{\infty} \frac{A^n}{n!}\,,\tag{2.37}$$

where $A^0 = \mathbb{1}$ is the identity operator. For exponentials such as eqn (2.37) the fact that linear operators do not generally commute must be taken into account, resulting in so-called *Baker–Campbell–Hausdorff* formulas. Simple examples for such formulas, assuming that $[A,[A,B]] = [B,[A,B]] = 0$, are

$$e^{A}\,e^{B} = e^{B}\,e^{A}\,e^{[A,B]}\,,\tag{2.38a}$$

$$e^{A}\,e^{B} = e^{A+B}\,e^{\frac{1}{2}[A,B]}\,.\tag{2.38b}$$

In cases where the operators A and B do not commute with their commutator $[A,B]$, the expressions contain infinite sums of nested commutators, for instance, in the famous *Hadamard lemma* of the Baker–Campbell–Hausdorff,

$$e^{A}\,B\,e^{-A} = B + [A,B] + \frac{1}{2!}\,[A,[A,B]] + \frac{1}{3!}\,[A,[A,[A,B]]] + \cdots\,.\tag{2.39}$$

Such constructions play a tremendously important role in quantum physics, in particular for the description of *symmetries*. We shall learn more about these concepts in Chapter 6, but we refer the interested reader to the extensive literature available on Lie-group theory, see, e.g., Warner (1983) and Bertlmann (2000).

2.4 Expectation Values and Variances

We are now in a position to combine the concepts we have previously introduced—wave functions, scalar products, and observables—to formulate the rules of quantum mechanics. Although we have claimed that the eigenvalues of self-adjoint operators correspond to individual measurement outcomes, there is generally no way of predicting *deterministically* which of these values is going to be realized in a single run of a measurement. However, based on the insights of Section 1.7.2 (recall Born's probability interpretation, Proposition 1.7.2) we are interested in making a *probabilistic* prediction for the outcomes of many individual measurements. So-called *expectation values*—quantum-mechanical averages—provide exactly such predictive tools.

2.4.1   Expectation Values of Operators

The construction of quantum-mechanical averages can most easily be understood from the analogy with classical averages. Consider a (classical) measurement with n different possible outcomes $x_1, x_2, \ldots, x_n$ that are all equally likely, i.e., each outcome occurs with probability $1/n$. For example, one may think of rolling fair (six-sided) dice and labelling the possible different outcomes by $x_i = i$ for $i \in \{1, 2, \ldots, 6\}$. In any single actual measurement, one of these outcomes will be obtained, but we have no means of predicting which one is realized in any particular measurement. However, we may repeat the measurement a number of times, say N times, obtaining N outcomes $\tilde{x}_1, \tilde{x}_2, \ldots, \tilde{x}_N$ with $\tilde{x}_j \in \{x_i\}_{i=1}^n \, \forall \, j \in \{1, 2, \ldots, N\}$, for which we can calculate an average value $\bar{x}^{(N)} = \frac{1}{N} \sum_{j=1}^N \tilde{x}_j$. For this average value—the *sample mean*—we can make a prediction: Assuming that the measurements are independent and their potential outcomes are identically distributed, the law of large numbers (see Ross, 2009, Chap. 8) implies that $\bar{x}^{(N)}$ converges to the *arithmetic mean* (or "expected value")

$$\langle \, x \, \rangle = \frac{1}{n} \sum_{i=1}^n x_i \,, \tag{2.40}$$

as N tends to infinity, $\lim_{N \to \infty} \bar{x}^{(N)} = \langle \, x \, \rangle$. As we will discuss in more detail in Section 24.1 in Part III, the sample mean is a so-called *unbiased estimator* of the expected value. When the individual measurements are repeated sufficiently many times[5], we can thus predict the mean value. This is also true if the possible values x_i are not equally likely, but are known (or assumed) to occur with probabilities ρ_i. In this case, we use the *weighted arithmetic mean*

$$\langle \, x \, \rangle = \sum_{i=1}^n \rho_i \, x_i \,, \tag{2.41}$$

where the probabilities ρ_i sum to 1. Similarly, the average of a function $f(x)$ can be found by replacing x_i by $f(x_i)$ in eqns (2.40) and (2.41). Now, suppose the set $\{x_i\}$ of different values is not a discrete set, but instead it is continuous. If the function $f(x)$ is (Riemann) integrable the weighted arithmetic mean generalizes to

$$\langle \, f \, \rangle = \int dx \, \rho(x) \, f(x) \,, \tag{2.42}$$

where $\rho(x)$ is an appropriately normalized probability distribution, i.e., $\int dx \rho(x) = 1$. Returning to quantum-mechanical considerations, we note that the probability density

---

[5]Technically, infinitely many times, but the probability for a deviation of $\bar{x}^{(N)}$ from $\langle \, x \, \rangle$ for finite N is exponentially suppressed in N and in the size of the deviation via Hoeffding's inequality, as we discuss in more detail in Section 24.1 in Part III.

$\rho = \psi^*\psi$ (see Definition 2.1) provides exactly such an object. Meanwhile, the function $f(x)$ is replaced by an observable—a self-adjoint operator A—that is inserted in the integral between the wave function and its conjugate. Hence, we write the quantum-mechanical weighted average as

$$\langle A \rangle = \int dx \, \psi^*(x) \, A \, \psi(x). \tag{2.43}$$

The resemblance with the right-hand side of the scalar product in eqn (2.25) is now uncanny, and, in accordance with our notation from eqn (2.27), we define the quantum-mechanical *expectation value*.

Definition 2.10   *The **expectation value** of an observable A in the state ψ is given by*

$$\langle A \rangle_\psi = \langle \psi | \, A \, | \psi \rangle.$$

The expectation value $\langle A \rangle_\psi$ represents the quantum-mechanical average for many repeated measurements of a quantity corresponding to the operator A on an ensemble of systems that have all been prepared in the state ψ. Even though we typically cannot deterministically predict the outcome of any of the individual measurements—only probabilities can be assigned—we may predict their mean value when we have knowledge of the quantum state. The expectation value has the following properties for all linear operators A and B, and for all $c_1, c_2 \in \mathbb{C}$.

$$\text{linearity} \qquad \langle c_1 A + c_2 B \rangle = c_1 \langle A \rangle + c_2 \langle B \rangle, \tag{2.44a}$$

$$\text{normalization} \qquad \langle \mathbb{1} \rangle = 1, \tag{2.44b}$$

$$\text{real if } A = A^\dagger \qquad \langle A \rangle \in \mathbb{R}, \tag{2.44c}$$

$$\text{imaginary if } A = -A^\dagger \qquad i \langle A \rangle \in \mathbb{R}, \tag{2.44d}$$

$$\text{non-negative if } A \geq 0 \qquad \langle A \rangle \geq 0. \tag{2.44e}$$

Let us illustrate the expectation value with some examples.

(i) **Expectation value of the position operator $\langle X \rangle$**

Consider an ensemble of identically prepared quantum-mechanical particles in one spatial dimension that are described by the state $\psi(x)$. The average position on the x axis (relative to the origin) is given by the expectation value of the operator X,

$$\langle X \rangle = \langle \psi | \, X \, | \psi \rangle = \int dx \, x \, |\psi(x)|^2. \tag{2.45}$$

This is a paradigmatic example for a classical average weighted by a probability distribution $|\psi(x)|^2$.

(ii) **Expectation value of the potential energy** $\langle\, V(x)\,\rangle$

Similarly, suppose the particles of the previous example are subject to the external potential $V(x)$. The expectation value for measurements of the potential energy is then given by

$$\langle\, V(x)\,\rangle = \langle\, \psi\,|\, V(x)\,|\,\psi\,\rangle \;=\; \int dx\, V(x)\,|\psi(x)|^2 \,. \tag{2.46}$$

(iii) **Expectation value of the momentum operator** $\langle\, P\,\rangle$

Finally, we wish to predict the average momentum of the particles, and compute the expectation value of the momentum operator $P = -i\hbar\nabla = -i\hbar\partial/\partial x$,

$$\langle\, P\,\rangle = -i\hbar \int dx\, \psi(x)^*\, \nabla\, \psi(x) \;=\; \int dp\, p\,|\tilde{\psi}(p)|^2 \,, \tag{2.47}$$

where $\tilde{\psi}(p)$ is the *Fourier transform* of the wave function in one dimension, which we define as

$$\tilde{\psi}(p) := \frac{1}{\sqrt{2\pi\hbar}} \int dx\, e^{-ipx/\hbar}\psi(x)\,. \tag{2.48}$$

Note that the occurrence of Planck's constant in the factor $1/\sqrt{2\pi\hbar}$ is simply due to the change of variables, from k to $p = \hbar k$ with respect to the usual Fourier transform. We see that the Fourier-transformed wave function $\tilde{\psi}(p)$ takes over the role of the wave function in momentum space, and $|\tilde{\psi}(p)|^2$ represents the probability density for the momenta. Let us quickly examine how the right-hand side of eqn (2.47) is obtained. First, we form the inverse Fourier transform to eqn (2.48), i.e.,

$$\psi(x) = \frac{1}{\sqrt{2\pi\hbar}} \int dp\, e^{ipx/\hbar}\, \tilde{\psi}(p)\,. \tag{2.49}$$

Inserting this expression into the left-hand side of the expectation value in eqn (2.47), one arrives at the expression

$$\langle\, P\,\rangle = \frac{1}{2\pi\hbar} \iiint dp\, dp'\, dx\; p\, e^{i\,(p-p')\,x/\hbar}\, \tilde{\psi}(p')\, \tilde{\psi}(p)\,. \tag{2.50}$$

At this point we recognize the *Dirac delta* distribution

$$\delta(p - p') = \frac{1}{2\pi\hbar} \int dx\, e^{i\,(p-p')\,x/\hbar}\,, \tag{2.51}$$

with the useful property

$$\int_{-\infty}^{\infty} dx'\, \delta(x - x')\, f(x') = f(x)\,. \tag{2.52}$$

With the help of the Dirac delta function we immediately arrive at the conclusion presented in eqn (2.47). However, the relationship between the position and momentum variables, and the role of the Fourier transform in it, will be of further interest to us in Chapter 3.

2.4.2 Uncertainty of Observables

With the formulation of quantum-mechanical expectation values we have gained an invaluable tool that allows us to make probabilistic predictions for the outcomes of a series of measurements on identically prepared quantum systems. But does the expectation value reveal any information about the outcome of an individual measurement?

We are generally not able to predict the outcome of such a single measurement with certainty, but we may yet hope that the average of all outcomes is close to the expectation value. Hence, we are interested in estimating how precise the predictions of the expectation values are with respect to such averages. At the same time, it is useful to be able to say how representative the expectation value is for individual outcomes. For this purpose one introduces the *variance*—the mean square deviation or mean square error—of an operator, to quantify the *uncertainty* about individual measurement outcomes. To understand this quantity, let us again begin with averages of classical ensembles.

Consider again the ensemble $\{x_i\}$ of classical measurement outcomes with probabilities ρ_i, whose weighted arithmetic mean $\langle x \rangle$ is given by eqn (2.41). Each outcome x_i

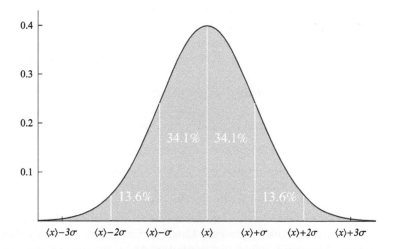

Fig. 2.2 Standard deviation of Gaussian distribution. The Gaussian probability distribution $\exp\left[-(x - \langle x \rangle)^2/(2\sigma^2)\right]/(\sigma\sqrt{2\pi})$ of the random variable x is centred around the expected value $\langle x \rangle$. The probability of finding the random variable within one standard deviation is 68.2%, while already 95.4% and 99.7% of the results lie within two and three standard deviations, respectively.

has a deviation $(x_i - \langle x \rangle)$ from the mean value $\langle x \rangle$. To quantify the average distance from the expected value we form the weighted arithmetic mean of the squared deviations, i.e.,

$$(\sigma(x))^2 = \langle (x_i - \langle x \rangle)^2 \rangle = \sum_{i=1}^{n} \rho_i (x_i - \langle x \rangle)^2. \qquad (2.53)$$

The quantity in eqn (2.53) is called the *variance* of the ensemble $\{(x_i, \rho_i)\}$, and its square root $\sigma(x)$—the *standard deviation*—is used as a figure of merit[6] for the precision with which a series of measurements is estimating the measured quantity. If $\sigma(x)$ is small, the x_i are found close to the expected value with high probability, while a large value of $\sigma(x)$ indicates that the x_i are distributed over a broad range, see Fig. 2.2.

For the quantum-mechanical analogue it is useful to define the deviation in terms of an operator, i.e., to every (self-adjoint) operator A we associate the observable $\overline{A}$, defined as

$$\overline{A} := A - \langle A \rangle. \qquad (2.54)$$

The quantum-mechanical variance is then given by the expectation value of the squared deviation operator

$$\langle \overline{A}^2 \rangle = \langle A^2 \rangle - \langle A \rangle^2. \qquad (2.55)$$

Note that we have suppressed the subscript indicating the particular quantum state in both eqns (2.54) and (2.55), and it is implicitly assumed that all expectation values in these expressions are evaluated with respect to the same quantum state. The right-hand side of eqn (2.55) immediately follows from the linearity of the expectation value, eqn (2.44a), since $\langle A \rangle \in \mathbb{C}$ (or $\mathbb{R}$ if A is Hermitian). As before, one typically considers the square root of the variance because it can be compared directly with the expectation value. We thus define the *uncertainty* of a quantum-mechanical observable.

Definition 2.11   *The **uncertainty** of an observable A in the state ψ is given by*

$$\Delta A = \sqrt{\langle A^2 \rangle_\psi - \langle A \rangle_\psi^2}.$$

Even though there are many similarities with the classical standard deviation, the quantum-mechanical uncertainty has conceptually quite different origins and it is customary to use the symbol Δ as a prefix for the observable, despite the possible confusion with the Laplacian that is featuring, for instance, in the Hamiltonian of eqn (2.5). After all, the two uses of the upper-case Greek letter Δ can be easily distinguished from the context. Moreover, using the symbol $\sigma(A)$ in analogy to the classical case

---

[6]The standard deviation informs us about properties of the distribution of individual measurement outcomes, but if (it is assumed that) the individual measurements are independent and their outcomes identically distributed, the standard deviation of the mean, which can be used to quantify the confidence in (and hence the precision of) the final estimate, is proportional to the standard deviation (of the sample), as we discuss in more detail in Section 24.1 in Part III.

could lead to a misinterpretation of the uncertainty. While the "classical" standard deviation σ is attributed to technical imperfections and systematic errors, issues that could in principle be avoided, the quantum-mechanical uncertainty of a given observable is determined solely by the quantum state that represents *all* information about the system. In other words, the uncertainty of an observable in a particular state cannot be reduced by more accurate measurement devices or better control of the system. It is a *genuine quantum feature*. However, this does not mean that the uncertainty of an observable is always non-zero.

Theorem 2.3 (Vanishing uncertainty)

The uncertainty of an observable A vanishes if, and only if, the state ψ of the system is an eigenstate of the observable A, i.e.,

$$A \,|\, \psi \,\rangle = a \,|\, \psi \,\rangle \quad \Leftrightarrow \quad \Delta A = 0 \,,$$

where a is the corresponding eigenvalue.

Proof Although we have not yet given a mathematical definition for the notation $|\,\psi\,\rangle$ we can nonetheless insert the symbolical expression $A\,|\,\psi\,\rangle = a\,|\,\psi\,\rangle$ into Definition 2.10 to prove the theorem. To this end we evaluate the expectation values of A and A^2 in an eigenstate of A,

$$\langle\, A \,\rangle_\psi = \langle\, \psi\,|\,A\,|\,\psi\,\rangle = \langle\, \psi\,|\,a\,|\,\psi\,\rangle = a\,\langle\,\psi|\psi\,\rangle = a \,, \tag{2.56a}$$

$$\langle\, A^2 \,\rangle_\psi = \langle\, \psi\,|\,A^2\,|\,\psi\,\rangle = \langle\, \psi\,|\,a^2\,|\,\psi\,\rangle = a^2\,\langle\,\psi|\psi\,\rangle = a^2 \,. \tag{2.56b}$$

Inserting into Definition 2.11 proves one direction ($\Rightarrow$) of Theorem 2.3. To show the other direction ($\Leftarrow$), we need to demonstrate that $\langle\, A^2 \,\rangle$ can only be equal to $\langle\, A \,\rangle^2$ for eigenstates of A. First, let us write $A\,|\,\psi\,\rangle = |\,\phi\,\rangle$, which must be a normalizable state in order for $\langle\, A^2 \,\rangle$ to exist. The uncertainty vanishes only if $\langle\,\phi|\phi\,\rangle = |\,\langle\,\psi|\phi\,\rangle\,|^2$. This, in turn, can only occur if $|\,\psi\,\rangle$ and $|\,\phi\,\rangle$ are proportional to each other, $a\,|\,\psi\,\rangle = |\,\phi\,\rangle = A\,|\,\psi\,\rangle$, which concludes the proof. $\square$

As mentioned in Definition 2.4, the eigenvalues of self-adjoint operators correspond to possible individual measurement outcomes. Hence, Theorem 2.3 can be understood in the following way. When the system is in an eigenstate ψ that corresponds to a particular measurement outcome a, then a series of measurements of the observable A will always result in that very same value, with certainty. Of course, this statement holds only provided that there are no technical imperfections that would lead to an additional classical variance. In the following we shall ignore possible technical imperfections and errors introduced by inadequate measurement procedures, and focus on quantum-mechanical uncertainties.

2.5   The Uncertainty Principle

We now revisit the uncertainty principle discussed in Section 1.6 and give it a mathematically precise meaning. As we have learned in Section 2.4.2, the uncertainty of an observable is the quantum-mechanical analogue of the classical standard deviation for the results of repeated measurements on identically prepared systems. However, the uncertainty is tied to the state of the quantum system, rather than the measurement procedure. The quantum state encodes information about possible measurement results, but also about the precision with which future outcomes can be predicted.

Theorem 2.3 reveals that some states provide particularly precise information about certain observables. Naively, we might assume that the problem of quantum-mechanical uncertainties can then be removed by simply preparing the system in a simultaneous eigenstate of all measurable quantities. As it turns out, this is not possible. Quantum mechanics does not allow for all observables to be measured simultaneously with arbitrary precision. Formally, this can be expressed in so-called *uncertainty relations*.

2.5.1   Uncertainty Relation for Observables

Let us take a closer look at the possibility of two observables with simultaneous eigenstates. Assume that the operators A and B do have some common eigenstate $|\psi\rangle$, $A|\psi\rangle = a|\psi\rangle$, and $B|\psi\rangle = b|\psi\rangle$, with $a, b \in \mathbb{R}$ (or $\mathbb{C}$ in case the operators are not Hermitian). This immediately implies that the commutator $[A, B]$ vanishes. Conversely, if $[A, B] = 0$, then A and B have common eigenstates. Consequently, the corresponding uncertainties ΔA and ΔB vanish for such states. But what happens when the operators do not commute? The answer can be phrased in the following theorem, first proven by *Erwin Schrödinger* (in 1930) following the derivation of a slightly weaker inequality (without the covariance term) by *Howard Percy Robertson* (in 1929).

Theorem 2.4 (Robertson–Schrödinger uncertainty relation)

*Let A and B be two Hermitian operators with domains $\mathcal{D}(A)$ and $\mathcal{D}(B)$, respectively, then for all states ψ for which $A\psi \in \mathcal{D}(B)$ and $B\psi \in \mathcal{D}(A)$ the following inequality, the **uncertainty relation**, holds*

$$(\Delta A)^2 (\Delta B)^2 \geq \left| \tfrac{1}{2} \langle \{A, B\} \rangle - \langle A \rangle \langle B \rangle \right|^2 + \left| \tfrac{1}{2i} \langle [A, B] \rangle \right|^2 ,$$

where $[\cdot, \cdot]$ is the commutator from Definition 2.9 and $\{A, B\} = AB + BA$ is called the anti-commutator.

The quantity $\tfrac{1}{2} \langle \{A, B\} \rangle_\psi - \langle A \rangle \langle B \rangle_\psi$ is called the *covariance* of the operators A and B in the state ψ. For operators whose commutator evaluates to a non-zero constant, the covariance can still be non-zero, but it often depends on the particular state and may change with time, as we shall see in an explicit example in the next section. Therefore, the uncertainty principle is often stated in its weaker form $\Delta A \, \Delta B \geq \frac{1}{2} |\langle [A,B] \rangle|$, which is usually referred to as Heisenberg uncertainty relation or Robertson uncertainty relation (1929). Here, we shall prove Theorem 2.4 in full generality.

Proof of Theorem 2.4   Recall that the variances can be written as the expectation values of the squared deviation operators $\overline{A}$ and $\overline{B}$ from eqn (2.54). We can thus rewrite the product of the variances as

$$(\Delta A)^2 \, (\Delta B)^2 = \langle \overline{A}^2 \rangle \, \langle \overline{B}^2 \rangle = \langle \overline{A}\psi | \overline{A}\psi \rangle \, \langle \overline{B}\psi | \overline{B}\psi \rangle \,, \qquad (2.57)$$

where, in the last step, we have used the fact that the operators A and B (and hence also the corresponding deviation operators) are Hermitian. The quantity of eqn (2.57) can now be bounded from below using the *Cauchy–Schwarz* inequality, i.e.,

$$\langle \overline{A}\psi | \overline{A}\psi \rangle \, \langle \overline{B}\psi | \overline{B}\psi \rangle \geq \left| \langle \overline{A}\psi | \overline{B}\psi \rangle \right|^2 \,. \qquad (2.58)$$

Since $\langle \overline{A}\psi | \overline{B}\psi \rangle = \langle \overline{B}\psi | \overline{A}\psi \rangle^*$ is a complex number, we may split it into its real and imaginary part,

$$\text{Re}\big(\langle \overline{A}\psi | \overline{B}\psi \rangle\big) = \frac{1}{2}\big(\langle \overline{A}\psi | \overline{B}\psi \rangle + \langle \overline{B}\psi | \overline{A}\psi \rangle\big) = \frac{1}{2} \langle \{A,B\} \rangle - \langle A \rangle \langle B \rangle \,, \qquad (2.59a)$$

$$\text{Im}\big(\langle \overline{A}\psi | \overline{B}\psi \rangle\big) = \frac{1}{2i}\big(\langle \overline{A}\psi | \overline{B}\psi \rangle - \langle \overline{B}\psi | \overline{A}\psi \rangle\big) = \frac{1}{2i} \langle [A,B] \rangle \,. \qquad (2.59b)$$

Finally, we write

$$\left| \langle \overline{A}\psi | \overline{B}\psi \rangle \right|^2 = \text{Re}^2\big(\langle \overline{A}\psi | \overline{B}\psi \rangle\big) + \text{Im}^2\big(\langle \overline{A}\psi | \overline{B}\psi \rangle\big) \,, \qquad (2.60)$$

and insert (2.59) to complete the proof. $\square$

Note that, because the commutator of two Hermitian operators is anti-Hermitian, see eqn (2.36d), the expectation value of $[A,B]$ is purely imaginary by virtue of eqn (2.44d), such that all terms on the right-hand side of eqn (2.60) are non-negative. The proof further relies on the fact that the operators A and B may subsequently be applied to the same state, that is, that the states $B\psi$ and $A\psi$ lie in the domains of A and B, respectively. This is certainly true for all *bounded* operators[7], for instance, operators that can be represented by matrices of finite rank. But it also applies to some unbounded operators, in particular position and momentum, as we shall discuss in the following section.

2.5.2   Position-Momentum Uncertainty

Let us now consider a situation with one spatial dimension and select the observables $A = X$ and $B = P$, corresponding to the position and linear momentum along

---

[7]More information on this matter will be given in Chapter 3.

that direction, respectively, for insertion into Theorem 2.2. As we shall see in Section 2.6.4, the covariance of position and momentum may vanish, but, as we have learned from Theorem 2.2, their commutator evaluates to a constant, $[X, P] = i\hbar$. The *Robertson–Schrödinger* uncertainty relation thus implies *Heisenberg's* uncertainty relation (recall eqn (1.39))

$$\Delta X \, \Delta P \geq \frac{\hbar}{2}. \tag{2.61}$$

Nonetheless, we should take a look at the fine print of the theorem. The wave functions ψ which lie in the domains of both X and P are square integrable over the chosen interval, for instance $\mathbb{R}$. Further, also $x\psi$ and $\nabla\psi$ need to be square integrable over the same interval. It can then easily be seen that also $\nabla(x\psi)$ is square integrable, such that the proof of Theorem 2.4 holds. This is not generally the case for arbitrary operators, for instance, for operators that correspond to an angle and the derivative with respect to this angle the uncertainty relation fails in this form (see, e.g., Hall, (2013), p. 245). However, as discussed in detail in Chisolm (2001) by recasting the Robertson–Schrödinger uncertainty relation for observables A and B in the form

$$(\Delta A)^2 \, (\Delta B)^2 \geq \mathrm{Re}^2(\langle \overline{A}\psi | \overline{B}\psi \rangle) + \mathrm{Im}^2(\langle \overline{A}\psi | \overline{B}\psi \rangle), \tag{2.62}$$

which follows directly from the proof of Theorem 2.4, one only needs to require that the quantum states one considers lie in the domains of both A and B, and one hence arrives at an inequality that is valid more generally.

The inequality of the uncertainty relation for position and momentum holds for *all* states in the domain of the corresponding operators, that is, for all states for which we can meaningfully speak of these quantities. However, for some states the inequality is saturated—it becomes an equality for Gaussian wave packets, which we shall discuss in the following section.

2.5.3 Uncertainty of Gaussian Wave Packets

Let us return to the notion of a wave packet that we have considered at the beginning of this chapter. Instead of the three-dimensional version of eqn (2.1) we shall keep only one spatial dimension for simplicity. At some fixed instant of time, without loss of generality we may pick $t_o = 0$, we consider a wave packet of Gaussian shape, represented by the wave function

$$\psi(x) = N \, e^{-\left(\frac{x-x_o}{2\sigma}\right)^2} e^{i p_o x / \hbar}. \tag{2.63}$$

The meaning of the real parameters N, σ, x_o, and p_o will become apparent from some simple calculations. For these computations we shall make use of two Gaussian-integral formulas, i.e.,

$$\int\limits_{-\infty}^{\infty} dx \, e^{-x^2} = \sqrt{\pi}, \qquad \int\limits_{-\infty}^{\infty} dx \, x^2 \, e^{-x^2} = \frac{\sqrt{\pi}}{2}. \tag{2.64}$$

We start with the normalization constant N by evaluating the scalar product

$$\langle \psi | \psi \rangle = |N|^2 \int_{-\infty}^{\infty} dx\, e^{-\frac{(x-x_o)^2}{2\sigma^2}} = |N|^2 \sqrt{2}\,\sigma \int_{-\infty}^{\infty} dy\, e^{-y^2} = |N|^2 \sqrt{2\pi}\,\sigma\,, \quad (2.65)$$

where we have performed a change of variables $x \to y = (x - x_o)/(\sqrt{2}\sigma)$ and then used eqn (2.64). We may thus choose the normalization constant to be $N = 1/\sqrt{\sqrt{2\pi}\,\sigma}$. Next, we consider the expectation value of the position operator

$$\langle X \rangle = \frac{1}{\sqrt{2\pi}\,\sigma} \int_{-\infty}^{\infty} dx\, x\, e^{-\frac{(x-x_o)^2}{2\sigma^2}} = \frac{1}{\sqrt{\pi}} \int_{-\infty}^{\infty} dy\, \left(\sqrt{2}\,\sigma\, y + x_o \right) e^{-y^2} = x_o\,, \quad (2.66)$$

where we have used the same change of variables as before. The integral in (2.66) can be evaluated by noting that e^{-y^2} is an even function, while $y e^{-y^2}$ is odd. Thus, changing variables from y to $-y$ reveals that the integral over $y e^{-y^2}$ evaluates to zero,

$$\int_{-\infty}^{\infty} dy\, y\, e^{-y^2} = 0\,. \quad (2.67)$$

One thus arrives at the position expectation value $\langle X \rangle = x_o$. For an ensemble of particles prepared in the state $\psi(x)$, the average position found at the time $t = t_o$ is at $x = x_o$. With the same techniques, in particular, the already familiar change of variables, as well as eqns (2.64) and (2.67) one easily finds the expectation value of X^2, i.e.,

$$\langle X^2 \rangle = \frac{1}{\sqrt{2\pi}\,\sigma} \int_{-\infty}^{\infty} dx\, x^2\, e^{-\frac{(x-x_o)^2}{2\sigma^2}} = \frac{1}{\sqrt{\pi}} \int_{-\infty}^{\infty} dy\, \left(\sqrt{2}\,\sigma\, y + x_o \right)^2 e^{-y^2} = \sigma^2 + x_o^2\,.$$

$$(2.68)$$

We may now construct the position uncertainty ΔX of Definition 2.11 from our previous results for $\langle X \rangle$ and $\langle X^2 \rangle$ and obtain

$$\Delta X = \sqrt{\langle X^2 \rangle - \langle X \rangle^2} = \sigma\,. \quad (2.69)$$

The parameter $\sigma = \Delta X$ is indeed equal to the uncertainty of the position measurement at $t = t_o$. In fact, we have already subsumed the results of this section in Fig. 2.2, which is a graphical representation of the probability density $|\psi(t_o, x)|^2$ corresponding to the wave packet in eqn (2.63). Up to this point the phase factor $\exp(i p_o x / \hbar)$ has not played a role in our computations. However, this will change when we evaluate the expectation value of the momentum operator $P = -i\hbar\nabla$ in the state of eqn (2.63). We find

$$\langle P \rangle = \frac{-i\hbar}{\sqrt{2\pi}\,\sigma} \int_{-\infty}^{\infty} dx\, e^{-\frac{(x-x_o)^2}{2\sigma^2}} \left[\frac{i p_o}{\hbar} - \frac{x - x_o}{2\sigma^2} \right] = \frac{-i\hbar}{\sqrt{\pi}} \int_{-\infty}^{\infty} dy\, e^{-y^2} \left(\frac{i p_o}{\hbar} - \frac{y}{\sqrt{2}\sigma} \right) = p_o\,,$$

$$(2.70)$$

where we have used the same substitution as before, along with eqns (2.67) and (2.64). The parameter p_o simply represents the average momentum of the wave packet(2.63). In a completely analogous computation one obtains the expectation value of P^2, i.e.,

$$\langle P^2 \rangle = \frac{-\hbar^2}{\sqrt{2\pi}\,\sigma} \int\limits_{-\infty}^{\infty} dx\, e^{-\frac{(x-x_o)^2}{2\sigma^2}} \left(\left[\frac{ip_o}{\hbar} - \frac{x-x_o}{2\sigma^2} \right]^2 - \frac{1}{2\sigma^2} \right) = p_o^2 + \frac{\hbar^2}{4\sigma^2}. \qquad (2.71)$$

Finally, we combine the expectation values of eqns (2.70) and (2.71) to express the momentum uncertainty

$$\Delta P = \sqrt{\langle P^2 \rangle - \langle P \rangle^2} = \frac{\hbar}{2\sigma}. \qquad (2.72)$$

At $t = t_o$ the product of the position and momentum uncertainties for the Gaussian wave packet thus equates to

$$\Delta X\, \Delta P = \frac{\hbar}{2}. \qquad (2.73)$$

For Gaussian wave packets the position-momentum uncertainty relation is tight—the right-hand side takes on its minimal value of $\hbar/2$. To fully appreciate the role of the Gaussian wave packet we now want to explore its representation in momentum space.

2.5.4   Wave Packets in Momentum Space

In the previous section, in eqn (2.47), we have already hinted at the role that the *Fourier transform* plays in the conversion between positions and momenta. The nature of this relation will be fully revealed when we discuss infinite-dimensional Hilbert spaces in Chapter 3. We may now nonetheless study the Fourier transform of the Gaussian wave packet(2.63) to examine the momentum dependence of the quantum state. We hence form

$$\tilde{\psi}(p) = \frac{1}{\sqrt{2\pi\hbar}} \int dx\, \psi(x)\, e^{-ipx/\hbar} = \frac{1}{\sqrt{(2\pi)^{3/2}\hbar\sigma}} \int dx\, \exp\left[-\frac{(x-x_o)^2}{4\sigma^2} - \frac{i(p-p_o)x}{\hbar} \right]. \qquad (2.74)$$

To evaluate the integral we first set $x' := [x_o - 2i(p-p_o)\sigma^2/\hbar]$, which completes the square in the exponent, and we extract all exponentials that do not depend on x. The remaining integral is then

$$\int dx\, \exp\left[-\frac{(x-x')^2}{4\sigma^2} \right] = 2\sigma \int dy\, e^{-y^2} = 2\sigma\sqrt{\pi}, \qquad (2.75)$$

where we have substituted $y = (x-x')/(2\sigma)$, before using eqn (2.64). Finally, we reinsert into (2.74) to find the *momentum-space* wave function of the Gaussian wave packet to be given by

$$\tilde{\psi}(p) = \sqrt{\frac{\sqrt{2}\sigma}{\sqrt{\pi}\hbar}}\, e^{-\frac{(p-p_o)^2\sigma^2}{\hbar^2}}\, e^{-i(p-p_o)x_o/\hbar}. \qquad (2.76)$$

As can be seen, a Gaussian wave packet of initial width σ in x space is likewise represented as a Gaussian wave packet in p-space, but this time of width $\hbar/(2\sigma)$, see Fig. 2.3.

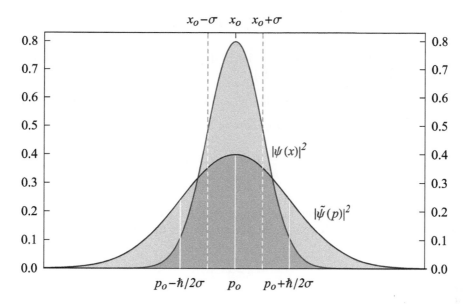

Fig. 2.3 Gaussian wave packet in position and momentum space. A wave packet $\psi(x)$, describing a quantum-mechanical particle localized to width σ, is Fourier transformed to a wave packet $\tilde{\psi}(p)$ with a spread proportional to $1/\sigma$ in momentum space. If the particle is well localized, i.e., σ is small, then the spread in momenta is large, or vice versa.

2.6 Time Evolution in Quantum Mechanics

The previous sections have supplied us with a framework to discuss quantum-mechanical systems, the states that describe them, the observables that represent measurable quantities, as well as the associated expectation values and uncertainties that allow us to relate these concepts to practical measurements and their outcome statistics. With this arsenal at hand, let us now return to the outset of this chapter, the Schrödinger equation, which governs the time evolution of quantum systems.

2.6.1 The Propagator

In all our considerations so far, we have omitted the role of time. For instance, we analysed the Gaussian wave packets of Sections 2.5.3 and 2.5.4 at a fixed instant of time. However, before returning to these specific examples in Section 2.6.4, we want to address this subject in the most general way. We may ask quite generally how the state $\psi(t, x)$ of a quantum system at a time t is obtained, if we are given the state $\psi(t_o, x)$ at some earlier time $t_o < t$. Since we require the probability to be conserved (recall Theorem 2.1) the wave functions at different times must be related by a transformation $U(t, t_o)$ that preserves the scalar product. But how do we construct such a transformation? The answer is provided by the Schrödinger equation (Proposition 2.1.1), i.e., we write

$$i\hbar \frac{\partial}{\partial t} U(t,t_o)\,\psi(t_o,\vec{x}) = H\,U(t,t_o)\,\psi(t_o,\vec{x})\,. \qquad (2.77)$$

When the Hamiltonian H does not explicitly depend on time, we can easily define a suitable time-evolution operator $U(t,t_o)$:

Definition 2.12   *For time-independent Hamiltonians H, the **time-shift operator**,*

*also called **propagator**, is defined by*

$$U(t,t_o) = e^{-\frac{i}{\hbar}H(t-t_o)}\,.$$

It can be immediately seen that the *propagator* satisfies eqn (2.77) and exhibits all the properties that we expect: It leaves the scalar product invariant, i.e.,

$$\langle\, U(t,t_o)\psi(t_o)|\,U(t,t_o)\phi(t_o)\,\rangle = \langle\, U^\dagger(t,t_o)\,U(t,t_o)\psi(t_o)|\,\phi(t_o)\,\rangle = \langle\,\psi(t_o)|\,\phi(t_o)\,\rangle\,, \qquad (2.78)$$

where we have used eqn (2.26) and the fact that the Hamiltonian is Hermitian. In other words, the propagator is an example for a *unitary* operator, which we will discuss in more detail in Chapter 3. The propagator further obeys the composition rule

$$U(t_3,t_2)\,U(t_2,t_1) = U(t_3,t_1)\,, \qquad (2.79)$$

which, in turn, immediately implies that $U(t,t) = \mathbb{1}$ and $U(t_o,t) = U^{-1}(t,t_o)$. The last statement suggests that the time evolution of a (closed) quantum-mechanical system is a reversible process. In general, the Hamiltonian need not be time-independent. In this case, one also obtains a unitary propagator, but it is no longer of the same form as in Definition 2.12, as we will discuss in more detail in Section 10.5.

2.6.2   Schrödinger Picture and Heisenberg Picture

We have noted that quantum-mechanical time evolution leaves the scalar product invariant. But what happens to the expectation values? To find out, let us assume that at some reference time t_o we have evaluated the expectation value of an observable A that does not explicitly depend on time. Using the notation of eqn (2.27) we can then explicitly write down the expression for the expectation value at time t as

$$\langle\, A(t)\,\rangle_\psi = \langle\,\psi(t_o)|\,U^\dagger(t,t_o)\,A\,U(t,t_o)|\,\psi(t_o)\,\rangle\,. \qquad (2.80)$$

Here one notices that the time-dependence can be assigned to either the states or the operators without altering the expectation value in any way. These two choices[8]

---

[8]In the presence of interactions, it becomes useful to define a third, commonly used convention for distributing the time-dependence between states and operators: the *interaction picture*, which we will discuss in Section 10.5.2 of Chapter 10.

are referred to as the *Schrödinger picture* and *Heisenberg picture*, respectively. In the Schrödinger picture, the quantum states depend on time,

$$\psi_{\mathrm{S}}(t) = U(t, t_o)\, \psi_{\mathrm{S}}(t_o)\,, \tag{2.81}$$

and the operators do not change with time, $A_{\mathrm{S}}(t) = A_{\mathrm{S}}(t_o)$. In the Heisenberg picture, on the other hand, the states are time independent, $\psi_{\mathrm{H}}(t) = \psi_{\mathrm{H}}(t_o)$, while the operators time evolve as

$$A_{\mathrm{H}}(t) = U^{\dagger}(t, t_o)\, A_{\mathrm{H}}(t_o)\, U(t, t_o)\,. \tag{2.82}$$

Physically, the two pictures describe the same situation, that is, in either case the expectation values at time t will be of the form of eqn (2.80). The subscripts for the Schrödinger-picture and Heisenberg-picture quantities that we have introduced for pedagogical reasons can be dropped in the following and it is typically sufficiently clear from the context which picture is used.

2.6.3 Time Evolution of Expectation Values

Being aware of the various ways to view time evolution, we may ask how the expectation value of an observable A changes in time when the operator explicitly depends on time. We simply take the derivative with respect to time of eqn (2.80) and, using the propagator from Definition 2.12, we find

$$\frac{d}{dt}\,\langle\, A(t)\,\rangle_{\psi} = \langle\, \psi(t_o)\,|\, U^{\dagger}(t, t_o) \left(\frac{i}{\hbar}\,[\,H\,, A\,] + \frac{\partial A}{\partial t}\right) U(t, t_o)\,|\, \psi(t_o)\,\rangle\,. \tag{2.83}$$

The time derivative of an observable's expectation value is hence determined by the expectation values of two operator-valued quantities, the commutator of the Hamiltonian with the observable, and the partial time derivative of the observable itself, i.e.,

$$\frac{d}{dt}\,\langle\, A\,\rangle = \frac{i}{\hbar}\,\langle\,[\,H\,, A\,]\,\rangle + \langle\, \frac{\partial}{\partial t} A\,\rangle\,. \tag{2.84}$$

Several observations can be made from this important result. First, note that, in general, the time derivative of the expectation value is itself time-dependent. Second, the time derivative is non-vanishing even if the observable does not explicitly depend on time. That is, if $\frac{\partial}{\partial t} A = 0$, the average value $\langle\, A\,\rangle$ may still change with time, and the time derivative then depends on the commutator $[\,H\,, A\,]$,

$$\frac{d}{dt}\,\langle\, A\,\rangle = \frac{i}{\hbar}\,\langle\,[\,H\,, A\,]\,\rangle\,. \tag{2.85}$$

If the commutator of the Hamiltonian H and A vanishes in addition to the partial derivative, the expectation value remains constant in time—the observable is *conserved* in time.

Theorem 2.5 (Conserved quantities)

An observable A that commutes with the Hamiltonian and does not explicitly depend on time corresponds to a conserved quantity,

$$[H, A] = \frac{\partial}{\partial t} A = 0 \quad \Rightarrow \quad \langle A \rangle = \text{const.}$$

For instance, consider a free particle in one spatial dimension with momentum operator P. The Hamiltonian for the free motion is given by

$$H = -\frac{\hbar^2}{2m} \Delta = \frac{P^2}{2m}, \tag{2.86}$$

which commutes with the momentum operator P. Since P does not explicitly depend on time either, the momentum of free particles is a conserved quantity. But what about other properties of free quantum-mechanical particles? For instance, its position, or the corresponding uncertainties. Such quantities will depend on the chosen wave function, a particularly interesting candidate of which is the Gaussian wave packet.

2.6.4   Time Evolution of Free Wave Packets

Intuitively, the first example that comes to mind when thinking of the wave function of a free quantum-mechanical particle is a Gaussian wave packet. As we have seen in Section 2.5.3, such a wave packet can be chosen to be initially localized at some location x_o, with an initial momentum p_o. Both quantities, position and momentum, have corresponding uncertainties, ΔX and ΔP, respectively, such that Heisenberg's uncertainty relation, eqn (2.61), is satisfied. In particular, the state can be chosen such that equality holds initially, $\Delta X \Delta P = \hbar/2$ at $t = 0$. For instance, we may have an initially well-localized particle with a large spread of momenta. Now, let us discover how the free wave packet behaves as time progresses.

We begin with the momentum uncertainty. We have noted that the momentum operator P commutes with the free Hamiltonian of eqn (2.86), and, by Theorem 2.5, the expectation value $\langle P \rangle$ is independent of time. The same holds for the operator P^2. Consequently, the momentum uncertainty in eqn (2.72) is also constant. Trivially, also the expectation value of the Hamiltonian, that is, the average energy, is conserved in time. Combining eqn (2.71) with (2.86) we immediately arrive at

$$\langle H \rangle = \frac{1}{2m} \langle P^2 \rangle = \frac{p_o^2}{2m} + \frac{\hbar^2}{8m\sigma^2} = E_{\text{kin}} + E_{\text{loc}}. \tag{2.87}$$

As can be seen from this expression, the average of the quantum-mechanical free energy of Gaussian wave packets contains two contributions. The first term is the kinetic energy $E_{\text{kin}} = p_o^2/(2m)$ of a classical particle with mass m and momentum p_o. The remaining term, $E_{\text{loc}} = \hbar^2/(8m\sigma^2)$, is a purely quantum-mechanical contribution that represents the energy that is needed to localize the wave packet. The better a particle is to be localized, the larger the required energy. On the other hand, E_{loc} is inversely proportional to the particle's mass, that is, particles of larger mass are easier localized. Consider, for example, an electron with mass $m_{\text{e}} \cong 0.5$ MeV$/c^2$ that is represented by a Gaussian wave packet of width $\sigma = 1$ Å. The *localization energy* is then roughly $E_{\text{loc}} \approx 1$ eV. In Chapter 7 we will see that this the same order of magnitude as the ionization energy of electrons in a hydrogen atom.

We proceed with the time evolution of the position degree of freedom, which will need a little bit more work. The expectation value $\langle X(t) \rangle_\psi$ is to be evaluated in the state given by eqn (2.63). To obtain this, we shall work in the Heisenberg picture (see Section 2.6.2), where the operator is assumed to carry the time-dependence, i.e., $X(t) = U^\dagger(t, t_o) X(t_o) U(t, t_o)$. Here, it is convenient to employ the Hadamard lemma of the Baker–Campbell–Hausdorff formula, eqn (2.39), which allows us to write

$$X(t) = e^{iH(t-t_o)/\hbar} X e^{-iH(t-t_o)/\hbar} = X + \tfrac{it}{\hbar}[H, X] + \tfrac{1}{2}\left(\tfrac{it}{\hbar}\right)^2 [H, [H, X]] + \dots . \tag{2.88}$$

The commutator of the Hamiltonian and the position operator evaluates to

$$[H, X] = \frac{1}{2m}[P^2, X] = -\frac{i\hbar}{m}P, \tag{2.89}$$

where we have used the canonical commutation relations (Theorem 2.2). All higher order commutators then vanish since $[P^2, P] = 0$, and, inserting the initial expectation values (2.66) and (2.70) we find

$$\langle X(t) \rangle = \langle X(0) \rangle + \frac{t}{m}\langle P(0) \rangle = x_o + \frac{p_o}{m}t, \tag{2.90}$$

that is, the centre of the wave packet moves away from its original position x_o with constant velocity $v_o = p_o/m$. Next, we study the shape of the wave packet by determining the position uncertainty at time t. To this end we make use of eqn (2.90) to compute

$$\langle X^2(t) \rangle = \left\langle \left(X(0) + \frac{P(0)t}{m}\right)^2 \right\rangle = \langle X^2(0) \rangle + \frac{t}{m}\langle 2X(0)P(0) - i\hbar \rangle + \frac{t^2}{m^2}\langle P^2(0) \rangle . \tag{2.91}$$

The expectation values $\langle X^2(0) \rangle$ and $\langle P^2(0) \rangle$ have already been evaluated in eqns (2.68) and (2.71), respectively, and a similar computation reveals

$$\langle 2X(0)P(0) - i\hbar \rangle = x_o p_o . \tag{2.92}$$

Assembling these individual results we get

$$\langle X^2(t) \rangle = \sigma^2 + \left(x_o + \frac{p_o t}{m}\right)^2 + \left(\frac{\hbar t}{2m\sigma}\right)^2 . \tag{2.93}$$

The time-evolved position uncertainty can then be written as

$$\Delta X(t) = \sigma \sqrt{1 + \left(\frac{\hbar t}{2m\sigma^2}\right)^2} = \sigma \sqrt{1 + \left(\frac{4E_{\text{loc}}t}{\hbar}\right)^2} . \tag{2.94}$$

We find that, although the momentum uncertainty and the energy of the wave packet remain constant, the position uncertainty increases with time. The wave packet loses its initial localization—it *spreads*—and the localization energy E_{loc} governs how rapidly this happens. An illustration of a spreading wave packet is shown in Fig. 2.4. It can be seen from the ratio $\Delta X(t)/\Delta X(0)$, that the wave packet increases in width quicker, the better it is localized initially.

Explicitly writing the product of the time-evolved variance of the position observable from eqn (2.94) with the time-independent value $(\Delta P)^2$ from eqn (2.72) we find the time-evolved uncertainty relation

$$\big(\Delta X(t)\big)^2 (\Delta P)^2 = \big(2\,E_{\text{loc}}\,t\big)^2 + \Big(\frac{\hbar}{2}\Big)^2 . \tag{2.95}$$

A quick comparison with the Robertson-Schrödinger uncertainty relation of Theorem 2.4 allows us to assign the time-dependence of the uncertainty relation to the covariance

$$\frac{1}{2}\big\langle\,\{\,X(t)\,,P(t)\,\}\,\big\rangle \,-\, \big\langle\,X(t)\,\big\rangle\big\langle\,P(t)\,\big\rangle = 2\,E_{\text{loc}}\,t , \tag{2.96}$$

while the time-independent term $(\hbar/2)^2$ in eqn (2.95) appears due to the canonical commutator $[X,P] = i\hbar$ (see Theorem 2.2). That the covariance evaluates exactly to the result in eqn (2.96)—recall that the right-hand side of the uncertainty relation only provides a lower bound for the product of the variances—may be seen directly by combining eqns (2.92) and (2.71), since we know that the Heisenberg-picture position

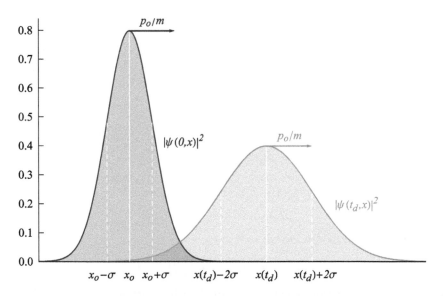

Fig. 2.4 Spreading of the free wave packet. A wave packet $\psi(x,0)$ (blue) with initial position x_o, and initial width σ, time evolves under the free Hamiltonian of eqn (2.86). After the time $t_d = 6m\sigma^2/\hbar$ the width of the wave packet has doubled, $\Delta X(t_d) = 2\sigma$. The peak of the wave packet propagates at fixed velocity p_o/m, which is here chosen to be $\hbar/(m\sigma)$. Thus, at time t_d the wave packet $\psi(x,t_d)$ (green) is centred around $x(t_d) = x_o + 6\sigma$.

operator is $X(t) = X(0) + P(0)t/m$ for the free time evolution, see (2.90). Consequently, the covariance of X and P vanishes initially, and increases as time progresses. Independently of the time evolution, the Robertson–Schrödinger uncertainty relation remains tight for the Gaussian wave packets, that is, the product of their position and momentum uncertainty is minimal at all times.

To better understand what brings about the wave packet's increase in width we again consider the initial state $\psi(0, x)$ from eqn (2.63). Now, recall from eqn (2.49) that the wave function can be written as the inverse Fourier transform of the momentum-space wave function $\tilde{\psi}(p)$. The Gaussian wave packet in eqn (2.63) can thus be understood as a superposition of plane waves $e^{-ikx} = e^{-ipx/\hbar}$ at time $t = 0$. Each of these plane waves has a fixed momentum, and a fixed energy $E = \hbar\omega$ with respect to the Hamiltonian of eqn (2.86), and, as time passes, they acquire phases $e^{-i\omega(k)t} = e^{-iE(p)t/\hbar}$, where $E = \hbar\omega$ and $p = \hbar k$, such that

$$\psi(t, x) = \frac{1}{\sqrt{2\pi\hbar}} \int dp\, e^{\frac{i}{\hbar}(px - E(p)t)}\, \tilde{\psi}(p) = \sqrt{\frac{\hbar}{2\pi}} \int dk\, e^{i(kx - \omega(k)t)}\, \tilde{\psi}(p(k)). \quad (2.97)$$

Plane waves propagate at the *phase velocity* $v_{\text{phase}} = \omega/k$, and inserting $\omega(k)$ from above immediately allows us to write

$$v_{\text{phase}} = \frac{\omega(k)}{k} = \frac{E(p)}{p} = \frac{p}{2m}. \quad (2.98)$$

Two important observations can be made from this expression. First, we note that the phase velocity is different for plane waves of different frequencies. This effect is called *dispersion* and it is the reason for the spreading of the wave packet, whose constituent plane waves are travelling at different speeds, changing the shape of the wave packet in the process. This is a typical feature found in quantum-mechanical particles with non-zero mass. On the other hand, for massless particles such as photons in the vacuum, the plane waves of different frequencies travel at the same speed, and dispersion occurs only when the light enters a medium.

The second observation we can make in eqn (2.98) is that the plane wave whose momentum matches the momentum p_o of the centre of the wave packet travels only at half of the expected speed. The phase velocity is hence not very useful for describing the propagation of wave packets. Instead, one uses the so-called *group velocity*. A wave packet $\psi(t, x)$ that represents a massive, quantum-mechanical particle, and whose Fourier transform $\tilde{\psi}(p(k))$ is centred around $p_o = \hbar k_o$, is moving at the velocity

$$v_{\text{group}} = \left.\frac{d\omega}{dk}\right|_{k=k_o} = \frac{p_o}{m} = 2v_{\text{phase}}. \quad (2.99)$$

This definition agrees with our previous calculations, and it also accounts for wave packets describing photons travelling in the vacuum, for which one finds

$$v_{\text{group}} = v_{\text{phase}} = c. \quad (2.100)$$

2.6.5   Energy-Time Uncertainty

As we have seen, for instance, in the exponents of the plane waves $e^{i(px-E(p)t)/\hbar}$, the duality that we have encountered between position and momentum is mirrored by the quantities energy and time. This duality can be expressed in terms of an uncertainty relation analogous to the position-momentum uncertainty of eqn (2.61). However, in deriving such an *energy-time uncertainty* relation we cannot immediately use the general Robertson–Schrödinger uncertainty relation of Theorem 2.4.

A slight diversion is required to find a suitable representative for the "time uncertainty" Δt, since time only appears as a parameter in quantum mechanics, but not as an observable. Instead, we measure the passage of time by the changes of other quantities, which themselves are observables. These, in turn, have associated uncertainties. Therefore, instead of assigning an uncertainty to time, one can ask how much time is needed to bring about a significant change in such an observable. Since the outcomes of successive, individual measurements of an observable do not typically yield the same results even without time evolution, we have to consider changes in the expectation values. Hence

Definition 2.13   *To every observable A with uncertainty ΔA, for which $\frac{d}{dt}\langle A \rangle \neq 0$, we associate a time Δt, during which the expectation value $\langle A \rangle$ changes by ΔA, i.e.,*

$$\Delta A = \left| \frac{d}{dt}\langle A \rangle \right| \Delta t .$$

The characteristic time Δt depends on both the observable that is chosen, and the state of the quantum system that serves as the "clock". With this definition in mind we return to the Robertson–Schrödinger uncertainty relation (Theorem 2.4). Taking one of the observables to be the Hamiltonian H, the operator associated with energy, and ignoring the non-negative covariance term of the right-hand side, we are left with the commutator $[A, H]$. Now, recall from eqn (2.85) that for observables which are not explicitly time-dependent, the commutator with the Hamiltonian determines the time evolution of the expectation value, such that

$$\Delta A \, \Delta H \geq \frac{1}{2} |\langle [A, H] \rangle| = \frac{\hbar}{2} \left| \frac{d}{dt}\langle A \rangle \right| \tag{2.101}$$

holds. Subsequently, we denote ΔH as ΔE, the *energy uncertainty* and insert Definition 2.13 to obtain the *energy-time uncertainty* relation

$$\Delta E \, \Delta t \geq \frac{\hbar}{2} . \tag{2.102}$$

Thus, whenever the energy of a quantum system is determined very precisely, the average values of all other observables change very slowly compared to their respective uncertainties. Conversely, if a system remains stationary, that is, the expectation values of all observables remain constant, then the energy of the system has a "sharp" value. Later on, in Chapter 4, we shall investigate this statement from the perspective of eigenvalue equations, but for now it will be useful to illustrate the energy-time uncertainty with a familiar example—the Gaussian wave packet.

For a free wave packet in one spatial dimension the position is a natural candidate to keep track of the passage of time. From eqns (2.90) and (2.94) we straightforwardly arrive at

$$\Delta t_x = \frac{\Delta X(t)}{\left|\frac{d}{dt}\langle X(t)\rangle\right|} = \frac{m\sigma}{|p_0|}\sqrt{1 + \left(\frac{4E_{\text{loc}}t}{\hbar}\right)^2}, \tag{2.103}$$

where we have assumed that $m > 0$. To determine the uncertainty of the free Hamiltonian in eqn (2.86) we still need to calculate $\langle P^4 \rangle$. This is best done by working in momentum space, that is, we compute

$$\langle P^4 \rangle = \int dp\, p^4\, |\tilde{\psi}(p)|^2 \tag{2.104}$$

as in eqn (2.47), with the momentum-space wave function $\tilde{\psi}(p)$ of eqn (2.76). Note that $\tilde{\psi}(p) = \tilde{\psi}(t, p)$ depends on time via a phase factor that cancels in eqn (2.104), such that $\langle P^4 \rangle$ is time-independent, as expected. A straightforward calculation, which we will leave as an exercise for the interested reader, then reveals

$$\Delta E = \frac{\hbar}{2} \frac{|p_0|}{m\sigma} \sqrt{1 + \frac{1}{2}\frac{E_{\text{loc}}}{E_{\text{kin}}}}, \tag{2.105}$$

where we have substituted the localization energy E_{loc} and the classical kinetic energy E_{kin} from eqn (2.87). Varying σ, whilst keeping all other parameters fixed, one can observe a trade-off between the quantities Δt and ΔE. As σ increases, the energy uncertainty becomes smaller, while the time that one needs to identify a significant change in the average position becomes larger. Combining Δt and ΔE one finally arrives at

$$\Delta E\, \Delta t = \frac{\hbar}{2} \sqrt{1 + \left(\frac{4E_{\text{loc}}t}{\hbar}\right)^2} \sqrt{1 + \frac{1}{2}\frac{E_{\text{loc}}}{E_{\text{kin}}}} \geq \frac{\hbar}{2}, \tag{2.106}$$

where it is apparent that the minimum is attained only in the limit where the particle is completely delocalized.

2.7 Recovering Classical Physics

The previous sections have provided us with the basic tools of Schrödinger's wave mechanics. In particular, we have seen how free particles can be described quantum-mechanically by wave functions that take the shape of Gaussian wave packets. Their time evolution is governed by the Hamiltonian via the Schrödinger equation and measurable quantities are represented by self-adjoint operators. Since quantum-mechanical systems can behave quite differently than classical systems, one may wonder how classical mechanics fits into this framework. It is the goal of this section to establish exactly this relationship between quantum physics and classical physics.

2.7.1 The Ehrenfest Theorem

A very intuitive approach to connect quantum mechanics and classical Newtonian mechanics is the inspection of the expectation values of familiar quantities and the comparison of the corresponding predictions of quantum theory with those of Newton's laws of motion. *Ehrenfest's theorem*, which we will derive in the following, delivers this desired comparison and is named after *Paul Ehrenfest*, who first formulated it (1927). To begin, let us consider a quantum-mechanical particle of mass $m > 0$ that is subject to an external potential $V(\vec{x})$, such that its motion is governed by the Hamiltonian

$$H = -\frac{\hbar^2}{2m}\Delta + V(\vec{x}) = \frac{\vec{P}^2}{2m} + V(\vec{x}). \tag{2.107}$$

The time evolution of the position operator $\vec{X}$ and the momentum operator $\vec{P}$, with component operators X_n and P_n, respectively, neither of which depends explicitly on time, is determined by their commutators with the Hamiltonian, see eqn (2.85). For the commutators $[H, X_n]$ we note that the X_n commute with the potential operator, and we can use the canonical commutation relations from Theorem 2.2 to obtain

$$[H, X_n] = \frac{1}{2m}[\vec{P}^2, X_n] = -\frac{i\hbar}{m}P_n. \tag{2.108}$$

Employing eqn (2.85) we hence recover in a more general setting what we have previously discovered for free wave packets, i.e.,

$$\frac{d}{dt}\langle \vec{X} \rangle = \frac{1}{m}\langle \vec{P} \rangle. \tag{2.109}$$

For the time evolution of the momentum operators $P_n = -i\hbar\nabla_n$ we use the product rule $\nabla V = (\nabla V) + V\nabla$, and find the commutators

$$[H, P_n] = -i\hbar[V(\vec{x}), \nabla_n] = i\hbar(\nabla_n V(\vec{x})), \tag{2.110}$$

which, subsequently, allow us to write

$$\frac{d}{dt}\langle \vec{P} \rangle = -\langle \vec{\nabla}V \rangle. \tag{2.111}$$

Identifying $-\vec{\nabla}V$ with the operator for a conservative force $\vec{F}$ we gather our results from eqns (2.109) and (2.111) to formulate Ehrenfest's theorem.

Theorem 2.6 (Ehrenfest theorem)

For quantum-mechanical particles, subject to a potential V that commutes with the momentum $\vec{P}$, the expectation values obey the classical Newtonian equations of motion, i.e.,

$$m\frac{d^2}{dt^2}\langle \vec{X} \rangle = \langle \vec{F} \rangle.$$

As expected, Planck's constant $\hbar$ does not appear in Theorem 2.6. In this sense we do recover a classical law of motion. However, keep in mind that the theorem applies to the average force $\langle\, F(\vec{x})\,\rangle$, but not necessarily to the force at the average position $F(\langle\,\vec{X}\,\rangle)$. These expressions only match if the second and all higher derivatives of the force vanish, i.e., $\nabla^n \vec{F} = 0$, $(n = 2, 3, \ldots)$. Such a situation occurs when the force is changing slowly as the position is varied, which is the case, e.g., for the free motion of eqn (2.86), where no force is present, or for the harmonic oscillator that we will encounter in Chapter 5. But it is not true in general, when the force may change more sharply with changing position.

2.7.2   Dirac's Rule

Ehrenfest's theorem establishes a connection between Newtonian and quantum mechanics. But a more general statement can be made by retracing our steps in the derivation of Theorem 2.6. The commutators that we have evaluated in eqns (2.109) and (2.111) were already specifically chosen to obtain the time evolution of the expectation values of the position and momentum operators, respectively. When looking for a more general expression that can be compared with classical physics we have to turn to eqn (2.84), and, indeed, a very similar expression exists in classical mechanics. There, observables on the phase space of N particles are represented by functions $f(t, q_m, p_n)$ that depend on time t, and the canonical coordinates q_m and p_n, with $(m, n = 1, \ldots, N)$. For these classical observables the time evolution is given by

$$\frac{d}{dt} f(t, q_m, p_n) = \{\, f\, , H\, \}_{\text{Poisson}} + \frac{\partial}{\partial t} f(t, q_m, p_n)\,, \qquad (2.112)$$

where H denotes Hamilton's principal function and $\{\, .\, ,\, .\, \}_{\text{Poisson}}$ is the *Poisson-bracket* (not to be confused with the anti-commutator appearing in Theorem 2.4). For two phase-space functions f and g the Poisson bracket is defined as

$$\{\, f\, , g\, \}_{\text{Poisson}} = \sum_{n=1}^{N} \left[\frac{\partial f}{\partial q_n} \frac{\partial g}{\partial p_n} - \frac{\partial f}{\partial p_n} \frac{\partial g}{\partial q_n} \right]. \qquad (2.113)$$

Inspection of eqn (2.112), and comparison with eqn (2.84) then reveals that the transition from classical mechanics to quantum mechanics requires the replacement of the Poisson bracket with the commutator and a factor of $-i/\hbar$,

$$\{\, .\, ,\, .\, \}_{\text{Poisson}} \qquad \rightarrow \qquad -\frac{i}{\hbar}\, [\, .\, ,\, .\,]\,, \qquad (2.114)$$

in addition to the quantization of the observables (see Section 2.1.2). The substitution in (2.114), which is called *Dirac's rule*, is quite helpful in outlining the parallels and differences between classical and quantum physics. It highlights the roles of Planck's constant, complex numbers, and the nature of observables as generally non-commuting operators in quantum mechanics. On the other hand, a certain familiarity of concepts from classical mechanics is preserved. Examples for the application of Dirac's rule that we have already encountered are the canonical commutation relations (Theorem 2.2), and we shall also discover the quantum analogue of the classical *Liouville equation* in Chapter 11 of Part II.

3

Mathematical Formalism of Quantum Mechanics

Let us now return to the mathematical foundation of our discussion of the previous chapter, where we have inspected Schrödinger's wave mechanics. It was already hinted that wave functions can be understood as part of a more general framework—*Hilbert spaces*—in terms of which all quantum physics can be described. The present chapter's aim is to introduce the basic concepts that form the mathematical backbone of quantum theory.

3.1 Hilbert Space

As we have established in Definitions 2.3 and 2.4, all available information about quantum-mechanical systems is encoded in quantum states, which are represented as vectors in so-called Hilbert spaces. Similarly, all measurable quantities correspond to self-adjoint operators that act on these spaces. To give mathematical meaning to these statements, we need to define the Hilbert space.

Definition 3.1 *A **Hilbert space** $\mathcal{H}$ is a complete inner-product space.*

The vector-space structure is required by the superposition principle, that is, up to normalization, any linear combination of state vectors is again a state vector. That the Hilbert space is an *inner-product space*, simply means that the vector space is equipped with an inner product $\langle\,.\,|\,.\,\rangle \in \mathbb{C}$, which satisfies (2.22). The elements of the Hilbert space—the vectors—represent the quantum states.

3.1.1 Norm and Completeness

The inner product $\langle\,\phi\,|\,\psi\,\rangle$ of two vectors $\phi, \psi \in \mathcal{H}$, encodes the probability amplitude to find the system in the state ϕ if it was prepared in the state ψ, and hence provides a notion of distance between the states. More precisely, the inner product induces a *norm* $\|.\|$, given by

$$\|\psi\| = \sqrt{\langle\,\psi\,|\,\psi\,\rangle}\,. \tag{3.1}$$

The function $d(\phi, \psi) = \|\phi - \psi\|$ then defines a distance between the states ϕ, and ψ, i.e., it is symmetric, $d(\phi, \psi) = d(\psi, \phi)$, it vanishes only if $\phi = \psi$, and it satisfies the triangle inequality $d(\phi, \psi) + d(\psi, \chi) \geq d(\phi, \chi)$ for all ϕ, ψ, and χ in the Hilbert space. The property of *completeness*, which we require for every Hilbert space, asserts that a series of vectors ψ_n that is absolutely convergent, i.e., for which the sum of the norms of the ψ_n converges,

$$\sum_{n=1}^{\infty} \|\psi_n\| < \infty, \qquad (3.2)$$

also converges to an element of the Hilbert space, i.e., there exists a $\psi \in \mathcal{H}$, such that

$$\lim_{m \to \infty} \sum_{n=1}^{m} \psi_n = \psi. \qquad (3.3)$$

The sequence $\{C_m\}_{m=1}^{\infty}$ of partial sums $C_m = \sum_{n=1}^{m} \psi_n$ of any such absolutely convergent series is called a *Cauchy sequence*. For any real number $\varepsilon > 0$ there exists an $N \in \mathbb{N}$, such that for all natural numbers $m, n > N$ the distance $d(C_m, C_n) < \varepsilon$. Completeness may hence be equivalently defined by demanding that every Cauchy sequence in a complete normed space converges to a limit in that space. Loosely speaking, completeness ensures that Hilbert spaces do not have any "holes".

However, one needs to be careful regarding the interpretation of the distance measure $d(\phi, \psi)$ as a measure of the similarity of two states in a physical sense. We have previously stated that physical states are represented by vectors in a Hilbert space (see Definition 2.3). To be precise, we should rephrase this in the following way.

Definition 3.2   *The **state** of a quantum-mechanical system is represented by a ray of vectors $\{e^{i\theta}\psi, \theta \in \mathbb{R}\}$ in a Hilbert space $\mathcal{H}$.*

That is to say, the overall (global) phase θ of a quantum state does not matter. The states $\psi' = e^{i\theta}\psi$ and ψ are physically equivalent and cannot be distinguished. Only relative phases in a superposition are of physical significance, e.g., for two orthogonal vectors $\psi, \phi \in \mathcal{H}$, with $\langle \psi | \phi \rangle = 0$, the phase $\varphi \in \mathbb{R}$ in the (unnormalized) state $\psi + e^{i\varphi}\phi$ can be determined. The measure $d(.,.)$ does not appropriately capture this distinction. In our example this means that $d(\psi, \psi') = d(\psi, \phi) = \sqrt{2}$, even though $|\langle \psi | \psi' \rangle| = 1$, while $\langle \psi | \phi \rangle = 0$. The invariance of the laws of quantum mechanics under a multiplication by a global phase factor $e^{i\theta}$ is a simple example for a *gauge symmetry*, similar to the invariance of the laws of nature under Lorentz transformations in the theory of special relativity. We shall explore the deeper role of gauge symmetries in quantum mechanics in Section 9.2.

3.1.2   Dimensionality of Hilbert Spaces

With the previous subtleties in mind, the definition of a Hilbert space provides an intuitive arena for our further analysis. In analogy to a classical phase space, the elements of the vector space, the vectors, represent the possible physical states. However, Hilbert spaces can be *finite- or infinite-dimensional* (even uncountably infinite-dimensional), while classical phase spaces have finite dimension for finite numbers of particles. For instance, the classical phase space for n particles in three spatial dimensions is $(6n)$-dimensional.

Typical Hilbert spaces that we will encounter here will be *separable spaces*[1], that is, Hilbert spaces with finite, or, at most, countably infinite dimensions. This means that they admit countable bases. With respect to such a basis $\{\psi_n\}$ all elements $\phi \in \mathcal{H}$ have a unique decomposition $\phi = \sum_n c_n \psi_n$, and ϕ may therefore be formally written as a column vector with (possibly infinitely many) components c_n,

$$\psi = \begin{pmatrix} c_1 \\ c_2 \\ \vdots \end{pmatrix}. \tag{3.4}$$

Conceptually, the only difference to vectors in a classical phase space is that the components may now be complex, i.e., $c_n \in \mathbb{C}$. If the chosen basis is an orthonormal basis, that is, if $\langle \psi_m | \psi_n \rangle = \delta_{mn}$ holds, the inner product $\langle \phi | \psi \rangle$ of two vectors $\psi = \sum_n c_n \psi_n$ and $\phi = \sum_n d_n \psi_n$ assumes the usual form

$$\langle \phi | \psi \rangle = \phi^\dagger \cdot \psi = (d_1^*, d_2^*, \dots) \begin{pmatrix} c_1 \\ c_2 \\ \vdots \end{pmatrix} = \sum_n d_n^* c_n, \tag{3.5}$$

apart from the additional complex conjugation of the components of the vector in the first slot of the scalar product, the "bra" (see Section 3.1.3). The normalization condition (2.8) now appears as a restriction on the complex components, i.e.,

$$\|\psi\| = \langle \psi | \psi \rangle = \sum_n c_n^* c_n = \sum_n |c_n|^2 = 1. \tag{3.6}$$

Despite the initial simplicity of describing the vectors in the Hilbert space independently of the Hilbert-space dimension, the dimensionality of the Hilbert space plays a role for the mathematical representation of the physical quantities that we want to measure, the observables, which are now operators acting on the vectors, rather than functions on phase space.

In a separable space we may also think of linear operators on the Hilbert space (recall Definition 2.6) simply as matrices with possible infinitely many entries, that map vectors in $\mathcal{H}$ unto vectors in $\mathcal{H}$. When the dimension of the Hilbert space is finite, the treatment is straightforward and linear algebra provides all necessary tools to formulate quantum mechanics, but matters are somewhat more complicated for infinite dimensions. For instance, not all linear operators A on infinite-dimensional,

---

[1] Not to be confused with the completely unrelated notion of *separable states*, which we will encounter in Part II.

normed vector spaces are *bounded*, that is, there may not exist a real number m, such that $\|A\psi\| < m\|\psi\|$ for all $\psi \in \mathcal{H}$. Such unbounded operators complicate the formulation of the rules of quantum mechanics in infinite dimensions.

To keep the discussion accessible, whilst conveying the conceptual ideas of quantum mechanics, we will restrict the discussion to finite-dimensional Hilbert spaces until the end of Section 3.2. Once we have established the relevant concepts and notation, we will revisit those facets of infinite-dimensional quantum systems that need a more subtle treatment in Section 3.3.

3.1.3   The Dual Hilbert Space and Dirac Notation

Before we discuss operators on Hilbert spaces and their properties in more detail, it is practical to consider the notation introduced in 1930 by *Paul Adrien Maurice Dirac* in his famous book *The Principles of Quantum Mechanics*. This so-called *Dirac notation*, also referred to as "*bra-ket*" notation, has proven to be very useful, and easy to handle. Consequently, it became the standard notation used throughout all areas of quantum physics. To fully appreciate the simplicity of this notation it is necessary to take a closer look at *dual Hilbert spaces*.

Let us consider a finite-dimensional Hilbert space $\mathcal{H}$ over the field of complex numbers $\mathbb{C}$. For any pair of vectors $\phi, \psi \in \mathcal{H}$, the inner product $\langle \phi | \psi \rangle$ provides a non-degenerate sesquilinear form, i.e., a map $\langle . | . \rangle : \mathcal{H} \times \mathcal{H} \to \mathbb{C}$ that is anti-linear in the first argument, and linear in the second, see eqns (2.22). Now, note that instead we could consider a fixed ϕ and view the inner product as a linear map $\phi(.) := \langle \phi | . \rangle : \mathcal{H} \to \mathbb{C}$—a *functional*—that takes any $\psi \in \mathcal{H}$ to a complex number $\phi(\psi)$. All of these maps $\phi(.)$ form a vector space—the *dual space* $\mathcal{H}^*$—which is a Hilbert space as well.

Definition 3.3   *The **dual Hilbert space** $\mathcal{H}^*$ is the space of all linear functionals over the Hilbert space $\mathcal{H}$.*

The elements of the dual space are called *covectors* and they have a one-to-one correspondence to vectors in the original Hilbert space. In other words, the spaces $\mathcal{H}$ and $\mathcal{H}^*$ are isometrically isomorphic, a fact known as the *Riesz–Fréchet theorem* (Fréchet, 1907; Riesz, 1907). The isomorphism between the vectors and covectors via the bracket of the inner product motivates the following notation. The elements ϕ of the dual Hilbert space $\mathcal{H}^*$ are denoted as $\langle \phi |$ ("bra"), while the elements ψ of the Hilbert space $\mathcal{H}$ are denoted as $| \psi \rangle$ ("ket"). Applying a covector $\langle \phi |$ to a vector $| \psi \rangle$ from the left hence naturally forms the inner product $\langle \phi | \psi \rangle$.

In terms of the matrix notation of eqn (3.4) a vector $| \chi \rangle$ may be understood as a column of components d_i with respect to a chosen basis $| \psi_i \rangle$. An immediate advantage of the Dirac notation is that the symbol $| . \rangle$ already provides the information that we are dealing with a vector in the Hilbert space, and we can hence simplify the notation of for the basis vectors from $| \psi_i \rangle$ to $| i \rangle$, such that

$$|\psi\rangle = \sum_i c_i \,|i\rangle = \begin{pmatrix} c_1 \\ c_2 \\ \vdots \end{pmatrix}. \tag{3.7}$$

Similarly, we can consider the dual basis $\{\langle j\,|\}$ in $\mathcal{H}^*$, where $\langle j\,|\,i\rangle = \delta_{ij}$, and represent the covector $\langle\phi\,|$ as a row of components g_j with respect to it, i.e.,

$$\langle\phi\,| = \sum_j g_j \,\langle j\,| = (g_1, \, g_2, \, \dots). \tag{3.8}$$

The isomorphism between $\mathcal{H}$ and $\mathcal{H}^*$ is then simply realized by Hermitian conjugation (transposition and complex conjugation), $|\phi\rangle = \langle\phi\,|^\dagger$ (compare to eqn (3.5) with $g_j = d_j^*$), and of course $\langle\phi\,| = |\phi\rangle^\dagger$. The inner product can then be viewed as standard matrix multiplication.

3.1.4   The Hilbert Space for Photon Polarization

Let us apply the new notation to an example, the polarization of a photon. Physically, we may first think of a beam of light that has been collimated by a lens. Along the beam, the electric and magnetic fields oscillate in directions perpendicular to the direction of propagation, and to each other. We may hence concentrate on one of the fields, for instance, the electric field, and describe its oscillation in the plane perpendicular to the direction of propagation. If the light is linearly polarized, the electric field oscillates along a fixed direction in this plane, for instance, horizontally or vertically (with respect to the laboratory). The Hilbert space of the polarization degree of freedom is two-dimensional, the Cartesian product of two complex planes, i.e., $\mathcal{H} = \mathbb{C} \times \mathbb{C} = \mathbb{C}^2$. The states for horizontal and vertical polarization, $|H\rangle$ and $|V\rangle$, respectively, form an orthonormal basis of $\mathbb{C}^2$, i.e.,

$$\langle H|H\rangle = \langle V|V\rangle = 1, \tag{3.9a}$$

$$\langle H|V\rangle = 0. \tag{3.9b}$$

Physically, this expresses the fact that a beam of photons prepared in the state $|H\rangle$, for instance, by sending an unpolarized beam through a polarization filter oriented horizontally with respect to the laboratory, will pass another horizontal filter undiminished. If the second filter is rotated ninety degrees to vertical orientation, the light will no longer pass through, and the screen behind the second filter will remain dark. An arbitrary polarization state $|\psi\rangle$ may be written as a two-component vector with respect to the two basis states $|H\rangle$ and $|V\rangle$, i.e.,

$$|\psi\rangle = \alpha\,|H\rangle + \beta\,|V\rangle = \begin{pmatrix} \alpha \\ \beta \end{pmatrix}, \tag{3.10}$$

where normalization requires $|\alpha|^2 + |\beta|^2 = 1$. Alternatively to the horizontal and vertical directions, we may select the ± 45-degree diagonals, with the corresponding orthonormal basis $\{|+\rangle, |-\rangle\}$. Written with respect to the basis $\{|H\rangle, |V\rangle\}$, this new basis reads

$$|\pm\rangle = \frac{1}{\sqrt{2}}(|H\rangle \pm |V\rangle) = \frac{1}{\sqrt{2}}\begin{pmatrix} 1 \\ \pm 1 \end{pmatrix}. \tag{3.11}$$

The horizontal and vertical components are either in phase ($|+\rangle$), or they have a relative phase shift of π ($|-\rangle$). On the other hand, if the relative phase shift is $\pm\pi/2$, then the magnitude of the electric-field vector remains constant, but instead it rotates in the plane perpendicular to the direction of propagation—the light is left-handed or right-handed *circularly* polarized. The corresponding basis states are

$$|L\rangle = \frac{1}{\sqrt{2}}\begin{pmatrix} 1 \\ +i \end{pmatrix}, \quad |R\rangle = \frac{1}{\sqrt{2}}\begin{pmatrix} 1 \\ -i \end{pmatrix}. \tag{3.12}$$

Physically, circular polarization can be obtained from a beam of linearly polarized light by a quarter-wave plate, i.e., an appropriately oriented, birefringent crystal of a thickness such that, for the chosen wavelength, the different propagation speeds of two perpendicular, linearly polarized components introduce a relative shift of a quarter wavelength. Using the scalar product, we may now evaluate some transition amplitudes. For instance, let us assume that a beam of circularly polarized photons was created in the state $|L\rangle$, which is consecutively sent through a polarization filter oriented horizontally. The probability for a photon from the beam to pass through the filter is then

$$|\langle H|L\rangle|^2 = |\frac{1}{\sqrt{2}}\begin{pmatrix} 1 & 0 \end{pmatrix} \cdot \begin{pmatrix} 1 \\ +i \end{pmatrix}|^2 = \frac{1}{2}. \tag{3.13}$$

Similarly, some further computations reveal

$$|\langle H|R\rangle|^2 = |\langle V|L\rangle|^2 = |\langle V|R\rangle|^2 = \frac{1}{2}, \tag{3.14}$$

which means that for circularly polarized light, the orientation of the polarization filter does not matter, 50% of the photons pass through the filter.

3.2 Operators on Finite-Dimensional Hilbert Spaces

On finite-dimensional Hilbert spaces all linear operators (recall Definition 2.6) can be represented as matrices with real or complex entries, that is, the action of an operator A on an n dimensional Hilbert space $\mathcal{H}$ that maps the vector $|\psi\rangle = \sum_i c_i |i\rangle \in \mathcal{H}$ to $|\phi\rangle = \sum_i d_i |i\rangle$ can be written as

$$A|\psi\rangle = \begin{pmatrix} A_{11} & \cdots & A_{1n} \\ \vdots & \ddots & \vdots \\ A_{n1} & \cdots & A_{nn} \end{pmatrix} \begin{pmatrix} c_1 \\ \vdots \\ c_n \end{pmatrix} = \begin{pmatrix} d_1 \\ \vdots \\ d_n \end{pmatrix} = |\phi\rangle, \tag{3.15}$$

with respect to the basis $\{\,|\,i\,\rangle\}$. The components A_{ij}, i.e., the matrix elements of A, may hence be written as

$$A_{ij} = \langle\, i\,|\, A\,|\, j\,\rangle\,. \tag{3.16}$$

With the decomposition of eqn (3.16) in mind, we may decompose the operator A using the so-called *tensor product* "$\otimes$", that is

$$A = \sum_{i,j} A_{ij}\,|\, i\,\rangle \otimes \langle\, j\,|\,. \tag{3.17}$$

The tensor product maps elements of two vector spaces V and W to a larger vector space $V \otimes W$, with $\dim(V \otimes W) = \dim(V)\dim(W)$, such that for all $v, v_1, v_2 \in V$, $w, w_1, w_2 \in W$, and for all $c \in \mathbb{C}$ holds that

$$(v_1 \otimes w) + (v_2 \otimes w) = (v_1 + v_2) \otimes w\,, \tag{3.18a}$$

$$(v \otimes w_1) + (v \otimes w_2) = v \otimes (w_1 + w_2)\,, \tag{3.18b}$$

$$c(v \otimes w) = (c\,v) \otimes w = v \otimes (c\,w)\,. \tag{3.18c}$$

For instance, for the vectors $|\,\psi\,\rangle \in \mathcal{H}$ and $|\,\phi\,\rangle \in \mathcal{H}$, we can form $|\,\psi\,\rangle \otimes |\,\phi\,\rangle \in \mathcal{H} \otimes \mathcal{H}$, which is then of the form

$$|\,\psi\,\rangle \otimes |\,\phi\,\rangle = \sum_{i,j} c_i\, d_j\,|\, i\,\rangle \otimes |\, j\,\rangle\,. \tag{3.19}$$

In other words, if $\{v_i \in V\}$ and $\{W_j \in W\}$ form bases of V and W, respectively, then $\{v_i \otimes w_j \in V \otimes W\}$ is a basis of $V \otimes W$. Moreover, for all linear operators M and N, acting on V and W, respectively, $M \otimes N$ is a linear operator acting on $V \otimes W$ according to

$$(M \otimes N)(v \otimes w) = (M\,v) \otimes (N\,w)\,. \tag{3.20}$$

Returning to our example operator A from eqn (3.17), where $|\,i\,\rangle \in \mathcal{H}$ and $\langle\, j\,| \in \mathcal{H}^*$, we note that eqn (3.16) can be interpreted as reading

$$A_{ij} = \left(\langle\, i\,| \otimes \mathbb{1}_{\mathcal{H}^*}\right) \left(\sum_{m,n} A_{mn}\,|\, m\,\rangle \otimes \langle\, n\,|\right) \left(\mathbb{1}_{\mathcal{H}} \otimes |\, j\,\rangle\right). \tag{3.21}$$

This notation appears to be rather cumbersome. Therefore, we rely again on the simplicity of the Dirac notation, which permits us to suppress the symbols for the tensor product and the identity operators, and we just write

$$\langle\, i\,|\, \sum_{m,n} A_{mn}\,|\, m\,\rangle\langle\, n\,|\,|\, j\,\rangle = \sum_{m,n} A_{mn}\,\langle\, i\,|\, m\,\rangle\,\langle\, n\,|\, j\,\rangle = \sum_{m,n} A_{mn}\,\delta_{im}\,\delta_{nj} = A_{ij}\,. \tag{3.22}$$

The simple rule of thumb is hence, "bra" applied to "ket" is a scalar, $\langle\,\psi\,|\,\phi\,\rangle \in \mathbb{C}$, while "ket" applied to "bra" yields a linear operator $|\,\psi\,\rangle\langle\,\phi\,| : \mathcal{H} \mapsto \mathcal{H}$. In this notation it is also easy to see that the adjoint operators are obtained by Hermitian conjugation, that is, transposition ($A_{ij} \to A_{ji}$), and complex conjugation ($A_{ij} \to A_{ij}^*$). Explicitly, we have

$$A^\dagger = \sum_{i,j} A_{ij}^\dagger \left(|i\rangle\langle j| \right)^\dagger = \sum_{i,j} A_{ij}^* |j\rangle\langle i| = \sum_{i,j} A_{ji}^* |i\rangle\langle j|, \qquad (3.23)$$

where we have used $|\psi\rangle^\dagger = \langle\psi|$ and $\langle\psi|^\dagger = |\psi\rangle$. All Hermitian operators (see Definition 2.8) on finite-dimensional Hilbert spaces are moreover necessarily self-adjoint.

3.2.1 Projectors

We now turn to a special type of linear operators called *projection operators*, or simply *projectors*. A projector P is a linear operator with the property

$$P^2 = P. \qquad (3.24)$$

In other words, the action of the operator does not change when applied more than once. More precisely, we shall limit our attention to *orthogonal* projectors, which are Hermitian, $P = P^\dagger$, in addition to the property of eqn (3.24). All such projectors can be written in terms of the notation introduced in the previous section, that is, every orthogonal projector can be written as

$$P = \sum_{i \in \mathcal{I}} |i\rangle\langle i|, \qquad (3.25)$$

with respect to an orthonormal basis $\{|i\rangle \,|\, i = 1, \ldots, \dim(\mathcal{H})\}$ of the Hilbert space $\mathcal{H}$, where $\mathcal{I} \subseteq \{1, \ldots, \dim(\mathcal{H})\}$. The operator P can hence be interpreted in the geometrical sense of projecting any vector in $\mathcal{H}$ to a vector in a linear subspace of $\mathcal{H}$ that is spanned by the $|i\rangle$ with $i \in \mathcal{I}$. The matrix representation of P with respect to the basis $\{|i\rangle\}$ is a diagonal matrix with entries 1 on the diagonal for every $i \in \mathcal{I}$ and entries 0 everywhere else. With respect to an arbitrary (orthogonal) basis this need not be the case, except when we project on the entire Hilbert space. In that case, that is, if $\mathcal{I} = \{1, \ldots, \dim(\mathcal{H})\}$, the $|i\rangle$ form a complete orthonormal system—a basis—and we have

$$\sum_{i=1}^{\dim(\mathcal{H})} |i\rangle\langle i| = \mathbb{1}_{\mathcal{H}}. \qquad (3.26)$$

The expression in eqn (3.26), which we will refer to as a *completeness relation* or a *resolution of the identity*, is independent of the choice of basis as long as the sum runs over the entire orthonormal system. It is further convenient to introduce a subscript for the projectors to avoid confusion with the momentum operator. For instance, projectors on individual states $|\psi\rangle$ may be written as

$$P_\psi = |\psi\rangle\langle\psi|. \qquad (3.27)$$

Physically, projection operators represent observables. For a quantum system prepared in the state $|\psi\rangle$ the probability of finding the system in some state $|\phi\rangle$ is given by the expectation value of the projector on $|\phi\rangle$,

$$\langle P_\phi \rangle_\psi = \langle |\phi\rangle\langle\phi| \rangle_\psi = \langle \psi|\phi\rangle \langle \phi|\psi\rangle = |\langle \phi|\psi\rangle|^2 . \tag{3.28}$$

After a measurement represented by a projector, the quantum system is described by a state that is an element of the linear subspace corresponding to the support of the projector. In a measurement corresponding to the observable P_ϕ of an ensemble of quantum systems prepared in the state $|\psi\rangle$ only the fraction corresponding to $|\langle \phi|\psi\rangle|^2$ of the particles of the ensemble remain after the measurement. The ensemble of these remaining quantum systems is then described by the state $|\phi\rangle$. Interventions of this type are called *projective measurements*. An example that intuitively illustrates this concept is a photon-polarization filter. Let us assume that a beam of photons prepared with the polarization state $|\psi\rangle$ from eqn (3.10) is sent through a horizontally oriented polarization filter. The intensity of the beam behind the filter, proportional to the number of photons passing through, will be reduced by the factor $|\alpha|^2 \leq 1$, but all photons that have passed the filter are now horizontally polarized. Projective measurements and their generalizations are discussed in more detail in Chapter 23.

The completeness relation of (3.26) is also very useful to switch between different bases. Given a state $|\psi\rangle$ and its decomposition with respect to an orthonormal basis $\{|i\rangle\}$, i.e., $|\psi\rangle = \sum_i c_i |i\rangle$, we may insert the identity operator $\mathbb{1}$ and decompose it into projectors of another basis $\{|\tilde{i}\rangle\}$ according to the completeness relation, i.e.,

$$|\psi\rangle = \sum_i c_i \mathbb{1} |i\rangle = \sum_i c_i \sum_j |\tilde{j}\rangle\langle\tilde{j}|i\rangle = \sum_j \tilde{c}_j |\tilde{j}\rangle , \tag{3.29}$$

where the new components are $\tilde{c}_j = \sum_i c_i \langle \tilde{j}|i\rangle$. For the photon-polarization state $|\psi\rangle$ from eqn (3.10) we can, e.g., switch from the basis $\{|H\rangle, |V\rangle\}$ to the basis $\{|L\rangle, |R\rangle\}$. We compute

$$|\psi\rangle = \alpha |H\rangle + \beta |V\rangle = (|L\rangle\langle L| + |R\rangle\langle R|)(\alpha |H\rangle + \beta |V\rangle)$$

$$= (\alpha \langle L|H\rangle + \beta \langle L|V\rangle)|L\rangle + (\alpha \langle R|H\rangle + \beta \langle R|V\rangle)|R\rangle$$

$$= \frac{\alpha - i\beta}{\sqrt{2}} |L\rangle + \frac{\alpha + i\beta}{\sqrt{2}} |R\rangle . \tag{3.30}$$

3.2.2   The Spectral Theorem

Beyond their role in projective measurements, projectors play an important conceptual role for the *spectral decomposition* of all other observables, which do not generally satisfy the property of eqn (3.24). We shall phrase this relationship in terms of the so-called *spectral theorem*, which holds, more generally, for all *normal* operators, i.e., for which $A^\top A = A A^\top$, and therefore, in particular, for all Hermitian operators. The latter are guaranteed to have a real spectrum, that is, the eigenvalues λ_j of any Hermitian operator A, which are determined by the *eigenvalue equation*

$$A |j\rangle = \lambda_j |j\rangle \tag{3.31}$$

are always real, $\lambda_j \in \mathbb{R}$ for all j. The eigenvectors of every normal operator, when appropriately normalized, form a complete orthonormal system—the *eigenbasis*. With this terminology we formulate the spectral theorem.

Theorem 3.1 (Spectral theorem)

Every normal operator A can be decomposed into projectors onto its eigenbasis $\{|j\rangle\}$, i.e.,

$$A = \sum_j \lambda_j |j\rangle\langle j|.$$

For a proof of this fundamental theorem of linear algebra we refer the interested reader to the book by Hoffman and Kunze (1971, Theorem 18, Chap. 8). To interpret the spectral theorem in terms of quantum mechanics, recall from Definition 2.4, that we associate possible measurement outcomes to the eigenvalues of the Hermitian operators representing observables. This association is based exactly on the relation with projectors that is provided by the spectral theorem. When a measurement of the observable A is performed, any single run will provide a definite measurement result j that corresponds to one of the eigenvalues λ_j of A and the measured individual quantum system is left in the state $|j\rangle$. The expectation value of A represents the average of many such measurements, that is, a weighted sum of different eigenvalues, where the weights are determined by the overlap of the initially prepared quantum state and the eigenstates $|j\rangle$,

$$\langle A\rangle_\psi = \langle\psi|\left(\sum_j \lambda_j |j\rangle\langle j|\right)|\psi\rangle = \sum_j \lambda_j |\langle j|\psi\rangle|^2. \tag{3.32}$$

The simplest non-trivial applications of the spectral theorem are the so-called *Pauli matrices*. These Hermitian 2×2 matrices are linear operators on $\mathbb{C}^2$, represented by

$$\sigma_x = \begin{pmatrix} 0 & 1 \\ 1 & 0 \end{pmatrix}, \quad \sigma_y = \begin{pmatrix} 0 & -i \\ i & 0 \end{pmatrix}, \quad \sigma_z = \begin{pmatrix} 1 & 0 \\ 0 & -1 \end{pmatrix}, \tag{3.33}$$

where the meaning of the subscripts x, y, and z will become apparent in Chapter 8, but they may be taken as mere labels for now. The eigenvalues of all of the Pauli matrices are ± 1, i.e., there are two possible measurement outcomes for each corresponding observable. But which measurements are described by these operators? Some elementary calculations reveal that the eigenstates of σ_x are exactly the linear-polarization states $|\pm\rangle$ of eqn (3.11). Similarly, the eigenstates of σ_y are the states of left- and right-circular polarization, $|L\rangle$ and $|R\rangle$, from eqn (3.12), while the eigenstates of σ_z are $|H\rangle$ and $|V\rangle$. According to Theorem 3.1 we may hence write σ_x, σ_y, and σ_z as

$$\sigma_x = (+1)|+\rangle\langle+| + (-1)|-\rangle\langle-|, \tag{3.34}$$

$$\sigma_y = (+1)|L\rangle\langle L| + (-1)|R\rangle\langle R|, \tag{3.35}$$

$$\sigma_z = (+1)|H\rangle\langle H| + (-1)|V\rangle\langle V|, \tag{3.36}$$

and the Pauli matrices can be associated to measurements of photon polarization.

An important corollary of the spectral theorem pertains to functions of Hermitian operators, for which one finds

$$f(A) = \sum_j f(\lambda_j) \, |\, j \, \rangle\langle\, j\, | \, . \tag{3.37}$$

An example for a function where this useful relation may be applied is the matrix exponential, see eqn (2.37), for instance, for the Hermitian operator σ_y we may form

$$\exp(-i\tfrac{\theta}{2}\sigma_y) = \exp(-i\tfrac{\theta}{2}(+1)) \, |\, L\, \rangle\langle\, L\, | \, + \, \exp(-i\tfrac{\theta}{2}(-1)) \, |\, R\, \rangle\langle\, R\, |$$

$$= \begin{pmatrix} \tfrac{1}{2}\big(e^{i\theta/2} + e^{-i\theta/2}\big) & \tfrac{i}{2}\big(e^{i\theta/2} - e^{-i\theta/2}\big) \\ \tfrac{-i}{2}\big(e^{i\theta/2} - e^{-i\theta/2}\big) & \tfrac{1}{2}\big(e^{i\theta/2} + e^{-i\theta/2}\big) \end{pmatrix} = \begin{pmatrix} \cos\big(\tfrac{\theta}{2}\big) & \sin\big(\tfrac{\theta}{2}\big) \\ -\sin\big(\tfrac{\theta}{2}\big) & \cos\big(\tfrac{\theta}{2}\big) \end{pmatrix} . \tag{3.38}$$

This operation represents a rotation of the linear polarization by the angle $\theta/2$, for instance, for $\theta = \pi$, the basis vector $|\, H\, \rangle$ is rotated to $\exp(-i\tfrac{\pi}{2})\, |\, H\, \rangle = |\, V\, \rangle$. We shall discuss the relationship between spatial rotations and the Pauli operators in more detail in Chapter 8.

3.2.3   Unitary Operators

The operator in eqn (3.38) is a perfect example for another important class of operations, so-called *unitary* operators. In a vector space over the field $\mathbb{C}$, these operators are the generalization of orthogonal transformations in a real vector space, that is, they preserve lengths and angles. In the Hilbert-space language that we use here, this means that unitary operations U leave the inner product invariant, $\langle\, U\psi\, |\, U\phi\, \rangle = \langle\, \psi\, |\, U^\dagger U\, |\, \phi\, \rangle = \langle\, \psi\, |\, \phi\, \rangle \; \forall \psi, \phi \in \mathcal{H}$. We shall phrase this requirement in the following definition.

Definition 3.4   *A linear operator $U : \mathcal{H} \to \mathcal{H}$ on a Hilbert space $\mathcal{H}$ is called*
                  unitary, *if $U^\dagger U = U U^\dagger = \mathbb{1}$.*

The significance of such operators in quantum mechanics lies in the fact that by preserving the inner product, probabilities are preserved, normalized states are mapped to normalized states. Unitary operators are therefore used to represent, for example, time evolution (recall the propagator in Definition 2.12), or symmetries such as rotations, which we will further discuss in Chapter 6. For all these cases eqn (3.38) can be a guiding example. That is, we may represent these unitary operators as exponentials of self-adjoint operators, simply note that for $A = A^\dagger$, the operator $U = e^{iA}$ is (formally) unitary, since $U^\dagger = e^{-iA^\dagger} = e^{-iA} = U^{-1}$, and one may use the spectral theorem, specifically eqn (3.37), for practical applications.

3.3   Infinite-Dimensional Hilbert Spaces

In this section we will turn our attention to some mathematical subtleties that arise for infinite-dimensional Hilbert spaces. Although the notation that we have previously introduced will be applicable with only some minor modifications, it is crucial to understand where certain features of finite-dimensional quantum mechanics require adjustment in infinite dimensions. Specifically, we shall scrutinize self-adjoint operators on infinite-dimensional Hilbert spaces, and their spectra.

3.3.1 Self-adjoint Operators on Infinite-Dimensional Hilbert Spaces

On a finite-dimensional Hilbert space every linear operator can be represented by a matrix. Determining whether or not a given operator is self-adjoint hence becomes a simple task. In infinite dimensions this task is not as straightforward. Not every operator that is formally identical to its adjoint is necessarily self-adjoint, recall Definition 2.8. In such cases, it may be possible to restrict the domain of the operator such that the obtained extension of the operator is self-adjoint. We will provide only a mathematically minimalist treatment of this topic here, stating a central theorem due to *Hermann Weyl* (1910) and *John von Neumann* (1929). For additional details and examples we direct the reader to Bonneau *et al.* (2001).

The task at hand is as follows. Given an operator A we are to determine whether it is self-adjoint or, failing that, if any *self-adjoint extension* exist. Fortunately, this generally difficult task has been solved and can be phrased in a concise theorem. To formulate it, let us establish some terminology. Let $(A, \mathcal{D}(A))$ be an operator on an (infinite-dimensional) Hilbert space $\mathcal{H}$. We call the domain of the operator *dense*, or, likewise, the operator *densely defined*, if for every $\psi \in \mathcal{H}$ there exists a sequence of vectors $\phi_n \in \mathcal{D}(A)$, which converges in norm to ψ. An operator is further called *closed*, if for any sequence ϕ_n that converges within the domain, i.e., $\lim_{n\to\infty} \phi_n = \phi$, with $\phi_n, \phi \in \mathcal{D}(A)$, it holds that $\lim_{n\to\infty} A\phi_n = \psi = A\phi$.

Theorem 3.2 (Weyl–von Neumann)

Let $(A, \mathcal{D}(A))$ be a densely defined, closed Hermitian operator with adjoint $(A^\dagger = A, \mathcal{D}(A^\dagger))$, and let $\mathcal{N}_\pm$ be the eigenspaces of $A^\dagger$ corresponding to eigenvalues $z_\pm$ with positive/negative imaginary parts, that is,

$$\mathcal{N}_\pm = \left\{ \psi \in \mathcal{D}(A^\dagger), \; A^\dagger \psi = z_\pm \psi, \; \mathrm{Im}(z_+) > 0, \; \mathrm{Im}(z_-) < 0 \right\}.$$

For the dimensions $n_\pm = \dim(\mathcal{N}_\pm)$ of these spaces, called the deficiency indices, *there are three possibilities:*

    (i) If $n_+ = n_- = 0$, the operator is self-adjoint.

    (ii) If $n_+ = n_- \geq 1$, the operator admits infinitely many self-adjoint extensions, parameterized by $n_\pm^2$ real parameters.

    (iii) If $n_+ \neq n_-$, the operator admits no self-adjoint extension.

The basis of this theorem, which we shall not prove here, is that eigenvalues of Hermitian operators need not be real in infinite dimensions. Operators for which this is the case do not qualify as corresponding to measurable quantities, but they may admit self-adjoint extensions which do not have this problem. Bear in mind that this still leaves the task of selecting a particular one of the (infinitely many) extensions as the "right" one. The choice is, in principle, arbitrary, but may be made based on other physical or aesthetical considerations. Let us now study some simple examples that illustrate the theorem, and which will be useful also in preparing the ground for the discussion in Chapter 4.

(i) Momentum operator on the real axis or semi-axis

Consider the momentum operator $P = -i\hbar\partial/\partial x$ on the whole real line, the Hilbert space is $\mathcal{H} = \mathfrak{L}_2(\mathbb{R})$. To apply Theorem 3.2, we select $z_\pm = \pm i p_0$, where $p_0 > 0$ is taken to be a constant of dimension momentum. We then aim to solve

$$P\psi_\pm(x) = -i\hbar\frac{\partial}{\partial x}\psi_\pm(x) = \pm i p_0 \psi_\pm(x) \,. \tag{3.39}$$

The solutions are quickly found to be $\psi_\pm = c_\pm e^{\mp p_0 x/\hbar}$, with constants $c_\pm \in \mathbb{C}$. Neither of these solutions is normalizable on $\mathbb{R}$. Consequently, these functions cannot be elements of $\mathcal{H}$, or its subset $\mathcal{D}(P^\dagger)$, and so we find $n_\pm = 0$. We conclude that, as previously claimed, the momentum operator is self-adjoint on the real line.

Now, suppose that the quantum-mechanical system is confined to only the positive half of the real line, $\mathcal{H} = \mathfrak{L}_2(\mathbb{R}_+)$. While $\psi_- = c_- e^{-p_0 x/\hbar}$ is normalizable on $[0, \infty]$, such that $n_- = 1$, the same is not true for ψ_+, and hence $n_+ = 0$. Quite surprisingly, we find that $P = -i\hbar\partial/\partial x$ is not self-adjoint on the real semi-axis, and does not admit any self-adjoint extension either. As far as the restriction to the half-space can be considered a reasonable physical model, the momentum operator does not correspond to a measurable quantity for particles confined in this way.

(ii) Free Hamiltonian on the real axis or semi-axis

Similar to the momentum operator, we may consider the free-particle Hamiltonian on both $\mathfrak{L}_2(\mathbb{R})$ and $\mathfrak{L}_2(\mathbb{R}_+)$. As before, we select a real, positive constant $k_0 > 0$ with appropriate dimensions, and solve

$$H \phi_\pm(x) = -\frac{\hbar^2}{2m}\frac{\partial^2}{\partial x^2}\phi_\pm(x) = \pm i k_0^2 \phi_\pm(x) \,. \tag{3.40}$$

The general solution to this second-order (partial) differential equation can easily be confirmed to be

$$\phi_\pm(x) = A_\pm e^{k_\pm x} + B_\pm e^{-k_\pm x} \,, \tag{3.41}$$

with $k_\pm = (1 \mp i)\sqrt{m}k_0/\hbar$ and constants $A_\pm, B_\pm \in \mathbb{C}$. The linearly independent solutions $e^{k_\pm x}$ and $e^{-k_\pm x}$ diverge as x approaches $+\infty$ and $-\infty$, respectively.

Consequently, there are no solutions on the whole real line, and $n_\pm = 0$. The free Hamiltonian is indeed self-adjoint for $\mathcal{H} = \mathfrak{L}_2(\mathbb{R})$.

On the other hand, for the positive real semi-axis, there is a one-dimensional space of normalizable solutions $\phi_\pm(x) = B_\pm\, e^{-k_\pm x}$, such that $n_+ = n_- = 1$. The operator $H = -\hbar^2/(2m)(\partial^2/\partial x^2)$ admits infinitely many self-adjoint extensions on $\mathcal{H} = \mathfrak{L}_2(\mathbb{R}_+)$. These extensions are parameterized by a single real parameter that relates to the boundary condition imposed on the wave functions at $x = 0$.

3.3.2   Continuous Spectra

The previous section has provided us with the tools to test the self-adjointness of Hermitian operators, and the existence of possible self-adjoint extension. Provided such an enterprise has been successful, we now wish to proceed along similar lines as we have done for finite-dimensional Hilbert spaces, by investigating the spectrum of self-adjoint operators on infinite-dimensional Hilbert spaces.

As we have argued, the momentum operator P is an example for a self-adjoint operator on the space $\mathfrak{L}_2(\mathbb{R})$. Consequently, the eigenvalues of $\big(P, \mathfrak{L}_2(\mathbb{R})\big)$ are real, but, within (relativistic) quantum theory, there is no reason to expect that these values may be labelled by a discrete index. In other words, it seems reasonable to assume that a single measurement of the momentum of a free particle may provide any real value—the spectrum is continuous. Let us proceed by naively writing an eigenvalue equation for the corresponding momentum eigenvectors, i.e.,

$$P\,|\,p\,\rangle = p\,|\,p\,\rangle\,. \tag{3.42}$$

Ignoring some technical difficulties for now, we may think of the objects $|\,p\,\rangle$ as elements of an uncountably infinite basis of the Hilbert space. Note that this does not imply that $\mathcal{H}$ is inseparable. The Hilbert space $\mathfrak{L}_2(\mathbb{R})$ is infinite-dimensional, but admits a countable basis (as we shall see in Chapter 4). However, the elements of such a basis need not be eigenvectors of P. Furthermore, with the use of the Dirac notation, we have written eqn (3.42) in a basis-independent way, where $|\,p\,\rangle$ represents an abstract vector, whose components with respect to some basis are yet to be specified. One such basis is yet another eigenbasis of an operator with a continuous spectrum, the position operator X, with

$$X\,|\,x\,\rangle = x\,|\,x\,\rangle\,. \tag{3.43}$$

Denoting the components of $|\,p\,\rangle$ with respect to this basis as $\psi_p(x) := \langle\,x\,|\,p\,\rangle$, and using the position representation $P = -i\hbar(\partial/\partial x)$ of the momentum operator, we have to solve

$$-i\hbar\frac{\partial}{\partial x}\psi_p(x) = p\,\psi_p(x)\,. \tag{3.44}$$

A straightforward computation gives the expression

$$\psi_p(x) = \langle x | p \rangle = \frac{1}{\sqrt{2\pi\hbar}} e^{ipx/\hbar},\tag{3.45}$$

where we have chosen the pre-factor $(1/\sqrt{2\pi\hbar})$ for later convenience. We observe, in the position representation, the momentum eigenstates are found to be plane waves. These solutions are not normalizable (unless we restrict to a finite interval), but they are nonetheless conceptually very valuable. For instance, in analogy to eqn (3.26) we are able to write the completeness relations as

$$\int dp\, | p \rangle\langle p | = \int dx\, | x \rangle\langle x | = \mathbb{1},\tag{3.46}$$

by replacing the sums by integrals. This is just an expression for the fact that the spectral theorem (Theorem 3.1) also holds for (bounded and unbounded) normal operators in infinite-dimensional Hilbert spaces (see, e.g., Hall, 2013, Chap. 8 and 10). Similarly, the spectral decomposition (see Theorem 3.1) of the momentum operator is expressed as

$$P = \int dp\, p\, | p \rangle\langle p |,\tag{3.47}$$

and likewise for the position operator. Inserting the identity in terms of the completeness relation, we can determine the components of any other state vector $| \psi \rangle$ with respect to the momentum basis, that is,

$$| \psi \rangle = \int dp\, | p \rangle\langle p | \psi \rangle.\tag{3.48}$$

The inner product $\langle p | \psi \rangle$ can then be evaluated in the position representation that we have previously encountered in eqn (2.21),

$$\langle p | \psi \rangle = \int_{-\infty}^{\infty} dx\, \psi_p^*(x)\, \psi(x) = \frac{1}{\sqrt{2\pi\hbar}} \int_{-\infty}^{\infty} dx\, e^{-ipx/\hbar}\, \psi(x).\tag{3.49}$$

The right-hand side of eqn (3.49) can immediately be recognized as the *Fourier transform* of the wave function $\psi(x)$. Hence, the wave function $\psi(x) = \langle x | \psi \rangle$ and its Fourier transform $\tilde{\psi}(p) = \langle p | \psi \rangle$ can be identified with the components of the state vector $| \psi \rangle$ with respect to the position and momentum bases, respectively. Although when using this terminology one should keep in mind that neither $| p \rangle$, nor $| x \rangle$, are elements of the Hilbert-space construction we have (thus far) introduced. None of these "basis vectors" can be normalized, and the elements are orthogonal only in the sense of the Dirac delta function of eqns (2.51) and (2.52), i.e.,

$$\langle p | p' \rangle = \int dx\, \langle p | x \rangle \langle x | p' \rangle = \frac{1}{2\pi\hbar} \int dx\, e^{-i(p-p')x/\hbar} = \delta(p - p'),\tag{3.50a}$$

$$\langle x | x' \rangle = \int dp\, \langle x | p \rangle \langle p | x' \rangle = \frac{1}{2\pi\hbar} \int dp\, e^{ip(x-x')/\hbar} = \delta(x - x').\tag{3.50b}$$

Although colloquially referred to as a function, the Dirac delta is actually not a function, but a generalization of this concept, a *distribution* or *functional,* as we shall briefly elaborate on in the following section.

3.3.3 Distributional Aspects of Quantum Mechanics

Distributions extend the concept of functions. Where the latter map numbers to numbers, distributions map *functions* to numbers. Distributions are defined with respect to a set Φ of *test functions* $f \in \Phi$, for which the map $\phi : \Phi \to \mathbb{R}$, which takes f to $\phi(f) = \langle \phi | f \rangle$, is well-defined. Loosely speaking, the more the space of test functions Φ is restricted, the fewer constraints are placed on the space of distributions $D(\Phi)$. We have already encountered examples for such objects in the form of continuous linear functionals over the Hilbert space $\mathcal{H}$. For square-integrable functions $\psi \in \mathcal{H} = \Phi$ the dual Hilbert space is the space of distributions, $D(\Phi) = \mathcal{H}^*$. Since the Hilbert space is self-dual, any square-integrable function ψ may be identified with a distribution $T_\psi = \langle \psi | \cdot \rangle \in \mathcal{H}^*$ via the integration of the scalar product,

$$T_\psi(\phi) = \langle \psi | \phi \rangle = \int_{-\infty}^{\infty} dx\, \psi^*(x)\, \phi(x)\,, \tag{3.51}$$

where $\phi \in \mathcal{H}$. In a slight abuse of notation, one may use the same symbol for the function ϕ and the corresponding functional. For Hilbert-space vectors we avoid this ambiguity by using the Dirac notation. For other distributions, for instance the Dirac delta, it is conventional to use the same symbol $\delta(x)$ for both objects, that is

$$\delta(x)[f] = \int_{-\infty}^{\infty} dx\, \delta(x)\, f(x) = f(0)\,. \tag{3.52}$$

The particular class of test functions that is interesting for many physical applications is the family of *Schwartz* functions. Without giving a detailed technical definition, let us simply remark that Schwartz functions are falling off at infinity more rapidly than any polynomial. They have the particularly useful property that their Fourier transform always exists and is in turn a Schwartz function. The corresponding distributions, i.e., the continuous linear functions over the Schwartz space, are called *tempered distributions*. All square-integrable functions can be understood as tempered distributions in the sense of eqn (3.51), and also the Dirac delta is a tempered distribution. In particular, the following hierarchy of the spaces $\Phi, \mathcal{H},$ and $D(\Phi)$, called *Gelfand tripple*, holds

$$\Phi \subseteq \mathcal{H} \subseteq D(\Phi)\,. \tag{3.53}$$

This inclusion is even dense for the case of the Schwartz space Φ and the Hilbert space $\mathcal{H} = \mathfrak{L}_2$. A prominent example for the dense inclusion of $\mathcal{H}$ in $D(\Phi)$ is the Dirac delta arising as the limit of ever more narrow Gaussian functions, that is,

$$\delta(x) = \lim_{\sigma \to 0} \frac{1}{\sqrt{2\pi}\sigma} e^{-\frac{x^2}{2\sigma^2}}\,. \tag{3.54}$$

The Gaussian functions on the right-hand side are square integrable, as we have explicitly confirmed in Chapter 2, but the Dirac delta is certainly not. Not even its square is well-defined. To prove eqn (3.54) we hence have to show that

$$\lim_{\sigma \to 0} \frac{1}{\sqrt{2\pi}\sigma} \int_{-\infty}^{\infty} dx \; e^{-\frac{x^2}{2\sigma^2}} \, f(x) = f(0) \tag{3.55}$$

for all Schwartz functions $f(x)$. To achieve this, we first substitute the Gaussian with its Fourier transform

$$\frac{1}{\sqrt{2\pi}\sigma} e^{-\frac{x^2}{2\sigma^2}} = \frac{1}{2\pi} \int_{-\infty}^{\infty} dk \; e^{-\frac{k^2\sigma^2}{2}} \, e^{-ikx} \,, \tag{3.56}$$

which allows us to evaluate the limit $\sigma \to 0$, and we find

$$\lim_{\sigma \to 0} \frac{1}{\sqrt{2\pi}\sigma} \int_{-\infty}^{\infty} dx \; e^{-\frac{x^2}{2\sigma^2}} \, f(x) = \frac{1}{2\pi} \int_{-\infty}^{\infty} dk \, dx \; f(x) \, e^{-ikx} = \frac{1}{\sqrt{2\pi}} \int_{-\infty}^{\infty} dk \; \tilde{f}(k) = f(0) \,,$$

$$\tag{3.57}$$

which concludes the proof.

The eigenfunctions of the position and momentum operators, albeit not elements of the actual Hilbert space, may thus be incorporated in the construction of a so-called *rigged* Hilbert space, that is, the pair $(\mathcal{H}, \Phi)$.

4

The Time-Independent Schrödinger Equation

With the mathematical techniques of the previous chapter at hand, we now return to the Schrödinger equation. In Chapter 2, we have used it to describe the dynamics of free quantum-mechanical particles. In this chapter, we wish to extend our analysis to interacting systems by considering Hamiltonians including non-trivial potentials. As we shall see, already simple choices of one-dimensional, time-independent potentials can lead to remarkably interesting solutions, describing bound states of quantum-mechanical particles and scattering phenomena such as the *tunnel effect*. We shall therefore keep the discussion focused on problems in one spatial dimension and without explicitly time-dependent interactions throughout this chapter. However, we will extend our analysis to three-dimensional problems in Chapter 7 and we will consider time-dependent interactions in Section 10.5 of Chapter 10.

4.1   Solving the Schrödinger Equation

The key technical step that is required right away is to side-step the time-dependence of the problem by introducing so-called *stationary states*.

4.1.1   Stationary States

Let us again consider the time-dependent Schrödinger equation of Proposition 2.1.1, including some non-zero, real, and time-independent potential $V(x)$, i.e.,

$$i\hbar \frac{\partial}{\partial t} \psi(t,x) = \left(-\frac{\hbar^2}{2m} \frac{\partial^2}{\partial x^2} + V(x) \right) \psi(t,x) = H \psi(t,x) . \qquad (4.1)$$

To obtain the solutions, we will make use of the method of *separation of variables*, that is, we will first restrict to solutions $\psi(t,x)$ that may be written as

$$\psi(t,x) = f(t) \psi(x), \qquad (4.2)$$

where $f(t)$ and $\psi(x)$ depend only on t and x, respectively. Next, we insert this ansatz into the time-dependent Schrödinger equation, and assuming, for now, that $\psi(t,x) \neq 0$, we may divide by $f(t)$ and $\psi(x)$ to arrive at

$$i\hbar \frac{1}{f(t)} \frac{\partial f(t)}{\partial t} = -\frac{\hbar^2}{2m} \frac{1}{\psi(x)} \frac{\partial^2 \psi(x)}{\partial x^2} + V(x). \qquad (4.3)$$

Note that while the left-hand side of eqn (4.3) depends only on t, the right-hand side just depends on x. From this observation one may conclude that either side must in fact be independent of the respective variable, and thus equal to a constant that we shall auspiciously call E. Considering only the left-hand side we may then turn the partial derivative $\partial f / \partial t$ into a total derivative df/dt and integrate

$$i\hbar \int \frac{df}{f(t)} = \int E \, dt \,, \tag{4.4}$$

which yields the logarithm $\ln(f) = -iEt/\hbar + \text{const.}$ that one can finally exponentiate to arrive at

$$f(t) = \text{const.} \, e^{-iEt/\hbar} \,. \tag{4.5}$$

The resulting wave functions $\psi(t, x)$ depend on time only in a trivial way, that is, only via a global phase $e^{-iEt/\hbar}$ that does not contribute to the probability density, i.e.,

$$|\psi(t, x)|^2 = e^{iEt/\hbar} \psi^*(x) \, e^{-iEt/\hbar} \psi(x) = |\psi(x)|^2 \,. \tag{4.6}$$

Similarly, expectation values are time independent in this case,

$$\langle A \rangle_{\psi(t,x)} = \int dx \, e^{iEt/\hbar} \psi^*(x) \, A \, e^{-iEt/\hbar} \psi(x) = \langle A \rangle_{\psi(x)} \,, \tag{4.7}$$

as long as the observable A does not explicitly depend on time t or on derivatives with respect to t. In this sense, the states $\psi(t, x) = f(t)\psi(x)$ are *stationary*.

Definition 4.1   *A state is called a **stationary state** if it is of the form*

$$\psi(t, x) = \psi(x) \, e^{-iEt/\hbar} \,.$$

Reinserting the stationary states into the Schrödinger equation we then get

$$i\hbar \frac{\partial}{\partial t} f(t) \, \psi(x) = E \, f(t) \, \psi(x) = \left(-\frac{\hbar^2}{2m} \frac{\partial^2}{\partial x^2} + V(x) \right) f(t) \, \psi(x) \,, \tag{4.8}$$

where we have absorbed the integration constant into $\psi(x)$. This remaining spatial part of the wave function must hence be an eigenfunction of the Hamiltonian H. The corresponding eigenvalue equation is referred to as the *time-independent Schrödinger equation*.

Theorem 4.1 (Time-independent Schrödinger equation)

For time-independent Hamiltonians $H - \frac{\hbar^2}{2m}\Delta + V(x)$, the spatial part $\psi(x)$ of a stationary state satisfies

$$H \, \psi(x) = E \, \psi(x) \,.$$

The time-independent Schrödinger equation is an eigenvalue equation, that is, the wave functions $\psi(x)$ are the eigenstates of the Hamiltonian H, and the energies E are the corresponding eigenvalues. Before we analyse this statement in terms of vectors in Hilbert spaces in the next section, let us briefly remark upon some consequences of Theorem 4.1.1, phrased in the following lemmata.

Lemma 4.1 *For normalizable solutions $\psi(x)$ of the time-independent Schrödinger equation the energy E is real and time-independent.*

Proof To see this, first note that the time-independence follows immediately from eqn (4.7). Since the potential was chosen to be real, $V^*(x) = V(x)$, it follows that the Hamiltonian is self-adjoint, such that

$$\langle\, H \,\rangle_\psi = \int dx\, \psi^* \, H \, \psi(x) \;=\; \int dx\, \psi^* \, E \, \psi(x) \;=\; E$$

$$= \int dx\, (H\,\psi)^* \, \psi(x) \;=\; \int dx\, E^* \, \psi^* \, \psi(x) \;=\; E^* \, . \tag{4.9}$$

Consequently, we have $E = \text{const.} \in \mathbb{R}$. The allowed values for E are determined by the specific form of the Hamiltonian (via the potential) and its domain (via restrictions placed on the allowed wave functions). □

Lemma 4.2 *The solutions $\psi(x)$ of the time-independent Schrödinger equation can always be chosen to be real.*

Proof For the proof, one simply forms the complex conjugate of the time-independent Schrödinger equation, that is $H^*\psi^* = E^*\psi^*$. Lemma 4.1 then immediately allows us to conclude that ψ^* is an eigenfunction of H with the same eigenvalue as ψ, which means that the same is true for $(\psi(x) + \psi^*(x))/2 = \text{Re}(\psi(x))$. □

Lemma 4.3 *For symmetric potentials, $V(x) = V(-x)$, the solutions $\psi(x)$ of the time-independent Schrödinger equation can always be chosen to be even, $\psi(x) = \psi(-x)$, or odd, $\psi(x) = -\psi(-x)$.*

Proof When $V(x) = V(-x)$, a change of coordinates from x to $-x$ transforms the time-independent Schrödinger equation to $H\psi(-x) = E\psi(-x)$. Therefore, if $\psi(x)$ is a solution with eigenvalue E, then so is $\psi(-x)$. In analogy to the proof of Lemma 4.2 we may thus form solutions $\psi_\pm(x) = \psi(x) \pm \psi(-x)$, which are even ("+") and odd ("−") solutions of the time-independent Schrödinger equation, respectively. □

4.1.2   The Schrödinger Equation as an Eigenvalue Problem

Before we continue with some paradigmatic examples for potentials that find application in quantum mechanics, it will be useful to cast the time-independent Schrödinger equation and the notion of stationary states into the Hilbert-space language that we have introduced in Chapter 3. In particular, let us make use of the Dirac notation to rewrite Theorem 4.1.1 as

$$H \, | \, \psi_n \rangle = E_n \, | \, \psi_n \rangle \, . \tag{4.10}$$

One may immediately recognize this as an eigenvalue equation, that is, the stationary states $| \, \psi_n \rangle$ are the eigenvectors of the Hamiltonian H with eigenvalues E_n. The subscripts n label the different eigenstates and eigenvalues. The collection of the eigenvalues, i.e., the set $\{E_n\}_n$, is referred to as the spectrum of H. Depending on the situation, the spectrum may be discrete and/or continuous. In this chapter we will encounter both cases. For scattered states, with energies that are larger than (in a sense that will be clear soon) the potential energy, the spectrum will be found to be continuous, as one would expect also from similar classical problems, such as the Kepler problem. For bound states with energies too small to escape the (classical) potential, on the other hand, the spectra are *quantized*. In either case, the stationary states, being eigenstates of H, have precisely defined energies. That is, every individual measurement of the energy of a stationary state $| \, \psi_n \rangle$ always gives the same result E_n, which therefore also equals the average outcome, such that

$$\langle H \rangle_{\psi_n} = \langle \psi_n | \, H \, | \psi_n \rangle = \langle \psi_n | \, E_n \, | \psi_n \rangle = E_n \, \langle \psi_n | \psi_n \rangle = E_n \, . \tag{4.11}$$

Consequently, the energy uncertainty vanishes for stationary states $\Delta E = 0$ (recall Theorem 2.3). In addition, the fact that the Hamiltonian is self-adjoint proves to be very useful indeed when taking advantage of the following theorem.

Theorem 4.2   *The eigenvalues of self-adjoint operators are real and the eigenvectors corresponding to different eigenvalues are orthogonal.*

Proof   We formulate the proof in terms of H and its eigenvectors and eigenvalues as a representative for any other self-adjoint operator. First, note that the expectation value in eqn (4.11) was obtained by acting with H on $| \, \psi_n \rangle$. One could as well apply the adjoint operator $H^\dagger = H$ to the covector $\langle \psi_n |$, i.e.,

$$\langle \psi_n | \, H \, | \psi_n \rangle = \langle H\psi_n | \psi_n \rangle = \langle \psi_n | \, E_n^* \, | \psi_n \rangle = E_n^* \langle \psi_n | \psi_n \rangle = E_n^* \, . \tag{4.12}$$

Since the results of eqns (4.11) and (4.12) must match, we can conclude that $E_n = E_n^*$, and hence that $E_n \in \mathbb{R}$. For the remaining part of the proof we consider

$$\begin{aligned}\langle \psi_m | \, H \, | \psi_n \rangle &= \langle H\psi_m | \psi_n \rangle = \langle \psi_m | \, E_m \, | \psi_n \rangle = E_m \, \langle \psi_m | \psi_n \rangle \\ &= \langle \psi_m | H\psi_n \rangle = \langle \psi_m | \, E_n \, | \psi_n \rangle = E_n \, \langle \psi_m | \psi_n \rangle \, .\end{aligned} \tag{4.13}$$

If $E_m \neq E_n$, the expression in eqn (4.13) must vanish, such that $\langle \psi_m | \psi_n \rangle = 0$. $\square$

If the eigenvectors are normalized, as we require, then they form an orthonormal basis of the Hilbert space, $\langle \psi_m | \psi_n \rangle = \delta_{mn}$, such that the completeness relation reads

$$\sum_n | \psi_n \rangle\langle \psi_n | = \mathbb{1} \, . \tag{4.14}$$

4.1.3 Expansion into Stationary States

Since the stationary states form an orthonormal basis of the Hilbert space, any state vector $| \psi \rangle$ in the Hilbert space may be expanded with respect to this basis as

$$| \psi \rangle = \sum_n c_n | \psi_n \rangle \, , \tag{4.15}$$

where the complex expansion coefficients are given by $c_n = \langle \psi_n | \psi \rangle$ and the normalization of $| \psi \rangle$ requires

$$\langle \psi | \psi \rangle = \sum_{m,n} c_m^* c_n \langle \psi_m | \psi_n \rangle = \sum_{m,n} c_m^* c_n \delta_{mn} = \sum_n |c_n|^2 = 1 \, . \tag{4.16}$$

The coefficients may be computed, for instance, in the position representation

$$c_n = \langle \psi_n | \psi \rangle = \int dx \, \langle \psi_n | x \rangle\langle x | \psi \rangle = \int dx \, \psi_n^*(x)\psi(x) \, , \tag{4.17}$$

where we have inserted the completeness relation for the position eigenstates from eqn (3.46). We can then immediately specify the general solution $| \psi(t) \rangle$ of the Schrödinger equation. We just recall that stationary states $| \psi_n \rangle$ transform under time evolution according to Definition 4.1 to find

$$| \psi(t) \rangle = \sum_n c_n e^{-iE_n t/\hbar} | \psi_n \rangle \, . \tag{4.18}$$

All solutions of the Schrödinger equation form a Hilbert space, where a general state vector can be expressed as a superposition of different energy eigenstates. Such superpositions are not time-independent. As time goes by, the components c_n acquire relative phases. In other words, superpositions of stationary states with different energies are no longer stationary states. Nonetheless, any individual measurement of the energy of the system always results in an outcome E_n, i.e., one of the eigenvalues of the Hamiltonian. The probability for the outcome E_n is given by $|c_n|^2$, such that the average energy is obtained as

$$\langle H \rangle_{\psi(t)} = \langle \psi(t) | H | \psi(t) \rangle = \sum_{m,n} c_m^* c_n \langle \psi_m | E_n | \psi_n \rangle = \sum_n |c_n|^2 E_n \, . \tag{4.19}$$

The coefficients c_n may thus be viewed as probability amplitudes for obtaining the nth eigenvalue as the measurement outcome.

4.1.4 Physical Interpretation of the Expansion Coefficients

This role of the expansion coefficients can be understood in much more general terms. Indeed, any state in the Hilbert space may be expanded with respect to the eigenstates $|\, n\,\rangle$ of any self-adjoint operator A with eigenvalues a_n. That is, for

$$A \,|\, n\,\rangle = a_n \,|\, n\,\rangle \,, \tag{4.20}$$

one may write $|\,\psi\,\rangle = \sum_n c_n \,|\, n\,\rangle$, and interpret $c_n = \langle\, n|\,\psi\,\rangle$ as the probability amplitude to obtain the result a_n in a measurement of the observable associated to the operator A. The probability[1] for the outcome a_n is then $|c_n|^2$ and the expectation value of A is simply

$$\langle\, A\,\rangle = \sum_n |c_n|^2 a_n \,. \tag{4.21}$$

When an individual measurement yields the result a_n, the system is said to be projected[2] into the state $|\, n\,\rangle$. If the system is prepared in an eigenstate of A to begin with, e.g., by a previous measurement of A, consecutive measurements of A leave the state unchanged and yield the same eigenvalue every time. In such a case, the uncertainty of A vanishes, $\Delta A = 0$.

The reduction step from $\sum_n c_n \,|\, n\,\rangle$ to $|\, n\,\rangle$, sometimes dubbed "collapse of the wave function", is the cause for some unease for some physicists, and attempts were (and are) therefore made to model this transition as a physical, albeit not directly observable process. We shall instead view the transition from the superposition to some fixed eigenstate as an update of our knowledge about the system, as we shall discuss in more detail in Section 13.4 of Part II.

4.2 Bound States

In this section we turn to bound-state problems, that is, we will consider solutions of the time-independent Schrödinger equation, which do not have enough energy to escape (at least classically) from the grasp of an attractive potential. As we shall learn, the quantum-mechanical solutions exhibit two features that are conceptually quite different from classical bound-state problems. First, the quantum-mechanical solutions are quantized, i.e., at most countably many bound-state energy levels are permitted, of which the solution with the lowest energy, the ground state, need not have vanishing energy. In this sense the quantum-mechanical bound states are more restricted than their classical counterparts. Second, quantum-mechanical particles may have non-vanishing probabilities to enter classically forbidden regions, a phenomenon referred to as the *tunnel effect*, which we will discuss in Section 4.3.

---

[1] Note that this is just the generalization of Born's probability interpretation (Proposition 1.7.2), i.e., when the operator in question is X, with $X \,|\, x\,\rangle = x \,|\, x\,\rangle$, the expansion coefficients are $\langle\, x|\,\psi\,\rangle = \psi(x)$, and $|\psi(x)|^2$ is the probability to obtain the measurement result x.

[2] Recall that the spectral decomposition allows us to write A as a weighted sum of projectors, $A = \sum_n a_n \,|\, n\,\rangle\langle\, n\,|$.

4.2.1 The Finite Potential Well

To begin our investigation of quantum-mechanical bound states, we consider a one-dimensional, finite potential well, i.e., a constant, finite, attractive potential in some region of space that sharply disappears outside of this region. Physically, one may think of this as a model of electrons in a metal, which they cannot leave, while the electrons are free to move within the confines of the atomic lattice. Other examples where a finite potential well serves as a useful model are neutrons in nuclei of heavy atoms (where the strong force is approximately constant in the bulk of baryons, but rapidly diminishes in strength when the distance from it is increased) or so-called quantum dots. To good approximation, the potential describing such situations is

$$V(x) = \begin{cases} 0 & x < -L/2 & \text{region I} \\ -V_0 & |x| \leq L/2 & \text{region II} \\ 0 & x > L/2 & \text{region III} \end{cases} \qquad (4.22)$$

where $L > 0$ and $V_0 > 0$ are real constants of dimensions length and energy, respectively, as illustrated in Fig. 4.1. Now, consider a particle that is subject to this potential. Since any contributions of the kinetic energy of the particle must be non-negative, the total energy E must be larger than the potential depth, i.e., $E \geq -V_0$. There are now two options, either $E \geq 0$ or $E < 0$. In the first case, which we shall consider in Section 4.3, the particle is not bound by the potential well but is scattered instead. For the second case, where $-V_0 \leq E < 0$, which we will discuss now, one would classically expect the particle to be strictly confined to region II, but free to move within it. On average, a classical particle at a fixed energy $E < 0$ should be found equally likely at any position within the well, but never outside of it. As we shall see, neither of these classical expectations is met for quantum-mechanical particles.

To determine the stationary bound states, we first separately consider the regions I, II, and III, where the potential is piecewise constant. In region I, the potential vanishes and the time-independent Schrödinger equation takes the form

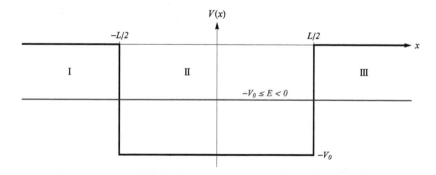

Fig. 4.1 Finite potential well. In a simple one-dimensional model, the attractive potential $V(x) = -V_0 < 0$ is constant in some finite interval of length L ($|x| \leq L/2$, region II), and vanishes everywhere else, that is, for $|x| > L/2$ (regions II and III).

$$\left(\frac{\partial^2}{\partial x^2} + \frac{2mE}{\hbar^2}\right)\psi(x) = \left(\frac{\partial^2}{\partial x^2} - \frac{2m|E|}{\hbar^2}\right)\psi(x) = 0.$$ (4.23)

The general solution of eqn (4.23) is of the form

$$\psi_{\mathrm{I}}(x) = A\,e^{kx} + B\,e^{-kx},$$ (4.24)

where $k = \sqrt{2m|E|}/\hbar > 0$. However, since $x < 0$ in region I, the solution e^{-kx} is not normalizable and we must therefore conclude that $B = 0$. We hence find $\psi_{\mathrm{I}}(x) = A e^{kx}$. In region II, we may write the time-independent Schrödinger equation as

$$\left(\frac{\partial^2}{\partial x^2} + \frac{2m(V_0 - |E|)}{\hbar^2}\right)\psi(x) = \left(\frac{\partial^2}{\partial x^2} + q^2\right)\psi(x) = 0.$$ (4.25)

We recognize this as the equation of motion of a classical harmonic oscillator, that is, we can write the general solution as

$$\psi_{\mathrm{II}}(x) = C\,e^{iqx} + D\,e^{-iqx} = (C+D)\cos(qx) + i(C-D)\sin(qx),$$ (4.26)

with $q = \sqrt{2m(V_0 - |E|)}/\hbar > 0$. In region III, the general solution is of the form of eqn (4.24), but this time $x > 0$, so we have to exclude e^{kx} as a viable option. Collecting these results, we may write the stationary-state wave function as

$$\psi(x) = \begin{cases} A\,e^{kx} & x < -L/2 & \text{region I} \\ C\,e^{iqx} + D\,e^{-iqx} & |x| \le L/2 & \text{region II} \\ F\,e^{-kx} & x > L/2 & \text{region III} \end{cases}.$$ (4.27)

Next, we make use of Lemma 4.3, which suggests that, since $V(x) = V(-x)$, we may pick all eigenvectors to be even or odd functions of x. For the wave functions in eqn (4.27), a parity transformation $x \to -x$ exchanges regions I and II, as well as the constants C and D. The even solutions ψ_+ must hence satisfy $A = F$ and $C = D$, while the constants for the odd solutions ψ_- are of the form $A = -F$ and $C = -D$. In summary, we then have

$$\psi_\pm(x) = \begin{cases} A\,e^{kx} & x < -L/2 & \text{region I} \\ C\left(e^{iqx} \pm e^{-iqx}\right) & |x| \le L/2 & \text{region II} \\ \pm A\,e^{-kx} & x > L/2 & \text{region III} \end{cases}.$$ (4.28)

The remaining constants A and C are then related by requiring the wave function to be *continuous* at the boundaries of the different regions. The conditions $\psi_{\mathrm{I}}(-L/2) = \psi_{\mathrm{II}}(-L/2)$ and $\psi_{\mathrm{II}}(L/2) = \psi_{\mathrm{III}}(L/2)$ provide the relation

$$A\,e^{-kL/2} = C\left(e^{-iqL/2} \pm e^{iqL/2}\right),$$ (4.29)

such that

$$\psi_\pm(x) = A\begin{cases} e^{kx} & x < -L/2 & \text{region I} \\ \pm\dfrac{e^{-kL/2}\left(e^{iqx} \pm e^{-iqx}\right)}{e^{iqL/2} \pm e^{-iqL/2}} & |x| \le L/2 & \text{region II} \\ \pm e^{-kx} & x > L/2 & \text{region III} \end{cases}.$$ (4.30)

Finally, the constant A can be determined from the normalization of the wave function, i.e.,

$$|A|^{-2} \int_{-\infty}^{\infty} dx\, |\psi_\pm(x)|^2 = \int_{-\infty}^{-L/2} dx\, e^{2kx} + \frac{2e^{-kL}}{1 \pm \cos(qL)} \int_{-L/2}^{L/2} dx\, (1 \pm \cos(2qx)) + \int_{L/2}^{\infty} dx\, e^{-2kx}$$

$$(4.31)$$

$$= 2\left[\frac{e^{2kx}}{2k}\right]_{-\infty}^{-L/2} + \frac{e^{-kL}}{1 \pm \cos(qL)} \left[x \pm \frac{\sin(2qx)}{2q}\right]_{-L/2}^{L/2} = e^{-kL}\left(\frac{1}{k} + \frac{1}{q}\frac{Lq \pm \sin(qL)}{1 \pm \cos(qL)}\right),$$

and one arrives at

$$A = e^{kL/2}\left(\frac{1}{k} + \frac{1}{q}\frac{Lq \pm \sin(qL)}{1 \pm \cos(qL)}\right)^{-1/2}. \qquad (4.32)$$

The stationary-state wave functions that we have obtained are now normalized solutions of the time-independent Schrödinger equation and are continuous everywhere. In particular, the probability density has no sudden "jumps". However, the first derivative $\psi'(x) \equiv d\psi(x)/dx$ may still have discontinuities at $x = \pm L/2$. This, in turn, would imply formally infinite second derivatives at these positions, resulting in infinite energies according to the time-independent Schrödinger equation. We shall therefore only be interested in functions with continuous first derivatives. It is convenient to phrase this restriction in terms of the logarithmic derivative $d\ln(\psi)/dx = \psi'/\psi$. For instance, at $x = -L/2$ we require $\psi'_I(-L/2)/\psi_I(-L/2) = \psi'_{II}(-L/2)/\psi_{II}(-L/2)$ and one finds

$$k = iq\,\frac{e^{-iqL/2} \mp e^{iqL/2}}{e^{-iqL/2} \pm e^{iqL/2}} = \begin{cases} +\tan(qL/2) & \text{for } \psi_+ \\ -\cot(qL/2) & \text{for } \psi_- \end{cases}. \qquad (4.33)$$

These equations implicitly determine the energy E, recall that $k = \sqrt{2m|E|}/\hbar$ and $q = \sqrt{2m(V_0 - |E|)}/\hbar$, but, unfortunately, yield no analytic solutions. Such *transcendental* equations may be solved numerically to within any desired precision, but much insight can be gained already from a graphical analysis. To this end, let us first call $z \equiv qL$ and write

$$(kL)^2 + z^2 = \frac{2m(V_0 - |E|)L^2}{\hbar^2} + \frac{2m|E|L^2}{\hbar^2} = \frac{2mV_0L^2}{\hbar^2} \equiv z_0^2. \qquad (4.34)$$

With this nomenclature, eqn (4.33) takes the form

$$\sqrt{\frac{z_0^2}{z^2} - 1} = \begin{cases} +\tan(z/2) & \text{for } \psi_+ \\ -\cot(z/2) & \text{for } \psi_- \end{cases}. \qquad (4.35)$$

For fixed values of L and V_0 the functions on the left- and right-hand side of eqn (4.35) can be plotted, and their intersections provide the allowed values of $z_n = q_n L$ ($n = 1, 2, \ldots$), and thus, the energies

$$E_n = \frac{\hbar^2 z_n^2}{2mL^2} - V_0, \qquad (4.36)$$

as illustrated in Fig. 4.2 (a).

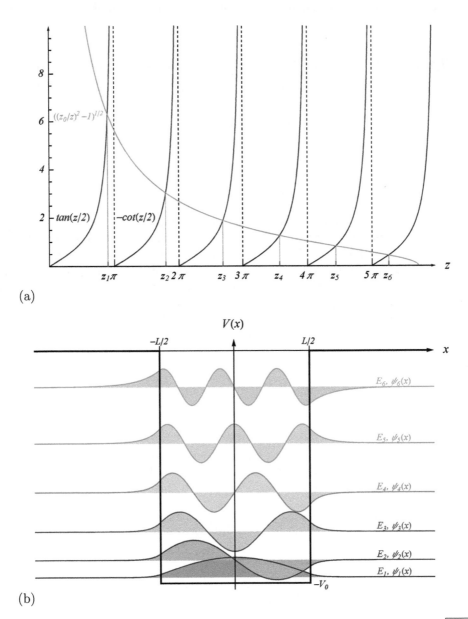

(a)

(b)

Fig. 4.2 Energy levels of the finite potential well. In (a), the functions $\sqrt{\frac{z_0^2}{z^2} - 1}$ (green), $\tan(z/2)$ (blue), and $-\cot(z/2)$ (purple) of the transcendental eqn (4.35) are plotted for the (purely) illustrative value $z_0 = 18$. The intersections yield the allowed values for z_n ($n = 1, 2, \ldots$). (b) shows the corresponding energy levels E_n and stationary-state wave functions ψ_n given by eqns (4.36) and (4.30), respectively.

The graphical approach allows us to make some elementary observations. First, note that the allowed energies are *quantized*, that is, only countably (and in this case finitely) many values are possible. Regardless of the choice of L and V_0, we have $z_n < z_{n+1}$ and $(n-1)\pi < z_n < n\pi$, and the solutions are alternatingly even (for odd n) and odd (for even n). The number of possible energy levels depends on the width and depth of the potential well, increasing the deeper and/or wider the well is. However, there is always[3] at least one (even) bound state when $L > 0$ and $V_0 > 0$. Another observation we can make at this point is that the quantum-mechanical particles with energies $E < 0$ are not strictly confined to the potential well. The probability to find the particle outside the well, that is, in regions I or II, decreases exponentially with the distance from the well, but is larger as E_n increases, as can be seen from the example in Fig. 4.2 (b).

Before we consider energies that are indeed large enough for the particle to permanently leave the well, i.e., scattering, we shall take a closer look at two extremal cases of potential wells. One is infinitely deep, the other is—in a sense—infinitely narrow.

4.2.2 The Infinite Potential Well

Let us now consider the limiting case of an infinitely deep potential well, that is, we take $|E|$ and V_0 to infinity, while keeping $\Delta E = V_0 - |E|$ finite. Ignoring any scattered states, one may regard this procedure as the equivalent of considering a potential $V(x)$ that is infinite for $|x| > L/2$ and vanishes for $|x| < L/2$. Formally, we shall here consider the limit $k \to \infty$, while q is kept finite. The first obvious consequence concerns regions I and II, where the solutions ψ_{I} and ψ_{II} must vanish identically. The particle is therefore now truly confined to the well. We thus drop the region label for the non-trivial solutions in region II and from eqn (4.30), including the normalization constant A from eqn (4.32), we find

$$\psi_{\pm}(x) = \pm\sqrt{\frac{2q}{gL \pm \sin(qL)}} \begin{cases} \cos(qx) & \text{for } \psi_+ \\ \sin(qx) & \text{for } \psi_- \end{cases}. \qquad (4.37)$$

On the other hand, when taking the limit $k \to \infty$ for the transcendental equations in (4.33), we find the conditions for the even and odd solutions to be

$$\psi_+ : \quad q\tan(qL/2) = k \to \infty \qquad \Rightarrow \qquad \frac{qL}{2} = \frac{2n+1}{2}\pi, \quad n = 1, 2, \ldots, \quad (4.38)$$

$$\psi_- : \quad -q\cot(qL/2) = k \to \infty \qquad \Rightarrow \qquad \frac{qL}{2} = n\pi, \quad n = 1, 2, \ldots. \quad (4.39)$$

For the even solutions, qL is an odd-integer multiple of π, for the odd solutions, an even-integer multiple of π. Taken together, we hence find $q_n L = n\pi$ ($n = 1, 2, \ldots$). This simplifies the stationary-state wave functions of eqn (4.37) to

---

[3]This is only true in the one-dimensional problem. As we shall see in Chapter 7, three-dimensional problems generally require a minimal potential depth to admit a bound state.

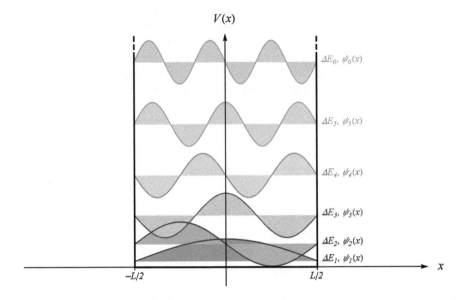

Fig. 4.3 Energy levels of the infinite potential well. The (first six) stationary-state wave functions $\psi_n(x)$ of the infinite potential well (from eqn (4.40)) and their respective energies ΔE_n (from eqn (4.41), as offset from the bottom of the well) are shown.

$$\psi_n(x) = (-1)^{n+1} \sqrt{\frac{2}{L}} \begin{cases} \cos(n\pi x/L) & \text{if } n \text{ is odd} \\ \sin(n\pi x/L) & \text{if } n \text{ is even} \end{cases}, \tag{4.40}$$

where the arbitrary phase factor $(-1)^{n+1}$ may be dropped. The energy levels of the infinite potential well are obtained from $q_n = \sqrt{2m\Delta E_n}/\hbar$ as

$$\Delta E_n = \frac{n^2\pi^2\hbar^2}{2mL^2}. \tag{4.41}$$

Both the stationary-state wave functions and the corresponding energy levels are illustrated in Fig. 4.3.

Note that the energies ΔE_n with respect to the bottom of the well grow as n^2, but the energy difference to the next energy level only grows linearly, i.e., $\Delta E_{n+1} - \Delta E_n = (2n+1)\pi^2\hbar^2/(2mL^2)$. Let us take a closer look at the stationary states that we have obtained. First, it is easy to see that all the $\psi_n(x)$ vanish at $x = \pm L/2$. So instead of starting from a finite well and raising the walls of the potential to infinity, we could have considered an infinite potential to begin with, but demand the boundary conditions $\psi(\pm L/2) = 0$ to arrive at the same solutions. The price one pays for taking the limit as we have described is a discontinuity in the first derivative of the wave function at the boundaries $x = \pm L/2$.

Before we continue with the next section, let us make some brief remarks on the self-adjointness of the Hamiltonian for the infinite potential well. Since we have limited the solutions solely to the region $|x| \leq L/2$, where the potential vanishes, we may simply ask if $H = -\hbar^2/(2m)\partial^2/\partial x^2$ is self-adjoint on the domain of square-integrable functions in the well, i.e., $\mathfrak{L}_2([-L/2, +L/2])$. To do this, we consider an example from Bonneau *et al.* (2001), i.e., the state

$$\psi(x) = \begin{cases} \sqrt{\frac{30}{L^5}} \left(\frac{L^2}{4} - x^2 \right) & \text{if } |x| \leq L/2 \\ 0 & \text{if } |x| > L/2 \end{cases} . \tag{4.42}$$

This state is normalized and continuous (i.e., it vanishes) at the boundaries. Since it is an even function, we can find an expansion in terms of the even basis states $\psi_{2l-1}(x)$ from eqn (4.40), that is $\psi(x) = \sum_{l=1}^{\infty} c_l \psi_{2l-1}(x)$ with

$$c_l = \langle \psi_{2l-1} | \psi \rangle = \sqrt{\frac{60}{L^6}} \int\limits_{-L/2}^{L/2} dx \left(\frac{L^2}{4} - x^2 \right) \cos\left(\frac{(2l-1)\pi x}{L} \right) = \frac{8\sqrt{15}}{\pi^3} \frac{(-1)^{l-1}}{(2l-1)^3} . \tag{4.43}$$

We then have two alternative ways to compute the average energy. On one hand, we may view ψ as the state that provides the energy ΔE_{2l-1} with probability $|c_l|^2$, such that

$$\langle H \rangle = \sum_{l=1}^{\infty} = |c_l|^2 \Delta E_{2l-1} = \frac{480\hbar^2}{\pi^4 m L^2} \sum_{l=1}^{\infty} \frac{1}{(2l-1)^4} = \frac{5\hbar^2}{mL^2} . \tag{4.44}$$

On the other hand, we can straightforwardly compute $H\psi(x) = \sqrt{30/L^5}\hbar^2/m = $ const., from which we find

$$\langle H \rangle = \langle \psi | H\psi \rangle = \frac{30\hbar^2}{mL^5} \int\limits_{-L/2}^{L/2} dx \left(\frac{L^2}{4} - x^2 \right) = \frac{5\hbar^2}{mL^2} , \tag{4.45}$$

and find that both results are consistent. Similarly, we now go on to compute $\langle H^2 \rangle$ and use the self-adjointness of H to find

$$\langle H^2 \rangle = \langle H\psi | H\psi \rangle = \int\limits_{-L/2}^{L/2} dx \frac{30\hbar^4}{m^2 L^5} = \frac{30\hbar^4}{m^2 L^4} . \tag{4.46}$$

However, if H is applied to $H\psi(x)$, which would be valid for a self-adjoint operator, one would obtain $H^2\psi(x) = 0$, and consequently $\langle H^2 \rangle = \langle \psi | H^2\psi \rangle = 0$. This is clearly nonsensical since it would imply $(\Delta E)^2 = \langle H^2 \rangle - \langle H \rangle^2 < 0$. But why does this happen? The reason is simply that H is *not* a self-adjoint operator for all square-integrable functions in $\mathfrak{L}_2([-L/2, L/2])$, but only for functions with appropriate boundary conditions. In other words, one needs to restrict the domain of H, for instance, to $\mathcal{D}_0(H) = \{\psi \in \mathfrak{L}_2([-L/2, L/2]), \psi(\pm L/2) = 0\}$.

The operator $(H, \mathcal{D}_0(H))$ is a self-adjoint extension of H, and not the only possible one. Indeed, recall Theorem 3.2 and the accompanying example, in particular eqns (3.41) and (3.40). For both k_+ and k_-, the linearly independent solutions $e^{k_\pm x}$ and $e^{-k_\pm x}$ are normalizable on $[-L/2, L/2]$, and the deficiency indices are hence $n_\pm = 2$. There are hence infinitely many self-adjoint extension, each one corresponding to particular boundary conditions, i.e., choices of $\psi(\pm L/2)$ and $\psi'(\pm L/2)$, which determine the energy spectrum. For more details, see Bonneau *et al.* (2001). Which of the possible self-adjoint extensions applies is a matter of physical restrictions (such as parity symmetry or positivity of the spectrum) and, ultimately, of matching the predictions with empirical data.

4.2.3   The Dirac-Delta Potential Well

Let us now consider a different limiting case of the finite potential well. Suppose the potential depth is kept fixed at $-V_0$, but the width of the well is taken to be arbitrarily small, that is, the potential is perfectly concentrated at some fixed point, e.g., at $x = 0$. Formally, we may take the limit $L \to 0$ of the solutions of eqn (4.30). Since the middle region II is effectively reduced to just a single point, we are left with the exponentially decreasing tails,

$$\psi_\pm(x) = A \begin{cases} e^{kx} & x < 0 \\ \pm e^{-kx} & x > 0 \end{cases}, \tag{4.47}$$

where the normalization constant can be shown to reduce to $A = \sqrt{k}$ using de l'Hôpital's rule. The antisymmetric solution is discontinuous, and we hence disregard it right away, such that we are left with the stationary-state wave function

$$\psi(x) = \sqrt{k}\, e^{-k|x|}, \tag{4.48}$$

as illustrated in Fig. 4.4. To model the limit $L \to 0$ for the potential itself, we replace the potential by a Dirac delta function,

$$V(x) = -V_0\, \delta(x). \tag{4.49}$$

To obtain the possible energies of the bound states, one would normally make use of the continuity of $\psi(x)$ and $\psi'(x)$. The latter can be easily found as

$$\psi'(x) = k^{3/2} \begin{cases} e^{kx} & x < 0 \\ -e^{-kx} & x > 0 \end{cases}. \tag{4.50}$$

Unfortunately, while the even solution is continuous at $x = 0$, the derivative of $\psi(x)$ is not. To take into account the non-vanishing potential at the origin, we therefore employ a method that exploits the useful properties of the delta function: We integrate the time-independent Schrödinger equation over a symmetric ϵ-interval around $x = 0$ and take the limit $\epsilon \to 0$, i.e., we compute

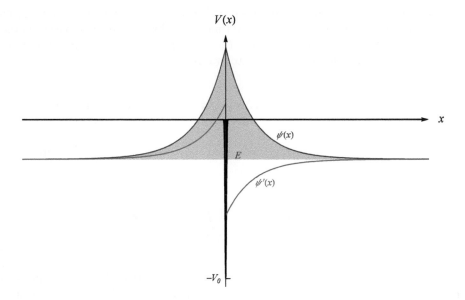

Fig. 4.4 Bound state of the delta potential. The (only allowed) bound state of the Dirac-delta potential is shown. The stationary-state wave function $\psi(x)$ (blue) is continuous, but its derivative $\psi'(x)$ (red) has a finite discontinuity at the origin. The delta potential (here illustrated by a black downward spike) allows us to deal with the discontinuity and obtain a finite bound-state energy E.

$$\lim_{\epsilon \to 0} \int_{-\epsilon}^{\epsilon} dx \left(-\frac{\hbar^2}{2m} \psi''(x) - V_0\, \delta(x)\, \psi(x) \right) = E \lim_{\epsilon \to 0} \int_{-\epsilon}^{\epsilon} dx\, \psi(x) \, . \qquad (4.51)$$

The right-hand side of eqn (4.51) vanishes because $\psi(x)$ is a continuous function. On the left-hand side, the integration over the delta function evaluates the wave function at $x = 0$, such that we get

$$\lim_{\epsilon \to 0} \left(\psi'(\epsilon) - \psi'(-\epsilon) \right) + \frac{2mV_0}{\hbar^2} \psi(0) = -k^{3/2} e^{-0} - k^{3/2} e^0 + k^{1/2} \frac{2mV_0}{\hbar^2} = 0 \, . \quad (4.52)$$

Finally, we note that the energy is related to k via $k^2 = -2mE/\hbar^2$ to arrive at

$$E = -\frac{mV_0^2}{2\hbar^2} \, . \qquad (4.53)$$

We hence find a single possible bound state is present even for a formally infinitely narrow potential, and its energy is determined by the strength V_0 of the potential. However, there is a price to be paid in terms of obtaining a solution with a discontinuous derivative. Previously, we have argued that this is undesirable, but let us now see where things can go wrong in practice.

Suppose that a free wave packet is prepared in the state $\psi(x) = \sqrt{k} \exp(-k|x|)$ of eqn (4.48). For instance, we may think of a situation where the delta potential arises due to an interaction that can be "switched off", taking V_0 to zero very rapidly, but leaving the wave function unperturbed. The time evolution of the wave packet is then simply governed by the free Hamiltonian $H = P^2/(2m)$. As one would expect, the expectation value of the momentum operator vanishes, i.e.,

$$\langle P \rangle_\psi = -i\hbar \int_{-\infty}^{\infty} dx \, \psi^*(x) \, \psi'(x) = -i\,\hbar\,k \left(\int_{-\infty}^{0} dx \, e^{2kx} - \int_{0}^{\infty} dx \, e^{-2kx} \right) = 0 \, . \qquad (4.54)$$

But, when we compute $\langle P^2 \rangle_\psi$ using the second derivative $\psi''(x) = k^{5/2} \exp(-k|x|)$, we get

$$\langle P^2 \rangle_\psi = -\hbar^2 \int_{-\infty}^{\infty} dx \, \psi^*(x) \, \psi''(x) = -\hbar^2 \, k^3 \left(\int_{-\infty}^{0} dx \, e^{2kx} + \int_{0}^{\infty} dx \, e^{-2kx} \right) = -\hbar^2 k^2 \, .$$
$$(4.55)$$

This clearly does not make any sense. The average kinetic energy $\langle P^2 \rangle /(2m)$ should be positive. So where did we go wrong? When the derivative was applied to the continuous function $\psi(x)$, everything still seemed to work out, but then the momentum operator was applied again, this time on a discontinuous function, which is not in the domain of the differential operator $-i\hbar\partial_x$, $\psi'(x) \notin \mathcal{D}(P)$. To check that this is the source of the problem, note that we can compute $\langle P^2 \rangle$ without differentiating any discontinuous functions, that is, via

$$\langle P^2 \rangle_\psi = \langle P\psi | P\psi \rangle = \int_{-\infty}^{\infty} dx \, \left(-i\hbar \, \psi'(x) \right)^* \left(-i\hbar \, \psi'(x) \right) = \hbar^2 k^2 \, . \qquad (4.56)$$

This last result looks reasonable, and one can verify it by working directly in momentum space. A quick computation reveals that the Fourier transform $\tilde{\psi}(p)$ is given by a *Cauchy–Lorentz* distribution (also sometimes called *Cauchy* distribution, *Lorentz* distribution, or *Breit–Wigner* distribution),

$$\tilde{\psi}(p) = \frac{1}{\sqrt{2\pi\hbar}} \int_{-\infty}^{\infty} dp \, \psi(x) \, e^{-i\,p\,x/\hbar} = \sqrt{\frac{2}{\pi\hbar k}} \frac{1}{1 + \left(\frac{p}{\hbar k} \right)^2} \, . \qquad (4.57)$$

The Cauchy–Lorentz distribution is a smooth continuous function, and, in particular, we can confirm that

$$\langle P^2 \rangle_\psi = \int_{-\infty}^{\infty} dp \, p^2 \, |\tilde{\psi}(p)|^2 = \frac{2}{\pi\hbar k} \int_{-\infty}^{\infty} dp \, p^2 \, \frac{1}{\left[1 + \left(\frac{p}{\hbar k} \right)^2 \right]^2} = \hbar^2 k^2 \, . \qquad (4.58)$$

Analogously to the example at the end of Section 4.2.2, we hence find that discontinuities in the derivative of the wave function can cause problems when applying differential operators. In the situations that we have encountered, these difficulties can be handled by paying attention to the domains of the operators H and P, respectively. We can therefore conclude that both the infinitely deep potential well and the infinitely narrow potential well can serve as useful approximations to the finite potential well, and we will consider applications of these models in the next section.

4.2.4 The Double Well and the Ammonia Molecule

In this section we want to apply the techniques we have discussed for solving the time-independent Schrödinger equation to a slightly more complicated bound-state problem with an application in chemistry: the ammonia molecule NH_3, a compound of three hydrogen atoms to which a single nitrogen atom is bound. At an atomic weight of 14 *amu*, the nitrogen atom is much heavier than the hydrogen atoms (1 *amu* each), and it is hence convenient to view it as a fixed point relative to which the three other atoms are arranged. With respect to the position of the nitrogen atom, there are two possible classical equilibrium configurations for the three hydrogen atoms, all on the left-hand side, or all on the right-hand side of the nitrogen atom, as illustrated in Fig. 4.5.

As an approximation to the potential depicted in Fig. 4.5, it proves useful to consider a double-well potential, that is, an infinite potential well, with a finite barrier in the middle. More specifically, for real parameters $a > L/2 > 0$, both of dimension length, we use the potential given by (see also Fig. 4.6)

$$V(x) = \begin{cases} V_0 & \text{for } |x| < a - L/2 \\ 0 & \text{for } a - L/2 \le |x| \le a + L/2 \\ \infty & \text{for } |x| > a + L/2 \end{cases} \qquad (4.59)$$

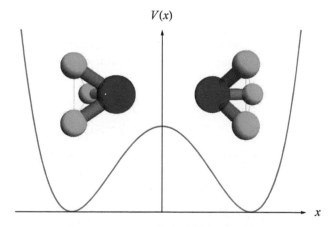

Fig. 4.5 Ammonia molecule. Classically, there are two equilibrium positions for the plane of the H_3-compound (illustrated in green) of ammonia with respect to the nitrogen atom (blue). A qualitative illustration of the one-dimensional potential $V(x)$ that governs the relative distance between N and H_3 is shown. The repulsive Coulomb interaction between the atomic nuclei creates a barrier at the centre, while the chemical bonds between the compounds cause a steep incline of the potential beyond the equilibrium positions.

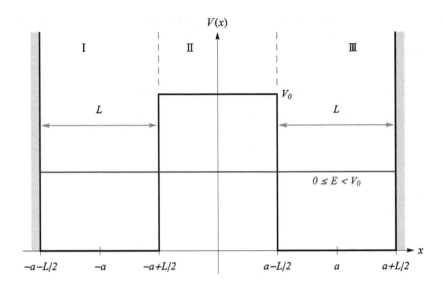

Fig. 4.6 Double-well potential. The relative distance between the nitrogen and tri-hydrogen compounds of ammonia are governed by a potential that is well approximated by a double well with infinitely high walls at $x = \pm(a+L/2)$. The left- and right-hand side regions I and III of width L are separated by a finite barrier of height $V_0 > 0$ (region I).

As before, the potential is symmetric with respect to the origin, $V(x) = V(-x)$, which suggests that a basis of even and odd functions can be selected for the stationary states (recall Lemma 4.3). With this in mind, we consider the regions I and III, where the potential vanishes. The general solutions of the time-dependent Schrödinger equation in these regions must be sine and cosine functions, similar to eqn (4.26). However, since $\cos(x) = \sin(x + \frac{\pi}{2})$, we may express these solutions solely as sine functions. With a little bit of forethought we write

$$\psi_{\text{I}}(x) = A \sin\left(k[a + \tfrac{L}{2} + x]\right), \tag{4.60a}$$

$$\psi_{\text{III}}(x) = \pm A \sin\left(k[a + \tfrac{L}{2} - x]\right), \tag{4.60b}$$

where $k = \sqrt{2mE/\hbar^2}$, the energy is positive, $E > 0$, and we have already included the boundary condition $\psi(x = \pm(a + L/2)) = 0$. In the central region II, the time-dependent Schrödinger equation takes the form

$$\left(\frac{\partial^2}{\partial x^2} - \frac{2m(V_0 - E)}{\hbar^2}\right) \psi_{\text{II}}(x) = \left(\frac{\partial^2}{\partial x^2} - q^2\right) \psi_{\text{II}}(x) = 0, \tag{4.61}$$

with $q = \sqrt{2m(V_0 - E)/\hbar^2}$. The even and odd solutions of eqn (4.61) are the hyperbolic functions

$$\psi_{\text{II}}(x) = B \begin{cases} \cosh(qx) & \text{even} \\ \sinh(qx) & \text{odd} \end{cases}. \tag{4.62}$$

Focusing first on the symmetric solutions, the continuity of the wave function at the boundary between regions I and II, i.e., $\psi_{\mathrm{I}}(-a + L/2) = \psi_{\mathrm{II}}(-a + L/2)$ provides the equation

$$A \sin(kL) = B \cosh\left(q[a - \tfrac{L}{2}]\right). \tag{4.63}$$

The continuity of the derivative of the even wave functions at the same position yields the condition

$$kA \cos(kL) = -qB \sinh\left(q[a - \tfrac{L}{2}]\right). \tag{4.64}$$

Combining eqns (4.63) and (4.64), continuity of the logarithmic derivative $\psi'(x)/\psi(x)$ of the symmetric solutions at $x = -a + L/2$ then results in the constraint

$$\tan(kL) = -\frac{k}{q} \coth\left(q[a - \tfrac{L}{2}]\right), \tag{4.65}$$

and, similarly, the antisymmetric solutions are restricted by the condition

$$\tan(kL) = -\frac{k}{q} \tanh\left(q[a - \tfrac{L}{2}]\right). \tag{4.66}$$

For arbitrary values of a, L, and V_0 these last two equations can be solved numerically to obtain the allowed energies. We shall instead consider a more specialized situation, to gain some analytical understanding. First, we restrict to energies that are *small* compared to the height of the central potential barrier, $E \ll V_0$, such that $q \approx \sqrt{2mV_0/\hbar^2} := q_0$ becomes independent of E at the leading order of the approximation. In addition, we assume a *broad barrier* in region II, i.e., $q(a - L/2) \gg 1$. With this, the transcendental equations, eqns (4.65) and (4.66), can be written as

$$\tan(kL) = -kL \frac{1 \pm 2e^{-q_0(2a-L)}}{q_0 L} = -kL\,\varepsilon_\pm. \tag{4.67}$$

Since $0 < \varepsilon_- < \varepsilon_+$, the graphical solution (see Fig. 4.7) of eqn (4.67) provides the energies $E_\pm = (\hbar k_\pm)^2/(2m)$ where $E_+ < E_-$. For small values $\varepsilon_\pm \ll 1$ it is easy to see that the first two solutions for $k_\pm L$ will be close to π. Expanding $\tan(z)$ around $z = \pi$, eqn (4.67) can be further approximated to $(kL - \pi) = -kL\,\varepsilon_\pm$, which in turn yields the energies

$$E_\pm = \frac{\pi^2 \hbar^2}{2mL^2}(1 - 2\varepsilon_\pm) \tag{4.68}$$

of the lowest two energy levels, with the corresponding wave functions $\psi_\pm$, such that

$$H\,\psi_\pm = E_\pm\,\psi_\pm. \tag{4.69}$$

Let us examine the energy levels more closely. The energy difference between the ground state and the first excited state decays exponentially with the width $(2a - L)$ of the potential barrier, i.e.,

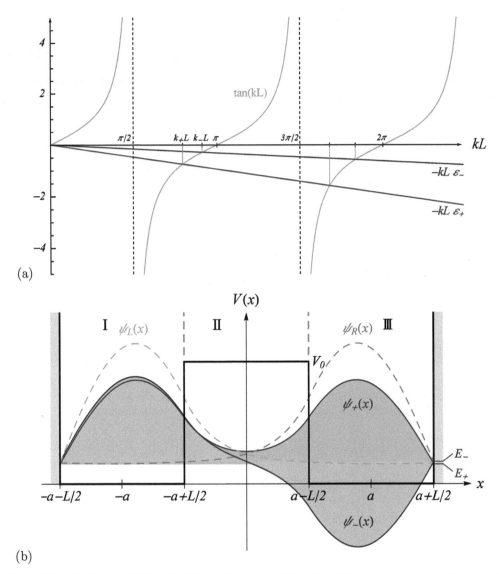

(a)

(b)

Fig. 4.7 Solutions of the double-well potential. (a) In the approximation of a broad barrier, the allowed values for k can be determined from the intersections of $\tan(kL)$ and the straight lines $-kL\,\varepsilon_\pm$ for the even (ε_+) and odd (ε_-) low-energy solutions. (b) The corresponding even (blue) and odd (purple) stationary states $\psi_+(x)$ and $\psi_-(x)$ with energies E_+ and E_-, respectively, are not localized to one of the sides of the double well, but their (non-stationary) superpositions $\psi_L = (\psi_+ + \psi_-)/\sqrt{2}$ and $\psi_R = (\psi_+ - \psi_-)/\sqrt{2}$ are mostly concentrated on the left- and right-hand side, respectively.

$$\Delta E = E_- - E_+ = \frac{\pi^2 \hbar^2}{mL^2}(\varepsilon_+ - \varepsilon_-) = \frac{\pi^2 \hbar^2}{mL^2} \frac{4e^{-q_0(2a-L)}}{q_0 L}. \qquad (4.70)$$

Indeed, when the barrier becomes very wide or high (which would correspond to the limit $q \to \infty$), we arrive at two separate infinite potential wells, each with an energy spectrum given by eqn (4.41). However, for finite height and width of the barrier, the energy levels are not degenerate. Intuitively this can be understood as a tunnelling effect, where the wave functions of the well on the left-hand side extend to the right-hand side of the well, and vice versa. Nonetheless, neither of the states $\psi_\pm$ corresponds to the classical equilibrium positions, since neither is localized on one side or the other. However, this can be quickly remedied by constructing the equally weighted superpositions

$$\psi_{L/R} = \frac{1}{\sqrt{2}} \left(\psi_+(x) \pm \psi_-(x) \right), \qquad (4.71)$$

as illustrated in Fig. 4.7. But, as the states $\psi_{L/R}$ are not stationary, their time evolution is non-trivial. That is, when the system is prepared in the state ψ_L at time $t = 0$, the state at time t is given by

$$\psi_L(t) = \frac{1}{\sqrt{2}} \left(e^{-iE_+ t/\hbar} \psi_+ + e^{-iE_- t/\hbar} \psi_- \right) = \frac{e^{-iE_+ t/\hbar}}{\sqrt{2}} \left(\psi_+ + e^{-i\Delta E t/\hbar} \psi_- \right). \quad (4.72)$$

In other words, a molecule initially in a classical equilibrium position oscillates between the left- and right-hand side at the Bohr frequency $\omega = \Delta E/\hbar$. This oscillation is the source of electromagnetic radiation that can be detected and the emitting systems can be identified via their specific frequencies. For instance, for ammonia the energy splitting is $\Delta E \approx 10^{-4}$ eV. The emitted radiation is hence in the microwave regime, $\nu = \omega/2\pi \approx 24$ GHz. Using spectroscopy to analyse light received from interstellar space, the presence of ammonia molecules in these regions can be verified (Cheung *et al.*, 1968).

4.3 Scattering and the Tunnel Effect

4.3.1 The Finite Potential Barrier

As we have seen in the case of the double well in Section 4.2.4, repulsive potential barriers do not fully prevent quantum-mechanical particles from entering classically forbidden regions. We have seen that, even if the energy is smaller than the height of the potential barrier, $E < V_0$, a particle first localized on the left-hand side may be found on the right-hand side after some time. To better understand this phenomenon, let us view it without the confines of the surrounding infinite potential well, i.e., we now consider the potential

$$V(x) = \begin{cases} V_0 & \text{for } |x| < L/2 \\ 0 & \text{for } |x| \geq L/2 \end{cases}, \qquad (4.73)$$

as illustrated in Fig. 4.8. The general solutions to the time-independent Schrödinger equation in regions I $(x \leq -L/2)$ and III $(x \geq L/2)$ are plane waves $e^{\pm ikx}$, where

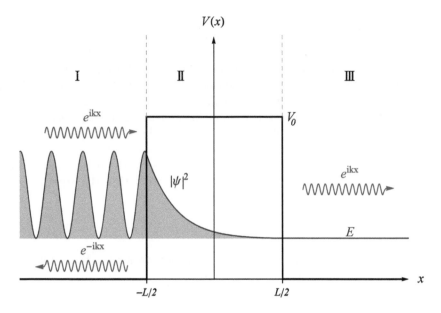

Fig. 4.8 Scattering from a potential wall The repulsive potential $V_0 > 0$ in region II is too large to allow classical particles with energies $E < V_0$ to pass from the left (region I) to the right (region III) of the barrier. The wave functions of quantum-mechanical particles, on the other hand, also permit non-zero amplitudes for the transmitted plane waves e^{ikx} in region III, in addition to the amplitudes of the ingoing (e^{ikx} in region I) and reflected (e^{-ikx}) plane waves.

$k = \sqrt{2mE/\hbar^2} > 0$. These wave functions carry momenta $\pm k$, which can easily be seen by an application of the momentum operator $-i\hbar\partial/\partial_x$, and they are propagating to the right ($+k$) and to the left ($-k$), respectively. If we assume, as we have previously, that a particle is prepared in region I, it seems reasonable to further require that no leftward-propagating plane waves are incident from the right-hand side of region III. We may then write the general solutions for a particle of energy $0 < E < V_0$ in all regions as

$$\psi(x) = \begin{cases} A\,e^{ikx} + B\,e^{-ikx} & \text{I} \\ C\,e^{qx} + D\,e^{-qx} & \text{II} \\ F\,e^{ikx} & \text{III} \end{cases}, \qquad (4.74)$$

where $q = \sqrt{2m(V_0 - E)/\hbar^2}$.

4.3.2   Reflection and Transmission

The continuity conditions for the wave function and its derivative at $x = \pm L/2$ supply a system of four linear equations for the five constants A, B, C, D, and F, i.e.,

$$\psi_{\mathrm{I}}(-\tfrac{L}{2}) = \psi_{\mathrm{II}}(-\tfrac{L}{2}): \qquad A\,e^{-ikl/2} + B\,e^{ikL/2} = C\,e^{-qL/2} + D\,e^{qL/2}, \qquad (4.75a)$$

$$\psi_{\mathrm{I}}'(-\tfrac{L}{2}) = \psi_{\mathrm{II}}'(-\tfrac{L}{2}): \quad ik\big(A\,e^{-ikl/2} - B\,e^{ikL/2}\big) = q\big(C\,e^{-qL/2} - D\,e^{qL/2}\big), \quad (4.75b)$$

$$\psi_{\mathrm{II}}(\tfrac{L}{2}) = \psi_{\mathrm{III}}(\tfrac{L}{2}): \qquad C\,e^{qL/2} + D\,e^{-qL/2} = F\,e^{ikL/2}, \qquad (4.75c)$$

$$\psi_{\mathrm{II}}'(\tfrac{L}{2}) = \psi_{\mathrm{III}}'(\tfrac{L}{2}): \qquad q\big(C\,e^{qL/2} - D\,e^{-qL/2}\big) = ik\,F\,e^{ikL/2}. \qquad (4.75d)$$

For square-integrable functions, the normalization condition would provide a fifth equation to determine all constants. Here, however, the plane-wave solutions in regions I and III are not normalizable. Nonetheless, one may still eliminate four out of the five constants and compare the amplitudes for the ingoing (A), reflected (B), and transmitted (F) plane waves. The quotient of eqns (4.75c) and (4.75d) yields the relation

$$D = C\,e^{qL}\,\frac{q - ik}{q + ik}. \qquad (4.76)$$

Inserting this expression back into eqn (4.75c), one immediately arrives at

$$F = C\,e^{-ikL/2}\,e^{qL/2}\,\frac{2q}{q + ik}. \qquad (4.77)$$

With this, eqns (4.75a) and (4.75b) become

$$A\,e^{-ikl/2} + B\,e^{ikL/2} = C\,e^{-qL/2}\left(1 + e^{2qL}\,\frac{q - ik}{q + ik}\right), \qquad (4.78a)$$

$$ik\big(A\,e^{-ikl/2} - B\,e^{ikL/2}\big) = q\,C\,e^{-qL/2}\left(1 - e^{2qL}\,\frac{q - ik}{q + ik}\right), \qquad (4.78b)$$

and, forming their quotient, a few lines of algebra reveal that A and B are related by

$$A = -B\,\frac{e^{ikL}}{q^2 + k^2}\left(q^2 - k^2 - 2iqk\,\coth(qL)\right). \qquad (4.79)$$

Now, since A and B are the amplitudes of the ingoing right-moving and reflected left-moving plane waves, respectively, we may consider $|B|^2/|A|^2$ as the probability for a reflection from the barrier. From eqn (4.79) we find the *reflection coefficient*

$$R = \frac{|B|^2}{|A|^2} = \left(1 + \frac{4q^2 k^2}{(q^2 + k^2)^2 \sinh^2(qL)}\right)^{-1}. \qquad (4.80)$$

The probability for the reflection is generally smaller than one, and the complementary probability of transmission $T = 1 - R$ must therefore be non-zero. That is, the *transmission coefficient* is given by

$$T = \frac{|F|^2}{|A|^2} = 1 - R = \left(1 + \frac{(q^2 + k^2)^2}{4q^2 k^2}\,\sinh^2(qL)\right)^{-1}. \qquad (4.81)$$

As we see, the coefficient that determines the probabilities for reflection and transmission is

$$\frac{|B|^2}{|F|^2} = \frac{(q^2 + k^2)^2}{4q^2 k^2}\,\sinh^2(qL) = \frac{V_0^2}{4E(V_0 - E)}\,\sinh^2\sqrt{\frac{2m(V_0 - E)L^2}{\hbar^2}}. \qquad (4.82)$$

4.3.3   Tunnelling and the Gamow Factor

The fact that the transmission coefficient for energies below the classical threshold is non-zero is referred to as the *tunnel effect*. In a typical situation, the energy of the particle is non-zero, but not close to V_0 either. In this regime, the transmission probability is governed by the argument qL of the hyperbolic sine. Indeed, for $qL \gg 1$, we have $\sinh(qL) \approx e^{qL}/2$ and, with the abbreviation $x = E/V_0$, we find

$$T(E) \approx \frac{4x(1-x)}{4x(1-x) + e^{2qL}/4} \approx 16x(1-x)\,e^{-2qL} = \exp\!\left(-2qL + \ln\!\left[16x(1-x)\right]\right),$$

$$(4.83)$$

where we have used the fact that $0 \leq 4x(1-x) \leq 1$ is much smaller than $e^{2qL} \gg 1$. Furthermore, the function $\left|\ln\!\left[16x(1-x)\right]\right|$ is much smaller than $qL \gg 1$ as long as x is not very close to either 0 or 1, and may therefore be neglected to arrive at

$$T(E) \approx \exp\!\left(-\frac{2L}{\hbar}\sqrt{2m(V_0 - E)}\right).$$

$$(4.84)$$

For fixed energy $E < V_0$, the transmission probability is hence exponentially suppressed with increasing barrier width. However, in many practical situations there is no particular reason to expect that the potential $V(x)$ has the shape of a rectangular barrier. It is therefore of interest to develop an expression for the transmission probability for more general potentials.

To this end, let us first consider not one, but two consecutive potential walls of the same width $L/2$, but with different heights, V_1 and V_2, respectively. For an incident particle of energy E, the probability to pass both barriers is the product $T(E, V_1)T(E, V_2)$ of the probabilities $T(E, V_1)$ and $T(E, V_2)$ of being transmitted through each of the individual barriers. Multiplying the corresponding expressions from eqn (4.83), the product of the exponential functions becomes a sum of the exponents, i.e.,

$$T(E) = T(E, V_1)T(E, V_2) \approx \exp\!\left(-2\left[\frac{L}{2}\frac{\sqrt{2m(V_1 - E)}}{\hbar} + \frac{L}{2}\frac{\sqrt{2m(V_2 - E)}}{\hbar}\right]\right).$$

$$(4.85)$$

We may then consider any continuous function $V(x)$ within an interval $[x_0, x_N]$ of length $L = x_N - x_0$ as being approximated by N rectangular potential barriers of finite widths L/N and heights $V_n = V(\frac{x_{n-1}+x_n}{2})$ $(n = 1, 2, \ldots, N)$, see Fig. 4.9. In the limit of an infinite fine-graining of this decomposition, that is, for the limit $N \to \infty$ such that $L = \text{const.}$, we arrive at an integral, the so-called *Gamow factor*

$$T(E) \approx \exp\!\left(-\frac{2}{\hbar}\int_{x_{\min}}^{x_{\max}} dx\sqrt{2m(V(x) - E)}\right),$$

$$(4.86)$$

named after *George Gamow*, who first developed a formula of this type in 1928 for the description of *alpha decay*, as illustrated in Fig. 4.10.

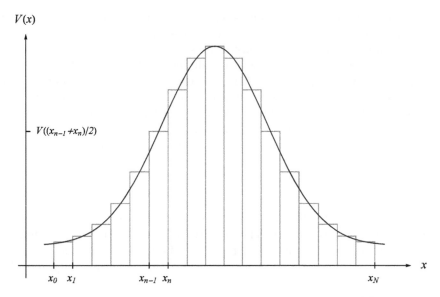

Fig. 4.9 Gamow factor. The transmission coefficient for a potential $V(x)$ may be computed by approximating the potential by a series of rectangular barriers. In the limit of infinitely many barriers of infinitesimal width, one arrives at an integral expression, eqn (4.86), the Gamow factor.

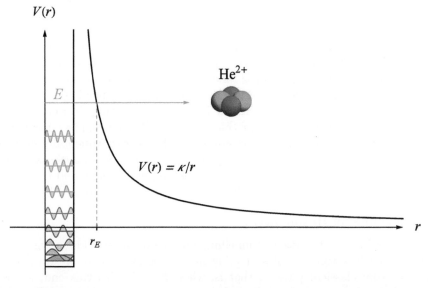

Fig. 4.10 Alpha decay. The strong interaction between the nucleons in a heavy atom can be modelled as a potential well. When four of these nucleons form a He^{2+} ion (an α particle) within the nucleus, they acquire the necessary energy to tunnel through the Coulomb barrier that repels positive charges outside.

This type of radioactive decay occurs when two protons and two neutrons within a heavy atomic nucleus (e.g., uranium) spontaneously form a stable He^{2+} compound. Whereas the remaining individual nucleons are still tightly bound in one of the lower energy eigenstates of the nucleus (recall the discussion in Section 4.2.1), the newly formed α-particle has gained some additional energy uniting the two protons and two neutrons. This boost of energy greatly increases the probability to tunnel through the Coulomb barrier, which separates the nucleus from other positively charged particles. Assuming the nucleus to be perfectly point-like and located at the origin, we may write the Coulomb potential for this one-dimensional problem as

$$V(r) = \frac{Qq_\alpha}{4\pi\epsilon_o}\frac{1}{r} = \frac{\kappa}{r}, \tag{4.87}$$

where Q and q_α are the electric charges of the nucleus and the α-particle, respectively, and ϵ_o is the vacuum permittivity. The potential of eqn (4.87) extends infinitely far in principle, but we may consider the α-particle to be free once it is far enough away at some $r > r_E = \kappa/E$ such that $V(r_E) = E$. The tunnelling probability given by the Gamow factor of eqn (4.86) is then

$$T(E) \approx \exp\left(-\frac{\sqrt{8m\kappa}}{\hbar}\int_0^{\kappa/E} dr\sqrt{\frac{1}{r} - \frac{E}{\kappa}}\right) = e^{-\sqrt{\frac{2m\pi^2\kappa^2}{\hbar^2 E}}}. \tag{4.88}$$

The tunnel effect is a characteristic quantum-mechanical phenomenon that can be used to explain many interesting observations. An example are electrons tunnelling through layers of insulating materials, for instance in semiconductors, an effect discovered by *Leo Esaki* in 1958, which led to the development of *tunnelling diodes* (also called Esaki diodes). In superconductors, that is, in certain metals at very low temperatures, two electrons may form what is called a Cooper pair. This pairing allows them to move practically unhindered through the bulk of atomic nuclei. When two pieces of superconducting material are separated by a thin insulating barrier, the Cooper pairs may collectively tunnel through this obstacle. This mechanism forms the basis for what is known as the *Josephson effect*, which was experimentally observed in 1960 by *Ivar Giaever* and theoretically explained by *Brian David Josephson* in 1962. For their respective discoveries of electronic tunnelling phenomena and the Josephson effect, Easki, Giaever, and Josephson shared the 1973 Nobel Prize in Physics.

4.3.4   Transmission Resonances

Having discussed cases of particles tunnelling into classically forbidden regions, let us now turn to the investigation of scattering in the regimes where we do not expect obstructions from classical physics, that is, when $E > V_0$. To this end, let us take a closer look at the range of validity of the transmission coefficient of eqn (4.81). Although we had assumed that $E < V_0$ initially, eqns (4.79)–(4.82) are well behaved when we transition to energies larger than the height of the potential barrier, i.e., where $E > V_0$. Simply note that, while $V_0^2/[4E(V_0-E)] = -V_0^2/[4E(E-V_0)]$ generates a sign flip, an additional sign switch is generated from squaring $\sinh\sqrt{-2m(E - V_0)/\hbar^2} =$

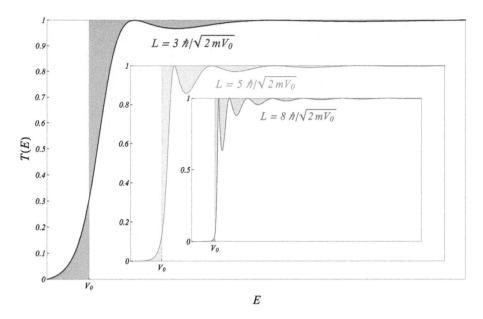

Fig. 4.11 Transmission coefficient. The transmission coefficient $T(E)$ for a potential barrier of height $V_0 > 0$ and width (in units $\hbar/\sqrt{2mV_0}$) $L = 2$ (blue), $L = 5$ (green), $L = 8$ (orange) is plotted as a function of the energy E of an incident plane wave. As the width of the barrier increases, the transmission probability for energies $E < V_0$ drastically decreases.

$\pm i \sin \sqrt{2m(E - V_0)/\hbar^2}$. We hence obtain a regular sine function in eqn (4.81) instead of the hyperbolic sine when $E > V_0$. Collecting these results for the transmission coefficient, we arrive at

$$T = \begin{cases} \left(1 + \dfrac{V_0^2}{4E(V_0-E)} \sinh^2 \sqrt{\dfrac{2m(V_0-E)L^2}{\hbar^2}}\right)^{-1} & \text{for } 0 < E < V_0 \,, \\[3mm] \left(1 + \dfrac{V_0^2}{4E(E-V_0)} \sin^2 \sqrt{\dfrac{2m(E-V_0)L^2}{\hbar^2}}\right)^{-1} & \text{for } V_0 < E \,. \end{cases} \tag{4.89}$$

Moreover, note that there is no longer any need to keep V_0 positive, i.e., we may consider scattering from an attractive potential, like the potential wells we have considered in Section 4.2.1, as long as $E > 0$. An immediate consequence of the sine appearing in $T(E)$ for $V_0 < E$ is an oscillatory character of the transmission probability, see Fig. 4.11.

We find that, while classical particles are always transmitted when $E > V_0$, the transmission probability for a quantum-mechanical particle has local minima when the argument of the sine function satisfies $(2n+1)\pi/2$ with $n \in \mathbb{N}$. The potential wall (or well for $V_0 < 0$) becomes transparent only at specific resonances, that is, when $qL = \sqrt{2m(E-V_0)L^2/\hbar^2} = n\pi$ for integer n, and hence only for energies

$$E_n = \frac{n^2\pi^2\hbar^2}{2mL^2} + V_0 \,. \tag{4.90}$$

Note that, apart from the constant offset V_0, these resonance energies coincide with the energy levels of the infinite potential well of eqn (4.41). To examine the resonance more closely, let us expand the coefficient $|B|^2/|F^2| = T(E)^{-1} - 1$ of eqn (4.82) in a Taylor series around the resonance energies. To this end, we compute the derivatives with respect to E, and evaluate them at $E = E_n$, i.e.,

$$\frac{\partial}{\partial E} \frac{V_0^2}{4E(E-V_0)} \sin^2 \sqrt{\frac{2m(E-V_0)L^2}{\hbar^2}} = \left(\frac{\partial}{\partial E} \frac{V_0^2}{4E(E-V_0)} \right) \sin^2 \sqrt{\frac{2m(E-V_0)L^2}{\hbar^2}}$$

$$+ \frac{V_0^2}{4E(E-V_0)} \sqrt{\frac{2mL^2}{\hbar^2(E-V_0)}} \sin \sqrt{\frac{2m(E-V_0)L^2}{\hbar^2}} \cos \sqrt{\frac{2m(E-V_0)L^2}{\hbar^2}}. \qquad (4.91)$$

Since the resonance condition demands that the sine functions vanish at $E = E_n$, the linear terms of the expansion disappear, i.e.,

$$\frac{\partial}{\partial E} \frac{V_0^2}{4E(E-V_0)} \sin^2 \sqrt{\frac{2m(E-V_0)L^2}{\hbar^2}} \bigg|_{E=E_n} = 0. \qquad (4.92)$$

This insight significantly simplifies the computation of the second derivative, which we immediately evaluate at $E = E_n$ to find only the cosine term remaining. Since $\cos^2 \sqrt{2m(E_n - V_0)L^2/\hbar^2} = 1$, we arrive at

$$\frac{\partial^2}{\partial^2 E} \frac{V_0^2}{4E(E-V_0)} \sin^2 \sqrt{\frac{2m(E-V_0)L^2}{\hbar^2}} \bigg|_{E=E_n} = \frac{1}{E_n} \frac{m^3 L^6 V_0^2}{n^4 \pi^4 \hbar^6}. \qquad (4.93)$$

Inserting into the Taylor expansion of $|B|^2/|F^2| = T(E)^{-1} - 1$, given by

$$\left(\frac{1}{T(E)} - 1 \right) = \left(\frac{1}{T(E_n)} - 1 \right) + \tfrac{1}{2}(E - E_n)^2 \frac{\partial^2}{\partial^2 E} \left(\frac{1}{T(E)} - 1 \right) \bigg|_{E=E_n} + \mathcal{O}((E - E_n)^2),$$
$$(4.94)$$

the transmission coefficient close to the resonance can be written as

$$T(E) = \frac{\frac{\Gamma_n^2}{4}}{\frac{\Gamma_n^2}{4} + (E - E_n)^2}, \qquad (4.95)$$

which we recognize as a Cauchy–Lorentz distribution, where

$$\Gamma_n = \left(\frac{2\hbar^2}{mL^2} \right)^{3/2} \frac{\sqrt{E_n}}{|V_0|} n^2 \pi^2 \qquad (4.96)$$

is the full width at the half height of the Cauchy–Lorentz distribution of the nth resonance, see Fig. 4.12. As n increases, the resonances become less pronounced and the Cauchy–Lorentz distributions become wider. The results of eqns (4.95) and (4.96) also apply when the particle is scattered from a potential well, i.e., for $V_0 < 0$. One only needs to keep in mind that $E_n = (n\pi\hbar)^2/(2mL^2) - |V_0| > 0$, and hence the transmission resonances occur for integers $n > \sqrt{2m|V_0|L^2/(\pi\hbar)^2}$. One can then observe that the resonances become more pronounced, that is, the half-widths Γ_n become smaller, as the depth $|V_0|$ of the potential well increases.

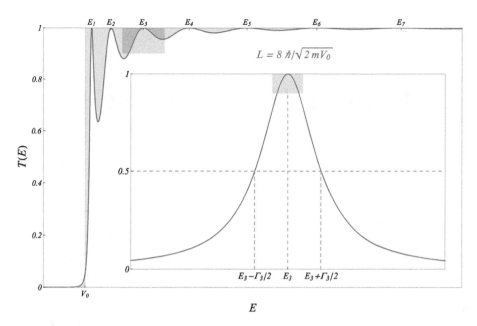

Fig. 4.12 Transmission resonances. Close to the resonances, the transmission coefficient $T(E)$, here shown for a potential barrier of height $V_0 > 0$ and width $L = 8\hbar/\sqrt{2mV_0}$, has the shape of a Cauchy–Lorentz distribution, see eqn (4.95).

The resonance of transmission can be nicely seen in the scattering of slow electrons in a noble gas (e.g., Ne, Ar, Xe) which has been studied independently by *Carl Ramsauer* (1921), as well as *John Sealy Townsend* and his student *Victor Albert Bailey* (1921). The probability for the electrons to collide with the gas particles, which classically should decrease monotonically for increasing energy, is observed to reach local minima for certain energies. This *Ramsauer–Townsend–Bailey*[4] effect is in complete agreement with the quantum-mechanical prediction of the transparency of the potential well for the resonance energies.

[4]This effect is sometimes called Ramsauer effect, Townsend effect, or Ramsauer–Townsend effect, but we felt it only fair to give credit also to Townsend's student V. A. Bailey.

5

The Quantum Harmonic Oscillator

As we have seen, piece-wise constant potentials can be handled well within the framework of Schrödinger's equation, and scattering from smooth potentials can be approximated using Gamow's formula in eqn (4.86). In this chapter we want to study the bound states of a particular smooth potential, the *harmonic oscillator*. The significance of the harmonic potential is twofold. Firstly, its widespread occurrence throughout all areas of quantum theory, from the description of the motion of ions trapped in magnetic fields, and the vibrations of molecular structures, to the oscillation of charges in superconductors or propagating electromagnetic fields, to name but a few. Secondly, understanding the algebraic techniques introduced for solving the Schrödinger equation of the harmonic potential will prove to be paramount also for discussing other quantities of interest, such as orbital angular momentum, which we will take a closer look at in Chapter 6. There is thus great pedagogical value in studying the harmonic oscillator. For now, it will nonetheless be most useful to think of the analogue of a classical pendulum. That is, a point-like particle of mass m that experiences a force proportional to the distance from the rest position (without loss of generality we may consider the origin), at least for small distances. The corresponding potential is given by

$$V(x) = \frac{m\omega^2}{2} x^2 \, .\tag{5.1}$$

The potential in eqn (5.1) is positive for all values of x, and we will therefore be interested in determining its bound states. In principle, we may go about this the same way as in Chapter 4 and solve the time-independent Schrödinger equation with the potential of eqn (5.1). And, indeed, we shall use this *analytic* method in Section 5.2 for the sake of completeness. Before we do so, however, we shall use the extremely powerful *algebraic* approach.

5.1 Algebraic Method

We begin again with the time-independent Schrödinger equation,

$$H\psi = \left(-\frac{\hbar^2}{2m} \frac{\partial^2}{\partial x^2} + \frac{m\omega^2}{2} x^2 \right) \psi = E\psi \, ,\tag{5.2}$$

where we have included the harmonic-oscillator potential from (5.1). In terms of the position and momentum operators X and P the Hamiltonian can be written as

$$H = \tfrac{1}{2m} \left[P^2 + (m\omega X)^2 \right] = \tfrac{1}{2m} \left[P^2 + \tilde{X}^2 \right]\tag{5.3}$$

where we have introduced the operator $\tilde{X} = m\omega X$.

5.1.1 Annihilation and Creation Operators

Our aim will now be to factorize the Hamiltonian. Preferably, we would like to express H as the square of some yet unknown operator. To this end, note that the commutator of $\tilde{X}$ and P evaluates to a constant, i.e., $[\tilde{X}, P] = im\omega\hbar$ (recall Theorem 2.2), and hence

$$(\tilde{X} + iP)(\tilde{X} - iP) = P^2 + \tilde{X}^2 - i[\tilde{X}, P] = P^2 + \tilde{X}^2 + m\omega\hbar. \qquad (5.4)$$

Equation (5.4) suggests that, up to a constant, the Hamiltonian in eqn (5.3) can be factorized by defining new operators a and $a^\dagger$ as:

Definition 5.1 *The **annihilation operator** a and the **creation operator** $a^\dagger$ are defined as*

$$a := \frac{1}{\sqrt{2m\omega\hbar}}(m\omega X + iP),$$

$$a^\dagger := \frac{1}{\sqrt{2m\omega\hbar}}(m\omega X - iP).$$

The names of these operators already signify their interpretation. They annihilate and create one quantum of energy $\hbar\omega$, respectively, as we shall soon substantiate. In terms of a and $a^\dagger$, the position and momentum operators may be written as

$$X = \sqrt{\frac{\hbar}{2m\omega}}\left(a + a^\dagger\right), \qquad (5.5a)$$

$$P = -i\sqrt{\frac{m\omega\hbar}{2}}\left(a - a^\dagger\right). \qquad (5.5b)$$

Next, we calculate the commutator of the creation and annihilation operators. Obviously, each operator commutes with itself,

$$[a, a] = \left[a^\dagger, a^\dagger\right] = 0. \qquad (5.6)$$

To obtain the commutator of a with $a^\dagger$, we first calculate the products $aa^\dagger$ and $a^\dagger a$,

$$aa^\dagger = \frac{1}{2m\omega\hbar}\left[(m\omega X)^2 - im\omega\,[X, P] + P^2\right], \qquad (5.7a)$$

$$a^\dagger a = \frac{1}{2m\omega\hbar}\left[(m\omega X)^2 + im\omega\,[X, P] + P^2\right]. \qquad (5.7b)$$

We hence find that the canonical commutator $[X, P] = i\hbar$ translates to the simple commutation relation

$$\left[a, a^\dagger\right] = 1. \qquad (5.8)$$

Comparing the Hamiltonian from eqn (5.3) with the expressions in (5.7), we can immediately write

$$H = \frac{1}{2m} \left(P^2 + (m\omega X)^2 \right) = \frac{\hbar\omega}{2} \left(a^\dagger a + a\, a^\dagger \right) . \tag{5.9}$$

Further making use of the commutator in eqn (5.8) we finally arrive at the factorized Hamiltonian

$$H = \hbar\omega(a^\dagger a + \tfrac{1}{2}) . \tag{5.10}$$

5.1.2   The Occupation-Number Operator

In other words, the Schrödinger equation from (5.2) can now be phrased as an eigenvalue equation for the operator $a^\dagger a$, that is,

$$a^\dagger a \psi = \left(\frac{E}{\hbar\omega} - \frac{1}{2} \right) \psi , \tag{5.11}$$

where the eigenvalues of $a^\dagger a$, up to an additive constant of $\frac{1}{2}$, are energies in units of $\hbar\omega$. Because of its prominent role and frequent occurrence, let us define this operator.

Definition 5.2   *The **occupation-number** (or particle-number) **operator** is defined as*

$$N := a^\dagger a .$$

For the following discussion it will be useful to consider the commutators of N with a and $a^\dagger$, respectively. It is easy to verify that these evaluate to

$$\left[N, a^\dagger \right] = a^\dagger , \tag{5.12a}$$

$$\left[N, a \right] = -a . \tag{5.12b}$$

Since the eigenstates of N are also eigenstates of the Hamiltonian, we may concentrate just on N and determine the solutions to the equation

$$N \, |\, \psi_n \,\rangle = n \, |\, \psi_n \,\rangle , \tag{5.13}$$

where n are the real eigenvalues of the Hermitian operator N and we have switched to the basis-independent Dirac notation $|\, \psi_n \,\rangle$ instead of the position representation $\psi_n(x)$ for now. When we apply $\langle\, \psi_n \,|$ to the right-hand side of eqn (5.13), we just retrieve the eigenvalue, $\langle\, \psi_n \,|\, N \,|\, \psi_n \,\rangle = n$. On the other hand, we can write

$$n = \langle\, \psi_n \,|\, N \,|\, \psi_n \,\rangle = \langle\, \psi_n \,|\, a^\dagger a \,|\, \psi_n \,\rangle = \langle\, a\psi_n \,|\, a\psi_n \,\rangle \geq 0. \tag{5.14}$$

We hence find that $n \geq 0$ by virtue of the positive definiteness of the scalar product, see eqn (2.22d).

5.1.3   The Ground State of the Harmonic Oscillator

The non-degeneracy of the scalar product, eqn (2.22e), further implies that for $n = 0$ we must have

$$a \, | \, \psi_0 \rangle = 0 \, , \tag{5.15}$$

that is, the ground state of the harmonic oscillator is annihilated by the annihilation operator. We may use this condition to determine the position representation $\psi_0(x)$ of the ground state, by inserting the annihilation operator from Definition 5.1 into the condition of eqn (5.15). This yields the differential equation

$$a \, \psi_0(x) = \frac{1}{\sqrt{2m\omega\hbar}} \Big(m\omega x + \hbar \frac{d}{dx} \Big) \psi_0(x) = 0 \, . \tag{5.16}$$

Using the methods of separation of variables we can integrate, i.e.,

$$\int \frac{d\psi_0}{\psi_0} = - \int dx \, \frac{m\omega}{\hbar} x \quad \Rightarrow \quad \ln \psi_0 = - \frac{m\omega}{2\hbar} x^2 + \ln \mathcal{N} \, , \tag{5.17}$$

where, with a little bit of forethought, we have written the integration constant as $\ln \mathcal{N}$. Up to the normalization constant $\mathcal{N}$, we hence find the harmonic-oscillator ground-state wave function to be given by

$$\psi_0(x) = \mathcal{N} e^{-\frac{x^2}{2\sigma_o^2}} \, , \tag{5.18}$$

where we have introduced the characteristic length $\sigma_o = \sqrt{\hbar/(m\omega)}$. We see that the ground state of the harmonic oscillator is a Gaussian distribution. The normalization condition $\int_{-\infty}^{\infty} dx \, |\psi_0(x)|^2 = 1$ can easily be evaluated using the useful formula for Gaussian integrals from eqn (2.64), obtaining

$$\mathcal{N} = \Big(\frac{m\omega}{\pi\hbar} \Big)^{\frac{1}{4}} \, . \tag{5.19}$$

5.1.4   Eigenstates of the Harmonic Oscillator

To determine all other eigenstates and eigenvalues of N we formulate the following three lemmas.

Lemma 5.1   *If $| \, \psi_n \rangle$ is an eigenstate of N with eigenvalue n, then also the state $a^\dagger \, | \, \psi_n \rangle$ is an eigenstate of N with eigenvalue $(n + 1)$.*

Proof   For the proof, we simply apply N to $a^\dagger \, | \, \psi_n \rangle$ and make use of the commutator from eqn (5.12a), that is,

$$N \, a^\dagger \, | \, \psi_n \rangle = \big(a^\dagger N + a^\dagger \big) \, | \, \psi_n \rangle = a^\dagger \, (N + 1) | \, \psi_n \rangle$$
$$= a^\dagger \, (n + 1) \, | \, \psi_n \rangle = (n + 1) \, a^\dagger \, | \, \psi_n \rangle \, . \tag{5.20}$$

$\square$

The state $a^\dagger |\psi_n\rangle$ is not yet normalized, but it is proportional to $|\psi_{n+1}\rangle$. To normalize the state we simply compute

$$\langle \psi_{n+1} | \psi_{n+1} \rangle \propto \langle \psi_n | aa^\dagger | \psi_n \rangle = \langle \psi_n | a^\dagger a + 1 | \psi_n \rangle = \langle \psi_n | N+1 | \psi_n \rangle$$
$$= \langle \psi_n | n+1 | \psi_n \rangle = (n+1) \langle \psi_n | \psi_n \rangle = n+1. \tag{5.21}$$

Consequently, the action of the creation operator can be written as

$$a^\dagger |\psi_n\rangle = \sqrt{n+1} \ |\psi_{n+1}\rangle . \tag{5.22}$$

Lemma 5.2   *If $|\psi_n\rangle$ is an eigenstate of N with eigenvalue $n \neq 0$, then also the state $a|\psi_n\rangle$ is an eigenstate of N with eigenvalue $(n-1)$.*

Proof   The lemma can be proven in complete analogy to Lemma 5.1, using the commutator of eqn (5.12b), i.e.,

$$N a|\psi_n\rangle = \big(a N - a\big)|\psi_n\rangle = a(N-1)|\psi_n\rangle = a(n-1)|\psi_n\rangle = (n-1)a|\psi_n\rangle. \tag{5.23}$$

$\square$

As before, we can use the normalization of the eigenstates to determine the exact action of the operator a. By calculating

$$\langle \psi_{n-1} | \psi_{n-1} \rangle \propto \langle \psi_n | a^\dagger a | \psi_n \rangle = \langle \psi_n | n | \psi_n \rangle = n \langle \psi_n | \psi_n \rangle = n, \tag{5.24}$$

we easily find

$$a|\psi_n\rangle = \sqrt{n}|\psi_{n-1}\rangle . \tag{5.25}$$

We have thus found that, starting from the ground state $|\psi_0\rangle$, an infinite number of eigenstates of N can be reached by applying creation operators $a^\dagger$. One may return to any of the previous states by the action of the annihilation operator a. But can *all* eigenstates of N be reached in this way?

Lemma 5.3   *All eigenstates $|\psi_n\rangle$ of N are labelled by non-negative integers, $n \in \mathbb{N}_0$.*

Proof   Let us assume that there exists an eigenstate $|\psi_m\rangle$ for some $m \in \mathbb{R}$, but $m \notin \mathbb{N}$. By Lemma 5.2, we may successively apply annihilation operators until we reach an eigenstate $|\psi_{m'}\rangle$ with $m' < 0$. Applying the operator a again, one then reaches the state $|\psi_{m'-1}\rangle$. Setting $n = m'-1$ in eqn (5.21) we find that $\||\psi_{m'-1}\rangle\|^2 = m' < 0$. Since the norm of any vector must be positive (unless it is the null vector, in which case it is zero) we arrive at a contradiction. The eigenvalues of N may hence not take on any non-integer values, and we must conclude that $n \in \mathbb{N}_0$. $\square$

This means that, indeed, all eigenstates $|\,n\,\rangle \equiv |\,\psi_n\,\rangle$ of N can be reached by the action of creation operators on the ground state. Using eqn (5.22) we can write these eigenstates as

$$|\,n\,\rangle = \frac{(a^\dagger)^n}{\sqrt{n!}}\,|\,0\,\rangle\,, \tag{5.26}$$

where $|\,0\,\rangle \equiv |\,\psi_0\,\rangle$ is the ground state. The eigenstates $|\,n\,\rangle$ form a ladder of alternating even and odd energy states with energies

$$E_n = \hbar\omega\left(n + \tfrac{1}{2}\right), \tag{5.27}$$

separated by one quantum of energy $\hbar\omega$ each, i.e., the energy levels are equally spaced. The creation and annihilation operators then "climb" and "descend" this energy ladder step by step, which is why they are also called *ladder operators*.

The ground-state wave function $\psi_0 = \langle x | \psi_0 \rangle$ is the Gaussian curve given by eqn (5.18). The explicit form of the excited state wave functions ψ_n will be calculated later on, but for now let us already reveal that they are proportional to a product of the ground state and a family of functions, the so-called *Hermite polynomials* H_n. The eigenstates of the harmonic oscillator are illustrated in Fig. 5.1.

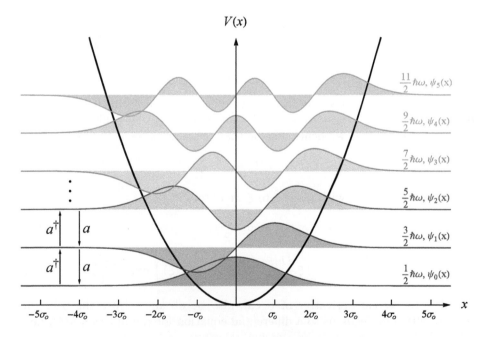

Fig. 5.1 Quantum harmonic oscillator. The eigenstates of the harmonic-oscillator potential $V(x) = \frac{\hbar\omega}{2}\left(\frac{x}{\sigma_o}\right)^2$, with the characteristic length $\sigma_o = \sqrt{\hbar/(m\omega)}$. The energy levels $E_n = \hbar\omega\left(n + \frac{1}{2}\right)$ are equally spaced. The ladder operators a and $a^\dagger$ allow descending and ascending this ladder of eigenstates.

5.2   Analytic Method

5.2.1   The Differential Equation of the Harmonic Oscillator

In this section we will contrast the algebraic approach of the previous section with the straightforward method of viewing the Schrödinger equation as a differential equation. That is, we will attempt to solve

$$H\,\psi(x) = \left(-\frac{\hbar^2}{2m}\frac{d^2}{dx^2} + \frac{m\omega^2}{2}x^2\right)\psi(x) = E\,\psi(x)\,. \tag{5.28}$$

Multiplying both sides of this equation by $2/(\hbar\omega)$ and introducing the dimensionless variables $\xi = \frac{x}{\sigma_o}$ and $K = 2E/(\hbar\omega)$, where we have again used the characteristic length $\sigma_o = \sqrt{\hbar/(m\omega)}$, we obtain the differential equation

$$\frac{d^2}{d\xi^2}\,\psi(\xi) = \left(\xi^2 - K\right)\psi(\xi)\,. \tag{5.29}$$

To get a better idea of what the solution could look like, we first consider the asymptotic behaviour of the solutions for large values of ξ. In the case where $\xi^2 \gg K$ and $\xi^2 \gg 1$, the term in eqn (5.29) that is proportional to K can be neglected such that

$$\frac{d^2}{d\xi^2}\,\psi(\xi) \approx \xi^2\,\psi(\xi)\,. \tag{5.30}$$

For this differential equation we then try the ansatz $\psi = \exp(-\xi^2/2)$, which leads to the correct result

$$\frac{d^2}{d\xi^2}\,\psi(\xi) = \left(\xi^2 - 1\right)\psi(\xi) \approx \xi^2\,\psi(\xi)\,. \tag{5.31}$$

To find the general solution we can then use a similar ansatz, differing from the previous one only insofar as we multiply the ansatz by a function of ξ. Using this method, known as *variation of constants*, we hence assume that the solution is of the form

$$\psi(\xi) = h(\xi)\,\exp(-\frac{\xi^2}{2})\,. \tag{5.32}$$

Differentiating eqn (5.32) with respect to ξ we obtain

$$\frac{d}{d\xi}\,\psi(\xi) = \left(\frac{d\,h(\xi)}{d\xi} - \xi\,h(\xi)\right)\exp(-\frac{\xi^2}{2})\,, \tag{5.33a}$$

$$\frac{d^2}{d\xi^2}\,\psi(\xi) = \left(\frac{d^2 h(\xi)}{d\xi^2} - 2\,\xi\,\frac{d\,h(\xi)}{d\xi} + (\xi^2 - 1)\,h(\xi)\right)\exp(-\frac{\xi^2}{2})\,. \tag{5.33b}$$

Meanwhile, the second derivative of ψ can also be obtained from the Schrödinger equation (5.29). This leads us to a differential equation for the function $h(\xi)$, i.e.,

$$\frac{d^2 h(\xi)}{d\xi^2} - 2\,\xi\,\frac{d\,h(\xi)}{d\xi} + (K-1)\,h(\xi) = 0\,. \tag{5.34}$$

At this point we use another practical method for solving differential equations, a *power-series ansatz*. That is, we write $h(\xi) = \sum_{m=0,1,\ldots} c_m\,\xi^m$ and differentiate,

$$\frac{dh(\xi)}{d\xi} = \sum_{m=0,1,\dots} m c_m \, \xi^{m-1} \,, \tag{5.35a}$$

$$\frac{d^2 h(\xi)}{d\xi^2} = \sum_{m=0,1,\dots} m(m-1) c_m \, \xi^{m-2} = \sum_{m=0,1,\dots} (m+2)(m+1) c_{m+2} \, \xi^m \,, \tag{5.35b}$$

where we have renamed the summation index in the last step. Substituting back into eqn (5.34), we arrive at the new equation

$$\sum_{m=0,1,\dots} \left((m+2)(m+1) c_{m+2} - m c_m + (K-1) c_m \right) \xi^m = 0 \,. \tag{5.36}$$

Since equality must hold for all values of ξ, the expression in parentheses must vanish identically and we arrive at the recursion relation

$$c_{m+2} = \frac{m - K + 1}{(m+1)(m+2)} \, c_m \tag{5.37}$$

for the coefficients of the power series. In order for the wave function in eqn (5.32) to be normalizable, the power series for $h(\xi)$ must terminate at a finite power of ξ. In other words, there must be some maximal value n such that $c_n \neq 0$ but $c_{n+2} = 0$. This implies that $n - K + 1 = 0$ and with $K = 2E/(\hbar\omega)$ we therefore arrive at the familiar relation

$$E_n = \hbar\omega \left(n + \tfrac{1}{2} \right) , \tag{5.38}$$

as we had previously found in eqn (5.27) using the algebraic approach.

5.2.2 The Hermite Polynomials

Further setting $(K - 1) = n$ in eqn (5.34) one can recognize the differential equation for the *Hermite polynomials* H_n, which we shall state here using the corresponding *formula of Rodrigues*, i.e.,

$$H_n(\xi) = (-1)^n \, e^{\xi^2} \left(\frac{d^n}{d\xi^n} e^{-\xi^2} \right) . \tag{5.39}$$

From eqn (5.39) one may straightforwardly verify the recursion relation

$$\frac{dH_n(\xi)}{d\xi} = 2\xi \, H_n(\xi) - H_{n+1}(\xi) = 2n \, H_{n-1}(\xi) \,, \tag{5.40}$$

which can be used to assert that the Hermite polynomials satisfy the differential equation (5.34). We will leave the explicit computation as an exercise problem for the interested reader. With the help of the Hermite polynomials we can now also give the exact expression for the eigenfunctions of the harmonic oscillator, that is,

$$\psi_n = \frac{1}{\sqrt{2^n (n!)}} \left(\frac{m\omega}{\pi\hbar} \right)^{\frac{1}{4}} H_n(\frac{x}{\sigma_o}) \, \exp(-\frac{x^2}{2\sigma_o^2}) \,, \tag{5.41}$$

where the normalization constant follows from the orthogonality relation of the Hermite polynomials,

$$\int_{-\infty}^{\infty} d\xi \, H_m(\xi) \, H_n(\xi) \, e^{-\xi^2} = \sqrt{\pi} \, 2^n \, (n!) \, \delta_{mn} \,. \tag{5.42}$$

As expected, the Hermite polynomials are even functions for even n, and odd functions for odd n. To give some examples we list the first few Hermite polynomials, i.e.,

$$H_0(x) = 1 \,, \quad H_1(x) = 2x \,, \quad H_2(x) = 4x^2 - 2$$
$$H_3(x) = 8x^3 - 12x \,, \quad H_4(x) = 16x^4 - 48x^2 + 12 \,. \tag{5.43}$$

The corresponding wave functions from eqn (5.41) are illustrated in Fig. 5.1. Now that we have substantiated our results from Section 5.1, we will work again with the ladder operators of the algebraic approach to examine the properties of the quantum harmonic oscillator in more detail.

5.3   Zero-Point Energy

Upon inspection of the energy levels $E_n = \hbar\omega(n + \frac{1}{2})$, one immediately notices that, contrary to the classical expectation, the lowest possible energy level of the harmonic oscillator is not zero but $E_0 = \frac{1}{2}\hbar\omega$, the *zero-point energy*. This can be understood in the following way. Suppose that one wishes to localize the particle in a small interval around the origin $x = 0$, allowing only small oscillations with $\langle P \rangle = 0$, i.e., in analogy to a classical pendulum at rest at the equilibrium position. By Heisenberg's uncertainty relation, eqn (2.61), the momentum uncertainty must grow as one makes this interval smaller. On the other hand, the ground state is an eigenfunction of the Hamiltonian, which, in turn, contains contributions from both the kinetic and the potential energy. The potential vanishes at the origin, $V(x = 0) = 0$, but the system must have some non-zero kinetic energy at $x = 0$, since we now have $(\Delta P)^2 = \langle P^2 \rangle = 2mE_{\text{kin}}$ (see Definition 2.11). A vanishing ground-state energy is therefore incompatible with the uncertainty principle.

5.3.1   Uncertainty Relation for the Harmonic Oscillator

In the light of this realization, let us examine the position-momentum uncertainty for eigenstates of the harmonic oscillator more closely. For the position uncertainty, we first use eqn (5.5a) to compute the expectation value

$$\langle X \rangle_n \propto \langle n \,|\, a + a^\dagger \,|\, n \rangle = \left(\sqrt{n} \, \langle n \,|\, n-1 \rangle + \sqrt{n+1} \, \langle n \,|\, n+1 \rangle \right) = 0 \,, \tag{5.44}$$

where one makes use of the orthogonality of the eigenstates $\langle m \,|\, n \rangle = \delta_{mn}$. Similarly, one finds

$$\langle X^2 \rangle_n = \frac{\hbar}{2m\omega} \langle n \,|\, \left(a + a^\dagger \right)^2 \,|\, n \rangle = \frac{\hbar}{2m\omega} \langle n \,|\, a^2 + aa^\dagger + a^\dagger a + (a^\dagger)^2 \,|\, n \rangle \,. \tag{5.45}$$

The expectation values of a^2 and $(a^\dagger)^2$ vanish identically, since $a^2 |n\rangle \propto |n-2\rangle$ and $(a^\dagger)|n\rangle \propto |n+2\rangle$. For the remaining terms we proceed by using the commutator from eqn (5.8) and write $a^\dagger a = N$ to obtain

$$\langle X^2 \rangle_n = \frac{\hbar}{2m\omega} \langle n | 2N+1 | n \rangle = \frac{\hbar}{m\omega}\left(n + \tfrac{1}{2}\right) = \sigma_o^2\left(n + \tfrac{1}{2}\right), \qquad (5.46)$$

where we have again used the characteristic length $\sigma_o = \sqrt{\hbar/(m\omega)}$ of the harmonic oscillator. Combining this result with eqn (5.44), the (squared) position uncertainty in the nth eigenstate is thus given by the expression

$$\left(\Delta X\right)^2 = \sigma_o^2\left(n + \tfrac{1}{2}\right), \qquad (5.47)$$

which vanishes in the classical limit, i.e., $\hbar \to 0$ or $m \to \infty$. Continuing with the momentum uncertainty, a computation similar to (5.44) reveals the momentum expectation value to be

$$\langle P \rangle_n \propto \langle n | a - a^\dagger | n \rangle = \left(\sqrt{n}\,\langle n|n-1\rangle - \sqrt{n+1}\,\langle n|n+1\rangle\right) = 0. \qquad (5.48)$$

Consequently, the (squared) momentum uncertainty is determined by the expectation value

$$\langle P^2 \rangle_n = \left(\Delta P\right)^2 = \frac{-m\omega\hbar}{2} \langle n | \left(a + a^\dagger\right)^2 | n \rangle = \frac{\hbar^2}{2\sigma_o^2} \langle n | 2N+1 | n \rangle = \frac{\hbar^2}{\sigma_o^2}\left(n + \tfrac{1}{2}\right). \qquad (5.49)$$

From eqns (5.47) and (5.49) we can then obtain the uncertainty relation

$$\Delta X \, \Delta P = \hbar\left(n + \tfrac{1}{2}\right) \geq \tfrac{\hbar}{2}, \qquad (5.50)$$

where equality holds exactly for the ground state, that is, for $n = 0$ we have $\Delta X \Delta P = \hbar/2$. The ground state is indeed the minimum-uncertainty energy eigenstate of the harmonic oscillator.

5.3.2 The Zero-Point Energy of the Harmonic Oscillator

To strengthen this connection between the minimum-energy state and the uncertainty relation, let us turn this argument on its head. That is, one can show that the minimal Heisenberg uncertainty also implies that the ground-state energy must be at least $\hbar\omega/2$. To see this, first note that the uncertainty relation $\left(\Delta X\right)^2\left(\Delta P\right)^2 \geq \hbar^2/4$ can be used to obtain the lower bound

$$\langle X^2 \rangle \geq \frac{\hbar^2}{4} \frac{1}{\langle P^2 \rangle - \langle P \rangle^2} + \langle X \rangle^2 \geq \frac{\hbar^2}{4\langle P^2 \rangle}. \qquad (5.51)$$

For the average energy $E = \langle H \rangle$ of the harmonic oscillator this means

$$E = \frac{\langle P \rangle^2}{2m} + \frac{m\omega^2 \langle X^2 \rangle}{2} \geq \frac{\langle P \rangle^2}{2m} + \frac{m\omega^2\hbar^2}{8\langle P^2 \rangle}. \qquad (5.52)$$

We may minimize the lower bound by differentiating the right-hand side with respect to $\langle P^2 \rangle$ and setting the result equal to zero,

$$\frac{1}{2m} + \frac{m\omega^2\hbar^2}{8\langle P^2 \rangle^2} = 0 \quad \Rightarrow \quad \langle P^2 \rangle = \frac{m\omega\hbar}{2}. \tag{5.53}$$

Inserting the result into the mean energy of eqn (5.52) we hence arrive at

$$E \geq \frac{1}{2m}\frac{m\omega\hbar}{2} + \frac{m\omega^2\hbar^2}{8}\frac{2}{m\omega\hbar} = \frac{\hbar\omega}{4} + \frac{\hbar\omega}{4} = \frac{\hbar\omega}{2}. \tag{5.54}$$

Indeed, the minimal energy that is compatible with the uncertainty principle is the ground-state energy, and, at the same time, the ground state ψ_0 of the harmonic oscillator is a minimal uncertainty state. Note that the ground state, illustrated in Fig. 5.1 is just a Gaussian wave packet of width σ_o, just as discussed in Section 2.5.3.

Theorem 5.1 (Zero-point energy)

The zero-point energy is the smallest possible energy of a physical system that is compatible with the uncertainty relation. It is the energy of its ground state.

5.4 Comparison with the Classical Oscillator

Having identified one important difference between the quantum harmonic oscillator and its classical counterpart, we shall devote this section to further compare the quantum-mechanical oscillator with predictions obtained from classical physics. The motion of a classical oscillator, e.g., a mass on a spring (see Fig. 5.2), is described by the function

$$x(t) = q_0 \sin(\omega t), \tag{5.55}$$

which is a solution to the classical equation of motion $(\partial^2/\partial t^2 + \omega^2)x(t) = 0$, where $q_0 = x(0)$ is a starting value at $t = 0$. The total energy of the system is constant and given by

$$E_{\text{class}} = \frac{m\omega^2}{2} q_0^2. \tag{5.56}$$

To make a comparison with quantum mechanics we want to identify a classical probability density $W_{\text{class}}(x)$. To do this, consider the derivative

$$\frac{dx(t)}{dt} = \omega q_0 \cos(\omega t) = \omega q_0 \sqrt{1 - \sin^2(\omega t)} = \frac{2\pi q_0}{T}\sqrt{1 - \left(\frac{x}{q_0}\right)^2}, \tag{5.57}$$

where we have substituted the period $T = 2\pi/\omega$. Rearranging eqn (5.57) and integrating both sides over one period of oscillation, we then find

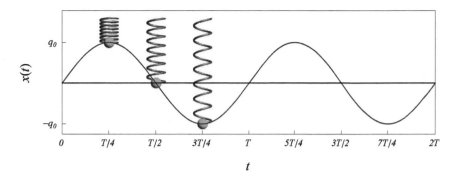

Fig. 5.2 Classical harmonic oscillator. For small displacements and neglecting the damping, a simple spring is a good example for a classical harmonic oscillator. The displacement of the mass from the equilibrium position $x = 0$ is described by the function in eqn (5.55). The oscillator returns to its initial configuration after a full period of duration $T = 2\pi/\omega$.

$$2 \int_{-q_0}^{q_0} \left(2\pi q_0 \sqrt{1 - \left(\tfrac{x}{q_0}\right)^2} \right)^{-1} dx = \frac{1}{T} \int_0^T dt = 1 \,, \tag{5.58}$$

where the factor of 2 on the left-hand side arises because the oscillator passes each position between $-q_0$ and q_0 twice during one full oscillation period. We can hence interpret the function

$$W_{\text{class}}(x) = \frac{1}{\pi q_0} \frac{1}{\sqrt{1 - \left(\tfrac{x}{q_0}\right)^2}} \tag{5.59}$$

as the probability density of the classical harmonic oscillator, see Fig. 5.3. Finally, we note that the starting value q_0, which just corresponds to the maximal distance from the equilibrium position of the classical oscillator, is fixed by the available energy of eqn (5.56). This gives a basis for comparing the classical and the quantum-mechanical oscillators on an equal footing. We simply set $E_n = \hbar\omega\left(n + \tfrac{1}{2}\right) = E_{\text{class}}$ and arrive at

$$q_0 = \sqrt{\frac{(2n+1)\hbar}{m\omega}} = \sqrt{2n+1}\,\sigma_o \,. \tag{5.60}$$

In the comparison of the quantum harmonic oscillator and its classical counterpart, which is illustrated in Fig. 5.3, we find the following: For the low-lying (small n) states, of course, the two probabilities for finding the particle differ considerably. However, for the highly energetic states (e.g., as shown for $n = 100$), the envelope of the quantum probability closely resembles the classical one, as postulated in Proposition 1.7.

5.5 The Three-Dimensional Harmonic Oscillator

Let us briefly inspect again the illustration of the classical oscillator in Fig. 5.2. The mass attached to the spring is oscillating along one direction that we have chosen to

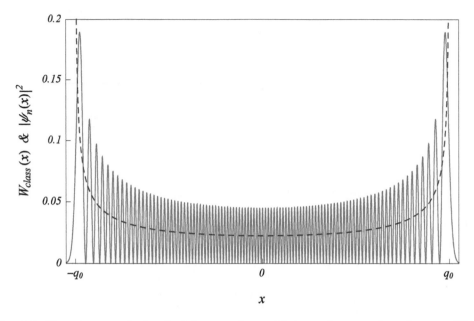

Fig. 5.3 Quantum and classical harmonic oscillators. A comparison for $n = 100$ of the probability densities $|\psi_n(x)|^2$ (solid orange curve) and $W_{\text{class}}(x)$ with $q_0/\sigma_o = \sqrt{2n+1}$ (dashed blue curve) of the quantum-mechanical and classical harmonic oscillator, respectively.

be the x direction[1]. One could easily attach springs in directions perpendicular to this direction as well. For small displacements the oscillations in all three directions then decouple. Similarly, we can consider a three-dimensional quantum-mechanical harmonic oscillator: After inspection of eqn (5.3), we may write the corresponding Hamiltonian in three dimensions as

$$H = \sum_{i=1}^{3} H_i = \sum_{i=1}^{3} \left(\frac{1}{2m} P_i^2 + \frac{m\omega_i^2}{2} X_i^2 \right), \tag{5.61}$$

where $i = 1, 2, 3$ labels the three spatial directions and $P_i = -i\hbar\partial/\partial x^i$. From the canonical commutation relations (Theorem 2.2) it easy to see that the Hamiltonian operators for different directions commute with each other, $[H_i, H_j] = 0 \ \forall i, j$. In other words, we have three completely independent quantum-mechanical harmonic oscillators. For simplicity we assume that the oscillator frequencies are identical for all three directions, $\omega_i \equiv \omega \ \forall i$. For each direction we may therefore introduce ladder operators in complete analogy to Definition 5.1, that is,

$$a_i := \frac{1}{\sqrt{2m\omega\hbar}} (m\omega X_i + iP_i), \tag{5.62a}$$

$$a_i^\dagger := \frac{1}{\sqrt{2m\omega\hbar}} (m\omega X_i - iP_i). \tag{5.62b}$$

---

[1]We may ignore the role of gravity for the sake of this argument and imagine the mass to be fixed to the centre of a spring whose ends are attached to some fixed points at either end.

The ladder operators defined in this way satisfy the commutation relations $[a_i, a_j^\dagger] = \delta_{ij}$ and $[a_i, a_j] = 0$, and the ith creation operator $a_i^\dagger$ creates an excitation of the ith oscillator. The Hamiltonian of eqn (5.61) can now be rewritten as

$$H = \hbar\omega \sum_{i=1}^{3} a_i^\dagger a_i + \tfrac{3}{2}\hbar\omega. \tag{5.63}$$

5.5.1 Eigenstates of the Three-Dimensional Harmonic Oscillator

Since the three oscillators decouple, the eigenstates of their Hamiltonian must be labelled by three integers, $n_i \in \mathbb{N}_0$. Denoting these eigenstates by $|\,n_1, n_2, n_3\,\rangle$ the time-independent Schrödinger equation can be written as

$$H\,|\,n_1, n_2, n_3\,\rangle = \hbar\omega\big(n_1 + n_2 + n_3 + \tfrac{3}{2}\big)\,|\,n_1, n_2, n_3\,\rangle. \tag{5.64}$$

The ground state is also easily identified, i.e., it is given by the state $|\,0,0,0\,\rangle$, which satisfies $a_i\,|\,0,0,0\,\rangle = 0$ for all i. However, when looking for the next energy eigenstate we have three options. The states $|\,1,0,0\,\rangle$, $|\,0,1,0\,\rangle$, and $|\,0,0,1\,\rangle$ all have the same energy $\tfrac{5}{2}\hbar\omega$. The spectrum of the Hamiltonian is *degenerate* and the eigenvectors are no longer uniquely determined by their energy alone. When the eigenvalues of one operator, in this case the eigenvalues $\hbar\omega(n + \tfrac{3}{2})$ of H with $n = n_1 + n_2 + n_3$, are not enough to label all the eigenstates, we can try to find an additional operator in the hope that the pair of the operators' eigenvalues will achieve such a labelling. However, this only makes sense if the operators commute and, consequently, permit a set of simultaneous eigenstates. In the example that we have chosen here, such an operator is given by the z component of the orbital angular-momentum operator, i.e.,

$$L_3 = X_1 P_2 - X_2 P_1. \tag{5.65}$$

Using the canonical commutation relations (Theorem 2.2), it is easy to verify that $[H, L_3] = 0$. We may also express L_3 via the harmonic-oscillator ladder operators by substituting $X_i = \sqrt{\hbar/(2m\omega)}\,(a_i + a_i^\dagger)$ and $P_i = -i\sqrt{m\omega\hbar/2}\,(a_i - a_i^\dagger)$ to obtain

$$L_3 = i\hbar\big(a_1 a_2^\dagger - a_2 a_1^\dagger\big). \tag{5.66}$$

Since $a_i\,|\,0,0,0\,\rangle = 0$ for all i, the ground state $|\,0,0,0\,\rangle$ is already an eigenstate of L_3 with eigenvalue $m = 0$, that is,

$$L_3\,|\,0,0,0\,\rangle = i\hbar\big(a_1 a_2^\dagger - a_2 a_1^\dagger\big)\,|\,0,0,0\,\rangle = 0. \tag{5.67}$$

The ground state is hence labelled by the pair $(n, m) = (0, 0)$. Similarly, one can write $|\,0,0,1\,\rangle = a_3^\dagger\,|\,0,0,0\,\rangle$ and note that $[L_3, a_3^\dagger] = 0$ to see that $L_3\,|\,0,0,1\,\rangle = 0$. The state $|\,0,0,1\,\rangle$ is therefore described by the labelling $(n, m) = (1, 0)$. The two other states, $|1,0,0\rangle$ and $|0,1,0\rangle$ are not eigenstates of L_3. Instead a short computation reveals

$$L_3\,|1,0,0\rangle = i\hbar\left(a_1 a_2^\dagger - a_2 a_1^\dagger\right)|1,0,0\rangle = i\hbar\,|0,1,0\rangle\,, \tag{5.68a}$$

$$L_3\,|0,1,0\rangle = i\hbar\left(a_1 a_2^\dagger - a_2 a_1^\dagger\right)|0,1,0\rangle = -i\hbar\,|1,0,0\rangle\,. \tag{5.68b}$$

Nonetheless, from eqn (5.68) one may quickly construct the superpositions $(|1,0,0\rangle \pm i\,|0,1,0\rangle)/\sqrt{2}$ which are eigenstates of L_3 with eigenvalues $m\hbar = \pm\hbar$, that is,

$$L_3\,\frac{1}{\sqrt{2}}\bigl(|1,0,0\rangle \pm i\,|0,1,0\rangle\bigr) = \pm\hbar\,\frac{1}{\sqrt{2}}\bigl(|1,0,0\rangle \pm i\,|0,1,0\rangle\bigr)\,. \tag{5.69}$$

The four eigenstates of lowest energy can therefore be labelled by pairs (n,m) with $n = 0,1$ and $m = -1,\ldots,+1$. The regularity of this labelling is no coincidence and we shall learn to understand this in a more general context in Chapter 6, where we will study the role of orbital angular momentum in the Schrödinger equation in more depth.

Before we do so, however, it us useful to give a mathematically more precise (and general) account of the description of the three coupled harmonic oscillators.

5.5.2 Systems of Multiple Degrees of Freedom

At the beginning of Section 5.5 we have argued that the oscillations along orthogonal directions can be considered independently. This warranted the definition of operators X_i, P_i, and $a_i, a_i^\dagger$ for each direction, commuting with the operators of the respective orthogonal directions. This is a procedure that is applied in general. To any system with N degrees of freedom described by generalized coordinates $q_1, q_2, \ldots, q_N$, one associates a set of commuting Hermitian[2] operators $\{Q_1, Q_2, \ldots, Q_N\}$, such that $[Q_i, Q_j] = 0\ \forall i,j = 1,\ldots,N$. These could correspond, for instance, to the positions of a single particle in N dimensions, or the position (along a single spatial dimension) of N different particles. In the latter case, we can describe each particle on its own if they are well-separated and have not interacted before. The position of the ith particle can then be described by a state vector $|\psi_i\rangle$ in a Hilbert space $\mathcal{H}_i$. At this point, we note that two qualitatively different assumptions are being made here: on the one hand, we can assume that the degrees of freedom in question are represented by operators that commute with each other. On the other hand, we can assume that the considered operators act on different Hilbert spaces and, as we shall describe shortly, we can combine these different Hilbert spaces in a tensor product. Whenever the operators in question are diagonalizable, in particular, for finite-dimensional Hilbert spaces and compact infinite-dimensional Hilbert spaces, these assumptions are equivalent, as we can see from the following theorem.

---

[2]Self-adjoint operators if the corresponding degree of freedom is continuous.

Theorem 5.2 (Joint eigenstates of commuting observables)

Any two observables represented by self-adjoint and diagonalizable operators A and B, that commute with each other, $[A, B] = 0$, have a common basis of joint eigenstates $\{|\psi_{m,n}\rangle\}$,

$$A | \psi_{m,n} \rangle = \alpha_m | \psi_{m,n} \rangle \,,$$

$$B | \psi_{m,n} \rangle = \beta_n | \psi_{m,n} \rangle \,.$$

Proof To prove Theorem 5.2, first note that if A and B are diagonalizable, each operator must have a set of eigenvectors and eigenvalues. Let us denote these by $|\alpha_m\rangle$ and $|\beta_n\rangle$, respectively, such that

$$A | \alpha_m \rangle = \alpha_m | \alpha_m \rangle \,, \tag{5.70a}$$

$$B | \beta_n \rangle = \beta_n | \beta_n \rangle \,. \tag{5.70b}$$

Now let us successively apply the operators A and B on an eigenvector of A, that is

$$B\,A\,|\,\alpha_m\,\rangle = B\,\alpha_m\,|\,\alpha_m\,\rangle = \alpha_m\,B\,|\,\alpha_m\,\rangle \overset{!}{=} A\,B\,|\,\alpha_m\,\rangle \,, \tag{5.71}$$

where we have used the fact that A and B commute in the last step. Comparing the last two expressions on the right-hand sides in each line we see that $B\,|\,\alpha_m\,\rangle$ must be an eigenvector of A with eigenvalue α_m. If the eigenvalue a_m is non-degenerate, that is, if there is only one eigenvector of A with this particular eigenvalue, then it follows immediately that $B\,|\,\alpha_m\,\rangle$ must be proportional to $|\,\alpha_m\,\rangle$. In other words, $|\,\alpha_m\,\rangle$ must be an eigenvector of B, with some eigenvalue β_n. We can then use the notation $|\,\psi_{m,n}\,\rangle$ for the joint eigenstate of A and B with eigenvalues α_m and β_n, respectively.

If the eigenvalue α_m is degenerate, that is, if there exist several (say, k) eigenstates $|\,\alpha_m^i\,\rangle$ $(i = 1, 2, \ldots, k)$ with the same eigenvalue, then we have to do a little more work. In this case, eqn (5.71) informs us that $B\,|\,\alpha_m^i\,\rangle$ is a linear combination of all the $|\,\alpha_m^j\,\rangle$. Although the $|\,\alpha_m^j\,\rangle$ may not be eigenstates of B, the action of B does not leave the subspace corresponding to the eigenvalue α_m of A. This means that this subspace can also be spanned by the eigenvectors of B or a subset of them, and all of these eigenvectors of B are eigenvectors of A to the eigenvalue α_m. Every vector $|\,\alpha_m^j\,\rangle$ can hence be written as a linear combination of (some of) the eigenstates $|\,\beta_n\,\rangle$ of B. We then just need to determine a basis of eigenstates of B within the subspace spanned by the $|\,\alpha_m^j\,\rangle$. Such a diagonalization must be possible since B is self-adjoint and we again denote these joint eigenstates by $|\,\psi_{m,m}\,\rangle$.

Repeating this procedure for all eigenvalues of A, we then find the desired basis $\{|\,\psi_{m,n}\,\rangle\}$ of joint eigenstates. Since the eigenvectors of either A or B already span the entire Hilbert space, the vectors $|\,\psi_{m,n}\,\rangle$ must also satisfy the completeness relation $\sum_{m,n} |\,\psi_{m,n}\,\rangle\langle\,\psi_{m,n}\,| = \mathbb{1}$. $\qquad\square$

So, if observables are represented by diagonalizable operators and commute, they have joint eigenstates. In comparison, recall again what happens with operators that do not commute. Not only do they not have joint eigenstates, from Theorem 2.4 one sees that operators that do not commute have non-trivial uncertainty relations. That is, if $[A, B] \neq 0$ then in general also $\Delta A \Delta B > 0$. With this in mind, we now focus again on commuting observables, in particular, on the case for observables Q_i $(i = 1, 2, \ldots, N)$ for different degrees of freedom. What Theorem 5.2 leaves unresolved is the following problem: If we are given the commuting observables Q_i with eigenstates $|\psi_n\rangle_i$ $(n = 1, 2, \ldots, d_i)$ spanning Hilbert spaces $\mathcal{H}_i$ with $\dim(\mathcal{H}_i) = d_i$, then how do we construct the joint eigenstates and the corresponding Hilbert space?

This is achieved via the *tensor product* that we have already encountered in Section 3.2. For two Hilbert spaces $\mathcal{H}_1$ and $\mathcal{H}_2$, the tensor product is just a bilinear map from the direct product of these spaces to the so-called tensor-product space $\mathcal{H} = \mathcal{H}_1 \otimes \mathcal{H}_2$, that is

$$\mathcal{H}_1 \times \mathcal{H}_2 \rightarrow \mathcal{H}_1 \otimes \mathcal{H}_2$$

$$(|\psi\rangle_1 \in \mathcal{H}_1, |\phi\rangle_2 \in \mathcal{H}_2) \mapsto |\psi\rangle_1 \otimes |\phi\rangle_2 , \tag{5.72}$$

where bilinearity is captured by the conditions in eqns (3.18). If $\{|\psi_m\rangle_1\}_{m=1,\ldots,d_1}$ and $\{|\phi_n\rangle_2\}_{n=1,\ldots,d_2}$ are bases of $\mathcal{H}_1$ and $\mathcal{H}_2$, respectively, then

$$\{|\psi_m, \phi_n\rangle = |\psi_m\rangle_1 \otimes |\phi_n\rangle_2\}_{\substack{m=1,\ldots,d_1 \\ n=1,\ldots,d_2}} \tag{5.73}$$

is a basis of the Hilbert space $\mathcal{H} = \mathcal{H}_1 \otimes \mathcal{H}_2$. Consequently, the dimension of the tensor-product space is the product of the dimensions of the subspaces,

$$\dim(\mathcal{H}) = \dim(\mathcal{H}_1) \dim(\mathcal{H}_2) . \tag{5.74}$$

It is interesting to note here that eqn (5.74) has far-reaching consequences. The dimension of the total Hilbert space grows exponentially with the number of degrees of freedom, whereas the dimension of a classical phase space only grows linearly with the number of degrees of freedom. The extra "space" of the quantum-mechanical description is exactly what is needed to capture superpositions, i.e., coherence (and interference) between different basis states. But this also means that it becomes very hard to fully describe quantum systems with many degrees of freedom on a (classical) computer. In terms of the description of the quantum states this manifests in the following way. Every quantum state $|\psi\rangle \in \mathcal{H}$ of the tensor-product space $\mathcal{H} = \mathcal{H}_1 \otimes \mathcal{H}_2$ can be written as

$$|\psi\rangle = \sum_{m,n} c_{m,n} |\psi_m\rangle_1 \otimes |\phi_n\rangle_2 \tag{5.75}$$

for some complex amplitudes $c_{m,n}$, but almost no state factorizes into a tensor product, i.e., most states cannot be written as $|\psi\rangle_1 \otimes |\phi\rangle_2$. This statement has many interesting consequences and leads to a number of fascinating features of quantum mechanics, which will keep us busy in Part II of this book. To continue, we now need to briefly learn how to do computations in a tensor-product space, e.g., calculating expectation values. For this, we need the scalar product, which is evaluated for each of the subspaces separately, that is, the scalar product between $|\psi\rangle_1 \otimes |\phi\rangle_2$ and $|\psi'\rangle_1 \otimes |\phi'\rangle_2$ is given by

$$_1\langle\psi'| \otimes {}_2\langle\phi' | \, |\psi\rangle_1 \otimes |\phi\rangle_2 = \langle\psi'|\psi\rangle \langle\phi'|\phi\rangle \in \mathbb{C}. \tag{5.76}$$

Furthermore, if A and B are linear operators on the subspaces $\mathcal{H}_1$ and $\mathcal{H}_2$, then $A \otimes B$ is a linear operator on the tensor-product space, such that

$$A \otimes B \, |\psi\rangle_1 \otimes |\phi\rangle_2 = A\,|\psi\rangle_1 \otimes B\,|\phi\rangle_2 , \tag{5.77}$$

which implies that, similarly, for products of operators we have

$$(A' \otimes B')(A \otimes B) = A'A \otimes B'B . \tag{5.78}$$

This also means that any known operator on a subspace can be trivially extended to the larger tensor-product space by a tensor product with the identity. If A and B are linear operators on $\mathcal{H}_1$ and $\mathcal{H}_2$, respectively, then $A \otimes \mathbb{1}$ and $\mathbb{1} \otimes B$ as well as the product of the latter operators are linear operators on $\mathcal{H}_1 \otimes \mathcal{H}_2$. If it is clear from the context which spaces the operators are acting on non-trivially, one can even use the same notation for A and $A \otimes \mathbb{1}$, or for B and $\mathbb{1} \otimes B$, respectively. This finally brings us back to systems of N degrees of freedom. Clearly, the operators $A \otimes \mathbb{1}$ and $\mathbb{1} \otimes B$ commute. By eqn (5.78) we have

$$AB = (A \otimes \mathbb{1})\,(\mathbb{1} \otimes B) = A \otimes B = (\mathbb{1} \otimes B)\,(A \otimes \mathbb{1}) = BA , \tag{5.79}$$

which, following Theorem 5.2, tells us that A and B have a common set of eigenstates. These eigenstates are nothing but the tensor products of the eigenvectors of the operators on the Hilbert spaces $\mathcal{H}_1$ and $\mathcal{H}_2$. In the notation of eqn (5.70) we just have

$$|\psi_{m,n}\rangle = |\alpha_m\rangle \otimes |\beta_n\rangle \equiv |\alpha_m, \beta_n\rangle \equiv |m,n\rangle . \tag{5.80}$$

Note that this is exactly what we have naively done already for the eigenstates of the three-dimensional harmonic oscillator in eqn (5.64), where we had assumed that the ladder operators a_i for different directions commute. However, since the Hilbert space of the harmonic oscillator is infinite-dimensional, and the operators we consider are not necessarily compact, we can actually not apply Theorem 5.2. Indeed, determining if the commutation of operators implies a tensor-product structure (or not) is at the heart of Tsirelson's problem, as we shall comment on again in Section 15.3.4. Therefore, in the case of non-diagonalizable operators on infinite-dimensional systems one usually assumes a tensor-product structure, which then implies commutation of the operators in question. In particular, we see that we have simply considered a tensor product of three harmonic oscillators. But with this construction, it is immediately obvious that the construction of the Hamiltonian and its eigenstates would have worked for any number of uncoupled oscillators. Indeed, such constructions play a vital role in quantum optics and quantum field theory, where even infinitely many oscillators are considered. The associated Hilbert spaces, so-called Fock spaces, can be constructed based on the same principles as we have discussed here, and we shall return to them in Chapter 25 of Part III.

What we have learned in this section also teaches us how to proceed with the treatment of wave mechanics in three spatial dimensions in the next chapters. Instead of a single position operator X, we can define commuting position operators X_i ($i = 1, 2, 3$) for each spatial direction. The joint eigenstates of these operators can then be written as $|x, y, z\rangle \equiv |x\rangle \otimes |y\rangle \otimes |z\rangle$, such that

$$X_1 \,|\, x, y, z \,\rangle = x \,|\, x, y, z \,\rangle \,, \tag{5.81a}$$

$$X_2 \,|\, x, y, z \,\rangle = y \,|\, x, y, z \,\rangle \,, \tag{5.81b}$$

$$X_3 \,|\, x, y, z \,\rangle = z \,|\, x, y, z \,\rangle \,. \tag{5.81c}$$

Indeed, keeping the tensor-product structure in mind, we can even collect the operators X_i into a vector operator

$$\vec{X} = X_1 \,\vec{e}_1 + X_2 \,\vec{e}_2 + X_3 \,\vec{e}_3 \,, \tag{5.82}$$

where $\vec{e}_i \in \mathbb{R}^3$, and the eigenvalues into a vector $\vec{x} = (x, y, z)^\top$ to write

$$\vec{X} \,|\, \vec{x} \,\rangle = \vec{x} \,|\, \vec{x} \,\rangle \,. \tag{5.83}$$

And of course, a similar treatment can be applied to the momentum operators P_i, obtaining

$$\vec{P} \,|\, \vec{p} \,\rangle = \vec{p} \,|\, \vec{p} \,\rangle \,. \tag{5.84}$$

With this, we are well-prepared for the following discussion of orbital angular momentum in Chapter 6 and of the three-dimensional Schrödinger equation in Chapter 7.

6
Orbital Angular Momentum

In Section 5.5 we have seen that another observable—*orbital angular momentum*—enters into quantum-mechanical considerations in three dimensions. This quantity, familiar already from classical mechanics, is crucial to understand the quantum-mechanical description of atoms and molecules. In this chapter we will therefore introduce the technical framework to discuss angular momentum in quantum physics, which will allow us to give a basic quantum-mechanical model for the hydrogen atom that replaces the postulates of Bohr's atom model (recall Propositions 1.4 and 1.5) in Chapter 7.

6.1 Angular Momentum and the Rotation Group

6.1.1 The Orbital Angular Momentum Operator

In classical mechanics, an object's tendency to rotate around some chosen point is described by the vector $\vec{L}$ of orbital angular momentum, given by

$$\vec{L} = \vec{x} \times \vec{p}, \tag{6.1}$$

in terms of the position $\vec{x}$ and momentum $\vec{p}$ of the object. For now, we shall simply refer to it as angular momentum, dropping the prefix "orbital", when there is no danger of confusion with internal angular momentum, which (classically) arises from the rotation of a composite body's constituents about its centre of mass. Note that the angular momentum depends on the choice of origin much like the linear momentum $\vec{p}$ depends on the choice of inertial observer. Nevertheless, both quantities are conserved in closed systems. To make the transition to quantum mechanics, we first express the classical angular-momentum vector in components, i.e.,

$$L_i = \varepsilon_{ijk}\, x_j\, p_k\,, \tag{6.2}$$

where summation over repeated indices is implied (Einstein summation convention) and

$$\varepsilon_{ijk} = \begin{cases} +1 & \text{if } (i,j,k) = (1,2,3), (3,1,2), (2,3,1) \\ -1 & \text{if } (i,j,k) = (1,3,2), (3,2,1), (2,1,3) \\ 0 & \text{otherwise} \end{cases} \tag{6.3}$$

is the completely antisymmetric Levi–Civita symbol. Subsequently, we replace the classical position and momentum variables x_i and p_i with their quantum-mechanical operator counterparts X_i and $P_i = -i\hbar\nabla_i$, respectively. To distinguish the classical variables in eqn (6.2) from the Hilbert-space operators, we could use lower-case and capital letters, respectively, hats ($\hat{L}_i$ instead of L_i), or other notation. Instead, we shall leave the distinction to the context of application, as is common in the literature, and write the angular-momentum operators as

$$L_i = \varepsilon_{ijk} X_j P_k \,. \tag{6.4}$$

The *commutation relations* for the angular-momentum operators then evaluate to

$$[L_i, L_j] = i\hbar\,\varepsilon_{ijk} L_k \,, \tag{6.5a}$$

$$[L_i, X_j] = i\hbar\,\varepsilon_{ijk} X_k \,, \tag{6.5b}$$

$$[L_i, P_j] = i\hbar\,\varepsilon_{ijk} P_k \,, \tag{6.5c}$$

where the first line, eqn (6.5a), defines the *Lie algebra* of the group SO(3), the rotation group in three dimensions. With the rotation group algebra at hand, we may in principle skip directly to Section 6.3, where we will continue with an algebraic treatment of the angular-momentum operator that is reminiscent to the harmonic oscillator in Chapter 5. Nonetheless, we invite the interested reader to continue with the brief excursion into group theory in the more mathematically oriented Section 6.1.2 to fully appreciate the relationship between the rotation group and angular momentum.

6.1.2 The Rotation Group in Three Dimensions

Intuitively, it is clear that rotations are somehow connected with the physical quantity angular momentum. To make this relationship more precise in the context of quantum mechanics we should first ask how rotations are represented in classical mechanics. Consider a vector $\vec{x} \in \mathbb{R}^3$. To rotate it, one can apply a real 3×3 matrix R. Since rotations preserve the lengths of vectors and angles between them, the Euclidean scalar product $\vec{x} \cdot \vec{y} = \vec{x}^\top \vec{y}$ must be rotationally invariant. To guarantee this, the matrix R must be orthogonal, $R^\top R = \mathbb{1}$, which also implies that $\det(R) = \pm 1$. In addition, rotations are required to preserve the "handedness" (also called "chirality") of any orthonormal basis. Put simply, there is no way to rotate one's left hand such that it looks exactly like one's right hand, only a mirror operation can achieve this. Mathematically, this is expressed as the requirement that $\det(R) = +1$. Now, it is simple to check that the rotation matrices form a group.

Definition 6.1   *A **group** is a set G that is closed under an operation "$\odot$", called the group multiplication, i.e., $g \odot g' \in G$ for all $g, g' \in G$. For all group elements g, g', and g'' the map "$\odot$" must further be associative, $g \odot (g' \odot g'') = (g \odot g') \odot g''$, admit an identity element $\mathbb{1}$, such that $\mathbb{1} \odot g = g \odot \mathbb{1} = g$, and contain an inverse g^{-1}, such that $g^{-1} \odot g = g \odot g^{-1} = \mathbb{1}$.*

For rotation matrices, the group multiplication is simply matrix multiplication and we will drop the symbol "$\odot$" from now on. The product of two rotation matrices is again a rotation matrix. The identity matrix is a trivial rotation, and the inverse of any rotation $R(\vec{\varphi})$ around the axis $\vec{\varphi} \in \mathbb{R}^3$ by an angle $|\vec{\varphi}|$ is $R(-\vec{\varphi}) = R^{\top}(\vec{\varphi})$. Products of orthogonal matrices are again orthogonal matrices and $\det(RR') = \det(R) \det(R')$. Orthogonal 3×3 matrices with determinant $+1$ hence form a group called the *special orthogonal group* SO(3), the rotation group in three dimensions, where "special" indicates that $\det(R) = +1$, and each element can be fully specified by three real parameters collected in the vector $\vec{\varphi}$. In compact form, we can write the group as

$$\text{SO}(3) = \left\{ R \in \text{GL}(3, \mathbb{R}) \,|\, R^T R = RR^T = \mathbb{1}, \det(R) = 1 \right\}, \tag{6.6}$$

where $\text{GL}(n, \mathbb{R})$ is the (general linear) group of $n \times n$ matrices with real entries. Since we now know how to describe rotations in $\mathbb{R}^3$, the crucial question is, how to describe three-dimensional rotations in the Hilbert space of states. In more mathematical terms, we are looking for a *representation* of the rotation group in the Hilbert space.

Definition 6.2   *A **representation** of a group G is a group homomorphism π from G to the group of linear operators on a vector space V, that is, a map $\pi : G \to V$, which respects the group structure, such that $\pi(gg') = \pi(g)\pi(g')$ for all $g, g' \in G$.*

In other words, orthogonal 3×3 matrices with determinant $+1$ are only one of the possible realizations—a representation—of the group SO(3), and we are interested in finding a representation on the Hilbert space. One restriction that we can make right away is that we are looking for a *unitary representation*, that is, one where every group element is mapped to a unitary operator on $\mathcal{H}$. At this point we could make an educated guess and verify that the proposed operators indeed represent rotations, and anyone who wishes to do so may jump directly to Section 6.2, in particular, to Theorem 6.2. However, it is both instructive and enlightening to find a constructive method to obtain a unitary representation of the rotation group. To achieve this, we will introduce some basic notions from the theory of Lie groups and Lie algebras, a topic on which extensive additional literature is available, see, for instance, the textbooks by Tung (1985) or Warner (1983).

6.1.3   Lie Groups and Lie Algebras

The rotation group is an example for a so-called *Lie group*, that is, a structure as laid out in Definition 6.1 that is also a differentiable manifold. Loosely speaking[1], this means that one is talking about an object that can be locally described as $\mathbb{R}^n$, where one can define coordinates, and functions of these coordinates. Globally, the object might not be similar to $\mathbb{R}^n$ at all, but any two different regions may be mapped onto each other using a set of *charts*. As an example, think of a sphere. Close to any chosen point, the sphere appears to be flat and can be described as $\mathbb{R}^2$. In other words, each point on the surface can be thought of as the origin of a tangent space, the space spanned by all vectors tangential to the surface in that point. If we stay close to this point, we can work in the tangent space with familiar concepts like functions and

---

[1] For a more detailed discussion, see Section 9.5.

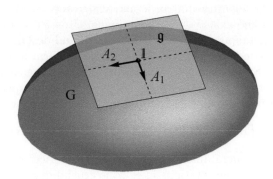

Fig. 6.1 Lie groups and Lie algebras. Illustration of a two-dimensional Lie group G. Every group element can be thought of as a point on the two-dimensional surface of the three-dimensional ellipsoid shown. The corresponding Lie algebra $\mathfrak{g}$ is the (also two-dimensional) tangent space at the point corresponding to the identity element $\mathbb{1}$ of G. The Lie algebra is spanned by the basis vectors A_1 and A_2.

derivatives. For Lie groups one uses a similar method. For a Lie group, the points on the manifold (in our example, the points on the sphere) are the group elements, and the tangent space close to a particular point, the identity element $\mathbb{1}$, is the *Lie algebra* $\mathfrak{g}$ corresponding to the Lie group G, as illustrated in Fig. 6.1.

Definition 6.3   *A (complex)* **Lie algebra** $\mathfrak{g}$ *is a vector space over the field* $\mathbb{C}$ *that is closed under a bilinear map* $[\,.\,,\,.\,] : \mathfrak{g} \times \mathfrak{g} \to \mathfrak{g}$, *a* **Lie bracket**, *which is antisymmetric,* $[A,B] = -[B,A]$, *and satisfies the Jacobi identity* $[A,[B,C]] + [C,[A,B]] + [B,[C,A]] = 0$, *for all* $A, B, C \in \mathfrak{g}$.

Definition 6.4   *A* **Lie-algebra representation** *is a homomorphism* $\phi : \mathfrak{g} \to \mathfrak{h}$, *i.e., a linear map between two Lie algebras* $\mathfrak{g}$ *and* $\mathfrak{h}$ *that preserves the structure of the Lie bracket,* $\phi([A,B]) = [\phi(A),\phi(B)] \; \forall \, A, B \in \mathfrak{g}$.

Much as with a (Lie) group, a Lie algebra is an abstract concept that may have different representations. However, a Lie algebra is a vector space, a concept somewhat more familiar to us by now. In particular, we know that we may choose a basis $\{A_i\}$ for a Lie algebra $\mathfrak{g}$. Since $\mathfrak{g}$ is closed under the Lie bracket, $[A_i, A_j]$ is a vector in $\mathfrak{g}$ that has a unique expansion with respect to the basis $\{A_i\}$, such that

$$[A_i, A_j] = f_{ijk} A_k, \tag{6.7}$$

where summation is implied for subscripts that are repeated twice (Einstein summation convention). The quantities f_{ijk} are called the *structure constants* of the Lie algebra, which, in turn, is completely determined by these constants. Fortunately, we have already encountered such a structure before in eqn (6.5a) and the commutator indeed has the properties required for a Lie bracket, see Definition 6.3. The commutator is clearly antisymmetric and satisfies the Jacobi identity, see eqns (2.36b) and (2.36f). We hence find that the angular-momentum operators $L_i = \varepsilon_{ijk} X_j P_k$ from eqn (6.4) span a Lie algebra. The elements of this algebra are vectors of this vector space, but also operators on the Hilbert space. In addition, any element can be written in terms of its components with respect to the Lie-algebra basis $\{L_i\}$ as $\sum_i \varphi_i L_i =: \vec{\varphi} \cdot \vec{L}$.

The task that we shall turn to next is establishing the connection between a Lie algebra and the corresponding Lie group, and more specifically, showing that the Lie algebra formed by the angular-momentum operators indeed is the Lie algebra of the rotation group SO(3). Without introducing the necessary aspects of differential geometry, for which we direct the curious reader to our brief treatment in Section 9.5 as well as the extensive literature such as Nakahara (1990), Bertlmann (2000), Sontz (2015), Frankel (1997), or Hou and Hou (1997), let us simply state the important relation between a Lie group and its Lie algebra in the following theorem.

Theorem 6.1 (Lie groups and Lie algebras)

For every Lie algebra representation there exists a unique Lie group representation if the Lie group is simply connected. If the group is connected, but not simply connected, a group representation still exists but may not be unique.

Theorem 6.1 is a condensed version of several more technical statements that we do not wish to analyse here in detail but we refer the interested reader to Warner (1983, Theorems 3.14, 3.16, 3.25, and 3.27).

For us, the remark on connectedness in Theorem 6.1 is, in fact, relevant, since SO(3) is connected, but not simply connected. This can be easily seen by noting that the group has the same parametrization as the inside of a sphere, i.e., a three-dimensional ball of radius π. A vector from the origin to any point in the ball defines a rotation axis, and the length of the vector determines the rotation angle. However, as rotations by π and $-\pi$ are equivalent, all points on the surface must be identified with their antipodal counterparts. Consequently, any curve through the ball that connects two opposite points on the surface is closed and cannot be contracted to a single point without breaking. The group is hence not simply connected, which means that different Lie-algebra elements may correspond to the same Lie-group element. For now, we will ignore this potential ambiguity, but we will revisit this issue in Chapter 8 (Section 8.2.5), when we discuss the related group SU(2), the double cover of SO(3).

6.2   Rotations in the Hilbert Space

6.2.1   Unitary Representations of the Rotation Group

Now, we shall be concerned with constructing a unitary Lie-group representation from a given Lie-algebra representation. This is achieved using the *exponential map* from eqn (2.37). Simply note that the exponential e^{iA} of any Hermitian operator $A^\dagger = A$ gives a formally unitary operator. We just need to make sure that the set of unitary operators

$$U(\vec{\varphi}) = e^{-i\vec{\varphi}\cdot\vec{L}/\hbar} \tag{6.8}$$

that we define by exponentiating the Lie-algebra elements $\vec{\varphi}\cdot\vec{L}/\hbar$, where we have included $\hbar$ for dimensional reasons, forms a group as specified in Definition 6.1. Indeed, the group defined in this way is closed, i.e., $U(\vec{\varphi})U(\vec{\varphi}')$ can be written as $U(\vec{\varphi}'')$ for some $\vec{\varphi}''$ by virtue of the Baker–Campbell–Hausdorff formula, see eqn (2.38), because the Lie algebra is closed under the commutator, see Definition 6.3. The identity element is obtained when $\vec{\varphi} = 0$, and the inverse of $U(\vec{\varphi})$ is given by $U(-\vec{\varphi})$.

It can also be easily checked that the Lie algebra spanned by the $\{L_i\}$ is the tangent space at the identity element of the group. The derivative of $U(\vec{\varphi})$ with respect to φ_k, evaluated at $\vec{\varphi} = 0$, gives $-iL_k/\hbar$. The tangent vectors of the group at $\mathbb{1}$ hence span the associated Lie algebra.

6.2.2   The Lie Algebra of the Rotation Group

Finally, all that is left to do is to show that the Lie algebra of the angular-momentum operators really is the Lie algebra of the rotation group. Let us show this by again considering the rotations $R(\vec{\varphi})$ in $\mathbb{R}^3$ from Section 6.1.2. We know that the usual 3×3 rotation matrices form a representation of the group SO(3). They can hence be written as $R(\vec{\varphi}) = \exp(-i\vec{\varphi}\cdot\vec{T})$, where $\vec{\varphi}\cdot\vec{T} := \sum_i \varphi_i T_i$, that is, as exponents of Hermitian 3×3 matrices T_i, given by

$$T_x = \begin{pmatrix} 0 & 0 & 0 \\ 0 & 0 & -i \\ 0 & i & 0 \end{pmatrix}, \quad T_y = \begin{pmatrix} 0 & 0 & i \\ 0 & 0 & 0 \\ -i & 0 & 0 \end{pmatrix}, \quad T_z = \begin{pmatrix} 0 & -i & 0 \\ i & 0 & 0 \\ 0 & 0 & 0 \end{pmatrix}. \tag{6.9}$$

More compactly, the components of these matrices can be written as $(T_i)_{jk} = -i\varepsilon_{ijk}$, and we can write the Lie algebra so(3) corresponding to the Lie group SO(3) as

$$\text{so}(3) = \left\{ T \in \text{GL}(3,\mathbb{C}) \,|\, T^\dagger = T,\, Tr(T) = 0 \right\}. \tag{6.10}$$

We will leave it as an exercise to verify that the matrices T_i satisfy the commutation relations

$$[T_i, T_j] = i\,\varepsilon_{ijk}\,T_k. \tag{6.11}$$

The structure constants are the same as for the angular-momentum operators $L_i/\hbar$, see eqn (6.5a). Since the Lie algebra is fully determined by these constants, we arrive at our conclusion that we can phrase in the following theorem.

Theorem 6.2 (Generators of rotations)

*The orbital angular-momentum operators $L_i/\hbar$ are the **generators of rotations** in Hilbert space, that is, three-dimensional rotations are represented on the Hilbert space by unitary operators*

$$U(\vec{\varphi}) = e^{-i\vec{\varphi}\cdot\vec{L}/\hbar}.$$

It is interesting to note that the matrices T_i from eqn (6.9) are traceless, that is $\mathrm{Tr}(T_i) = 0$, where the trace of an operator is defined as $\mathrm{Tr}(A) = \sum_i \langle \psi_i | A | \psi_i \rangle$. For the trace one may choose any orthonormal basis $\{| \psi_i \rangle\}$, in particular, one may choose the eigenbasis of A. The trace is hence just the sum of the eigenvalues of A. The property $\mathrm{Tr}(T_i) = 0$ can then easily be seen to be a consequence of the requirement that proper rotations $R(\vec{\varphi})$ should have $\det(R) = 1$, since $\det(e^A) = e^{\mathrm{Tr}(A)}$. Similarly, to represent proper rotations, the operators $U(\vec{\varphi})$ are required to have $\det(U) = 1$, which implies $\mathrm{Tr}(L_i) = 0$. We can hence already see that the eigenvalues of the operators L_i, which we will study in more detail in Section 6.3, must sum to zero.

6.2.3   Rotation of the Wave Function

Let us quickly illustrate Theorem 6.2 by considering the action of the unitary operator $U(\vec{\varphi})$ on a state vector $\psi(\vec{x})$ when $|\vec{\varphi}| \ll 1$. Expanding the exponential and keeping only the linear-order terms, we get

$$U(\vec{\varphi})\,\psi(\vec{x}) \approx \left(\mathbb{1} - \tfrac{i}{\hbar}\,\vec{\varphi}\cdot\vec{L}\right)\psi(\vec{x}) = \left(\mathbb{1} - \tfrac{i}{\hbar}\,\vec{\varphi}\cdot(\vec{X}\times\tfrac{\hbar}{i}\vec{\nabla})\right)\psi(\vec{x}). \qquad (6.12)$$

By rearranging this expression using the simple vector identity $\vec{a}\cdot(\vec{b}\times\vec{c}) = (\vec{a}\times\vec{b})\cdot\vec{c}$, we get

$$U(\vec{\varphi})\,\psi(\vec{x}) \approx \left(\mathbb{1} - (\vec{\varphi}\times\vec{x})\cdot\vec{\nabla}\right)\psi(\vec{x}). \qquad (6.13)$$

This last expression can be recognized as the leading-order terms of the Taylor expansion of $\psi(\vec{x} - (\vec{\varphi}\times\vec{x}))$ for small $|\vec{\varphi}|$, where $\vec{x} - (\vec{\varphi}\times\vec{x})$ in turn is the leading-order expansion of $\vec{x}' = R^{-1}(\vec{\varphi})\vec{x}$, i.e.,

$$U(\vec{\varphi})\,\psi(\vec{x}) \approx \psi\big(\vec{x} - (\vec{\varphi}\times\vec{x})\big) \approx \psi(\vec{x}'). \qquad (6.14)$$

This nicely illustrates a general transformation property.

Theorem 6.3 (Rotation of wave function)

*The wave function $\psi(\vec{x})$ transforms like a **scalar field**, i.e.,*

$$U(\vec{\varphi})\,\psi(\vec{x}) = \psi\big(R^{-1}(\vec{\varphi})\vec{x}\big).$$

A rotation of our coordinate system translates to a counter-rotation of the argument of the scalar field $\psi(\vec{x})$. Since we have just sketched a proof for this statement here for small $|\vec{\varphi}|$, an explicit proof for all $\vec{\varphi}$ will be given in Section 6.4 to supplement the group-theoretic arguments presented so far.

6.2.4 Rotation of Operators

Since we now know how the wave functions transform under rotations, let us find out how operators are changed by such operations. Consider some (linear) operator A acting on the state $\psi(\vec{x}) \in \mathcal{H}$, such that $A\psi(\vec{x}) = \phi(\vec{x})$. We apply the unitary rotation operator $U(\vec{\varphi})$ to $\phi(\vec{x})$ and insert the identity $\mathbb{1} = U^\dagger U$, such that

$$U\,\phi(\vec{x}) = U\,A\,\psi(\vec{x}) = U\,A\,U^\dagger\,U\,\psi(\vec{x}) = U\,A\,U^\dagger\,\psi\big(R^{-1}(\vec{\varphi})\vec{x}\big) \overset{!}{=} \phi\big(R^{-1}(\vec{\varphi})\vec{x}\big). \tag{6.15}$$

We can then just read off the operator A' in the rotated system, since $A'\psi(\vec{x}') = \phi(\vec{x}')$, i.e.,

$$A' = U\,A\,U^\dagger = A - \tfrac{i}{\hbar}\vec{\varphi}\cdot[\vec{L},A] + \mathcal{O}(|\vec{\varphi}|^2), \tag{6.16}$$

where we have performed a similar expansion as in eqn (6.12). To study the transformation of operators under rotations in more detail, let us consider some examples.

6.2.5 Rotation of Vector Operators

Consider a set of three Hermitian operators A_i ($i = x, y, z$) that can be seen as the components of a vector operator $\vec{A}$. Examples for such operators are $\vec{X}$, $\vec{P}$, and $\vec{L}$, for which we know the commutators with the angular-momentum operators from (6.5). For instance, in analogy to eqn (6.16), we can write the jth component of the operator $\vec{X}'$ in the rotated frame as

$$X_j' = U\,X\,U^\dagger = X_j - \tfrac{i}{\hbar}\vec{\varphi}\cdot[\vec{L},X_j] + \mathcal{O}(|\vec{\varphi}|^2). \tag{6.17}$$

Inspecting the second term in the expansion more closely, we can insert the commutator from eqn (6.5b), to find

$$-\tfrac{i}{\hbar}\varphi_i\,[L_i,X_j] = \varphi_i\,\varepsilon_{ijk}\,A_k = i\,\varphi_i\,(T_i)_{jk}\,A_k = i\,\big[(\vec{\varphi}\cdot\vec{T})\vec{A}\big]_j, \tag{6.18}$$

where summation is implied over the repeated indices (Einstein summation convention). In eqn (6.18) we can recognize the expansion

$$R^{-1}(\vec{\varphi}) = \exp\big(i\vec{\varphi}\cdot\vec{T}\big) = \mathbb{1} + i\vec{\varphi}\cdot\vec{T} + \mathcal{O}(|\vec{\varphi}|^2), \tag{6.19}$$

which indicates that the operator in the rotated frame is $\vec{X}' = R^{-1}(\vec{\varphi})\vec{X}$. Indeed, this can be easily confirmed. Inserting $U^\dagger U = \mathbb{1}$ as in eqn (6.15), the expectation value of $\vec{X}$ can be written as

$$\langle \vec{X} \rangle_\psi = \int d^3x\, \psi(\vec{x})\, \vec{X} \psi(\vec{x}) = \int d^3x\, \psi(R^{-1}\vec{x})\, U\vec{X} U^\dagger \psi(R^{-1}\vec{x}) \qquad (6.20)$$

$$= \int d^3x\, \psi(R^{-1}\vec{x})\, U\vec{x} U^\dagger \psi(R^{-1}\vec{x}) = \int d^3x'\, \psi(\vec{x}')\, U R\vec{x}' U^\dagger\, \psi(\vec{x}')\,.$$

In the last step we have substituted $\vec{x} = R\vec{x}'$, where $d^3x = d^3x'$ since the Jacobian determinant is just $\det(R) = 1$. From this we can deduce that $\vec{x}' = U(R\vec{x}')U^\dagger$, which confirms that the position operator transforms as $U(\vec{\varphi})\vec{X} U^\dagger(\vec{\varphi}) = R^{-1}(\vec{\varphi})\vec{X}$. At the same time, we can see that, via the Hadamard lemma of eqn (2.39), this transformation property is completely determined by the commutation relation with the angular-momentum operator in eqn (6.5b), which brings us to the following conclusion.

Theorem 6.4 (Rotation of vector operators)

Every vector-valued operator $\vec{A}$ that satisfies the commutation relations $[L_i, A_j] = i\hbar\varepsilon_{ijk} A_k$ for all its components A_j, transforms under rotations as

$$U(\vec{\varphi})\vec{A} U^\dagger(\vec{\varphi}) = R^{-1}(\vec{\varphi})\vec{A}\,.$$

6.2.6 Rotation of Scalar Operators

Many scalar-valued operators, such as $H = \vec{P}^2/(2m)$, can be written as functions of vector operators, e.g., $A = A(\vec{X}, \vec{P})$. To form a scalar, such operators typically depend on scalar products of vector operators, such as $\vec{X}^2 = \vec{X} \cdot \vec{X}$. Since the Euclidean scalar product is invariant under rotations, we have

$$U(\vec{\varphi})\, \vec{X} \cdot \vec{X} U^\dagger(\vec{\varphi}) = \big(R^{-1}(\vec{\varphi})\vec{X}\big) \cdot \big(R^{-1}(\vec{\varphi})\vec{X}\big) = \vec{X} \cdot \vec{X}\,. \qquad (6.21)$$

Scalar operators such as $H = \vec{P}^2/(2m)$, $\vec{X}^2$, and $\vec{L}^2$ hence commute with the angular-momentum operators L_i. In general, all operators A for which $[L_i, A] = 0\ \forall\ i$ are invariant under rotations.

Lemma 6.1 (Rotationally invariant operators)

Every operator A that commutes with the angular-momentum operator $\vec{L}$ is invariant under rotations,

$$[\vec{L}, A] = 0 \qquad \Leftrightarrow \qquad U(\vec{\varphi})A U^\dagger(\vec{\varphi}) = A\,.$$

Now that we have a clear idea of the connection between angular momentum and rotations, and how to represent either on the Hilbert space, it is time to study the spectrum of the angular-momentum operators.

6.3 Angular Momentum Eigenstates and Eigenvalues

Returning again to the Lie algebra of the rotation group SO(3),

$$[L_i, L_j] = i\hbar\varepsilon_{ijk}L_k, \tag{6.22}$$

we also notice that different components of the angular-momentum operator, e.g., L_x and L_y, do not commute with each other, i.e., they are incompatible observables. By remembering the general uncertainty relation for operators A and B from Theorem 2.4, where we ignore the additional term due to the covariance, we can immediately deduce the uncertainty relation

$$\Delta L_x \cdot \Delta L_y \geq \tfrac{\hbar}{2}|\langle L_z\rangle|. \tag{6.23}$$

In the context of an experiment, this translates to the statement that different components of the angular-momentum operator $\vec{L}$ cannot be measured "simultaneously". The measurement of one component changes the possible outcomes for a consecutive measurement of the other component for the same system. In a theoretical framework this is expressed through the fact that these operators do not have common eigenfunctions. Nonetheless, as we have just argued in the example of the last section, the operator $\vec{L}^2$ commutes with all components L_i of $\vec{L}$, i.e.,

$$[\vec{L}^2, L_i] = 0. \tag{6.24}$$

The operator $\vec{L}^2$ and any fixed component—in the absence of any interactions or other considerations that break rotational symmetry we may select L_z without loss of generality—have common eigenfunctions. It will be our next task to determine these eigenfunctions $f_{\lambda,m}$ and the corresponding eigenvalues $\hbar^2\lambda$ and $\hbar m$ of $\vec{L}^2$ and L_z, respectively, such that

$$\vec{L}^2 f_{\lambda,m} = \hbar^2\lambda f_{\lambda,m} \quad \text{and} \quad L_z f_{\lambda,m} = \hbar m f_{\lambda,m}, \tag{6.25}$$

where we have included factors of $\hbar^2$ and $\hbar$, respectively, such that λ and m are dimensionless.

6.3.1 Angular Momentum Ladder Operators

In complete analogy to the algebraic approach to the harmonic-oscillator problem (Section 5.1) we use the technique of *ladder operators*.

Definition 6.5 *The angular momentum **ladder operators** $L_\pm$ are defined as*

$$L_\pm := L_x \pm i L_y,$$

*where L_+ and L_- are called the **raising operator** and **lowering operator**, respectively.*

With Definition 6.5 at hand it is easy to see that the commutation relations defining the Lie algebra of the rotation group in eqn (6.22) imply the relations

$$[L_z, L_\pm] = \pm \hbar L_\pm, \tag{6.26a}$$

$$[\vec{L}^2, L_\pm] = 0, \tag{6.26b}$$

$$[L_+, L_-] = 2 \hbar L_z. \tag{6.26c}$$

Similar to the Lemmas 5.1 and 5.2, formulated for the harmonic-oscillator ladder operators, we can make the following statement for the angular-momentum ladder operators.

Lemma 6.2 *If $f_{\lambda,m}$ is an eigenfunction of $\vec{L}^2$ and L_z then the function $L_\pm f_{\lambda,m}$ is also an eigenfunction of $\vec{L}^2$ and L_z.*

Proof We shall prove this in two steps. First, simply note that since $\vec{L}^2$ and $L_\pm$ commute, see eqn (6.26b), it trivially follows from eqn (6.25) that

$$\vec{L}^2 L_\pm f_{\lambda,m} = L_\pm \vec{L}^2 f_{\lambda,m} = \hbar^2 \lambda L_\pm f_{\lambda,m}, \tag{6.27}$$

i.e., $L_\pm f_{\lambda,m}$ is also an eigenfunction of $\vec{L}^2$ with eigenvalue $\hbar^2 \lambda$. To show that $L_\pm f_{\lambda,m}$ is an eigenfunction of L_z, we add zero and use the commutator from eqn (6.26a), i.e.,

$$L_z L_\pm f_{\lambda,m} = L_\pm L_z f_{\lambda,m} + [L_z, L_\pm] f_{\lambda,m} = \hbar(m \pm 1) L_\pm f_{\lambda,m}. \tag{6.28}$$

We see that $L_\pm f_{\lambda,\mu}$ is an eigenfunction of L_z with eigenvalue $(\mu \pm \hbar)$, which proves the lemma. □

Starting from some eigenvalue μ, the ladder operators L_+ and L_- switch between the possible eigenvalues of L_z, "climbing" and "descending" the "ladder" of eigenvalues, respectively. However, the number of rungs on this ladder depends on λ, that is, the eigenvalues of L_z are bounded, $\sqrt{\lambda} > |m| \geq 0$.

To see this, simply note that, since the operators L_i are Hermitian, they have real eigenvalues, which suggests that the eigenvalues of L_i^2 and $\vec{L}^2 = \sum_i L_i^2$ must be positive. Consequently, for a fixed eigenfunction $f_{\lambda,m}$, the eigenvalue $\hbar^2 \lambda$ is larger or equal than the eigenvalue of the squared component L_z^2.

6.3.2 Angular Momentum Eigenvalues

To make the last observation of the previous subsection more precise, let us switch to the Dirac notation (see Section 3.1.3) and denote the functions $f_{\lambda,m}$ as vectors $| \lambda, m \rangle$ in a Hilbert space. We are then interested in normalizing the eigenvectors obtained by acting with $L_\pm$ as described in Lemma 6.2. To do this, the following short computation will be useful,

$$L_\pm L_\mp = \left(L_x \pm iL_y\right)\left(L_x \mp iL_y\right) = L_x^2 + L_y^2 \mp i\left[L_x, L_y\right] = \vec{L}^{\,2} - L_z^2 \pm \hbar L_z\,. \tag{6.29}$$

With this, we compute the norm

$$\|L_\pm \,|\,\lambda, m\,\rangle\|^2 = \langle\,\lambda, m\,|\,L_\mp L_\pm\,|\,\lambda, m\,\rangle = \langle\,\lambda, m\,|\,\vec{L}^{\,2} - L_z^2 \mp \hbar L_z\,|\,\lambda, m\,\rangle$$
$$= \hbar^2\big(\lambda - m(m \pm 1)\big)\,. \tag{6.30}$$

Since the norm must be positive definite, see eqn (2.22d), we see that for a fixed λ, there must be a maximal value $m_{\max}$ for m, let us denote it as $m_{\max} = l$, such that

$$\lambda = l(l+1)\,. \tag{6.31}$$

This means that λ is fully determined by l, the *azimuthal quantum number*, and we can label our eigenstates as $|\,l, m\,\rangle$ instead of $|\,l(l+1), m\,\rangle$ for brevity, where m is called the *magnetic quantum number*. From eqn (6.30) one can further deduce that the magnetic quantum number also has a minimal value $m_{\min} = -l$ when the azimuthal quantum number is fixed. Remembering that the scalar product is non-degenerate, i.e., the norm can only vanish for a null vector, we hence find

$$L_+ \,|\,l, m = +l\,\rangle = L_- \,|\,l, m = -l\,\rangle = 0\,. \tag{6.32}$$

Starting from the state $|\,l, -l\,\rangle$, the raising operator L_+ then shifts upwards on this discrete ladder of eigenstates, while the lowering operator L_- shifts downwards. By inspection of eqn (6.30) we can write

$$L_\pm \,|\,l, m\,\rangle = \hbar\sqrt{l(l+1) - m(m \pm 1)}\,|\,l, m \pm 1\,\rangle\,. \tag{6.33}$$

6.3.3   Angular Momentum Eigenstates

With each application of L_+ or L_-, the angular momentum $\hbar m$ along the z direction increases or decreases by one unit of $\hbar$, respectively, until the topmost state $|\,l, +l\,\rangle$ or the bottommost state $|\,l, -l\,\rangle$ is reached after $2l$ steps. For a fixed eigenvalue of $\vec{L}^{\,2}$ one hence finds $N = 2l + 1$ eigenstates with different eigenvalues of L_z. But are there any more?

Suppose there was a continuum of eigenstates of L_z with arbitrary eigenvalues in the interval between $m = -l$ and $m = +l$ on the real line. If that was the case for some m, one could apply the operator L_- until reaching some m' between $-l$ and $-l + 1$. According to Lemma 6.2, an additional application would then yield another eigenstate $|\,l, m''\,\rangle$ with $-l - 1 < m'' < -l$. Following eqn (6.30), this state would have a negative norm, which cannot occur for any valid state. For a fixed eigenvalue of $\vec{L}^{\,2}$, only the previously discussed $N = 2l$ eigenstates are allowed.

From the previous statement we can also infer that $l = (N-1)/2 \geq 0$ can in principle only take on integer values $(0, 1, 2, \ldots)$ or half-integer values $(\frac{1}{2}, \frac{3}{2}, \ldots)$. However, as we shall see in Section 6.4, the operator $\vec{L}^2$ for the orbital angular momentum only has eigenvalues corresponding to integer azimuthal quantum numbers, $l = 0, 1, 2, \ldots$. Nonetheless, taking into account *spin*, which we will do in Chapter 8, half-integer values are possible for the *total angular momentum* observable $\vec{J}^2$, but more on this later. For now, let us collect our observations in the following theorem.

Theorem 6.5 (Angular momentum eigenstates)

The orbital angular-momentum operators $\vec{L}^2$ and L_z have a discrete common set of eigenstates $|\, l, m \,\rangle$ with eigenvalues $\hbar^2 l(l+1)$ and $\hbar m$, respectively, i.e.,

$$\vec{L}^2 \,|\, l, m \,\rangle = \hbar^2 l(l+1) \,|\, l, m \,\rangle \,,$$
$$L_z \,|\, l, m \,\rangle = \hbar m \,|\, l, m \,\rangle \,,$$

*where the **azimuthal quantum number** l and the **magnetic quantum number** m take on values*

$$l = 0, 1, 2, \ldots, \qquad and \qquad m = -l, -l+1, \ldots, l-1, l \,.$$

Before we turn to explicitly determining the wave functions of the states $|\, l, m \,\rangle$, some brief remarks about the magnetic quantum number are in order. First, note that the eigenvalues of L_z sum to zero, and we would have got the same result had we considered L_x or L_y instead. This is nothing but the requirement that $\mathrm{Tr}(L_i) = 0$, such that $\det\big(U(\vec{\varphi})\big) = 1$.

The second remark concerns the fact that $\vec{L}^2$ and L_z commute. This means that these quantities are simultaneously measurable. For instance, for an eigenstate $|\, l, m \,\rangle$ with $l > 0$, a measurement of $\vec{L}^2$ will show that the angular momentum of the system is a vector of length $\hbar\sqrt{l(l+1)}$ and the z component of this vector is $\hbar m$. However, we cannot specify the other components exactly. Even when m takes its maximal (or minimal) value $m = +l$ (or $m = -l$), the z component does not account for the entire length of the vector. Any measurement results can only determine the angular-momentum vector up to an arbitrary rotation around the z axis (or any other reference direction), see Fig. 6.2. Another way to express this fact lies in the uncertainty principle. If the system is in an eigenstate of L_z (and $l \neq 0$) then the observables L_x and L_y have non-zero uncertainties. For $m \neq 0$ this can easily be seen from the non-vanishing commutator in Theorem 2.4, which yields the uncertainty relation

$$\Delta L_x \, \Delta L_y \geq \tfrac{\hbar}{2} |\langle L_z \rangle| = \frac{\hbar^2}{2} m \,. \tag{6.34}$$

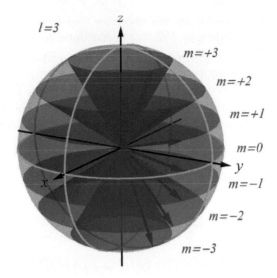

Fig. 6.2 Orbital angular momentum. In an eigenstate $|l, m\rangle$ of $\vec{L}^2$ and L_z, the angular-momentum vector has length $|\vec{L}| = \hbar\sqrt{l(l+1)}$ and the component in the z direction is $\hbar|m| < |\vec{L}|$. This defines a direction along a cone, as illustrated here for $l = 3$. The other components of the angular momentum are unspecified, with uncertainties $\Delta L_x = \Delta L_y = \hbar\sqrt{(l(l+1) - m^2)/2}$.

The uncertainties are even non-zero when $m = 0$ as long as $l \neq 0$. In general one finds

$$\Delta L_x \, \Delta L_y = \frac{\hbar^2}{2}\left(l(l+1) - m^2\right) \tag{6.35}$$

for a state $|l, m\rangle$, but in order to show this we must first find the eigenfunctions of $\vec{L}^2$ and L_z, which we shall do in the next section.

6.4 Angular Momentum Eigenfunctions

In this section, we are going to determine the wave functions that correspond to the simultaneous eigenstates $|l, m\rangle$ of $\vec{L}^2$ and L_z.

6.4.1 Spherical Polar Coordinates

This is most conveniently done using spherical polar coordinates (r, ϑ, φ), which are related to the usual Cartesian coordinates (x, y, z) via

$$x = r \sin\vartheta \cos\varphi, \tag{6.36a}$$
$$y = r \sin\vartheta \sin\varphi, \tag{6.36b}$$
$$z = r \cos\vartheta, \tag{6.36c}$$

see Fig. 6.3. It will now be our task to express the angular-momentum operators

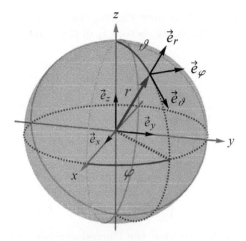

Fig. 6.3 Spherical polar coordinates. Spherical polar coordinates (r, ϑ, φ) allow to specify any point in $\mathbb{R}^3$ by the radial distance from the origin, the radius r, a polar angle ϑ measured from a fixed reference direction called the zenith, which is here chosen to be the z axis, and an azimuthal angle φ in a plane (here the x–y plane) orthogonal to the zenith. However, unlike the basis vectors $\vec{e}_x$, $\vec{e}_y$, and $\vec{e}_z$ of the Cartesian coordinate system, the directions of the basis vectors $\vec{e}_r$, $\vec{e}_\vartheta$, and $\vec{e}_\varphi$ of the spherical coordinate system depend on the position, see eqn (6.42).

using the coordinates (r, ϑ, φ). The first step towards this goal is to transform the nabla operator

$$\vec{\nabla} = \vec{e}_x \frac{\partial}{\partial x} + \vec{e}_y \frac{\partial}{\partial y} + \vec{e}_z \frac{\partial}{\partial z} \qquad (6.37)$$

by transforming its components, the partial derivatives $\partial/\partial x^i$, where $(x^1, x^2, x^3) = (x, y, z)$, as well as the basis vectors $\vec{e}_i$, where $i = x, y, z$ into spherical polar coordinates. The partial derivatives are easily transformed using

$$\frac{\partial}{\partial x^i} = \frac{\partial x'^j}{\partial x^i} \frac{\partial}{\partial x'^j} , \qquad (6.38)$$

where $(x'^1, x'^2, x'^3) = (r, \vartheta, \varphi)$, so that we have

$$\frac{\partial}{\partial x} = \frac{\partial r}{\partial x} \frac{\partial}{\partial r} + \frac{\partial \vartheta}{\partial x} \frac{\partial}{\partial \vartheta} + \frac{\partial \varphi}{\partial x} \frac{\partial}{\partial \varphi}$$

$$= \sin\vartheta \cos\varphi \frac{\partial}{\partial r} + \frac{1}{r} \cos\vartheta \cos\varphi \frac{\partial}{\partial \vartheta} - \frac{1}{r} \frac{\sin\varphi}{\sin\vartheta} \frac{\partial}{\partial \varphi} , \qquad (6.39a)$$

$$\frac{\partial}{\partial y} = \sin\vartheta \sin\varphi \frac{\partial}{\partial r} + \frac{1}{r} \cos\vartheta \sin\varphi \frac{\partial}{\partial \vartheta} + \frac{1}{r} \frac{\cos\varphi}{\sin\vartheta} \frac{\partial}{\partial \varphi} , \qquad (6.39b)$$

$$\frac{\partial}{\partial z} = \cos\vartheta \frac{\partial}{\partial r} - \frac{1}{r} \sin\vartheta \frac{\partial}{\partial \vartheta} . \qquad (6.39c)$$

To determine the basis vectors of the new coordinate system, let us first consider the position vector $\vec{r}$. In terms of the basis vectors $\vec{e}_i$ of the Cartesian coordinate system, we can write it as

$$\vec{r} = x\,\vec{e}_x + y\,\vec{e}_y + z\,\vec{e}_z. \tag{6.40}$$

Now, the direction specified by $\vec{e}_r$ is the direction along which the coordinate r increases, while the other coordinates ϑ and φ remain constant. This means that $\vec{e}_r$ is proportional to the partial derivative $\partial\vec{r}/\partial r$, which we just have to normalize to obtain a unit vector, i.e.,

$$\vec{e}_r = \frac{\frac{\partial\vec{r}}{\partial r}}{\left|\frac{\partial\vec{r}}{\partial r}\right|} = \frac{\vec{e}_x\frac{\partial x}{\partial r} + \vec{e}_y\frac{\partial y}{\partial r} + \vec{e}_z\frac{\partial z}{\partial r}}{\sqrt{\left(\frac{\partial x}{\partial r}\right)^2 + \left(\frac{\partial y}{\partial r}\right)^2 + \left(\frac{\partial z}{\partial r}\right)^2}}, \tag{6.41}$$

where we have made use of the fact that the Cartesian basis vectors do not depend on the coordinates. Following the same logic, one finds $\vec{e}_\vartheta = \frac{\partial\vec{r}}{\partial\vartheta}/|\frac{\partial\vec{r}}{\partial\vartheta}|$ and $\vec{e}_\varphi = \frac{\partial\vec{r}}{\partial\varphi}/|\frac{\partial\vec{r}}{\partial\varphi}|$. From eqn (6.36) we then arrive at

$$\vec{e}_r = \sin\vartheta\,\cos\varphi\,\vec{e}_x + \sin\vartheta\,\sin\varphi\,\vec{e}_y + \cos\vartheta\,\vec{e}_z, \tag{6.42a}$$

$$\vec{e}_\vartheta = \cos\vartheta\,\cos\varphi\,\vec{e}_x + \cos\vartheta\,\sin\varphi\,\vec{e}_y - \sin\vartheta\,\vec{e}_z, \tag{6.42b}$$

$$\vec{e}_\varphi = -\sin\varphi\,\vec{e}_x + \cos\varphi\,\vec{e}_y. \tag{6.42c}$$

From eqn (6.42) we can easily compute the Cartesian basis vectors with respect to the basis of the spherical coordinate system, which yields

$$\vec{e}_x = \sin\vartheta\,\cos\varphi\,\vec{e}_r + \cos\vartheta\,\cos\varphi\,\vec{e}_\vartheta - \sin\varphi\,\vec{e}_\varphi, \tag{6.43a}$$

$$\vec{e}_y = \sin\vartheta\,\sin\varphi\,\vec{e}_r + \cos\vartheta\,\sin\varphi\,\vec{e}_\vartheta + \cos\varphi\,\vec{e}_\varphi, \tag{6.43b}$$

$$\vec{e}_z = \cos\vartheta\,\vec{e}_r - \sin\vartheta\,\vec{e}_\vartheta. \tag{6.43c}$$

Finally, we combine the partial derivatives of eqn (6.39) with the Cartesian basis vectors of eqn (6.43) to form the nabla-operator of eqn (6.37), and arrive at

$$\vec{\nabla} = \vec{e}_r\frac{\partial}{\partial r} + \vec{e}_\vartheta\frac{1}{r}\frac{\partial}{\partial\vartheta} + \vec{e}_\varphi\frac{1}{r\sin\vartheta}\frac{\partial}{\partial\varphi}. \tag{6.44}$$

6.4.2   Angular Momentum Operators in Spherical Coordinates

We now want to express the components L_x, L_y, and L_z of the angular-momentum operator with respect to the Cartesian basis in terms of spherical polar coordinates. Recalling that the angular-momentum operator is defined as

$$\vec{L} = \vec{x} \times \vec{p} = \frac{\hbar}{i}\,\vec{x} \times \vec{\nabla}, \tag{6.45}$$

we can insert the transformed nabla operator from eqn (6.44). Further comparing eqn (6.36) and the basis vector $\vec{e}_r$ from eqn (6.42a), we note that the position vector can be written as $\vec{x} = r\,\vec{e}_r$, which allows us to write $\vec{L}$ with respect to the basis $\{\vec{e}_r, \vec{e}_\vartheta, \vec{e}_\varphi\}$. Finally, we can expand the basis vectors of the spherical coordinate system in terms of their Cartesian counterparts $\{\vec{e}_x, \vec{e}_y, \vec{e}_z\}$ using eqn (6.42) and arrive at

$$L_x = \frac{\hbar}{i} \left(-\sin\varphi\, \frac{\partial}{\partial\vartheta} - \cos\varphi\, \cot\vartheta\, \frac{\partial}{\partial\varphi} \right), \tag{6.46a}$$

$$L_y = \frac{\hbar}{i} \left(\cos\varphi\, \frac{\partial}{\partial\vartheta} - \sin\varphi\, \cot\vartheta\, \frac{\partial}{\partial\varphi} \right), \tag{6.46b}$$

$$L_z = \frac{\hbar}{i} \frac{\partial}{\partial\varphi}. \tag{6.46c}$$

Alternatively, we could have obtained the same result by transforming the partial derivatives $\partial/\partial x^i$ and expressing $\vec{x}$ via r, ϑ, and φ, but we wanted to explicitly present the apparatus of spherical polar coordinates, especially the form of the nabla operator and how one can calculate it. Before we move on to the remaining angular-momentum operators, we can now supply a simple proof of Theorem 6.3.

Proof of Theorem 6.3   Consider the operator $\exp(-i\alpha L_z/\hbar)$ and its action on a wave function expressed in terms of spherical polar coordinates, i.e., $\psi(\vec{x}) = \psi(r, \vartheta, \varphi)$. Since L_z only acts non-trivially on the coordinate φ, we will suppress the other coordinates, writing

$$\exp(-i\alpha L_z/\hbar)\psi(\varphi) = \exp\!\left(-\alpha\frac{\partial}{\partial\varphi}\right)\psi(\varphi) = \sum_{n=0}^{\infty} \frac{(-\alpha)^n}{n!} \frac{\partial^n}{\partial\varphi^n} \psi(\varphi). \tag{6.47}$$

The right-hand side of eqn (6.47) can now be recognized as the Taylor expansion of the function ψ around the point $(\varphi - \alpha)$, i.e.,

$$\exp(-i\alpha L_z/\hbar)\psi(\varphi) = \sum_{n=0}^{\infty} \frac{\big((\varphi - \alpha) - \varphi\big)^n}{n!} \frac{\partial^n \psi(\varphi)}{\partial\varphi^n} = \psi(\varphi - \alpha) = \psi\big(R^{-1}(\vec{\alpha})\vec{x}\big), \tag{6.48}$$

where $\vec{\alpha} = (0, 0, \alpha)^{\mathrm{T}}$. The operator L_z hence generates rotations around the z axis. The reason that this can be easily seen is because the spherical coordinate system was chosen with the z axis as its zenith, see Fig. 6.3. Had we chosen a different axis as the zenith, say, the x axis or y axis, then we could as easily confirm that L_x and L_y generate rotations about these axes, respectively, which proves Theorem 6.3.   □

Next, we turn to the transformation of the ladder operators $L_\pm$ (see Definition 6.5) into spherical coordinates using eqn (6.46), which yields

$$L_\pm = \hbar\, e^{\pm i\varphi} \left(\pm\frac{\partial}{\partial\vartheta} + i\cot\vartheta\, \frac{\partial}{\partial\varphi} \right), \tag{6.49}$$

while a similar computation for the squared angular-momentum operator results in

$$\vec{L}^{\,2} = -\hbar^2 \left(\frac{1}{\sin\vartheta} \frac{\partial}{\partial\vartheta} \sin\vartheta\, \frac{\partial}{\partial\vartheta} + \frac{1}{\sin^2\vartheta} \frac{\partial^2}{\partial\varphi^2} \right). \tag{6.50}$$

6.4.3 The Spherical Harmonics

Denoting the wave functions for the coordinates ϑ and φ as $Y_{lm}(\vartheta, \varphi) = \langle \vartheta, \varphi | l, m \rangle$, we can rewrite the eigenvalue equations for the operators $\vec{L}^2$ and L_z from Theorem 6.5 using eqns (6.50) and (6.46c), and obtain

$$\left(\frac{1}{\sin \vartheta} \frac{\partial}{\partial \vartheta} \sin \vartheta \frac{\partial}{\partial \vartheta} + \frac{1}{\sin^2 \vartheta} \frac{\partial^2}{\partial \varphi^2} \right) Y_{lm} = -l(l+1) Y_{lm} \qquad (6.51a)$$

$$\frac{\partial}{\partial \varphi} Y_{lm} = i m Y_{lm} . \qquad (6.51b)$$

Although there may be some potential for confusion with the position operator for the y direction, we want to stick to the widely used convention of denoting the sought after solutions, which will turn out to be the *spherical harmonics*, by $Y_{lm}(\vartheta, \varphi)$, trusting that the double subscripts will alleviate the potential for confusion. In order to determine the solutions of the partial differential equations in (6.51), we use a *separation ansatz*, the same method we have used in eqn (4.2) in Section 4.1.1, that is, we try a solution of the form

$$Y_{lm}(\vartheta, \varphi) = P(\vartheta) \Phi(\varphi) . \qquad (6.52)$$

Inserting this ansatz into eqn (6.51b), we immediately find

$$\Phi(\varphi) = e^{i m \varphi} . \qquad (6.53)$$

Since the wave function must be continuous, we have to require that $\Phi(\varphi + 2\pi) = \Phi(\varphi)$. This is only satisfied if the magnetic quantum number m (and thus also l) is an integer (eliminating the possibility for half-integer numbers), as we have claimed in Theorem 6.5, i.e., $l = 0, 1, 2, \ldots$ and $m = -l, -l+1, \ldots, l-1, l$. The solutions that we are interested in are hence of the form $Y_{lm}(\vartheta, \varphi) = e^{i m \varphi} P(\vartheta)$, which we insert into the eigenvalue equation (6.51a) to get

$$\left(\frac{1}{\sin \vartheta} \frac{\partial}{\partial \vartheta} \sin \vartheta \frac{\partial}{\partial \vartheta} - \frac{m^2}{\sin^2 \vartheta} + l(l+1) \right) e^{i m \varphi} P(\vartheta) = 0 . \qquad (6.54)$$

We can now eliminate $e^{i m \varphi}$, since it is non-vanishing for all m and for all φ. With a simple change of variables, $\xi = \cos \vartheta$, where $\frac{\partial}{\partial \vartheta} = \frac{\partial \xi}{\partial \vartheta} \frac{\partial}{\partial \xi} = -\sin \vartheta \frac{\partial}{\partial \xi}$, eqn (6.54) becomes

$$\left((1 - \xi^2) \frac{d^2}{d\xi^2} - 2\xi \frac{d}{d\xi} + l(l+1) - \frac{m^2}{1 - \xi^2} \right) P_{lm}(\xi) = 0 , \qquad (6.55)$$

where we have added the subscripts to P_{lm} since the solutions still depend on l and m, and we have replaced the partial derivatives by total derivatives because only one variable remains. The differential equation (6.55) is the so-called *general Legendre equation*. It is solved by the *associated Legendre functions*, also called associated Legendre polynomials[2], for which we make the ansatz

---

[2]Technically speaking, they are only polynomials if m is even, but the term became accustomed and is also used for odd values of m.

$$P_{lm}(\xi) = \left(1 - \xi^2\right)^{\frac{m}{2}} \frac{d^m}{d\xi^m} P_l(\xi) \,. \tag{6.56}$$

Inserting eqn (6.56) into eqn (6.55), we can straightforwardly differentiate and use the simple rule

$$\xi \frac{d^k}{d\xi^k} = \frac{d^k}{d\xi^k} \xi - k \frac{d^{k-1}}{d\xi^{k-1}} \tag{6.57}$$

to arrive at the differential equation

$$\left(\left(1 - \xi^2\right) \frac{d^2}{d\xi^2} - 2\xi \frac{d}{d\xi} + l\left(l + 1\right) \right) P_l(\xi) = 0 \,. \tag{6.58}$$

This differential equation, called Legendre equation, is solved by the (ordinary) Legendre polynomials P_l, which, in turn can be obtained via the corresponding *formula of Rodrigues*, i.e.,

$$P_l(\xi) = \frac{1}{2^l \, l!} \frac{d^l}{d\xi^l} \left(\xi^2 - 1\right)^l \,. \tag{6.59}$$

We will leave it as an exercise to verify that the Legendre polynomials P_l indeed satisfy eqn (6.58). Note that for $m = 0$ the associated Legendre polynomials just reduce to the regular Legendre polynomials, which are polynomials of the order l in $\xi = \cos\vartheta$, and the first few are

$$P_0 = 1, \quad P_1 = \cos\vartheta, \quad P_2 = \tfrac{1}{2}\left(3\cos^2\vartheta - 1\right). \tag{6.60}$$

The associated functions P_{lm} on the other hand, are polynomials of order $(l - m)$, multiplied by a factor $\sqrt{(1 - \xi^2)^m} = \sin^m\theta$, and have $(l - m)$ roots in the interval $[-1, +1]$. To give an impression of their form, let us also list the first few of these functions, i.e.,

$$P_{0,0} = 1, \quad P_{1,0} = \cos\vartheta, \quad P_{1,1} = \sin\vartheta \,. \tag{6.61}$$

Both the regular and the associated Legendre polynomials form orthogonal systems on the interval $[-1, +1]$, that is, we may write

$$\int_{-1}^{+1} d\xi \, P_l(\xi) \, P_{l'}(\xi) = \frac{2}{2l + 1} \delta_{ll'} \,, \tag{6.62}$$

and the similar orthogonality relation for the associated functions is

$$\int_{-1}^{+1} d\xi \, P_{lm}(\xi) \, P_{l'm}(\xi) = \frac{2}{2l + 1} \frac{(l + m)!}{(l - m)!} \delta_{ll'} \,. \tag{6.63}$$

Finally, we can combine the solutions $\Phi(\varphi) = e^{im\varphi}$ and $P_{lm}(\vartheta)$ into the *spherical harmonics* $Y_{lm}(\vartheta, \varphi)$ given by

$$Y_{lm}(\vartheta, \varphi) = (-1)^{\frac{(m+|m|)}{2}} \left[\frac{2l+1}{4\pi} \frac{(l-|m|)!}{(l+|m|)!} \right]^{\frac{1}{2}} e^{im\varphi} P_{l|m|}(\cos\vartheta) , \qquad (6.64)$$

where we have included an appropriate normalization factor. The spherical harmonics Y_{lm} are the desired eigenfunctions of $\vec{L}^2$ and L_z, satisfying the equations

$$\vec{L}^2 Y_{lm} = \hbar^2 l(l+1) Y_{lm} , \qquad (6.65a)$$

$$L_z Y_{lm} = \hbar m Y_{lm} , \qquad (6.65b)$$

$$L_{\pm} Y_{lm} = \hbar \sqrt{l(l+1) - m(m\pm 1))} Y_{lm\pm 1} , \qquad (6.65c)$$

as stated in Theorem 6.5 and eqn (6.33), respectively. With the normalization constant chosen in eqn (6.64), the spherical harmonics form a complete orthonormal system on the unit sphere S_2,

$$\int_{S_2} d\Omega\, Y_{lm}^*(\vartheta,\varphi)\, Y_{l'm'}(\vartheta,\varphi) = \int_0^{2\pi} d\varphi \int_0^{\pi} d\vartheta\, \sin\vartheta\, Y_{lm}^*(\vartheta,\varphi)\, Y_{l'm'}(\vartheta,\varphi) = \delta_{ll'}\,\delta_{mm'} .$$

$$(6.66)$$

The integration is carried out over the unit sphere, that is, where the integral over the full solid angle is $\int d\Omega = 4\pi$. Practically speaking, eqn (6.66) means that all functions that are square integrable on the surface of a sphere of finite, non-zero radius can be expanded in terms of these special functions on said surface. The closure relations for the spherical harmonics that are relevant for us here read

$$\sum_{l=0}^{\infty} \sum_{m=-l}^{+l} Y_{lm}^*(\vartheta,\varphi)\, Y_{lm}(\vartheta',\varphi') = \frac{1}{\sin\vartheta}\, \delta(\vartheta - \vartheta')\,\delta(\varphi - \varphi') , \qquad (6.67)$$

$$\sum_{m=-l}^{+l} Y_{lm}^*(\vartheta,\varphi)\, Y_{lm}(\vartheta,\varphi) = \frac{2l+1}{4\pi} . \qquad (6.68)$$

Alternatively, since we know that $Y_{lm}(\vartheta,\varphi) = \langle \vartheta,\varphi | l,m \rangle$ are the wave functions of the states $| l,m \rangle$ with respect to the coordinates ϑ and φ, we can write

$$\int_{S_2} d\Omega\, | \vartheta,\varphi \rangle \langle \vartheta,\varphi | = \sum_{m=-l}^{+l} \sum_{l=0}^{\infty} | l,m \rangle \langle l,m | = \mathbb{1} , \qquad (6.69)$$

where $\langle \vartheta,\varphi | \vartheta',\varphi' \rangle = \frac{1}{\sin\vartheta}\delta(\vartheta - \vartheta')\delta(\varphi - \varphi')$ and $\langle l,m | l',m' \rangle = \delta_{ll'}\delta_{mm'}$. To illustrate the spherical harmonics, let us explicitly write down the first few of these functions,

$$Y_{0,0} = \sqrt{\frac{1}{4\pi}} , \quad Y_{1,0} = \sqrt{\frac{3}{4\pi}} \cos\vartheta , \quad Y_{1,\pm 1} = \mp\sqrt{\frac{3}{8\pi}} \sin\vartheta\, e^{\pm i\varphi} , \qquad (6.70)$$

which are also depicted in Fig. 6.4.

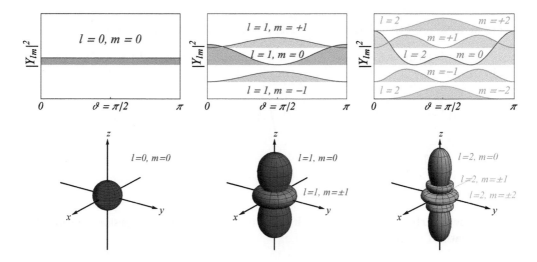

Fig. 6.4 Spherical harmonics. The illustration shows $|Y_{lm}|^2$, the modulus squared of the spherical harmonics, the common eigenfunctions of $\vec{L}^2$ and L_z, for $l = 0, 1, 2$ and $m = -l, \ldots, +l$, as functions of the polar angle ϑ on top, and $\frac{\vec{r}}{|\vec{r}|}|Y_{lm}|^2$ as functions of ϑ and φ below. Note that the amplitudes $Y_{lm}(\vartheta, \varphi)$ also depend on φ, but the probabilities are independent of φ, and hence invariant under rotations around the z axis.

6.4.4 Uncertainty of Angular-Momentum Operators

Finally, let us put what we have learned to practical use, and compute the uncertainty of L_x for a simultaneous eigenstate of $\vec{L}^2$ and L_z. First, we calculate the expectation value $\langle L_x \rangle_{l,m}$ using the operator L_x in spherical polar coordinates from eqn (6.46a), i.e.,

$$\langle L_x \rangle_{l,m} = \int\limits_{S_2} d\Omega \, Y_{lm}^*(\vartheta, \varphi) \frac{\hbar}{i} \left(-\sin\varphi \frac{\partial}{\partial\vartheta} - \cos\varphi \cot\vartheta \frac{\partial}{\partial\varphi} \right) Y_{lm}(\vartheta, \varphi) \qquad (6.71)$$

$$= -\frac{\hbar}{i} \frac{2l+1}{4\pi} \frac{(l-|m|)!}{(l+|m|)!} \int\limits_{S_2} d\Omega \, P_{l|m|}(\cos\vartheta) \left(\sin\varphi \frac{\partial}{\partial\vartheta} + im\cos\varphi \cot\vartheta \right) P_{l|m|}(\cos\vartheta).$$

Since the integrals of $\sin\varphi$ and $\cos\varphi$ over the full period from 0 to 2π vanish, it is easy to see that $\langle L_x \rangle_{l,m} = 0$. We must hence have $(\Delta L_x)^2 = \langle L_x^2 \rangle_{l,m}$, and, similar to eqn (6.71) we can write

$$\langle L_x^2 \rangle_{l,m} = \hbar^2 \frac{2l+1}{4\pi} \frac{(l-|m|)!}{(l+|m|)!} \int\limits_{S_2} d\Omega \left| \left(\sin\varphi \frac{\partial}{\partial\vartheta} + im\cos\varphi \cot\vartheta \right) P_{l|m|}(\cos\vartheta) \right|^2 \qquad (6.72)$$

$$= \hbar^2 \frac{2l+1}{4} \frac{(l-|m|)!}{(l+|m|)!} \int\limits_0^\pi d\vartheta \, \sin\vartheta \left[\left(\frac{\partial}{\partial\vartheta} P_{l|m|}(\cos\vartheta) \right)^2 + m^2 \cot^2\vartheta P_{l|m|}^2(\cos\vartheta) \right],$$

where we have used that $\int\limits_{0}^{2\pi} d\varphi \sin^2\varphi = \int\limits_{0}^{2\pi} d\varphi \cos^2\varphi = \pi$. To further evaluate the integral, we use the substitution $\xi = \cos\vartheta$, where $d\xi = -\sin\vartheta d\vartheta$ and $\partial/\partial\vartheta = -\sqrt{1-\xi^2}\partial/\partial\xi$, to obtain

$$\int\limits_{0}^{\pi} d\vartheta \, \sin\vartheta \left(\frac{\partial}{\partial\vartheta} P_{l|m|}(\cos\vartheta)\right)^2 = \int\limits_{-1}^{+1} d\xi \, (1-\xi^2)\left(\frac{\partial}{\partial\xi} P_{l|m|}(\xi)\right)^2 , \tag{6.73a}$$

$$\int\limits_{0}^{\pi} d\vartheta \, \sin\vartheta \, m^2 \cot^2\vartheta P_{l|m|}^2(\cos\vartheta) = \int\limits_{-1}^{+1} d\xi \, \frac{m^2 \, \xi^2}{1-\xi^2} \, P_{l|m|}^2(\xi) . \tag{6.73b}$$

Having now nearly arrived at the result, we employ some useful recursion relations for the associated Legendre polynomials, i.e.,

$$\frac{1}{2}\left[P_{lm+1}(\xi) - (l+m)(l-m+1) P_{lm-1}(\xi)\right] = \sqrt{1-\xi^2}\frac{d}{d\xi} P_{lm}(\xi) , \tag{6.74a}$$

$$\frac{1}{2}\left[P_{lm+1}(\xi) + (l+m)(l-m+1) P_{lm-1}(\xi)\right] = \frac{m\xi P_{lm}(\xi)}{\sqrt{1-\xi^2}} , \tag{6.74b}$$

which can both be proven using the Legendre equation (6.58), along with eqns (6.59), (6.56), and (6.57). With this, we can easily combine the integrals of eqn (6.73) and evaluate them using the orthogonality relation (6.63) of the associated Legendre polynomials. We then arrive at the result

$$(\Delta L_x)^2_{l,m} = \langle L_x^2 \rangle_{l,m} = \frac{\hbar^2}{2}\left(l(l+1) - m^2\right) . \tag{6.75}$$

Since a similar calculations gives the same result for $(\Delta L_y)^2_{l,m}$, we recover the uncertainty relation of eqn (6.35).

After this extensive look at orbital angular momentum, its eigenvalues and eigenfunctions $Y_{lm}(\vartheta, \varphi)$, we now want to include the coordinate r into our considerations and consider the three-dimensional Schrödinger equation in Chapter 7.

7

The Three-Dimensional Schrödinger Equation

In this chapter, we want to discuss the time-independent Schrödinger equation in three dimensions. In many respects, the task ahead of us is similar to that in Chapter 4 in the sense that it will be our goal to determine the bound states for some potentials of interest. In particular, we will consider the Coulomb potential, which describes the interaction between electrically charged particles, like electrons and protons. We shall hence demonstrate how quantum mechanics provides a model for the formation of atoms and thus the basis for chemistry. Besides this essential task, we will discuss general techniques to determine whether bound states exist and how one may obtain bounds on the ground-state energy that are applicable also to other three-dimensional problems.

7.1 The Radial Schrödinger Equation

We now consider the time-independent Schrödinger equation in three dimensions, that is,

$$H\,\psi(\vec{x}) = \left(-\frac{\hbar^2}{2m}\,\Delta\,+\,V(\vec{x})\right)\psi(\vec{x})\,=\,E\,\psi(\vec{x})\,, \qquad (7.1)$$

where the operator Δ is the Laplacian (or Laplace operator)

$$\Delta = \vec{\nabla}^2 \,=\, \frac{\partial^2}{\partial x^2}\,+\,\frac{\partial^2}{\partial y^2}\,+\,\frac{\partial^2}{\partial z^2}\,. \qquad (7.2)$$

7.1.1 Angular Momentum in the Schrödinger Equation

In general, the Hamiltonian may depend on the coordinates in a complicated fashion via the potential $V(\vec{x})$. However, for many physically relevant scenarios, it is reasonable to expect a spherically symmetric situation in which the potential only depends on r, i.e., $V = V(r)$. We now restrict our discussion to such scenarios, where, as it will turn out, the knowledge that we have obtained in Chapter 6 about the (orbital) angular-momentum operator $\vec{L}$ will be very useful. To see this, we first transform the Laplacian of eqn (7.2) to spherical polar coordinates by inserting eqns (6.39a)–(6.39c) before differentiating according to the Leibnitz rule. This leads us to

$$\Delta = \frac{1}{r^2}\frac{\partial}{\partial r}\left(r^2\frac{\partial}{\partial r}\right) + \frac{1}{r^2\sin\vartheta}\frac{\partial}{\partial\vartheta}\left(\sin\vartheta\frac{\partial}{\partial\vartheta}\right) + \frac{1}{r^2\sin^2\vartheta}\frac{\partial^2}{\partial\varphi^2}\,. \tag{7.3}$$

Comparing eqn (7.3) to the squared angular-momentum operator $\vec{L}^{\,2}$ in spherical polar coordinates, see eqn (6.50), we recognize that we can rewrite the kinetic part of the Hamiltonian of eqn (7.1) as

$$-\frac{\hbar^2}{2m}\Delta = -\frac{\hbar^2}{2m}\frac{1}{r^2}\frac{\partial}{\partial r}\left(r^2\frac{\partial}{\partial r}\right) + \frac{\vec{L}^{\,2}}{2mr^2}\,. \tag{7.4}$$

The Schrödinger equation is then of the form

$$\left(-\frac{\hbar^2}{2m}\frac{1}{r^2}\frac{\partial}{\partial r}\left(r^2\frac{\partial}{\partial r}\right) + \frac{\vec{L}^{\,2}}{2mr^2} + V(r)\right)\psi(\vec{x}) = E\,\psi(\vec{x})\,, \tag{7.5}$$

and we can employ an already familiar separation ansatz. That is, we assume that the wave function can be written as the product of a function that only depends on the radial coordinate r, the so-called *radial wave function $R(r)$*, and a function containing all the angle-dependencies of the wave function. Since we already know the spherical harmonics to be the eigenfunctions of the squared angular-momentum operator, see eqn (6.65a), we can identify the latter function with $Y_{lm}(\vartheta,\varphi)$, such that

$$\psi(r,\vartheta,\varphi) = R(r)\,Y_{lm}(\vartheta,\varphi)\,. \tag{7.6}$$

The Schrödinger equation is thus reduced to a differential equation for just the radial component $R(r)$, i.e.,

$$\left(-\frac{\hbar^2}{2m}\frac{1}{r^2}\frac{\partial}{\partial r}\left(r^2\frac{\partial}{\partial r}\right) + \frac{\hbar^2\,l\,(l+1)}{2mr^2} + V(r)\right)R_l(r) = E\,R_l(r)\,, \tag{7.7}$$

where we have included a subscript on the radial wave function $R_l(r)$, since the solution depends on the azimuthal quantum number l.

7.1.2   Reduced Wave Function and Effective Potential

To simplify the differential equation (7.7), we follow three simple steps. The first step is to rewrite the differential operator on the left-hand side according to

$$\frac{1}{r^2}\frac{\partial}{\partial r}\left(r^2\frac{\partial}{\partial r}\right) = \frac{\partial^2}{\partial r^2} + \frac{2}{r}\frac{\partial}{\partial r}\,. \tag{7.8}$$

Next we replace the function $R_l(r)$ by the so-called *reduced wave function $u_l(r)$*, which is defined in the following way.

Definition 7.1   *The **reduced wave function** $u_l(r)$ is defined as*

$$u_l(r) := r\,R_l(r)\,.$$

Differentiating the reduced wave function twice with respect to r we get

$$\frac{\partial^2}{\partial r^2} u_l = \frac{\partial^2}{\partial r^2}(r\,R_l) = \frac{\partial}{\partial r}\left(r\,\frac{\partial R_l}{\partial r} + R_l\right) = \left(r\,\frac{\partial^2}{\partial r^2} + 2\,\frac{\partial}{\partial r}\right)R_l\,. \tag{7.9}$$

Comparing this result to that of eqn (7.8), we immediately see that the radial Schrödinger equation of (7.7), when multiplied by r, can be reformulated for the reduced wave function as

$$\left(-\frac{\hbar^2}{2m}\frac{\partial^2}{\partial r^2} + \frac{\hbar^2\,l\,(l+1)}{2mr^2} + V(r)\right)u_l(r) = E\,u_l(r)\,. \tag{7.10}$$

The second-order differential equation that we have obtained for u_l looks remarkably like the one-dimensional Schrödinger equation, if it were not for the term containing the azimuthal quantum number. In the third step we therefore define an effective potential $V_{\text{eff.}}(r)$ to bring the equation to a more appealing form.

Definition 7.2   *The **effective potential** $V_{\text{eff.}}(r)$ is defined as*

$$V_{\text{eff.}}(r) := \frac{\hbar^2\,l\,(l+1)}{2mr^2} + V(r)\,.$$

For spherically symmetric problems, the Schrödinger equation in three dimensions, a second-order partial differential equation, has thus been turned into a one-dimensional ordinary differential equation for the reduced wave function, i.e.,

$$\frac{d^2}{dr^2}u_l(r) + \frac{2m}{\hbar^2}\left(E_l - V_{\text{eff.}}(r)\right)u_l(r) = 0\,. \tag{7.11}$$

This differential equation takes the form of the ordinary one-dimensional Schrödinger equation, except for the term $\hbar^2\,l\,(l+1)/2mr^2$ in the effective potential. This additional term represents a repulsive potential, called the *centrifugal term*, in analogy to the classical (fictitious) force experienced in a rotating reference frame. We indicate this dependence on the angular momentum by adding a subscript l to the energy E_l.

7.2   Bound States in Three Dimensions

7.2.1   Normalization of the Radial Wave Function

Having formulated the Schrödinger equation in three dimensions, we can now in principle proceed as in Chapter 4 and determine the bound states for some given potential $V(r)$. Since such bound states $\psi(r,\vartheta,\varphi)$ must be normalized, they must satisfy

$$\int_{S_2} d\Omega \int_0^\infty dr\, r^2\, |\psi(r,\vartheta,\varphi)|^2 = 1\,. \tag{7.12}$$

Inserting the wave function from eqn (7.6) and using the normalization of the spherical harmonics, see eqn (6.66), we can easily transform the normalization condition to

$$\int\limits_0^\infty dr\, r^2\, |R_l(r)|^2 = 1\, , \tag{7.13}$$

which translates to the reduced wave function as

$$\int\limits_0^\infty dr\, |u_l(r)|^2 = 1\, . \tag{7.14}$$

The radial wave function $R(r) = \frac{u_l(r)}{r}$ is only normalizable if it is bounded at the origin. Consequently, the reduced wave function is required to vanish at the origin

$$u_l(r = 0) = 0\, . \tag{7.15}$$

This constraint imposes the character of an odd function on the reduced wave function. That is, if negative values were allowed for r, the function $u_l(r)$ would be reminiscent of one of the (odd) excited bound states of a one-dimensional potential. However, as we have seen in Chapter 4, not every one-dimensional potential features an excited state, and we can take this as an indication that not every three-dimensional potential allows for a ground state. Interestingly, this is a feature that appears only once three spatial dimensions are reached, which we will phrase in a theorem following Yang and de Llano (1989), and Chadan *et al.* (2003):

Theorem 7.1 (Existence of bound states)

For an attractive potential $V(\vec{x})$ in n spatial dimensions satisfying $\int_{\mathbb{R}^n} d^n x\, V(\vec{x}) < 0$, the Hamiltonian $H = -\frac{\hbar^2}{2m}\Delta + V(\vec{x})$ is guaranteed to admit a bound state with energy $E_{\mathrm{g}} < 0$ only if $n \leq 2$.

7.2.2 Rayleigh–Ritz Variational Principle

To prove Theorem 7.1, we will employ what is called the *Rayleigh–Ritz variational principle*, which is a simple method to establish an upper bound on the ground-state energy of any Hamiltonian. It is based on the following straightforward lemma.

Lemma 7.1 (Upper bound on ground-state energy)

For arbitrary (not necessarily normalized) states ψ in the domain of a Hamiltonian H, the ground-state energy E_{g} of H satisfies

$$E_{\mathrm{g}} \leq \frac{\langle\, \psi\, |\, H\, |\, \psi\, \rangle}{\langle\, \psi|\, \psi\, \rangle}.$$

Proof of Lemma 7.1 To see that Lemma 7.1 indeed holds true, consider $\{|\,n\,\rangle\}_n$ to be the basis of eigenstates of H, such that

$$H\,|\,n\,\rangle = E_n\,|\,n\,\rangle\,, \tag{7.16}$$

where $E_g = E_1 \leq E_2, E_3, \ldots$ is the ground-state energy. When we then compute the expectation value of H for some arbitrary state $|\,\psi\,\rangle$, which we will assume need not be normalized, we can insert the completeness relation for the basis $\{|\,n\,\rangle\}_n$ to find

$$\langle\,\psi\,|\,H\,|\,\psi\,\rangle = \sum_n \langle\,\psi\,|\,H\,|\,n\,\rangle\langle\,n\,|\,\psi\,\rangle = \sum_n E_n|\langle\,n\,|\,\psi\,\rangle|^2 \leq E_g \sum_n |\langle\,n\,|\,\psi\,\rangle|^2 = E_g\,\langle\,\psi\,|\,\psi\,\rangle\,. \tag{7.17}$$

With this, Lemma 7.1 is proven, and any state ψ will give an upper bound on the ground-state energy. However, to obtain a useful bound, we want to do better. $\quad\square$

Proof of Theorem 7.1 Suppose ψ_ϵ is a family of normalizable test wave functions that are parameterized by ϵ, then

$$E(\epsilon) = \frac{\langle\,\psi_\epsilon\,|\,H\,|\,\psi_\epsilon\,\rangle}{\langle\,\psi_\epsilon\,|\,\psi_\epsilon\,\rangle} \geq E_g \tag{7.18}$$

for all values of ϵ. Using standard variational methods one can then determine

$$\min_\epsilon E(\epsilon) \geq E_g \tag{7.19}$$

to arrive at a better upper bound on the ground-state energy. In Section 7.4.2 we will use this method to bound the ground-state energy of the hydrogen atom.

For now, let us apply it for the more general task of proving Theorem 7.1, as laid out by Yang and de Llano (1989). Consider the family of states given by $\psi_\epsilon(r) = \exp(-(r+1)^\epsilon)$, for $r^2 = \sum_{i=1,\ldots,n} x_i^2$. For $\epsilon > 0$ these states are normalizable, since

$$\langle\,\psi_\epsilon\,|\,\psi_\epsilon\,\rangle = \int_{\mathbb{R}^n} d^n\!x\,|\psi_\epsilon(r)|^2 \propto \int_0^\infty dr\,r^{n-1}e^{-2(r+1)^\epsilon} \leq \int_0^\infty dr\,(r+1)^{n-1}e^{-2(r+1)^\epsilon}$$

$$= \frac{1}{\epsilon}\int_1^\infty dh\,h^{(n-\epsilon)/\epsilon}e^{-2h} < \infty\,, \tag{7.20}$$

where we have used the substitution $h = (r+1)^\epsilon$. We then consider the expectation value of the kinetic-energy operator $\vec{P}^2/(2m) = -\frac{\hbar^2}{2m}\Delta$, where we can use that $\vec{P}$ is Hermitian and

$$\vec{P}\psi_\epsilon(r) = i\hbar\vec{\nabla}\psi_\epsilon(r) = -i\hbar\,\epsilon\,(r+1)^{\epsilon-1}e^{-(r+1)^\epsilon}\frac{\vec{x}}{r} \tag{7.21}$$

to calculate

$$\langle\, \psi_\epsilon \,|\, \vec{P}^{\,2} \,|\, \psi_\epsilon \,\rangle = \int_{\mathbb{R}^n} d^n \vec{x} \,|\vec{P}\psi_\epsilon(r)|^2 \propto \epsilon^2 \int\limits_0^\infty dr\, r^{n-1}\,(r+1)^{2\epsilon-2}\,e^{-2(r+1)^\epsilon}$$

$$\leq \epsilon^2 \int\limits_0^\infty dr\,(r+1)^{2\epsilon+n-3}\,e^{-2(r+1)^\epsilon} = \epsilon \int\limits_1^\infty dh\, h^{1+(n-2)/\epsilon}\,e^{-2h}. \qquad (7.22)$$

The kinetic energy is always positive, and as long as $n \leq 2$, the integral in eqn (7.22) approaches the value 0 from above in the limit $\epsilon \to 0^+$. When n is larger or equal than 2, however, the expression diverges for vanishing ϵ. Meanwhile, in the limit of $\epsilon \to 0$, the expectation value of the potential energy $V(\vec{x})$ evaluates to

$$\lim_{\epsilon \to 0}\langle\, \psi_\epsilon \,|\, V(\vec{x}) \,|\, \psi_\epsilon \,\rangle = \lim_{\epsilon \to 0}\int_{\mathbb{R}^n} d^n \vec{x}\, V(\vec{x})\,|\psi_\epsilon(r)|^2 = e^{-2}\int_{\mathbb{R}^n} d^n \vec{x}\, V(\vec{x}) < 0\,, \qquad (7.23)$$

where we have assumed that the spatial integral of the potential is negative, as we have required in Theorem 7.1. We see that, as ϵ approaches zero from above, the potential energy approaches a negative constant, while the kinetic energy becomes smaller and smaller, such that, eventually, $\langle\, \psi_\epsilon \,|\, H \,|\, \psi_\epsilon \,\rangle$ becomes negative for some finite $\epsilon > 0$. By Lemma 7.1 it is then clear that the ground-state energy is bounded from above by that negative value. In other words, a ground state with energy $E_{\mathrm{g}} < 0$ certainly exists for $n \leq 2$, which proves Theorem 7.1.   □

The proof relies on the spatial dimension being smaller than or equal to two, but in principle, this does not exclude the possibility that also all attractive potentials in three dimensions have a ground state. To show that this is not the case, we shall study a particular example in the next section.

7.3   The Spherical Potential Well

In this section we will take a brief look at the three-dimensional spherical potential well illustrated in Fig. 7.1, which is a generalization of the finite potential well in Section 4.2.1. That is, the potential that we consider has a finite negative value $-V_0 < 0$ within some sphere of radius r_o, and vanishes outside, such that

$$V(r) = \begin{cases} -V_0 & r < r_o & \text{region I} \\ 0 & r \geq r_o & \text{region II} \end{cases}. \qquad (7.24)$$

7.3.1   General Solutions for the Spherical Potential Well

We then proceed as we have done already a few times throughout Chapter 4, and determine the solutions of the time-independent Schrödinger equation in the different regions, where we require continuity of the wave function and its derivative at region boundaries to obtain the allowed bound-state energy eigenvalues E, where $-V_0 \leq E < 0$. In region I, the radial Schrödinger equation from eqn (7.10) becomes

$$\left(\frac{\partial^2}{\partial r^2} - \frac{l\,(l+1)}{r^2} + \kappa^2 \right) u_l(r) = 0\,, \qquad (7.25)$$

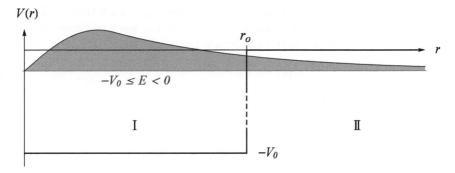

Fig. 7.1 Spherical potential well. The spherical potential well is the generalization of the finite potential well to three dimensions. The potential is attractive for $r < r_o$ (region I), and vanishes for $r \geq r_o$ (region II). Whether a ground-state wave function exists as shown, depends on the depth and width of the potential well.

where $\kappa = \sqrt{2m(V_0 - |E|)}/\hbar$. We then make the ansatz $u_l(r) = \sqrt{r}\,J(r)$ and differentiate, i.e.,

$$\frac{\partial}{\partial r}u_l(r) = \tfrac{1}{2}r^{-1/2}J(r) + r^{1/2}\frac{\partial}{\partial r}J(r)\,, \tag{7.26a}$$

$$\frac{\partial^2}{\partial r^2}u_l(r) = -\tfrac{1}{4}r^{-3/4}J(r) + r^{-1/2}\frac{\partial}{\partial r}J(r) + r^{1/2}\frac{\partial^2}{\partial r^2}J(r)\,. \tag{7.26b}$$

With the simple substitution $\tilde{r} = \kappa r$ we then find that the functions J must satisfy

$$\left(\tilde{r}^2\frac{\partial^2}{\partial \tilde{r}^2} + \tilde{r}\frac{\partial}{\partial \tilde{r}} + (\tilde{r}^2 - \alpha^2)\right)J(\tilde{r}) = 0\,, \tag{7.27}$$

with $\alpha = (2l+1)/2$. This is known as the *Bessel differential equation* and the linearly independent solutions $J_\alpha(x)$ and $Y_\alpha(x)$ for $\alpha \in \mathbb{C}$ and $x \in \mathbb{R}$ are called the *Bessel functions of the first and second kind*, respectively, given by

$$J_\alpha(x) = \sum_{m=0}^{\infty}\frac{(-1)^m}{m!\,\Gamma(m+\alpha+1)}\left(\frac{x}{2}\right)^{2m+\alpha}\,, \tag{7.28a}$$

$$Y_\alpha(x) = \frac{J_\alpha(x)\,\cos(\pi\alpha) - J_{-\alpha}(x)}{\sin(\pi\alpha)}\,. \tag{7.28b}$$

The function Γ in the denominator of eqn (7.28a) is the *gamma function*

$$\Gamma(t) = \int_0^{\infty}dx\,x^{t-1}\,e^{-x}\,, \tag{7.29}$$

which generalizes the notion of the factorial function to complex arguments (except non-positive integers), i.e., for positive integers n, one has $\Gamma(n) = (n-1)!$. However, the

Bessel functions of the second kind are divergent at the origin, such that $\sqrt{r}\, Y_\alpha(\kappa r) \neq 0$ as $r \to 0$. According to our constraint from eqn (7.15) we can hence discard the functions Y_α, and the general solution in region I is of the form

$$u_l^{\mathrm{I}}(r) = A\sqrt{r}\, J_{l+1/2}(\kappa r), \tag{7.30}$$

where A is a normalization constant.

In region II, the radial Schrödinger equation from (7.25) is modified only by replacing κ^2 with $-q^2 = -2m|E|/\hbar^2$. It is therefore easy to see that the solutions in region II can be obtained by replacing the argument κr of the Bessel functions by iqr. The resulting Bessel functions $J_\alpha(ix)$ and $Y_\alpha(ix)$ with purely imaginary arguments are related to the *modified Bessel functions* of the first and second kind, $I_\alpha(x)$ and $K_\alpha(x)$, via $J_\alpha(ix) = i^\alpha I_\alpha(x)$ and $Y_\alpha(ix) = i e^{i\pi\alpha/2} I_\alpha(x) - \frac{2}{\pi} e^{-i\pi\alpha/2} K_\alpha(x)$. In turn, using eqn (7.28) the modified Bessel functions can then be written as

$$I_\alpha(x) := i^{-\alpha} J_\alpha(ix) = \sum_{m=0}^{\infty} \frac{1}{m!\,\Gamma(m+\alpha+1)} \left(\frac{x}{2}\right)^{2m+\alpha}, \tag{7.31a}$$

$$K_\alpha(x) = \frac{\pi}{2} \frac{I_{-\alpha}(x) - I_\alpha(x)}{\sin(\pi\alpha)}. \tag{7.31b}$$

Unfortunately, neither the functions $\sqrt{r}\, I_\alpha(qr)$ nor $\sqrt{r}\, K_\alpha(qr)$ are normalizable in region II. It is therefore more convenient to write the general solution in region II in terms of the linear combinations $H_\alpha^{(1)} = J_\alpha + iY_\alpha$ and $H_\alpha^{(2)} = J_\alpha - iY_\alpha$, known as the *Hankel functions* of the first and second kind. Out of these two functions, $H_\alpha^{(1)}$ is normalizable in region II, while $H_\alpha^{(2)}$ is not. We hence find

$$u_l^{\mathrm{II}}(r) = B\sqrt{r}\, H_{l+1/2}^{(1)}(qr). \tag{7.32}$$

To obtain the bound states of the spherical potential well, one then has to match the solutions u_l^{I} and u_l^{II} and their derivatives at $r = r_o$.

7.3.2 The Ground State of the Spherical Potential Well

Here we will only be concerned with a more basic question, that is, what is the condition for at least one bound state, the ground state, to exist at all? For this task, we only have to consider the states with $l = 0$. If there is no bound state with azimuthal quantum number $l = 0$, then the potential does not admit a ground state for any other value of l either. Concentrating on $l = 0$, we can see that the radial Schrödinger equations in regions I and II is solved by

$$u_{l=0}(r) = \begin{cases} C\left(e^{i\kappa r} - e^{-i\kappa r}\right) & r < r_o \quad \text{region I} \\ D\,e^{-qr} & r \geq r_o \quad \text{region II} \end{cases}, \tag{7.33}$$

where we have already ensured that $u_{l=0}(r=0) = 0$. The continuity of the reduced wave functions and its derivative then lead us to the constraint

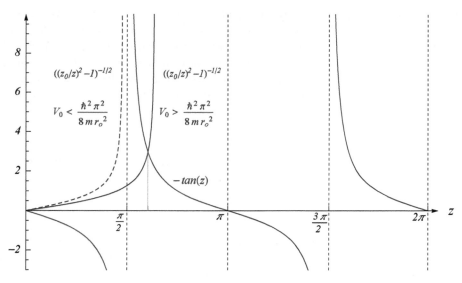

Fig. 7.2 Transcendental equation for the spherical potential well. The spherical potential well is an example of a three-dimensional potential that only admits bound states under certain conditions. Only when the well has a minimum depth $V_0 > \hbar^2\pi^2/(8mr_o^2)$ (relative to its width r_o), is there at least one bound state.

$$\frac{\kappa}{q} = -\tan(\kappa r_o), \tag{7.34}$$

which we can rewrite as the transcendental equation

$$\left(\frac{z_o^2}{z^2} - 1\right)^{-1/2} = -\tan(z) \tag{7.35}$$

for $z = \kappa r_o \geq 0$ and $z_o = \sqrt{2mV_0 r_o^2}/\hbar \geq 0$. As can be seen in Fig. 7.2, the function on the left-hand side of eqn (7.35) has an asymptote at $z = z_o$ and is non-negative, whereas the right-hand side is negative between $z = 0$ and the asymptote $z = \frac{\pi}{2}$. For the functions to have at least one intersection, we must hence have $z_o > \frac{\pi}{2}$. In other words, the spherical potential well only admits a ground state when the potential has a minimum strength, i.e., when

$$V_0 > \frac{\hbar^2\pi^2}{8mr_o^2}. \tag{7.36}$$

7.4 The Coulomb Potential and the Stability of Matter

7.4.1 The Coulomb Potential

For the remainder of this chapter, we will focus on a potential of particular interest, the *Coulomb potential* which describes the electrostatic interaction of charged particles. Specifically, we want to determine the energy levels of the simplest atomic system, the *hydrogen atom*, consisting only of a proton of mass $m_{\mathrm{p}} \approx 938.27$ MeV$/c^2$ and (elementary) charge $q_{\mathrm{e}} \approx 1.602 \times 10^{-19}$ C and an electron of mass $m_{\mathrm{e}} \approx 0.5$ MeV$/c^2$ and charge $-q_{\mathrm{e}}$. Since $m_{\mathrm{e}} \ll m_{\mathrm{p}}$ we can (for now) think of the electron as being subjected to a potential $V(r)$ caused by the proton at rest at $r = 0$. In Gaussian cgs (centimetre-gram-second) units, which are most natural for describing electric and magnetic fields on the same footing (see, e.g., Ohanian, 1988), the Coulomb potential is given by

$$V(r) = -\frac{q_{\mathrm{e}}^2}{r} . \tag{7.37}$$

If one wishes to be more precise, as we shall be later on, one can introduce a reduced mass for the joint atomic system. In this chapter, we will further disregard the spin of the involved particles, which requires additional attention and will be discussed in Chapter 8.

It is further customary (and instructive) to introduce constants that are typical characteristics of such a system to further simplify and analyse the problem at hand. The first of which will be the *fine-structure constant* α, given by

$$\alpha = \frac{q_{\mathrm{e}}^2}{\hbar c} \approx \frac{1}{137} . \tag{7.38}$$

The fine-structure constant is a fundamental constant, introduced by *Arnold Sommerfeld* (1940), whose numerical value does not depend on the choice of units, and which therefore can be displayed in numerous ways, e.g., in units where $\hbar = c = 1$ one has $\alpha = q_{\mathrm{e}}^2$ and thus the Coulomb potential is simply $V(r) = -\alpha/r$. This is also the reason for its many different physical interpretations, the most common of which is as the coupling constant of the electromagnetic interaction, i.e., the strength at which photons couple to the electric charge. Another interpretation, relevant for our calculations here, is that the fine-structure constant can be seen as the ratio of the typical electron velocity in a bound state of the hydrogen atom to the speed of light. Taking this view, the smallness of α gives us the justification for initially neglecting effects predicted by the theory of special relativity, which we shall revisit in Chapter 8.

Classically, we can think of the potential as causing the electron to experience a force $-\vec{\nabla}V(r) = -\frac{q_{\mathrm{e}}}{r^2}\frac{\vec{x}}{r}$ directed towards the origin. For zero angular momentum, the potential is negative everywhere and approaches 0 as $r \to \infty$. If the azimuthal quantum number is non-zero, then the centrifugal term in Definition 7.2 diverges as $r \to 0$ and the effective potential is negative only for r larger than some finite value. Nonetheless, for classical particles of finite angular momentum the Coulomb potential admits a continuum of bound states with negative energies.

At first glance, the (classical) Coulomb problem hence appears to be similar to the classical Kepler problem, that is, bound states of a gravitational potential (see, e.g., Landau and Lifshitz, 1971, pp. 287–300). However, using classical methods one immediately runs into a vexing dilemma. The bound states of classical particles are realized as circular or elliptic orbits and the particles on such orbits are continuously accelerated by the gravitational or, here, electrostatic force. At the same time, electro-dynamics tells us that accelerated charges emit electromagnetic radiation. In a classical model of the hydrogen atom, the accelerated electron would hence immediately radi-ate away its energy and plummet into the proton in the nucleus. The hydrogen atom would not be stable, and even if it was, its spectrum would be continuous. It is one of the great successes of quantum mechanics to solve this puzzle and explain the stability of matter and correctly predict atomic absorption and emission spectra.

Before we show this quantum-mechanical solution, we want to take extra care and first show that a ground state indeed exists. Although one may arrive at this insight also by directly solving the Schrödinger equation (as we shall do in Section 7.5), it is a priori not guaranteed that a ground state exists for any given potential (recall the results of Section 7.1). The techniques that we discuss in this section are therefore very useful to establish the existence of a ground state before getting into a lengthy calculation.

7.4.2 An Upper Bound on the ground-state energy of the H-Atom

First, let us make sure that the Coulomb potential admits a bound state at all by determining an upper bound on the ground-state energy using the Rayleigh–Ritz vari-ational principle from Section 7.2.2. We then choose a family of (unnormalized) test wave functions

$$\psi_\epsilon(r, \vartheta, \varphi) = e^{-r/\epsilon} \tag{7.39}$$

parameterized by ϵ. For the normalization we then find

$$\langle \psi_\epsilon | \psi_\epsilon \rangle = \int_{S_2} d\Omega \int_0^\infty dr\, r^2\, |\psi_\epsilon|^2 = 4\pi \int_0^\infty dr\, r^2\, e^{-2r/\epsilon} = \pi\epsilon^3, \tag{7.40}$$

where the last step of the calculation can be carried out by twice using integration by parts. We then wish to compute the expectation value of the Hamiltonian $H = \vec{P}^2/(2m) - q_e^2/r$, since we may assume $l = 0$ for the ground state. For the kinetic energy term $\vec{P}^2/(2m)$ we recall the expression for the Laplacian in spherical polar coordinates from eqn (7.3), of which we only need the first term because our test wave functions do not depend on ϑ and φ. We hence have

$$\langle \frac{\vec{P}^2}{2m} \rangle = -\frac{2\pi\hbar^2}{m} \int_0^\infty dr\, e^{-r/\epsilon} \frac{\partial}{\partial r}\left(r^2 \frac{\partial}{\partial r}\right) e^{-r/\epsilon} = \frac{2\pi\hbar^2}{m\epsilon} \int_0^\infty dr\, r^2 e^{-2r/\epsilon}\left(2r - \frac{r^2}{\epsilon}\right) = \frac{\pi\hbar^2\epsilon}{2m}, \tag{7.41}$$

where we have again used integration by parts. Similarly, the potential energy term gives

$$\langle V(r) \rangle = -4\pi q_e^2 \int_0^\infty dr\, r^2\, \frac{1}{r} e^{-2r/\epsilon} = -\pi\, q_e^2\, \epsilon^2. \tag{7.42}$$

Combining eqns (7.40)–(7.42), we obtain the bound

$$E(\epsilon) = \frac{\langle \frac{\vec{P}^2}{2m} + V(r) \rangle}{\langle \psi_\epsilon | \psi_\epsilon \rangle} = \frac{\hbar^2}{2m\epsilon^2} - \frac{q_e^2}{\epsilon} \geq E_g. \tag{7.43}$$

Finally, we can minimize over the values of ϵ by setting

$$\frac{\partial E(\epsilon)}{\partial \epsilon} = -\frac{\hbar^2}{m\epsilon^3} + \frac{q_e^2}{\epsilon^2} \overset{!}{=} 0, \tag{7.44}$$

from which it is easy to see that we have $\epsilon_{min} = \hbar^2/(mq_e^2)$ at the minimum. Consequently, we find

$$\min_{\epsilon} E(\epsilon) = E(\epsilon_{min}) = -\frac{mq_e^4}{2\hbar^2} \geq E_g, \tag{7.45}$$

which informs us that the Coulomb potential indeed admits a bound state with negative energy, although we still have not concluded that the ground-state energy is finite. It will be the aim of the next section to establish such a lower bound. In anticipation of the results of Section 7.5, we shall nonetheless give away some of the results to come by noting that the upper bound of eqn (7.45) indeed already is tight. That is, the ground-state energy of the hydrogen atom will turn out to be exactly

$$E_g = -\frac{mq_e^4}{2\hbar^2} = \frac{q_e^2}{2r_B}, \tag{7.46}$$

where $r_B = \hbar^2/(mq_e^2)$ is called the *Bohr radius*. The reason for this apparent coincidence is that the ground-state wave function happens to be already contained within the family of test functions we have chosen in eqn (7.39). Before we confirm this explicitly, let us briefly discuss some techniques to obtain lower bounds on the ground-state energy.

7.4.3   A Lower Bound from Heisenberg's Uncertainty

Historically, a predecessor of the full quantum-mechanical solution of the Coulomb problem incarnated in the form of Bohr's atom model (recall Section 1.4). While ultimately incorrect, Bohr's model was nevertheless a milestone on the road to quantum mechanics and it was indeed correct regarding some aspects. Here we therefore want to present a "derivation" of a lower bound on the ground-state energy of the hydrogen atom that is inspired by Bohr's model to illustrate how some subtle assumptions can lead us astray. With this lesson learned we then inspect our assumptions more thoroughly in Section 7.4.4 before we proceed with the proper quantum-mechanical solution of the Coulomb problem in Section 7.5.

Let us hence assume that, much like in Bohr's model, the hydrogen atom has a ground state in which it has a "size" (in a sense we yet have to make more precise) that is characterized by a radius r_o. Since one may assume that $l = 0$ for the ground state, we expect a spherically symmetric wave function, such that $\langle X \rangle = \langle Y \rangle = \langle Z \rangle = 0$

and also $\langle P_x \rangle = \langle P_y \rangle = \langle P_z \rangle = 0$. At the same time, the expectation values $\langle X^2 \rangle = \langle Y^2 \rangle = \langle Z^2 \rangle$ and $\langle P_x^2 \rangle = \langle P_y^2 \rangle = \langle P_z^2 \rangle$ for such spherically symmetric functions can generally be non-zero and we hence have

$$\langle r^2 \rangle = \langle X^2 + Y^2 + Z^2 \rangle = 3 \langle X^2 \rangle . \tag{7.47}$$

We can then make use of Heisenberg's uncertainty relation in three dimensions, which can be obtained from eqn (2.61) and which reads

$$\left(\Delta \vec{X} \right)^2 \left(\Delta \vec{P} \right)^2 \geq \frac{9\hbar^2}{4} . \tag{7.48}$$

Here, we can use our previous assumption of spherical symmetry to obtain

$$\left(\Delta \vec{X} \right)^2 = \langle \vec{X}^2 \rangle - \langle \vec{X} \rangle^2 = \langle r^2 \rangle , \tag{7.49a}$$

$$\left(\Delta \vec{P} \right)^2 = \langle \vec{P}^2 \rangle - \langle \vec{P} \rangle^2 = 3 \langle P_x^2 \rangle . \tag{7.49b}$$

We may then combine eqns (7.49a) and (7.49b) and Heisenberg's uncertainty relation to bound the kinetic energy $\langle \vec{P}^2 \rangle / (2m)$, by noting

$$\langle P_x^2 \rangle \geq \frac{3\hbar^2}{4 \langle r^2 \rangle} . \tag{7.50}$$

We just need to make some reasonable assumptions about the wave function that stabilizes the system. In Bohr's atom model, the electrons are restricted to classical trajectories at some fixed radius. Let us therefore choose a wave function for which $|\psi|^2$ is proportional to a delta function $\delta(r - r_o)$, such that $\langle V(r) \rangle = -q_e^2 / r_o$ and

$$\langle r^2 \rangle = r_o^2 . \tag{7.51}$$

For the total average energy we then find

$$\bar{E} = \frac{\langle \vec{P}^2 \rangle}{2m} + \langle V(r) \rangle = \frac{3 \langle P_x^2 \rangle}{2m} + \langle V(r) \rangle \geq \frac{3}{2m} \frac{3\hbar^2}{4 r_o^2} - \frac{q_e^2}{r_o} := E_{\min} . \tag{7.52}$$

The interpretation of this approach is the following: Classically, the accelerated electron is expected to fall towards $r = 0$ as it radiates away its energy. As r_o becomes smaller, the potential energy decreases, and $\langle V(r) \rangle \to -\infty$ as $r_o \to 0$. This is where the Heisenberg uncertainty may be of help, telling us that the kinetic energy must increase as r_o goes to zero, hopefully balancing the decrease in potential energy to stabilize the system. Now, since we must assume that r_o is the radius at which the energy becomes minimal, we must have

$$\frac{\partial E_{\min}}{\partial r_o} = -2 \frac{9\hbar^2}{4 m r_o^3} + \frac{q_e^2}{r_o^2} \stackrel{!}{=} 0 . \tag{7.53}$$

From this we then easily find $r_o = 9\hbar^2 / (4 m q_e^2) = \frac{9}{4} r_{\mathrm{B}}$, and consequently

$$E_{\mathrm{g}} \geq -\frac{2}{9} \frac{m q_e^4}{\hbar^2} = -\frac{4}{9} \frac{q_e^2}{2 r_{\mathrm{B}}} . \tag{7.54}$$

At first glance it hence appears that we can derive a lower bound on the ground-state energy using Heisenberg's uncertainty principle. In fact, we can even do a little bit better. While the delta function $\delta(r - r_o)$ fits well with Bohr's picture of stable classical trajectories, it seems very restrictive that the electron can only be found at a fixed radius, but not at other positions within the atom. Suppose we therefore replace the choice of a delta function that we have made before with a wave function that is in some sense centred around a spherical shell around a characteristic value r_o, such that $\langle r \rangle = r_o$ and $\langle V(r) \rangle = -q_e^2/r_o$. But let us also assume that the electron can in principle be found anywhere within this radius r_o, such that $\Delta r = r_o$ as well. From this we immediately obtain

$$\langle r^2 \rangle = (\Delta r)^2 + \langle r \rangle^2 = 2 r_o^2 . \tag{7.55}$$

Inserting this result into eqn (7.52) instead of eqn (7.51), we obtain different values, $r_o = \frac{9}{8} r_{\mathrm{B}}$ and

$$E_{\mathrm{g}} \geq -\frac{8}{9} \frac{q_e^2}{2 r_{\mathrm{B}}} . \tag{7.56}$$

Since $8/9 \approx 1$, this result is indeed rather close to the actual ground-state energy from eqn (7.46). However, note that this lower bound is not compatible with the upper bound $E_{\mathrm{g}} \leq mq_e^4/(2\hbar^2) = q_e^2/(2r_{\mathrm{B}})$ that we have derived in the last section. So what went wrong? Motivated by Bohr's atom model of stable circular orbits, we have naively made some unwarranted assumptions about the existence and shape of the ground-state wave function. In other words, we cannot conclude that the stability of the system was derived from our model, if we assume right from the start that a stabilizing wave function exists.

7.4.4   A Lower Bound: Sobolev Inequalities

Although our previous choices of wave functions were evidently incorrect, we may still wonder whether the Heisenberg uncertainty relation is able to stabilize the system. That is, we may ask if the uncertainty principle guarantees a finite lower bound on the ground-state energy for any spherically symmetric wave function. A simple counterexample shows that the answer is no. Consider a family of spherically symmetric wave functions $\psi_\rho(r)$, given by

$$\psi_\rho(r) = \begin{cases} \sqrt{\frac{3}{4\pi\rho^3}} & r \leq \rho \\ 0 & r > \rho \end{cases} . \tag{7.57}$$

A particle described by the wave function $\psi_\rho(r)$ is confined entirely within a ball of radius ρ, with a constant probability of finding the particle anywhere within, but vanishing probability to find it outside. Suppose now we describe a particle as a superposition $\psi = (\psi_\epsilon + \psi_R)/\sqrt{N}$ of two such wave functions, concentrated within radii ϵ and R, respectively, and let N be a normalization constant. If we further assume that $R \gg \epsilon$, then the normalization condition

$$\int_{\mathbb{R}^3} d^3\!x \, |\psi|^2 = 4\pi \int_0^\infty dr \, r^2 \frac{1}{N} (\psi_\epsilon + \psi_R)^2 = \frac{3}{N} \left(\frac{1}{\epsilon^3} + \frac{2}{\sqrt{\epsilon^3 R^3}} \right) \int_0^\epsilon dr \, r^2 + \frac{3}{N} \frac{1}{R^3} \int_0^R dr \, r^2 ,$$
$$\tag{7.58}$$

tells us that $N = 2\left(1 + \sqrt{(\epsilon/R)^3}\right) \approx 2$. With this information, we can immediately compute the expectation values

$$\langle\, r^2 \,\rangle \approx \frac{4\pi}{2} \int\limits_0^\infty dr\, r^4 \left(\,\psi_\epsilon + \psi_R\right)^2 = \frac{3}{10}\left(\epsilon^2 + R^2 + \frac{\epsilon^{7/2}}{R^{3/2}}\right) \approx \frac{3R^2}{10}\,, \tag{7.59a}$$

$$\langle\, \frac{1}{r} \,\rangle \approx \frac{4\pi}{2} \int\limits_0^\infty dr\, r \left(\,\psi_\epsilon + \psi_R\right)^2 = \frac{3}{4}\left(\frac{1}{\epsilon} + \frac{1}{R} + \frac{\epsilon^{1/2}}{R^{3/2}}\right) \approx \frac{3}{4\epsilon}\,, \tag{7.59b}$$

and insert them into eqns (7.52) and (7.50) to bound the ground-state energy, i.e.,

$$E_{\mathrm{g}} \geq \frac{9\hbar^2}{8m\,\langle\, r^2 \,\rangle} - q_{\mathrm{e}}^2\,\langle\, \frac{1}{r} \,\rangle \approx \frac{15\hbar^2}{4mR^2} - \frac{3q_{\mathrm{e}}^2}{4\epsilon}\,. \tag{7.60}$$

Unfortunately, we cannot use this expression for a minimization procedure as in the previous section. If we keep R fixed, and consider the limit of $\epsilon \to 0$, the kinetic energy term approaches a constant, $15\hbar^2/(4mR^2)$, whereas the potential energy diverges, $\lim_{\epsilon \to 0} = -\infty$. Although we had speculated that the Heisenberg uncertainty would guarantee that the increase in kinetic energy would balance the decrease in potential energy as we approach $r = 0$, we see that this is not the case for all wave functions. The Heisenberg uncertainty alone is not sufficient to ensure the stability of matter. Of course we will later find that the ground state of the hydrogen atom is not of the same form as the wave function ψ in our example here, but there is no grounds for ruling out such a shape a priori.

If we want to show that a ground state exists without explicitly solving the Schrödinger equation, we hence need a relation which predicts a stronger increase of the kinetic energy during the "fall into the centre" than can be inferred from Heisenberg's uncertainty relation. For the system to be stabilized, the total energy must remain constant. A relation that provides the desired behaviour is given by a so-called *Sobolev inequality*.

Theorem 7.2 (Sobolev inequality)

Spherically symmetric wave functions $\psi(r)$ satisfy the Sobolev inequality

$$\int_{\mathbb{R}^3} d^3\vec{x}\, |\vec{\nabla}\psi|^2 \geq \mu_F \left(\int_{\mathbb{R}^3} d^3\vec{x}\, |\psi|^6\right)^{1/3},$$

where $\mu_F = 3(\pi/2)^{4/3} \approx 5.4779$.

Proof For the proof of this particular example of a Sobolev inequality we will follow the calculation of Grosse and Martin (1997, p. 155). More details on the intricacies of the formal proof of a more general inequality can be found also in Glaser *et al.* (1976, p. 169). For additional information on Sobolev spaces and inequalities for norms thereon see also Adams and Fournier (2003, Chap. 3 and p. 101). Indeed, one may show that the inequality in Theorem 7.2 also holds for wave functions that are not spherically symmetric, but since we are here only interested in the (spherically symmetric) ground state, we shall restrict the proof (and hence the theorem) to the simpler case of rotational symmetry. As we will see, the value μ_F arises as the infimum over all spherically symmetric wave functions $\psi(r)$ of the functional

$$F[\psi] = \frac{\int_{\mathbb{R}^3} d^3\vec{x}\,|\vec{\nabla}\psi|^2}{\left(\int_{\mathbb{R}^3} d^3\vec{x}\,|\psi|^6\right)^{1/3}} = (4\pi)^{2/3}\frac{\int_0^\infty dr\,r^2\,|\partial_r\psi(r)|^2}{\left(\int_0^\infty dr\,r^2\,|\psi(r)|^6\right)^{1/3}} = (4\pi)^{2/3}\frac{I_N}{I_D^{1/3}}, \qquad (7.61)$$

where we have used the expression for the nabla operator in spherical polar coordinates from eqn (6.44), and we have introduced abbreviations for the integrals I_N in the numerator, and I_D in the denominator, respectively. Next, we substitute $q = \ln(r)$ and $\phi = \sqrt{r}\psi$ to rewrite the two integrals. That is, we calculate

$$I_N = \int_0^\infty dr\,r^2\left(\frac{1}{\sqrt{r}}\frac{\partial\phi}{\partial r} - \frac{1}{2}\frac{\phi}{r^{3/2}}\right)^2 = \int_{-\infty}^\infty dq\left((\partial_q\phi)^2 + \frac{1}{4}\phi^2\right), \qquad (7.62)$$

where we have eliminated the cross-terms using integration by parts and $\phi(r=0) = \phi(r=\infty) = 0$, and we have assumed that ϕ is real-valued without loss of generality (recall Lemma 4.2). The integral in the denominator simply gives

$$I_D = \int_0^\infty dr\,r^2\left(\frac{\phi}{\sqrt{r}}\right)^6 = \int_{-\infty}^\infty dq\,\phi^6. \qquad (7.63)$$

The functional F can now be minimized using standard variational methods by setting $\delta F = 0$. Straightforwardly, we compute

$$\delta I_N = \frac{\partial I_N}{\partial\phi}\delta\phi + \frac{\partial I_N}{\partial(\partial_q\phi)}\delta(\partial_q\phi) = \int_{-\infty}^\infty dq\left(2(\partial_q\phi)\,\delta(\partial_q\phi) + \frac{1}{2}\phi\,\delta\phi\right) \qquad (7.64)$$

$$= \int_{-\infty}^\infty dq\left(2\partial_q\big[(\partial_q\phi)\,\delta\phi\big] - 2(\partial_q^2\phi)\,\delta\phi + \frac{1}{2}\phi\,\delta\phi\right) = \int_{-\infty}^\infty dq\left(\frac{1}{2}\phi - 2(\partial_q^2\phi)\right)\delta\phi,$$

where the term $2\partial_q\big[(\partial_q\phi)\,\delta\phi\big]$ vanishes after integrating and evaluating at the boundary $q = \pm\infty$. The variation of the second integral is much simpler and gives

$$\delta I_D = 6\int_{-\infty}^\infty dq\,\phi^5\,\delta\phi. \qquad (7.65)$$

Combining eqns (7.64) and (7.65) one then arrives at the relation

$$\delta F = (4\pi)^{2/3} \Big(\frac{\delta I_N}{(I_D)^{1/3}} - \frac{1}{3} \frac{I_N}{(I_D)^{4/3}} \delta I_D \Big) \tag{7.66}$$

$$= 2 \frac{(4\pi)^{2/3}}{(I_D)^{1/3}} \int_{-\infty}^{\infty} dq \, \Big(\frac{\phi}{4} - \partial_q^2 \phi - \frac{I_N}{I_D} \phi^5 \Big) \delta\phi \overset{!}{=} 0 \,. \tag{7.67}$$

We hence find the second-order differential equation

$$\partial_q^2 \phi = \frac{\phi}{4} - \frac{I_N}{I_D} \phi^5 \tag{7.68}$$

with the boundary condition $\phi(q = \pm\infty) = \partial_q \phi(q = \pm\infty) = 0$. It can be easily checked that eqn (7.68) is solved by $\phi(q) = k/\sqrt{\cosh(q)}$, for an arbitrary (real) constant k, for which we obtain $I_N = 3\pi k^2/8$ and $I_D = \pi k^6/2$. Inserting these results into the functional form, eqn (7.61), we find $F[\phi] = (4\pi)^{2/3} I_N / I_D^{1/3} = 3(\pi/2)^{4/3} = \mu_F$, as stated in Theorem 7.2. $\qquad\square$

With the Sobolev inequality of Theorem 7.2 proven, we now return to our problematic example wave function $\psi \approx (\psi_\epsilon + \psi_R)/\sqrt{2}$ with ψ_ϵ and ψ_R as in eqn (7.57). Instead of using the Heisenberg uncertainty relation to bound the kinetic-energy term, we now employ the Sobolev inequality and first compute

$$\frac{\langle \vec{P}^2 \rangle}{\hbar^2} = \int_{\mathbb{R}^3} d^3\vec{x} \, |\vec{\nabla}\psi|^2 \geq \int_{\mathbb{R}^3} d^3\vec{x} \, |\psi|^6 \approx 4\pi \int_0^\infty dr \, r^2 \frac{1}{2^3} (\psi_\epsilon + \psi_R)^6 \approx \frac{9}{8(4\pi)^2 \epsilon^6} \,. \tag{7.69}$$

For the (average) total energy we therefore find

$$E = \frac{\hbar^2}{2m} \int_{\mathbb{R}^3} d^3\vec{x} \, |\vec{\nabla}\psi|^2 + \langle V(r) \rangle \geq \frac{\hbar^2 \mu_F}{2m} \Big(\int_{\mathbb{R}^3} d^3\vec{x} \, |\psi|^6 \Big)^{1/3} - q_e^2 \langle \frac{1}{r} \rangle \tag{7.70}$$

$$\approx \frac{\hbar^2 \mu_F}{2m} \Big(\frac{9}{8(4\pi)^2} \Big)^{1/3} \frac{1}{\epsilon^2} - \frac{3q_e^2}{4\epsilon} := E_{\min} \,,$$

where we have inserted our previous result for $\langle \frac{1}{r} \rangle$ from eqn (7.59b). Having made use of the Sobolev inequality, we can proceed as usual with the minimization procedure. Setting $\partial E_{\min}/\partial\epsilon = 0$, we arrive at the bound

$$E \geq -\frac{9}{16\mu_F} \Big(\frac{8(4\pi)^2}{9} \Big)^{1/3} \frac{m q_e^4}{2\hbar^2} = -\Big(\frac{3}{2\pi^2} \Big)^{1/3} \frac{m q_e^4}{2\hbar^2} \,. \tag{7.71}$$

The Sobolev inequality has hence allowed us to establish a lower bound even for a wave function for which the Heisenberg uncertainty relation alone could not guarantee that the electron would not "fall into the centre". However, we once again find that the lower bound we obtain is not in fact compatible with the upper bound from eqn (7.45). For the hydrogen atom this is not much of a problem. Indeed, in the next section we shall determine the energy levels exactly. Nonetheless, the techniques that we have discussed here for deriving upper and lower bounds on the ground-state energy can be quite useful when this is not possible.

7.5   The Hydrogen Atom

7.5.1   The Radial Schrödinger Equation for the Hydrogen Atom

Finally, let us turn to the task of solving the radial Schrödinger equation for the Coulomb potential $V(r) = -q_e^2/r$. Inserting into the differential equation for the reduced wave functions in eqn (7.10) we have

$$\left(-\frac{\hbar^2}{2m} \frac{\partial^2}{\partial r^2} + \frac{\hbar^2 \, l \, (l+1)}{2mr^2} - \frac{q_e^2}{r} - E \right) u_l(r) = 0 \,. \tag{7.72}$$

Since we are only interested in the bound-state solutions, we may set $E = -|E|$ and use $k = \sqrt{2m|E|}/\hbar > 0$, just as for the finite potential well in Section 4.2.1. In addition, we perform a change of variables, $r \to \rho = kr$ to rewrite eqn (7.72) as

$$\left(k^2 \frac{\partial^2}{\partial \rho^2} - k^2 \frac{l \, (l+1)}{\rho^2} + \frac{2mq_e^2}{\hbar^2} \frac{k}{\rho} - k^2 \right) u_l = 0 \,. \tag{7.73}$$

Introducing the abbreviation $\rho_o = 2mq_e^2/(k\hbar^2)$, the radial Schrödinger equation hence becomes

$$\left(\frac{\partial^2}{\partial \rho^2} - \frac{l \, (l+1)}{\rho^2} + \frac{\rho_o}{\rho} - 1 \right) u_l(r) = 0 \,. \tag{7.74}$$

In a first step towards solving this differential equation for u_l, we make the ansatz

$$u_l(\rho) = \rho^{l+1} \, e^{-\rho} \, L(\rho) \,. \tag{7.75}$$

With a little work differentiating this product of functions, we can transform eqn (7.74) into a differential equation for the function $L(\rho)$, which is of the form

$$\left(\rho \frac{\partial^2}{\partial \rho^2} + 2(l + 1 - \rho) \frac{\partial}{\partial \rho} + (\rho_o - 2l - 2) \right) L(\rho) = 0 \,. \tag{7.76}$$

It is now convenient to make a power-series ansatz for the function $L(\rho)$, that is, we assume the function $L(\rho)$ may be written as

$$L(\rho) = \sum_{j=0} a_j \, \rho^j \,. \tag{7.77}$$

We differentiate twice and insert the results back into eqn (7.76) to obtain

$$\rho \sum_{j=2} j \, (j-1) \, a_j \, \rho^{j-2} + 2(l+1-\rho) \sum_{j=1} j \, a_j \, \rho^{j-1} + (\rho_o - 2l - 2) \sum_{j=0} a_j \, \rho^j \tag{7.78}$$

$$= \sum_{j=0} \rho^j \Big(a_{j+1}(j+1)(j+2l+2) + a_j \big(\rho_o - 2(j+l+1) \big) \Big) = 0,$$

where we have performed some relabelling of the summation indices and ordered the terms by the powers of ρ. Since the differential equation must be true for all values of ρ, eqn (7.78) provides a recursion relation for the coefficients a_j, i.e.,

$$a_{j+1} = a_j \frac{2j + 2l + 1 - \rho_o}{(j+1)(j + 2l + 2)} \qquad \text{for } j = 0, 1, 2, \ldots . \qquad (7.79)$$

In principle, our desired solutions $L(\rho)$ are determined by this recursion relation, although it is not straightforward to give a closed expression for $L(\rho)$ with this information. We shall therefore have to invest a little bit more work to obtain the bound-state solutions of the hydrogen atom, and we shall do so in Section 7.5.3. However, before we do so, we can make another crucial observation to determine the energy levels of the hydrogen atom.

7.5.2   The Energy Levels of the Hydrogen Atom

For large j, the recursion relation in eqn (7.79) approximately reduces to $a_{j+1} \approx (2/j)a_j$. If we compare this with the exponential function

$$e^{2\rho} = \sum_j \frac{2^j}{j!} \rho^j = \sum_j b_j \rho^j , \qquad (7.80)$$

we see that here the recursion relation of the power-series coefficients b_j for large $j \gg 1$ is

$$b_{j+1} = \frac{2^{j+1}}{(j+1)!} = \frac{2}{j+1} \frac{2^j}{j!} = \frac{2}{j+1} b_j \approx \frac{2}{j} b_j . \qquad (7.81)$$

In other words, if the power series for $L(\rho)$ continues indefinitely, the function will grow exponentially with ρ. This, in turn, means that if the wave functions that we seek are to be normalizable, the power series must terminate eventually. For each wave function there must exist some finite value $j = n'$ such that $a_{n'} \neq 0$ but $a_{n'+1} = 0$. The recursion relation (7.79) then tells us that these values $n' \in \mathbb{N}_0$, called the *radial quantum numbers*, satisfy

$$\rho_o = 2\left(n' + l + 1\right) := 2n , \qquad (7.82)$$

where we have introduced yet another quantum number, $n = n' + l + 1$. If we now recall that $\rho_o = 2mq_e^2/(k\hbar^2)$ where $k = \sqrt{-2mE}/\hbar$ we can express the allowed energy levels in terms of this *principal quantum number* n, that is, we find the *Bohr formula*

$$E_n = -\frac{mq_e^4}{2\hbar^2 n^2} . \qquad (7.83)$$

Since the radial quantum number n' and the azimuthal quantum number l can take values $0, 1, 2, \ldots$, we discover from eqn (7.82) that the principal quantum number may take values $n = 1, 2, 3, \ldots$. Vice versa, since $n' = n - l - 1$, the azimuthal quantum number is bounded from above, $l \leq n - 1$. In addition, we recall from Theorem 6.5 that each value of l allows for $2l + 1$ different values of the magnetic quantum number

m. Since neither l nor m enter into E_n in eqn (7.83), we see that the degeneracy of the energy levels for a given principal quantum number n is

$$\sum_{l=0}^{n-1} (2l + 1) = n^2 \ . \tag{7.84}$$

The only non-degenerate energy level is the lowest level obtained for $n = 1$, where $l = 0$, and we find the ground-state energy

$$E_{\mathrm{g}} = E_1 = -\frac{m\, q_{\mathrm{e}}^4}{2\hbar^2} = -\tfrac{1}{2} m\, c^2\, \alpha^2 \approx -13.6\, eV \ , \tag{7.85}$$

just as we had revealed earlier in eqn (7.46). Its absolute value, $E_{\mathrm{I}} = -E_{\mathrm{g}}$ is also called the *ionization energy* of the hydrogen atom. Alternatively, we can also express the energy in terms of the Bohr radius r_{B}, i.e., $E_{\mathrm{g}} = -q_{\mathrm{e}}^2/(2r_{\mathrm{B}})$. This characteristic length of the hydrogen atom is given by

$$r_{\mathrm{B}} = \frac{\hbar^2}{m\, q_{\mathrm{e}}^2} = \frac{1}{\alpha}\frac{\hbar}{m\, c} = \frac{1}{\alpha}\lambda_{\mathrm{C}} \approx 0.53\,\mathring{A} \ , \tag{7.86}$$

where λ_{C} is the Compton wavelength of the electron when $m = m_{\mathrm{e}}$ is the electron mass. Recall that one may introduce a reduced mass m_{r} for this problem (see, e.g., Greiner, 2000, p. 387), that is,

$$m_{\mathrm{r}} = \frac{m_{\mathrm{e}}}{1 + \frac{m_{\mathrm{e}}}{m_{\mathrm{p}}}} \approx m_{\mathrm{e}}\left(1 - \frac{1}{1836}\right), \tag{7.87}$$

where m_{p} is the mass of the proton and $m \approx m_{\mathrm{e}}$.

Before we move on to calculating the form of the bound-state wave functions, let us make a few brief remarks. First, the fact that there is degeneracy with respect to the angular momentum is an interesting property of the $1/r$ potential. It hints at an additional symmetry other than the rotational invariance. This symmetry also has a counterpart in the classical theory, where it is the *Laplace–Runge–Lenz* vector that is a constant of motion in this kind of potential (see, e.g., the discussion in Jackiw, 1995, p. 233).

Second, note that the treatment of the hydrogen atom has been completely non-relativistic in this section. When we employ the correct relativistic description in Chapter 8, including the relativistic energy-momentum relation and spin, we will find an additional splitting of the now degenerate levels. Such effects include the two possible spin levels for each set of quantum numbers n, l, and m (see Section 8.3), as well as an interaction of the magnetic moments caused by the spin and orbital angular momentum of the electron—the spin-orbit interaction, which we will discuss in Section 10.2 of Chapter 10. These, and other corrections, lead to the *fine structure* of the hydrogen atom. In addition, one may include the interaction of the electron with the spin of the proton in the nucleus to obtain the *hyperfine structure*, and the so-called *Lamb shift*, an effect arising from quantum electrodynamics. All these effects can be better understood in terms of the Dirac equation, which is the relativistic counterpart of the Schrödinger equation, although these effects can also be treated within the framework provided by the Schrödinger equation using perturbative methods to get satisfying results (see Chapter 10). With such perturbative methods we can also treat the influences of exterior fields which cause the *Zeeman effect* (see Section 10.3) and the *Stark effect* (see Section 10.4) for external magnetic and electric fields, respectively.

7.5.3 The Laguerre Polynomials

At last, we also want to determine the radial wave functions for the hydrogen atom explicitly. Returning to eqn (7.76) we substitute $\xi = 2\rho$ to rewrite the differential equation as

$$\left(\xi \frac{\partial^2}{\partial \xi^2} + (\beta + 1 - \xi) \frac{\partial}{\partial \xi} + (\gamma - \beta)\right) L^{\beta}_{\gamma}(\xi) = 0, \tag{7.88}$$

where $\beta = 2l + 1$ and $\gamma = n' + \beta = n + l$, and we have inserted the radial quantum number $n' = n - l - 1$ from before. Equation (7.88) can be recognized as the *general Laguerre* differential equation for the *associated Laguerre polynomials* L^{β}_{γ}, given by

$$L^{\beta}_{\gamma}(\xi) = \frac{\partial^{\beta}}{\partial \xi^{\beta}} L_{\gamma}(\xi), \tag{7.89}$$

where L_{γ} are the (regular) *Laguerre polynomials*. With the information from eqn (7.89), we can evaluate the terms in eqn (7.88), that is,

$$\xi \frac{\partial^2}{\partial \xi^2} L^{\beta}_{\gamma}(\xi) = \frac{\partial^{\beta}}{\partial \xi^{\beta}} \left(\xi \frac{\partial^2}{\partial \xi^2} - \beta \frac{\partial}{\partial \xi}\right) L_{\gamma}(\xi), \tag{7.90a}$$

$$(\beta + 1 - \xi) \frac{\partial}{\partial \xi} L^{\beta}_{\gamma}(\xi) = \frac{\partial^{\beta}}{\partial \xi^{\beta}} \left((\beta + 1 - \xi) \frac{\partial}{\partial \xi} + \beta\right) L_{\gamma}(\xi), \tag{7.90b}$$

to find that the functions $L_{\gamma}(\xi)$ must satisfy the *Laguerre differential equation*

$$\left(\xi \frac{\partial^2}{\partial \xi^2} + (1 - \xi) \frac{\partial}{\partial \xi} + \gamma\right) L_{\gamma}(\xi) = 0. \tag{7.91}$$

The solutions to this differential equation can be written as

$$L_{\gamma}(\xi) = e^{\xi} \frac{\partial^{\gamma}}{\partial \xi^{\gamma}} e^{-\xi} \xi^{\gamma}. \tag{7.92}$$

To see that the functions defined in eqn (7.92) indeed solve eqn (7.91), we compute

$$\xi \frac{\partial^2}{\partial \xi^2} L_{\gamma}(\xi) = \gamma e^{\xi} \frac{\partial^{\gamma-1}}{\partial \xi^{\gamma-1}} e^{-\xi} \left[-(\gamma - 1)\xi^{\gamma-2} - (\gamma - 1)\xi^{\gamma-1}\right], \tag{7.93a}$$

$$(1 - \xi) \frac{\partial}{\partial \xi} L_{\gamma}(\xi) = \gamma e^{\xi} \frac{\partial^{\gamma-1}}{\partial \xi^{\gamma-1}} e^{-\xi} \left[(\gamma - 1)\xi^{\gamma-2} - \xi^{\gamma-1} + \xi^{\gamma}\right], \tag{7.93b}$$

$$\gamma L_{\gamma}(\xi) = \gamma e^{\xi} \frac{\partial^{\gamma-1}}{\partial \xi^{\gamma-1}} e^{-\xi} \left[\gamma \xi^{\gamma-1} - \xi^{\gamma}\right]. \tag{7.93c}$$

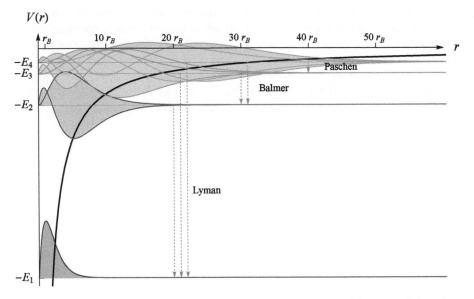

Fig. 7.3 Hydrogen atom. The reduced radial wave functions $u(r) = r R_{nl}(r)$ and energy levels E_n of the Coulomb potential $V(r) = -q_e^2/r = 2E_1 r_B/r$ (shown in black) are shown for $n = 1, 2, 3, 4$ (bottom to top) and $l = 0, \ldots, n-1$, shifted by their respective energies $E_n = -E_1/n^2$. The vertical arrows indicate the corresponding transitions of the Lyman, Balmer, and Paschen series.

Combining this result with our previous ansatz from eqn (7.75), we hence find that, up to a normalization constant $\mathcal{N}$, the radial wave functions for the hydrogen atom are labelled by the quantum numbers n and l, and they are given by

$$R_{nl}(r) = \frac{u_l}{r} = \mathcal{N} \left(\frac{r}{n\, r_B} \right)^l e^{-\frac{r}{n r_B}} L_{n+l}^{2l+1}\left(\frac{2r}{n\, r_B}\right), \tag{7.94}$$

where we have also used the Bohr formula to write $k = \sqrt{-2mE_n}/\hbar = 1/(nr_B)$. For an illustration of the radial solutions see Fig. 7.3. Noting that the associated Laguerre polynomials satisfy the orthogonality relation

$$\int\limits_0^\infty d\xi\, \xi^{\beta+1}\, e^{-\xi}\, L_\gamma^\beta(\xi)\, L_{\gamma'}^{\beta'}(\xi) = \frac{(2\gamma - \beta + 1)(\gamma!)^3}{(\gamma - \beta)!}\, \delta_{\beta\beta'}\, \delta_{\gamma\gamma'}, \tag{7.95}$$

we find the missing normalization constant

$$\mathcal{N} = -\frac{2^{l+1}}{n^2} \sqrt{\frac{(n-l-1)!}{(r_B(n+l)!)^3}}, \tag{7.96}$$

where we have made the arbitrary choice of including an overall minus sign for later convenience. The complete solutions of the three-dimensional Schrödinger equation for the Coulomb potential is then obtained by combining the radial solution with the spherical harmonics from eqn (6.64). The bound-state wave functions

$$\psi_{n,l,m}(r,\vartheta,\varphi) = R_{nl}(r)\,Y_{lm}(\vartheta,\varphi) \tag{7.97}$$

are hence labelled by the quantum numbers

$$n = 1, 2, \ldots, \qquad \text{(principal quantum number)}, \tag{7.98a}$$
$$l = 0, \ldots, n-1, \qquad \text{(azimuthal quantum number)}, \tag{7.98b}$$
$$m = -l, \ldots, l, \qquad \text{(magnetic quantum number)}, \tag{7.98c}$$

but the energy levels (as far as we know at this point), only depend on the principal quantum number n via the Bohr formula of eqn (7.83). In particular, we can note that the energy-level structure determines the spectroscopic properties of the atom. When an electron transitions from an excited state with energy E_m to another bound state with smaller energy $E_n < E_m$, a photon carrying the energy difference is emitted (see, e.g., the discussion in Section 10.5.3). We can easily see that the corresponding frequency $\nu_{mn} = \omega_{mn}/(2\pi)$ of this photon is given by the *Rydberg formula*

$$\nu_{mn} = \frac{1}{2\pi\hbar}(E_m - E_n) = \frac{m_e q_e^4}{4\pi\hbar^3}\left(\frac{1}{n^2} - \frac{1}{m^2}\right) = c\mathcal{R}\left(\frac{1}{n^2} - \frac{1}{m^2}\right), \tag{7.99}$$

where $\mathcal{R} = m_e q^4/(8\epsilon_o^2 h^3 c) = 1,097 \times 10^5\,\mathrm{cm}^{-1}$ is called the *Rydberg constant*. Conversely, a photon of frequency ν_{mn} is absorbed during a transition of an electron from E_n to E_m. To get a more exact expression for the transition frequencies, the electron mass has to be replaced by the reduced mass of eqn (7.87)[1]. Nonetheless, even without the corrections that we are yet to discover in the next chapters, the Rydberg formula already gives a very accurate account of the absorption and emission spectra of the hydrogen atom. For historical reasons, the spectral lines corresponding to frequencies ν_{mn} of transitions to fixed final levels n are grouped together in spectral series. For instance the transitions towards the ground state ($n = 1$) produce the *Lyman series*. The transitions towards the excited states $n = 2, 3, 4, 5, 6$ are called *Balmer*, *Paschen*, *Brackett*, *Pfund*, and *Humphreys* series, respectively, while the series for $n > 6$ remain unnamed.

In principle, this energy-level structure is the foundation of modern chemistry and molecular physics since it is no problem to generalize our treatment here to atomic nuclei with more than one proton (if we ignore the interaction between the nucleons). However, some work still lies ahead of us to discuss how additional electrons populate the available energy levels and we still have to make some modifications to this atom model even for just a single electron. Most importantly, we have to properly take into account the electron spin that we will discuss in Chapter 8, and, as we shall see in Chapter 10, the relativistic treatment of the hydrogen atom introduces additional corrections—the *fine structure*—to the picture that we have developed so far. Although

---

[1]To express this difference, the corrected Rydberg constant for the hydrogen atom can be denoted by R_{H}, whereas the symbol R_∞ can be used instead of $\mathcal{R}$ to emphasize that R_∞ arises from R_{H} by taking the mass of the nucleus to infinity.

these considerations will slightly modify the energy levels of the hydrogen atom, in particular, by removing some of the degeneracies, the basic structure remains unchanged, and we can hence proceed with some remarks about the physical properties of the hydrogen atom in Section 7.5.4.

7.5.4   Properties of the Hydrogen Atom

With the knowledge we have obtained about the bound-state wave functions of the hydrogen atom, we are now in a position to make statements about its radial structure. In particular, we can follow up on some of the previous guesswork from Section 7.4 regarding the location of the electron within the atom. In principle, the wave functions $\psi_{n,l,m}$ of eqn (7.97) provide all the information to compute expectation values for arbitrary observables, e.g., the mean radius $\langle r \rangle_{n,l,m}$. For fixed values of n, l, and m such a calculation can be carried out in a straightforward way, although appropriate computer software might come in useful for larger quantum numbers.

Here, however, we want to present a method that allows us to compute such expectation values for all values of (n, l, m) in a concise and systematic way without having to deal with complicated integrals of the associated Laguerre polynomials. In the approach that we will discuss now, we shall make use of the so-called *Kramers–Pasternack* recursion relation between expectation values of different powers of (r/r_{B}) for eigenstates of the hydrogen-atom Hamiltonian, first developed[2] by *Simon Pasternack* (1937) and independently by *Hendrik A. Kramers* (1938). Here, we will follow the more recent treatment in Armstrong (1971) and Balasubramanian (2000).

Theorem 7.3 (Kramers–Pasternack recursion relation)

For the hydrogen eigenstates $\psi_{n,l,m}$ of eqn (7.97) *the expectation values* $\langle \left(\frac{r}{r_B}\right)^q \rangle_{n,l,m} =: \langle \zeta^q \rangle$ *satisfy the recursion relation*

$$4(q+1)\,\langle \zeta^q \rangle - 4n^2(2q+1)\,\langle \zeta^{q-1} \rangle + n^2 q\big[(2l+1)^2 - q^2\big]\,\langle \zeta^{q-2} \rangle = 0\,.$$

Proof   To prove Theorem 7.3, we first note that the expectation values of powers of r only depend on the radial part of the wave function. We may hence immediately write an integral for the (real) reduced wave function $u(r) = R_{nl}(r)$ similar to eqn (7.14)

$$\langle \left(\frac{r}{r_{\mathrm{B}}}\right)^q \rangle_{n,,l,m} = \int_0^\infty dr \left(\frac{r}{r_{\mathrm{B}}}\right)^q |u(r)|^2 = r_{\mathrm{B}} \int_0^\infty d\zeta \, \zeta^q \, u^2 =: \langle \zeta^q \rangle \,, \qquad (7.100)$$

where we have substituted $\zeta = r/r_{\mathrm{B}}$. On the other hand, inserting Bohr's formula of eqn (7.83) into the Schrödinger equation (7.72) for the reduced wave function and again changing the variable r to ζ we arrive at the relation

---

[2]For a historical overview of their development surrounding these relations and an alternative derivation we refer to Szymanski and Freericks, (2021).

$$\frac{\partial^2}{\partial \zeta^2} u = \left(\frac{l(l+1)}{\zeta^2} - \frac{2}{\zeta} + \frac{1}{n^2} \right) u. \tag{7.101}$$

In other words, the second derivative $u'' = \partial^2 u / \partial \zeta^2$ can be written as a linear combination of functions that are proportional to u up to some power of ζ. This can be used to calculate the integral

$$r_{\text{B}} \int_0^\infty d\zeta\, \zeta^q\, u''\, u = l(l+1) \langle \zeta^{q-2} \rangle - 2 \langle \zeta^{q-1} \rangle + \frac{1}{n^2} \langle \zeta^q \rangle . \tag{7.102}$$

The strategy is then to use several steps of integration by parts to evaluate the left-hand side of eqn (7.102) and compare the result to the right-hand side to obtain the desired recursion relation. We begin by differentiating $u\zeta^q$ and integrating u'' to compute

$$r_{\text{B}} \int_0^\infty d\zeta\, \zeta^q\, u''\, u = -r_{\text{B}} \int_0^\infty d\zeta\, (\zeta^q u)'\, u' = -r_{\text{B}} q \int_0^\infty d\zeta\, \zeta^{q-1} u\, u' - r_{\text{B}} \int_0^\infty d\zeta\, \zeta^q u'\, u', \tag{7.103}$$

where the boundary term $r_{\text{B}} \left[\zeta^q u u' \right]_0^\infty$ vanishes. We then evaluate the remaining two integrals on the right-hand side separately. For the first one, simply note that $u u' = (u^2/2)'$ before the integration by parts to obtain

$$r_{\text{B}} \int_0^\infty d\zeta\, \zeta^{q-1} u\, u' = \frac{r_{\text{B}}}{2} \int_0^\infty d\zeta\, \zeta^{q-1} (u^2)' = -r_{\text{B}} \frac{(q-1)}{2} \int_0^\infty d\zeta\, \zeta^{q-2} u^2 = -\frac{(q-1)}{2} \langle \zeta^{q-2} \rangle , \tag{7.104}$$

where the boundary term again disappears because u must fall off stronger than any polynomial as ζ approaches infinity. For the second integral in eqn (7.103) we integrate ζ^q and differentiate $(u')^2$ to get

$$r_{\text{B}} \int_0^\infty d\zeta\, \zeta^q u'\, u' = -\frac{2r_{\text{B}}}{q+1} \int_0^\infty d\zeta\, \zeta^{q+1} u'\, u'' = -\frac{2r_{\text{B}}}{q+1} \int_0^\infty d\zeta\, \zeta^{q+1} \left(\frac{l(l+1)}{\zeta^2} - \frac{2}{\zeta} + \frac{1}{n^2} \right) u u', \tag{7.105}$$

where we have inserted the radial Schrödinger equation in the form of (7.101). It is then easy to observe that every term on the right-hand side of eqn (7.105) is of the same form as eqn (7.104) for some value of q. Combining the results of eqns (7.103)–(7.105) we have

$$r_{\text{B}} \int_0^\infty d\zeta\, \zeta^q\, u''\, u = (q-1)\left(\frac{q}{2} - \frac{l(l+1)}{q+1} \right) \langle \zeta^{q-2} \rangle + \frac{2q}{q+1} \langle \zeta^{q-1} \rangle - \frac{1}{n^2} \langle \zeta^q \rangle . \tag{7.106}$$

The final comparison with the calculation in eqn (7.102) and some simplification then provides the relation stated in Theorem 7.3. $\qquad\square$

Theorem 7.3 can now be used to compute the expectation values $\langle (r/r_{\mathrm{B}})^q \rangle_{n,l,m}$. We start with the case $q = 0$, for which $\langle \zeta^q \rangle = 1$ due to the normalization and the coefficient of $\langle \zeta^{q-2} \rangle$ vanishes in the recursion relation, which provides an expression for $\langle \zeta^{-1} \rangle$ that we can rewrite to obtain

$$\langle n, l, m \,|\, \frac{1}{r} \,|\, n, l, m \rangle = \frac{1}{r_{\mathrm{B}} n^2} \,. \tag{7.107}$$

For the negative values of q we still need to compute $\langle \zeta^{-2} \rangle$ to be able to determine the remaining expectation values. We shall do so when we have need of these particular expectation values in Section 10.2. Nonetheless, we should note already here that the values of $\langle \zeta^q \rangle$ are not finite for all values of $q < 0$. When $q = -(2l + 1)$, the coefficient of $\langle \zeta^{q-2} \rangle$ again vanishes and the Kramers–Pasternack relation therefore reaches the limit of its usefulness to predict further expectation values. On the other hand, inspection of eqn (7.94) reveals that R_{nl} is proportional to r^l, and therefore none of the expectation values for $q \leq -(2l+1)$ are finite. Turning to the expectation values for positive q, we start with $q = 1$, which provides

$$\langle n, l, m \,|\, r \,|\, n, l, m \rangle = r_{\mathrm{B}} \frac{3n^2 - l(l+1)}{2} \tag{7.108}$$

when inserting $\langle \zeta^0 \rangle = 1$ and $\langle \zeta^{-1} \rangle$ from eqn (7.107). Similarly, we can continue with $q = 2$ to obtain

$$\langle n, l, m \,|\, r^2 \,|\, n, l, m \rangle = r_{\mathrm{B}}^2 \frac{n^2(5n^2 - 3l(l+1) + 1)}{2} \,, \tag{7.109}$$

and in principle one may proceed with higher powers of r. Clearly, this technique for systematically calculating expectation values applies more generally to eigenstates of Hamiltonians other than the one studied here, for example, when considering the fully relativistic treatment (see, e.g., Andrae, 1997, and Suslov, 2010). For now, we shall be content with the results found so far. Most importantly, we may associate the result of eqn (7.108) with an effective size of the hydrogen atom, which grows as n^2 with increasing principal quantum number (recall that $l \leq n - 1$.). For the ground state $n = l = m = 0$, we find $\langle r \rangle_{0,0,0} = 3r_{\mathrm{B}}/2$, which confirms the role of the Bohr radius as a characteristic length of the hydrogen atom. With growing average distance of the electron from the nucleus, the electron also becomes more "smeared out", that is, the radial uncertainty is found to be

$$\Delta r = \frac{r_{\mathrm{B}}}{2} \sqrt{(n^2 + 1)^2 - [l(l+1)]^2 - 1} \,, \tag{7.110}$$

by combining eqns (7.108) and (7.109).

8
Spin and Atomic Structure

Throughout Chapters 2 to 6 we have formulated the quantum-mechanical formalism without specific reference to the type of quantum-mechanical system we wish to describe. In Chapter 7, we departed from this generic approach when discussing the interaction between electric charges. More specifically, we have been interested in finding the bound states of an electron in the electric field of a proton. By including the Coulomb interaction into the Hamiltonian, electrostatics was hence easily incorporated into the quantum-mechanical language. In this chapter we shall try to include also some of the dynamical aspects of electromagnetism, in particular the interaction of quantum-mechanical systems with magnetic fields. As in the historical development of quantum mechanics, this will lead us to the concept of *spin*.

8.1 The Magnetic Dipole Moment

8.1.1 Classical Magnetic Dipoles

To include the interaction with magnetic fields into our formalism, let us briefly return to classical physics. A particle with electric charge q, mass m_{q}, and velocity $\vec{v}$ in the presence of electric and magnetic fields, $\vec{E}(t, \vec{x})$ and $\vec{B}(t, \vec{x})$, respectively, is subject to the *Lorentz force*

$$\vec{F}_{\mathrm{L}} = q\left(\vec{E} + \frac{1}{c}\vec{v} \times \vec{B}\right), \tag{8.1}$$

where the factor $\frac{1}{c}$ is due to the choice of Gaussian cgs units, which we will use throughout this chapter. For more information on this convention and the conversion to other units, see, e.g., the textbooks by Ohanian (1988) and Jackson (1999). When incorporating the Lorentz force into quantum mechanics and considering only the electric field, we can simply write $\vec{E} = -\vec{\nabla}\phi$. For the electric field generated by a point-like charge we identify ϕ with the Coulomb potential from eqn (7.37). For the magnetic field, we have to be a little more inventive since the Lorentz force is in general not conservative. The Lorentz force therefore cannot be fully represented by just adding another potential term to the Hamiltonian. What is required is the so-called *minimal substitution* as we will discuss in Section 9. For now, let us try a different approach. From classical electrodynamics we know that, while an electric monopole (a charge as described before) is subject to the Lorentz force, a magnetic dipole[1] is subject to a

---

[1]Recall that, as far as we know empirically, magnetic monopoles do not exist in nature. For a discussion, see Section 9.6.3 or the available literature, e.g., Bertlmann (2000, p. 297).

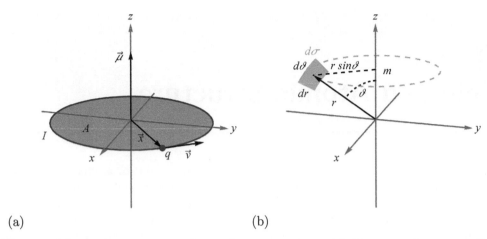

(a)   (b)

Fig. 8.1 Magnetic moment and angular momentum. (a) The magnetic moment $\vec{\mu}$ of a wire loop enclosing the surface A with a current I has magnitude IA/c (in Gaussian cgs units), and can be thought of as a particle of charge q and mass m_{q} with orbital velocity $\vec{v}$. The angular momentum $\vec{L} = m_{\mathrm{q}}\vec{x} \times \vec{v}$ of such a classical particle is therefore proportional to the magnetic moment. (b) At fixed r and ϑ, the infinitesimal area element orthogonal to the azimuthal direction is $d\sigma = r\,dr\,d\vartheta$ and its distance from the z axis is $r\sin\vartheta$.

force $\vec{F}_{\mathrm{mag}} = -\vec{\nabla}V_{\mathrm{mag}}$ when placed into a magnetic field. The corresponding potential energy is given by

$$V_{\mathrm{mag}}(\vec{x}) = -\vec{\mu} \cdot \vec{B}\,, \tag{8.2}$$

where $\vec{\mu}$ is the *magnetic moment* of the dipole. When the dipole moment and the magnetic field are parallel, then the potential energy is minimal, while the potential energy is maximal, when they are antiparallel. A simple example for a magnetic dipole from classical physics is an ideally thin loop of wire carrying a current I that encircles an area A as illustrated in Fig. 8.1 (a). The magnetic dipole moment $\vec{\mu}$ of this wire loop is orthogonal to the surface A, pointing upwards if the direction of the current is counter-clockwise (right-hand rule), and its magnitude is just $|\vec{\mu}| = IA/c$ (in Gaussian cgs units), see, e.g., Ohanian (1988) and Jackson (1999). Classically, we can think of the electric current as being caused by a point-like charge q rotating on a circular orbit with velocity $\vec{v}$ and radius $r = |\vec{x}|$. In that case we have $I = q|\vec{v}|/(2\pi r)$ and $A = r^2\pi$. On the other hand, according to eqn (6.1), the particle with mass m_{q} has angular momentum $\vec{L} = m_{\mathrm{q}}\vec{x} \times \vec{v}$, which is aligned with its magnetic moment, $\vec{L} \propto \vec{\mu}$ and $|\vec{L}| = m_{\mathrm{q}}vr$. From our example we can then easily see that

$$\vec{\mu} = \frac{q}{2m_{\mathrm{q}}c}\,\vec{L} = \gamma\,\vec{L}\,, \tag{8.3}$$

where $\gamma = q/(2m_{\mathrm{q}}c)$ is called the *gyromagnetic ratio* of the particle under consideration.

8.1.2 The Magnetic Dipole Moment of the Hydrogen Atom

The question is now whether we can apply this observation of classical electromagnetism to a quantum-mechanical system. As we have found in the previous chapter, the electron is delocalized within the hydrogen atom in the sense that the probability density given by the modulus squared of the radial wave function is non-zero nearly everywhere (see Fig. 7.3). Meanwhile, recall that the probability density is associated to a conserved probability current given by Definition 2.2, e.g., for an electron of mass m_e we have

$$\vec{j}(t, \vec{x}) := \frac{\hbar}{2 m_e i} \left[\psi^* \vec{\nabla} \psi - \psi \vec{\nabla} \psi^* \right]. \tag{8.4}$$

For an electron with charge $q = -q_e$ such a probability current is also an electric current, and we shall next compute this current for an eigenstate $\psi_{n,l,m}(r, \vartheta, \varphi)$ of the hydrogen atom. Since we have determined these stationary states with respect to spherical polar coordinates, it also appears natural to compute the components of the probability current with respect to the basis vectors $\vec{e}_r$, $\vec{e}_\vartheta$, and $\vec{e}_\varphi$. This is easily achieved by using the operator $\vec{\nabla}$ in spherical polar coordinates. From eqn (6.44) we know it is of the form

$$\vec{\nabla} = \vec{e}_r \frac{\partial}{\partial r} + \vec{e}_\vartheta \frac{1}{r} \frac{\partial}{\partial \vartheta} + \vec{e}_\varphi \frac{1}{r \sin \vartheta} \frac{\partial}{\partial \varphi}. \tag{8.5}$$

One can immediately see that the currents in the r direction and ϑ direction vanish, since the parts of the wave function which depend on r and ϑ, that is, the radial wave function $R_{nl}(r)$ from eqn (7.94) and the associated Legendre polynomials $P_{l|m|}(\cos \vartheta)$ featuring in eqn (6.64) are real-valued. Consequently, we have $\psi^* \frac{\partial}{\partial r} \psi = \psi \frac{\partial}{\partial r} \psi^*$ and $\psi^* \frac{\partial}{\partial \vartheta} \psi = \psi \frac{\partial}{\partial \vartheta} \psi^*$ and hence $j_r = j_\vartheta = 0$ for the electrons in the hydrogen atom. For the third component, note that the spherical harmonics depend on φ via the function $\Phi(\varphi) = e^{im\varphi}$ and therefore one finds

$$Y_{lm}^*(\vartheta, \varphi) \frac{\partial}{\partial \varphi} Y_{lm}(\vartheta, \varphi) = - Y_{lm}(\vartheta, \varphi) \frac{\partial}{\partial \varphi} Y_{lm}^*(\vartheta, \varphi) = im \left| Y_{lm}(\vartheta, \varphi) \right|^2, \tag{8.6}$$

where m is the magnetic quantum number (not to be confused with the mass m_e of the electron). The probability current j_φ in the azimuthal direction is then just

$$j_\varphi = \frac{\hbar m}{m_e} \frac{1}{r \sin \vartheta} \left| \psi_{n,l,m}(r, \vartheta, \varphi) \right|^2. \tag{8.7}$$

This expression is remarkable intuitive, the azimuthal current is directly proportional to the magnetic quantum number m. Now consider the current at fixed values of r and ϑ passing through an infinitesimal area element $d\sigma$ that is orthogonal to the current, see Fig. 8.1 (b). The electric current through $d\sigma$ is

$$dI_\varphi = - q_e \, j_\varphi \, d\sigma \tag{8.8}$$

and the area enclosed by this circular current loop is now just

$$A = (r \sin \vartheta)^2 \pi \,. \tag{8.9}$$

According to our previous observation for the classical wire loop, the magnetic moment along the z direction corresponding to the current through $d\sigma$ must therefore be

$$d\mu_z = \frac{A}{c} dI_\varphi = -\frac{q_e\, r^2 \sin^2 \vartheta\, \pi\, j_\varphi}{c} d\sigma = -\frac{q_e \hbar m \pi}{m_e c}\, r\, \sin \vartheta\, |\psi_{n,l,m}(r,\vartheta,\varphi)|^2\, d\sigma \,. \tag{8.10}$$

The total magnetic moment in the z direction is then obtained by summing up all the infinitesimal elements, that is, by integrating and noting that $d\sigma = r\,dr\,d\vartheta$ we arrive at

$$\mu_z = \int d\mu_z = -\frac{q_e \hbar m \pi}{m_e c} \int d\sigma\, r\, \sin \vartheta\, |\psi_{n,l,m}(r,\vartheta,\varphi)|^2 = -\frac{q_e \hbar m}{2 m_e c} = -m\mu_B \,, \tag{8.11}$$

where we have used the fact that $|\psi_{n,l,m}(r,\vartheta,\varphi)|^2$ is independent of φ and normalized. The ratio between $|\mu_z|$ for the electron and the magnetic quantum number is a constant called the *Bohr magneton* which, in Gaussian cgs units, is given by

$$\mu_B = \frac{q_e \hbar}{2 m_e c} \,. \tag{8.12}$$

Since m is an integer between $-l$ and $+l$, the size of the magnetic moment of the hydrogen atom (due to orbital angular momentum) is given in multiples of the Bohr magneton, and magnetic moments of other objects are often specified in units of μ_B. Now, remember from eqn (6.65b) that $\hbar m$ is just the eigenvalue of L_z,

$$L_z \psi_{n,l,m} = \hbar m\, \psi_{n,l,m} \,. \tag{8.13}$$

It hence becomes clear that the operator for the magnetic moment in the z direction is just proportional to the angular-momentum operator L_z,

$$\mu_z = -\frac{q_e}{2 m_e c} L_z = \gamma_e\, L_z \,, \tag{8.14}$$

where γ_e is just the gyromagnetic ratio (compare to eqn (8.3)) for the orbital angular momentum of the electron. On the one hand, we recover what we would have expected from classical electrodynamics. The orbital angular momentum of the electron couples to the magnetic field like its classical counterpart. To emphasize that this could have been different, and may still be different in other cases, one introduces the dimensionless *g-factor* for the electron orbital angular momentum,

$$g_l = \frac{|\mu_z|/\mu_B}{|L_z|/\hbar} = 1 \,, \tag{8.15}$$

i.e., the ratio of the size of the magnetic moment in units of the Bohr magneton to the orbital angular momentum in units of $\hbar$. On the other hand, our computation indicates a behaviour that is in strong contrast to that of classical magnetic dipoles. We find that the magnetic moment along the z direction is quantized. A measurement of μ_z, regardless of the preparation of the atom, will only give results in integer multiples of the Bohr magneton. This observation deserves a more detailed discussion to which we will return when discussing the *Stern–Gerlach* experiment in Section 8.2.1.

Let us further note that obtaining a relation for the z component of the angular momentum is indeed enough to conclude that $\vec{\mu} = \gamma \vec{L}$. Since all components of the angular-momentum operator commute with $\vec{L}^2$, we are free to choose a reference direction with respect to which we define the magnetic quantum number. When no magnetic field is present, this choice is arbitrary and has no physical significance. Recall that the energy levels given by the Bohr formula in eqn (7.83) are independent of the quantum number m. However, when a (homogeneous) magnetic field is present it defines a preferred direction. Orienting our coordinates system such that the z axis aligns with the magnetic field, eqn (8.14) then provides the magnetic moment that determines the shift in potential energy according to eqn (8.2). The effect of such an external magnetic field on the energy levels will be further discussed in Section 10.3.

8.1.3 Magnetic Dipoles in External Magnetic Fields

Irrespective of the changes to the bound-state energies, an atom with a non-zero magnetic dipole moment in a homogeneous magnetic field will experience a torque $\vec{\tau} = \vec{\mu} \times \vec{B}$, the co-called *Larmor torque*, causing the precession of $\vec{\mu}$ around the direction of $\vec{B}$ with the *Larmor frequency*

$$\omega = -\gamma\,|\vec{B}|\,, \tag{8.16}$$

where the gyromagnetic ratio γ is negative for the negatively charged electron that is causing the magnetic moment. To see this, let us do a quick quantum-mechanical computation. Forgetting for a moment about its internal structure, let us treat the hydrogen atom as a single quantum-mechanical particle with mass m_{H} in a constant homogeneous magnetic field $\vec{B}$. Without loss of generality, we may choose coordinates such that the z axis aligns with the magnetic field, such that $\vec{B} = B_z \vec{e}_z$ and $B_z = |\vec{B}|$. The Hamiltonian for this particle is then just

$$H = \frac{\vec{P}^2}{2m_{\mathrm{H}}} - \vec{\mu}\cdot\vec{B} = \frac{\vec{P}^2}{2m_{\mathrm{H}}} - \gamma\,\vec{L}\cdot\vec{B}\,. \tag{8.17}$$

We then want to ask how the magnetic moment $\vec{\mu}$, which is now an operator, is changing with time. Working in the Heisenberg picture (see Section 2.6.2), we have

$$\vec{\mu}(t) = e^{iH(t-t_o)/\hbar}\,\vec{\mu}(t_o)\,e^{-iH(t-t_o)/\hbar}\,. \tag{8.18}$$

In full analogy to eqn (2.85) we then find

$$\frac{d}{dt}\,\vec{\mu}(t) = \frac{i}{\hbar}\,[\,H\,,\vec{\mu}(t)\,] = \frac{i\gamma}{\hbar}\,[\,H\,,\vec{L}\,]\,. \tag{8.19}$$

To evaluate the commutator, note that the kinetic energy $\vec{P}^2/(2m_{\mathrm{H}})$ commutes with all components of the angular-momentum operator (see the comment below eqn (6.21)), but from eqn (8.17) we find

$$[H, L_k] = -\gamma \sum_j [L_j, L_k] B_j = -i\hbar\gamma \sum_{j,l} \epsilon_{jkl} L_l B_j = -i\hbar (\vec{\mu} \times \vec{B})_k, \quad (8.20)$$

where we have inserted the Lie-algebra relation (6.5a) for the angular-momentum operators and used the definition of the cross product. Inserting back into eqn (8.19) we arrive at

$$\frac{d}{dt}\vec{\mu}(t) = \gamma\,\vec{\mu}(t) \times \vec{B}. \quad (8.21)$$

With our previous assumptions that the magnetic field is constant and oriented along the z axis, this differential equation reduces to the coupled differential equations $\dot{\mu}_x = \gamma B_z \mu_y$ and $\dot{\mu}_y = -\gamma B_z \mu_x$, where the dot represents the derivative with respect to time and the third component trivially gives $\dot{\mu}_z = 0$. The z component of the magnetic moment hence remains unchanged, but differentiating the other components once more and inserting the original first-order differential equations we obtain

$$\left(\frac{d^2}{dt^2} + \gamma^2 B_z^2\right)\mu_{x/y} = \left(\frac{d^2}{dt^2} + \omega^2\right)\mu_{x/y} = 0. \quad (8.22)$$

The components of the magnetic moment that are perpendicular to the homogeneous magnetic field satisfy the differential equation of a classical harmonic oscillator with frequency $\omega = -\gamma|\vec{B}|$, where we have chosen the negative sign in front of the square root because $\gamma < 0$ for a negative "spinning" charge. This *Larmor precession* finds use in many practical applications and lies at the heart of important phenomena such as nuclear magnetic resonance.

With the inclusion of (homogeneous) magnetic fields, the picture seems to fit together nicely. The bound states of the hydrogen atom are labelled by three quantum numbers, n, l, and m, of which the principal quantum number n determines the energy levels (for now disregarding small relativistic corrections, the fine structure, that we will take a closer look at in Section 10.2) and hence the spectral lines for absorbed and emitted photons. When placing the atom in an external magnetic field (without loss of generality we may assume it is oriented along the z direction), the angular momentum of the electron, represented by the azimuthal quantum number l causes a magnetic moment, which precesses around the direction of the magnetic field. The z component of the angular momentum, represented by the magnetic quantum number m further determines the splitting of each spectral line (given by n) into $2l+1$ different lines. We shall study this *Zeeman splitting* or *Zeeman effect* for weak magnetic fields in Section 10.3. It hence seems that by the beginning of the 1920s, the behaviour of the hydrogen atom had been fully understood. That this is not so was revealed by the ground-breaking Stern–Gerlach experiment, which we will discuss in the next section.

8.2   Spin

In 1922, at a time the hydrogen atom was thought to be completely understood in terms of Bohr's atom model, two assistants at the University of Frankfurt, *Otto Stern* and *Walther Gerlach*, performed an experiment (1922*a*, 1922*b*, 1922*c*) which showed that electrons carry an intrinsic angular momentum, the spin, which is quantized in two distinct levels. It was one of the most important experiments conducted in the twentieth century, convincing the community of the quantized nature of quantum-mechanical measurement results, and further leading to many interesting experimental and theoretical applications.

8.2.1   The Stern–Gerlach Experiment

In the Stern–Gerlach experiment, the magnetic moments of silver atoms are measured by sending a beam of atoms through an inhomogeneous magnetic field and detecting them on a screen, as illustrated in Fig. 8.2. An oven ejecting the atoms along with some collimating gratings is used to produce this narrow beam of silver atoms, which we assume are initially travelling along the y direction. The particular magnet that is used to create the inhomogeneous field—a *Stern–Gerlach magnet*—consists of two differently shaped parts. The bottom "south-pole" part of the magnet has a horseshoe profile, while the top "north-pole" part has the profile of a downward pointing wedge, The south–north direction defines the spatial orientation of the Stern–Gerlach magnet along which the field strength changes. In the experiment, this direction is aligned with the z axis, which corresponds to a measurement of the component of the magnetic moment along the z axis. As we have established, the magnetic moment of magnetic dipoles couples to the magnetic field. If the field is homogeneous, the dipoles experience a torque that causes the magnetic moment to precess, as discussed in Section 8.1.3. In an inhomogeneous field such as that produced by the Stern–Gerlach magnet, the force

$$\vec{F}_{\mathrm{mag}} = \vec{\nabla}\left(\vec{\mu}\cdot\vec{B}\right) \tag{8.23}$$

also changes the direction of motion of the particles. If $\vec{\mu}\cdot\vec{B} > 0$ (< 0), the force is directed towards increasing (decreasing) field strength, while atoms for which $\vec{\mu}\cdot\vec{B} = 0$ are not deflected at all. In the experiment, this causes the particles to change direction along the z axis when passing through the Stern–Gerlach magnet. The angle of deflection for each atom depends on the z component of its magnetic moment, which hence determines the point of impact on the screen. The silver atoms are prepared such that the overall orbital angular momentum of their electrons combines to zero. The magnetic moments of the silver atoms are therefore determined by any intrinsic angular momentum that may be carried by the electrons. For reasons that will become apparent a little later, it is in fact only the intrinsic angular momentum—*the spin*—of the single valence electron in the outermost shell (opening up an s-orbital with $l = 0$ in the terminology that will be established in Section 8.3.4) that is responsible for the magnetic moment of these silver atoms.

The classical expectation of the outcome of the experiment is the following. Assuming that the electrons indeed carry some spin, the orientations of the magnetic

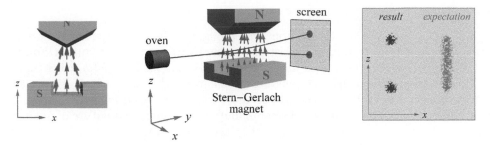

Fig. 8.2 Stern–Gerlach experiment. In the experimental setup to measure the intrinsic magnetic moment of electrons, a beam of silver atoms ejected from an oven along the y direction is sent through a Stern–Gerlach magnet oriented along the z direction. The inhomogeneous magnetic field created by the Stern–Gerlach magnet deflects the atoms along the z direction according to the z component of their magnetic moments, which is caused by the intrinsic magnetic moment (spin) of their single valence electrons. Contrary to the classical expectation, which suggests a continuous distribution, the electron spin and hence the distribution on the screen is quantized in two discrete values.

moments of the individual atoms in an unpolarized beam are distributed randomly. This means that the angle between the atom's magnetic moment and the orientation of the Stern–Gerlach apparatus is expected to vary continuously between 0 and π. One hence expects to detect the atoms along a continuous vertical line, with the topmost and bottommost detections corresponding to the largest and smallest z components of the detected magnetic moments, respectively. The larger the intrinsic angular momentum of the electrons is, the farther apart are the maximum and minimum. Contrary to this classical expectation, the experiment resulted in two distinct, concentrated distributions of detections, see Fig. 8.2. Half of the particles end up in the upper spot ("spin up"), while the other half ends up in the lower spot ("spin down"), with no detections in between.

The Stern–Gerlach experiment hence managed to demonstrate two important features of quantum theory. First, it confirmed that the electrons carry intrinsic angular momentum, the spin. Second, the experiment showed, very convincingly, that spin (along a given direction) is quantized in two distinct levels. Measurements of the spin along any axis can only ever give one of two results, spin up or spin down. This last observation is so crucial for the understanding of the principles of quantum mechanics that we will return to discuss it again in Section 8.2.4.

8.2.2 Spin $\frac{1}{2}$

For now, let us stay with the first insight provided by the Stern–Gerlach experiment. The electrons carry some angular momentum that is not captured by the orbital angular-momentum operator $\vec{L}$ and the quantum numbers[2] l and m_l. This means

---

[2]To distinguish the magnetic quantum number m for the orbital angular momentum from the magnetic quantum number for spin, one adds the subscript "l" to m and writes m_l.

that we must have some additional angular-momentum operator $\vec{S}$, with operator-valued components S_i for $i = x, y, z$, along with a set of two new quantum numbers, the *spin quantum number s*, and the *magnetic spin quantum number m_s*, which label the corresponding eigenstates. In complete analogy to Theorem 6.5 one can write

$$\vec{S}^2 \, | \, s, m_s \, \rangle = \hbar^2 s(s+1) \, | \, s, m_s \, \rangle \, , \tag{8.24a}$$

$$S_z \, | \, s, m_s \, \rangle = \hbar m_s \, | \, s, m_s \, \rangle \, , \tag{8.24b}$$

where m_s can take on one of the $(2s+1)$ values between $-s, -s+1, \ldots, s-1, s$. We further know from the Stern–Gerlach experiment that the magnetic spin quantum number can only take on two different values, i.e., $2s + 1 = 2$, we hence find

$$s = \frac{1}{2} \qquad \text{and} \qquad m_s = \pm \frac{1}{2}. \tag{8.25}$$

Particles with $s = \frac{1}{2}$, such as the silver atoms in the experiment are called spin-$\frac{1}{2}$ particles in reference to their spin quantum number. Note that the value $\frac{1}{2}$ does not refer to the magnitude of the spin vector, which is $|\vec{S}| = \hbar\sqrt{s(s+1)}$ as indicated in eqn (8.24a). The existence of objects with spin $\frac{1}{2}$ is deeply connected to the representation theory of the rotation group (recall the discussion in Section 6.1), which we shall briefly comment on in the next section and in Section 8.3.3. The crucial feature of spin angular momentum, which sets it apart from orbital angular momentum, is that spin angular momentum cannot be removed by a change of reference frame. The orbital angular momentum vanishes in the (momentary) rest-frame of a particle, which follows directly from the definition of $\vec{L}$ in eqn (6.1). The total angular momentum, on the other hand, remains non-zero for particles with non-zero spin, which can be proven within relativistic quantum field theory (see, e.g., Peskin and Schroeder, 1995, p. 60). Spin thus represents the angular momentum of a particle in its momentary rest frame. Every elementary particle carries a characteristic spin value. For (pseudo)scalar particles such as certain mesons (see, e.g., the discussion in Chapter 26 of Part III) the spin is zero, while vector particles like the photon have spin 1. In addition, a large group of particles in the standard model of particle physics—all so-called *leptons*—are spin-$\frac{1}{2}$ particles.

In particular, the (valence) electrons in the Stern–Gerlach experiment are spin-$\frac{1}{2}$ particles. However, spin is not measured directly in this experiment, but the magnetic moment is. Here the Stern–Gerlach setup provides yet another surprise. We have inferred the value $s = \frac{1}{2}$ from the observation of two possible measurement outcomes and the fact that we are describing an angular momentum observable, but the experiment allows to determine the two possible values of the magnetic moment corresponding to $m_s = \pm\frac{1}{2}$. From eqn (8.3) one would expect to find that $\vec{\mu} = \gamma \vec{S}$, but this is not the case. Instead one finds that the spin angular momentum of the electrons contributes to the magnetic moment (roughly) twice as much as the electron's orbital angular momentum. This so-called *anomalous magnetic moment of the electron* is captured by setting the *g*-factor g_s for the angular momentum associated to the electron spin to 2, such that

$$\vec{\mu} = -\frac{q_e}{2m_e c}\left(g_l \vec{L} + g_s \vec{S}\right) = \gamma_e\left(\vec{L} + 2\vec{S}\right). \tag{8.26}$$

The value $g_s = 2$ is a consequence of the quantum-mechanical description of spin, as we shall discover in Section 9. The derivation of this constant, and, indeed, of the existence of spin, does not require the theory of (special) relativity, but can be obtained by coupling electromagnetism to a linearized[3] Schrödinger equation, a procedure due to *Jean-Marc Lévy-Leblond* (1967) that is accessibly explained in Greiner (2000, Sec. 13.1). Nonetheless, when applying relativistic quantum field theory, that is, within quantum electrodynamics, the g-factor for the spin of the electron is not exactly equal to two, but receives small corrections of the order of the fine-structure constant α (see, for instance, Peskin and Schroeder, 1995, Secs. 6.2 and 6.3).

8.2.3   Mathematical Formulation of Spin

Let us now find a suitable mathematical formulation for spin $\frac{1}{2}$. The fact that the measurement of the spin only has two potential outcomes, spin up "↑" and spin down "↓", means that the corresponding Hilbert space is two-dimensional. As the canonical orthonormal basis states we may chose the (simultaneous) eigenstates of $\vec{S}^2$ and S_z, i.e.,

$$\left| s = \tfrac{1}{2}, m_s = +\tfrac{1}{2} \right\rangle = \left| \uparrow \right\rangle = \begin{pmatrix} 1 \\ 0 \end{pmatrix}, \qquad \left| s = \tfrac{1}{2}, m_s = -\tfrac{1}{2} \right\rangle = \left| \downarrow \right\rangle = \begin{pmatrix} 0 \\ 1 \end{pmatrix}, \tag{8.27}$$

such that $\langle \uparrow | \uparrow \rangle = \langle \downarrow | \downarrow \rangle = 1$ and $\langle \uparrow | \downarrow \rangle = 0$. Any other possible quantum state of the spin degree of freedom of a spin-$\frac{1}{2}$ particle can be described as a superposition of these two basis states, that is, a general state $| \psi \rangle$ will be of the form

$$| \psi \rangle = \alpha \, | \uparrow \rangle + \beta \, | \downarrow \rangle, \tag{8.28}$$

where α and β are complex amplitudes satisfying $|\alpha|^2 + |\beta|^2 = 1$ such that the state is normalized. In other words, the Hilbert space for the spin degree of freedom is $\mathcal{H} = \mathbb{C}^2$. With respect to this choice of basis, we can then determine the representations of the spin operators S_i which satisfy eqn (8.24). These are just proportional to the *Pauli matrices* that we have already encountered in Section 3.2.2. Specifically, the components of the vector-valued operator $\vec{S}$ are

$$S_i = \frac{\hbar}{2}\,\sigma_i, \tag{8.29}$$

where the operators σ_i for $i = x, y, z$ are the 2×2 matrices

---

[3] Instead of a single differential equation containing a first-order (partial) derivative with respect to time and second-order (partial) derivatives with respect to the spatial coordinates, one can rewrite the Schrödinger equation as a system of two coupled differential equations containing only linear terms, i.e., first-order (partial) derivatives for all coordinates. The two equations correspond to the two components of the spinor describing spin-$\frac{1}{2}$ particles, see Section 8.2.3.

$$\sigma_x = \begin{pmatrix} 0 & 1 \\ 1 & 0 \end{pmatrix}, \quad \sigma_y = \begin{pmatrix} 0 & -i \\ i & 0 \end{pmatrix}, \quad \sigma_z = \begin{pmatrix} 1 & 0 \\ 0 & -1 \end{pmatrix}. \tag{8.30}$$

One can then easily see that

$$S_z \, | \tfrac{1}{2}, +\tfrac{1}{2} \rangle = \frac{\hbar}{2} \begin{pmatrix} 1 & 0 \\ 0 & -1 \end{pmatrix} \begin{pmatrix} 1 \\ 0 \end{pmatrix} = +\frac{\hbar}{2} \, | \tfrac{1}{2}, +\tfrac{1}{2} \rangle, \tag{8.31}$$

$$S_z \, | \tfrac{1}{2}, -\tfrac{1}{2} \rangle = \frac{\hbar}{2} \begin{pmatrix} 1 & 0 \\ 0 & -1 \end{pmatrix} \begin{pmatrix} 0 \\ 1 \end{pmatrix} = -\frac{\hbar}{2} \, | \tfrac{1}{2}, -\tfrac{1}{2} \rangle. \tag{8.32}$$

Moreover, since all Pauli matrices square to the identity, $\sigma_i^2 = \mathbb{1}$ for all i, we further find

$$\vec{S}^{\,2} = \frac{\hbar^2}{4} \sum_i \sigma_i^2 = \hbar^2 \frac{3}{4} \mathbb{1} = \hbar^2 \tfrac{1}{2} (\tfrac{1}{2} + 1) \mathbb{1}, \tag{8.33}$$

from which we can trivially confirm that the components of the spin operator all commute with $\vec{S}^{\,2}$,

$$[\vec{S}^{\,2}, S_i] = 0. \tag{8.34}$$

This justifies our assumption of the existence of simultaneous eigenstates of $\vec{S}^{\,2}$ and S_z a posteriori. As for the orbital angular momentum (see Definition 6.5), one can define ladder operators that switch between the two basis states, i.e., we define

$$S_\pm := S_x \pm i S_y = \frac{\hbar}{2} (\sigma_x \pm i \sigma_y) = \hbar \sigma_\pm, \tag{8.35}$$

with the corresponding Pauli ladder operators $\sigma_\pm$, such that

$$\sigma_+ \, | \downarrow \rangle = \begin{pmatrix} 0 & 1 \\ 0 & 0 \end{pmatrix} \begin{pmatrix} 0 \\ 1 \end{pmatrix} = \begin{pmatrix} 1 \\ 0 \end{pmatrix} = | \uparrow \rangle, \tag{8.36}$$

$$\sigma_- \, | \uparrow \rangle = \begin{pmatrix} 0 & 0 \\ 1 & 0 \end{pmatrix} \begin{pmatrix} 1 \\ 0 \end{pmatrix} = \begin{pmatrix} 0 \\ 1 \end{pmatrix} = | \downarrow \rangle, \tag{8.37}$$

while the other combinations vanish, $\sigma_+ \, | \uparrow \rangle = \sigma_- \, | \downarrow \rangle = 0$. Inspecting this calculation, we see that we have just recovered a formula analogous to eqn (6.33), that is

$$S_\pm \, | s, m_s \rangle = \hbar \sqrt{s(s+1) - m_s(m_s \pm 1)} \, | s, m_s \pm 1 \rangle. \tag{8.38}$$

And much like in (6.26) we recover the commutation relations

$$[S_z, S_\pm] = \pm \hbar S_\pm, \tag{8.39a}$$

$$[\vec{S}^2, S_\pm] = 0, \tag{8.39b}$$

$$[S_+, S_-] = 2 \hbar S_z. \tag{8.39c}$$

Finally, we can inspect the commutation relations of the spin-operator components with each other. For the Pauli matrices one finds the relation

$$[\sigma_i, \sigma_j] = 2 i \epsilon_{ijk} \sigma_k. \tag{8.40}$$

This implies that the spin operators for different directions satisfy the commutation relations given by

$$[S_i, S_j] = i \hbar \epsilon_{ijk} S_k, \tag{8.41}$$

which we recognize as the defining relation (recall Section 6.1) for the *Lie algebra* of the three-dimensional rotation group from eqn (6.5a). A direct practical consequence of this observation is that different components of the spin do not commute. Similar to the orbital angular-momentum vector (operator) $\vec{L}$, the components of the spin vector (operator) $\vec{S}$ therefore cannot be measured simultaneously. Moreover, learning that the spin component along the z axis is $m_s = +\frac{1}{2}$ or $-\frac{1}{2}$ does not fully determine the direction of the spin vector, see Fig. 8.3 and also compare to Fig. 6.2.

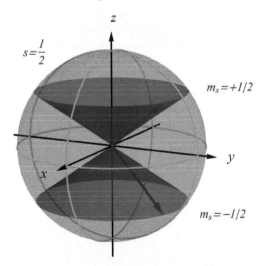

Fig. 8.3 Spin $\frac{1}{2}$. In an eigenstate $|s = \frac{1}{2}, m_s = \pm\frac{1}{2}\rangle$ of $\vec{S}^2$ and S_z, the spin vector has length $|\vec{S}| = \hbar\sqrt{s(s+1)} = \hbar\sqrt{3/4}$ and the component in the z direction is $\hbar/2$ (spin up) or $-\hbar/2$ (spin down). The remaining components S_x and S_y of the spin are unspecified, with uncertainties $\Delta S_x = \Delta S_y = \hbar/2$. The spin vector can hence be thought of as being confined to a cone given by these constraints.

8.2.4 Spin Measurements

To follow up on the previous observation that the spin operators for orthogonal directions do not commute and are therefore not simultaneously measurable, let us consider a series of consecutive spin measurements. Practically, we can think of these as sending a particle beam through a series of differently oriented Stern–Gerlach apparatuses. To begin, consider two measurements along the same direction, e.g., both along the z direction. When the initially unpolarized beam passes the first device, each particle is deflected either upwards or downwards, each with a probability of $1/2$. The particles of the beam exiting the first device upwards or downwards are now prepared in the states $|\uparrow\rangle$ and $|\downarrow\rangle$, respectively. We then let each beam exiting the first device pass through a second Stern–Gerlach magnet, again oriented along the z direction. This time, the direction in which each particle is deflected is no longer random. All of the particles that were in the beam prepared in the state $|\uparrow\rangle$ by the first device exit the second device in the upper beam, they again give the result "spin up" and stay in the state $|\uparrow\rangle$. Similarly, all the particles in the lower beam, prepared in the state $|\downarrow\rangle$, again give the result "spin down" and stay in the state $|\downarrow\rangle$, see Fig. 8.4.

Now consider a situation where the second measurement along the z direction is replaced by a measurement along the x direction. As we have established, the first measurement prepares the particles in eigenstates of S_z, randomly producing the results "spin up" or "spin down" along the z direction. For the second measurement along an orthogonal direction, the results are again completely random. Fifty per cent of the particles will give the result "spin up" along the x axis, while the other fifty per cent result in "spin down" along that axis, as illustrated in Fig. 8.4. From a classical point of view, one could expect that, having learned the component of the spin along one direction, a second measurement might provide knowledge also about the component along an orthogonal direction, thus narrowing down the precise orientation of the spin of the particle. However, this is not the case. There is no such memory effect. Were we to measure the same particle first along the z direction, then along the y direction,

(a) (b)

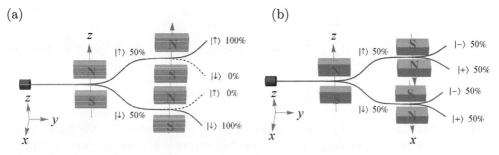

Fig. 8.4 Spin measurements. A Stern–Gerlach apparatus prepares the particles passing through it in one of two eigenstates associated to the direction of the device, for instance, the z direction, as shown here. If the particle already is in an eigenstate of the spin observable for that direction, e.g., $|\uparrow\rangle$ for a measurement of S_z, then a second measurement along the same direction, gives the same answer. However, for a second measurement along an orthogonal direction the result is completely random.

and again along the z direction, the third result would be completely independent of the outcomes of the two before.

Spin measurements are hence a paradigmatic example for a *projective measurement*, which we have already briefly discussed in Section 3.2.1 and which will be discussed in more mathematical detail in Chapter 23. In other words, the measurement of the spin along the z direction is represented by the observable S_z, which can be written in spectral decomposition (recall Theorem 3.1) as

$$S_z = \tfrac{\hbar}{2} \, |\uparrow\rangle\langle\uparrow| - \tfrac{\hbar}{2} \, |\uparrow\rangle\langle\uparrow| \,. \tag{8.42}$$

For a particle prepared in the state $|\,\psi\,\rangle$, the measurement of the spin component along this axis gives the results $\uparrow$ $(m_s = +\tfrac{\hbar}{2})$ or $\downarrow$ $(m_s = -\tfrac{\hbar}{2})$ with probabilities

$$p(\uparrow) = \langle\psi|\uparrow\rangle\langle\uparrow|\psi\rangle = |\langle\psi|\uparrow\rangle|^2 \,, \tag{8.43a}$$

$$p(\downarrow) = \langle\psi|\downarrow\rangle\langle\downarrow|\psi\rangle = |\langle\psi|\downarrow\rangle|^2 \,. \tag{8.43b}$$

After the measurement, the particle is left in the state $|\uparrow\rangle$ or $|\downarrow\rangle$, if the result was $\uparrow$ or $\downarrow$, respectively. Since the spin observables corresponding to measurements along different directions do not commute, they have no common eigenstates. Consequently, consecutive measurements along different directions change the states of the particles and the results of each measurement are a priori undetermined, in particular, when the directions are orthogonal. To further analyse this, we compute the uncertainties of S_x and S_y in an eigenstate of S_z, where

$$\Delta S_i = \sqrt{\langle S_i^2 \rangle - \langle S_i \rangle^2} \,. \tag{8.44}$$

The expectation values of S_x and S_y in the states $|\tfrac{1}{2}, \pm\tfrac{1}{2}\rangle$ are both zero, i.e.,

$$\langle S_x \rangle = \frac{\hbar}{2} \, \langle \tfrac{1}{2}, \pm\tfrac{1}{2} | \, \sigma_x \, | \tfrac{1}{2}, \pm\tfrac{1}{2} \rangle = 0 = \langle S_y \rangle \,. \tag{8.45}$$

On the other hand, since $\sigma_i^2 = \mathbb{1}$, one easily finds $\langle S_i^2 \rangle = \hbar^2/4$ and therefore we immediately obtain

$$\Delta S_x = \Delta S_y = \tfrac{\hbar}{2} \,. \tag{8.46}$$

In other words, for a spin-$\tfrac{1}{2}$ particle prepared in an eigenstate of S_z, the outcome of a measurement of the spin along an orthogonal direction is completely undetermined, as we have also seen in the illustration of Fig. 8.3.

8.2.5   Spinors and the Relation of SO(3) and SU(2)

In addition to what we have learned in the last section, the commutation relation from eqn (8.41) tells us that the spin operators really are angular-momentum operators in every respect. In particular, the operators S_i form a representation of the Lie algebra of the rotation group. Following on from Theorem 6.2 we must therefore conclude that rotations in the Hilbert space $\mathbb{C}^2$ are of the form

$$U(\vec{\varphi}) = e^{-i\vec{\varphi}\cdot\vec{S}/\hbar}. \tag{8.47}$$

It is crucial to note that the transformation behaviour of the two-component vectors $|\psi\rangle \in \mathbb{C}^2$ is qualitatively different than that of vectors in $\mathbb{R}^n$. For instance, recall from Section 6.2.2 that vectors in $\mathbb{R}^3$ are rotated by matrices $R(\vec{\varphi}) = \exp(-i\vec{\varphi}\cdot\vec{T}) \in$ SO(3), i.e., the R are orthogonal 3×3 matrices with determinant one. Vectors in n spatial dimensions hence transform under rotations via elements of the group SO(n), in particular, also when $n = 2$.

The elements of the vector space $\mathbb{C}^2$, on the other hand, are not spatial vectors in the sense of, say, describing the position of a particle in a plane. They are abstract vectors in the sense of elements of a Hilbert space, and spatial rotations of these objects are not represented by orthogonal matrices, but *unitary* 2×2 matrices $e^{-i\vec{\varphi}\cdot\vec{S}/\hbar}$. Since $\mathrm{Tr}(S_i) = 0$, which can be easily verified from eqn (8.30), the matrices $e^{-i\vec{\varphi}\cdot\vec{S}/\hbar}$ also have determinant one (remember the discussion following Theorem 6.2) and are hence elements of the group SU(2). To capture this difference to spatial vectors and scalars (recall that scalars are rotationally invariant, but scalar fields transform according to Theorem 6.3) it is customary to use the term *spinor* for such objects.

Theorem 8.1 (Rotation of spinors)

Vectors $|\psi\rangle \in \mathbb{C}^2$ that transform under rotations as

$$U(\vec{\varphi})\,|\psi\rangle = e^{-i\vec{\varphi}\cdot\vec{S}/\hbar}\,|\psi\rangle,$$

where $U(\vec{\varphi}) \in$ SU(2), are called **spinors**.

In analogy to the group SO(3) defined in eqn (6.6), we define the group SU(2) as

$$\mathrm{SU}(2) = \left\{ U \in \mathrm{GL}(2,\mathbb{C}) \,|\, U^\dagger U = UU^\dagger = \mathbb{1},\ \det(U) = 1 \right\}. \tag{8.48}$$

The corresponding Lie algebra su(2) is then defined as

$$\mathrm{su}(2) = \left\{ A \in \mathrm{GL}(2,\mathbb{C}) \,|\, A = A^\dagger,\ \mathrm{Tr}(A) = 0 \right\}. \tag{8.49}$$

The most general 2×2 matrix that satisfies the constraints of the Lie algebra su(2) is determined by three real parameters, a_1, a_2, and a_3, which can be collected into a three-component vector $\vec{a}$ such that

$$A = \begin{pmatrix} a_3 & a_1 - ia_2 \\ a_1 + ia_2 & -a_3 \end{pmatrix} = \sum_i a_i \sigma_i = \vec{a}\cdot\vec{\sigma}. \tag{8.50}$$

In other words, we find that the Lie algebra su(2) is three-dimensional[4] and the Pauli matrices σ_i from eqn (8.30) form a basis. We can now see the similarity of the Lie algebras su(2) and so(3) from eqn (6.10), that is, they are both three-dimensional vector spaces, consist of traceless, Hermitian matrices, and the respective basis elements T_i and $\sigma_i/2$ satisfy the same commutation relations, see eqns (6.11) and (8.40), respectively. This means that replacing T_i with $\sigma_i/2$ is a Lie-algebra homomorphism (see Definition 6.4).

There is, however, a difference between the two corresponding groups that is important. As remarked right after Theorem 6.1, the group SO(3) is connected, but not simply connected. On the other hand, the group SU(2) is both. Following the discussion in Sexl and Urbantke (2001), one first notes that the conditions of unitarity and unimodularity (i.e., that $\det(U) = 1$) leave three free parameters that can be thought of as spherical coordinates on the three-dimensional surface of a ball in $\mathbb{R}^4$. Other than with the parametrization of SO(3), antipodal points on this surface[5] are not identified. Consequently, any closed curves can be represented as loops on the surface, which can be contracted to individual points. The group SU(2) is hence also simply connected. According to Theorem 6.1 we can therefore conclude that the Lie-group representation of the algebra su(2) is unique, but that of so(3) is not. This last statement can be made more precise by exploiting the intimate connection between the groups SU(2) and SO(3). To do this, it is useful to first find a simpler expression for the SU(2) elements $U(\vec{\varphi})$.

Lemma 8.1 (SU(2) rotations) *Any SU(2) rotation $U(\vec{\varphi})$ around an axis $\vec{\varphi} = \varphi\vec{n}$ and an angle $\varphi = |\vec{\varphi}|$ can be written as*

$$U(\vec{\varphi}) = e^{-i\vec{\varphi}\cdot\vec{S}/\hbar} = \mathbb{1}\cos\left(\tfrac{\varphi}{2}\right) - i\,\vec{n}\cdot\vec{\sigma}\,\sin\left(\tfrac{\varphi}{2}\right).$$

Proof Considering the Euclidean scalar product of the unit vector $\vec{n}$ and the vector of the Pauli operators $\vec{\sigma}$, one notes that

$$(\vec{n}\cdot\vec{\sigma})^2 = (n_x\sigma_x + n_y\sigma_y + n_z\sigma_z)^2 = |\vec{n}|^2\mathbb{1} + n_xn_y(\sigma_x\sigma_y + \sigma_y\sigma_x)$$
$$+ n_xn_z(\sigma_x\sigma_z + \sigma_z\sigma_x) + n_yn_z(\sigma_y\sigma_z + \sigma_z\sigma_y) = \mathbb{1}\,, \tag{8.51}$$

where we have used the fact that $\sigma_i\sigma_j = -\sigma_j\sigma_i$ for $i \neq j$, which follows from the commutation relations of the Pauli matrices in eqn (8.40), and the normalization $|\vec{n}|^2 = 1$. We thus see that all even powers of $\vec{n}\cdot\vec{\sigma}$ give the identity, while all odd powers are just $\vec{n}\cdot\vec{\sigma}$. This means we can split the sum in the series expansion of $e^{-i\vec{\varphi}\cdot\vec{S}/\hbar}$ and get

---

[4]Recall from Definition 6.3 that a Lie algebra is a vector space.

[5]In contrast, SO(3) can be visualized as the inside of a three-dimensional ball with a two-dimensional surface whose antipodal points are identified.

$$e^{-i\vec{\varphi}\cdot\vec{S}/\hbar} = \sum_{k\in\mathbb{N}_0}\frac{\left(-i\frac{\varphi}{2}\right)^k}{k!}(\vec{n}\cdot\vec{\sigma})^k = \sum_{l\in\mathbb{N}_0}\frac{\left(-i\frac{\varphi}{2}\right)^{2l}}{(2l)!}(\vec{n}\cdot\vec{\sigma})^{2l} + \sum_{l\in\mathbb{N}_0}\frac{\left(-i\frac{\varphi}{2}\right)^{2l+1}}{(2l+1)!}(\vec{n}\cdot\vec{\sigma})^{2l+1}$$

$$= \mathbb{1}\sum_{l\in\mathbb{N}_0}(-1)^l\frac{\left(\frac{\varphi}{2}\right)^{2l}}{(2l)!} - i\,\vec{n}\cdot\vec{\sigma}\sum_{l\in\mathbb{N}_0}(-1)^l\frac{\left(\frac{\varphi}{2}\right)^{2l+1}}{(2l+1)!}\,. \tag{8.52}$$

Since we can now simply recognize the series expansions of $\cos\left(\frac{\varphi}{2}\right)$ and $\sin\left(\frac{\varphi}{2}\right)$, respectively, we arrive at the relation stated in Lemma 8.1. $\qquad\square$

With Lemma 8.1 proven, we have all the tools together to formulate the precise relation between SU(2) and SO(3).

Theorem 8.2 (Relation of SU(2) and SO(3))

The group SU(2) *is the universal double cover of* SO(3),

$$\mathrm{SU}(2)/\{\mathbb{1},-\mathbb{1}\} \cong \mathrm{SO}(3),$$

that is, for every $R \in \mathrm{SO}(3)$ *there exist two elements* $\pm U \in \mathrm{SU}(2)$ *such that*

$$U^\dagger\,\vec{\sigma}\,U = R\,\vec{\sigma}\,.$$

Proof First we can note that the components of every vector $\vec{a} \in \mathbb{R}^3$ can be used to define a Hermitian 2×2 matrix $A = \vec{a}\cdot\vec{\sigma}$ as in eqn (8.50). The matrix A is then an observable, whose expectation value we can compute in a reference frame rotated around an axis $\vec{\varphi}$ by an angle $\varphi = |\vec{\varphi}|$, i.e.,

$$\langle\psi\,|\,U^\dagger(\vec{\varphi})AU(\vec{\varphi})\,|\,\psi\rangle = \sum_i a_i\,\langle\psi\,|\,U^\dagger(\vec{\varphi})\sigma_i U(\vec{\varphi})\,|\,\psi\rangle\,. \tag{8.53}$$

To simplify the computation, we can then choose a coordinate system in which the z axis is aligned with the direction of $\vec{\varphi}$, such that $n_x = n_y = 0$ and $n_z = 1$. We can then apply the SU(2) rotation $U(\vec{\varphi})$ in the form of Lemma 8.1 to get

$$U^\dagger A U = \left(\mathbb{1}\cos\left(\tfrac{\varphi}{2}\right) + i\,\sigma_z\sin\left(\tfrac{\varphi}{2}\right)\right)\vec{a}\cdot\vec{\sigma}\left(\mathbb{1}\cos\left(\tfrac{\varphi}{2}\right) - i\,\sigma_z\sin\left(\tfrac{\varphi}{2}\right)\right) \tag{8.54}$$

$$= \begin{pmatrix} e^{i\varphi/2} & 0 \\ 0 & e^{-i\varphi/2} \end{pmatrix}\begin{pmatrix} a_3 & a_1 - ia_2 \\ a_1 + ia_2 & -a_3 \end{pmatrix}\begin{pmatrix} e^{-i\varphi/2} & 0 \\ 0 & e^{i\varphi/2} \end{pmatrix} = \begin{pmatrix} a_3' & a_1' - ia_2' \\ a_1' + ia_2' & -a_3' \end{pmatrix},$$

where we have inserted from eqn (8.50), and the right-hand side of eqn (8.54) can be considered as the quantity $\vec{a}'\cdot\vec{\sigma}$ for the counter-rotated vector given by

$$\vec{a}' = \begin{pmatrix} \cos\varphi & \sin\varphi & 0 \\ -\sin\varphi & \cos\varphi & 0 \\ 0 & 0 & 1 \end{pmatrix}\begin{pmatrix} a_1 \\ a_2 \\ a_3 \end{pmatrix} = R^{-1}(\vec{\varphi})\,\vec{a}\,. \tag{8.55}$$

Likewise, we could rotate the basis of Pauli operators instead of the vector $\vec{a}$, that is

$$\left(R^{-1}(\vec{\varphi})\,\vec{a}\right) \cdot \vec{\sigma} = \vec{a} \cdot \left(R(\vec{\varphi})\vec{\sigma}\right), \tag{8.56}$$

while the left-hand side of eqn (8.54) can be written as

$$U^{\dagger} A U = U^{\dagger} \vec{a} \cdot \vec{\sigma} U = \sum_i a_i U^{\dagger} \sigma_i U. \tag{8.57}$$

Since these arguments must hold independently of the choice of $\vec{a}$ and the chosen coordinate system, we arrive at the result

$$U^{\dagger}(\vec{\varphi})\,\vec{\sigma}\,U(\vec{\varphi}) = R(\vec{\varphi})\,\vec{\sigma}, \tag{8.58}$$

as stated in Theorem 8.2. Looking back at our calculation, we see that for every rotation $R(\vec{\varphi}) = \exp(-i\vec{\varphi} \cdot \vec{T}) \in \mathrm{SO}(3)$, rotating a vector $\vec{a}$, there exist two matrices, $U(\vec{\varphi}) = \exp(-\frac{i}{2}\vec{\varphi} \cdot \vec{\sigma})$ and $-U(\vec{\varphi})$ that rotate the matrix representation A of $\vec{a}$. $\square$

8.3   The Atomic Structure—Revisited

The last section has taught us about the existence of the spin degree of freedom and how to describe it. However, as we have also learned previously, spin is not the only degree of freedom to consider: In particular, we require a combined description of orbital angular momentum and spin.

8.3.1   Total Angular Momentum

As we have discussed in Section 5.5.2, the correct quantum-mechanical description of a system with several degrees of freedom is the tensor product of the respective Hilbert spaces. Here, we have a two-dimensional Hilbert space $\mathcal{H}_{\mathrm{S}} = \mathbb{C}^2$ for the spin, and the infinite-dimensional Hilbert space $\mathcal{H}_{\mathrm{E,L}} = \mathcal{L}_2(\mathbb{R}^3)$ for the energy and orbital angular momentum of the particle. Since $\{|\,s=\frac{1}{2}, m_s\,\rangle\}_{m_s}$ and $\{|\,n,l,m_l\,\rangle\}_{n,l,m_l}$ are bases of the two Hilbert spaces in question, a basis for the joint space is spanned by the states

$$|\,n,l,m_l\,\rangle \otimes |\,s=\tfrac{1}{2}, m_s\,\rangle. \tag{8.59}$$

When representing the states $|\,n,l,m_l\,\rangle$ with respect to the basis $\{|\,r,\vartheta,\varphi\,\rangle\}_{r,\vartheta,\varphi}$ of $\mathcal{H}_{\mathrm{E,L}}$, we just find the bound-state wave functions

$$\psi_{n,l,m_l}(r,\vartheta,\varphi) = \langle\,r,\vartheta,\varphi\,|\,n,l,m_l\,\rangle \tag{8.60}$$

from eqn (7.97), i.e., the wave functions are the components of the vectors $|\,n,l,m_l\,\rangle$ with respect to the basis $\{|\,r,\vartheta,\varphi\,\rangle\}_{r,\vartheta,\varphi}$, as we have explained in Section 3.3.2. Of course, we could also choose a different basis, e.g., $\{|\,\vec{x}\,\rangle\}_{\vec{x}}$ and write the components of any vector $|\,\psi\,\rangle \in \mathcal{H}_{\mathrm{E,L}}$ with respect to this basis as $\psi(\vec{x}) = \langle\,\vec{x}\,|\,\psi\,\rangle$. This quantity, which we have referred to as the wave function corresponding to the state $|\,\psi\,\rangle$, transforms under rotations like a scalar field, see Theorem 6.3. Similarly, we can collect the components of a vector $|\,\phi\,\rangle \in \mathcal{H}_{\mathrm{S}}$ with respect to a chosen basis (such as $\{|\,s = \frac{1}{2}, m_s = \pm\frac{1}{2}\,\rangle\}$) in a two-component column vector

$$|\,\phi\,\rangle = \begin{pmatrix} \langle\, s = \frac{1}{2}, m_s = +\frac{1}{2}\,|\,\phi\,\rangle \\ \langle\, s = \frac{1}{2}, m_s = -\frac{1}{2}\,|\,\phi\,\rangle \end{pmatrix} = \begin{pmatrix} \phi_+ \\ \phi_- \end{pmatrix}, \tag{8.61}$$

which we have learned transforms under rotations like a spinor as stated in Theorem 8.1. Combining these two observations, we can choose a basis of the joint space $\mathcal{H} = \mathcal{H}_{\mathrm{E,L}} \otimes \mathcal{H}_{\mathrm{S}}$, for instance $\{|\,\vec{x}\,\rangle \otimes |\,s = \frac{1}{2}, m_s\,\rangle\}_{\vec{x}, m_s = \pm 1/2}$, and collect components of any vector $|\,\Psi\,\rangle \in \mathcal{H}$ with respect to this basis in the two-component object

$$|\,\Psi(\vec{x})\,\rangle = \begin{pmatrix} \langle\,\vec{x}\,| \otimes \langle\,\frac{1}{2}, +\frac{1}{2}\,|\,\Psi\,\rangle \\ \langle\,\vec{x}\,| \otimes \langle\,\frac{1}{2}, -\frac{1}{2}\,|\,\Psi\,\rangle \end{pmatrix} = \begin{pmatrix} \Psi_+(\vec{x}) \\ \Psi_-(\vec{x}) \end{pmatrix}. \tag{8.62}$$

In this way we obtain a two-component spinor, i.e., a spinor whose components are functions of $\vec{x}$. In the literature, especially in particle physics, it is often customary to write $\Psi(\vec{x})$ rather than $|\,\Psi(\vec{x})\,\rangle$ when it is clear what type of object is being considered (see, e.g., Peskin and Schroeder, 1995, Chap. 3).

When applying spatial rotations to spinor fields, the transformation of the two-component spinor is generated by the spin observable $\vec{S}$, as stated in Theorem 8.1, whereas each component of the spinor transforms like a scalar field, for which rotations are generated by the orbital angular-momentum operator $\vec{L}$. We thus come to the following conclusion:

Theorem 8.3 (Total angular momentum)

*The **total angular momentum** operators $J_i = L_i + S_i$ are the generators of rotations in Hilbert space, that is, three-dimensional rotations are represented on the Hilbert space by unitary operators*

$$U(\vec{\varphi}) = e^{-i\vec{\varphi}\cdot\vec{J}/\hbar}.$$

Since orbital angular momentum and spin correspond to two independent degrees of freedom, the extensions of the operators $\vec{L}$ and $\vec{S}$ to the larger Hilbert space $\mathcal{H}_{\mathrm{E,L}} \otimes \mathcal{H}_{\mathrm{S}}$ are given by $\vec{L} \otimes \mathbb{1}$ and $\mathbb{1} \otimes \vec{S}$, respectively. The total angular-momentum operators are hence just

$$\vec{J} = \vec{L} \otimes \mathbb{1} + \mathbb{1} \otimes \vec{S} \equiv \vec{L} + \vec{S}, \tag{8.63}$$

where we can omit the identity operators and the tensor-product symbols as a simplification of the notation. Of course this construction is not unique to orbital angular momentum and spin. For example, we could consider a total angular-momentum operator for the combination of two scalar particles, A and B, with orbital angular momenta $\vec{L}_A$ and $\vec{L}_B$, respectively, or two particles with spin. In any such case, rotations on the joint Hilbert space are generated by the respective total angular-momentum operators. Moreover, the construction ensures that the angular-momentum commutation relations of $\vec{L}$ and $\vec{S}$ carry over to $\vec{J}$. That is, from eqns (6.5a) and (8.41) it immediately follows that

$$[J_i, J_j] = i\hbar\,\epsilon_{ijk}\,J_k\,.\tag{8.64}$$

These commutation relations also immediately imply that $\vec{J}^2$ commutes with each of its component operators, that is, in complete analogy to eqns (6.24) and (8.34) we find

$$[\vec{J}^2, J_i] = 0\,.\tag{8.65}$$

In addition, eqns (6.24) and (8.34) also imply that the components J_i also commute with both $\vec{L}^2$ and $\vec{S}^2$, since L_i and S_j are operators on different parts of the tensor product and commute trivially. We can thus write

$$[\vec{L}^2, J_i] = [\vec{S}^2, J_i] = [\vec{J}^2, J_i] = 0\,.\tag{8.66}$$

Furthermore, since $\vec{L}^2$ and $\vec{S}^2$ commute with all components of $\vec{J}$, they automatically also satisfy

$$[\vec{L}^2, \vec{J}^2] = [\vec{S}^2, \vec{J}^2] = [\vec{S}^2, \vec{L}^2] = 0\,.\tag{8.67}$$

By virtue of Theorem 5.2 it can hence be concluded that the operators $\vec{L}^2$, $\vec{S}^2$, $\vec{J}^2$ and any single chosen component of $\vec{J}$, which is conventionally chosen to be J_z, commute with each other and therefore have a set of common eigenstates $|\,l, s, j, m_j\,\rangle$ satisfying

$$\vec{L}^2\,|\,l, s, j, m_j\,\rangle = \hbar^2\,l(l+1)\,|\,l, s, j, m_j\,\rangle\,,\tag{8.68a}$$

$$\vec{S}^2\,|\,l, s, j, m_j\,\rangle = \hbar^2\,s(s+1)\,|\,l, s, j, m_j\,\rangle\,,\tag{8.68b}$$

$$\vec{J}^2\,|\,l, s, j, m_j\,\rangle = \hbar^2\,j(j+1)\,|\,l, s, j, m_j\,\rangle\,,\tag{8.68c}$$

$$J_z\,|\,l, s, j, m_j\,\rangle = \hbar m_j\,|\,l, s, j, m_j\,\rangle\,.\tag{8.68d}$$

Here, j is referred to as the *total angular-momentum quantum number* and m_j is the associated magnetic quantum number, which we will elaborate on in the next section. In stating the relations of (8.68) we have omitted the principal quantum number n on purpose to illustrate that the joint Hilbert space of two angular-momentum degrees of freedom (for fixed l and s) can be spanned by the vectors $|\,l, s, j, m_j\,\rangle$ rather than by the vectors $|\,l, m_l; s, m_s\,\rangle$, which are the joint eigenstates of $\vec{L}^2$, L_z, $\vec{S}^2$, and S_z. Mathematically, this statement is independent of the choice of the Hamiltonian of the system. However, in practice there can be a strong physical motivation to prefer one set of basis vectors over the other.

As will be discussed in more detail in Section 10.2, the complete Hamiltonian of the hydrogen atom contains a term proportional to $\vec{S} \cdot \vec{L}$. This so-called *spin-orbit* interaction originates from the fact that the positively charged proton in the nucleus is seen to be moving in the instantaneous rest frame of the electron, where it causes a magnetic field. In its rest frame, the magnetic dipole moment of the electron is proportional to its spin, and couples to this magnetic field. The components of $\vec{L}$, on the other hand, can be interpreted as quantifying the circular motion of the electron[6], leading to the spin-orbit term proportional to $\vec{S} \cdot \vec{L}$, as we will discuss in more detail in Section 10.2.2. In the presence of this term, it is no longer the case that the operators L_z and S_z commute with the Hamiltonian, that is,

$$[\,\vec{S} \cdot \vec{L}\,, L_z\,] = i\hbar \left(L_x S_y - L_y S_x \right), \tag{8.69a}$$

$$[\,\vec{S} \cdot \vec{L}\,, S_z\,] = i\hbar \left(L_y S_x - L_x S_y \right). \tag{8.69b}$$

Consequently, the set of operators $\{H, \vec{L}^2, L_z, \vec{S}^2, S_z\}$ does not admit a common set of eigenvectors. In other words, m_l and m_s are no longer "good" quantum numbers. Conversely, it is easy to see from eqn (8.69) that $J_z = L_z + S_z$ still commutes with the Hamiltonian. In addition, J_z commutes with $\vec{J}^2$, $\vec{L}^2$, and $\vec{S}^2$, see eqn (8.66). It is therefore desirable to switch to the eigenstates $|\, l, s, j, m_j \,\rangle$ of these operators and we will discuss in the next section how to relate these coupled eigenstates to the already known states $|\, l, m_l; s, m_s \,\rangle$.

8.3.2 Addition of Angular Momenta

In the last section we have learned that it can be of (physical) interest to relate the easily constructed *product basis*

$$|\, l, m_l \,\rangle \otimes |\, s, m_s \,\rangle \equiv |\, l, m_l; s, m_s \,\rangle, \tag{8.70}$$

i.e., the joint eigenstates of $\vec{L}^2$ and L_z, as well as of $\vec{S}^2$ and S_z, to the *coupled basis* $|\, l, s, j, m_j \,\rangle$, the common eigenstates of $\vec{L}^2$ and $\vec{S}^2$, as well as of $\vec{J}^2$ and J_z, where $J_i = L_i + S_i$. Since both of these families of states are eigenstates of $\vec{L}^2$ and $\vec{S}^2$, the quantum numbers l and s are fixed, and we must have a relation of the form

$$|\, l, s, j, m_j \,\rangle = \sum_{m_l=-l}^{+l} \sum_{m_s=-s}^{+s} C_{l,m_l;s,m_s}^{(j,m_j)} |\, l, m_l; s, m_s \,\rangle. \tag{8.71}$$

Our task is hence to determine the complex amplitudes $C_{l,m_l;s,m_s}^{(j,m_j)}$, which are generally known as the *Clebsch–Gordan coefficients* for the angular momenta l and s. These coefficients can be obtained for any two (types of) angular momenta, not just for coupling orbital angular momentum and spin. The general method for obtaining these

---

[6]Recall the discussion in Section 8.1.2.

coefficients follows the same procure irrespective of the type of angular momenta. We will therefore now go through the derivation of the Clebsch–Gordan coefficients for the examples $l = 0, s = \frac{1}{2}$ and $l = 1, s = \frac{1}{2}$ for the first two values of the azimuthal quantum number and a particle with spin $\frac{1}{2}$, and we leave any further values as exercises for the interested reader or to be looked up in the (plentifully) available literature, e.g., Cohen-Tannoudji *et al.* (1991, p. 1009) or Hagiwara *et al.* (2002).

To start, we fix some pair of quantum numbers l and s, which corresponds to fixing the eigenvalues of the operators $\vec{L}^2$ and $\vec{S}^2$. There are $2l+1$ possible values for $m_l = -l, -l+1, \ldots, +l$ for each l, and $2s+1$ possible values for $m_s = -s, \ldots, +s$ for each value of s. This means we are operating in a $(2l+1)(2s+1)$-dimensional subspace of the vector space $\mathcal{H}_L \otimes \mathcal{H}_S$. The next observation we can make concerns the minimally and maximally allowed values of the quantum number j, given the choices for l and s. The maximal value $j_{\max}$ is achieved when the two angular momenta are parallel, while the minimal value $j_{\min}$ is obtained when they are antiparallel, i.e.,

$$j_{\min} = |l - s|, \qquad \text{and} \quad j_{\max} = l + s. \tag{8.72}$$

Let us consider a trivial example. When $l = 0$ and $s = \frac{1}{2}$, which describes, e.g., the ground state of the hydrogen atom, the two possible states of the product basis are labelled by $m_s = \pm\frac{1}{2}$. For the coupled basis we note that $j_{\min} = j_{\max} = \frac{1}{2}$, only one value of j is allowed and the corresponding total magnetic quantum number can take on values $m_j = -j, -j+1 = \pm j = \pm\frac{1}{2}$. We must therefore have

$$| \, l = 0, s = \tfrac{1}{2}, j = \tfrac{1}{2}, m_j = \pm\tfrac{1}{2} \, \rangle = | \, l = 0, m_l = 0; s = \tfrac{1}{2}, m_s = \pm\tfrac{1}{2} \, \rangle. \tag{8.73}$$

In this simple example it was intuitively clear how to assign the two states of the two pairs to each other, but in the general case we need a more systematic approach. To start, one notes that the sum on the right-hand side of eqn (8.71) cannot contain contributions from arbitrary values of m_l and m_s. Applying the operator $J_z = L_z + S_z$ on both sides, we see that we can only select terms such that

$$m_j = m_l + m_s. \tag{8.74}$$

In our first example, this was clearly satisfied. Let us now apply this insight to a slightly more complicated example for $l = 1$ and $s = \frac{1}{2}$. In this case, the minimal value of the total angular-momentum quantum number is $j_{\min} = 1 - \frac{1}{2} = \frac{1}{2}$, while the maximal value is $j_{\max} = 1 + \frac{1}{2} = \frac{3}{2}$. Clearly, we now have more options. For instance, the state $| \, l, s, j = \frac{3}{2}, m_j = \frac{1}{2} \, \rangle$ will be some superposition of product-basis states such that $m_l + m_s = \frac{1}{2}$, which is the case for $m_l = 0$ with $m_s = \frac{1}{2}$ and also for $m_l = 1$ with $m_s = -\frac{1}{2}$. Since we currently do not have a guess for the respective Clebsch–Gordan coefficients, we have to start somewhere else. The trick is to start with a state for which we have only a single choice. Take for instance, the "topmost" state of the ladder of eigenstates. When $j = \frac{3}{2}$ and $m_j = \frac{3}{2}$ we are forced to select $m_l = 1$ and $m_s = \frac{1}{2}$, that is, we find

$$| \, l = 1, s = \tfrac{1}{2}, j = \tfrac{3}{2}, m_j = \tfrac{3}{2} \, \rangle = | \, l = 1, m_l = 1; s = \tfrac{1}{2}, m_s = \tfrac{1}{2} \, \rangle. \tag{8.75}$$

To make a statement about the previously mentioned state with $j = \frac{3}{2}, m_j = \frac{1}{2}$ we have to lower the quantum number m_j by one. Fortunately, we know an operator that does exactly this. In complete analogy to the ladder operators for the orbital angular momentum in Definition 6.5 and the ladder operators for spin $\frac{1}{2}$ in eqn (8.35) we define the *total angular momentum ladder operators*

$$J_{\pm} = J_x \pm J_y = L_{\pm} + S_{\pm} \,, \tag{8.76}$$

and as in eqn (6.33) the commutation relations of the angular-momentum observables imply the relations

$$J_{\pm} \, | \, l, s, j, m_j \, \rangle = \hbar \sqrt{j(j+1) - m_j(m_j \pm 1)} \, | \, l, s, j, m_j \pm 1 \, \rangle \,. \tag{8.77}$$

On the other hand, we know that $L_{\pm}$ and $S_{\pm}$ act on the product basis according to

$$L_{\pm} \, | \, l, m_l; s, m_s \, \rangle = \hbar \sqrt{l(l+1) - m_l(m_l \pm 1)} \, | \, l, m_l \pm 1; s, m_s \, \rangle \,, \tag{8.78a}$$

$$S_{\pm} \, | \, l, m_l; s, m_s \, \rangle = \hbar \sqrt{s(s+1) - m_s(m_s \pm 1)} \, | \, l, m_l; s, m_s \pm 1 \, \rangle \,. \tag{8.78b}$$

Returning to our example, we now apply J_- to the state $| \, l = 1, s = \frac{1}{2}, j = \frac{3}{2}, m_j = \frac{3}{2} \, \rangle$ from eqn (8.75) and for ease of notation we will suppress the fixed quantum numbers $l = 1$ and $s = \frac{1}{2}$ in this computation. We then obtain

$$J_- \, | \, j = \tfrac{3}{2}, m_j = \tfrac{3}{2} \, \rangle = \hbar \sqrt{\tfrac{3}{2} \tfrac{5}{2} - \tfrac{3}{2} \tfrac{1}{2}} \, | \, j = \tfrac{3}{2}, m_j = \tfrac{1}{2} \, \rangle = \hbar \sqrt{3} \, | \, j = \tfrac{3}{2}, m_j = \tfrac{1}{2} \, \rangle$$

$$= L_- \, | \, m_l = 1; m_s = \tfrac{1}{2} \, \rangle + S_- \, | \, m_l = 1; m_s = \tfrac{1}{2} \, \rangle$$

$$= \hbar \sqrt{1(1+1)} \, | \, m_l = 0; m_s = \tfrac{1}{2} \, \rangle + \sqrt{\tfrac{3}{4} + \tfrac{1}{4}} \, | \, m_l = 1; m_s = -\tfrac{1}{2} \, \rangle \,. \tag{8.79}$$

By comparing the first and third line of eqn (8.79) and dividing by $\hbar\sqrt{3}$ we then arrive at

$$| \, j = \tfrac{3}{2}, m_j = \tfrac{1}{2} \, \rangle = \sqrt{\tfrac{2}{3}} \, | \, m_l = 0; m_s = \tfrac{1}{2} \, \rangle + \sqrt{\tfrac{1}{3}} \, | \, m_l = 1; m_s = -\tfrac{1}{2} \, \rangle \,. \tag{8.80}$$

To obtain the next state, we apply the lowering operator J_- to eqn (8.80), that is,

$$J_- \, | \, j = \tfrac{3}{2}, m_j = \tfrac{1}{2} \, \rangle = \hbar \sqrt{\tfrac{15}{4} + \tfrac{1}{4}} \, | \, j = \tfrac{3}{2}, m_j = -\tfrac{1}{2} \, \rangle = 2\hbar \, | \, j = \tfrac{3}{2}, m_j = -\tfrac{1}{2} \, \rangle \tag{8.81}$$

$$= \hbar \sqrt{\tfrac{4}{3}} \, | \, m_l = -1; m_s = \tfrac{1}{2} \, \rangle + \hbar \sqrt{\tfrac{8}{3}} \, | \, m_l = 0; m_s = -\tfrac{1}{2} \, \rangle \,,$$

from which we can conclude that the normalized state $| \, j = \frac{3}{2}, m_j = -\frac{1}{2} \, \rangle$ is given by

$$| \, j = \tfrac{3}{2}, m_j = -\tfrac{1}{2} \, \rangle = \sqrt{\tfrac{1}{3}} \, | \, m_l = -1; m_s = \tfrac{1}{2} \, \rangle + \sqrt{\tfrac{2}{3}} \, | \, m_l = 0; m_s = -\tfrac{1}{2} \, \rangle \,. \tag{8.82}$$

A third application of J_-, with another round of normalization finally gives the last state with $j = \frac{3}{2}$, i.e.,

$$| \, j = \tfrac{3}{2}, m_j = -\tfrac{3}{2} \, \rangle = | \, m_l = -1; m_s = -\tfrac{1}{2} \, \rangle \,. \tag{8.83}$$

With eqns (8.75), (8.80), (8.82), (8.83) we have expressed four out of the six states of the coupled basis. To find the Clebsch–Gordan coefficients for the missing states

$|\, j = \tfrac{1}{2}, m_j = \pm\tfrac{1}{2}\,\rangle$, we start again with the "uppermost" state $|\, j = \tfrac{1}{2}, m_j = j\,\rangle$ and write it as a general linear combination of the allowed states (that is, for which $m_l + m_s = m_j$) of the product basis. In other words, we have

$$|\, j = \tfrac{1}{2}, m_j = \tfrac{1}{2}\,\rangle = \alpha\,|\, m_l = 1; m_s = -\tfrac{1}{2}\,\rangle + \beta\,|\, m_l = 0; m_s = \tfrac{1}{2}\,\rangle\,. \qquad (8.84)$$

This state must be orthogonal to the already known state $|\, j = \tfrac{3}{2}, m_j = \tfrac{1}{2}\,\rangle$, which must of course also be composed of the same product-basis states, but with different probability amplitudes. This allows us to compute

$$\langle\, j = \tfrac{3}{2}, m_j = \tfrac{1}{2}\,|\, j = \tfrac{1}{2}, m_j = \tfrac{1}{2}\,\rangle = \sqrt{\tfrac{1}{3}}\,\alpha + \sqrt{\tfrac{2}{3}}\,\beta = 0\,. \qquad (8.85)$$

Since we do not have to care about the overall phase, any choice of α and β for which this is satisfied will do. For instance, we can set $\alpha = \sqrt{\tfrac{2}{3}}$ and $\beta = -\sqrt{\tfrac{1}{3}}$, and hence we find the state decomposition

$$|\, j = \tfrac{1}{2}, m_j = \tfrac{1}{2}\,\rangle = \sqrt{\tfrac{2}{3}}\,|\, m_l = 1; m_s = -\tfrac{1}{2}\,\rangle - \sqrt{\tfrac{1}{3}}\,|\, m_l = 0; m_s = \tfrac{1}{2}\,\rangle\,. \qquad (8.86)$$

Now, as before, we can apply the lowering operator J_- and normalize the result to determine the last missing state of the coupled basis, for which we find

$$|\, j = \tfrac{1}{2}, m_j = -\tfrac{1}{2}\,\rangle = \sqrt{\tfrac{1}{3}}\,|\, m_l = 0; m_s = -\tfrac{1}{2}\,\rangle - \sqrt{\tfrac{2}{3}}\,|\, m_l = -1; m_s = \tfrac{1}{2}\,\rangle\,. \qquad (8.87)$$

This completes our example for $l = 1$ and $s = \tfrac{1}{2}$, but we have gained some valuable insights that allow us to determine the Clebsch–Gordan coefficients for two arbitrary angular momenta. For simplicity, let us still denote the corresponding quantum numbers as l and s, with magnetic quantum numbers m_l and m_s, respectively. In general, one can then start from the topmost state $|\, j = j_{\text{max}}, m_j = j_{\text{max}}\,\rangle$ of the coupled basis, where $j_{\text{max}} = l + s$ for which

$$|\, j = j_{\text{max}}, m_j = j_{\text{max}}\,\rangle = |\, l, m_l = l; s, m_s = s\,\rangle\,, \qquad (8.88)$$

before applying the lowering operator J_- (and normalizing) $(2j_{\text{max}} + 1)$ times, until one reaches the state $|\, j = j_{\text{max}}, m_j = -j_{\text{max}}\,\rangle$. To lower the value of j from $j = j_{\text{max}}$ to $j = j_{\text{max}} - 1$, we then write $|\, j = j_{\text{max}} - 1, m_j = j_{\text{max}} - 1\,\rangle$ as the linear combination

$$|\, j = j_{\text{max}} - 1, m_j = j_{\text{max}} - 1\,\rangle = \alpha\,|\, m_l = l - 1; m_s = s\,\rangle + \beta\,|\, m_l = l; m_s = s - 1\,\rangle\,, \qquad (8.89)$$

and use the fact that the state $|\, j = j_{\text{max}} - 1, m_j = j_{\text{max}} - 1\,\rangle$ is orthogonal to the (already known) state $|\, j = j_{\text{max}}, m_j = j_{\text{max}} - 1\,\rangle$ to determine the coefficients α and β. Then one applies the lowering operator again and continues the procedure in the same fashion as before.

For our continuing discussion of atomic energy-level structures, we can be content that there is a reliable procedure to determine all of the joint eigenstates of H, $\vec{L}^{\,2}$, $\vec{S}^{\,2}$, $\vec{J}^{\,2}$, and J_z. With or without the spin-orbit interaction, the corresponding quantum numbers n, l, s, j, and m_j are "good" quantum numbers, where of course $s = \frac{1}{2}$ is fixed. In any case, the existence of the spin of the electron has doubled our state space. Instead of the $2l + 1$ states previously encountered for some fixed n and l we now have twice as many possible states that can be occupied by a single electron, e.g., in an electrically neutral hydrogen atom. But can we also learn something about the electronic structure of other, heavier nuclei? As we shall discuss next, this is indeed possible, but requires some considerations about multi-electron wave functions.

8.3.3 Indistinguishable Particles and Pauli Principle

When only a single electron is present, the model for the hydrogen atom can be adapted to describe atoms with higher mass and electric charge by simply changing the parameters for the (reduced) mass and charge used in the Schrödinger equation. Additional nucleons (protons and neutrons) may be considered by multiplying q_e^2 in the Coulomb potential of eqn (7.37) by the effective nuclear charge Z, i.e., the number of protons in the nucleus. The proton mass m_p also has to be replaced by the product of A and m_p in the expression for the reduced mass of eqn (7.87), where A is the atomic mass number, i.e., the total number of protons and neutrons of which the nucleus is composed. Of course we have not provided a model for the interaction of the nucleons with each other, and we must hence assume that the nucleus is stable for a given number of protons and neutrons as we proceed.

However, it is highly desirable to describe, in particular, the electrically neutral configurations of the atoms in question. In principle, this would require solving the Schrödinger equation for several electrons. This task, as it turns out, is practically not feasible (three-body problem). Nonetheless, remarkably precise predictions can be made for the (ground-state) configurations of the electrons in heavier atoms by assuming that electrons are not interacting and compensating for their Coulomb repulsion with a modified (lower) nuclear charge. The starting point for further investigation is hence the assumption that N electrons bound to an atomic nucleus may be in any quantum state

$$|\psi\rangle_{Ne^-} \in \mathcal{H}_{Ne^-} = \mathcal{H}_{1e^-}^{\otimes N} \tag{8.90}$$

in the Hilbert space corresponding to N copies of the single-electron Hilbert space $\mathcal{H}_{1e^-}$. The basis vectors of each of the Hilbert spaces $\mathcal{H}_{1e^-}$ can be labelled by the previously established quantum numbers n, l, m_l, and s, m_s, i.e., $\{|\,n, l, m_l, m_s\,\rangle\}$ (since $s = \frac{1}{2}$ is fixed). For N particles, one may thus select a basis of $\mathcal{H}_{Ne^-}$ with basis vectors denoted as

$$|\, n^{(1)}, l^{(1)}, m_l^{(1)}, m_s^{(1)}; n^{(2)}, l^{(2)}, m_l^{(2)}, m_s^{(2)}; \ldots; n^{(N)}, l^{(N)}, m_l^{(N)}, m_s^{(N)} \,\rangle\,. \tag{8.91}$$

Alternatively one may chose a basis for the coupled angular momenta, replacing the quantum numbers $m_l^{(i)}$ and $m_s^{(i)}$ (for all i) by j and m_j, as explained in Section 8.3.2.

What one is left to explain is the way in which a number of electrons populate the available state space, in particular, the energy levels given by Bohr's formula (7.83). For instance, can one expect that the electrons of an atom in its ground state can all be found in the ground state with $n = 0$? The answer is no. And attempting to do so,

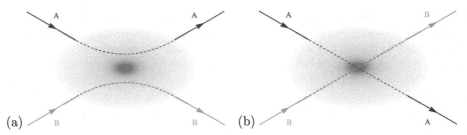

Fig. 8.5 Distinguishability. In a classical scattering experiment, e.g., in a game of billiards, one can assign labels to otherwise indistinguishable objects and track those labels throughout the experiment. For quantum-mechanical objects, this is not possible. If the particles initially labelled by A and B are identical, then one cannot say whether the particles exchanged positions or not.

we encounter an important restriction, *Pauli's exclusion principle*, which arises from the connection between spin and particle statistics. To understand this restriction, it is paramount to first think about the notion of distinguishability.

In classical physics, any two particles can in principle be distinguished by their trajectories. Pictorially speaking, we can put a label on a particle by simply tracking its motion throughout any interaction. For quantum-mechanical particles, we do not have this option. First and foremost, we cannot rely on the notion of a well-defined trajectory. As long as one considers only a single-particle theory, this question clearly does not arise, but when two (or more) particles are present and we want to describe their interaction we have to take this into account. Let us study this question using a simple scattering experiment as an example. Suppose two particles of the same type, e.g., electrons, approach each other before being scattered as illustrated in Fig. 8.5. Initially, we assign the labels "A" and "B" to the particles in the upper and lower path, respectively. Since we cannot continuously monitor the two quantum-mechanical particles throughout their interaction, we don't know if the particles crossed paths or not. What can be observed is that two particles interact and that two particles emerge. But if the particles are truly indistinguishable it does not matter which particle is in which path. All physical quantities of interest are independent of these labels.

Let us try to capture these ideas in a mathematically more precise way. The joint quantum state $|\psi\rangle_{AB}$ of two particles with Hilbert spaces $\mathcal{H}_A$ and $\mathcal{H}_B$ is given by

$$|\psi\rangle_{AB} = \sum_{m,n} c_{mn} |m\rangle_A \otimes |n\rangle_B , \qquad (8.92)$$

where $\{|m\rangle_A\}_m \{|n\rangle_B\}_n$ are bases of $\mathcal{H}_A$ and $\mathcal{H}_B$, respectively. If the two particles are of the same type, their Hilbert spaces are isomorphic, $\mathcal{H}_A \cong \mathcal{H}_B$, and we can exchange the states of the two subspaces. That is, we can define an exchange operator.

Definition 8.1   *The **parity operator** $\hat{P}_{AB}$ for the exchange of two particles with Hilbert spaces $\mathcal{H}_A \cong \mathcal{H}_B$ acts on the joint space $\mathcal{H}_A \otimes \mathcal{H}_B$ as*

$$\hat{P}_{AB} |m\rangle_A \otimes |n\rangle_B = |n\rangle_A \otimes |m\rangle_B .$$

For instance, when two particles without spin are being considered, the position representation of their joint state is the wave function $\psi(\vec{x}, \vec{y})$, where $|\psi(\vec{x}, \vec{y})|^2 d^3x\, d^3y$ is the probability to find the particles A and B within infinitesimal volume elements around the positions $\vec{x}$ and $\vec{y}$, respectively. The action of the parity operator is then just

$$\hat{P}_{AB} \, |\psi\rangle = \hat{P}_{AB} \iint d^3x\, d^3y\; \psi(\vec{x}, \vec{y}) \, |\vec{x}\rangle_A |\vec{y}\rangle_B = \iint d^3x\, d^3y\; \psi(\vec{y}, \vec{x}) \, |\vec{x}\rangle_A |\vec{y}\rangle_B \,, \tag{8.93}$$

where we have shifted the exchange of $\vec{x}$ and $\vec{y}$ from the ket vectors to the argument of the wave function and we have dropped the tensor-product symbol "$\otimes$" for ease of notation. When the particles also have spin then the action of the parity operator extends to the total Hilbert space consisting of the spatial degrees of freedom and the spin, such that the state

$$|\psi\rangle = \sum_{m_A=-s_A}^{s_A} \sum_{m_B=-s_B}^{s_B} \iint d^3x\, d^3y\; \psi_{m_A, m_B}(\vec{x}, \vec{y}) \, |\vec{x}, m_A\rangle_A |\vec{y}, m_B\rangle_B \,, \tag{8.94}$$

is transformed by the parity operator as

$$\hat{P}_{AB} |\psi\rangle = \sum_{m_A=-s_A}^{s_A} \sum_{m_B=-s_B}^{s_B} \iint d^3x\, d^3y\; \psi_{m_B, m_A}(\vec{y}, \vec{x}) \, |\vec{x}, m_A\rangle_A |\vec{y}, m_B\rangle_B \,. \tag{8.95}$$

In other words, the total parity operator can be understood as a tensor product of parity operators for each degree of freedom, e.g., $\hat{P}_{AB} = \hat{P}_{AB}^{\text{pos.}} \otimes \hat{P}_{AB}^{\text{spin}}$. When the two particles in question have non-zero spin quantum numbers s_A and s_B it is useful to switch to the coupled basis as we have discussed in the previous Section 8.3.2. For instance, for two spin-$\frac{1}{2}$ particles, i.e., for $s_A = s_B = \frac{1}{2}$, we denote the azimuthal quantum number for the total angular momentum as s and the corresponding magnetic number as m and write

$$|\psi\rangle = \sum_{s=0,1} \sum_{m=-s}^{+s} \iint d^3x\, d^3y\; \psi_{s,m}(\vec{y}, \vec{x}) \, |\vec{x}, \vec{y}, s, m\rangle \,. \tag{8.96}$$

We will leave it as an exercise to the interested reader to use the method explained in the previous section to verify that the *triplet* of coupled-basis states for $s = 1$ are then related to the product-basis states according to

$$|s=1, m=-1\rangle = |m_A = -\tfrac{1}{2}, m_B = -\tfrac{1}{2}\rangle = |\uparrow\uparrow\rangle \,, \tag{8.97a}$$

$$|s=1, m=0\rangle = \tfrac{1}{\sqrt{2}}\left(|m_A = \tfrac{1}{2}, m_B = -\tfrac{1}{2}\rangle + |m_A = -\tfrac{1}{2}, m_B = \tfrac{1}{2}\rangle \right)$$

$$= \tfrac{1}{\sqrt{2}}\left(|\uparrow\downarrow\rangle + |\downarrow\uparrow\rangle \right), \tag{8.97b}$$

$$|s=1, m=+1\rangle = |m_A = \tfrac{1}{2}, m_B = \tfrac{1}{2}\rangle = |\downarrow\downarrow\rangle \,, \tag{8.97c}$$

while the *singlet* state for $s = 0$ is given by

$$| \, s = 0, m = 0 \, \rangle = \tfrac{1}{\sqrt{2}} \Big(| \, m_A = \tfrac{1}{2}, m_B = -\tfrac{1}{2} \, \rangle - | \, m_A = -\tfrac{1}{2}, m_B = \tfrac{1}{2} \, \rangle \Big)$$

$$= \tfrac{1}{\sqrt{2}} \Big(| \uparrow \downarrow \rangle - | \downarrow \uparrow \rangle \Big) . \tag{8.98}$$

The triplet states of eqn (8.97) are all symmetric under the exchange of the two particles, while the singlet state in eqn (8.98) is antisymmetric, and we can therefore compactly write

$$\hat{P}_{AB}^{\text{spin}} \, | \, s, m \, \rangle = (-1)^{s+1} \, | \, s, m \, \rangle . \tag{8.99}$$

With this, we can further rewrite eqn (8.95) as

$$\hat{P}_{AB} \, | \, \psi \, \rangle = \sum_{s=0,1} \sum_{m=-s}^{+s} \iint d^3x \, d^3y \, (-1)^{s+1} \psi_{s,m}(\vec{y}, \vec{x}) \, | \, \vec{x}, \vec{y}, s, m \, \rangle . \tag{8.100}$$

For the singlet and triplet states, the joint state of two indistinguishable particles must give the same physical predictions when the particles are exchanged. In other words, the parity operation must be a *symmetry*. This is so crucial that we will promote this to our definition for identical particles.

Definition 8.2   *Two particles are **identical** (or indistinguishable) if all physical predictions for all joint states $| \, \psi \, \rangle$ of the two particles are left invariant by the exchange of the two particles.*

This means that the exchange of any two indistinguishable particles can, at most, result in an overall phase, i.e., for some real γ one must have

$$\hat{P}_{AB} \, | \, \psi \, \rangle = e^{i\gamma} \, | \, \psi \, \rangle . \tag{8.101}$$

However, since a second application of the parity operator returns the original state, formally we may write $\hat{P}_{AB}^2 = \mathbb{1}$, the phase factor $e^{i\gamma}$ can only be $+1$ or -1. According to Definition 8.2, the joint states of any two identical particles must satisfy

$$\hat{P}_{AB} \, | \, \psi \, \rangle = \pm \, | \, \psi \, \rangle . \tag{8.102}$$

We can therefore conclude that only two types of indistinguishable particles can occur in nature.

Definition 8.3   *Particles whose multi-particle wave functions are **symmetric** with respect to the exchange of any two of the constituent particles are called **bosons**. Particles whose multi-particle wave functions are **antisymmetric** under the exchange of any two of the constituent particles are called **fermions**.*

Systems of non-interacting bosons and fermions in thermal equilibrium with a heat bath satisfy *Bose–Einstein statistics* and *Fermi–Dirac statistics*, respectively, which significantly deviate from classical Maxwell–Boltzmann statistics at low temperatures. Examples for bosons include the fundamental force exchange particles such as photons, gauge bosons ($W^\pm$ and Z^0), and gluons, but also the infamous Higgs boson, as well as composite particles such as α particles, π mesons ("pions"), and kaons[7]. Examples for fundamental fermions are electrons and positrons ($e^\pm$), and their second and third generation counterparts, the muons and tauons, as well as their respective neutrinos, but also quarks. Fermions that are composed of other elementary particles are, for instance, the nucleons (protons and neutrons). A most interesting distinction between bosons and fermions is captured by the *spin-statistics theorem*, which arises from quantum field theory. Since we lack the tools to prove this theorem here, we shall simply state it here without proof, but we refer the interested reader to Pauli (1940), Schwinger (1951), and Peskin and Schroeder (1995, p. 52).

Theorem 8.4 (Spin-statistics theorem)

*All **bosons** have **integer spin**. All **fermions** have **half-integer spin**.*

The wave function $\psi(\vec{x}, \vec{y}) = \langle\, \vec{x}, \vec{y}|\, \psi\,\rangle$ for two spin-0 particles hence satisfies $\psi(\vec{x}, \vec{y}) = \psi(\vec{y}, \vec{x})$. At this point it is important to note that the operator $\hat{P}_{AB}$ for the exchange of two particles, although being called parity operator, is different from the spatial parity operator $\hat{P}$, which corresponds to a spatial reflection. That is, the action of $\hat{P}$ is to flip the sign of (one or all of) the spatial coordinates of each single particle, i.e.,

$$\hat{P} \int d^3x\; \psi(\vec{x})\,|\,\vec{x}\,\rangle = \int d^3x\; \psi(\vec{x})\,|-\vec{x}\,\rangle = \int d^3x\; \psi(-\vec{x})\,|\,\vec{x}\,\rangle. \qquad (8.103)$$

Or, in short, $\hat{P}\psi(\vec{x}) = \vec{\psi}(-\vec{x})$. Now, a particle can both be a boson, e.g., with spin 0, and at the same time be antisymmetric under a spatial parity transformation, such that $\hat{P}\psi(\vec{x}) = -\psi(\vec{x})$. For example, the π mesons are such *pseudoscalar* particles.

With this, we return to the parity operator for particle exchange, and consider the spatial wave function of two spin-$\frac{1}{2}$ particles. According to Theorem 8.3.3, these are fermions, and the total spatial wave function

$$\langle\, \vec{x}, \vec{y}|\, \psi\,\rangle = \psi_{00}(\vec{x}, \vec{y})\,|\,s = 0, m = 0\,\rangle + \sum_{m=-1}^{+1} \psi_{1,m}(\vec{x}, \vec{y})\,|\,s = 1, m\,\rangle, \qquad (8.104)$$

must hence be antisymmetric with respect to the exchange of the two particles. Therefore, $\psi_{00}(\vec{x}, \vec{y})$ must be symmetric, $\psi_{00}(\vec{x}, \vec{y}) = \psi_{00}(\vec{y}, \vec{x})$ since the singlet of eqn (8.98) is antisymmetric, while $\psi_{1,m}(\vec{x}, \vec{y}) = -\psi_{1,m}(\vec{y}, \vec{x})$ must be antisymmetric, since all the triplet states of eqn (8.97) are symmetric. This means that the spatial and spin wave

---

[7]We will discuss kaons in more detail in Chapter 26 of Part III.

functions are correlated. In the product basis the wave function of two fermions must be of the form

$$\psi_{m_1,m_2}(\vec{x},\vec{y}) = -\psi_{m_2,m_1}(\vec{y},\vec{x}). \tag{8.105}$$

This implies that the wave function for two fermions with the same quantum numbers satisfy $\psi_{m,m}(\vec{x},\vec{x}) = 0$. This observation is typically phrased as the so-called *Pauli exclusion principle* or simply *Pauli principle*.

Theorem 8.5 (Pauli principle)

No two fermions can simultaneously occupy the same single-particle quantum state. In particular, two fermions in the same spin state cannot be in the same location.

Returning to the question of the atomic structure, we must conclude from Theorem 8.3.3 that each electronic state $| n,l,m_l,m_s \rangle$ of the single-particle Hilbert space $\mathcal{H}_{1e^-}$ can be occupied by at most a single electron. In other words, for states in an N-electron Hilbert space $\mathcal{H}_{1e^-}^{\otimes N}$, such as in eqn (8.91), the label (n,l,m_l,m_s) cannot be repeated in the same ket-vector. That is, $(n^{(i)}, l^{(i)}, m_l^{(i)}, m_s^{(i)}) \neq (n^{(j)}, l^{(j)}, m_l^{(j)}, m_s^{(j)})$ for $i \neq j$. This means that, even in the atomic ground state, some of the electrons of heavier neutral atoms will be found in excited states with principal quantum numbers $n > 1$.

8.3.4   Electronic Orbitals

As we have learned in Section 7.5, the atomic energy levels are determined by the principal quantum number n. For each value of n, there are $n-1$ different values of the azimuthal quantum number l, each of which in turn allows for $2l+1$ different values of the magnetic number m_l. Together, these three quantum numbers (n,l,m_l), more precisely, the wave function described by these numbers, are referred to as an electronic *orbital*. Furthermore, the previous section has taught us that for each choice of these three quantum numbers, two different spin states with magnetic spin quantum numbers $m_s = \pm\frac{1}{2}$ exist. Each configuration (n,l,m_l,m_s) is called a *spin orbital*. According to Pauli's exclusion principle (Theorem 8.3.3) each orbital can be occupied by two electrons, which implies that each fixed energy level (i.e., a given value of n) hence allows for $2\sum_{l=0,\dots,n-1}(2l+1) = 2n^2$ electrons to be bound to the atomic nucleus.

For a spectroscopic description of atomic configurations, not all of these quantum numbers are needed (provided that differences from fine-structure corrections that we will discuss in Section 10.2 are not resolved). If no external magnetic field determining a preferred direction is present, the magnetic quantum numbers m_l and m_s (and m_j) are not strictly speaking "good" quantum numbers. That is, without a magnetic field coupling to the magnetic moments (see Section 8.1) of the electrons there is no specific reason to expect that the electrons are more likely to be found in an eigenstate of L_z (or S_z, J_z), rather than of, e.g., L_y. In a solid, on the other hand, one may face the situation that there are preferred axes determined by the lattice structure of the solid, but no preferred directions along these axes. In this case there may not be a reason to distinguish, e.g., $m = +1$ from $m = -1$, whereas $|m|$ may still be a useful quantum number. Consequently, a somewhat superficial but practical spectroscopic description of the electrons in a particular atom may consist only of a list of pairs (n, l) of principal and azimuthal quantum numbers, with the corresponding occupation numbers as superscripts. Each such combination (n, l), referred to as a *shell*, can contain up to $2(2l + 1)$ electrons. For historical reasons (see, e.g., Griffiths, 1995, p. 190, or Levine, 2000, p. 144), an alphabetic code is used for the quantum number l instead of numerical values, labelling the corresponding orbitals and shells as

$$
\begin{array}{lll}
\text{s} & \text{``}sharp\text{''} & l = 0 \\[4pt]
\text{p} & \text{``}principal\text{''} & l = 1 \\[4pt]
\text{d} & \text{``}difffuse\text{''} & l = 2 \\[4pt]
\text{f} & \text{``}fundamental\text{''} & l = 3
\end{array}
$$

and alphabetically from there on, e.g., g, h, i, k for $l = 4, 5, 6, 7$, and so on (``j'' is not used to avoid confusion with ``i''). For example, the electronic occupation of an atom with ten electrons in the lowest possible configuration can be written as

$$
1\text{s}^2\, 2\text{s}^2\, 2\text{p}^6 \,, \tag{8.106}
$$

i.e., two electrons with $(n, l) = (0, 0)$, two with $(n, l) = (1, 0)$, and six (corresponding to $m_l = 0, \pm 1$) with $(n, l) = (2, 1)$. When illustrating these orbitals, one often chooses wave functions that are linear combinations of the solutions with fixed n, l, and $|m|$. Inspecting the wave functions $\psi_{n,l,m}$ in eqn (7.97), one notes that the radial part $R_{nl}(r)$ shown in eqn (7.94) is real, while the spherical harmonics Y_{lm} from eqn (6.64) are proportional to $e^{i\,m\,\varphi}$. With our phase convention, the linear combinations

$$
\psi^{+}_{n,l,|m|} = \frac{1}{\sqrt{2}}\left(\psi_{n,l,m} + (-1)^m \psi_{n,l,-m}\right), \tag{8.107a}
$$

$$
\psi^{-}_{n,l,|m|} = -\frac{i}{\sqrt{2}}\left(\psi_{n,l,m} - (-1)^m \psi_{n,l,-m}\right) \tag{8.107b}
$$

are hence real valued functions, which are still orthonormal in the sense of eqn (6.66). The corresponding probability densities $|\psi^{\pm}_{n,l,|m|}|^2$ up to and including $n = 3$ are illustrated in Fig. 8.6. To keep the terminology clear in the following, let us collect the nomenclature for the single-electron configurations. That is, a single (non-relativistic) electron may occupy

$$
\begin{array}{lll}
\text{``}energy\ levels\text{''} & \text{labelled by} & n, \\[6pt]
\text{``}shells\text{''} & \text{labelled by} & (n, l), \\[6pt]
\text{``}orbitals\text{''} & \text{labelled by} & (n, l, m_l), \\[6pt]
\text{``}spin\ orbitals\text{''} & \text{labelled by} & (n, l, m_l, m_s).
\end{array}
$$

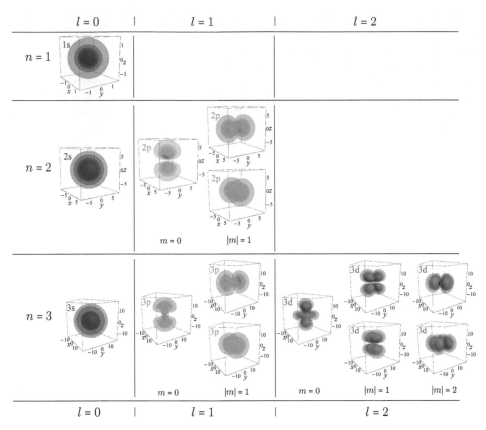

Fig. 8.6 Electronic orbitals. The probability densities $|\psi^{\pm}_{n,l,|m|}(\vec{x})|^2$ for the electronic orbitals with principal quantum numbers $n = 1, 2, 3$ and azimuthal quantum numbers $l = 0, \ldots, n-1$ are shown.

Here, it is interesting to note that when we consider the $2l+1$ probability densities of a shell with fixed values of (n, l) (and hence different values m_l), averaging over the orbitals of that shell with uniform probability $1/(2l+1)$ results in a spherically symmetric function (hence the name "shell"). That is, when no information about the component of the spin in z direction (or any other specific direction) is known, we obtain a probability density that depends only on r and is given by

$$\frac{1}{2l+1} \sum_{m=-l}^{+l} |\psi_{n,l,m}|^2 = \frac{1}{4\pi} |R_{nl}(r)|^2. \tag{8.108}$$

This can be easily verified using the closure relation for the spherical harmonics in eqn (6.68). The observant reader may have noticed that we have innocently introduced the notion of probabilistic mixtures in this way. In other words, we have considered the average over the probability densities $|\psi|^2$, rather than, e.g., trying to average over the wave functions themselves. In Chapter 11 we will return to this problem and establish that this is indeed the right way to describe probabilistic mixtures. Moreover, we will provide a careful mathematical treatment of this topic, which is of great conceptual importance for the combination of quantum mechanics with thermodynamics and information theory.

Turning our attention back on the ten-electron configuration in eqn (8.106), it turns out that this is indeed the ground-state configuration of the noble gas neon (Ne, atomic number 10). This seems reasonable from what we have established so far, since we have assigned electrons to the ten available states (quantum numbers) with the lowest energy as given by Bohr's formula in eqn (7.83). Nevertheless, we should be careful with statements about situations with several electrons since we have technically only solved the Schrödinger equation for a single electron. If the electrons were each only interacting with the nucleus of the atom via the Coulomb interaction, and not with each other, then we could naively assume that the order in which available energy levels are populated in the atomic ground state is determined only by the principal quantum number n. That is, the first two electrons populate the ground state with $n = 1$, and the next $8 = 2n^2$ electrons may fill up the orbitals corresponding to $n = 2$ in any order, and so on.

However, both empirical evidence and theoretical arguments suggest that this is typically not the case. A different rule was first[8] deduced by *Charles Janet* in 1927–1929 (see Janet, 1929, and the debate in Stewart, 2010), and later empirically established by *Erwin Madelung* (1936, p. 359), and suggests that the order in which electrons fill up the energy levels also depends on their orbital angular momentum. This rule, which *Vsevolod M. Klechkovsky* (1962) first supported with theoretical arguments based on the statistical *Thomas–Fermi model* (see Thomas, 1927; Fermi, 1927; and Wong, 1979), is called the *Janet–Madelung–Klechkovsky* rule (or sometimes only Madelung rule), and can be stated in the following way.

Proposition 8.1 (Janet–Madelung–Klechkovsky rule)

(i) *In neutral atoms in their ground state, electrons first occupy shells (n, l) with lower values of $n + l$ before occupying shells with higher values of $n + l$.*

(ii) *If two shells have the same value of $n + l$, the one with the lower value of n is filled up first.*

The Janet–Madelung–Klechkovsky rule, for brevity we will just refer to it as Madelung rule in the following, applies for many neutral atoms in their ground state and is often considered as the foundation for understanding the structure of the periodic table of elements, as illustrated in Fig. 8.7. In particular, it explains why the 4s shell $(n + l = 4 + 0 = 4)$ is (typically) populated before the 3d orbital $(n + l = 3 + 2 = 5)$.

---

[8]The rule has been reported to have been known to Erwin Madelung as early as 1926 (see Goudsmit and Richards, 1964, p. 670).

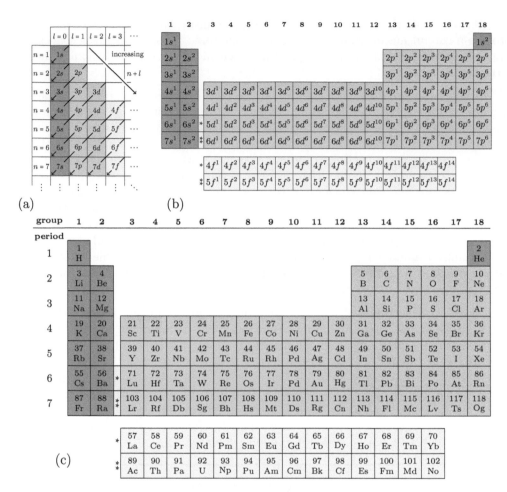

Fig. 8.7 Madelung rule. According to Proposition 8.3.4, electrons in the ground-state configuration of neutral atoms first populate available shells (n, l) with the smallest values of $n + l$. That is, the order in which the shells are filled is 1s, 2s, 2p, 3s, 3p, 4s, 3d, and so on, as indicated by the arrows in (a). The pattern in (b) is obtained by subsequently listing the configurations of neutral atoms with up to 118 electrons according to the Madelung rule. A new row is started, when a new s-shell is populated, creating a pattern where configurations in the same column have the same number of electrons in their outermost (occupied shell with largest value $n + l$) shell. For reasons of convention and comparison with the actual periodic table in (c), the f shells have been listed below the main table. Similarly, the configuration where (the only) two electrons occupy the 1s shell, corresponding to the element helium (He), has been moved to the very right. In (c), the periodic table is shown listing the 118 currently known named elements (see Karol *et al.*, 2016*a* and 2016*b*), which are arranged in eighteen main groups (columns), seven periods (rows), plus the lanthanides (row marked with one asterisk *) and actinides (row marked with two asterisks). The colour coding has been chosen to match the structure in (a) for easier comparison, even though the Madelung rule does not strictly hold for all elements.

At the same time, one should not overstate the significance of the explanatory power of the Madelung rule, given that there are a number of well-known exceptions (Meek and Allen, 2002). For instance, in the ground state of neutral copper atoms the 3d orbital is indeed populated before the 4s orbital. And although one can theoretically argue for its validity (Wong, 1979), the Madelung rule is not a proven statement (which is reassuring, since it does not always hold). Consequently, we have here phrased it as a proposition that should be viewed as an *a posteriori* justification of empirical facts. Nonetheless, one should not be tempted to think that there is any lack of precision in the predictions made by quantum mechanics in this regard.

As argued by Friedrich (2004) in reply to Scerri (1998, 2004), only a few bound-state problems in quantum mechanics can be solved analytically, including such prominent examples as square potential wells (see Section 4.2), the harmonic oscillator (see Chapter 5), and the problem of a single (non-relativistic) electron in a Coulomb potential. For more complicated potentials, including those arising from the interaction of multiple particles, we cannot give closed formulas for the corresponding bound-state wave functions. However, we can nonetheless treat such cases approximately, meaning that methods exist to calculate quantities of interest, e.g., the energy levels of atoms and ions, to arbitrary precision. In other words, while one may lack a certain intuition that would be provided by a closed, formal solution, this does not prevent one from making accurate predictions. We shall encounter some tools to obtain such predictions using perturbative methods in Chapter 10.

Taking the Madelung rule as a tentative description of the order in which electrons fill the available shells (n, l), one can of course also ask how the electrons populate the available states within each shell. This we will briefly discuss in the next section.

8.3.5 Term Symbols and Hund's Rules

As we have seen with the Madelung rule, not only the principal quantum number, but also the orbital angular momentum determines the energy levels in atoms with multiple electrons. In particular, we have learned in which order the different shells (n, l) of neutral atoms in their ground state are populated by electrons. Now we would like to determine how the individual electrons populate the states in a given shell. This behaviour is influenced not only by the orbital angular momentum of the single-electron orbitals (n, l, m_l), but by the total angular momentum. As we have discussed in Section 8.3.2, different angular momenta combine in specific ways. In particular, the spins and orbital angular momenta of the electrons bound to a nucleus interact with each other, leading to a set of instructions called Hund's rules that determine the order in which the microstates in a given shell are occupied, as we will explain now.

Let us start with a simple example. Consider a neutral helium atom with the configuration $1s^2$, that is, we have two electrons in the shell $(n, l) = (1, 0)$. The two electrons are hence assumed to have the quantum numbers $l_1 = l_2 = 0$ and $s_1 = s_2 = 0$. According to the rules discussed in Section 8.3.2, the orbital angular momenta are combined into a total orbital angular momentum with quantum number L ranging from $L_{\min} = |l_1 - l_2|$ to $L_{\max} = l_1 + l_2$, and in this example we hence have $L = 0$.

Similarly, the spins combine to a total spin quantum number S, which could take the values $S = 0$ or $S = 1$, if considered on its own. In turn, the quantum numbers L and S combine to a total angular momentum quantum number J with associated magnetic quantum number $M_J = -J, \ldots, +J$.

Now let us take a step back. The 1s shell has two possible single-electron states, $|\, n, l, m_l, m_s \,\rangle = |\, 1, 0, 0, \pm\frac{1}{2} \,\rangle$, but following the Pauli principle (Theorem 8.3.3), the two electrons may not occupy the same state. In other words, within the 1s shell the Hilbert space for a single electron is two-dimensional, an electron can occupy one of two states. But when a second electron is present, one of them must occupy one state, while the other occupies an orthogonal state, there is no freedom left to choose, and the Hilbert space of two electrons in 1s is one-dimensional. This means that the total spin quantum number $S = 1$ cannot occur in this state, since it would lead to $J = 1$, and a three-dimensional Hilbert space with $M_J = -1, 0, +1$.

The two electrons in the filled 1s shell must hence be in a joint state $|\, L, S, J, M_J \,\rangle = |\, 0, 0, 0, 0 \,\rangle$. With respect to the product basis we can hence write the state as $|\, L, M_L \,\rangle \otimes |\, S, M_S \,\rangle$ where the orbital angular momentum state $|\, L, M_L \,\rangle = |\, l_1, m_{l_1} \,\rangle \, |\, l_2, m_{l_2} \,\rangle = |\, 0, 0 \,\rangle \, |\, 0, 0 \,\rangle$ is symmetric with respect to the exchange of the electrons, and the total spin state is

$$|\, S = 0, M_S = 0 \,\rangle = \tfrac{1}{\sqrt{2}} \big(|\, m_{s_1} = \tfrac{1}{2}; m_{s_2} = -\tfrac{1}{2} \,\rangle - |\, m_{s_1} = -\tfrac{1}{2}; m_{s_2} = \tfrac{1}{2} \,\rangle \big), \quad (8.109)$$

i.e., the antisymmetric spin-singlet state of eqn (8.98). The total two-electron wave function is hence antisymmetric, as required by the spin-statistics connection (see Definition 8.3).

The fact that the total angular momenta vanishes in the previous example is no coincidence and has to do with the fact that the shell is completely filled. Indeed, the same argument can be made for any shell (n, l) with maximal occupancy $N_{\max} = 2(2l + 1)$. The corresponding Hilbert space dimension for a single electron is $N_{\max}$, but for k electrons the Pauli principle limits the dimension to

$$N = \binom{N_{\max}}{k}. \quad (8.110)$$

For any completely filled shell this leaves only one possible state to distribute the $N_{\max}$ electrons, i.e., one Hilbert space dimension, which leads to the conclusion that the total angular momentum of any full shell vanishes.

When the shell is not fully occupied, the order in which the available states are populated is not determined by this observation, and we hence need additional guidance. For this case, *Hund's rules* (1927) (see also, e.g., Engel and Reid, 2006, p. 477) provide the required instructions.

Proposition 8.2 (Hund's rules)

For a neutral atom with multiple electrons, the ground-state electronic configuration is determined in the following way. Start by populating the shells (n, l) according to the Madelung rule (Proposition 8.3.4). Determine the outermost shell, i.e., the shell with highest value of $n+l$ (and highest n if there is more than one) that is occupied by $k > 0$ electrons. For fixed n and l, determine the states $|L, S, J, M_J\rangle$ of k electrons (see Section 8.3.2) allowed by the Pauli principle (Theorem 8.3.3). Out of these states, the ground state is the state with

(i) maximal value of the total spin S,

(ii) maximal value of the total orbital angular momentum L, given the maximal S.

(iii) For fixed S and L, the ground state has maximal total angular momentum J if the shell is more than half full, $k > N_{\max}/2$, and minimal J if it is less than (or exactly) half full, i.e., when $k \leq N_{\max}/2$.

As with the Madelung rule, Hund's rules are to be understood as empirical observations, rather than proven theorems. Nonetheless, Proposition 8.3.5 is fairly accurate for light atoms, where the spin-orbit interaction is weak, and L and S represent "good" quantum numbers. The rules, especially the first one, are sometimes phrased in terms of individual electrons occupying single-electron (spin) orbitals (n, l, m_l, m_s) in particular orders. For instance, (i) may be loosely stated as "spins are as parallel as possible". However, as discussed in Campbell (1991), this is not accurate, since such statements refer to the uncoupled basis of single-electron states with quantum numbers l, m_l, and m_s of each electron, whereas Hund's rules are formulated for quantum numbers L, S, and J, i.e., total angular momenta of the joint system.

To better illustrate this, let us consider an example. Suppose we wish to determine the electronic ground-state configuration of a (neutral) carbon atom (C, atomic number 6). According to the Madelung rule, the ground state configuration in spectroscopic notation is $1s^2 2s^2 2p^2$, where the 1s and 2s shells are completely full and can hence be disregarded. The p orbital, however, is only occupied by $k = 2$ electrons, but would have space for $N_{\max} = 6$. In this p shell, the two electrons have azimuthal quantum numbers $l_1 = l_2 = 1$ and (of course) spin quantum numbers $s_1 = s_2 = \frac{1}{2}$. In principle, these could combine to the total orbital angular momentum and total spin quantum numbers $L = 0, 1, 2$ and $S = 0, 1$, respectively, but not all combinations $J = |L - S|, \ldots, L + S$ are possible for fermions. The Pauli principle allows only $N = \binom{6}{2} = 15$ states. Take, for example, the total angular momentum $J = 3$. It could only be obtained from the available quantum numbers by combining $L = 2$ and $S = 1$. For $J = 3$, there would then exist a state with magnetic quantum number $M_J = M_L + M_S$, such that $|J = 3, M_J = 3\rangle = |L = 2, M_L = 2\rangle |S = 1, M_S = 1\rangle$, with

$$|L = 2, M_L = 2\rangle = |l_1 = 1, m_{l_1} = 1\rangle |l_2 = 1, m_{l_2} = 1\rangle, \tag{8.111a}$$

$$|S = 1, M_S = 1\rangle = |s_1 = \tfrac{1}{2}, m_{m_s} = \tfrac{1}{2}\rangle |s_2 = \tfrac{1}{2}, m_{s_2} = \tfrac{1}{2}\rangle. \tag{8.111b}$$

Since both of the states in eqns (8.111) are fully symmetric with respect to the exchange of the two electrons, the Pauli principle would be violated. After going through all the possible combinations of L and S in this example (which we leave as an exercise for the interested reader), one finds that three pairs (L, S) are allowed, as listed in the following table:

L	S	J	M_J	#
0	0	0	0	1
1	1	0	0	1
1	1	1	-1,0,+1	3
1	1	2	-2,-1,0,+1,+2	5
2	0	2	-2,-1,0,+1,+2	5

As we see, there are a total of $1 + 1 + 3 + 5 + 5 = 15$ allowed states, as expected, but these are labelled by quantum numbers corresponding to total (orbital) angular momentum and total spin of the joint system, rather than those corresponding to angular momenta of the individual electrons. To simplify the terminology, it is customary to introduce the following nomenclature for the combinations of quantum numbers: Multiple electrons within a shell (n, l) occupy

"terms"	labelled by	(L, S),
"levels"	labelled by	(L, S, J),
"states"	labelled by	(L, S, J, M_J).

For the terms, one can then apply a similar spectroscopic notation as before for the orbitals of single electrons (see Section 8.3.4). That is, the total orbital angular momentum quantum numbers L are labelled by capital letters S ($L = 0$), P ($L = 1$), D ($L = 2$), F ($L = 3$), and alphabetically from there on (again, the letter J is not used). These letters are now supplemented by a superscript index for the total spin multiplicity $2S + 1$ on the left-hand side, while the total angular momentum quantum number J is placed as a subscript on the right.

In our example of two electrons in the 2p shell, we hence have the allowed terms $(L, S) = (0, 0)$, $(1, 1)$, and $(2, 0)$, denoted as ^{1}S, ^{3}P, and ^{1}D, respectively. Following Hund's first rule, (i) in Proposition 8.3.5, the term with the maximal value of S is $(L, S) = (1, 1)$ or ^{3}P. Since there is only one term with $S = 1$ here, we do not need to invoke (ii) in this example, since only $L = 1$ is possible. The third rule, (iii), then suggests that the ground state is one of the states in the level with minimal J within the term $(L, S) = (1, 1)$, since $k = 2 \leq N_{\max}/2 = 3$. More specifically, this is the case here for $J = 0$, which fixes the ground state to be ^{3}P$_0$. In general, different magnetic quantum numbers for the total angular momentum are still possible for the ground states of arbitrary atoms, but these are degenerate in the absence of an external magnetic field (and disregarding the magnetic interaction with the nuclear spin). For our example, there is no such degeneracy because $J = 0$.

In the ground state of a (neutral) carbon atom, the two electrons in the outermost shell are hence in the state $|L, S, J, M_J\rangle = |1, 1, 0, 0\rangle$. But what does this state look like in the uncoupled basis of single-electron orbitals? Following the procedure laid out in Section 8.3.2, one finds

$$|L = 1, S = 1, J = 0, M_J = 0\rangle = \tfrac{1}{\sqrt{3}}(|L = 1, M_L = -1\rangle|S = 1, M_S = 1\rangle$$

$$- |L = 1, M_L = 0\rangle|S = 1, M_S = 0\rangle$$

$$+ |L = 1, M_L = 1\rangle|S = 1, M_S = -1\rangle),$$

$$(8.112)$$

where the states with $L = 1$ and $M_L = -1, 0, +1$ are combinations of two orbital angular momenta $l_1 = l_2 = 1$, for which one in turn finds the decomposition into the product-basis states $|l_1, m_{l_1}\rangle|l_2, m_{l_2}\rangle$, i.e.,

$$|L = 1, M_L = -1\rangle = \tfrac{1}{\sqrt{2}}(|1, -1\rangle|1, 0\rangle - |1, 0\rangle|1, -1\rangle), \qquad (8.113a)$$

$$|L = 1, M_L = 0\rangle = \tfrac{1}{\sqrt{2}}(|1, -1\rangle|1, 1\rangle - |1, 1\rangle|1, -1\rangle), \qquad (8.113b)$$

$$|L = 1, M_L = 1\rangle = \tfrac{1}{\sqrt{2}}(|1, 0\rangle|1, 1\rangle - |1, 1\rangle|1, 0\rangle). \qquad (8.113c)$$

These coupled orbital angular momentum states are hence all antisymmetric with respect to the exchange of the electrons, whereas the three spin states for adding $s_1 = \tfrac{1}{2}$ and $s_2 = \tfrac{1}{2}$ to $S = 1$ are given by the fully symmetric triplet states of eqn (8.97). Overall, we see that the joint state $|L = 1, S = 1, J = 0, M_J = 0\rangle$ of the two electrons is antisymmetric, as required. Moreover, we see that the individual spin states of the two electrons are not well defined. The ground state is a superposition of states $|\uparrow\uparrow\rangle$, $|\downarrow\downarrow\rangle$, $|\uparrow\downarrow\rangle$, and $|\downarrow\uparrow\rangle$. Speaking of parallel or antiparallel spins is therefore meaningless in this context, although this may be suggested by graphical representations of term schemes found in the literature (see Campbell, 1991 and references therein for a discussion).

9
Electromagnetism in Quantum Mechanics

9.1 The Pauli Equation

After the introduction of spin via the Stern–Gerlach experiment, and exploring the consequences of this additional property for the atomic structure, let us now return to the problem discussed at the very beginning of this chapter. How can the interaction of quantum-mechanical particles with electromagnetic fields be incorporated into the Schrödinger equation? And how can spin be included in such a description? To answer these questions, we return to the Lorentz force, that is, a (classical) particle with charge q, mass m_{q}, and velocity $\vec{v}$ is subject to the force

$$\vec{F}_{\mathrm{L}} = q\left(\vec{E} + \frac{1}{c}\vec{v}\times\vec{B}\right) \tag{9.1}$$

in the presence of an electric field $\vec{E}(t,\vec{x})$ and a magnetic field $\vec{B}(t,\vec{x})$. In other words, the equation of motion of such a particle is $m_{\mathrm{q}}\frac{d^2\vec{x}}{dt^2} = \vec{F}_{\mathrm{L}}$. A convenient way to arrive at the corresponding quantum-mechanical equation of motion is to switch from the Newtonian to the Hamiltonian form, and promote the Hamiltonian function $H(\vec{x},\vec{p},t)$ to an operator $\hat{H}(\vec{X},\vec{P},t)$ in the canonical way.

9.1.1 Hamiltonian for the Interaction with the Electromagnetic Field

The classical Hamilton function $H(\vec{x},\vec{p},t)$ satisfies the equations

$$\frac{dx^i}{dt} = \frac{\partial H}{\partial p_i}, \quad \frac{dp^i}{dt} = -\frac{\partial H}{\partial x_i}, \tag{9.2}$$

for $\vec{x} = (x^i)$ and $\vec{p} = (p^i)$. As we will show now, the function H for which Hamilton's equations of motion correspond to Netwon's with the Lorentz force on the right-hand side is given by

$$H = \frac{1}{2m}\left(\vec{p} - \frac{q}{c}\vec{A}\right)^2 + q\,\phi, \tag{9.3}$$

where ϕ is the (scalar) electric potential and $\vec{A}$ is the magnetic vector potential such that

$$\vec{E} = -\vec{\nabla}\phi - \frac{1}{c}\frac{\partial\vec{A}}{\partial t}, \quad \vec{B} = \vec{\nabla}\times\vec{A}. \tag{9.4}$$

For the Hamiltonian function of eqn (9.3), Hamilton's equations (9.2) become

$$\frac{dx^i}{dt} = \frac{1}{m}(p^i - \tfrac{q}{c}A^i), \tag{9.5a}$$

$$\frac{dp^i}{dt} = \frac{q}{mc}\sum_{j=1}^{3}(p^j - \tfrac{q}{c}A^j)\frac{\partial A_j}{\partial x_i} - q\frac{\partial \phi}{\partial x_i} = \frac{q}{c}\sum_{j=1}^{3}\frac{dx^j}{dt}\frac{\partial A_j}{\partial x_i} - q\frac{\partial \phi}{\partial x_i}, \tag{9.5b}$$

where eqn (9.5a) has been inserted in the second step of eqn (9.5b). We can then take the derivative with respect to time t on both sides of eqn (9.5a),

$$\frac{d^2x^i}{dt^2} = \frac{1}{m}\Big(\frac{dp^i}{dt} - \frac{q}{c}\frac{dA^i}{dt}\Big) = \frac{1}{m}\Big(\frac{dp^i}{dt} - \frac{q}{c}\sum_{j=1}^{3}\frac{\partial A^i}{\partial x^j}\frac{dx^j}{dt} - \frac{q}{c}\frac{\partial A^i}{\partial t}\Big). \tag{9.6}$$

Here we can insert $\frac{dp^i}{dt}$ from eqn (9.5b) and multiply both sides by m to obtain

$$m\frac{d^2x^i}{dt^2} = \frac{q}{c}\sum_{j=1}^{3}\frac{dx^j}{dt}\Big(\frac{\partial A_j}{\partial x_i} - \frac{\partial A^i}{\partial x^j}\Big) - q\frac{\partial \phi}{\partial x_i} - \frac{q}{c}\frac{\partial A^i}{\partial t}. \tag{9.7}$$

Finally, we recognize $\vec{E}$ from eqn (9.4) and by setting $\frac{dx^j}{dt} \equiv v^j$ and noting that

$$\big(\vec{v} \times \vec{B}\big)^i = \big(\vec{v} \times (\vec{\nabla} \times \vec{A})\big)^i = \sum_{j=1}^{3} v^j\Big(\frac{\partial A_j}{\partial x_i} - \frac{\partial A^i}{\partial x^j}\Big), \tag{9.8}$$

we arrive at the equation of motion

$$m\frac{d^2x^i}{dt^2} = \frac{q}{c}\,\vec{v} \times \vec{B} + q\vec{E}. \tag{9.9}$$

For a charged (scalar) particle of mass m in the presence of electromagnetic fields we can hence promote the Hamilton function $H(\vec{x}, \vec{p}, t)$ of eqn (9.3) to the Hamiltonian (operator)

$$\hat{H} = \frac{1}{2m}\Big(\vec{P} - \frac{q}{c}\hat{\vec{A}}\Big)^2 + q\,\hat{\phi}, \tag{9.10}$$

where $\hat{\vec{A}} = A(\hat{\vec{X}}, t)$ and $\hat{\phi} = \phi(\hat{\vec{X}}, t)$ and we immediately drop the hats on $\vec{A}$ and ϕ again in a slight abuse of notation. In the absence of a magnetic potential $\vec{A}$, we just recover the Hamiltonian that we have previously used in the Schrödinger equation. When the vector potential is non-zero, $\vec{A} \neq 0$, we simply replace the momentum (operator) $\vec{P}$, i.e.,

$$\vec{P} \longmapsto \vec{P} - \tfrac{q}{c}\vec{A}, \tag{9.11}$$

a transformation called *minimal substitution* or *minimal coupling*.

In the case of non-vanishing $\vec{A}$, we of course expect the minimally coupled Hamiltonian to reproduce what we have previously learned about the interaction of quantum-mechanical particles and their spin with magnetic fields. Before we worry about the inclusion of spin, let us first check that our approach is consistent with the observations made about magnetic moments earlier on in this chapter. That is, we expect a term $-\vec{\mu} \cdot \vec{B}$ to appear in the Hamiltonian. Expressing the first term in eqn (9.10) in the position representation, we have

$$\frac{1}{2m}\left(\vec{P} - \frac{q}{c}\vec{A}\right)^2 = \frac{1}{2m}\left(-i\hbar\vec{\nabla} - \frac{q}{c}\vec{A}\right)^2 = -\frac{\hbar^2}{2m}\vec{\nabla}^2 + \frac{iq\hbar}{2mc}\left(\vec{\nabla} \cdot \vec{A} + \vec{A} \cdot \vec{\nabla}\right) + \frac{q^2}{2mc^2}\vec{A}^2. \tag{9.12}$$

For the electromagnetic potentials there is a *gauge freedom* (or *gauge symmetry*), as we shall elaborate on more in Section 9.2, and we may hence use the *Coulomb gauge* $\vec{\nabla} \cdot \vec{A} = 0$ to write

$$\vec{\nabla} \cdot \vec{A} + \vec{A} \cdot \vec{\nabla} = \sum_i \left((\nabla_i A^i) + 2A^i \nabla_i\right) = 2\vec{A} \cdot \vec{\nabla}. \tag{9.13}$$

Inserting this back into the Hamiltonian, we have

$$H = -\frac{\hbar^2}{2m}\vec{\nabla}^2 + \frac{iq\hbar}{mc}\vec{A} \cdot \vec{\nabla} + \frac{q^2}{2mc^2}\vec{A}^2. \tag{9.14}$$

9.1.2   Paramagnetic and Diamagnetic Contributions

Besides the usual kinetic term, we hence find two contributions to the Hamiltonian in eqn (9.14), where $\frac{iq\hbar}{mc}\vec{A} \cdot \vec{\nabla}$ is called the *paramagnetic* contribution, while $\frac{q^2}{2mc^2}\vec{A}^2$ is referred to as the *diamagnetic* contribution. To better understand their role, consider a constant, homogeneous magnetic field $\vec{B}$. Without loss of generality, we may assume this field is pointing along the z direction, i.e., $\vec{B} = B\vec{e}_z$. As illustrated in Fig. 9.1, a vector potential

$$\vec{A} = -\frac{1}{2}\vec{r} \times \vec{B} \tag{9.15}$$

gives rise to exactly such a magnetic field. To see this, we can calculate the ith component of $\vec{B} = \vec{\nabla} \times \vec{A}$ for this choice, i.e.,

$$(\vec{\nabla} \times \vec{A})^i = -\frac{1}{2}\left(\vec{\nabla} \times (\vec{r} \times \vec{B})\right)^i = -\frac{1}{2}\varepsilon^i_{\ jk}\nabla^j \varepsilon^k_{\ lm}r^l B^m,$$

$$= -\frac{1}{2}\left(\delta^i_{\ l}\delta_{jm} - \delta^i_{\ m}\delta_{jl}\right)\nabla^j r^l B^m = -\frac{1}{2}\left(\nabla_m r^i B^m - \nabla_l r^l B^i\right). \tag{9.16}$$

where summation over indices appearing twice (once as a superscript and once as a subscript) is implied (Einstein summation convention) and $\varepsilon^i_{\ jk}$ is the completely antisymmetric Levi–Civita symbol from 6.3. Since the components of the magnetic field are constants, $B^m = B\delta^m_3$, the derivatives only act on the components of $\vec{r} =$

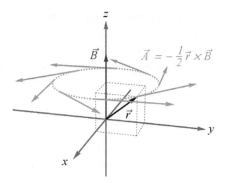

Fig. 9.1 Vector potential for static magnetic field. A constant, homogeneous magnetic field, e.g., along the z direction, $\vec{B} = B\vec{e}_z$, can be described by the vector potential $\vec{A} = -\frac{1}{2}\vec{r} \times \vec{B}$ with $\vec{r} = (x, y, z)^{\top}$.

$(x, y, z)^{\top}$, for which we have $\nabla_m r^i = \delta^i{}_m$ and $\nabla_l r^l = \delta^l{}_l = 3$, and we therefore have $(\vec{\nabla} \times \vec{A})^i = -\frac{1}{2}(\delta^i{}_m B^m - 3B^i) = B^i$, as claimed before. Let us then evaluate the potential energy contributions to the Hamiltonian of eqn (9.14) for the vector potential of eqn (9.15). For the paramagnetic contribution we find

$$\frac{iq\hbar}{mc}\vec{A} \cdot \vec{\nabla} = -\frac{iq\hbar}{2mc}\varepsilon^j{}_{kl}\, r^k B^l \nabla_j. \tag{9.17}$$

Once more, we make use of the fact that the magnetic field is constant and thus commutes with the derivative. We further exploit the antisymmetry of the Levi–Civita symbol, $\varepsilon^j{}_{kl} = -\varepsilon_{lk}{}^j$, to write

$$\frac{iq\hbar}{mc}\vec{A} \cdot \vec{\nabla} = -\frac{q}{2mc}\varepsilon_{lk}{}^j r^k(-i\hbar\nabla_j)B^l = -\frac{q}{2mc}\vec{L} \cdot \vec{B} = -\vec{\mu} \cdot \vec{B}, \tag{9.18}$$

where we have inserted the position representation of the momentum operator $\vec{P} = -i\hbar\vec{\nabla}$, and, subsequently, the orbital angular momentum $\vec{L} = \vec{r} \times \vec{P}$. We thus find that the paramagnetic term is responsible for the coupling of the magnetic field to the magnetic moment $\vec{\mu}$ that we have discussed in Section 8.1.

With this desired result at hand, let us now inspect the diamagnetic term, for which the constant magnetic field $\vec{B} = B\,\vec{e}_z$ results in a contribution

$$\frac{q^2}{2mc^2}\vec{A}^2 = \frac{q^2}{8mc^2}(\vec{r} \times \vec{B})^2 = \frac{q^2}{8mc^2}\varepsilon^i{}_{jk}\, r^j B^k \varepsilon_{imn}\, r^m B^n \tag{9.19}$$

$$= \frac{q^2}{8mc^2}\big(\delta_{jm}\delta_{kn} - \delta_{jn}\delta_{km}\big)r^j B^k r^m B^n = = \frac{q^2}{8mc^2}\big(\vec{r}^2\vec{B}^2 - (\vec{r} \cdot \vec{B})^2\big).$$

For a field oriented in the z direction, the diamagnetic contribution (as an operator in the position representation) then reduces to $\frac{q^2}{8mc^2}B^2(x^2 + y^2)$. Since we have not previously taken into account this term in our description of the hydrogen atom, let us

estimate its size for the H atom. Assuming that the typical length scale for electrons
$(q = q_{\rm e})$ bound to the hydrogen atom is the Bohr radius, $\langle\, x^2 + y^2\,\rangle \approx r_{\rm B} = \hbar^2/(mq_{\rm e}^2)$,
and that the typical orbital angular momentum along the z axis is $\langle\, L_z\,\rangle \approx \hbar$, we can
evaluate the relative size[1] of the diamagnetic and paramagnetic contributions as

$$\frac{\frac{q_{\rm e}^2}{8m_{\rm e}c^2}B^2 r_{\rm B}^2}{\frac{q_{\rm e}}{2m_{\rm e}c}\langle\, L_z\,\rangle B} = \frac{\hbar^3}{4m_{\rm e}^2 c\, q_{\rm e}^3}B \approx 10^{-6}\frac{B}{\rm T}\,. \tag{9.20}$$

For the description of atomic spectra, the paramagnetic term is hence dominant and
the diagmagnetic term only becomes relevant for extremely strong external magnetic
fields, as would occur, for instance, in a neutron star. However, note that the diamagnetic contribution may be significant in other physical contexts, e.g., in a synchrotron
for which $x^2 + y^2 \ll r_{\rm B}$.

Thus ignoring the diamagnetic contribution, we find the Hamiltonian in the pres-
ence of a constant magnetic field $|\vec{B}| = B_z \equiv B$ in the z direction is given by

$$H = H_o - \frac{q}{2mc}B\,L_z = H_o - \gamma B\,L_z, \tag{9.21}$$

where $H_o = \vec{P}^2/(2m)$ is the Hamiltonian in the absence of (external) fields, e.g., for
the hydrogen atom we have (in position representation)

$$H_o\,\psi_{n,l,m_l} = E_n\,\psi_{n,l,m_l}. \tag{9.22}$$

As we have discussed in Section 8.1.3, external magnetic fields lead to a Larmor torque,
causing a precession of the magnetic moment around the direction of the external field
with Larmor frequency $\omega_{\rm L} = -\gamma B$. In addition, we now see that the additional term
lifts the degeneracy of the energy levels. That is, we now have

$$H\,\psi_{n,l,m_l} = \big(E_n - \gamma\hbar m_l B\big)\psi_{n,l,m_l} = \big(E_n - \hbar\omega_{\rm L}m_l\big)\psi_{n,l,m_l}. \tag{9.23}$$

Whereas the energy levels were determined only by the principal quantum number n
before, now we have $2l + 1$ different energy levels for each given azimuthal quantum
number l, see Fig. 9.2. Since l is a positive integer, this means that one expects to
find an odd number of spectral lines for each fixed pair n, l. At least, that would be
(and indeed was) the case if one did not know about spin. Recall from Section 8.2.2
that the spin of charged particles also couples to external magnetic fields, and it does
so in an *anomalous* way, i.e., twice as strongly as orbital angular momentum. We will
discuss this so-called Zeeman effect in more detail in Section 10.3.

9.1.3   The Stern–Gerlach Term

Having rewritten the Schrödinger equation for charged spinless particles in terms of
couplings to the electromagnetic potentials, we are of course interested in incorporating

---

[1]Note that we are using Gaussian cgs units, and that the conversion of the appropriate unit (Gauss)
of the magnetic field strength to Tesla, is $1{\rm T}= 10^4{\rm G}$.

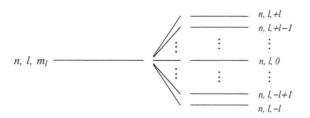

Fig. 9.2 Zeeman effect. The energy levels E_n for fixed azimuthal quantum number l split into $2l + 1$ contributions according to their magnetic quantum number $m_l = -l, -1 + 1, \ldots, l - 1, l$ in the presence of an external magnetic field.

spin in a similarly elegant way. To this end, let us re-examine the term $\left(\vec{P} - \frac{q}{c}\vec{A}\right)^2$. Since this term gave rise to the coupling between orbital angular momentum and the magnetic field, let us try to incorporate the Pauli operators from eqn (8.30) in some fashion. In particular, note that the term we consider can be regarded as the inner product of two vectors, $\vec{a} \cdot \vec{b}$. Here, we can attempt a naive ansatz to include the Pauli matrices and write

$$(\vec{a} \cdot \vec{\sigma})(\vec{b} \cdot \vec{\sigma}) = \sum_{i,j} a_i b_j \sigma_i \sigma_j = \sum_{i,j} a_i b_j (\delta_{ij} + i\varepsilon_{ijk}\sigma_k) = \vec{a} \cdot \vec{b} + i\,\vec{\sigma}(\vec{a} \times \vec{b}), \quad (9.24)$$

where we have made use of the commutation relations from eqn (8.40). In our case we have $\vec{a} = \vec{b} = \vec{P} - \frac{q}{c}\vec{A}$, and we can therefore write

$$\frac{1}{2m}\left((\vec{P} - \frac{q}{c}\vec{A})\vec{\sigma}\right)^2 = \frac{1}{2m}\left(\vec{P} - \frac{q}{c}\vec{A}\right)^2 + \frac{i}{2m}\left((\vec{P} - \frac{q}{c}\vec{A}) \times (\vec{P} - \frac{q}{c}\vec{A})\right)\vec{\sigma}$$

$$= \frac{1}{2m}\left(\vec{P} - \frac{q}{c}\vec{A}\right)^2 - \frac{\hbar q}{2mc}(\vec{\nabla} \times \vec{A} + \vec{A} \times \vec{\nabla})\vec{\sigma}. \quad (9.25)$$

Inspection of the ith component of the second term on the right-hand side reveals

$$\left(\vec{\nabla} \times \vec{A} + \vec{A} \times \vec{\nabla}\right)^i = \varepsilon^i{}_{jk}\left(\nabla^j A^k + A^j \nabla^k\right) = \varepsilon^i{}_{jk}\left((\nabla^j A^k) + A^k \nabla^j + A^j \nabla^k\right)$$

$$= B^i + \varepsilon^i{}_{jk}\left(A^k \nabla^j + A^j \nabla^k\right) = B^i, \quad (9.26)$$

since $\varepsilon^i{}_{jk}$ and $(A^k \nabla^j + A^j \nabla^k)$ are antisymmetric and symmetric with respect to the exchange of the summation indices j and k, respectively, the corresponding sum vanishes. We thus have

$$\frac{1}{2m}\left((\vec{P} - \frac{q}{c}\vec{A})\vec{\sigma}\right)^2 = \frac{1}{2m}\left(\vec{P} - \frac{q}{c}\vec{A}\right)^2 - \frac{q}{mc}\vec{B} \cdot \frac{\hbar}{2}\vec{\sigma}$$

$$= \frac{1}{2m}\left(\vec{P} - \frac{q}{c}\vec{A}\right)^2 - 2\gamma\vec{B} \cdot \vec{S}. \quad (9.27)$$

That is, the direct inclusion of the vector of Pauli matrices along with the minimal substitution has led directly to the *Stern–Gerlach term* proportional to $\vec{B} \cdot \vec{S}$ with the correct pre-factor for the anomalous magnetic moment of the electron. The full (non relativistic) equation of motion for a charged spin-$\frac{1}{2}$ particle, known as the *Pauli equation*, can hence be written in the compact form

$$\left[\frac{1}{2m}\left((\vec{P} - \frac{q}{c}\vec{A})\vec{\sigma}\right)^2 + q\,\phi\right]\psi = i\hbar\frac{\partial}{\partial t}\psi, \qquad (9.28)$$

where ψ is a *spinor field*, i.e., a vector in the Hilbert space $\mathfrak{L}_2(\mathbb{R}^3) \otimes \mathbb{C}^2$. In other words, we can think of ψ as a two-component spinor whose components are square-integrable functions (see Section 8.2.5). The Pauli equation, suggested by *Wolfgang Pauli* in 1927, can be considered to be a middle ground between the Schrödinger equation, its limiting case for vanishing magnetic fields, and the Dirac equation, from which the Pauli equation itself arises in the non-relativistic limit (see, e.g., Greiner, 2000, Sec. 2.3). In particular, the splitting into electric and magnetic fields is observer dependent and the Pauli equation is hence not Lorentz covariant. However, it features a different important symmetry, *gauge symmetry*.

9.2   Gauge Symmetries in Quantum Mechanics

The equations of motion of classical electromagnetism, i.e., Maxwell's equations, are gauge invariant. This means that there is a certain freedom of choosing the gauge potential $\vec{A}$ and the electric potential ϕ without changing the underlying physics. More specifically, given solutions $\vec{A}$ and ϕ one may select an arbitrary (differentiable) *gauge function* $f(\vec{x}, t)$ and define the transformed quantities

$$\vec{A}' = \vec{A} + \vec{\nabla}f \quad \text{and} \quad \phi' = \phi - \frac{1}{c}\frac{\partial f}{\partial t}. \qquad (9.29)$$

Evaluating the corresponding electric and magnetic fields, one obtains

$$\vec{E}' = -\vec{\nabla}\phi' - \frac{1}{c}\frac{\partial \vec{A}'}{\partial t} = -\vec{\nabla}(\phi - \frac{1}{c}\frac{\partial f}{\partial t}) - \frac{1}{c}\frac{\partial}{\partial t}(\vec{A} + \vec{\nabla}f)$$

$$= -\vec{\nabla}\phi - \frac{1}{c}\frac{\partial \vec{A}}{\partial t} + \frac{1}{c}\vec{\nabla}\frac{\partial}{\partial t}f - \frac{1}{c}\frac{\partial}{\partial t}\vec{\nabla}f = -\vec{\nabla}\phi - \frac{1}{c}\frac{\partial \vec{A}}{\partial t} = \vec{E}, \qquad (9.30)$$

as well as

$$\vec{B}' = \vec{\nabla} \times \vec{A}' = \vec{\nabla} \times \vec{A} + \vec{\nabla} \times \vec{\nabla}f = \vec{\nabla} \times \vec{A} = \vec{B}. \qquad (9.31)$$

The electric and magnetic fields, and hence Maxwell's equations, are hence invariant under a *local gauge transformation*. Here, "local" refers to the fact that f is not a mere constant, but a function of the coordinates $\vec{x}$ and t. *But what about the Pauli equation?* If we are to naively insert the transformed fields into the Hamiltonian, where we can work without the explicit inclusion of the Pauli matrices since $[\sigma_i, \nabla_j f] = 0$, we have

$$H' = \frac{1}{2m}\left(\vec{P} - \frac{q}{c}\vec{A}'\right)^2 + q\,\phi' = \left[\frac{1}{2m}\left(\vec{P} - \frac{q}{c}\vec{A}\right)^2 + q\,\phi\right]$$

$$- \left[\frac{1}{2m}\left((\vec{P} - \frac{q}{c}\vec{A})\vec{\nabla}f + \vec{\nabla}f(\vec{P} - \frac{q}{c}\vec{A})\right) - \frac{q^2}{2mc^2}(\vec{\nabla}f)^2 + \frac{q}{c}\frac{\partial f}{\partial t}\right], \qquad (9.32)$$

that is, a number of extra terms appear. However, there is one more freedom aside from the gauge transformation of the electromagnetic field: that of the wave function. We know that quantum-mechanical predictions are invariant under a global phase shift, $\psi \to e^{i\lambda}\psi$, i.e., a U(1) gauge transformation. Let us therefore attempt to promote this global gauge symmetry to a local one by defining the transformation

$$\psi'(\vec{x}, t) = e^{i\lambda(\vec{x},t)}\psi(\vec{x}, t) \tag{9.33}$$

for a function $\lambda(\vec{x}, t)$ that is yet to be determined. We may again disregard the Pauli operators and, setting the scalar potential ϕ aside for a moment, the kinetic term of the Hamiltonian applied to ψ' gives

$$\frac{1}{2m}\left(\vec{P} - \frac{q}{c}\vec{A}'\right)^2 \psi'(\vec{x}, t) = \frac{1}{2m}\left(-\hbar^2\Delta + \frac{i\hbar q}{c}(\vec{\nabla}\vec{A}') + 2\frac{i\hbar q}{c}\vec{A}'\vec{\nabla} + \frac{q^2}{c^2}\vec{A}'^2\right)\psi'(\vec{x}, t). \tag{9.34}$$

We then have to take into account the gauge freedom when differentiating ψ, that is,

$$\vec{\nabla}\psi' = \vec{\nabla}e^{i\lambda(\vec{x},t)}\psi = e^{i\lambda}\left[i(\vec{\nabla}\lambda) + \vec{\nabla}\right]\psi, \tag{9.35}$$

$$\Delta\psi' = \vec{\nabla}\vec{\nabla}\psi' = e^{i\lambda}\left[i(\vec{\nabla}\lambda) - (\vec{\nabla}\lambda)^2 + 2i(\vec{\nabla}\lambda)\vec{\nabla} + \Delta\right]\psi. \tag{9.36}$$

Inserting these derivatives in eqn (9.34), and combining with the missing term $q\phi'\psi' = \left(q\phi - \frac{q}{c}\left(\frac{\partial f}{\partial t}\right)\right)e^{i\lambda}\psi$, the right-hand side of the minimally coupled Schrödinger equation becomes

$$H'\psi' = e^{i\lambda}\left[\frac{1}{2m}\left(\vec{P} - \frac{q}{c}\vec{A}'\right)^2 + q\phi\right]\psi + e^{i\lambda}\left[-i\hbar^2\left((\Delta\lambda) - \frac{q}{\hbar c}(\Delta f)\right)\right. \tag{9.37}$$

$$\left. + \hbar^2\left((\vec{\nabla}\lambda) - \frac{q}{\hbar c}(\vec{\nabla}f)\right)^2 - 2\hbar\left((\vec{\nabla}\lambda) - \frac{q}{\hbar c}(\vec{\nabla}f)\right)\left(\frac{q}{c}\vec{A} + i\hbar\vec{\nabla}\right) - \frac{q}{c}\left(\frac{\partial f}{\partial t}\right)\right]\psi,$$

whereas the left-hand side is

$$i\hbar\frac{\partial}{\partial t}\psi' = e^{i\lambda}\left[i\hbar\frac{\partial}{\partial t} - \hbar\left(\frac{\partial\lambda}{\partial t}\right)\right]\psi. \tag{9.38}$$

We see that the minimally coupled Schrödinger equation, or equivalently the Pauli equation, is gauge invariant for the choice $\lambda(\vec{x}, t) = \frac{q}{\hbar c}f(\vec{x}, t)$. The gauge symmetry of classical electrodynamics thus becomes a *local U(1) symmetry of quantum mechanics*. In other words, Maxwell's equations and the Pauli equation are invariant under the *gauge transformation*

$$\vec{A} \longmapsto \vec{A}' = \vec{A} + \vec{\nabla}f(\vec{x}, t), \tag{9.39a}$$

$$\phi \longmapsto \phi' = \phi - \frac{1}{c}\frac{\partial f(\vec{x}, t)}{\partial t}, \tag{9.39b}$$

$$\psi \longmapsto \psi' = \exp\left(i\frac{q}{\hbar c}f(\vec{x}, t)\right)\psi. \tag{9.39c}$$

9.3   The Aharonov–Bohm effect

The gauge freedom of electromagnetism may lead one to the conclusion that the gauge potentials $\vec{A}$ and ϕ are physically irrelevant, and only non-zero electric and magnetic fields matter. Such a sweeping conclusion, however, turns out to be too naive, as is beautifully captured by the Aharonov–Bohm effect, which was first predicted by *Werner Ehrenberg* and *Raymond Eldred Siday* (1949), but only rose to prominence with the independent discovery by *Yakir Aharonov* and *David Bohm* (1959, 1961). Consider the following experimental setup: Suppose a charged particle with electric charge q, e.g., an electron, is emitted from a source at position $\vec{x}$ and sent through a double slit and is detected on a screen at position $\vec{y}$. As we recall from Section 1.7, repeating such an experiment many times results in an interference pattern on the screen.

We can think of this interference pattern as a probability density for finding the particle at different screen positions $\vec{y}$ in any single run of the experiment. The probability $p(\vec{y})$ for detecting a particle at a given position $\vec{y}$ is given by the modulus squared of the particle wave function, $p(\vec{y}) = |\psi(\vec{y})|^2$. For simplicity, we can think of $\psi(\vec{y})$ as being the superposition of two contributions, $\psi(\vec{y}) = c_1\psi_1(\vec{y}) + c_2\psi_2(\vec{y})$, with ψ_1 and ψ_2 corresponding to the propagation through slits 1 and 2, respectively. The particular value of $\psi(\vec{y})$ depends on the relative phase between $\psi_1(\vec{y})$ and $\psi_2(\vec{y})$. So far, we have described the usual double-slit experiment as in Section 1.7. To discover the Aharonov–Bohm effect, let us modify this picture by introducing a very thin solenoid behind the double slit that generates a magnetic field perpendicular to the plane of the setup, as illustrated in Fig. 9.3. We assume that the solenoid is sufficiently long, thin and well shielded that, although it is threaded by a finite magnetic field $\vec{B}$ on the

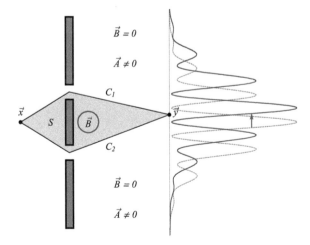

Fig. 9.3 Setup for Aharonov–Bohm effect. Double-slit experiment with additional solenoid. The generated magnetic field $\vec{B}$ is perpendicular to the plane shown and non-zero only within the solenoid (red circle), but vanishes outside. The vector potential is non-zero everywhere, see, e.g., Fig. 9.1. The phase shift due to the resulting Aharonov–Bohm effect shifts the interference pattern.

inside, outside of the solenoid, where the electrons pass by, the magnetic field vanishes identically, $\vec{B} = 0$. However, note that because $\vec{B} = \vec{\nabla} \times \vec{A}$, the gauge potential $\vec{A}$ is generally non-zero also in the area that the electrons pass through.

On the one hand, one could now take the stance that, since both electric and magnetic field strengths vanish along the possible paths of the electron, the latter is not subject to a deflecting Lorentz force according to eqn (8.1). Therefore, the interference pattern should remain unaltered, irrespective of the size of $\vec{B}$ within the area that the electrons never enter. On the other hand, let us consider two possible paths, C_1 and C_2, that an electron could take from the source at $\vec{x}$ to a particular point $\vec{y}$ on the screen, such that C_1 and C_2 pass through slits 1 and 2, respectively, and enclose the region of the non-zero magnetic field between each other. To uncover the influence of the non-zero gauge potential $\vec{A}$ along different paths, we have to understand how to describe translations along these paths. As we shall see, in quantum mechanics, spatial translations are generated by the momentum operator much in the same way that spatial rotations are generated by angular-momentum operators (recall Theorem 8.3).

Theorem 9.1 (Generators of translations)

*The momentum operators P_i are the **generators of translations** in Hilbert space, that is, translations by vectors $\vec{a} \in \mathbb{R}^3$ are represented on the Hilbert space by unitary operators*

$$U(\vec{a}) = \exp\left(\tfrac{-i\vec{a}\cdot\vec{P}}{\hbar}\right).$$

To see this, let us consider the action of the (unitary) operator $\exp\left(-i\vec{a} \cdot \vec{P}/\hbar\right)$ and commute the position operator $\vec{X}$ past it using the Hadamard lemma of eqn (2.39). That is, we write

$$\vec{X} \exp\left(-i\vec{a} \cdot \vec{P}/\hbar\right)$$
$$= \exp\left(-i\vec{a} \cdot \vec{P}/\hbar\right) \left(\vec{X} + \tfrac{i}{\hbar}[\vec{a}\cdot\vec{P},\vec{X}] + \tfrac{1}{2}\left(\tfrac{i}{\hbar}\right)^2 [\vec{a}\cdot\vec{P},[\vec{a}\cdot\vec{P},\vec{X}]] + \ldots\right)$$
$$= \exp\left(-i\vec{a} \cdot \vec{P}/\hbar\right)(\vec{X} + \vec{a}), \tag{9.40}$$

where we have used the canonical commutation relations (Theorem 2.2) to find that

$$\tfrac{i}{\hbar}[\vec{a}\cdot\vec{P},X_i] = \tfrac{i}{\hbar}\sum_j a_j\,[P_j\,,X_i] = \sum_j a_j \delta_{ij} = a_i \tag{9.41}$$

is a constant. Therefore, on the right-hand side of the first line of eqn (9.40) every commutator except the first vanishes. The result of eqn (9.40) implies that if $|\,\vec{x}\,\rangle$ is an eigenstate of $\vec{X}$ with eigenvalue $\vec{x}$, then $\exp\left(-i\vec{a}\cdot\vec{P}/\hbar\right)|\,\vec{x}\,\rangle = |\,\vec{x}+\vec{a}\,\rangle$ is an eigenstate of $\vec{X}$ with eigenvalue $\vec{x} + \vec{a}$. However, as we know from eqn (9.11), gauge symmetry demands that we use the minimally substituted momentum operator, implying that translations are indeed generated by $\vec{P} - \frac{q}{c}\vec{A}$ with $\vec{A} = \vec{A}(\vec{x}, t)$. For an infinitesimal translation from $\vec{x}$ to $\vec{x} + d\vec{x}$, any state vector $\psi(\vec{x}, t)$ will thus pick up a phase factor, i.e.,

$$\psi(\vec{x}, t) \rightarrow e^{i\frac{q}{\hbar c}\vec{A}(\vec{x}, t)d\vec{x}}\, \psi(\vec{x}, t) = e^{i\varphi}\, \psi(\vec{x}, t)\,. \tag{9.42}$$

For any particular path C, the corresponding phase that is picked up is then obtained by integration, i.e.,

$$\varphi = \frac{q}{\hbar c}\int\limits_{C} \vec{A}(\vec{x}, t)d\vec{x}\,. \tag{9.43}$$

Moreover, when we evaluate the relative phase between the two paths C_1 and C_2, we can make use of the fact that reversing the direction that C_2 is traversed just corresponds to a sign flip for the line integral, and the composition of C_1 and (the reversed curve) C_2 is a closed curve enclosing the surface S. We thus find

$$\Delta\varphi = \frac{q}{\hbar c}\int\limits_{C_1}\vec{A}(\vec{x}, t)d\vec{x} - \frac{q}{\hbar c}\int\limits_{C_2}\vec{A}(\vec{x}, t)d\vec{x} = \frac{q}{\hbar c}\oint\vec{A}(\vec{x}, t)d\vec{x} = \frac{q}{\hbar c}\iint\limits_{S}\vec{B}d^2x = \Phi_{\mathrm{B}},$$
$$\tag{9.44}$$

where we have turned the line integral over the closed curve into a surface integral over S. The additional[2] relative phase between the two paths is thus given exactly by the magnetic flux Φ_{B} through the surface enclosed by the two paths. Even though the electrons pass through a region of non-zero magnetic fields, the field strength $\vec{B}$ in a region the particles cannot enter determines the phase shift, and hence a shift of the entire interference pattern. One way of interpreting this result is that the particles indeed interact with the non-zero gauge potential, even though its precise value at any position can be altered by way of a physically inconsequential gauge transformation.

Although certain aspects of the interpretation of the Aharonov–Bohm effect are still the subject of ongoing discussions (Vaidman, 2012), the existence of the effect itself is uncontroversial and has been confirmed in various experiments as early as 1960 by Chambers. A notable experimental refinement quite beautifully capturing and unequivocally confirming the Aharonov–Bohm effect was performed by Tonomura *et al.* and by Osakabe *et al.* in 1986. There, the magnetic field is realized by a toroidal magnet that is covered by a thin layer of superconducting material and another layer of copper, preventing the electron beams from entering the magnet. Meanwhile, the magnetic field is strictly confined to within the torus by the *Meissner–Ochsenfeld effect* (Meissner and Ochsenfeld, 1933) of the superconducting covering. The interference is then observed for two electron beams, one passing through, the other outside the torus.

---

[2]Note that the original interference pattern when no gauge fields are present is already encoded in the coefficients c_1 and c_2 and the wave functions $\psi_1(\vec{y})$ and $\psi_2(\vec{y})$, respectively.

9.4   Geometric Phases

So far we have discussed the Aharonov–Bohm effect on physical grounds but it also has an interesting topological or geometric interpretation, which we will discuss in this section. To do so, let us first introduce the concept of geometric phases more generally.

In 1984 *Michael Berry* published a paper (Berry, 1984) on the existence of a pure geometric phase in quantum systems, which had an enormous impact on physics. This seminal paper inspired a wave of research activity in the 1980s and 1990s to interpret physics in terms of topological or geometrical concepts. In particular, the mathematician *Barry Simon*, getting to know Berry's work at an early stage, was able to formulate the geometric phase as a holonomy in a line bundle since the adiabatic theorem naturally defines a connection in such a bundle (Simon, 1983), as we shall elaborate on in the following. In the following, the condition of adiabaticity was relaxed by *Yakir Aharonov* and *Jeeva Anandan* (1987), before *Joseph Samuel* and *Rajendra Bhandari* constructed a geometrical phase even for non-cyclic evolutions (1988). In this context, the earlier work of *Shivaramakrishnan Pancharatnam* (1956) was rediscovered, who had described effects due to a geometric phase by investigating polarized light already in the 1950s[3].

In the 1990s and thereafter, geometric phases have also played a role in quantum computing and quantum information (Zanardi and Rasetti, 1999; Jones *et al.*, 2000; Duan *et al.*, 2001; Fujii, 2001; Durstberger, 2003), they can be defined for mixed states[4] (Uhlmann, 1986; Uhlmann, 1991; Sjöqvist *et al.*, 2000; Durstberger, 2005) and for off-diagonal elements (Manini and Pistolesi, 2000; Filipp and Sjöqvist, 2003). As the concept of a geometric phase is quite profound it is no surprise that it also entered into other areas of physics such as gauge theories, solid-state physics and condensed-matter physics (Shapere and Wilczek, 1989; Nakahara, 1990; Bertlmann, 2000). Of course, there is a huge amount of literature available in this field, and we highly recommend the following additional articles (Wang, 1990; Fernandez *et al.*, 1992; Mironova *et al.*, 2013), reviews, and books (Anandan, 1992; Anandan *et al.*, 1997; Durstberger, 2003; Chruściński and Jamiołkowski, 2004; Durstberger, 2005; Filipp, 2006; Mironova, 2010; Basler, 2014), and references therein, where one may also find literature pertinent to experimental demonstrations of geometric phases.

9.4.1   Holonomy

Let us first consider an example in elementary geometry to get a feeling for how a geometric phase can arise. Let us imagine an oscillating pendulum resting on the north pole of a sphere. Next, let us consider the pendulum being moved along a closed path along the sphere: Starting from the north pole, the pendulum is first moved down to the equator along a meridian and then some distance along the equator until reaching another meridian enclosing an azimuthal angle φ with the first meridian. If we now transport the pendulum back to the north pole along the second meridian, seemingly

---

[3]Shivaramakrishnan Pancharatnam (1934–1969) made significant contributions in optics but died rather young at an age of 35. Guided by his advisor *Chandrashekhara Venkata Raman* he studied interference patterns produced by light waves in crystal plates and discovered what is nowadays known as the geometric phase.

[4]We have not yet discussed mixed states, but shall do so in Chapter 11 of Part II.

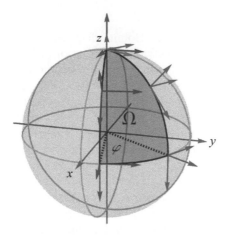

Fig. 9.4 Holonomy. Parallel transport of a vector on a sphere. An orthogonal triad of vectors (red) initially at the north pole of the sphere is transported along a meridian down to the equator, along the equator, and back to the north pole along another meridian at an azimuthal angle φ with respect to the first meridian. Although the path of the triad along the sphere is closed, enclosing a solid angle Ω, the orientation of the final triad (orange) differs from the orientation of the original triad.

nothing has changed. But still, something *has* changed, namely the plane, wherein the pendulum oscillates, see Fig. 9.4. This phenomenon can be described in terms of the *parallel transport* of a vector on a sphere. After completing the closed path one observes that the vector defining the plane of oscillation of the pendulum has changed its direction by an angle φ although the vector is kept parallel to its previous direction for every infinitesimal segment of the path throughout, as illustrated in Fig. 9.4. The solid angle Ω enclosed by the closed path in this example is

$$\Omega = \int\limits_{0}^{\varphi} d\varphi' \int\limits_{0}^{\pi/2} d\vartheta \, \sin\vartheta = \varphi \,. \tag{9.45}$$

The phenomenon where a system is (abstractly speaking) "moved" along a closed path but does not return to its initial state is called a *holonomy*[5]. The holonomic process was already known to Carl Friedrich Gauss and is nowadays described by the so-called *Hannay angle* (Hannay, 1985). The parallel transport on a sphere getting into trouble is related to the so-called *hairy ball theorem*, first derived (for the 2-sphere) by *Henri Poincaré* in 1885, and sometimes colloquially described as "*You can't comb the hair on a coconut without creating a cowlick*" (compare, e.g., similar formulations in Richeson, 2008, Chap. 19).

---

[5] *Holonomy* is the expression used in mathematics, whereas in physics, e.g., in mechanics, the same phenomenon is sometimes called an *anholonomy*. We stick to the mathematical convention and use the term holonomy.

Obviously, the discussed holonomy gets its geometric interpretation due to the curvature of the sphere. Curved space can be described as a Riemannian manifold, where the coordinate system cannot be given by a constant metric tensor g_{ik} everywhere. For a sphere with radius R, for example, we have (in abstract index notation)

$$
g_{ik} = \begin{pmatrix} 1 & 0 & 0 \\ 0 & R^2 & 0 \\ 0 & 0 & R^2 \sin^2 \theta \end{pmatrix} .
\tag{9.46}
$$

9.4.2 The Berry Phase

Michael Berry showed in a series of papers (1984, 1987, 1990) that a quantum system which evolves adiabatically and cyclically picks up a phase factor on pure geometrical grounds. This geometric phase arises in addition to the familiar dynamical phase which is governed by the Hamiltonian of the system. Berry proved that this geometric phase cannot be gauged away, and, consequently, it has a physical—a measurable—effect. In the following we will derive this phase.

Consider a quantum system whose Hamiltonian $H(\lambda(t))$ depends on an external (potentially multidimensional) parameter $\lambda(t)$ which itself depends on time, for example, a time-dependent magnetic field $\vec{B}(t)$. As we know, the time evolution of the system is determined by the time-dependent Schrödinger equation

$$
H(\lambda(t)) \, | \, \psi(t) \, \rangle = i \, \hbar \, \frac{\partial}{\partial t} \, | \, \psi(t) \, \rangle \ .
\tag{9.47}
$$

Going forward, we assume that the energy spectrum of H is discrete, the eigenvalues are not degenerate, and no level crossing occurs during the evolution (see, e.g., the discussion in Section 10.1.4). We further assume that at $t = 0$ the system is in its nth energy eigenstate, which we denote as $| \, \psi(0) \, \rangle = | \, n, \lambda(0) \, \rangle$, such that we have

$$
H(\lambda(0)) \, | \, n, \lambda(0) \, \rangle = E_n(\lambda(0)) \, | \, n, \lambda(0) \, \rangle \ .
\tag{9.48}
$$

But what about the state $| \, \psi(t) \, \rangle$ at a later time $t > 0$?

To analyse this situation in more detail, let us assume that the time evolution of the system is *adiabatic*, which means $\lambda(t)$ varies slowly, and we perform a cycle, i.e., the evolution is described by a closed loop in parameter space: $\lambda(0) \equiv \lambda(t = T)$ for some $T > 0$. If the time evolution is sufficiently slow so that the *adiabatic theorem* (which we will discuss in more detail later in this section) applies, the system remains in its nth energy eigenstate throughout the evolution, i.e.,

$$
H(\lambda(t)) \, | \, n, \lambda(t) \, \rangle = E_n(\lambda(t)) \, | \, n, \lambda(t) \, \rangle \ .
\tag{9.49}
$$

Then, besides the *dynamical phase*

$$
\theta_n(t) = -\tfrac{1}{\hbar} \int_0^t E_n(\lambda(t')) \, dt' \ ,
\tag{9.50}
$$

we have to introduce an additional phase $\gamma_n(t)$ in order to satisfy the Schrödinger equation (9.47), as we shall see in the following. Our ansatz for the time evolution of the state therefore is

$$| \psi(t) \rangle = e^{i\gamma_n(t)} e^{i\theta_n(t)} | n, \lambda(t) \rangle \, . \tag{9.51}$$

The phase $\gamma_n(t)$ is established by the simultaneous validity of eqns (9.47) and (9.49). The left-hand side of the Schrödinger equation (9.47) is simply

$$H\big(\lambda(t)\big) | \psi(t) \rangle = E_n\big(\lambda(t)\big) | \psi(t) \rangle \, , \tag{9.52}$$

while the inserting our ansatz from eqn (9.51) into the right-hand side of eqn (9.47) gives

$$i\hbar \frac{\partial}{\partial t} | \psi(t) \rangle = \left[-\hbar \frac{\partial}{\partial t} \gamma_n(t) + E_n\big(\lambda(t)\big) \right] | \psi(t) \rangle + e^{i\gamma_n(t)} e^{i\theta_n(t)} i\hbar \frac{\partial}{\partial t} | n, \lambda(t) \rangle \, . \tag{9.53}$$

Equating the right-hand sides eqns (9.52) and (9.53), and applying $\langle \psi(t) |$ from the left, we then find

$$\frac{\partial}{\partial t} \gamma_n(t) = i \langle n, \lambda(t) | \frac{\partial}{\partial t} | n, \lambda(t) \rangle \, . \tag{9.54}$$

Integration of eqn (9.54) then gives

$$\gamma_n(t) = i \int_0^t \langle n, \lambda(t') | \frac{\partial}{\partial t'} | n, \lambda(t') \rangle \, dt' = i \int_{\lambda(0)}^{\lambda(t)} \langle n, \lambda | \frac{\partial}{\partial \lambda} | n, \lambda \rangle \, d\lambda \, . \tag{9.55}$$

Here we note that, if λ is not a single parameter but a set of parameters that can be collected in a vector $\vec{\lambda} = (\lambda_1, \lambda_2, \ldots)^\top$, we should formally replace the partial derivative with respect to λ by the corresponding nabla operator, i.e., the vector of partial derivatives $\vec{\nabla}_\lambda$ with components $\partial/\partial\lambda_i$ and integrate over the line element $d\vec{\lambda}$, but to keep the expressions more compact in the following we will implicitly assume this and stay with the notation λ, $d\lambda$ and $\partial/\partial\lambda$ as in eqn (9.55). For a closed loop $\lambda(0) = \lambda(T)$ we obtain

$$\gamma_n = i \oint \langle n, \lambda | \frac{\partial}{\partial \lambda} | n, \lambda \rangle \, d\lambda \neq 0 \, , \tag{9.56}$$

which does not vanish in general, since the integrand is not necessarily a total differential. The phase γ_n is called *Berry phase*. To summarize, let us state the result of our derivation in the following theorem.

Theorem 9.2 (Berry phase)

Consider a quantum system governed by a Hamiltonian $H(\lambda(t))$ that depends on an external time-dependent parameter $\lambda(t)$ but has a discrete and non-degenerate spectrum. For adiabatic and cyclic time evolution free of energy-level crossings, the system state at time t is given by

$$|\psi(t)\rangle = e^{i\gamma_n(t)}\, e^{i\theta_n(t)}\, |n, \lambda(t)\rangle \,,$$

such that, besides the dynamical phase

$$\theta_n(t) = -\tfrac{1}{\hbar} \int\limits_0^t E_n\big(\lambda(t')\big)\, dt' \,,$$

the system picks up a phase of purely geometric nature—the Berry phase:

$$\gamma_n(t) = i \int\limits_0^t \langle n, \lambda(t') | \tfrac{\partial}{\partial t'} | n, \lambda(t')\rangle\, dt' \,,$$

$$= i \oint \langle n, \lambda | \tfrac{\partial}{\partial \lambda} | n, \lambda\rangle\, d\lambda \,.$$

Let us now discuss some important properties of the Berry phase, which we shall state in terms of the following corollaries.

Corollary 9.1 (Berry's phase is real)

The Berry phase is a real quantity, $\gamma_n(t) \in \mathbb{R}$.

Proof For the proof let us simply consider the normalization of the states, that is, from $\langle n, \lambda | n, \lambda\rangle = 1$ it follows that $\frac{\partial}{\partial d\lambda} \langle n, \lambda | n, \lambda\rangle = 0$. Carrying out the partial derivative, we have

$$0 = \left(\tfrac{\partial}{\partial \lambda} \langle n, \lambda |\right) |n, \lambda\rangle + \langle n, \lambda | \tfrac{\partial}{\partial \lambda} |n, \lambda\rangle$$

$$= \left(\langle n, \lambda | \tfrac{\partial}{\partial \lambda} |n, \lambda\rangle\right)^* + \langle n, \lambda | \tfrac{\partial}{\partial \lambda} |n, \lambda\rangle$$

$$= 2\,\mathrm{Re}\left(\langle n, \lambda | \tfrac{\partial}{\partial \lambda} |n, \lambda\rangle\right),. \tag{9.57}$$

Consequently, $\langle n, \lambda | \tfrac{\partial}{\partial \lambda} |n, \lambda\rangle$ is purely imaginary, and hence γ_n is real. $\qquad\square$

Corollary 9.2 (Berry potential)

The Berry phase is the integral $\gamma_n = \oint A_n \, d\lambda$ of a U(1) gauge potential

$$A_n = i \left\langle n, \lambda \left| \frac{\partial}{\partial \lambda} \right| n, \lambda \right\rangle .$$

Proof Under a unitary gauge transformation with one-parameter gauge group U(1) the eigenstates transform as

$$| n, \lambda \rangle \longmapsto | n, \lambda \rangle^{\mathsf{g}} = e^{i\alpha_n(\lambda)} | n, \lambda \rangle , \qquad (9.58)$$

where $\alpha_n(\lambda)$ is a *gauge function*, which is a single-valued real function. The new gauge-transformed eigenstates $| n, \lambda \rangle^{g}$ also form a basis of eigenstates for the Hamiltonian $H\big(\lambda(t)\big)$. Inserting the gauge transformation (9.58) of the eigenstates into the definition of the Berry phase Theorem 9.2, we have

$$\gamma_n \longmapsto \gamma_n^{\mathsf{g}} = i \oint \left\langle n, \lambda \left| e^{-i\alpha_n(\lambda)} \frac{\partial}{\partial \lambda} e^{i\alpha_n(\lambda)} \right| n, \lambda \right\rangle d\lambda$$

$$= i \oint \left\langle n, \lambda \left| \left(\frac{\partial}{\partial \lambda} + i \frac{\partial}{\partial \lambda} \alpha_n(\lambda) \right) \right| n, \lambda \right\rangle d\lambda, \qquad (9.59)$$

which provides the familiar gauge transformation from eqn (9.39a) for A_n, i.e.,

$$A_n(\lambda) \longmapsto A_n^{\mathsf{g}}(\lambda) = A_n(\lambda) - \frac{\partial}{\partial \lambda} \alpha_n(\lambda) , \qquad (9.60)$$

and we thus indeed recognize A_n as a gauge potential. □

Corollary 9.3 (Gauge invariance)

The Berry phase γ_n is gauge invariant.

Proof As we have seen in eqn (9.59), the Berry phase transforms under a gauge transformation as

$$\gamma_n \longmapsto \gamma_n^{\mathsf{g}} = \oint A_n \, d\lambda - \oint \frac{\partial}{\partial \lambda} \alpha_n(\lambda) \, d\lambda . \qquad (9.61)$$

The second term, however, vanishes

$$\oint \frac{\partial}{\partial \lambda} \alpha_n(\lambda) \, d\lambda = \alpha_n(\lambda(T)) - \alpha_n(\lambda(0)) = 0 , \qquad (9.62)$$

since we consider a closed loop for which we have $\alpha_n(\lambda(T)) = \alpha_n(\lambda(0))$ and $\alpha(\lambda)$ is a single-valued function. We thus find $\gamma_n^{\mathsf{g}} = \gamma_n$. □

Remark about the adiabatic theorem: Adiabatic processes in mechanics and early quantum mechanics have been first studied by *Paul Ehrenfest* in 1916, followed up by mathematically more rigorous studies by *Max Born* and *Vladimir Aleksandrovich Fock* (1928). In the 1950s *Tosio Kato* (1950) reconsidered adiabatic evolutions in quantum mechanics and presented an exact proof of the theorem. Nowadays the adiabatic theorem is treated accessibly in standard textbooks like Griffiths (1995, Chap. 10.1) or Messiah (1969, p. 744).

The exact proof of the adiabatic theorem is quite involved, nevertheless, we want to argue why our ansatz (9.51) represents an adiabatic process. Quite generally, the time-evolved state is a superposition of *all* possible eigenstates $|\,n, \lambda(t)\,\rangle$,

$$|\,\psi(t)\,\rangle = \sum_n e^{i\gamma_n(t)}\, e^{i\theta_n(t)}\, |\,n, \lambda(t)\,\rangle \,, \qquad (9.63)$$

and not just a specific energy eigenstate like in the ansatz of eqn (9.51). When we consider the time evolution according to the time-dependent Schrödinger equation (9.47), we can insert the expression from (9.63) instead of our ansatz (9.51), and in full analogy to the calculation in eqns (9.52) and (9.53) we obtain the relation

$$\sum_m \left(\frac{\partial}{\partial t}\gamma_m(t)\right) e^{i\gamma_m(t)} e^{i\theta_m(t)} |\,m, \lambda(t)\,\rangle = i\sum_m e^{i\gamma_m(t)} e^{i\theta_m(t)} \frac{\partial}{\partial t} |\,m, \lambda(t)\,\rangle . \qquad (9.64)$$

We can now apply $e^{i\gamma_n(t)} e^{i\theta_n(t)} \langle\, n, \lambda(t)\,|$ from the left, before using $\langle\, n, \lambda\,|\,m, \lambda\,\rangle = \delta_{nm}$, to arrive at

$$\frac{\partial}{\partial t}\gamma_n(t) = \sum_m e^{i(\gamma_m - \gamma_n)}\, e^{i(\theta_m - \theta_n)}\, i\,\langle\, n, \lambda(t)\,|\,\frac{\partial}{\partial t}\,|\,m, \lambda(t)\,\rangle \qquad (9.65)$$

$$= i\,\langle\, n, \lambda(t)\,|\,\frac{\partial}{\partial t}\,|\,n, \lambda(t)\,\rangle + \sum_{m\neq n} e^{i(\gamma_m - \gamma_n)}\, e^{i(\theta_m - \theta_n)}\, i\,\langle\, n, \lambda\,|\,\frac{\partial}{\partial t}\,|\,m, \lambda\,\rangle \,.$$

If we now inspect the phase factor $e^{i(\theta_m - \theta_n)}$, and insert the expressions for θ_n and θ_m from Theorem 9.2, we see that it is of the form

$$e^{i(\theta_m - \theta_n)} = \exp\left(-\frac{i}{\hbar}\int_0^t [E_m(t') - E_n(t')]dt'\right). \qquad (9.66)$$

If the energy levels do not cross during the time evolution, as we had assumed, then the sign of the integral in the exponent is fixed by the choice of m and n for all $t > 0$. What is more, the inverse of the (minimal) difference between the energy eigenvalues sets an intrinsic timescale. If the evolution of the system is slow compared to this timescale, then the phase factor oscillates very rapidly and the integration over t that is required to obtain γ_n from eqn (9.65) will effectively eliminate all contributions from $m \neq n$. Precisely this occurs in the adiabatic regime, in which case we arrive at the result that we have previously obtained from our ansatz (9.51) for the Berry phase.

9.4.3 The Aharonov–Anandan Phase

In 1987 *Yakir Aharonov* and *Jeeva Anandan* presented an important generalization of the Berry phase. They considered time evolutions of a quantum system which are cyclic but not adiabatic. In such a case, the system does not remain in an eigenstate of the Hamiltonian and the geometric phase cannot be associated to the parameter space of the Hamiltonian. For practical applications this is an important generalization of the assumptions made for the derivation of the Berry phase since the adiabatic condition is never exactly fulfilled in real-world experiments.

In order to find an appropriate state space for such a setting we recall that the state space of a quantum system is only determined up to a phase. A shift by a global phase $\alpha \in \mathbb{R}$ results in the same probabilities, the same predictions, the same physics, i.e., $|\psi\rangle \sim e^{i\alpha}|\psi\rangle$ are *equivalent*. But we have to take care here! It is only the global phase that can be neglected. The phase difference in a superposition of states $|\psi\rangle = |\psi_1\rangle + e^{i\beta}|\psi_2\rangle$ certainly cannot be disregarded—it describes a different state.

Projective Hilbert space: Physical states are defined as *equivalence classes* of vectors in the Hilbert space $\mathcal{H}$. The space built up of these equivalence classes is called a *projective Hilbert space* $\mathcal{P}$. Let us denote the set of normalized non-zero states in $\mathcal{H}$ as $\mathcal{N}_0$, i.e.,

$$\mathcal{N}_0 = \left\{ |\psi\rangle \in \mathcal{H} \,\middle|\, \langle\psi|\psi\rangle = 1 \right\}. \tag{9.67}$$

Then we can define a *projective map* Π from the Hilbert space $\mathcal{N}_0$ to the projective Hilbert space $\mathcal{P}$ by

$$\Pi : \quad \mathcal{N}_0 \longrightarrow \mathcal{P}, \tag{9.68a}$$

$$\Pi(|\psi\rangle) = \left\{ |\phi\rangle \in \mathcal{N}_0 \,\middle|\, |\phi\rangle = c|\psi\rangle \,,\; c \in \mathbb{C} \right\} = \mathcal{P}. \tag{9.68b}$$

Thus the projective Hilbert space contains all states of $\mathcal{N}_0$ where the states differing just by a phase factor are identified. This describes a *ray*, which is a representative of $\mathcal{P} = \mathcal{N}_0 / \sim$, where $\sim$ denotes the equivalence relation. Thus the projective space is the space of rays (recall Definition 3.2), which are the physically relevant quantities.

An *equivalence relation* can be expressed mathematically more precisely in terms of the following proposition.

Proposition 9.1 (Equivalence relation)

The state vector $|\phi\rangle$ is equivalent to the vector $|\psi\rangle$ iff $|\phi\rangle$ is an element of the projective space $\Pi(|\psi\rangle) = \mathcal{P}$,

$$|\phi\rangle \sim |\psi\rangle \quad \Longleftrightarrow \quad |\phi\rangle \in \Pi(|\psi\rangle) = \mathcal{P}.$$

Every element of this space represents a distinct state since every element is actually a different ray (disregarding any gauge symmetries of the system).

However, in adopting the idea of projective spaces, we have to be careful since a projective space is *not a vector space* any more. In particular, it has no zero element.

An equivalence class of states can be represented by the projection operator

$$| \psi \rangle \langle \psi | \, , \tag{9.69}$$

which is the *outer product* or *tensor product* of two vectors[6] of dimension m and n, resulting in an $m \times n$ matrix; recall our discussion about operators in Section 3.2. For example, for two-component vectors, like that describing the spin of a spin-$\frac{1}{2}$ particle, or more abstractly, a qubit, we have

$$\begin{pmatrix} a \\ b \end{pmatrix} \begin{pmatrix} c & d \end{pmatrix} = \begin{pmatrix} ac & ad \\ bc & bd \end{pmatrix} \, . \tag{9.70}$$

Note that the outer/tensor product should not be confused with the exterior product of two vectors, which results in another vector. Physically, the operator (9.69) can be interpreted as the density matrix for the state $| \psi \rangle$, and indeed, for the entire equivalence class $\mathcal{P}$, as will be discussed in detail in Chapter 11.

Now, let us consider a cyclic evolution of the state $| \psi(t) \rangle$ with a period T, such that $| \psi(T) \rangle = e^{i\Phi} | \psi(0) \rangle$. This evolution will trace out an arbitrary curve C in the Hilbert space $\mathcal{N}_0$, but the projection of this curve will give a closed curve $\tilde{C} = \Pi(C)$ in $\mathcal{P}$ since the projection operators coincide for $t = 0$ and $t = T$, i.e.,

$$| \psi(T) \rangle \langle \psi(T) | = e^{i\Phi} | \psi(0) \rangle \langle \psi(0) | e^{-i\Phi} = | \psi(0) \rangle \langle \psi(0) | \, , \tag{9.71}$$

as illustrated in Fig. 9.5.

To reach any point along the curve C in the Hilbert space $\mathcal{N}_0$ without changing the corresponding curve $\tilde{C}$ in the projective space $\mathcal{P}$, we define a state $| \phi(t) \rangle \in \mathcal{N}_0$ such that

$$| \phi(t) \rangle = e^{-if(t)} | \psi(t) \rangle \quad \text{and} \quad f(T) - f(0) = \Phi \, . \tag{9.72}$$

Then the state lies in the projective space, i.e., $| \phi(t) \rangle \in \mathcal{P}$

$$| \phi(T) \rangle = e^{-if(T)} | \psi(T) \rangle = e^{-if(T)} e^{i\Phi} | \psi(0) \rangle$$

$$= e^{-i(f(T)-f(0))} e^{i\Phi} | \phi(0) \rangle = | \phi(0) \rangle \, . \tag{9.73}$$

The time-dependence of the phase can then be obtained from the Schrödinger equation in (9.47), that is, by differentiating eqn (9.72) and applying $\langle \phi(t) |$ from the left we obtain

---

[6]More precisely, of a vector and a covector, see eqn (3.17) and the discussion thereafter.

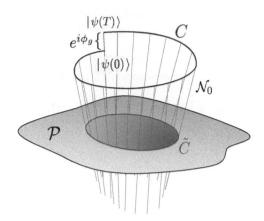

Fig. 9.5 Projective Hilbert space. The space of rays in $\mathcal{N}_0$. The rays denote states that differ by a U(1) phase factor. After a cycle the curve C in the Hilbert space $\mathcal{N}_0$ is open and differs by the geometric phase factor $e^{i\phi_g}$, whereas the corresponding curve $\tilde{C}$ in the projective space $\mathcal{P}$ is closed.

$$\frac{\partial}{\partial t} f(t) = -\tfrac{1}{\hbar} \langle \psi(t) | H | \psi(t) \rangle + i \langle \phi(t) | \frac{\partial}{\partial t} | \phi(t) \rangle \;. \tag{9.74}$$

Integrating eqn (9.74) in the interval $t \in [0, T]$ gives

$$f(T) - f(0) = -\tfrac{1}{\hbar} \int_0^T \langle \psi(t) | H | \psi(t) \rangle\, dt + i \int_0^T \langle \phi(t) | \frac{\partial}{\partial t} | \phi(t) \rangle\, dt \;, \tag{9.75}$$

where the first term is the *dynamical phase*

$$\theta = -\tfrac{1}{\hbar} \int_0^T \langle \psi(t) | H | \psi(t) \rangle\, dt \;, \tag{9.76}$$

and the second term represents the geometric or *Aharonov–Anandan phase*

$$\phi_g := i \int_0^T \langle \phi(t) | \frac{\partial}{\partial t} | \phi(t) \rangle\, dt = -\theta + \Phi \;. \tag{9.77}$$

The phase ϕ_g is purely geometric and does not depend on the Hamiltonian. It is solely determined by the geometry of the closed curve $\tilde{C}$ in the projective space $\mathcal{P}$ and uniquely defined modulo 2π. In the derivation we did not rely on the adiabatic condition, we only made use of the cyclicity of the time evolution of the system. Hence the Aharonov–Anandan phase ϕ_g is valid for arbitrary state vectors. In the adiabatic limit, of course, the Aharonov–Anandan phase converges to the Berry phase.

9.4.4 Spin $\frac{1}{2}$ in Adiabatically Rotating Magnetic Field

Next, let us study an important example for a typical physical problem where the Berry phase plays a role, a spin-$\frac{1}{2}$ particle moving in a rotating external magnetic field $\vec{B}(t)$. We will assume that the rotation of the field is sufficiently slow so that the adiabatic condition is satisfied, and we consider the rotation axis to be inclined with respect to the z axis by an angle δ, such that

$$
\vec{B}(t) = B_0 \begin{pmatrix} \sin\delta \, \cos\omega t \\ \sin\delta \, \sin\omega t \\ \cos\delta \end{pmatrix} , \tag{9.78}
$$

with ω the angular frequency of the rotation and B_0 the field strength. When the field rotates slowly enough, the particle's spin will be "pinned" to the direction of the magnetic field $\vec{B}(t)$ at any time t. That means the spin, which is assumed to be in an eigenstate of $H(0)$ at the beginning (at $t = 0$), will be in an eigenstate of $H(t)$ at any later time $t > 0$.

To make the example more specific, let us take the particle to be an electron. From eqn (8.17) the interaction Hamiltonian for the spin with the magnetic field is given by

$$
H(t) = \mu_B \, \vec{\sigma} \cdot \vec{B}(t) = \mu_B B_0 \begin{pmatrix} \cos\delta & e^{-i\omega t} \sin\delta \\ e^{i\omega t} \cos\delta & -\cos\delta \end{pmatrix} , \tag{9.79}
$$

where we have inserted the magnetic moment $\vec{\mu} = -\frac{q_e}{2m_e c} g_s \vec{S}$ with the electron g-factor $g_s = 2$ from eqn (8.26), and the spin operator $\vec{S} = \frac{\hbar}{2}\vec{\sigma}$ from eqn (8.29) with the vector of Pauli operators given by $\vec{\sigma} = (\sigma_x, \sigma_y, \sigma_z)^\top$. Finally, the constant μ_B denotes the Bohr magneton from eqn (8.12). For this interaction Hamiltonian, the eigenvalue equation

$$
H(t) \, | \, n(t) \, \rangle = E_n \, | \, n(t) \, \rangle \tag{9.80}
$$

is satisfied by the normalized eigenstates

$$
| \, n_+(t) \, \rangle = \begin{pmatrix} \cos\frac{\delta}{2} \\ e^{i\omega t} \sin\frac{\delta}{2} \end{pmatrix} , \tag{9.81a}
$$

$$
| \, n_-(t) \, \rangle = \begin{pmatrix} -\sin\frac{\delta}{2} \\ e^{i\omega t} \cos\frac{\delta}{2} \end{pmatrix} , \tag{9.81b}
$$

with the corresponding energy eigenvalues

$$
E_\pm = \pm \mu_B B_0 . \tag{9.82}
$$

The states $| \, n_+(t) \, \rangle$ and $| \, n_-(t) \, \rangle$ denote the eigenstates of spin up and spin down along the direction of the magnetic field $\vec{B}$.

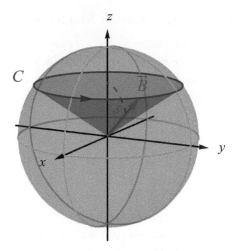

Fig. 9.6 Parameter space. A magnetic field $\vec{B}(t)$, inclined at an angle δ with respect to the z axis, is (slowly) rotating around the z axis, tracing out a closed curve C on the sphere S^2.

The Hamiltonian depends on $\vec{B}(t)$, which is in turn determined by the polar coordinates δ, $\varphi(t) = \omega t$, and $r = B_0$. Therefore the parameter space of the Hamiltonian can be identified with a sphere S^2. The magnetic field $\vec{B}(t)$ traces out a closed curve C on S^2, which is the intersection of a cone with angle δ and the sphere S^2, as illustrated in Fig. 9.6.

To calculate the Berry phase we need the derivative in parameter space, which in our case is the nabla operator in spherical polar coordinates, as in eqn (6.44) we have

$$\vec{\nabla} = \vec{e}_r \frac{\partial}{\partial r} + \vec{e}_\vartheta \frac{1}{r} \frac{\partial}{\partial \vartheta} + \vec{e}_\varphi \frac{1}{r \sin \vartheta} \frac{\partial}{\partial \varphi} \,, \tag{9.83}$$

with $\vec{e}_r, \vec{e}_\vartheta, \vec{e}_\varphi$ as in eqns (6.42) of Section 6.4. Applied to the eigenstates (9.81a) and (9.81b) we find

$$\vec{\nabla} \, | \, n_+(t) \, \rangle = \frac{1}{r} \begin{pmatrix} -\frac{1}{2} \sin \frac{\delta}{2} \\ \frac{1}{2} e^{i\omega t} \cos \frac{\delta}{2} \end{pmatrix} \vec{e}_\vartheta + \frac{1}{r \sin \delta} \begin{pmatrix} 0 \\ i e^{i\omega t} \sin \frac{\delta}{2} \end{pmatrix} \vec{e}_\varphi \,, \tag{9.84a}$$

$$\vec{\nabla} \, | \, n_-(t) \, \rangle = \frac{1}{r} \begin{pmatrix} -\frac{1}{2} \cos \frac{\delta}{2} \\ -\frac{1}{2} e^{i\omega t} \sin \frac{\delta}{2} \end{pmatrix} \vec{e}_\vartheta + \frac{1}{r \sin \delta} \begin{pmatrix} 0 \\ i e^{i\omega t} \cos \frac{\delta}{2} \end{pmatrix} \vec{e}_\varphi \,. \tag{9.84b}$$

Applying the corresponding covectors $\langle \, n_+(t) \, |$ and $\langle \, n_-(t) \, |$, respectively, gives

$$\langle\, n_+(t)\,|\,\vec{\nabla}\,|\,n_+(t)\,\rangle \;=\; i\,\frac{\sin^2\frac{\delta}{2}}{r\sin\delta}\,\vec{e}_\varphi\,, \tag{9.85a}$$

$$\langle\, n_-(t)\,|\,\vec{\nabla}\,|\,n_-(t)\,\rangle \;=\; i\,\frac{\cos^2\frac{\delta}{2}}{r\sin\delta}\,\vec{e}_\varphi\,. \tag{9.85b}$$

We then integrate along the closed curve $C(r=\text{const.}, \delta=\text{const.}, \varphi\in[0,2\pi])$, which leads to

$$\oint \langle\, n_\pm\,|\,\vec{\nabla}\,|\,n_\pm\,\rangle\, r\sin\delta\, d\varphi\, \vec{e}_\varphi \;=\; i\pi(1\pm\cos\delta)\,, \tag{9.86}$$

so that the Berry phase from eqn (9.56) is finally found to have the explicit form

$$\gamma_\pm(C) \;=\; -\,\pi\,(1\mp\cos\delta)\,. \tag{9.87}$$

Geometrical interpretation of the Berry phase: The interesting result is that the calculated Berry phase depends only on the angle δ and *not* on the angular frequency ω or the magnetic field strength B_0. This is why the Berry phase is a *purely* geometric phase. Furthermore, there exists a relation between the Berry phase and the solid angle of the area enclosed by the curve C in the parameter space. The solid angle is calculated as

$$\Omega \;=\; \int d\Omega \;=\; \int_0^{2\pi} d\varphi \int_0^{\delta} \sin\delta\, d\delta \;=\; \int_0^{2\pi} d\varphi\, (1-\cos\delta(\varphi))\,, \tag{9.88}$$

where $\delta(\varphi)$ parametrizes the curve C, which is constant in our case. Thus we have

$$\Omega \;=\; 2\pi\,(1-\cos\delta)\,, \tag{9.89}$$

implying the relation

$$\gamma_\pm(C) \;=\; \mp\,\frac{1}{2}\,\Omega(C)\,. \tag{9.90}$$

The Berry phase is equal to the half of the solid angle of the area enclosed by the curve C, an interesting fact indeed.

The dynamical phase for a rotation, given by

$$\theta_\pm(T) \;=\; -\tfrac{1}{\hbar}\int_0^T E_\pm(t)\, dt \;=\; \mp\mu_{\mathrm{B}} B_0 T\,, \tag{9.91}$$

clearly depends on the period $T=\frac{2\pi}{\omega}$. After one rotation, where $B(T)=B(0)$, the energy eigenstate can be expressed as

$$|\,n_\pm(T)\,\rangle \;=\; e^{-i\pi(1\mp\cos\delta)}\, e^{\mp\mu_{\mathrm{B}} B_0 T}\, |\,n_\pm(0)\,\rangle\,, \tag{9.92}$$

i.e., the state picks up a geometric phase in addition to the dynamical phase.

9.5 A Rush through Differential Geometry and Topology

Differential geometry and topology are important and elegant tools in theoretical physics. They are the proper modern mathematical framework to formulate theories such as gauge theories and gravitation. Within differential geometry one may formulate Maxwell equations, non-Abelian Yang–Mills theories, and even the gravitational equations in a very compact and clear way. But also the physical phenomena we consider in our book, like the Aharonov–Bohm effect or the Berry phase, can be formulated purely from the point of view of geometry and topology. To provide the necessary mathematical background, we will first provide a brief overview of basic concepts in differential geometry and topology, but for a more extensive mathematical treatment we refer to comprehensive books such as Nakahara (1990), Frankel (1997), Hou and Hou (1997), Bertlmann (2000), and Sontz (2015).

9.5.1 Differential Geometry

The basic concepts of differential geometry are differential manifolds and differential forms with their associated operations.

Differentiable manifold: A differentiable manifold M ($\dim M = m$) is some smooth surface—a topological space—which looks Euclidean locally but is not necessarily Euclidean globally. Via *homeomorphisms*—one-to-one mappings between sets such that both the function and its inverse are continuous—to $\mathbb{R}^m$ one may associate an open set of coordinates to each point of M and require the transition from one set to another to be smooth (infinitely differentiable). Specifically, one may define *charts* $\{(U_\alpha, \varphi_\alpha)\}$, where $U_\alpha \subset M$ is an open subset of M with $\cup_\alpha U_\alpha = M$ and $\varphi_\alpha \to R_\alpha$ is a homeomorphic map into the open set $R_\alpha \subset \mathbb{R}^m$.

Tangent space: A *tangent vector* X on a manifold M is a differential operator $X = X^i \frac{\partial}{\partial x^i}$ given by its action on a function $f(x^i(t))$ over M in local coordinates, i.e.,

$$Xf = \frac{d}{dt} f(x^i(t)) = \frac{d}{dt} x^i(t) \frac{\partial}{\partial x^i} f = X^i \frac{\partial}{\partial x^i} f , \qquad (9.93)$$

where summation over indices appearing twice (once as a superscript and once as a subscript) is implied, i.e., we use the Einstein summation convention. The tangent vectors X span a vector space, the *tangent space* $T_x(M)$ ($\dim T_x(M) = \dim M = m$) for which one may choose the basis $\{\frac{\partial}{\partial x^i} \equiv \partial_i\}$, $i = 1, \ldots, m$.

Cotangent space: Contravariant vectors are dual to covariant (tangent) vectors, where duality is defined via the inner product

$$\left(dx^i, \frac{\partial}{\partial x^j} \right) = \delta^i{}_j . \qquad (9.94)$$

The set of differentials $\{dx^i\}$, $i = 1, \ldots, m$ forms a basis for the *cotangent space* $T_x^*(M)$, the dual space to $T_x(M)$. For example, the differential

$$df = \partial_i f(x) \, dx^i = \frac{\partial}{\partial x^j} f(x) \, dx^i \in T_x^*(M) \qquad (9.95)$$

is a covariant vector and as such an element of the cotangent space $T_x^*(M)$. The action of df on X is defined by the inner product

$$(df, X) = (\frac{\partial}{\partial x^i} f \, dx^i, X^j \frac{\partial}{\partial x^j}) = X^i \frac{\partial}{\partial x^i} f = X f . \tag{9.96}$$

Quite generally, an element of $T_x^*(M)$ is also called a 1-*form* $\omega = \omega_i dx^i \in T_x^*(M)$.

Tensors: *Tensors of type* (a, b) are constructed by mapping a elements of $T_x(M)$ and b elements of $T_x^*(M)$ into $\mathbb{R}$, i.e.,

$$T_b^a = \underbrace{T_x(M) \otimes \cdots \otimes T_x(M)}_{a \text{ factors}} \otimes \underbrace{T_x^*(M) \otimes \cdots \otimes T_x^*(M)}_{b \text{ factors}} . \tag{9.97}$$

An element of T_b^a is a *mixed tensor* with contravariant rank a and covariant rank b, in terms of local coordinates

$$T(x) = T^{i_1 \cdots i_a}{}_{j_1 \cdots j_b}(x) \frac{\partial}{\partial x^{i_1}} \cdots \frac{\partial}{\partial x^{i_a}} dx^{j_1} \cdots dx^{j_b} \in T_b^a . \tag{9.98}$$

Differential forms: In order to generalize the 1-forms of $T_x^*(M)$ we introduce the *wedge product* "$\wedge$", an antisymmetrized tensor product, i.e.,

$$dx^\mu \wedge dx^\nu = dx^\mu \otimes dx^\nu - dx^\nu \otimes dx^\mu . \tag{9.99}$$

From now on we choose Greek letters as indices since these are commonly used in many physical applications, in particular, when referring to the $3 + 1$ (spatial + temporal) coordinates of a spacetime with Minkowskian metric. In this notation, the wedge product implies $dx^\mu \wedge dx^\nu = - dx^\nu \wedge dx^\mu$ and $dx^\mu \wedge dx^\mu = 0$.

   Analogously, the higher wedge products are defined as totally antisymmetric tensors, e.g.,

$$dx^\mu \wedge dx^\nu \wedge dx^\sigma = dx^\mu \otimes dx^\nu \otimes dx^\sigma + dx^\nu \otimes dx^\sigma \otimes dx^\mu$$

$$+ dx^\sigma \otimes dx^\mu \otimes dx^\nu - dx^\mu \otimes dx^\sigma \otimes dx^\nu$$

$$- dx^\sigma \otimes dx^\nu \otimes dx^\mu - dx^\nu \otimes dx^\mu \otimes dx^\sigma, \tag{9.100}$$

$$dx^\mu \wedge dx^\nu \wedge dx^\sigma \wedge dx^\rho = dx^\mu \otimes dx^\nu \otimes dx^\sigma \otimes dx^\rho$$

$$+ \text{ all } \begin{pmatrix} +1 & \text{even} \\ -1 & \text{odd} \end{pmatrix} \text{ permutations,}$$

$$\text{etc.} \tag{9.101}$$

In this way we obtain totally antisymmetric tensors of rank: $0, 1, 2, \ldots, m = \dim M$. If the rank $p > m = \dim M$ exceeds the dimension of the manifold the wedge product vanishes.

With the wedge product at hand, *p-forms* are defined by

$$0\text{-form}\quad \omega = \omega(x)$$
$$1\text{-form}\quad \omega = \omega_\mu(x)\, dx^\mu$$
$$2\text{-form}\quad \omega = \frac{1}{2!}\omega_{\mu\nu}(x)\, dx^\mu \wedge dx^\nu$$
$$\cdots \cdots$$
$$p\text{-form}\quad \omega = \frac{1}{p!}\omega_{\mu_1 \ldots \mu_p}(x)\, dx^{\mu_1} \wedge \ldots \wedge dx^{\mu_p}\,, \tag{9.102}$$

where $\omega_{\mu_1\ldots\mu_p}(x)$ expresses a totally antisymmetric, covariant tensor field of rank $p \leq m$; in particular, ω vanishes for $p > m$, $\omega = 0$. The set of all p-forms is denoted by Λ^p. It is a vector space of dimension

$$\dim(\Lambda^p) = \binom{m}{p} = \frac{m!}{p!(m-p)!}\,. \tag{9.103}$$

Exterior derivative: We differentiate the forms by introducing the *exterior derivative*

$$d = \frac{\partial}{\partial x^\mu} dx^\mu \tag{9.104}$$

acting on a *p*-form in the following way

$$d\omega = \frac{1}{p!}\frac{\partial}{\partial x^\nu}\omega_{\mu_1\ldots\mu_p}(x)\, dx^\nu \wedge dx^{\mu_1} \wedge \ldots \wedge dx^{\mu_p}. \tag{9.105}$$

The exterior derivative is a map $d : \Lambda^p \to \Lambda^{p+1}$ which transforms p-forms into $(p+1)$-forms. It is *nilpotent*, satisfying the important property $d^2 = 0$. In addition, the exterior derivative d obeys the *anti-derivation rule*

$$d(\alpha_p\beta_q) = (d\alpha_p)\beta_q + (-)^p\alpha_p d\beta_q, \tag{9.106}$$

where we have indicated the degrees of the forms by the respective subscripts.

If $d\omega_p = 0$ the form ω_p is called *closed*, if $\omega_p = d\alpha_{p-1}$ it is called *exact*. Note that

$$\text{exact} \quad \overset{\Rightarrow}{\not\Leftarrow} \quad \text{closed}. \tag{9.107}$$

Indeed, the relationship between the properties closedness and exactness is more intricate, as can be formulated in terms of the following lemma.

Lemma 9.1 (Poincaré's lemma)

Any closed form, $d\omega = 0$, can be expressed as a locally exact form, $\omega = d\alpha$, but is not necessarily globally exact.

For a full proof of Poincaré's lemma we refer the interested reader to Bertlmann (2000, p. 328). Here, let us illustrate the lemma using an example of a closed form that is not exact. Let us consider the manifold $M = \mathbb{R}^2 \setminus \{0\}$, and the 1-form $\omega = \omega_i dx^i$, with components

$$\omega_1 = \frac{-y}{r^2} \quad \text{and} \quad \omega_2 = \frac{x}{r^2}, \tag{9.108}$$

where we have chosen the coordinates $x^1 = x$, and $x^2 = y$ with $r^2 = x^2 + y^2$. This 1-form ω is well defined on M. It is closed since $d\omega = \frac{\partial}{\partial x^i} \frac{x^i}{r^2} dx^1 dx^2 = 0$. But ω is *not* exact. Exactness would mean that there is a function $f(x, y)$ such that $\omega = df = d\left(\arctan \frac{y}{x}\right)$. However, $f(x, y)$ is not continuous (hence not differentiable) everywhere on the manifold $M = \mathbb{R}^2 \setminus \{0\}$. Therefore $\omega \neq df$ *globally*, ω is *not* exact.

As another example, consider the 1-form $\omega = \omega_{\text{Re}} + i\,\omega_{\text{Im}} = \frac{dz}{z} = d \ln z$, where

$$z = x + iy, \qquad dz = dx + idy, \qquad \frac{dz}{z} = \frac{xdx + ydy + i(-ydx + xdy)}{r^2}, \tag{9.109}$$

and we can directly read off the real-valued and purely imaginary components of ω_{Re} and ω_{Im}, respectively. We then have $d\omega = 0$, ω is closed, since $d^2 = 0$ or $d\omega_{\text{Re}} = d\omega_{\text{Im}} = 0$ explicitly. However, if we consider the same manifold as in the previous example, the logarithm $\ln z$ is *not* defined on the whole manifold $M = \mathbb{R}^2 \setminus \{0\}$, hence $\omega \neq d \ln z$ *globally* and ω is *not* exact. On the other hand, if we cut the plane along the negative x axis, so if we consider the manifold $M = \mathbb{R}^2 \setminus \{(x, y) | x \leq 0, y = 0\}$, where $\omega = d \ln z$ is well defined, we find that ω is exact.

Interior product: The *interior product* i_X is a contracted multiplication of a p-form $\omega_p \in \Lambda^p$ with a vector field $X \in T_x(M)$ and acts according to

$$i_X \omega_p = \frac{1}{p!} \sum_{s=1}^{p} X^{\mu_s} \omega_{\mu_1 \ldots \mu_s \ldots \mu_p} (-)^{s-1} dx^{\mu_1} \wedge \ldots \widehat{dx^{\mu_s}} \ldots \wedge dx^{\mu_p}$$

$$= \frac{1}{(p-1)!} X^{\nu} \omega_{\nu \mu_2 \ldots \mu_p} dx^{\mu_2} \wedge \ldots \wedge dx^{\mu_p}, \tag{9.110}$$

where the symbol $\widehat{}$ means that the corresponding quantity has been omitted. Note that the minus sign stems from the anti-derivative property of i_X, clearly $i_X^2 = 0$. Viewed as a mapping, the operation $i_X : \Lambda^p \to \Lambda^{p-1}$ can be considered as a mapping in the "opposite direction" to the exterior derivative d. That is, we can proceed with the contraction of a p-form to arrive at a 0-form

$$i_{X_p} \ldots i_{X_1} \omega_p = \omega_{\mu_1 \ldots \mu_p} X^{\mu_1} \ldots X^{\mu_p} = \omega(X_1, \ldots, X_p). \tag{9.111}$$

Integration of differential forms: The integration of differential forms on a manifold M is the inverse operation to the exterior derivative. The differential forms automatically provide an integration measure or volume element together with the correct transformation properties. More specifically, let us choose an *orienting m-form* on a compact, orientable manifold M

$$\omega \; = \; \frac{1}{m!} \, \varepsilon_{\mu_1 \ldots \mu_m} \, dx^{\mu_1} \wedge \ldots \wedge dx^{\mu_m} \; = \; dx^1 dx^2 \ldots dx^m \; \equiv \; dV_x \, , \qquad (9.112)$$

with $\varepsilon_{\mu_1 \ldots \mu_m}$ the *totally antisymmetric ε-tensor* or *Levi–Civita tensor*:

$$\varepsilon_{\mu_1 \ldots \mu_m} \; = \; \begin{cases} 1 \text{ for any even permutations of } 1, \ldots, m \\ 0 \text{ if two indices are equal} \\ -1 \text{ for any odd permutations of } 1, \ldots, m. \end{cases} \qquad (9.113)$$

This provides an oriented volume element dV_x.

For any smooth function f on M we may then define an m-form given by $f\omega$. We define the integral of such an m-form in the domain of a chart $(U_\alpha, \varphi_\alpha)$ with coordinates $(x^1, x^2, \ldots, x^m)$ by

$$\int_{U_\alpha} f\omega \; = \; \int_{\varphi_\alpha(U_\alpha)} f\big(\varphi_\alpha^{-1}(x)\big) \, dx^1 \ldots dx^m \; = \; \int_{R_\alpha} f(x) \, dV_x. \qquad (9.114)$$

To extend the integral of the m-form to the whole manifold we have to consider an *atlas of charts* (a set of compatible charts covering M)

$$\int_M f\omega \; = \; \sum_\alpha \int_{U_\alpha} f_\alpha \omega \, , \qquad (9.115)$$

where $f = \sum_\alpha f_\alpha$ and f_α has its support on the chart $(U_\alpha, \varphi_\alpha)$. With these notions at hand, we can formulate an important theorem.

Theorem 9.3 (Stokes' theorem)

Let M be an orientable compact manifold with dimension $\dim(M) = m$ and non-empty boundary ∂M, and let $\omega \in \Lambda^{m-1}$ be an $(m-1)$-form, then

$$\int_M d\omega \; = \; \int_{\partial M} \omega \, .$$

Stokes' theorem expresses an interesting connection between information about a quantity that is available in the whole manifold and the information accessible at the boundary ∂M of the manifold.

As an example in two dimensions, consider a disk $M = \{(x_1, x_2) | x_1^2 + x_2^2 \leq 1\}, \partial M = \{(x_1, x_2) | x_1^2 + x_2^2 = 1\}$. We then define the forms

$$\omega = \omega_1(x)dx^1 + \omega_2(x)dx^2 \equiv \vec{A}(x)d\vec{x} \,, \tag{9.116}$$

$$d\omega = \frac{\partial}{\partial x^i}\omega_j(x)\, dx^i \wedge dx^j = \frac{1}{2}(\partial_i A_j - \partial_j A_i)\, dx^i dx^j$$

$$\equiv \frac{1}{2}\varepsilon_{ijk}B^k\, dx^i dx^j = B^k df_k = \vec{B}d\vec{f} \tag{9.117}$$

with the curl $B^k = \varepsilon^{kij}\partial_i A_j$ or $\vec{B} = \vec{\nabla} \times \vec{A}$ multiplied by an area element $df_k = \frac{1}{2}\varepsilon_{kij}dx^i dx^j$. Then we recover Stokes' theorem in the familiar notation, i.e., we get *Stokes' theorem in two dimensions,*

$$\int_M d\omega = \int_{\text{disk}} (\vec{\nabla} \times \vec{A})\, d\vec{f} = \int_{\text{circle}} \vec{A}\, d\vec{x} = \int_{\partial M} \omega \,, \tag{9.118}$$

where $\vec{A}$ is the vector potential and $\vec{B}$ the magnetic field. The integral of the magnetic flux through the disk equals the integral of the vector potential along the boundary.

The three-dimensional analogue to the previous example is to consider a ball, that is, we define the forms

$$\omega = \frac{1}{2}\omega_{ij}(x)\, dx^i \wedge dx^j \equiv \frac{1}{2}\varepsilon_{ijk}E^k(x)\, dx^i \wedge dx^j = E^k df_k = \vec{E}d\vec{f} \,, \tag{9.119}$$

$$d\omega = \frac{1}{2}\partial_\ell \omega_{ij}\, dx^\ell \wedge dx^i \wedge dx^j = \frac{1}{2}\varepsilon_{ijk}\partial_\ell E^k\, dx^\ell \wedge dx^i \wedge dx^j$$

$$= \frac{1}{2}\partial_\ell E^k \varepsilon_{ijk}\varepsilon^{\ell ij}\, dx^1 dx^2 dx^3 = \partial_k E^k\, dV = \vec{\nabla}\vec{E}\, dV \,, \tag{9.120}$$

where $\vec{E}$ is the electric field. In this case, we recover what we usually call the *Gauss law* or Gauss' divergence theorem, i.e.,

$$\int_M d\omega = \int_{\text{ball}} \vec{\nabla}\vec{E}\, dV = \int_{\text{sphere}} \vec{E}\, d\vec{f} = \int_{\partial M} \omega \,. \tag{9.121}$$

Hodge star operation: There is a duality—in the form of an isomorphism—between the vector spaces Λ^p and Λ^{m-p} (with $\dim(\Lambda^{m-p}) = \dim(\Lambda^p)$) given by the *Hodge $*$ operation*: $\Lambda^p \overset{*}{\longrightarrow} \Lambda^{m-p}$. The star transforms p-forms into $(m-p)$-forms and is defined by

$$*\omega_p = \frac{1}{p!(m-p)!}\, \omega_{\mu_1 \dots \mu_p}\, \varepsilon^{\mu_1 \dots \mu_p}{}_{\mu_{p+1} \dots \mu_m}\, dx^{\mu_{p+1}} \wedge \dots \wedge dx^{\mu_m} \,, \tag{9.122}$$

where the indices of the Levi–Civita tensor are raised and lowered by the metric tensor $g^{\mu\nu}$ (in m dimensions)

$$\varepsilon^{\mu_1 \dots \mu_p}{}_{\mu_{p+1} \dots \mu_m} = g^{\mu_1 \nu_1} \dots g^{\mu_p \nu_p}\, \varepsilon_{\nu_1 \dots \nu_p \mu_{p+1} \dots \mu_m} \,. \tag{9.123}$$

When the Hodge star is applied twice we have

$$
** = \begin{cases} (-)^{p(m-p)} \\[2mm] (-)^{p(m-p)+1} \end{cases} \quad \text{or} \quad *^{-1} = \begin{cases} (-)^{p(m-p)}* & \text{Euclidean} \\[2mm] (-)^{p(m-p)+1}* & \text{Minkowskian.} \end{cases} \tag{9.124}
$$

Push forward: Consider a smooth map $f : M \to N$ between two manifolds M and N. It induces a map between the tangent spaces of the manifolds which is called, for this reason, *tangent map* or *push forward* $f_* : T_x(M) \to T_{f(x)}(N)$. Its action on a vector field $X \in T_x(M)$ is given by

$$
f_* X[g] = X[gf] \,, \tag{9.125}
$$

where $f_* X \in T_{f(x)}(N)$, the map $g : N \to \mathbb{R}$ and consequently $gf : M \to \mathbb{R}$. In terms of the components of a vector $X^\mu \frac{\partial}{\partial x^\mu}$ we can write

$$
f_* X^\nu = X^\mu \frac{\partial y^\nu}{\partial x^\mu} \,, \tag{9.126}
$$

where x^μ and y^ν are the respective coordinates of the charts. Thus the components X^μ and $f_* X^\nu$ are related by the Jacobian $\frac{\partial y^\nu}{\partial x^\mu}$ of the map f.

Pullback: The smooth map $f : M \to N$ also induces a map between the dual spaces—the cotangent spaces—which is called the *pullback* $f^* : T_{f(x)}^*(N) \to T_x^*(M)$. Choosing a vector $X = X^\nu \frac{\partial}{\partial x^\nu} \in T_x(M)$ and a 1-form $\omega = \omega_\mu dy^\mu \in T_{f(x)}^*(N)$ the pullback $f^* \omega = f^* \omega_\mu dx^\mu \in T_x^*(M)$ is defined by

$$
(f^* \omega, X) = (\omega, f_* X) \,, \tag{9.127}
$$

or $f^* \omega(X) = \omega(f_* X)$ in contraction notation. Then we find the components of the pullback as

$$
f^* \omega_\mu(x) = \omega_\nu\big(y(x)\big) \frac{\partial y^\nu}{\partial x^\mu} \,. \tag{9.128}
$$

Again the two components ω_ν and $f^* \omega_\mu$ are related by the Jacobian of the map f. In local coordinates f is defined by $y^\nu(x^\mu)$ with $\mu = 1, \ldots, m = \dim(M)$ and $\nu = 1, \ldots, n = \dim(N)$.

It is straightforward to generalize the pullback f^* for p-forms $\omega_p \in \Lambda^p(N)$. In this case, the pullback map $f^* : \Lambda_{f(x)}^p(N) \to \Lambda_x^p(M)$ is defined by (in the notation of eqn (9.111))

$$
f^* \omega(X_1, \ldots, X_p) = \omega(f_* X_1, \ldots, f_* X_p) \,, \tag{9.129}
$$

and explicitly we have

$$
f^* \omega = \frac{1}{p!} f^* \omega_{\mu_1 \ldots \mu_p} \, dx^{\mu_1} \wedge \ldots \wedge dx^{\mu_p} \in \Lambda^p(M) \,, \tag{9.130a}
$$

$$
f^* \omega_{\mu_1 \ldots \mu_p}(x) = \omega_{\nu_1 \ldots \nu_p}(y(x)) \frac{\partial y^{\nu_1}}{\partial x^{\mu_1}} \cdots \frac{\partial y^{\nu_p}}{\partial x^{\mu_p}} \,. \tag{9.130b}
$$

9.5.2 Fibre Bundles

In topology a fibre bundle is a space that locally looks like a product space but which may globally not be one. The essential quantities of a bundle are the so-called fibres. Pictorially speaking, the way these fibres "twist" when moving around in the space reveals the topological structure of the bundle and as such the features of the physical situation that the bundle represents, in our case, the Aharanov-Bohm effect, the Berry phase, and the Dirac monopole.

The *bundle setup*, denoted by (E, Π, M, F, G), is the following:

i) *E total space* (topological space),

ii) *M base space* (topological space),

iii) $\Pi : E \to M$ *projection* (surjection),

iv) *F standard fibre* (topological space). It is homeomorphic (continuous and invertible) to all inverse images $\Pi^{-1}(x) = F_x$ with $x \in M$; the F_x are called fibres,

v) *G structure group*: A group of homeomorphisms of the fibre F,

vi) *Local trivialization*: A covering of open sets $\{U_\alpha\}$ of M together with homeomorphisms ϕ_α, i.e., $\forall U_\alpha \exists \phi_\alpha : \Pi^{-1}(U_\alpha) \to U_\alpha \times F$ such that $\Pi\phi_\alpha^{-1}(x, f) = x$ with $x \in U_\alpha$, $f \in F$. The homeomorphism ϕ_α maps $\Pi^{-1}(U_\alpha) \subset E$ onto the direct product space $U_\alpha \times F$ and is called a *local trivialization*.

A fibre bundle is illustrated in Fig. 9.7 for the example of a Möbius strip consisting of line segments, the fibres F. The whole strip represents the total fibre space E which is projected down by Π onto the base space $M = S^1$. The homeomorphism ϕ_α "untwists" the fibre space $\Pi^{-1}(U_\alpha)$ onto the product space $U_\alpha \times F$, the local trivialization.

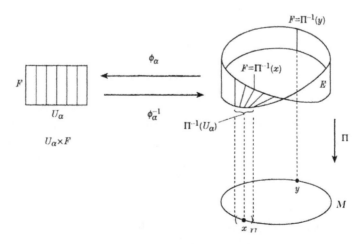

Fig. 9.7 Fibre bundle. Illustration by a Möbius strip. F denotes the fibre and E the total space, Π is the projection and M the base space. ϕ_α denotes a homeomorphic map from the bundle space $\Pi^{-1}(U_\alpha)$ onto the product space $U_\alpha \times F$. Reproduced from Fig. 2.38 in Reinhold A. Bertlmann, *Anomalies in Quantum Field Theory* (2000), with permission from Oxford University Press.

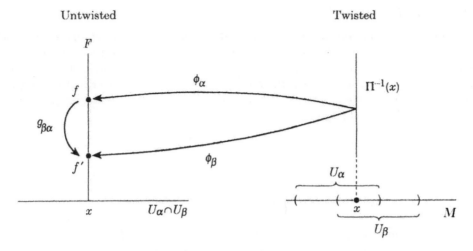

Fig. 9.8 Transition function. The transition function $g_{\beta\alpha}$ from one fibre point f to another $f' = g_{\beta\alpha}f$ untwists the bundle. Reproduced from Fig. 2.39 in Reinhold A. Bertlmann, *Anomalies in Quantum Field Theory* (2000), with permission from Oxford University Press.

Global bundle structure—transition functions: The global structure of the bundle is determined by the structure group G which arises by transitioning from one set of *local bundle coordinates* (U_α, ϕ_α) to another set (U_β, ϕ_β). The homeomorphic map in the overlap $U_\alpha \cap U_\beta \neq \emptyset$ is called *transition function*

$$g_{\alpha\beta} := \phi_\alpha \cdot \phi_\beta^{-1} : \quad (U_\alpha \cap U_\beta) \times F \longrightarrow (U_\alpha \cap U_\beta) \times F. \tag{9.131}$$

As illustrated in Fig. 9.8, the transition function maps one fibre point $f \in F$ corresponding to the local trivialization (U_α, ϕ_α) to another point $f' \in F$ corresponding to (U_β, ϕ_β),

$$f' = g_{\beta\alpha}f. \tag{9.132}$$

The transition functions form the structure group $G = \{g_{\beta\alpha}\}$, they tell us how the fibres are "glued" together and give us the full information about the global structure of the bundle. The transition functions satisfy the *consistency conditions*

$$
\begin{aligned}
g_{\alpha\alpha}(x) &= \mathbb{1}, & x &\in U_\alpha \\
g_{\alpha\beta}(x) &= g_{\beta\alpha}^{-1}(x), & x &\in U_\alpha \cap U_\beta \\
g_{\alpha\beta}(x)g_{\beta\gamma}(x) &= g_{\alpha\gamma}(x), & x &\in U_\alpha \cap U_\beta \cap U_\gamma\,,
\end{aligned}
\tag{9.133}
$$

and they are *not* unique but can be transformed by a homeomorphism h_α that belongs to the structure group $g'_{\alpha\beta}(x) = h_\alpha^{-1}(x)g_{\alpha\beta}(x)h_\beta(x)$. That means the bundles E with structure group elements $\{g_{\alpha\beta}\}$ and E' with $\{g'_{\alpha\beta}\}$ are topologically equivalent.

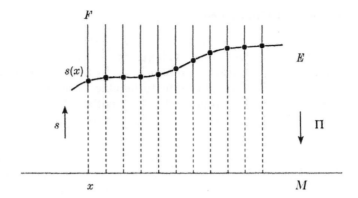

Fig. 9.9 Section of bundle. A section $s(x) : M \rightarrow E$ of a fibre bundle E. Reproduced from Fig. 2.41 in Reinhold A. Bertlmann, *Anomalies in Quantum Field Theory* (2000), with permission from Oxford University Press.

Section: A (global) *section* of a bundle E is a continuous map

$$s : M \rightarrow E \quad \text{with } \Pi s(x) = x, \ \ \forall x \in M. \tag{9.134}$$

We can always define a *local section* over an open subset $U \subset M$ of the manifold, however, a global section might not exist, see Fig. 9.9. The section transforms under a change of the local coordinates according to $s_\beta(x) = s_\alpha(x) \cdot g_{\alpha\beta}(x)$.

Examples for different types of fibre bundles include:

Principal bundle: The fibre F_x is identical with the structure group G, i.e., $F_x \equiv G$. The principal bundle is denoted by $P(M, G)$ and plays an important role in physics, specifically in gauge theories.

Vector bundle: The fibre F_x is a vector space.

Line bundle: The fibre F_x is a one-dimensional vector space, a line.

Associated bundle: The structure group G has a certain representation ρ on the vector space F. It is termed *associated to the principal bundle* if its transition functions $h_{\alpha\beta}$ are the images under ρ of the corresponding transition functions of the principal bundle: $h_{\alpha\beta} = \rho(g_{\alpha\beta})$.

Tangent bundle: The fibre F_x is the tangent space of the base manifold M: $F_x \equiv T_x(M)$.

Cotangent bundle: When the fibre F_x is identical to the cotangent space of the base manifold M: $F_x \equiv T_x^*(M)$.

9.5.3 Connection and Curvature

The concepts connection and curvature are of great importance for fibre bundles. The connection allows for parallel transport of the fibres and also induces a curvature in the bundle. More specifically, we can reach each fibre point by multiplication with a group element, recall eqn (9.132). But we also want to be able to move in the (pictorially speaking) horizontal direction, i.e., from one fibre to another. To this end the concept of a connection is introduced. It also describes how vectors are parallel transported.

Connection: Let us then study a principal bundle $P(M, G)$ with a Lie group G (recall Section 6.1.3), e.g., U(N), SU(N), GL($m, \mathbb{R}$), SO(d), or similar. Consider the tangent space $T_u(P)$ to a fibre point $u \in P$, then we can split the tangent space into two complimentary subspaces, called the vertical and horizontal subspaces, as illustrated in Fig. 9.10.

The *vertical subspace* $V_u(P)$ is defined by

$$V_u(P) := \{ X \in T_u(P) \,|\, \Pi_* X = 0 \} , \tag{9.135}$$

i.e., it consists of all tangent vectors at u for which the push forward vanishes, i.e., it is the kernel of the tangent map. Meanwhile, a *horizontal subspace* $H_u(P)$ is defined relative to the vertical subspace in such a way that the *direct sum* of the vertical and

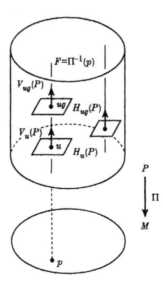

Fig. 9.10 Vertical and horizontal subspaces. The tangent space $T_u(P)$ of the fibre bundle $P(M, G)$ is decomposed into the vertical subspace $V_u(P)$ and horizontal subspace $H_u(P)$. One may move along the fibre via the right-action of the group G on P, $R_g u = ug$. Reproduced from Fig. 2.43 in Reinhold A. Bertlmann, *Anomalies in Quantum Field Theory* (2000), with permission from Oxford University Press.

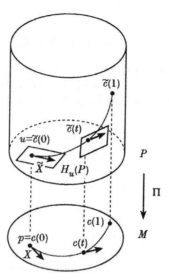

Fig. 9.11 Parallel transport. A tangent vector X in a base manifold M is lifted into the horizontal subspace $\widetilde{X} \in H_u(P)$. The fibre point u is parallel-transported along the base curve c. Reproduced from Fig. 2.44 in Reinhold A. Bertlmann, *Anomalies in Quantum Field Theory* (2000), with permission from Oxford University Press.

horizontal subspaces yields the tangent space $V_u(P) \oplus H_u(P) = T_u(P)$. While the vertical subspace is uniquely determined by the mentioned condition, there is still room to choose different horizontal subspaces, which can be used to define a connection.

A *connection* assigns a horizontal subspace $H_u(P) :\ u \in P \mapsto H_u(P) \subset T_u(P)$ to each fibre point $u \in P$ in a smooth way, i.e., such that

$$\text{i)}\ \ T_u(P) = V_u(P) \oplus H_u(P)\,, \tag{9.136a}$$

$$\text{ii)}\ \ R_{g*}H_u(P) = H_{ug}(P)\,, \tag{9.136b}$$

where R_{g*} denotes the push forward of the right-action $R_g u = ug$ of the structure group G on the bundle P; and $g \in G$, $u \in P$.

Horizontal lift: Tangent vector X can be "lifted" from a base manifold M into the horizontal subspace $H_u(P)$ of the fibre bundle. The *horizontal lift* $\widetilde{X} \in H_u(P)$ of a tangent vector $X \in T_p(M)$, illustrated in Fig. 9.11, is defined by

$$\text{i)}\ \ \Pi_* \widetilde{X} = X, \tag{9.137a}$$

$$\text{ii)}\ \ \widetilde{X}^V = 0 \qquad \text{vertical component}\,. \tag{9.137b}$$

Similarly we can lift curves $c(t)\colon t \in [0,1] \to M$ from the base manifold M into the bundle P, as illustrated in Fig. 9.11. The *lifted curve* $\widetilde{c}(t)\colon t \in [0,1] \to P$ has to satisfy

$$\text{i) } \Pi\widetilde{c} = c, \tag{9.138a}$$

$$\text{ii) tangent vector to } \widetilde{c}\colon \widetilde{X} \in H_{\widetilde{c}(t)}(P). \tag{9.138b}$$

If a curve $c(t) \in M$ in the base manifold M is closed, $c(1) = c(0)$, the horizontal lift $\widetilde{c}(t)$ *need not* be closed, $\widetilde{c}(1) = \widetilde{c}(0) \cdot h$, with $h \in G$. If the lifted curve is not closed, we call it a *holonomy* and $H = \{h\} \subset G$ the *holonomy group*.

Connection 1-form: A quantity of great physical significance, and indeed, usefulness, is the *connection 1-form ω*. It is a Lie-algebra valued 1-form on the bundle P, i.e., a tensor product of a 1-form and an element of a Lie algebra $\omega \in T^* \otimes \mathrm{Lie}G$. The basis elements of $\mathrm{Lie}G = \{T^a\}$, called generators T^a of the Lie group $G = \{g = e^{\alpha^a T_a}\}$, satisfy the commutation relation $[T_a, T_b] = f_{abc}T_c$ with f_{abc} the totally antisymmetric *structure constants* of the algebra, recall Section 6.1.3.

The connection 1-form ω is defined by the projection of the tangent space $T_u(P)$ onto the vertical subspace $V_u(P) \simeq \mathrm{Lie}G$, i.e., $\omega : T_u(P) \to V_u(P)$. Then the horizontal subspace $H_u(P)$ is given by the kernel of the map ω

$$H_u(P) = \{X \in T_u(P) \,|\, \omega(X) = 0\}. \tag{9.139}$$

In physics we work with local coordinates. Locally the connection 1-form ω represents a Yang–Mills gauge potential $\mathcal{A}$ which arises from the pullback of a local section

$$s_\alpha : U_\alpha \to \Pi^{-1}(U_\alpha) \tag{9.140}$$

$$\text{then} \quad \mathcal{A}_\alpha = s_\alpha^* \omega \in \Lambda^1(U_\alpha) \otimes \mathrm{Lie}G. \tag{9.141}$$

The connection 1-form ω is uniquely defined on the bundle, therefore we have the identity $\omega|_{U_\alpha} \equiv \omega|_{U_\beta}$ in the overlap of two neighbourhoods U_α, U_β with $U_\alpha \cap U_\beta \neq \emptyset$. As a consequence the Yang–Mills potential obeys the following *compatibility condition* in the overlap, which is nothing but the familiar *gauge transformation law for Yang–Mills potentials*

$$\mathcal{A}_\beta = g_{\alpha\beta}^{-1}\mathcal{A}_\alpha g_{\alpha\beta} + g_{\alpha\beta}^{-1}dg_{\alpha\beta} \quad \text{on } U_\alpha \cap U_\beta, \tag{9.142}$$

with $g_{\alpha\beta}$ the transition function—the gauge element in physics—from U_α to U_β.

For example, in electrodynamics, which is a U(1) gauge theory, the algebra is trivial, $g = e^{\alpha(x)}$, and we obtain, written as 1-form,

$$\mathcal{A}^g(x) = \mathcal{A}(x) + d\alpha(x). \tag{9.143}$$

Whereas for SU(3) Yang–Mills theories, with gauge $g(x) = e^{\alpha^a(x)T_a} \in SU(3)$, we find

$$\mathcal{A}^g(x) = \mathcal{A}_\mu^{a\,g}(x)T_a dx^\mu = g^{-1}(x)\mathcal{A}_\mu^a(x)T_a dx^\mu g(x) + g^{-1}(x)\partial_\mu dx^\mu g(x)$$

$$= g^{-1}\mathcal{A}g + g^{-1}dg. \tag{9.144}$$

Curvature 2-form: First we define the *exterior covariant derivative D* on a q-form φ via the relation

$$D\,\varphi(X_1, \ldots, X_{q+1}) = d_p\,\varphi(X_1^{\mathrm{H}}, \ldots, X_{q+1}^{\mathrm{H}}), \tag{9.145}$$

where $X_i^{\mathrm{H}} \in H_u(P)$ refers to the horizontal component of $X_i \in T_u(P)$ determined by the connection, and $d_p\varphi = d_p\varphi^a \otimes T_a$ is the exterior derivative on the bundle P, evaluated on the horizontal components of the vector fields. Then the *curvature 2-form* Ω on the bundle P is defined as the covariant derivative of the connection 1-form ω

$$\Omega = D\omega = d_p\omega + \omega^2 \in \Lambda^2(P) \otimes \mathrm{Lie}G. \tag{9.146}$$

In our notation ω^2 means the wedge product of ω with itself, which is half of the commutator: $\omega^2 = \omega \wedge \omega = \frac{1}{2}[\omega, \omega]$. Note, the commutator $[\omega, \omega]$ does *not* vanish due to the commutation relation of the Lie algebra.

The local representation provides the Yang–Mills field strength $\mathcal{F}$

$$\mathcal{F} = s^*\Omega = s^*d_p\omega + s^*(\omega\omega) \in \Lambda^2(U) \otimes \mathrm{Lie}G$$

$$= ds^*\omega + s^*\omega s^*\omega = d\mathcal{A} + \mathcal{A}^2, \tag{9.147}$$

where s^* is the pullback of a local section $s : U \to \Pi^{-1}(U)$ on a chart $U \subset M$.

The *compatibility condition* in the overlap of the charts $U_\alpha \cap U_\beta$ is given by

$$\mathcal{F}_\beta = g_{\alpha\beta}^{-1}\mathcal{F}_\alpha g_{\alpha\beta} \quad \text{on } U_\alpha \cap U_\beta. \tag{9.148}$$

This is precisely the *gauge transformation of the field strength* in physics

$$\mathcal{F}^g = g^{-1}\mathcal{F}g. \tag{9.149}$$

For U(1) electrodynamics we have

$$\mathcal{F} = d\mathcal{A} = \partial_\mu\mathcal{A}_\nu\,dx^\mu \wedge dx^\nu = \frac{1}{2}\,\mathcal{F}_{\mu\nu}\,dx^\mu \wedge dx^\nu, \tag{9.150a}$$

with the electromagnetic field-strength tensor $\mathcal{F}_{\mu\nu} = \partial_\mu\mathcal{A}_\nu - \partial_\nu\mathcal{A}_\mu$, whereas for a Yang–Mills theory with $\mathcal{A} = \mathcal{A}_\mu^a T_a dx^\mu$ we find

$$\mathcal{F} = d\mathcal{A} + \mathcal{A}^2 = \partial_\mu\mathcal{A}_\nu\,dx^\mu \wedge dx^\nu + \mathcal{A}_\mu\mathcal{A}_\nu\,dx^\mu \wedge dx^\nu$$

$$= \frac{1}{2}\,\mathcal{F}_{\mu\nu}\,dx^\mu \wedge dx^\nu \tag{9.151a}$$

with the corresponding field-strength tensor $\mathcal{F}_{\mu\nu} = \partial_\mu\mathcal{A}_\nu - \partial_\nu\mathcal{A}_\mu + [\mathcal{A}_\mu, \mathcal{A}_\nu]$.

9.6 Topological Interpretation of Physical Effects

The physics behind the Aharonov–Bohm effect, the Berry phase and the Dirac monopole (which we will discuss in Section 9.6.3) can also be understood from a purely mathematical—or more precisely, geometrical—perspective. We can describe these phenomena elegantly within topology, within the fibre bundle formalism. For further literature, we refer the interested reader to Nakahara (1990), Bertlmann (2000), and Sontz (2015).

9.6.1 Aharonov–Bohm Effect and Topology

Recalling the setup of the Aharonov–Bohm experiment (Section 9.3), the double slit with the solenoid in the middle (Fig. 9.3), the natural choice for a base manifold M is a two-dimensional plane with the origin $\{0\}$ extracted, or, alternatively, cutting out a disk $\{D^2\}$ corresponding to the diameter of the solenoid,

$$M = \mathbb{R}^2 \setminus \{0\}. \tag{9.152}$$

The gauge group of electromagnetism is U(1). It supplies the *structure group* U(1) of the bundle, where all possible group elements build up the *fibre*. So we construct the *principal bundle* $P(M,\text{U}(1))$. The wave function $\psi(x)$ describing the paths of the electron corresponds to a *section* of the *associated vector bundle*, a line bundle.

The gauge potential $\vec{A}$ in the Aharonov–Bohm experiment can be written as a 1-form

$$A = A_i dx^i = \frac{\Phi_\text{M}}{2\pi r^2}(-ydx + xdy) = \frac{\Phi_\text{M}}{2\pi}\,d\varphi\,, \tag{9.153}$$

with $\frac{e}{\hbar c}\Phi_\text{M} = \Phi_\text{B}$ the magnetic flux (9.44). It represents the local *connection 1-form* of the bundle, which determines the parallel transport of the fibre

$$\mathcal{A} = iA. \tag{9.154}$$

The field strength $F = dA$ corresponds to the local curvature 2-form

$$\mathcal{F} = iF = d\mathcal{A}. \tag{9.155}$$

In the case of the Aharonov–Bohm effect, we have $\mathcal{F} = 0$, so $\mathcal{A}$ is closed but not exact (recall the example in eqn (9.109) of Section 9.5.1). Thus we cannot gauge away the potential on the entire manifold.

Holonomy: Generally, a fibre point is *parallel-transported* along a curve σ in the base manifold M by the phase factor containing the connection (let us here choose $q_\text{e} = \hbar = c = 1$ for simplicity)

$$\exp\left[-\int_\sigma \mathcal{A}\right] \overset{\text{closed curve in } M}{\longrightarrow} \exp\left[-i\oint A\right] = \exp\left[-i\,\Phi_\text{M}\right]. \tag{9.156}$$

For a closed curve in M (9.152), a circle around the origin, the lifted curve in the fibre space U(1) is not closed any more. We get a *holonomy* in the fibre space. Classifying the curves in M by the winding numbers n we obtain the *holonomy group*

$$H = \left\{h = e^{-in\Phi_\text{M}}\right\}, \quad n \in \mathbb{Z}. \tag{9.157}$$

It characterizes the maps of the closed curves in M into the fibre space U(1).

We also detect this holonomy by considering the covariant derivative for the parallel transport of wave function. Then we have

$$\mathcal{D}\psi = (d + \mathcal{A})\psi = 0,\qquad(9.158)$$

where $\mathcal{D} = d + \mathcal{A}$ denotes the covariant derivative. Using polar coordinates for the calculation, we have

$$\mathcal{D}\psi = \left(d + i\frac{\Phi_{\mathrm{M}}}{2\pi}\,d\varphi\right)\psi = 0,\qquad(9.159)$$

and we easily get the solution

$$\psi(\varphi) = \text{const} \cdot \exp\left[-i\Phi_{\mathrm{M}}\,\frac{\varphi}{2\pi}\right].\qquad(9.160)$$

Along a circle S^1 in the base manifold, given by $\varphi : 0 \to 2\pi$ (winding number $n = +1$), the wave function—a fibre point—is transported as

$$\psi(0) \mapsto \psi(2\pi) = \psi(0)\cdot e^{-i\Phi_{\mathrm{M}}}.\qquad(9.161)$$

The corresponding fibre curve does not close: A *holonomy* occurs. So, geometrically the Aharonov–Bohm effect occurs due to a holonomy in the fibre space.

9.6.2 Berry Phase and Topology

Next we want to present Barry Simon's (1983) fibre-bundle perspective of the Berry phase, which we have discussed in Section 9.4.2. In order to do so, we have to construct a bundle that corresponds to the setup where the Berry phase occurs: The *base manifold* M is identified with the parameter space $M = \{\lambda\}$. At each point of the parameter space the eigenstate of the system is defined up to a complex phase $e^{i\alpha} \in \mathrm{U}(1)$ by the eigenvalue equation (9.49). Therefore an eigenstate $|n,\lambda\rangle$ is completely characterized by the pair $(\lambda, e^{i\alpha})$. Thus, we construct a *principal bundle* $P(M,\mathrm{U}(1))$ with the *structure group* $\mathrm{U}(1)$.

The *section* of the *associated line bundle* fixes the phase of the eigenstate, i.e., the fibre point of the bundle. The time evolution of an eigenstate generates a curve in the bundle space P, which corresponds to a curve in the base space M, i.e., the base curve is lifted into the bundle space. This *horizontal lift* is defined by the connection 1-form.

In Berry's case of an adiabatic evolution governed by the parameter λ, we can identify the $\mathrm{U}(1)$ gauge potential A with

$$A(\lambda) = \langle n,\lambda | \frac{\partial}{\partial\lambda} | n,\lambda\rangle,\qquad(9.162)$$

and rewrite it as *connection 1-form*, called the *Berry connection*

$$\mathcal{A} = iA\,d\lambda = i\langle n,\lambda | \frac{\partial}{\partial\lambda} | n,\lambda\rangle\,d\lambda = i\langle n,\lambda | d | n,\lambda\rangle,\qquad(9.163)$$

where $d = \frac{\partial}{\partial\lambda}\,d\lambda$ denotes the exterior derivative in the parameter space λ.

Then the *curvature 2-form*, the *Berry curvature*, is given by (recall the parameter λ is multi-dimensional)

$$\mathcal{F} = d\mathcal{A} = i\frac{\partial}{\partial \lambda^i}\langle n,\lambda \,|\, \frac{\partial}{\partial \lambda^j}\,|\, n,\lambda\rangle \, d\lambda^i \wedge d\lambda^j \ . \tag{9.164}$$

Finally, the *holonomy group* $H = \{h = e^{i\gamma}\} \subset \mathrm{U}(1)$ is determined by integrating the connection 1-form along a closed curve C, a cycle $\lambda(0) = \lambda(T)$

$$\gamma = \oint_C \mathcal{A} = i\oint_C \langle n,\lambda\,|\,\frac{\partial}{\partial \lambda}\,|\,n,\lambda\rangle \, d\lambda \equiv \gamma_n \ , \tag{9.165}$$

which provides exactly the Berry phase of eqn (9.56). So the Berry phase has the geometric interpretation of a holonomy in fibre space.

Applying Stokes' Theorem 9.3 we get the Berry phase

$$\gamma = \oint_C \mathcal{A} = \int_{S, \partial S = C} d\mathcal{A} = \int_S \mathcal{F}\ , \tag{9.166}$$

which, in topology, is called the *first Chern number*, a number that characterizes the topology of the bundle.

We also recover the correct *compatibility conditions*, when one fibre point is transformed into another $|\,n,\lambda'\,\rangle = e^{i\alpha(\lambda)}\,|\,n,\lambda\rangle$, that implies $\mathcal{A}' = \mathcal{A} - d\alpha$, the correct U(1) gauge transformation.

9.6.3 Dirac Monopole and Topology

Finally, we also want to discuss the so-called Dirac monopole which is another simple but topologically non-trivial example of a gauge theory. In classical electrodynamics there is a strong symmetry between the electric field $\vec{E}$ and the magnetic field $\vec{B}$. However, a magnetic charge—*magnetic monopole*—is absent in the familiar Maxwell equations. It is nevertheless quite instructive to take a look at the possibility of such a charge from a theoretical point of view, which we will do in this section. Assuming the existence of such a magnetic charge g with density ρ_M, we would have

$$\vec{\nabla}\vec{B}(\vec{x}) = \rho_M(\vec{x}) = 4\pi g\, \delta^3(\vec{x}) \neq 0\ , \tag{9.167}$$

and the symmetry between $\vec{E}$ and $\vec{B}$ would be perfect. It was *Paul Dirac* (1931) who discovered that, as a consequence of quantum mechanics, any potential magnetic charge (if it were to exist) must be quantized in units of $\frac{1}{2}q_{\mathrm{e}}$. If we choose units where $q_{\mathrm{e}} = \hbar = c = 1$, the magnetic charge is quantized according to the rule

$$g = \frac{n}{2}, \quad n = 1, 2, \dots . \tag{9.168}$$

This is a very remarkable outcome of quantum physics, but it can also be understood from a purely geometrical perspective by considering the topology of the corresponding fibre bundle, as discussed in Wu and Yang (1975), Eguchi *et al.* (1980), and Bertlmann (2000).

To solve the monopole problem topologically we consider a sphere S^2 and construct two different gauge potentials leading to the same magnetic field. One potential, A_+, is valid in the upper hemisphere H_+ the other, A_-, in the lower hemisphere H_-. In the overlap region, the equator, the potentials must be related by a gauge transformation.

Let us write the two potentials in terms of differential forms using spherical polar coordinates, such that we have the *gauge-potential 1-forms*

$$A_\pm = \frac{n}{2} (\pm 1 - \cos\theta) \, d\varphi \quad \text{on} \quad U_\pm = \mathbb{R}^3_\pm \setminus \{0\} , \qquad (9.169)$$

with the gauge transformation

$$A_+ = A_- + n \, d\varphi \quad \text{on} \quad U_+ \cap U_- = \mathbb{R}^2 \setminus \{0\} . \qquad (9.170)$$

Applying the exterior derivative d provides the *field-strength 2-form*

$$F = dA_\pm = \frac{n}{2} \sin\theta \, d\theta \wedge d\varphi . \qquad (9.171)$$

Clearly F is closed $dF = d^2 A_\pm = 0$ but *not* exact since dA_+ is defined only *locally* on U_+, and dA_- on U_-.

Topology: Now let us construct a fibre bundle that mirrors the physics of the Dirac monopole. The space where the Dirac monopole is defined is $\mathbb{R}^3$ but removing the origin $\{0\}$ where the monopole is assumed to rest. Topologically, this space is equivalent (homotopic) to the sphere S^2. Therefore the *base manifold* M for our bundle is the sphere S^2,

$$M = \mathbb{R}^3 \setminus \{0\} \sim S^2 , \qquad (9.172)$$

which we cover by two hemispheres, H_+ and H_-, the *charts*, see Fig. 9.12.

The gauge group U(1) of electrodynamics supplies the *structure group* U(1) for the bundle. The set of all possible U(1) elements establishes the *fibre*. The fibres are identical to the structure group itself, therefore we construct a *principal bundle* $P(S^2, \mathrm{U}(1))$. The wave function $\psi(x)$ corresponds to a *section* of the associated line bundle. *Locally* the bundle thus takes the form

$$
\begin{array}{ccc}
H_+ \times \mathrm{U}(1) & & (\theta, \varphi, e^{i\alpha_+}) \\
& \text{with bundle coordinates} & \\
H_- \times \mathrm{U}(1) & & (\theta, \varphi, e^{i\alpha_-})
\end{array}
\qquad (9.173)
$$

and with $0 \le \theta < \pi$ and $0 \le \varphi < 2\pi$.

Dirac monopole

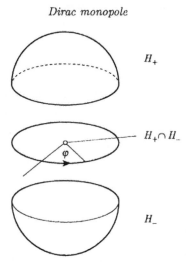

Fig. 9.12 Dirac monopole. The sphere S^2 is covered by two hemispheres H_+, H_- with the equator $H_+ \cap H_- = S^1$. Reproduced from Fig. 6.2 in Reinhold A. Bertlmann, *Anomalies in Quantum Field Theory* (2000), with permission from Oxford University Press.

The *transition functions* h connect fibres from one chart to another,

$$e^{i\alpha_-} = h_{-+} e^{i\alpha_+}, \tag{9.174a}$$

$$\psi_- = h_{-+} \psi_+. \tag{9.174b}$$

They represent elements of the U(1) structure group and are functions of φ defined along the equator $(\theta = \pi/2)\ H_+ \cap H_- = S^1,\ h_{+-}(\varphi) \in$ U(1). Intuitively speaking, they must "glue" the fibres together when we complete a full turn at the equator. Then the sole choice is

$$h_{-+} = e^{in\varphi}, \tag{9.175}$$

with $n = 1, 2, \ldots$ being an integer. This is the topological origin for Dirac's quantization condition (9.168) for the magnetic charge.

The Lie-algebra valued *connection 1-form* of the bundle is locally represented by the gauge potentials

$$\mathcal{A}_\pm = iA_\pm, \tag{9.176}$$

which are defined on the hemispheres $H_\pm$. According to our definitions we just have a Lie-algebra factor i (for our Abelian case U(1)), when $A_\pm$ is given by eqn (9.169).

The *compatibility condition*

$$\mathcal{A}_+ = h_{-+}^{-1} \mathcal{A}_- h_{-+} + h_{-+}^{-1} dh_{-+} \tag{9.177}$$

expresses the gauge transformation law and gives

$$\mathcal{A}_+ = \mathcal{A}_- + in\,d\varphi \tag{9.178}$$

in agreement with eqn (9.170).

The Lie-algebra valued *curvature 2-form* of the bundle, given by

$$\mathcal{F} = d\mathcal{A}_\pm = iF, \tag{9.179}$$

represents the gauge-field strength F from eqn (9.171). As already mentioned, $\mathcal{F}$ is closed, $d\mathcal{F} = 0$, but not exact, and the Abelian $\mathcal{F}$ is gauge invariant.

Winding number and monopole charge: Consider again the transition functions of the bundle $h_{-+}(\varphi) \in U(1)$, eqn (9.175). They define a map from the equator of the base manifold into the structure group

$$h_{-+} : H_+ \cap H_- = S^1 \to U(1). \tag{9.180}$$

When φ varies from $0 \to 2\pi$, the group U(1) is wound around n times. Therefore the number n is called *winding number* and can be expressed quite generally by the integral (Bertlmann, 2000)

$$n = \frac{1}{2\pi i} \int_0^{2\pi} d\varphi \; h_{-+}^{-1}(\varphi) \frac{d}{d\varphi} h_{-+}(\varphi) = \frac{1}{2\pi i} \int_0^{2\pi} d\varphi \cdot in, \tag{9.181}$$

where n represents the quantized magnetic monopole charge $g = \frac{n}{2}$ (in units $q_e = \hbar = c = 1$).

These winding numbers—*monopole charges*—are typical numbers of the topology of the bundle and are also determined by the so-called *first Chern number*

$$C_1 = \frac{i}{2\pi} \int_{S^2} \mathrm{tr}\,\mathcal{F}. \tag{9.182}$$

The explicit evaluation gives

$$-C_1 = \frac{1}{2\pi} \int_{S^2} F = \frac{1}{2\pi} \left[\int_{H_+} dA_+ + \int_{H_-} dA_- \right] \tag{9.183}$$

$$= \frac{1}{2\pi} \left[\int_{\partial H_+} A_+ + \int_{\partial H_-} A_- \right] = \frac{1}{2\pi} \int_{S^1} (A_+ - A_-) = \frac{1}{2\pi} \int_0^{2\pi} d\varphi \cdot n = n,$$

where we have used Stokes' Theorem 9.3.

Résumé: There is a remarkable correspondence between physics and geometry. In fact, fibre bundles appear as the natural geometric setup for describing the physics of gauge theories. Assuming some internal symmetry group—the gauge symmetry—the fibre bundle automatically supplies a purely *geometric* interpretation for the concepts encountered in a gauge theory, specifically for the gauge potential, the field strength, the gauge element, and gauge transformations. This we have demonstrated in detail for the Aharonov–Bohm effect, the Berry phase and the Dirac monopole, where the bundle formalism provides a geometrical-topological reason for the quantized charge of the magnetic monopole. Furthermore, bundles also provide deep, geometric explanations for other physical phenomena such as instantons and anomalies of non-Abelian gauge theories, see (Bertlmann, 2000).

Aide-memoire	
Physics	**Geometry**
gauge theory	principal fibre bundle
space-time	base manifold
all possible gauge elements	fibre
gauge group	structure group
gauge element	transition function
gauge potential	connection
gauge field strength	curvature
gauge transformations	compatibility conditions
wave function	section of fibre bundle
wave-function phase factor	parallel transport of fibre
Aharonov–Bohm-effect Berry phase	holonomy in fibre space
magnetic monopole charge	winding number, Chern number C_1

10
Perturbative Methods in Quantum Mechanics

Unfortunately, apart from a few simple examples, the Schrödinger equation[1] is generally not exactly solvable and we therefore have to rely upon approximative methods to deal with more realistic situations. Such methods include *variational methods* such as the Rayleigh—Ritz variational principle that we have encountered in Section 7.2.2 or the Hartree–Fock method (see, e.g., Levine, 2000, Chap. 11.1, and references therein), approximations such as the *WKB* method[2], and perturbative methods, some of which we shall briefly review here. In short, perturbation theory attempts to describe the effects of "small" perturbations of quantities of interest on the predictions of an exactly solvable model. Here, this quantity of interest is the Hamiltonian of the quantum system, while the perturbations are typically terms describing interactions. In Section 10.1 we will consider the simplest case, time-independent perturbations, which result in corrections to the energy eigenvalues and the corresponding eigenstates, as we shall illustrate for some prominent examples, the fine structure of the hydrogen atom, the Zeeman effect, as well as the Stark effect, in Sections 10.2, 10.3, and 10.4, respectively, before turning to time-dependent perturbations in Section 10.5, which influence the transition probabilities between unperturbed eigenstates.

10.1 Time-Independent Perturbation Theory

10.1.1 Rayleigh–Schrödinger Perturbation Theory

The premise of time-independent perturbation theory is that a system described by Hamiltonian H_0 is "slightly" perturbed by an interaction represented by a second, constant Hamiltonian H_{int}, such that the total Hamiltonian is time-independent and given by

$$H = H_0 + \lambda H_{\text{int}}, \qquad (10.1)$$

where the coupling constant λ is assumed to be small (in a way that we shall make more technically precise shortly). It is further assumed that the Schrödinger equation for the unperturbed Hamiltonian H_0 is exactly solvable, and H_0 thus has well-known eigenvalues $E_n^{(0)}$ and associated eigenvectors $|\, n^{(0)} \,\rangle$, i.e.,

---

[1] As well as its generalizations such as the Pauli equation, see Section 9.1.3

[2] Named after *Gregor Wentzel*, *Hendrik Kramers*, and *Léon Brillouin*, see Hall, (2013), Chap. 15 for an introduction.

$$H_0 \, | \, n^{(0)} \, \rangle = E_n^{(0)} \, | \, n^{(0)} \, \rangle \, , \tag{10.2}$$

where the eigenstates are chosen to be normalized

$$\langle \, m^{(0)} | \, n^{(0)} \, \rangle = \delta_{mn} \, . \tag{10.3}$$

The aim is then to approximately solve the Schrödinger equation for the total Hamiltonian H, i.e., to approximately determine the eigenvalues E_n and eigenvectors $| \, n \, \rangle$ of the total Hamiltonian H

$$H \, | \, n \, \rangle = E_n \, | \, n \, \rangle \, . \tag{10.4}$$

To achieve this, the eigenvalues and eigenvectors are expanded into a power series in the real parameter λ, that is,

$$E_n = E_n^{(0)} + \lambda \, E_n^{(1)} + \lambda^2 \, E_n^{(2)} + \mathcal{O}(\lambda^3), \tag{10.5a}$$

$$| \, n \, \rangle = | \, n^{(0)} \, \rangle + \lambda \, | \, n^{(1)} \, \rangle + \lambda^2 \, | \, n^{(2)} \, \rangle + \mathcal{O}(\lambda^3), \tag{10.5b}$$

where $\mathcal{O}(x)$ denotes terms such that $\mathcal{O}(x)/x$ is finite as $x \to 0$. In writing down such perturbative expansions, some assumptions are made implicitly: First, it is tacitly assumed that $E_n(\lambda)$ and $| \, n(\lambda) \, \rangle$ are continuously differentiable functions (indeed, C^∞) of λ and that the series expansions in eqn (10.5) indeed converges. Second, the limit $\lambda \to 0$ needs to reproduce the spectrum of the unperturbed Hamiltonian, i.e., $E_n(\lambda = 0) = E_n^{(0)}$ and $| \, n(\lambda = 0) \, \rangle = | \, n^{(0)} \, \rangle$. In short, this approach may not always be applicable. Indeed, there are prominent examples, e.g., in particle physics, where, despite providing incredibly accurate predictions that match empirical results to within extremely small margins of error, perturbative expansions can be shown to yield divergent series. In fact, it may even happen that individual terms are formally infinite and require the introduction of cut-offs (see, for instance, the discussion in Peskin and Schroeder, 1995, Chap. 1). Such issues have been the subject of an entire branch of theoretical research that can be subsumed under the term *renormalization theory*. Here, we will not further discuss questions of convergence but refer the interested reader to Peskin and Schroeder (1995, Part II) and references therein.

Despite potential theoretical issues in the context of renormalization, the method of perturbation theory has been successfully applied in many situations throughout quantum mechanics, some of which we will discuss in Sections 10.3, 10.4, and 10.2 in more detail. The general method was first developed in the context of quantum mechanics by *Erwin Schrödinger* (1926c), who himself referenced previous work by *Rayleigh* (Strutt, 2011), and this approach is thus referred to as the *Rayleigh–Schrödinger* perturbation method. To study the particular mechanisms of this approach, it is useful to separately consider the non-degenerate and degenerate cases, as we shall do here in Sections 10.1.2 and 10.1.3, respectively.

10.1.2 Non-Degenerate Perturbation Theory

The simplest cases of time-independent perturbation theory involve Hamiltonians for which the unperturbed eigenvalues are non-degenerate, i.e.,

$$E_m^{(0)} \neq E_n^{(0)} \quad \text{for} \quad m \neq n. \tag{10.6}$$

We will therefore start by considering non-degenerate unperturbed Hamiltonians, and return to the more general case of *degenerate perturbation theory* in Section 10.1.3.

Using this assumption, we insert the perturbative expansions in eqn (10.5) into the eigenvalue equation (10.4) for the total Hamiltonian,

$$\left(H_0 + \lambda H_{\text{int}} \right) \left(| n^{(0)} \rangle + \lambda | n^{(1)} \rangle + \lambda^2 | n^{(2)} \rangle + \mathcal{O}(\lambda^3) \right) \tag{10.7}$$
$$= \left(E_n^{(0)} + \lambda E_n^{(1)} + \lambda^2 E_n^{(2)} + \mathcal{O}(\lambda^3) \right) \left(| n^{(0)} \rangle + \lambda | n^{(1)} \rangle + \lambda^2 | n^{(2)} \rangle + \mathcal{O}(\lambda^3) \right).$$

By comparing the coefficients for different powers of λ we obtain the following relations:

$$\lambda^0 : H_0 \, | n^{(0)} \rangle = E_n^{(0)} \, | n^{(0)} \rangle \, , \tag{10.8a}$$

$$\lambda^1 : H_0 \, | n^{(1)} \rangle + H_{\text{int}} \, | n^{(0)} \rangle = E_n^{(0)} \, | n^{(1)} \rangle + E_n^{(1)} \, | n^{(0)} \rangle \, , \tag{10.8b}$$

$$\lambda^2 : H_0 \, | n^{(2)} \rangle + H_{\text{int}} \, | n^{(1)} \rangle = E_n^{(0)} \, | n^{(2)} \rangle + E_n^{(1)} \, | n^{(1)} \rangle + E_n^{(2)} \, | n^{(0)} \rangle \, . \tag{10.8c}$$

Similarly, by demanding that the sought-after eigenstates of the total Hamiltonian are normalized, $\langle n | n \rangle = 1$, and inserting from eqn (10.5b), we have

$$\langle n | n \rangle = \left(\langle n^{(0)} | + \lambda \langle n^{(1)} | + \lambda^2 \langle n^{(2)} | + \mathcal{O}(\lambda^3) \right) \left(| n^{(0)} \rangle + \lambda | n^{(1)} \rangle + \lambda^2 | n^{(2)} \rangle + \mathcal{O}(\lambda^3) \right)$$
$$= \langle n^{(0)} | n^{(0)} \rangle + \lambda \left(\langle n^{(0)} | n^{(1)} \rangle + \langle n^{(1)} | n^{(0)} \rangle \right)$$
$$+ \lambda^2 \left(\langle n^{(1)} | n^{(1)} \rangle + \langle n^{(2)} | n^{(0)} \rangle + \langle n^{(0)} | n^{(2)} \rangle \right) + \mathcal{O}(\lambda^3). \tag{10.9}$$

Considering the coefficients up to and including second order in λ, we thus obtain the conditions

$$\langle n^{(0)} | n^{(0)} \rangle = 1 \, , \tag{10.10a}$$

$$\langle n^{(0)} | n^{(1)} \rangle = - \langle n^{(1)} | n^{(0)} \rangle \, , \tag{10.10b}$$

$$\langle n^{(1)} | n^{(1)} \rangle = - \langle n^{(0)} | n^{(2)} \rangle - \langle n^{(2)} | n^{(0)} \rangle . \tag{10.10c}$$

First-order corrections: With this information at hand, we can apply $\langle n^{(0)} |$ from the left to both sides of eqn (10.8b), which results in

$$\langle n^{(0)} | H_0 | n^{(1)} \rangle + \langle n^{(0)} | H_{\text{int}} | n^{(0)} \rangle = E_n^{(0)} \langle n^{(0)} | n^{(1)} \rangle + E_n^{(1)} \langle n^{(0)} | n^{(0)} \rangle . \tag{10.11}$$

The first term on the left-hand side yields $E_n^{(0)} \langle n^{(0)} | n^{(1)} \rangle$ and thus cancels with the first term on the right-hand side. Inserting the normalization condition from eqn (10.10a) then lets us arrive at the expression for the first-order correction to the energy eigenvalue.

Theorem 10.1 (First-order corrections (non-degenerate))

The first-order corrections to the (non-degenerate) eigenvalues $E_n^{(0)}$ are given by the expected values of the perturbing interaction Hamiltonian H_{int} with respect to the corresponding unperturbed eigenstates $|\, n^{(0)} \,\rangle$,

$$E_n^{(1)} = \langle\, n^{(0)} \,|\, H_{\text{int}} \,|\, n^{(0)} \,\rangle\,.$$

To obtain the first-order correction to the eigenvector $|\, n^{(0)} \,\rangle$ itself, we first note that an expansion of $\langle\, m \,|\, n \,\rangle = \delta_{mn}$ in analogy to eqn (10.9) implies that $\langle\, m^{(0)} \,|\, n^{(0)} \,\rangle = \delta_{mn}$, and we must hence be able to expand $|\, n^{(1)} \,\rangle$ with respect to the basis $\{|\, m^{(0)} \,\rangle\}_m$, such that

$$|\, n^{(1)} \,\rangle = \sum_m \langle\, m^{(0)} \,|\, n^{(1)} \,\rangle \,|\, m^{(0)} \,\rangle\,. \tag{10.12}$$

To determine the coefficients $\langle\, m^{(0)} \,|\, n^{(1)} \,\rangle$, we then apply $\langle\, m^{(0)} \,|$ from the left to eqn (10.8b), and arrive at

$$\langle\, m^{(0)} \,|\, H_0 \,|\, n^{(1)} \,\rangle + \langle\, m^{(0)} \,|\, H_{\text{int}} \,|\, n^{(0)} \,\rangle = E_n^{(0)} \langle\, m^{(0)} \,|\, n^{(1)} \,\rangle + E_n^{(1)} \langle\, m^{(0)} \,|\, n^{(0)} \,\rangle\,. \tag{10.13}$$

Since $\langle\, m^{(0)} \,|\, H_0 = E_m^{(0)} \langle\, m^{(0)} \,|$ and $\langle\, m^{(0)} \,|\, n^{(0)} \,\rangle = \delta_{mn}$, we thus have

$$\langle\, m^{(0)} \,|\, n^{(1)} \,\rangle \left(E_n^{(0)} - E_m^{(0)} \right) = \langle\, m^{(0)} \,|\, H_{\text{int}} \,|\, n^{(0)} \,\rangle \tag{10.14}$$

whenever $m \neq n$, which implies $E_m^{(0)} \neq E_n^{(0)}$ in the non-degenerate case. This, in turn, allows us to write

$$|\, n^{(1)} \,\rangle = \sum_{m \neq n} \frac{\langle\, m^{(0)} \,|\, H_{\text{int}} \,|\, n^{(0)} \,\rangle}{E_n^{(0)} - E_m^{(0)}} \,|\, m^{(0)} \,\rangle + \langle\, n^{(0)} \,|\, n^{(1)} \,\rangle \,|\, n^{(0)} \,\rangle\,. \tag{10.15}$$

Second-order corrections: To obtain the higher-order corrections, we can proceed in a similar fashion. Applying $\langle\, n^{(0)} \,|$ from the left to both sides of eqn (10.8c), we have

$$\langle\, n^{(0)} \,|\, H_0 \,|\, n^{(2)} \,\rangle + \langle\, n^{(0)} \,|\, H_{\text{int}} \,|\, n^{(1)} \,\rangle = E_n^{(0)} \langle\, n^{(0)} \,|\, n^{(2)} \,\rangle + E_n^{(1)} \langle\, n^{(0)} \,|\, n^{(1)} \,\rangle$$
$$+ E_n^{(2)} \langle\, n^{(0)} \,|\, n^{(0)} \,\rangle\,. \tag{10.16}$$

Using $\langle\, n^{(0)} \,|\, H_0 = E_n^{(0)} \langle\, n^{(0)} \,|$ and $\langle\, n^{(0)} \,|\, n^{(0)} \,\rangle = 1$, and inserting for $|\, n^{(1)} \,\rangle$ from eqn (10.15) we find

$$E_n^{(2)} = \langle\, n^{(0)} \,|\, H_{\text{int}} \,|\, n^{(1)} \,\rangle - E_n^{(1)} \langle\, n^{(0)} \,|\, n^{(1)} \,\rangle$$

$$= \langle\, n^{(0)} \,|\, H_{\text{int}} \left(\sum_{m \neq n} \frac{\langle\, m^{(0)} |H_{\text{int}}| n^{(0)} \,\rangle}{E_n^{(0)} - E_m^{(0)}} \,|\, m^{(0)} \,\rangle + \langle\, n^{(0)} \,|\, n^{(1)} \,\rangle \,|\, n^{(0)} \,\rangle \right) - E_n^{(1)} \langle\, n^{(0)} \,|\, n^{(1)} \,\rangle$$

$$= \sum_{m \neq n} \frac{|\langle\, m^{(0)} |H_{\text{int}}| n^{(0)} \,\rangle|^2}{E_n^{(0)} - E_m^{(0)}} + \left(\langle\, n^{(0)} \,|\, H_{\text{int}} \,|\, n^{(0)} \,\rangle - E_n^{(1)} \right) \langle\, n^{(0)} \,|\, n^{(1)} \,\rangle\,. \tag{10.17}$$

In the last step, we then recognize that the term in parenthesis vanishes by inserting from Theorem 10.1 and we hence arrive at the second-order correction to the (non-degenerate) energy eigenvalues.

Theorem 10.2 (Second-order corrections (non-degenerate))

The second-order corrections to the (non-degenerate) energy eigenvalues $E_n^{(0)}$ *are given by*

$$E_n^{(2)} = \sum_{m \neq n} \frac{|\langle m^{(0)} \, | \, H_{\text{int}} \, | \, n^{(0)} \, \rangle |^2}{E_n^{(0)} - E_m^{(0)}}.$$

Up to and including second-order corrections in λ, we thus have

$$E_n = E_n^{(0)} + \lambda \, \langle n^{(0)} \, | \, H_{\text{int}} \, | \, n^{(0)} \, \rangle + \lambda^2 \sum_{m \neq n} \frac{|\langle m^{(0)} \, | \, H_{\text{int}} \, | \, n^{(0)} \, \rangle |^2}{E_n^{(0)} - E_m^{(0)}} + \mathcal{O}(\lambda^3).$$
$$(10.18)$$

In writing down such an expression, it is imperative to keep in mind the limitations of the validity of such perturbative expansions. Obviously, λ is not a free parameter entirely and is restricted to cases where $\lambda \ll 1$. But, indeed, one has to be somewhat stricter than this and take into account the relative size of all correction terms. That is, even for any fixed "small" value of λ such that $\lambda \ll 1$, it is important to note that the approximation by the power series is only valid as long as, order by order, the respective correction terms are much smaller than the corresponding preceding lower-order terms, provided the former do not vanish identically. In other words, it is assumed that

$$|\lambda^k E_n^{(k)}| \gg |\lambda^{k+1} E_n^{(k+1)}|$$
$$(10.19)$$

unless $E_n^{(k)} = 0$. In other words, if e.g., the second-order correction $\lambda^2 E_n^{(2)}$ were to be of a magnitude comparable to the first-order correction $\lambda E_n^{(1)}$, then by what virtue could we cut off the power expansion after inclusion of the terms linear in λ? Similarly, it is hence assumed that all following corrections, here captured by $\mathcal{O}(\lambda^3)$ are negligible compared to the first- and second-order corrections. However, it may well be that corrections at some order vanish identically, e.g., $E_n^{(1)} = 0$, while $E_n^{(2)} \neq 0$. In addition, one should also be careful in considering the size of the corrections relative not just to their respective unperturbed eigenvalues, but with respect to the unperturbed energy gaps. That is, perturbative expansions become problematic when the perturbations are of the order of magnitude of the energy gaps of the unperturbed system. To better understand why this is the case, let us first consider situations where these gaps vanish, i.e., degenerate energy levels, before coming back to the phenomenon of so-called *avoided crossings* in Section 10.1.4.

10.1.3 Degenerate Perturbation Theory

Let us now turn to the case where (some of) the unperturbed energy levels are degenerate. That is, we now assume that there are k_n orthogonal eigenstates $|\, n_i^{(0)} \,\rangle$ for $i = 1, \ldots, k_n$ associated to the same unperturbed eigenvalue $E_n^{(0)}$, such that

$$H_0 \,|\, n_i^{(0)} \,\rangle \;=\; E_n^{(0)} \,|\, n_i^{(0)} \,\rangle \quad \forall\, i = 1, \ldots, k_n \,. \tag{10.20}$$

To begin, we are interested in the first-order corrections $E_{n_i}^{(1)}$ to the unperturbed eigenvalues. Naively, we could make an ansatz along the lines of Theorem 10.1, that is, we could try

$$E_{n_i}^{(1)} \;=\; \langle\, n_i^{(0)} \,|\, H_{\mathrm{int}} \,|\, n_i^{(0)} \,\rangle \,. \tag{10.21}$$

However, this would not be unambiguous because every superposition of vectors $|\, n_i^{(0)} \,\rangle$ with the same n but different i again results in an eigenstate of the unperturbed Hamiltonian with the same unperturbed eigenvalue $E_n^{(0)}$. Consequently, there are in principle infinitely many inequivalent ways of making an ansatz like in eqn (10.21). Moreover, inspecting the first-order corrections to the eigenstates in the non-degenerate case, eqn (10.15), we see that the denominator $E_n^{(0)} - E_m^{(0)}$ in the sum is not well defined if we were to replace $n \to n_i$ and $m \to m_j$ and sum over all m and j except for the case when $m = n$ and $i = j$. To get out of this conundrum, let us reconsider how we obtained eqn (10.15) in the first place: We applied $\langle\, m^{(0)} \,|$ from the left to eqn (10.8b). Let us try this step again, but with the replacement $n \to n_i$ and $m \to n_j$, such that we have

$$\langle\, n_j^{(0)} \,|\, H_0 \,|\, n_i^{(1)} \,\rangle + \langle\, n_j^{(0)} \,|\, H_{\mathrm{int}} \,|\, n_i^{(0)} \,\rangle = E_n^{(0)} \,\langle\, n_j^{(0)} \,|\, n_i^{(1)} \,\rangle + E_{n_i}^{(1)} \,\langle\, n_j^{(0)} \,|\, n_i^{(0)} \,\rangle \,. \tag{10.22}$$

Since $\langle\, n_j^{(0)} \,|\, H_0 = E_n^{(0)} \,\langle\, n_j^{(0)} \,|$, the first terms on each side cancel, and since we may still assume $\langle\, n_j^{(0)} \,|\, n_i^{(0)} \,\rangle = \delta_{ij}$, we have

$$E_{n_i}^{(1)} \,\delta_{ij} \;=\; \langle\, n_j^{(0)} \,|\, H_{\mathrm{int}} \,|\, n_i^{(0)} \,\rangle \,. \tag{10.23}$$

In other words, for $i = j$ we have exactly the usual condition $E_{n_i}^{(1)} = \langle\, n_i^{(0)} \,|\, H_{\mathrm{int}} \,|\, n_i^{(0)} \,\rangle$ for the linear-order correction to the energy eigenvalue, but in addition we have the constraint $\langle\, n_j^{(0)} \,|\, H_{\mathrm{int}} \,|\, n_i^{(0)} \,\rangle = 0$ for all $i \neq j$. This can easily be satisfied if the unperturbed basis states in the degenerate subspace are chosen such that they are eigenstates of the perturbing Hamiltonian H_{int}. In absence of the perturbation there is no particular reason to select one particular basis in the degenerate subspace over any other basis, but the presence of H_{int} means that there is now a preferred basis, in other words, we have to diagonalize the perturbation in the degenerate subspace.

> **Theorem 10.3 (First-order corrections (degenerate))**
>
> *The first-order corrections to the k_n degenerate energy eigenvalues $E_{n_i}^{(0)}$ with corresponding unperturbed eigenstates $| n_i^{(0)} \rangle$ for $i = 1, \dots, k_n$ are given by the eigenvalues of the $k_n \times k_n$ matrix H_{int_n} with components*
>
> $$(H_{\mathrm{int}_n})_{ij} = \langle n_i^{(0)} | H_{\mathrm{int}} | n_j^{(0)} \rangle.$$

The associated simultaneous eigenstates $| \tilde{n}_i^{(0)} \rangle$ of H_0 and H_{int_n} then satisfy

$$\left(H_0 + \lambda H_{\mathrm{int}_n} \right) | \tilde{n}_i^{(0)} \rangle = \left(E_n^{(0)} + \lambda E_{n_i}^{(1)} \right) | \tilde{n}_i^{(0)} \rangle. \tag{10.24}$$

However, we note that H_0 and H_{int} generally do not have to have joint eigenstates, i.e., in general, $[H_0, H_{\mathrm{int}}] \neq 0$ and thus $H_{\mathrm{int}} \neq \bigoplus_n H_{\mathrm{int}_n}$. Nevertheless, the effect of the perturbation in a degenerate eigenspace is thus to provide corrections $E_{n_1}^{(1)}, \dots, E_{n_{k_n}}^{(1)}$ that are in general different from each other, lifting the degeneracy.

In combination with our previous appeal to caution in the non-degenerate case, i.e., to be careful when the leading-order perturbations of non-degenerate levels are comparable in size to the unperturbed energy gaps, one may hence wonder if the opposite effect may also occur: Can a perturbation create degeneracy between non-degenerate levels? Perhaps surprisingly, the answer is quite categorically: *no*—that is, not unless the perturbation is represented by an operator that is diagonal with respect to the unperturbed eigenstates. Since the last clause is really quite specific, i.e., relating only to cases where the eigenstates of the unperturbed and perturbing Hamiltonians coincide to begin with, it can quite generically be said that non-degenerate energy levels do not cross, as we shall explain in the following section.

10.1.4 Avoided Crossings

Following the detailed discussion in Cohen-Tannoudji *et al.*, (1999, p. 406), let us consider a two-level system with unperturbed Hamiltonian $H_0 = \mathrm{diag}\{E_0^{(0)}, E_1^{(0)}\}$ that is perturbed by a potential

$$V = \begin{pmatrix} V_{00} & V_{01} \\ V_{01}^* & V_{11} \end{pmatrix}, \tag{10.25}$$

where we make no assumption about the size of the matrix elements $V_{ij} \in \mathbb{R}$, in particular, no requirements arising from justifying the truncation of power expansions. The total Hamiltonian thus is of the form

$$H = H_0 + V = \begin{pmatrix} E_0^{(0)} + V_{00} & V_{01} \\ V_{01}^* & E_1^{(0)} + V_{11} \end{pmatrix}. \tag{10.26}$$

Using the notation $\tilde{E}_0 = E_0^{(0)} + V_{00}$ and $\tilde{E}_1 = E_1^{(0)} + V_{11}$, the (exact) eigenvalues of

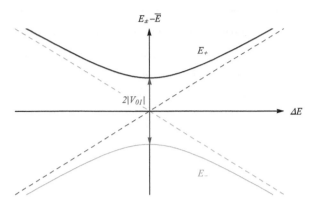

Fig. 10.1 Avoided crossing. The difference between the perturbed energies E_+ (blue, top) and E_- (green, bottom) and their average $\overline{E}$ is plotted against the difference $\Delta E = (\tilde{E}_0 + \tilde{E}_1)/2$. The diagonal and antidiagonal, dashed lines correspond to the limit $V_{01} \to 0$, and hence to $E_+ = +\Delta E$ (dashed, blue) and $E_- = -\Delta E$ (dashed, green).

H are then easily found to be

$$E_\pm = \tfrac{1}{2}\big(\tilde{E}_0 + \tilde{E}_1\big) \pm \sqrt{\tfrac{1}{4}\big(\tilde{E}_0 - \tilde{E}_1\big)^2 + |V_{01}|^2}. \tag{10.27}$$

Aside from the off-diagonal element of the perturbation, the new eigenvalues are governed by two quantities,

$$\overline{E} = \tfrac{1}{2}\big(\tilde{E}_0 + \tilde{E}_1\big) = \tfrac{1}{2}\big(E_0^{(0)} + E_1^{(0)}\big) + \tfrac{1}{2}\big(V_{00} + V_{11}\big), \tag{10.28a}$$

$$\Delta E = \tfrac{1}{2}\big(\tilde{E}_0 - \tilde{E}_1\big) = \tfrac{1}{2}\big(E_0^{(0)} - E_1^{(0)}\big) + \tfrac{1}{2}\big(V_{00} - V_{11}\big), \tag{10.28b}$$

that is, we can rewrite eqn (10.27) as

$$E_\pm = \overline{E} \pm \sqrt{\Delta E^2 + |V_{01}|^2}. \tag{10.29}$$

For any fixed value of V_{01} we can then simply plot these energies, or rather, the difference between the perturbed energies $E_\pm$ and their average $\overline{E} = (E_+ + E_-)/2$ as a function of ΔE, as illustrated in Fig. 10.1. We see that, for any non-zero value of V_{01} there is a separation of energies, regardless of the splitting ΔE. The diagonal elements of V with respect to the unperturbed energy eigenbasis do not matter for this qualitative statement, since their sum $V_{00} + V_{11}$ just corresponds to a shift of the horizontal axis, whereas their difference shifts the vertical axis. The energy levels thus only meet in the case that both $V_{01} = 0$ and $\Delta E = 0$. In particular, this means that any coupling between the two levels, represented by V_{01}, leads to a non-degeneracy that cannot be undone irrespective of the specific values of the original energies. In other words, two energy levels coupled by a perturbation do not cross and one hence speaks of "avoided crossings", see, for instance, the discussion in Garrison and Chiao (2008, Sec. 12.1) in the context of the so-called Jaynes–Cummings model (which we will discuss in more detail in Section 22.4 in Part III).

The occurrence of avoided crossings, or perhaps one should say the absence of crossings, going back to seminal work by *John von Neumann* and *Eugene Wigner* (1993), has here been discussed for a two-level example, which turns out to be a relevant case, e.g., in the Jaynes–Cummings model that we will present in Section 22.4. However, the effect can be understood more generally in terms of the observation that the (initially distinct) eigenvalues of an observable parameterized by a set of k parameters can only intersect on a manifold of dimension $k-2$ (if no additional symmetries of the perturbing Hamiltonian are taken into account), as is discussed in more detail in, e.g., Landau and Lifshitz (1977, p. 304). In the example before we may assume a real value for V_{01} without loss of generality and we can hence vary the three parameters V_{01}, V_{00}, V_{11} of the perturbation V. However, for any fixed value of $E_0^{(0)} - E_1^{(0)}$, there is only a one-dimensional manifold in the parameter space with coordinates (V_{01}, V_{00}, V_{11}), where both the conditions $V_{01} = 0$ and $\Delta E = 0$ are met, i.e., for the line given by the set of points $\{(0, V_{00}, V_{00} + E_0^{(0)} - E_1^{(0)})|V_{00} \in \mathbb{R}\}$.

Loosely speaking, the effect of avoided crossings can be summarized in the following way. For a system with two energy eigenstates, one with lower and one with higher energy, perturbations will generally not lead to an exchange of the populations of the two levels. A particle originally in the eigenstate with lower energy will remain in the lower-energy eigenstate (even if the specific value of the corresponding energy has changed), while a particle in the higher-energy state will remain there.

10.2   The Fine Structure of the Hydrogen Atom

With the basic framework of time-independent perturbation theory in place, let us now come to an application that can be considered to be an impressive success of early quantum theory, the prediction of the electronic fine structure of the hydrogen atom. In Section 7.5, we have studied the Coulomb problem for the hydrogen atom, consisting of a single electron in the Coulomb potential created by a single proton at the coordinate origin, in the context of non-relativistic quantum mechanics. More specifically, we have determined the exact solutions of the stationary Schrödinger equation in three spatial dimensions, represented by the Hamiltonian

$$H_0 = \frac{\vec{P}^2}{2m} - \frac{q_e^2}{r}, \tag{10.30}$$

where $\vec{P}$ is the three-dimensional momentum operator, m is the reduced mass[3] of the electron and proton, see eqn (7.87), $q_e > 0$ is the elementary charge in Gaussian cgs units and r is the (operator) for the radial coordinate. Note that in the context of the perturbative calculations that are to follow, we have adopted the notation H_0 rather than H as in eqn (7.37). The (unperturbed) eigenvalues of this Hamiltonian are the energies given by Bohr's formula from eqn (7.83),

$$E_n^{(0)} = - \frac{q_e^2}{2\, r_{\mathrm{B}}\, n^2}, \tag{10.31}$$

---

[3]Note that we have dropped the subscript r for the reduced mass as compared to eqn (7.87) for ease of notation.

expressed via the Bohr radius $r_{\mathrm{B}} = \hbar^2/(m q_{\mathrm{e}}^2)$ from eqn (7.86). As we have seen, the unperturbed energy levels depend only on the principal quantum number $n = 1, 2, 3, \ldots$, but the corresponding eigenstates $|\, n, l, s, m_l, m_s \,\rangle$ are also labelled by the azimuthal quantum number $l = 0, 1, \ldots, n-1$, the spin quantum number $s = \frac{1}{2}$ (for the electron we consider here), and the corresponding magnetic quantum numbers $m_l = -l, -l+1, \ldots, l-1, l$ and $m_s = \pm\frac{1}{2}$. As we have discussed in Section 8.3, within each subspace of fixed l and s, we may instead switch to a basis of eigenstates $|\, n, l, s, j, m_j \,\rangle$ labelled by the corresponding total angular momentum quantum numbers $j = |l - s|, |l - s| + 1, \ldots, l + s$ and $m_j = -j, -j+1, \ldots, j-1, j$, but for the bare Hamiltonian H_0 and its eigenvalues, this is of no consequence, the unperturbed eigenvalues $E_n^{(0)}$ are degenerate.

As we shall see next, this changes when perturbations are taken into account, and we shall begin by examining the so-called *fine-structure* corrections, which were first measured by *Albert Michelson* and *Edward Morley* (1887) and later underpinned by theoretical work by *Arnold Sommerfeld* (see, e.g., Sommerfeld, 1940), *Charles G. Darwin*[4] (see, e.g., Darwin, 1928), as well as *Werner Heisenberg* and *Pascual Jordan* (1926). Although each of the fine-structure correction terms can be heuristically justified at least in good approximation, their exact form follows from the fully relativistic treatment of the electromagnetic interaction between the electron (field) and the electromagnetic field generated by the proton charge. That is, the correct Lorentz-covariant equation of motion generalizing the Schrödinger equation is the Dirac equation (see, e.g., Peskin and Schroeder, 1995, Chap. 3). The Dirac equation describes the electron-positron field in terms of a four-component spinor[5] and can be coupled to the electromagnetic field via the minimal substitution from eqn (9.11) just like we have done to obtain the Pauli equation from eqn (9.28). Indeed, in the non-relativistic, low-energy regime, the Dirac equation decouples into two equations for independent two-component spinors and thus reduces to the Pauli equation. However, the Dirac equation now allows one to take into account also corrections to the fully non-relativistic description, specifically, by expanding the minimally coupled Dirac Hamiltonian in a power series in $(mc^2)^{-1}$ and considering weak electromagnetic fields, repeated application of so-called (unitary) Foldy–Wouthuysen transformations, as is shown in detail in Greiner (2000, Sec. 11.1). For a point-like nucleus creating a spherically symmetric field configuration, as is the case in the Coulomb problem, one thus arrives at the full Hamiltonian

$$H = H_0 + H_{\mathrm{fine-structure}}, \tag{10.32}$$

where the unperturbed Hamiltonian is

$$H_0 = \frac{\vec{P}^{\,2}}{2m} - \frac{q_{\mathrm{e}}^2}{r}, \tag{10.33}$$

---

[4]The grandson of the famous naturalist and biologist *Charles Robert Darwin*.

[5]Meaning, it is an object that first looks like a vector with four components, but which does not transform like a vector under the action of rotations or Lorentz transformation (recall Theorems 6.4 and 8.1), but rather transforms like a direct sum of representations corresponding to two-component spinors, see Friis (2010, Sec. 3.2.2) for a compact treatment.

and the fine-structure perturbation term has three contributions,

$$H_{\text{fine-structure}} = H_{\text{rel.kin.}} + H_{\text{spin-orbit}} + H_{\text{Darwin}}, \tag{10.34}$$

where

$$H_{\text{rel.kin.}} = -\frac{\vec{P}^4}{8m^3c^2}, \tag{10.35a}$$

$$H_{\text{spin-orbit}} = \frac{q_e^2}{2m^2c^2}\frac{\vec{S}\cdot\vec{L}}{r^3}, \tag{10.35b}$$

$$H_{\text{Darwin}} = \frac{\pi}{2}\frac{q_e^2\hbar^2}{m^2c^2}\,\delta^3(\vec{x}). \tag{10.35c}$$

In the following, we shall discuss each of these contributions in turn and calculate their respective corrections to the unperturbed energy levels $E_n^{(0)}$, before combining the results in a compact formula.

10.2.1 Relativistic Correction to the Kinetic Energy

The first correction can be understood as the first-order relativistic correction to the non-relativistic kinetic energy. That is, the relativistic energy-momentum relation is

$$E = \sqrt{m^2c^4 + \vec{p}^2c^2}. \tag{10.36}$$

Expanding the square root in the last expression for small $\vec{p}^2/(mc)^2$, we have

$$E = mc^2\sqrt{1 + \frac{\vec{p}^2}{m^2c^2}} = mc^2\left[1 + \frac{\vec{p}^2}{2m^2c^2} - \frac{\vec{p}^4}{8m^4c^4} + \mathcal{O}([\frac{\vec{p}^2}{m^2c^2}]^3)\right]$$

$$\approx mc^2 + \frac{\vec{p}^2}{2m} - \frac{\vec{p}^4}{8m^3c^2}. \tag{10.37}$$

By treating the momentum as an operator, one thus arrives at the fine-structure perturbation for the relativistic kinetic energy

$$H_{\text{rel.kin.}} = -\frac{\vec{P}^4}{8m^3c^2}. \tag{10.38}$$

Now, since the unperturbed eigenstates of H_0 have degenerate energy levels labelled only by the principal quantum number n, we have to proceed with caution. Technically, we have to diagonalize the projections of the perturbing operator on the subspaces of fixed n. In each such subspace, we are currently using a basis of vectors $|n, l, m_l, s, m_s\rangle$ for fixed n but with $l = 0, 1, \ldots, n-1$, $m_l = -l, -l+1, \ldots, +l$, $s = \frac{1}{2}$ and $m_s = \pm\frac{1}{2}$. By inspecting the components L_i of the orbital angular-momentum operator from eqn (6.46) and the squared orbital angular-momentum operator $\vec{L}^2$ from eqn (6.50), and noting that $[\,\vec{L}^{\,2}, L_i\,] = 0$ for $i = x, y, z$, it is obvious from eqn (7.4) that the operator $\vec{P}^{\,2}$ commutes with all of them,

$$[\,\vec{P}^{\,2}, \vec{L}^{\,2}\,] = [\,\vec{P}^{\,2}, L_i\,] = 0. \tag{10.39}$$

In addition, $\vec{P}^{\,2}$ commutes trivially with $\vec{S}^{\,2}$ and with all components S_i of the spin operator,

$$[\,\vec{P}^{\,2}, \vec{S}^{\,2}\,] = [\,\vec{P}^{\,2}, S_i\,] = 0. \tag{10.40}$$

The perturbation proportional to $\vec{P}^{\,4}$ is thus diagonal with respect to the unperturbed basis $\{\,|\,n, l, m_l, s, m_s\,\rangle\,\}_{l, m_l, s, m_s}$ within the subspace[6] of any fixed n. Despite the degeneracy of the unperturbed energy levels, we can therefore proceed to calculate the leading-order corrections using the machinery of non-degenerate perturbation theory, in particular, Theorem 10.1. Moreover, we can simplify the calculation a little by noting that $\vec{P}^{\,4}$ does not act on the spin degree of freedom at all, and we can hence determine the correction using only $|\,n, l, m_l\,\rangle$. In addition, we can rewrite $H_{\text{rel.kin.}}$ using the identity $\vec{P}^{\,2}/(2m) = H_0 + \frac{q_e^2}{r}$ obtained from eqn (10.33), i.e.,

$$H_{\text{rel.kin.}} = -\frac{\vec{P}^{\,4}}{8m^3c^2} = -\frac{1}{2mc^2}\left(\frac{\vec{P}^{\,2}}{2m}\right)^2 = -\frac{1}{2mc^2}\left(H_0 + \frac{q_e^2}{r}\right)^2$$

$$= -\frac{1}{2mc^2}\left(H_0^2 + H_0\frac{q_e^2}{r} + \frac{q_e^2}{r}H_0 + \frac{q_e^4}{r^2}\right). \tag{10.41}$$

With this, we can now calculate

$$\langle\,n, l, m_l\,|\,H_{\text{rel.kin.}}\,|\,n, l, m_l\,\rangle = -\frac{1}{2mc^2}\Big[\langle\,n, l, m_l\,|\,H_0^2\,|\,n, l, m_l\,\rangle$$

$$+ \langle\,n, l, m_l\,|\,H_0\frac{q_e^2}{r}\,|\,n, l, m_l\,\rangle + \langle\,n, l, m_l\,|\,\frac{q_e^2}{r}H_0\,|\,n, l, m_l\,\rangle + \langle\,n, l, m_l\,|\,\frac{q_e^4}{r^2}\,|\,n, l, m_l\,\rangle\Big]$$

$$= -\frac{1}{2mc^2}\Big[E_n^{(0)\,2} + 2E_n^{(0)}q_e^2\,\langle\,n, l, m_l\,|\,\frac{1}{r}\,|\,n, l, m_l\,\rangle + q_e^4\,\langle\,n, l, m_l\,|\,\frac{1}{r^2}\,|\,n, l, m_l\,\rangle\Big], \tag{10.42}$$

where we have used the eigenvalue equation of the unperturbed Hamiltonian (and its adjoint). We are thus left with calculating the expectation values of $\frac{1}{r}$ and $\frac{1}{r^2}$ for the eigenstates of the hydrogen atom. Luckily, we already have found the expression for $\frac{1}{r}$ using the Kramers–Pasternack relation (Theorem 7.3), and from eqns (7.107) and (10.31) we have

$$\langle\,n, l, m_l\,|\,\frac{1}{r}\,|\,n, l, m_l\,\rangle = \frac{1}{r_{\text{B}}n^2} = -\frac{2E_n^{(0)}}{q_e^2}. \tag{10.43}$$

Unfortunately, we cannot apply the same method to determine $\langle\,\frac{1}{r^2}\,\rangle_{n, l, m_l}$ since the pre-factor of the corresponding term in Theorem 7.3 vanishes identically for $q = 0$.

---

[6]Note that the perturbation does not necessarily have to commute with H_0.

To obtain $\langle \frac{1}{r^2} \rangle_{n,l,m_l}$, we turn to the radial Schrödinger equation for the hydrogen atom from eqn (7.72), which takes the form

$$\left(-\frac{\hbar^2}{2m} \frac{\partial^2}{\partial r^2} + \frac{\hbar^2 l(l+1)}{2mr^2} - \frac{q_e^2}{r} \right) u_{nl}(r) = E_n^{(0)} u_{nl}(r) , \qquad (10.44)$$

where we have explicitly included the dependence of the reduced wave function $u_{nl}(r) = rR_{nl}(r)$ on the principal quantum number n. Now, with a change of variables from r to $y = \frac{r}{r_B}$, such that $\frac{\partial}{\partial r} = \frac{1}{r_B} \frac{\partial}{\partial y}$, this equation becomes

$$\left(-\frac{\partial^2}{\partial y^2} + \frac{l(l+1)}{y^2} - \frac{2}{y} \right) u_{nl} = \frac{2\hbar^2 E_n^{(0)}}{mq_e^4} u_{nl}(r) = \epsilon\, u_{nl} , \qquad (10.45)$$

with the dimensionless quantity $\epsilon = \frac{2\hbar^2 E_n^{(0)}}{mq_e^4} = -\frac{1}{n^2}$. This now has the form of an eigenvalue equation for the operator $\tilde{H} := -\frac{\partial^2}{\partial y^2} + \frac{l(l+1)}{y^2} - \frac{2}{y}$, i.e., $\tilde{H} u_{nl} = \epsilon u_{nl}$. Differentiating both sides of this equation with respect to the azimuthal quantum number l we find

$$\frac{\partial \tilde{H}}{\partial l} u_{nl} + \tilde{H} \frac{\partial u_{nl}}{\partial l} = \frac{\partial \epsilon}{\partial l} u_{nl} + \epsilon \frac{\partial u_{nl}}{\partial l}. \qquad (10.46)$$

We can now multiply by $u_{nl}^*(r)$ and take the integral over r, formally, we apply the map $(u_{nl}, \cdot) := \int_o^\infty u_{nl}^*(r) \cdot$, and obtain

$$\left(u_{nl}, \frac{\partial \tilde{H}}{\partial l} u_{nl} \right) + \left(u_{nl}, \tilde{H} \frac{\partial u_{nl}}{\partial l} \right) = \frac{\partial \epsilon}{\partial l} \left(u_{nl}, u_{nl} \right) + \epsilon \left(u_{nl}, \frac{\partial u_{nl}}{\partial l} \right). \qquad (10.47)$$

Since $\tilde{H}$ is self-adjoint, we can transform the second term on the left-hand side as

$$\left(u_{nl}, \tilde{H} \frac{\partial u_{nl}}{\partial l} \right) = \left(\tilde{H} u_{nl}, \frac{\partial u_{nl}}{\partial l} \right) = \epsilon \left(u_{nl}, \frac{\partial u_{nl}}{\partial l} \right), \qquad (10.48)$$

which cancels with the second term on the right-hand side of eqn (10.47). Moreover, because normalization demands that $(u_{nl}, u_{nl}) = 1$, see eqn (7.14), we arrive at

$$\frac{\partial \epsilon}{\partial l} = \left(u_{nl}, \frac{\partial \tilde{H}}{\partial l} u_{nl} \right). \qquad (10.49)$$

Both of the involved derivatives can be easily calculated and yield

$$\frac{\partial \epsilon}{\partial l} - \frac{\partial}{\partial l} \frac{1}{n^2} = -\frac{\partial}{\partial l} \frac{1}{(n'+l+1)^2} = \frac{1}{(n'+l+1)^3} = \frac{1}{n^3}, \qquad (10.50)$$

where we have inserted the radial quantum number $n' = n' + l + 1$ from eqn (7.82), and

$$\frac{\partial \tilde{H}}{\partial l} = \frac{\partial}{\partial l} \left(-\frac{\partial^2}{\partial y^2} + \frac{l(l+1)}{y^2} - \frac{2}{y} \right) = \frac{2l+1}{y^2}, \qquad (10.51)$$

respectively. Inserting the last two results into eqn (10.49), we thus have

$$\left(u_{nl}, \frac{1}{y^2} u_{nl}\right) = \frac{2}{n^3 (2l + 1)}. \tag{10.52}$$

And since $\left(u_{nl}, \frac{1}{y^2} u_{nl}\right) = r_{\mathrm{B}}^2 \left\langle \frac{1}{r^2} \right\rangle_{nl}$, we arrive at

$$\langle n, l, m_l \mid \frac{1}{r^2} \mid n, l, m_l \rangle = \frac{2}{r_{\mathrm{B}}^2 n^3 (2l + 1)} = \frac{8n E_n^{(0)\,2}}{q_{\mathrm{e}}^4 (2l + 1)}. \tag{10.53}$$

At last, we can insert this result into eqn (10.42) together with $\langle \frac{1}{r} \rangle_{nl}$ from eqn (10.43), and calculate

$$\langle n, l, m_l \mid H_{\mathrm{rel.kin.}} \mid n, l, m_l \rangle = -\frac{1}{2mc^2} \left[E_n^{(0)\,2} + 2 E_n^{(0)} q_{\mathrm{e}}^2 \left(-\frac{2 E_n^{(0)}}{q_{\mathrm{e}}^2} \right) + q_{\mathrm{e}}^4 \left(\frac{8n E_n^{(0)\,2}}{q_{\mathrm{e}}^4 (2l + 1)} \right) \right],$$

$$= \frac{E_n^{(0)\,2}}{2mc^2} \left[3 - \frac{4n}{l + \frac{1}{2}} \right]. \tag{10.54}$$

10.2.2   Spin-Orbit Correction

Let us now proceed with the second perturbation arising from the Dirac equation, the so-called *spin-orbit* term,

$$H_{\mathrm{spin-orbit}} = \frac{q_{\mathrm{e}}^2}{2m^2 c^2} \frac{\vec{S} \cdot \vec{L}}{r^3}. \tag{10.55}$$

This perturbation can be heuristically understood when considering the electron and proton as point-like particles in relative motion. From the point of view of the electron rest frame, the moving proton charge is responsible for a magnetic field $\vec{B}$ that the magnetic dipole moment of the electron spin—which persists in the electron's rest frame—couples to. Using the law of Ampère–Biot–Savart, e.g., as discussed in Jackson (1999, p. 176) or Ohanian (1988, p. 256), the magnetic field experienced in the rest frame of the electron at position $\vec{x}$ for a proton of mass m and charge q_{e} moving with velocity $\vec{v}$ is (again, in Gaussian cgs units)

$$\vec{B} = \frac{q_{\mathrm{e}}}{c} \frac{\vec{v} \times \vec{x}}{r^3} = -\frac{q_{\mathrm{e}}}{mc} \frac{\vec{L}}{r^3}, \tag{10.56}$$

where we have inserted the orbital angular momentum $\vec{L} = \vec{x} \times \vec{p} = -m\vec{v} \times \vec{x}$ from eqn (6.1). Meanwhile, the magnetic moment of the electron spin from eqn (8.26), including the g-factor $g_s \approx 2$ for the anomalous magnetic moment, is

$$\vec{\mu}_{\mathrm{s}} = \frac{q_{\mathrm{e}}}{mc} \vec{S}. \tag{10.57}$$

Combining eqns (10.56) and (10.57) in the interaction term $-\vec{\mu} \cdot \vec{B}$ from eqn (8.2), and comparing with the spin-orbit term in eqn (10.55), we note that we come up a factor $\frac{1}{2}$ short. The reason for this discrepancy lies in the non-inertial motion of the electron. In taking the point of view of the electron, we have formally performed a Lorentz transformation to its rest frame. However, since the motion is non-inertial, the rest frame changes and the corresponding Lorentz boosts in non-colinear directions do not commute. In particular, the effect of performing a Lorentz boost $L(p^\mu)$ from the rest-frame four-momentum $k^\mu = (mc, 0)^\top$ to p^μ, followed by an arbitrary Lorentz transformation Λ, and an (inverse) boost $L^{-1}(\Lambda p^\mu)$ back to the new rest frame is a transformation

$$W(\Lambda, p^\mu) = L^{-1}(\Lambda p^\mu)\, \Lambda\, L(p^\mu) \tag{10.58}$$

called a *Wigner rotation*. For a more detailed discussion, see, e.g., Sexl and Urbantke (2001, Sec. 9.4) or Friis (2010, Sec. 3.2.4). That it is truly a rotation can be seen from the fact that it leaves the rest-frame four-momentum k^μ invariant. The net effect of this rotation of the reference frame when the electron's motion around the nucleus is treated as continuous is the so-called *Thomas precession*, which was first calculated by *Llewellyn Thomas* (1926) for the spin-orbit interaction in the hydrogen atom, where it leads exactly to the missing factor of $\frac{1}{2}$, the *Thomas half*.

Let us now turn to the perturbative calculation of the energy correction resulting from $H_{\text{spin-orbit}}$. As in the previous subsection, we first have to determine to what extent we have to apply the tools of degenerate perturbation theory, since the unperturbed energies $E_n^{(0)}$ are degenerate and only depend on n. However, unlike for the correction Hamiltonian $H_{\text{rel.kin.}}$, the "uncoupled" basis $\{|\,n, l, s, m_l, m_s\,\rangle\}$ for any fixed n is only useful as long as spin and orbital angular momentum are not coupled—as the name of the basis suggests. But this is exactly what the operator $\vec{S} \cdot \vec{L}$ does. This means for any fixed n we should switch to the "coupled" or total angular momentum basis $|\,n, l, s, j, m_j\,\rangle$ from eqn (8.71), which is a joint eigenbasis of the operators $\vec{L}^2$, $\vec{S}^2$, $\vec{J}^2$, and $\vec{J}_z$, see eqn (8.68).

These four operators commute with each other, see eqns (8.66) and (8.67), but they also all commute with $\vec{S} \cdot \vec{L}$, which we can easily see by rearranging $\vec{J}^2 = (\vec{S} + \vec{L})^2 = \vec{S}^2 + \vec{L}^2 + 2\vec{S} \cdot \vec{L}$ to obtain

$$\vec{S} \cdot \vec{L} = \tfrac{1}{2}(\vec{J}^2 - \vec{S}^2 - \vec{L}^2). \tag{10.59}$$

Indeed, the mentioned operators also all commute with the unperturbed Hamiltonian H_0, which can be seen by comparing eqn (7.4) and eqn (6.46), and we can hence write

$$\vec{S} \cdot \vec{L}\,|\,n, l, s, j, m_j\,\rangle = \frac{1}{2}(\vec{J}^2 - \vec{S}^2 - \vec{L}^2)\,|\,n, l, s, j, m_j\,\rangle \tag{10.60}$$

$$= \frac{\hbar^2}{2}\left(j(j+1) - s(s+1) - l(l+1)\right)|\,n, l, s, j, m_j\,\rangle,$$

where we have used eqn (8.68).

We can now evaluate the perturbative correction arising from the spin-orbit term,

$$\langle n, l, s, j, m_j \,|\, H_{\text{spin}-\text{orbit}} \,|\, n, l, s, j, m_j \,\rangle = \frac{q_{\text{e}}^2}{2m^2c^2} \,\langle n, l, s, j, m_j \,|\, \frac{\vec{S} \cdot \vec{L}}{r^3} \,|\, n, l, s, j, m_j \,\rangle$$

$$= \frac{q_{\text{e}}^2 \hbar^2}{4m^2c^2} \left(j(j+1) - s(s+1) - l(l+1) \right) \langle n, l, s, j, m_j \,|\, \frac{1}{r^3} \,|\, n, l, s, j, m_j \,\rangle. \quad (10.61)$$

Once more we are thus in a position where we have to calculate expectation values of powers of $1/r$, specifically, the third power for the hydrogen atom. We may once again drop the dependence on the spin, and employ the Kramers–Pasternack relation from Theorem 7.3. In particular, since these relations do not depend on m_l, we may work with the coupled basis, and simply set $q = -1$ in Theorem 7.3 to obtain the relation

$$r_{\text{B}} l(l+1) \langle n, l, s, j, m_j \,|\, \frac{1}{r^3} \,|\, n, l, s, j, m_j \,\rangle = \langle n, l, s, j, m_j \,|\, \frac{1}{r^2} \,|\, n, l, s, j, m_j \,\rangle. \quad (10.62)$$

For $l \neq 0$ we can hence express $\langle \frac{1}{r^3} \rangle_{n,l}$ via $\langle \frac{1}{r^2} \rangle_{n,l}$ by inserting from eqn (10.53), which results in

$$\langle \tfrac{1}{r^3} \rangle_{n,l} = \frac{4n E_n^{(0)\,2}}{r_{\text{B}}\, q_{\text{e}}^4\, l(l+1)(l+\frac{1}{2})} \quad \text{for } l \neq 0. \quad (10.63)$$

Finally inserting into eqn (10.61) with $s = \frac{1}{2}$, we arrive at

$$\langle n, l, s, j, m_j \,|\, H_{\text{spin}-\text{orbit}} \,|\, n, l, s, j, m_j \,\rangle = \frac{n\, E_n^{(0)\,2}}{mc^2}\, \frac{j(j+1) - l(l+1) - \frac{3}{4}}{l(l+1)(l+\frac{1}{2})} \quad (10.64)$$

for $l \neq 0$, whereas the application of $\vec{L}$ on any state with $l = 0$ trivially gives $\vec{L} \,|\, n, l = 0, s, j, m_j \,\rangle = 0$. For non-zero values of l, the quantum number j may then have one of two values, $j = l + \frac{1}{2}$ and $j = l - \frac{1}{2}$, because $s = \frac{1}{2}$. Distinguishing all of these cases, we can thus write

$$\langle\, H_{\text{spin}-\text{orbit}} \,\rangle_{n,l,j} = \frac{n\, E_n^{(0)\,2}}{mc^2} \times \begin{cases} \frac{1}{(l+1)(l+\frac{1}{2})} & \text{for } l \neq 0,\ j = l + \frac{1}{2}, \\ 0 & \text{for } l = 0,\ j = s = \frac{1}{2},, \\ \frac{-1}{l(l+\frac{1}{2})} & \text{for } j = l - \frac{1}{2}, \end{cases} \quad (10.65)$$

which will be useful later on when we collect the corrections from all fine-structure terms.

10.2.3 The Darwin Correction

Let us now turn to the final fine-structure term,

$$H_{\text{Darwin}} = \frac{\pi}{2} \frac{q_{\text{e}}^2 \hbar^2}{m^2c^2}\, \delta^3(\vec{x}), \quad (10.66)$$

named after *Charles Galton Darwin* in recognition of his contributions to the investigation of the fine structure (see, e.g., Darwin, 1928). Heuristically, the Darwin term can be understood to originate from what is sometimes referred to as "Zitterbewegung" or smearing out of the electron position when trying to confine it to a small region. In this case, when the electron position is confined to within a distance of the nucleus of the order of one Compton wavelength $2\pi\lambda_{\rm C} = \frac{\hbar}{mc}$, the position-momentum uncertainty relation implies a momentum uncertainty that results in energies comparable to the electron rest mass mc^2. In this regime, (virtual) pair creation effects become relevant, effectively smearing out the position of the electron interacting with the electric field of the nucleus. This can be modelled by a smearing function, e.g., a Gaussian of suitable width, which provides a correction term of the form of eqn (10.66), which is off by a factor that only depends on the choice of the smearing function (see, e.g., Cohen-Tannoudji *et al.*, 1991, p. 1212), thus giving support to this intuition behind the effect.

To evaluate the (exact) correction, we then first note that the three-dimensional delta function at the origin eliminates contributions from all but the solutions with $l = 0$ (and hence also $m_l = 0$). The corresponding energy levels are still twice degenerate ($m_s = \pm\frac{1}{2}$), but since $[H_{\rm Darwin}, \vec{S}^{\,2}] = [H_{\rm Darwin}, S_z] = 0$, we can ignore the spin for the calculation of the perturbative corrections. Nevertheless, we switch to the calculation in the coupled (total angular momentum) basis as mandated by the presence of the spin-orbit interaction. We thus have

$$\langle\, n, l, s, j, m_j \,|\, H_{\rm Darwin} \,|\, n, l, s, j, m_j \,\rangle = \begin{cases} \frac{\pi}{2}\frac{q_{\rm e}^2\hbar^2}{m^2c^2}\,|\psi_{n,l=0,m_l=0}(r=0)|^2 & \text{for } l = 0 \\ 0 & \text{for } l \neq 0 \end{cases}.$$

$$(10.67)$$

Inserting the (unperturbed) hydrogen wave functions ψ_{n,l,m_l} from eqn (7.97), and noting that $|Y_{0,0}|^2 = \frac{1}{4\pi}$ from eqn (6.70), we are left with determining $|R_{n,0}(r=0)|^2$. From eqns (7.94) and (7.96) we have

$$|R_{n,0}(r=0)|^2 = \frac{4}{n^4}\frac{(n-1)!}{(r_{\rm B}\,n!)^3}\,[L_n^1(0)]^2.$$

$$(10.68)$$

To determine the value of the associated Laguerre polynomials at the origin, we recall the definitions from eqn (7.89) and eqn (7.92) and consider the following limit,

$$L_n^1(0) = \lim_{\xi\to 0}\frac{\partial}{\partial\xi}L_n(\xi) = \lim_{\xi\to 0}\frac{\partial}{\partial\xi}e^\xi\frac{\partial^n}{\partial\xi^n}e^{-\xi}\,\xi^n.$$

$$(10.69)$$

In this form, it is not entirely straightforward to take the limit $\xi \to 0$, so let us write the exponential functions as a power series and simply carry out the partial derivatives,

$$\frac{\partial}{\partial\xi}e^\xi\frac{\partial^n}{\partial\xi^n}e^{-\xi}\,\xi^n = \frac{\partial}{\partial\xi}\sum_{k'=0}^{\infty}\frac{\xi^{k'}}{k'!}\frac{\partial^n}{\partial\xi^n}\sum_{k=0}^{\infty}\frac{(-1)^k}{k!}\xi^{k+n}$$

$$= \frac{\partial}{\partial\xi}\sum_{k,k'=0}^{\infty}\frac{(-1)^k}{k!k'!}\,(n+k)(n+k-1)\ldots(k+1)\,\xi^{k+k'}$$

$$= \sum_{k,k'=0}^{\infty}\frac{(-1)^k\,(n+k)!(k+k')}{(k!)^2k'!}\,\xi^{k+k'-1}.$$

$$(10.70)$$

Now, as we take the limit $\xi \to 0$, the only surviving terms in the sum occur for the index pairs $k = 0$, $k' = 1$ and $k = 1$, $k' = 0$, such that

$$L_n^1(0) = n! - (n+1)! = n! - (n+1)n! = -n\,n!\,. \tag{10.71}$$

Inserting into eqn (10.68) we have

$$|\psi_{n,l=0,m_l=0}(r=0)|^2 = \frac{1}{4\pi}\frac{4}{n^4}\frac{(n-1)!}{(r_{\mathrm{B}}\,n!)^3}\,n^2(n!)^2 = \frac{1}{\pi\,n^2\,r_{\mathrm{B}}^3}\frac{(n-1)!}{n!} = \frac{1}{\pi\,n^3\,r_{\mathrm{B}}^3}, \tag{10.72}$$

and we finally obtain the Darwin correction by combining the result with eqn (10.67),

$$\langle\,n,l=0,s,j,m_j\,|\,H_{\mathrm{Darwin}}\,|\,n,l=0,s,j,m_j\,\rangle = \frac{\pi}{2}\frac{q_{\mathrm{e}}^2\hbar^2}{m^2c^2}\frac{1}{\pi\,n^3\,r_{\mathrm{B}}^3} = \frac{2n\,E_n^{(0)\,2}}{mc^2}, \tag{10.73}$$

where we have again used $r_{\mathrm{B}} = \frac{\hbar^2}{m\,q_{\mathrm{e}}^2}$ and $E_n^{(0)} = -\frac{q_{\mathrm{e}}^2}{2\,r_{\mathrm{B}}\,n^2}$ from eqn (10.31). To summarize, the effect of the Darwin correction is thus

$$\langle\,n,l,s,j,m_j\,|\,H_{\mathrm{Darwin}}\,|\,n,l,s,j,m_j\,\rangle = \begin{cases} \frac{2n\,E_n^{(0)\,2}}{mc^2} & \text{for } l=0 \\ 0 & \text{for } l\neq 0 \end{cases}. \tag{10.74}$$

10.2.4 Combined Fine-Structure Correction

With all three corrections at hand, we first note a convenient coincidence: While the Darwin correction from eqn (10.74) vanishes for $l \neq 0$, the spin-orbit correction from eqn (10.65) vanishes identically for $l = 0$. Yet, for $j = l + \frac{1}{2}$, the limit $l \to 0$ of the spin-orbit correction exactly matches the contribution from the Darwin term. Since we have $j = \frac{1}{2}$ for $l = 0$, we can therefore combine these two terms into the simple expression

$$\langle\,H_{\mathrm{spin-orbit}} + H_{\mathrm{Darwin}}\,\rangle_{n,l,j} = \frac{n\,E_n^{(0)\,2}}{mc^2} \times \begin{cases} \frac{1}{(l+1)(l+\frac{1}{2})} & \text{for } j = l+\frac{1}{2}, \\ \frac{-1}{l(l+\frac{1}{2})} & \text{for } j = l-\frac{1}{2}, \end{cases}. \tag{10.75}$$

To put this together with the relativistic correction to the kinetic energy from eqn (10.54), let us evaluate the cases for $j = l + \frac{1}{2}$ and $j = l - \frac{1}{2}$ separately. In the first case we have

$$\langle\,H_{\mathrm{fine-structure}}\,\rangle_{n,l,j=l+\frac{1}{2}} = \frac{E_n^{(0)\,2}}{2mc^2}\left[3 - \frac{4n}{l+\frac{1}{2}} + \frac{2n}{(l+1)(l+\frac{1}{2})}\right]$$

$$= \frac{E_n^{(0)\,2}}{2mc^2}\left[3 - \frac{4n}{l+\frac{1}{2}}\frac{l+1-\frac{1}{2}}{(l+1)}\right]$$

$$= \frac{E_n^{(0)\,2}}{2mc^2}\left[3 - \frac{4n}{j+\frac{1}{2}}\right], \tag{10.76}$$

where we have inserted $l + \frac{1}{2} = j$ in the last step. For the second case, we find

$$\langle H_{\text{fine-structure}} \rangle_{n,l,j=l-\frac{1}{2}} = \frac{E_n^{(0)\,2}}{2mc^2}\left[3 - \frac{4n}{l + \frac{1}{2}} - \frac{2n}{l(l + \frac{1}{2})} \right]$$

$$= \frac{E_n^{(0)\,2}}{2mc^2}\left[3 - \frac{4n}{l + \frac{1}{2}} \frac{l + \frac{1}{2}}{l} \right]$$

$$= \frac{E_n^{(0)\,2}}{2mc^2}\left[3 - \frac{4n}{j + \frac{1}{2}} \right], \tag{10.77}$$

where we have set $l = j + \frac{1}{2}$. We thus see that, remarkably, both cases result in the same formula and the combined fine-structure correction hence only depends on the principal quantum number n and the total angular-momentum quantum number j, i.e.,

$$\langle H_{\text{fine-structure}} \rangle_{n,j} = \frac{E_n^{(0)\,2}}{2mc^2}\left[3 - \frac{4n}{j + \frac{1}{2}} \right]. \tag{10.78}$$

We observe here that, since $l \leq n - 1$ and thus $j \leq l + 1/2 \leq n - 1/2$, we have $4n/(j+1/2) \geq 4n/n = 4$, such that the fine-structure corrections are always negative, i.e., the unperturbed energies are lowered for all n and j. For fixed n, the levels are lowered more strongly the smaller j is, with the $j = \frac{1}{2}$ levels lowered the most, as illustrated in Fig. 10.2. The largest correction (in absolute terms) arising from the fine structure is obtained for the ground state, $n = 1$ and $j = \frac{1}{2}$, in which case the relative shift is

$$\left| \frac{\langle H_{\text{fine-structure}} \rangle_{n=1,l=1/2}}{E_{n=1}^{(0)}} \right| = \left| \frac{E_{n=1}^{(0)}}{2mc^2} \right| = \frac{\alpha^2}{4} \approx 1.3 \times 10^{-5}, \tag{10.79}$$

where we have expressed the ground-state energy using the fine-structure constant $\alpha \approx \frac{1}{137}$ from eqn (7.38) as in eqn (7.85).

Before we move on to discuss additional perturbations due to external fields, let us remark that there are two effects that we have not yet included here, and which we will not discuss here at length either, but which nevertheless need to be mentioned: the *Lamb shift* and the *hyperfine structure*. The atomic hyperfine structure describes corrections that arise from the interaction between the nucleus and the electron(s), more specifically, the interaction of the nuclear magnetic dipole moment with the magnetic field due to the electron motion, as well as between the electric quadrupole moment of the nucleus and the charge distribution in the atom, see Cohen-Tannoudji *et al.* (1991, Chap. XII) for a more detailed discussion or Griffiths (1982) for a pedagogical derivation. The hyperfine corrections are of the order of 10^{-6} eV, and thus about two orders of magnitude below the fine structure, see Griffiths (1982, p. 701). Despite the small size of the hyperfine corrections, the former can be experimentally resolved with high precision (Hellwig *et al.*, 1970) and are of fundamental importance, for instance,

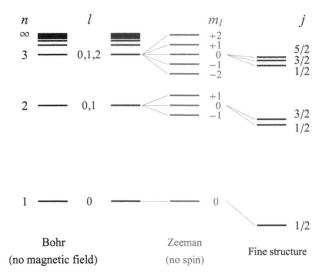

Fig. 10.2 Fine structure. The fine-structure splitting of the degenerate unperturbed energy levels $E_n^{(0)}$ of the Bohr model for the hydrogen atom is shown (not to scale) for the first few energy levels. While the Zeeman effect (see the previous discussion in Section 9.1.2, especially Fig. 9.2) suggests a splitting of the degenerate energy levels into tuplets of $2n - 2$ lines according to the magnetic quantum number m_l in the presence of an external magnetic field, the fine-structure corrections split the energy levels according to the total angular-momentum quantum number j without any external fields. The effect of additional external fields on the fine structure is discussed in Sections 10.3 and 10.4, respectively.

the famous 21 cm line of neutral hydrogen is important in radio astronomy (Ewen and Purcell, 1951). The hyperfine splitting also plays a practical role in the development of quantum technologies, more precisely, hyperfine sublevels are used, e.g., for storage of quantum information in trapped-ion processors, see, for example Leibfried *et al.* (2003) and Schindler *et al.* (2013).

The Lamb shift, on the other hand, is a correction that does not follow simply from the Dirac equation but collects effects from quantum electrodynamics, i.e., which requires a full quantum-field theory treatment, see for example the contributions discussed in Ryder (1996, p. 342) and Peskin and Schroeder (1995, p. 253). First observed by *Willis Lamb* and *Robert Retherford* in 1947 and first theoretically derived in full detail by *Hans Bethe* (1947), this effect leads to a splitting of the fine-structure level $n = 2$, $j = 1/2$, or, in other words, a rise of the level $2s_{1/2}$ (for $(n, l, j) = (2, 0, 1/2)$) above the level $2p_{1/2}$ (for $(n, l, j) = (2, 1, 1/2)$) by about 4.372×10^{-6} eV (or 1057 MHz), and so is of the same order of magnitude as the hyperfine splitting.

10.3 The Zeeman Effect

In this section, we will consider corrections to the energy levels of the hydrogen atom that are caused by (time-independent) external magnetic fields. The resulting splitting of the energy levels is referred to as the *Zeeman effect* that we have already encountered in Section 9.1.2. The corresponding effect for static electric fields—called the *Stark effect*—is considered in Section 10.4.

Let us consider a neutral hydrogen atom in the presence of a static, homogeneous external magnetic field $\vec{B}$. As discussed in Section 8.1, the magnetic dipole moment $\vec{\mu}$ of the atom couples to the magnetic field via an interaction term of the form

$$H_{\text{Zeeman}} = -\vec{\mu} \cdot \vec{B}. \tag{10.80}$$

If we ignore any contributions from the nuclear magnetic moment, we can use the magnetic moment of the electron from eqn (8.26), which contains contributions from the electron spin as well as from its orbital angular momentum, i.e.,

$$\vec{\mu} = -\frac{q_e}{2m_e c}\left(\vec{L} + g_s \vec{S}\right). \tag{10.81}$$

Now, since we know that the complete Hamiltonian for the hydrogen atom contains the spin-orbit interaction from eqn (10.55), the (magnetic quantum numbers for) spin and orbital angular momentum are not separately conserved but coupled to the conserved total angular momentum $\vec{J} = \vec{L} + \vec{S}$. At the same time, the external magnetic field couples with different strength to $\vec{L}$ and $\vec{S}$, and we must therefore take into account the strength of the magnetic field relative to the size of the spin-orbit coupling.

10.3.1 Weak Field—Anomalous Zeeman Effect

When the magnetic field is too weak to break the coupling between the electron's spin and orbital angular momentum, we can consider both to be precessing around the direction of the total angular momentum $\vec{J}$ such that the components orthogonal to $\vec{J}$ are averaged out. In this case we can therefore replace $\vec{L}$ and $\vec{S}$ in eqn (10.81) by the components parallel to $\vec{J}$, which are given by $\frac{\vec{L}\cdot\vec{J}}{\vec{J}^2}\vec{J}$ and $\frac{\vec{S}\cdot\vec{J}}{\vec{J}^2}\vec{J}$, respectively. Moreover, squaring $\vec{S} = \vec{J} - \vec{L}$ and $\vec{L} = \vec{J} - \vec{S}$, we can write

$$\vec{L}\cdot\vec{J} = \tfrac{1}{2}\left(\vec{J}^2 + \vec{L}^2 - \vec{S}^2\right), \tag{10.82a}$$

$$\vec{S}\cdot\vec{J} = \tfrac{1}{2}\left(\vec{J}^2 - \vec{L}^2 + \vec{S}^2\right). \tag{10.82b}$$

With this, we can rewrite the Zeeman term from eqn (10.80) as

$$H_{\text{Zeeman}} = \frac{q_e}{2m_e c}\left(1 + \frac{(g_s-1)\left[\vec{J}^2 - \vec{L}^2 + \vec{S}^2\right]}{2\vec{J}^2}\right)\vec{J}\cdot\vec{B}. \tag{10.83}$$

For this perturbing Hamiltonian we have the commutator $[H_{\text{Zeeman}}, H] = 0$ with $H = H_0 + H_{\text{fine-structure}}$, and we thus see that, despite the $(2j+1)$-fold degeneracy of the unperturbed energies (including the fine-structure corrections from eqn (10.78)), the unperturbed basis $|\,n,l,s,j,m_j\,\rangle$ can be used to evaluate the Zeeman corrections. Assuming without loss of generality that the magnetic field is oriented along the z direction, $\vec{B} = B\vec{e}_z$, we then have

$$\langle\,H_{\text{Zeeman}}\,\rangle_{l,j,m_j} = \langle\,n,l,s,j,m_j\,|\,H_{\text{Zeeman}}\,|\,n,l,s,j,m_j\,\rangle = \mu_{\text{B}}\,g_j\,m_j\,B, \qquad (10.84)$$

where we have used eqn (8.68), $\mu_{\text{B}} = \frac{q_e\hbar}{2m_e c}$ is the Bohr magneton from eqn (8.12), and the *Landé g-factor* (Landé, 1921) for the total angular momentum is given by

$$g_j = \left(1 + \frac{(g_s-1)(j(j+1) - l(l+1) + s(s+1))}{2j(j+1)}\right), \qquad (10.85)$$

with $s = \frac{1}{2}$ for a neutral hydrogen atom with a single electron. This splitting according to different quantum numbers l and m_j, in addition to n and j for the fine structure, in the presence of a weak (compared to the spin-orbit interaction) magnetic field is illustrated in Fig. 10.3 for the neutral hydrogen atom. There, the transitions from $n > 1$ to $n = 1$ give rise to a line spectrum called the Lyman series, with Greek letters enumerating transitions from different n, and we consider the Lyman-α lines for transitions between the energy levels $n = 1$ and $n = 2$. Taking into account the *selection rules* for electric dipole transitions of a single electron between two energy levels, which require $\Delta l = \pm 1$ and $\Delta m_j = -1, 0, +1$ (see, e.g., the detailed discussion in Cohen-Tannoudji *et al.*, 1991, Complement A XIII), one thus finds that each of the two lines in the fine-structure dublet, corresponding to the transitions[7] $2p_{1/2} \to 1s_{1/2}$ and[8] $2p_{3/2} \to 1s_{1/2}$, is split into another triplet of lines corresponding to transitions with $\Delta m_j = -1, 0, +1$. For fixed Δj and Δl (note that $\Delta l = -1$ is fixed for all lines in this example), all possible (according to the dipole selection rules) transitions with the same Δm_j give rise to the same spectral lines, since the energy differences for fixed g_j only depend on Δm_j.

The effect of the splitting of spectral lines in the presence of an external magnetic field is named after *Pieter Zeeman*, who first discovered this effect in 1896 (Zeeman, 1897), and who was subsequently awarded the 1902 Nobel Prize in Physics for his discovery. The prefix "anomalous", typically referring to the effect for non-zero net spin, is an artefact of the time before the discovery of the electron spin, at which point the splitting only due to the magnetic quantum number m_l was not sufficient to explain the more complex structure of line spectra that was observed in experiments (Preston, 1899).

10.3.2   Strong Field—Paschen–Back Effect

When the magnetic field is much stronger than the spin-orbit coupling, one enters the regime of the Paschen–Back effect, named after *Friedrich Paschen* and *Ernst Back*

---

[7]Here we use the spectroscopic notation where the first number specifies n, the letter represents l using the labelling discussed in Section 8.3.4, and the subscript indicates the total angular momentum quantum number j.

[8]Note that the transition from $2s_{1/2}$ to $1s_{1/2}$ is "dipole forbidden" since $\Delta l = 0$.

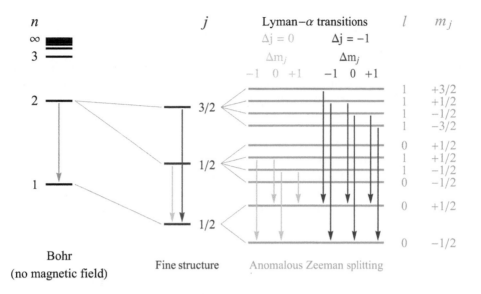

Fig. 10.3 Anomalous Zeeman effect. In the presence of a weak magnetic field (compared to the strength of the spin-orbit coupling), the fine-structure levels of the neutral hydrogen atom, determined by n and j, further split up into an even number of Zeeman levels, labelled by the magnetic quantum number $m_j = -j, -j+1, \ldots, j-1, j$ of the total angular momentum according to eqn (10.84), and by the azimuthal quantum number l via the Landé g-factor of eqn (10.85). For the Lyman-α line for the transition between $n = 2$ and $n = 1$, this leads to a further splitting of the fine-structure dublet (here with $\Delta l = -1$, $\Delta j = 0, -1$) into two triples of spectral lines for the "dipole allowed" transitions with $\Delta m_j = 0, \pm 1$ (for a discussion of these selection rules see Section 10.4.1).

(Paschen and Back, 1921). In this case, we are not treating the Zeeman term as a perturbation, for sufficiently strong fields this would simply not be justified. But by assuming the field to be oriented along the z axis, which we may do without loss of generality, we have $\vec{B} = B\,\vec{e}_z$, and

$$H_{\text{Zeeman}} = -\vec{\mu} \cdot \vec{B} = \frac{q_e B}{2m_e c}\left(L_z + g_s S_z\right),\qquad(10.86)$$

which commutes with the Coulomb Hamiltonian. In other words, the eigenstates $|\,n, l, s, m_l, m_s\,\rangle$ of the (unperturbed) hydrogen atom are also eigenstates of H_{Zeeman} and the eigenvalues $E_n^{(0)}$ from eqn (10.31) are thus modified by the addition of the term

$$E_{\text{Zeeman}} = \mu_{\text{B}}\,B\left(m_l + g_s m_s\right),\qquad(10.87)$$

where we have used Theorem 6.5 and eqn (8.24b), and the Bohr magneton $\mu_{\text{B}} = \frac{q_e \hbar}{2m_e c}$ from eqn (8.12). The total energies are thus

$$E^{(0)}_{n,m_l,m_s} = -\frac{q_e^2}{2\,r_B\,n^2} + \mu_B\,B\,(m_l + g_s m_s).\tag{10.88}$$

But what about the fine-structure corrections? In the strong-field regime, the roles of the Zeeman term and the fine-structure terms are reversed with respect to the weak-field situation, the fine-structure Hamiltonian is now a perturbation of the Zeeman energy levels.

For the corrections arising from the relativistic kinetic energy and from the Darwin term we need not worry, since both corresponding Hamiltonians (from eqns (10.38) and (10.66)) are diagonal with respect to both the coupled and uncoupled bases. The corrections from eqns (10.54) and (10.74) thus remain unchanged. In particular, in the case where $l = 0$, the spin-orbit term vanishes identically (when applied to the corresponding eigenstates), and we can thus also obtain the combined fine-structure correction by setting $j = \frac{1}{2}$ in eqn (10.78), i.e.,

$$\langle H_{\text{fine-structure}}\rangle_{n,l=0} = \frac{2n\,E_n^{(0)\,2}}{mc^2}\left[\frac{3}{4n} - 1\right] \quad \text{for } l = 0.\tag{10.89}$$

When $l \neq 0$, the spin-orbit term generally does not vanish, and is not diagonal with respect to the basis of states $|\,n,l,s,m_l,m_s\,\rangle$. Indeed, we see that one would have to employ degenerate perturbation theory here since there are different combinations of m_l and m_s that would give rise to the same values $m_l + g_s m_s$, at least in very good approximation, when setting $g_s \approx 2$. However, for a strong magnetic field, it is reasonable to expect that $\vec{S}$ and $\vec{L}$ precess about the direction of $\vec{B}$ and their components orthogonal to $\vec{B}$ hence vanish in a time average. We may hence replace $\vec{S}\cdot\vec{L}$ by $S_z L_z$, for which $|\,n,l,s,m_l,m_s\,\rangle$ are eigenstates after all, and we have

$$\langle n,l,s,m_l,m_s\,|\,S_z\,L_z\,|\,n,l,s,m_l,m_s\rangle = \hbar^2\,m_s m_l.\tag{10.90}$$

Since the calculation of $\langle\frac{1}{r^3}\rangle_{n,l}$ is unaffected by any of this, we can insert from eqn (10.63) and for $H_{\text{spin-orbit}}$ from eqn (10.55) to obtain

$$\langle H_{\text{spin-orbit}}\rangle_{n,l,m_l,m_s} = \frac{2n\,E_n^{(0)\,2}}{mc^2}\frac{m_s m_l}{l(l+1)(l+\frac{1}{2})}.\tag{10.91}$$

Combining this with the other fine-structure corrections from eqns (10.54) and (10.74), we then have

$$\langle H_{\text{fine-structure}}\rangle_{n,l,m_l,m_s} = \frac{2n\,E_n^{(0)\,2}}{mc^2} \times \begin{cases} \left[\frac{3}{4n} - \frac{l(l+1) - m_s m_l}{l(l+1)(l+\frac{1}{2})}\right] & \text{for } l \neq 0 \\ \left[\frac{3}{4n} - 1\right] & \text{for } l = 0 \end{cases}.\tag{10.92}$$

It is easy to see that the correction for $l = 0$ is always negative. When $l \neq 0$, we observe that because $m_s \leq \frac{1}{2}$ and $m_l \leq l$, we have

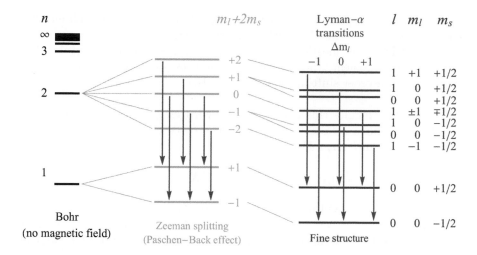

Fig. 10.4 Paschen–Back effect. In the presence of a strong magnetic field (compared to the strength of the spin-orbit coupling), the Zeeman corrections (not shown to scale) split the Bohr energy levels according to the values of $m_l + g_s m_s$. As a result, the α-line of the Lyman series is split into a triple (each with two possible transitions, see Section 10.4.1 for a discussion of the dipole selection rules) with $\Delta m_l = 0, \pm 1$. The fine-structure corrections (also not shown to scale) then further split up the Zeeman levels, depending on the quantum numbers l, m_l, and m_s according to eqn (10.92). This leads to five distinct Lyman-α lines (taking into account the selection rules for allowed electronic dipole transitions, here, $\Delta l = -1$ and $\Delta m_s = 0$), two each with $\Delta m_l = +1$ and $\Delta m_l = -1$, where the differences in m_s entering via the product $m_s m_l$ in eqn (10.92) causes the splitting into pairs, and a single line for the two transitions with $\Delta m_l = 0$.

$$\frac{l(l+1) - m_s m_l}{l(l+1)(l+\frac{1}{2})} \geq \frac{l(l+1) - \frac{l}{2}}{l(l+1)(l+\frac{1}{2})} = \frac{1}{l+1}, \tag{10.93}$$

which in turn is larger or equal than $\frac{1}{n}$ since $l \leq n - 1$. We therefore see that, in the strong-field regime, the fine-structure corrections shift all Zeeman levels towards smaller energies, as illustrated in Fig. 10.4 for the energy levels $n = 1$ and $n = 2$ of the neutral hydrogen atom.

In the intermediate regime, where the influence of the magnetic field is comparable to the fine-structure terms, both have to be treated as perturbations of the Coulomb Hamiltonian. A discussion of this regime for selected examples can be found in Griffiths (1982, Sec. 6.4.3).

10.4 The Stark Effect for the Hydrogen Atom

The Stark effect is the electric analogue to the Zeeman effect, that is, an atom is considered to be placed in a static, homogeneous external electric field $\vec{E}$. As a consequence, atoms in configurations that carry an (average) electric dipole moment, like (certain energy levels of) the hydrogen atom, will be subject to a splitting of their energy levels proportional to the external field, with the proportionality factor depending on the specific energy levels in question. Even for those atoms and energy levels without a permanent dipole moment, the external field induces an electric dipole moment, resulting in a second-order shift of the energy levels.

Here, we will examine this effect for the hydrogen atom with a Hamiltonian as in eqn (10.30), which, in addition to the fine-structure terms from eqn (10.35), is perturbed by the *Stark term* H_{Stark}

$$H_{\text{Stark}} = -\vec{\mathcal{E}}\,\vec{\mu}_{\text{el}} = -q\,\vec{\mathcal{E}}\,\vec{X} = q_{\text{e}}\,\mathcal{E}\,Z\,, \tag{10.94}$$

where $q_{\text{e}} > 0$ is the elementary charge, $\vec{\mathcal{E}}$ denotes the external electric field[9] which, without loss of generality, we choose to be along the z direction, with Z denoting the corresponding position operator, $\mathcal{E} = |\vec{\mathcal{E}}|$, and $\vec{\mu}_{\text{el}}$ is the electrical dipole moment.

We are now interested in calculating the energy corrections of first and second order following Theorems 10.1, 10.2, and 10.3, respectively. As discussed before for the Zeeman effect, one needs to distinguish between the weak-field and strong-field cases relative to the magnitude of the fine-structure corrections, but in both cases the unperturbed energy levels are generally degenerate. In the weak-field case, we consider the perturbation through the Stark Hamiltonian H_{Stark} to be much smaller than the fine-structure corrections and the unperturbed energy levels are determined by the principal quantum number n and the total angular momentum quantum number j. For each fixed pair of n and j, there are at most two different values of l, since $s = \frac{1}{2}$, $l = 0, 1, \ldots, n$, and $j = |l - s|, \ldots, l + s$, such that $l = j \pm \frac{1}{2}$ for $j \neq n - \frac{1}{2}$ and $l = n - 1$ for $j = n - \frac{1}{2}$. For every j, we further have the magnetic quantum numbers $m_j = -j, -j + 1, \ldots, j - 1, j$, and following Theorem 10.3 we thus have to consider a matrix $H_{\text{int}_{n,j}}$ whose components are given by

$$\langle\, n, l, s, j, m_j \,|\, H_{\text{Stark}} \,|\, n, l', s, j, m'_j \,\rangle = q_{\text{e}}\,\mathcal{E}\,\langle\, n, l, s, j, m_j \,|\, Z \,|\, n, l', s, j, m'_j \,\rangle, \tag{10.95}$$

for $l, l' = j \pm \frac{1}{2}$ and $m_j, m'_j = -j, \ldots, +j$. Each matrix element can further be written in terms of matrix elements with respect to the uncoupled basis. From eqn (8.71) we have

$$|\, n, l, s, j, m_j \,\rangle = C^{(j,m_j)}_{l, m_l = m_j + \frac{1}{2}; s, m_s = -\frac{1}{2}}\,|\, n, l, m_l = m_j + \tfrac{1}{2}; s = \tfrac{1}{2}, m_s = -\tfrac{1}{2}\,\rangle$$

$$+ C^{(j,m_j)}_{l, m_l = m_j - \frac{1}{2}; s, m_s = +\frac{1}{2}}\,|\, n, l, m_l = m_j - \tfrac{1}{2}; s = \tfrac{1}{2}, m_s = +\tfrac{1}{2}\,\rangle, \tag{10.96}$$

where the $C^{(j,m_j)}_{l, m_l; s, m_s}$ are the Clebsch–Gordan coefficients and one must always have $m_j = m_l + m_s$. In the strong-field regime, the fine structure can be (initially) neglected and the unperturbed energy levels are determined only by n alone, which means the

---

[9]We use a calligraphic letter here for easier distinction between the electric field strength $\mathcal{E} = |\vec{\mathcal{E}}|$ and the energy levels of the atom.

degeneracy includes n different values of $l = 0, 1, \ldots, n-1$, with $2l+1$ values of $m_l = -l, \ldots, +l$, and two possible values $m_s = \pm\frac{1}{2}$. In either of the two regimes, we thus require calculating a number of matrix elements of the form

$$\langle\, n, l, m_l; s, m_s \mid Z \mid n, l', m_l'; s, m_s' \,\rangle . \tag{10.97}$$

Fortunately, the calculation is greatly simplified by the fact that many of these matrix elements vanish, which we will show in the following.

To begin, let us consider the commutator $[L_z, Z] = 0$ from eqn (6.5b). We can of course formally write the matrix elements of this trivial operator with respect to the uncoupled basis, i.e.,

$$
\begin{aligned}
0 &= \langle\, n, l, m_l; s, m_s \mid [L_z, Z] \mid n, l', m_l'; s, m_s' \,\rangle \\
&= \langle\, n, l, m_l; s, m_s \mid L_z Z \mid n, l', m_l'; s, m_s' \,\rangle - \langle\, n, l, m_l; s, m_s \mid L_z Z \mid n, l', m_l'; s, m_s' \,\rangle \\
&= (m_l - m_l')\, \langle\, n, l, m_l; s, m_s \mid Z \mid n, l', m_l'; s, m_s' \,\rangle . \tag{10.98}
\end{aligned}
$$

As we easily see, this implies that

$$\langle\, n, l, m_l; s, m_s \mid Z \mid n, l', m_l'; s, m_s' \,\rangle = 0 \quad \text{for } \ m_l \neq m_l' . \tag{10.99}$$

Since we also trivially have $[S_z, Z] = 0$, and this in turn implies that $[J_z, Z] = [L_z, Z] + [S_z, Z] = 0$, we can set $m_l = m_l'$, $m_s = m_s'$ and $m_j = m_j'$ in all matrix elements we consider in this section. However, the non-trivial combinations of l and l' are quite limited. To see this, it is most instructive to turn to a much more general result known as the *Wigner–Eckart theorem*.

10.4.1 The Wigner–Eckart Theorem

Loosely speaking, the Wigner–Eckart theorem makes a statement about matrix elements of so-called *spherical tensor operators*—operators with a particular kind of transformation behaviour under spatial rotations—with respect to the eigenstates of angular-momentum operators $\vec{J}^{\,2}$ and J_z. Let us first define the operators under consideration.

Definition 10.1 *An operator $\hat{T}_q^{(k)}$ with $k \in \mathbb{N}_0$ and $q = -k, -k+1, \ldots, k-1, k$ that satisfies*

$$
\begin{aligned}
[J_z, \hat{T}_q^{(k)}] &= \hbar\, q\, \hat{T}_q^{(k)}, \\
[J_\pm, \hat{T}_q^{(k)}] &= \hbar\, \sqrt{k(k+1) - q(q \pm 1)}\, \hat{T}_{q\pm1}^{(k)},
\end{aligned}
$$

*is called a rank-k **spherical tensor operator** with components labelled by q.*

Although we have used slightly different notation to capture a broader class of operators (namely, for all k) with this definition, we have indeed already encountered a number of Cartesian tensor operators with rank $k = 1$. i.e., vector operators such as $\vec{X}$, $\vec{P}$, $\vec{L}$, $\vec{S}$, and $\vec{J}$. Indeed, all of these operators are of the form $\vec{V} = (V_i)$ with components V_i that satisfy the commutation relations

$$[J_i, V_j] = i\hbar\,\varepsilon_{ijk}\,V_k, \tag{10.100}$$

as we recall from eqns (6.5), (8.41), and (8.64). For each of these Cartesian vector operators $\vec{V}$ we can define a spherical vector operator $\hat{T}_q^{(1)}$ with components labelled by $q = 0, \pm 1$ via

$$\hat{T}_{q=0}^{(1)} = V_z, \tag{10.101a}$$

$$\hat{T}_{q=\pm 1}^{(1)} = \mp \tfrac{1}{\sqrt{2}}\left(V_x \pm i\,V_y\right). \tag{10.101b}$$

It can then be checked straightforwardly that the commutation relations in eqn (10.100) imply exactly the commutation relations in Definition 10.1. Specifically, for $q = 0$ we simply have

$$[J_z, \hat{T}_{q=0}^{(1)}] = [J_z, V_z] = 0, \tag{10.102}$$

while the commutator of J_z with $\hat{T}_{q=\pm 1}^{(1)}$ gives

$$[J_z, \hat{T}_{q=\pm 1}^{(1)}] = \mp \tfrac{1}{\sqrt{2}}\left([J_z, V_x] \pm i\,[J_z, V_y]\right) = \mp \tfrac{1}{\sqrt{2}}\left(i\hbar V_y \pm i(-i\hbar V_x)\right) = \pm\hbar\hat{T}_{q=\pm 1}^{(1)}. \tag{10.103}$$

Similarly, we can evaluate the commutators with $J_\pm = J_x \pm iJ_y$, that is,

$$[J_\pm, \hat{T}_{q=0}^{(1)}] = [J_x, V_z] \pm i\,[J_y, V_z] = \mp \hbar V_x - i\hbar V_y = \sqrt{2}\,\hbar\hat{T}_{q=\pm 1}^{(1)}, \tag{10.104}$$

$$[J_\pm, \hat{T}_{q=+1}^{(1)}] = -\tfrac{1}{\sqrt{2}}\left(i\,[J_x, V_y] \pm i\,[J_y, V_x]\right) = \tfrac{\hbar}{\sqrt{2}}\left(V_z \mp V_z\right) = \hbar\sqrt{2 - (1 \pm 1)}\,\hat{T}_0^{(1)}. \tag{10.105}$$

With a calculation of $[J_\pm, \hat{T}_{q=-1}^{(1)}]$ in complete analogy to (10.105), we can then summarize the commutation relations compactly as in Definition 10.1.

We thus see that all the vector operators that we have encountered here can easily be turned into spherical tensor operators of rank 1. Indeed, we have also already dealt with spherical tensor operators of arbitrary rank, the spherical harmonics from Sec. 6.4.3. Comparing the relations in eqns (6.65b) and (6.65c) with those in Definition 10.1, we see that interpreting the spherical harmonics $Y_{lm}(\vartheta, \varphi)$ as operators in the position representation[10] with respect to the coordinates ϑ and φ, means that Y_{lm} is a spherical tensor operator with rank $k = l$ and components labelled by $q = m$.

With this we now come to the formulation of an important theorem.

---

[10] In other words, we can express the spherical harmonics as functions of the vector operator $\vec{x}/r$ using the relation between Cartesian and spherical polar coordinates in eqn (6.36).

Theorem 10.4 (Wigner–Eckart Theorem)

For any rank-k spherical tensor operator $\hat{T}_q^{(k)}$, the matrix elements with respect to the eigenstates $|\,j,m\,\rangle$ of $\vec{J}^{\,2}$ and J_z are given by

$$\langle\,j',m'\,|\,\hat{T}_q^{(k)}\,|\,j,m\,\rangle = C_{j,m;k,q}^{(j',m')}\,\langle\,j'\,\|\,\hat{T}^{(k)}\,\|\,j\,\rangle\,,$$

where $C_{j,m;k,q}^{(j',m')} = \langle\,j,m;k,q\,|\,j',m'\,\rangle$ is the Clebsch–Gordan coefficient for adding the angular momentum quantum numbers j and k to the total angular momentum quantum number j', with m, q, and m' the associated magnetic quantum numbers, and $\langle\,j'\,\|\,\hat{T}^{(k)}\,\|\,j\,\rangle$ is a quantity that is independent of m, q, and m'.

Proof Let us begin the proof of the Wigner–Eckart theorem by inspecting the matrix element $\langle\,j',m'\,|\,[J_\pm,\hat{T}_q^{(k)}]\,|\,j,m\,\rangle$. On the one hand, we can use the commutation relation in Definition 10.1 to write

$$\langle\,j',m'\,|\,[J_\pm,\hat{T}_q^{(k)}]\,|\,j,m\,\rangle = \hbar\sqrt{k(k+1)-q(q\pm 1)}\,\langle\,j',m'\,|\,\hat{T}_{q\pm 1}^{(k)}\,|\,j,m\,\rangle\,. \quad (10.106)$$

On the other hand, we can expand the commutator and apply the ladder operators $J_\pm$ to the ket and bra, respectively, recalling that $J_\pm^\dagger = J_\mp$, we have

$$\langle\,j',m'\,|\,[J_\pm,\hat{T}_q^{(k)}]\,|\,j,m\,\rangle = \langle\,j',m'\,|\,J_\pm\hat{T}_q^{(k)} - \hat{T}_q^{(k)}J_\pm\,|\,j,m\,\rangle \quad (10.107)$$

$$= \hbar\sqrt{j'(j'+1)-m'(m'\mp 1)}\,\langle\,j',m'\mp 1\,|\,\hat{T}_q^{(k)}\,|\,j,m\,\rangle$$

$$- \hbar\sqrt{j(j+1)-m(m\pm 1)}\,\langle\,j',m'\,|\,\hat{T}_q^{(k)}\,|\,j,m\pm 1\,\rangle\,,$$

where we have used eqn (8.77). Equating the right-hand sides of eqns (10.106) and (10.107), we see that we have obtained a recursion relation for the desired matrix elements with fixed j', j, and k, which we can write in a compact form using the shorthand notation $C_\pm(j,m) := \sqrt{j(j+1)-m(m\pm 1)}$, i.e.,

$$C_\mp(j',m')\,\langle\,j',m'\mp 1\,|\,\hat{T}_q^{(k)}\,|\,j,m\,\rangle = C_\pm(j,m)\,\langle\,j',m'\,|\,\hat{T}_q^{(k)}\,|\,j,m\pm 1\,\rangle \quad (10.108)$$

$$+ C_\pm(k,q)\,\langle\,j',m'\,|\,\hat{T}_{q\pm 1}^{(k)}\,|\,j,m\,\rangle\,.$$

We can do the same for the Clebsch–Gordan coefficients. starting with the expansion of the total angular momentum state $|\,j',m'\,\rangle$ into the basis of uncoupled states $|\,j_1,m_1\,\rangle\otimes|\,j_2,m_2\,\rangle = |\,j_1,m_1;j_2,m_2\,\rangle$, i.e.,

$$|\,j',m'\,\rangle = \sum_{m_1,m_2} |\,j_1,m_1;j_2,m_2\,\rangle\!\langle\,j_1,m_1;j_2,m_2\,|\,j',m'\,\rangle$$

$$= \sum_{m_1,m_2} C_{j_1,m_1;j_2,m_2}^{(j',m')}\,|\,j_1,m_1;j_2,m_2\,\rangle\,. \quad (10.109)$$

We then apply the ladder operators $J_\pm = J_{1,\pm} + J_{2,\pm}$ to both sides, where the left-hand side gives

$$
\begin{aligned}
J_\pm \,|\, j', m' \,\rangle &= \hbar C_\pm(j'; m') \,|\, j', m' \pm 1 \,\rangle \\
&= \hbar \sum_{m_1, m_2} C_\pm(j', m') \, C^{(j', m' \pm 1)}_{j_1, m_1; j_2, m_2} \,|\, j_1, m_1; j_2, m_2 \,\rangle .
\end{aligned} \tag{10.110}
$$

Meanwhile, applying the ladder operators to each of the subsystems corresponding to the uncoupled angular momenta gives the result

$$
\begin{aligned}
J_\pm \,|\, j', m' \,\rangle &= \sum_{m_1, m_2} C^{(j', m')}_{j_1, m_1; j_2, m_2} \big(J_{1,\pm} + J_{2,\pm} \big) \,|\, j_1, m_1; j_2, m_2 \,\rangle \\
&= \hbar \sum_{m_1, m_2} C^{(j', m')}_{j_1, m_1; j_2, m_2} \Big[C_\pm(j_1, m_1) \,|\, j_1, m_1 \pm 1; j_2, m_2 \,\rangle + C_\pm(j_2, m_2) \,|\, j_1, m_1; j_2, m_2 \pm 1 \,\rangle \Big] \\
&= \hbar \sum_{m_1, m_2} \Big[C_\pm(j_1, m_1 \mp 1) C^{(j', m')}_{j_1, m_1 \mp 1; j_2, m_2} + C_\pm(j_2, m_2 \mp 1) C^{(j', m')}_{j_1, m_1; j_2, m_2 \mp 1} \Big] \,|\, j_1, m_1; j_2, m_2 \,\rangle .
\end{aligned} \tag{10.111}
$$

Comparing the coefficients for $|\, j_1, m_1; j_2, m_2 \,\rangle$ in the last lines of eqn (10.110) and eqn (10.111), we find the recursion relation for the Clebsch–Gordan coefficients,

$$
\begin{aligned}
C_\pm(j', m') \, C^{(j', m' \pm 1)}_{j_1, m_1; j_2, m_2} &= C_\pm(j_1, m_1 \mp 1) C^{(j', m')}_{j_1, m_1 \mp 1; j_2, m_2} \\
&\quad + C_\pm(j_2, m_2 \mp 1) C^{(j', m')}_{j_1, m_1; j_2, m_2 \mp 1}.
\end{aligned} \tag{10.112}
$$

Finally, we note that $C_\pm(j_i, m_i \mp 1) = C_\mp(j_i, m_i)$ for $i = 1, 2$, and exchanging $\pm \leftrightarrow \mp$ in all expressions, we have

$$
C_\mp(j', m') \, C^{(j', m' \mp 1)}_{j_1, m_1; j_2, m_2} = C_\pm(j_1, m_1) \, C^{(j', m')}_{j_1, m_1 \pm 1; j_2, m_2} + C_\pm(j_2, m_2) \, C^{(j', m')}_{j_1, m_1; j_2, m_2 \pm 1}. \tag{10.113}
$$

With the identification $j_1 = j$, $j_2 = k$, and $m_1 = m$, $m_2 = q$, we recover the exact same recursion relation for the Clebsch–Gordan coefficients as we have for the matrix elements in eqn (10.108). Consequently, the matrix elements $\langle\, j', m' \,|\, \hat{T}^{(k)}_q \,|\, j, m \,\rangle$ must be proportional to the Clebsch–Gordan coefficients $C^{(j', m')}_{j, m; k, q}$, and the proportionality factor must be a constant that is independent of m', m, and q, exactly as claimed in Theorem 10.4. For more details on this argument, see, e.g., Sakurai and Napolitano (2017, p. 253). $\qquad \square$

Selection rules: Let us now contemplate some of the consequences of Theorem 10.4. First, it permits us to use our knowledge of adding angular momenta to conclude that certain matrix elements must vanish, which is of great relevance for spectroscopy in determining the *selection rules* for dipole (and more generally, multipole) transitions. In particular, the dipole operator is a vector operator whose components can be written as linear combinations of the operators $\hat{T}^{(1)}_q$ for $q = 0, \pm 1$. Consequently, a dipole transition between two states $|n, l, m_l\rangle$ and $|n', l', m'_l\rangle$ is said to be forbidden exactly when $\langle n', l', m'_l | \hat{T}_q^{(1)} | n, l, m_l\rangle = 0$ for all q. These matrix elements, in turn, are proportional to the Clebsch–Gordan coefficients $C^{(l', m'_l)}_{l, m_l; k=1, q} = \langle l, m_l; 1, q | l', m'\rangle$. Since $q = 0, \pm 1$ and the magnetic quantum numbers of the uncoupled basis must add up to that of the coupled basis to obtain a non-zero coefficient, $m'_l = m_l + q$, we have the selection rule

$$\Delta m_l = m'_l - m_l = 0, \pm 1. \tag{10.114}$$

Moreover, recall from eqn (8.72) that the corresponding total azimuthal quantum number is bounded by $|l - k| \leq l' \leq l + k$, with $k = 1$ in this case, so that

$$\Delta l = l' - l = 0, \pm 1, \tag{10.115}$$

except when $l = 0$, in which case we must have $l' = 1$ and $\Delta l = +1$, and similarly when $l' = 0$, we must have $\Delta l = -1$. The exact same arguments apply when we consider the coupled basis and the corresponding matrix elements for transitions from $|n, l, s, j, m_j\rangle$ to $|n', l', s, j', m'_j\rangle$, and we hence have

$$\Delta j = j' - j = 0, \pm 1, \qquad \Delta m_j = m'_j - m_j = 0, \pm 1. \tag{10.116}$$

Finally, we note that since $[S_z, X] = [S_z, Y] = [S_z, Z] = 0$, we also have $\Delta m_s = 0$ for electric dipole transitions. This does not mean that other transitions are not possible, just that the latter would have to correspond to higher-order terms in a multipole expansion and would hence be far less likely to occur (see, e.g., Condon and Shortley, 1970, Chap. IV). Such transitions are hence referred to as "dipole forbidden". In Figs. 10.3 and 10.4 we have therefore only included the "dipole allowed" transitions giving rise to the Lyman-α lines.

10.4.2 First-Order Stark Effect

For our discussion of the Stark effect, we are only interested in matrix elements of the operator Z, for which we have already concluded that $\Delta m_j = \Delta m_l = \Delta m_s = 0$, and from the dipole selection rules we further have $\Delta j, \Delta l = 0, \pm 1$. In addition, we can make one more simplification before we turn to the calculation of the Stark corrections themselves: The rules so far allow us to restrict to matrix elements of the form

$$\langle n, l, m_l; s, m_s | Z | n, l', m_l; s, m_s\rangle = \langle n, l, m_l | Z | n, l', m_l\rangle, \tag{10.117}$$

where we have eliminated the spin since it clearly has no bearing on this problem. Closer inspection of the remaining matrix element gives

$$\langle n, l, m_l | Z | n, l', m_l\rangle = \int_{S_2} d\Omega \int_0^\infty dr\, r^2\, \psi^*_{n, l, m_l}\, z\, \psi_{n, l', m_l}$$

$$= \sqrt{\tfrac{4\pi}{3}} \int_0^\infty dr\, r^3\, R^*_{nl} R_{nl'} \int_{S_2} d\Omega\, Y^*_{l, m_l}\, Y_{1, 0}\, Y_{l', m_l}, \tag{10.118}$$

where we have inserted $z = r \cos \vartheta = r \sqrt{\frac{4\pi}{3}} Y_{1,0}(\vartheta)$ from eqns (6.36) and (6.70). The integral over the unit sphere further simplifies because of the matching magnetic quantum numbers, in which case the shape of the spherical harmonics in eqn (6.64) implies that the integrand has no dependence on φ such that we have

$$\int_{S_2} d\Omega \, Y^*_{l,m_l} \, Y_{1,0} \, Y_{l',m_l} \propto \int_0^\pi d\vartheta \, \sin\vartheta \, P^*_{l|m_l|}(\cos\vartheta) \, P_{10}(\cos\vartheta) \, P_{l'|m_l|}(\cos\vartheta)$$

$$= \int_{-1}^{+1} d\xi \, P^*_{l|m_l|}(\xi) \, P_{10}(\xi) \, P_{l'|m_l|}(\xi), \tag{10.119}$$

where we have substituted $\cos\vartheta = \xi$ in the last step. The associated Legendre functions $P_{lm}(\xi)$ have even or odd parity, depending on the combination of quantum numbers, that is, they satisfy $P_{lm}(-\xi) = (-1)^{l+m} P_{lm}(\xi)$, which one can see directly from eqns (6.56) and (6.59). We can thus easily see that the integral in eqn (10.119) vanishes whenever $l' - l + 1 = \Delta l + 1$ is odd. This is the case, in particular, when $\Delta l = 0$, and we can hence restrict to matrix elements where $\Delta l = \pm 1$.

As a consequence of these restrictions on the non-zero matrix elements, we see that there is no first-order perturbative correction for the ground state ($n = 1$), because there is only one possible value of the azimuthal quantum number ($l = 0$) and hence one necessarily has $l' = l$ in that case.

Let us now examine the general form of the first-order Stark corrections for all other energy levels, starting with the weak-field effect, where the unperturbed energies are fixed by n and j. Since $s = \frac{1}{2}$, any given value of j can only arise from two possible values of the azimuthal quantum number, $l = j \pm \frac{1}{2}$, unless $j = n - \frac{1}{2}$, in which case only $l = n - 1$ is possible. In the latter case, there is no first-order Stark correction, for the same reason that the corresponding correction vanishes for the ground state, which corresponds to $n = 1$, $j = n - \frac{1}{2} = \frac{1}{2}$.

In the case where $j \neq n - \frac{1}{2}$, there are two possible values of l, and the requirement $\Delta l = \pm 1$ allows for first-order corrections, that we can calculate as the eigenvalues of the matrix $H_{\text{int}_{n,j}}$ for fixed n and j. But since we must have $\Delta m_j = 0$ for all non-zero entries of this matrix, we find that $H_{\text{int}_{n,j}}$ is block-diagonal, with non-trivial 2×2 blocks $H_{\text{int}_{n,j,m_j}}$ labelled by m_j. These 2×2 blocks have vanishing diagonal elements, and their off-diagonal elements are given by $\langle n, j, l = j - \frac{1}{2}, m_j \, | \, H_{\text{Stark}} \, | \, n, j, l = j + \frac{1}{2}, m_j \rangle$ and its complex conjugate, respectively. The eigenvalues of these 2×2 blocks, and thus the first-order Stark corrections in the weak-field regime are

$$E^{(1)}_{n,j,m_j} = \pm q_e \mathcal{E} \begin{cases} |\langle n, j, l = j - \frac{1}{2}, m_j \, | \, Z \, | \, n, j, l = j + \frac{1}{2}, m_j \rangle| & \text{if } j \neq n - \frac{1}{2} \\ 0 & \text{if } j = n - \frac{1}{2} \end{cases}.$$

$$\tag{10.120}$$

We can calculate these off-diagonal matrix elements via the uncoupled basis states $|\,n, l, m_l; s, m_s\,\rangle$ using eqn (10.96), for which we find

$$\langle\,n, j, l = j - \tfrac{1}{2}, m_j\,|\,Z\,|\,n, j, l = j + \tfrac{1}{2}, m_j\,\rangle =$$

$$= \sum_{m_l = m_j \pm \frac{1}{2}} C^{(j,m_j)\,*}_{l=j-1/2, m_l; s, m_s = m_j - m_l}\; C^{(j,m_j)}_{l=j+1/2, m_l; s, m_s = m_j - m_l}$$

$$\times\;\langle\,n, l = j - \tfrac{1}{2}, m_l\,|\,Z\,|\,n, l = j + \tfrac{1}{2}, m_l\,\rangle. \tag{10.121}$$

Whenever $m_j = +j$, only the term with $m_l = j - \tfrac{1}{2}$ contributes to the sum in eqn (10.121), since $m_l = j + \tfrac{1}{2}$ would exceed the maximal value allowed by $l = j - \tfrac{1}{2}$, and since $m_l = j - \tfrac{1}{2}$ and $m_s = \tfrac{1}{2}$ is the only allowed combination of magnetic quantum numbers that adds up to $m_j = j$, the first Clebsch–Gordan coefficient in eqn (10.121) is $C^{(j, m_j = j)}_{l=j-1/2, m_l = j-1/2; s, m_s = 1/2} = 1$. Similarly only $m_l = j + \tfrac{1}{2}$ contributes to the sum when $m_j = -j$ and we have $C^{(j, m_j = -j)}_{l=j+1/2, m_l = j+1/2; s, m_s = -1/2} = 1$.

The evaluation of the remaining matrix elements $\langle\,n, l = j - \tfrac{1}{2}, m_l\,|\,Z\,|\,n, l = j + \tfrac{1}{2}, m_l\,\rangle$ can be carried out by inserting for the radial wave functions from eqn (7.94) and for the spherical harmonics from eqn (6.64). Meanwhile, the Clebsch–Gordan coefficients can be calculated using the procedure discussed in Section 8.3.2, or they can be simply looked up (see, e.g., Hagiwara, 2002) but are nowadays included as standard functions in many scientific software distributions. In addition, we can limit our calculations to the case of non-negative m_j, since the corrections have the symmetry $E^{(1)}_{n,j,-m_j} = E^{(1)}_{n,j,m_j}$. To see this, we simply note that switching the sign of m_j is equivalent to switching the sign of the coordinate z, and so the matrix elements $\langle\,n, j, l = j - \tfrac{1}{2}, m_j\,|\,Z\,|\,n, j, l = j + \tfrac{1}{2}, m_j\,\rangle$ changes their sign under the transformation $m_j \to -m_j$, which does not affect the eigenvalues of the relevant 2×2 matrix. We thus find that, whenever $j \neq n - \tfrac{1}{2}$, the Stark effect in the weak-field regime results in a splitting of the fine-structure levels into pairs determined by fixed values of $|m_j|$, as illustrated in Fig. 10.5.

As a hands-on example, let us calculate the first-order correction for the first excited state of the hydrogen atom with non-trivial (first-order) corrections, which is the case for $n = 2$, $j = \tfrac{1}{2}$. For this example, we have the Clebsch–Gordan coefficients

$$C^{(j=\frac{1}{2}, m_j=\pm\frac{1}{2})}_{l=0, m_l=0; s=\frac{1}{2}, m_s=\pm\frac{1}{2}} = 1 \tag{10.122a}$$

$$C^{(\frac{1}{2}, m_j=\pm\frac{1}{2})}_{l=1, m_l=0; s=\frac{1}{2}, m_s=\pm\frac{1}{2}} = \mp\frac{1}{\sqrt{3}} \tag{10.122b}$$

from eqn (8.73), as well as eqns (8.86) and (8.87), respectively. Meanwhile, the relevant radial integral for the matrix element $\langle\,n = 2, l = 0, m_l = 0\,|\,Z\,|\,n = 2, l = 1, m_l = 0\,\rangle$ gives

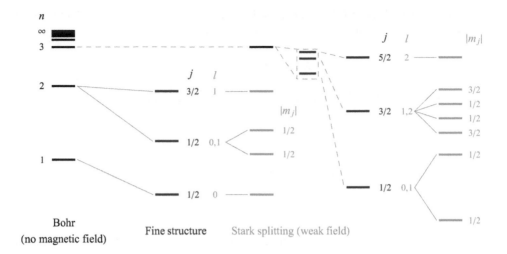

Bohr (no magnetic field)   **Fine structure**   Stark splitting (weak field)

Fig. 10.5 Stark effect (weak field). In the presence of a weak electric field (compared to the strength of the spin-orbit coupling), the fine-structure levels of the neutral hydrogen atom, determined by n and j, receive first-order corrections whenever $j \neq n - \frac{1}{2}$, which causes the degenerate levels to further split up into pairs corresponding to fixed $|m_j|$. Note that the energy levels and their splitting are not shown to scale.

$$\int_0^\infty dr\, r^3 R_{20}^* R_{21} = -\frac{1}{8\sqrt{3}} \int_0^\infty dr\, \frac{r^4}{r_B^4}\left(\frac{r}{r_B} - 2\right) e^{-r/r_B} = -3\sqrt{3}\, r_B, \qquad (10.123)$$

and the integral over the unit sphere evaluates to

$$\int_{S_2} d\Omega\, Y_{0,0}^* \cos\vartheta\, Y_{1,0} = \frac{\sqrt{3}}{2} \int_0^\pi d\vartheta\, \sin\vartheta \cos^2\vartheta = \frac{1}{\sqrt{3}}. \qquad (10.124)$$

We thus find the correction

$$E_{n=2,j=1/2,m_j=\pm1/2}^{(1)} = \pm\sqrt{3}\, q_e\, \mathcal{E}\, r_B. \qquad (10.125)$$

Strong-field Stark effect: When we turn to the strong-field Stark effect, where the corrections due to the strength of the electric field are much larger than the fine-structure corrections, the unperturbed energy levels are determined only by n, but are degenerate. That is, even ignoring spin, which plays no role here, for every n there are n different values of l, and $2l + 1$ different values of m_l for each l. As before, we can restrict to matrix elements with matching values of m_l, which means that the matrix H_{int_n} of corrections in the degenerate subspace is block-diagonal and each block is labelled by a fixed value of m_l, and of size $(n - |m_l|) \times (n - |m_l|)$ since $m_l \leq l$ and there are n possible values of l. Moreover, as before, $\pm m_l$ give the same corrections.

Therefore, we can calculate corrections separately for fixed values of n and $|m_l|$ by diagonalizing matrices $H_{\text{int}_{n,m_l}}$ with components

$$(H_{\text{int}_{n,m_l}})_{ll'} = q_e \mathcal{E} \left(\delta_{l+1,l'} + \delta_{l,l'+1} \right) \langle n, l, m_l \mid Z \mid n, l', m_l \rangle. \tag{10.126}$$

For instance, for our previous example where $n = 2$, we only have two possible values for l, i.e., $l = 0$ and $l = 1$. We thus have two states with $m_l = \pm 1$ which do not receive a correction, because these magnetic quantum numbers are only compatible with a single value of l, which is $l = 1$, but for $m_l = 0$ we have a 2×2 matrix with off-diagonal elements $\langle n = 2, l = 0, m_l = 0 \mid Z \mid n = 2, l' = 1, m_l = 0 \rangle = \langle n = 2, l = 1, m_l = 0 \mid Z \mid n = 2, l' = 0, m_l = 0 \rangle$. Inserting from eqns (10.123) and (10.124), we thus find the strong-field correction

$$E^{(1)}_{n=2,m_l} = \begin{cases} \pm 3\, q_e\, \mathcal{E}\, r_{\text{B}} & \text{if } m_l = 0 \\ 0 & \text{if } |m_l| = 1 \end{cases} \tag{10.127}$$

and thus a split into three distinct energy levels, which are in turn further modified by the fine-structure corrections. Of the latter, the corrections for the relativistic kinetic energy and the Darwin term commute with H_{Stark} and the results hence carry over, but the corrections of the spin-orbit term have to be reevaluated, which we will not do here for the sake of brevity.

Instead, let us turn to another, perhaps conceptually and technically more interesting aspect of the Stark effect, which manifests in the calculation of the second-order correction for the ground state.

10.4.3 Second-Order Stark Effect

As we have observed, for some of the eigenstates of the hydrogen atom, there are no first-order perturbative corrections from the Stark effect, and the first and most prominent example of such a state is the ground state, $n = 1$. This does not mean that the presence of the external electric field has no effect on these energy levels, just that the onset of the effect is only at second order in this case. In principle, higher-order corrections can of course also be calculated for states which receive non-zero first-order corrections, but we will here focus on the ground state and the strong-field case to illustrate the challenges and interesting features of such calculations for a comparably simple example. In particular, the ground state forces the azimuthal quantum number to vanish, $l = 0$ and $m_l = 0$, and since the spin does not play a role in the calculation of the Stark shift, we can hence leave aside potential complications due to degeneracies or distinctions between the weak-field and strong-field regime and focus on the second-order correction for a fixed single state $|n^{(0)}\rangle = |n = 1, l = 0, m_l = 0\rangle$. Following Theorem 10.2, the task at hand is thus to calculate an infinite sum of the form

$$E^{(2)}_n = \sum_{m \neq n} \frac{|\langle m^{(0)} \mid H_{\text{int}} \mid n^{(0)} \rangle|^2}{E^{(0)}_n - E^{(0)}_m}. \tag{10.128}$$

With what we have previously learned about the involved matrix elements, in particular, the selection rules $\Delta l = +1$ and $\Delta m_l = 0$ that apply here and inserting for the unperturbed energies from eqn (10.31), this sum can be written as[11]

$$E_{n=1}^{(2)} = -2\, r_{\mathrm{B}}\, \mathcal{E}^2 \sum_{k \neq 1} \frac{|\langle k, l=1, m_l=0\,|\, Z\,|\, n=1, l=0, m_l=0 \rangle|^2}{1 - \frac{1}{k^2}}. \qquad (10.129)$$

Unfortunately, despite these simplifications, the calculation of the (still infinitely many) remaining matrix elements is a tedious affair (which we shall not go through here explicitly). An alternative approach that makes it easier to carry out this particular calculation, albeit at the expense of first having to solve the Schrödinger equation in parabolic coordinates, is discussed in Landau and Lifshitz (1977, § 77). However, for the purpose of calculating the second-order correction for a specific state (here, the ground state), there is a convenient work-around, discussed in Shankar (1994, p. 462), Ohanian (1990, Chap. 10) or Merzbacher (1998, p. 461), that we can employ here. The key ingredient is to determine an operator $\widehat{\Omega}$, whose commutator with the unperturbed Hamiltonian has the same effect on the unperturbed eigenstates as the perturbation, i.e.,

$$H_{\mathrm{int}}\,|\,n^{(0)}\rangle = [\widehat{\Omega}, H_0]\,|\,n^{(0)}\rangle\,. \qquad (10.130)$$

With an operator with this property at hand, we can rewrite the general second-order correction from eqn (10.128) in the following way:

$$\begin{aligned}
E_n^{(2)} &= \sum_{m \neq n} \frac{\langle n^{(0)}\,|\,H_{\mathrm{int}}\,|\,m^{(0)}\rangle \langle m^{(0)}\,|\,H_{\mathrm{int}}\,|\,n^{(0)}\rangle}{E_n^{(0)} - E_m^{(0)}} \\[2mm]
&= \sum_{m \neq n} \frac{\langle n^{(0)}\,|\,H_{\mathrm{int}}\,|\,m^{(0)}\rangle \langle m^{(0)}\,|\,[\widehat{\Omega}, H_0]\,|\,n^{(0)}\rangle}{E_n^{(0)} - E_m^{(0)}} \\[2mm]
&= \sum_{m \neq n} \frac{\langle n^{(0)}\,|\,H_{\mathrm{int}}\,|\,m^{(0)}\rangle \left(\langle m^{(0)}\,|\,\widehat{\Omega}\,H_0\,|\,n^{(0)}\rangle - \langle m^{(0)}\,|\,H_0\,\widehat{\Omega}\,|\,n^{(0)}\rangle\right)}{E_n^{(0)} - E_m^{(0)}} \\[2mm]
&= \sum_{m \neq n} \frac{\langle n^{(0)}\,|\,H_{\mathrm{int}}\,|\,m^{(0)}\rangle \left(E_n^{(0)} - E_m^{(0)}\right) \langle m^{(0)}\,|\,\widehat{\Omega}\,|\,n^{(0)}\rangle}{E_n^{(0)} - E_m^{(0)}} \\[2mm]
&= \sum_{m \neq n} \langle n^{(0)}\,|\,H_{\mathrm{int}}\,|\,m^{(0)}\rangle\langle m^{(0)}\,|\,\widehat{\Omega}\,|\,n^{(0)}\rangle\,. \qquad (10.131)
\end{aligned}$$

In the last line, we note that we (almost) have a sum over projectors on a complete basis, that is,

$$\sum_{m \neq n} |\,m^{(0)}\rangle\langle m^{(0)}\,| = \mathbb{1} - |\,n^{(0)}\rangle\langle n^{(0)}\,|, \qquad (10.132)$$

---

[11] Actually, as we shall see shortly, this is not entirely correct, since the sum in eqn (10.129) is neglecting the continuum spectrum of the hydrogen atom which is implicitly included in the expression in eqn (10.128).

and we can thus express the second-order correction as

$$E_n^{(2)} = \langle n^{(0)} | H_{\text{int}} \widehat{\Omega} | n^{(0)} \rangle - \langle n^{(0)} | H_{\text{int}} | n^{(0)} \rangle \langle n^{(0)} | \widehat{\Omega} | n^{(0)} \rangle . \qquad (10.133)$$

Provided that we find suitable operators $\widehat{\Omega}$, this is a significant reduction in computational effort, from calculating potentially infinitely many to just three matrix elements. However, two important observations should be made here about this method. First, there may not be an operator $\widehat{\Omega}$ for every pair H_0 and H_{int} that satisfies eqn (10.130), and less so one that does so for all n. Nevertheless, as long as eqn (10.130) holds for the particular eigenvalue n we are interested in, the calculation goes through and we can easily evaluate $E_n^{(2)}$ for that specific energy level.

The second remark has no direct effect on the calculation itself, but goes on to highlight an interesting aspect of perturbative methods in infinite-dimensional Hilbert spaces. The expression in Theorem 10.2 contains a sum over all unperturbed eigenvalues of a given operator H_0, but for infinite-dimensional Hilbert spaces, the Hamiltonian can have a discrete as well as a continuous part of the spectrum. For some cases, like the harmonic oscillator (see Chapter 5) or the infinite potential well (see Section 4.2.2) the spectrum is entirely discrete, but for others, like the finite potential well (see Section 4.2.1) and, in particular, the Coulomb potential, one encounters a discrete spectrum of bound states with energies $E_n \leq 0$ and a continuum spectrum of states with energies $E > 0$. For the finite potential well, these continuum solutions are simply the eigenstates of the momentum operator.

In the case of the hydrogen atom, we have not explicitly derived these solutions, but one may do so by noting that for any sign of E the Schrödinger equation for the Coulomb Hamiltonian(7.72) leads to the same differential equation as in eqn (7.88), except that n in $\gamma = n + l$ is replaced by $\nu = \frac{q_e^2}{\hbar} \sqrt{\frac{m}{-2E}}$. However, for $E > 0$, the quantity ν is not a positive integer and, indeed, not even a real number. In this case, eqn (7.88) can be recognized as Kummer's differential equation, whose solutions $_1F_1(a, b, \xi)$ are called the confluent hypergeometric functions (of the first kind[12]) with $a = l + 1 - \nu$ and $b = 2l + 2$. As with the eigenstates of the momentum operator, the corresponding solutions of the Schrödinger equation, denoted here as $| E, l, m_l \rangle$ based on these functions are not square-integrable, but they can be "normalized" in the sense that $\langle E, l, m | E', l', m' \rangle = \delta_{ll'} \delta_{mm'} \delta(E - E')$, and they are certainly part of the rigged Hilbert space (see Section 3.3.3).

The sum over all unperturbed eigenstates in Theorem 10.2 must hence be read as a sum over all contributions from the discrete spectrum as well as an integral over the continuum spectrum, which we would have had to calculate in addition to the infinite sum arising from the discrete eigenstates. Likewise, the resolution of the identity in eqn (10.132) must hence be understood as containing contributions from both the discrete and continuous part of the spectrum to cover all allowed energies, both negative and positive. In this way, the method employing the operator $\widehat{\Omega}$ as in

---

[12] As the name suggests, there is also a second kind, but the physically relevant solutions correspond to the functions known as the hypergeometric functions of the first kind.

eqn (10.133) conveniently takes care also of any contributions that may arise from the eigenstates of the continuum spectrum without having to even write down any of them.

Returning now to the matter of calculating the second-order Stark shift for the hydrogen ground state, the corresponding operator $\widehat{\Omega}$ is

$$\widehat{\Omega} = -\frac{\mathcal{E}}{q_e}\left(\frac{r^2}{2} + r_{\mathrm{B}}\, r\right)\cos\vartheta\,. \tag{10.134}$$

To check that this operator satisfies eqn (10.130), we first write $H_{\mathrm{int}} = q_e\,\mathcal{E}\,r\,\cos\vartheta$, and applying it to the ground-state wave function $\psi_{1,0,0} = R_{1,0}(r)\,Y_{0,0} = \frac{1}{\sqrt{\pi r_{\mathrm{B}}^3}}e^{-r/r_{\mathrm{B}}}$, where we have inserted $Y_{0,0} = \sqrt{\frac{1}{4\pi}}$ and $R_{1,0} = \left(2/\sqrt{r_{\mathrm{B}}^3}\right)e^{-r/r_{\mathrm{B}}}$ from eqns (6.70) and (7.94), respectively, we have

$$H_{\mathrm{int}}\,\psi_{1,0,0} = \frac{q_e\,\mathcal{E}}{\sqrt{\pi r_{\mathrm{B}}}}\,\frac{r}{r_{\mathrm{B}}}\,\cos\vartheta\,e^{-r/r_{\mathrm{B}}}\,. \tag{10.135}$$

For the first term of the commutator, we can write

$$\widehat{\Omega}\,H_0\,\psi_{1,0,0} = \widehat{\Omega}\,E_1^{(0)}\,\psi_{1,0,0} = \frac{q_e\,\mathcal{E}}{\sqrt{\pi r_{\mathrm{B}}}}\left(\frac{r^2}{4r_{\mathrm{B}}^2} + \frac{r}{2r_{\mathrm{B}}}\right)\cos\vartheta\;e^{-r/r_{\mathrm{B}}}\,. \tag{10.136}$$

In order to evaluate the remaining term, we write the unperturbed Hamiltonian as $H_0 = -\frac{\hbar^2}{2m}\Delta - \frac{q_e^2}{r}$ and insert the Laplacian in spherical polar coordinates from eqn (7.3). A short calculation that we will leave for the interested reader to check then gives

$$H_0\,\widehat{\Omega}\,\psi_{1,0,0} = \frac{q_e\,\mathcal{E}}{\sqrt{\pi r_{\mathrm{B}}}}\left(\frac{r^2}{4r_{\mathrm{B}}^2} - \frac{r}{2r_{\mathrm{B}}}\right)\cos\vartheta\;e^{-r/r_{\mathrm{B}}}\,. \tag{10.137}$$

Combining these results, we can thus confirm that $\widehat{\Omega}$ has the desired property.

For the hydrogen ground state, we further have

$$\langle\,n^{(0)}\,|\,H_{\mathrm{int}}\,|\,n^{(0)}\,\rangle = q_e\,\mathcal{E}\,\langle\,1,0,0\,|\,Z\,|\,1,0,0\,\rangle = 0 \tag{10.138}$$

because of the already encountered selection rule $\Delta l = \pm 1$. The second-order correction can therefore be evaluated just from a single matrix element,

$$E_n^{(2)} = \langle 1,0,0 | H_{\text{int}} \, \widehat{\Omega} | 1,0,0 \rangle = -\mathcal{E}^2 \langle 1,0,0 | \left(\frac{r^3}{2} + r_{\text{B}} \, r^2 \right) \cos^2 \vartheta \, | 1,0,0 \rangle$$

$$= -\mathcal{E}^2 \int\limits_0^\infty dr \, r^2 \int\limits_0^\pi d\vartheta \, \sin \vartheta \int\limits_0^{2\pi} d\varphi \, |\psi_{1,0,0}|^2 \left(\frac{r^3}{2} + r_{\text{B}} \, r^2 \right) \cos^2 \vartheta$$

$$= -\frac{2\,\mathcal{E}^2}{r_{\text{B}}^3} \int\limits_0^\infty dr \, e^{-2r/r_{\text{B}}} \left(\frac{r^5}{2} + r_{\text{B}} \, r^4 \right) \int\limits_0^\pi d\vartheta \, \sin \vartheta \, \cos^2 \vartheta$$

$$= -\mathcal{E}^2 \, r_{\text{B}}^3 \, \frac{1}{3} \int\limits_0^\infty dr' \, e^{-r'} \left(\frac{r'^5}{2^5} + \frac{r'^4}{2^3} \right) = -\mathcal{E}^2 \, r_{\text{B}}^3 \, \frac{9}{4}, \tag{10.139}$$

as in Shankar (1994, eqn (17.2.43)) or, obtained using a different method in Griffiths (1982, p. 291).

With this example, we conclude our discussion of the Stark effect and of time-independent perturbation theory, but for further information on this topic we refer the interested reader to a wealth of literature, for instance Cohen-Tannoudji *et al.* (1991, Complement E XII), Landau and Lifshitz (1977, Chap. VI), Griffiths (1982, Chap. 6), or Shankar (1994, Chap. 17).

10.5 Time-Dependent Perturbation Theory

10.5.1 Time-Dependent Hamiltonians

Let us now briefly turn to the time-dependent case, where we have a Hamiltonian of the form

$$H = H_0 + H_{\text{int}}(t). \tag{10.140}$$

Here, the unperturbed Hamiltonian H_0 is typically assumed not to be explicitly time-dependent[13], while the interaction Hamiltonian $H_{\text{int}}(t)$ may now explicitly depend on time and the coupling constant has been absorbed into the definition of H_{int}. We will assume that the interaction starts at some particular time t_o such that $H_{\text{int}}(t) = 0$ for $t < t_o$. As in the time-independent case, it is further assumed that the (now time-dependent) solutions $|\psi^{(0)}(t)\rangle$ to the Schrödinger equation are fully known in the absence of the interaction term, i.e.,

$$i\hbar \frac{\partial}{\partial t} |\psi^{(0)}(t)\rangle = H_0 |\psi^{(0)}(t)\rangle \qquad \text{for } t < t_o, \tag{10.141}$$

while the general solutions $|\psi(t)\rangle$ satisfying

$$i\hbar \frac{\partial}{\partial t} |\psi(t)\rangle = \left[H_0 + H_{\text{int}}(t) \right] |\psi(t)\rangle \tag{10.142}$$

are unknown.

---

[13]In cases where it is, expressions such as $e^{\pm i H_0 t/\hbar}$ need to be replaced with the corresponding time-ordered integral, which we shall discuss shortly.

The first subtlety that we encounter in dealing with this problem lies in the observation that the time evolution under time-dependent Hamiltonians is not as straightforward any more as for time-independent Hamiltonians that we have discussed in Section 2.6. The propagator $U(t, t_o)$ that maps the Schrödinger-picture state $|\psi(t_o)\rangle$ at time t_o to the corresponding state $|\psi(t)\rangle = U(t, t_o)|\psi(t_o)\rangle$ at a later time t now has a more complicated form. To determine this form explicitly, we insert the state $|\psi(t)\rangle$ into the Schrödinger equation to obtain

$$i\hbar \frac{\partial}{\partial t} U(t, t_o)|\psi(t_o)\rangle = H U(t, t_o)|\psi(t_o)\rangle \,. \tag{10.143}$$

Since this must hold for any ψ and for any t_o, this implies

$$i\hbar \frac{\partial}{\partial t} U(t, t_o) = H U(t, t_o)\,. \tag{10.144}$$

Making a power-series ansatz for the propagator, we see that it must be of the form

$$U(t, t_o) = 1 + \frac{-i}{\hbar} \int_{t_o}^{t} dt_1\, H(t_1) + \left(\frac{-i}{\hbar}\right)^2 \int_{t_o}^{t} dt_1 \int_{t_o}^{t_1} dt_2\, H(t_1)\, H(t_2)$$

$$+ \left(\frac{-i}{\hbar}\right)^3 \int_{t_o}^{t} dt_1 \int_{t_o}^{t_1} dt_2 \int_{t_o}^{t_2} dt_3\, H(t_1)\, H(t_2)\, H(t_3) + \dots, \tag{10.145}$$

because differentiation of any term in the sum always results in the previous term multiplied by $-iH(t)/\hbar$, such that eqn (10.144) holds with the initial condition $U(t_o, t_o) = 1$. This expression, known as a *Born-von Neumann series*, can further be simplified by noting that the integration limits imply $t_1 \geq t_2 \geq t_3 \geq \dots t_o$ and so the operators in the integrand are time ordered in the sense that operators further to the left are evaluated at later times than operators to their right. We can formalize this ordering by introducing a *time-ordering* symbol $\mathcal{T}$. For any two time-dependent operators $A(t)$ and $B(t)$ we can define it via

$$\mathcal{T}\{A(t_1)\, B(t_2)\} = \begin{cases} A(t_1)\, B(t_2) & \text{if } t_1 > t_2, \\ B(t_2)\, A(t_1) & \text{if } t_1 < t_2 \end{cases}, \tag{10.146}$$

and similarly for products of three or more operators. Replacing the integrand in each term in eqn (10.145) by its time-ordered equivalent, we do not have to keep track of the ordering in the integration limits and can evaluate each integral in the limits t_o to t instead. However, doing so we get the same contribution from each of the $n!$ permutations of the integration variables $t_1, t_2, \dots, t_n$ in the nth term (starting with $n = 0$), where only one contribution (corresponding to the correct time order) contributed before (cf. Peskin and Schroeder, 1995, p. 84). We therefore have to include a pre-factor $1/n!$, and arrive at

$$U(t,t_o) = 1 + \frac{-i}{\hbar} \int_{t_o}^{t} dt_1 \, H(t_1) + \frac{1}{2}\left(\frac{-i}{\hbar}\right)^2 \int_{t_o}^{t} dt_1 \int_{t_o}^{t} dt_2 \, \mathcal{T}\{H(t_1)\,H(t_2)\}$$

$$+ \frac{1}{3!}\left(\frac{-i}{\hbar}\right)^3 \int_{t_o}^{t} dt_1 \int_{t_o}^{t} dt_2 \int_{t_o}^{t} dt_3 \, \mathcal{T}\{H(t_1)\,H(t_2)\,H(t_3)\} + \dots$$

$$= \sum_{n=0}^{\infty} \frac{1}{n!}\left(\frac{-i}{\hbar}\right)^n \int_{t_o}^{t} dt_1 \int_{t_o}^{t} dt_2 \dots \int_{t_o}^{t} dt_n \, \mathcal{T}\{H(t_1)\,H(t_2)\dots H(t_n)\}. \quad (10.147)$$

We can write this expression more compactly as a time-ordered exponential

$$U(t,t_o) = \mathcal{T}\left\{\exp\left(-\frac{i}{\hbar}\int_{t_o}^{t} dt' \, H(t')\right)\right\}, \quad (10.148)$$

and we obviously recover the usual form $U(t,t_o) = \exp(-iH(t-t_o)/\hbar)$ from Definition 2.12 when the Hamiltonian is time independent. It is also obvious from this form that the propagator for any (time-dependent) Hamiltonian satisfies the composition rule

$$U(t_3,t_2)\,U(t_2,t_1) = U(t_3,t_1) \qquad \text{for all } t_3 \geq t_2 \geq t_1. \quad (10.149)$$

10.5.2 The Interaction Picture

In the case of the perturbative calculations we are interested in here, the situation one is typically facing is a time-independent unperturbed Hamiltonian H_0 and a time-dependent interaction term $H_{\text{int}}(t)$ corresponding, e.g., to some time-dependent external driving fields. To proceed in such situations, it is generally useful to switch to the so-called *interaction picture*, which can in some sense be considered to be a middle ground between the Schrödinger and Heisenberg pictures discussed in Section 2.6.2.

In the interaction picture, we begin with the Schrödinger-picture state $|\psi_{\text{S}}(t)\rangle$, which is a solution to the Schrödinger equation in (10.142) for the full time-dependent Hamiltonian $H = H_0 + H_{\text{int}}(t)$, and which hence time evolves according to

$$|\psi_{\text{S}}(t)\rangle = U(t,t_o)\,|\psi_{\text{S}}(t_o)\rangle, \quad (10.150)$$

where the propagator $U(t,t_o)$ is given by the time-ordered integral in eqn (10.148). We then define the interaction-picture state $|\psi_{\text{I}}(t)\rangle$ by applying an additional inverse "free" propagator, i.e.,

$$|\psi_{\text{I}}(t)\rangle = e^{iH_0 t/\hbar}\,|\psi_{\text{S}}(t)\rangle. \quad (10.151)$$

Now, if the free Hamiltonian H_0 commutes with the total Hamiltonian H, in our case this means $[H_0, H_{\text{int}}(t)] = 0$ for all t, then this definition means that the time evolution of $|\psi_{\text{I}}(t)\rangle$ is governed by the (time-ordered exponential of) $H_{\text{int}}(t)$ only. In general, one finds that the time evolution of $|\psi_{\mathrm{I}}(t)\rangle$ is governed by the interaction-picture version $H_{\mathrm{int,I}}$ of H_{int}, given by

$$H_{\mathrm{int,I}}(t) = e^{iH_0 t/\hbar}\, H_{\mathrm{int}}(t)\, e^{-iH_0 t/\hbar}\,. \tag{10.152}$$

We can then use these definitions to determine the form of the Schrödinger equation in the interaction picture. Starting from the left-hand side of the Schrödinger equation (in the Schrödinger-picture) in (10.142), we substitute $|\psi_{\mathrm{S}}(t)\rangle = \exp(-iH_0 t/\hbar)\,|\psi_{\mathrm{I}}(t)\rangle$ and differentiate to obtain

$$i\,\hbar\,\frac{\partial}{\partial t}\,|\psi_{\mathrm{S}}(t)\rangle = i\,\hbar\,\frac{\partial}{\partial t}\,e^{-iH_0 t/\hbar}\,|\psi_{\mathrm{I}}(t)\rangle$$

$$= H_0\, e^{-iH_0 t/\hbar}\,|\psi_{\mathrm{I}}(t)\rangle + i\,\hbar\, e^{-iH_0 t/\hbar}\,\frac{\partial}{\partial t}\,|\psi_{\mathrm{I}}(t)\rangle\,, \tag{10.153}$$

while the right-hand side of eqn (10.142) gives

$$i\hbar\,\frac{\partial}{\partial t}\,|\psi_{\mathrm{S}}(t)\rangle = \big[H_0 + H_{\mathrm{int}}(t)\big]\,|\psi_{\mathrm{S}}(t)\rangle = \big[H_0 + H_{\mathrm{int}}(t)\big]\,e^{-iH_0 t/\hbar}\,|\psi_{\mathrm{I}}(t)\rangle\,. \tag{10.154}$$

Comparing the two expressions, we note that the terms $H_0 \exp(-iH_0 t/\hbar)\,|\psi_{\mathrm{I}}(t)\rangle$ cancel, and applying $\exp(iH_0 t/\hbar)$ from the left to the remaining terms and inserting from eqn (10.152) we find

$$i\,\hbar\,\frac{\partial}{\partial t}\,|\psi_{\mathrm{I}}(t)\rangle = e^{iH_0 t/\hbar}\, H_{\mathrm{int}}(t)\, e^{-iH_0 t/\hbar}\,|\psi_{\mathrm{I}}(t)\rangle = H_{\mathrm{int,I}}(t)\,|\psi_{\mathrm{I}}(t)\rangle\,. \tag{10.155}$$

That is, the interaction-picture state $|\psi_{\mathrm{I}}(t)\rangle$ satisfies the Schrödinger equation with respect to the interaction-picture interaction Hamiltonian $H_{\mathrm{int,I}}(t)$. Since $H_{\mathrm{int,I}}(t)$ is time-dependent, the solution to this differential equation can again be expressed in terms of a time-ordered exponential integral, as discussed in Section 10.5.1, i.e.,

$$|\psi_{\mathrm{I}}(t)\rangle = \mathcal{T}\left\{\exp\left(-\frac{i}{\hbar}\int_{t_o}^{t} dt'\, H_{\mathrm{int,I}}(t')\right)\right\}\,|\psi_{\mathrm{I}}(t_o)\rangle\,. \tag{10.156}$$

Let us now consider the situation where the system is initially prepared such that it is in an eigenstate $|n\rangle$ of the unperturbed Hamiltonian H_0 with $H_0\,|n\rangle = E_n\,|n\rangle$ prior to the interaction, and we are interested in determining the transition amplitude to some other unperturbed eigenstate $|m\rangle$ as a function of t. Up to a global phase factor that we can get rid of without loss of generality by setting $t_o = 0$, we thus have the interaction-picture initial state $|\psi_{\mathrm{I}}(t_o)\rangle = |n\rangle$. We can then write the desired transition amplitude as

$$\langle m | \psi_{\mathrm{I}}(t) \rangle = \langle m | \mathcal{T} \left\{ \exp \left(-\frac{i}{\hbar} \int_{t_o}^{t} dt'\, H_{\mathrm{int,I}}(t') \right) \right\} | n \rangle$$

$$= \langle m | n \rangle - \frac{i}{\hbar} \int_{t_o}^{t} dt'\, \langle m | H_{\mathrm{int,I}}(t') | n \rangle \qquad (10.157)$$

$$+ \tfrac{1}{2} \left(\frac{-i}{\hbar} \right)^2 \int_{t_o}^{t} dt' \int_{t_o}^{t} dt''\, \langle m | \mathcal{T}\{ H_{\mathrm{int,I}}(t')\, H_{\mathrm{int,I}}(t'') \} | n \rangle + \dots .$$

To first order in the perturbation, the transition amplitude between two eigenstates is thus given by

$$\langle m | \psi_{\mathrm{I}}(t) \rangle \approx \delta_{mn} - \frac{i}{\hbar} \int_{t_o}^{t} dt'\, \langle m | e^{i H_0 t'/\hbar}\, H_{\mathrm{int}}(t')\, e^{-i H_0 t'/\hbar} | n \rangle$$

$$= \delta_{mn} - \frac{i}{\hbar} \int_{t_o}^{t} dt'\, e^{\frac{i}{\hbar}(E_m - E_n)t'}\, \langle m | H_{\mathrm{int}}(t') | n \rangle , \qquad (10.158)$$

and the corresponding first-order transition probability for two orthogonal eigenstates is

$$P_{n \to m}^{(1)} = \left| \frac{i}{\hbar} \int_{t_o}^{t} dt'\, e^{\frac{i}{\hbar}(E_m - E_n)t'}\, \langle m | H_{\mathrm{int}}(t') | n \rangle \right|^2 . \qquad (10.159)$$

In the next section, we will discuss this expression for some special cases that are of particular interest.

10.5.3   Fermi's Golden Rule

In the final section of this chapter, we want to take a closer look at some particularly relevant cases of time-dependent interactions, firstly, a step function, i.e., an interaction Hamiltonian that is constant after it has been "switched on", and secondly, a sinusoidal perturbation. In treating these cases, we will discover three expressions for transition rates under these perturbations that we will jointly refer to as different versions of *Fermi's golden rule*[14].

Constant perturbation: For the first special case, we can write the interaction as

$$H_{\mathrm{int}}(t) = V_{\mathrm{int}}\, \Theta(t), \qquad (10.160)$$

where V_{int} is a time-independent operator and $\Theta(t)$ is the step function given by

$$\Theta(t) = \begin{cases} 0 & \text{if } t < 0 \\ 1 & \text{if } t > 0 \end{cases} . \qquad (10.161)$$

---

[14]Although the basic ingredients for these expressions can be traced back to work by Paul Dirac (1927), the name "golden rule" was coined by *Enrico Fermi* (see Fermi, 1950, eqn VIII.2).

The time-dependence of the problem hence resides entirely in the discontinuous change in the Hamiltonian at $t = t_o = 0$. Inserting into the transition probability from eqn (10.159) and using the shorthand $\omega_{mn} \equiv (E_m - E_n)/\hbar$, we have

$$P^{(1)}_{n \to m} = \frac{1}{\hbar^2} \left| \int_0^t dt' \, e^{i\omega_{mn}t'} \langle m | V_{\text{int}} | n \rangle \right|^2 = \frac{1}{\hbar^2} \left| \frac{-i}{\omega_{mn}} \left(e^{i\omega_{mn}t} - 1 \right) \langle m | V_{\text{int}} | n \rangle \right|^2$$

$$= \frac{1}{\hbar^2} \left| \frac{-i e^{i\omega_{mn}t/2}}{\omega_{mn}} \left(e^{i\omega_{mn}t/2} - e^{-i\omega_{mn}t/2} \right) \langle m | V_{\text{int}} | n \rangle \right|^2$$

$$= \frac{1}{\hbar^2} \left[\frac{\sin(\omega_{mn}t/2)}{\omega_{mn}/2} \right]^2 |\langle m | V_{\text{int}} | n \rangle|^2 . \tag{10.162}$$

At this point it is useful to briefly take a closer look at the one-parameter family of functions of the form

$$\delta_t(\omega) = \frac{\sin^2(\omega t)}{\pi \omega^2 t}. \tag{10.163}$$

For any test function $f(\omega)$ (see, e.g., the discussion in Section 3.3.3), we have

$$\lim_{t \to \infty} \int_{-\infty}^{\infty} d\omega \, \delta_t(\omega) \, f(\omega) = \lim_{t \to \infty} \int_{-\infty}^{\infty} d\omega \, \frac{\sin^2(\omega t)}{\pi \omega^2 t} f(\omega) = \lim_{t \to \infty} \int_{-\infty}^{\infty} dx \, \frac{\sin^2(x)}{\pi x^2} f\left(\frac{x}{t}\right)$$

$$= \int_{-\infty}^{\infty} dx \, \frac{\sin^2(x)}{\pi x^2} f(0) = f(0), \tag{10.164}$$

where we have substituted $x = \omega t$ and the used the integral $\int_{-\infty}^{\infty} dx \, \frac{\sin^2(x)}{x^2} = \pi$. We thus recognize $\delta_t(\omega)$ as a representation of the Dirac delta distribution, i.e., formally

$$\lim_{t \to \infty} \delta_t(\omega) = \delta(\omega). \tag{10.165}$$

The comparison of eqn (10.163) with the transition probability in eqn (10.162) thus suggests that the *transition rate* per unit time, which we will denote as $\Gamma^{(1)}_{n \to m} \equiv P^{(1)}_{n \to m}/t$, is proportional to the Dirac delta in the limit of large times t. That is,

$$\lim_{t \to \infty} \frac{1}{t \, \hbar^2} \left[\frac{\sin(\omega_{mn}/2)}{\omega_{mn}/2} \right]^2 = \lim_{t \to \infty} \frac{\pi}{\hbar^2} \delta_t\left(\frac{\omega_{mn}}{2} \right) = \lim_{t \to \infty} \frac{\pi}{\hbar^2} \delta_t\left(\frac{E_m - E_n}{2\hbar} \right)$$

$$= \frac{\pi}{\hbar^2} \delta\left(\frac{E_m - E_n}{2\hbar} \right) = \frac{2\pi}{\hbar} \delta(E_m - E_n), \tag{10.166}$$

where we have used the property $\delta(ax) = \frac{1}{a}\delta(x)$ of the Dirac delta distribution, which one can easily check by integrating the distribution with a test function and substituting $y = ax$, i.e.,

$$\int\limits_{-\infty}^{\infty} dx\, f(x)\, \delta(ax) = \tfrac{1}{a} \int\limits_{-\infty}^{\infty} dy\, f\!\left(\tfrac{y}{a}\right) \delta(y) = \tfrac{1}{a}\, f(0) = \int\limits_{-\infty}^{\infty} dx\, f(x)\, \tfrac{1}{a}\, \delta(x). \qquad (10.167)$$

With this, we can formulate the following Theorem:

Theorem 10.5 (Fermi's golden rule (i))

To first order in the perturbation, the long-time limit of the transition rate (per unit time) between two unperturbed eigenstates $|\,n\,\rangle$ and $|\,m\,\rangle$ under a constant (step-function) perturbation $H_{\text{int}}(t) = V_{\text{int}}\,\Theta(t)$ is given by

$$\lim_{t\to\infty} \Gamma^{(1)}_{n\to m} = \frac{2\pi}{\hbar} \delta\big(E_m - E_n\big)\, |\langle\, m\,|\, V_{\text{int}}\,|\, n\,\rangle|^2\,.$$

Scattering into a set of eigenstates: The identification with the Dirac delta strictly only makes sense if we assume that the energies of the final states form a continuum such that the integration over the delta function is well defined, while we can still use the expressions in eqn (10.162) for discrete spectra. At the same time, we note that the proportionality factor $\frac{2\pi}{\hbar}$ in eqn (10.166) arises from an integration over all of $\mathbb{R}$, which may not correspond exactly to the spectrum considered. To better understand the specific conditions under which the approximation of the function $\delta_t\!\left(\frac{E_m - E_n}{2\hbar}\right)$ using the delta function as in Theorem 10.5 is justified, we should therefore take a closer look at the set of final states into which the perturbation "scatters" the initial state. For instance, we might think of a set of plane waves with fixed wave number but whose wave vectors lie within a certain range of angles.

Here, we denote the density of states as a function of the corresponding target energy by $\rho(E_m)$, such that $\rho(E_m)dE_m$ is the number of states within an energy interval of width dE_m. We can then determine the transition rate into the target set described by the density $\rho(E_m)$ starting from an initial state $|\,n\,\rangle$ as

$$\lim_{t\to\infty} \sum_m \Gamma_{n\to m} = \lim_{t\to\infty} \int dE_m\, \rho(E_m)\, \Gamma_{n\to m}. \qquad (10.168)$$

For a constant perturbation, we can then insert from Theorem 10.5 to obtain the transition rate to first order in the perturbation and in the long-time limit.

Theorem 10.6 (Fermi's golden rule (ii))

To first order in the perturbation, the long-time limit of the transition rate (per unit time) from an unperturbed eigenstate $|\,n\,\rangle$ into a group of eigenstates $|\,m\,\rangle$ under a constant (step-function) perturbation $H_{\mathrm{int}}(t) = V_{\mathrm{int}}\,\Theta(t)$ is given by

$$\lim_{t\to\infty}\int dE_m\,\rho(E_m)\,\Gamma^{(1)}_{n\to m} \;=\; \frac{2\pi}{\hbar}\,\rho(E_n)\,|\langle\,E_n\,|\,V_{\mathrm{int}}\,|\,n\,\rangle|^2\,,$$

where $|\,E_n\,\rangle$ denotes an eigenstate with energy E_n.

We note that this second version of Fermi's golden rule assumes that the matrix element $\langle\,E_n\,|\,V_{\mathrm{int}}\,|\,n\,\rangle$ is the same for all continuum states (all states in the target set) with the same energy, which we have denoted by $|\,E_n\,\rangle$ here to emphasize that they are not (necessarily) from the same set as the initial state $|\,n\,\rangle$. Nevertheless, we see again that transitions induced by a constant potential conserve energy, at least approximately, in the sense of Theorems 10.5 and 10.6.

Validity of the approximation: Let us now briefly examine the simplifications we have made here. Using the delta function as in Theorem 10.5 to approximate the transition rate proportional to the function $\delta_t\!\left(\frac{E_m-E_n}{2\hbar}\right)$, and subsequently integrating over it, requires two assumptions: First, we must require that the width ΔE of the density $\rho(E_m)$ of the considered set of final states is much bigger than the width of the function $\delta_t\!\left(\frac{E_m-E_n}{2\hbar}\right)$. As a representative of the latter we can consider the width of the central peak of $\delta_t\!\left(\frac{E_m-E_n}{2\hbar}\right)$ between the first local minima at $E_n \pm \frac{2\pi\hbar}{t}$, as illustrated in Fig. 10.6. The first requirement can thus be formulated as

$$\Delta E \gg \tfrac{2\pi\hbar}{t}. \tag{10.169}$$

At the same time, we must assume that sufficiently many eigenstates lie close to each other in the range covered by the central peak of $\delta_t\!\left(\frac{E_m-E_n}{2\hbar}\right)$. If we denote the (average) distance of the energy levels within the target set in the proximity of the target energy as δE, then we can write this requirement as

$$\delta E \ll \tfrac{2\pi\hbar}{t}. \tag{10.170}$$

For further discussions of the validity of this approach see Cohen-Tannoudji *et al.* (1991, Sec. XIII C).

Sinusoidal perturbation: With the limitations of the approximation in mind, let us now turn to a second example, a periodic perturbation represented by a sinusoidal function. The interaction Hamiltonian for such a perturbation can be written as

$$H_{\mathrm{int}}(t) = V_{\mathrm{int}}\,\sin(\omega t)\,\Theta(t), \tag{10.171}$$

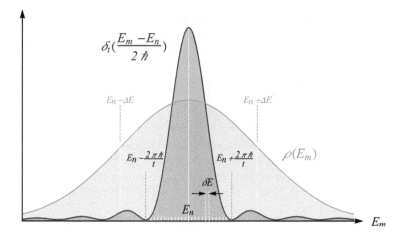

Fig. 10.6 Validity of replacing δ_t with Dirac delta. Replacing the function $\delta_t\left(\frac{E_m - E_n}{2\hbar}\right)$ with a Dirac delta is a good approximation if the width ΔE of the density function $\rho(E_m)$ of target states is much bigger than the distance between the first-order local minima of δ_t at $E_n \pm \frac{2\pi\hbar}{t}$, while the distance δE between the individual relevant energy levels in the target set is much smaller than $\frac{2\pi\hbar}{t}$.

where $\Theta(t)$ is the step function from eqn (10.161) and V_{int} is a time-independent operator as before. We now calculate the first-order transition probability from eqn (10.159)
for this interaction Hamiltonian and obtain the expression

$$P^{(1)}_{n \to m} = \frac{1}{\hbar^2} \left| \int_0^t dt'\, e^{i\omega_{mn}t'}\, \sin(\omega t')\, \langle m \,|\, V_{\text{int}} \,|\, n \rangle \right|^2 \tag{10.172}$$

$$= \frac{1}{4\hbar^2} \left| \frac{-i}{\omega_{mn} + \omega}\left(e^{i(\omega_{mn}+\omega)t} - 1\right) - \frac{-i}{\omega_{mn} - \omega}\left(e^{i(\omega_{mn}-\omega)t} - 1\right) \right|^2 |\langle m \,|\, V_{\text{int}} \,|\, n \rangle|^2$$

$$= \frac{1}{4\hbar^2} \left| \frac{e^{i\omega_{mn}^{+}t/2}\sin\left(\frac{\omega_{mn}^{+}t}{2}\right)}{\omega_{mn}^{+}/2} - \frac{e^{i\omega_{mn}^{-}t/2}\sin\left(\frac{\omega_{mn}^{-}t}{2}\right)}{\omega_{mn}^{+}/2} \right|^2 |\langle m \,|\, V_{\text{int}} \,|\, n \rangle|^2,$$

where we have used the shorthand $\omega_{mn}^{\pm} = \omega_{mn} \pm \omega$ in the last step. When we inspect
the modulus square of the sum of sine functions, we note that squares of the two terms
give expressions just like in the last line of eqn (10.162), except with the substitutions
$\omega_{mn} \to \omega_{mn}^{\pm}$. In the long-time limit, these terms will thus be proportional to the Dirac
delta distributions $\delta\left(\frac{\omega_{mn}^{\pm}}{2}\right)$ and will thus only give a contribution when $\omega_{mn} = \pm\omega$.
Meanwhile, the sum of the cross terms in the modulus squared is proportional to

$$\text{Re}\left(\frac{e^{i\omega_{mn}^{+}t/2}e^{-i\omega_{mn}^{-}t/2}\sin\left(\frac{\omega_{mn}^{+}t}{2}\right)\sin\left(\frac{\omega_{mn}^{-}t}{2}\right)}{\omega_{mn}^{+}\,\omega_{mn}^{-}}\right) = \text{Re}\left(\frac{e^{i\omega t}\sin\left(\frac{\omega_{mn}^{+}t}{2}\right)\sin\left(\frac{\omega_{mn}^{-}t}{2}\right)}{\omega_{mn}^2 - \omega^2}\right)$$

$$= \frac{\cos(\omega t)\sin\left(\frac{\omega_{mn}^{+}t}{2}\right)\sin\left(\frac{\omega_{mn}^{-}t}{2}\right)}{\omega_{mn}^2 - \omega^2}. \tag{10.173}$$

When $\omega_{mn} \neq \pm\omega$, this contribution is suppressed by the growing denominator since the sine and cosine functions are bounded by ± 1. In the regime where the Dirac delta functions give a non-vanishing contribution, that is, when $\omega_{mn} = \pm\omega$, we have

$$\lim_{\omega_{mn} \to \pm\omega} \frac{\cos(\omega t)\, \sin\left(\frac{\omega_{mn}^+ t}{2}\right) \sin\left(\frac{\omega_{mn}^- t}{2}\right)}{\omega_{mn}^2 - \omega^2} = \frac{t}{8\omega} \sin(2\omega t). \tag{10.174}$$

In this case, the corresponding contribution to the transition rate is proportional to $\sin(2\omega t)/\omega$ and thus suppressed in the magnitude of the frequency of the perturbation. For all practical purposes we can therefore disregard the cross terms. Proceeding as we have for the constant perturbation, we can thus consider the long-time limit of the transition rate using a calculation analogous to eqn (10.166), and obtain the following result:

Theorem 10.7 (Fermi's golden rule (iii))

To first order in the perturbation, the long-time limit of the transition rate (per unit time) between two unperturbed eigenstates $|\, n \,\rangle$ and $|\, m \,\rangle$ under a periodic perturbation $H_{\text{int}}(t) = V_{\text{int}} \sin(\omega t)\, \Theta(t)$ is given by

$$\lim_{t \to \infty} \Gamma_{n \to m}^{(1)} = \frac{\pi}{2\hbar} \left[\delta\big(E_m - E_n - \hbar\omega\big) + \delta\big(E_m - E_n + \hbar\omega\big) \right] |\langle\, m \,|\, V_{\text{int}} \,|\, n \,\rangle|^2 \,.$$

We see that the periodic perturbation, for instance, by an oscillating electromagnetic field, leads to transitions between states that differ in energy exactly by $\hbar\omega$, the energy of photons corresponding to the driving frequency. In very good approximation, transitions between discrete energy levels such as in the hydrogen atom thus only occur if the driving frequency is in *resonance* with the corresponding energy gap. When the final-state energy E_m is smaller than the energy E_n of the initial state, what we have described here is therefore simply the process of *stimulated emission*: An external drive at frequency $\hbar\omega$ leads to the release of a photon of frequency $\hbar\omega$. Similarly, when $E_m > E_n$, a photon with energy $\hbar\omega$ is absorbed from the field. The particular transitions within the atomic structure are then determined by the remaining matrix element $|\langle\, m \,|\, V_{\text{int}} \,|\, n \,\rangle|^2$, following the selection rules discussed in Section 10.4.1, e.g., when V_{int} corresponds to, say, the operator $\vec{X}$.

What is interesting to note here is that the results of this section seem to suggest that there is no reason to expect radiation from *spontaneous emission*. In the absence of any external fields, stationary states should remain stable and no transitions should occur. This would be the case if all external fields were classical entities themselves, but the fact that, in particular, the electromagnetic field is a quantum field (which we will discuss in more detail in Section 25.1), and hence features a non-zero vacuum energy even in the absence of excitations (photons), means that there is never really no field at all. In this sense, spontaneous emission is never really spontaneous but can be thought of as stimulated emission catalyzed by ever-present quantum fields.

The perturbative treatment of the solutions of the Schrödinger equation, including time-independent corrections to the energy levels, and time-dependent interactions with external fields, thus leads to a detailed and accurate description of the atomic structure and the corresponding emission and absorption spectra. The description of the structure of matter and of the interactions between its components and the surrounding quantum fields thus provided by quantum mechanics, be it on the level of solid-state physics, atomic or subatomic physics, has greatly extended our understanding of the mechanisms that we encounter in nature. However, one of the most notable frontiers of quantum science today lies in the abstraction of quantum-mechanical principles to quantum information theory, which has paved the way for the advent of fascinating technologies such as quantum computing and quantum communication.

In Parts II and III, we aim to provide an introduction to quantum information theory that supplies a perspective on the origins of quantum information but also covers a range of methods and techniques employed in these fields.

Part II

Entanglement and Non-Locality

In Part II we turn to quantum information, an area of quantum theory that emerged in the late 1980s and 1990s and received much attention since, in particular, because it opened up new directions for thinking about disciplines of science—information theory, communication theory, and computation theory—that had themselves only risen to prominence in the twentieth century. In all of these disciplines, the combination with ideas from quantum mechanics opened up new possibilities for quantum-information processing, quantum communication, and quantum computation. Moreover, an important catalyst for the development of quantum information came from work by John Bell on the foundation of quantum mechanics. His now famous Bell inequalities and their experimental violation, dubbed non-locality, connect a basic feature of quantum states called entanglement to what were up to this point commonly thought to be purely philosophical questions about reality. Subsequently, the mathematical and physical features of entanglement became a fast-growing and fruitful research topic, with entanglement taking centre stage as the basic ingredient of quantum information theory, where two parties — in the popular terminology called Alice and Bob — are communicating via quantum channels. And the zoo of fascinating features, theoretical challenges, and counter-intuitive possibilities keeps on expanding when higher dimensions and more parties are involved.

In our treatment here we first introduce the mathematical framework of the density-matrix formalism, before we turn to the physical and interpretational questions surrounding hidden-variable theories, contextuality, and the Kochen–Specker theorem. Then we explain the Einstein–Podolsky–Rosen paradox, which is the starting point for studying Bell inequalities. Several types of Bell inequalities are derived explicitly and analysed regarding their relevance for quantum physics, before we summarize key experiments which are unequivocally in agreement with quantum-mechanical predictions.

Next we focus on entanglement and its applications, foremost quantum teleportation as a building block of quantum-information processing protocols. We then return to the more mathematical problems of quantum information, delving more deeply into entanglement theory. There, we first consider the bipartite scenario, for which we discuss methods for entanglement detection, entanglement witnesses, entanglement measures, as well as the geometric features of separability, entanglement, and non-locality in the Hilbert–Schmidt space of quantum states. Finally, we turn to higher-dimensional quantum systems, like qutrits and qudits, as well as to multipartite systems, like the Greenberger–Horne–Zeilinger states, and explain their quantum characteristics. In particular, we discuss approaches and challenges to the characterization, detection, and quantification of entanglement in such systems.

11
Density Matrices

In Part I, we have focused on the description of quantum systems as state vectors in a Hilbert space (recall Definition 2.3), or as wave functions, when considering the position representation. In a sense that will become more precise over the course of Part II, such quantum states represent the most complete information—knowing exactly which state vector describes a system—that we can have about the system. However, a premise of (quantum) information theory is that one does not always have (access to) complete information. This first chapter of Part II therefore introduces a description of quantum states that is more suitable to represent our knowledge about physical systems within the context of quantum information theory: *density matrices*, also called *density operators*, which conceptually take the role of the state vectors discussed so far, as they encode all the (accessible) information about a quantum system.

As we will see, states that are described by state vectors $|\psi\rangle$ in Hilbert space—so-called *pure states*—are idealized descriptions that cannot characterize statistical (incoherent) mixtures, which one encounters in experiments, i.e., in nature. These *mixed states* are represented by *density matrices*[1], which are hence of fundamental importance for the theory of quantum information and quantum communication. Here we will discuss the most important properties of density matrices and give some physical examples, including cases of entangled states, which are crucial for quantum information. Further details about the density-matrix formalism can be found, e.g., in Thirring (1980), Nielsen and Chuang (2000), Breuer and Petruccione (2002), and Wilde (2017).

11.1 Pure States

Let us begin by considering the possibility of representing a statistical mixture of state vectors as another state vector. To this end, we have to first take a closer look at what we mean by "statistical mixture". Suppose we know that a particular quantum system can be in one of two different orthogonal quantum states, $|\psi\rangle$ or $|\phi\rangle$, for instance, representing the transitions of a particle through the two apertures of a double slit, respectively. If we know the *relative phase* φ between the two contributions, if there is *coherence*, then we can of course assign a *coherent superposition* $\alpha|\psi\rangle + e^{i\varphi}|\phi\rangle$, where $|\alpha^2|$ and $|\beta|^2 = 1 - |\alpha^2|$ are the probabilities for finding the system in the states

---

[1]Remark for experts: It is possible to find a vector representation for every given quantum-mechanical state, even those represented by a density matrix. This can be done via the so-called *purification* (see Section 15.1.4) or the more general GNS (Gelfand–Neumark–Segal) construction (Gelfand and Naimark, 1943; Segal, 1947). This vector representation need not, however, be of any practical form and dealing with the concept of a density matrix is therefore inevitable.

$|\psi\rangle$ and $|\phi\rangle$, respectively. But how do we describe the system if we do not know the relative phase?

Simply fixing the phase to some particular value will generally not match the experimental observations since the value of the phase influences the outcomes of measurements in other measurement bases. In the example of the double slit this will correspond to a shift of the interference pattern. A naive approach would be to simply average over the phase, but since the integral over the phase vanishes,

$$\int_0^{2\pi} d\varphi\, e^{i\varphi} = 0 \,, \tag{11.1}$$

this clearly also does not work. We therefore have to look for a different description for incoherent mixtures.

To get an idea of how to approach this problem, we start again with a system described by a state vector—a *pure state* $|\psi\rangle$—and we consider the expectation value of an observable A for this state,

$$\langle\, A\,\rangle_\psi = \langle\psi|\, A\,|\psi\rangle = \langle\psi|\sum_n |n\rangle\langle n|\, A\,|\psi\rangle = \sum_n \langle n|\, A\,|\psi\rangle\langle\psi|\,|n\rangle\,. \tag{11.2}$$

Here, we have inserted a complete orthonormal system $\sum_n |n\rangle\langle n| = \mathbb{1}$. Let us then make the following definition:

Definition 11.1   *The **trace of an operator** D is given by*

$$\mathrm{Tr}(D) := \sum_n \langle n|\, D\,|n\rangle \,,$$

where the set $\{\,|n\rangle\,\}_n$ forms a complete orthonormal system.

With this, we see that we can write the pure-state expectation value from eqn (11.2) as

$$\langle\, A\,\rangle_\psi = \mathrm{Tr}\big(A\,|\psi\rangle\langle\psi|\big). \tag{11.3}$$

Note that the order of the operator A and the projector on the state in the trace does not matter, since operators in the trace can generally be cyclically permuted. That is, for arbitrary operators A and B we have

$$\mathrm{Tr}(AB) = \sum_n \langle n|\, AB\,|n\rangle = \sum_n \langle n|\, A \sum_m |m\rangle\langle m|\, B\,|n\rangle \sum_{m,n} \langle n|\, A\,|m\rangle\,\langle m|\, B\,|n\rangle$$

$$= \sum_{m,n} \langle m|\, B\,|n\rangle\,\langle n|\, A\,|m\rangle = \sum_m \langle m|\, B \sum_n |n\rangle\langle n|\, A\,|m\rangle$$

$$= \sum_m \langle m|\, BA\,|m\rangle = \mathrm{Tr}(BA). \tag{11.4}$$

We see that the expected value for a pure state $|\psi\rangle$ can be cast in terms of the projection operator $|\psi\rangle\langle\psi|$. We can also recall from eqn (9.69) that a projection operator is the correct representation for an equivalence class of states related by a global phase shift. This motivates the following definition:

Definition 11.2 *The **density matrix** ρ of a pure state $|\psi\rangle$ is given by*

$$\rho := |\psi\rangle\langle\psi| \ .$$

This density matrix has the following properties:

$$\text{(i)} \qquad \rho^\dagger = \rho \qquad \text{Hermitian,} \tag{11.5a}$$

$$\text{(ii)} \qquad \text{Tr}(\rho) = 1 \qquad \text{normalized,} \tag{11.5b}$$

$$\text{(iii)} \qquad \rho \geq 0 \qquad \text{positive semidefinite,} \tag{11.5c}$$

$$\text{(iv)} \qquad \rho^2 = \rho \qquad \text{projector.} \tag{11.5d}$$

Properties (i) and (iv) follow immediately from Definition 11.2, and property (ii) can be verified using the definition of the trace operation for an arbitrary operator D. Specifically, let us consider the operator $D = |\psi\rangle\langle\phi|$ representing an arbitrary matrix element of any more general (linear) operator, and calculate its trace, then we get

$$\text{Tr}(D) = \sum_n \langle n|\psi\rangle\langle\phi|n\rangle = \sum_n \langle\phi\underbrace{|n\rangle\langle n|}_{1}\psi\rangle = \langle\phi|\psi\rangle \ . \tag{11.6}$$

Consequently, we have $\text{Tr}(|\psi\rangle\langle\psi|) = \langle\psi|\psi\rangle = 1$, the state is normalized.

Property (iii) means that the eigenvalues of ρ are not negative, or in other words, for any vector $|\varphi\rangle$ we have

$$\langle\varphi|\rho|\varphi\rangle = \langle\varphi|\psi\rangle\langle\psi|\varphi\rangle = |\langle\varphi|\psi\rangle|^2 \geq 0 \ . \tag{11.7}$$

Indeed, this is nothing but the Born rule for the probability of obtaining a measurement outcome corresponding to the state $|\varphi\rangle$, and positivity of the density operator is thus clearly an important property because probabilities are always positive or zero (not negative). We have covered that density operators established by Definition 11.2 reproduce the expectation value of an observable using the trace over their product. Moreover, what we shall see shortly is that this definition of the expectation value also carries over to the description of more general mixed quantum states. Let us therefore formulate the following theorem, which we will prove in Section 11.2:

Theorem 11.1 (Expectation value of operator)

The expectation value of an observable A in a state, represented by a density matrix ρ, is given by

$$\langle A \rangle_\rho = \text{Tr}(A\rho)$$

Now let us consider an **ensemble of individual systems**. If all individual systems that we are investigating are in one and the same quantum state then the ensemble can be described by a pure state. As we know, we can still only make probabilistic statements about the outcomes of individual measurements of the observable for the whole ensemble of identically prepared systems. Let the system be in the state $|\psi\rangle$ which we can expand with respect to the eigenstates of a Hermitian operator A, i.e.,

$$|\psi\rangle = \sum_n c_n |n\rangle, \quad \text{with} \quad c_n = \langle n|\psi\rangle \quad \text{and} \quad A|n\rangle = a_n |n\rangle. \tag{11.8}$$

The expectation value is then given by

$$\langle A\rangle_\psi = \sum_n |c_n|^2 a_n = \sum_n \frac{N_n}{N} a_n, \tag{11.9}$$

where $|c_n|^2$ is the probability to measure the eigenvalue a_n in any given measurement. It corresponds to the fraction N_n/N, the frequency with which the eigenvalue a_n occurs, where N_n is the number of times this eigenvalue has been observed when all N individual systems of the ensemble are measured.

The state is characterized by a density matrix of the form of Definition 11.2, with the properties (i)–(iv) from eqns (11.5a)–(11.5d), where we can combine properties (ii) and (iv) to conclude that a pure-state density matrix satisfies

$$\text{Tr}(\rho^2) = 1. \tag{11.10}$$

11.2 Mixed States

Let us next study the situation where not all of the N individual systems of the ensemble are in the same state, i.e., N_i systems are in the state $|\psi_i\rangle$, respectively, such that $\sum_i N_i = N$. The probability p_i to find an individual system of the ensemble to be described by the state $|\psi_i\rangle$ is then given by

$$p_i = \frac{N_i}{N}, \quad \text{where} \quad \sum_i p_i = 1. \tag{11.11}$$

We can thus write down the *mixed state* of the ensemble as a convex sum, i.e., a weighted sum with $\sum_i p_i = 1$ of pure-state density matrices, i.e.,

$$\rho_{\text{mixed}} = \sum_i p_i \rho_i^{\text{pure}} = \sum_i p_i |\psi_i\rangle\langle\psi_i|. \tag{11.12}$$

The expectation value is again given by Theorem 11.1

$$\langle A\rangle_{\rho_{\text{mixed}}} = \text{Tr}(A\rho_{\text{mix}}), \tag{11.13}$$

where we can express the *expectation value of the mixed state* as a convex sum of expectation values of its constituent pure states

$$\langle A\rangle_{\rho_{\text{mixed}}} = \sum_i p_i \langle\psi_i| A |\psi_i\rangle. \tag{11.14}$$

Proof of Theorem 11.1

$$\text{Tr}(A\,\rho_{\text{mixed}}) = \text{Tr}\left(A\sum_i p_i\,|\psi_i\rangle\langle\psi_i|\right) = \sum_n\sum_i p_i\,\langle n|\,A\,|\psi_i\rangle\langle\psi_i|\,n\rangle$$

$$= \sum_i p_i\,\langle\psi_i|\,A\underbrace{\sum_n|n\rangle\langle n|}_{1}\,|\psi_i\rangle = \sum_i p_i\,\langle\psi_i|\,A\,|\psi_i\rangle\,. \qquad (11.15)$$

$\square$

As a consequence, we can again phrase the *Born rule* for the probability of obtaining a measurement outcome corresponding to the eigenstate $|n\rangle$ of an observable as

$$p(n) = \langle n|\,\rho_{\text{mixed}}\,|n\rangle = \sum_i p_i\,\langle n|\psi_i\rangle\langle\psi_i|n\rangle = \sum_i p_i|\langle n|\psi_i\rangle|^2, \qquad (11.16)$$

in analogy to eqn (11.7), but now featuring a convex sum of the transition probabilities $|\langle n|\psi_i\rangle|^2$. Moreover, the properties (i)–(iii) from eqns (11.5a)–(11.5c) are still valid for mixed states, and we can thus raise them to the status of defining a general density operator.

Definition 11.3 *The **density operator** ρ describing the state of a quantum system is a Hermitian ($\rho^\dagger = \rho$), positive semidefinite ($\rho \geq 0$), and normalized [$\text{Tr}(\rho) = 1$] operator.*

In particular, this definition implies that every convex sum $\rho = \sum_i p_i\,|\psi_i\rangle\langle\psi_i|$ representing a probabilistic mixture of normalized but not necessarily orthogonal pure states $|\psi_i\rangle$ with associated probabilities $p_i \geq 0$ with $\sum_i p_i = 1$ is a valid density operator. At the same time, given a valid density operator, we can always find a decomposition into a convex sum of projectors on orthogonal pure states. That is, since ρ is Hermitian, it can be diagonalized and the eigenvectors form an orthonormal basis, and since density operators are positive (semidefinite) and normalized, their eigenvalues must be non-negative and sum to one, respectively. Consequently, any density operator satisfying properties (i)–(iii) from eqns (11.5a)–(11.5c) can be written as a convex sum of projectors onto its eigenstates, i.e., orthogonal pure states. Let us summarize these observations in the following theorem:

Theorem 11.2 (Decomposition of density operators)

Every density matrix ρ given by a Hermitian ($\rho^\dagger = \rho$), positive semidefinite ($\rho \geq 0$), and normalized [$\text{Tr}(\rho) = 1$] operator can be decomposed as a convex sum of projectors,

$$\rho = \sum_n p_n\,|\psi_n\rangle\langle\psi_n|$$

with $0 \leq p_n \leq 1$ and $\sum_n p_n = 1$ for some orthonormal basis $\{|\psi_n\rangle\}_n$.

However, property (iv) from eqn (11.5d) no longer holds in general, that is, using the decomposition into a orthonormal basis from Theorem 11.2, we have

$$\rho_{\text{mixed}}^2 = \sum_i \sum_j p_i \, p_j \, |\psi_i\rangle \underbrace{\langle \psi_i \, | \, \psi_j \rangle}_{\delta_{ij}} \langle \psi_j | = \sum_i p_i^2 \, |\psi_i\rangle\langle\psi_i| \neq \rho_{\text{mixed}} \, . \quad (11.17)$$

Moreover, we can then easily calculate the trace of ρ_{mixed}^2 by noting that the trace is independent of the chosen orthonormal basis. From eqn (11.17) we thus immediately have

$$\text{Tr}(\rho_{\text{mixed}}^2) = \sum_j \langle \psi_j | \sum_i p_i^2 \, |\psi_i\rangle\langle\psi_i| \, |\psi_j\rangle = \sum_i p_i^2 < \sum_i p_i = 1, \quad (11.18)$$

which, in contrast to pure states, is no longer one, but smaller than one since $0 \leq p_i \leq 1$ and therefore $p_i^2 \leq p_i$.

We conclude that the trace of ρ^2 is a reasonable measure for the mixedness of a density matrix, since it is equal to one for pure states and strictly smaller than one for mixed states. For a *maximally mixed state* ρ_{mix} for which all probabilities p_i have the same value $\frac{1}{d}$ in a given dimension d, we have $\rho_{\text{mix}} = \frac{1}{d}\mathbb{1}_d$, and we hence find

$$\text{Tr}(\rho_{\text{mix}}^2) = \frac{1}{d} > 0 \, . \quad (11.19)$$

The quantity

$$P(\rho) = \text{Tr}(\rho^2), \quad (11.20)$$

with the range $\frac{1}{d} \leq P(\rho) \leq 1$ is called *purity*, and the gap to the maximal value of one, i.e.,

$$S_{\text{L}}(\rho) = 1 - P(\rho) \quad (11.21)$$

is called the *mixedness* or *linear entropy* of the system, which we will discuss in more detail in Chapter 20.

11.3   Time Evolution of Density Matrices

Now that we have argued that density operators are the correct way of replacing pure-state vectors as the general description of our knowledge of quantum systems, let us proceed by discussing how these objects evolve in time. That is, we now want to determine the correct equation of motion for density matrices describing closed quantum systems. To do this, let us assume that at some given time t we are given the density operator $\rho = \sum_i p_i \, |\psi_i\rangle\langle\psi_i|$ of a mixed state from eqn (11.12), where the probabilities p_i are fixed but the pure states $|\psi_i\rangle$ depend on time. Using the time-dependent Schrödinger equation from Proposition 2.1.1 and its Hermitian conjugate

$$i\hbar \, \frac{\partial}{\partial t} \, |\psi\rangle = H \, |\psi\rangle \quad \xrightarrow{\dagger} \quad -i\hbar \, \frac{\partial}{\partial t} \, \langle \psi | = \langle \psi | \, H \, , \quad (11.22)$$

we can then multiply the density matrix ρ by $i\hbar$ and differentiate with respect to time. Noting that $\frac{\partial}{\partial t} p_i = 0$ we obtain

$$i\hbar \frac{\partial}{\partial t} \rho = i\hbar \sum_i p_i \left[\left(\frac{\partial}{\partial t} | \psi_i \rangle \right) \langle \psi_i | + | \psi_i \rangle \left(\frac{\partial}{\partial t} \langle \psi_i | \right) \right]$$

$$= i\hbar \sum_i p_i \left[\left(-\tfrac{i}{\hbar} H | \psi_i \rangle \right) \langle \psi_i | + | \psi_i \rangle \left(\tfrac{i}{\hbar} \langle \psi_i | H \right) \right]$$

$$= \sum_i p_i \left[H | \psi_i \rangle\langle \psi_i | - | \psi_i \rangle\langle \psi_i | H \right] = [H, \rho]. \qquad (11.23)$$

Let us summarize this observation in the following theorem:

Theorem 11.3 (von Neumann equation)

The time evolution of a density matrix describing a closed quantum system is determined by the von Neumann equation

$$i\hbar \tfrac{\partial}{\partial t} \rho = [H, \rho].$$

Some remarks on this equation of motion are now in order. First, the von Neumann equation, Theorem 11.3, is the quantum-mechanical analogue of the classical *Liouville equation* in statistical mechanics

$$\frac{\partial}{\partial t} \rho(t) = \{H, \rho\}, \qquad (11.24)$$

where $\rho(p, q, t)$ represents the classical probability density of a statistical ensemble in the phase space with coordinates (p, q), H is the Hamilton function, and $\{.,.\} = \frac{\partial}{\partial q}\frac{\partial}{\partial p} - \frac{\partial}{\partial p}\frac{\partial}{\partial q}$ denotes the Poisson bracket. This analogy nicely demonstrates Dirac's quantization rule, which suggests that the Poisson bracket $\{.,.\}$ is replaced by the commutator $\frac{-i}{\hbar}[.,.]$, in addition to promoting all observables to operators, as we recall from Section 2.7.2. For example, for the canonically conjugate variables q and p one has

$$\{q, p\} = 1 \quad \longrightarrow \quad -\frac{i}{\hbar} [\hat{q}, \hat{p}] = 1. \qquad (11.25)$$

Second, note that we are considering a closed quantum system, which manifests here in the assumption that probabilities are conserved, $\frac{\partial}{\partial t} p_i = 0$ for all i. The resulting dynamics are hence governed exclusively by the Hamiltonian H of the system. In general, however, quantum systems need not be closed and interact with their environment. This leads to the phenomenon referred to as *decoherence*, mathematically reflected in the appearance of additional terms on the right-hand side of the von Neumann equation, as we will discuss in detail in Chapter 22. The commutator $[H, \rho]$

thus represents the *unitary* aspect of the dynamics. Indeed, the time evolution of density matrices can also be written in terms of the *unitary propagator* $U(t, t_o)$, given by Definition 2.12 in case of time-independent Hamiltonians, or by the more general expression in eqn (10.148) when the Hamiltonian depends on time. In any case, the propagator allows us to relate the density matrix at a later time t to the density matrix at some earlier time t_o via the relation

$$\rho(t) = U(t, t_o) \, \rho(t_o) \, U^\dagger(t, t_o) \,. \tag{11.26}$$

Furthermore, this helps us to prove, for instance, that the purity $P(\rho) = \text{Tr}(\rho^2)$ and the mixedness $S_\text{L}(\rho) = 1 - P(\rho)$ of a density matrix describing a closed quantum system are time independent.

Corollary 11.1 (Time-independence of purity/mixedness)

The purity $P(\rho) = \text{Tr}(\rho^2)$ and mixedness $S_\text{L}(\rho) = 1 - P(\rho)$ of a quantum state ρ describing a closed quantum system are time independent.

Proof   Using the relation for the time evolution of $\rho(t)$ from eqn (11.26), we calculate

$$\text{Tr}(\rho^2(t)) = \text{Tr}(U \, \rho(t_0) \, \underbrace{U^\dagger \, U}_{1} \, \rho(t_0) \, U^\dagger) = \text{Tr}(\rho(t_0) \, \rho(t_0) \, \underbrace{U^\dagger \, U}_{1}) = \text{Tr}(\rho^2(t_0)) \,,$$
$$\tag{11.27}$$

where we used the cyclicity of the trace operation in the second-to-last step, see eqn (11.4). □

We can even make this observation somewhat more general by noting that unitary evolution leaves the eigenvalues of any operator that the evolution is applied to invariant. Consequently, all functions of a density operator that depend only on its eigenvalues are also invariant under time evolution (or any other unitary transformation of the system). For the specific case of the trace of ρ^2, we already saw in eqn (11.18) that the function in question is just a sum of the squares of the eigenvalues, but recalling this general property will be useful for us later on, for instance, when discussing entropies in Chapter 20.

Before we delve more deeply into the mathematical and geometric aspects that the description of the states of quantum systems as density operators opens up (see Section 15.5), let us first consider some examples for density matrices for simple but highly relevant physical situations that we have already encountered earlier on in this book. Quantum systems in thermal equilibrium, which we will discuss in Section 11.4, and two-level quantum systems such as spin-$\frac{1}{2}$ particles or photon polarization, for which we will examine the appropriate density operators in Section 11.5.

11.4 Density Matrices for Quantum Systems in Thermal Equilibrium

With what we have learned about density operators so far, we are now in a position to consider an important family of mixed quantum states: *thermal states*, which describe quantum systems that are in thermal equilibrium with a heat reservoir at (some) temperature T. For such a system with Hamiltonian H, we denote the density operator by the symbol τ, which is given by Messiah (1969) and Thirring (1980)

$$\tau(\beta) = \frac{e^{-\beta H}}{\mathcal{Z}}, \tag{11.28}$$

where $\beta = \frac{1}{k_\mathrm{B} T}$ is the inverse temperature and k_B denotes Boltzmann's constant. The *partition function* is defined by

$$\mathcal{Z} = \mathrm{Tr}\!\left(e^{-\beta H}\right). \tag{11.29}$$

Writing the thermal-state density operator with respect to the eigenstates $|\,n\,\rangle$ with energies E_n of the Hamiltonian, such that $H = \sum_n E_n |\,n\,\rangle\!\langle\,n\,|$, we have

$$\tau(\beta) = \mathcal{Z}^{-1} \sum_n e^{-\beta E_n} |\,n\,\rangle\!\langle\,n\,|, \tag{11.30}$$

and applying Born's rule for density operators from eqn (11.16), we see that the probability for finding the system in the nth energy eigenstate is

$$p(n) = \langle\,n\,|\,\tau(\beta)\,|\,n\,\rangle = \mathcal{Z}^{-1} e^{-\beta E_n}, \tag{11.31}$$

in accordance with the classical Boltzmann distribution of thermodynamics, compare, e.g., eqn (1.7). Similar to eqn (11.30), we can also take the trace appearing in the definition of the partition function in eqn (11.29) with respect to the energy eigenstates, obtaining

$$\mathcal{Z} = \mathrm{Tr}\!\left(e^{-\beta H}\right) = \sum_n \langle\,n\,|\,e^{-\beta H}\,|\,n\,\rangle = \sum_n e^{-\beta E_n}. \tag{11.32}$$

From $\mathcal{Z}$ we can then calculate the mean energy of the system as a function of the temperature. Starting with the definition of the mean energy (see Theorem 11.1),

$$\langle\,H\,\rangle = \mathrm{Tr}\!\left(H\rho\right), \tag{11.33}$$

we note that $\langle\,H\,\rangle$ can be deduced from the partition function (11.29) as follows. Differentiating $\mathcal{Z}$ with respect to β, we have

$$-\frac{\partial \mathcal{Z}}{\partial \beta} = \mathrm{Tr}\!\left(H e^{-\beta H}\right). \tag{11.34}$$

We then simply multiply by $\mathcal{Z}^{-1}$ to obtain

$$-\mathcal{Z}^{-1}\frac{\partial\mathcal{Z}}{\partial\beta} = \mathcal{Z}^{-1}\operatorname{Tr}\big(He^{-\beta H}\big) = \operatorname{Tr}\big(H\mathcal{Z}^{-1}e^{-\beta H}\big) = \operatorname{Tr}\big(H\tau(\beta)\big) = \langle\, H\,\rangle\,,$$
(11.35)

which we can compactly summarize as a logarithmic derivative

$$\langle\, H\,\rangle = -\frac{\partial}{\partial\beta}\ln\mathcal{Z}\,.$$
(11.36)

We are now prepared to make our thermal-state example more specific and consider the case of a harmonic oscillator in thermodynamic equilibrium with a heat bath at temperature T. As we shall see in Section 25.1.2, where we consider the quantized electromagnetic field, this corresponds to the situation encountered, e.g., Planck's experiment on black-body radiation where the oscillating atoms, by emitting and absorbing photons, are in thermal equilibrium with the photons inside the cavity, as we recall from our discussion at the very beginning of the book in Section 1.1.3. The corresponding quantum state for the oscillator is a mixed state represented by the density operator in eqn (11.28) with the oscillator Hamiltonian $H = \hbar\omega(a^\dagger a + \frac{1}{2})$ from eqn (5.10) in Section 5.1, whose eigenstates are given by the occupation-number states $|\,n\,\rangle$, that is, $H\,|\,n\,\rangle = \hbar\omega(n+\frac{1}{2})\,|\,n\,\rangle$. From eqn (11.32) we can then easily calculate the partition function explicitly,

$$\mathcal{Z} = \sum_n e^{-\beta E_n} = \sum_{n=0}^{\infty} e^{-\beta\hbar\omega(n+\frac{1}{2})} = e^{-\beta\hbar\omega/2}\sum_{n=0}^{\infty}\big(e^{-\beta\hbar\omega}\big)^n\,.$$
(11.37)

Using the summation formula $\sum_{n=0}^{\infty} y^n = \frac{1}{1-y}$ with $|y| < 1$, we can rewrite this as

$$\mathcal{Z} = \frac{e^{-\beta\hbar\omega/2}}{1 - e^{-\beta\hbar\omega}} = \frac{1}{2\,\sinh\big(\frac{\hbar\omega}{2k_{\mathrm{B}}T}\big)}\,.$$
(11.38)

With the partition function at hand, we can calculate the average energy for the harmonic oscillator at thermal equilibrium. Starting from the partition function (11.38), we first consider its logarithm

$$\ln(\mathcal{Z}) = -\beta\frac{\hbar\omega}{2} - \ln\big(1 - e^{-\beta\hbar\omega}\big)\,,$$
(11.39)

before taking the partial derivative with respect to β, i.e.,

$$-\frac{\partial}{\partial\beta}\ln(\mathcal{Z}) = \frac{\hbar\omega}{2} + \frac{\hbar\omega\,e^{-\beta\hbar\omega}}{1 - e^{-\beta\hbar\omega}}\,,$$
(11.40)

to arrive at the *mean energy of the harmonic oscillator* in thermal equilibrium

$$\langle\, H\,\rangle = \frac{\hbar\omega}{2} + \frac{\hbar\omega}{e^{\hbar\omega/k_{\mathrm{B}}T} - 1} = \frac{\hbar\omega}{2}\coth\big(\frac{\hbar\omega}{2k_{\mathrm{B}}T}\big)\,.$$
(11.41)

The first term $\frac{\hbar\omega}{2}$ in the middle expression of eqn (11.41) is constant and represents the ground-state energy of the harmonic oscillator but the second term depends on the temperature T and provides precisely Planck's law from Theorem 1.3 in Section 1.1.2.

When we now consider the limits of low and high temperature, we find the desired physical properties. At *very low temperatures*, where $k_B T \ll \hbar\omega$, the oscillator remains in its ground state

$$\langle H \rangle \approx \frac{\hbar\omega}{2} \,, \tag{11.42}$$

and at *very high temperatures*, where $k_B T \gg \hbar\omega$, the mean energy is given by the classical Maxwell–Boltzmann statistics

$$\langle H \rangle \approx k_B T \,. \tag{11.43}$$

Next we calculate the mean energy as a function of the average excitation number $\bar{n}$. Firstly, we apply again the oscillator Hamiltonian $H = \hbar\omega(a^\dagger a + \frac{1}{2})$ to the occupation-number states $|n\rangle$, that is, we use the relation $a^\dagger a |n\rangle = n|n\rangle$ to obtain

$$\langle H \rangle = \text{Tr}\big(H\,\tau(\beta)\big) = \sum_{n=0}^{\infty} \hbar\omega(n+\tfrac{1}{2})\langle n|\,\tau\,|n\rangle = \sum_{n=0}^{\infty} \hbar\omega(n+\tfrac{1}{2})\,p(n), \tag{11.44}$$

where $p(n) = \langle n|\,\tau\,|n\rangle$ is the probability of finding the oscillator in its nth occupation-number state (or energy eigenstate) at thermal equilibrium, see eqn (11.31), and τ is expanded into its occupation-number states as

$$\tau = \sum_{n=0}^{\infty} p(n)\,|n\rangle\langle n| \,. \tag{11.45}$$

Combining eqn (11.31) with the expression for the partition function in eqn (11.38) and setting $E_n = \hbar\omega(n+\frac{1}{2})$, we have

$$p(n) = \mathcal{Z}^{-1} e^{-\beta E_n} = \big(1 - e^{-\beta\hbar\omega}\big) e^{\beta\hbar\omega/2} e^{-\beta\hbar\omega(n+\frac{1}{2})} = e^{-\beta\hbar\omega n}\big(1 - e^{-\beta\hbar\omega}\big) \,. \tag{11.46}$$

Secondly, we calculate the average photon number

$$\bar{n} = \langle N \rangle = \sum_{n=0}^{\infty} p(n)\, n \quad \text{with} \quad \sum_{n=0}^{\infty} p(n) = 1 \,. \tag{11.47}$$

Now using the explicit expression (11.46) for $p(n)$ we get

$$\bar{n} = \big(1 - e^{-\beta\hbar\omega}\big) \sum_{n=0}^{\infty} n\, e^{-\beta\hbar\omega n} = -\frac{1}{\hbar\omega}\big(1 - e^{-\beta\hbar\omega}\big)\frac{\partial}{\partial\beta}\sum_{n=0}^{\infty}\big(e^{-\beta\hbar\omega}\big)^n, \tag{11.48}$$

where we have replaced the factor n in the sum by a partial derivative with respect to β. We can then once more apply the formula $\sum_{n=0}^{\infty} y^n = \frac{1}{1-y}$ with $|y| < 1$ to evaluate the sum, and calculate

$$\bar{n} = -\frac{1}{\hbar\omega}\big(1 - e^{-\beta\hbar\omega}\big)\frac{\partial}{\partial\beta}\frac{1}{1 - e^{-\beta\hbar\omega}} = \frac{e^{-\beta\hbar\omega}}{1 - e^{-\beta\hbar\omega}} \,. \tag{11.49}$$

Thus we finally find the *average photon number* of the harmonic oscillator in thermal equilibrium with a heat bath

$$\overline{n} = \frac{1}{e^{\hbar\omega/k_{\mathrm{B}}T} - 1}, \tag{11.50}$$

which is precisely the formula (1.9) we derived by considering two-level atoms that absorb and emit (spontaneously) photons in a cavity, which leads to Planck's law for black-body radiation, see eqns (1.17) and (1.18) in Section 1.1.3.

Thus the mean energy and the density matrix of the harmonic-oscillator thermal state can be expressed in terms of the average photon number $\overline{n}$ as

$$\langle H \rangle = \hbar\omega \left(\overline{n} + \tfrac{1}{2} \right), \tag{11.51}$$

$$\tau(\beta) = \frac{1}{\overline{n}+1} \sum_{n=0}^{\infty} \left(\frac{\overline{n}}{\overline{n}+1} \right)^{n} |n\rangle\langle n| . \tag{11.52}$$

11.5   Density Matrices for Two-Level Quantum Systems

To obtain a more hands-on intuition for the utility of the density-operator formalism, let us now consider another paradigmatic example in the form of quantum systems described by two-dimensional Hilbert spaces. Such two-level systems—referred to as *quantum bits*, or *qubits* in the language of quantum information theory—are the simplest and most widely used carriers of quantum information and are therefore often employed as basic units of quantum information. Physically, qubits can be encoded in a variety of ways, including (but not limited to) the spin degree of freedom of a spin-$\tfrac{1}{2}$ particle (see Section 8.2.3), the polarization of a single photon (see our discussion in Section 3.1.4), two different paths that a single quantum particle can take (e.g., in an interferometer), two particular electronic energy levels in an atom, two different positions in a double well (see Section 4.2.4), or even more exotic degrees of freedom such as the "strangeness" of kaons (which we will elaborate on in Section 26.1).

11.5.1   Pure and Mixed States of a Single Qubit

Mathematically, all of these cases can be treated on the same footing, that is, the effective pure-state Hilbert space for the two levels can be identified with $\mathcal{H} = \mathbb{C}^2$. State vectors $|\psi\rangle \in \mathbb{C}^2$ representing pure states can be written as

$$|\psi\rangle = \alpha |0\rangle + \beta |1\rangle , \tag{11.53}$$

where $\alpha, \beta \in \mathbb{C}$ with $|\alpha|^2 + |\beta|^2 = 1$, and we use the notation $|0\rangle, |1\rangle$ to denote the two states forming the *computational basis*,

$$|0\rangle = \begin{pmatrix} 1 \\ 0 \end{pmatrix}, \qquad |1\rangle = \begin{pmatrix} 0 \\ 1 \end{pmatrix}, \tag{11.54}$$

rather than previous notation that is more specific to the particular physical realization of the qubit such as $|\uparrow\rangle, |\downarrow\rangle$ or $|H\rangle, |V\rangle$ for spin $\tfrac{1}{2}$ or photon polarization, respectively. The density operator for a general pure single-qubit state is then of the form

$$\rho_{\text{pure}} = |\psi\rangle\langle\psi| = \begin{pmatrix} |\alpha|^2 & \alpha\beta^* \\ \alpha^*\beta & |\beta|^2 \end{pmatrix}. \tag{11.55}$$

For a more general *mixed* single-qubit state, we have an operator of the form

$$\rho = \begin{pmatrix} \rho_{00} & \rho_{01} \\ \rho_{10} & \rho_{11} \end{pmatrix}, \tag{11.56}$$

with the constraint $\rho_{10} = \rho_{01}^*$ for the off-diagonal elements since ρ is Hermitian, $\rho^\dagger = \rho$, while the diagonal elements must satisfy $\rho_{00}, \rho_{11} \geq 0$ because the probabilities $\rho_{00} = p(0) = \langle 0|\rho|0\rangle$ and $\rho_{11} = p(1) = \langle 1|\rho|1\rangle$ are non-negative, and normalization implies $\text{Tr}(\rho) = \rho_{00} + \rho_{11} = 1$ or $\rho_{11} = 1 - \rho_{00}$. Since the probabilities for measurement outcomes in any basis must be non-negative, we also have the condition $\rho \geq 0$, the density operator is positive semidefinite. Consequently, the eigenvalues of ρ must be non-negative, which results in the condition

$$\tfrac{1}{2}\left(1 \pm \sqrt{(1 - 2\rho_{00})^2 + 4|\rho_{01}|^2}\right) \geq 0. \tag{11.57}$$

Since these conditions still are fairly abstract, we would like to get a better intuition of what constitutes the physical difference between a pure and a mixed quantum state. We recall:

A *pure* state is a *coherent* superposition of states. Unless the particular pure state considered is already an element of the chosen basis, the state's density operator features off-diagonal elements that are responsible for the coherence and contain the information about the relative phases between the basis states.

A *mixed* state is an *incoherent* superposition—a *mixture*—of states. There are generally also off-diagonal elements, but they are typically smaller, constrained by (11.57), representing a reduced phase-information content.

To make these statements more precise, let us ask: What differences between pure and mixed states can we observe when considering the respective predictions for the outcomes of measurements in different bases, physically represented, e.g., by spin measurements using a Stern–Gerlach apparatus as in Section 8.2.1?

We now compare the maximally mixed single-qubit state

$$\rho_{\text{mix}} = \tfrac{1}{2}\left(|0\rangle\langle 0| + |1\rangle\langle 1|\right) = \frac{1}{2}\begin{pmatrix} 1 & 0 \\ 0 & 1 \end{pmatrix} = \tfrac{1}{2}\mathbb{1}_2, \tag{11.58}$$

to a particular pure state given by

$$\rho_{\text{pure}} = \frac{1}{2} \begin{pmatrix} 1 & 1 \\ 1 & 1 \end{pmatrix}, \tag{11.59}$$

where one may easily check the purity by noting that $\rho_{\text{pure}}^2 = \rho_{\text{pure}}$.

Let us then first consider a measurement in the computational basis, which can in principle be represented by any observable that is diagonal in the basis $\{|0\rangle, |1\rangle\}$, but a typical example (e.g., for a spin measurement along the z direction) is the Pauli operator

$$\sigma_z = |0\rangle\langle 0| - |1\rangle\langle 1| = \begin{pmatrix} 1 & 0 \\ 0 & -1 \end{pmatrix}. \tag{11.60}$$

The expectation values of this operator in the states ρ_{mix} and ρ_{pure} from eqns (11.58) and (11.59), respectively, are

$$\langle \sigma_z \rangle_{\text{mix}} = \text{Tr}(\sigma_z \rho_{\text{mix}}) = \text{Tr}\left[\frac{1}{2} \begin{pmatrix} 1 & 0 \\ 0 & -1 \end{pmatrix} \begin{pmatrix} 1 & 0 \\ 0 & 1 \end{pmatrix} \right] = 0, \tag{11.61a}$$

$$\langle \sigma_z \rangle_{\text{pure}} = \text{Tr}(\sigma_z \rho_{\text{pure}}) = \text{Tr}\left[\frac{1}{2} \begin{pmatrix} 1 & 0 \\ 0 & -1 \end{pmatrix} \begin{pmatrix} 1 & 1 \\ 1 & 1 \end{pmatrix} \right] = 0. \tag{11.61b}$$

That means in both cases, for the chosen mixed and pure quantum states, the probability for either measurement outcome, 0 (or $\uparrow$ for spin) and 1 (or $\downarrow$ for spin), is 50%, such that the expectation value of σ_z vanishes. With respect to measurements in this particular basis, both states are hence equivalent. This is also reflected in the expectation values of the projectors P_0 and P_1

$$P_0 = |0\rangle\langle 0| = \begin{pmatrix} 1 & 0 \\ 0 & 0 \end{pmatrix}, \qquad P_1 = |1\rangle\langle 1| = \begin{pmatrix} 0 & 0 \\ 0 & 1 \end{pmatrix}, \tag{11.62}$$

for which we obtain the expectation values

$$\langle P_0 \rangle_{\text{mix}} = \text{Tr}(P_0 \rho_{\text{mix}}) = \text{Tr}\left[\frac{1}{2} \begin{pmatrix} 1 & 0 \\ 0 & 0 \end{pmatrix} \begin{pmatrix} 1 & 0 \\ 0 & 1 \end{pmatrix} \right] = \frac{1}{2} \equiv \langle P_1 \rangle_{\text{mix}}, \tag{11.63a}$$

$$\langle P_0 \rangle_{\text{pure}} = \text{Tr}(P_0 \rho_{\text{pure}}) = \text{Tr}\left[\frac{1}{2} \begin{pmatrix} 1 & 0 \\ 0 & 0 \end{pmatrix} \begin{pmatrix} 1 & 1 \\ 1 & 1 \end{pmatrix} \right] = \frac{1}{2} \equiv \langle P_1 \rangle_{\text{pure}}. \tag{11.63b}$$

If, instead, we now consider a measurement with respect to a different basis, for instance, the basis $\{|\pm\rangle = \frac{1}{\sqrt{2}}(|0\rangle \pm |1\rangle)\}$ corresponding to the eigenstates of σ_x, i.e., spin along the x direction, we have the projectors

$$P_+ = |+\rangle\langle+| = \frac{1}{2}\begin{pmatrix} 1 & 1 \\ 1 & 1 \end{pmatrix}, \tag{11.64a}$$

$$P_- = |-\rangle\langle-| = \frac{1}{2}\begin{pmatrix} 1 & -1 \\ -1 & 1 \end{pmatrix}. \tag{11.64b}$$

In this case we obtain the expectation values

$$\langle P_+ \rangle_{\mathrm{mix}} = \mathrm{Tr}(P_+ \rho_{\mathrm{mix}}) = \mathrm{Tr}\left[\frac{1}{4}\begin{pmatrix} 1 & 1 \\ 1 & 1 \end{pmatrix}\begin{pmatrix} 1 & 0 \\ 0 & 1 \end{pmatrix}\right] = \frac{1}{2} \equiv \langle P_- \rangle_{\mathrm{mix}}, \tag{11.65}$$

whereas the considered pure state yields

$$\langle P_+ \rangle_{\mathrm{pure}} = \mathrm{Tr}(P_+ \rho_{\mathrm{pure}}) = \mathrm{Tr}\left[\frac{1}{4}\begin{pmatrix} 1 & 1 \\ 1 & 1 \end{pmatrix}\begin{pmatrix} 1 & 1 \\ 1 & 1 \end{pmatrix}\right] = 1, \tag{11.66a}$$

$$\langle P_- \rangle_{\mathrm{pure}} = \mathrm{Tr}(P_- \rho_{\mathrm{pure}}) = \mathrm{Tr}\left[\frac{1}{4}\begin{pmatrix} 1 & -1 \\ -1 & 1 \end{pmatrix}\begin{pmatrix} 1 & 1 \\ 1 & 1 \end{pmatrix}\right] = 0. \tag{11.66b}$$

We observe that a measurement in this new basis provides either outcome ("+" or "−") with the same probability for the maximally mixed state, but for the pure state ρ_{pure} from eqn (11.59) one of the two outcomes ('+' in this case) occurs with certainty. This indeed points to a more general principle:

Résumé: Pure and mixed quantum states have different measurement characteristics. For a pure state there always exists an "optimal" test—a projective measurement with respect to a basis that contains the considered pure state—such that one outcome occurs in 100% of cases, with probability one, whereas for a mixed quantum state no such test exists.

Remark I: The expectation value of all Pauli operators σ_i vanishes for the maximally mixed state. A spin-$\frac{1}{2}$ particle described by such a state is unpolarized since all spin directions are equivalent. In particular, we have

$$\langle \sigma_i \rangle_{\mathrm{mix}} = \mathrm{Tr}(\sigma_i \rho_{\mathrm{mix}}) = \frac{1}{2}\mathrm{Tr}(\sigma_i) = 0, \tag{11.67}$$

since the Pauli operators from eqn (8.30) are traceless, and since a spin measurement along any direction $\vec{n}$ can be represented by a linear combination of Pauli operators, specifically, by the operator $\vec{n}\vec{\sigma}$, we have

$$\langle \vec{n} \cdot \vec{\sigma} \rangle_{\mathrm{mix}} = \sum_{i=x,y,z} n_i \langle \sigma_i \rangle_{\mathrm{mix}} = 0. \tag{11.68}$$

Remark II: Further note that the decomposition (11.58) into projectors $|0\rangle\langle0|$ and $|1\rangle\langle1|$ is by no means unique, we can obtain the maximally mixed state $\rho_{\mathrm{mix}} = \frac{1}{2}\mathbb{1}_2$

in many different ways. Indeed, because the maximally mixed state $\rho_{\text{mix}} = \frac{1}{2}\mathbb{1}_2$ is proportional to the identity, which is invariant under unitary transformation, $\hat{U}\mathbb{1}\hat{U}^\dagger = \mathbb{1}$, we could have chosen an equally weighted mixture of projectors onto the states of any orthonormal basis. We will further discuss this in the context of general mixed states in Section 11.5.2.

11.5.2   The Bloch Decomposition

Quite generally the density matrix of a qubit will be a 2×2 matrix that can be written as linear combination of the identity $\mathbb{1}_2$ and the three Pauli matrices σ_x, σ_y, and σ_z.

Theorem 11.4 (Bloch decomposition)

Every single-qubit density operator ρ can be written as

$$\rho = \tfrac{1}{2}\big(\mathbb{1}_2 + \vec{a} \cdot \vec{\sigma}\big) = \tfrac{1}{2}\Big(\mathbb{1}_2 + \sum_{i=x,y,z} a_i\,\sigma_i\Big),$$

*where $\vec{a} \in \mathbb{R}^3$, with $|\vec{a}| \leq 1$ is called the **Bloch vector** of the system.*

Proof   For the proof, let us begin by considering an arbitrary pure single-qubit state $|\psi\rangle$ as in eqn (11.53), which we can parameterize as

$$|\psi\rangle = \cos\big(\tfrac{\vartheta}{2}\big)\,|0\rangle + e^{i\varphi}\,\sin\big(\tfrac{\vartheta}{2}\big)\,|1\rangle \tag{11.69}$$

without loss of generality, that is, up to an irrelevant global phase. Forming the projector onto this state, we have

$$|\psi\rangle\langle\psi| = \begin{pmatrix} \cos^2\big(\tfrac{\vartheta}{2}\big) & e^{-i\varphi}\,\sin\big(\tfrac{\vartheta}{2}\big)\cos\big(\tfrac{\vartheta}{2}\big) \\ e^{i\varphi}\,\sin\big(\tfrac{\vartheta}{2}\big)\cos\big(\tfrac{\vartheta}{2}\big) & \sin^2\big(\tfrac{\vartheta}{2}\big) \end{pmatrix}. \tag{11.70}$$

Writing the Pauli operators in their standard representation with respect to the computational basis,

$$\sigma_x = \begin{pmatrix} 0 & 1 \\ 1 & 0 \end{pmatrix}, \quad \sigma_y = \begin{pmatrix} 0 & -i \\ i & 0 \end{pmatrix}, \quad \sigma_z = \begin{pmatrix} 1 & 0 \\ 0 & -1 \end{pmatrix}, \tag{11.71}$$

we can then compare the matrix in eqn (11.70) with the expression for ρ from the Bloch decomposition in Theorem 11.4, which is of the form

$$\tfrac{1}{2}\big(\mathbb{1}_2 + \vec{a} \cdot \vec{\sigma}\big) = \tfrac{1}{2}\begin{pmatrix} 1 + a_z & a_x - ia_y \\ a_x + ia_y & 1 - a_z \end{pmatrix}. \tag{11.72}$$

Comparing the matrix elements of eqns (11.70) and (11.72), we immediately obtain the coefficients a_i as

$$a_x = \sin(\vartheta)\cos(\varphi)\,, \tag{11.73a}$$

$$a_y = \sin(\vartheta)\sin(\varphi)\,, \tag{11.73b}$$

$$a_z = \cos(\vartheta)\,. \tag{11.73c}$$

Since these coefficients satisfy $\sum_i |a_i|^2 = 1$ we can conclude that every *pure* single-qubit state can be written in Bloch decomposition for a Bloch vector of unit length, i.e., with $|\vec{a}| = 1$. Now we already know from Theorem 11.2 that every mixed-state density operator can be decomposed as a convex sum of pure states, and for each pure state $|\psi_n\rangle$ in such a decomposition, we can find a normalized Bloch vector $\vec{a}_n$. For an arbitrary single-qubit mixed state ρ, we can therefore always find a Bloch decomposition of the form

$$\rho = \sum_n p_n\,|\psi_n\rangle\langle\psi_n| = \sum_n p_n\,\tfrac{1}{2}\big(\mathbb{1}_2 + \vec{a}_n \cdot \vec{\sigma}\big) = \tfrac{1}{2}\big(\mathbb{1}_2 + \sum_n p_n\,\vec{a}_n \cdot \vec{\sigma}\big), \tag{11.74}$$

where we have used the fact that $\sum_n p_n = 1$ in the last step. Finally, we can define $\vec{a} := \sum_n p_n\,\vec{a}_n$, and since the probabilities satisfy $p_n \leq 1$ and sum to one, and because we also have $|\vec{a}_n| = 1$, we must have $|\vec{a}| \leq 1$. $\square$

The coefficients a_i of the Bloch vector $\vec{a}$, illustrated in Fig. 11.1, can be calculated as the expectation values of the Pauli matrices

$$\vec{a} = \mathrm{Tr}(\vec{\sigma}\rho) = \langle\vec{\sigma}\rangle\,. \tag{11.75}$$

Consequently, the Bloch vector can be interpreted as the (average) spin vector for a spin-$\frac{1}{2}$ particle. All single-qubit (or spin-$\frac{1}{2}$) density matrices lie on or within the so-called Bloch sphere (with radius $|\vec{a}| = 1$) and are determined by the Bloch vector $\vec{a}$, see Fig. 11.1. The length of the Bloch vector tells us something about the mixedness, the polarization of an ensemble, i.e., of a beam of spin $\frac{1}{2}$ particles, e.g., electrons or neutrons. We say the beam is polarized if $a_i = 1$ and completely unpolarized if $a_i = 0$, $\forall i = 1, 2, 3$. This means that pure and mixed states can be characterized via the Bloch vector in the following way:

$$\text{pure state} \quad \rho^2 = \rho \quad \Rightarrow \quad |\vec{a}| = 1\,, \tag{11.76}$$

$$\text{mixed state} \quad \rho^2 \neq \rho \quad \Rightarrow \quad |\vec{a}| < 1\,. \tag{11.77}$$

In particular, the maximally mixed state is obtained for $\vec{a} = 0$. For general mixed states we can easily calculate the mixedness of eqn (11.21) as

$$S_{\mathrm{L}}(\rho) = 1 - \mathrm{Tr}\Big[\tfrac{1}{4}\big(\mathbb{1}_2 + \sum_i a_i\sigma_i\big)^2\Big] = 1 - \tfrac{1}{4}\mathrm{Tr}\Big[\mathbb{1}_2 + 2\sum_i a_i\sigma_i + \sum_{i,j} a_i a_j \sigma_i \sigma_j\Big]$$

$$= 1 - \tfrac{1}{2}\big(1 + \sum_i a_i^2\big) = \tfrac{1}{2}\big(1 - |\vec{a}|^2\big), \tag{11.78}$$

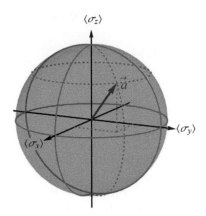

Fig. 11.1 Bloch sphere of a two-level quantum system. The density operator of any two-level quantum system, or *qubit*, can be fully described by a Bloch vector $\vec{a} \in \mathbb{R}^3$ with $|\vec{a}| \leq 1$, i.e., as a point within a unit sphere, the *Bloch sphere*. Pure states are represented as points on the surface, while mixed states lie within. Using the expectation values $\langle \sigma_i \rangle$ of the three Pauli matrices as coordinate axes, the computational-basis states $|0\rangle$ and $|1\rangle$, the eigenstates of σ_z, lie on the north and south poles of the Bloch sphere, respectively, while the maximally mixed state $\rho_{\mathrm{mix}} = \frac{1}{2}\mathbb{1}_2$ resides at the origin with $\vec{a} = 0$.

where we have used the fact that $\mathrm{Tr}(\sigma_i) = 0$ and $\mathrm{Tr}(\sigma_i \sigma_j) = 2\delta_{ij}$. The latter identity follows from the observation that all Pauli operators square to the identity, but products of two different Pauli operators are always again proportional to a Pauli operator.

Let us remark here that the choice of the parameterized state $|\psi\rangle$ in eqn (11.69) was not entirely arbitrary, indeed, it is easy to check that this state is a spin eigenstate with eigenvalue $+1$ for the spin operator along the direction $\vec{a}$ given by (11.73). That is, using the notation $|\psi(\vartheta, \varphi)\rangle \equiv |+\vec{a}\rangle$, we have

$$\vec{a} \cdot \vec{\sigma}\, |\psi(\vartheta, \varphi)\rangle = + |\psi(\vartheta, \varphi)\rangle \,. \tag{11.79}$$

Similarly, a corresponding orthogonal eigenstate $|-\vec{a}\rangle$ with eigenvalue -1 can be constructed,

$$|-\vec{a}\rangle = -\sin\left(\tfrac{\vartheta}{2}\right) |0\rangle + e^{i\varphi} \cos\left(\tfrac{\vartheta}{2}\right) |1\rangle \,, \tag{11.80}$$

such that

$$\vec{a} \cdot \vec{\sigma}\, |\pm \vec{a}\rangle = \pm |\pm \vec{a}\rangle \,. \tag{11.81}$$

For $\vartheta = 0$ we obtain the computational basis $\{|0\rangle, |1\rangle\}$, while for $\vartheta = \pi/2$ and $\varphi = 0$ we have an orthonormal basis consisting of the states $|+\rangle$ and $-|-\rangle$ with $|\pm\rangle \tfrac{1}{2}(|0\rangle \pm |1\rangle)$ and $\sigma_x |\pm\rangle = \pm |\pm\rangle$. The additional sign in front of the vector $|-\rangle$ is due to the fact that a rotation of the Bloch vector in $\mathbb{R}^3$ is achieved by a 3×3 rotation matrix in SO(3). As we recall from Section 8.2.5, such transformations are represented on the two-dimensional spinor Hilbert space by 2×2 matrices in

SU(2), which have determinant $+1$. In our example, a rotation by the angle $\pi/2$ around the y axis, which would rotate the z axis into the x axis, is represented by matrix $U_y(\pi/2) = \mathbb{1} - i\sin(\pi/4)\sigma_y$, see eqn (8.1), such that $U_y(\pi/2)\,|\,0\,\rangle = |\,+\,\rangle$ and $U_y(\pi/2)\,|\,1\,\rangle = -\,|\,-\,\rangle$. In other words, we *cannot* transform the basis $\{|\,0\,\rangle,|\,1\,\rangle\}$ into the basis $\{|\,\pm\,\rangle\}$ via a unitary in SU(2), i.e., with determinant $+1$, but since all orthonormal bases in a fixed dimension can be transformed into each other unitarily, we can find some unitary that does this. Here, this would be the so-called *Hadamard gate*, represented by the matrix

$$\frac{1}{\sqrt{2}}\begin{pmatrix} 1 & 1 \\ 1 & -1 \end{pmatrix}, \tag{11.82}$$

which has determinant -1, and which is of further relevance in the theory of quantum computation (see, e.g., Nielsen and Chuang, 2000).

Inspection of the pure-state density operator $|\,\vec{a}\,\rangle\langle\,\vec{a}\,|$ further shows that the sum of the diagonal elements, i.e., the probability to measure spin up and spin down in direction $\vec{a}$, is of course one. In the notation tailored to the description of spin we can write $\rho_{\uparrow\uparrow} + \rho_{\downarrow\downarrow} = \cos^2\!\left(\frac{\vartheta}{2}\right) + \sin^2\!\left(\frac{\vartheta}{2}\right) = 1$. Meanwhile, the difference of the diagonal elements $\rho_{\uparrow\uparrow} - \rho_{\downarrow\downarrow} = \cos^2\!\left(\frac{\vartheta}{2}\right) - \sin^2\!\left(\frac{\vartheta}{2}\right) = \cos(\vartheta) = \langle\,+\vec{a}\,|\,\sigma_z\,|\,+\vec{a}\,\rangle$ gives the expectation value of σ_z in the state $|\,+\vec{a}\,\rangle$. Therefore the diagonal elements describe the *longitudinal polarization* $\langle\,\sigma_z\,\rangle$.

The off-diagonal elements of the density matrix, on the other hand, represent the *transversal polarization* $|\rho_{\uparrow\downarrow}| = |\rho_{\downarrow\uparrow}| = \frac{1}{2}\sin(\vartheta) = \frac{1}{2}|\,\langle\,\vec{\sigma}\,\rangle_\perp\,|$, where $\langle\,\vec{\sigma}\,\rangle_\perp$ is the projection of $\langle\,\vec{\sigma}\,\rangle$ onto the x–y plane and the argument (the angle between $\langle\,\vec{\sigma}\,\rangle_\perp$ and the x axis) is φ.

Remark: As we had remarked already previously, the maximally mixed state ρ_{mix} represents a statistical mixture of various states $|\,+\vec{n}\,\rangle$ which are equally probable for all directions. In addition to the mixtures of orthogonal states that we have considered earlier, for instance, the decomposition (11.58), the maximally mixed state can also be seen as an equally weighted mixture of more than two states, e.g., of the three states $|\,\vec{a}_i\,\rangle$ given by

$$|\,\vec{a}_1\,\rangle = |\,0\,\rangle\,, \tag{11.83a}$$

$$|\,\vec{a}_2\,\rangle = \cos\!\left(\tfrac{2\pi}{6}\right)|\,0\,\rangle + \sin\!\left(\tfrac{2\pi}{6}\right)|\,1\,\rangle\,, \tag{11.83b}$$

$$|\,\vec{a}_3\,\rangle = \cos\!\left(\tfrac{2\pi}{6}\right)|\,0\,\rangle - \sin\!\left(\tfrac{2\pi}{6}\right)|\,1\,\rangle\,, \tag{11.83c}$$

which form an equilateral triangle in the x–z plane of the Bloch sphere. Recalling that $\cos\!\left(\frac{\pi}{3}\right) = \frac{1}{2}$ and $\sin\!\left(\frac{\pi}{3}\right) = \frac{\sqrt{3}}{2}$, the mixture

$$\rho_{\mathrm{mix}} = \sum_{i=1}^{3} \tfrac{1}{3}\,|\,\vec{a}_i\,\rangle\langle\,\vec{a}_i\,| = \frac{1}{2}\begin{pmatrix} 1 & 0 \\ 0 & 1 \end{pmatrix} = \frac{1}{2}\,\mathbb{1} \tag{11.84}$$

again results in the maximally mixed state. This leads us to the following proposition.

Proposition 11.1 (Decomposition of mixed density matrix)
There are various, in general infinitely many, mixtures of quantum states that lead to the same mixed-state density matrix.

Physical predictions depend only on the density matrix and not on how it is structured. That means we cannot distinguish the several types of statistical mixtures leading to the same mixed density matrix. We therefore have to consider different decompositions as different expressions of one and the same incomplete information we have about the system. In the course of this book this insight will lead us to formulate the concept of entropy for quantum systems. Entropy measures the degree of uncertainty about a quantum system. We will elaborate on this issue, which is of central importance in quantum information, in a dedicated chapter, Chapter 20 of Part III. Here, let us just quote *Walter Thirring* with regards to the role of entropy in quantum systems: *"The entropy measures how much is missing from maximal information."* (Translated from Thirring, 1980.)

11.5.3 Spin $\frac{1}{2}$ in an External Magnetic Field

With the geometric picture provided by the Bloch sphere, let us now consider a physical example for the time evolution of a density operator: a spin-$\frac{1}{2}$ particle placed into an external static and homogenous magnetic field. As we recall from our treatment in Sections 8.1.3 and 8.2.2, especially eqn (8.26), when the motion of the particle can be disregarded, this situation can be described by the Hamiltonian

$$H = -\vec{\mu} \cdot \vec{B} = -\gamma g\, \vec{S} \cdot \vec{B}, \qquad (11.85)$$

where the magnetic dipole moment $\vec{\mu}$ is determined by the *gyromagnetic ratio* $\gamma = q/(2m_q c)$ for a particle of electric charge q and mass m_q and g denotes the corresponding g-factor for spin. For instance, for an electron, we have $q = -q_e < 0$, $m = m_e$, and the g-factor is $g = g_s \approx 2$ corresponding to the anomalous magnetic moment of the electron (small corrections from quantum field theory are ignored), such that with $\vec{S} = \frac{\hbar}{2}\vec{\sigma}$ we have $\vec{\mu} = -\mu_B \vec{\sigma}$, where $\mu_B = \frac{q_e \hbar}{2m_e c}$ is the Bohr magneton from eqn (8.12). Without loss of generality we choose the magnetic field to be oriented along the z direction, $\vec{B} = B\,\vec{e}_z$, which allows us to rewrite the Hamiltonian as

$$H = -\frac{\gamma\, g\, \hbar\, B}{2}\sigma_z = -\frac{\hbar\,\omega_L}{2}\sigma_z\,, \qquad (11.86)$$

where $\omega_L = \gamma\, g\, B$ denotes the *Larmor frequency*. This is the angular frequency of the precession of the magnetic moment of the particle about the external magnetic field, recall Section 8.1.3. The Hamilton in eqn (11.86) yields the following energy levels:

$$E = \begin{cases} +\frac{\hbar\omega_L}{2} & \text{for spin } \downarrow \\[2mm] -\frac{\hbar\omega_L}{2} & \text{for spin } \uparrow\,. \end{cases} \qquad (11.87)$$

Now we calculate the time evolution of the spin-$\frac{1}{2}$ particle in the magnetic field. Following Theorem 11.3, we need the commutator

$$[\sigma_z, \rho] = 2 \begin{pmatrix} 0 & \rho_{01} \\ -\rho_{10} & 0 \end{pmatrix} \tag{11.88}$$

to evaluate the von Neumann equation, where we have used the notation $\rho_{ij} = \langle i|\rho|j \rangle$ for the density-matrix components, as in eqn (11.56). The von Neumann equation thus reads

$$\frac{\partial}{\partial t} \begin{pmatrix} \rho_{00} & \rho_{01} \\ \rho_{10} & \rho_{11} \end{pmatrix} = i\,\omega_{\mathrm{L}} \begin{pmatrix} 0 & \rho_{01} \\ -\rho_{10} & 0 \end{pmatrix} . \tag{11.89}$$

The solutions show that the spin is rotating with frequency ω_{L} around the z axis

$$\rho_{00} = \rho_{00}(0) \quad \text{constant} \quad \text{and} \quad \rho_{11} = \rho_{11}(0) \quad \text{constant} \tag{11.90}$$

$$\rho_{01}(t) = e^{i\omega_{\mathrm{L}}t}\,\rho_{01}(0) \quad \text{and} \quad \rho_{10}(t) = e^{-i\omega_{\mathrm{L}}t}\,\rho_{10}(0)\,. \tag{11.91}$$

This is even more obvious when viewing this in the Bloch picture, where it is easy to see that we have

$$a_x(t) = \tfrac{1}{2}\big(\rho_{01} + \rho_{10}\big) = \cos(\omega_{\mathrm{L}}t)\,, \tag{11.92a}$$

$$a_y(t) = \tfrac{i}{2}\big(\rho_{01} - \rho_{10}\big) = -\sin(\omega_{\mathrm{L}}t)\,, \tag{11.92b}$$

$$a_z(t) = a_z(0) \quad \text{constant}\,. \tag{11.92c}$$

11.6 Geometry of the State Space

Although the Bloch decomposition discussed in Section 11.5.2 is extremely successful in providing a geometric picture for the (three-dimensional) state space of a single qubit, it is quite obvious that for higher-dimensional quantum systems we cannot generally expect to find a picture that is at the same time as geometrically simple and as informative about the underlying physics. Nevertheless, it is possible to describe quantum systems of arbitrary dimension, including those partitioned into two or more subsystems, by a generalized Bloch decomposition, as we will discuss in more detail in Section 15.2.3 and Chapter 17. This turns out to be convenient for many calculations and can even provide an intuitive picture of the Hilbert-space geometry for some cases (e.g., for a three-dimensional system, a *qutrit*, see Eltschka *et al.*, 2021).

Here we want to focus on a view of the state space that is less informative regarding properties of individual density operators but provides a remarkably simple geometrical picture of *all* density matrices.

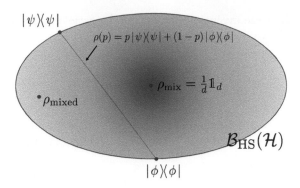

Fig. 11.2 Geometry of quantum states. The set of all density operators on $\mathcal{H}$ forms a convex subset of the Hilbert–Schmidt space $\mathcal{B}_{HS}(\mathcal{H})$. The boundary of this set is formed by pure states $|\psi\rangle$, while all mixed states, for instance, mixtures $p\,|\psi\rangle\langle\psi|+(1-p)\,|\phi\rangle\langle\phi|$ of two pure states $|\psi\rangle$ and $|\phi\rangle$ lie inside of the convex set. The maximally mixed state $\rho_{\text{mix}} = \frac{1}{d}\mathbb{1}_d$ lies at the centre of this set in the set that all states with the same purity (in particular, all pure states with purity one) have the same distance from it.

As we recall from Theorem 11.2, every density matrix can be viewed as a convex sum of pure states, and indeed, arbitrarily many equivalent decompositions of this kind exist for any mixed state (see Proposition 11.1). At the same time, arbitrary convex combinations of mixed states must again result in valid (mixed-state) density operators. This suggests that density operators (in any fixed dimension) form a *convex set*, illustrated in Fig. 11.2. Pure states form the surface or boundary of this set, since no pure state can be obtained via convex combinations of other states, while each mixed state is represented by a point in the bulk or interior of this convex set, and vice versa, every point in the set can be understood as a state in the form of a density operator.

This convex state space is part of a Hilbert space called the Hilbert–Schmidt space. To define this space, we first define the *Hilbert–Schmidt inner product* as follows:

Definition 11.4 *The **Hilbert–Schmidt inner product** of two bounded linear operators A and B on a Hilbert space $\mathcal{H}$ is given by*

$$(A, B)_{HS} := \mathrm{Tr}\left(A^\dagger B\right)\ \ .$$

The Hilbert–Schmidt inner product induces a norm via the relation $\|A\|_{HS}^2 = (A, A)_{HS}$, the *Hilbert–Schmidt norm* is thus given by

$$\|A\|_{HS} = \sqrt{\mathrm{Tr}\left(A^\dagger A\right)}. \tag{11.93}$$

In some cases it can be useful to understand this norm as a special case of a more general family of matrix norms, the so-called *Schatten p-norms*, named after *Robert Schatten* (1960). Restricting our discussion here to finite-dimensional Hilbert spaces

$\mathcal{H}_1$ and $\mathcal{H}_2$, for any linear operator from $\mathcal{H}_1$ to $\mathcal{H}_2$ we can always find a singular-value decomposition

$$A = \sum_i s_i \,|\, u_i \,\rangle\langle\, v_i \,|, \tag{11.94}$$

for some orthonormal bases $\{|\, u_i \,\rangle\}_i$ and $\{|\, v_j \,\rangle\}_j$ of $\mathcal{H}_2$ and $\mathcal{H}_1$, respectively. The Schatten p-norm of A is then

$$\|A\|_p := \left(\sum_i s_i^p \right)^{1/p}. \tag{11.95}$$

We can then calculate the Hilbert–Schmidt norm in terms of the singular values of A,

$$\|A\|_{\mathrm{HS}} = \sqrt{\mathrm{Tr}\big(A^\dagger A\big)} = \sqrt{\mathrm{Tr}\big(\sum_{i,j} s_i s_j \,|\, u_i \,\rangle\langle\, v_i \,|\, v_j \,\rangle\langle\, u_j \,|\big)}$$

$$= \sqrt{\mathrm{Tr}\big(\sum_i s_i^2 \,|\, u_i \,\rangle\langle\, u_i \,|\big)} = \sqrt{\sum_i s_i^2}, \tag{11.96}$$

to find that the Hilbert–Schmidt norm is the Schatten p-norm for $p = 2$. We will encounter another important member of this family, the *trace norm* (corresponding to $p = 1$) later on, in Section 23.4. Here, we now use the Hilbert–Schmidt norm to define the Hilbert–Schmidt space:

Definition 11.5 *The **Hilbert–Schmidt space** $\mathcal{B}_{\mathrm{HS}}(\mathcal{H})$ is the space of all bounded linear operators on $\mathcal{H}$ with finite Hilbert–Schmidt norm.*

The Hilbert–Schmidt space is complete with respect to the Hilbert–Schmidt inner product and hence forms a Hilbert space. Moreover, the Hilbert–Schmidt space is isometrically isomorphic to the tensor product $\mathcal{H}^* \otimes \mathcal{H}$ of the dual Hilbert space $\mathcal{H}^*$ and of $\mathcal{H}$ itself (Conway, 1990, p. 268).

Moreover, we can use the Hilbert–Schmidt norm to define a distance measure between operators.

Definition 11.6 *The **Hilbert–Schmidt distance** $D_{\mathrm{HS}}(A, B)$ between two bounded linear operators A and B is given by*

$$D_{\mathrm{HS}}(A, B) := \|A - B\|_{\mathrm{HS}}\ .$$

The Hilbert–Schmidt distance is the metric on the Hilbert–Schmidt space induced by the Hilbert–Schmidt inner product via the Hilbert–Schmidt norm, recall that all inner-product spaces are also normed vectors spaces, which in turn must be metric spaces, a well-known fact of linear algebra (see, e.g., Hoffman and Kunze 1971). In particular, for all density operators ρ_1, ρ_2, and ρ_3 the Hilbert–Schmidt distance has the following properties of a metric:

(i) $\qquad\qquad D_{\text{HS}}(\rho_1, \rho_2) \geq 0$ $\qquad\qquad\qquad$ non-negative, $\qquad\qquad$ (11.97a)

(ii) $\qquad D_{\text{HS}}(\rho_1, \rho_2) = D_{\text{HS}}(\rho_2, \rho_1)$ $\qquad\qquad$ symmetric, $\qquad\qquad$ (11.97b)

(iii) $D_{\text{HS}}(\rho_1, \rho_3) \leq D_{\text{HS}}(\rho_1, \rho_2) + D_{\text{HS}}(\rho_2, \rho_3)$ triangle inequality, $\qquad$ (11.97c)

(iv) $\qquad D_{\text{HS}}(\rho_1, \rho_2) = 0 \;\Leftrightarrow\; \rho_1 = \rho_2$ $\qquad$ identity of indiscernibles. (11.97d)

With the Hilbert–Schmidt metric at hand, let us take a closer look at some examples. First, let us examine the distance between two pure states $|\psi\rangle$ and $|\phi\rangle$. To this end, we first note that for any choice of these two vectors, we can decompose $|\phi\rangle$ into a component parallel to $|\psi\rangle$ and a component $|\psi^{\perp}\rangle$ orthogonal to $|\psi\rangle$, such that $|\phi\rangle = \alpha|\psi\rangle + \sqrt{1-|\alpha|^2}\,|\psi^{\perp}\rangle$ with $\alpha = \langle\psi|\phi\rangle$, where we can choose the coefficient $\sqrt{1-|\alpha|^2}$ to be real and non-negative since any relative phase can be absorbed into the definition of $|\psi^{\perp}\rangle$. With this we calculate

$$|\psi\rangle\langle\psi| - |\phi\rangle\langle\phi| = (1-|\alpha|^2)|\psi\rangle\langle\psi| - \sqrt{1-|\alpha|^2}\left(\alpha|\psi\rangle\langle\psi^{\perp}| + \alpha^*|\psi^{\perp}\rangle\langle\psi|\right)$$
$$- (1-|\alpha|^2)|\psi^{\perp}\rangle\langle\psi^{\perp}| \,. \qquad (11.98)$$

Since $\langle\psi|\psi^{\perp}\rangle = 0$, the square of this matrix has the simple form

$$\left(|\psi\rangle\langle\psi| - |\phi\rangle\langle\phi|\right)^2 = (1-|\alpha|^2)|\psi\rangle\langle\psi| + (1-|\alpha|^2)|\psi^{\perp}\rangle\langle\psi^{\perp}| \,, \qquad (11.99)$$

that is, it is a diagonal matrix with two identical and non-negative eigenvalues $(1-|\alpha|^2)$. With this, we evaluate the Hilbert–Schmidt distance,

$$D_{\text{HS}}(|\psi\rangle\langle\psi|, |\phi\rangle\langle\phi|) = \||\psi\rangle\langle\psi| - |\phi\rangle\langle\phi|\|_{\text{HS}} = \sqrt{\text{Tr}\left[\left(|\psi\rangle\langle\psi| - |\phi\rangle\langle\phi|\right)^2\right]}$$
$$= \sqrt{2(1-|\alpha|^2)} = \sqrt{2(1-|\langle\psi|\phi\rangle|^2)}\,, \qquad (11.100)$$

which turns out to be related to the overlap or *fidelity* $|\langle\psi|\phi\rangle|^2$ of the involved pure states. The fidelity is also often used as a measure of distance between quantum states, but the generalization of the pure-state fidelity to mixed states in terms of the *Uhlmann fidelity*, which we will discuss in Section 24.2.2, generally does not relate to the Hilbert–Schmidt distance in the same way as it does for pure states in eqn (11.100).

We can also easily evaluate the Hilbert–Schmidt distance of an arbitrary state ρ to the maximally mixed state $\rho_{\text{mix}} = \frac{1}{d}\mathbb{1}_d$ in a d-dimensional Hilbert space. For this case we find

$$D_{\text{HS}}(\rho, \rho_{\text{mix}}) = \sqrt{\text{Tr}\left[(\rho - \rho_{\text{mix}})^2\right]} = \sqrt{\text{Tr}\left[\rho^2 - 2\frac{1}{d}\rho + \frac{1}{d^2}\mathbb{1}_d\right]}$$
$$= \sqrt{\text{Tr}\left[\rho^2\right] - \frac{2}{d} + \frac{1}{d}} = \sqrt{\text{Tr}\left[\rho^2\right] - \frac{1}{d}} = \sqrt{P(\rho) - \frac{1}{d}}\,, \qquad (11.101)$$

where $P(\rho)$ is the purity from eqn (11.20) with $\frac{1}{d} \leq P(\rho) \leq 1$. We thus see that states with the same purity have the same distance from the maximally mixed state. In particular, all pure states have the maximal distance $D_{\text{HS}}(|\psi\rangle, \rho_{\text{mix}}) = \sqrt{1 - (1/d)}$ from the maximally mixed state, which we can thus regard as forming the centre of the state space.

As a final example that brings together all of the previously obtained intuition about the geometry of the state space, let us focus on single qubits in the Bloch picture. Using the Bloch decomposition from Theorem 11.4, we can represent two arbitrary single-qubit states ρ_A and ρ_B by their Bloch vectors $\vec{a}$ and $\vec{b}$, respectively, and we have

$$\rho_A - \rho_B = \tfrac{1}{2}\big(\mathbb{1}_2 + \vec{a}\cdot\vec{\sigma}\big) - \tfrac{1}{2}\big(\mathbb{1}_2 + \vec{b}\cdot\vec{\sigma}\big) = \tfrac{1}{2}\big(\vec{a} - \vec{b}\big)\cdot\vec{\sigma}. \tag{11.102}$$

For an arbitrary vector $\vec{r} \in \mathbb{R}^3$ we also have

$$\text{Tr}\big(\vec{r}\cdot\vec{\sigma}\big)^2 = \text{Tr}\Big(\sum_{i,j} r_i r_j \sigma_i \sigma_j\Big) = \sum_{i,j} r_i r_j \, \text{Tr}\big(\sigma_i \sigma_j\big) = 2\sum_i r_i^2 = 2\,|\vec{r}|^2, \tag{11.103}$$

where we have used $\text{Tr}(\sigma_i \sigma_j) = 2\delta_{ij}$. The Hilbert–Schmidt distance then simply evaluates to

$$D_{\text{HS}}(\rho_A, \rho_B) = \sqrt{\text{Tr}\big[\big(\rho_A - \rho_B\big)^2\big]} = \tfrac{1}{2}\sqrt{\text{Tr}\big(\big[(\vec{a} - \vec{b})\cdot\vec{\sigma}\big]^2\big)} = \tfrac{1}{\sqrt{2}}|\vec{a} - \vec{b}|. \tag{11.104}$$

We thus see, up to a pre-factor of $\frac{1}{\sqrt{2}}$, the Hilbert–Schmidt distance for two qubits is just the Euclidean distance between the respective points on the Bloch sphere. Indeed, as we shall discuss in more detail in Chapter 17, one can find a *generalized Bloch decomposition* (Bertlmann and Krammer, 2008*a*) with Bloch vectors $\vec{a} \in \mathbb{R}^{d^2-1}$ for all dimensions d. In this case, the expression for the Hilbert–Schmidt distance between two arbitrary states described by the (generalized) Bloch vectors $\vec{a}$ and $\vec{b}$ is as in eqn (11.104), modified only by a pre-factor that depends on the particular choice of generalization of the Pauli matrices to higher dimensions.

11.7 Density Matrices for Bipartite Quantum Systems

Finally, let us briefly consider density operators of composite systems in preparation for a more detailed mathematical analysis of the distinction between *separability* and *entanglement* in Chapter 15, which will also be conceptually highly relevant already for our earlier discussion of *non-locality* (the violation of Bell-type inequalities, see Chapter 13) and *quantum teleportation* (see Chapter 14). Here, we simply want to establish the most basic notation and examples to illustrate how composite systems are described within the density-operator formalism.

Let us consider two quantum systems (for instance, two spin-$\frac{1}{2}$ particles) at two different (potentially far apart) locations, where they are controlled by observers Alice and Bob, respectively. Although the particles are separated in this way, we may treat them as one composite quantum system. The density matrices describing such a composite quantum system are defined in the Hilbert–Schmidt space $\mathcal{B}_{\mathrm{HS}}(\mathcal{H}_A \otimes \mathcal{H}_B)$, the space of all bounded linear operators (with finite Hilbert–Schmidt norm, see Definition 11.5) on the bipartite tensor-product space $\mathcal{H} = \mathcal{H}_A \otimes \mathcal{H}_B$ associated with the systems controlled by Alice and Bob, respectively. This Hilbert–Schmidt space can also be considered as an algebra $\mathcal{A}$ which we will refer to in some cases in the following, e.g., in Sections 14.3.3 and 16.4.1. The tensor-product space $\mathcal{H}$ has dimension $d_A \times d_B$, where $d_A = \dim(\mathcal{H}_A)$ and $d_B = \dim(\mathcal{H}_B)$.

To provide some hands-on examples, we now focus on the simplest example of a pair of two-dimensional systems, two qubits, with $d_A = d_B = 2$, realized in the examples below as two spin-$\frac{1}{2}$ particles, such that we use the notation $|\uparrow\rangle \equiv |0\rangle$ and $|\downarrow\rangle \equiv |1\rangle$. Indeed, we have already discussed such a composite system in Section 8.3.3, in particular, we studied the tensor products of two state vectors and focused on the spin triplet and singlet states in eqns (8.97) and (8.98), respectively.

As a basis for the tensor-product space $\mathcal{H}$, we can consider the *standard-product basis* or *two-qubit computational basis*, which is given by

$$|\uparrow\uparrow\rangle \equiv |\uparrow\rangle \otimes |\uparrow\rangle, \qquad |\downarrow\downarrow\rangle \equiv |\downarrow\rangle \otimes |\downarrow\rangle \tag{11.105a}$$

$$|\uparrow\downarrow\rangle \equiv |\uparrow\rangle \otimes |\downarrow\rangle, \qquad |\downarrow\uparrow\rangle \equiv |\downarrow\rangle \otimes |\uparrow\rangle. \tag{11.105b}$$

Alternatively, we can choose, e.g., a coupled basis of joint eigenstates of the (square) total spin operator $\vec{S}^{\,2}$ with $\vec{S} = \vec{S}_A + \vec{S}_B$ and its z component S_z, i.e., the totally symmetric triplet and totally antisymmetric singlet states from eqns (8.97) and (8.98), respectively. Both bases have the two basis vectors $|\uparrow\uparrow\rangle$ and $|\downarrow\downarrow\rangle$ in common. However, we can also construct a basis where all basis vectors are completely symmetric or antisymmetric with respect to the exchange of A and B but which have no basis vectors in common with the product basis in eqn (11.105), the so-called *Bell states*

$$|\psi^{\pm}\rangle = \tfrac{1}{\sqrt{2}}\left(|\uparrow\downarrow\rangle \pm |\downarrow\uparrow\rangle\right), \tag{11.106a}$$

$$|\phi^{\pm}\rangle = \tfrac{1}{\sqrt{2}}\left(|\uparrow\uparrow\rangle \pm |\downarrow\downarrow\rangle\right). \tag{11.106b}$$

Whereas states (11.105) are product states and thus an example of so-called *separable states*, the four Bell states (11.106) are examples for *maximally entangled* states. We are going to discuss the notions of separability and entanglement, along with their mathematical and physical implications, more generally in Section 15.2. Here, let us just make an important remark on the mathematical structure encountered for pure states: Every pure state that is not decomposable into product states (with respect to a fixed assignment of two subsystems, here A and B) is an *entangled state*.

The main intuition behind the concept of entanglement is that it is a form of correlation that is, in a sense, stronger than any form of classical correlation. We can clearly see from the decomposition of the Bell states that measurements of the spin along the z direction of Alice and Bob are perfectly correlated (in the case of $|\phi^{\mp}\rangle$) or anti-correlated (in the case of $|\psi^{\mp}\rangle$).

Rules for tensor products: Before we take a closer look at the density operators corresponding to these states, let us first state some of the rules for applying tensor products. This will be useful in the following chapters, where we will extensively work with tensor products of spin states. For the mathematical details of tensor products, we refer to our discussion in Section 3.2 or to the mathematical literature, e.g., to Hungerford (1974). Here we just want to recall some important rules for calculating tensor products. For matrices A, B, C, D, vectors x, y, z, and scalars a, b, c, d the following rules hold:

$$(A \otimes B)(C \otimes D) = AC \otimes BD, \qquad (11.107a)$$

$$(A \otimes B)(x \otimes y) = Ax \otimes By, \qquad (11.107b)$$

$$(x + y) \otimes z = x \otimes z + y \otimes z, \qquad (11.107c)$$

$$z \otimes (x + y) = z \otimes x + z \otimes y, \qquad (11.107d)$$

$$ax \otimes by = ab(x \otimes y). \qquad (11.107e)$$

For the tensor product of matrices with a matrix U we have

$$\begin{pmatrix} A & B \\ C & D \end{pmatrix} \otimes U = \begin{pmatrix} A \otimes U & B \otimes U \\ C \otimes U & D \otimes U \end{pmatrix}, \qquad (11.108)$$

which we can specialize to a matrix of scalars tensored with a matrix U:

$$\begin{pmatrix} a & b \\ c & d \end{pmatrix} \otimes U = \begin{pmatrix} aU & bU \\ cU & dU \end{pmatrix}. \qquad (11.109)$$

For complex conjugation the rule is

$$(A \otimes B)^* = A^* \otimes B^*. \qquad (11.110)$$

Typical examples for spin states: Let us now take a closer look at the density matrices for the previously mentioned states. Beginning with the state vectors for the pure product states, we have, for instance,

$$|\uparrow\uparrow\rangle = |\uparrow\rangle \otimes |\uparrow\rangle = \begin{pmatrix} 1 \\ 0 \end{pmatrix} \otimes \begin{pmatrix} 1 \\ 0 \end{pmatrix} = \begin{pmatrix} 1 \\ 0 \\ 0 \\ 0 \end{pmatrix}, \qquad (11.111a)$$

$$|\uparrow\downarrow\rangle = |\uparrow\rangle \otimes |\downarrow\rangle = \begin{pmatrix} 1 \\ 0 \end{pmatrix} \otimes \begin{pmatrix} 0 \\ 1 \end{pmatrix} = \begin{pmatrix} 0 \\ 1 \\ 0 \\ 0 \end{pmatrix}. \qquad (11.111b)$$

With this in mind, we then form the projectors for the product states, which we can also write in components with respect to the product/computational basis:

$$
\rho_{\uparrow\uparrow} = |\uparrow\uparrow\rangle\langle\uparrow\uparrow| = \begin{pmatrix} 1 \\ 0 \\ 0 \\ 0 \end{pmatrix} \big(1, 0, 0, 0\big) = \begin{pmatrix} 1 & 0 & 0 & 0 \\ 0 & 0 & 0 & 0 \\ 0 & 0 & 0 & 0 \\ 0 & 0 & 0 & 0 \end{pmatrix}, \qquad (11.112a)
$$

$$
\rho_{\uparrow\downarrow} = |\uparrow\downarrow\rangle\langle\uparrow\downarrow| = \begin{pmatrix} 0 \\ 1 \\ 0 \\ 0 \end{pmatrix} \big(0, 1, 0, 0\big) = \begin{pmatrix} 0 & 0 & 0 & 0 \\ 0 & 1 & 0 & 0 \\ 0 & 0 & 0 & 0 \\ 0 & 0 & 0 & 0 \end{pmatrix}, \qquad (11.112b)
$$

$$
\rho_{\downarrow\uparrow} = |\downarrow\uparrow\rangle\langle\downarrow\uparrow| = \begin{pmatrix} 0 \\ 0 \\ 1 \\ 0 \end{pmatrix} \big(0, 0, 1, 0\big) = \begin{pmatrix} 0 & 0 & 0 & 0 \\ 0 & 0 & 0 & 0 \\ 0 & 0 & 1 & 0 \\ 0 & 0 & 0 & 0 \end{pmatrix}, \qquad (11.112c)
$$

$$
\rho_{\downarrow\downarrow} = |\downarrow\downarrow\rangle\langle\downarrow\downarrow| = \begin{pmatrix} 0 \\ 0 \\ 0 \\ 1 \end{pmatrix} \big(0, 0, 0, 1\big) = \begin{pmatrix} 0 & 0 & 0 & 0 \\ 0 & 0 & 0 & 0 \\ 0 & 0 & 0 & 0 \\ 0 & 0 & 0 & 1 \end{pmatrix}. \qquad (11.112d)
$$

In case of the entangled Bell states we have

$$
\rho^{\pm} = |\psi^{\pm}\rangle\langle\psi^{\pm}| = \tfrac{1}{2} \begin{pmatrix} 0 \\ 1 \\ \pm1 \\ 0 \end{pmatrix} \big(0, 1, \pm1, 0\big) = \tfrac{1}{2} \begin{pmatrix} 0 & 0 & 0 & 0 \\ 0 & 1 & \pm1 & 0 \\ 0 & \pm1 & 1 & 0 \\ 0 & 0 & 0 & 0 \end{pmatrix}, \qquad (11.113a)
$$

$$
\omega^{\pm} = |\phi^{\pm}\rangle\langle\phi^{\pm}| = \tfrac{1}{2} \begin{pmatrix} 1 \\ 0 \\ 0 \\ \pm1 \end{pmatrix} \big(1, 0, 0, \pm1\big) = \tfrac{1}{2} \begin{pmatrix} 1 & 0 & 0 & \pm1 \\ 0 & 0 & 0 & 0 \\ 0 & 0 & 0 & 0 \\ \pm1 & 0 & 0 & 1 \end{pmatrix}. \qquad (11.113b)
$$

Finally, we take a look at the tensor product of two Pauli operators,

$$\sigma_x \otimes \sigma_y = \begin{pmatrix} 0 & 1 \\ 1 & 0 \end{pmatrix} \otimes \begin{pmatrix} 0 & -i \\ i & 0 \end{pmatrix} = \begin{pmatrix} 0 & 0 & 0 & -i \\ 0 & 0 & i & 0 \\ 0 & -i & 0 & 0 \\ i & 0 & 0 & 0 \end{pmatrix} \neq \sigma_y \otimes \sigma_x \, . \quad (11.114)$$

Résumé: Statistical mixtures of quantum states are described by density matrices, which form elements of the Hilbert–Schmidt space. One and the same density matrix can be expressed in terms of different mixtures of pure states. The time evolution of density operators is determined by the von Neumann equation, which is the quantum-mechanical analogue of the classical Liouville equation in statistical mechanics.

We have studied the density matrix of a harmonic oscillator in thermodynamic equilibrium with a heat reservoir and explicitly derived the mean energy of the oscillator featuring in Planck's law.

We have also discussed the important example of spin $\frac{1}{2}$ (single qubits) in the Bloch decomposition and examined composite systems of two spins, in particular, product states and entangled states. More specifically, we concentrated on Bell states as they are of fundamental importance in quantum information theory.

Finally, we have presented some important rules for calculating tensor products of matrices, vectors, and scalars with the typical examples for spin, the standard product basis of a composite system and the Bell basis.

12
Hidden-Variable Theories

The idea of hidden variables is as old as quantum theory itself. As we shall discuss in this chapter, many of the founders of quantum mechanics, like *Albert Einstein*, *Erwin Schrödinger*, *Louis de Broglie*, and others, were not satisfied with the intrinsic indeterminism of quantum theory. It was believed, for instance by *David Bohm* and (perhaps most importantly) by *John Bell* that the predictions of quantum mechanics could perhaps be derived from a deterministic, yet unknown and hidden, "sub-quantum" theory which can be described by so-called hidden variables. Here, we discuss the historical development of these and related ideas at the foundation of quantum theory, which have formed a conceptual starting point for many aspects of modern quantum information theory, as we shall see. A review of the hidden-variable theory à la de Broglie–Bohm can be found in Holland (1993) and Cushing (1994).

12.1   Historical Overview and Hidden-Variable Basics

The historical origin of hidden-variable theories lies in the famous *Bohr–Einstein debate*, an ongoing exchange of differing views on the interpretation of the wave function between *Niels Bohr* and *Albert Einstein*, in particular, the debate during the course of the fifth Solvay Conference in 1927, where Bohr refuted Einstein's objections to quantum mechanics (Bohr, 1983). Nevertheless, Einstein remained unsatisfied and his dissatisfaction with the statistical interpretation of Max Born became enshrined in his now famous dictum "He [God] does not play dice!" (Einstein, 1926). Einstein was convinced that quantum mechanics is an incomplete theory that misses what physics should represent, namely "a reality in time and space, free from spooky actions at a distance." (Einstein, 1947).

In 1935—in a by now renowned article of *Albert Einstein*, *Boris Podolsky*, and *Nathan Rosen*, in short EPR (1935)—Einstein thought to have found a proof for the *incompleteness* of quantum mechanics. Bohr's reply followed immediately (Bohr, 1935) with an opposite statement "Quantum mechanics is complete!" He did not question the logical reasoning of EPR but rather their starting point, Einstein's conception of a *physical reality*. However, the dispute Einstein contra Bohr was considered as rather philosophical; the quantum-mechanical formalism worked perfectly well anyhow, so why bother? And the debate disappeared for the following decades.

It took nearly thirty years for John Stewart Bell to stir up the debate with his celebrated papers "On the Einstein–Podolsky–Rosen paradox" (Bell, 1964) and "On the problem of hidden variables in quantum mechanics" (Bell, 1966). These papers eventually caused a dramatic change since Bell was able to move the purely philosophical debate of Einstein and Bohr into the realm of experimentally testable science. With the help of Bell's inequalities we are now able to distinguish between hidden-variable theories and quantum mechanics, in theory and in experiment.

Hidden-variable basics: Hidden-variable theories as well as quantum mechanics describe ensembles of individual systems. Whereas the orthodox (Copenhagen) doctrine in quantum mechanics tells us that measured properties, e.g., the spin of a particle along some direction, have no definite values before measurement, hidden-variable theories postulate that the properties of individual systems do *have* pre-existing values, which are revealed by the act of measurement.

What are the features of a hidden-variable theory quite generally? Let us consider an ensemble of individual systems, each of which is prepared in a state $| \psi \rangle$, which is described by a set of observables

$$A, B, C, \ldots \quad . \tag{12.1}$$

A hidden-variable theory (HVT) assigns to each individual system a set of values corresponding to the observables of eqn (12.1), i.e., one of the eigenvalues

$$v(A), v(B), v(C), \ldots \quad , \tag{12.2}$$

of the corresponding operator such that a measurement of observable A in an individual system gives the numerical value $v(A)$.

The hidden-variable theory now provides rules for how the values (12.2) should be distributed over all individual systems of the ensemble that is given by the state $| \psi \rangle$. Of course, these rules must be such that the statistical distribution of the results agrees with quantum mechanics. The states, specified by the quantum-mechanical state vector *and* by an additional hidden variable which determines individual measurement outcomes as in classical statistical mechanics, are called *dispersion free*.

If, in particular, a functional relation such as

$$f(A, B, C, \ldots) = 0 \tag{12.3}$$

is satisfied by a set of mutually commuting observables $A, B, C, \ldots$ then the same relation must hold for the values in the individual systems

$$f\big(v(A), v(B), v(C), \ldots\big) = 0 . \tag{12.4}$$

Amazingly, just by relying on conditions (12.3) and (12.4) one can construct so-called *no-go theorems* that allow us to rule out certain hidden-variable models.

12.2 Von Neumann and Additivity of Measurement Values

12.2.1 Von Neumann's Assumption

John Bell started his investigation "On the problem of hidden variables in quantum mechanics" by criticizing von Neumann. Already in 1932 *John von Neumann* had formulated a proof that dispersion-free states, and thus hidden variables, are incompatible with quantum mechanics. Due to von Neumann's significant reputation in the physics community his proof was widely accepted. Nevertheless, Bell examined von Neumann's proof carefully and critically. The essential point was the following.

Let us consider three mutually commuting operators, A, B, and C with a functional relation as in eqn (12.3), explicitly, let us suppose the operators satisfy the condition

$$C = A + B .\tag{12.5}$$

Then it follows that the corresponding attached values obtained from measurements of the individual systems must also satisfy

$$v(C) = v(A) + v(B) ,\tag{12.6}$$

since the operators are supposed to commute.

Now, von Neumann's assumption was to impose condition (12.6) on a hidden-variable theory also for *non-commuting* operators. Interestingly, the mathematician and philosopher *Grete Hermann* had raised this objection to von Neumann's assumption already in 1935 (Hermann, 1935a; Hermann, 1935b) but she was ignored. Also *Simon Kochen* and *Ernst Specker* (Conway and Kochen, 2002), when reading von Neumann's proof in 1961, had their doubts about the additivity (12.6) for non-commuting operators. And finally John Bell simply claimed[1], "This is false!"

Why? Consider the following example. Suppose we consider the measurement of the magnetic dipole moment of a spin-$\frac{1}{2}$ particle. A measurement corresponding to the operator σ_x then requires a suitably oriented Stern–Gerlach apparatus (see Section 8.2.1. The measurement for σ_y demands a different orientation, and again a different one for $(\sigma_x + \sigma_y)$. The operators do not commute, they cannot be measured simultaneously, thus there is no reason for imposing the additivity relation (12.6). Perhaps von Neumann's erroneous assumption was based on the quantum-mechanical expectation values,

$$\langle \psi | A + B | \psi \rangle = \langle \psi | A | \psi \rangle + \langle \psi | B | \psi \rangle ,\tag{12.7}$$

that is, additivity holds for the averages, irrespective of whether A and B commute or not.

Bell then proceeded to construct a simple two-dimensional example demonstrating that von Neumann's additivity assumption (12.6) is not reasonable for non-commuting observables.

---

[1]Private communication to R. A. Bertlmann.

12.2.2 Bell's Two-Dimensional Hidden-Variable Model

To begin, recall that a spin measurement along some direction $\vec{n}$ is represented by the observable $\vec{\sigma} \cdot \vec{n}$ whose eigenstates $|\pm\vec{n}\rangle$ satisfy

$$\vec{\sigma} \cdot \vec{n} \, |\pm\vec{n}\rangle \;=\; \pm\,|\pm\vec{n}\rangle \;. \tag{12.8}$$

At the same time, we note that arbitrary observables A and B on a two-dimensional Hilbert space can be written in terms of the Pauli matrices as

$$A \;=\; a_0 \,\mathbb{1} + \vec{a} \cdot \vec{\sigma}\,, \qquad B \;=\; b_0 \,\mathbb{1} + \vec{b} \cdot \vec{\sigma}\,, \tag{12.9}$$

where the a_i are real constants, because such operators must be Hermitian. For instance for A, the choices of a_0 and a_z allow us to obtain arbitrary real diagonal elements $\langle 0|A|0\rangle = a_0 + a_z$ and $\langle 1|A|1\rangle = a_0 - a_z$, while the most general off-diagonal elements can be obtained from combinations of a_x and a_y, i.e., $\langle 0|A|1\rangle = (\langle 1|A|0\rangle)^* = a_x - ia_y$. The operators A and B commute if and only if $\vec{a}$ and $\vec{b}$ are parallel or antiparallel. The allowed values for an individual system are the eigenvalues of A and B given by

$$v(A) \;=\; a_0 \pm |\vec{a}|\,, \qquad v(B) \;=\; b_0 \pm |\vec{b}|\,. \tag{12.10}$$

Therefore, if the observables commute—let us first choose the vectors $\vec{a}$ and $\vec{b}$ to be parallel—we obtain

$$
\begin{aligned}
v(A+B) &= v\big((a_0 + b_0)\mathbb{1} + (\vec{a}+\vec{b})\cdot\vec{\sigma}\big) \\
&= a_0 + b_0 \pm |\vec{a}+\vec{b}| \\
&= v(A) + v(B)\,.
\end{aligned} \tag{12.11}
$$

The corresponding attached values are additive since $|\vec{a}+\vec{b}| = |\vec{a}|+|\vec{b}|$. If $\vec{a}$ and $\vec{b}$ are antiparallel we have $|\vec{a}+\vec{b}| = |\vec{a}|-|\vec{b}|$ for $|\vec{a}| > |\vec{b}|$, which leads to the same additivity result (12.11) since in this case $v(B) = b_0 \mp |\vec{b}|$.

If, however, the observables are non-commuting—that is, if the vectors are not (anti)parallel—then we have $|\vec{a}+\vec{b}| \neq |\vec{a}|+|\vec{b}|$ and consequently

$$v(A+B) \;\neq\; v(A) + v(B)\,. \tag{12.12}$$

Bell went on to show the invalidity of von Neumann's conclusion by formulating a hidden-variable model that reproduces the quantum-mechanical predictions in this system. Consider the expectation value, for such a mean value we can distribute the single values in the following way

$$
\begin{aligned}
v(A) &= a_0 + |\vec{a}| \quad &&\text{for} \quad &(\vec{\lambda}+\vec{n})\cdot\vec{a} > 0 \\
v(A) &= a_0 - |\vec{a}| \quad &&\text{for} \quad &(\vec{\lambda}+\vec{n})\cdot\vec{a} < 0\,,
\end{aligned} \tag{12.13}
$$

where $\vec{\lambda}$ denotes some random unit vector, the hidden variable. Then the mean value of (12.13) over a uniform distribution of directions $\vec{\lambda}$ provides the quantum-mechanical result as required

$$\bar{v}(A) \; = \; \frac{1}{4\pi} \int d\Omega(\vec{\lambda}) \, v(A) \; = \; a_0 + \vec{a} \cdot \vec{n} \; = \; \langle +\vec{n} | \, A \, | +\vec{n} \rangle \,, \qquad (12.14)$$

and, of course, with the additivity property

$$\langle +\vec{n} | \, A + B \, | +\vec{n} \rangle \; = \; \langle +\vec{n} | \, A \, | +\vec{n} \rangle + \langle +\vec{n} | \, B \, | +\vec{n} \rangle \,. \qquad (12.15)$$

This is a simple two-dimensional hidden-variable model for quantum mechanics that demonstrates that von Neumann's assumption in eqn (12.6) is not justified.

12.3   Contextuality

Let us now consider a theorem of high importance for the mathematical formalism of quantum mechanics: *Gleason's theorem*. It can be viewed as a derivation of the Born rule from fundamental assumptions about probabilities for obtaining certain measurement outcomes. According to the Born rule, the probability p_i for obtaining a measurement outcome i associated with a projector P_i for a system described by the density operator ρ is given by the trace of $P_i\rho$, i.e., $p_i = \mathrm{Tr}(P_i\rho)$. Since we are considering a projector, we can think of a measurement corresponding to P_i as an elementary yes–no proposition, i.e., restricting the measured outcomes to the eigenvalues 0 or 1 means that the system is checked for having a certain property. For instance, a measurement corresponding to the projector $| \uparrow \rangle\langle \uparrow |$ can be viewed as asking if the system has the property of spin up, i.e., the z component of its spin being oriented parallel to the z axis, or not. Here it is important to note that the assignment of p_i only depends on the system state ρ and the mathematical representation P_i of the measured property, not on the larger *context* of the measurement. In particular, if P_i is a rank-one projector such that $P_i = | \psi_i \rangle\langle \psi_i |$, then it does not matter for the assignment of p_i which orthonormal basis the vector $| \psi_i \rangle$ is part of.

A theorem first established by *Andrew M. Gleason* in 1957 provides a converse statement to the Born rule in the sense that it shows that all functions f that associate probabilities to outcomes of measurements of quantities represented by projectors P_i, such that $0 \leq f(P_i) \leq 1$ for all i and $\sum_i f(P_i) = 1$ if $\sum_i P_i = \mathbb{1}$, must arise from applying the Born rule to the product of the respective projector and some density operator, i.e., there exists a ρ such that $f(P_i) = \mathrm{Tr}(P_i\rho)$. Here, as before, it is assumed that the assignment of a probability $f(P_i)$ to the projector P_i depends only on the specific P_i, and not on any potential context. Leaving the proof to be looked up by the interested reader in Gleason (1957), let us here state the theorem as follows.

Theorem 12.1 (Gleason's theorem)

In a Hilbert space $\mathcal{H}$ of $\dim(\mathcal{H}) > 2$ the only probability measures of the state associated with a linear subspace $\mathcal{V}$ of the Hilbert space are of the form $\mathrm{Tr}\big[P(\mathcal{V})\rho\big]$, where $P(\mathcal{V})$ is the projection operator onto $\mathcal{V}$ and ρ the density matrix of the system.

A property of Theorem 12.1 is certainly that the probability associated to the sum of commuting projections matches the sum of the probabilities associated with those projections individually. However, $\mathrm{Tr}[P(\mathcal{V})\rho]$ cannot have values restricted just to 0 or 1 for all projection operators $P(\mathcal{V})$, and consequently the state represented by ρ cannot be dispersion free for each observable.

Bell, on the other hand, established his own corollary which is more directed to hidden-variable theories.

Corollary 12.1 (Bell's corollary)

Consider a state space $\mathcal{V}$. If $\dim(\mathcal{V}) > 2$ then the additivity requirement for expectation values of commuting operators cannot be met for dispersion-free states.

Corollary 12.1 states that for $\dim(\mathcal{V}) > 2$ it is in general impossible to assign a definite value for each observable in each individual quantum system. Note that this is not in conflict with Bell's hidden-variable model (see Section 12.2), which has only two dimensions.

Thus Bell pointed to another class of hidden-variable models, where the results may depend on different settings of the apparatus—the context. Such models are called *contextual* and may agree with quantum mechanics. Corollary 12.1, on the other hand, states that all *non-contextual* hidden-variable theories are in conflict with quantum mechanics (for Hilbert-space dimensions larger than two). Hence the essential feature for the difference between hidden-variable theories and quantum mechanics is *contextuality*.

In 1967, *Simon Kochen* and *Ernst Specker* published their famous paper on "The problem of hidden variables in quantum mechanics", where they established their no-go theorem that non-contextual hidden-variable theories are *incompatible* with quantum mechanics (Kochen and Specker, 1967). But before we discuss this topic we want to consider an even simpler statement by *Asher Peres* about the incompatibility of certain premises with quantum mechanics (Peres, 1990).

12.4   Statements Incompatible with Quantum Mechanics

As *Asher Peres* discovered in 1990, there are "Incompatible results of quantum measurements". To make this more precise, let us formulate the following assumptions:

(i) **Measurement outcomes for the measurement of an observable are independent of the choice of other types of measurements.**

   The results of the measurement of an observable corresponding to an operator A depend only on A and on the state of the quantum system being measured. The total state is specified by the quantum-mechanical state (the density operator) and by any hidden variable one may invent.

(ii) **Measurement outcomes for the measurement of an observable corresponding to a tensor product of operators match the product of the individual measurement results.**

In particular, if another operator B commutes with A, the result of a measurement of the observable corresponding to their product AB is equal to the product of the results of the separate measurements of A and B.

Corollary 12.2 (Peres' corollary)

The two assumptions (i) and (ii) are incompatible with quantum mechanics.

Proof   To prove Corollary 12.2 we consider two spin-$\frac{1}{2}$ particles in a spin singlet state, specifically, the antisymmetric singlet state from eqn (8.98) in Section 8.3.3 with spin total spin quantum number $s = 0$ and magnetic spin quantum number $m_s = 0$,

$$| s = 0, m = 0 \rangle = \tfrac{1}{\sqrt{2}} \left(| \uparrow \rangle \otimes | \downarrow \rangle - | \downarrow \rangle \otimes | \uparrow \rangle \right) . \tag{12.16}$$

We further assume that two experimenters, Alice and Bob, each control one of the two particles and independently perform measurements on them at two distant locations. If only Alice measures the spin of her particle, say along the x direction, and Bob remains inactive, the corresponding observable is $\sigma_x^A \otimes \mathbb{1}_x^B$ and the result x_A will be

$$\sigma_x^A \otimes \mathbb{1}_x^B \quad \longrightarrow \quad x_A = \pm 1 . \tag{12.17}$$

Analogously, if only Bob measures along the x direction and Alice remains inactive, his outcome x_B will be

$$\mathbb{1}_x^A \otimes \sigma_x^B \quad \longrightarrow \quad x_B = \pm 1 . \tag{12.18}$$

The same holds for measurements along the y and z directions with the results $y_{A,B} = \pm 1$ and $z_{A,B} = \pm 1$, respectively. If both measure simultaneously, the result for the joint measurement of the two spins in the spin singlet state is

$$\langle \sigma_x^A \otimes \sigma_x^B \rangle = \langle s = 0, m = 0 | \sigma_x^A \otimes \sigma_x^B | s = 0, m = 0 \rangle = -1 . \tag{12.19}$$

According to assumption (ii) the result of the joint measurement is the product of the single measurement results, i.e., for any measurement direction we have

$$\begin{aligned} x_A \cdot x_B &= -1 , \\ y_A \cdot y_B &= -1 , \\ z_A \cdot z_B &= -1 . \end{aligned} \tag{12.20}$$

Consider now a joint measurement in different directions. The result of a separate measurement of Alice along x direction is $x_A = \pm 1$ and of Bob along y direction is $y_B = \pm 1$. According to assumption (i) the result x_A of Alice is independent of whether Bob measures along the x direction or along the y direction. For the joint measurement, we therefore get the result

$$\sigma_x^A \otimes \sigma_y^B \quad \longrightarrow \quad x_A \cdot y_B \,. \tag{12.21}$$

We notice that the commutator

$$\left[\sigma_x^A \otimes \sigma_y^B \,,\, \sigma_y^A \otimes \sigma_x^B \right] = 0 \tag{12.22}$$

vanishes, so that we can measure both operators $\sigma_x^A \otimes \sigma_y^B$ and $\sigma_y^A \otimes \sigma_x^B$ without mutual disturbance. Of course, we cannot determine the results for the operator $\sigma_y^A \otimes \sigma_x^B$ together with $\sigma_x^A \otimes \mathbb{1}_x^B$ or $\mathbb{1}_y^A \otimes \sigma_y^B$ separately, without disturbance, since the former does not commute with either of the latter two. By the same argument the result of the joint measurement with exchanged choices of directions is

$$\sigma_y^A \otimes \sigma_x^B \quad \longrightarrow \quad y_A \cdot x_B \,. \tag{12.23}$$

Next we study the product of such operators, that is,

$$\sigma_x^A \otimes \sigma_y^B \cdot \sigma_y^A \otimes \sigma_x^B = \sigma_z^A \otimes \sigma_z^B \,, \tag{12.24}$$

for which the expectation value in the spin singlet state (12.16) is

$$\langle \sigma_z^A \otimes \sigma_z^B \rangle = -1 \,. \tag{12.25}$$

But then we get

$$x_A \cdot y_B \cdot y_A \cdot x_B = \langle \sigma_x^A \otimes \sigma_y^B \cdot \sigma_y^A \otimes \sigma_x^B \rangle = -1 \,, \tag{12.26}$$

which is in contradiction to the products

$$x_A \cdot x_B = y_A \cdot y_B = -1 \,, \tag{12.27}$$

that we had before in eqn (12.20). □

Thus we have to conclude that, within any hidden-variable model, the prediction of an individual measurement result for an observable corresponding to an operator A does not only depend on A and on the state of the system but also on the choice of other measurements that may possibly be performed. We therefore again recover the notion of *contextuality* that we have discussed before.

We also remark here that the dependence on the quantum state is not essential as we will see in the next section.

12.5   The Kochen–Specker Theorem

Simon Kochen and *Ernst Specker* had also read the proof about the non-existence of hidden variables in von Neumann's book (von Neumann, 1932) and were dissatisfied with its assumption, namely with the additivity of the values for non-commuting operators, eqn (12.6). They constructed a counterexample by considering a spin-1 system.

12.5.1   Kochen–Specker Theorem for Spin-1 System

Let us consider a spin-1 system. From Section 8.3.1, specifically from eqn (8.68c), we recall that this means that we have a spin quantum number $s = 1$ corresponding to the eigenvalue $\hbar^2 s(s+1)$ of the square $\vec{S}^2 = S_x^2 + S_y^2 + S_z^2$ of a spin operator $\vec{S} = (S_x, S_y, S_z)^\top$. We can thus write

$$S_x^2 + S_y^2 + S_z^2 = s(s+1) = 2, \tag{12.28}$$

where we have switched to units where $\hbar = 1$ for simplicity, which we will use in the following. We have also dropped the symbol for the identity operator in the expressions in the middle and on the right-hand side, keeping in mind that we are considering an operator on a three-dimensional Hilbert space spanned by three associated spin eigenstates with magnetic quantum numbers $m_s = -1, 0, +1$ for the component of $\vec{S}$ along any fixed direction, usually considered to be the z direction. Consequently, the eigenvalues of the squared components S_i^2 $(i = x, y, z)$, which are the observables we are considering now, are m_s^2 and evaluate to 0 or 1. Moreover, the commutation relations for the S_i from eqn (8.64) imply that, although the components themselves do not commute, their squares do, $[S_i^2, S_j^2] = 0$ for all i and j. Therefore, it is possible to measure S_x^2, S_y^2, and S_z^2 simultaneously and the three resulting measurement outcomes must give two results of 1 and one result of 0, in some order, to satisfy eqn (12.28).

We have previously formulated our considerations for a coordinate system based on the x, y, and z directions, but all of the arguments are true for any orthogonal triad of directions. In particular, denoting the spin operator along any given direction $\vec{d}$ by $S_d = \vec{d} \cdot \vec{S}$, we have $[S_d^2, S_{d'}^2] = 0$, for $\vec{d} \perp \vec{d}'$, all squared operators commute if the directions are orthogonal. If we now choose a set of directions given by different orthogonal triads, the corresponding set of observables consist of the squared spin components obeying relation (12.28). Since these observables mutually commute, the same relation (12.28) must hold for the measurement outcomes (0 or 1) of the individual systems. Therefore two of the values must be 1 and the third one 0.

Based on these observations, Kochen and Specker provided the following no-go theorem, proving that non-contextual theories are incompatible with quantum mechanics.

Theorem 12.2 (Kochen–Specker theorem for spin-1 system)

For a spin-1 system there is a set of directions for which it is impossible to assign the values 0 and 1 to the directions in a way that is consistent with the constraint $S_x^2 + S_y^2 + S_z^2 = s(s+1) = 2$ for every orthogonal triad.

Proof   For the proof of the Kochen–Specker theorem we will proceed in two steps. First, we will present an argument by *Robert Clifton* (1993), which is based on (and useful for understanding) the construction previously used by Kochen and Specker.

The proof by Clifton is simpler but relies on a statistical argument, whereas the original proof by Kochen and Specker (which we will present thereafter following Kochen and Specker, 1967; Conway and Kochen, 2002) is more elaborate but arrives at a logical contradiction without invoking probabilities.

To begin, let us recall the constraints of a spin-1 system. If D is any set of directions then a prediction function for the set $\{S_d^2 \mid d \in D\}$ is provided by a function $v : D \to \{0, 1\}$ such that

(i) For any orthogonal frame (x, y, z) the function v takes the value 0 exactly once.

(ii) For any orthogonal pair (d, d') the function v takes the value 0 at most once.

The proof can then be formulated in a compact and illustrative way by representing these conditions graphically. In order to do this, we simply set up the following rules: Every set D of directions is represented as a graph that consists of vertices and edges connecting (some of) the vertices. For a chosen direction the corresponding squared spin component that is measured is represented by a vertex. Vertices are coloured blue • if the assigned value is 1 and red • if it is zero. Vertices corresponding to orthogonal directions are connected by an edge. The constraints (i) and (ii) then translate to the following conditions:

(i) Two vertices connected by an edge can be blue. If one vertex on an edge is red the other one must be blue.

(ii) In every triangle formed by three edges, two vertices must be blue and one vertex must be red.

Now let us consider a set of eight directions $D = \{\vec{a}_i\}_{i=0,\dots,7}$ given by the unit vectors

$$\vec{a}_0 = \frac{1}{\sqrt{3}}\begin{pmatrix} 1 \\ 1 \\ 1 \end{pmatrix}, \quad \vec{a}_1 = \frac{1}{\sqrt{2}}\begin{pmatrix} 0 \\ 1 \\ -1 \end{pmatrix}, \quad \vec{a}_2 = \frac{1}{\sqrt{2}}\begin{pmatrix} 1 \\ -1 \\ 0 \end{pmatrix}, \quad \vec{a}_3 = \frac{1}{\sqrt{2}}\begin{pmatrix} 0 \\ 1 \\ 1 \end{pmatrix}, \quad \cdot$$

$$\vec{a}_4 = \frac{1}{\sqrt{2}}\begin{pmatrix} 1 \\ 1 \\ 0 \end{pmatrix}, \quad \vec{a}_5 = \begin{pmatrix} 1 \\ 0 \\ 0 \end{pmatrix}, \quad \vec{a}_6 = \begin{pmatrix} 0 \\ 0 \\ 1 \end{pmatrix}, \quad \vec{a}_7 = \frac{1}{\sqrt{3}}\begin{pmatrix} 1 \\ -1 \\ 1 \end{pmatrix}, \quad (12.29)$$

which yield the graph Γ_1 of Fig. 12.1. That is, vectors $\vec{a}_1$, $\vec{a}_3$, and $\vec{a}_5$, as well as vectors $\vec{a}_2$, $\vec{a}_4$, and $\vec{a}_6$ form orthogonal triads, and additionally $\vec{a}_0$ is orthogonal to $\vec{a}_1$ and $\vec{a}_2$, while $\vec{a}_7$ is orthogonal to $\vec{a}_3$ and $\vec{a}_4$. Finally, the two vectors $\vec{a}_5$ and $\vec{a}_6$ are orthogonal to each other.

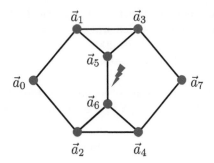

Fig. 12.1 Graph Γ_1 of eight spin directions. The directions given by eqn (12.29) are chosen by Clifton (1993) to arrive at a contradiction.

The directions given by $\vec{a}_0$ and $\vec{a}_7$ are not orthogonal, the overlap of the corresponding unit vectors evaluates to

$$\left| \frac{\vec{a}_0}{|\vec{a}_0|} \cdot \frac{\vec{a}_7}{|\vec{a}_7|} \right|^2 = \frac{1}{9} \,, \tag{12.30}$$

which means that there are quantum states, e.g., the eigenstate with eigenvalue 0 of the spin along the direction $\vec{a}_0$, for which there is a non-zero probability that the measurement of both squared spin components gives the result 0. The quantum-mechanical prediction is thus that if many measurements of the spin along the direction $\vec{a}_7$ are carried out on the individual particles of an ensemble prepared in the eigenstate with eigenvalue 0 for the direction $\vec{a}_0$, a statistically significant fraction of them will also give the result 0.

On the other hand, if we assume

$$v(\vec{a}_0) = v(\vec{a}_7) = 0 \,, \tag{12.31}$$

which means that both vertices are to be coloured in red, then, according to rule (i) we must colour the adjacent vertices in blue,

$$v(\vec{a}_1) = v(\vec{a}_2) = v(\vec{a}_3) = v(\vec{a}_4) = 1 \,, \tag{12.32}$$

in which case rule (ii) implies that the remaining vertices $\vec{a}_5$ and $\vec{a}_6$ should be coloured red,

$$v(\vec{a}_5) = v(\vec{a}_6) = 0 \,. \tag{12.33}$$

But this is in contradiction to rule (i).

With this construction we have thus obtained a contradiction between the probabilities assigned to the outcomes of observables of a hidden-variable theory and of quantum mechanics. However, the original argument by Kochen and Specker, although somewhat more complicated, could do "better" in the sense that they could show that

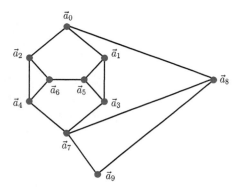

Fig. 12.2 Graph Γ_2 of ten spin directions. A set of eight directions corresponding to the geometry of Γ_1 in Fig. 12.1 is supplemented by two additional directions, $\vec{a}_8$ and $\vec{a}_9$, which exist if the angle between $\vec{a}_0$ and $\vec{a}_9$ is smaller than $\pi/10$. The colouring is chosen consistently with the rules (i) and (ii), under the assumption that $\vec{a}_0$ has been assigned the colour red. To avoid a contradiction, this means, in particular, that $\vec{a}_7$ must be blue and $\vec{a}_9$ must be red.

there is a set of 117 directions represented by the graph Γ_3 in Fig. 12.3 for which one directly arrives at logical contradiction without invoking measurement statistics.

To construct the graph Γ_3, let us first consider another graph, Γ_2 shown in Fig. 12.2, which is a simple extension of the graph Γ_1 in Fig. 12.1, obtained by adding two directions, $\vec{a}_8$ and $\vec{a}_9$, that are orthogonal to each other. Moreover, let $\vec{a}_9$ be orthogonal to $\vec{a}_7$, and $\vec{a}_8$ be orthogonal to both $\vec{a}_0$ and $\vec{a}_7$. A possible choice of such directions is

$$\vec{a}_8 = \frac{1}{\sqrt{2}} \begin{pmatrix} 1 \\ 0 \\ -1 \end{pmatrix}, \quad \text{and} \quad \vec{a}_9 = \frac{1}{3} \begin{pmatrix} 2 \\ 1 \\ 2 \end{pmatrix}. \tag{12.34}$$

By construction, the vectors $\vec{a}_0$, $\vec{a}_7$, and $\vec{a}_9$ all lie in a plane orthogonal to $\vec{a}_8$, and the angle between $\vec{a}_0$ and $\vec{a}_9$ is

$$\arccos(\vec{a}_9 \cdot \vec{a}_9) = \frac{5}{3\sqrt{3}} < \tfrac{\pi}{10}, \tag{12.35}$$

but it can also be shown (see, e.g., Kochen and Specker, 1967 or Redhead, 1987, p. 126) that there exists a choice of directions $\{\vec{a}_i\}_{i=0,\dots,9}$ satisfying the orthogonality constraints of Γ_2 in Fig. 12.2 such that the angle between $\vec{a}_0$ and $\vec{a}_9$ is exactly $\tfrac{\pi}{10}$.

Now that we know that such a choice of ten directions is possible, and ignoring the colouring for now, let us embed the graph Γ_2 in a more complicated structure. To this end, we consider an orthogonal triad $(\vec{p}, \vec{q}, \vec{r})$ and without loss of generality we make the identification $\vec{r} = \vec{a}_0$ and $\vec{q} = \vec{a}_8$, as shown in the graph in Fig. 12.3 (a). Since $\vec{p}$ is also orthogonal to $\vec{q} = \vec{a}_8$, it must lie in the plane spanned by $\vec{r} = \vec{a}_0$ and $\vec{a}_9$, which means that we can rotate $\vec{a}_0$ by $\tfrac{\pi}{2}$ around the axis parallel to $\vec{a}_8$ to obtain $\vec{p}$. If we

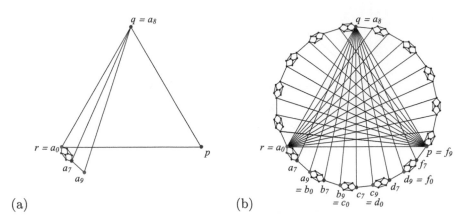

Fig. 12.3 Graph Γ_3**—117 directions of the spin.** For a set of directions represented by the graph in (b), Kochen and Specker were able to construct a logical contradiction between (non-contextual) hidden-variable models and quantum mechanics, now known as the Kochen–Specker theorem.

choose the vectors in Γ_2 such that the angle between $\vec{a}_0$ and $\vec{a}_9$ is exactly $\frac{\pi}{10}$, then this means we can connect $\vec{r} = \vec{a}_0$ and $\vec{p}$ by five interlocking copies of Γ_2, where the point corresponding to $\vec{a}_9$ of the previous copy is identified with the point corresponding to $\vec{a}_0$ of the next, but all five copies share the same vector $\vec{a}_8 = q$. With the same construction we can connect $\vec{p}$ with $\vec{q}$, as well as $\vec{q}$ and $\vec{r}$ by five interlocking graphs Γ_2 each, whose corresponding points $\vec{a}_8$ are identified with $\vec{r}$ and $\vec{p}$, respectively, as illustrated by the graph Γ_3 in Fig. 12.3 (b).

Let us now try to colour the graph Γ_3. Since the directions $\vec{p}$, $\vec{q}$, and $\vec{r}$ are mutually orthogonal, one of the three corresponding vertices must be coloured red according to rule (ii), and without loss of generality we assume it is $\vec{r}$. In order to avoid a contradiction as in Fig. 12.1, we cannot colour both $\vec{a}_0$ and $\vec{a}_7$ of any given graph Γ_2 in red, but if we choose (as we do here) $\vec{a}_0 = \vec{r}$ of the first instance of Γ_2 to be red, then following the rules (i) and (ii), we find that $\vec{a}_7$ must be blue, while $\vec{a}_9$ must be red, and similarly for the following interlocking copies of Γ_2, we must have red vertices $\vec{b}_0 = \vec{a}_9$, $\vec{c}_0 = \vec{b}_9$, $\vec{d}_0 = \vec{c}_9$, and $\vec{f}_0 = \vec{d}_9$. However, for the fifth copy, we arrive at a contradiction, since we have the identification $\vec{f}_9 = \vec{p}$, and $\vec{p}$ is already coloured blue by construction. $\qquad \square$

Thus there exists a quantum system, the spin-1 system, and a set of measurements represented by the graph Γ_3 in Fig. 12.3 (b), which cannot be described by a non-contextual hidden-variable theory.

The proof of Kochen and Specker requires a set of 117 directions, a number which is obtained by counting the vectors in the fifteen interlocking graphs Γ_2 and subtracting the directions that have been identified. Since 1967 other sets of directions with fewer vectors that cannot be consistently coloured have been found. The present records are 31 directions discovered by *John Conway* and *Simon Kochen* (Kochen, 2000; Conway and Kochen, 2002) and independently by *Sixia Yu* and *Choo-Hiap Oh* (2012). Furthermore a proof with thirty-three rays for dimension three and with only twenty-four rays for dimension four has been given by *Asher Peres* (1991).

12.5.2   Peres' Nonet for Two Qubits

When regarding higher-dimensional Hilbert spaces proofs of statements like that of Kochen and Specker turn out to be quite simple. Let us start with an example in four dimension due to *Asher Peres* (1990). There, the relevant observables are represented by Pauli matrices σ_i with $(i = x, y, z)$ from eqn (11.71), or tensor products thereof for two independent spin-$\frac{1}{2}$ particles (i.e., for two qubits). To simplify the labelling of the operators for the two qubits, we use the notation $\sigma_x \equiv X$, $\sigma_y \equiv Y$, $\sigma_z \equiv Z$ frequently used in quantum information, and we add subscripts to identify the subsystem that an operator acts upon non-trivially, i.e.,

$$
\begin{aligned}
X_1 &= \sigma_x \otimes \mathbb{1} & Y_1 &= \sigma_y \otimes \mathbb{1} & Z_1 &= \sigma_z \otimes \mathbb{1} \,, \\
X_2 &= \mathbb{1} \otimes \sigma_x & Y_2 &= \mathbb{1} \otimes \sigma_y & Z_2 &= \mathbb{1} \otimes \sigma_z \,,
\end{aligned}
\tag{12.36}
$$

which implies that products of operators with no common indices commute, e.g., in terms of the Pauli matrices the expression $X_1 Y_2$ stands for the tensor product

$$
X_1 Y_2 = \sigma_x \otimes \mathbb{1} \cdot \mathbb{1} \otimes \sigma_y = \sigma_x \otimes \sigma_y = Y_2 X_1 \,.
\tag{12.37}
$$

We then focus on a specific set of nine observables, arranged in groups of three, which form a nonet (or square), as shown in Fig. 12.4. Each observable A in Fig. 12.4 squares to the identity, that is, $A^2 = \mathbb{1}$, since $\sigma_i^2 = \mathbb{1}$. Therefore the eigenvalues of each are $A \to \pm 1$, which means that the values attached to each individual system must be

$$
v(A) = \pm 1 \,.
\tag{12.38}
$$

Moreover, whenever two observables A and B commute, a non-contextual hidden-variable model should allow us to assign the values $v(A)$, $v(B)$ and $v(AB)$ such that

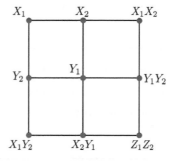

Fig. 12.4 Peres' nonet or square. Nine spin operators acting on the four-dimensional Hilbert space of a two spin-$\frac{1}{2}$ system are arranged such that the operators in each row and in each column mutually commute.

$$v(AB) = v(A) \cdot v(B) \quad \text{for} \quad [A, B] = 0. \tag{12.39}$$

Now we notice that the observables in each row and in each column commute amongst each other and are thus simultaneously measurable, but operators belonging to different rows or columns do not necessarily commute. Let us therefore consider the functional relation of the products of the observables in each row and in each column. For each of the rows we obtain the identity, i.e.,

$$\begin{aligned} X_1 \cdot X_2 \cdot X_1 X_2 &= \mathbb{1} \,, \\ Y_2 \cdot Y_1 \cdot Y_1 Y_2 &= \mathbb{1} \,, \\ X_1 Y_2 \cdot X_2 Y_1 \cdot Z_1 Z_2 &= \mathbb{1} \,. \end{aligned} \tag{12.40}$$

The product of the observables in the first and second column gives the identity as well,

$$\begin{aligned} X_1 \cdot Y_2 \cdot X_1 Y_2 &= \mathbb{1} \,, \\ X_2 \cdot Y_1 \cdot X_2 Y_1 &= \mathbb{1} \,, \end{aligned} \tag{12.41}$$

but in the third column we get the identity with a pre-factor of -1,

$$X_1 X_2 \cdot Y_1 Y_2 \cdot Z_1 Z_2 = -\mathbb{1} \,. \tag{12.42}$$

Since the observables of each row and of each column mutually commute, the relations (12.40) and (12.42) must also hold for the values of the individual systems. In particular, the products of the three values assigned to the operators in all rows and in the first two columns should give $+1$, while the product in the third column should give -1. However, focusing on the third row and third column the relations of the individual values are

$$v(X_1 Y_2) \, v(X_2 Y_1) \, v(Z_1 Z_2) = v(X_1) \, v(Y_2) \, v(X_2) \, v(Y_1) \, v(Z_1) \, v(Z_2) = 1 \,, \tag{12.43}$$

$$v(X_1 X_2) \, v(Y_1 Y_2) \, v(Z_1 Z_2) = v(X_1) \, v(Y_2) \, v(X_2) \, v(Y_1) \, v(Z_1) \, v(Z_2) = -1 \,, \tag{12.44}$$

which is *impossible* to satisfy!

We can then formulate the Peres construction for a system of two qubits as a type of Kochen–Specker theorem:

Theorem 12.3 (Kochen–Specker theorem for two qubits)

For a two-qubit system there is no way to assign definite values to each observable of Peres' nonet in a way that is consistent with the functional identities.

This means that, within any hidden-variable theory compatible with quantum mechanics the assignment of the result of the measurement of an observable, e.g., of $Z_1 Z_2 \to v(Z_1 Z_2)$, depends on the choice of other simultaneous measurements—*contextuality*.

12.5.3 Mermin's Pentagram for Three Qubits

Finally, let us consider an instructive example in eight dimensions constructed by *David Mermin* (1993). There, three independent spin-$\frac{1}{2}$ particles or three qubits are considered, and the relevant observables are again given by (tensor products of) Pauli matrices in the quantum-information notation that we have used in Section 12.5.2. In Mermin's example, the tensor products of eqn (12.36) become extended by operators on the subspace of a third qubit,

$$
\begin{aligned}
X_1 &= \sigma_x \otimes \mathbb{1} \otimes \mathbb{1} & Y_1 &= \sigma_y \otimes \mathbb{1} \otimes \mathbb{1} & Z_1 &= \sigma_z \otimes \mathbb{1} \otimes \mathbb{1}\,, \\
X_2 &= \mathbb{1} \otimes \sigma_x \otimes \mathbb{1} & Y_2 &= \mathbb{1} \otimes \sigma_y \otimes \mathbb{1} & Z_2 &= \mathbb{1} \otimes \sigma_z \otimes \mathbb{1}\,, \\
X_3 &= \mathbb{1} \otimes \mathbb{1} \otimes \sigma_x & Y_3 &= \mathbb{1} \otimes \mathbb{1} \otimes \sigma_y & Z_3 &= \mathbb{1} \otimes \mathbb{1} \otimes \sigma_z\,.
\end{aligned}
\tag{12.45}
$$

Then ten spin observables are arranged in five groups of four operators such that the graph obtained by representing each operator by a vertex and connecting vertices pertaining to operators within the same group by an edge is a five-pointed star—a pentagram, as shown in Fig. 12.5. As in the previous example of Peres, each observable A in Fig. 12.5 gives the identity when squared. Consequently, the possible eigenvalues are again ± 1, and so are the values associated to each individual system $v(A) = \pm 1$.

The four observables on each of the five rays of Mermin's pentagram mutually commute, while the product of the observables on each line except for the horizontal line gives the identity $\mathbb{1}$; while the horizontal line gives $-\mathbb{1}$.

All relations must also be satisfied by the values of the individual systems. Therefore the product of all the values associated to all five lines must be $(-1) \cdot (+1)^4 = -1$. This last condition, however, results in a *contradiction* since each of the ten observables appears twice in the product (each observable lies on the intersection of two lines), thus

$$
\left(\left(v(A) = \pm 1 \right)^2 \right)^{10} = 1\,.
\tag{12.46}
$$

Therefore, it is impossible to assign definite values to all ten observables in a way that is consistent with their functional relations.

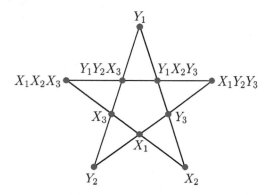

Fig. 12.5 Mermin's pentagram. Ten three-qubit observables are chosen such that five groups of four observables represented by the lines in a pentagram mutually commute.

We also want to draw attention to the horizontal line of Mermin's pentagram, which provides a direct link between the Kochen–Specker theorem and the Greenberger–Horne–Zeilinger theorem, which we will discuss in Section 18.2.

To summarize, contextuality is an essential feature of quantum mechanics and has also become an important issue in the research of quantum systems (Cabello, 2016; Cañas *et al.*, 2014; Gühne *et al.*, 2014).

Let us therefore formulate the Kochen–Specker theorem here quite generally in the following way:

Theorem 12.4 (Kochen–Specker theorem general)

In a Hilbert space $\mathcal{H}$ of $\dim(\mathcal{H}) > 2$ it is impossible to assign values to all physical observables while simultaneously preserving the functional relations between them.

Résumé: Hidden-variable theories describe an ensemble of individual systems. A hidden-variable theory postulates that the properties of the individual systems do have pre-existing values, revealed by the act of measurement, which are distributed over all individual systems of the ensemble according to a given rule.

Gleason and afterwards Bell proved that in a Hilbert space $\mathcal{H}$ with $\dim(\mathcal{H}) > 2$ it is in general impossible to assign a definite value for each observable in each individual (dispersion-free) system.

Hidden-variable models, where the results may depend on different possible settings of the apparatus, are called contextual. According to the proofs by Gleason and Bell, hidden-variable theories are in conflict with quantum mechanics.

This leads to the celebrated Kochen–Specker theorem which states that non-contextual hidden-variable theories are *incompatible* with quantum mechanics (for a Hilbert space with $\dim(\mathcal{H}) > 2$).

The essential difference between hidden-variable theories and quantum mechanics hence is contextuality!

13
Bell Inequalities

As we have seen in Chapter 12, hidden-variable models, which attempt to postulate rules that deterministically assign measurement outcomes to measurements of quantum-mechanical systems, must be constrained to be *contextual* in order to be compatible with quantum mechanics. In a paper published in 1964 *John Bell* demonstrated that there exist empirical tests in the form of what is nowadays called a *Bell inequality*, which allow us to experimentally distinguish between the predictions of certain types of hidden-variable models and quantum mechanics. But to discuss such Bell inequalities, which we will do in this chapter, we should start with the paradox formulated by *Albert Einstein*, *Boris Podolsky*, and *Nathan Rosen* (1935) that attempted to show that quantum mechanics is incomplete.

13.1 The EPR Paradox

13.1.1 The EPR Criteria

Einstein was convinced that quantum mechanics, the theory to which he contributed so much, was not the final theory of the microscopic world (see, e.g., his letters to Max Born: Einstein, 1952). The element that he seems to have missed in the theory was that of physical reality. In the article "Can Quantum-Mechanical Description of Physical Reality Be Considered Complete?" Einstein, Podolsky, and Rosen (1935), or EPR for short, constructed a gedanken experiment in order to demonstrate that quantum mechanics is—according to their definition—an incomplete theory. Let us go through their argumentation in detail. For *completeness* of a theory EPR define:

> *Every element of the physical reality must have a counterpart in the theory.*

A main concern of the EPR article is the concept of physical reality, which can be applied to both the macroscopic and to the microscopic world. EPR thus formulate their premises with great care and extreme generality:

1. **Criterion for physical reality.**
   "If, without in any way disturbing a system, we can predict with certainty (i.e., with probability equal to one) the value of a physical quantity, then there exists an element of physical reality corresponding to this physical quantity."

2. **There is local causality (no action at a distance).**

3. **The quantum-mechanical predictions are correct**
   when we consider a certain system consisting of two spatially separated particles.

13.1.2 The EPR Paradox—Aharonov–Bohm Scenario

In their article, EPR considered a system of two particles and examined their complementary quantities position and momentum. In this respect we do not follow their original work here. In view of the Bell inequalities and the following experiments, which we present in Section 13.2, we discuss the spin version of the EPR scenario advocated by *Yakir Aharonov* and *David Bohm* (Bohm, 1952; Bohm and Aharonov, 1957).

There, one considers a system of two spin-$\frac{1}{2}$ particles that we will label particles 1 and 2, respectively, which are prepared in the antisymmetric spin singlet state from eqn (8.98) with spin total spin quantum number $s = 0$ and magnetic spin quantum number $m_s = 0$, which coincides with the antisymmetric Bell state from eqn (11.106), i.e., $|\, s = 0, m = 0 \,\rangle \equiv |\, \psi^- \,\rangle$. The particles move freely in opposite directions and measurements are performed along various spin directions by two Stern–Gerlach magnets on the left-hand side and on the right-hand side, as illustrated in Fig. 13.1. We also assume—and this was most reasonable for EPR—that if the two measurements are made at two distant locations the orientation of one magnet does not influence the result obtained with the other.

Quantum-mechanically, the spin of such a system, which will be the relevant degree of freedom for our considerations here, is described by the state vector (recall Section 8.3.2)

$$|\, \psi^- \,\rangle \;=\; \frac{1}{\sqrt{2}} \big(|\, +\vec{a} \,\rangle_1 \otimes |\, -\vec{a} \,\rangle_2 \,-\, |\, -\vec{a} \,\rangle_1 \otimes |\, +\vec{a} \,\rangle_2 \big) \,, \qquad (13.1)$$

where the vectors $|\, \pm\vec{a} \,\rangle_1$ and $|\, \pm\vec{a} \,\rangle_2$ correspond to the states of the particles 1 and 2, respectively, having spin "up" ($+$) or "down" ($-$) along a given direction $\vec{a}$. As we have already seen in eqn (11.81), the latter states are the eigenvectors of the operators $\vec{a} \cdot \vec{\sigma} = a_x \sigma_x + a_y \sigma_y + a_z \sigma_z$ with eigenvalues ± 1, i.e.,

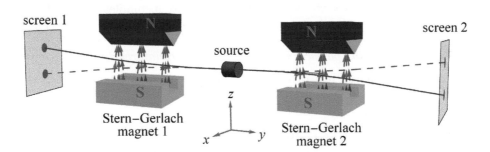

Fig. 13.1 EPR–Aharonov–Bohm setup. A system of two spin-$\frac{1}{2}$ particles is prepared in a singlet state $|\, \psi^- \,\rangle$. One particle is propagating to the left-hand side while the other is propagating to the right-hand side. The spins of both particles are measured by two Stern–Gerlach magnets oriented along the same direction $\vec{a}$. In this setting, when particle 1 is detected on screen 1 in the location corresponding to spin "up", particle 2 will be found in the location on screen 2 corresponding to spin "down", and vice versa.

$$\vec{a} \cdot \vec{\sigma} \, | \pm \vec{a} \rangle_{1,2} = \pm \, | \pm \vec{a} \rangle_{1,2} \, . \tag{13.2}$$

The fact that the spin singlet can be written in this form with respect to an arbitrary direction $\vec{a}$ and not just for the z direction is a direct consequence of the fact that the state has a total spin quantum number $s = 0$, and is hence invariant under rotations. Any such three-dimensional rotation $R(\vec{\varphi}) \in$ SO(3) is represented on the space of the pair of two-component spinors as the tensor product of two unitaries $U(\vec{\varphi}) \otimes U(\vec{\varphi})$ with $U(\vec{\varphi}) \in$ SU(2) as in Lemma 8.1. And since we can always find a rotation $R(\vec{\varphi})$ that rotates the z axis to a desired direction $\vec{a}$ such that $U(\vec{\varphi})| \uparrow \rangle = | + \vec{a} \rangle$ and $U(\vec{\varphi})| \downarrow \rangle = | - \vec{a} \rangle$ the equivalence of the states $| \psi^- \rangle$ in eqns (13.1) and (11.106) follows.

Suppose now we measure the spin of particle 1 along the z direction. The outcome is not predetermined by the singlet state $| \psi^- \rangle$ and on average we get the result "up" $(+)$ in 50% of the cases and "down" $(-)$ in the other 50% of cases. Nevertheless the state $| \psi^- \rangle$ allows us to predict:

If particle 1 gives the result "up" $(+)$ then particle 2 will give the result "down" $(-)$ with certainty—provided that particle 2 is also measured along the z direction.

The prediction is made without disturbing particle 2 in any way because the particles can be assumed to be arbitrarily far apart. Therefore, to avoid invoking any *"action at a distance"*, EPR have to attribute an *"element of physical reality"* to the spin of particle 2, which predetermines the outcome of the measurement. Such predetermined values (here spin "up" or "down")—such real properties—fixed in advance of observation, are not included in the quantum-state description $| \psi^- \rangle$ of eqn (13.1), and are not contained in the quantum-mechanical formalism. According to EPR one is therefore forced to conclude:

The quantum-mechanical formalism is incomplete!

Moreover, if one now considers a measurement of the spin of particle 1 along the x direction (or any other direction for that matter), an argument along the same lines leads one to conclude that the spin of particle 2 also has simultaneous reality along the x direction, in fact along any direction, in advance of observation. Vice versa, reality also must be attributed to the spin properties of particle 1.

This certainly is not the case in the quantum-mechanical formalism. Not only does the particular quantum state considered not assign an appropriate counterpart to the element of physical reality corresponding to the spin component of either particle along any fixed direction, the non-commutation of the spin operators for different directions means that *no* quantum state can simultaneously reflect the elements of reality corresponding to these different spin components. Thus an extended theory—a deterministic hidden-variable theory—is required if one takes the point of view of EPR. There, a quantum state would be specified more completely by a parameter (or set of parameters), the hidden variable, reflecting the predetermination à la EPR.

The idea is that the quantum-mechanical description via the spin singlet state may actually only be an approximate description for the "true" physical state which possesses definite values for the spin along x, y, and z axes, described by so-called hidden variables. Such a more complete theory should then contain all "elements of reality". Furthermore, there must be some mechanism acting on these variables to give rise to the observed effects of non-commuting quantum observables. Such a theory is called a hidden-variable theory (HVT). In the mentioned experiment an HVT is indistinguishable from quantum mechanics, and as we recall from Chapter 12 such theories must necessarily be contextual[1], but it turns out that there are more serious challenges to the idea of hidden variables. These will lead us to Bell inequalities. But before we discuss these issues we want to mention the few but important replies to the article of EPR.

13.1.3   Bohr's Reply to EPR

Niels Bohr immediately wrote an extensive reply to the EPR article, with the same title, in the same journal, in the same year (Bohr, 1935). His basic message was:

Quantum mechanics is complete!

In his article Bohr did not question the EPR reasoning once all premises were accepted. Instead he attacked the premises of EPR, heavily criticizing EPR's "criterion for physical reality" which he felt contained "an essential ambiguity as regards the meaning of the expression, 'without in any way disturbing a system'."

Bohr advocated a completely different point of view. His principle of *"complementarity"* served as a key to understand his view.

In the phenomena concerned we are not dealing with an incomplete description characterized by the arbitrary picking out of different elements of physical reality at the cost of sacrifying [sacrificing] other such elements, but with a rational discrimination between essentially different experimental arrangements and procedures ...

Bohr continued:

In fact, the renunciation in each experimental arrangement of the one or the other of two aspects of description of physical phenomena—the combination of which characterizes the method of classical physics, and which therefore in this sense may be considered as complementary to one another—depends essentially on the impossibility, in the field of quantum theory, of accurately controlling the reaction of the object on the measuring instruments.

In another passage Bohr explained explicitly the ambiguity of the expression "without in any way disturbing a system":

---

[1]Note that this was not known at the time that EPR (1935) published their paper.

There is essentially the question of an influence on the very conditions which define the possible types of predictions regarding the future behavior of the system ... this description ... may be characterized as a rational utilization of all possibilities of unambiguous interpretation of measurements, compatible with the finite and uncontrollable interaction between the objects and the measuring instruments in the field of quantum theory.

In short, Bohr argues as follows: In quantum theory there exists a finite quantum action in the sense of a finite interaction between measured object and measuring instrument. But this also implies a finite disturbance of the measured object, which is entirely unpredictable and cannot be eliminated in principle. This limits our information about the object, and we cannot attribute a physical reality to an object independently of the measurement. So for Bohr *quantum particles do not have any definite properties in advance of observation.*

Aage Petersen, an assistant of Bohr, characterized Bohr's philosophy using the following words (Petersen, 1963):

When asked whether the algorithm of quantum mechanics could be considered as somehow mirroring an underlying quantum world, Bohr would answer: "There is no quantum world. There is only an abstract quantum physical description. It is wrong to think that the task of physics is to find out how nature is. Physics concerns what we can say about nature".

What a different point of view! Einstein was convinced throughout his life that particles must have properties whether they are measured or not. Applying now Bohr's position to the spin experiment, we have to argue in the following way:

Suppose we measure the spin of particle 1 in the z direction, then the spin of particle 1 is real. Only if particle 2 is also measured will its spin be real too. The same argumentation holds for a measurement along the x direction or any other direction. Bohr argues that we cannot in principle measure the spin of particle 2 (or 1) simultaneously along the z direction and along the x direction—they are *complementary* quantities described by the *complementary operators* σ_z and σ_x—therefore the spin of particle 2 (or 1) cannot have simultaneous reality in both directions. For this reason *quantum mechanics is complete and there is no paradox.*

13.1.4 Schrödinger's Reply to EPR

In 1935 *Erwin Schrödinger* wrote his by now legendary trilogy of articles (in German) entitled *"Die gegenwärtige Situation in der Quantenmechanik"* ("The current situation in quantum mechanics", Schrödinger, 1935a), which was also published by the Cambridge Philosophical Society (Schrödinger, 1935c; Schrödinger, 1936). His papers were intended as a reply to the article of EPR. Schrödinger was the first to realize that the EPR paradox was connected to what he called *entanglement*, *Verschränkung* in German, and he described the phenomenon already in our modern terms of quantum information. For him entanglement was not just *one* but *the* characteristic trait of quantum mechanics:

When two systems, of which we know the states by their respective representatives, enter into temporary physical interaction due to known forces between them, and when after a time of mutual influence the systems separate again, then they can no longer be described in the same way as before, viz. by endowing each of them with a representative of its own. I would not call that one but rather the characteristic trait of quantum mechanics, the one that enforces its entire departure from classical lines of thought. By the interaction the two representatives have become entangled.

Schrödinger continued:

Of either system, taken separately, all previous knowledge may be entirely lost, leaving us but one privilege: to restrict the experiments to one only of the two systems. After reestablishing one representative by observation, the other one can be inferred simultaneously. ... Another way of expressing the peculiar situation is: the best possible knowledge of a whole does not necessarily include the best possible knowledge of all its parts, even though they may be entirely separated and therefore virtually capable of being "best possibly known", i.e. of possessing, each of them, a representative of its own. The lack of knowledge is by no means due to the interaction being insufficiently known—at least not in the way that it could possibly be known more completely—it is due to the interaction itself.

This is precisely how we see the EPR situation nowadays in quantum information, and finally, also Schrödinger, when referring directly to the article of EPR, had to confess that the EPR situation was rather disconcerting and indeed quite paradoxical:

Attention has recently [Einstein, Podolsky, Rosen, 1935] *been called to the obvious but very disconcerting fact that even though we restrict the disentangling measurements to one system, the representative obtained for the other system is by no means independent of the particular choice of observations which we select for that purpose and which by the way are entirely arbitrary. It is rather discomforting that the theory should allow a system to be steered or piloted into one or the other type of state at the experimenter's mercy in spite of his having no access to it. This paper* [Schrödinger, 1935c] *does not aim at a solution of the paradox, it rather adds to it, if possible.*

It is precisely this discomforting feature of the theory, that we can steer the type of a state in spite of having no access to it—the characteristics of entanglement—which is utilized in quantum information with great success. Schrödinger himself, unfortunately, did not pursue his discovery further.

It is remarkable in connection with this EPR debate that neither Einstein, nor Bohr, nor Schrödinger referred to the mathematical work of *John von Neumann* (1932) who had previously published a proof that hidden-variable theories were incompatible with quantum mechanics. In hindsight, we of course now know (see Section 12.2) that this proof relied on incorrect assumptions. Nevertheless, EPR, Bohr, and Schrödinger just relied on physical or "philosophical" arguments in their debate.

In fact, the Bohr–Einstein debate was considered by many in the physics community as being purely "philosophical" since it did not affect the calculations within the extremely successful quantum-mechanical formalism. Therefore this debate on the foundations of quantum mechanics disappeared for the next decades and the "Copenhagen interpretation", Bohr's point of view, became the dominant standard interpretation. About thirty years later, however, the situation changed dramatically when *John Bell* published his analysis "On the Einstein–Podolsky–Rosen paradox" (Bell, 1964) leading to the so-called *Bell Inequalities* or *Bell Theorem*. Bell moved this issue from purely "philosophical speculations" onto experimental grounds. As a result of Bell's theorem we are now able to *experimentally* distinguish between quantum mechanics and local realistic theories with hidden variables, and it triggered a new development in physics, namely that of quantum information, quantum communication, and quantum computing.

13.2 Bell Inequalities—Theory

13.2.1 The Setup

Bell's starting point was the argumentation of EPR, which he considered in the Aharonov–Bohm spin version (Bohm, 1951; Bohm and Aharonov, 1957) discussed already. The paradox of EPR served as an argument that quantum mechanics is an incomplete theory and that it should be supplemented by additional parameters, the hidden variables. These additional variables would (supposedly) restore causality and locality in the theory. In his analysis "On the Einstein–Podolsky–Rosen paradox" Bell (1964) discovered a surprising and profound feature of quantum mechanics. He discovered that for any hidden-variable theory that was *realistic* in the sense that a hidden variable λ determines the measurement outcomes, the requirement of *locality* is *incompatible* with the statistical predictions of quantum mechanics. He phrased the requirement that created the essential difficulty in the following way (Bell, 1964):

> *The result of a measurement on one system be unaffected by operations on a distant system with which it has interacted in the past.*

Let us go through Bell's argument step by step. We again consider a setup where a pair of spin-$\frac{1}{2}$ particles is prepared in a spin singlet state $|\psi^-\rangle$ from eqn (13.1), and the individual particles propagate freely in opposite directions—an EPR–Aharonov–Bohm setup, see Fig. 13.2.

The spin measurement on one side, customarily called "Alice" in quantum information, realized by a Stern–Gerlach magnet oriented along some direction given by a unit vector $\vec{a}$, is described by the operator $\vec{a} \cdot \vec{\sigma}$ with possible measurement outcomes ± 1. Meanwhile, a similar spin measurement along the direction $\vec{b}$ (also with $|\vec{b}| = 1$) is carried out on the other side, Bob's side, which is represented by $\vec{b} \cdot \vec{\sigma}$ with possible outcomes ± 1 as well. If Alice and Bob measure along any fixed matching direction, knowing the outcome of Alice allows us to predict with certainty and in advance the outcome for Bob, it thus seems like the result must be predetermined.

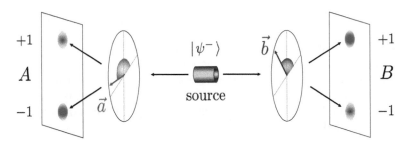

Fig. 13.2 Schematic EPR–Aharonov–Bohm setup. A pair of spin $\frac{1}{2}$ particles is prepared in a spin singlet state and propagates freely in opposite directions to the measuring stations called Alice and Bob. Alice measures the spin in direction $\vec{a}$, whereas Bob measures simultaneously in direction $\vec{b}$.

This predetermination can be cast in terms of an additional variable λ, on which the measurement outcome may depend in addition to the respective measurement directions. In such an extended theory we denote the measurement results of Alice and Bob by

$$A(\vec{a},\lambda) = \pm 1,0 \qquad \text{and} \qquad B(\vec{b},\lambda) = \pm 1,0 \,, \qquad (13.3)$$

respectively. To be more general, in particular, in expectation of the statistical argument that we are about to make, we also include the possible result 0 for imperfect measurements, i.e., what we actually require is

$$|A| \leq 1 \qquad \text{and} \qquad |B| \leq 1 \,. \qquad (13.4)$$

This permits us to take into account situations where, e.g., a measurement result is inconclusive or a particle is lost, that is, if only either Alice or Bob register a particle. Then the expectation value of the joint spin measurement of Alice and Bob can be decomposed in the following way:

Proposition 13.1 (Expectation value—Bell's locality hypothesis)
$$E(\vec{a},\vec{b}) = \int d\lambda \, \rho(\lambda) \, A(\vec{a},\lambda) \, B(\vec{b},\lambda) \,.$$

The choice of the product $A(\vec{a},\lambda) \, B(\vec{b},\lambda)$ in the expectation value (Proposition 13.1), where A does not depend on Bob's settings $\vec{b}$, and B does not depend on Alice's setting $\vec{a}$, is called *Bell's locality hypothesis*. It is in some sense an obvious definition from the point of view of a physicist or engineer, and *must not* be confused with other locality definitions, like local interactions or locality in quantum field theory.

The function $\rho(\lambda)$ represents some probability distribution for the variable λ, and is thus assumed to be normalized

$$\int d\lambda \, \rho(\lambda) = 1 \,. \qquad (13.5)$$

Crucially, it does not depend on the measurement settings $\vec{a}$ and $\vec{b}$, which can be chosen freely or randomly by the observers. This is essential!

13.2.2 The CHSH Inequality

Now, with the preliminaries of Section 13.2.1 at hand, it is quite easy to derive an inequality that is well adapted to real experiments with imperfect sources and measurement apparatuses. We start by considering the difference between the expectation values $E(\vec{a}, \vec{b})$ and $E(\vec{a}, \vec{b}')$ with the same direction $\vec{a}$ for Alice but different directions $\vec{b}$ and $\vec{b}'$ for Bob,

$$E(\vec{a}, \vec{b}) - E(\vec{a}, \vec{b}') = \int d\lambda\, \rho(\lambda)\, A(\vec{a}, \lambda)\, B(\vec{b}, \lambda) - \int d\lambda\, \rho(\lambda)\, A(\vec{a}, \lambda)\, B(\vec{b}', \lambda)$$

$$\pm \int d\lambda\, \rho(\lambda) A(\vec{a}, \lambda) B(\vec{b}, \lambda) A(\vec{a}', \lambda) B(\vec{b}', \lambda) \mp \int d\lambda\, \rho(\lambda) A(\vec{a}, \lambda) B(\vec{b}, \lambda) A(\vec{a}', \lambda) B(\vec{b}', \lambda),$$

$$= \int d\lambda\, \rho(\lambda)\, A(\vec{a}, \lambda)\, B(\vec{b}, \lambda) \big(1 \pm A(\vec{a}', \lambda)\, B(\vec{b}', \lambda) \big)$$

$$- \int d\lambda\, \rho(\lambda)\, A(\vec{a}, \lambda)\, B(\vec{b}', \lambda) \big(1 \pm A(\vec{a}', \lambda)\, B(\vec{b}, \lambda) \big), \tag{13.6}$$

where we have inserted the last two terms which add to zero in the first step. Now we can take the modulus of this expression and use the triangle inequality $|x+y| \leq |x|+|y|$, i.e.,

$$\big| E(\vec{a}, \vec{b}) - E(\vec{a}, \vec{b}') \big| \leq \left| \int d\lambda\, \rho(\lambda)\, A(\vec{a}, \lambda)\, B(\vec{b}, \lambda) \big(1 \pm A(\vec{a}', \lambda)\, B(\vec{b}', \lambda) \big) \right|$$

$$+ \left| \int d\lambda\, \rho(\lambda)\, A(\vec{a}, \lambda)\, B(\vec{b}', \lambda) \big(1 \pm A(\vec{a}', \lambda)\, B(\vec{b}, \lambda) \big) \right|. \tag{13.7}$$

Then we inspect the individual terms on the right-hand side, and note that since $|AB| \leq 1$ we have $1 \pm A(\vec{a}', \lambda)\, B(\vec{b}', \lambda) \geq 0$, and further $\rho(\lambda) \geq 0$, such that we can write

$$\left| \int d\lambda\, \rho(\lambda)\, A(\vec{a}, \lambda)\, B(\vec{b}, \lambda) \big(1 \pm A(\vec{a}', \lambda)\, B(\vec{b}', \lambda) \big) \right|$$

$$= \int d\lambda\, \rho(\lambda)\, \big| A(\vec{a}, \lambda)\, B(\vec{b}, \lambda) \big| \, \big(1 \pm A(\vec{a}', \lambda)\, B(\vec{b}', \lambda) \big)$$

$$\leq \int d\lambda\, \rho(\lambda)\, \big(1 \pm A(\vec{a}', \lambda)\, B(\vec{b}', \lambda) \big)$$

$$\leq \int d\lambda\, \rho(\lambda) \pm \int d\lambda\, \rho(\lambda)\, A(\vec{a}', \lambda)\, B(\vec{b}', \lambda) = 1 \pm E(\vec{a}', \vec{b}'). \tag{13.8}$$

Similarly, we must also have

$$\left| \int d\lambda\, \rho(\lambda)\, A(\vec{a}, \lambda)\, B(\vec{b}', \lambda) \big(1 \pm A(\vec{a}', \lambda)\, B(\vec{b}, \lambda) \big) \right| \leq 1 \pm E(\vec{a}', \vec{b}), \tag{13.9}$$

and we therefore obtain the inequality

$$\left| E(\vec{a}, \vec{b}) - E(\vec{a}, \vec{b}') \right| \leq 2 \pm \left(E(\vec{a}', \vec{b}') + E(\vec{a}', \vec{b}) \right). \tag{13.10}$$

Since the last inequality holds regardless of the sign in front of the parenthesis on the right-hand side, we can choose the sign such that the right-hand side becomes minimal, that is

$$\left| E(\vec{a}, \vec{b}) - E(\vec{a}, \vec{b}') \right| \leq 2 - \left| E(\vec{a}', \vec{b}') + E(\vec{a}', \vec{b}) \right|. \tag{13.11}$$

Moving the absolute values to the left-hand side, we thus arrive at the CHSH inequality:

Proposition 13.2 (CHSH inequality)

$$S_{\text{CHSH}} := \left| E(\vec{a}, \vec{b}) - E(\vec{a}, \vec{b}') \right| + \left| E(\vec{a}', \vec{b}') + E(\vec{a}', \vec{b}) \right| \leq 2.$$

Proposition 13.2 is the familiar *CHSH inequality*, named after *John Clauser, Michael Horne, Abner Shimony,* and *Richard Holt,* who published it in 1969. Bell presented his own, more general derivation of the CHSH inequality at the International School of Physics "Enrico Fermi" in Varenna at Lake Como in 1970 (Bell, 1971) and in his "Bertlmann's socks" paper in 1980 (Bell, 1981).

For any hidden-variable theory that satisfies the condition of *reality*, i.e., which assigns definite outcomes $A(\vec{a}, \lambda)$ and $B(\vec{b}, \lambda)$ to the measurements performed by Alice and Bob, and which does so while respecting Bell's *locality* hypothesis in the sense that A is independent of $\vec{b}$ and B independent of $\vec{a}$ and the expectation values are given by Proposition 13.1, the CHSH inequality must hold.

With this in mind, let us then calculate the quantum-mechanical expectation value for the joint measurement when the system is given by the spin singlet state $|\psi^-\rangle$ from eqn (13.1). When we apply the operators

$$\vec{a} \cdot \vec{\sigma} = a_x \sigma_x + a_y \sigma_y + a_z \sigma_z = \begin{pmatrix} a_z & a_x - ia_y \\ a_x + ia_y & -a_z \end{pmatrix} \text{ and } \vec{b} \cdot \vec{\sigma} = \begin{pmatrix} b_z & b_x - ib_y \\ b_x + ib_y & -b_z \end{pmatrix}$$

$$\tag{13.12}$$

to the computational basis states $|0\rangle = |\uparrow\rangle$ and $|1\rangle = |\downarrow\rangle$ we have

$$\vec{a} \cdot \vec{\sigma} |0\rangle = a_z |0\rangle + (a_x + ia_y)|1\rangle, \tag{13.13a}$$

$$\vec{a} \cdot \vec{\sigma} |1\rangle = (a_x - ia_y)|0\rangle - a_z |1\rangle, \tag{13.13b}$$

$$\vec{b} \cdot \vec{\sigma} |0\rangle = b_z |0\rangle + (b_x + ib_y)|1\rangle, \tag{13.13c}$$

$$\vec{b} \cdot \vec{\sigma} |1\rangle = (b_x - ib_y)|0\rangle - b_z |1\rangle. \tag{13.13d}$$

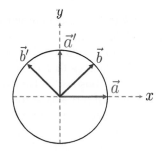

Fig. 13.3 Bell angles for maximal CHSH violation. For the so-called "Bell angles" $(\alpha, \beta, \alpha', \beta') = (0, \frac{\pi}{4}, 2\frac{\pi}{4}, 3\frac{\pi}{4})$, such that $\theta(\vec{a}, \vec{b}) = \theta(\vec{a}', \vec{b}) = \theta(\vec{a}', \vec{b}') = \frac{\pi}{4}$ and $\theta(\vec{a}, \vec{b}') = \frac{3\pi}{4}$, the quantum-mechanical Bell–CHSH parameter $S_{\mathrm{CHSH}}^{\mathrm{QM}}$ takes its maximal value $2\sqrt{2}$.

And so we arrive at

$$E(\vec{a}, \vec{b}) = \langle \Psi^- | \vec{a} \cdot \vec{\sigma} \otimes \vec{b} \cdot \vec{\sigma} | \Psi^- \rangle = \tfrac{1}{2}\big(\langle 01 | - \langle 10 | \big) \vec{a} \cdot \vec{\sigma} \otimes \vec{b} \cdot \vec{\sigma} \big(| 01 \rangle - | 10 \rangle \big)$$

$$= \tfrac{1}{2}\big(\langle 01 | - \langle 10 | \big) \Big[\big(a_z | 0 \rangle + (a_x + i a_y) | 1 \rangle \big) \big((b_x - i b_y) | 0 \rangle - b_z | 1 \rangle \big)$$

$$\quad - \big((a_x - i a_y) | 0 \rangle - a_z | 1 \rangle \big) \big(b_z | 0 \rangle + (b_x + i b_y) | 1 \rangle \big) \Big]$$

$$= \tfrac{1}{2}\big(-a_z b_z - (a_x - i a_y)(b_x + i b_y) - (a_x + i a_y)(b_x - i b_y) - a_z b_z \big)$$

$$= -a_x b_x - a_y b_y - a_z b_z = -\vec{a} \cdot \vec{b} \,. \tag{13.14}$$

Now we select measurement directions in the plane perpendicular to the direction of propagation, without loss of generality we choose the x–y plane, and describe the unit vectors $\vec{a}$ and $\vec{b}$ by their respective angles α and β relative to the x axis. In this case we have

$$E(\vec{a}, \vec{b}) = -\vec{a} \cdot \vec{b} = -\cos(\alpha - \beta) \,. \tag{13.15}$$

Then we know that for the choice of the *Bell angles* $(\alpha, \beta, \alpha', \beta') = (0, \frac{\pi}{4}, 2\frac{\pi}{4}, 3\frac{\pi}{4})$, illustrated in Fig. 13.3, the CHSH inequality (Proposition 13.2) is (maximally) violated

$$S_{\mathrm{CHSH}}^{\mathrm{QM}} = 2\sqrt{2} = 2.828 > 2 \,. \tag{13.16}$$

We thus see that local realistic hidden-variable theories and, in fact, any local realistic theories in the sense of the assumptions in Section 13.2.1 are incompatible with the predictions of quantum mechanics.

Ultimately the question of course is which theory is supported by empirical observations and in the case of the CHSH inequality, corresponding experimental tests have been performed, as we will discuss in detail in Section 13.3. Without taking away from this discussion, let us nevertheless remark here that in tests of the CHSH inequality using polarization-entangled photons, e.g., by *Anton Zeilinger's* group, there are subtle differences in the setup. In particular, translating the Bell angles from the Bloch-sphere vectors $\vec{a}$ and $\vec{b}$ (which coincide with the spatial orientations of the Stern–Gerlach apparatuses used for spin measurements) to angles α and β for the orientation of polarization filters one incurs a factor of 2 in the expectation value, i.e., the expectation value (13.15) changes to $E(\vec{a}, \vec{b}) = -\cos[2(\alpha - \beta)]$. This can be seen as a simple consequence of the bosonic nature of photons, that is, although we recall from Section 3.1.4 that the Hilbert space for photon polarization is $\mathcal{H} = \mathbb{C}^2$, the same as that for spin $\frac{1}{2}$, the polarization vector is a vector in $\mathbb{R}^3$ that happens to be confined to the plane perpendicular to the photon propagation since the latter is travelling at the speed of light (see Section 25.1.1). Polarization hence transforms under rotations like a vector, not like a spinor: The orthogonal polarization states $|H\rangle$ and $|V\rangle$ are connected by spatial rotation by $\frac{\pi}{2}$, while the transformation of $|\uparrow\rangle$ to $|\downarrow\rangle$ corresponds to a spatial rotation by an angle π.

13.2.3 Bell's Inequality

Having established the more practical CHSH inequality, let us nevertheless take a look at the inequality originally derived by Bell (Bell, 1964), which follows directly from the CHSH inequality. One just assumes perfect anti-correlation,

$$E(\vec{a}, \vec{b} = \vec{a}) = -1,\tag{13.17}$$

and reduces the possible measurement settings to only three different orientations, which is the minimal choice, e.g., by equating $\vec{a}' = \vec{b}'$. Then the CHSH inequality (Proposition 13.2) becomes

---

Proposition 13.3 (Bell's inequality)

$$S_{\text{Bell}} := |E(\vec{a}, \vec{b}) - E(\vec{a}, \vec{b}')| - E(\vec{b}', \vec{b}) \leq 1.$$

---

which is *Bell's original inequality.*

Bell's inequality (Proposition 13.3) is violated maximally for the quantum-mechanical expectation value (13.15) of the Bell state $|\psi^-\rangle$ and the choice $(\alpha, \beta, \beta') = (0, 2\frac{\pi}{3}, \frac{\pi}{3})$, in which case one obtains

$$S_{\text{Bell}}^{\text{QM}} = \frac{3}{2} = 1.5 > 1.\tag{13.18}$$

13.2.4 Wigner's Inequality

Next we discuss another type of Bell inequality which is used in experiments. It has a very simple form and has been derived by *Eugene P. Wigner* (1970). Instead of focusing on expectation values he considered directly probabilities whose estimates are proportional to the number of clicks observed with a detector. In terms of probabilities $P(\vec{a}, A; \vec{b}, B)$ for joint-measurement outcomes $A \in \{\uparrow, \downarrow\}$ and $B \in \{\uparrow, \downarrow\}$ of Alice and

Bob for measurement settings $\vec{a}$ and $\vec{b}$, respectively, the expectation value can be expressed as

$$E(\vec{a},\vec{b}) = P(\vec{a},\uparrow;\vec{b},\uparrow) + P(\vec{a},\downarrow;\vec{b},\downarrow) - P(\vec{a},\uparrow;\vec{b},\downarrow) - P(\vec{a},\downarrow;\vec{b},\uparrow), \quad (13.19)$$

and assuming that $P(\vec{a},\uparrow;\vec{b},\uparrow) \equiv P(\vec{a},\downarrow;\vec{b},\downarrow)$ and $P(\vec{a},\uparrow;\vec{b},\downarrow) \equiv P(\vec{a},\downarrow;\vec{b},\uparrow)$ together with $\sum_{A,B} P(\vec{a},A;\vec{b},B) = 1$ the expectation value becomes

$$E(\vec{a},\vec{b}) = 4\,P(\vec{a},\uparrow;\vec{b},\uparrow) - 1\,. \quad (13.20)$$

Inserting expression (13.20) into Bell's inequality (Proposition 13.3) yields *Wigner's inequality* for the joint probabilities, where Alice measures spin $\uparrow$ along direction $\vec{a}$ and Bob also $\uparrow$ along direction $\vec{b}$ (from now on we drop the spin notation $\uparrow$ in the formulae)

$$P(\vec{a},\vec{b}) \leq P(\vec{a},\vec{b}') + P(\vec{b}',\vec{b})\,, \quad (13.21)$$

which we rewrite as

Proposition 13.4 (Wigner inequality)

$$S_{\text{Wigner}} := P(\vec{a},\vec{b}) - P(\vec{a},\vec{b}') - P(\vec{b}',\vec{b}) \leq 0\,.$$

For the Bell state $|\psi^-\rangle$ the quantum-mechanical prediction for the probability gives

$$P(\vec{a},\vec{b}) = |(\langle\vec{a}| \otimes \langle\vec{b}|)\,|\psi^-\rangle|^2 = \frac{1}{2}\sin^2\frac{1}{2}(\alpha - \beta)\,, \quad (13.22)$$

with $|\vec{a}\rangle$ and $|\vec{b}\rangle$ as in eqn (11.81), which leads to a maximal violation of Wigner's inequality from Proposition 13.4,

$$S_{\text{Wigner}}^{\text{QM}} = \frac{1}{8} = 0.125 > 0\,, \quad (13.23)$$

for $(\alpha,\beta,\beta') = (0,2\frac{\pi}{3},\frac{\pi}{3})$, the same choice as in Bell's original inequality.

13.2.5 Clauser–Horne Inequality

As a last inequality to consider here, we want to present the *Clauser–Horne inequality* (Clauser and Horne, 1974), which is also often used in experiment. It relies on weaker assumptions and is very well suited for photon experiments with absorptive analysers. Clauser and Horne work with relative counting rates, i.e., the number of particles registered by the detectors. More precisely, the quantity $N(\vec{a},\vec{b})$ is the rate of simultaneous events, the *coincidence rate*, for the photon detectors of Alice and Bob after the photons passed the corresponding polarizers oriented along the direction $\vec{a}$ and $\vec{b}$, respectively. The relative rate $N(\vec{a}, \vec{b})/N = P(\vec{a}, \vec{b})$, where N represents the number of all events registered in the absence of the polarizers, corresponds to the joint probability $P(\vec{a}, \vec{b})$ in the limit of infinitely many events, which is a well-justified approximation in practice. If one polarizer is removed, say on Bob's side, then expression $N_A(\vec{a})/N = P_A(\vec{a})$ stands for the single probability at Alice's (or the correspondingly at Bob's) side.

We can then start from the purely algebraic inequality

$$S := xy - xy' + x'y + x'y' - x' - y \leq 0, \tag{13.24}$$

which holds for any real numbers x, x', y, and y' in the interval $[0, 1]$. To see this, note that we can bring all quantities to the left-hand side and group them together in different ways,

$$S = y(x - 1) + x'(y - 1) + y'(x' - x), \tag{13.25a}$$

$$= x'(y' - 1) + y(x' - 1) + x(y - y'), \tag{13.25b}$$

$$= x'(y - 1) + y(x' - 1) + (x - x')(y - y'), \tag{13.25c}$$

where it is easy to see that the first two terms in each line of the right-hand side of eqn (13.25) are smaller than or equal to zero. The remaining term in the first line is non-positive when $x' \leq x$, and in the second line the last term is non-positive for $y \leq y'$, while the last term in the third line is negative when both $x' > x$ and $y > y'$.

With the inequality (13.24) thus proven, we turn to the formulation of the probabilities within any hidden-variable model. Let $P_A(\vec{a}|\lambda)$ and $P_B(\vec{b}|\lambda)$ denote the conditional probabilities of obtaining clicks in the detectors of Alice and Bob oriented along the directions $\vec{a}$ and $\vec{b}$, respectively, provided that the hidden variable has the value λ, and let $\rho(\lambda)$ be the normalized distribution of the hidden variable. Then the individual probabilities for observing clicks on Alice's and Bob's sides are

$$P_A(\vec{a}) = \int d\lambda\, \rho(\lambda)\, P_A(\vec{a}|\lambda), \quad \text{and} \quad P_B(\vec{b}) = \int d\lambda\, \rho(\lambda)\, P_B(\vec{b}|\lambda), \tag{13.26}$$

while the joint probability $P(\vec{a}, \vec{b})$ for obtaining clicks on both sides is given by

$$P(\vec{a}, \vec{b}) = \int d\lambda\, \rho(\lambda)\, P_A(\vec{a}|\lambda)\, P_B(\vec{b}|\lambda). \tag{13.27}$$

It is now easy to derive the Clauser–Horne inequality for probabilities by equating $P_A(\vec{a}|\lambda) = x$, $P_A(\vec{a}'|\lambda) = x'$, $P_B(\vec{b}|\lambda) = y$, $P_B(\vec{b}'|\lambda) = y'$.

Proposition 13.5 (Clauser–Horne inequality)

$$S_{\mathrm{CH}} := P(\vec{a}, \vec{b}) - P(\vec{a}, \vec{b}') + P(\vec{a}', \vec{b}) + P(\vec{a}', \vec{b}') - P_A(\vec{a}') - P_B(\vec{b}) \leq 0.$$

The Clauser–Horne inequality (Proposition 13.5) has been used by *Alain Aspect* and his collaborators *Jean Dalibard* and *Gérard Roger* in their time-flip experiment (1982*a*), which we will discuss in Section 13.3. The two-photon state produced in this experiment was the symmetrical Bell state $| \phi^+ \rangle = \frac{1}{\sqrt{2}} (| R \rangle \otimes | L \rangle + | L \rangle \otimes | R \rangle) = \frac{1}{\sqrt{2}} (| H \rangle \otimes | H \rangle + | V \rangle \otimes | V \rangle)$, where $| R \rangle, | L \rangle$ denote the right- and left-handed circularly polarized photons and $| H \rangle, | V \rangle$ the horizontally and vertically polarized ones.

In case of polarization-entangled photons in the state $| \phi^+ \rangle$ the quantum-mechanical probability to detect a linearly polarized photon with an angle α on Alice's side, and simultaneously another linearly polarized photon with angle β on Bob's side, is given by

$$P(\vec{a}, \vec{b}) = \left| \left[(\langle H | \cos(\alpha) + \langle V | \sin(\alpha)) \otimes (\langle H | \cos(\beta) + \langle V | \sin(\beta)) \right] | \phi^+ \rangle \right|^2$$
$$= \frac{1}{2} \cos^2(\alpha - \beta) \,. \tag{13.28}$$

Now choosing the angles $(\alpha, \beta, \alpha', \beta') = (0, \frac{\pi}{8}, 2\frac{\pi}{8}, 3\frac{\pi}{8})$ the quantum-mechanical probabilities (13.28) violate the Clauser–Horne inequality (Proposition 13.5) maximally, i.e.,

$$S_{\mathrm{CH}}^{\mathrm{QM}} = \frac{\sqrt{2} - 1}{2} = 0.207 \, > \, 0 \,. \tag{13.29}$$

Further literature can be found in the books *Quantum [Un]speakables* (Bertlmann and Zeilinger (eds.), 2002) and *Quantum [Un]speakables II* (Bertlmann and Zeilinger (eds.), 2017), and in the review article Brunner *et al.* (2014).

What are the conclusions? The combination of Bell's locality and his vision of reality are *incompatible* with quantum mechanics! The essence of all Bell inequalities can be summarized in the following theorem:

Theorem 13.1 (Bell's theorem, 1964)

There are physical situations, where all *local realistic theories are incompatible with quantum mechanics.*

In his seminal work in 1964, John Bell realized the far-reaching consequences of a realistic theory as an extension to quantum mechanics and expressed it in the following way:

In a theory in which parameters are added to quantum mechanics to determine the results of individual measurements, without changing the statistical predictions, there must be a mechanism whereby the setting of one measuring device can influence the reading of another instrument, however remote. Moreover, the signal involved must propagate instantaneously, so that such a theory could not be Lorentz invariant.

He continued and stressed the crucial point in such EPR-type experiments:

Experiments ... , in which the settings are changed during the flight of the particles, are crucial.

Thus it is of utmost importance *not* to allow some mutual report by the exchange of signals with velocity less than or equal to the speed of light. Experiments where this condition is met are discussed in Section 13.3.

Résumé: EPR argued that quantum mechanics is an incomplete theory. A logical consequence from such a conclusion is that the theory should be supplemented by additional parameters—hidden variables predetermining the outcome of a measurement—that would restore causality and locality. Bell discovered that the requirement of locality—Bell's locality hypothesis in the expectation value of the joint measurements of Alice and Bob—is *incompatible* with the predictions of quantum mechanics. The incompatibility is demonstrated via a Bell inequality. Besides Bell's original inequality several other so-called Bell inequalities do exist, including the inequalities by Clauser, Horne, Shimony, and Holt, by Clauser and Horne, as well as by Wigner.

13.3   Bell Inequalities—Experiments

After Bell had published his article introducing his inequality there was practically no immediate interest in this field. It was the "dark era" of the foundations of quantum mechanics. *Wolfgang Pauli*—referring to *Otto Stern*—wrote to *Max Born* in 1954 (Pauli, 1999):

One should no more rack one's brain about the problem of whether something one cannot know anything about exists at all, than about the ancient question of how many angels are able to sit on the point of a needle.

Nevertheless, interest in empirical tests of Bell's inequality started to grow, culminating in first experiments in the 1970s and subsequent developments to the present day, as we shall elaborate on in the following.

13.3.1   First-Generation Experiments in the 1970s

The first who got interested in the subject in the late sixties was John Clauser, at the time a young graduate student at Columbia University. When he studied Bell's paper (1964) presenting a bound for all hidden-variable theories, he was astounded by its result. As a true experimentalist he wanted to see empirical evidence for the predicted violation of the bound on local realistic models. So he planned to perform an experiment. Although experiments of this type were not appreciated at that time—we warmly recommended reading Clauser's recollections (Clauser, 2002) in this regard—he prepared the experimental setup. He sent an abstract describing the proposed experiment to the Spring Meeting of the American Physical Society (Clauser, 1969). Soon afterwards, Abner Shimony called him and told him that he and his student Michael Horne had had the same idea. So they joined forces and, together with Richard Holt, at the

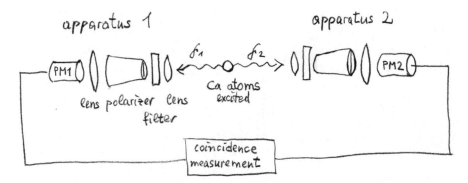

Fig. 13.4 Schematic experimental setup. Illustration of the experiment of (Clauser and Freedman, 1972). By a special procedure excited $^{40}_{20}Ca$ atoms emit two photons γ_1 and γ_2 in an atomic cascade. These emitted photons, quantum-mechanically in the symmetric Bell state $|\phi^+\rangle$, are observed by symmetrically placed optical systems: lenses, filters, rotatable and removable polarizers, and photon detectors (photon multipliers). The coincidence rate for two photons at different angles of linear polarization is recorded.

time a PhD student of *Francis Pipkin* at Harvard, they wrote the famous CHSH article (Clauser *et al.*, 1969), where they proposed the CHSH inequality, which was more suitable to experiments than Bell's original inequality.

Clauser carried out the experiment in 1972 together with *Stuart Freedman* (1972), then a graduate student at Berkeley, who received his PhD with this experiment (Freedman, 1972*b*). As pointed out in the CHSH paper (Clauser *et al.*, 1969), pairs of photons emitted in an atomic radiative cascade would be suitable for a Bell inequality test. Clauser and Freedman chose calcium atoms pumped by a special procedure, where the excited atoms emitted the desired photon pairs, see Fig. 13.4 for the schematic setup. These emitted photons, quantum-mechanically in the symmetric Bell state $|\phi^+\rangle = \frac{1}{\sqrt{2}}(|R\rangle \otimes |L\rangle + |L\rangle \otimes |R\rangle) = \frac{1}{\sqrt{2}}(|H\rangle \otimes |H\rangle + |V\rangle \otimes |V\rangle)$, were observed by symmetrically placed optical systems: lenses, filters, rotatable and removable polarizers, and photon detectors (photon multipliers). The coincidence rate for two photons at different angles of linear polarization was finally measured.

Clauser and Freedman measured the correlation in the linear polarization of the two emitted photons in the calcium cascade $6\,^1S_0 \to 4\,^1P_1 \to 4\,^1S_0$. Here the level notation of atomic spectroscopy $n\,^{(2S+1)}L_J$ is used, where n denotes the principal quantum number, S the spin, L the orbital angular momentum, and $J = L + S$ the total angular momentum, see Fig. 13.5.

More precisely, by resonance absorption of a (2275 Å) photon the calcium atoms were excited to the $3d4p\,^1P_1$ energy level (s, p, d denote the atomic subshells occupied by the electron; the upper index counts the number of electrons that are placed into the subshell, where the maximal number is given by $2l(l + 1)$ with l the azimuthal quantum number). About 7% of the atoms decayed to the $4p^2\,^1S_0$ level, from which

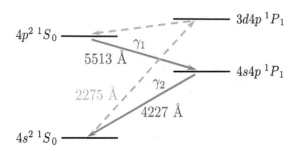

Fig. 13.5 Relevant energy levels of calcium $^{40}_{20}Ca$. By resonance absorption of a (2275 Å) photon the calcium atoms were excited to the $3d4p\,^1P_1$ energy level. The atoms decayed to the $4p^2\,^1S_0$ level, from which they cascaded through the $4s4p\,^1P_1$ intermediate state to the ground state $4s^2\,^1S_0$ by emitting two photons γ_1 (5513 Å) and γ_2 (4227 Å).

they cascaded through the $4s4p\,^1P_1$ intermediate state to the ground state $4s^2\,^1S_0$ by emitting two photons γ_1(5513 Å) and γ_2(4227 Å), see Fig. 13.5.

The coincidence rate $R(\varphi)$ for two-photon detection was measured as a function of the angle φ between the planes of linear polarization produced by the inserted polarizers, see Fig. 13.4. In the ideal quantum-mechanical case we have

$$\frac{R(\varphi)}{R_0} = \frac{1}{4}\left(1 + \cos(2\varphi)\right), \qquad (13.30)$$

where R_0 represents the coincidence rate with both polarizers removed. Of course, in the real experiment the transmittance of the polarizers and the lenses must be taken into account, which reduces the coincidence rate (13.30) by about a factor of 0.85, see Clauser and Freedman (1972).

The signals were very weak at that time: A measurement lasted for about 200 hours. For comparison with theoretical predictions a very practical inequality, where just two angles enter, was used. It was derived by Freedman (1972a, 1972b).

Proposition 13.6 (Freedman inequality)

$$S_{\text{Freedman}} = \left|\frac{R(67,5°)}{R_0} - \frac{R(22,5°)}{R_0}\right| - \frac{1}{4} \le 0.$$

The outcome of the experiment is well known. Clauser and Freedman obtained a clear violation of the inequality (Proposition 13.6),

$$S_{\text{Freedman}}^{\text{CF exper}} = 0.050 \pm 0.008, \qquad (13.31)$$

very much in accordance with quantum mechanics, i.e.,

$$S_{\text{Freedman}}^{\text{QM}} = 0.044\,. \tag{13.32}$$

At the same time, Francis Pipkin and Richard Holt performed a similar experiment at Harvard, using the radiative cascade $9\,{}^1P_1 \to 7\,{}^3S_1 \to 6\,{}^3P_0$ in the mercury isotope ${}_{80}^{198}$Hg. The outcome was just contrary, the results agreed with a hidden-variable theory and disagreed with quantum mechanics (Holt and Pipkin, 1974). But Clauser repeated the Holt–Pipkin experiment investigating the same cascade in mercury ${}_{80}^{202}$Hg and applying the same excitation technique as Holt and Pipkin. His result, however, violated the Bell inequality significantly, in agreement with quantum mechanics (Clauser, 1976).

In 1976 *Edward Fry* and his student *Randall Thompson* (1976) independently set up an experiment in Houston. As in Clauser's experiment the linear-polarization correlation was measured between the two photons emitted from a radiative cascade in mercury atoms whose atomic levels had been excited by lasers. In particular, they focused on the $7\,{}^3S_1 \to 6\,{}^3P_1 \to 6\,{}^1S_0$ cascade of ${}_{80}^{200}Hg$. Due to the much better signals with improved lasers they could collect enough data in 80 minutes. This data was used to evaluate Freedman's version (Proposition 13.6) of the Bell inequality. The result was Fry and Thompson (1976)

$$S_{\text{Freedman}}^{\text{FT exper}} = 0.046 \pm 0.014\,, \tag{13.33}$$

in clear violation of the inequality $S_{\text{Freedman}} \le 0$ (Proposition 13.6) and in excellent agreement with the quantum-mechanical prediction $S_{\text{Freedman}}^{\text{QM}} = 0.044 \pm 0.007$ (13.32).

Thus at that time, there was already convincing empirical evidence that hidden-variable theories did not work but quantum mechanics was correct. However, the experiments were not perfect yet, the analysers were static, only a small amount of photon pairs were registered, etc. Several loopholes were still open: the detection-efficiency or fair-sampling loophole, and the communication-absence or freedom-of-choice loophole, just to mention some important ones. Closing these loopholes thus remained as a challenge for future experiments.

13.3.2 Second-Generation Experiments in the 1980s

In the late 1970s and beginning of the 1980s, the general atmosphere in the physics community was still such that it could be summarized as:

Quantum mechanics works very well, so don't worry!

Alain Aspect, however, when reading Bell's inequality paper (Bell, 1964), was so strongly impressed that he immediately decided to do his *"thèse d'état"* on this fascinating topic. Aspect's idea and final goal was to include variable analysers.

Together with his collaborators *Philippe Grangier*, *Gérard Roger*, and *Jean Dalibard* he performed a whole series of experiments (Aspect, 1976; Aspect *et al.*, 1981;

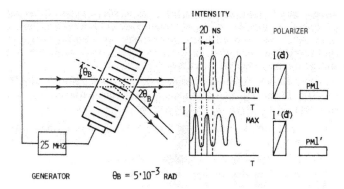

Fig. 13.6 Optical switch. The incident photons are switched by acousto-optical interaction with ultrasonic standing waves in water. The photons are either transmitted, when the amplitude of the standing wave is minimal, or deflected, when it is maximal. The intensity of the transmitted and deflected photons is shown qualitatively. $I(\vec{a})$ and $I'(\vec{a'})$ are different polarizers at different orientations; PM1, PM1' are photomultipliers. From Fig. 5 in Bertlmann (1990), adapted from Fig. 3 in Aspect *et al.* (1982*a*), copyrighted by the American Physical Society, with permission of Alain Aspect.

Aspect *et al.*, 1982*a*; Aspect *et al.*, 1982*b*; Aspect and Grangier, 1985) with gradually improving designs, approaching an "ideal" setup configuration step by step. Like Clauser, they chose the radiative cascade $6\ {}^1S_0 \to 4\ {}^1P_1 \to 4\ {}^1S_0$ in calcium, see Fig. 13.5, but irradiated the beam of calcium atoms (typical density: 3×10^{10} atoms/cm^3) at 90° by a krypton ion laser (406 nm) and a rhodamine dye laser (581 nm). The such excited Ca atoms emitted photon pairs (λ_1, λ_2) in opposite directions forming the Bell state $|\phi^+\rangle$. For comparison with theory the Clauser–Horne inequality (Proposition 13.5) was used, which was violated significantly in each experiment.

The biggest advance was achieved together with Jean Dalibard and Gérard Roger in the final time-flip experiment (Aspect *et al.*, 1982*a*), where a clever acousto-optical switch mechanism was incorporated. They used the possibility of scattering photons on ultrasonic standing waves in water. The incident photons were either transmitted straight on into a polarizer with a given orientation or they were Bragg-deflected into another polarizer with a different orientation. Transmission occurred when the amplitude of the standing wave was minimal, and deflection when it was maximal, see Fig. 13.6. This switching worked very rapidly, 10 ns for switching between the two different polarizers.

It worked such that the switching time between the polarizers (10 ns), as well as the lifetime of the photon cascade (5 ns), was much smaller than the time of flight (40 n) of the photon pair from the source to the analysers, which were located at a distance of 12 m, see Fig. 13.7. This implied a space-like separation of the event intervals. However, the time-flipping mechanism was still not ideal, i.e., truly random, but *quasi-periodic*, as they called it. The mean for two runs which lasted about two hours yielded the following result

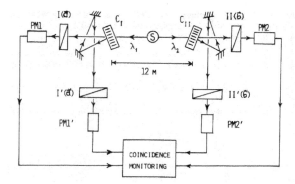

Fig. 13.7 Setup of the Aspect–Dalibard–Roger experiment (Aspect *et al.*, 1982*a*). S is the photon source, irradiated calcium atoms which emit correlated photon pairs in opposite direction. C_I and C_{II} are the optical switches which are followed by the polarizers I, I' and II, II' with different orientations. Behind the polarizers the photomultipliers PM are placed. A coincidence monitoring provides the counting rates of the photon pairs, which are used to calculate the Clauser–Horne inequality (Proposition 13.5). From Fig. 6 in Bertlmann (1990), adapted from Fig. 2 in Aspect *et al.* (1982*a*), copyrighted by the American Physical Society, with permission of Alain Aspect.

$$S_{\text{CH}}^{\text{ADR exp}} = 0.101 \pm 0.020 \,,\tag{13.34}$$

in very good agreement with the quantum-mechanical prediction

$$S_{\text{CH}}^{\text{QM corr}} = 0.113 \pm 0.005 \,,\tag{13.35}$$

that had been adapted for the experiment (compared with the ideal value $S_{\text{CH}}^{\text{QM}} = 0.207$ from eqn (13.29)).

This time-flip experiment of Aspect *et al.* received much attention in the physics community and also in popular science. It represents a turning point, the physics community began to realize that there was something essential in it. Research started into a new direction and flourished, initiating a lot of activity in what is nowadays called the field of quantum information and quantum communication.

13.3.3   Third-Generation Experiments in the 1990s

In the 1990s, the spirit towards foundations in quantum mechanics totally changed since quantum information gained increasing interest, Bell inequalities and quantum entanglement were the basis of these developments.

Meanwhile, the technical facilities including the electronics and lasers also improved considerably. Most important was the invention of a new source for producing entangled photon pairs: *spontaneous parametric down conversion* (SPDC), where a non-linear crystal is pumped with a laser and the pump photons are converted into two photons—in accordance with the conservation of energy and momentum, see Fig 13.8. The vertically and horizontally polarized photons originating from the

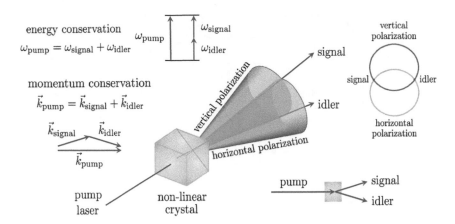

Fig. 13.8 Spontaneous parametric down conversion. A non-linear crystal is pumped by a laser. Some incident photons of frequency ω_{pump} are converted to two photons with lower frequencies ω_{signal} and ω_{idler}. Due to dispersion in the crystal, i.e., dependence of the refractive index on the frequency, polarization, and angle of the beams, the conservation of energy and momentum only allows phase matching for specific combinations of frequencies. In type-II SPDC, the horizontally and vertically polarized photons of the created pairs are confined to two cones and for each photon pair the positions on the cones are symmetric and the polarizations are anti-correlated, if the photon on one cone is vertically polarized then the corresponding photon on the other cone is horizontally polarized. Along the intersections of the two cones, the polarizations are perpendicular but undetermined for each photon: The photons are entangled in polarization.

down-conversion process propagate along two different cones. Photon pairs propagating along the two lines where the cones intersect are entangled in their polarization, as shown in Fig. 13.8.

Such an EPR source was used by Zeilinger's group, when they performed their Bell experiment in 1998 (Weihs *et al.*, 1998). Zeilinger's student *Gregor Weihs* obtained his PhD (1998) with this experiment, the setup of which is shown in Fig. 13.9. Their ambitious goal was to construct an ultra-fast and truly random setting of the analysers at each side of Alice and Bob, such that strict Einstein locality—no mutual influence between the two observers Alice and Bob (according to the laws of relativity)—could be enforced in the experiment. The ultra-fast and random setting was achieved, on the one hand, by an electro-optic modulator that changed the orientation of the polarization of the propagating photons, and on the other hand, by a random-number generator that controlled the modulator, see Fig. 13.9. For the random-number generator a 50:50 beam splitter was used.

The data was compared with the CHSH inequality (Proposition 13.2) and the experimental result was

$$S_{\text{CHSH}}^{\text{WZ exp}} = 2.73 \pm 0.02 \,, \tag{13.36}$$

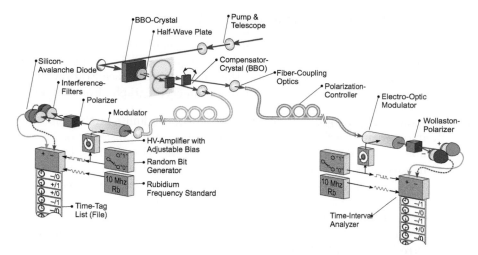

Fig. 13.9 The timing experiment (Weihs *et al.*, 1998). The EPR source is a so-called BBO (beta-barium borate) crystal pumped by a laser, the outgoing photons are vertically and horizontally polarized on two different cones and in the overlap region they are entangled. These entangled photons are separately led to the measurement stations of Alice and Bob via optical fibres. During the photon propagation the orientations of polarizations are changed by an electro-optic modulator which is driven by a random-number generator on each side. In this way the strict Einstein locality condition—no mutual influence between the two observers Alice and Bob—is achieved in the experiment. Figure © Gregor Weihs.

which corresponded to a violation of the inequality by 30 standard deviations. Due to this high-efficiency photon source the measurement could be performed already in three to four minutes. It was *the* experiment that truly included the vital time factor that John Bell had insisted upon so strongly[2].

At about the same time other groups (Brendel *et al.*, 1992; Tapster *et al.*, 1994) investigated energy-correlated photon pairs to test Bell inequalities. A record was set by the group of *Nicolas Gisin* (Tittel *et al.*, 1998), by using energy-time entangled photon pairs in optical fibres. They managed to separate their observers Alice and Bob by more than 10 km and could show that this distance had practically no effect on the entanglement of the photons. The investigated Bell inequalities were violated by 16 standard deviations.

A historical description of the exciting time of "rebuilding the foundations" of quantum mechanics is given in the book *The Quantum Dissidents* (Freire Jr., 2015) and some more reflections on Bell's theorem can be found in Cushing and McMullin (1989).

---

[2]For the experiments described above the Nobel Prize in Physics 2022 was awarded to Alain Aspect, John F. Clauser, and Anton Zeilinger on the grounds of "experiments with entangled photons, establishing the violation of Bell inequalities and pioneering quantum information science". Their results paved the way for new technologies based on quantum information.

13.3.4 Fourth-Generation Experiments After 2000

In the new millennium, a race started to achieve records in entanglement-based long-distance quantum communication. The vision was (and still is) to install a global network, in particular via satellites or via the International Space Station, that can provide access to secure communication via quantum cryptography from any location.

It was again Zeilinger's group that pushed the distance limits further and further. Firstly, in an open-air experiment in the city of Vienna over a distance of 7.8 km the group Resch *et al.* (2005) could violate a CHSH inequality (Proposition 13.2) by more than 13 standard deviations. Secondly, setting a new world record at that time, the group Ursin *et al.* (2007) successfully carried out an open-air Bell experiment over 144 km between the two Canary Islands La Palma and Tenerife. In 2020 the limit was pushed even further by the group of Zeilinger's former student *Jian-Wei Pan* in an experiment demonstrating "Entanglement-based secure quantum cryptography over 1.120 kilometres" (Yin *et al.*, 2020).

Furthermore, an impressive cosmic Bell experiment with polarization-entangled photons, a "Cosmic Bell test using random measurement settings from high-redshift quasars" (Rauch *et al.*, 2018) was carried out, whose light was emitted billions of years ago. The experiment ensures locality. A statistically significant violation of Bell's inequality by 9.3 standard deviations was observed. This experiment pushes back to at least ≈ 7.8 Gyr ago, the most recent time by which any local-realist influences could have exploited the "freedom-of-choice" loophole to account for the observed Bell violation. Any such mechanism is practically excluded, extending almost from the Big Bang to today.

As in every experiment, the Bell-type experiment requires assumptions that provide loopholes that allow, at least in principle, for a local realistic theory to explain the experimental data. The most significant loopholes are the following.

Loopholes:

- **Locality loophole**
  The locality loophole exists if the measurement settings or result of Alice could in principle be communicated to Bob in time in order to influence his measurement result. It can be closed by a space-like separation—strict Einstein locality—of the local measurements of Alice and Bob. That means the measurement settings of Alice and Bob must be chosen independently and sufficiently rapidly such that no physical signal—limited to propagate at the speed of light —can travel from Alice to Bob in time to pass any relevant information about the chosen setting or the measurement result.

- **Freedom-of-choice loophole**
  The freedom-of-choice loophole represents the requirement that the setting choices must be "free" or truly random, such that there is no possible interdependency between the choice of the settings and the properties of the system being measured. It can be closed by generating the settings independently at the locations

of Alice and Bob and by separating the settings space-like from the creation point of the particles.

- **Fair-sampling loophole or detection loophole**
  The fair-sampling loophole refers to the possibility that under local realism it is conceivable that a sub-ensemble of the emitted pair of particles violate a Bell inequality, while the total ensemble does not. It can be closed by detecting the particles with sufficiently high efficiency.

To close the locality loophole is sometimes considered to be most important. The first who tackled this loophole was Alain Aspect (1982*a*) in his time-flip experiment where the polarizer orientations were changed rapidly, but nevertheless not truly randomly (see Section 13.3.2). The Zeilinger group (Weihs *et al.*, 1998) followed with their experiment using an ultra-fast and truly random switch, thus closing the locality loophole (see Section 13.3.3).

The experimental group Scheidl *et al.* (2010) excluded the freedom-of-choice loophole and the group Giustina *et al.* (2013) closed the fair-sampling loophole. More precisely, by using an inequality à la *Eberhard* (1993*a*), their results of violating this inequality were valid without assuming that the sample of measured photons accurately represented the entire ensemble. Furthermore, the groups Rowe *et al.* (2001) and Matsukevich *et al.* (2008) closed the detection loophole by working with ion traps.

In this connection we want to refer to the Bell inequality tests of the group of *Helmut Rauch* (Hasegawa *et al.*, 2003; Bartosik *et al.*, 2009; Hasegawa *et al.*, 2011) who worked with neutron interferometry. These neutron-interferometer experiments are of particular interest since in this case the quantum correlations are explored in the degrees of freedom of a single particle, the neutron. Physically, this can be interpreted more as a test of contextuality rather than of (spatial) non-locality.

However, up to 2015, all these loopholes had only been closed separately. To close all three mentioned loopholes simultaneously in a single experiment with high statistical significance demanded great effort. Nevertheless, this was achieved by the research groups led by Anton Zeilinger (Giustina *et al.*, 2015) in Vienna, *Sae Woo Nam* (Shalm *et al.*, 2015) in Boulder and *Ronald Hanson* (Hensen *et al.*, 2015) in Delft. Whereas the experiments by the Vienna group and the Boulder group worked with photons in the familiar EPR–Bell-type setup, see Fig. 13.2, the Delft group used quite a different scheme that entangled the electron spins of remote nitrogen-vacancy centres (a kind of artificial atom embedded in a diamond crystal) which were placed at different locations.

Let us discuss the experiment of the Vienna group (Giustina *et al.*, 2015) in more detail. The experimental design, illustrated in Fig. 13.10, became much more refined as compared to the former setups, see Figs 13.7 and 13.9. The whole setup, the photon source and measurement stations of Alice and Bob depicted in Fig. 13.10 (a), was located in the sub-basement of the Vienna Hofburg palace. To create the entangled photon pair spontaneous parametric down conversion in a non-linear crystal (as described in Section 13.3.3) was used, see Fig. 13.10 (b). The photons were coupled

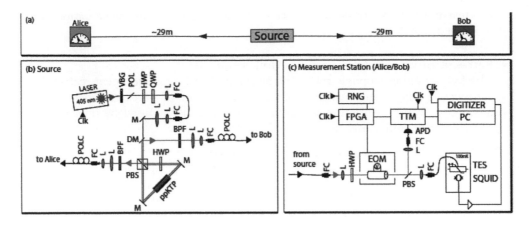

Fig. 13.10 (a) **Setup of the experiment** (Giustina *et al.*, 2015). From the source, two polarization-entangled photons propagate to the two identically constructed and spatially separated measurement stations Alice and Bob (distance ≈ 58 m). (b) Source: As source a type-II spontaneous parametric down conversion process in a periodically poled crystal (ppKTP) is used. The crystal is pumped with a 405 nm pulsed diode laser, whose light is filtered. The ppKTP crystal is pumped from both sides to create polarization entanglement. Each photon pair is split at the polarizing beam splitter (PBS) and collected into two different single-mode fibres leading to the measurement stations Alice and Bob. (c) Measurements stations: In each station, one of the two linear-polarization directions is selected, as controlled by an electro-optical modulator (EOM), which acts as a switchable polarization rotator in front of a PBS plate. Adapted electronics (FPGA) sample the output of a random-number generator (RNG) to trigger the switching of the EOM. The output of the PBS is delivered via a fibre to the transition-edge sensor (TES). The signal of the TES is amplified by a superconducting quantum interference device (SQUID), additional electronics, digitized, and recorded together with the setting choices. Synchronization is carried out with clock input (Clk). Figure from Giustina *et al.* (2015, Fig. 1).

into fibres and sent to the two distant stations, Alice and Bob. There the polarization measurements were performed in one of the two setting angles, as illustrated in Fig. 13.10 (c). While the photons propagated, the choice of the setting in each station was made by a random-number generator placed there. The measurement was carried out with the help of a fast electro-optical modulator that transmitted the particular photon polarization chosen by the random-number generator. The photons were observed in a single-photon detector and the signal was finally amplified by a series of amplifiers, see Fig. 13.10 (c).

The photon and setting data stored locally at the locations of Alice and Bob were collected by a computer which evaluated the Bell inequality. To close the loopholes a specific space-time configuration of the experiment was chosen.

To tackle the fair-sampling loophole a Clauser–Horne–Eberhard inequality for the probabilities of the outcomes was used, which did not rely on the fair-sampling assumption. The inequality was developed by Bierhorst (2015) and Kofler *et al.* (2016) and could be violated with high efficiencies. It needed only one detector on each side and limited the probabilities in the following way:

Proposition 13.7 (Clauser–Horne–Eberhard Inequality)

$$S_{\text{CHE}} =: P_{++}(\vec{a}, \vec{b}) - P_{+0}(\vec{a}, \vec{b}') - P_{0+}(\vec{a}', \vec{b}) - P_{++}(\vec{a}', \vec{b}') \leq 0 .$$

In the notation of Proposition 13.7, the subscripts "+" and "0" in the first (Alice) and second (Bob) position stand for a detection and no detection, respectively, when the polarization directions are chosen as $\vec{a}$ on Alice's side and as $\vec{b}'$ on Bob's side.

The inequality (Proposition 13.7) can be violated by using Eberhard states

$$| \psi_{\text{E}} \rangle = \frac{1}{\sqrt{1 + r^2}} (| V \rangle \otimes | H \rangle + r | H \rangle \otimes | V \rangle) , \qquad (13.37)$$

a one-parameter family of two-photon states that include a product state for $r = 0$ and a maximally entangled state for $r = 1$. In the experiment, the state was chosen with $r \approx -2.9$ and measured at angles $\vec{a} = 94.4°, \vec{a}' = 62.4°, \vec{b} = -6.5°$, and $\vec{b}' = 25.5°$ for about 3510 seconds. The estimates for the probabilities obtained in this way provided an experimental S value of

$$S_{\text{CHE}}^{\text{exper}} = 7.27 \cdot 10^{-6} , \qquad (13.38)$$

which is positive and definitely violates the Clauser–Horne–Eberhard inequality (Proposition 13.7). Quantum mechanics, for comparison, predicts an optimal S value of

$$S_{\text{CHE}}^{\text{QM}} = 4 \cdot 10^{-5} . \qquad (13.39)$$

The group also computed the statistical significance of their experimental result and found that under local realism the probability of observing the measured value (13.38) did not exceed $3.74 \cdot 10^{-31}$ corresponding to a confidence corresponding to 11.5 standard deviations.

By closing all three mentioned loopholes simultaneously the experiments of Giustina *et al.* (2015), Shalm *et al.* (2015), and Hensen *et al.* (2015) thus provide the experimental cornerstone for the following statement:

Local realistic theories are incompatible with quantum mechanics and with nature!

13.4 Interpretations of Quantum Mechanics

In the course of the development of Bell inequalities the question of formulating an interpretation of the quantum-mechanical formalism had reappeared and eventually became an active research subject. In this regard, it is perhaps tempting to express the hope that such an interpretation might bring about an intuitive "understanding" of quantum mechanics. Some might say that "understanding" quantum mechanics on the grounds of our day-to-day experience and our preemptive—perhaps too naive—notions of reality is an impossible task that already the founders of quantum theory had their great difficulties with, and in all honesty, so do we. Before getting into such a discussion, it might be prudent to first agree on what we mean by "understanding" in the first place. Clearly, the mathematical framework of quantum mechanics is uncontroversial and so are the predictions that we can make when using it. In this sense we can safely rely on the mathematical formalism. But what may be a point of contention are the predictions that the formalism *does not* allow us to make, in particular, regarding the question *why* certain measurement results are realized in a given experiment and not others, that is, the infamous *measurement problem*. As we now know, attempts to establish deterministic hidden-variable models to ascribe this problem to our mere ignorance are severely restricted, if not altogether thwarted by the requirement of contextuality (see Section 12.3) and by numerous convincing experimental refutations of local realism by Bell inequality violations (see Section 13.3). But is there a middle ground between the realist views of the likes of Einstein or Bell and retreating entirely to an attitude once summarized by David Mermin (see Hardy and Spekkens, 2010, p. 76) as "Shut up and calculate!"? Or is the formalism—mathematics in itself, the geometry, topology, symmetries, etc.—perhaps more real than we think? Then we certainly would have to expand our vision of reality.

In this section, we want to briefly touch upon the issue of interpretations of quantum mechanics by describing two opposite positions to "understand" quantum mechanics, the position of the *realists* and the position of the proponents of *information*. For the sake of brevity, we will largely leave aside other interpretations and approaches, like the *many-worlds interpretation* (see, e.g., the anthology Saunders *et al.*, 2010 for a collection of essays on this topic), *Bohmian mechanics* (see Bohm, 1952, or for instance, the more recent books by Holland, 1993 or Dürr and Teufel, 2009), *quantum Darwinism* (see Zurek, 2009), works advocating objective wave-function collapse (see, e.g., the review Bassi *et al.*, 2013 and references therein) and others (an overview of different interpretations can be found in Cabello, 2017), leaving these to be discovered and independently assessed by interested readers on their own.

Realism goes back to the founders of quantum mechanics like Albert Einstein and Erwin Schrödinger, followed by David Bohm, John Bell, and others. The approach we refer to here as "information" or "informationist", on the other hand, is a more modern point of view, for which we focus on the work of *Časlav Brukner* and Anton Zeilinger. A more extensive treatment of the contrast between realism and information can be found in "Real or not real that is the question ..." by Bertlmann (2020). The general question "What is real?" has been discussed on a historical and more popular level in the book Becker (2018).

Let us concentrate first on John Bell's view[3] of the world, since he was one of the most outstanding figures in this interpretational debate who in some sense triggered the whole issue. Bell was the "true" realist. "Everything has definite properties!" was his credo. This position led him to one of the most profound discoveries in physics, to Bell's theorem, but also to further deep insights in quantum field theory, like the "Anomalies" (Bell and Jackiw, 1969), for an overview of this field, see the book Bertlmann (2000).

On the other hand, for Zeilinger (1999) and Brukner–Zeilinger (1999, 2001, 2003) quantum mechanics is a theory of information, a mathematical representation of what an observer has to know in order to calculate the probabilities for the outcomes of measurements.

13.4.1 Realism

According to John Bell (1976, 2004) a physical theory should consist of *physical* quantities—*real* entities. He called them *beables*—not observables!— that refer to the ontology of the theory, to what *is*, to what *exists*. The physical quantities—the ontology—are represented by both, the mathematical quantities *and* the dynamics. The dynamics determines how the physical quantities evolve, either deterministic or probabilistic, by precise, unambiguous equations.

As we know, the "Copenhagen Interpretation" of quantum mechanics does not meet these demands, whereas "Bohm's theory" (Bohm, 1952) does. Bohm's theory was not appreciated by the physics community: neither by Einstein, writing in a letter (Einstein, 1952) to his friend Max Born "That way seems too cheap to me"; nor by Wolfgang Pauli, who strongly disliked theories with hidden parameters which produce no observable effects, see, in particular, Pauli's 1951 letter to David Bohm (Meyenn and Brown, 1996, p. 436). Pauli even called such theories "artificial metaphysics" (Pauli, 1953). Bell, however, was very much impressed by Bohm's work and often remarked, "I saw the impossible thing done", and even more, he continued: "In every quantum mechanics course you should learn Bohm's model!"

Bell also had reservations about the prominent mathematical book of John von Neumann about the foundations of quantum mechanics (von Neumann, 1932). Bell's point of view was more physical than mathematical. Firstly, Bell criticized von Neumann to impose false assumptions in his proof that hidden-variable theories are incompatible with quantum mechanics, as we recall from our discussion in Chapter 12. It was the starting point for Bell's famous paper "On the problem of hidden variables in quantum mechanics" (Bell, 1966), where he pointed to the feature of *contextuality* and stated:

All non-contextual hidden-variable theories [for $\dim(\mathcal{H}) > 2$*] are in conflict with quantum mechanics!*

---

[3]John Bell's views as reported here, including quotations attributed to him, are (unless otherwise stated) based on the personal recollections of Reinhold Bertlmann and private communications between him and John Bell.

This statement has been proven independently by Simon Kochen and Ernst Specker (1967, 2002) and is known as *Kochen–Specker theorem*, recall Chapter 12.

Secondly, Bell particularly disliked von Neumann's description of "projective measurements", where the quantum state "jumps" from one state into another, also sometimes called "collapse of the wave function". As a mathematical operation such a projection functions extremely well, and agrees with the experimental outcomes. But does it correspond to a real physical change of the system? And in what sense is the apparatus involved which most definitely interacts with the system. In this regard the quantum-mechanical formalism makes no statement.

At the same time, we are inclined to venture a guess here that a major part of the physics community would not ascribe physical *reality* to the fundamental Schrödinger wave function (or a density operator for that matter), the "jumping" wave function if one so wishes. Instead, it rather represents a kind of information, a point of view that is becoming increasingly important and that we would like to discuss in the following.

For Bell, however, it was clear that quantum mechanics had to be completed with variables that refer to the *real* properties of the involved objects. He was absolutely convinced that *realism* is the right position to be held by a scientist. He believed that the experimental results are predetermined and not just induced by the measurement process. To put it in John Bell's own words as spoken during an interview he gave in the late 1980s (see Bell 1990, 13:44):

John Bell's confession:

Oh, I'm a realist and I think that idealism is a kind of ... it's a kind of ... I think it's an artificial position which scientists fall into when they discuss the meaning of their subject and they find that they don't know what it means. I think that in actual daily practice all scientists are realists, they believe that the world is really there, that it is not a creation of their mind. They feel that there are things there to be discovered, not a world to be invented but a world to be discovered. So I think that realism is a natural position for a scientist and in this debate about the meaning of quantum mechanics I do not know any good arguments against realism.

For Bell—and for Schrödinger too—the concept of "quantum jumps" was a relict, a leftover from Bohr's old quantum theory which should not be necessary in a complete, consistent theory. Bell was convinced that one has to complete quantum mechanics with additional variables, the hidden variables, which refer to the real properties of the system, and which finally led to the inequalities, named after him.

An apology for the realism of the world was presented by Bell in his delightful article "Bertlmann's socks and the nature of reality" (Bell, 1980; Bell, 1981), where he elucidated with sharpness and wit—typical characteristics of Bell—that an EPR-like situation of (classically correlated) socks—which he illustrated in a cartoon shown in Fig. 13.11—leads to a Bell inequality that is broken by *quantum socks*.

Les chaussettes
de M. Bertlmann
et la nature
de la réalité

Fondation Hugot
juin 17 1980

pink

not
pink

Fig. 13.11 Bell's cartoon of Monsieur Bertlmann. Illustration in the CERN preprint Ref.TH.2926-CERN *"Bertlmann's socks and the nature of reality"* of John Bell from 18 July 1980 (Bell, 1980), published as Fig. 1 in Journal de Physique **42** (C2), 41–61 (1981). The article is based on an invited lecture John Bell gave at le Colloque sur les *"Implications conceptuelles de la physique quantique"*, organized by the Foundation Hugot du Collège de France, 17 June 1980.

Bell's original phrasing:

The philosopher in the street, who has not suffered a course in quantum mechanics, is quite unimpressed by Einstein–Podolsky–Rosen correlations. He can point to many examples of similar correlations in everyday life. The case of Bertlmann's socks is often cited. Dr Bertlmann likes to wear two socks of different colours. Which colour he will have on a given foot on a given day is quite unpredictable. But when you see (on Fig. 13.11) that the first sock is pink you can be already sure that the second sock will not be pink. Observation of the first, and experience of Bertlmann, gives immediate information about the second. There is no accounting for tastes, but apart from that there is no mystery here. And is not the EPR business just the same?

Then Bell proceeds and demonstrates that the EPR business is *not* the same! Why? Assuming that the colour of a sock survives one thousand washing cycles at 45°, or at 90°, or at 0°, one can easily derive a Bell inequality by relying just on classical set theory. Testing the two members of a pair of socks, where the pairing is à la Bertlmann, denoting $P(0° \,|\, 45°)$ the probability of being able to pass the test at 0° and not able at 45°, etc . . . , then the *Bell inequality for socks* is expressed by the following proposition:

Proposition 13.8 (Bell inequality for socks)

$$P(0° \,|\, 45°) + P(45° \,|\, 90°) \geq P(0° \,|\, 90°) \,.$$

The inequality, however, is not respected by *quantum socks*. The quantum-mechanical probability for magnets passing through a non-parallel Stern–Gerlach magnet—which is a realization of quantum socks—is given by

$$P(\text{up}, \text{up}) = P(\text{down}, \text{down}) = \frac{1}{2}\sin^2\frac{\alpha - \beta}{2},\qquad(13.40)$$

where α and β are the angles of the orientations of the Stern–Gerlach magnets at Alice's and Bob's side, respectively.

Then clearly the Bell inequality for socks, Proposition 13.8, is violated,

$$\frac{1}{2}\sin^2 22, 5° + \frac{1}{2}\sin^2 22, 5° \not\geq \frac{1}{2}\sin^2 45°,$$

$$0, 1464 \not\geq 0, 2500.\qquad(13.41)$$

Résumé: The EPR correlations are such that the result of the experiment on one side immediately allows us to predict the outcome on the other side, whenever the analysers are parallel. Following EPR, to avoid invoking an instantaneous action at a distance we are obliged to assume that the results on both sides are predetermined, i.e., they are elements of reality. But this would imply—according to Bell—that there is a conflict with quantum mechanics for non-parallel magnets—there is a form of *non-locality*! Intervention on one side would immediately influence the other.

Interestingly, Bell did hold on to the hidden-variable program, despite the great success of (ordinary) quantum mechanics. He was not discouraged by the outcome of the Bell-type experiments but rather puzzled. For him "The situation was very intriguing that at the foundation of all that impressive success [of quantum mechanics] there are these great doubts", as he once remarked.

Bell was deeply disturbed by this supposedly non-local feature of quantum mechanics since for him it was equivalent to a "breaking of Lorentz invariance" in an extended theory for quantum mechanics, a conclusion he could hardly accept. He often remarked: "It's a great puzzle to me ... behind the scenes something is going faster than the speed of light."

Let us emphasize at this point that there is a fundamental difference in the correlations between *Bertlmann's socks* and *quantum socks*, which is, perhaps due to the amusing character of the analogy, often overlooked.

Correlations:

- **Bertlmann's socks**

  They exhibit classical correlations. Observation of the left sock as being "pink" gives information about the colour of the right sock, "not-pink". But this observation on the left *does not influence* (in the sense of an action) the colour of the right sock. The colour of the socks is predetermined (by Bertlmann in the morning) and is real! There is no mystery here.

Fig. 13.12 Conclusion. Modified sketch from Fig. 8 in Bertlmann, *Bell's Theorem and the Nature of Reality*, Found. Phys. **20**, 1191–1212 (1990); used with permission of Springer Nature BV conveyed through Copyright Clearance Center, Inc. The paper was dedicated to John Bell in 1988 on occasion of his 60th birthday (Bertlmann, 1988; Bertlmann, 1990).

- **Quantum socks**

  The quantum socks, in contrast, show EPR–Bell correlations, quantum correlations due to the entanglement of their joint state. Before the measurement there is only the quantum-mechanical state, somehow neutral between the two possibilities "pink" and "not-pink". Then the decision between the possibilities is made for *both* distant socks by looking (measuring) just *one* of them.

It is this "spooky action at a distance", this seeming non-locality, which excited physicists like Bell, leading to illustrations such as in a cartoon (Fig. 13.12) in the conclusions of the reply to Bell's "Bertlmann's socks" paper, titled "Bell's theorem and the nature of reality" (Bertlmann 1988,1990), which satirizes the (apparently) immediate determination of events at a distant system by events at a nearby system. And Bell remarked (if there is no realism) "It's a mystery if looking at one sock makes the sock pink and the other one not-pink at the same time." Further details on Bell's views can be found in the reminiscences of Reinhold Bertlmann (2002, 2014, 2015, 2017).

13.4.2 Information

A completely different view of the meaning of a quantum state (given by a wave function or, more generally, a density matrix) is that it represents information. "Information about what?" Bell repeatedly remarked. The answer of Anton Zeilinger (1999) and Brukner–Zeilinger (1999, 2001, 2003) is "Information about possible future experimental outcomes!". Thus the quantum states is not about reality, about physical elements (like spin or polarization) that exist prior to or independently of a measurement. It is about *knowledge* or *information*.

This view certainly sheds a different light on non-locality discussed so far. In the case of an entangled state share between two parties Alice and Bob, a measurement by Alice has an instantaneous effect on her ability to predict the outcome of a measurement Bob could make. The *ability to predict* is a kind of *knowledge* that we can use synonymously with *information*. This is the kind of non-locality we observe. It circumvents the vision of an instantaneous evolution due to a *physical* action. So there is no breaking of Lorentz symmetry—which Bell worried about—no contradiction with special relativity.

According to Zeilinger (1999) and Brukner–Zeilinger (1999, 2001, 2003) information is the most fundamental concept in quantum physics. The physical description of a system is nothing but a set of propositions together with their truth values, *"true"* or *"false"*. Any proposition assigned to a quantum system is based on observation and represents our knowledge, i.e., the *information*. Their understanding of information is very much influenced by the "Ur" hypothesis of *Carl Friedrich von Weizsäcker* (1985) and by the "It from bit" idea of *John Archibald Wheeler* (1989). Only a few propositions form the basis of this programme:

Propositions:

- **Information content**
  The information content of a quantum system is finite!

- **Elementary system—one bit**
  An elementary system carries only one bit!

- N **elementary systems—N bits** N elementary systems carry N bits!

Then Brukner and Zeilinger describe the correlation content of composite systems, the correlations of constituents, i.e., the joint properties, and quantitatively define the measure of information, which is a sum of squared probabilities.

Amazingly, relying on these few information-theoretical assumptions Brukner and Zeilinger are able to derive the characteristic features of quantum mechanics like: coherence-interference, complementarity, the "true" randomness, the quantum evolution equation in the form of the von Neumann equation (i.e., the equivalence of the Schrödinger equation), and most importantly entanglement, very much in the sense of Erwin Schrödinger (1935a):

Maximal knowledge of a total system does not necessarily include total knowledge of all its parts, not even when these are fully separated from each other and at the moment are not influencing each other at all.

Exactly in this sense the information content in the correlations of a composite system is larger in the case of entanglement than in the case of separability.

For Brukner and Zeilinger the wave function that describes the state of a system is just the mathematical representation of our knowledge about the system. In a measurement the abrupt collapse of the wave function corresponds merely to our sudden change of knowledge and does not correspond to a real physical process. Therefore no "spooky action at a distance" is involved. According to Brukner and Zeilinger (2003):

When a measurement is performed, our knowledge of the system changes, and therefore its representation, the quantum state, also changes. In agreement with the new knowledge, it instantaneously changes all its components, even those which describe our knowledge in the regions of space quite distant from the site of measurement.

Brukner–Zeilinger's idea of understanding the collapse of the wave function as an update of knowledge goes back to Schrödinger (1935*c*):

The abrupt change by measurement ... is the most interesting part of the entire theory. It is exactly that point which requires breaking with naive realism. For this reason, the psi-function cannot take the place of the model or of something real. Not because we can't expect a real object or a model to change abruptly and unexpectedly, but because from a realist point of view, observation is a natural process like any other which can't cause a disruption of the course of nature.

Furthermore, Brukner's and Zeilinger's view is very much in the sense of *Shimon Malin* "What are quantum states?" (Malin, 2006), who argues against an ontic interpretation of quantum states but shows that an epistemic interpretation is not an appropriate alternative. Instead, Malin takes the following view:

Quantum states as representing the available knowledge about the potentialities of a quantum system from the perspective of a particular point in space. Unlike ordinary knowledge, which requires a knower, available knowledge can be assumed to be present regardless of a knower.

And Časlav Brukner has formulated such a view more precisely in his article "On the quantum measurement problem" (Brukner, 2016):

The quantum state is a representation of knowledge necessary for a hypothetical observer—respecting her experimental capabilities—to compute probabilities of outcomes of all possible future experiments.

Investigations along these lines have been carried out by *Borivoje Dakić* and Brukner (2011, 2016), concentrating on "The essence of entanglement" by Časlav Brukner, Marek Żukowski, and Anton Zeilinger (2001).

Also in the famous "Wigner's-friend-type experiments" (Wigner, 1961; Wigner, 1963; Deutsch, 1985)—for literature on this topic, see, e.g., Baumann *et al.* (2018), Baumann and Wolf (2018), Baumann and Brukner (2020), DeBrota *et al.* (2020), Baumann (2021), Guérin *et al.* (2021), Lostaglio and Bowles (2021), Renes (2021), and Xu *et al.* (2021)—which are directly linked to the measurement problem, the notion of realism—facts that exist independently of an observer—cannot be upheld. Brukner, when investigating this more closely, concludes (2016):

Measurement records— "facts"—coexisting for both Wigner and his friend ... can have a meaning only relative to the observer; there are no "facts of the world per se".

That's a strong statement against realism, indeed—one that should be viewed also in context of the result of *Matthew F. Pusey, Jonathan Barrett,* and *Terry Rudolph* (2012) who proved a no-go theorem, the *Pusey–Barrett–Rudolph theorem*, which they summarize as: "... any model in which a quantum state represents mere information about an underlying physical state of the system, and in which systems that are prepared independently have independent physical states, must make predictions that contradict those of quantum theory." Together with further results in this direction by *Roger Colbeck* and *Renato Renner* (2012, 2017), this can be seen as a blow to the view to regard the quantum state as "mere" information about reality, or as a further argument against realism—perhaps there is no underlying ontology.

Finally, we want to draw attention to several information-theoretic studies. Firstly, to the work of *Lucien Hardy* (2001) who derives the characteristics of quantum mechanics from five axioms, secondly, to the approaches of *Rob Clifton, Jeffrey Bub,* and *Hans Halvorson* (2003), *Lluis Masanes* and *Markus P. Müller* (2011), as well as by *Giulio Chiribella, Giacomo Mauro D'Ariano,* and *Paolo Perinotti* (2011), who recover quantum mechanics by information-theoretic constraints. Secondly, to work on quantum Bayesiansm, specifically to "QBism" advocated by *Christopher Fuchs* and *Rüdiger Schack* (2013, 2015) as well as by *David Mermin* (2014, 2017), and last but not least to "Relational quantum mechanics" of *Carlo Rovelli* (1996, see also the exchange with Brukner and *Jacques Pienaar,* Di Biagio and Rovelli, 2022; Di Biagio and Rovelli, 2021; Brukner, 2021; Pienaar, 2021). A discussion of (some of) these works as well as some by Brukner and Zeilinger (1999, 2001, 2003) can also be found in the Bachelor theses of *Ferdinand Horvath* (2013) and *Christoph Regner* (2015).

Résumé: An alternative view to realism is the information-theoretic approach towards "understanding" quantum mechanics. Here, we have focused on the work of Brukner and Zeilinger and leave other interpretations to be discovered by the reader. For Brukner and Zeilinger quantum mechanics is a theory of information, a mathematical representation of what an observer has to know in order to calculate the probabilities for the outcomes of measurements. For them the wave function that describes the state of a system is just the mathematical representation of our knowledge about the system. The so-called "collapse of the wave function" in a measurement process corresponds merely to the sudden change of knowledge and does not represent a real physical process. Therefore, the problem of "breaking of Lorentz invariance", when considering the *non-locality* in entangled quantum systems, does not occur here. Brukner's and Zeilinger's view actually emphasizes the importance of mathematics in describing and communicating with nature—a view we share too.

14
Quantum Teleportation

Quantum teleportation is a communication protocol invented by *Charles Bennett* and his collaborators *Gilles Brassard, Claude Crépeau, Richard Jozsa, Asher Peres*, and *William K. Wootters* in 1993 that enshrines one of the most exciting issues in quantum mechanics into a practically useful procedure that can be considered to form the basis for modern quantum information and quantum communication. The protocol is a special kind of information transfer from Alice to Bob, borrowing its name from science fiction and sometimes even referred to as *beaming* in popular culture. However, the image invoked by the latter term is quite misleading, there is *no* transfer of matter or energy involved in quantum teleportation beyond that associated entirely with classical communication. Alice's particle is not being physically transported to Bob, only the particle's quantum state is being transferred. But, as we will see, this is enough from an information-theoretic point of view. Let us see how it works.

14.1 Quantum Teleportation

14.1.1 The Teleportation Protocol

In the teleportation protocol we again consider two observers, Alice and Bob. The aim of quantum teleportation is the transmission of a known or unknown (unspecified) quantum state—for instance, one qubit, which we will consider for now—from one physical system at Alice's location onto another system at Bob's location by exploiting the correlations between an EPR pair. The basis for such a phenomenon is the observation that, quantum-mechanically, the involved qubits can be (maximally) entangled in different ways.

Specifically, suppose that Alice receives a message represented by the quantum state $| \psi \rangle_1$ of a quantum system that we label 1, and she wishes to send this message (state) to Bob. This could be done by sending the quantum system directly to Bob using a suitable channel. However, here we will consider another option, illustrated in Fig. 14.1. There, Alice keeps the particular quantum system 1 under her control but in addition Alice and Bob share a bipartite quantum system consisting of subsystems 2 and 3 in a maximally entangled state $| \psi^- \rangle_{23}$, such that Alice receives system 2, while Bob receives system 3. This can be achieved by a an EPR source under the control of one of the observers or even a third party, which produces, for example, two entangled photons, one of which is sent to Alice and the other to Bob (much like in a Bell test, see Chapter 13).

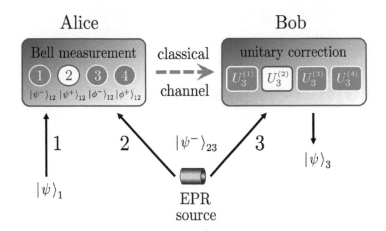

Fig. 14.1 Quantum teleportation scheme. The state $|\psi\rangle$, at first carried by system 1 at Alice's location is transferred to system 3 at Bob's location via a previously shared pair of entangled systems 2 and 3. This is done by entangling systems 1 and 2 via a Bell measurement and transmitting the information of the outcome (e.g., outcome 2 corresponding to the state $|\psi^+\rangle_{12}$) of this measurement to Bob via a classical channel. Based on this information, Bob can apply a (unitary) correction operation, e.g., $U_3^{(2)}$ in the displayed example, to recover the state $|\psi\rangle_3$.

Let us now phrase this setup in terms of the mathematical formalism of quantum mechanics. The quantum states, described by vectors, are elements of a tensor product of three Hilbert spaces, $\mathcal{H}_{\text{tot}} = \mathcal{H}_1 \otimes \mathcal{H}_2 \otimes \mathcal{H}_3$, and we assume for now that each of the subspaces $\mathcal{H}_i$ for $i = 1, 2, 3$ has dimension $d = 2$. The spaces $\mathcal{H}_1$ and $\mathcal{H}_2$ are controlled by Alice, while $\mathcal{H}_3$ is controlled by Bob. The state $|\psi\rangle_1$ of Alice's first quantum system can generally be written as

$$|\psi\rangle_1 = \alpha\,|0\rangle_1 + \beta\,|1\rangle_1\,,\tag{14.1}$$

with $\alpha, \beta \in \mathbb{C}$ and $|\alpha|^2 + |\beta|^2 = 1$. This could, for example, be physically realized by the polarization of a photon, in which case one could choose, for instance, horizontal and vertical linear polarization as the computational basis, $|0\rangle \equiv |H\rangle$ and $|1\rangle \equiv |V\rangle$. Meanwhile, Alice knows that her second quantum system $\mathcal{H}_2$ is entangled with Bob's system $\mathcal{H}_3$ by a source of entangled photons, the EPR source, such that the total state restricted to $\mathcal{H}_2 \otimes \mathcal{H}_3$ is maximally entangled and given by the antisymmetric Bell state

$$|\psi^-\rangle_{23} = \tfrac{1}{\sqrt{2}}\big(|0\rangle_2 \otimes |1\rangle_3 - |1\rangle_2 \otimes |0\rangle_3\big)\,,\tag{14.2}$$

while the joint state of all three quantum systems is

$$|\psi\rangle_{123} = |\psi\rangle_1 \otimes |\psi^-\rangle_{23} = \tfrac{\alpha}{\sqrt{2}}\big(|0\rangle_1|0\rangle_2|1\rangle_3 - |0\rangle_1|1\rangle_2|0\rangle_3\big)$$
$$+ \tfrac{\beta}{\sqrt{2}}\big(|1\rangle_1|0\rangle_2|1\rangle_3 - |1\rangle_1|1\rangle_2|0\rangle_3\big)\,,\tag{14.3}$$

where we have dropped the tensor-product symbols in the last step for ease of notation.

In writing the decomposition in eqn (14.3), we have made the choice of representing the state of systems 1 and 2 in terms of the two-qubit computational basis $\{|00\rangle_{12}, |01\rangle_{12}, |10\rangle_{12}, |11\rangle_{12}\}$. However, we may as well utilize the fact that the four Bell states $|\psi^{\pm}\rangle$ and $|\phi^{\pm}\rangle$, given by

$$|\psi^{\pm}\rangle = \tfrac{1}{\sqrt{2}}(|01\rangle \pm |10\rangle)\,, \tag{14.4a}$$

$$|\phi^{\pm}\rangle = \tfrac{1}{\sqrt{2}}(|00\rangle \pm |11\rangle)\,, \tag{14.4b}$$

form a complete orthonormal system for two qubits, which allows for any state in $\mathcal{H}_1 \otimes \mathcal{H}_2$ to be expanded in terms of the Bell states. Rewriting the total state (14.3) in such an expansion, we have

$$
\begin{aligned}
|\psi\rangle_{123} = \tfrac{1}{2\sqrt{2}}\Big[&- \big(|0\rangle_1 |1\rangle_2 - |1\rangle_1 |0\rangle_2\big)\big(\alpha|0\rangle + \beta|1\rangle\big) \\
&- \big(|0\rangle_1 |1\rangle_2 + |1\rangle_1 |0\rangle_2\big)\big(\alpha|0\rangle - \beta|1\rangle\big) \\
&+ \big(|0\rangle_1 |0\rangle_2 - |1\rangle_1 |1\rangle_2\big)\big(\beta|0\rangle + \alpha|1\rangle\big) \\
&- \big(|0\rangle_1 |0\rangle_2 + |1\rangle_1 |1\rangle_2\big)\big(\beta|0\rangle - \alpha|1\rangle\big)\Big],
\end{aligned} \tag{14.5}
$$

which suggests that qubits 1 and 2 can be considered to be in a superposition of being entangled in four different ways, corresponding to projections onto the four Bell states. In each of the four cases, the remaining state of system 3 is related to the original state in eqn (14.1) by a unitary transformation. Thus, up to some local unitary transformations $U_3^{(i)}$ on system 3, the expansion of eqn (14.5) can be rewritten more compactly as

$$
\begin{aligned}
|\psi\rangle_{123} = \tfrac{1}{2}\Big[&- |\psi^-\rangle_{12}\, U_3^{(1)}\, |\psi\rangle_3 - |\psi^+\rangle_{12}\, U_3^{(2)}\, |\psi\rangle_3 \\
&+ |\phi^-\rangle_{12}\, U_3^{(3)}\, |\psi\rangle_3 - |\phi^+\rangle_{12}\, U_3^{(4)}\, |\psi\rangle_3\,,
\end{aligned} \tag{14.6}
$$

where the unitary transformations are given by the identity and the Pauli matrices, respectively, i.e.,

$$
U_3^{(1)} = \mathbb{1} = \begin{pmatrix} 1 & 0 \\ 0 & 1 \end{pmatrix}, \quad U_3^{(2)} = \sigma_z = \begin{pmatrix} 1 & 0 \\ 0 & -1 \end{pmatrix},
$$

$$
U_3^{(3)} = \sigma_x = \begin{pmatrix} 0 & 1 \\ 1 & 0 \end{pmatrix}, \quad U_3^{(4)} = i\,\sigma_y = \begin{pmatrix} 0 & 1 \\ -1 & 0 \end{pmatrix}. \tag{14.7}
$$

Alice then performs a so-called *Bell measurement*, a projective measurement in the basis of maximally entangled Bell states of subsystems 1 and 2. Quantum-mechanically, this can be described by projection operators onto the four states of eqn (14.4). For instance, applying the projector $P_{12}^- = |\psi^-\rangle\langle\psi^-|_{12}$ onto the Bell state $|\psi^-\rangle_{12}$ to the joint state of eqn (14.6) we have

$$|\psi^-\rangle\langle\psi^-|_{12} \otimes \mathbb{1}_3 \, |\psi\rangle_{123} = -\tfrac{1}{2}\,|\psi^-\rangle_{12} \otimes (\alpha|0\rangle_3 + \beta|1\rangle_3) = -\tfrac{1}{2}\,|\psi^-\rangle_{12}|\psi\rangle_3, \tag{14.8}$$

with the result that the quantum state of system 3 on Bob's side is left in the state initially carried by system 1.

However, Bob does not know onto which of the four Bell states the total state was projected; this happens randomly and all four outcomes have the same probability,

$$P(\psi^-) = \mathrm{Tr}\big(P_{12}^-\,|\psi\rangle\langle\psi|_{123}\big) = \tfrac{1}{4} = P(\psi^+) = P(\phi^-) = P(\phi^+), \tag{14.9}$$

independently of the particular initial state $|\psi\rangle_1$ of system 1 that is to be teleported. Consequently, Bob requires additional classical information from Alice (e.g., via mobile phone) to decide which of the four unitary operations $U_3^{(1)}$, $U_3^{(2)}$, $U_3^{(3)}$, or $U_3^{(4)}$ he has to perform in order to recover the state $|\psi\rangle_3$. Without access to this information, the state that Bob ascribes to system 3 is a probabilistic mixture of the four possible states $U_3^{(i)}|\psi\rangle_3$ with equal probabilities, which evaluates to

$$\rho_3 = \sum_{i=1}^{4} \tfrac{1}{4} U_3^{(i)}|\psi\rangle\langle\psi|_3 U_3^{(i)\dagger}$$

$$= \frac{1}{4}\left[\begin{pmatrix} |\alpha|^2 & \alpha\beta^* \\ \alpha^*\beta & |\beta|^2 \end{pmatrix} + \begin{pmatrix} |\alpha|^2 & -\alpha\beta^* \\ -\alpha^*\beta & |\beta|^2 \end{pmatrix} + \begin{pmatrix} |\beta|^2 & \beta\alpha^* \\ \beta^*\alpha & |\alpha|^2 \end{pmatrix} + \begin{pmatrix} |\beta|^2 & -\beta\alpha^* \\ -\beta^*\alpha & |\alpha|^2 \end{pmatrix} \right]$$

$$= \frac{1}{2}\begin{pmatrix} |\alpha|^2 + |\beta|^2 & 0 \\ 0 & |\alpha|^2 + |\beta|^2 \end{pmatrix} = \frac{1}{2}\begin{pmatrix} 1 & 0 \\ 0 & 1 \end{pmatrix} = \rho_{\mathrm{mix}}, \tag{14.10}$$

i.e., a maximally mixed state as in eqn (11.19) carrying no information about the teleported state $|\psi\rangle_1$. But once Bob obtains the information about Alice's measurement outcome, which can be encoded in two classical bits, he can apply the corresponding correction operation and find his system 3 in exactly the same state that system 1 was in before.

With the main ingredients of the protocol established, let us remark that all the involved experimental procedures, the distribution of entangled pairs, the Bell measurement, and the correction operations can nowadays be easily implemented experimentally and a series of experiments demonstrating teleportation have been performed, as we will discuss in more detail in Section 14.2.

But what makes quantum teleportation so special as a form of communicating information? Before returning to this question, let us make some important remarks and observations about the protocol.

14.1.2 Remarks on the Teleportation Protocol

First, we observe that the details of the quantum state to be teleported do not matter. The state can be left unspecified and can even be entirely unknown without influencing the protocol. In particular, since the protocol works in the same way for any initial pure state, it also works for any mixed quantum state ρ_1 that Alice might want to teleport. Consequently, the protocol also goes through if the quantum system 1 that initially carries the state to be teleported is a subsystem of larger quantum system, i.e., it may itself be entangled with another quantum system. This is the basis for the phenomenon of *entanglement swapping* and the subsequent construction of *quantum repeaters* (Briegel *et al.*, 1998), which we will discuss in Section 14.3.1.

Second, note that the initial state of the quantum-information carrier system 1 is transferred to system 3 during the protocol and the final state of system 1 after the Bell measurement no longer carries any information about its initial state $|\psi\rangle_1$. That is, once systems 1 and 2 have been randomly projected into one of the Bell states, the respective state does not depend on the parameters α and β any more. Thus, the teleportation protocol really transfers the information about the initial state of system 1 to system 3 and does not create a copy of the initial state. Indeed, it is generally impossible to (perfectly) duplicate an unknown quantum state, which is expressed in the so-called *no-cloning theorem* (Wootters and Zurek, 1982) that we will discuss in more detail in Section 21.5. In many applications, in particular in those employing photons, the question of the information content in system 1 does not naturally come up though, since the incoming photon in channel 1 is destroyed during the teleportation process.

Third, there is *no superluminal transfer of information*. Although this may be a trope (implicitly) suggested by popular science or at least pop-culture references to teleportation, the measurement on Alice's systems does not instantaneously lead to any physical effect on Bob's system, at least when taking an information-theoretic stance as that outlined in Section 13.4.2, while a realist point of view as described in Section 13.4.1 might lead one to assume that the transmission occurs *instantaneously* via EPR correlations and that all information must instantaneously be present at Bob's side. However, in order for Bob to obtain any meaningful results from the system under his control he needs additional information from Alice, which is transmitted by a classical channel. Therefore, even if one were to take a realist perspective, special relativity is not violated, since the (classical) signal cannot travel faster than the speed of light. From an information-centric perspective, it is enough to say that Alice's knowledge of the state of Bob's system changes (is updated, when using the language of Bayesian inference) when she observes the measurement outcome, but her knowledge must be sent to Bob (at most at the speed of light) in order to update his knowledge of the state he controls. However, aside from keeping these considerations in mind, the teleportation process does not underly any distance restrictions in principle—it can occur over any distances.

Finally, we should mention that although we have presented the protocol for qubits here, one may teleport the state of any quantum system in principle. In particular, let us consider the state $|\psi\rangle_1$ of a d-dimensional quantum system—a *qudit* in the language of quantum information theory. To apply the teleportation protocol, we simply assume that Alice and Bob first share a maximally entangled two-qudit state, e.g., one of the states

$$| \phi_{mn} \rangle_{23} = \frac{1}{\sqrt{d}} \sum_{k=0}^{d-1} e^{\frac{2\pi i}{d} nk} | k \rangle_2 | (k+m)_{\mathrm{mod}(d)} \rangle_3 \, , \qquad (14.11)$$

with $m, n = 0, \ldots, d-1$, for instance $| \psi_{00} \rangle = \frac{1}{\sqrt{2}} \sum_{k=0}^{d-1} | k \rangle_2 | k \rangle_3$. Alice then has to perform a Bell measurement, but instead of carrying out this measurement with respect to the basis consisting of the four two-qubit Bell states, the measurement has to be performed with respect to the basis $\{| \psi_{mn} \rangle_{12}\}_{m,n}$ of maximally entangled states in eqn (14.11). For such a measurement there are d^2 possible outcomes, the correct one of which will have to be communicated to Bob classically, and Bob then has to apply one of the corresponding d^2 correction unitaries to retrieve the state $| \psi \rangle_1$ (see, e.g., Bennett *et al.*, 1993 for more details).

With these observations giving some context to the teleportation protocol, let us now turn to some milestones in the experimental demonstration of this effect, before we examine quantum teleportation in the context of more general quantum communication protocols in Section 14.3.

14.2   Experiments on Quantum Teleportation

14.2.1   Milestones of Experimental Teleportation

The first experiment where a quantum state independent from the source was teleported from Alice to Bob was carried out by the team of *Anton Zeilinger* in Innsbruck in 1997. In the years after, several other groups reported on successful demonstrations that refined various aspects of teleportation using photon polarization (Boschi *et al.*, 1998; Kim *et al.*, 2001), optical coherence (Furusawa *et al.*, 1998), and nuclear magnetic resonance (Nielsen *et al.*, 1998). Many practically useful protocols have been tested using photons, for instance, for long-distance high-fidelity communication (Pan *et al.*, 2001; Marcikic *et al.*, 2003; Ma *et al.*, 2012a) or recently for chip-to-chip teleportation with applications to integrated photonic quantum technologies (Llewellyn *et al.*, 2020). A particularly appealing variant of the experiment was carried out by *Rupert Ursin* and collaborators in 2004, who teleported a quantum state across the river Danube in Vienna, see Fig. 14.2.

In 2012, a notable record for quantum teleportation across the longest distance was set by Zeilinger's group (Ursin *et al.*, 2007; Ma *et al.*, 2012a), who achieved teleportation over a distance of 143 km (or 89.5 mi) between the two Canary Islands of La Palma and Tenerife off the Atlantic coast of North-Africa. Conceptually, this already represents a significant step for potential space applications, since this corresponds to the minimum distance between the ground stations on Earth and orbiting satellites, see Fig. 14.3. But this should only be a portent of developments to come. Indeed, more recently, in 2017, the previous distance record was significantly surpassed by a demonstration of quantum teleportation between the ground and actual satellites in orbit at distances of up to 1400 km by the group of *Jian-Wei Pan* (Ren *et al.*, 2017).

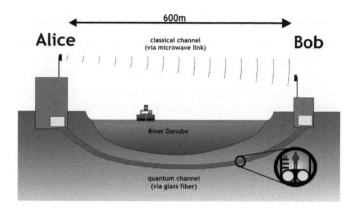

Fig. 14.2 Quantum teleportation under the river Danube. The experiment was performed by a group around Rupert Ursin and Anton Zeilinger in Ursin *et al.* (2004), where EPR pairs, created at Alice's location on one side of the Danube, were shared via a glass fibre to teleport quantum states across the river at a distance of approximately 600 m. Figure used with permission of Anton Zeilinger and Markus Aspelmeyer © OEAW.

In parallel to advances in photonic setups, there have been significant advances in realizing teleportation with massive particles, like atoms and ions, for which experiments have been carried out successfully by several groups (Barrett *et al.*, 2004; Riebe *et al.*, 2004; Olmschenk *et al.*, 2009; Nölleke *et al.*, 2013; Pfaff *et al.*, 2014).

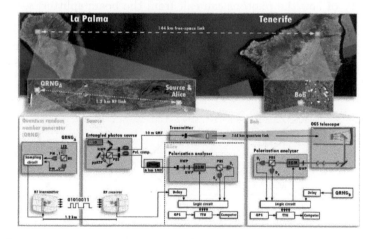

Fig. 14.3 Quantum teleportation between La Palma and Tenerife. Successful teleportation over a distance of 144 km (or 89.5 mi) was achieved by Ursin, Zeilinger and collaborators as reported in (Ursin *et al.*, 2007). With a similar setup shown here, enhanced by the addition of quantum random-number generators, the violation of a Bell inequality without the locality and freedom-of-choice loopholes was demonstrated in (Scheidl *et al.*, 2010). Image from Fig. 1 in Scheidl et al., "Violation of local realism with freedom of choice", Proc. Natl. Acad. Sci. U.S.A. **107** (46), 19708–19713, (2010).

14.2.2   Experimental Bell-state Measurements via a Beam Splitter

A particular detail of the experimental implementation using photon pairs entangled in their polarization that we have not discussed yet is the realization of the Bell measurement. To achieve this, Alice inserts a beam splitter into the channels $(1, 2)$ of the incoming particles. As the name suggests, a beam splitter is a (typically optical, when referring to photons) device that splits a beam of particles into two. It is a crucial part of most interferometers. In the case of photons, such a device can be realized by a half-silvered mirror or from two joined triangular prisms, and we here consider the transmittivity to be such that 50% of the particles in an incident beam are transmitted and 50% reflected. Let us describe this situation quantum-mechanically in more detail.

A 50:50 standard beam splitter has two spatial input modes (u, v) and two spatial output modes (x, y) (Weinfurter, 1994; Braunstein and Mann, 1995; Pan, 1999). Quantum-mechanically, we can describe the action of the beam splitter on the input modes as

$$|u\rangle \rightarrow \frac{1}{\sqrt{2}}(i\,|x\rangle + |y\rangle), \quad |v\rangle \rightarrow \frac{1}{\sqrt{2}}(|x\rangle + i\,|y\rangle). \tag{14.12}$$

The phase-flip factor $i = e^{i\frac{\pi}{2}}$ is due to the reflection at the semi-transparent mirror.

Suppose now that two otherwise indistinguishable photons, photon 1 with a given polarization state in the input mode u and photon 2 with another polarization state in mode v, are incident on the beam splitter. Each photon has the same probability $p = \frac{1}{2}$ to be transmitted or reflected, therefore there are four possibilities with the same probability of how the photons can exit from the beam splitter, see Fig. 14.4.

Classically, the particles are indistinguishable and thus no interferences occur. We can predict that in two cases the particles are found in different output channels with probability $p = \frac{1}{2}$ and in the other two cases the particles exit in the same output channel, with probability $p = \frac{1}{4}$ for each of the two, see Fig. 14.4.

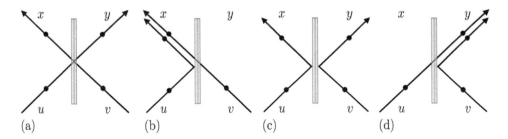

Fig. 14.4 Photon pair incident on a beam splitter. Two indistinguishable incident photons simultaneously hit a beam splitter from different sides, labelled as the input channels u and v. There exist four possibilities, illustrated in (a), (b), (c), and (d), respectively, how the particles may leave from the beam splitter. As it is not possible to decide which of the incident photons arrives at a given output channel, x or y, we obtain coherent superpositions corresponding to the states for the different possibilities, resulting in a maximally entangled Bell state.

Quantum-mechanically, however, the indistinguishability of the particles means that we have to consider a superposition of the amplitudes for these possibilities. We start with input state

$$|\psi_{\text{in}}\rangle = (\alpha\,|\,H\,\rangle_1 + \beta\,|\,V\,\rangle_1)\otimes|\,u\,\rangle_1 \otimes (\gamma\,|\,H\,\rangle_2 + \delta\,|\,V\,\rangle_2)\otimes|\,v\,\rangle_2\,, \quad (14.13)$$

where photon 1 enters in the spatial mode u and photon 2 in the spatial mode v, and the pairs $\alpha,\beta \in \mathbb{C}$ and $\gamma,\delta \in \mathbb{C}$ with $|\alpha|^2 + |\beta|^2 = 1 = |\gamma|^2 + |\delta|^2$ determine the polarization states of the two photons, respectively, which is reflected in the notation $|\,H\,\rangle \equiv |\,0\,\rangle$ and $|\,V\,\rangle \equiv |\,1\,\rangle$.

As the photons pass through the beam splitter the spatial modes are unitarily transformed according to eqn (14.12). If the two particles were distinguishable we would hence obtain the output state

$$|\psi_{\text{out}}\rangle_{12} = \frac{1}{\sqrt{2}}(\alpha\,|\,H\,\rangle_1 + \beta\,|\,V\,\rangle_1)\otimes(i\,|\,x\,\rangle_1 + |\,y\,\rangle_1)$$

$$\otimes \frac{1}{\sqrt{2}}(\gamma\,|\,H\,\rangle_2 + \delta\,|\,V\,\rangle_2)\otimes(|\,x\,\rangle_2 + i\,|\,y\,\rangle_2)\,. \quad (14.14)$$

However, since the photons are indistinguishable and have to obey Bose–Einstein statistics the final two-photon state (the polarization and the spatial part) must be symmetric under exchange of the labels 1 and 2. Denoting the state of eqn (14.14) with exchanged particle labels as $|\psi_{\text{out}}\rangle_{21}$, we have

$$|\psi_{\text{out}}\rangle_{21} = \frac{1}{\sqrt{2}}(\alpha\,|\,H\,\rangle_2 + \beta\,|\,V\,\rangle_2)\otimes(i\,|\,x\,\rangle_2 + |\,y\,\rangle_2)$$

$$\otimes \frac{1}{\sqrt{2}}(\gamma\,|\,H\,\rangle_1 + \delta\,|\,V\,\rangle_1)\otimes(|\,x\,\rangle_1 + i\,|\,y\,\rangle_1)\,, \quad (14.15)$$

and the final outgoing state is given by an equally weighted superposition of these two possibilities of labelling the particles (we will return to the discussion of symmetrization for quantum states of light in Section 22.4.2), that is

$$|\psi_{\text{out}}\rangle = \frac{1}{\sqrt{2}}(|\psi_{\text{out}}\rangle_{12} + |\psi_{\text{out}}\rangle_{21})\,. \quad (14.16)$$

Restructuring eqn (14.16) by separating and collecting the polarization and the spatial part we finally arrive at

$$|\psi_{\text{out}}\rangle = \frac{1}{2\sqrt{2}}\Big\{(\alpha\gamma + \beta\delta)\,(|\,H\,\rangle_1|\,H\,\rangle_2 + |\,V\,\rangle_1|\,V\,\rangle_2)\,i\,(|\,x\,\rangle_1|\,x\,\rangle_2 + |\,y\,\rangle_1|\,y\,\rangle_2)$$

$$+ (\alpha\gamma - \beta\delta)\,(|\,H\,\rangle_1|\,H\,\rangle_2 - |\,V\,\rangle_1|\,V\,\rangle_2)\,i\,(|\,x\,\rangle_1|\,x\,\rangle_2 + |\,y\,\rangle_1|\,y\,\rangle_2)$$

$$+ (\alpha\delta + \beta\gamma)\,(|\,H\,\rangle_1|\,V\,\rangle_2 + |\,V\,\rangle_1|\,H\,\rangle_2)\,i\,(|\,x\,\rangle_1|\,x\,\rangle_2 + |\,y\,\rangle_1|\,y\,\rangle_2)$$

$$+ (\alpha\delta - \beta\gamma)\,(|\,H\,\rangle_1|\,V\,\rangle_2 - |\,V\,\rangle_1|\,H\,\rangle_2)\,(|\,y\,\rangle_1|\,x\,\rangle_2 - |\,x\,\rangle_1|\,y\,\rangle_2)\Big\}\,,$$

$$(14.17)$$

where we insert the Bell states from (14.4) in the linear-polarization basis to arrive at

$$|\psi_{\text{out}}\rangle = \frac{1}{2}\Big\{ (\alpha\gamma + \beta\delta)\,|\phi^+\rangle_{12}\, i\,(|x\rangle_1|x\rangle_2 + |y\rangle_1|y\rangle_2)$$
$$+ (\alpha\gamma - \beta\delta)\,|\phi^-\rangle_{12}\, i\,(|x\rangle_1|x\rangle_2 + |y\rangle_1|y\rangle_2)$$
$$+ (\alpha\delta + \beta\gamma)\,|\psi^+\rangle_{12}\, i\,(|x\rangle_1|x\rangle_2 + |y\rangle_1|y\rangle_2)$$
$$+ (\alpha\delta - \beta\gamma)\,|\psi^-\rangle_{12}\, (|y\rangle_1|x\rangle_2 - |x\rangle_1|y\rangle_2) \Big\}. \tag{14.18}$$

We find that the final output state (14.18) is a superposition of all four Bell states for the polarization, each in a product state with certain two-photon states for the spatial output modes. However, for three of the four possibilities—whenever the polarization state is symmetric with respect to the exchange of the particle labels—the two photons always exit in the same output mode, either both in x or both in y.

For the antisymmetric Bell state $|\psi^-\rangle_{12} = \frac{1}{\sqrt{2}}(|H\rangle_1|V\rangle_2 - |V\rangle_1|H\rangle_2)$, on the other hand, the output state of the spatial modes must also be antisymmetric for the bosonic two-photon system, as we see in the last line of eqn (14.18). Therefore we can identify the polarization Bell state $|\psi^-\rangle_{12}$ uniquely if the two photons are detected simultaneously on different sides of the beam splitter, see Fig. 14.5 (a). Meanwhile, the symmetric Bell state $|\psi^+\rangle_{12} = \frac{1}{\sqrt{2}}(|H\rangle_1|V\rangle_2 + |V\rangle_1|H\rangle_2)$, where the two photons appear on the same side, can be identified with the help of two additional polarizing beam splitters (PBS, also called two-channel polarizers), one inserted into each of the output channels. A PBS is a beam splitter that fully transmits one polarization and fully reflects photons with the orthogonal polarization, e.g., transmitting horizontally polarized photons and reflecting vertically polarized ones. In this way, two photons exiting from different sides of either of the PBSs identify the polarization state $|\psi^+\rangle_{12}$, see Fig. 14.5 (b).

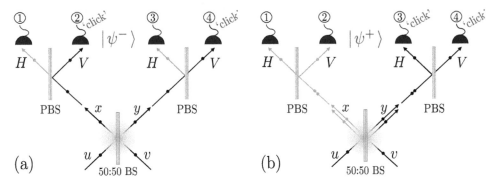

Fig. 14.5 Bell-state detection on a beam splitter. A 50:50 beam splitter (BS), two polarizing beam splitters (PBS), and four photon detectors can be arranged such that certain click patterns uniquely identify two out of the four Bell states. Simultaneous clicks of (a) detectors 1 and 3, or of 2 and 4 identify the antisymmetric Bell states $|\psi^-\rangle$, whereas (b) simultaneous clicks in detectors 1 and 2, or in 3 and 4 identify the state $|\psi^+\rangle$. The remaining two Bell states $|\phi^\pm\rangle$ both result in two photons arriving at the same detector, and hence cannot be distinguished in this setup.

The two remaining Bell states $|\phi^{\pm}\rangle_{12} = \frac{1}{\sqrt{2}}(|H\rangle_1|H\rangle_2 \pm |V\rangle_1|V\rangle_2)$, for which both photons exit the first beam splitter in the same spatial mode and both photons carry the same polarization $|H\rangle_1|H\rangle_2$ or $|V\rangle_1|V\rangle_2$, cannot be identified by this arrangement. Nevertheless, beam splitters provide a convenient splitting of the final state into the Bell states such that an observation of specific click patterns in suitable photon detectors—projections onto particular combinations of spatial output modes—allow us to identify two of the four Bell states.

Résumé: Quantum teleportation is the transmission of an unspecified quantum state from one system at Alice's location onto another system at Bob's location via a previously shared EPR pair. Alice performs a joint Bell measurement of the incoming particle and her half of the EPR pair. In terms of Alice's knowledge of the state of Bob's remaining particle the state of the second EPR particle is updated *instantaneously* via EPR correlations. However, in order for Bob to read out a meaningful result from his data he needs additional information from Alice (which Bell state she measured), transmitted by a classical channel. There is *no* transfer of matter or energy involved. Alice's particle is not being physically transported to Bob, only the particle's quantum state is being transferred. Experimentally, quantum teleportation has been successfully carried out over large distances in space (e.g., up to 144 km between ground stations on Earth, and up to 1400 km between satellites and the ground) by several experimental groups.

14.3   Primitives of Quantum Communication

Since its inception in the 1990s, the teleportation protocol has motivated and inspired numerous communication protocols that explore the possibilities offered by communication based on the exchange of quantum systems. The field of quantum communication has become vast since then, and covering it in its entirety would fill books in its own right, but we nevertheless want to outline a few of the most basic quantum communication protocols in this section.

14.3.1   Entanglement Swapping

Another striking quantum phenomenon that is closely related to the teleportation of individual quantum states is so-called *entanglement swapping* (Żukowski *et al.*, 1993). It can be interpreted as the teleportation of a qubit that is itself part of an EPR pair, and therefore does not itself have well-defined polarization properties. As illustrated in Fig. 14.6, the setup for entanglement swapping hence consists of four quantum systems, labelled 1, 2, 3, and 4, respectively, of which the pairs of systems 1 and 2 as well as 3 and 4 are initially in maximally entangled Bell states. By performing a Bell measurement on systems 2 and 3, which can be considered as entangling these systems, the entanglement is swapped to systems 1 and 4. Experimentally entanglement swapping has been first demonstrated by the team of Anton Zeilinger, including *Jian-Wei Pan*, *Dik Bouwmeester*, and *Harald Weinfurter* (1998) and nowadays represents a standard procedure in quantum-information processing (Bertlmann and Zeilinger, 2002).

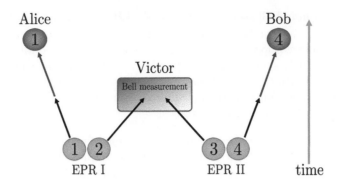

Fig. 14.6 Entanglement swapping. Two pairs of entangled quantum systems (e.g., photons) are shared by the sources EPR I and EPR II. When Victor entangles two systems from different pairs, i.e., performs a Bell measurement on systems 2 and 3, the remaining two systems 1 and 4 controlled by Alice and Bob, respectively, become entangled—the entanglement has been swapped.

In quantum mechanics the setup is described by a tensor product of four Hilbert spaces $\mathcal{H}_1 \otimes \mathcal{H}_2 \otimes \mathcal{H}_3 \otimes \mathcal{H}_4$ of equal dimensions. The entanglement-swapping phenomenon then illustrates the different possibilities for partitioning a four-fold tensor product into pairs of bipartite Hilbert spaces, in particular, $1, 2|3, 4$ or $1, 4|2, 3$. Let's discuss the formalism in detail.

In an experimental setup corresponding to the sketch in Fig. 14.6 we have two EPR sources, EPR I and EPR II, each emitting a pair of entangled photons into the channels $(1, 2)$ and $(3, 4)$ and we assume that both pairs are produced in the Bell state $|\psi^-\rangle$. Consequently, the overall initial state can be described by a tensor product with entanglement between subsystems 1 and 2 as well as between 3 and 4, and is given by

$$
\begin{aligned}
|\psi\rangle_{1234} &= |\psi^-\rangle_{12} \otimes |\psi^-\rangle_{34}, \\
&= \tfrac{1}{2}\big(|H\rangle_1|V\rangle_2|H\rangle_3|V\rangle_4 - |V\rangle_1|H\rangle_2|H\rangle_3|V\rangle_4 \\
&\quad - |H\rangle_1|V\rangle_2|V\rangle_3|H\rangle_4 + |V\rangle_1|H\rangle_2|V\rangle_3|H\rangle_4\big).
\end{aligned} \tag{14.19}
$$

As in the case of teleportation we may think of the total state as being entangled in different ways, i.e., we can re-express the state (14.19) as a superposition of Bell states. Inserting

$$
\begin{aligned}
|H\rangle|V\rangle &= \tfrac{1}{\sqrt{2}}\big(|\psi^+\rangle + |\psi^-\rangle\big), \quad |H\rangle|H\rangle = \tfrac{1}{\sqrt{2}}\big(|\phi^+\rangle + |\phi^-\rangle\big), \\
|V\rangle|H\rangle &= \tfrac{1}{\sqrt{2}}\big(|\psi^+\rangle - |\psi^-\rangle\big), \quad |V\rangle|V\rangle = \tfrac{1}{\sqrt{2}}\big(|\phi^+\rangle - |\phi^-\rangle\big),
\end{aligned} \tag{14.20}
$$

for the channel pairs $(1, 4)$ and $(2, 3)$ in the expansion (14.19) we find the four-photon state is a superposition of four different pairs of entangled two-photon states in the channels $(1, 4)$ and $(2, 3)$,

$$|\psi\rangle_{1234} = \tfrac{1}{2}\left(|\psi^+\rangle_{14} \otimes |\psi^+\rangle_{23} - |\psi^-\rangle_{14} \otimes |\psi^-\rangle_{23}\right.$$
$$\left. - |\phi^+\rangle_{14} \otimes |\phi^+\rangle_{23} + |\phi^-\rangle_{14} \otimes |\phi^-\rangle_{23}\right). \qquad (14.21)$$

Now Victor performs a Bell measurement in channels $(2,3)$, and let us suppose that in our case the projection happens to be onto the Bell state $|\psi^-\rangle_{23}$

$$|\psi^-\rangle\langle\psi^-|_{23} \otimes \mathbb{1}_{14} \, |\psi\rangle_{1234} = -\tfrac{1}{2}|\psi^-\rangle_{23} \otimes |\psi^-\rangle_{14}, \qquad (14.22)$$

then Victor concludes that the photons in channels $(1,4)$ are now in the same Bell state $|\psi^-\rangle_{14}$—the entanglement has been swapped—which can be measured afterwards by Alice and Bob, e.g., by violating a Bell inequality. Here it is interesting to remark that when Alice and Bob perform their measurement after Victor's Bell measurement, they will be able to test a Bell inequality for an entangled pair of photons in channels $(1,4)$ that never directly interacted!

From a technological perspective, it is also important to note that, together with *entanglement distillation* protocols—procedures to obtain fewer, more strongly entangled states from many shared copies of less entangled states (see Section 16.1.3)—entanglement swapping is the working principle behind the idea of quantum repeaters, first envisioned by *Hans J. Briegel, Wolfgang Dür, Juan Ignacio Cirac*, and *Peter Zoller* in 1998. There, the aim is to use multiple repeater stations locally producing Bell pairs and performing Bell measurements to create long-distance entanglement via entanglement swapping, while compensating for the effects of noise and loss using entanglement distillation.

14.3.2 The Formalism of Isometries

To demonstrate what happens quantum-mechanically within a sequence of measurements the mathematical formalism of isometries (introduced in Uhlmann; Thirring *et al.*; Bertlmann *et al.* is quite powerful and convenient to handle. Isometries are distance-preserving mappings from one Hilbert space to another, valid in any dimension d and therefore useful for characterizing entangled states quite generally. Let us discuss this formalism in more detail for the case of teleportation, where we have a tensor product of three Hilbert spaces $\mathcal{H}_1 \otimes \mathcal{H}_2 \otimes \mathcal{H}_3$ of equal dimensions.

Teleportation:

A maximally entangled state $|\psi\rangle_{23} \in \mathcal{H}_2 \otimes \mathcal{H}_3$ defines a map[1] that is an anti-linear isometry $\tilde{I}_{32}$ between the vectors from one tensor-product factor to the other (Uhlmann, 2003; Bertlmann *et al.*, 2013). More precisely, the anti-linear isometry $\tilde{I}_{32}$ is a bijective map from an orthogonal basis $\{|\varphi_i\rangle_2\}$ in $\mathcal{H}_2$ into a corresponding orthogonal basis $\{|\kappa_i\rangle_3\}$ in $\mathcal{H}_3$. In our notation the first index of $\tilde{I}_{32}$ corresponds to the range of the

---

[1]For instance, for the state $|\phi^+\rangle = \frac{1}{\sqrt{d}}\sum_{k=0}^{d-1}|k\rangle|k\rangle$ one obtains the Choi isomorphism that we will discuss in more detail in Section 21.4.1.

map and the second one to the domain. The map is such that the components of the entangled state

$$|\psi\rangle_{23} = \frac{1}{\sqrt{d}} \sum_{i=1}^{d} |\varphi_i\rangle_2 \otimes |\kappa_i\rangle_3 \tag{14.23}$$

are related by

$$|\tilde{I}_{32}\,\varphi_{i\,2}\rangle_3 = |\kappa_i\rangle_3 \quad \text{and} \quad |\tilde{I}_{32}\,U_2\,\varphi_{i\,2}\rangle_3 = |U_3^*\,\tilde{I}_{32}\,\varphi_{i\,2}\rangle_3 = U_3^*\,|\kappa_i\rangle_3\,, \tag{14.24}$$

for every unitary operator U. Notice our choice of notation for the transpose, adjoint and complex conjugate of an operator: $(U_{ij})^\top = U_{ji}$, $(U_{ij})^\dagger = (U_{ji})^*$, $(U_{ij})^* = U_{ij}^*$.

Of course, the formalism also allows us to formulate a definition where the roles of the two Hilbert spaces are exchanged, which certainly has no impact on the physics. Then the anti-linear isometry $\tilde{I}_{23}$ defines the map from the basis $\{|\kappa_i\rangle_3\}$ in $\mathcal{H}_3$ to the basis $\{|\varphi_i\rangle_2\}$ in $\mathcal{H}_2$, where the components are related by

$$|\tilde{I}_{23}\,\kappa_{i\,3}\rangle_2 = |\varphi_i\rangle_2 \quad \text{and} \quad |\tilde{I}_{23}\,U_3\,\kappa_{i\,3}\rangle_2 = |U_2^*\,\tilde{I}_{23}\,\kappa_{i\,3}\rangle_2 = U_2^*\,|\varphi_i\rangle_2\,. \tag{14.25}$$

Clearly, the isometry (14.25) is the inverse map of (14.24): $\tilde{I}_{23} = (\tilde{I}_{32})^{-1}$.

The possibility to transfer Alice's incoming state in $\mathcal{H}_1$ into a state of $\mathcal{H}_3$ for Bob uses the fact that an isometry I_{31} between these two spaces was taken for granted. Now Alice chooses an isometry $\tilde{I}_{21}$, which corresponds to choosing a maximally entangled state $|\psi\rangle_{12} \in \mathcal{H}_1 \otimes \mathcal{H}_2$, such that the isometry relation

$$\tilde{I}_{32} \circ \tilde{I}_{21} = I_{31} \tag{14.26}$$

is satisfied. We denote the anti-linear isometry by a tilde, while the linear isometry has no tilde, and the composition of two maps is denoted by the symbol $\circ$. The operation $\circ$ means that the map $\tilde{I}_{21}$ is followed by $\tilde{I}_{32}$. Note that the operation $\circ$ is a purely mathematical composition of maps and need not correspond to the time ordering of the physical processes, which is given by the order of successive projection operators (see Section 14.3.3).

Expressed with respect to $\{\varphi_i\}$, an orthonormal basis of one tensor-product factor, the state vector can be written as

$$|\psi\rangle_{12} = \frac{1}{\sqrt{d}} \sum_{i=1}^{d} |\varphi_i\rangle_1 \otimes |\tilde{I}_{21}\,\varphi_{i\,1}\rangle_2\,. \tag{14.27}$$

A measurement by Alice in $\mathcal{H}_1 \otimes \mathcal{H}_2$ with the outcome of an entangled state (14.27) produces the desired state in $\mathcal{H}_3$, i.e., the incoming state $|\phi\rangle_1 \in \mathcal{H}_1$ of Alice has been teleported to Bob: $|\phi\rangle_3 \in \mathcal{H}_3$, see Fig. 14.7. Mathematically, this is described by the linear isometry I_{31} (14.26).

The outcome of other maximally entangled states, orthogonal to the first one, corresponds to a unitary transformation of the form $U_{12} = U_1 \otimes \mathbb{1}_2$ in $\mathcal{H}_1 \otimes \mathcal{H}_2$, which

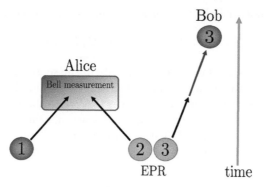

Fig. 14.7 Quantum teleportation. A pair of entangled photons is emitted into the channels $(2,3)$ by an EPR source. Independently, a photon in an arbitrary state in channel 1 arrives at Alice's side together with one of the photons of the EPR pair in channel 2. When Alice performs a Bell measurement in channels $(1,2)$ the state of photon 1 is teleported to photon 3 at Bob's side. However, which of the four Bell states Alice has obtained as measurement outcome has to be communicated to Bob in a classical way so that Bob can perform the appropriate unitary transformation to find his photon in the state of Alice's initially incoming photon.

produces a unique unitary transformation U_3 in $\mathcal{H}_3$ that Bob can perform to obtain the desired state, see eqn (14.6). Thus Alice just has to tell Bob her measurement outcome via some classical channel. The measurements of Alice produce the following results for Bob:

$$|\psi\rangle\langle\psi|_{12} \otimes \mathbb{1}_3 \, |\phi\rangle_1 \otimes |\psi\rangle_{23} = \tfrac{1}{d}|\psi\rangle_{12} \otimes |\phi\rangle_3 \,, \tag{14.28}$$

$$U_{12} \, |\psi\rangle\langle\psi|_{12} \, U_{12}^{\dagger} \otimes \mathbb{1}_3 \, |\phi\rangle_1 \otimes |\psi\rangle_{23} = \tfrac{1}{d} U_{12}|\psi\rangle_{12} \otimes U_3|\phi\rangle_3 \,, \tag{14.29}$$

with the unitary transformation $U_{12} = U_1 \otimes \mathbb{1}_2$ in $\mathcal{H}_1 \otimes \mathcal{H}_2$. The corresponding unitary transformations U_3 at Bob's side are given explicitly in eqn (14.7).

The state vectors can be expressed with respect to an orthonormal basis for a fixed isometry, e.g., choosing $\tilde{I}_{32}$ in the following way

$$|\psi\rangle_{23} = \frac{1}{\sqrt{d}} \sum_{i=1}^{d} |\varphi_i\rangle_2 \otimes |\varphi_i\rangle_3 \,, \tag{14.30}$$

$$|\phi\rangle_{1\,\text{or}\,3} = \sum_{i=1}^{d} \alpha_i \, |\varphi_i\rangle_{1\,\text{or}\,3} \,. \tag{14.31}$$

The anti-linear isometries $\tilde{I}_{21}$ and $\tilde{I}_{32}$ corresponding to the two given maximally entangled states combine to the linear isometry $I_{31} = \tilde{I}_{32} \circ \tilde{I}_{21}$ defining the map in eqn (14.31).

Changing to another Bell state in Alice's measurement corresponds to

$$\tilde{I}_{32} \circ \tilde{I}_{21} \, U_1 \;=\; \tilde{I}_{32} \circ U_2^* \, \tilde{I}_{21} \;=\; U_3 \, \tilde{I}_{32} \circ \tilde{I}_{21} \;=\; U_3 \, I_{31} \;. \tag{14.32}$$

Résumé: If Alice measures the same Bell state in the sense of relation (14.26) between $\mathcal{H}_1$ and $\mathcal{H}_2$ as there was between $\mathcal{H}_2$ and $\mathcal{H}_3$, as determined by the EPR source, she knows that her measurement left Bob's system in $\mathcal{H}_3$ in the state $|\phi\rangle_3$ which is Alice's incoming state. If Alice finds a different Bell state, which is given by $U_{12}|\psi\rangle_{12}$ (and $U_{12} = U_1 \otimes \mathbb{1}_2$) since all other Bell states are connected by local unitary transformations, then Bob will obtain the state vector $U_3|\phi\rangle_3$, where the unitary transformation U_3 is determined by U_{12}.

Entanglement swapping:

In complete analogy to the case of teleportation we can discuss entanglement swapping within the formalism of isometries. In this case the starting point is to consider two maximally entangled pure states combined in a tensor product $|\psi\rangle_{12} \otimes |\psi\rangle_{34}$. Expressed with respect to an orthonormal basis of one tensor-product factor, the entangled states are given by eqn (14.27) and describe, e.g., two pairs of EPR photons. These propagate into different directions and at the interaction point of two of them, say photons 2 and 3, Victor performs a Bell measurement, i.e., a measurement with respect to an orthogonal set of maximally entangled states. For two qubits, these are usually chosen as the Bell states (14.4). But the isometry results hold more generally, the four tensor-product factors just have to be of the same dimension d, which can be arbitrary.

Thus, analogously to the case of teleportation discussed before, the effect of the projection corresponding to Victor's measurement on the state is

$$|\psi\rangle\langle\psi|_{23} \otimes \mathbb{1}_{14} \, |\psi\rangle_{12} \otimes |\psi\rangle_{34} \;=\; \tfrac{1}{d} \, |\psi\rangle_{23} \otimes |\psi\rangle_{14} \;, \tag{14.33}$$

where $|\psi\rangle_{14} \in \mathcal{H}_1 \otimes \mathcal{H}_4$ is a maximally entangled state that can be expressed with respect to an orthonormal basis $\{\varphi_i\}$ of one tensor-product factor

$$|\psi\rangle_{14} \;=\; \frac{1}{\sqrt{d}} \sum_{i=1}^{d} |\varphi_i\rangle_1 \otimes |\tilde{I}_{41}\,\varphi_{i\,1}\rangle_4 \;. \tag{14.34}$$

This means the entangled state (14.34) is determined by the anti-linear isometry $\tilde{I}_{41}$ satisfying the relation

$$\tilde{I}_{41} \;=\; \tilde{I}_{43} \circ \tilde{I}_{32} \circ \tilde{I}_{21} \;. \tag{14.35}$$

The other isometries, $\tilde{I}_{43}$, $\tilde{I}_{32}$, and $\tilde{I}_{21}$ represent the other maximally entangled states.

To summarize, after the Bell measurement of photons 2 and 3 resulting in a definite entangled state, the photons 1 and 4 become entangled and are left in the state corresponding to the outcome of the Bell measurement, see Fig. 14.6, which is described by the anti-linear isometry (14.35).

Résumé: There are two pairs of entangled photons in channels $(1,2)$ and $(3,4)$ emitted by different sources. Victor entangles two photons, i.e., performs a Bell measurement on channels $(2,3)$. The remaining two photons in channels $(1,4)$ then become entangled, they are projected onto the same entangled state—the entanglement has been swapped—which can be verified in measurements carried out by Alice and Bob.

14.3.3 Delayed-Choice Entanglement Swapping

A fascinating aspect of entanglement swapping manifests itself in a variant of this protocol called *delayed-choice entanglement swapping*, first formulated by Asher Peres in 2000. For those viewing the projection of a quantum system into a definite state during a measurement as a physical process (rather than an update of the observer's knowledge, see the discussion in Section 13.4), such delayed-choice phenomena appear to be quite paradoxical since such quantum effects can mimic an influence of future actions on past events. Experimentally this phenomenon was first demonstrated by *Xiao-Song Ma, Johannes Kofler*, and *Anton Zeilinger* (2012*b*). In this experiment the order of successive measurements is reversed as compared to entanglement swapping. Alice and Bob measure first and record their data, then at a later time Victor is free to choose a projection of his two states onto an entangled or separable state, i.e., to measure them jointly or individually. The outcome of the measurements of Alice, Bob, and Victor is recorded and compared to the previous case where Victor measured before Alice and Bob. It turns out that the joined probability for the outcome of these two cases is the same. Particularly, in case of delayed-choice scenarios it is Victor's measurement that decides the context and determines the interpretation of Alice and Bob's data. Alice and Bob can sort their already recorded data, according to Victor's later choice and his results, in such a way that they can verify having measured either entangled or separable states. For a review, see Ma *et al.* (2016).

In terms of the mathematical formalism of quantum mechanics, however, this effect is not paradoxical at all and can be traced back to the commutativity of the projection operators that are involved in the corresponding measurement process (Bertlmann *et al.*, 2013). How does this work in detail? Let us consider a sequence of measurements where the results are obtained with a certain probability. Then we ask the question:

To what extent does the probability of the measurement outcome depend on the chronological order of the measurements?

Within classical physics, we can consider time-invariant states. For these, the measurement outcomes are clearly independent of the chosen time, thus independent of the order in which the measurements are carried out.

In quantum physics, however, we may consider a sequence of events at successive times corresponding to a sequence of outcomes of commuting or non-commuting projectors. Such sequences are called *quantum histories* in the literature (Gell-Mann and Hartle, 1993; Griffiths, 1996; Griffiths, 1998). They determine the outcome of the measurements and thus the chronological order of the measurements.

Consider a quantum state that we now represent by a density matrix ρ defined on some larger algebra $\mathcal{A}$ since this description is more convenient for our following considerations. We then successively perform several measurements on the system represented by this state. We start by considering two measurements, M_1 and M_2, that we describe by sets of projectors $\{P_1^{(i)} = \left| \psi^i \middle\rangle \middle\langle \psi^i \right|_1\}_i$ and $\{Q_2^{(j)} = \left| \phi^j \middle\rangle \middle\langle \phi^j \right|_2\}_j$ on some states of the subalgebras $\mathcal{A}_1, \mathcal{A}_2 \subset \mathcal{A}$, respectively.

Successively performing the measurements M_1 and M_2 we obtain[2]

$$\rho \xrightarrow{M_1} \rho^{(i)} = \tfrac{1}{w_i} P_1^{(i)} \rho\, P_1^{(i)} \xrightarrow{M_2} \rho^{(ji)} = \tfrac{1}{w_{ji}} Q_2^{(j)} P_1^{(i)} \rho\, P_1^{(i)} Q_2^{(j)}, \qquad (14.36)$$

where $w_i = \mathrm{Tr}\big(P_1^{(i)} \rho\, P_1^{(i)}\big) = \mathrm{Tr}\big(P_1^{(i)} \rho\big)$ is the probability to obtain the state $\rho^{(i)}$ after the first measurement M_1, and similarly

$$w_{ji} = \mathrm{Tr}\big(Q_2^{(j)} P_1^{(i)} \rho\, P_1^{(i)} Q_2^{(j)}\big) = \mathrm{Tr}\big(P_1^{(i)} Q_2^{(j)} P_1^{(i)} \rho\big) \qquad (14.37)$$

is the probability to obtain the state $\rho^{(ji)}$ after the second measurement M_2.

Now we interchange the order of measurements, $M_1 \longleftrightarrow M_2$. Then the state

$$\rho^{(ij)} = \tfrac{1}{w_{ij}} P_1^{(i)} Q_2^{(j)} \rho\, Q_2^{(j)} P_1^{(i)} \qquad (14.38)$$

occurs with probability $w_{ij} = \mathrm{Tr}\big(P_1^{(i)} Q_2^{(j)} \rho\big)$. In the case of commuting projection operators we can see immediately that there is no dependence on the order of measurement operations, i.e.,

$$\big[P_1^{(i)}, Q_2^{(j)} \big] = 0 \quad \Longrightarrow \quad w_{ij} = w_{ji}. \qquad (14.39)$$

The probabilities of the measurement outcomes, which can be determined from the protocols of the measurements, are the same! Actually, it is enough to examine whether the condition $\mathrm{Tr}\big[\rho\, \big(Q_2^{(j)} P_1^{(i)} Q_2^{(j)} - P_1^{(i)} Q_2^{(j)} P_1^{(i)}\big)\big] = 0$ holds.

Let us next turn to the experiment of Zeilinger's group (Ma *et al.*, 2012*b*), specifically, we begin with the case of entanglement swapping as depicted in Fig. 14.6. There we start with a tensor product of four matrix algebras $\mathcal{A}_{\text{tot}} = \mathcal{A}_1 \otimes \mathcal{A}_2 \otimes \mathcal{A}_3 \otimes \mathcal{A}_4$ of equal dimensions; $\mathcal{A}_1$ and $\mathcal{A}_4$ correspond to systems controlled by Alice and Bob, respectively, and the subalgebra $\mathcal{A}_2 \otimes \mathcal{A}_3$ refers to systems under Victor's control. At the beginning the state described by the density matrix $\rho \in \mathcal{A}_{\text{tot}}$ is separable with respect to the bipartition $\mathcal{A}_1 | \mathcal{A}_4$ but by a suitable manipulation—a Bell measurement—on $\mathcal{A}_2 \otimes \mathcal{A}_3$ the state over $\mathcal{A}_1 \otimes \mathcal{A}_4$ becomes entangled.

Thus initially we have the two Bell states ρ_{12} and ρ_{34} and first Victor performs a Bell measurement $M_1 = \{Q_{23}^{(i)}\}$ that projects onto some maximally entangled state $\rho_{23}^{(i)} \equiv Q_{23}^{(i)}$ on $\mathcal{A}_2 \otimes \mathcal{A}_3$. Then we obtain

$$\rho = \rho_{12} \otimes \rho_{34} \xrightarrow{M_1} \rho^{(i)} = \tfrac{1}{w_i} Q_{23}^{(i)} \rho_{12} \otimes \rho_{34}\, Q_{23}^{(i)} = \rho_{14}^{(i)} \otimes \rho_{23}^{(i)}, \qquad (14.40)$$

---

[2]For a more detailed discussion of quantum measurements, see Chapter 23.

where each of the maximally entangled states $\rho_{14}^{(i)}$ on $\mathcal{A}_1 \otimes \mathcal{A}_4$ occur with probability $w_i = \frac{1}{4}$ and the isometry relation

$$\tilde{I}_{41}^i = \tilde{I}_{43} \circ \tilde{I}_{32}^i \circ \tilde{I}_{21} \tag{14.41}$$

between the algebras is satisfied. The state $\left(\frac{1}{w_i} Q_{23}^{(i)} \rho\, Q_{23}^{(i)}\right)_{\mathcal{A}_1 \otimes \mathcal{A}_4}$ after the projection is thus pure and maximally entangled on $\mathcal{A}_1 \otimes \mathcal{A}_4$. Entanglement has been swapped.

The second measurement $M_2 = \{P_{14}^{(ab)} = P_1^{(a)} \otimes P_4^{(b)}\}$ is performed by Alice and Bob. They measure the incoming states at $\mathcal{A}_1$ and $\mathcal{A}_4$ by projecting on some states $|\psi^{(a)}\rangle_1$ and $|\psi^{(b)}\rangle_4$. Denoting the corresponding projectors by $P_1^{(a)} = |\psi^{(a)}\rangle\langle\psi^{(a)}|_1$ and $P_4^{(b)} = |\psi^{(b)}\rangle\langle\psi^{(b)}|_4$, respectively, the state turns into

$$\rho^{(i)} \xrightarrow{M_2} \rho^{(iab)} = \frac{1}{w_{iab}} P_{14}^{(ab)} Q_{23}^{(i)} \rho\, Q_{23}^{(i)} P_{14}^{(ab)} = \rho_1^{(a)} \otimes \rho_{23}^{(i)} \otimes \rho_4^{(b)}. \tag{14.42}$$

The probability $w_{iab} = \text{Tr}\big(\rho\, Q_{23}^{(i)} P_{14}^{(ab)} Q_{23}^{(i)}\big)$ of finding the state $\rho^{(iab)}$ is determined by the isometry $\tilde{I}_{41}^i$, i.e., it depends on the particular entangled state labelled by i.

Next we study the reversed order of measurements, i.e., the case of delayed-choice entanglement swapping, illustrated in Fig. 14.8. The first measurement is now $M_2 = \{P_{14}^{(ab)} = P_1^{(a)} \otimes P_4^{(b)}\}$ performed by Alice and Bob. They measure their incoming states by projecting on some states $|\psi^{(a)}\rangle_1$ and $|\psi^{(b)}\rangle_4$. Then the total state ρ turns into

$$\rho \xrightarrow{M_2} \rho^{(ab)} = \frac{1}{w_{ab}} P_{14}^{(ab)} \rho_{12} \otimes \rho_{34}\, P_{14}^{(ab)} = \rho_1^{(a)} \otimes \rho_2^{(a)} \otimes \rho_3^{(b)} \otimes \rho_4^{(b)}, \tag{14.43}$$

with $w_{ab} = \text{Tr}\big(P_{14}^{(ab)} \rho_{12} \otimes \rho_{34}\big)$, which is a separable and even pure state on the subalgebra $\mathcal{A}_1 \otimes \mathcal{A}_4$.

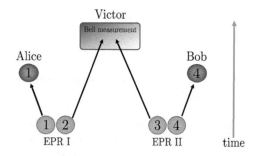

Fig. 14.8 Delayed-choice entanglement swapping. There are two EPR sources I and II and the order of measurements is reversed with respect to the usual entanglement-swapping protocol (compare with Fig. 14.6). Alice and Bob measure first in channels $(1, 4)$ and record their data and later on Victor freely chooses to project his state in channels $(2, 3)$ either onto an entangled or onto a separable state, corresponding to a measurement with respect to Bell-state basis or the computational basis, respectively. When comparing this case with entanglement swapping of Fig. 14.6 it turns out that the joint probability for the outcome of these two cases is the same due to the commutativity of the projection operators, i.e., the results are independent of the order of the successive measurements.

In a second measurement Victor may choose freely to project onto an entangled state or onto a separable state on $\mathcal{A}_2 \otimes \mathcal{A}_3$. Let us first discuss the measurement $M_1 = \{Q_{23}^{(i)}\}$ onto an entangled state. There, the above state changes into

$$\rho^{(ab)} \xrightarrow{M_1} \rho^{(abi)} = \frac{1}{w_{abi}} Q_{23}^{(i)} P_{14}^{(ab)} \rho_{12} \otimes \rho_{34} P_{14}^{(ab)} Q_{23}^{(i)} = \rho_1^a \otimes \rho_{23}^{(i)} \otimes \rho_4^b, \qquad (14.44)$$

which remains separable on $\mathcal{A}_1 \otimes \mathcal{A}_4$. The entangled state $\rho_{23}^{(i)}$ clearly depends on the isometry $\tilde{I}_{32}^i(a, b)$ and on the particular outcomes (a, b).

Since the projection operators—the measurement operators—commute

$$\left[P_{14}^{(ab)}, Q_{23}^{(i)} \right] = 0 \qquad \Longrightarrow \qquad w_{abi} = w_{iab}, \qquad (14.45)$$

the probability $w_{abi} = \text{Tr}\left(\rho \, P_{14}^{(ab)} Q_{23}^{(i)} P_{14}^{(ab)} \right)$ for the occurrence of the final state $\rho^{(abi)}$ is the same as before, in fact, $\rho^{(abi)} = \rho^{(iab)}$ and the results are independent of the order of the measurements. This is valid for all projection operators $P_{14}^{(ab)}$ independently of the particular pair (a, b). Therefore different quantum histories lead to the same result.

In the case that Victor projects onto a separable state on $\mathcal{A}_2 \otimes \mathcal{A}_3$ we have measurement operators $\widehat{M}_1 = \{P_{23}^{(j)}\}$, where the projectors $P_{23}^{(j)} = \left| \phi^{(j)} \right\rangle\!\left\langle \phi^{(j)} \right|_{23}$ project onto the separable states $\left| \phi^{1,2,3,4} \right\rangle_{23} = \{ |0\rangle_2 \otimes |0\rangle_3, |1\rangle_2 \otimes |1\rangle_3, |0\rangle_2 \otimes |1\rangle_3, |1\rangle_2 \otimes |0\rangle_3 \}$. Then the commutator of the projectors vanishes,

$$\left[P_{14}^{(ab)}, P_{23}^{(j)} \right] = 0 \qquad \Longrightarrow \qquad w_{abj} = w_{jab}, \qquad (14.46)$$

and again we find independence of the order of the measurements.

Finally, one can also construct an experimental setup that depends on the order of the successive measurements corresponding to non-commutative projection operators. In this case entanglement swapping is used to teleport a quantum state from Alice to Bob—in fact, it is not so much about entanglement swapping but rather about double and triple teleportation—where Bob now *does have* the possibility to examine the non-commutativity within the quantum history. For further literature on this topic, we refer to Bertlmann *et al.* (2013).

Résumé: In delayed-choice entanglement swapping the order of the measurements of Victor—who entangles two photons of the two pairs—and Alice and Bob—who measure the remaining two photons—is reversed. As in the usual entanglement-swapping procedure, there are two EPR sources emitting entangled photons in channels $(1, 2)$ and $(3, 4)$. Alice and Bob measure first in channels $(1, 4)$ and record their data and later on Victor freely chooses whether to project his quantum state in channels $(2, 3)$ onto an entangled or onto a separable state. When comparing this case with the entanglement-swapping protocol depicted in Fig. 14.6 it turns out that the respective joint probabilities for the outcomes of these two cases are the same. This is due to the commutativity of the projection operators, i.e., there is independence of the order of successive measurements. This phenomenon may appear to be quite paradoxical since it seems to mimic an influence of future actions on past events.

Having discussed a number of variants of the teleportation protocol, let us next examine in which way quantum teleportation provides advantageous alternatives to classical communication protocols.

14.3.4   Quantum Teleportation versus Classical Information Transfer

As we have mentioned before, Alice also has other options (besides teleportation) of sending a qubit of quantum information that do not require knowing which specific state is sent. In particular, Alice could simply send the quantum system encoding the qubit directly by using a suitable quantum channel. For instance, a photon (where, e.g., the polarization can encode a qubit) could be sent via an optical link or through a glass fibre. Surely, there are some losses in such a channel, but the same is true for the channel that distributes the entangled pairs in the teleportation scheme. So what are the advantages of using teleportation?

Leaving potential interpretational issues related to the EPR paradox (see Section 13.1) aside, there are two key advantages of using teleportation over just sending the qubit carrying the quantum information along some channel.

First, recall that the protocol works independently of the distance between the parties, so long as the quality of the shared Bell pair is high enough. That is, qualitatively speaking, the lower the fidelity[3] of the shared entangled state is with a pure, maximally entangled state, the lower the fidelity of the output of the teleportation protocol with the desired state $|\psi\rangle$. So, even though one would also have to distribute these entangled pairs using imperfect channels, the fidelity of the Bell states could be improved beforehand, e.g., by entanglement distillation, while covering in principle arbitrary distances via lossy and noisy channels by using quantum repeaters. Once a sufficiently "good" Bell pair has been established, the quantum-state transfer can be carried out perfectly.

The second crucial practical difference to teleportation lies in the amount of information that could be intercepted by a third party. Consider the following scenario: Suppose the teleportation protocol is used as described before, but there is an eavesdropper who we will call Eve who intercepts system 3 as well as the classical signal (the two bits) sent from Alice to Bob, and sends a replacement quantum system as well as (copies of the) two classical bits to Bob. A priori we must assume that there is no way for Bob to know that the classical bits have been copied, and indeed, if Bob where to consult Alice, she could confirm the bit values she sent.

However, the situation is different for the qubit, system 3. If Eve makes a measurement of qubit 3 before sending it on to Bob, or if she sends a completely independent qubit, then there will not be any entanglement between Alice and Bob. In particular, quantum mechanics does not allow Alice to simply make a copy of the qubit (this is called the no-cloning theorem, which we will discuss in Section 21.5). Consequently, Bob will generally not receive the correct state after completing the teleportation protocol. But if Bob does not know which state he is supposed to receive, then Alice and

---

[3]See Section 24.2.2 for a definition of the fidelity that is suitable for mixed states.

Bob either have to communicate beforehand which states are being sent (defeating the point of sending this information later on), or they can find out only a posteriori that an eavesdropper has gained access to the state. However, the fact that the teleportation protocol is based on shared entanglement offers a way around this problem, as we will see next.

A simple modification to the teleportation protocol allows checking for the presence of an eavesdropper without revealing any information about the teleported state, while still perfectly transferring this state, and without either Alice or Bob having to know any details about the state $|\psi\rangle_1$ before, during, or after the teleportation.

To see how this works, recall that the entangled state $|\psi^-\rangle$ can be used to violate a Bell inequality, e.g., the CHSH inequality (see Chapter 13). Alice and Bob can hence proceed in the following way. Before attempting to teleport the state $|\psi\rangle_1$, they share a number of copies of $|\psi^-\rangle$. Then, Alice decides which of these copies to use for the joint Bell measurement with $|\psi\rangle_1$. The remaining qubits are used for a Bell test (e.g., she can make measurements along some directions $\vec{a}$ and $\vec{a}'$ as described in Section 13.2.2). Alice then publicly communicates to Bob which of the qubits she measured along the directions $\vec{a}$ or $\vec{a}'$, and what the outcomes of these measurements were. Bob can use this information to make corresponding measurements along directions $\vec{b}$ and $\vec{b}'$, and check the violation of the CHSH inequality for those copies.

If Alice and Bob are confident that the Bell inequality is violated sufficiently strongly, they can conclude that only a small fraction of qubits might have been intercepted (since this would break entanglement). They can then be reasonably sure (depending on the observed violation of the Bell inequality) that the remaining qubits have also not been intercepted (since Eve has no way of knowing beforehand which qubits she can and which she cannot intercept without alerting Alice and Bob to her presence), and proceed with the teleportation protocol (i.e., by sending the classical bits that allow Bob to perform the correction operations). If Eve has intercepted too many of the qubits used for the Bell inequality, Alice and Bob will not find a large enough violation and can abort the protocol before sending the classical bits. Recall, without these, all the previously sent qubits contain no information about $|\psi\rangle_1$.

One drawback of this method is that it requires Alice and Bob to share a number of Bell pairs whose constituents need to be locally stored (in some quantum memory) until it is determined which copy Alice uses for teleportation. This represents quite a technological challenge, since storing the involved quantum states for an extended period of time without them being degraded by noise and environmental effects is difficult. At the same time, one is in practice often interested in transmitting classical information instead of quantum information in the form of a qubit. We will therefore briefly discuss some entanglement-based protocols for quantum communication related to quantum teleportation: the *dense-coding* protocol[4] and the *Ekert-91 protocol*, as well as its (famous but not entanglement-based) precursor, the *BB84* protocol.

---

[4]Sometimes called "super-dense coding" in the literature (see, e.g., Nielsen and Chuang, 2000, Sec. 2.3.

14.3.5   The Dense-Coding Protocol

In many respects, the teleportation protocol represents one of the most important primitives for quantum-information processing, from its role in entanglement swapping and quantum repeaters (see Section 14.3.1), to so-called *gate teleportation*, which is the basic principle behind measurement-based quantum computation (Briegel and Raussendorf, 2001; Raussendorf and Briegel, 2001). In this one-way computation scheme, unitary transformations—called *quantum gates* in the context of quantum computation—are only applied locally to individual qubits, but the effects of these unitaries can be teleported onto other qubits via pre-shared entanglement in the form of (certain) so-called graph states by performing measurements on the initial qubits and applying suitable corrections to the target qubits (for a more detailed discussion of this approach, see the review by Briegel *et al.*, 2009 or Appendix A.5 in Friis *et al.*, 2017*b* for a compact summary of the basics). Here, however, we briefly want to explore an entanglement-based protocol for transmitting classical bits, the dense-coding protocol.

In dense coding, the goal is for Alice to send classical bits to Bob by encoding them in qubits. As it turns out, one cannot just send one classical bit using one qubit, it is possible to send *two* bits using *one* qubit. This is achieved, loosely speaking, by "inverting" the teleportation protocol, i.e., by exchanging the operations performed by Alice and Bob.

To formulate this more precisely, let us first slightly reformulate the teleportation protocol in the language of quantum computation by recasting the Bell measurement in terms of the application of certain quantum gates, followed by measurements in the computational basis. That is, a measurement with respect to the basis of the maximally entangled Bell states can be thought of as a suitable unitary that rotates the Bell basis $\{|\,\psi^{\pm}\,\rangle, |\,\phi^{\pm}\,\rangle\}$ to the basis $\{|\,m\,\rangle\,|\,n\,\rangle\}_{m,n=0,1}$, and then simply measuring with respect to the latter basis.

Such a unitary transformation can further be decomposed into a combination of a controlled NOT (or CNOT) transformation U_{CNOT} that acts jointly on both qubits, followed by a particular local unitary H, a Hadamard gate, acting on only one of the qubits. The CNOT gate can be written as

$$U_{\mathrm{CNOT}} = |\,0\,\rangle\langle\,0\,| \otimes \mathbb{1} + |\,1\,\rangle\langle\,1\,| \otimes \sigma_x, \tag{14.47}$$

where $\sigma_x = |\,0\,\rangle\langle\,1\,| + |\,1\,\rangle\langle\,0\,|$ is the usual Pauli matrix with respect to our chosen basis $\{|\,0\,\rangle, |\,1\,\rangle\}$. Thus, if the state of the first qubit (the "control") is $|\,1\,\rangle$, the chosen ("computational") basis states $|\,0\,\rangle$ and $|\,1\,\rangle$ of the second (target) qubit are "flipped"[5] to $|\,1\,\rangle$ and $|\,0\,\rangle$, respectively, otherwise, if the first qubit is in the state $|\,0\,\rangle$, the second qubit is left unchanged. Meanwhile, the Hadamard gate can be written with respect to the computational basis as

---

[5]This corresponds to a logical negation, a "NOT" operation, recall that $\sigma_x\,|\,0\,\rangle = |\,1\,\rangle$ and $\sigma_x\,|\,1\,\rangle = |\,0\,\rangle$.

$$H = \tfrac{1}{\sqrt{2}} \begin{pmatrix} 1 & 1 \\ 1 & -1 \end{pmatrix}, \tag{14.48}$$

and exchanges the eigenbases $\{|0\rangle, |1\rangle\}$ and $\{|\pm\rangle = \tfrac{1}{\sqrt{2}}(|0\rangle \pm |1\rangle)\}$ of σ_z and σ_x, respectively, i.e., $H|0\rangle = |+\rangle$, $H|+\rangle = |0\rangle$, and $H|1\rangle = |-\rangle$, $H|-\rangle = |1\rangle$. Applying the CNOT gate to the first two qubits of the joint state $|\psi\rangle_{123} = |\psi\rangle_1 \otimes |\psi^-\rangle_{23}$ of all three qubits in the teleportation protocol before the Bell measurement, eqn (14.3), we have

$$
\begin{aligned}
U_{\text{CNOT},\,12}\,|\psi\rangle_{123} &= \frac{\alpha}{\sqrt{2}} \left(U_{\text{CNOT},\,12}\,|0\rangle_1 |0\rangle_2 |1\rangle_3 - U_{\text{CNOT},\,12}\,|0\rangle_1 |1\rangle_2 |0\rangle_3 \right) \\
&\quad + \frac{\beta}{\sqrt{2}} \left(U_{\text{CNOT},\,12}\,|1\rangle_1 |0\rangle_2 |1\rangle_3 - U_{\text{CNOT},\,12}\,|1\rangle_1 |1\rangle_2 |0\rangle_3 \right) \\
&= \frac{\alpha}{\sqrt{2}} \left(|0\rangle_1 |0\rangle_2 |1\rangle_3 - |0\rangle_1 |1\rangle_2 |0\rangle_3 \right) \\
&\quad + \frac{\beta}{\sqrt{2}} \left(|1\rangle_1 |1\rangle_2 |1\rangle_3 - |1\rangle_1 |0\rangle_2 |0\rangle_3 \right).
\end{aligned}
\tag{14.49}
$$

Subsequently applying the Hadamard gate to the first qubit, we further obtain

$$
\begin{aligned}
H_1\,U_{\text{CNOT},\,12}\,|\psi\rangle_{123} &= \frac{\alpha}{\sqrt{2}} \left(|+\rangle_1 |0\rangle_2 |1\rangle_3 - |+\rangle_1 |1\rangle_2 |0\rangle_3 \right) \\
&\quad + \frac{\beta}{\sqrt{2}} \left(|-\rangle_1 |1\rangle_2 |1\rangle_3 - |-\rangle_1 |0\rangle_2 |0\rangle_3 \right),
\end{aligned}
\tag{14.50}
$$

which we can rewrite as

$$
\begin{aligned}
H_1\,U_{\text{CNOT},\,12}\,|\psi\rangle_{123} &= \frac{\alpha}{2} \left([|0\rangle_1 + |1\rangle_1]\,|0\rangle_2 |1\rangle_3 - [|0\rangle_1 + |1\rangle_1]\,|1\rangle_2 |0\rangle_3 \right) \\
&\quad + \frac{\beta}{2} \left([|0\rangle_1 - |1\rangle_1]\,|1\rangle_2 |1\rangle_3 - [|0\rangle_1 - |1\rangle_1]\,|0\rangle_2 |0\rangle_3 \right) \\
&= \frac{1}{2} \Big[|0\rangle_1 |0\rangle_2 \left(\alpha|1\rangle_3 - \beta|0\rangle_3 \right) - |0\rangle_1 |1\rangle_2 \left(\alpha|0\rangle_3 - \beta|1\rangle_3 \right) \\
&\quad + |1\rangle_1 |0\rangle_2 \left(\alpha|1\rangle_3 + \beta|0\rangle_3 \right) - |1\rangle_1 |1\rangle_2 \left(\alpha|0\rangle_3 + \beta|1\rangle_3 \right) \Big].
\end{aligned}
$$

Again adopting the quantum-information notation $\sigma_x = X$ and $\sigma_z = Z$ with subscripts added to denote the particular qubit that the operation is acting on, we can then more compactly write the state prior to Alice's measurement as

$$
H_1\,U_{\text{CNOT},\,12}\,|\psi\rangle_{123} = \sum_{b_1, b_2 = 0,1} (-1)^{b_2}\,|b_1\rangle_1\,|b_2\rangle_2\,Z^{b_1+1}\,X^{b_2+1}\,|\psi\rangle_3. \tag{14.51}
$$

We thus see that the teleportation protocol can be phrased as the following simple instructions: After applying $U_{\text{CNOT},\,12}$ and H_1, Alice just measures in the computational

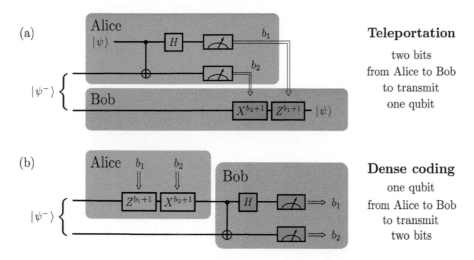

Fig. 14.9 Quantum teleportation and dense coding. The circuit diagrams for teleportation (a) and dense coding (b) are shown. Horizontal single lines represent qubits, double lines represent classical bits. The vertical line represents a controlled NOT gate $U_{\mathrm{CNOT,12}}$ from eqn (14.47) between two qubits, with the filled circle marking the control qubit and the cross-hair indicating the target qubit. Labelled boxes represent single-qubit gates corresponding to the labels, including the Hadamard gate H from eqn (14.48) as well as Pauli X and Z gates. Rectangular boxes with dials and pointer represent measurement in the computational basis of the respective qubit.

basis and sends the resulting two bit values b_1 and b_2 to Bob, who applies the operation $Z^{b_1+1} X^{b_2+1}$ to retrieve the single-qubit state $|\psi\rangle_3$. The corresponding quantum circuit is drawn in Fig. 14.9 (a).

To arrive at the dense-coding protocol, we now simply exchange the operations carried out by Alice and Bob. Alice selects two bit values b_1 and b_2 and applies the operations Z^{b_1+1} and X^{b_2+1} to her half of a shared Bell pair $|\psi^-\rangle_{12}$ and then send her single qubit to Bob. Bob then applies the CNOT gate on both qubits, followed by a Hadamard gate H on the qubit received from Alice. A measurement in the two-qubit computational basis then results in the outcome (b_1, b_2), as shown in Fig. 14.9 (b). Let us quickly check this with a calculation:

$$b_1 = 0,\, b_2 = 0:\ H_1 U_{\mathrm{CNOT,12}} X_1 Z_1 |\psi^-\rangle = H_1 U_{\mathrm{CNOT,12}} \tfrac{1}{\sqrt{2}}\big(|1\rangle_1 |1\rangle_2 + |0\rangle_1 |0\rangle_2\big)$$

$$= H_1 \tfrac{1}{\sqrt{2}}\big(|1\rangle_1 |0\rangle_2 + |0\rangle_1 |0\rangle_2\big) = H_1 |+\rangle_1 |0\rangle_2 = |0\rangle_1 |0\rangle_2, \tag{14.52a}$$

$$b_1 = 0,\, b_2 = 1:\ H_1 U_{\mathrm{CNOT,12}} Z_1 |\psi^-\rangle = H_1 U_{\mathrm{CNOT,12}} \tfrac{1}{\sqrt{2}}\big(|0\rangle_1 |1\rangle_2 + |1\rangle_1 |0\rangle_2\big)$$

$$= H_1 \tfrac{1}{\sqrt{2}}\big(|0\rangle_1 |1\rangle_2 + |1\rangle_1 |1\rangle_2\big) = H_1 |+\rangle_1 |1\rangle_2 = |0\rangle_1 |1\rangle_2, \tag{14.52b}$$

$b_1 = 1, \, b_2 = 0: \; H_1 U_{\text{CNOT},12} X_1 \, | \psi^- \rangle = H_1 U_{\text{CNOT},12} \frac{1}{\sqrt{2}} \big(| 1 \rangle_1 | 1 \rangle_2 - | 0 \rangle_1 | 0 \rangle_2 \big)$

$$= H_1 \frac{1}{\sqrt{2}} \big(| 1 \rangle_1 | 0 \rangle_2 - | 0 \rangle_1 | 0 \rangle_2 \big) = -H_1 | - \rangle_1 | 0 \rangle_2 = - | 1 \rangle_1 | 0 \rangle_2 \,,$$

$$(14.52c)$$

$b_1 = 1, \, b_2 = 1: \; H_1 U_{\text{CNOT},12} \, | \psi^- \rangle = H_1 U_{\text{CNOT},12} \frac{1}{\sqrt{2}} \big(| 0 \rangle_1 | 1 \rangle_2 - | 1 \rangle_1 | 0 \rangle_2 \big)$

$$= H_1 \frac{1}{\sqrt{2}} \big(| 0 \rangle_1 | 1 \rangle_2 - | 1 \rangle_1 | 1 \rangle_2 \big) = H_1 | - \rangle_1 | 1 \rangle_2 = | 1 \rangle_1 | 1 \rangle_2 \,.$$

$$(14.52d)$$

By pre-sharing an entangled Bell pair, a single qubit can thus be used to communicate two classical bits via dense coding, whereas teleportation allows the communication of a single qubit by sending two classical bits. In the next section, we will see that pre-shared entanglement can also be used to establish secure classical communication, but to properly appreciate these ideas we will have to step back and inspect the task of *quantum key distribution* more generally.

14.4 Quantum Key Distribution

We now come to an area where quantum-information processing is directly applied to problems that have become part of everyday life: the secure transmission of information. The field of study dealing with such problems is *quantum cryptography* and is generally concerned with the encryption, transmission, and decryption of information using quantum (and classical) channels (see, e.g., Gisin *et al.*, 2002 for a review). This can involve tasks where the participants do not trust each other, such as bit commitment or "coin flipping". However, here we want to focus on communication between two parties that trust each other, but want their communication to stay secret with respect to any (untrusted) third parties. This can be achieved by sharing a secret key and hence requires a secure protocol for key distribution.

The security of such key-distribution protocols is based on the *one-time pad*, illustrated in Fig. 14.10. That is, in order to encrypt or decrypt a message encoded in a bit string of length n, the message is added (modulo 2) bit-by-bit to the bits of a secret random key of equal length. As long as the key is truly random, is used only once, and kept completely secret, this encryption cannot be broken.

```
        Message: 0 1 1 0 1 0 0 0 0 1 1 0 0 1 0 1 0 1 1 0 1 1 0 0 0 1 1 0 1 1 0 0 0 1 1 0 1 1 1 1
            Key: 1 0 0 1 0 1 0 1 1 1 0 1 0 1 0 0 1 1 1 0 0 1 0 1 0 1 0 1 0 1 0 1 1 0 0 1 0 0 1
                 ↓ ↓ ↓
Encrypted Message: 1 1 1 1 1 1 0 1 1 0 1 1 0 1 0 1 1 0 0 1 1 1 1 0 1 1 0 0 0 1 1 0 1 0 1 0 0 1 1 0
            Key: 1 0 0 1 0 1 0 1 1 1 0 1 0 1 0 0 1 1 1 0 0 1 0 1 0 1 0 1 0 1 0 1 1 0 0 1 0 0 1
                 ↓ ↓ ↓
Decrypted Message: 0 1 1 0 1 0 0 0 0 1 1 0 0 1 0 1 0 1 1 0 1 1 0 0 0 1 1 0 1 1 0 0 0 1 1 0 1 1 1 1
```

Fig. 14.10 One-time pad. A message in the form of a bit string can be securely encrypted and decrypted using a share secret key of the same length as the bit string by adding (modulo 2) the bit values of the message and key bit by bit.

The question that quantum cryptography is therefore most concerned with is the creation and secure distribution of random keys. Here, we will take a closer look at the two most paradigmatic quantum key-distribution (QKD) protocols, developed by *Charles Bennett* and *Gilles Brassard* (1984, republished in 2014), and by *Artur Ekert* (1991), respectively.

14.4.1 The BB 84 Protocol

The first fully worked out example for a quantum key-distribution protocol is due to Charles Bennett and Gilles Brassard (1984, republished in 2014) and the acronym BB84 is typically used to refer to it. The protocol is based on the idea of "conjugate coding" by *Stephen Wiesner*, whose pioneering work on cryptography in the 1960s was not appreciated at the time[6], but who eventually published his manuscript in 1983 after being encouraged to do so by Bennett (see Rogers, 2010, p. 31). Bennet and Brassard then built on Wiesner's work to develop their protocol.

The premise of the BB84 protocol is that Alice is able to prepare random bit strings as well as quantum states of n qubits and that she can send quantum as well as classical information to Bob, while Bob is able to measure qubits in bases of his choosing and send classical information to Alice. The protocol then proceeds with the following steps:

(1) First, Alice generates two (random) bit strings of length n,

$$a = a_1, a_2, a_3, \ldots, a_n \qquad \text{and} \qquad b = b_1, b_2, b_3, \ldots, b_n . \qquad (14.53)$$

(2) Alice then uses these bit strings to generate an n-qubit state

$$| \psi \rangle = \bigotimes_{i=1}^{n} | \psi_{a_i, b_i} \rangle = | \psi_{a_1, b_1} \rangle | \psi_{a_2, b_2} \rangle \ldots | \psi_{a_n, b_n} \rangle , \qquad (14.54)$$

where $a_i \in \{0, 1\}$ determines the single-qubit basis that is chosen for the ith qubit (computational basis, i.e., the eigenbasis of σ_z if $a_i = 0$ or the eigenbasis of σ_x if $a_i = 1$), whereas $b_i \in \{0, 1\}$ determines the particular basis state within a given basis according to the following table:

$\| \psi_{a_i, b_i} \rangle$	$a_i = 0$	$a_i = 1$
$b_i = 0$	$\| 0 \rangle$	$\| + \rangle$
$b_i = 1$	$\| 1 \rangle$	$\| - \rangle$

(3) Bob receives a string of qubits and randomly chooses to perform one of two measurements on them: Either he measures in the computational basis, or he measures in

---

[6]Wiesner's ideas were independently developed in the 1980s under the term "oblivious transfer" (Rabin, 1981; Even *et al.*, 1985).

the eigenbasis of σ_x. Whenever Bob happens to measure a qubit in a basis corresponding to the basis it was prepared in (which happens in 50% of cases on average), he obtains a "correct" bit value b_i as measurement outcome. For instance, if he measures a qubit prepared in the state $|0\rangle$ in the computational basis, he obtains outcome 0. Whenever Bob guesses the basis incorrectly, he still has a chance of 50% of obtaining the "correct" outcome, since both outcomes are equally likely. In total, Bob's outcomes will thus be correct in 75% of the cases on average. However, Bob does not know which qubit was prepared in which basis, so he does not know which of his bits are correct and which are not.

(4) After performing the measurements, Bob sends his choices of measurement bases (but not his measurement results!) to Alice via a public classical channel.

(5) Alice can now check which of the qubits Bob measured in the correct basis and send this information to Bob. With this, Bob now has a key of (on average) $n/2$ bits.

Now, there might have been an eavesdropper (Eve) who could have intercepted, copied, and sent on any classical information as well as intercepted and replaced any qubits exchanged between Alice and Bob. How can Alice and Bob check if this is the case?

Suppose Eve intercepts the qubits sent from Alice to Bob. Eve cannot copy the information of the quantum state (see the no-cloning theorem that we will discuss in Section 21.5), but she can measure the qubits in a basis of her choice and then prepare replacement qubits that are sent to Bob. However, Eve is in the same initial position as Bob, she (initially) does not know which basis the qubits were prepared in. She can perform measurements on the qubits or store them in a quantum memory until she finds out more, but she will have to send some qubits to Bob, otherwise he will not continue with step **(4)**. If Eve measures the qubits in randomly chosen bases (computational or eigenbasis of σ_x), 50% (on average) will give the correct result, and she can hence prepare the qubits sent to Bob in the corresponding correct state. However, for the remaining 50%, Eve's measurement basis does not match the measurement basis Alice used. Therefore, if Bob measures a qubit in what Alice (and later also Bob) believes to be the correct basis, the result will be random (only correct half the time) if Eve prepared it in the wrong basis (which she does half of the time). Therefore, each bit in the key of step **(5)** has a 25% chance of being incorrect if Eve intercepted it. Alice and Bob can thus sacrifice a few (say k) of the key bits and publicly compare them to check if an eavesdropper was listening in.

Every bit of key whose corresponding qubit was intercepted is incorrect (on average) 25% of the time. Conversely, this means that the each bit is correct 75% of the time. And the probability that k bits are correct is thus $\left(\frac{3}{4}\right)^k$. If Alice and Bob compare k bits of the key, the probability to find a mismatch (and hence to conclude that an eavesdropper was present) is $p = 1 - \left(\frac{3}{4}\right)^k$. To illustrate this probability of detecting an eavesdropper, suppose that Alice and Bob are willing to leave a chance of 0.001

that an eavesdropper is present, in other words, $p = 0.999$. Then, since $\left(\frac{3}{4}\right)^k = 1 - p$ we have

$$k = \frac{\ln(1-p)}{\ln(3/4)} = \frac{\ln(0.001)}{\ln(3/4)} = 24.0118 , \tag{14.55}$$

which means that already $k = 25$ bits are enough. And with 50 bits the chance is already less than one in a million.

Résumé: The BB84 quantum key-distribution protocol is based only on the properties of conjugate observables (here, one might also refer to "mutually unbiased bases": no matter which basis vector in which basis is chosen, the probabilities to obtain any of the outcomes in the respective other basis are the same) and the fact that quantum information cannot be freely copied, i.e., the no-cloning theorem (see Section 21.5).

However, there are also other types of quantum key-distribution protocols based on entanglement, whose first and most well-known representative, the Ekert-91 protocol, we will consider next.

14.4.2 The Ekert-91 Protocol

As we have seen in the previous section, quantum key distribution can be achieved entirely based on sending and receiving single quantum systems. However, as shown by Artur Ekert in 1991 secure keys can also be distributed using protocols based on entangled quantum systems. The fascinating aspect of this possibility lies in the observation that the security of this protocol is guaranteed by Bell non-locality, i.e., the fact that no eavesdropper can "fake" the violation of a Bell inequality (see Chapter 13). The Ekert-91 protocol works in the following way:

(1) One of the parties (or a third party) distributes a sequence of particle pairs in the Bell state $|\Psi^-\rangle_{AB} = \frac{1}{\sqrt{2}}\left(|0\rangle_A|1\rangle_B - |1\rangle_A|0\rangle_B\right)$ to the two parties such that Alice and Bob each receive one particle of each entangled pair.

(2) For each particle they receive, Alice and Bob then each secretly choose a measurement direction (on the Bloch sphere, see Section 11.5.2) from the sets $\{\vec{a}_1, \vec{a}_2, \vec{a}_3\}$ and $\{\vec{b}_1, \vec{b}_2, \vec{b}_3\}$, respectively, where

$$\vec{a}_1 = \begin{pmatrix} 1 \\ 0 \\ 0 \end{pmatrix} , \quad \vec{a}_2 = \vec{b}_1 = \frac{1}{\sqrt{2}} \begin{pmatrix} 1 \\ 0 \\ 1 \end{pmatrix} , \quad \vec{a}_3 = \vec{b}_2 = \begin{pmatrix} 0 \\ 0 \\ 1 \end{pmatrix} , \quad \vec{b}_3 = \frac{1}{\sqrt{2}} \begin{pmatrix} -1 \\ 0 \\ 1 \end{pmatrix} \tag{14.56}$$

and measure along the chosen directions, as illustrated in Fig. 14.11.

(3) Alice and Bob then publicly announce the sequences of their chosen measurement directions, $(n_1^A, n_2^A, \ldots)$ and $(n_1^B, n_2^B, \ldots)$, respectively, where $n_j^{A/B} \in \{1, 2, 3\}$ label the respective directions, but they keep the measurement outcomes secret. Since the

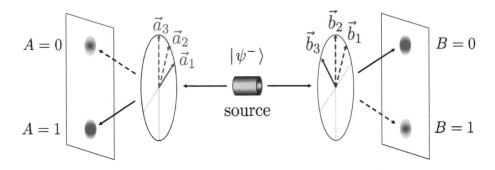

Fig. 14.11 Ekert-91 protocol. Schematic of the setup for the Ekert-91 protocol. A source distributes entangled states $|\psi^-\rangle$ shared by Alice and Bob, who each randomly select a measurement direction and obtain outcomes to which they assign bit values 0 and 1.

measurement outcomes for measurements along the same direction must be perfectly anti-correlated for the state $|\Psi^-\rangle_{AB}$, as we recall from eqn (13.1), Alice and Bob now know that whenever $n_j^A = 2$, $n_j^B = 1$ or $n_j^A = 3$, $n_j^B = 2$, the corresponding bit values A_j and B_j must have opposite values. By simply flipping the corresponding bits B_j Alice and Bob have thus obtained a shared secret key.

There are nine possible combinations of measurement directions, two of which are useful for generating the key. Therefore, exchanging n entangled pairs leads to $\frac{2n}{9}$ bits of key on average. However, (some of) the remaining outcomes are not without use. The outcomes $n_j^A = 1$, $n_j^B = 1$, $n_j^A = 1$, $n_j^B = 3$, $n_j^A = 3$, $n_j^B = 1$, and $n_j^A = 3$, $n_j^B = 3$ correspond exactly to the measurement directions used for a maximal violation of the CHSH inequality (see Section 13.2.2). Since these outcomes are not used to generate a key, Alice and Bob can just publicly announce these outcomes so that both parties can verify the violation of the CHSH inequality. If they observe a value significantly below the expected value of $2\sqrt{2}$, Alice and Bob conclude that an eavesdropper has intercepted (some of) the qubits and they discard the key.

The Ekert-91 protocol was first experimentally demonstrated by the group of Zeilinger (Ursin *et al.*, 2007) in a free-space experiment between the Canary Islands La Palma and Tenerife, while notable implementations of the BB84 protocol over long distances were realized via fibre by Hiskett *et al.* (2006) and via the Canary-island free-space link by Schmitt-Manderbach *et al.* (2007).

Résumé: The Ekert-91 protocol thus provides a strategy for generating a secret key in such a way that the security of the protocol depends on the fact that Eve cannot intercept information without breaking the entanglement between Alice and Bob that is necessary for the Bell inequality violation.

The protocols that we have described in Section 14.4 both allow for detecting an eavesdropper, but practical limitations mean that even without an eavesdropper interfering with the protocol, errors can lead to mismatched keys, and eavesdroppers may "hide" behind such small errors. This means that practical quantum key-distribution protocols involve additional steps such as information reconciliation (correcting partially mismatched keys) and privacy amplification (reducing the amount of information a potential eavesdropper might have obtained about the key).

In addition, there are known additional attacks on cryptographic protocols that can exploit particular features of the implementation. For instance, in the BB84 protocol, single qubits are sent from Alice to Bob. These qubits are typically encoded in single photons. However, in many cases light sources used only probabilistically produce single photons. Sometimes, two photons are emitted and there is no way for Alice or Bob to know when this may have happened. If Eve is able to split off the additional photons, she might get partial information about the key. Such a "photon-number splitting attack" is just an example for an ever expanding range of conceivable attacks and countermeasures which are investigated in the field of quantum cryptography.

In this book, we will not further explore these cryptographic problems, but simply note that the distribution protocols discussed are not the end of the story and refer to the extensive literature on the subject (e.g., Bennett and Brassard, 2014, or the reviews Scarani *et al.*, 2009; Pirandola *et al.*, 2020).

15
Entanglement and Separability

As its name suggests, this chapter will revolve around the notion of entanglement (*Verschränkung* in German), a term used by Erwin Schrödinger 1935*a*, already identified back then as a crucial characteristic of quantum mechanics. Today, entanglement is centrally important in quantum information and beyond. Although many interesting mathematical questions surrounding entanglement remain open (some of which we will come across here), it should be emphasized that there is nothing mysterious per se about the existence and generation of entanglement. Top put it minimalistically, entanglement is a form of correlation that arises as a consequence of the superposition principle. If we accept the central tenet of quantum mechanics that generally it is complex probability amplitudes that are additive, not real-valued probabilities, then the consequence is that quantum systems can be correlated in a way that classical systems cannot. Nevertheless, over the years the notion of entanglement has been at the heart of debates regarding the interpretation of quantum mechanics (for instance, via the characterization of entanglement as prompting "spooky action at a distance" by realists, as discussed in Chapter 13.4). In this chapter, we will focus on the mathematical aspects of entanglement: its definition for pure and mixed states, some necessary and sufficient separability criteria based on positive maps, the relation of entanglement and Bell non-locality (see Chapter 13), and a geometric view on entanglement and separability. We also illustrate these concepts for several types of two-qubit and two-qutrit states in the Hilbert–Schmidt space of density matrices.

In addition, the following Chapters 16, 17, and 18 will provide further room for discussing features of entanglement detection and quantification as well as entanglement in high-dimensional and multipartite systems. Nevertheless, the topic has grown vastly over the past three decades and we therefore refer to the excellent reviews on the topic for further reading, in particular Bruß (2002), Plenio and Virmani (2007), Gühne and Tóth (2009), Horodecki *et al.* (2009), Eltschka and Siewert (2014), and Friis *et al.* (2019).

15.1 Composite Quantum Systems

15.1.1 Entanglement and Separability for Pure States

The basis for our discussion of entanglement in this chapter is to consider two quantum systems A and B, typically considered to be controlled by observers called Alice and Bob, who may be (arbitrarily) far apart from each other. If the quantum systems A and B are independent and Alice and Bob each have full information about their respective systems, the latter can be represented by pure states, represented by vectors in Hilbert space,

$$|\varphi\rangle = \sum_m a_m |m\rangle \in \mathcal{H}_A \quad \text{and} \quad |\phi\rangle = \sum_n b_n |n\rangle \in \mathcal{H}_B, \tag{15.1}$$

where $\mathcal{H}_A$ with dimension $\dim(\mathcal{H}_A) = d_A$ denotes the Hilbert space of Alice and $\mathcal{H}_B$ with $\dim(\mathcal{H}_B) = d_B$ the Hilbert space of Bob, $a_m, b_n \in \mathbb{C}$ with $\sum_m |a_m|^2 = \sum_n |b_n|^2 = 1$, and the sets $\{|m\rangle\}_m$ and $\{|n\rangle\}_n$ form complete orthonormal systems (bases) of the respective Hilbert spaces.

The state space $\mathcal{H}$ of the combined system is given by the tensor-product space

$$\mathcal{H} = \mathcal{H}_A \otimes \mathcal{H}_B. \tag{15.2}$$

Quite generally, a vector in $\mathcal{H}$ can be expanded as

$$|\psi\rangle_{AB} = \sum_{m,n} c_{mn} |m\rangle_A \otimes |n\rangle_B, \tag{15.3}$$

where the coefficient matrix cannot in general be factorized

$$c_{mn} \neq a_m \cdot b_n, \tag{15.4}$$

which physically corresponds to two interacting systems in a joint entangled state. The special case, when the coefficient matrix factorizes,

$$c_{mn} = a_m \cdot b_n, \tag{15.5}$$

describes two independent quantum systems—a *product state* $|\psi\rangle_{AB} = |\varphi\rangle_A \otimes |\phi\rangle_B$. In this case, the joint state is separable and belongs to a subset of the possible states in $\mathcal{H}$ (15.2). Such product states feature no correlation between Alice and Bob, no measurements that Alice or Bob could perform on their systems could be used to make useful predictions about the measurement outcomes of the respective other system if the state of the latter is unknown. As we have already seen for the example of the Bell states, the situation is different for entangled states, where the observers can predict certain measurement outcomes of the respective other party with certainty (see Section 13.1). For pure states, we will raise this idea to the level of a definition of separability and entanglement:

Definition 15.1 *Any pure state $|\psi\rangle_{AB}$ that can be written as $|\psi\rangle_{AB} = |\varphi\rangle_A \otimes |\phi\rangle_B$ for some vectors $|\varphi\rangle_A$ and $|\phi\rangle_B$ is called **separable** with respect to the bipartition $A|B$, i.e., the Hilbert-space splitting $\mathcal{H} = \mathcal{H}_A \otimes \mathcal{H}_B$.*

Conversely, we define entanglement as "not being separable":

Definition 15.2 *Any pure state $|\psi\rangle_{AB}$ that is not separable is called **entangled**:*

$$|\psi\rangle_{AB} \neq |\varphi\rangle_A \otimes |\phi\rangle_B.$$

The simplicity of these definitions not withstanding, how do we decide if a given pure state $|\psi\rangle_{AB}$ is separable or entangled? This, as it turns out, is not difficult to answer, but in order to do so, it is useful to turn to the so-called Schmidt-decomposition theorem, which we will do next.

15.1.2   The Schmidt Decomposition

In order to characterize the quantum states of a composite system the following theorem, first formulated by *Erhard Schmidt* in 1907, is of great value.

Theorem 15.1 (Schmidt-decomposition theorem)

Any bipartite pure state $|\psi\rangle \in \mathcal{H}_A \otimes \mathcal{H}_B$ can be written as

$$|\Psi\rangle_{AB} = \sum_{i=1}^{k} \lambda_i \, |\chi_i\rangle_A \, |\eta_i\rangle_B$$

where $\{|\chi_i\rangle_A \in \mathcal{H}_A\}$ and $\{|\eta_i\rangle_B \in \mathcal{H}_B\}$ are orthonormal bases called "Schmidt base" of $|\Psi\rangle_{AB}$ in $\mathcal{H}_A$ and $\mathcal{H}_B$, respectively, the "Schmidt coefficients" $\lambda_i \in \mathbb{R}$ are real and non-negative, $\lambda_i \geq 0$, and at most $k \leq \min\{\dim(\mathcal{H}_A), \dim(\mathcal{H}_B)\}$ coefficients are non-zero.

Before we prove this theorem, let us examine what it means and why this is a big deal. Recall that when we first wrote a general basis decomposition for a (pure) bipartite quantum state in eqn (15.3), we had to use two different indices m and n to label the basis elements of the product basis $\{|m\rangle_A \otimes |n\rangle_B\}_{m,n}$. Naively, it therefore seems like we might need a lot of information—a number of complex coefficients in the amount of the product of the local dimensions—to fully describe arbitrary joint states, and consequently, their local properties (their reduced states). However, the Schmidt-decomposition theorem tells us that we can fully describe the state (and its local properties) using only real, non-negative numbers, and only as many as the smaller of the local dimensions at most. As we will see, this is a very powerful statement. But before we see what the consequences are, let us prove the theorem.

Proof   We begin again with the general form of the pure state $|\psi\rangle_{AB}$ from eqn (15.3). For any local Hilbert-space dimensions $\dim(\mathcal{H}_A) = d_A$ and $\dim(\mathcal{H}_B) = d_B$, the $d_A \times d_B$ coefficient matrix $c = (c_{mn})$ can be written in *singular-value decomposition* (a well-known statement from linear algebra), i.e., we can write the matrix as

$$c = U\,\Lambda^{\mathrm{diag}}\,V^{\top}, \tag{15.6}$$

where $U = (U_{mk})$ and $V = (V_{nl})$ are $d_A \times d_A$ and $d_B \times d_B$ complex unitary matrices, respectively, and Λ^{diag} is a diagonal $d_A \times d_B$ matrix with non-negative diagonal elements $\lambda_i := \Lambda_{ii}^{\mathrm{diag}} \geq 0$. In components we may thus write $c_{mn} = \sum_i U_{mi}\,\Lambda_{ii}^{\mathrm{diag}}\,V_{in}^{\top} = \sum_i U_{mi}\,\lambda_i\,V_{ni}$. Thus we get

$$|\psi\rangle_{AB} = \sum_{m,n} c_{mn} |m\rangle_A \otimes |n\rangle_B = \sum_{m,n,i} U_{mi} \lambda_i V_{ni} |m\rangle_A \otimes |n\rangle_B \ . \qquad (15.7)$$

Defining the vectors

$$|\chi_i\rangle_A := \sum_m U_{mi} |m\rangle_A \quad \text{and} \quad |\eta_i\rangle_B := \sum_n V_{ni} |n\rangle_B \ , \qquad (15.8)$$

which form complete orthonormal bases of $\mathcal{H}_A$ and $\mathcal{H}_B$, respectively, we arrive at

$$|\psi\rangle_{AB} = \sum_i^k \lambda_i |\chi_i\rangle_A \otimes |\eta_i\rangle_B \ . \qquad (15.9)$$

Moreover, since the values of the index i are bounded from above by the smaller of the two dimensions, $k \leq \min\{d_A, d_B\}$, we arrive at the statement of the theorem. $\qquad \square$

We also have the freedom to choose the unitaries U and V such that the Schmidt coefficients are ordered, $\lambda_1 \geq \lambda_2 \geq \ldots \geq \lambda_k > 0$. The number of non-vanishing Schmidt coefficients λ_i is called *Schmidt number* N_S, or *Schmidt rank*[1], of the state $|\psi\rangle_{AB}$, and N_S is uniquely defined for $|\psi\rangle_{AB}$, i.e., it does not depend on the particular Schmidt bases.

From the proof of the Schmidt-decomposition theorem it is also clear that the Schmidt number N_S is invariant under local unitary transformations U_A and U_B on the subspaces $\mathcal{H}_A$ and $\mathcal{H}_B$, respectively. Consequently, the following corollaries hold.

Corollary 15.1 (Local unitary invariance)

A product state cannot be transformed into an entangled state—and vice versa—by local unitary transformations of the subsystems A and B in the Hilbert spaces $\mathcal{H}_A$ and $\mathcal{H}_B$, respectively.

Corollary 15.2 (Entanglement, separability, and Schmidt number)

$$|\psi\rangle_{AB} \in \mathcal{H}_A \otimes \mathcal{H}_B \ \text{is entangled iff } N_S > 1$$
$$\text{is separable iff } N_S = 1$$

When $d_A = d_B = d$, a state $|\psi\rangle_{AB}$ is called *maximally entangled* if all Schmidt coefficients are equal $\lambda_i = \frac{1}{d}, \forall i$, which also implies a maximal Schmidt number

---
[1]This terminology will become clear when we discuss reduced states of the subsystems in the following.

$N_S = k = d$. Note that the reverse is not true; we can have states with $N_S = k = d$ for which, say, one of the Schmidt coefficients is (arbitrarily) close to one, while the others are all non-zero but sufficiently small (so that the state is normalized, $\sum_i \lambda_i^2 = 1$), and, for reasons that will become apparent soon, we do not consider such states to be maximally (or even strongly) entangled.

15.1.3   Subsystems and Reduced States

The Schmidt-decomposition theorem (Theorem 15.1) is also very useful to prove important properties of quantum states of composite systems and their subsystems. In order to properly explore these properties, we first need to establish how we can formally describe the subsystems of a composite system. To this end, we introduce the *partial-trace operation*.

Definition 15.3   *Let ρ be a (linear) operator acting on a Hilbert space $\mathcal{H}_A \otimes \mathcal{H}_B$. The **partial trace** of ρ over subsystem B is given by*

$$\mathrm{Tr}_B(\rho) := \sum_n {}_B\langle n \mid \rho \mid n \rangle_B \,,$$

where the set $\{\mid n \rangle_B\}_n$ is an orthonormal basis of $\mathcal{H}_B$.

Here, we have used the subscripts on the (co)vectors ${}_B\langle n \mid$ and $\mid n \rangle_B$ as a shorthand to indicate identity operators acting on the (all) other subsystem(s), i.e., it is often useful to keep in mind that a more precise way to express the partial trace is

$$\mathrm{Tr}_B(\rho) := \sum_n \left(\mathbb{1}_A \otimes {}_B\langle n \mid\right) \rho \left(\mathbb{1}_A \otimes \mid n \rangle_B\right), \tag{15.10}$$

when the operator in the argument of the partial trace is a density operator (as we have already hinted at by using the symbol ρ), the result of this operation is the *reduced density operator* or *reduced state* of the remaining subsystem(s).

Definition 15.4   *Let ρ_{AB} be the density operator of a bipartite system with Hilbert space $\mathcal{H}_A \otimes \mathcal{H}_B$. The **reduced states** of subsystems A and B are given by $\rho_A := \mathrm{Tr}_B(\rho_{AB})$ and $\rho_B := \mathrm{Tr}_A(\rho_{AB})$, respectively.*

The properties of Hermiticity ($\rho^\dagger = \rho$), normalization [$\mathrm{Tr}(\rho) = 1$], and positivity ($\rho \geq 0$) for the reduced states to represent proper density operators (see Chapter 11) follow immediately from this definition if ρ_{AB} satisfies them as well. We obviously have $\rho_A^\dagger = \rho_A$ and $\rho_B^\dagger = \rho_B$, and normalization follows since we can calculate the trace of the joint system with respect to a basis of product states $\mid m \rangle_A \otimes \mid n \rangle_B$. For positivity, we need to ensure that for any $\mid \psi \rangle_A \in \mathcal{H}_A$ we have ${}_A\langle \psi \mid \rho_A \mid \psi \rangle_A \geq 0$, but this is clear since we can write any such expression as a sum of non-negative terms due to the positivity of ρ_{AB}, i.e.,

$$_A\langle\psi|\rho_A|\psi\rangle_A = \sum_n {}_A\langle\psi|\otimes {}_B\langle n|\rho|\psi\rangle_A\otimes|n\rangle_B \geq 0. \qquad (15.11)$$

We thus observe that Definition 15.4 results in well-defined density operators for the subsystems, but of course this leaves out the question *why should we choose this particular way of obtaining reduced-state density operators* from the joint state. The simple answer to this question is: because the partial trace is the unique function that maps the joint state ρ_{AB} to subsystem density operators ρ_A and ρ_B such that the measurement statistics are correctly represented for all possible measurements on the subsystems (see, e.g., the discussion in Nielsen and Chuang, 2000, p. 107). More precisely, this definition ensures that for all observables O_A and O_B in the algebras of observables for subsystems A and B, respectively, we have

$$\mathrm{Tr}(O_A\,\rho_{AB}) = \mathrm{Tr}(O_A\,\rho_A), \quad \text{and} \quad \mathrm{Tr}(O_B\,\rho_{AB}) = \mathrm{Tr}(O_B\,\rho_B). \qquad (15.12)$$

Indeed, for the sake of generality we should promote satisfaction of the conditions in (15.12) to the definition of the reduced states, which turns out to be necessary for situations where one wishes to define subsystems in the absence of a tensor-product structure, e.g., for modes of fermionic quantum fields, see (Friis *et al.*, 2013a; Friis, 2016).

With what we now know about the description of subsystems, let us return to the Schmidt-decomposition theorem and see how it aids us in making statements about the reduced states.

Lemma 15.1 (Eigenvalues of subsystems)

If a composite system with subsystems A and B is in a pure state $|\psi\rangle_{AB}$ described by the density matrix $\rho_{AB} = |\psi\rangle\langle\psi|_{AB}$ then the reduced density matrices ρ_A and ρ_B have the same eigenvalues, which are given by the squared Schmidt coefficients of $|\psi\rangle_{AB}$, and are diagonal in regards to the respective Schmidt bases,

$$\rho_A = \sum_{i=1}^{k}\lambda_i^2\,|\chi_i\rangle\langle\chi_i|_A, \quad \text{and} \quad \rho_B = \sum_{i=1}^{k}\lambda_i^2\,|\eta_i\rangle\langle\eta_i|_B.$$

Proof To prove this, we simply write the state $|\psi\rangle_{AB}$ in Schmidt decomposition according to Theorem 15.1, form the projector $\rho_{AB} = |\psi\rangle\langle\psi|_{AB}$ and calculate the reduced states. In evaluating the partial traces we have the freedom to choose any local bases of subsystems A and B, respectively, which means that we may use the Schmidt bases $\{|\chi_i\rangle_A\}_i$ and $\{|\eta_i\rangle_B\}_i$. Starting with subsystem A we then have

$$\rho_A = \text{Tr}_B\left(|\psi\rangle\langle\psi|_{AB}\right) = \text{Tr}_B\left(\sum_{i,j}\lambda_i\lambda_j|\chi_i\rangle\langle\chi_j|_A \otimes |\eta_i\rangle\langle\eta_j|_B\right)$$

$$= \sum_k {}_B\langle\eta_k|\left(\sum_{i,j}\lambda_i\lambda_j|\chi_i\rangle\langle\chi_j|_A \otimes |\eta_i\rangle\langle\eta_j|_B\right)|\eta_k\rangle_B$$

$$= \sum_{i,j,k}\lambda_i\lambda_j\,|\chi_i\rangle\langle\chi_j|_A\,{}_B\langle\eta_k|\eta_i\rangle_B\,{}_B\langle\eta_j|\eta_k\rangle_B = \sum_k\lambda_k^2\,|\chi_k\rangle\langle\chi_k|_A\,, \quad (15.13)$$

and in complete analogy we obtain the corresponding expression for ρ_B when carrying out the partial trace over subsystem A with respect to the local basis $\{|\chi_i\rangle_A\}_i$.   □

Let us now make some remarks about this result.

Remark I: Firstly, we can now appreciate why $N_S \equiv k$ is called the "Schmidt rank". It is the rank of the reduced states, the number of non-zero eigenvalues. In particular, using the shorthand $p_i = \lambda_i^2$, we can write the reduced states with respect to the Schmidt bases as

$$\rho_A = \begin{pmatrix} p_1 & 0 & 0 & \cdots & 0 & & & \\ 0 & p_2 & 0 & \cdots & 0 & & & \\ 0 & 0 & p_3 & \cdots & 0 & & & \\ \vdots & \vdots & \vdots & \ddots & \vdots & & & \\ 0 & 0 & 0 & \cdots & p_k & & & \\ & & & & & 0 & \cdots & 0 \\ & & & & & \vdots & \ddots & \vdots \\ & & & & & 0 & \cdots & 0 \end{pmatrix}, \quad \text{and} \quad \rho_B = \begin{pmatrix} p_1 & 0 & 0 & \cdots & 0 & & & \\ 0 & p_2 & 0 & \cdots & 0 & & & \\ 0 & 0 & p_3 & \cdots & 0 & & & \\ \vdots & \vdots & \vdots & \ddots & \vdots & & & \\ 0 & 0 & 0 & \cdots & p_k & & & \\ & & & & & 0 & \cdots & 0 \\ & & & & & \vdots & \ddots & \vdots \\ & & & & & 0 & \cdots & 0 \end{pmatrix}, \quad (15.14)$$

i.e., in block-diagonal form, with a diagonal matrix $\text{diag}\{p_1, p_2, \ldots, p_k\}$ and remaining $(d_A - k) \times (d_A - k)$ and $(d_B - k) \times (d_B - k)$ blocks with all zeros.

Remark II: Second, we notice that the subsystems are in a mixed state even though the composite system is in a pure state. In particular, if $d_A = d_B = d$ and the composite system is maximally entangled the reduced density matrices are maximally mixed, i.e., proportional to the identity matrix

$$\rho_A = \tfrac{1}{d}\mathbb{1}_A\,, \quad \rho_B = \tfrac{1}{d}\mathbb{1}_B\,. \quad (15.15)$$

In this case the *purity* (recall Corollary 11.1) of the subsystems evaluates to

$$P(\rho_A) = \text{Tr}(\rho_A^2) = P(\rho_B) = \tfrac{1}{d} \quad (15.16)$$

and the *mixedness* is just $S_{\mathrm{L}}(\rho) = 1 - P(\rho) = 1 - \frac{1}{d}$. An interesting remark on the side here is that, since the eigenvalues of the reduced states are the same for any overall pure state, all functions of the eigenvalues such as the purity and mixedness, but also other entropic functions that we will encounter later, must give the same result for both subsystems. However, this is generally not the case when the overall state is mixed.

As an example, consider the Bell state

$$\rho^- = |\psi^-\rangle\langle\psi^-| = \frac{1}{2}\begin{pmatrix} 0 & 0 & 0 & 0 \\ 0 & 1 & -1 & 0 \\ 0 & -1 & 1 & 0 \\ 0 & 0 & 0 & 0 \end{pmatrix}, \tag{15.17}$$

for which the reduced density matrices are $\rho_A = \rho_B = \frac{1}{2}\mathbb{1}_2$ with a maximal mixedness of $S_{\mathrm{L}} = 1 - \mathrm{Tr}(\rho\rho_A^2) = 1 - \mathrm{Tr}(\rho\rho_B^2) = \frac{1}{2}$.

Choosing, on the other hand, a pure but not maximally entangled state like

$$|\psi\rangle = \tfrac{1}{\sqrt{3}}\left(|00\rangle + |01\rangle + |11\rangle\right), \tag{15.18}$$

and writing it with respect to the computational basis $\{|0\rangle, |1\rangle\}$, we obtain the density matrix

$$\rho = |\psi\rangle\langle\psi| = \frac{1}{3}\begin{pmatrix} 1 & 1 & 0 & 1 \\ 1 & 1 & 0 & 1 \\ 0 & 0 & 0 & 0 \\ 1 & 1 & 0 & 1 \end{pmatrix}, \tag{15.19}$$

and, in turn, the reduced density matrices

$$\rho_A = \tfrac{1}{3}\left(2\,|0\rangle\langle0| + |1\rangle\langle0| + |0\rangle\langle1| + |1\rangle\langle1|\right) = \tfrac{1}{3}\begin{pmatrix} 2 & 1 \\ 1 & 1 \end{pmatrix},$$

$$\rho_B = \tfrac{1}{3}\left(|0\rangle\langle0| + |1\rangle\langle0| + |0\rangle\langle1| + 2\,|1\rangle\langle1|\right) = \tfrac{1}{3}\begin{pmatrix} 1 & 1 \\ 1 & 2 \end{pmatrix}. \tag{15.20}$$

Both reduced density matrices have the same eigenvalues and thus the same purity $P(\rho_A) = P(\rho_B) = \frac{7}{9}$, respectively, the same mixedness $S_{\mathrm{L}} = 1 - \mathrm{Tr}(\rho_A^2) = 1 - \mathrm{Tr}(\rho_B^2) = \frac{2}{9}$.

Remark III: If the composite state $\rho = |\psi\rangle\langle\psi|$ is pure, it is a product state if and only if the reduced density matrices ρ_A and ρ_B are pure states.

For example, if Alice's spin is definitely up and Bob's down then the composite state represents a pure state

$$
\rho_{\uparrow\downarrow} = \rho_{\uparrow}^{(A)} \otimes \rho_{\downarrow}^{(B)} = \begin{pmatrix} 1 & 0 \\ 0 & 0 \end{pmatrix} \otimes \begin{pmatrix} 0 & 0 \\ 0 & 1 \end{pmatrix} = \begin{pmatrix} 0 & 0 & 0 & 0 \\ 0 & 1 & 0 & 0 \\ 0 & 0 & 0 & 0 \\ 0 & 0 & 0 & 0 \end{pmatrix}. \tag{15.21}
$$

Conversely, this means that every pure state whose subsystems are in mixed states must be entangled!

15.1.4 Purification of Quantum States

In quantum communication and quantum information in general, the notion of the purification of a quantum state plays an important role. Once again, the Schmidt-decomposition theorem (Theorem 15.1) is very helpful to understand this quantum feature. In a nutshell, purification means that we can associate a pure state on a larger Hilbert space with any given mixed state. Let us see how this works.

Suppose we have a quantum system A at Alice's side represented by the mixed density matrix ρ_A. Then we can always introduce another system B, not necessarily corresponding to a physical one, and define a pure state $|\psi\rangle_{AB}$ for the joint system of A and B, with corresponding density operator

$$
\rho_{AB} = |\psi\rangle\langle\psi|_{AB} , \tag{15.22}
$$

such that the pure state ρ_{AB} reduces to ρ_A, when we concentrate on subsystem A alone,

$$
\rho_A = \mathrm{Tr}_B(\rho_{AB}) = \mathrm{Tr}_B(|\psi\rangle\langle\psi|_{AB}) . \tag{15.23}
$$

Let us phrase this important fact in terms of theorem:

Theorem 15.2 (Purification)

For any mixed state on the Hilbert space $\mathcal{H}_A$ described by a density operator $\rho \in \mathcal{B}_{\mathrm{HS}}(\mathcal{H}_A)$ there exists a (non-unique) pure state $|\psi\rangle_{AB}$ in a larger Hilbert space $\mathcal{H}_A \otimes \mathcal{H}_B$ with $\dim(\mathcal{H}_B) \le \dim(\mathcal{H}_A)$ such that

$$
\rho_A = \mathrm{Tr}_B(|\psi\rangle\langle\psi|_{AB}) .
$$

*A state $|\psi\rangle_{AB}$ with this property is called a **purification** of ρ_A.*

Proof We start by writing the mixed state ρ_A with respect to its eigenbasis $\{|\,\psi_i\,\rangle_A\}_i$

$$\rho_A = \sum_i p_i \,|\,\psi_i\,\rangle\langle\,\psi_i\,|_A \;. \tag{15.24}$$

To purify ρ_A we introduce a system B with a Hilbert space of the same dimension, and choose an orthonormal basis $\{|\,\varphi_i\,\rangle_B \in \mathcal{H}_B\}_i$. Then we define a pure state for the joint system of A and B as

$$|\,\psi\,\rangle_{AB} = \sum_i \sqrt{p_i}\,|\,\psi_i\,\rangle_A \otimes |\,\varphi_i\,\rangle_B\,, \tag{15.25}$$

mirroring the Schmidt-decomposition theorem. By inspecting the proof of Theorem 15.1 we see that the proof that we are looking for here follows from the former proof by setting $\sqrt{p_i} = \lambda_i$. □

The fact that such a purification always exists means that we can always regard mixed states as an expression of our lack of knowledge about the system, in particular, of its entanglement with other (unknown, uncontrolled, or simply unspecified) quantum systems, and we will return to this statement in Section 21.1.

15.2   Entanglement and Separability

With our preliminary definitions and backbone of results for pure states at hand, we now turn to the heart of the matter: entanglement. Specifically, in this section we focus on features of bipartite entanglement, considering pairs of two quantum systems A and B—the situation of Alice and Bob—whose joint state we describe by a density matrix ρ_{AB} in a Hilbert–Schmidt space $\mathcal{B}_{\mathrm{HS}}(\mathcal{H}_A \otimes \mathcal{H}_B)$. The subsystems A and B are fully specified by the reduced density matrices $\rho_A = \mathrm{Tr}_B(\rho_{AB})$ and $\rho_B = \mathrm{Tr}_A(\rho_{AB})$.

15.2.1   Entanglement and Separability for Mixed States

As we have mentioned, entanglement is a form of correlation. Let us therefore attempt to characterize the features of quantum states according to the correlations of the measurement results of Alice and Bob. For example, if the total quantum state is a *product state*

$$\rho_{AB} = \rho_A \otimes \rho_B\,, \tag{15.26}$$

then the subsystems are entirely *uncorrelated*. As in the case of pure states (recall Definition 15.1), we classify mixed product states as separable states. However, opposed to the pure-state case, a mixed state does not have to be entangled if it is not a product state. Indeed, we can think of many examples of (classically) correlated systems that are by no means entangled. For instance, imagine that we prepare two pairs of quantum systems A and B and distribute them to Alice and Bob, the first system in the state $|\,0\,\rangle_A|\,1\,\rangle_B$, and the second one in the state $|\,1\,\rangle_A|\,0\,\rangle_B$, but we do not tell Alice and Bob which pair is in which state. They will be forced to describe the joint state of each pair by the density operator

$$\rho_{AB}^{c.c.} = \tfrac{1}{2}\Big(|0\rangle\langle 0|_A \otimes |1\rangle\langle 1|_B + |1\rangle\langle 1|_A \otimes |0\rangle\langle 0|_B \Big). \tag{15.27}$$

As we will discuss in more detail in Section 15.2.2, such a state is called *classically correlated* and despite not being in product form, it is not entangled either. For mixed states we will therefore have to take into account this possibility in our definition of entanglement and separability.

Definition 15.5   *Any mixed state ρ_{AB} in $\mathcal{B}_{\mathrm{HS}}(\mathcal{H}_A \otimes \mathcal{H}_B)$ that can be written as*

$$\rho_{AB} = \sum_i p_i \, \rho_i^{(A)} \otimes \rho_i^{(B)}$$

*for some density operators $\rho_i^{(A)}$ in $\mathcal{B}_{\mathrm{HS}}(\mathcal{H}_A)$ and $\rho_i^{(B)}$ in $\mathcal{B}_{\mathrm{HS}}(\mathcal{H}_B)$, and for some probability distribution $\{p_i\}_i$ with $0 \leq p_i \leq 1$ and $\sum_i p_i = 1$, is called **separable** with respect to the bipartition $A|B$.*

Again, we define entanglement as the contrapositive statement of not being separable:

Definition 15.6   *Any mixed state ρ_{AB} that is not separable is called **entangled**:*

$$\rho_{AB} \neq \sum_i p_i \, \rho_i^{(A)} \otimes \rho_i^{(B)}.$$

Mathematically, we can thus classify all quantum states into separable and entangled states, and we can define the corresponding sets of states.

Definition 15.7   *The **set of separable states** is defined as*

$$\mathcal{S} := \Big\{ \rho_{AB} = \sum_i p_i \, \rho_i^{(A)} \otimes \rho_i^{(B)} \,\big|\, 0 \leq p_i \leq 1, \ \sum_i p_i = 1 \Big\}.$$

The set of entangled states is just the complement of $\mathcal{S}$ in the space of all density operators in the Hilbert–Schmidt space $\mathcal{B}_{\mathrm{HS}}(\mathcal{H}_A \otimes \mathcal{H}_B)$.

Definition 15.8   *The **set of entangled states** is defined as the complement $\mathcal{S}^{\mathrm{C}}$ of the set of separable states $\mathcal{S}$, such that $\mathcal{S} \cup \mathcal{S}^{\mathrm{C}}$ is the set of all density operators (Hermitian, positive semidefinite, normalized) in $\mathcal{B}_{\mathrm{HS}}(\mathcal{H}_A \otimes \mathcal{H}_B)$.*

Geometrically, the set of separable states is convex:

> **Lemma 15.2 (Convexity of $\mathcal{S}$)**
>
> If $\rho_1, \rho_2 \in \mathcal{S}$ then for any $0 \leq p \leq 1$: $\quad \rho := p\,\rho_1 + (1-p)\,\rho_2 \in \mathcal{S}$.

Proof Let us choose two states that are, by definition, separable, and given by

$$\rho_1 = \sum_{i=1}^{m} p_i^{(1)} \rho_i^{(A)} \otimes \rho_i^{(B)}, \quad \rho_2 = \sum_{j=1}^{n} p_j^{(2)} \omega_j^{(A)} \otimes \omega_j^{(B)}, \tag{15.28}$$

then we form their convex sum

$$\rho = p\,\rho_1 + (1-p)\,\rho_2 = p \sum_{i=1}^{m} p_i^{(1)} \rho_i^{(A)} \otimes \rho_i^{(B)} + (1-p) \sum_{j=1}^{n} p_j^{(2)} \omega_j^{(A)} \otimes \omega_j^{(B)}$$

$$= \sum_{k=1}^{m+n} \tilde{p}_k \, \tilde{\rho}_k^{(A)} \otimes \tilde{\rho}_k^{(B)} \in \mathcal{S}, \tag{15.29}$$

where $\tilde{p}_k = p\,p_k^{(1)}$ and $\tilde{\rho}_k^{(A/B)} = \rho_k^{(A/B)}$ for $k = 1, \ldots, m$, while $\tilde{p}_k = (1-p)\,p_{k-m}^{(2)}$ and $\tilde{\rho}_k^{(A/B)} = \omega_{k-m}^{(A/B)}$ for $k = m+1, \ldots, m+n$. Clearly, the operators $\tilde{\rho}_k^{(A/B)}$ are valid density operators for all k, and $0 \leq \tilde{p}_k \leq 1$ with

$$\sum_{k} \tilde{p}_k = p \sum_{i=1}^{m} p_i^{(1)} + (1-p) \sum_{j=1}^{n} p_j^{(2)} = p + (1-p) = 1, \tag{15.30}$$

which proves the statement of Lemma 15.2. $\qquad\qquad\qquad\qquad\qquad\qquad \square$

As we have seen before in Section 15.5 the set of all quantum states as a whole is also convex. This means that the convex set $\mathcal{S}$ of separable states is enclosed by the set of entangled states. Nevertheless, since the boundary of the set of all states is formed by pure states (see Fig. 11.2), and we know that there exist pure, separable states (namely, product states $|\psi\rangle_A \otimes |\phi\rangle_B$), for which the set $\mathcal{S}$ touches the boundary of the set of all states. This geometric structure is schematically illustrated in Fig. 15.1.

15.2.2 Quantum Correlations versus Classical Correlations

With the central definitions for entanglement and separability of mixed states in place, let us now examine the (sometimes subtle, sometimes more pronounced) differences between states from these sets in terms of the correlations they exhibit. According to *Reinhard Werner* (1989) we have to differentiate between *uncorrelated* states, *classically correlated* states, and *quantum correlations* (or *EPR correlations*).

Any completely *uncorrelated state* is a product state in the form of eqn (15.26) because the expectation values for the joint measurement of any local observables A and B of Alice's and Bob's subsystems always factorize,

$$\langle A \otimes B \rangle = \mathrm{Tr}(\rho_{AB}\, A \otimes B) = \mathrm{Tr}\big[(\rho_A \otimes \rho_B)\,(A \otimes B)\big] = \mathrm{Tr}(\rho_A A)\,\mathrm{Tr}(\rho_B B). \tag{15.31}$$

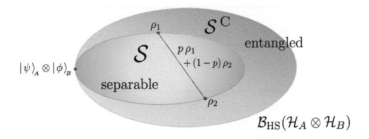

Fig. 15.1 Geometry of entanglement and separability. Schematic illustration of the convex set of all quantum states (as a subset of the Hilbert–Schmidt space $\mathcal{B}_{\mathrm{HS}}(\mathcal{H}_A \otimes \mathcal{H}_B)$) and the convex subset $\mathcal{S}$ of separable states (in blue), which is enclosed by the entangled states $\mathcal{S}^{\mathrm{C}}$ (in orange). Convex combinations $p\rho_1 + (1-p)\rho_2$ of separable states ρ_1 and ρ_2 are again separable states. The points where the set of separable states touch the boundary of the set of all states correspond to pure product states $|\psi\rangle_A \otimes |\phi\rangle_B$.

Physically such an uncorrelated state can be prepared by Alice and Bob by setting up devices that operate independently and produce the states ρ_A and ρ_B, respectively. Then the results of joint measurements of the two observables A and B factorize in the sense that the mean value of the joint measurement is the product of the mean values of the individual measurements, which corresponds to the classical rule of multiplying probabilities.

For *classically correlated states* we may think, for instance[2], of the following physical situation. Alice and Bob each have a device with a switch that can be set in different positions i with $i = 1, ..., n$. For each setting of the switch the devices prepare the states $\rho_i^{(A)}$ and $\rho_i^{(B)}$, respectively. Now suppose there also is a random-number generator which produces numbers $i = 1, ..., n$ with probability p_i. Before the measurement, a number i is randomly selected and the switches of the two devices are set according to this number. Then the expectation value of the joint measurement of the observables A and B will be given by the weighted sum of factorized expectation values

$$\langle A \otimes B \rangle = \sum_{i=1}^{n} p_i \operatorname{Tr}\!\big(\rho_i^{(A)} A\big)\, \operatorname{Tr}\!\big(\rho_i^{(B)} B\big)$$

$$= \sum_{i=1}^{n} p_i \operatorname{Tr}\!\big[\big(\rho_i^{(A)} \otimes \rho_i^{(B)}\big)(A \otimes B)\big] = \operatorname{Tr}\!\big(\rho_{AB}\, A \otimes B\big), \qquad (15.32)$$

where ρ_{AB} is a separable state as in Definition 15.7. Such a general separable state is called classically correlated. Here, the term "correlated" is due to the fact that the expectation value no longer factorizes but is expressed as a weighted sum like eqn (15.32), but the correlation is "classical" because the preparation of the correlations (with the help of a random-number generator) would work equally well with any

---

[2]Recall also the example in eqn (15.27).

classical systems with n different settings. However, note that since we can prepare the same state in many different ways (recall Proposition 11.1) classical correlation does not necessarily mean a preparation of the state exactly as described before but just that its statistical properties can be reproduced by a classical mechanism.

Of course, our definition for separability, Definition 15.5, contains both uncorrelated and classically correlated states since the convex sum may contain any number of product states, including the case of a single, uncorrelated product state.

States that are neither uncorrelated nor exclusively classically correlated are called *quantum correlated* or *EPR correlated* to emphasize their crucial role in the violation of a Bell inequality (as we will see in Section 15.3) and in the EPR paradox. But this brings us to the question: What specific quality of quantum correlations is it that brings about this special role? In what sense are quantum correlations different from classical correlations?

Let us compare the classically correlated (and thus separable) state $\rho_{AB}^{\text{c.c.}}$ from eqn (15.27) to the Bell state $|\psi^-\rangle_{AB} = \frac{1}{\sqrt{2}}(|0\rangle_A|1\rangle_B - |1\rangle_A|0\rangle_B)$. Suppose that both Alice and Bob make a measurement with respect to the computational basis (i.e., of σ_z, along the z direction on the Bloch sphere). Then for both states, whenever Alice obtains the outcome 0, Bob obtains the outcome 1, and vice versa. Phrasing this in terms of joint probabilities $p(m_A, n_B)$ of Alice obtaining outcome m and Bob obtaining outcome n, we thus have

$$p(0_A, 1_B) = p(1_A, 0_B) = 1, \tag{15.33a}$$

$$p(0_A, 0_B) = p(1_A, 1_B) = 0, \tag{15.33b}$$

perfect anti-correlation for both the Bell state and the classically correlated state.

Now, suppose that both Alice and Bob measure along different directions, say, with respect to the x bases, $\{|\pm\rangle_{A/B} = \frac{1}{\sqrt{2}}(|0\rangle_{A/B} \pm |1\rangle_{A/B})\}$. From eqn (13.1) we already know that the Bell state $|\psi^-\rangle_{AB}$ looks exactly the same in this new basis, and we thus have

$$p(+_A, -_B)_{\psi^-} = p(-_A, +_B)_{\psi^-} = 1, \tag{15.34a}$$

$$p(+_A, +_B)_{\psi^-} = p(-_A, -_B)_{\psi^-} = 0. \tag{15.34b}$$

Meanwhile, for the classically correlated state, we can calculate

$$p(m_A, n_B)_{\text{c.c.}} = {}_A\langle m|\,{}_B\langle n| \rho_{AB}^{\text{c.c.}} |m\rangle_A|n\rangle_B = \tfrac{1}{2}\Big(|\langle mn|01\rangle|^2 + |\langle mn|10\rangle|^2\Big), \tag{15.35}$$

where the inner products ${}_{AB}\langle mn|01\rangle_{AB}$ factorize and we have

$$|\langle mn|01\rangle|^2 = |\langle m|0\rangle|^2\,|\langle n|1\rangle|^2\,, \quad |\langle mn|10\rangle|^2 = |\langle m|1\rangle|^2\,|\langle n|0\rangle|^2\,. \tag{15.36a}$$

Since we further have $|\langle\pm|0\rangle|^2 = |\langle\pm|1\rangle|^2 = \tfrac{1}{2}$, we thus arrive at

$$p(+_A, -_B)_{\text{c.c.}} = p(-_A, +_B)_{\text{c.c.}} = p(+_A, +_B)_{\text{c.c.}} = p(-_A, -_B)_{\text{c.c.}} = \tfrac{1}{4} . \qquad (15.37)$$

The classically correlated state shows no correlations whatsoever with respect to the new bases. Indeed, no other choice of measurement directions for Alice and Bob will show correlations (of the same strength, here, perfect anti-correlations) for the state $\rho_{AB}^{\text{c.c.}}$. The Bell state, on the other hand, shows perfect anti-correlations in many, in fact, infinitely many different pairs of bases of Alice and Bob, as long as they choose the same measurement directions.

For other maximally entangled states (e.g., other Bell states), analogue statements can be made: As long as the choices of bases for Alice and Bob are correlated (in a way determined by choice of the particular entangled state, see, e.g., the discussion of the Werner states, Section 15.3.3), they will also find perfectly (anti-)correlated outcomes. Note, however, that this does not mean that maximally entangled states are correlated for arbitrary pairs of bases for Alice and Bob. For instance, if Alice were to measure along the z direction (computational basis) and Bob along the x direction, we would not find any correlations.

We conclude that entangled states can be "more strongly" correlated than classical (separable) states in the sense that quantum correlations can appear in (infinitely) many different bases corresponding, in general, to non-compatible (non-commuting) observables (e.g., σ_z and σ_x).

In the following, we are aiming to establish a quantitative description of entanglement and correlations. For this purpose, it is useful to extend the single-qubit Bloch decomposition from Section 11.5.2 to pairs of quantum systems, which we will do next.

15.2.3 Bloch Decomposition for Two Qubits

In this section, we extend the Bloch picture that we have discussed for single qubits in Section 11.5.2 to the case of two qubits, leaving the more general case of qudits (d-dimensional quantum systems) to be discussed in Section 17.3.

As we recall from Section 11.5.2, any matrix in Hilbert–Schmidt space of dimension two can be decomposed into a matrix basis $\{A_i\}_i$. The basis fulfils the following properties:

(i) The basis contains the identity matrix $\mathbb{1}$ and $2 \times 2 - 1 = 3$ matrices which are traceless, $\text{Tr}(\Gamma_i) = 0$.

(ii) The matrices of the basis $\{\Gamma_i\}_i$ are orthogonal, i.e., $\text{Tr}(\Gamma_i^\dagger \Gamma_j) = N_\Gamma \delta_{ij}$ with N_Γ some normalization constant.

In dimension two, we can conveniently choose the Pauli matrices $\{\sigma_1, \sigma_2, \sigma_3\}$, where we have switched to the labelling $1, 2, 3$ rather than x, y, z for ease of notation when generalizing to composite (and later higher-dimensional) quantum systems. This leads to the already familiar *Bloch decomposition*

$$\rho = \tfrac{1}{2} \left(\mathbb{1}_2 + \vec{a} \cdot \vec{\sigma} \right) . \qquad (15.38)$$

The normalization $\frac{1}{2}$ is chosen such that $\text{Tr}(\rho) = 1$. In the term $\vec{a} \cdot \vec{\sigma} = a_1\sigma_1 + a_2\sigma_2 + a_3\sigma_3$ the *Bloch vector* $\vec{a} = \text{Tr}(\rho\vec{\sigma})$ is real, three-dimensional and such that its norm is $|\vec{a}| \leq 1$. When considering a qubit, for $|\vec{a}| < 1$, the state ρ is mixed, with $\vec{a} = 0$ representing the maximally mixed state $\rho_{\text{mix}} = \frac{1}{2}\mathbb{1}_2$, while cases where $|\vec{a}| = 1$, corresponding to the boundary of the Bloch sphere, represent pure states.

For two qubits a general state ρ_{AB} can also be decomposed into a sum of tensor products of Pauli operators (see, e.g., Bertlmann *et al.* 2002; Bertlmann and Krammer, 2008*a*)

$$\rho_{AB} = \frac{1}{4}\left[\mathbb{1}_A \otimes \mathbb{1}_B + \sum_i a_i\, \sigma_i^{(A)} \otimes \mathbb{1}_B + \sum_j b_j\, \mathbb{1}_A \otimes \sigma_j^{(B)} + \sum_{i,j} t_{ij}\, \sigma_i^{(A)} \otimes \sigma_j^{(B)} \right].$$

$$(15.39)$$

Here, the coefficients a_i and b_j are the components of the Bloch vectors $\vec{a}$ and $\vec{b}$ of the individual qubits, which we can confirm with a short calculation by tracing over one of the subsystems,

$$\rho_A = \text{Tr}_B(\rho_{AB}) = \frac{1}{4}\left[\mathbb{1}_A\,\text{Tr}_B(\mathbb{1}_B) + \sum_i a_i\, \sigma_i^{(A)}\,\text{Tr}_B(\mathbb{1}_B) \right.$$

$$\left. + \sum_j b_j\, \mathbb{1}_A\,\text{Tr}_B(\sigma_j^{(B)}) + \sum_{i,j} t_{ij}\, \sigma_i^{(A)}\,\text{Tr}_B(\sigma_j^{(B)}) \right]$$

$$= \frac{1}{4}\left(\mathbb{1}_A\,2 + \sum_i a_i\, \sigma_i^{(A)}\,2 \right) = \frac{1}{2}\left(\mathbb{1}_A + \sum_i a_i\, \sigma_i^{(A)} \right), \qquad (15.40)$$

and similarly for the reduced state $\rho_B = \text{Tr}_A(\rho_{AB})$ of subsystem B. Consequently, we must have $|\vec{a}| \leq 1$ and $|\vec{b}| \leq 1$ in order to obtain well-defined reduced density matrices for Alice and Bob. The real coefficients t_{ij} are the components of the so-called *correlation tensor* $T = (t_{ij})$, which must also be real-valued since they arise as expectation values of Hermitian operators,

$$t_{ij} = \text{Tr}\left(\rho_{AB}\, \sigma_i^{(A)} \otimes \sigma_j^{(B)} \right). \qquad (15.41)$$

However, note that the coefficients a_i, b_j, and t_{ij} cannot be chosen completely independently because they are jointly (non-trivially) constrained by the positivity of ρ_{AB}.

In terms of the two-qubit Bloch decomposition, a product state has the form

$$\rho_{AB} = \frac{1}{4}\left(\mathbb{1}_A \otimes \mathbb{1}_B + \sum_i a_i\, \sigma_i^{(A)} \otimes \mathbb{1}_B + \sum_j b_j\, \mathbb{1}_A \otimes \sigma_j^{(B)} + \sum_{i,j} a_i b_j\, \sigma_i^{(A)} \otimes \sigma_j^{(B)} \right)$$

$$= \frac{1}{2}\left(\mathbb{1}_A + \sum_i a_i\, \sigma_i^{(A)} \right) \otimes \frac{1}{2}\left(\mathbb{1}_B + \sum_j b_j\, \sigma_j^{(B)} \right), \qquad (15.42)$$

i.e., the matrix $t_{ij} = a_i b_j$ factorizes into the product of the separate Bloch-vector contributions of Alice and Bob. Any separable state $\rho_{AB} \in \mathcal{S}$ (pure or mixed) can be written as a convex combination of expressions that are of the form of eqn (15.42).

As an example for a product state, consider the state $|01\rangle\langle01|$, which can be written in Bloch decomposition as

$$|01\rangle\langle01| = \tfrac{1}{4}\left(\mathbb{1}_A \otimes \mathbb{1}_B + \sigma_3^{(A)} \otimes \mathbb{1}_B - \mathbb{1}_A \otimes \sigma_3^{(B)} - \sigma_3^{(A)} \otimes \sigma_3^{(B)}\right). \quad (15.43)$$

For the maximally entangled Bell state $\rho^- = |\psi^-\rangle\langle\psi^-|$, on the other hand, we have

$$\rho^- = \tfrac{1}{4}\left(\mathbb{1}_A \otimes \mathbb{1}_B - \sum_i t_{ii}\, \sigma_i^{(A)} \otimes \sigma_i^{(B)}\right), \quad (15.44)$$

the local Bloch vectors vanish, $\vec{a} = \vec{b} = 0$ since the reduced states are maximally mixed, and the correlation tensor is diagonal, $T = \mathrm{diag}\{-1, -1, -1\}$.

15.2.4   The Peres–Horodecki Criterion

The question of separability versus entanglement, i.e., determining for any given density operator of any bipartite system whether it is separable or entangled, is one of the most important, difficult[3] and generally open problems of quantum information theory. We will discuss this question in more detail in Section 16.3—but for "lower" dimensions there exist necessary and sufficient separability criteria, which turn out to be very practical.

Such a useful criterion is the so-called *positive partial transposition* (PPT) *criterion*, independently developed by *Asher Peres* (1996) as well as by *Michał, Paweł, and Ryszard Horodecki* (1996 *a*), but we need a crucial definition first:

Definition 15.9   *Let ρ_{AB} be a density operator of a bipartite quantum system. For any choice of local bases $\{|i\rangle_A\}_i$ and $\{|k\rangle_B\}_k$, in terms of which ρ_{AB} takes the form*

$$\rho_{AB} = \sum_{i,j,k,l} \rho_{ijkl} |i\rangle\langle j|_A \otimes |k\rangle\langle l|_B$$

*the **partial transpose** of subsystem B is defined as*

$$\rho_{AB}^{\mathsf{T}_B} = \sum_{i,j,k,l} \rho_{ijlk} |i\rangle\langle j|_A \otimes |k\rangle\langle l|_B.$$

And of course we can similarly define a partial transpose of subsystem A as

$$\rho_{AB}^{\mathsf{T}_A} = \sum_{i,j,k,l} \rho_{jikl} |i\rangle\langle j|_A \otimes |k\rangle\langle l|_B. \quad (15.45)$$

When we write a given density matrix ρ_{AB} in the Hilbert–Schmidt space $\mathcal{B}_{\mathrm{HS}}(\mathcal{H}_A \otimes \mathcal{H}_B)$ in its Bloch-decomposition form (15.39) then we can alternatively obtain the partial transposition of either of the subsystems by transposing the off-diagonal elements of

---

[3]For general density operators, the separability problem is known to be an NP-hard problem (Gurvits, 2003; Gurvits, 2004), even when relaxing the problem such that one allows a margin of error that is inversely polynomial (compared with inverse exponential error margins in Gurvits' approach) in the dimension of the system (Gharibian, 2010).

the respective Pauli matrices: $(\sigma_i)_{kl}^\top = (\sigma_i)_{lk}$. With these preliminaries established, we can then formulate the PPT criterion.

Theorem 15.3 (PPT criterion)

$$\textit{If} \quad \rho_{AB}^{\mathsf{T}_A} \ngeq 0 \quad \textit{or} \quad \rho_{AB}^{\mathsf{T}_B} \ngeq 0 \quad \Longrightarrow \quad \rho_{AB} \textit{ entangled.}$$

To put Theorem 15.3 in words, whenever one finds that the partial transpose of a bipartite density operator is no longer a positive semidefinite operator, i.e., when the partial transpose has one or more negative eigenvalues, one may conclude that the state is entangled.

Proof To prove this statement, we simply apply the partial transposition to a separable state as in Definition 15.5, that is,

$$\rho_{AB}^{\mathsf{T}_B} = \sum_i p_i \, \rho_i^{(A)} \otimes \left(\rho_i^{(B)}\right)^\top . \tag{15.46}$$

Since $\left(\rho_i^{(B)}\right)^\top \geq 0$ is a positive semidefinite operator, normalized operator, it must again be valid density operator, and therefore $\rho_{AB}^{\mathsf{T}_B}$ must also be a valid density operator, in particular, a positive semidefinite operator. Therefore, the partial transpose of a separable state cannot result in negative eigenvalues. Conversely, any negative eigenvalue implies entanglement. □

We call a state that remains positive under partial transposition a *PPT state*. If, however, at least one of the eigenvalues of the partially transposed matrix is negative we call the state an *NPT state* (negative partial transposition), and it has to be entangled. However, the converse is generally not true in arbitrary dimensions, i.e., finding a positive partial transpose generally does not imply that a state is necessarily separable. In arbitrary dimensions the PPT criterion is only necessary but not sufficient for separability. Nevertheless, the PPT criterion turns out to be necessary *and* sufficient for separability when the dimension of the joint system is smaller than or equal to six.

Theorem 15.4 (PPT criterion for $d \leq 6$)

For $\rho_{AB} \in \mathcal{B}_{\mathrm{HS}}(\mathcal{H}_A \otimes \mathcal{H}_B)$ with $\dim(\mathcal{H}_A) \times \dim(\mathcal{H}_B) \leq 6$,

$$\rho_{AB}^{\mathsf{T}_A} \geq 0 \quad \textit{and} \quad \rho_{AB}^{\mathsf{T}_B} \geq 0 \quad \Longleftrightarrow \quad \rho_{AB} \textit{ separable.}$$

We will discuss the proof of this stronger statement in Section 15.4.2 in the context of a more general family of separability criteria based positive maps.

In dimensions other than 2×2, 2×3 or 3×2 the PPT criterion is generally not tight. There exist entangled states that remain positive semidefinite under partial transposition, *PPT entangled states*. Despite the fact that such states are entangled, they can be considered to be weakly entangled in the sense that they are not distillable (Horodecki *et al.*, 1998), that is, local operations and classical communication (LOCC, see Section 16.1.2) on multiple copies of PPT entangled states cannot result in (a smaller number of) maximally entangled Bell pairs, entanglement distillation (discussed in more detail in Section 17.3) does not work for such states. At the same time it is not known if all *NPT states* are distillable (Pankowski *et al.*, 2010), although it is known that finite numbers of copies are not sufficient for distilling arbitrary NPT states (Watrous, 2004). States that cannot be distilled to maximally entangled states are called *bound-entangled states*.

Before discussing an example for the application of the PPT criterion for two qubits, which we will do in Section 15.3.3, let us briefly examine the relevance of our definition of entanglement for the violation of Bell inequalities.

15.3 Entanglement and Non-Locality

15.3.1 Separable States Cannot Violate a Bell Inequality

Now an interesting question in relation to our adopted definition of separability versus entanglement is whether or not the correlations represented by a bipartite quantum state allow for a hidden-variable model. As we have seen in Section 13.2.2, Bell inequalities can be violated by quantum mechanics, or more specifically, by an entangled state. Clearly, one can now wonder what specific feature enables this violation and it goes without saying that one intuitively suspects this violation to have something to do with entanglement. Indeed, in 1989 Reinhard Werner was able to show that all separable (at most classically correlated) states admit a hidden-variable model and hence satisfy all Bell inequalities.

Theorem 15.5 (Separability and Non-Locality)

Every separable state is compatible with a local hidden-variable model.

Proof To see this, consider two local measurements represented by Hermitian operators A and B independently carried out by Alice and Bob, respectively, on their subsystems, which we first assume to be represented by a product state $\rho_{AB} = \rho^{(A)} \otimes \rho^{(B)}$. From eqn (15.31), we then recall that this implies that the expectation value factorizes, $E(A, B) = \langle A \otimes B \rangle = \text{Tr}(\rho^{(A)} A)\, \text{Tr}(\rho^{(B)} B)$. Writing the observables in their spectral decompositions $A = \sum_m \mu_m |m\rangle\langle m|_A$ and $B = \sum_n \nu_n |n\rangle\langle n|_B$ with respect to their eigenvalues μ_m and ν_n (the possible measurement outcomes), we further have

$$\text{Tr}(\rho^{(A)} A) = \sum_m \mu_m \,_A\langle m | \rho^{(A)} | m \rangle_A = \sum_m \mu_m\, p^{(A)}(m), \tag{15.47a}$$

$$\text{Tr}(\rho^{(B)} B) = \sum_n \nu_n \,_B\langle n | \rho^{(B)} | n \rangle_B = \sum_n \nu_n\, p^{(B)}(n). \tag{15.47b}$$

Consequently, we can write the joint expectation value as

$$E(A, B) = \langle A \otimes B \rangle = \sum_{m,n} \mu_m \nu_n \, p^{(A)}(m) \, p^{(B)}(n) \,. \tag{15.48}$$

Now, if we have an arbitrary separable state of the form $\rho_{AB} = \sum_i p_i \, \rho_i^{(A)} \otimes \rho_i^{(B)}$, we can just define $p_i^{(A)}(m) := {}_A\langle m \,|\, \rho_i^{(A)} \,|\, m \rangle_A$ and $p_i^{(B)}(n) := {}_B\langle n \,|\, \rho_i^{(B)} \,|\, n \rangle_B$ to obtain

$$E(A, B) = \langle A \otimes B \rangle = \sum_{m,n,i} \mu_m \nu_n \, p_i \, p_i^{(A)}(m) \, p_i^{(B)}(n) \,. \tag{15.49}$$

Then consider the following simple local hidden-variable model. Let

$$\lambda = \{\lambda_{A_1}, \lambda_{A_2}, \dots, \lambda_{B_1}, \lambda_{B_2}, \dots\} \tag{15.50}$$

be a list of all possible measurement outcomes $\lambda_{A_j} \in \{\mu_m^j\}_m$ and $\lambda_{B_k} \in \{\nu_n^k\}_n$ for all possible measurement settings A_j and B_k of Alice and Bob, respectively, such that $A(j, \lambda) = \lambda_{A_j}$ and $B(k, \lambda) = \lambda_{B_k}$ for all i and k. Alice's choice of measurement direction has no influence over Bob's choice of measurement direction, and vice versa. As the normalized distribution of the hidden variable we take a product of individual distributions, $\rho(\lambda) = p_{\lambda_{A_1}} p_{\lambda_{A_2}} \cdots p_{\lambda_{B_1}} p_{\lambda_{B_2}} \cdots$, so that

$$\sum_{\lambda} \rho(\lambda) = \sum_{\lambda_{A_1}} \sum_{\lambda_{A_2}} \cdots \sum_{\lambda_{B_1}} \sum_{\lambda_{B_2}} \cdots p_{\lambda_{A_1}} p_{\lambda_{A_2}} \cdots p_{\lambda_{B_1}} p_{\lambda_{B_2}} \cdots = 1 \,. \tag{15.51}$$

In that case, for the jth and kth measurement settings chosen by Alice and Bob, we would get the expected value

$$E(A_j, B_k) = \sum_{\lambda} \rho(\lambda) \, A(j, \lambda) \, B(k, \lambda) = \sum_{\lambda_{A_j}} \sum_{\lambda_{B_k}} p_{\lambda_{A_j}} p_{\lambda_{B_k}} \lambda_{A_j} \lambda_{B_k} \,. \tag{15.52}$$

If we introduce an additional element of randomness, i.e., by introducing a set of probability distributions $\{\rho_i(\lambda) = p_{\lambda_{A_1}}^{(i)} p_{\lambda_{A_2}}^{(i)} \cdots p_{\lambda_{B_1}}^{(i)} p_{\lambda_{B_2}}^{(i)} \cdots\}_i$ each of which occurs with probability p_i, then the expectation value further becomes

$$E(A_j, B_k) = \sum_i \sum_{\lambda} \rho_i(\lambda) \, A(j, \lambda) \, B(k, \lambda) = \sum_i \sum_{\lambda_{A_j}} \sum_{\lambda_{B_k}} p_i \, p_{\lambda_{A_j}}^{(i)} p_{\lambda_{B_k}}^{(i)} \lambda_{A_j} \lambda_{B_k} \,, \tag{15.53}$$

which is exactly of the form of eqn (15.49). The correlations of every separable state can thus be explained by a local hidden-variable model. $\qquad\square$

Conversely this means that every state violating a Bell inequality, i.e., not admitting a hidden-variable model, must be quantum correlated. However, Werner (Werner, 1989) also discovered that the reverse statement is not necessarily true, that is, the inability of a state to allow one to violate a given Bell inequality (say, the CHSH inequality) does not mean that the state is separable, as we will see for ourselves in Section 15.3.3. In particular, it is still unclear to this day if entanglement and non-locality actually coincide because it is not known if any entangled state can violate some (generalized) Bell inequality or not.

In the following, we will therefore restrict our analysis of the relationship between entanglement and non-locality to the specific case of the CHSH inequality and focus on the question if, and if so how strongly, a given quantum state can violate the CHSH inequality.

15.3.2 The CHSH-Operator Criterion

As we have seen, all separable states admit a hidden-variable model (see Theorem 15.5) and thus cannot violate any Bell inequality, in particular, not the CHSH inequality (see Section 13.2.2). However, we yet do not know if, and if so how strongly, a given entangled state can violate the CHSH inequality. Taking the violation of the CHSH inequality as our guide post for quantifying non-locality we therefore first rephrase it in terms of the expectation value of an operator

$$\mathcal{B}_{\text{CHSH}}(\vec{a}, \vec{a}', \vec{b}, \vec{b}') = \vec{a} \cdot \vec{\sigma}^{(A)} \otimes (\vec{b} - \vec{b}') \cdot \vec{\sigma}^{(B)} + \vec{a}' \cdot \vec{\sigma}^{(A)} \otimes (\vec{b} + \vec{b}') \cdot \vec{\sigma}^{(B)} \,, \quad (15.54)$$

which depends on two pairs of chosen measurement directions represented by unit vectors $\vec{a}$, $\vec{a}'$ and $\vec{b}$, $\vec{b}'$, respectively, such that the CHSH inequality can be written as

$$\overline{\mathcal{B}}_{\text{CHSH}} := (\rho, \mathcal{B}_{\text{CHSH}})_{\text{HS}} = \text{Tr}(\rho \, \mathcal{B}_{\text{CHSH}}) \leq 2 \,, \quad (15.55)$$

where we have used the definition of the Hilbert–Schmidt inner product (Definition 11.4).

Now, we know that quantum states exist for which the quantity $\overline{\mathcal{B}}_{\text{CHSH}}$ takes the value $2\sqrt{2}$ for certain combinations of directions $\vec{a}$, $\vec{a}'$, $\vec{b}$, and $\vec{b}'$, but for an arbitrary set of directions the value of $\overline{\mathcal{B}}_{\text{CHSH}}$ will generally be smaller. More generally, many quantum states will only allow us to reach smaller values no matter which directions we choose. So how can we tell how large $\overline{\mathcal{B}}_{\text{CHSH}}$ can become for a given quantum state without having to try out all combinations of directions? More concretely, how can we determine the maximum value

$$\overline{\mathcal{B}}_{\text{CHSH}}^{\text{max}} := \max_{\vec{a}, \vec{b}, \vec{a}', \vec{b}'} \text{Tr}(\rho \, \mathcal{B}_{\text{CHSH}}(\vec{a}, \vec{a}', \vec{b}, \vec{b}')), \quad (15.56)$$

that the left-hand side of the CHSH inequality—which we will refer to as the *Bell–CHSH parameter* $\overline{\mathcal{B}}_{\text{CHSH}}$ from now on—may take for any choice of measurement directions? In this regard, another theorem by *Ryszard*, *Paweł*, and *Michał Horodecki* (1995) proves to be very powerful and greatly simplifies the problem.

Theorem 15.6 (CHSH–Bell operator criterion)

For any two-qubit density operator ρ, the maximally possible value $\overline{\mathcal{B}}_{\text{CHSH}}^{\max}$ of the Bell–CHSH parameter for any choice of measurement directions is given by

$$\overline{\mathcal{B}}_{\text{CHSH}}^{\max} = 2\sqrt{\mu_1 + \mu_2}\,,$$

where μ_1 and μ_2 are the two largest eigenvalues of the matrix $M_\rho = T_\rho^\top T_\rho$, and $T_\rho = (t_{ij})$ is the correlation tensor of the state ρ in the two-qubit Bloch decomposition, $t_{ij} = \text{Tr}\big(\rho\,\sigma_i^{(A)} \otimes \sigma_j^{(B)}\big).$

Proof We will carry out the proof in several independent steps. First, we note that the Bell–CHSH parameter can be written as a sum of two (Euclidean) inner products, $\vec{a}\cdot T_\rho(\vec{b} - \vec{b}')$ and $\vec{a}' \cdot T_\rho(\vec{b} + \vec{b}')$ of vectors in $\mathbb{R}^3$. To see this, we just write the state ρ in the two-qubit Bloch decomposition from eqn (15.39) and calculate

$$\text{Tr}(\rho\,\mathcal{B}_{\text{CHSH}}) = \tfrac{1}{4}\Big[\text{Tr}\big(\mathbb{1}_A \otimes \mathbb{1}_B \mathcal{B}_{\text{CHSH}}\big) + \sum_i a_i\,\text{Tr}\big(\sigma_i^{(A)} \otimes \mathbb{1}_B\,\mathcal{B}_{\text{CHSH}}\big)$$

$$+ \sum_j b_j\,\text{Tr}\big(\mathbb{1}_A \otimes \sigma_j^{(B)}\,\mathcal{B}_{\text{CHSH}}\big) + \sum_{i,j} t_{ij}\,\text{Tr}\big(\sigma_i^{(A)} \otimes \sigma_j^{(B)}\,\mathcal{B}_{\text{CHSH}}\big)\Big]$$

$$= \tfrac{1}{4}\sum_{i,j} t_{ij}\sum_{m,n}\big[a_m(b_n - b_n') + a_m'(b_n + b_n')\big]\,\text{Tr}\big(\sigma_i^{(A)} \otimes \sigma_j^{(B)}\,\sigma_m^{(A)} \otimes \sigma_n^{(B)}\big),$$

$$(15.57)$$

where we have used the fact that $\text{Tr}(\sigma_i) = 0$ for all i, which eliminates all terms in the Bloch decomposition that feature $\mathbb{1}_A$ or $\mathbb{1}_B$ when multiplying by $\sigma_m^{(A)} \otimes \sigma_n^{(B)}$ (from $\mathcal{B}_{\text{CHSH}}$) and evaluating the trace. In the expression we have left, we now use $\text{Tr}(\sigma_i\sigma_m) = 2\delta_{im}$ (and similarly for the indices j and n), such that

$$\text{Tr}(\rho\,\mathcal{B}_{\text{CHSH}}) = \sum_{i,j,m,n} t_{ij}\big[a_m(b_n - b_n') + a_m'(b_n + b_n')\big]\delta_{im}\delta_{jn}$$

$$= \sum_{m,n}\big[a_m\,t_{mn}\,(b_n - b_n') + a_m'\,t_{mn}\,(b_n + b_n')\big]$$

$$= \vec{a}\cdot T_\rho(\vec{b} - \vec{b}') + \vec{a}' \cdot T_\rho(\vec{b} + \vec{b}')\,. \qquad (15.58)$$

Now we note that the vectors $(\vec{b} - \vec{b}')$ and $(\vec{b} + \vec{b}')$ are orthogonal whenever $\vec{b}$ and $\vec{b}'$ are not parallel, and assuming the latter, we can define corresponding unit $\vec{c}$ and $\vec{c}'$ vectors as

$$\vec{c} = 2\sin(\theta)\,(\vec{b} - \vec{b}')\,, \quad \text{and} \quad \vec{c}' = 2\cos(\theta)\,(\vec{b} + \vec{b}')\,, \qquad (15.59)$$

where the angle between $\vec{b}$ and $\vec{b}'$ is 2θ, such that $\vec{b} \cdot \vec{b}' = \cos(2\theta)$. With this, we can write

$$\text{Tr}(\rho \, \mathcal{B}_{\text{CHSH}}) = 2 \sin(\theta) \, \vec{a} \cdot T_\rho \vec{c} + 2 \cos(\theta) \, \vec{a}' \cdot T_\rho \vec{c}' . \qquad (15.60)$$

The maximization over the four directions $\vec{a}$, $\vec{a}'$, $\vec{b}$, and $\vec{b}'$ has thus become a maximization over $\vec{a}$, $\vec{a}'$, $\vec{c}$, $\vec{c}'$, and the angle θ. The expressions $\vec{a} \cdot T_\rho \vec{c}$ and $\vec{a}' \cdot T_\rho \vec{c}'$ are maximal when $\vec{a}$ is parallel to $T_\rho \vec{c}$ and $\vec{a}'$ is parallel to $T_\rho \vec{c}'$,

$$\vec{a}_{\text{max}} = \frac{T_\rho \vec{c}}{|T_\rho \vec{c}|} , \quad \text{and} \quad \vec{a}'_{\text{max}} = \frac{T_\rho \vec{c}'}{|T_\rho \vec{c}'|} , \qquad (15.61)$$

and we thus have

$$\max_{\vec{a},\vec{b},\vec{a}',\vec{b}'} \text{Tr}(\rho \, \mathcal{B}_{\text{CHSH}}(\vec{a},\vec{a}',\vec{b},\vec{b}')) = 2 \max_{\theta,\vec{c},\vec{c}'} \left[\sin(\theta) \, |T_\rho \vec{c}| + \cos(\theta) \, |T_\rho \vec{c}'| \right]. \qquad (15.62)$$

Differentiating with respect to θ immediately shows that the maximum over θ is taken for the angle $\theta = \theta_{\text{max}}$ for which

$$\cos(\theta_{\text{max}}) \, |T_\rho \vec{c}| = \sin(\theta_{\text{max}}) \, |T_\rho \vec{c}'| , \qquad (15.63)$$

and for which both $\cos(\theta_{\text{max}}) > 0$ and $\sin(\theta_{\text{max}}) > 0$. With this, we can write

$$\max_\theta \left[\sin(\theta) \, |T_\rho \vec{c}| + \cos(\theta) \, |T_\rho \vec{c}'| \right] = \sin(\theta_{\text{max}}) \, |T_\rho \vec{c}| + \cos(\theta_{\text{max}}) \, |T_\rho \vec{c}'|$$

$$= \sqrt{\left(\sin(\theta_{\text{max}}) \, |T_\rho \vec{c}| + \cos(\theta_{\text{max}}) \, |T_\rho \vec{c}'| \right)^2}$$

$$= \sqrt{\sin^2(\theta_{\text{max}}) \, |T_\rho \vec{c}|^2 + 2 \sin(\theta_{\text{max}}) \, |T_\rho \vec{c}| \, \cos(\theta_{\text{max}}) \, |T_\rho \vec{c}'| + \cos^2(\theta_{\text{max}}) \, |T_\rho \vec{c}'|^2}$$

$$= \sqrt{\left[\sin^2(\theta_{\text{max}}) + \cos^2(\theta_{\text{max}}) \right] |T_\rho \vec{c}|^2 + \left[\sin^2(\theta_{\text{max}}) + \cos^2(\theta_{\text{max}}) \right] |T_\rho \vec{c}'|^2}$$

$$= \sqrt{|T_\rho \vec{c}|^2 + |T_\rho \vec{c}'|^2} , \qquad (15.64)$$

where we have inserted the condition from eqn (15.63) in the second-to-last step. Finally, we can carry out the maximization over $\vec{c}$ and $\vec{c}'$ and insert into eqn (15.62), i.e.,

$$\mathcal{B}_{\text{CHSH}}^{\text{max}} = 2 \max_{\vec{c},\vec{c}'} \sqrt{|T_\rho \vec{c}|^2 + |T_\rho \vec{c}'|^2} . \qquad (15.65)$$

Since $|T_\rho \vec{c}|^2 = \vec{c}^{\,\top} T_\rho^\top T_\rho \vec{c}$ and $\vec{c}$ is a unit vector, the maximum of $|T_\rho \vec{c}|^2$ is just the largest eigenvalue μ_1 of $T_\rho^\top T_\rho = M_\rho$, and since $\vec{c}'$ must be orthogonal to $\vec{c}$, this leaves the second largest eigenvalue μ_2 for the maximum of $|T_\rho \vec{c}'|^2$, and we thus arrive at the result

$$\mathcal{B}_{\text{CHSH}}^{\text{max}} = 2 \sqrt{\mu_1 + \mu_2} . \qquad (15.66)$$

$\square$

The CHSH criterion from Theorem 15.6 gives us a simple method to determine, loosely speaking, how "non-local" a given (two-qubit) quantum state can be. And, perhaps not surprisingly, for the four Bell states with density operators $\rho^{\pm}$ and $\omega^{\pm}$ from eqns (11.113a) and (11.113b), respectively, one finds[4] the correlation tensors

$$
T_{\rho_{\pm}} = \begin{pmatrix} \pm 1 & 0 & 0 \\ 0 & \pm 1 & 0 \\ 0 & 0 & -1 \end{pmatrix} \quad \text{and} \quad T_{\omega_{\pm}} = \begin{pmatrix} \pm 1 & 0 & 0 \\ 0 & \mp 1 & 0 \\ 0 & 0 & 1 \end{pmatrix}, \tag{15.67}
$$

such that the maximal Bell–CHSH parameter for all four states is $\overline{\mathcal{B}}_{\text{CHSH}}^{\max} = 2\sqrt{2}$, while all non-maximally entangled states give smaller values for $\overline{\mathcal{B}}_{\text{CHSH}}^{\max}$. In the next section, we will consider a family of states for which the CHSH criterion can be used to nicely demonstrate the differences between entanglement and non-locality.

15.3.3 Werner States

In 1989 Reinhard Werner discovered that certain mixed entangled states may also satisfy the CHSH inequality for all choices of measurement directions, i.e., admits a local hidden-variable model. This was a great surprise, suggesting that entanglement and non-locality are not necessarily the same but *different* concepts!

The one-parameter family of states that Werner used to formulate this contrast, now called the *Werner states*, are formed as a mixture of the pure, maximally entangled state $|\psi^{-}\rangle$, represented by the density operator ρ^{-}, and the two-qubit maximally mixed state $\frac{1}{2}\mathbb{1}_2 \otimes \frac{1}{2}\mathbb{1}_2 = \frac{1}{4}\mathbb{1}_4$, given by

$$
\rho_{\text{Werner}}(\alpha) = \alpha\,\rho^{-} + \frac{1-\alpha}{4}\,\mathbb{1}_4 = \frac{1}{4}\begin{pmatrix} 1-\alpha & 0 & 0 & 0 \\ 0 & 1+\alpha & -2\alpha & 0 \\ 0 & -2\alpha & 1+\alpha & 0 \\ 0 & 0 & 0 & 1-\alpha \end{pmatrix}, \tag{15.68}
$$

or in two-qubit Bloch decomposition

$$
\rho_{\text{Werner}}(\alpha) = \frac{1}{4}\left(\mathbb{1}_2 \otimes \mathbb{1}_2 - \alpha \sum_i \sigma_i \otimes \sigma_i\right). \tag{15.69}
$$

For parameter values $\alpha \in [0,1]$ it is clear that we have a convex sum of two density operators and hence a valid state that corresponds to a setup that produces a pure entangled state accompanied by white noise due to imperfections. However, by diagonalizing the central 2×2 block of the matrix in eqn (15.68), we see that the eigenvalues of $\rho_{\text{Werner}}(\alpha)$ are $(1-\alpha)/4$ (thrice degenerate) and $(1+3\alpha)/4$, and we thus see that

---

[4]We leave the short calculation of the components $t_{ij} = \text{Tr}\big(\rho\,\sigma_i^{(A)} \otimes \sigma_j^{(B)}\big)$ as an exercise for the interested reader.

$\rho_{\text{Werner}}(\alpha)$ is positive semidefinite (and hence a valid density operator) in the larger parameter range $\alpha \in [-\frac{1}{3}, 1]$.

Separability of Werner states: From the definition of the Werner states it is obvious that the parameter value $\alpha = 1$ must yield an entangled state ρ^-, while one obtains a separable (even product) state for $\alpha = 0$. All convex combinations of these two possibilities form a line in the state space, ranging from the pure state ρ^- at the boundary of the set of density operators to the maximally mixed state lying within the set of separable states $\mathcal{S}$. Consequently there must be some parameter value at which this line crosses from the set $\mathcal{S}^C$ of entangled states into the separable set $\mathcal{S}$ (see, e.g., Fig. 15.1). To determine what exactly this value is we apply the PPT criterion (Theorem 15.3). To this end, we determine the partial transposition

$$\rho_{\text{Werner}}^{\mathsf{T}_B} = (\mathbb{1}_A \otimes \mathsf{T}_B)\,\rho_{\text{Werner}} = \frac{1}{4}\begin{pmatrix} 1-\alpha & 0 & 0 & -2\alpha \\ 0 & 1+\alpha & 0 & 0 \\ 0 & 0 & 1+\alpha & 0 \\ -2\alpha & 0 & 0 & 1-\alpha \end{pmatrix}, \qquad (15.70)$$

which has eigenvalues $\lambda_{1,2,3} = \frac{1+\alpha}{4}$ and $\lambda_4 = \frac{1-3\alpha}{4}$. Therefore the states are separable (i.e., all eigenvalues remain positive) for $0 \le \alpha \le \frac{1}{3}$ and entangled for $\alpha > \frac{1}{3}$. We also observe that for negative values of α, that is, for $-\frac{1}{3} \le \alpha < 0$ the eigenvalues remain positive, i.e., the states remain separable. Beyond this lower limit for α the states become unphysical, i.e. $\rho < 0$, as we have seen.

Non-locality of Werner states: In order to find the Werner states that may violate the CHSH inequality we use the CHSH–Bell operator criterion (Theorem 15.6). We choose the Bloch decomposition (15.69) to directly read off the correlation matrix

$$T = (t_{ij}) = \begin{pmatrix} -\alpha & 0 & 0 \\ 0 & -\alpha & 0 \\ 0 & 0 & -\alpha \end{pmatrix}. \qquad (15.71)$$

All eigenvalues of the product of the correlation matrices $(t_{ij})^T (t_{ij})$ are hence equal to α^2 and the maximal violation of the CHSH inequality ist thus given by $\overline{\mathcal{B}}_{\text{CHSH}}^{\text{max}} = 2\sqrt{2\alpha^2} > 2$, implying that for all $\alpha > \frac{1}{\sqrt{2}} \approx 0.7071$ the CHSH inequality is violated.

Now, this leaves an interval $[\frac{1}{3}, \frac{1}{\sqrt{2}}]$ where the Werner states are entangled but still do not violate the CHSH inequality for any choice of measurement directions, from which one might naively conclude that entanglement and non-locality are distinct. Yet, other options can still be explored. In particular, it is possible to construct other Bell-type inequalities besides CHSH for which entangled Werner states in the region in question display non-locality, and indeed examples for such inequalities could be shown to be violated already for values $\alpha > 0.7056$ (Vértesi, 2008). For even smaller values, work by Hirsch *et al.* (2017) using a known bound (Acín *et al.*, 2006) based on Grothendieck's constant demonstrated that no Bell inequality employing only projective measurements can be violated by Werner states with $\alpha \leq 0.682$. But also this is not the end of the story, since one may consider more general measurements based on *positive operator-valued measures* (POVMs, see Chapter 23) for which it is in principle possible to go beyond what is possible with projective measurement alone. Taking into account such POVMs, Hirsch *et al.* (2017) were able to improve upon previous bounds (Barrett, 2002; Oszmaniec *et al.*, 2017) and showed that the correlations of Werner states with $\alpha < 0.4547$ are compatible with local hidden-variable models. Thus, there is currently only a small but nevertheless noticeable region of uncertainty between the parameter values 0.4547 and 0.7056 for which it is not clear if entanglement and non-locality coincide (on the single-copy level) or not, but there is also a region of states characterized by values of α between $\frac{1}{3}$ and 0.4547, which are certainly entangled but cannot violate any Bell inequality as long as only a single copy is considered.

Non-locality and distillability: When more than one copy of a given state is available, things become even more complex. In such a situation it is clear that all distillable states can be used to obtain a pure, maximally entangled state from some number of copies (see Section 16.1.3). Since the latter can be used to violate a Bell inequality, it is therefore obvious that all distillable states are asymptotically (in the number of copies) non-local. However, in higher dimensions one may find states that are undistillable for any number of copies, so-called *bound-entangled* states, and it is known that all PPT entangled states are bound entangled (Horodecki *et al.*, 1998), even if it is still unknown if all NPT states are distillable asymptotically (Pankowski *et al.*, 2010), despite the fact that it is known that there exist NPT states that are not distillable for any finite number of copies (Watrous, 2004). Regardless of the potential (in)equivalence of a negative partial transposition and distillability it was conjectured by Asher Peres in 1996 that no PPT state, and more generally, no bound-entangled state may violate a Bell inequality. Interestingly, Peres' conjecture was disproven by *Tamás Vértesi* and *Nicolas Brunner* (2014): In general (in higher dimensions), one may in fact find PPT states (which are hence undistillable/bound entangled) that can violate certain Bell inequalities (Dür, 2001; Vértesi and Brunner, 2012 and 2014) or so-called steering criteria (Moroder *et al.*, 2014). Hence, the relationship of non-locality with entanglement, and with the PPT criterion is generally more complicated.

We will return to the discussion of bound entanglement in Section 17.3, and a geometric illustration of the discussed features of the Werner states in the Hilbert–Schmidt space will be presented in Section 15.5.

15.3.4　Tsirelson's Bound

Let us now more closely examine the value $2\sqrt{2}$ that we have observed can be produced when evaluating the expectation values of CHSH inequality for maximally entangled two-qubit states. In particular, one may wonder if perhaps other quantum systems or observables could result in even higher values. Naively, we see that the CHSH inequality consists of four expectation values, each of which is bounded from above by $+1$, which suggests a (trivial) upper bound of four for the CHSH parameter. One might therefore wonder if it is possible to find correlations that are (in this sense) even "stronger" than those of the maximally entangled (two-qubit) state $|\psi^-\rangle$. Such questions have first been considered, or at least, published, by *Boris Tsirelson* (Tsirelson, 1980; Tsirelson, 1987), who showed that $2\sqrt{2}$ is indeed the maximal violation of the CHSH inequality that is possible within quantum mechanics.

Theorem 15.7 (Tsirelson's Bound)

The maximal value that can be obtained for the Bell–CHSH parameter by any quantum system is $2\sqrt{2}$.

Proof Using the CHSH criterion (Theorem 15.6), we have already seen that $2\sqrt{2}$ can be achieved for maximally entangled two-qubit states when carrying out measurements for Alice and Bob that are represented by unit vector on the Bloch sphere, i.e., for single-qubit observables $A(\vec{a}) = \vec{a} \cdot \vec{\sigma}^{(A)} \otimes \mathbb{1}_B$ and $B(\vec{b}) = \mathbb{1}_A \otimes \vec{b} \cdot \vec{\sigma}^{(B)}$. Now, let us relax this assumption and assume, just as in the derivation of the CHSH inequality in Section 13.2.2, that $A(\vec{a})$ and $B(\vec{b})$ are arbitrary self-adjoint operators with eigenvalues ± 1, such that $A^2 = \mathbb{1}_A$ and $B^2 = \mathbb{1}_B$. As we will see shortly, any eigenvalues in the interval $[-1, +1]$ would not change the result either, since this would imply $A^2 \leq \mathbb{1}_A$ and $B^2 \leq \mathbb{1}_B$. Further using the shorthand $A \equiv A(\vec{a})$, $A' \equiv A(\vec{a}')$, $B \equiv B(\vec{b})$, and $B' = B(\vec{b}')$, we see that locality implies

$$[A, B] = [A, B'] = [A', B] = [A', B'] = 0. \tag{15.72}$$

With this notation, we can redefine the operator $\mathcal{B}_{\mathrm{CHSH}}(\vec{a}, \vec{a}', \vec{b}, \vec{b}')$ from eqn (15.54) as

$$\mathcal{B}_{\mathrm{CHSH}}(\vec{a}, \vec{a}', \vec{b}, \vec{b}') = AB - AB' + A'B + A'B. \tag{15.73}$$

Squaring this operator, we further have

$$\mathcal{B}_{\mathrm{CHSH}}^2 = (AB - AB' + A'B + A'B)(AB - AB' + A'B + A'B) \tag{15.74}$$

$$= 4\,\mathbb{1}_A \otimes \mathbb{1}_B + AA'[B, B'] - A'A[B, B'] = 4\,\mathbb{1}_{AB} + [A, A'][B, B'],$$

where we have used the commutation relations from eqn (15.72) and the fact that the operators square to the identity. Now, since the operator $\mathcal{B}_{\mathrm{CHSH}}$ is a Hermitian operator with non-negative variance, $\Delta \mathcal{B}_{\mathrm{CHSH}}^2 = \langle \mathcal{B}_{\mathrm{CHSH}}^2 \rangle - \langle \mathcal{B}_{\mathrm{CHSH}} \rangle^2 \geq 0$, we have

$$\langle \mathcal{B}_{\mathrm{CHSH}} \rangle^2 \leq \langle \mathcal{B}_{\mathrm{CHSH}}^2 \rangle = \langle 4\,\mathbb{1}_{AB} + [A, A'][B, B'] \rangle = 4 + \langle [A, A'][B, B'] \rangle$$

$$\leq 4 + \| [A, A'][B, B'] \| = 4 + \| [A, A'] \| \| [B, B'] \|$$

$$\leq 4 + \|A\| \|A'\| \|B\| \|B'\| = 8, \tag{15.75}$$

where $\|A\| = \sup\{\frac{\|A|\psi\rangle\|}{\||\psi\rangle\|} \,|\, |\psi\rangle \in \mathcal{H},\, |\psi\rangle \neq 0\}$ is the (usual) operator norm (in contrast to other matrix norms we have discussed in Chapter 11). With $\langle \mathcal{B}_{\mathrm{CHSH}} \rangle^2$ bounded from above by the value 8, we thus arrive at the conclusion that the maximal value for $|\langle \mathcal{B}_{\text{CHSH}} \rangle|$ that is allowed by quantum mechanics is

$$|\langle \mathcal{B}_{\text{CHSH}} \rangle| \leq 2\sqrt{2}.$$ (15.76)

$\square$

Tsirelson's problem: Now, an interesting aspect of this formulation of the problem is the fact that it was assumed that the operators A and A' commute with the operators B and B', rather than outright assuming a tensor-product structure. Since we have also implicitly assumed that the operators are diagonalizable, this difference does not matter here, as we recall from Theorem 5.2. However, in general one may wonder if there are situations (outside the realm of applicability of Theorem 5.2, i.e., in infinite-dimensional systems) where the sets of correlations arising from commuting operators need not match the set of correlations arising from tensor products of operators. This issue, known as *Tsirelson's problem* (see, e.g., Fritz, 2012) was claimed to be solved by *Zhengfeng Ji, Anand Natarajan, Thomas Vidick, John Wright*, and *Henry Yuen* (2020), with a negative answer. If the result holds up, correlations of tensor-product observables form a strict subset of correlations arising from commuting operators.

Realizability of correlations within quantum mechanics: Although the specific bound in Theorem 15.7 pertains to the CHSH inequality, today the term *Tsirelson bound* is applied for any Bell-type inequality when referring to the maximal value achievable within quantum mechanics. However, as Tsirelson's problem showcases, it is not straightforward to even decide which correlations may or may not be realized within quantum mechanics, and thus to identify the corresponding bounds. A framework for addressing this realizability issue for Bell-like inequalities was put forward in terms of the so-called *Navascués–Pironio–Acín hierarchy*, cast as an infinite family of semidefinite programmes (SDPs) by *Miguel Navascués, Stefano Pironio*, and *Antonio Acín* (2007, 2008).

Beyond-quantum correlations: Thus, even staying within quantum mechanics, a number of challenges are encountered in this area. But this has of course not stopped people from pushing ever deeper onwards. The question which properties of physical theories are responsible for limiting the achievable correlations within the respective theory such that the resulting violations of Bell-like inequalities are bounded by Tsirelson's bound has given rise to an active field of research in quantum foundations. In particular, it was shown by *Boris Tsirelson* and *Leonid Aleksandrovich Khalfin* (1985) as well as independently by *Peter Rastall* (1985) that correlations limited only by relativistic causality can indeed achieve violations of the CHSH inequality up to the value of 4, but this result only became widely known upon its rediscovery by *Sandu Popescu* and *Daniel Rohrlich* in 1994, who formulated it in terms of what are nowadays called *Popescu–Rohrlich* (or simply PR) boxes producing the correlations in question.

15.3.5   Hidden Non-Locality

Whereas Reinhard Werner (recall Section 15.3.3) demonstrated that mixed entangled states may not necessarily be able to violate a Bell inequality—thus showing that Bell inequality violation is not a faithful measure for entanglement—it was *Nicolas Gisin* (1996) who showed that some quantum states initially satisfying a given Bell inequality can lead to a violation after certain local selective measurements, i.e., local filtering operations which amplify the entanglement of the original state. In this way the non-local character of the quantum system, previously hidden, is revealed (in this context see also Popescu, 1995).

To understand this phenomenon we follow the presentation in (Friis *et al.*, 2017*a*) and consider a class of quantum states which arises as a mixture of pure entangled states ρ_θ with a particular mixed separable state ρ_{top}. The entangled states are defined as

$$| \psi_\theta \rangle = \sin(\theta) | 01 \rangle + \cos(\theta) | 10 \rangle , \qquad (15.77)$$

for $0 < \theta < \frac{\pi}{2}$, with the corresponding density matrix in the Bloch representation

$$\rho_\theta = \tfrac{1}{4} \Big[\mathbb{1} \otimes \mathbb{1} - \cos(2\theta) \left(\sigma_z \otimes \mathbb{1} - \mathbb{1} \otimes \sigma_z \right)$$
$$+ \sin(2\theta) \left(\sigma_x \otimes \sigma_x + \sigma_y \otimes \sigma_y \right) - \sigma_z \otimes \sigma_z \Big] . \qquad (15.78)$$

Meanwhile, the mixed separable state is given by

$$\rho_{\text{top}} = \tfrac{1}{2} \left(| 00 \rangle\langle 00 | + | 11 \rangle\langle 11 | \right) = \tfrac{1}{4} \left(\mathbb{1} \otimes \mathbb{1} + \sigma_z \otimes \sigma_z \right) , \qquad (15.79)$$

which is located at the top corner of the double pyramid of separable states shown in Fig. 15.3. The *Gisin states* ρ_{Gisin} (Gisin, 1996) are then given by

$$\rho_{\text{Gisin}}(\lambda, \theta) = \lambda \rho_\theta + (1 - \lambda) \rho_{\text{top}}$$
$$= \tfrac{1}{4} \Big[\mathbb{1} \otimes \mathbb{1} - \lambda \cos(2\theta) \left(\sigma_z \otimes \mathbb{1} - \mathbb{1} \otimes \sigma_z \right)$$
$$+ \lambda \sin(2\theta) \left(\sigma_x \otimes \sigma_x + \sigma_y \otimes \sigma_y \right) + (1 - 2\lambda) \sigma_z \otimes \sigma_z \Big] , \qquad (15.80)$$

with the parameter range $0 \le \lambda \le 1$. Note that the Gisin states are in general not Weyl states (which we will discuss in Section 15.5), i.e., they are generally not locally maximally mixed, the local Bloch vectors do not vanish for the whole parameter range. Only the subset for which $\theta = \pi/4$ can be represented in the tetrahedron of Weyl states (shown in Fig. 15.3) as a straight line on the surface of the tetrahedron that is connecting the state ρ_{top} at the top of the separable double pyramid with the maximally entangled state $| \psi^+ \rangle$.

Since we are considering a family of two-qubit states, we can determine the parameter region for which the Gisin states are entangled with the help of the PPT criterion, Theorem 15.3, from which one can easily infer that the Gisin states are entangled if and only if

$$1 - \lambda < \lambda \sin(2\theta) . \qquad (15.81)$$

To quantify the entanglement, one may use an entanglement monotone called the concurrence C, which we will discuss in detail in Section 16.2.2. For the Gisin state the calculation of this quantity gives

$$C[\rho_{\text{Gisin}}] = \max\left\{0, (\lambda \sin(2\theta) + \lambda - 1)\right\}. \tag{15.82}$$

In contrast, we can also calculate the parameter range for which $\rho_{\text{Gisin}}(\lambda, \theta)$ is non-local according to the CHSH–Bell operator criterion of Theorem 15.6. To this end we have to consider the correlation tensor t_{ij} for ρ_{Gisin} to calculate the maximal value $\overline{\mathcal{B}}_{\text{CHSH}}^{\max} = 2\sqrt{\mu_1 + \mu_2}$ of the Bell–CHSH parameter, where μ_1 and μ_2 are the two largest eigenvalues of the matrix $M_\rho = T_\rho^\top T_\rho$, and $T_\rho = (t_{ij})$ denotes the correlation tensor for a given state. Reading off the matrix elements t_{ij} from the Bloch decomposition in eqn (15.80) and determining the eigenvalues of the correlation tensor, the maximally possible expectation value of the CHSH–Bell operator is (see, e.g., Friis *et al.*, 2017*a*)

$$\overline{\mathcal{B}}_{\text{CHSH}}^{\max}(\rho_{\text{Gisin}}) = 2 \max\left\{ \sqrt{\lambda^2 \sin^2(2\theta) + (1 - 2\lambda)^2}, \ \sqrt{2}\lambda \sin(2\theta) \right\}. \tag{15.83}$$

For $\overline{\mathcal{B}}_{\text{CHSH}}^{\max}(\rho_{\text{Gisin}}) > 2$ (recall Theorem 15.6) the Gisin states are hence definitely non-local. What is crucial here is to note that the parameter region for which the Gisin states are entangled is strictly larger than (and includes) the region of non-locality. An illustration of these regions of entanglement and non-locality is given in Friis *et al.* (2017*a*).

Gisin's filtering procedure: A most interesting feature of the Gisin states is revealed when a local filtering procedure is applied (Gisin, 1996). That is, suppose that after sharing the state $\rho_{\text{Gisin}}(\lambda, \theta)$ for some θ between 0 and $\frac{\pi}{4}$, Alice and Bob locally amplify their qubit states $|0\rangle_A$ and $|1\rangle_B$, respectively. In that case we have $\sin(\theta) < \cos(\theta)$, and thus the component of $|01\rangle$ increases with respect to that of $|10\rangle$ in $|\psi_\theta\rangle$, which effectively moves the state closer to the maximally entangled state $|\psi^+\rangle$. Likewise, if $\frac{\pi}{4} < \theta < \frac{\pi}{2}$, amplifying $|1\rangle_A$ and $|0\rangle_B$, respectively, will have the same effect.

Mathematically, these filtering operations are represented by a family of local, completely positive, and trace-nonincreasing maps $\mathcal{F}_\theta$, parameterized by θ

$$\mathcal{F}_\theta : \quad \rho \longrightarrow \mathcal{F}_\theta[\rho] = F_\theta \rho F_\theta^\dagger. \tag{15.84}$$

The operators F_θ are the so-called *Kraus operators* for this operation (for a general definition, see Chapter 21.3 in Part III) and satisfy the relation $F_\theta^\dagger F_\theta \leq \mathbb{1}$. In our case they are of the form

$$F_\theta = \begin{cases} F_0(\theta) \otimes F_1(\theta) & \text{if } 0 < \theta \leq \frac{\pi}{4} \\ F_1^{-1}(\theta) \otimes F_0^{-1}(\theta) & \text{if } \frac{\pi}{4} < \theta < \frac{\pi}{2} \end{cases}, \tag{15.85}$$

where the local operations are

$$F_0(\theta) = \begin{pmatrix} 1 & 0 \\ 0 & \sqrt{\tan(\theta)} \end{pmatrix} \quad \text{and} \quad F_1(\theta) = \begin{pmatrix} \sqrt{\tan(\theta)} & 0 \\ 0 & 1 \end{pmatrix}. \tag{15.86}$$

The probability for successful filtering can be computed as

$$\text{Tr}\big(\mathcal{F}_\theta[\rho]\big) = \begin{cases} N(\lambda,\theta)\,\cot(\theta) & \text{if } 0 < \theta < \frac{\pi}{4} \\ N(\lambda,\theta)\,\tan(\theta) & \text{if } \frac{\pi}{4} < \theta < \frac{\pi}{2} \end{cases}, \tag{15.87}$$

with the normalization $N(\lambda,\theta) = \lambda\sin(2\theta) + 1 - \lambda$. For any pair of parameters λ and θ we then get the normalized quantum state after the filtering procedure,

$$\rho_{\text{filtered}}(\lambda,\theta) = \frac{\mathcal{F}_\theta(\rho)}{\text{Tr}\big(\mathcal{F}_\theta[\rho]\big)} = \frac{1}{N(\lambda,\theta)}\Big[\lambda\sin(2\theta)\,\rho^+ + (1-\lambda)\,\rho_{\text{top}}\Big] \tag{15.88}$$

$$= \frac{1}{4}\bigg[\mathbb{1}\otimes\mathbb{1} + \frac{\lambda\sin(2\theta)}{N(\lambda,\theta)}\,(\sigma_x\otimes\sigma_x + \sigma_y\otimes\sigma_y) + \frac{1-\lambda-\lambda\sin(2\theta)}{N(\lambda,\theta)}\,\sigma_z\otimes\sigma_z\bigg].$$

The *filtered Gisin state* now fully lies within the set of Weyl states. In fact, the set of filtered and unfiltered Gisin states coincides for $\theta = \pi/4$, and can be represented by a line from ρ_{top} to $\rho^+ = |\psi^+\rangle\langle\psi^+|$ (as we will see in Fig. 15.3).

For the concurrence of the filtered Gisin states from eqn (15.88) one finds

$$C[\rho_{\text{filtered}}] = \max\left\{0, \frac{\lambda\sin(2\theta) - 1 + \lambda}{\lambda\sin(2\theta) + 1 - \lambda}\right\}. \tag{15.89}$$

The filtered state $\rho_{\text{filtered}}(\lambda,\theta)$ is entangled if (and only if) $\lambda\sin(2\theta) > (1-\lambda)$. This means the filtering is not able to[5] entangle initially separable states. However, due to the denominator $\lambda\sin(2\theta) + 1 - \lambda \le 1$, all already entangled states become more entangled. This is possible despite the fact that the filtering operation is local because the part of the initial quantum state that does not pass the filters is disregarded. If we were to complete the (trace-nonincreasing) quantum operation $\mathcal{F}_\theta$ from (15.84) to a (trace-preserving) quantum channel $\overline{\mathcal{F}}_\theta$ with Kraus operators F_θ and $\overline{F}_\theta = (\mathbb{1} - F_\theta^\dagger F_\theta)^{1/2}$, then we would see that the entanglement would not increase on average.

Now turning to non-locality, we study the effect of the filtering procedure. As before we calculate the maximally possible expectation value of the CHSH inequality from Theorem 15.6, which yields

$$\overline{\mathcal{B}}_{\text{CHSH}}^{\max}(\rho_{\text{filtered}}) = \frac{2}{N(\lambda,\theta)}\max\left\{\sqrt{\lambda^2\sin^2(2\theta) + \big[1-\lambda-\lambda\sin(2\theta)\big]^2}, \sqrt{2}\lambda\sin(2\theta)\right\}. \tag{15.90}$$

Focusing on the region where $\rho_{\text{filtered}}(\lambda,\theta)$ is entangled, i.e., for $\lambda > \big[1 + \sin(2\theta)\big]^{-1}$, the condition $\overline{\mathcal{B}}_{\text{CHSH}}^{\max}(\rho_{\text{filtered}}) > 2$ provides the region of non-locality for the filtered Gisin states

$$\lambda > \frac{1}{(\sqrt{2}-1)\sin(2\theta) + 1}. \tag{15.91}$$

---

[5] If it could, this would be rather worrying, since it is a local filtering operation and local operations cannot create entanglement, see the discussion in Section 16.1.2.

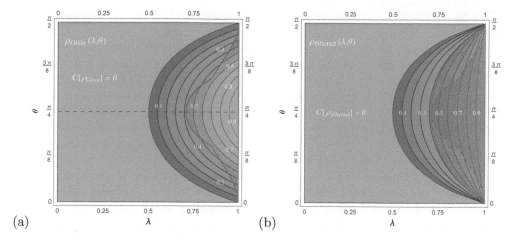

Fig. 15.2 Gisin states. The parameter regions (λ, θ) of separability (blue), entanglement (green), and non-locality (orange) are shown for the unfiltered Gisin states ρ_{Gisin} of eqn (15.80) in panel (a) and for the filtered Gisin states $\rho_{\text{filtered}}(\lambda, \theta)$ from eqn (15.88) in (b). The separable states lie in the blue regions on the respective left-hand sides. In the green entangled regions on the right-hand side the contour lines of the concurrence $C[\rho_{\text{Gisin}}]$ and $C[\rho_{\text{filtered}}]$ from eqns (15.82) and (15.89) are drawn in (a) and (b), respectively, for values of the concurrence from 0.1 to 0.9 in steps of 0.1. The solid orange lines on the right-hand sides delimit the regions of unfiltered and filtered Gisin states, respectively, that can violate the CHSH inequality, and the dashed orange line in (b) indicates the boundary for the unfiltered Gisin states for comparison. The horizontal dashed red line in (a) indicates those Gisin states that are also Weyl states. (a) and (b) modified from Figs. 2 and 3, respectively, in *Geometry of two-qubit states with negative conditional entropy*, Friis *et al.*, J. Phys. A: Math. Theor. **50**, 125301 (2017) © IOP Publishing. Reproduced with permission. All rights reserved.

As shown in Fig. 15.2 (b) the non-local parameter region for $\rho_{\text{filtered}}(\lambda, \theta)$ includes the entire region of non-locality of the unfiltered state, but is strictly larger. Some previously local (entangled) states become more strongly entangled and even non-local due to the filtering. The amplification of entanglement hence reveals the hidden non-locality of some of the Gisin states, while others remain local. Although this separation may be attributed to the choice of filtering operation, it should be remarked that not every entangled state can become non-local under local filtering operations (see Hirsch *et al.*, 2016).

Résumé: As discovered by Nicolas Gisin, there are quantum states that initially satisfy a Bell inequality, i.e., they are local, but lead to a violation of the inequality after local filtering operations that amplify the entanglement of the initial state. In this way the non-local character of the quantum state, previously hidden, is revealed.

15.4 Separability Criteria from Positive Maps

Although we have by now obtained a clearer view of the differences between entanglement, separability, classical correlations, and non-locality, we nevertheless find ourselves facing the question: *Given some density operator, how can we determine if the quantum state is entangled or not?* As we have seen, finding a Bell inequality that is violated by the quantum state in question unambiguously identifies the state as being entangled, but finding no such inequality, or even finding a hidden-variable model compatible with the state does not necessarily mean that the state is separable. All the while, such methods are generally difficult to check on a case-by-case basis. The PPT criterion does provide a test that can be straightforwardly applied to any density operator, but is also known not to catch all entangled states (in dimensions larger than 2×3). However, the PPT criterion can be seen as a special case of a larger family of criteria based on *positive but not completely positive* maps, which we will discuss in this section. These criteria still do not fully resolve the separability problem, and indeed there are a number of other, generally inequivalent entanglement-detection criteria, ranging from *entanglement witnesses* (see Section 16.3), over methods such as the *computable cross-norm or realignment* criterion (see Section 17.4.1) to approaches based on *entropic uncertainty relations* (see Section 17.4.2). Nevertheless, positive-map criteria provide us with a first systematic approach to entanglement detection.

15.4.1   The Positive-Map Theorem

Positive (but not completely positive) maps play an important role in the determination of the separability range of composite quantum systems. To proceed let us first clarify what these objects are in terms of the following definitions.

Definition 15.10 *A **linear map** $\Lambda : L(\mathcal{H}_1) \to L(\mathcal{H}_2)$ from the space $L(\mathcal{H}_1)$ of linear operators on the Hilbert space $\mathcal{H}_1$ to the space $L(\mathcal{H}_2)$ of linear operators on the Hilbert space $\mathcal{H}_2$ is called **positive** if it maps all positive operators $A \in L(\mathcal{H}_1)$ onto positive ones on $L(\mathcal{H}_2)$,*

$$\Lambda(A) \geq 0 \quad \forall\, A \geq 0 \,.$$

Definition 15.11 *A positive map $\Lambda^{\mathrm{CP}} : L(\mathcal{H}_1) \to L(\mathcal{H}_2)$ is **completely positive** if it remains positive under composition with the identity $\mathbb{1}_d$ on the space $L(\mathcal{H}_d)$ of all $d \times d$ matrices, that is, if the map*

$$\Lambda^{\mathrm{CP}} \otimes \mathbb{1}_d : \; L(\mathcal{H}_1 \otimes \mathcal{H}_d) \longrightarrow L(\mathcal{H}_2 \otimes \mathcal{H}_d)$$

is still a positive map $\forall\, d = 2, 3, 4, \ldots$.

Thus, a completely positive map satisfies $\Lambda^{\mathrm{CP}} \otimes \mathbb{1}_d \geq 0$, that is, $(\Lambda^{\mathrm{CP}} \otimes \mathbb{1}_d)[A] \geq 0$ for all operators $A \in L(\mathcal{H}_1 \otimes \mathcal{H}_d)$ that are themselves positive, $A \geq 0$. Physically, this extended positive map $\Lambda^{\mathrm{CP}} \otimes \mathbb{1}_d \geq 0$ corresponds to an operation on the combined system of Alice and Bob that only acts non-trivially locally on Alice's subsystem without influencing Bob's side. Perhaps it comes at a surprise to some readers that there are indeed positive maps that are not completely positive, but we have already encountered a rather prominent candidate for such a map before.

Partial transposition: An example of a *positive but not completely positive map* is the transposition $\rho^{\mathsf{T}} = \mathsf{T}[\rho]$ defined with respect to a fixed basis $\{|\,j\,\rangle\}_j$ as

$$\rho^{\mathsf{T}} = \mathsf{T}[\rho] = \mathsf{T}\Big(\sum_{j,k} \rho_{jk} \,|\,j\,\rangle\langle\,k\,|\Big) = \sum_{j,k} \rho_{kj} \,|\,j\,\rangle\langle\,k\,|\,. \tag{15.92}$$

In Theorem 15.3 we already made use of the fact that, despite T being a positive map, the partial transposition $\mathbb{1} \otimes \mathsf{T}$ is not necessarily positive, and with the Werner states (recall Section 15.3.3) we have also seen examples of states (represented by positive operators) for which the partial transpose results in operators with a negative eigenvalue. Thus, the partial transposition is not positive (semidefinite),

$$\mathbb{1}_A \otimes \mathsf{T}_B \not\geq 0\,, \tag{15.93}$$

i.e., T_B is a positive but not a completely positive map.

Such positive but not completely positive maps turn out to be crucial for the detection of entanglement, and a central statement in this endeavour is the following theorem for positive maps, established by *Michał, Paweł*, and *Ryszard Horodecki* (1996*a*, see also Horodecki *et al.*, 1996*a*; Horodecki *et al.*, 2009).

Theorem 15.8 (Positive-map theorem)

A bipartite state ρ_{AB} is separable if and only if

$$(\mathbb{1}_A \otimes \Lambda_B)[\rho_{AB}] \geq 0 \quad \textit{for all} \quad \Lambda \geq 0\,.$$

Proof   The necessary condition can be proven very quickly. Let us assume that ρ_{AB} is separable, i.e., it can be expressed as $\rho_{AB} = \sum_i p_i\, \rho_i^{(A)} \otimes \rho_i^{(B)}$. Applying the map $\mathbb{1}_A \otimes \Lambda_B$ gives

$$(\mathbb{1}_A \otimes \Lambda_B)[\rho_{AB}] = \sum_i p_i\, \rho_i^{(A)} \otimes \Lambda(\rho_i^{(B)})\,. \tag{15.94}$$

Since Λ is positive, $\Lambda \geq 0$, then $\Lambda(\rho_i^{(B)}) \geq 0$ is positive as well, and therefore $(\mathbb{1}_A \otimes \Lambda_B)[\rho_{AB}] \geq 0$ since the convex combination (i.e., linear combination with non-negative probability weights) of (tensor products of) positive operators is positive.

To prove the sufficient condition, i.e., that $(\mathbb{1}_A \otimes \Lambda_B)[\rho_{AB}] \geq 0 \ \forall \Lambda \geq 0$ implies that ρ_{AB} is separable, we need to rely on some techniques and statements that we will only prove and discuss in detail later in this book, but which we will formulate here in a way that is convenient for the proof of Theorem 15.8.

(i) A bipartite state ρ_{AB} is separable *iff* $\mathrm{Tr}(W\rho_{AB}) \geq 0$ for all Hermitian operators $W = W^\dagger$ for which $\mathrm{Tr}(W\rho_{\mathrm{sep}}) \geq 0$ for all separable states ρ_{sep}.

(ii) Every positive map $\Lambda \geq 0$ from $L(\mathcal{H}_1)$ to $L(\mathcal{H}_2)$ is in one-to-one correspondence with a Hermitian operator $\rho_\Lambda = \rho_\Lambda^\dagger$ that satisfies $\text{Tr}(\rho_\Lambda P \otimes Q) \geq 0$ for all projectors P and Q on $\mathcal{H}_1$ and $\mathcal{H}_2$, respectively.

Statement (i) follows directly from the *entanglement-witness theorem* (Theorem 16.8) and the Hermitian operators in question are called *entanglement witnesses*. Also in the case of the entanglement-witness theorem one direction is easily proven, clearly knowing that $\text{Tr}(W\rho_{\text{sep}}) \geq 0$ for all separable states implies $\text{Tr}(W\rho_{AB}) \geq 0$ if ρ_{AB} is separable, but to show that ρ_{AB} cannot be entangled if $\text{Tr}(W\rho_{AB}) \geq 0$ for all witnesses W requires some more work, as we will see in Chapter 16.

The second statement, (ii), refers to the so-called *Choi–Jamiołkowski isomorphism* between linear maps from one Hilbert space to another and operators on the tensor-product space of the two Hilbert spaces, which we will discuss in detail in Section 21.4.1. More specifically, we will formulate (and prove) the relation between positive maps Λ and their corresponding "*Choi states*" ρ_Λ in the form of Lemma 21.3 in Section 21.4.1. Although the original proof in Horodecki *et al.* (1996*a*) employs Jamiołkowski's version of the isomorphism, we will forego the latter's appeal of being basis-independent in favour of a slightly more compact proof of the positive-map theorem using the *Choi isomorphism*.

With these preliminaries out of the way, let us sketch the proof. Suppose Λ_B is a positive map from $L(\mathcal{H}_B)$ to $L(\mathcal{H}_{B'})$ with $d_{B'} \leq d_A$, and $(\mathbb{1}_A \otimes \Lambda_B)[\rho_{AB}] \geq 0$, then its trace must be positive, and the same must be true for the trace of the product with another positive operator, which can be chosen to be a projector $P_{00} = |\phi_{00}\rangle\langle\phi_{00}|$ with

$$|\phi_{00}\rangle = \frac{1}{\sqrt{d_A}} \sum_{k=0}^{d_A-1} |k\rangle_A \otimes |k\rangle_{B'} \in \mathcal{H}_A \otimes \mathcal{H}_{B'}, \tag{15.95}$$

as in eqn (14.11) for $m = n = 0$. We thus see that $(\mathbb{1}_A \otimes \Lambda_B)[\rho_{AB}] \geq 0$ implies

$$\text{Tr}\Big[(\mathbb{1}_A \otimes \Lambda_B)[\rho_{AB}] P_{00}\Big] = ((\mathbb{1}_A \otimes \Lambda_B)[\rho_{AB}]^\dagger, P_{00})_{\text{HS}} \geq 0, \tag{15.96}$$

where $(A, B)_{\text{HS}} = \text{Tr}(A^\dagger B)$ denotes the Hilbert–Schmidt inner product from Definition 11.4. Now, since Λ is positive but not necessarily completely positive, $(\mathbb{1}_A \otimes \Lambda_B)$ may not be a positive map for general arguments (we have only assumed here that it is when applied to ρ_{AB}). But since any positive map preserves Hermiticity, the same is true for $(\mathbb{1}_A \otimes \Lambda_B)$, and consequently we have $(\mathbb{1}_A \otimes \Lambda_B)[\rho]^\dagger = (\mathbb{1}_A \otimes \Lambda_B)[\rho^\dagger]$, as we will show in Section 21.4.1. By passing from the map $(\mathbb{1}_A \otimes \Lambda_B)$ from $L(\mathcal{H}_A \otimes \mathcal{H}_B)$ to $L(\mathcal{H}_A \otimes \mathcal{H}_{B'})$ to the adjoint map $(\mathbb{1}_A \otimes \Lambda_{B'})$ from $L(\mathcal{H}_A \otimes \mathcal{H}_{B'})$ to $L(\mathcal{H}_A \otimes \mathcal{H}_B)$ we thus obtain the equivalent (to (15.96)) condition

$$(\rho_{AB}, (\mathbb{1}_A \otimes \Lambda_{B'})[P_{00}])_{\text{HS}} = \text{Tr}\Big[\rho_{AB}(\mathbb{1}_A \otimes \Lambda_{B'})[P_{00}]\Big] \geq 0. \tag{15.97}$$

But as we will see in Section 21.4.1, $(\mathbb{1}_A \otimes \Lambda_{B'})[P_{00}]$ is the *Choi state* ρ_Λ of the positive map $\Lambda_{B'}$. Thus, $(\mathbb{1}_A \otimes \Lambda_B)[\rho_{AB}] \geq 0 \ \forall \Lambda \geq 0$ implies $\text{Tr}(\rho_{AB}\rho_\Lambda) \geq 0$ for all ρ_Λ.

According to statement (ii) this operator is Hermitian and has a non-negative trace with all product projectors (see Lemma 21.3), in particular, with all separable states. Moreover, since there is a positive map Λ for every Hermitian operator that satisfies $\mathrm{Tr}(\rho_\Lambda P \otimes Q) \geq 0$, there is a positive map Λ for every entanglement witness from statement (i). Consequently, we have shown that $(\mathbb{1}_A \otimes \Lambda_B)[\rho_{AB}] \geq 0 \ \forall \Lambda \geq 0$ is equivalent to saying that $\mathrm{Tr}(W \rho_{AB}) \geq 0$ for all entanglement witnesses, and the state ρ_{AB} must hence be separable. $\qquad\square$

Theorem 15.8 thus establishes an interesting connection between positive maps and entanglement witnesses, and the statement can be turned around to assert:

Theorem 15.9 (Entanglement theorem)

A bipartite state ρ_{AB} is entangled if and only if $\exists$ positive map $\Lambda \geq 0$ such that

$$(\mathbb{1}_A \otimes \Lambda_B)[\rho_{AB}] \not\geq 0.$$

Thus, Theorem 15.9 can be understood as a generalization of the PPT criterion, Theorem 15.3: If the partial application (i.e., to one of the subsystems only) of *any* positive (but not completely positive) map Λ to a bipartite state ρ_{AB} results in an operator with a negative eigenvalue, then the state cannot be separable (according to Theorem 15.8) and must be entangled. However, if the density operator of a given state remains positive when applying a particular positive map to a subsystem, one may not in general conclude that the state is separable. This is only the case in some special cases, one of which we will discuss next.

15.4.2 Proof of the PPT Criterion in Dimension ≤ 6

A special case in which a positive but not completely positive map provides a necessary and sufficient separability criterion is the PPT criterion in dimensions smaller or equal to 6, i.e., when the bipartite quantum system consists of two qubits (dimension 2×2) or one qubit and one qutrit (dimension 2×3). To show that the positivity of the partial transposition is sufficient for separability in this case, and thus to provide a proof for Theorem 15.4, we need a theorem based on well-known results in mathematical physics by *Erling Størmer* (1963) and *Stanisław Lech Woronowicz* (Woronowicz, 1976):

Theorem 15.10 (Størmer–Woronowicz theorem)

Any positive map $\Lambda : L(\mathcal{H}_1) \to L(\mathcal{H}_2)$ with $d_1 = \dim(\mathcal{H}_1) = 2 = \dim(\mathcal{H}_2) = d_2$, or $d_1 = 2$, $d_2 = 3$, or $d_1 = 3$, $d_2 = 2$ can be decomposed as

$$\Lambda = \Lambda_1^{\mathrm{CP}} + \Lambda_2^{\mathrm{CP}} \circ \top,$$

where Λ_1^{CP} and Λ_2^{CP} are completely positive maps and $\top$ is the transposition.

The proof of the Størmer–Woronowicz theorem is somewhat non-trivial and going through the arguments presented in Størmer (1963) and Woronowicz (1976) in detail is a task that we will leave to the interested reader, but at least for the case of dimension 2×3, a somewhat simplified version of the proof can be found in Aubrun and Szarek (2015). Taking the Størmer–Woronowicz theorem as given, let us now see how it can be employed to prove Theorem 15.4.

Proof of Theorem 15.4   Suppose we restrict the dimension of the joint Hilbert space to $d \leq 2 \times 3$ as required in Theorem 15.10 and there is a state ρ_{AB} with $(\mathbb{1}_A \otimes \top_B)\rho_{AB} \geq 0$. Since Λ_1^{CP} and Λ_2^{CP} are completely positive maps we have

$$(\mathbb{1}_A \otimes \Lambda_{1,B}^{\mathrm{CP}})\rho_{AB} + (\mathbb{1}_A \otimes \Lambda_{2,B}^{\mathrm{CP}})(\mathbb{1}_A \otimes \top_B)\rho_{AB} \geq 0 \,,$$

$$\Rightarrow \quad (\mathbb{1}_A \otimes \Lambda_{1,B}^{\mathrm{CP}})\rho_{AB} + (\mathbb{1}_A \otimes \Lambda_{2,B}^{\mathrm{CP}} \circ \top_B)\rho_{AB} \geq 0 \,,$$

$$\Rightarrow \quad \left(\mathbb{1}_A \otimes [\Lambda_{1,B}^{\mathrm{CP}} + \Lambda_{2,B}^{\mathrm{CP}} \circ \top_B]\right)\rho_{AB} \geq 0 \,. \tag{15.98}$$

Since we have restricted to dimensions $(d_A, d_B) = (2,2)$, $(2,3)$, or $(3,2)$, and the Størmer–Woronowicz theorem (Theorem 15.10) implies that every positive map from $L(\mathcal{H}_B)$ to $L(\mathcal{H}_{B'})$ with $d_{B'} \leq d_A$ can be decomposed in this way. We hence have

$$(\mathbb{1}_A \otimes \Lambda_B)\rho_{AB} \geq 0 \quad \forall \, \Lambda \geq 0 \,. \tag{15.99}$$

The positive-map theorem (Theorem 15.8) then says that ρ_{AB} must necessarily be separable, which completes the proof. $\square$

However, let us remark that already when $(d_A, d_B) = (3,3)$ covering all relevant positive maps would require maps to spaces $L(\mathcal{H}_{B'})$ with $d_{B'} = 3$, for which the Størmer–Woronowicz theorem does not hold. Indeed, not all positive maps between Hilbert spaces of higher dimensions are decomposable in the desired way.

15.4.3   The Reduction Criterion

Let us now consider another example of a positive but not completely positive map Λ in the form

$$\Lambda : L(\mathcal{H}) \longrightarrow L(\mathcal{H}) \tag{15.100}$$

$$\rho \longmapsto \Lambda[\rho] = \mathrm{Tr}(\rho)\,\mathbb{1} - \rho \,.$$

If $\rho \geq 0$, we can write it in spectral decomposition as $\rho = \sum_n p_n \,|\, n \,\rangle\langle\, n \,|$ with $p_n \geq 0$ and take the trace with respect to the eigenbasis $\{|\, n \,\rangle\}_n$ of ρ, such that

$$\Lambda[\rho] = \mathrm{Tr}(\rho)\,\mathbb{1} - \rho = \sum_m p_m \sum_n |\, n \,\rangle\langle\, n \,| - \sum_n p_n \,|\, n \,\rangle\langle\, n \,|$$

$$= \sum_n \Big(\sum_m p_m - p_n\Big)\,|\, n \,\rangle\langle\, n \,| = \sum_n \Big(\sum_{m \neq n} p_m\Big)\,|\, n \,\rangle\langle\, n \,| \geq 0 \,. \tag{15.101}$$

If we apply the positive map to only one of the subsystems of a bipartite quantum state ρ_{AB}, the trace becomes a partial trace over the respective subsystem, leaving an identity operator in its place and the reduced state for the other subsystem, so we have

$$(\mathbb{1}_A \otimes \Lambda_B)\,\rho_{AB} = \mathrm{Tr}_B(\rho_{AB}) \otimes \mathbb{1}_B - \rho_{AB} = \rho_A \otimes \mathbb{1}_B - \rho_{AB}\,. \tag{15.102}$$

And of course we can also exchange the roles of the subsystems, such that finding a negative eigenvalue for $(\mathbb{1}_A \otimes \Lambda_B)\,\rho_{AB}$ or $(\Lambda_A \otimes \mathbb{1}_B)\,\rho_{AB}$ implies that ρ_{AB} is entangled, which is captured in the so-called reduction criterion first developed by Michał and Paweł Horodecki (1999) and shortly after independently discovered by *Nicolas Cerf*, *Christoph Adami*, and *Robert Gingrich* (1999):

Theorem 15.11 (Reduction criterion)

If $\ \rho_A \otimes \mathbb{1}_B - \rho_{AB} \ngeq 0\ $ *or* $\ \mathbb{1}_A \otimes \rho_B - \rho_{AB} \ngeq 0 \ \Longrightarrow \ \rho_{AB}$ *entangled.*

The reduction criterion thus provides another example for a positive map that, when applied to one of the subsystems, can be used to detect entanglement. However, it is not more powerful than the PPT criterion. When one of the subsystems is two-dimensional, i.e., when we have a $2 \times n$-dimensional joint Hilbert space, the reduction criterion is equivalent to the PPT criterion.

To see this, we recall that we can write any two-qubit system in terms of a generalized Bloch decomposition as in eqn (15.39). As we will discuss later on in Chapter 17, we can also find a similar decomposition for higher-dimensional systems where the Pauli operators are replaced by other traceless Hermitian operators. For a bipartite system of which one subsystem is a qubit, we can forget about the details of the other subsystem and consider the action of the map Λ from eqn (15.100) on the Pauli operators σ_i and the identity operator pertaining to the qubit subsystem, for which we find

$$\Lambda[\mathbb{1}_2] = \mathrm{Tr}(\mathbb{1}_2)\,\mathbb{1}_2 - \mathbb{1}_2 = 2\,\mathbb{1}_2 - \mathbb{1}_2 = \mathbb{1}_2, \tag{15.103}$$

$$\Lambda[\sigma_i] = \mathrm{Tr}(\sigma_i)\,\mathbb{1}_2 - \sigma_i = 0\,\mathbb{1}_2 - \sigma_i = -\sigma_i. \tag{15.104}$$

The effect of the map is thus to flip the sign of all Bloch-vector components of the target qubit. In comparison, the transpose T, carried out with respect to the computational basis leaves $\mathbb{1}_2$, as well as σ_x and σ_z invariant, but flips the sign of σ_y. However, we can combine the transposition with a rotation around the y axis about the angle π. From Lemma 8.1 we have $U_y(\pi) = -i\sigma_y$, and since $\sigma_y\sigma_x = -\sigma_x\sigma_y$ and $\sigma_y\sigma_z = -\sigma_z\sigma_y$, we can write

$$U_y(\pi)\,\mathsf{T}[\sigma_i]\,U_y^\dagger(\pi) = -\sigma_i = \Lambda[\sigma_i]. \tag{15.105}$$

We thus see that, up to a local unitary, the reduction map is equivalent to the transposition when applied to a single qubit. Local unitary transformations do not change the assignment of a state to the set of separable or entangled states, and we thus see that the reduction criterion is equivalent to the PPT criterion when one of the subsystems is a qubit. In particular, it is a necessary and sufficient separability criterion if the other subsystem is of dimension 2 or 3, according to Theorem 15.4.

However, in other dimensions the reduction criterion is strictly weaker than the PPT criterion. This follows from the fact that the reduction map is decomposable à la Størmer–Woronowicz (Theorem 15.10) in any dimension (Horodecki and Horodecki, 1999).

Lemma 15.3 (Decomposability of reduction map)

The reduction map $\Lambda[\rho] = \mathrm{Tr}(\rho)\,\mathbb{1} - \rho$ is decomposable, i.e.,

$$\Lambda = \mathsf{T} \circ \Lambda^{\mathrm{CP}}\,,$$

where Λ^{CP} is completely positive and T is the transposition.

Proof To see that the map is decomposable, we make use of the fact that a map Λ is completely positive (CP) if and only if the corresponding Choi state $\rho_\Lambda = (\mathbb{1}_A \otimes \Lambda_B)[P_{00}]$ is positive. One direction of this statement follows directly from the definition (Def. 15.11) of complete positivity, which implies that $\rho_\Lambda \geq 0$ if Λ is CP. For the reverse direction, we can invoke Theorem 21.1, the proof of which tells us that every positive operator ρ_Λ on a bipartite Hilbert space can be interpreted as the Choi state of a map Λ that admits what is called a *Kraus decomposition*, for which Lemma 21.2 in turn ensures that Λ must be CP.

With this information at hand, we take a closer look at the Choi state of the reduction map, which gives

$$\rho_\Lambda = \big(\mathbb{1}_A \otimes \Lambda_B\big)\big[P_{00}\big] = \rho_A \otimes \mathbb{1}_B - |\phi_{00}\rangle\langle\phi_{00}| = \tfrac{1}{d_A}\mathbb{1}_A \otimes \mathbb{1}_B - |\phi_{00}\rangle\langle\phi_{00}|\,,\tag{15.106}$$

where $P_{00} = |\phi_{00}\rangle\langle\phi_{00}|$ is the projector on the maximally entangled state $|\phi_{00}\rangle = \frac{1}{\sqrt{d_A}}\sum_i |ii\rangle$, and we hence noted that the reduced state ρ_A of subsystem A is the maximally mixed state $\frac{1}{d_A}\mathbb{1}_A$. Next, let us consider the partial transpose of the Choi state ρ_Λ,

$$\rho_\Lambda^{\mathsf{T}_B} = (\mathbb{1}_A \otimes \mathsf{T}_B)[\rho_\Lambda] = \tfrac{1}{d_A}\mathbb{1}_A \otimes \mathbb{1}_B - (\mathbb{1}_A \otimes \mathsf{T}_B)\big[|\phi_{00}\rangle\langle\phi_{00}|\big]$$

$$= \tfrac{1}{d_A}\mathbb{1}_A \otimes \mathbb{1}_B - \tfrac{1}{d_A}\sum_{i,j} |i\rangle\langle j| \otimes \big(|i\rangle\langle j|\big)^{\mathsf{T}} = \tfrac{1}{d_A}\mathbb{1}_A \otimes \mathbb{1}_B - \tfrac{1}{d_A}\sum_{i,j} |i\rangle\langle j| \otimes |j\rangle\langle i|\,.\tag{15.107}$$

Closer inspection of the sum on the right-hand side reveals that it represents a unitary operator

$$U_{\mathrm{SWAP}} := \sum_{i,j} |i\rangle\langle j| \otimes |j\rangle\langle i|\tag{15.108}$$

that exchanges any two states of the two subsystems, i.e., $U_{\text{SWAP}} \, | \, \psi \, \rangle \otimes | \, \phi \, \rangle = | \, \phi \, \rangle \otimes | \, \psi \, \rangle$. The eigenvectors of U_{SWAP} are hence symmetric (with eigenvalue $+1$) or antisymmetric states (with eigenvalue -1) with respect to the exchange of the two subsystems. Consequently, the operator $\rho_\Lambda^{T_B}$ has non-negative eigenvalues and is thus positive. From the statement at the beginning of the proof we can then conclude that $\mathsf{T}\Lambda$ is a CP map Λ^{CP}, and, conversely, that Λ can be composed into a CP map and the transposition T as stated in Lemma 15.3. $\qquad\qquad\qquad\qquad\qquad\qquad\qquad\qquad\qquad\qquad\qquad\quad$ $\square$

Since the reduction map is decomposable, it follows that the only way that its application to one subsystems can yield a negative eigenvalue is if the partial transposition is non-positive. Thus, every state whose entanglement is detected via a positive (but not completely positive) map that is decomposable is also detected by the PPT criterion. The reduction criterion can nevertheless be useful for practical purposes since it was shown (Horodecki and Horodecki, 1999) that every entangled state detected by the reduction criterion is distillable (see Section 16.1.3).

15.4.4 Isotropic States

Although the reduction criterion is not superior to the PPT criterion in general, it can nevertheless sometimes provide a practical way of detecting entangled states, in particular, for the family of the so-called *isotropic states*.

Definition 15.12 *An **isotropic state** $\rho_{\text{iso}} \in \mathcal{B}_{\text{HS}}(\mathcal{H}_A \otimes \mathcal{H}_B)$ with $d_A = d_B = d$ is*

$$\text{of the form } \rho_{\text{iso}}(\alpha) := \alpha \, | \, \phi_{00} \, \rangle\!\langle \, \phi_{00} \, | + \tfrac{1-\alpha}{d^2} \, \mathbb{1}_d \otimes \mathbb{1}_d, \text{ where}$$

$$\text{the real parameter } \alpha \text{ satisfies } -\tfrac{1}{d^2-1} \leq \alpha \leq 1 \text{ and } | \, \phi_{00} \, \rangle \text{ is the}$$

$$\text{maximally entangled state from eqn (15.95).}$$

The isotropic states form a one-parameter family of quantum states. For $0 \leq \alpha \leq 1$ they are a proper mixture of the maximally entangled state $| \, \phi_{00} \, \rangle$ and the maximally mixed state $\frac{1}{d^2}\mathbb{1}_d \otimes \mathbb{1}_d$ similar to the Werner states we encountered in Section 15.3.3, but instead of two qubits, we now have two d-dimensional quantum systems. For the negative parameter range $-\frac{1}{d^2-1} \leq \alpha < 0$ the operator $\rho_{\text{iso}}(\alpha)$ is not a proper mixture of the two aforementioned states, but nevertheless a valid state, i.e., a positive semidefinite operator. To see this, simply note that since $| \, \phi_{00} \, \rangle\!\langle \, \phi_{00} \, |$ is a pure state (i.e., represented by a rank-one projector), it has one eigenvalue $+1$ while all other eigenvalues vanish. Meanwhile, the identity operator $\mathbb{1}_d \otimes \mathbb{1}_d$ takes the same form in any basis, including an orthonormal basis containing the state $| \, \phi_{00} \, \rangle$, so we immediately see that the eigenvalues of $\rho_{\text{iso}}(\alpha)$ are $\frac{1-\alpha}{d^2}$ (with degeneracy $d^2 - 1$) and a single eigenvalue $\alpha + \frac{1-\alpha}{d^2}$. While the former are clearly all non-negative as long as $\alpha \leq 1$, the non-negativity of the latter single eigenvalue provides the range of allowed values for $\alpha \geq -\frac{1}{d^2-1}$.

States of this family are called *isotropic* in reference to their particular symmetry: They are invariant under all local unitaries of the form $U \otimes U^*$ (Horodecki and Horodecki, 1999).

> **Lemma 15.4 (Symmetry of isotropic states)**
>
> *A state $\rho \in \mathcal{B}_{\mathrm{HS}}(\mathcal{H}_d \otimes \mathcal{H}_d)$ is invariant under the transformation*
>
> $$\rho \longmapsto (U \otimes U^*)\rho(U \otimes U^*)^\dagger$$
>
> *for all $U \in U(d)$ **iff** it is an isotropic state, $\rho = \rho_{\mathrm{iso}}(\alpha)$ for some α.*

Proof Again, one direction of the proof is straightforward: $\mathbb{1} \otimes \mathbb{1}$ is obviously invariant under all unitaries. We can then note that for any local, linear operator $A = \sum_{j,k} A_{jk} \,|\, j \,\rangle\langle\, k \,|$, the maximally entangled state $|\, \phi_{00} \,\rangle$ satisfies

$$A \otimes \mathbb{1} \,|\, \phi_{00} \,\rangle = \sum_{j,k,l} A_{jk} \,|\, j \,\rangle\langle\, k \,|\, \otimes |\, l \,\rangle\langle\, l \,|\, \tfrac{1}{\sqrt{d}} \sum_m |\, m \,\rangle \otimes |\, m \,\rangle = \sum_{j,m} A_{jm} \,|\, j \,\rangle \otimes |\, m \,\rangle, \tag{15.109}$$

$$\mathbb{1} \otimes A^\top \,|\, \phi_{00} \,\rangle = \sum_{j,k,l} |\, l \,\rangle\langle\, l \,|\, \otimes A_{kj} \,|\, j \,\rangle\langle\, k \,|\, \tfrac{1}{\sqrt{d}} \sum_m |\, m \,\rangle \otimes |\, m \,\rangle = \sum_{j,m} A_{mj} \,|\, m \,\rangle \otimes |\, j \,\rangle, \tag{15.110}$$

such that one finds $A \otimes \mathbb{1} \,|\, \phi_{00} \,\rangle = \mathbb{1} \otimes A^\top \,|\, \phi_{00} \,\rangle$. Consequently, we have

$$U \otimes U^* \,|\, \phi_{00} \,\rangle\langle\, \phi_{00} \,|\, (U \otimes U^*)^\dagger = (\mathbb{1} \otimes U^*)(U \otimes \mathbb{1}) \,|\, \phi_{00} \,\rangle\langle\, \phi_{00} \,|\, (U \otimes \mathbb{1})^\dagger (\mathbb{1} \otimes U^*)^\dagger$$

$$= \mathbb{1} \otimes (UU^\dagger)^* \,|\, \phi_{00} \,\rangle\langle\, \phi_{00} \,|\, \mathbb{1} \otimes (UU^\dagger)^* = |\, \phi_{00} \,\rangle\langle\, \phi_{00} \,|, \tag{15.111}$$

so the isotropic states must be invariant under all transformations $U \otimes U^*$.

For the reverse direction, let us write a general bipartite density operator as $\rho = \sum_{m,n,m',n'} \rho_{mn,m'n'} \,|\, m \,\rangle\langle\, n \,|\, \otimes |\, m' \,\rangle\langle\, n' \,|$. If such a state is to be invariant under all $U \otimes U^*$, then it must be invariant for $U = \mathbb{1} - 2 \,|\, m_0 \,\rangle\langle\, m_0 \,|$ for all m_0, which are unitaries that each flip the sign of one particular basis vector. This eliminates all matrix elements except those of the form $\rho_{mm,nn}$, $\rho_{mn,mn}$, and $\rho_{mn,nm}$. Similarly, consider the unitaries $U = \mathbb{1} - (1 - i) \,|\, m_0 \,\rangle\langle\, m_0 \,|$, for which invariance under $U \otimes U^*$ implies that all $\rho_{mn,nm}$ must vanish except when $m = n$. When we make the same argument for two-element permutations of the form

$$U = \mathbb{1} - |\, m_0 \,\rangle\langle\, m_0 \,|\, - |\, n_0 \,\rangle\langle\, n_0 \,|\, + |\, m_0 \,\rangle\langle\, n_0 \,|\, + |\, n_0 \,\rangle\langle\, m_0 \,| \tag{15.112}$$

for all m_0 and n_0 we conclude that all elements of the form $\rho_{mm,nn}$ must be equal when $m \neq n$, all elements of the form $\rho_{mn,mn}$ must be equal when $m \neq n$, and also all $\rho_{mm,mm}$ must be equal. Any $U \otimes U^*$-invariant state must therefore be of the form

$$\rho = \alpha_1 \sum_{m \neq n} |\, mm \,\rangle\langle\, nn \,|\, + \alpha_2 \sum_{m \neq n} |\, mn \,\rangle\langle\, mn \,|\, + \alpha_3 \sum_m |\, mm \,\rangle\langle\, mm \,|, \tag{15.113}$$

and the α_i must be real since the density operator is Hermitian. However, when we replace the permutation in eqn (15.112) with a two-level rotation of the form

$$U = \mathbb{1} - (1 - \cos(\varphi))(|m_0\rangle\langle m_0| + |n_0\rangle\langle n_0|) + \sin(\varphi)(|m_0\rangle\langle n_0| - |n_0\rangle\langle m_0|)$$
$$\text{(15.114)}$$

we see that the last term in eqn (15.113) is not invariant on its own. The parameter α_3 must be linearly dependent on the other two, and the normalization condition $\mathrm{Tr}(\rho) = 1$ leaves only one free parameter. All $U \otimes U^*$-invariant states thus belong to a one-parameter family of states on which these states depend linearly. As we have seen previously, the family of isotropic states matches these conditions and all states invariant under $U \otimes U^*$ are therefore contained in the set of all $\rho_{\mathrm{iso}}(\alpha)$. $\qquad\square$

Separability of isotropic states: With these properties of the isotropic states established, let us now determine their range of separability via the reduction criterion of Theorem 15.11. Firstly, we calculate the reduced density matrix of Alice

$$\rho_{\mathrm{iso}}^{(A)}(\alpha) = \mathrm{Tr}_B(\rho_{\mathrm{iso}}(\alpha)) = \alpha\,\mathrm{Tr}_B(|\phi_{00}\rangle\langle\phi_{00}|) + \tfrac{1-\alpha}{d^2}\mathrm{Tr}_B(\mathbb{1}_d \otimes \mathbb{1}_d)$$
$$= \tfrac{\alpha}{d}\mathbb{1}_d + \tfrac{1-\alpha}{d}\mathbb{1}_d = \tfrac{1}{d}\mathbb{1}_d\,,\tag{15.115}$$

where we have made use of the fact that the reductions of the maximally entangled state $|\phi_{00}\rangle\langle\phi_{00}|$ give the maximally mixed state $\frac{1}{d}\mathbb{1}_d$ of the respective subsystem. For the reduction criterion we then calculate

$$\rho_{\mathrm{iso}}^{(A)}(\alpha) \otimes \mathbb{1}_d - \rho_{\mathrm{iso}}(\alpha) = \tfrac{1}{d}\mathbb{1}_d \otimes \mathbb{1}_d - \alpha|\phi_{00}\rangle\langle\phi_{00}| - \tfrac{1-\alpha}{d^2}\mathbb{1}_d \otimes \mathbb{1}_d$$
$$= \tfrac{\alpha+d-1}{d^2}\mathbb{1}_d \otimes \mathbb{1}_d - \alpha|\phi_{00}\rangle\langle\phi_{00}|_{\mathrm{diag}}\,.\tag{15.116}$$

As before, when we determined the range of α via the positivity of ρ_{iso}, we can make use of the fact that $|\phi_{00}\rangle\langle\phi_{00}|$ is a pure state that can be represented as a diagonal matrix with one eigenvalue 1 and all others equal to 0, and decompose the identity in an orthonormal basis that contains $|\phi_{00}\rangle$. Therefore the eigenvalues of expression (15.116) are

$$\lambda_1 = \tfrac{\alpha(1-d^2)+d-1}{d^2}\,,\qquad \lambda_2, ..., \lambda_d = \tfrac{\alpha+d-1}{d^2}\,.\tag{15.117}$$

The degenerate eigenvalues λ_i for $i = 2, \ldots, d$ are all positive in the entire range of α, but the eigenvalue λ_1 becomes negative when $\alpha > \frac{1}{d+1}$, which implies

$$\frac{1}{d+1} < \alpha \leq 1 \qquad\Rightarrow\qquad \rho_{\mathrm{iso}}(\alpha)\ \ \text{entangled}\,.\tag{15.118}$$

For the remaining parameter range, the reduction criterion alone does not specify if the states are separable or entangled, since the criterion is generally only necessary and sufficient for separability in 2×2 dimensions (or for 2×3 dimensions, which is irrelevant for the isotropic states since we have dimension $d \times d$ here).

However, in the case of the isotropic states this is not a shortcoming of the reduction criterion, indeed, the PPT criterion gives the same answer. Calculating the partial transpose of $\rho_{\mathrm{iso}}(\alpha)$ we find

$$\rho_{\mathrm{iso}}(\alpha)^{\mathsf{T}_B} = \frac{\alpha}{d} \sum_{m,n} |m\rangle\langle n| \otimes \left(|m\rangle\langle n|\right)^{\mathsf{T}} + \frac{1-\alpha}{d^2} \mathbb{1}_d \otimes \mathbb{1}_d^{\mathsf{T}}$$

$$= \frac{\alpha}{d} \sum_{m,n} |m\rangle\langle n| \otimes |n\rangle\langle m| + \frac{1-\alpha}{d^2} \mathbb{1}_d \otimes \mathbb{1}_d . \tag{15.119}$$

Careful inspection of this expression shows that $\rho_{\mathrm{iso}}(\alpha)^{\mathsf{T}_B}$ can be decomposed into a direct sum of a $d \times d$ diagonal matrix with entries $\frac{\alpha}{d} + \frac{1-\alpha}{d^2}$, corresponding to the cases where $m = n$ in the sum in eqn (15.119), and $(d-1)d/2$ blocks of size 2×2 with diagonal entries $\frac{1-\alpha}{d^2}$ and off-diagonal entries $\frac{\alpha}{d}$ for every pair (m,n) with $m < n$. These 2×2 matrices have eigenvalues $(1 \pm \alpha(d \mp 1))/d^2$ one of which is negative if $\alpha > 1/(d+1)$. We thus see that the PPT criterion is equivalent to the reduction criterion for all isotropic states.

To conclusively determine if the isotropic states for $\alpha \leq 1/(d+1)$ are separable (or not) it is useful to introduce a frequently used quantity in quantum information: The *singlet fraction* or *maximal fidelity* of a state ρ with respect to any maximally entangled, pure state $|\psi_{\mathrm{ent}}^{\mathrm{max}}\rangle$, which is defined as

$$\mathcal{F}_{\mathrm{max}}(\rho) := \max_{\psi_{\mathrm{ent}}^{\mathrm{max}}} \mathcal{F}\left(\rho, |\psi_{\mathrm{ent}}^{\mathrm{max}}\rangle\right) = \max_{\psi_{\mathrm{ent}}^{\mathrm{max}}} \langle \psi_{\mathrm{ent}}^{\mathrm{max}} | \rho | \psi_{\mathrm{ent}}^{\mathrm{max}} \rangle , \tag{15.120}$$

with $0 \leq \mathcal{F} \leq 1$. Physically, the fidelity $\mathcal{F}\left(\rho, |\psi_{\mathrm{ent}}^{\mathrm{max}}\rangle\right)$ expresses the probability that the state resulting from a projective measurement of ρ is $|\psi_{\mathrm{ent}}^{\mathrm{max}}\rangle$, suitably generalizing the notion of the overlap $|\langle\psi|\phi\rangle|^2$ of two pure states, and representing a special case of the more general *Uhlmann fidelity* of two arbitrary mixed states that we will encounter in Section 24.2.2.

Here, we are interested in the fidelity of a general isotropic state $\rho_{\mathrm{iso}}(\alpha)$ with the maximally entangled state $|\psi_{\mathrm{ent}}^{\mathrm{max}}\rangle \equiv |\phi_{00}\rangle$, for which the maximum is obtained in eqn (15.120) for the case of the isotropic states. The singlet fraction is therefore given by

$$\mathcal{F}\left(\rho, |\phi_{00}\rangle\right) = \langle \phi_{00} | \rho_{\mathrm{iso}}(\alpha) | \phi_{00} \rangle = \frac{1 + \alpha(d^2 - 1)}{d^2} . \tag{15.121}$$

We see that there is a one-to-one correspondence between the parameter α and the singlet fraction,

$$\alpha = \frac{d^2 \mathcal{F} - 1}{d^2 - 1} , \tag{15.122}$$

where we have abbreviated the singlet fraction as $\mathcal{F} \equiv \mathcal{F}(\rho, |\phi_{00}\rangle)$, and we will continue to use this simpler notation for the remainder of this chapter. Parameterizing ρ_{iso} in terms the fidelity $\mathcal{F}$ we then get

$$\rho_{\text{iso}}(\mathcal{F}) = \frac{d^2}{d^2 - 1} \left[\left(\mathcal{F} - \frac{1}{d^2} \right) | \phi_{00} \rangle\langle \phi_{00} | + \frac{1 - \mathcal{F}}{d^2} \mathbb{1}_d \otimes \mathbb{1}_d \right]. \tag{15.123}$$

Inserting α from eqn (15.122) into eqn (15.118) we find

$$\frac{1}{d} < \mathcal{F} \leq 1 \qquad \Rightarrow \qquad \rho_{\text{iso}}(\alpha) \text{ entangled .} \tag{15.124}$$

To understand how this is helpful to determine the full range of separability for the isotropic states, we note that since $| \phi_{00} \rangle$ is an isotropic state, we have $U \otimes U^* | \phi_{00} \rangle = | \phi_{00} \rangle$ for all unitaries U, and for any state ρ we therefore have

$$\mathcal{F}(\rho, | \phi_{00} \rangle) = \mathcal{F}(\rho, U \otimes U^* | \phi_{00} \rangle). \tag{15.125}$$

What is more, from Lemma 15.4 we see that for *any* (not necessarily isotropic) state $\rho \in \mathcal{B}_{\text{HS}}(\mathcal{H}_d \otimes \mathcal{H}_d)$, the state $\tilde{\rho}$ obtained by the so-called *twirling operation*

$$\tilde{\rho} = \int dU \, U \otimes U^* \rho \, (U \otimes U^*)^\dagger, \tag{15.126}$$

corresponding to an equally weighted (according to the Haar measure of the unitary group $U(\text{d})$) mixture of all states subject to the local unitary transformation $U \otimes U^*$, must be an isotropic state, since it is invariant under all $U \otimes U^*$. Combining this with eqn (15.125) we have

$$\mathcal{F}(\tilde{\rho}, | \phi_{00} \rangle) = \int dU \, \langle \phi_{00} | \, U \otimes U^* \rho \, (U \otimes U^*)^\dagger \, | \phi_{00} \rangle = \mathcal{F}(\rho, U^\dagger \otimes U^{*\dagger} | \phi_{00} \rangle)$$

$$= \mathcal{F}(\rho, | \phi_{00} \rangle). \tag{15.127}$$

All of this must be true also if we consider the (obviously) separable product state $\rho = | 0 \rangle\langle 0 | \otimes | \psi \rangle\langle \psi |$ with $| \psi \rangle = a | 0 \rangle + \sqrt{1 - a^2} | 1 \rangle$ for $0 \leq a \leq 1$, where $| 0 \rangle$ and $| 1 \rangle$ can be the first two basis states in any d-dimensional Hilbert space. In that case we have

$$\mathcal{F}(\rho, | \phi_{00} \rangle) = \frac{a^2}{d} . \tag{15.128}$$

But since this product state is separable and the corresponding isotropic state $\tilde{\rho}$ is obtained via local unitaries from it, also $\tilde{\rho}$ must be separable for any value of a. In other words, for any $\mathcal{F} \leq \frac{1}{d}$, the corresponding isotropic state can be obtained by twirling a product state, and must hence be separable.

We can therefore finally conclude that the reduction criterion and the PPT criterion are not just equivalent for isotropic states, but they are also necessary and sufficient for the separability of isotropic states,

$$\frac{1}{d} < \mathcal{F} \leq 1 \qquad \Longleftrightarrow \qquad \rho_{\text{iso}}(\alpha) \text{ entangled ,} \tag{15.129}$$

$$0 \leq \mathcal{F} \leq \frac{1}{d} \qquad \Longleftrightarrow \qquad \rho_{\text{iso}}(\alpha) \text{ separable .}$$

Note that for infinite dimensions, as $d \to \infty$, the set of separable isotropic states is of measure zero, essentially all isotropic states are entangled.

15.5 Geometry of Two-Qubit Quantum States

With what we have learned in the previous sections about entanglement, its description and its detection, let us now re-examine the geometric picture we have discussed in Section 15.2.1: The set of quantum states is convex and contains the set of separable states as a convex subset. We now want to describe this structure in more detail for the case of two qubits, where necessary and sufficient conditions for separability are available.

15.5.1 Weyl States and Their Geometric Representation

In 2×2 dimensions, the case of two qubits, we can use the previously established criteria to clearly characterize different types of correlations: separability and entanglement, locality and non-locality. Moreover, despite the fact that the Hilbert space is a complex four-dimensional vector space it turns out that one can provide an intuitive three-dimensional visualization of the state-space geometry when restricting to the set $\mathcal{W}$ of *locally maximally mixed states*, also called *Weyl states*.

Definition 15.13 *A **Weyl state** $\rho_{\mathrm{Weyl}} \in \mathcal{B}_{\mathrm{HS}}(\mathcal{H}_2 \otimes \mathcal{H}_2)$ is a two-qubit state whose marginals are maximally mixed, $\rho_A = \mathrm{Tr}_B(\rho_{\mathrm{Weyl}}) = \frac{1}{2}\mathbb{1}_2 = \rho_B$.*

Using the two-qubit Bloch decomposition from eqn (15.39), this just translates to the condition of vanishing Bloch vectors $a_i = b_j = 0 \;\; \forall i, j$. The set $\mathcal{W}$ contains all the maximally entangled states, in particular, the Bell states $\rho^{\pm}$ and $\omega^{\pm}$ from eqns (11.113a) and (11.113b), respectively, as well as the completely uncorrelated, maximally mixed state proportional to the identity in four dimensions $\rho_{\mathrm{mix}} = \frac{1}{4}\mathbb{1}_4 = \frac{1}{4}\mathbb{1}_A \otimes \mathbb{1}_B$. Consequently, the set $\mathcal{W}$ of Weyl states contains all convex combinations of maximally entangled and maximally mixed two-qubit states, including the one-parameter families of states that we have discussed before, the Werner states ρ_{Werner} from Section 15.3.3 and the two-qubit isotropic states ρ_{iso} from the previous section.

All states in $\mathcal{W}$ can be written in Bloch decomposition as

$$\rho = \frac{1}{4}\left[\mathbb{1}_A \otimes \mathbb{1}_B + \sum_{i,j} t_{ij}\, \sigma_i^{(A)} \otimes \sigma_j^{(B)} \right], \qquad (15.130)$$

and are thus fully determined by their correlation tensors $T = (t_{ij})$. We can now use a trick to further compress the information that we are interested in. That is, we are interested in capturing the geometry of entanglement versus separability. For this purpose, states with the same "amount" of entanglement can be identified. We have not yet talked about the quantification of entanglement in detail (but will do so in Chapter 16), but for our purposes here, it is enough to recall Corollary 15.1 and the related discussion, where we have noted that local unitary transformations do not change if or how strongly a state is entangled. This insight can be leveraged to obtain a more compact description of the Weyl states. If we apply a local unitary $U \otimes V$ to a Weyl state in generalized Bloch decomposition, we have

$$U \otimes V \rho U^\dagger \otimes V^\dagger = \frac{1}{4}\Big[\mathbb{1}_A \otimes \mathbb{1}_B + \sum_{i,j} t_{ij}\, U\sigma_i^{(A)}U^\dagger \otimes V\sigma_j^{(B)}V^\dagger \Big], \qquad (15.131)$$

because the unitaries leave the identity term invariant. In particular, let us consider the local unitaries U and V with determinant $+1$. In other words, we assume that U and V belong to the group SU(2), the group of unitary 2×2 matrices with $\det(U) = +1$, see Section 8.2.5.

We now want to use the special relationship between the group SU(2) and the group SO(3) of rotations in three dimensions, i.e., the group of real, orthogonal 3×3 matrices R with $\det(R) = 1$: From Theorem 8.2 we know that for every $R \in$ SO(3) there exists a[6] unitary $U \in$ SU(2) such that $U\vec{\sigma}U^\dagger = R\vec{\sigma}$. Without loss of generality we can associate the rotation matrices $R_1^\top$ and R_2 with the local unitaries U and V, respectively, and use their relation for the Weyl states by writing

$$\sum_{i,j} t_{ij}\, U\sigma_i U^\dagger \otimes V\sigma_j V^\dagger = \sum_{i,j} t_{ij}\, (R_1^\top \vec{\sigma})_i \otimes (R_2\vec{\sigma})_j$$

$$= \sum_{i,j} t_{ij} \Big(\sum_l (R_1^\top)_{il}\sigma_l\Big) \otimes \Big(\sum_k (R_2)_{jk}\sigma_k\Big) = \sum_{i,j,k,l} (R_1^\top)_{il}\, t_{ij}\, (R_2)_{jk}\sigma_l \otimes \sigma_k$$

$$= \sum_{k,l} \sum_{i,j} (R_1)_{li}\, t_{ij}\, (R_2)_{jk}\sigma_l \otimes \sigma_k = \sum_{k,l} (R_1 T R_2)_{lk}\, \sigma_l \otimes \sigma_k. \qquad (15.132)$$

Then we note that the correlation tensor T is a real 3×3 matrix, and can therefore be brought to diagonal form by orthogonal transformations, that is, we can make use of the singular-value decomposition (recall Section 15.1) of T (recall Section 15.1): $\exists O_1, O_2 \in O(3):\quad O_1 T O_2 = \Sigma = \mathrm{diag}\{s_i\}$. Now, the matrices O_1 and O_2 for a given T are orthogonal 3×3 matrices, but need not have determinant 1, that is, they are in $O(3)$ but need not be in $SO(3)$. Being orthogonal just implies that $\det(O_{1,2}) = \pm 1$, but for the connection to SU(2) to work, we need matrices in SO(3). This can be arranged. Since $\det(AB) = \det(A)\det(B)$ and $\det(-\mathbb{1}_3) = -1$, this means that for any given $O \in O(3)$ we can define $R := -\mathbb{1}_3 O \in$ SO(3). In fact, instead of $-\mathbb{1}_3$ we could also have multiplied by other diagonal 3×3 matrices with determinant -1, such as $\mathrm{diag}\{-1, +1, +1\}$, $\mathrm{diag}\{+1, -1, +1\}$, or $\mathrm{diag}\{+1, +1, -1\}$, to obtain an element of SO(3). This means that for any $O \in O(3)$ there exists a diagonal 3×3 matrix D with diagonal entries ± 1 and $\det(D) = \det(O)$ such that $DO, OD \in$ SO(3). In practice, this means that we can get a statement similar to the singular-value decomposition, but with matrices in the group SO(3), which results in a diagonal matrix with real but not necessarily non-negative entries:

$$\exists R_1, R_2 \in SO(3):\quad R_1 T R_2 = D_1 O_1 T O_2 D_2 = D_1 \Sigma D_2 = \tilde{T} = \mathrm{diag}\{\tilde{t}_1, \tilde{t}_2, \tilde{t}_3\}.$$

If we combine this with the statement of eqn (15.132), we see that the correlation tensor of any state can be brought to diagonal form by local unitaries: There exist unitaries U and V, corresponding to $R_1^\top$ and R_2, such that

---

[6]Actually, there are two unitaries for every $R \in$ SO(3): U and $-U$. The group SU(2) is the universal double cover of SO(3).

$$\sum_{k,l}(R_1 T R_2)_{lk}\, \sigma_l \otimes \sigma_k \; = \; \sum_{k,l}\tilde{t}_{lk}\sigma_l \otimes \sigma_k \; = \; \sum_{k,l}\delta_{lk}\tilde{t}_k\sigma_l \otimes \sigma_k \; = \; \sum_{k}\tilde{t}_k\sigma_k \otimes \sigma_k\,.$$

$$(15.133)$$

This is true for any two-qubit state, but for Weyl states the Bloch vectors vanish, so we see that up to local unitaries, every Weyl states can be written as

$$\rho = \tfrac{1}{4}\Big[\mathbb{1}_A \otimes \mathbb{1}_B + \sum_i \tilde{t}_i\, \sigma_i^{(A)} \otimes \sigma_i^{(B)}\Big].$$

$$(15.134)$$

Thus, for any Weyl state, we can describe the properties important for the characterization of entanglement using only three real parameters, just the right number for making an illustration by interpreting the $\tilde{r}_i$ as the components of a vector in $\mathbb{R}^3$.

The vector components $\tilde{t}_i$ all lie in the interval $-1 \leq \tilde{t}_i \leq +1$ since they are obtained as expectation values of (tensor products of) Pauli operators. Additionally, they are constrained by the positivity of ρ. Inserting for the Pauli matrices from eqn (11.71), we have (up to local unitaries) the form

$$\rho \cong \tfrac{1}{4}\begin{pmatrix} 1+\tilde{t}_3 & 0 & 0 & \tilde{t}_1-\tilde{t}_2 \\ 0 & 1-\tilde{t}_3 & \tilde{t}_1+\tilde{t}_2 & 0 \\ 0 & \tilde{t}_1+\tilde{t}_2 & 1-\tilde{t}_3 & 0 \\ \tilde{t}_1-\tilde{t}_2 & 0 & 0 & 1+\tilde{t}_3 \end{pmatrix},$$

$$(15.135)$$

whose eigenvalues can easily be computed for the two independent two-by-two blocks

$$\tfrac{1}{4}\begin{pmatrix} 1+\tilde{t}_3 & \tilde{t}_1-\tilde{t}_2 \\ \tilde{t}_1-\tilde{t}_2 & 1+\tilde{t}_3 \end{pmatrix} \quad \text{and} \quad \tfrac{1}{4}\begin{pmatrix} 1-\tilde{t}_3 & \tilde{t}_1+\tilde{t}_2 \\ \tilde{t}_1+\tilde{t}_2 & 1-\tilde{t}_3 \end{pmatrix},$$

$$(15.136)$$

with the result

$$\lambda = \tfrac{1}{4}\begin{cases} 1+\tilde{t}_1-\tilde{t}_2+\tilde{t}_3 \\ 1+\tilde{t}_1+\tilde{t}_2-\tilde{t}_3 \\ 1-\tilde{t}_1-\tilde{t}_2-\tilde{t}_3 \\ 1-\tilde{t}_1+\tilde{t}_2+\tilde{t}_3 \end{cases}.$$

$$(15.137)$$

For each of these four eigenvalues we require $\lambda \geq 0$, with equality ($\lambda = 0$) defining four planes that form a tetrahedron illustrated in Fig. 15.3 (a). All valid Weyl states are represented by points within the convex space enclosed by these four planes, and the four corners of the tetrahedron correspond to the four Bell states,

$$|\phi^\pm\rangle = \begin{pmatrix} \tilde{t}_1 \\ \tilde{t}_2 \\ \tilde{t}_3 \end{pmatrix} = \begin{pmatrix} \pm 1 \\ \mp 1 \\ 1 \end{pmatrix}, \qquad |\psi^\pm\rangle = \begin{pmatrix} \tilde{t}_1 \\ \tilde{t}_2 \\ \tilde{t}_3 \end{pmatrix} = \begin{pmatrix} \pm 1 \\ \pm 1 \\ -1 \end{pmatrix},$$

$$(15.138)$$

while the maximally mixed state $\rho_{\mathrm{mix}} = \tfrac{1}{4}\mathbb{1}_4$ is located at the origin $(\tilde{t}_1,\tilde{t}_2,\tilde{t}_3) = (0,0,0)$.

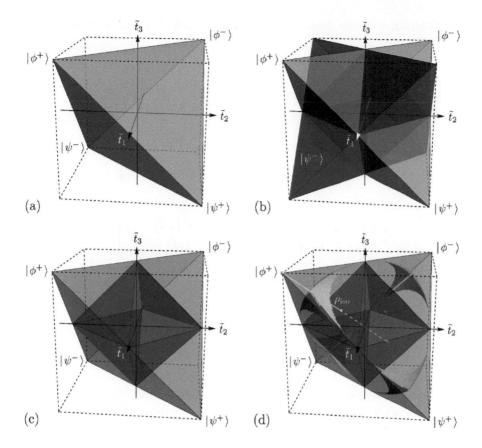

Fig. 15.3 Tetrahedron of Weyl states $\mathcal{W}$. All states in the set $\mathcal{W}$ can (up to local unitaries) be geometrically represented as points within a tetrahedron spanned by the four Bell states $|\psi^{\pm}\rangle$, $|\phi^{\pm}\rangle$. The singular values $(\tilde{t}_1, \tilde{t}_2, \tilde{t}_3)$ of the correlation tensor serve as coordinates. (b) The positivity of the partial transpose defines another tetrahedron (blue) whose intersection with the positivity tetrahedron of the original states forms a double pyramid (c), the set $\mathcal{S}$ of separable states, while all points in the remaining corners outside (green) represent entangled states. The maximally mixed state ρ_{mix} is located at the origin. (d) All local states that cannot violate the CHSH–Bell inequality (15.55) lie within the dark-yellow "parachutes" drawn outside the double pyramid, which include also some entangled states. The red line indicates the Werner states ρ_{Werner} of eqn (15.68), while the yellow line represents the family of isotropic two-qubit states ρ_{iso} from Section 15.4.4, with the solid and dashed parts of each line indicating the entangled and separable states of each family, respectively.

15.5.2 Entanglement and Separability of Weyl States

Within the tetrahedron of Weyl states, the region of separability can be determined by the PPT criterion (Theorem 15.3), since the latter is necessary and sufficient for two qubits (Theorem 15.4). It hence allows us to identify the bipartite quantum states as separable if the partial transposition of their density operator yields a positive operator. Starting with the density operator in eqn (15.135) we calculate the partial transpose

$$
\rho^{\mathsf{T}_B} \cong \frac{1}{4}
\begin{pmatrix}
1 + \tilde{t}_3 & 0 & 0 & \tilde{t}_1 + \tilde{t}_2 \\
0 & 1 - \tilde{t}_3 & \tilde{t}_1 - \tilde{t}_2 & 0 \\
0 & \tilde{t}_1 - \tilde{t}_2 & 1 - \tilde{t}_3 & 0 \\
\tilde{t}_1 + \tilde{t}_2 & 0 & 0 & 1 + \tilde{t}_3
\end{pmatrix}.
\tag{15.139}
$$

Again, the eigenvalues can easily be computed for two independent two-by-two blocks, given by

$$
\frac{1}{4}
\begin{pmatrix}
1 + \tilde{t}_3 & \tilde{t}_1 + \tilde{t}_2 \\
\tilde{t}_1 + \tilde{t}_2 & 1 + \tilde{t}_3
\end{pmatrix}
\quad \text{and} \quad
\frac{1}{4}
\begin{pmatrix}
1 - \tilde{t}_3 & \tilde{t}_1 - \tilde{t}_2 \\
\tilde{t}_1 - \tilde{t}_2 & 1 - \tilde{t}_3
\end{pmatrix},
\tag{15.140}
$$

from which we obtain the four eigenvalues

$$
\mu = \frac{1}{4}
\begin{cases}
1 + \tilde{t}_1 + \tilde{t}_2 + \tilde{t}_3 \\
1 + \tilde{t}_1 - \tilde{t}_2 - \tilde{t}_3 \\
1 - \tilde{t}_1 + \tilde{t}_2 - \tilde{t}_3 \\
1 - \tilde{t}_1 - \tilde{t}_2 + \tilde{t}_3
\end{cases}.
\tag{15.141}
$$

Once more the positivity of the four eigenvalues defines four planes that enclose a tetrahedron. Geometrically, the partial transposition corresponds to a $\frac{\pi}{2}$ rotation of the original positivity tetrahedron around the $\tilde{t}_3$ axis. The intersection of this ro-tated set of states $(\mathbb{1}_A \otimes \mathsf{T}_B)\,\rho$—the rotated tetrahedron—with the original tetrahedron of the set $\mathcal{W}$ of Weyl states has the form of a *double pyramid* with corners at $(\tilde{t}_1, \tilde{t}_2, \tilde{t}_3) = (\pm 1, 0, 0)$, $(0, \pm 1, 0)$ and $(0, 0, \pm 1)$, representing the set of separable states (see Fig. 15.3, as discussed, e.g., in Horodecki and Horodecki, 1996; Vollbrecht and Werner, 2001; Bertlmann *et al.*, 2002).

In addition to the distinction between separability and entanglement, we can use the CHSH criterion from Theorem 15.6 to numerically identify the states which can violate the CHSH–Bell inequality (15.55), at least for the case where only projective measurements are considered. All states that lie within the dark-yellow parachutes drawn outside the double pyramid of separability are entangled but do not violate the CHSH inequality for any measurement directions $\vec{a}$, $\vec{a}'$, $\vec{b}$, and $\vec{b}'$.

Moreover, Fig. 15.3 (d) shows the Werner states $\rho_{\text{Werner}}(\alpha)$ from eqn (15.68) in red, and the isotropic two-qubit states ρ_{iso} from Section 15.4.4 in yellow, with the separable and entangled regions of these one-parameter families of states indicated as dashed and solid lines respectively. Each of these families connects a maximally entangled state for the parameter value $\alpha = 1$, $|\psi^-\rangle$ in the case of the Werner states, and $|\phi_{00}\rangle \equiv |\phi^+\rangle$ in case of the isotropic states, with the maximally mixed state at the origin (in both cases for the parameter $\alpha = 0$), but both families continue onwards towards the respective opposite surfaces of the separable double pyramid for $\alpha = -\frac{1}{3}$. It shows nicely how the states change from maximally mixed and separable—within the blue double pyramid $(\alpha \leq \frac{1}{3})$—to local and mixed entangled—within the dark-yellow parachutes $(\alpha \leq 1/\sqrt{2})$—and finally to non-local and mixed entangled states, ending at pure, maximally entangled states.

The distance from the set of separable states, or (here) conversely, from the closest of the four Bell states presents itself as an intuitive descriptor of the entanglement contained in the respective Weyl states. Indeed, the idea translates beyond Weyl states, and one can easily see that all separable two-qubit states have a maximal fidelity of $\frac{1}{2}$ with any of the four Bell states, or indeed any maximally entangled pure state, i.e., the *singlet fraction* $\mathcal{F}_{\max}(\rho)$ from eqn (15.120) is bounded from above by $\frac{1}{2}$ for any separable state. To see this we inspect the generalized Bloch decomposition from eqn (15.39) for the four Bell states, whose Bloch vectors vanish and whose correlation tensors, given by eqn (15.67), are diagonal with eigenvalues ± 1. Since the Pauli matrices are traceless but square to the identity, we can define bit values $b_i \in \{0, 1\}$ that define each of the four Bell states $\rho_{\mathrm{Bell}} = |\psi_{\mathrm{Bell}}\rangle\langle\psi_{\mathrm{Bell}}|$ such that we can write

$$\mathcal{F}(\rho, \rho_{\mathrm{Bell}}) = \langle\psi_{\mathrm{Bell}}| \rho |\psi_{\mathrm{Bell}}\rangle = \mathrm{Tr}(\rho\,\rho_{\mathrm{Bell}}) \tag{15.142}$$

$$= \tfrac{1}{4}\Big[\mathrm{Tr}(\mathbb{1}\,\rho) + \sum_i (-1)^{b_i}\,\mathrm{Tr}\big(\rho\,\sigma_i^{(A)} \otimes \sigma_i^{(B)}\big)\Big] \leq \tfrac{1}{4}\Big[1 + \sum_i |\langle\sigma_i^{(A)} \otimes \sigma_i^{(B)}\rangle|\Big].$$

For any pure separable state $|\psi\rangle_A |\phi\rangle_B$, we have $\langle\sigma_i^{(A)} \otimes \sigma_i^{(B)}\rangle = \langle\sigma_i^{(A)}\rangle\,\langle\sigma_i^{(B)}\rangle$ such that

$$\sum_i |\langle\sigma_i^{(A)} \otimes \sigma_i^{(B)}\rangle| = \sum_i |\langle\sigma_i^{(A)}\rangle||\langle\sigma_i^{(B)}\rangle| \leq \sqrt{\sum_i |\langle\sigma_i^{(A)}\rangle|^2}\sqrt{\sum_j |\langle\sigma_j^{(B)}\rangle|^2} \leq 1, \tag{15.143}$$

where we have used the Cauchy–Schwarz inequality and the fact that Bloch vectors are bounded in norm by 1 for any state. For any pure separable state, the fidelity with any of the Bell states is thus bounded from above by $\frac{1}{2}$. The same is true for the fidelity with any maximally entangled (pure) two-qubit state since any of these can be obtained from the Bell states via local unitary operations, which would only correspond to a rotation of the Pauli matrices in (15.143). Since mixed separable state can be written as a convex combination of pure separable states, and because the Bell states form a subset of all maximally entangled two-qubit states, we have $\mathcal{F}(\rho_{\mathrm{sep}}, \rho_{\mathrm{Bell}}) \leq \mathcal{F}_{\max}(\rho_{\mathrm{sep}}) \leq \frac{1}{2}$. We can therefore conclude that

$$\mathcal{F}_{\max}(\rho) > \tfrac{1}{2} \quad\Longrightarrow\quad \rho \text{ entangled}. \tag{15.144}$$

Note that the reverse is not true in general (Verstraete and Verschelde, 2002); there are entangled two-qubit states that are not in the family of isotropic (or Werner) states for which the singlet fraction satisfies $\mathcal{F}_{\max}(\rho) \leq \frac{1}{2}$. We will come back to the idea of using distances to the set of separable states as a means of entanglement quantification in Section 16.2.3, while the idea of detecting entanglement using expectation values of particular operators (so-called "entanglement witnesses") will be examined more closely in Section 16.3.

Résumé: We have discussed how pure and mixed quantum states can be classified as being either separable or entangled, local and non-local, introducing a generalized Bloch decomposition in the process. We have formulated practically relevant conditions that are necessary for the separability of quantum states, including the PPT criterion or the reduction criterion, but which are generally only sufficient for special cases such as low dimensions (2×2 and 2×3) or restricted families of quantum states (such as the Werner or isotropic states). With the positive-map theorem, we have also formulated criteria that are generally necessary and sufficient, but hardly more practical than the very definition of separability itself.

We have also more closely inspected the interesting geometry of bipartite states in the Hilbert–Schmidt space of density matrices. There the tetrahedron of Weyl states, a simplex spanned by the four Bell states, has been investigated in detail. It contains the double pyramid of separable states, where the maximally mixed state is located at the origin, as well as the families of the Werner and (two-qubit) isotropic states. All local states, satisfying the CHSH inequality (when restricting to projective measurements), are enclosed by surfaces reminiscent of parachutes outside of the set of separable states.

16
Quantification and Conversion of Entanglement

In this chapter we aim to further deepen our understanding of entanglement and separability in bipartite quantum systems by studying methods for quantifying entanglement in Section 16.1. Although the resulting *entanglement measures* do not capture all properties of quantum states, especially with regards to the potential conversion of states into each other via local operations and classical communication (LOCC), some of them can nevertheless be useful for assessing the expected performance of certain quantum-information processing tasks, at least in the asymptotic sense when infinitely many copies are processed. At the same time, some measures allow us to appreciate a richer geometrical structure of the state space in contrast to the purely dichotomic distinction between entanglement and separability that we have made in the previous chapter. Based on these insights we will discuss so-called *entanglement witnesses* in Section 16.3 which provide means of detecting entanglement that are both practically measurable and can be related to certain measures of entanglement. In Section 15.3.5 we then take a closer look at the effect of state conversion via (non-deterministic) LOCC on non-locality. Finally, we consider situations not restricted to local operations and discuss the distinction between entanglement and separability from a purely mathematical point of view as a choice of factorization of a tensor-product space, where different factorizations are related by global unitary transformations, which we illustrate with several examples in Section 16.4.1.

16.1 Quantifying Entanglement

So far we have discussed the problem of deciding whether or not a given quantum state, represented by a density operator, is separable or entangled. This in itself is already a hard problem. Nevertheless, this information alone is not always enough to decide how useful a quantum system is for performing particular quantum information-processing tasks, such as, e.g., quantum teleportation (see Section 14.1) or entanglement swapping (see Section 14.3.1). In practice a quantum state is not always an ideal pure state but is mixed due to the interaction with the environment or simply because of a lack of information. It is therefore of great interest to know to what extent a quantum state still remains entangled enough to perform certain tasks—generally speaking, how strongly entangled the state is. That is, we would like to know what we can do with the entanglement contained in the state, and what we cannot do. The amount of entanglement can be quantified by so-called *entanglement measures*.

To identify suitable measures, we have to ask ourselves what we aim to capture with a measure of entanglement. We could resort to just saying that we would like to have some sort of hierarchy between entangled states, so that, given two states, we can say things like "state ρ_1 is more entangled than state ρ_2", but ultimately it would be better if "more entangled" also generally meant "more useful for tasks that require entanglement", for instance, for converting "more entangled" to "less entangled" states. Unfortunately, the situation is much more complicated than this, as we will briefly discuss in this chapter. There are many different types of entanglement measures with different properties, but none that have all desired properties: Some have a clear operational meaning but are extremely difficult to calculate, some can be easily calculated, but lack a clear operational meaning. Some tell us how "expensive" a given state is to prepare, others tell us how we can use the entanglement contained in a state. Generally, most measures are inequivalent. In particular, as far as is known today, there is no unique order (hierarchy) between entangled states. Mathematically, some of the questions involved are very challenging and are open problems. In this chapter, we want to get a qualitative overview of some important measures and apply them to simple examples, where possible, and we refer the interested reader to the extensive further literature on the subject (e.g., Plenio and Virmani, 2007; Gühne and Tóth, 2009; Horodecki *et al.*, 2009; Eltschka and Siewert, 2014).

16.1.1 Entropy of Entanglement—Quantifying Pure-State Entanglement

Once again we begin our examination with the pure-state case for which one can easily define a very useful measure that can be calculated for all pure states of the Hilbert space. The basis for this measure of entanglement is the Schmidt-decomposition theorem (Theorem 15.1) from Section 15.1.2, from which we have learned that for any pure bipartite state $|\psi\rangle_{AB} \in \mathcal{H}_{AB} = \mathcal{H}_A \otimes \mathcal{H}_B$ with density operator $\rho_{AB} = |\psi\rangle\langle\psi|_{AB}$ the two reduced states $\rho_A = \mathrm{Tr}_B(\rho_{AB})$ and $\rho_B = \mathrm{Tr}_A(\rho_{AB})$ have the same eigenvalues (Lemma 15.1), and the state is entangled if and only if more than one of these eigenvalue is non-zero (Corollary 15.2). In other words, if the reduced states are pure, then the pure bipartite state is separable, and if the reduced states are (maximally) mixed, then the joint state is (maximally) entangled. We can then extend this idea to provide a continuous measure of entanglement for the joint state by considering a continuous measure of the mixedness of the reduced states. In principle this could be the *mixedness* or *linear entropy* from eqn (11.21) but it is more common to use a different entropy measure instead, the *von Neumann entropy* of a density matrix ρ.

Definition 16.1 *The **von Neumann entropy** of a density matrix ρ is defined as*

$$S(\rho) := -\mathrm{Tr}\big(\rho\,\log(\rho)\big).$$

Here we have left the base of the logarithm unspecified for now, depending on the context it is customary to use either the natural logarithm ln (base e), the logarithm to base 2, $\log_2$, or to use the logarithm to base d in the case of matching subsystem dimensions $d_A = d_B = d$. As we will discuss in more detail in Section 20.1 the entropy $S(\rho)$ can be associated with the lack of information about the quantum state ρ, and thus represents a measure of its mixedness. When calculating the von Neumann entropy we have to take the trace of a matrix, which is a procedure independent of the chosen basis. Therefore we can always choose an eigenbasis such that

$$S(\rho) = -\operatorname{Tr}(\rho \log(\rho)) = -\sum_i p_i \log(p_i), \qquad (16.1)$$

where the set $\{p_i\}_i$ comprises the eigenvalues of the (density) matrix ρ. With this, one can define the following pure-state entanglement measure (Bennett *et al.*, 1996*a*; Bennett *et al.*, 1996*c*):

Definition 16.2   *The **entropy of entanglement** of a pure state $\rho_{AB} = |\psi\rangle\langle\psi|_{AB}$*

is defined as $\mathcal{E}(\rho_{AB}) := S(\rho_A) = S(\rho_B)$.

For a pure product state, the reduced states are also pure states with eigenvalues 0 and 1, such that $S(\rho_{\text{pure}}) = 0$ and hence $\mathcal{E}(|\psi\rangle_A |\phi\rangle_B) = 0$. For a maximally entangled state—for instance, the Bell states $\rho^\pm$ and $\omega^\pm$ from eqns (11.113a) and (11.113b), respectively, for two qubits with local dimensions $d_A = d_B = 2$—the reduced states of both subsystems are maximally mixed. For general but matching local dimensions $d_A = d_B = d$, examples for maximally entangled states are the states $|\phi_{mn}\rangle$ from eqn (14.11), and the maximally mixed reduced states are given by $\rho_A = \rho_B = \frac{1}{d}\mathbb{1}_d$, with the von Neumann entropy

$$S(\rho_A) = S(\rho_B) = -\sum_{i=1}^{d} \tfrac{1}{d}\log(\tfrac{1}{d}) = \tfrac{1}{d}\sum_{i=1}^{d}\big(\log(d) - \log(1)\big) = \log(d). \qquad (16.2)$$

Thus, if we choose the logarithm to base d, we get the maximal value $\log_d(d) = 1$ and hence the entanglement range $0 \leq \mathcal{E} \leq 1$. We also know that the measure $\mathcal{E}$ is *continuous* in this range in the sense that for any two (pure) states ρ_1 and ρ_2 whose Hilbert–Schmidt distance vanishes, $D_{\text{HS}}(\rho_1, \rho_2) \to 0$ it follows from eqn (11.97d) that $\rho_1 - \rho_2 \to 0$ and hence that $\mathcal{E}(\rho_1) - \mathcal{E}(\rho_2) \to 0$.

Further using Corollary 15.1 of the Schmidt-decomposition theorem (Theorem 15.1), we immediately see that the entropy of entanglement is *invariant under local unitary transformations*,

$$\mathcal{E}(U \otimes V \rho_{AB} U^\dagger \otimes V^\dagger) := S(U\rho_A U^\dagger) = S(\rho_A) = \mathcal{E}(\rho_{AB}), \qquad (16.3)$$

since the von Neumann entropy, being a function of the eigenvalues only, is invariant under unitaries. One can make an even more general statement when considering arbitrary local operations and classical communication (LOCC), as we will see in Section 16.1.2, for any probabilistic LOCC transformation that takes a pure state $|\psi\rangle_{AB}$ as input and results in a state $|\psi_j\rangle_{AB}$ from an ensemble of pure output states with probability p_j, in which case the entropy of entanglement must decrease on average

$$\mathcal{E}(|\psi\rangle_{AB}) \geq \sum_i p_i\, \mathcal{E}(|\phi_i\rangle_{AB}). \qquad (16.4)$$

Another important property of the entropy of entanglement is its *additivity*. That is, for any two pure states $|\psi\rangle_{AB} \in \mathcal{H}_{AB} = \mathcal{H}_A \otimes \mathcal{H}_B$ and $|\phi\rangle_{A'B'} \in \mathcal{H}_{A'B'} = \mathcal{H}_{A'} \otimes \mathcal{H}_{B'}$ we can calculate the entropy of entanglement for the product state $|\psi\rangle_{AB} \otimes |\phi\rangle_{A'B'}$ with respect to the bipartition $AA'|BB'$ and find

$$\mathcal{E}\big(|\psi\rangle_{AB} \otimes |\phi\rangle_{A'B'}\big) = \mathcal{E}\big(|\psi\rangle_{AB}\big) + \mathcal{E}\big(|\phi\rangle_{A'B'}\big). \tag{16.5}$$

This can be seen in the following way: For general product states we have

$$\mathrm{Tr}_{BB'}(\rho_{AB} \otimes \rho_{A'B'}) = \mathrm{Tr}_B(\rho_{AB}) \otimes \mathrm{Tr}_{B'}(\rho_{A'B'}) = \rho_A \otimes \rho_{A'}. \tag{16.6}$$

Now let λ_i and μ_i be the eigenvalues of ρ_A and $\rho_{A'}$, respectively, then the eigenvalues of the product state $\rho_A \otimes \rho_{A'}$ are just the products $\lambda_i \mu_j$ and

$$S(\rho_A \otimes \rho_{A'}) = -\sum_{i,j} \lambda_i \mu_j \log(\lambda_i \mu_j) = -\sum_{i,j} \lambda_i \mu_j \big(\log(\lambda_i) + \log(\mu_j)\big) \tag{16.7}$$

$$= -\sum_j \mu_j \sum_i \lambda_i \log(\lambda_i) - \sum_i \lambda_i \sum_j \mu_j \log(\mu_j) = S(\rho_A) + S(\rho_{A'}).$$

The additivity of eqn (16.5) follows, and as a consequence N copies of any state $|\psi\rangle_{AB}$ are exactly N times as entangled (as measured by $\mathcal{E}$) as a single copy.

As we have seen in eqn (16.3), the additivity relation must also hold if we perform local unitaries jointly on A and A' or on B and B', since this has no effect on the entanglement with respect to the bipartition $AA'|BB'$,

$$\mathcal{E}\big(U_{AA'} \otimes V_{BB'} |\psi\rangle_{AB} \otimes |\phi\rangle_{A'B'}\big) = \mathcal{E}\big(|\psi\rangle_{AB} \otimes |\phi\rangle_{A'B'}\big) = \mathcal{E}\big(|\psi\rangle_{AB}\big) + \mathcal{E}\big(|\phi\rangle_{A'B'}\big). \tag{16.8}$$

At the same time, it is clear that applying a joint unitary on A and A' generally changes the reduced states of A and A'. While the entanglement with respect to the bipartition $AA'|BB'$ stays the same, this gives the opportunity to distribute entanglement differently, e.g., entanglement between A and B could increase, if entanglement between A' and B' decreases such that their sum remains constant according to eqn (16.8).

This observation forms the basic principle behind procedures collectively known as *entanglement distillation*, a type of LOCC protocols to concentrate the entanglement of many weakly entangled states into fewer copies of strongly entangled states. As we will discuss in Section 16.1.3, the entropy of entanglement derives a unique preferred position among all ways of quantifying entanglement from such state-conversion protocols. However, to properly understand the significance of these protocols we first have to take a closer look at the paradigm of LOCC, which we will do in Section 16.1.2.

16.1.2   LOCC and Majorization

A key aspect to understanding entanglement and its quantification is to appreciate entanglement as a resource that cannot be created or increased using only local operations and classical communication.

Definition 16.3   *In a scenario of two or more observers sharing quantum states,* **LOCC** *(local operations and classical communication) refers to any set of interventions that includes only local manipulation (applying local unitary or even non-unitary local maps) and measurements (each observer can only manipulate the subsystem under their direct control), assisted by the exchange of classical information.*

In other words, entanglement can be viewed as a resource for quantum communication between distant labs. Quantum systems may be more difficult to control than classical systems, but today one can nevertheless prepare such systems very well locally. Here, we can take this to mean that when one has direct control over (both parts of) a bipartite quantum system in a single lab, one can also "easily" create entanglement. In particular, one can create maximally entangled pure states, at least in good approximation. However, sending the entangled systems to (or between) distant observers via noisy channels is difficult: Noise reduces the coherence and hence the entanglement of shared states (see, e.g., the discussion in Chapters 22 and 21), and creating entanglement between distant locations cannot be achieved by any local manipulations of the parties at these remote locations. In practice it is therefore important to know what can and cannot be achieved with a given quantum state using only LOCC.

For instance, we might be interested in using an LOCC protocol to concentrate (or "distil") entanglement, i.e., to create fewer copies of maximally entangled states from many copies of less entangled states, or to prepare more copies of a desired (non-maximally) entangled state using (fewer) copies of maximally entangled states. And, indeed, there are many other conceivable ways to distribute entanglement. In this context it is customary to distinguish between *deterministic* protocols, where a series of steps results in a particular outcome with probability 1, and probabilistic protocols, where the desired outcome is only obtained with a probability smaller than 1. When considering protocols of the latter type, one refers to the set of *stochastic local operations and classical communication*, or SLOCC for short.

Which operations are LOCC? For instance, Alice may measure her quantum system, communicate the measurement outcome to Bob via a classical channel, and Bob may then perform a unitary or measurement on his subsystem depending on Alice's outcome. This can include many rounds of measurements and adjustments, can include generalized operations and measurements (which we will discuss in Chapters 21 and 23, respectively), and of course can involve sending information via classical signals between Alice and Bob. But while it is quite easy to construct *some* LOCC protocol, or to say if a given protocol is LOCC, the description of the entire set of LOCC is notoriously difficult (see, e.g., the review Chitambar *et al.*, 2014). This brings us to the crucial question:

How do we know whether a particular transformation is possible via LOCC?

Suppose we want to convert a copy of $|\psi\rangle = \frac{1}{2}(\sqrt{3}|01\rangle + |10\rangle)$ into a copy of $|\Phi^+\rangle = \frac{1}{\sqrt{2}}(|00\rangle + |11\rangle)$ using local unitaries. This is not possible: The two states have different Schmidt coefficients (see Section 15.1.2), for $|\psi\rangle$ we have $\lambda_1 = \frac{\sqrt{3}}{2}$, and $\lambda_2 = \frac{1}{2}$ but for the maximally entangled state $|\Phi^+\rangle$ we have $\lambda_1' = \lambda_2' = \frac{1}{\sqrt{2}}$. Since the squares of the Schmidt coefficients are the eigenvalues of the reduced states (see Lemma 15.1), and the latter are left invariant by local unitaries, the conversion is not possible. But could this be done using more general LOCC operations? Even though we have not yet learned what such operations might look like, it turns out there is a simple way to check if (deterministic or probabilistic) conversion via LOCC is possible: *Michael Nielsen*'s majorization theorem (1999) for deterministic transformations. To formulate this theorem and to understand how it arises, we first have to talk about *majorization*.

Majorization is a mathematical relation that expresses that one set is more disordered than another. Given two vectors $\vec{r}$ and $\vec{r}\,'$ in $\mathbb{R}^d$, we say that $\vec{r}$ *is majorized* by $\vec{r}\,'$, denoted as $\vec{r} \preceq \vec{r}\,'$, if $\vec{r}$ can be obtained by mixing together permutations of the components of $\vec{r}\,'$. More formally, let us make the following definition:

Definition 16.4   *A vector $\vec{r} \in \mathbb{R}^d$ **is majorized by** $\vec{r}\,' \in \mathbb{R}^d$, $\vec{r} \preceq \vec{r}\,'$, if there exists a probability distribution $\{p_i\}_i$ and permutation matrices Π_i such that*
$$\vec{r} = \sum_i p_i\, \Pi_i\, \vec{r}\,'.$$

For instance, if the two vectors have non-negative entries that sum to one, then we have probability distributions and it becomes intuitively clear how randomly mixing together different distributions increases the disorder. The theory of majorization is a well-established area of mathematics and in the following we will refer to some of its central results without proof, but the curious reader is directed to the textbooks by Alberti and Uhlmann (1982) and Marshall *et al.* (2011), or the pedagogical review by Nielsen and Vidal (2001) on majorization in entanglement theory. The gap to quantum mechanics is bridged by three important mathematical results. First, the *König–Birkhoff–von Neumann theorem*[1], which tells us that the set of doubly stochastic matrices (whose rows and columns sum to one) is a convex polytope whose vertices are the permutation matrices, such that every doubly stochastic matrix can be represented as a convex sum of permutation matrices. Second, using this theorem, one can prove the *Schur–Horn theorem* named after *Issai Schur* (1923) and *Alfred Horn* (1954), which asserts that for any pair $\vec{r} = (r_i)$ and $\vec{r}\,' = (r_j')$ satisfying $\vec{r} \preceq \vec{r}\,'$ there exists a unitary matrix $U = (U_{ij})$ such that $r_i = \sum_j |U_{ij}|^2 r_j'$. Since U is unitary, the matrix $D = (D_{ij})$ with $D_{ij} = |U_{ij}|^2$ is doubly stochastic. This is relevant for quantum mechanics because the entries of $\vec{r}$ and $\vec{r}\,'$ can be interpreted as the eigenvalues

---

[1] First formulated and proven by *Dénes König* (see König, 1936, Chap. XIV, Sec. 3), but typically attributed to the (independent) work by *Garrett Birkhoff* (1946) and *John von Neumann* (1953), and thus sometimes referred to as *Birkhoff's theorem* (e.g., in Marshall *et al.*, 2011, p. 30), not to be confused with George David Birkhoff's (the father of Garrett Birkhoff) famous theorems in general relativity or ergodic theory.

of Hermitian operators R and R', which brings us to the third important result, the generalization of the Schur–Horn theorem, commonly attributed to *Armin Uhlmann* (1971, 1972, 1973, see also Nielsen and Vidal, 2001), which states that $\vec{r}(R) \preceq \vec{r}(R')$ if and only if there exists a probability distribution $\{p_i\}_i$ and unitary matrices U_i such that

$$R = \sum_i p_i\, U_i\, R'\, U_i^\dagger. \tag{16.9}$$

It now becomes evident that, for two density operators, a majorization relation between them means that one can be obtained from the other by applying random unitaries, thus mixing it. Aside from this conceptual significance of the definition of majorization, it also provides a practical way for checking if two states can be related in this way: The majorization condition of Definition 16.4 can be recast as a set of d linear conditions by ordering the components of $\vec{r}$ and $\vec{r}'$ in non-increasing order to obtain the vectors $\vec{r}^\downarrow$ and $\vec{r}'^\downarrow$, such that $r_1^\downarrow \geq r_2^\downarrow \geq \ldots \geq r_d^\downarrow$ and $r_1'^\downarrow \geq r_2'^\downarrow \geq \ldots \geq r_d'^\downarrow$. Then we have

$$\vec{r} \preceq \vec{r}' \quad \Leftrightarrow \quad \sum_{i=1}^{k} r_i^\downarrow \leq \sum_{i=1}^{k} r_i'^\downarrow \ \ \forall \ k = 1, 2, \ldots, d. \tag{16.10}$$

Theorem 16.1 (Nielsen's majorization theorem)

A pure state $|\psi\rangle$ with Schmidt coefficients $\{\lambda_i\}_i$ can be converted to another pure state $|\phi\rangle$ with Schmidt coefficients $\{\lambda_j'\}_j$ with unit probability via LOCC if and only if the vector $\vec{r}(\psi) = (\lambda_i^2)$ of squared Schmidt coefficients of $|\psi\rangle$ is majorized by the vector $\vec{r}(\phi) = (\lambda_i'^2)$ of squared Schmidt coefficients of the target state

$$\vec{r}(\psi) \preceq \vec{r}(\phi).$$

Shortly after the publication of Nielsen's original majorization theorem, generalizations to probabilistic transformations were provided by *Guifré Vidal* (1999), as well as by *Daniel Jonathan* and *Martin Plenio* (1999), and we focus here on the latter, most general variant (Jonathan and Plenio, 1999):

Theorem 16.2 (Probabilistic LOCC conversion)

A pure state $|\psi\rangle$ can be converted to (one of) the pure states $|\phi_i\rangle$ of an ensemble $\{|\phi_i\rangle\}_i$ with probability p_i via LOCC if and only if

$$\vec{r}(\psi) \preceq \sum_i p_i\, \vec{r}(\phi_i).$$

Proof   Clearly, Nielsen's version, Theorem 16.1, follows from the more general Theorem 16.1, so we focus on proving the latter. Starting with one direction of the proof, we want to first show that whenever the majorization relation is satisfied, an LOCC protocol exists that achieves the conversion. Following Definition 16.4, this means we can assume there exists a probability distribution $\{q_j\}_j$ along with a set of permutation operators Π_j such that

$$\vec{r}(\psi) = \sum_{i,j} p_i\, q_j\, \Pi_j\, \vec{r}(\phi_i). \tag{16.11}$$

Now, since $|\psi\rangle$ and $|\phi_i\rangle$ are pure states, it follows from Lemma 15.1 that the components r_n and $r_{n,i}$ of $\vec{r}(\psi)$ and $\vec{r}(\phi_i)$ are the eigenvalues of the reduced states $\rho_A = \mathrm{Tr}_B(|\psi\rangle\langle\psi|)$ and $\rho_{A,i} = \mathrm{Tr}_B(|\phi_i\rangle\langle\phi_i|)$, respectively. The operators ρ_A and $\rho_{A,i}$ need not be diagonal with respect to the same bases, but we can always find a set of unitaries $\{V_i\}_i$ such that all $\tilde{\rho}_{A,i} = V_i \rho_{A,i} V_i^\dagger$ are diagonal with respect to the eigenbasis $\{|n\rangle\}_n$ of ρ_A, i.e.,

$$\rho_A = \sum_{n=1}^{k} r_n\, |n\rangle\langle n| \quad \text{and} \quad \tilde{\rho}_{A,i} = \sum_{n} r_{n,i}\, |n\rangle\langle n|, \tag{16.12}$$

with $k \leq d = \dim(\mathcal{H})$ for $|\psi\rangle \in \mathcal{H} \otimes \mathcal{H}$, $r_n > 0$ for all $n = 1, \ldots, k$ and $r_n = 0$ for $k < n < d$ and it follows from eqn (16.11) that for each i at most k of the eigenvalues $r_{n,i}$ are non-zero. We can thus rewrite eqn (16.11) as the equivalent statement

$$\rho_A = \sum_{i,j} p_i\, q_j\, \Pi_j\, \tilde{\rho}_{A,i}\, \Pi_j^\dagger, \tag{16.13}$$

which, in essence, corresponds to Uhlmann's majorization theorem from eqn (16.9).

We then define a set of operators $\{M_{ij}\}_{i,j}$ labelled by two indices i and j and define them implicitly via the relation

$$M_{ij}\sqrt{\rho_A} = \sqrt{p_i q_j}\, \sqrt{\tilde{\rho}_{A,i}}\, \Pi_j^\dagger. \tag{16.14}$$

In the next step, we multiply the left-hand side by its adjoint from the left before inserting the right-hand side,

$$\left(M_{ij}\sqrt{\rho_A}\right)^\dagger M_{ij}\sqrt{\rho_A} = \sqrt{\rho_A}\, M_{ij}^\dagger\, M_{ij}\, \sqrt{\rho_A} = p_i\, q_j\, \Pi_j\, \tilde{\rho}_{A,i}\, \Pi_j^\dagger. \tag{16.15}$$

If we now sum over i and j we recognize the right-hand side as eqn (16.13) such that we have

$$\sqrt{\rho_A} \sum_{i,j} M_{ij}^\dagger\, M_{ij}\, \sqrt{\rho_A} = \sum_{i,j} p_i\, q_j\, \Pi_j\, \tilde{\rho}_{A,i}\, \Pi_j^\dagger = \rho_A, \tag{16.16}$$

from which we can deduce that $\sum_{i,j} M_{ij}^\dagger M_{ij} = \mathbb{1}$. To be precise, one should note here that eqn (16.14) only defines the operators M_{ij} in the subspace spanned by the eigenvectors of ρ_A with non-zero eigenvalues r_n. From eqn (16.13) it is clear that all $\Pi_j \, \tilde{\rho}_{A,\,i} \, \Pi_j^\dagger$ have their support in this subspace as well. The condition in eqn (16.16) thus only fixes $\sum_{i,j} M_{ij}^\dagger \, M_{ij} = \mathbb{1}_k$ to the identity $\mathbb{1}_k = \sum_{n=1}^k |\, n \, \rangle \langle \, n \,|$ in this k-dimensional subspace, but we can then simply complete the set $\{M_{ij}\}_{i,j}$ to satisfy $\sum_{i,j} M_{ij}^\dagger \, M_{ij} = \mathbb{1}$ by adding one additional operator $M_{00} = \mathbb{1}_d - \mathbb{1}_k = M_{00}^\dagger M_{00}$.

At the same time, we observe that the operators $E_{ij} := M_{ij}^\dagger \, M_{ij}$ are positive semidefinite, which we can see by taking a look at the matrix elements of M_{ij} with respect to the eigenbasis $\{|\, n \, \rangle\}_n$ of ρ_A and $\tilde{\rho}_{A,\,i}$. The left-hand side of eqn (16.14) gives

$$\langle \, k \,|\, M_{ij} \, \sqrt{\rho_A} \,|\, l \, \rangle \;=\; \langle \, k \,|\, M_{ij} \sum_n \sqrt{r_n} \,|\, n \, \rangle\langle \, n \,|\, l \, \rangle \;=\; \sqrt{r_l} \, \langle \, k \,|\, M_{ij} \,|\, l \, \rangle \qquad (16.17)$$

while the right-hand side is

$$\langle \, k \,|\, \sqrt{p_i \, q_j} \, \sqrt{\tilde{\rho}_{A,\,i}} \, \Pi_j^\dagger \,|\, l \, \rangle \;=\; \sum_n \sqrt{p_i \, q_j \, r_{n,i}} \, \langle \, k \,|\, n \, \rangle\langle \, n \,|\, \Pi_j^\dagger \,|\, l \, \rangle \;=\; \sqrt{p_i \, q_j \, r_{k,i}} \, \langle \, k \,|\, \Pi_j^\dagger \,|\, l \, \rangle \,,$$

such that we have $\sqrt{r_l} \, \langle \, k \,|\, M_{ij} \,|\, l \, \rangle = \sqrt{p_i \, q_j} \, \sqrt{r_{k,i}} \, \langle \, k \,|\, \Pi_j^\dagger \,|\, l \, \rangle$. With this, we calculate the matrix elements of $E_{ij} := M_{ij}^\dagger \, M_{ij}$ via

$$\sqrt{r_m r_n} \, \langle \, m \,|\, E_{ij} \,|\, n \, \rangle \;=\; \sqrt{r_m r_n} \, \langle \, m \,|\, M_{ij}^\dagger \sum_k |\, k \, \rangle\langle \, k \,|\, M_{ij} \,|\, n \, \rangle \qquad (16.18)$$

$$=\; \langle \, m \,|\, p_i \, q_j \, \Pi_j \sum_k r_{k,i} \,|\, k \, \rangle\langle \, k \,|\, \Pi_j^\dagger \,|\, n \, \rangle \;=\; \langle \, m \,|\, p_i \, q_j \, \Pi_j \, \tilde{\rho}_{A,\,i} \, \Pi_j^\dagger \,|\, n \, \rangle \,.$$

Since $\tilde{\rho}_{A,\,i}$ is diagonal in the chosen basis, $\Pi_j \, \tilde{\rho}_{A,\,i} \, \Pi_j^\dagger$ must also be diagonal because it just corresponds to a permutation of diagonal elements of $\tilde{\rho}_{A,\,i}$. Consequently, for all n for which $r_n \neq 0$ we have $\langle \, m \,|\, E_{ij} \,|\, n \, \rangle = \delta_{mn} \frac{p_i \, q_j}{r_n} \, \langle \, n \,|\, \Pi_j \, \tilde{\rho}_{A,\,i} \, \Pi_j^\dagger \,|\, n \, \rangle$, i.e., E_{ij} is diagonal in the eigenbasis of ρ_A with non-negative entries since $p_{ij} := p_i \, q_j \geq 0$ and $\Pi_j \, \tilde{\rho}_{A,\,i} \, \Pi_j^\dagger$ is a (positive semidefinite) density operator. For all m and n for which $r_m = 0$ or $r_n = 0$, the condition in eqn (16.18) is trivial, but as we have discussed below eqn (16.16) we then have $\langle \, m \,|\, E_{ij} \,|\, n \, \rangle = \langle \, m \,|\, E_{00} \,|\, n \, \rangle = \delta_{mn}$. We can thus conclude that $E_{ij} \geq 0$.

We thus have set of positive operators $\{E_{ij}\}_{i,j}$ that sums to the identity. This, as we shall learn in Chapter 23, is the mathematical description of a general measurement (a so-called POVM). The measurement outcome labelled by i and j occurs with probability $p_{ij} = \text{Tr}\big(E_{ij} \, \rho_A\big) = \sum_n r_n \, \langle \, n \,|\, E_{ij} \,|\, n \, \rangle = p_i q_j$ and results in the post-measurement state

$$\tfrac{1}{p_{ij}} M_{ij} \rho_A M_{ij}^\dagger \;=\; \tfrac{1}{p_{ij}} p_i \, q_j \, \sqrt{\tilde{\rho}_{A,\,i}} \, \Pi_j^\dagger \Pi_j \, \sqrt{\tilde{\rho}_{A,\,i}} \;=\; \tilde{\rho}_{A,\,i} \;=\; V_i \rho_{A,\,i} V_i^\dagger \,, \qquad (16.19)$$

as we can easily see by inserting from eqn (16.14). By replacing M_{ij} with $M_{ij} V_i^\dagger$ via a local unitary we can thus obtain the desired output states with the desired probabilities. Since this measurement is carried out locally on one of the subsystems, it constitutes an LOCC procedure, thus constructively showing that such a protocol exists whenever the majorization condition in Theorem 16.2 is met.

For the reverse direction of the proof we have to show that any LOCC transformation of the pure state $|\psi\rangle$ that results in one of the pure states $|\phi_i\rangle$ with probability p_i satisfies the stated majorization relation. This is greatly simplified by leveraging results by Lo and Popescu (1999), which shows that any such LOCC protocol is equivalent to one where a single projective measurement is carried out by one of the two parties. That is, for any such protocol there are operators M_i with $M_i^\dagger M_i \geq 0$ satisfying $\sum_i M_i^\dagger M_i = \mathbb{1}$ such that

$$p_i = \text{Tr}\big(M_i^\dagger M_i \otimes \mathbb{1}_B |\psi\rangle\langle\psi|\big) \quad \text{and} \quad |\phi_i\rangle = \tfrac{1}{\sqrt{p_i}} M_i \otimes \mathbb{1}_B |\psi\rangle, \qquad (16.20)$$

where we have chosen the measurement to be performed on subsystem A without loss of generality. Now, since such a local measurement cannot have an instantaneous effect on subsystem B, its reduced state should be unaffected by the measurement on A and thus satisfy

$$\rho_B = \text{Tr}_A\big(|\psi\rangle\langle\psi|\big) = \text{Tr}_A\big(M_i^\dagger M_i \otimes \mathbb{1}_B |\psi\rangle\langle\psi|\big)$$
$$= \text{Tr}_A\Big(\sum_i p_i |\phi_i\rangle\langle\phi_i|\Big) = \sum_i p_i \rho_{B,i}, \qquad (16.21)$$

where we have used eqn (16.20) and defined $\rho_{B,i} := \text{Tr}_A\big(|\phi_i\rangle\langle\phi_i|\big)$.

Here we can finally make use of the general majorization formula: For all Hermitian operators ρ_1 and ρ_2 we have

$$\vec{r}(\rho_1 + \rho_2) \preceq \vec{r}(\rho_1) + \vec{r}(\rho_1), \qquad (16.22)$$

where $\vec{r}(\rho)$ with components $r_n(\rho)$ denotes the vector of eigenvalues of ρ in non-increasing order. To see that this is true, let $\{|n\rangle\}_n$ be the eigenvectors of $\rho_1 + \rho_2$ and let $P = \sum_{n=1}^k |n\rangle\langle n|$ be the projector onto the subspace of the largest k of them, then

$$\sum_{n=1}^k r_n(\rho_1 + \rho_2) = \text{Tr}\big[(\rho_1 + \rho_2) P\big] = \text{Tr}(\rho_1 P) + \text{Tr}(\rho_2 P) \leq \sum_{n=1}^k r_n(\rho_1) + \sum_{n=1}^k r_n(\rho_2), \qquad (16.23)$$

where the last inequality follows from *Ky Fan's maximum principle* (see Nielsen and Vidal, 2001, p. 82). The majorization relation of Theorem 16.2 then follows immediately from eqn (16.21) since $|\psi\rangle$ and $|\phi_i\rangle$ are pure and thus $\vec{r}(\psi) = \vec{r}(\rho_B)$ and $\vec{r}(\phi_i) = \vec{r}(\rho_{B,i})$ (via Lemma 15.1). $\qquad\square$

Majorization relations thus express the disorder in the reduced states in a way that closely relates to the possibility of converting one pure state into an ensemble of others via LOCC, in particular, without creating entanglement. Moreover, this (partial) order established in the state space via a majorization relation is preserved by the von Neumann entropy.

Theorem 16.3 (Majorization and entropy of entanglement)

For any pair of bipartite pure states $|\psi\rangle$ and $|\phi\rangle$:

$$\vec{r}(\psi) \preceq \sum_i p_i \, \vec{r}(\phi_i) \quad \Longrightarrow \quad \mathcal{E}(|\psi\rangle) \geq \sum_i p_i \, \mathcal{E}(|\phi_i\rangle).$$

Proof Using the notation from the proof of Theorem 16.2, the entropy of entanglement of $|\psi\rangle$ is just the von Neumann entropy of the reduced state $\rho_A = \mathrm{Tr}_B(|\psi\rangle\langle\psi|)$. The majorization relation then means that we can use Uhlmann's majorization theorem from eqn (16.9) to write ρ_A in a similar form as in eqn (16.13), such that

$$\mathcal{E}(|\psi\rangle) = S(\rho_A) = S\Big(\sum_{i,j} p_i \, q_j \, U_j \, \rho_{A,i} \, U_j^\dagger\Big), \tag{16.24}$$

where $\rho_{A,i} = \mathrm{Tr}_B(|\phi_i\rangle\langle\phi_i|)$ and the U_j are unitaries. We can then use the fact that the von Neumann entropy is concave, $S(\sum_i p_i \rho_i) \geq \sum_i p_i S(\rho_i)$, as we will show in Section 20.1. We thus have

$$\mathcal{E}(|\psi\rangle) \geq \sum_{i,j} p_i \, q_j \, S(U_j \, \rho_{A,i} \, U_j^\dagger) = \sum_{i,j} p_i \, q_j \, S(\rho_{A,i}) = \sum_i p_i \, \mathcal{E}(|\phi_i\rangle), \tag{16.25}$$

because the von Neumann entropy is invariant under unitary transformations. $\square$

As a special case of Theorem 16.3 it follows that for any pair of pure states $|\psi\rangle$ and $|\phi\rangle$ for which $|\psi\rangle$ can be converted to $|\phi\rangle$ deterministically via LOCC, we have $\vec{r}(\psi) \preceq \vec{r}(\phi)$ and thus $\mathcal{E}(|\psi\rangle) \geq \mathcal{E}(|\phi\rangle)$. Entanglement cannot be created or increased under LOCC and this is correctly captured by the entropy of entanglement via the von Neumann entropy. However, this does not (yet) single out the entropy of entanglement, since the partial order established by the majorization relation is preserved for any Schur-convex function of the reduced states.

At the same time it is crucial to note that the converse ("$\Longleftarrow$") of Theorem 16.3 is not true in general: The ordering established by the majorization relation is only partial, some states are *incomparable* and cannot be converted into each other at all, for instance for pure states with vectors of squared Schmidt coefficients $\vec{r} = (\frac{1}{2}, \frac{1}{4}, \frac{1}{4})$ and $\vec{r}' = (\frac{4}{10}, \frac{4}{10}, \frac{2}{10})$, neither majorizes the other. Such an incomparability (via LOCC) of two pure states is not captured by the entropy of entanglement. This puts somewhat of a dent into the nice picture that has presented itself so far, but this problem can at least be partially ameliorated by the possibility of *entanglement catalysis* (Jonathan and Plenio, 1999). For some incomparable states $|\psi\rangle$ and $|\phi\rangle$, a deterministic transformation can be achieved (or the probability for the conversion in a probabilistic protocol can be increased) by adding a catalyst, i.e., a second entangled state $|\chi\rangle$ that returns to the same state at the end, which makes the transformation $|\psi\rangle \otimes |\chi\rangle \longrightarrow |\phi\rangle \otimes |\chi\rangle$ possible.

However, a perhaps more serious issue with the picture where single copies of pure states are converted into each other via LOCC is the observation (Vidal *et al.*, 2000) that there are some examples where the conversion from one pure state to another is possible with probability 1, but the probability (that is, even when allowing probabilistic LOCC transformations rather than deterministic ones) to convert to a pure state that is arbitrarily close (e.g., in Hilbert–Schmidt distance, see Definition 11.6, or in the trace distance which we will discuss in Section 23.4) is strictly 0. In general, this suggests that it makes sense to relax the idea of exact conversion (getting exactly the right state) and the notion of only converting single copies, as we will elaborate on in the next section.

16.1.3   The Asymptotic Setting—Cost and Distillation of Entanglement

As we have discussed, the conversion of individual copies of bipartite pure states into each other is not always possible via LOCC and even if it is, discontinuities might appear in the sense that the conversion might be possible and impossible respectively for two target states that are arbitrarily close to each other. It therefore makes sense to consider *asymptotically exact state conversion.* That is, transforming a large number N of copies of some state ρ, given by $\rho^{\otimes N}$, to a state $\tilde{\rho}^{(M)}$ that approximates $\tilde{\rho}^{\otimes M}$ well for some desired target state $\tilde{\rho}$ for some large M, such that, in the limit $N \to \infty$ and for fixed rate $r = M/N$ the approximation becomes arbitrarily good. The optimal achievable rate of conversion between two states can then be used as a measure of the relative entanglement content. In particular, for any fixed state ρ one can ask how many copies of maximally entangled two-qubit states are needed per copy asymptotically for preparing the state approximately via LOCC, and, likewise, one may ask how many copies of maximally entangled two-qubit states can be obtained per copy from ρ via LOCC. This leads one to two important measures of entanglement: The *entanglement cost* and the *entanglement of distillation.*

Entanglement cost: For a given state ρ, the entanglement cost $E_{\mathrm{C}}(\rho)$ quantifies the optimal rate of asymptotically exact conversion from (many copies of) maximally entangled two-qubit states (sometimes referred to as "ebits" in this context) to states that approximate (many copies of) ρ such that the conversion becomes exact for infinitely many copies. Denoting by Λ a (trace-preserving) LOCC protocol and using the shorthand $\omega^+ = |\phi^+\rangle\langle\phi^+|$ for the maximally entangled two-qubit Bell state ω^+ from eqn (11.113b), the entanglement cost can formally be defined as (see, e.g., Plenio and Virmani, 2007)

$$E_{\mathrm{C}}(\rho) = \inf\left\{r : \lim_{M \to \infty}\left[\inf_{\Lambda \in \mathrm{LOCC}} D\big(\rho^{\otimes M}, \Lambda\big[\omega^{+\otimes rM}\big]\big)\right] = 0\right\}, \qquad (16.26)$$

where $D(\rho, \sigma)$ is a suitable distance measure between two quantum states ρ and σ, e.g., the trace distance (see Section 23.4). Loosely speaking, the entanglement cost hence quantifies how "expensive" it is to create ρ via LOCC in terms of the number of needed Bell pairs (per copy). A trivial consequence of this definition is that the entanglement cost must be non-zero for any entangled state since $E_{\mathrm{C}}(\rho) = 0$ implies that the state can be created via LOCC without sharing entanglement, and therefore cannot be entangled to begin with.

While this is a reasonably motivated quantifier for entanglement, it is generally not known how to calculate the entanglement cost in practice except for the pure-state case, which we will come back to shortly. In addition, using a number of pure, highly entangled states to create a less entangled mixed state on purpose is typically not a task of high practical relevance. Instead, one more often worries about the reverse process. Consider, e.g., the following situation: Alice and Bob want to communicate with each other via quantum teleportation (see Chapter 14). This requires pre-shared entanglement, e.g., in the form of the two-qubit singlet state $\rho^- = |\psi^-\rangle\langle\psi^-|$ or the triplet state $\omega^+ = |\phi^+\rangle\langle\phi^+|$ of which one subsystem is distributed to Alice and one to Bob. Unfortunately, the transmission of quantum systems realistically occurs via noisy and imperfect channels such that Alice and Bob do not end up with a pure singlet state $|\psi^-\rangle$ but rather some mixed state ρ.

Now the question is, can they by any means regain a maximally entangled state? In many cases the answer is yes, they can!

Entanglement of distillation: If multiple copies of mixed entangled states are shared between Alice and Bob, then there exist LOCC protocols—so-called *entanglement distillation* protocols (Bennett *et al.*, 1996*a*; Bennett *et al.*, 1996*c*)—designed to recover a smaller number of more strongly entangled states from a larger number of less entangled states. Ideally, these more strongly entangled states are "almost" maximally entangled singlet states. Here, the word "almost" is to be understood in the sense of asymptotically exact state conversion as before, i.e., from a (large but finite) number N of mixed states ρ we can "distil" a smaller number M of maximally entangled states $\tilde{\rho} = \omega^+$ in the sense that $\tilde{\rho}^{(M)}$ approximates $\tilde{\rho}^{\otimes M}$ well, and the conversion can be made exact asymptotically as the number of copies increases. The maximal possible distillation rate is called *entanglement of distillation* or sometimes *distillable entanglement* and, in analogy to the entanglement cost, it can be formalized as (see, e.g., Plenio and Virmani, 2007)

$$E_{\mathrm{D}}(\rho) = \sup\left\{r : \lim_{M\to\infty}\left[\inf_{\Lambda\,\in\,\mathrm{LOCC}} D\big(\Lambda\big[\rho^{\otimes M}\big], \omega^{+\otimes rM}\big)\right] = 0\right\}. \qquad (16.27)$$

Distillation of entanglement can be viewed as the converse process of formation, see Fig. 16.1. The value of the distillable entanglement must generally be smaller than the entanglement cost since LOCC transformations cannot increase entanglement

$$E_{\mathrm{D}}(\rho) \leq E_{\mathrm{C}}(\rho). \qquad (16.28)$$

Converting N copies of some state ρ to M copies of a maximally entangled state ω^+ via LOCC and then converting these M copies back to M' copies of ω^+ generally means $M' \leq N$. This irreversibility in the formation of states can be seen as a consequence of loss of information about the decomposition of the state.

Entropy of entanglement and asymptotically exact conversion: As with the entanglement cost, there is generally no known method of calculating this measure for general mixed states. However, in their seminal paper in 1996*a Charles Bennett,*

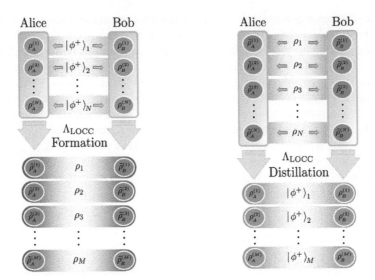

Fig. 16.1 Formation and distillation of entanglement. In the formation of entanglement, the entanglement cost is the minimum number of maximally entangled pairs $|\phi^+\rangle$ which are needed (per copy) to prepare a state ρ via LOCC operations in the limit of processing infinitely many copies. The converse process to formation is called distillation of entanglement. Alice and Bob perform LOCC operations to recover a smaller number of "almost" maximally entangled states from a given number of non-maximally entangled states.

Herbert J. Bernstein, Sandu Popescu and *Benjamin Schumacher* showed that in the setting of asymptotically exact conversion the majorization relations determining the possibility of LOCC transformation of N copies of a state $|\psi\rangle$ to M copies of a state $|\phi\rangle$ all collapse to single inequality for the von Neumann entropy of the reduced states $\rho_A = \mathrm{Tr}_B(|\psi\rangle\langle\psi|)$ and $\rho'_A = \mathrm{Tr}_B(|\phi\rangle\langle\phi|)$, i.e., the asymptotically exact conversion is possible *if and only if*

$$N\,S(\rho_A) \leq M\,S(\rho'_A)\,. \tag{16.29}$$

In the process of formation of a state from maximally entangled states, we have $|\psi\rangle = |\phi^+\rangle$ and $S(\rho_A) = S(\frac{1}{2}\mathbb{1}_2) = \log_2(2) = 1$, where we have used the logarithm to base 2 in the von Neumann entropy, while $|\phi\rangle$ is the target state with $S(\rho'_A) = \mathcal{E}(|\phi\rangle)$. The number of maximally entangled states needed per copy of $|\phi\rangle$ to form $|\phi\rangle$ via LOCC then satisfies $\frac{N}{M} \geq \mathcal{E}(|\phi\rangle)$ and the infimum over all rates in eqn (16.26) thus means $E_{\mathrm{C}}(|\phi\rangle) = \mathcal{E}(|\phi\rangle)$, the entanglement cost for pure states is exactly the entropy of entanglement. Conversely, for distillation processes the target state $|\phi\rangle$ is $|\phi^+\rangle$ with $S(\rho'_A) = 1$ and the number of distillable copies of $|\phi^+\rangle$ per copy of the initial state $|\psi\rangle$ thus satisfies $\frac{M}{N} \leq \mathcal{E}(|\psi\rangle)$. The supremum in the definition of the distillable entanglement in eqn (16.27) then implies $E_{\mathrm{D}}(|\phi\rangle) = \mathcal{E}(|\psi\rangle)$. For pure states we therefore see that the entropy of entanglement is the (unique) measure of entanglement that satisfies

$$E_{\mathrm{D}}(|\psi\rangle) = \mathcal{E}(|\psi\rangle) = E_{\mathrm{C}}(|\psi\rangle)\,. \tag{16.30}$$

Entanglement distillation: After the discussion of pure-state distillation in Bennett *et al.* (1996*a*), the first entanglement distillation protocols for mixed states were introduced by *Charles Bennett, Gilles Brassard, Sandu Popescu, Benjamin Schumacher, John Smolin* and *William K. Wooters* (in 1996*b*). To illustrate how such protocols work, let us briefly go through the simplest one for two qubits in Bennett *et al.* (1996*b*). The protocol operates on the premise that one has a (large) number of two-qubit input states ρ_{AB} available whose singlet fraction $\mathcal{F}_{\max}(\rho)$—the maximal fidelity with any maximally entangled state from eqn (15.120)—exceeds $\frac{1}{2}$ and which are hence entangled. Without loss of generality we assume that maximally entangled state for which this value $\mathcal{F}_{\max} > \frac{1}{2}$ is obtained is $|\psi^-\rangle$, otherwise the state ρ_{AB} can be brought to this form by local unitaries, and we use the shorthand $\mathcal{F} = \mathcal{F}(\rho_{AB}, |\psi^-\rangle)$ in the following. The distillation protocol then consists of the following steps:

(1) First, Alice and Bob apply a local unitary $U \otimes U$ with U a random unitary in $SU(2)$ to all of the two-qubit input states ρ_{AB}. The resulting state is invariant under all local unitaries of the form $U \otimes U$ and must therefore be in the family of *Werner states* $\rho_{\mathrm{Werner}}(\alpha)$, which, as we recall from Section 15.3.3 take the form

$$\rho_{\mathrm{Werner}}(\alpha) = \alpha\, |\psi^-\rangle\langle\psi^-| + \tfrac{1-\alpha}{4}\, \mathbb{1}_4\,. \tag{16.31}$$

That the Werner states are $U \otimes U$-invariant follows from the rotational invariance of the singlet $|\psi^-\rangle = |l = 0, m_l = 0\rangle$, i.e., it being the eigenstate of the orbital angular momentum operators $\vec{L}^2$ and L_z with respective quantum numbers $l = 0$ and $m_l = 0$, see Section 8.3.2. That all $U \otimes U$-invariant states must be of this form can be shown in similar fashion as the proof of the $U \otimes U^*$-invariance of the isotropic states in Lemma 15.4, but we will leave the details of this proof as an exercise for the interested reader. Here, we focus again on the singlet fraction, which can be expressed via α as

$$\mathcal{F}\big(\rho_{\mathrm{Werner}}(\alpha), |\psi^-\rangle\langle\psi^-|\big) = \langle\psi^-|\,\rho_{\mathrm{Werner}}(\alpha)\,|\psi^-\rangle = \alpha + \tfrac{1-\alpha}{4} = \tfrac{3\alpha-1}{4}, \tag{16.32}$$

which allows us to reparameterize the state via $\mathcal{F}$,

$$\rho_{\mathrm{Werner}}(\mathcal{F}) = \mathcal{F}\, |\psi^-\rangle\langle\psi^-| + \tfrac{1-\mathcal{F}}{3}\big(|\phi^+\rangle\langle\phi^+| + |\phi^-\rangle\langle\phi^-| + |\psi^+\rangle\langle\psi^+|\big). \tag{16.33}$$

(2) In the second step, we apply a local unitary $\sigma_y^{(A)} \otimes \mathbb{1}_B$ to all shared states, which results in the state

$$\rho^{(2)} := \sigma_y^{(A)} \otimes \mathbb{1}_B\, \rho_{\mathrm{Werner}}\, \sigma_y^{(A)} \otimes \mathbb{1}_B$$

$$= \mathcal{F}\, |\phi^+\rangle\langle\phi^+| + \tfrac{1-\mathcal{F}}{3}\big(|\psi^-\rangle\langle\psi^-| + |\psi^+\rangle\langle\psi^+| + |\phi^-\rangle\langle\phi^-|\big). \tag{16.34}$$

(3) In the third and crucial step we then take pairs of states $\rho^{(2)}$ of which we label Alice's subsystems A_1 and A_2 and Bob's B_1 and B_2, respectively. Each observer locally performs an entangling two-qubit gate, a controlled NOT or simply CNOT gate as in eqn (14.47) on the two qubits under their respective control. Subsequently, Alice and

Bob each measure their respective qubits in the second copy, A_2 and B_2 in the computational basis, communicate with each other on the results. If both results match, either both giving the result 0 or the result 1, they keep the first copy, otherwise they discard it. To describe the effect of this operation, we first calculate

$$\mathbb{1} \otimes \langle 0 | \, U_{\text{CNOT}} = \mathbb{1} \otimes \langle 0 | \left(|0\rangle\langle 0| \otimes \mathbb{1} + |1\rangle\langle 1| \otimes \sigma_x \right)$$

$$= |0\rangle\langle 0| \otimes \langle 0| + |1\rangle\langle 1| \otimes \langle 1|, \qquad (16.35)$$

before we define the operator

$$M_{00} := {}_{A_2 B_2}\langle 00 | \, U_{\text{CNOT}, B_1 B_2} \, U_{\text{CNOT}, A_1 A_2} = \sum_{m,n=0,1} |mn\rangle\langle mn|_{A_1 B_1} \otimes {}_{A_2 B_2}\langle mn|. \qquad (16.36)$$

With this operator at hand, it is easy to see that

$$M_{00} \, |\phi^+\rangle_{A_1 B_1} \otimes |\phi^+\rangle_{A_2 B_2} = M_{00} \, |\phi^-\rangle_{A_1 B_1} \otimes |\phi^-\rangle_{A_2 B_2} = \tfrac{1}{\sqrt{2}} \, |\phi^+\rangle_{A_1 B_1}, \quad (16.37)$$

$$M_{00} \, |\phi^+\rangle_{A_1 B_1} \otimes |\phi^-\rangle_{A_2 B_2} = M_{00} \, |\phi^-\rangle_{A_1 B_1} \otimes |\phi^+\rangle_{A_2 B_2} = \tfrac{1}{\sqrt{2}} \, |\phi^-\rangle_{A_1 B_1}, \quad (16.38)$$

$$M_{00} \, |\psi^+\rangle_{A_1 B_1} \otimes |\psi^+\rangle_{A_2 B_2} = M_{00} \, |\psi^-\rangle_{A_1 B_1} \otimes |\psi^-\rangle_{A_2 B_2} = \tfrac{1}{\sqrt{2}} \, |\psi^+\rangle_{A_1 B_1}, \quad (16.39)$$

$$M_{00} \, |\psi^+\rangle_{A_1 B_1} \otimes |\psi^-\rangle_{A_2 B_2} = M_{00} \, |\psi^-\rangle_{A_1 B_1} \otimes |\psi^+\rangle_{A_2 B_2} = \tfrac{1}{\sqrt{2}} \, |\psi^-\rangle_{A_1 B_1}, \quad (16.40)$$

while applying M_{00} to tensor products of $|\phi^\pm\rangle$ and $|\psi^\pm\rangle$ gives zero. We hence find

$$M_{00} \, \rho^{(2)\otimes 2} \, M_{00}^\dagger = \tfrac{1}{2} \Big[\mathcal{F}^2 + \tfrac{(1-\mathcal{F})^2}{9} \Big] \, |\phi^+\rangle\langle\phi^+| + \tfrac{\mathcal{F}(1-\mathcal{F})}{3} \, |\phi^-\rangle\langle\phi^-|$$

$$+ \tfrac{(1-\mathcal{F})^2}{9} \left(|\psi^-\rangle\langle\psi^-| + |\psi^+\rangle\langle\psi^+| \right), \qquad (16.41)$$

$$\text{Tr}\big(M_{00} \, \rho^{(2)\otimes 2} \, M_{00}^\dagger \big) = \tfrac{1}{18} \big(8\mathcal{F}^2 - 4\mathcal{F} + 5 \big). \qquad (16.42)$$

The outcome where both A_2 and B_2 are found in the state $|1\rangle$ gives the exact same result, and by applying the operation $\sigma_y^{(A_1)} \otimes \mathbb{1}_{B_1}$ from step **(2)** again to the remaining copy, which converts the state $|\phi^+\rangle$ back to $|\psi^-\rangle$ (up to an irrelevant global phase), we thus obtain the state

$$\rho^{(3)} := \sigma_y^{(A_1)} \otimes \mathbb{1}_{B_1} \frac{M_{00} \, \rho^{(2)\otimes 2} \, M_{00}^\dagger}{\text{Tr}\big(M_{00} \, \rho^{(2)\otimes 2} \, M_{00}^\dagger \big)} \, \sigma_y^{(A_1)} \otimes \mathbb{1}_{B_1}, \qquad (16.43)$$

with probability $\big(8\mathcal{F}^2 - 4\mathcal{F} + 5\big)/9 > \tfrac{1}{2}$ for $\mathcal{F} > \tfrac{1}{2}$. Moreover, this state now has a fidelity with $|\psi^-\rangle$ given by

$$\mathcal{F}' := \mathcal{F}\big(\rho^{(3)}, |\psi^-\rangle\langle\psi^-|\big) = \frac{10\mathcal{F}^2 - 2\mathcal{F} + 1}{8\mathcal{F}^2 - 4\mathcal{F} + 5} > \mathcal{F} \qquad (16.44)$$

for $\tfrac{1}{2} < \mathcal{F} < 1$. We are thus left with fewer copies of more strongly entangled states for which the procedure can be repeated.

As the fidelity approaches $\mathcal{F} \to 1$, the yield goes to zero, but this can be circumvented by extending the number of copies whose qubits are used as control qubits (here A_1 and B_1) for CNOT operations for the single target pair (here A_2 and B_2), see Bennett *et al.* (1996*b*). Further note that this protocol works only if $\mathcal{F} > \frac{1}{2}$ but since there exist entangled states with $\mathcal{F} \leq \frac{1}{2}$ (Verstraete and Verschelde, 2002), this does not yet imply that all two-qubit states are distillable, which was shown independently by Horodecki *et al.* (1997).

Distillability versus bound entanglement: The fact that all two-qubit states are distillable also follows from another theorem by *Michał* and *Paweł Horodecki* (1999):

Theorem 16.4 (Distillation theorem)

A bipartite quantum state can be distilled if and only if its entanglement is detected by the reduction criterion (Theorem 15.11).

For the proof we refer the interested reader to Horodecki and Horodecki (1999).

From this distillation theorem we can make a few important observations. As we recall from Lemma 15.3, the reduction map is decomposable à la Størmer–Woronowicz (Theorem 15.10) in any dimension (Horodecki and Horodecki, 1999), and so the violation of the PPT criterion is necessary to violate the reduction criterion and hence for distillability. In other words, PPT entangled states are not distillable and are therefore so-called *bound-entangled* states.

Corollary 16.1 (PPT bound entanglement)

All PPT entangled states are bound entangled, i.e., they cannot be distilled.

This corollary can also be attributed to the observation that the positivity of the partial transpose is *tensor-product stable* (or simply "tensor stable") meaning that if $\rho_{A_1 B_1}$ is PPT, then so is $\rho_{A_1 B_2} \otimes \rho_{A_2 B_2}$, or indeed any $\rho_{AB}^{\otimes N}$ when considering bipartitions into subsystems $(A_1, A_2, \ldots, A_N)$ and $(B_1, B_2, \ldots, B_N)$, and no LOCC protocol can therefore bring about an NPT state like $|\psi^-\rangle$ (Horodecki *et al.*, 1998).

However, on a more encouraging note, Theorem 16.4 also allows us to conclude that all entangled isotropic states and all entangled two-qubit states are "free entangled", i.e., distillable, since the reduction criterion is equivalent to the PPT criterion for these states and its violation is even necessary and sufficient for detecting entanglement in this case.

Corollary 16.2 (Distillability of isotropic and two-qubit states)

All entangled isotropic states and all entangled two-qubit states are distillable.

Combining our observations so far, we note that there is still a considerable gap in the possibility for distillation in higher-dimensional Hilbert spaces. There could be states that either are PPT entangled or that are NPT entangled but do not violate the reduction criterion (NPT bound entanglement), and, remarkably, such states exist (Horodecki *et al.*, 1998). However, although it is known that there exist NPT states that are not distillable for any finite number of copies (Watrous, 2004), it is still unknown if all NPT states are distillable asymptotically (Pankowski *et al.*, 2010). As far as the entanglement of distillation $E_\mathrm{D}(\rho)$ is concerned this means that

$$E_\mathrm{D}(\rho) = 0 \quad \not\Rightarrow \quad \rho \text{ separable} , \qquad (16.45)$$

the distillable entanglement can be zero for entangled states. We will discuss this issue further by investigating the regions of bound entanglement for the particular case of two qutrits (3×3 dimensions) in Section 17.3.

16.2   Entanglement Measures for Mixed States

We have seen so far that we can define operationally well-defined entanglement measures—the entropy of entanglement, the entanglement cost, and the distillable entanglement—that coincide and can also be straightforwardly calculated for pure states. For mixed states, the entropy of entanglement itself no longer carries any significance since the entropy of the reduced states may originate from entanglement of the joint state or its mixedness. The other two measures are well-defined and maintain their operational meaning, but in general we have no practical way of calculating them for a given density matrix. But when looking for an alternative, we can use what we have learned so far to formulate some basic requirements along with some optional desired properties for an entanglement measure $E(\rho)$ for mixed states.

16.2.1   Requirements for Entanglement Measures and Monotones

Quite generally, and ideally, we are looking for maps E from a bipartite Hilbert–Schmidt space $\mathcal{B}_\mathrm{HS}(\mathcal{H}_A \otimes \mathcal{H}_B)$ to the non-negative real numbers,

$$\begin{aligned} E : \ \mathcal{B}_\mathrm{HS}(\mathcal{H}_A \otimes \mathcal{H}_B) &\longrightarrow \mathbb{R}_+ \\ \rho &\longmapsto E(\rho) \end{aligned} \qquad (16.46)$$

where one can include a normalization factor such that $E(\rho) \in [0, \log(d)]$ for a $d \times d$-dimensional Hilbert space. The minimal requirements for such a measure are then:

(i) $E(\rho) = 0$ if ρ is **separable**,

(ii) $E(\rho) = \log(d)$ if ρ is **maximally entangled**,

(iii) **Non-increasing under LOCC on average**: For any LOCC protocol that takes ρ to ρ_i with probability p_i: $\sum_i p_i E(\rho_i) \leq E(\rho)$,

(iv) **Reduces to entropy of entanglement for pure states**: $E(|\psi\rangle\langle\psi|) = \mathcal{E}(|\psi\rangle)$.

By design, these requirements are chosen such that they are satisfied by the entanglement cost E_C and the distillable entanglement E_D. Following the convention used in, e.g., Plenio and Virmani (2007), a function E as in eqn (16.46) that satisfies (i)–(iii) but not (iv) is called an *entanglement monotone*, whereas the term *entanglement measure* is reserved for maps E that satisfy (i), (ii), and (iv) but instead of (iii) only the weaker requirement of not increasing under deterministic LOCC operations is made, i.e., (iii) only needs to hold when $p_i = 1$ for some i. In particular, this means that both entanglement measures and entanglement monotones must be *invariant under* (reversible) *local unitary transformations*, $E\big((U_A \otimes U_B)\,\rho\,(U_A \otimes U_B)^\dagger\big) = E(\rho)$. In addition to these basic requirements, there are number of more restricting properties that one would ideally like an a priori hypothetical, ideal entanglement measure to have, some of the most common of these are:

(v) $E(\rho) = 0 \iff \rho$ is **separable**,

(vi) **Asymptotic continuity**: $\dfrac{E(\rho_1) - E(\rho_2)}{\log(d)} \longrightarrow 0$ for $\|\rho_1 - \rho_2\|_{\mathrm{HS}} \longrightarrow 0$,

(vii) **Convexity**: $E\big(\sum_i p_i\, \rho_i\big) \le \sum_i p_i\, E(\rho_i)$,

(viii) **Additivity**: $E(\rho_1 \otimes \rho_2) = E(\rho_1) + E(\rho_2)$ for all ρ_1 and ρ_2,

(ix) **Additivity on copies**: $E(\rho^{\otimes n}) = n\, E(\rho)$ for all ρ,

(x) **Subadditivity**: $E(\rho_1 \otimes \rho_2) \le E(\rho_1) + E(\rho_2)$.

First, let us remark that the three additivity requirements are not independent since, clearly, (full) additivity (viii) implies additivity on copies (ix) and subadditivity (x), whereas the properties of additivity on copies and subadditivity do not imply full additivity, or each other in general. However, for any convex entanglement monotone additivity on copies implies full additivity (Fukuda and Wolf, 2007). Nevertheless we can already see that these additional requirements are not all met by all of the measures we have encountered so far and for a long time it was unclear if they could all be met by any single measure at all.

The distillable entanglement, on the one hand, vanishes for bound-entangled states and so $E_D = 0$ does not imply separability as would be required in (v), and there is some evidence (Shor *et al.*, 2001) that distillable entanglement may neither be fully additive (viii) nor convex (vii), despite the fact that it is additive on copies[2] (ix) by definition.

On the other hand, the entanglement cost is convex (Donald *et al.*, 2002) and since it is additive on copies (ix) by construction, it is also fully additive, while strict positivity for entangled states [$\Rightarrow$ in (v)] has been proven in Yang *et al.* (2005). However, it is not known for either the entanglement cost or the distillable entanglement whether or not they are asymptotically continuous in general (see, e.g., Table 3.4 in Christandl, 2006), but they are asymptotically continuous for pure states.

---

[2]Note that, if the distillable entanglement was convex, full additivity would follow via the result of Fukuda and Wolf (2007).

Squashed entanglement: However, there is an entanglement measure that does indeed satisfy (v)–(viii), the so-called *squashed entanglement* E_{Sq}, defined as

$$E_{\text{Sq}}(\rho_{AB}) = \inf\Big\{S(A:B|C)| \text{ with } \rho_{ABC} : \text{Tr}_C(\rho_{ABC}) = \rho_{AB}\Big\}, \qquad (16.47)$$

where $S(A:B|C)$ is the *quantum conditional mutual information* given by

$$S(A:B|C) = S(\rho_{AC}) + S(\rho_{BC}) - S(\rho_{ABC}) - S(\rho_C) \qquad (16.48)$$

with $S(\rho)$ the von Neumann entropy (see Definition 16.1) and the reduced states $\rho_{AC} = \text{Tr}_B(\rho_{ABC})$, $\rho_{BC} = \text{Tr}_A(\rho_{ABC})$, and $\rho_C = \text{Tr}_{AB}(\rho_{ABC})$.

As we will discuss in more detail in Section 20.4.2, the mutual information $S(A\!:\!B)$ can be understood as a measure of correlations that does not distinguish between classical and quantum correlations, but by considering the conditioning on a third system C and minimizing over all choices of C, one is—as *Matthias Christandl* and *Andreas Winter* put it (2004)—*"squashing out"* the purely classical correlations. Without going more into detail here with regards to the interpretation of the quantity $S(A:B|C)$, let us simply report that $S(A:B|C)$ was first considered in the context of quantum theory by *Nicolas Cerf* and *Christoph Adami* (1996b), followed by a series of investigations by *Robert Tucci* between 1999 and 2002, as well as by *Patrick Hayden, Richard Jozsa, Denes Petz*, and *Andreas Winter* in 2004 exploring the connection to entanglement.

But only with the breakthrough by Matthias Christandl and Andreas Winter in 2003 (published in 2004), which also established the name "squashed entanglement", was E_{Sq} shown to be a true entanglement measure (in particular, non-increasing under LOCC), that is also convex (vii) and additive (viii). Following work by *Robert Alicki* and *Mark Fannes* (2003) showed that squashed entanglement is asymptotically continuous (vi), while *Fernando Brandão, Matthias Christandl* and *Jon Yard* (2011) finally showed strict positivity for entangled states (v).

From its definition it is also easy to see that $E_{\text{Sq}}(\rho_{AB})$ reduces to the entropy of entanglement for pure states $\rho_{AB} = |\psi\rangle\langle\psi|$ since this implies that ρ_{ABC} is a product state of the form $\rho_{ABC} = \rho_{AB} \otimes \rho_C$ and $S(\rho_1 \otimes \rho_2) = S(\rho_1) + S(\rho_2)$, as we shall see in Section 20.1. Moreover, it was also shown in Christandl and Winter (2004) that E_{Sq} is a lower bound to the entanglement cost, as well as an upper bound to the distillable entanglement,

$$E_{\text{D}} \leq E_{\text{Sq}} \leq E_{\text{C}}. \qquad (16.49)$$

Because E_{Sq} is non-zero also for bound-entangled states one generally has $E_{\text{D}} \neq E_{\text{Sq}}$, and there are also examples (see, e.g., Christandl, 2006, Sec. 4.3.2) showing that generally $E_{\text{Sq}} \neq E_{\text{C}}$, which means that the inequality signs in (16.49) are generally strict. In particular this means that, despite its abstractly appealing properties, the squashed entanglement lacks the useful operational interpretation of E_{D} and E_{C}. In addition, the minimization in its definition means that calculating E_{Sq} is generically extremely difficult. In fact much like for entanglement cost (and other measures that we are yet to encounter), calculating the squashed entanglement is NP hard (Huang, 2014).

For more detailed discussions of these properties, their relations, and the zoo of entanglement measures that arise from these considerations we refer the interested reader to Christandl (2006), Plenio and Virmani (2007), and Horodecki *et al.* (2009). In the following, we will discuss some selected measures that either help us to better understand entanglement conceptually or that provide some practical way of actually being calculated.

16.2.2 Entanglement of Formation and Concurrence

As we have remarked at the beginning of this section, the entropy of entanglement works very well for pure states but is in itself useless for mixed states. To illustrate this, consider, for example, the two density operators

$$\rho_{\text{ent}} = \tfrac{1}{4}\big(\,|\,00\,\rangle\langle\,00\,| + |\,00\,\rangle\langle\,11\,| + |\,11\,\rangle\langle\,00\,| + |\,11\,\rangle\langle\,11\,|\,\big) = |\,\phi^+\,\rangle\langle\,\phi^+\,|\,, \quad (16.50)$$

$$\rho_{\text{sep}} = \tfrac{1}{2}\big(\,|\,00\,\rangle\langle\,00\,| + |\,11\,\rangle\langle\,11\,|\,\big) = \tfrac{1}{2}\big(\,|\,\phi^+\,\rangle\langle\,\phi^+\,| + |\,\phi^-\,\rangle\langle\,\phi^-\,|\,\big)\,. \quad (16.51)$$

The first state ρ_{ent} is entangled; it is the Bell state $|\,\phi^+\,\rangle$, the upper left tip of the tetrahedron in Fig. 15.3. The second state ρ_{sep} is a mixture of two Bell states and is separable, which is clear from the fact that it has a separable decomposition, and in the tetrahedron of Weyl states in Fig. 15.3; it is located on the top of the double pyramid. For both of the states ρ_{ent} and ρ_{sep}, the trace over Bob's subspace provides the same reduced density matrix for Alice, $\rho_A = \text{Tr}_B(\rho_{\text{ent}}) = \text{Tr}_B(\rho_{\text{sep}})$, and thus the same entropy of entanglement, if we were to simply adopt Definition 16.2 for mixed states as well.

Fortunately, the definition can be extended to mixed states in suitable way and the resulting entanglement measure E_F is called *entanglement of formation* (EOF) and we will now introduce it in several steps. As the name suggests, the quantity E_F is connected to the formation of a quantum state, but, as we shall see, without resorting to the asymptotic regime of infinitely many copies. For pure states, we simply equate the EOF with the entropy of entanglement:

(1) **EOF for pure states** ρ_{pure}: $\quad E_F(\rho_{\text{pure}}) = \mathcal{E}(\rho_{\text{pure}})$.

Now let us consider a particular ensemble of pure states ρ_{pure}^i, labelled by i, each of which is produced by a source with probability p_i. The amount of pure-state entanglement, quantified by the entropy of entanglement, that is needed per copy to prepare this ensemble is just the convex sum of the individual values $\mathcal{E}(\rho_{\text{pure}}^i)$.

(2) **EOF of pure-state ensemble** $\{p_i, \rho_{\text{pure}}^i\}$: $\quad E_F\big(\{p_i, \rho_{\text{pure}}^i\}\big) = \sum_i p_i\,\mathcal{E}(\rho_{\text{pure}}^i)$.

Finally, for any given mixed quantum state ρ, there are (generally infinitely) many different pure-state ensembles into which the state could be decomposed and some of these will overestimate how much entanglement is needed to form the state. Consider, for example, the maximally mixed two-qubit state $\rho_{\text{mix}} = \frac{1}{4}\mathbb{1}$. On the one hand, it can be prepared as an equal mixture of the four Bell states

$$\rho_{\text{mix}} = \tfrac{1}{4}\left(\,|\,\phi^+\,\rangle\langle\,\phi^+\,|\,+\,|\,\phi^-\,\rangle\langle\,\phi^-\,|\,+\,|\,\psi^+\,\rangle\langle\,\psi^+\,|\,+\,|\,\psi^-\,\rangle\langle\,\psi^-\,|\,\right). \qquad (16.52)$$

The entropy of entanglement is maximal for each of the Bell states, $\mathcal{E}(\rho_{\text{Bell}}) = \log(2)$, therefore the EOF of such an ensemble is also maximal. On the other hand, we can also prepare the state ρ_{mix} as an equal mixture of four orthogonal product states,

$$\rho_{\text{mix}} = \tfrac{1}{4}\left(\,|\,00\,\rangle\langle\,00\,|\,+\,|\,01\,\rangle\langle\,01\,|\,+\,|\,01\,\rangle\langle\,01\,|\,+\,|\,11\,\rangle\langle\,11\,|\,\right), \qquad (16.53)$$

and it is clear that this ensemble requires no entanglement to be prepared. Therefore the EOF for such an ensemble vanishes. For any fixed density operator ρ one therefore introduces a minimization (technically, an infimum) over all pure-state decompositions $\mathcal{D}(\rho)$, i.e., where $\mathcal{D}(\rho)$ is the set of all sets $\{p_i, \rho^i_{\text{pure}}\}_i$ such that $\rho = \sum_i p_i \rho^i_{\text{pure}}$ with $\sum_i p_i = 1$ and $0 \leq p_i \leq 1$. Such a construction is known as a *convex-roof construction* (Uhlmann, 1998), i.e., the entanglement of formation is the convex roof of the entropy of entanglement, and indeed many relevant quantities[3] in quantum information can be defined in this way for mixed states starting from the respective pure-state quantities. A specific decomposition which realizes the infimum is called an *optimal* decomposition of ρ.

(3) **EOF for mixed states** ρ: $\qquad E_{\text{F}}(\rho) = \inf_{\mathcal{D}(\rho)} \sum_i p_i\,\mathcal{E}(\rho^i_{\text{pure}}).$

Properties and interpretation of the EOF: The EOF defined in this way was introduced by *Charles Bennett, David Di Vincenzo, John Smolin*, and *William Wootters* (1996c). Since the entanglement of formation reduces to the entropy of entanglement for pure states by construction, it trivially satisfies properties (i), (ii), and (iv). Moreover, Bennett *et al.* (1996c) showed that the EOF does not increase under LOCC on average (iii), and hence is a proper entanglement measure.

Of the additional properties that one might desire for an entanglement measure it is quite easy to see that the EOF must be non-zero for any entangled state (v) since all decompositions of an entangled mixed state must contain (with non-zero probability weight) at least one entangled pure state (with non-zero entropy of entanglement) by definition. In addition, the original paper by Bennett *et al.* demonstrated that the EOF is continuous (vi), convex (vii), and subadditive (x). However, full additivity[4], although conjectured to hold for the EOF (see, e.g., the discussion in Plenio and Virmani, 2007) and supported by evidence for certain special cases (see, e.g., Benatti and Narnhofer, 2001; Vollbrecht and Werner, 2001; Vidal *et al.*, 2002), remained unproven for a long time.

---

[3]For instance, for certain situations, the so-called quantum Fisher information can be obtained as a convex roof of the variance, see eqn (24.72).

[4]Or additivity on copies (ix), which would be equivalent via the result of Fukuda and Wolf (2007) since the EOF is convex.

The particular significance of the question of additivity of E_F arises from the fact that it was shown in Hayden *et al.* (2001) that the entanglement cost is the so-called *regularization* of the entanglement of formation, that is, the entanglement cost arises as the limiting case of the EOF when infinitely many copies of ρ are jointly processed,

$$E_C(\rho) = \lim_{n \to \infty} \tfrac{1}{n} E_F(\rho^{\otimes n}). \qquad (16.54)$$

If the entanglement of formation would have turned out to be fully additive (viii), or even just additive on copies (ix), then it would have been equal to the entanglement cost. However, building on the equivalence of certain major additivity problems in quantum information, proven by *Peter Shor* in 2004, and on work by *Patrick Hayden* and *Andreas Winter* (2008), it was shown by *Matthew Hastings* in 2009 that the entanglement of formation is not additive in general. Together with the convexity of E_F (see Section 16.2.2), the result of *Motohisa Fukuda* and *Michael Wolf* (2007) implies that the entanglement of formation can also not be additive on copies and hence does not match the entanglement cost in general.

Consequently, the entanglement of formation is indeed *not* the minimal cost to prepare a given mixed state via LOCC, it is generally larger than then entanglement cost, such that the chain of inequalities from (16.49) is extended to

$$E_D \leq E_{Sq} \leq E_C \leq E_F. \qquad (16.55)$$

But the EOF nevertheless expresses the cost of preparing a quantum state via a *specific protocol*. That is, the state ρ can be created by preparing a number of identical systems in the pure states corresponding to the optimal decomposition, whose respective minimal preparation cost via LOCC (their entanglement cost) is given by their entropy of entanglement, and then "forgetting" (erasing the information) which system is in which state. The average entanglement cost per copy is then just the entanglement of formation. What is interesting to note here is that the inequivalence of the entanglement cost and the entanglement of formation means that one can in general save some of the cost of creating an entangled state by jointly processing multiple copies, i.e., the cost per copy is lower when creating n copies rather than just one.

Bounds on the EOF: The main advantage of the EOF with respect to the entanglement cost stems from the observation that, despite also being as difficult to calculate exactly in general (see, e.g., Huang, 2014), it can be calculated for certain special cases[5] and there are some known bounds on the EOF that can be easily evaluated for more general situations. For instance, already the team around Bennett (Bennett *et al.*, 1996*a*; Bennett *et al.*, 1996*c*) found a very simple formula, a lower bound, for the case of two qubits (dimension 2×2) that we will phrase in terms of the following theorem:

---

[5]In Section 25.4 we will encounter another simple situation, so-called symmetric two-mode Gaussian states for continuous-variable states on infinite-dimensional Hilbert spaces, for which the EOF can also be computed explicitly (Giedke *et al.*, 2003).

Theorem 16.5 (EOF bound)

The entanglement of formation of a two-qubit state ρ is bounded from below by

$$E_{\mathrm{F}}(\rho) \geq h\big(\mathcal{F}_{\max}(\rho)\big) \,,$$

where $\mathcal{F}_{\max}(\rho)$ is the singlet fraction from eqn (15.120), and the function $h\big(\mathcal{F}_{\max}(\rho)\big)$ is defined by

$$h\big(\mathcal{F}_{\max}\big) := \begin{cases} H_{\mathrm{bin}}\big(\tfrac{1}{2} + \sqrt{\mathcal{F}_{\max}(1 - \mathcal{F}_{\max})}\big) & \text{for} \quad \mathcal{F}_{\max} \geq 1/2 \\ 0 & \text{for} \quad \mathcal{F}_{\max} < 1/2, \end{cases}$$

with the binary entropy function

$$H_{\mathrm{bin}}(p) = -p \log_2(p) - (1 - p) \log_2(1 - p).$$

Here, the function $H_{\mathrm{bin}}(p)$ is just the *Shannon entropy* $H(\{p, 1 - p\})$ from classical information theory for the *Bernoulli distribution* $\{p, 1 - p\}$, which we will discuss in more detail in Section 19.2. For general mixed states ρ the function $h\big(\mathcal{F}_{\max}(\rho)\big)$ is only a lower bound to the EOF but for pure states and mixtures of Bell states, which are diagonal in the Bell basis, the bound is tight, i.e.,

$$E_{\mathrm{F}}(\rho_{\mathrm{pure}}) = h\big(\mathcal{F}_{\max}(\rho_{\mathrm{pure}})\big), \qquad E_{\mathrm{F}}(\rho_{\mathrm{mix}}^{\mathrm{Bell}}) = h\big(\mathcal{F}_{\max}(\rho_{\mathrm{mix}}^{\mathrm{Bell}})\big), \qquad (16.56)$$

and the EOF can in principle be easily calculated via the binary entropy function $H_{\mathrm{bin}}(p)$ in Theorem 16.5. However, we shall not go through this calculation or the proof of Theorem 16.5 here, since an exact formula for the entanglement of formation of two qubits was found in a series of works by *Scott Hill* and *William Wootters* (1997), and Wootters (1998 and 2001), expressing the EOF as a function of another quantity, the *concurrence*, which we will discuss shortly. For higher-dimensional systems other methods can be used to bound the entanglement of formation and we will look into these in more detail in Section 17.4.2.

Concurrence: The definition of EOF for a general mixed quantum state as the infimum of the entropy of entanglement over all possible ensembles is quite plausible and theoretically appealing. As we have seen from the example in Theorem 16.5, there are also lower bounds available for the EOF, which are typically saturated only in special cases (e.g., for pure states and mixtures of Bell states). In general, however, it is hard to calculate its exact value since it is difficult to determine all possible ensembles of pure states which represent a given mixed state. Nevertheless, for two-qubit states, results by William Wootters (1998, 2001) provide a method to calculate the exact value of EOF in terms of another quantity, the so-called *concurrence*. In fact, the concurrence itself serves as an entanglement monotone which turns out to be quite practical.

Let us start with some definitions. The *spin-flip operation* on a single-qubit pure state $|\psi\rangle$ is represented by the Pauli matrix σ_y

$$|\psi\rangle \longrightarrow |\tilde{\psi}\rangle = \sigma_y |\psi^*\rangle \,, \tag{16.57}$$

where the asterisk $*$ denotes complex conjugation with respect to the computational basis (the eigenbasis of σ_z). The operation is called *spin-flip* since it flips the spin of a state when interpreting the computational basis as the eigenstates of the spin-z operator, i.e.,

$$\sigma_y |\uparrow\rangle = i |\downarrow\rangle \,, \quad \sigma_y |\downarrow\rangle = -i |\uparrow\rangle \,, \tag{16.58}$$

with $|\uparrow\rangle \equiv |0\rangle, |\downarrow\rangle \equiv |1\rangle$ in the usual quantum-information notation. The spin-flip operation on a two-qubit density matrix can be expressed as

$$\rho \longrightarrow \tilde{\rho} = (\sigma_y \otimes \sigma_y) \rho^* (\sigma_y \otimes \sigma_y) \,. \tag{16.59}$$

With this, we can formulate the following two theorems (Hill and Wootters, 1997; Wootters, 1998 and 2001):

Theorem 16.6 (Concurrence)

The concurrence C of a two-qubit pure state $|\psi\rangle$ is given by

$$C(|\psi\rangle) = |\langle \psi | \tilde{\psi} \rangle|,$$

and the concurrence of a general two-qubit mixed state ρ is given by

$$C(\rho) = \max\{0, \lambda_1 - \lambda_2 - \lambda_3 - \lambda_4\},$$

where the λ_i are the square roots of the eigenvalues of the matrix $\rho\tilde{\rho}$ in decreasing order.

Theorem 16.7 (EOF—concurrence)

The entanglement of formation E_{F} of a general two-qubit state ρ is a function of the concurrence C

$$E_{\mathrm{F}}(\rho) = E_{\mathrm{F}}(C(\rho)) = H_{\mathrm{bin}}\left(\tfrac{1}{2}(1 + \sqrt{1 - C^2}\,)\right),$$

where H_{bin} is the binary entropy function from Theorem 16.5.

As the concurrence C increases from 0 to 1, the EOF E_{F} also increases monotonically from 0 to 1. Thus the concurrence itself is a valid entanglement monotone for two qubits that satisfies the properties (i)–(iii), but which does not reduce to the entropy of entanglement (iv). The concurrence is therefore sometimes used instead of the EOF, despite the fact that the former derives its meaning entirely via its relation to the latter. In particular, although there are some approaches towards extending the concurrence to higher dimensions (see, e.g., Audenaert *et al.*, 2001; Badziąg *et al.*, 2002; Rungta *et al.*, 2001), there is generally no unique definition for the concurrence beyond two qubits.

Tangle: However, a useful entanglement measure can be obtained when starting from the square of the pure-state concurrence, that is, for two-qubit pure states $|\psi\rangle$ we can define the *tangle* $\tau(|\psi\rangle)$ as

$$\tau(|\psi\rangle) = C^2(|\psi\rangle). \tag{16.60}$$

Writing the state as $|\psi\rangle = \sum_{m,n} c_{mn} |mn\rangle$ with $c_{mn} \in \mathbb{C}$ with respect to the computational basis, one can easily check that the tangle takes the form

$$\tau(|\psi\rangle) = |\langle\psi|\sigma_y \otimes \sigma_y|\psi^*\rangle|^2 = 4|c_{00}c_{11} - c_{01}c_{10}|^2. \tag{16.61}$$

Similarly, we can calculate the reduced state of Alice's qubit with respect to this basis decomposition,

$$\rho_A = \mathrm{Tr}_B(|\psi\rangle\langle\psi|) = \mathrm{Tr}_B\Big(\sum_{\substack{m,n\\m',n'}} c^*_{m'n'}c_{mn}|mn\rangle\langle m'n'|\Big) = \sum_{m,n,m'} c^*_{m'n}c_{mn}|m\rangle\langle m'|. \tag{16.62}$$

Since the normalization implies $\sum_{m,n} c^*_{mn}c_{mn} = 1$ we can further write the *linear entropy* from eqn (11.21) as

$$S_{\mathrm{L}}(\rho_A) = 1 - \mathrm{Tr}(\rho_A^2) = \Big(\sum_{m,n} c^*_{mn}c_{mn}\Big)^2 - \sum_{\substack{m,n\\m',n'}} c^*_{mn}c_{mm'}c^*_{n'm'}c_{n'n}$$

$$= \tfrac{1}{2}\sum_{\substack{m,n\\m',n'}} |c_{mn}c_{m'n'} - c_{mn'}c_{m'n}|^2, \tag{16.63}$$

where the last step requires some relabelling of summation indices to combine matching terms. For qubits, $m, n, m', n' \in \{0,1\}$, the sum contains four equal non-vanishing terms for $(m, n, m', n') = (0,0,1,1), (1,1,0,0), (0,1,1,0,$ and $(1,0,0,1)$ such that we have

$$S_{\mathrm{L}}(\rho_A) = 2|c_{00}c_{11} - c_{01}c_{10}|^2 = \tfrac{1}{2}\tau(|\psi\rangle). \tag{16.64}$$

For pure two-qubit states the tangle is hence just twice[6] the linear entropy of the reduced state, or, equivalently, the *linear entropy of entanglement* $\mathcal{E}_{\mathrm{L}}(|\psi\rangle) = S_{\mathrm{L}}(\rho_A)$, as the analogue of Definition 16.2 for the linear entropy. This means that one may extend the definition of the tangle to mixed states via a convex-roof construction over

---

[6]Note that conventions differ for the linear entropy and some authors define linear entropy with an additional pre-factor of 2, see, e.g., (Osborne and Verstraete, 2006).

the linear entropies of the subsystems (Osborne and Verstraete, 2006) in exactly the same way (up to a factor 2) as the entanglement of formation is constructed as a convex roof over the von Neumann entropies of the reduced states,

$$\tau(\rho) := 2 \inf_{\mathcal{D}(\rho)} \sum_i p_i \mathcal{E}_{\mathrm{L}}(\rho_{\mathrm{pure}}^i). \qquad (16.65)$$

This definition now naturally extends to bipartite systems of any dimension. But one should note that the tangle is generally not identical to the square of the suitable mixed-state extensions of the concurrence beyond two qubits (see, e.g., Osborne, 2005).

Monogamy of entanglement: Nonetheless, a very useful feature of the tangle emerges for more than two parties in the sense that it allows one to capture the so-called *monogamy of entanglement*. For three qubits A, B, and C the tangle satisfies the *Coffman–Kundu–Wootters* (CKW) inequality

$$\tau(\rho_{AB}) + \tau(\rho_{AC}) \leq \tau(\rho_{A(BC)}), \qquad (16.66)$$

where $\rho_{AB} = \mathrm{Tr}_C(\rho_{ABC})$, $\rho_{AC} = \mathrm{Tr}_B(\rho_{ABC})$, and $\tau(\rho_{A(BC)})$ is the tangle with respect to the bipartition $A|BC$. The CKW inequality was first proven by *Valerie Coffman, Joydip Kundu*, and *William Wootters* (2000), and subsequently extended to an arbitrary number of qubits by *Frank Verstraete* and *Tobias Osborne* (2006). Loosely speaking, the observation behind this phenomenon is that a quantum system can generally not be entangled arbitrarily strongly with multiple other quantum systems at the same time, in particular, a given qubit cannot be maximally entangled with more than one qubit at a time. Any gain in entanglement between qubits A and C must be compensated by a reduction in entanglement between A and B. The monogamy inequalities using the tangle do not generally hold beyond qubits (see, e.g., Ou, 2007), but possible extensions to qudits relying on other measures of entanglement have been proposed (Kim *et al.*, 2009), for instance squashed entanglement satisfies monogamy constraints in arbitrary dimensions (Koashi and Winter, 2004) and monogamy can be restored when restricting to specific sets of states and particular entanglement quantifiers, e.g., for Gaussian states in infinite-dimensional systems (Hiroshima *et al.*, 2007).

Example—concurrence and EOF of Werner states: To illustrate the connection between the concurrence and the entanglement of formation, let us now determine these quantities for the Werner states from eqn (15.68). For the concurrence we first have to calculate the spin-flipped Werner states $\tilde{\rho}_\alpha$ by applying the operation in (16.59). However, the Werner states only have real components with respect to the computational basis, and are hence unaffected by the complex conjugation (with respect to this basis). Moreover, as we recall from the discussion following eqn (16.31), the Werner states are invariant under all local unitaries of the form $U \otimes U$, and therefore in particular when $U = \sigma_y$. Therefore one finds that the spin-flipped states coincide with the original states $\tilde{\rho}_{\mathrm{Werner}}(\alpha) \equiv \rho_{\mathrm{Werner}}(\alpha)$.

Next we need the square roots of the eigenvalues of the matrix $\rho_{\mathrm{Werner}} \tilde{\rho}_{\mathrm{Werner}} = \rho_{\mathrm{Werner}}^2$, which are nothing but the eigenvalues of the Werner states $\rho_{\mathrm{Werner}}(\alpha)$

$$\lambda_1 = \frac{1 + 3\alpha}{4}, \quad \lambda_2 = \lambda_3 = \lambda_4 = \frac{1 - \alpha}{4}. \tag{16.67}$$

Subtracting the eigenvalues in decreasing order gives

$$\lambda_1 - \lambda_2 - \lambda_3 - \lambda_4 = \frac{3\alpha - 1}{2}, \tag{16.68}$$

so that, according to Theorem 16.6, we find the concurrence of the Werner states

$$C[\rho_{\text{Werner}}(\alpha)] = \begin{cases} 0 & \text{for} \quad -1/3 \le \alpha \le 1/3 \\ \frac{3\alpha - 1}{2} & \text{for} \quad 1/3 < \alpha \le 1. \end{cases} \tag{16.69}$$

The result for the concurrence is in agreement with our findings in Section 15.3.3. Further inserting the concurrence into the formula for the EOF from Theorem 16.7 we find

$$E_{\text{F}}(\rho_{\text{Werner}}) = \frac{2 + \gamma}{4} \log_2\left(\frac{4}{2 + \gamma}\right) + \frac{2 - \gamma}{4} \log_2\left(\frac{4}{2 - \gamma}\right), \tag{16.70}$$

with $\gamma = \sqrt{3 + 6\alpha - 9\alpha^2}$. The comparison of the expression in (16.69) with (16.70) shows that the concurrence has a much simpler form than the EOF, which makes it a little bit easier to work with.

16.2.3   Entanglement Measures Based on Distance

To round off our discussion of different entanglement measures, let us briefly take a look at geometric entanglement measures based on the distance of entangled states to suitable sets of states, e.g., to the set of separable states[7]. Let $D(\rho_1, \rho_2)$ denote a *distance function* of two states ρ_1 and ρ_2, which will be specified shortly. Then we can define a *distance measure* $\mathcal{D}(\rho_{\text{ent}})$ of an entangled state ρ_{ent} as the minimal distance of ρ_{ent} to the set $\mathcal{S}$ of separable states

$$\mathcal{D}(\rho_{\text{ent}}) := \min_{\rho \in \mathcal{S}} D(\rho_{\text{ent}}, \rho). \tag{16.71}$$

One now has several options to construct such a distance function, two of which we will now describe in more detail.

Relative entropy of entanglement is a prominent entanglement measure based on a quantity called the *quantum relative entropy* $S(\rho_1 \| \rho_2)$, which we will discuss more deeply in Section 20.3 in the context of quantum entropies, and which is given by

$$S(\rho_1 \| \rho_2) = \text{Tr}(\rho_1 [\log_2(\rho_1) - \log_2(\rho_2)]) \tag{16.72}$$

for two density matrices ρ_1 and ρ_2. What we notice first about $S(\rho_1 \| \rho_2)$ is that it is a strange distance measure, it is *not symmetric*, i.e., in general $S(\rho_1 \| \rho_2) \neq S(\rho_2 \| \rho_1)$.

---

[7]More generally, one can consider distances to sets that are closed under LOCC, e.g., the set of undistillable states or the set of PPT states (see, for instance, the discussion in Verstraete *et al.*, 2002*a*), but we shall not go into this discussion here and refer the interested reader to Horodecki *et al.* (2009, Sec. XV.C.1).

We can nevertheless use it here as a one-sided distance, e.g., from a bipartite state ρ_{AB} to the product state of its marginals, $\rho_A \otimes \rho_B$. As we will see in Lemma 20.2, this yields $S(\rho \| \rho_A \otimes \rho_B) = S(\rho_A) + S(\rho_B) - S(\rho_{AB})$, which gives rise to a measure of correlations called the quantum mutual information sometimes denoted as $S(A:B)$ or as $\mathcal{I}(\rho_{AB})$

$$S(A:B) = \mathcal{I}(\rho_{AB}) = S(\rho_A) + S(\rho_B) - S(\rho_{AB}). \tag{16.73}$$

The mutual information is a measure of total correlations, i.e., it does not distinguish between classical correlations and entanglement, and can thus be non-zero for separable states, as long as they are classically correlated. However, even for a maximally classically correlated state of the form $\rho_{\mathrm{MCC}} = \frac{1}{d} \sum_{m=1}^{d} | mm \rangle\langle mm |$ for which $S(\rho_{\mathrm{MCC}}) = \log(d)$ and both marginals are maximally mixed, $\rho_A = \rho_B = \frac{1}{d} \sum_{m=1}^{d} | m \rangle\langle m |$ with $S(\rho_A) = S(\rho_B) = \log(d)$ such that $S(A:B) = \mathcal{I}(\rho_{\mathrm{MCC}}) = \log(d)$. Any larger values of the mutual information hence detect entanglement. One way to quantify entanglement using the mutual information is to consider the *minimal average mutual information* over all decompositions. Indeed, with a factor of $\frac{1}{2}$ in front, it is easy to see that this gives exactly the entanglement of formation, because the entropy of the pure states vanishes, such that $\mathcal{I}(| \psi_i \rangle) = S(\rho_A) + S(\rho_B) = 2S(\rho_A)$ and we hence have

$$\frac{1}{2} \inf_{\mathcal{D}(\rho)} \sum_i p_i \, \mathcal{I}(| \psi_i \rangle) = E_{\mathrm{F}}(\rho). \tag{16.74}$$

When we now replace the "distance" to the product state by the minimal distance to any separable state as originally mentioned, we obtain the *relative entropy of entanglement*

$$E_{\mathrm{RE}}(\rho) := \inf_{\rho_{\mathrm{sep}} \in \mathcal{S}} S(\rho \| \rho_{\mathrm{sep}}), \tag{16.75}$$

which was introduced in a series of works by *Vlatko Vedral, Martin Plenio,* and collaborators (Vedral *et al.*, 1997*a*; Vedral *et al.*, 1997*b*; Vedral and Plenio, 1998; Vedral, 2002). In particular, it was shown in Vedral *et al.* (1997*b*) and Vedral and Plenio (1998) that $E_{\mathrm{RE}}(\rho)$ indeed satisfies all the required properties (i)–(iv) for a proper entanglement measure and is also strictly positive for entangled states (v), convex (vii), and subadditive (x). In addition, it was shown in Donald and Horodecki (1999) that the relative entropy of entanglement is asymptotically continuous (vi), and one finds that its value is bounded from below by the distillable entanglement (Vedral, 2002) and from above by the entanglement of formation (Vedral and Plenio, 1998), such that

$$E_{\mathrm{D}}(\rho) \leq E_{\mathrm{RE}}(\rho) \leq E_{\mathrm{F}}(\rho). \tag{16.76}$$

However, the relative entropy of entanglement is unfortunately not additive (Vollbrecht and Werner, 2001), i.e., neither (viii) nor (ix) hold.

Example—quantum relative entropy of Werner states: Let us again illustrate this measure via the Werner states $\rho_{\mathrm{Werner}}(\alpha)$ of eqn (15.68), whose eigenvalues λ_i are $(1 - \alpha)/4$ (thrice degenerate) and $(1 + 3\alpha)/4$. From geometric considerations (see

Section 15.5) we know that the closest separable state to any entangled Werner state must again be a Werner state, more specifically the Werner state $\rho_{\text{Werner}}(\alpha = \frac{1}{3})$ for $\alpha = \frac{1}{3}$ at the boundary between separability and entanglement. Denoting the eigenvalues of $\rho_{\text{Werner}}(\alpha = \frac{1}{3})$ as μ_i, we have $\mu_1 = \frac{1}{2}$ and $\mu_2 = \mu_3 = \mu_4 = \frac{1}{6}$. Then we find for the relative entropy of entanglement

$$
\begin{aligned}
E_{\text{RE}}(\rho_{\text{Werner}}) &= S\big(\rho_{\text{Werner}}(\alpha)\|\rho_{\text{Werner}}(\alpha = \tfrac{1}{3})\big) = \sum_i \lambda_i \big[\log_2(\lambda_i) - \log_2(\mu_i)\big] \\
&= \tfrac{3(1-\alpha)}{4}\left[\log_2\!\big(\tfrac{1-\alpha}{4}\big) + \log_2(6)\right] + \tfrac{1+3\alpha}{4}\left[\log_2\!\big(\tfrac{1+3\alpha}{4}\big) + 1\right] \\
&= c\log_2(c) + (1-c)\log_2(1-c) + 1\,,
\end{aligned}
\tag{16.77}
$$

with $c = (1+3\alpha)/4$. We can then easily check that this gives the expected results in the limits $\alpha \to \frac{1}{3}$ and $\alpha \to 1$, corresponding to $c \to \frac{1}{2}$ and $c \to 1$, respectively, for which we get

$$
E_{\text{RE}}\big(\rho_{\text{Werner}}(\alpha \to \tfrac{1}{3})\big) \longrightarrow 0 \quad \text{and} \quad E_{\text{RE}}\big(\rho_{\text{Werner}}(\alpha \to 1)\big) \longrightarrow 1\,. \tag{16.78}
$$

Entanglement measures based on metric distances: As we have mentioned, the (quantum) relative entropy is not an actual distance measure, but there are indeed a number of distance measures available for quantum states that are indeed also metrics on the Hilbert–Schmidt space, in particular, the Hilbert–Schmidt distance $D_{\text{HS}}(\rho_1, \rho_2)$ from Definition 11.6, but also the *trace distance* $D_{\text{tr}}(\rho_1, \rho_2)$ which we will discuss in Section 23.4, as well as the *Bures distance* D_{B} (Bures, 1969), which we will take a closer look at in Section 24.2.2 in the context of discussing the Uhlmann fidelity, to which the Bures distance is closely related. For any of these measures, we can formally define a minimum distance to the set of separable states using eqn (16.71), e.g., for the Hilbert–Schmidt distance we have the *Hilbert–Schmidt measure* (Witte and Trucks, 1999)

$$
\mathcal{D}_{\text{HS}}(\rho) = \min_{\rho_{\text{sep}} \in \mathcal{S}} D_{\text{HS}}(\rho, \rho_{\text{sep}}) = D_{\text{HS}}(\rho, \rho_0)\,, \tag{16.79}
$$

where ρ_0 denotes the closest separable state. By construction all geometric measures of this kind are non-zero if and only if ρ is entangled (v), they can be normalized (ii), and they are also invariant under local unitary operations. The measure $\mathcal{D}_{\text{B}}$ arising from the Bures distance also satisfies (iii), that is, it is also non-increasing under LOCC on average (see Vedral and Plenio, 1998) while the quantity $\mathcal{D}_{\text{Tr}}$ based on the trace distance is non-increasing under deterministic LOCC (see, e.g., Bengtsson and Życzkowski, 2006, p. 430), and both are also convex, even though properties (vi) and (ix) do not hold, that is, neither quantity is asymptotically continuous or additive on copies (see, for instance Bengtsson and Życzkowski, 2006, Table 16.2). Therefore, while they do not reduce to the entropy of entanglement (property (iv)), they still qualify as entanglement monotones. However, for the Hilbert–Schmidt "measure" $\mathcal{D}_{\text{HS}}$ it is not even clear if it is a monotone in the sense of (iii), since a stronger monotonicity property that would have implied that $\mathcal{D}_{\text{HS}}$ is non-increasing under LOCC was shown not to hold by *Masanao Ozawa* (2000). Nevertheless, the latter can serve as a geometric measure for the proximity of a given state to unentangled states that can provide some intuition. The advantage of the Hilbert–Schmidt measure lies in its easy calculation and it plays an important role in connection with entanglement witnesses, as we will discuss in Section 16.3.3.

Example—Hilbert–Schmidt measure of Werner states: Let us again investigate the Werner states as an example. Here it is convenient to start from their Bloch decomposition in eqn (15.69),

$$\rho_{\text{Werner}}(\alpha) = \tfrac{1}{4}\big(\mathbb{1} \otimes \mathbb{1} - \alpha \sum_i \sigma_i \otimes \sigma_i\big) \tag{16.80}$$

and we recall from Section 15.3.3 that they are separable for $-\tfrac{1}{3} \le \alpha \le \tfrac{1}{3}$ and entangled for $\tfrac{1}{3} < \alpha \le 1$. Again we can make use of geometric arguments from our discussion in Section 15.5.1 that the closest separable state is obtained for $\alpha = \tfrac{1}{3}$, and given by

$$\rho_0 = \rho_{\text{Werner}}(\alpha = \tfrac{1}{3}) = \tfrac{1}{4}\big(\mathbb{1} \otimes \mathbb{1} - \tfrac{1}{3}\sum_i \sigma_i \otimes \sigma_i\big) . \tag{16.81}$$

We then evaluate the difference between ρ_0 and an entangled state $\rho_{\text{Werner}}(\alpha > \tfrac{1}{3})$

$$\rho_0 - \rho_{\text{Werner}}(\alpha > \tfrac{1}{3}) = \tfrac{1}{4}(\alpha - \tfrac{1}{3})\sum_i \sigma_i \otimes \sigma_i, \qquad \text{for} \qquad \tfrac{1}{3} < \alpha \le 1 . \tag{16.82}$$

For the Hilbert–Schmidt norm $\|\sum_i \sigma_i \otimes \sigma_i\|_{\text{HS}}$ we find

$$\Big\|\sum_i \sigma_i \otimes \sigma_i\Big\|_{\text{HS}} = \sqrt{\text{Tr}\big(\sum_{i,j} \sigma_i\sigma_j \otimes \sigma_i\sigma_j\big)} = \sqrt{\sum_i \text{Tr}\big(\sigma_i^2 \otimes \sigma_i^2\big)} = 2\sqrt{3} . \tag{16.83}$$

Then the Hilbert–Schmidt distance evaluates to

$$\|\rho_0 - \rho_{\text{Werner}}(\alpha > \tfrac{1}{3})\|_{\text{HS}} = \tfrac{\sqrt{3}}{2}(\alpha - \tfrac{1}{3}), \qquad \text{for} \qquad \tfrac{1}{3} < \alpha \le 1 . \tag{16.84}$$

Now we can normalize this quantity to satisfy property (ii), which makes it easier to compare the Hilbert–Schmidt measure with other entanglement measures. Noting that $\|\rho_0 - \rho_{\alpha=1}\|_{\text{HS}} = \tfrac{1}{\sqrt{3}}$ we simply multiply by $\sqrt{3}$ to obtain

$$\mathcal{D}_{\text{HS}}[\rho_{\text{Werner}}(\alpha > \tfrac{1}{3})] = \frac{3\alpha - 1}{2} . \tag{16.85}$$

Remarkably, this coincides with the concurrence for Werner states from eqn (16.69),

$$\mathcal{D}_{\text{HS}}[\rho_{\text{Werner}}(\alpha > \tfrac{1}{3})] \equiv C[\rho_{\text{Werner}}(\alpha > \tfrac{1}{3})] . \tag{16.86}$$

Finally, when comparing to several other measures we have so far evaluated for the Werner states, we observe the following hierarchy

$$\mathcal{D}_{\text{HS}}(\rho_{\text{Werner}}) = C(\rho_{\text{Werner}}) \ge E_{\text{F}}(\rho_{\text{Werner}}) \ge E_{\text{RE}}(\rho_{\text{Werner}}) . \tag{16.87}$$

The corresponding curves in dependence of α are depicted in Fig. 16.2.

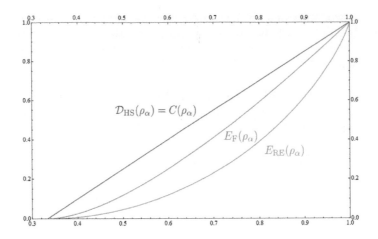

Fig. 16.2 Comparison of entanglement measures and monotones for the Werner states. The Hilbert–Schmidt measure $\mathcal{D}_{\mathrm{HS}}(\rho_{\mathrm{Werner}})$, concurrence $C(\rho_{\mathrm{Werner}})$ (both in blue), the entanglement of formation $E_{\mathrm{F}}(\rho_{\mathrm{Werner}})$ (orange), and the relative entropy of entanglement $E_{\mathrm{RE}}(\rho_{\mathrm{Werner}})$ (green) are plotted for the Werner states in the entangled parameter range $\frac{1}{3} \leq \alpha \leq 1$. For these particular states, the Hilbert–Schmidt distance to the separable states coincides with the concurrence, $\mathcal{D}_{\mathrm{HS}}(\rho_{\mathrm{Werner}}) = C(\rho_{\mathrm{Werner}})$.

16.2.4   Prominent Entanglement Monotones—Negativities

In addition to the entanglement monotones arising from geometric considerations, another important family of entanglement monotones is based directly on the PPT criterion (Theorem 15.3). These *negativity* measures were introduced by *Guifré Vidal* and *Reinhard Werner* (2002) and quantify—loosely speaking—how much a given state fails to be positive under partial transposition.

Definition 16.5   *The negativity $\mathcal{N}(\rho)$ of a bipartite state ρ is given by*

$$\mathcal{N}(\rho) := \tfrac{1}{2} \sum_i \left(|\lambda_i| - \lambda_i \right),$$

where the $\lambda_i \in [-\frac{1}{2}, 1]$ are the eigenvalues of ρ^{T_B}.

Thus, the negativity is the modulus of the sum of the at most $(d_A - 1)(d_B - 1)$ negative eigenvalues of the partial transposition (Sanpera *et al.*, 1998; Rana, 2013), where as usual, $d_A = \dim(\mathcal{H}_A)$ and $d_B = \dim(\mathcal{H}_B)$. Alternatively, the negativity can be defined in terms of the trace norm $\| \cdot \|_{\mathrm{tr}}$, i.e., $\mathcal{N}(\rho) = \frac{1}{2}(\|\rho^{\mathsf{T}_B}\|_{\mathrm{tr}} - 1)$, where $\|\rho\|_{\mathrm{tr}} = \mathrm{Tr}\sqrt{\rho^\dagger \rho}$, see Section 23.4. By virtue of the PPT criterion, the negativity vanishes for separable states (i) and if one were to include a pre-factor of 2, then it would also give the value 1 for maximally entangled two-qubit states, thus satisfying (ii) in principle. It was also shown in Vidal and Werner (2002) that $\mathcal{N}(\rho)$ is monotonic under LOCC, property (iii) and convex (vii), but the negativity does not reduce to the entropy of entanglement for pure states (iv) and it is not additive (on copies or otherwise, properties (ix) and (viii)). The latter issue can be amended by defining the so-called *logarithmic negativity* $E_\mathcal{N}$ as

$$E_\mathcal{N}(\rho) = \log_2 \|\rho^{T_B}\|_{\mathrm{tr}} = \log_2\big(2\mathcal{N}(\rho) + 1\big), \qquad (16.88)$$

which is *additive*, but still does not reduce to the entropy of entanglement in the pure-state case, and is not (or, no longer) convex (Plenio, 2005). Clearly, because of their relation to the PPT criterion neither of the negativities is able to capture bound entanglement, i.e., there are entangled states for which both $\mathcal{N}(\rho)$ and $E_\mathcal{N}(\rho)$ vanish. Nonetheless, the negativity measures are widely used because of their computational simplicity. Finally, it is interesting to note that one can relate the entanglement of formation and the negativity measures. It was shown in Verstraete *et al.* (2001*a*) that for any two-qubit state with given concurrence C the negativity is bounded by

$$\tfrac{1}{2}\big(\sqrt{(1-C)^2 + C^2} - (1-C)\big) \leq \mathcal{N} \leq \tfrac{1}{2}C. \qquad (16.89)$$

When we apply the negativity to our previous example of the Werner states, we immediately see from eqn (15.70) that the negativity for the entangled Werner states is just $\mathcal{N}[\rho_{\mathrm{Werner}}(\alpha > \tfrac{1}{3})] = (3\alpha - 1)/4$ which is exactly one half of the value of the concurrence for the Werner states from eqn (16.69), and we thus see that the upper bound on the negativity in (16.89) is tight for Werner states.

Résumé: To quantify the amount of entanglement in a given state—an important resource to perform certain tasks in quantum-information processing—there exist several entanglement measures and monotones. Among these measures, the entropy of entanglement stands out as the unique pure-state entanglement measure. Unfortunately, when it comes to mixed states, every single one of the measures proposed to date has some drawbacks, lacking in certain desirable qualities or simply being impossible to calculate. Nevertheless, many of these measures can be calculated explicitly for the case of two qubits, and there exist some useful bounds for some of the most important measures, the entanglement of formation (EOF) and the entanglement of distillation, as we shall see in more detail in Section 17.4.2. The EOF can be viewed as an upper bound on the minimum number of maximally entangled pairs needed to prepare a quantum state via LOCC, while the distillable entanglement represents the entanglement that can be obtained from the converse process, where Alice and Bob perform LOCC to recover a smaller number of "almost" maximally entangled singlet states from a given number of mixed entangled states. Not all states are distillable in this way, but there exist theorems telling us (at least for certain cases) which quantum states can be distilled (Theorem 16.4) and which cannot (Corollary 16.1). Finally, there are geometric approaches to entanglement quantification based on distance measures between an entangled state and suitable sets of states, e.g., the set of separable states. Entanglement monotones and measures such as the Hilbert–Schmidt measure and the relative entropy of entanglement differ in their properties depending on the choice of distance measure, but offer connections to practical methods for entanglement detection, as we will see in Section 16.3. To provide some examples, we have calculated many of these measures for the family of Werner states.

16.3   Entanglement Witnesses

Having discussed possible entanglement measures to quantify entanglement, and having encountered many of the challenges that come along with this task, we now want to turn to more practical means of detecting entangled states. Such a tool that enables us to decide whether a given quantum state is entangled or not is provided by so-called *entanglement witnesses*. Such objects represent necessary and sufficient criteria for identifying entanglement. The great value of entanglement witnesses lies in their duality of theoretical appeal arising from their clear geometrical meaning paired with experimental practicality originating in their role as measurable observables.

16.3.1   Entanglement-Witness Theorem

Let us start with the following theorem, which can be found in various forms (in, e.g., Horodecki *et al.*, 1996*a*; Terhal, 2000; Bertlmann *et al.*, 2002).

Theorem 16.8 (Entanglement-witness theorem)

*A quantum state ρ_{ent} is entangled if and only if $\exists$ a Hermitian operator W, called an **entanglement witness**, such that*

$$(\rho_{\text{ent}}, \mathrm{W})_{\text{HS}} = \text{Tr}(\mathrm{W}\,\rho_{\text{ent}}) < 0\,, \qquad \rho_{\text{ent}} \in \mathcal{S}^{\text{C}}$$
$$(\rho, \mathrm{W})_{\text{HS}} = \text{Tr}(\mathrm{W}\,\rho) \geq 0\,, \qquad \forall \rho \in \mathcal{S}\,.$$

Here, as before, $\mathcal{S}$ denotes the set of separable states and $\mathcal{S}^{\text{C}}$ its complement, such that $\mathcal{S} \cup \mathcal{S}^{\text{C}}$ is the set of all density operators in the total Hilbert–Schmidt space, see Definition 15.7. The entanglement-witness theorem (Theorem 16.8) can be derived via the *Hahn–Banach theorem* familiar from functional analysis. We want to carry out the proof within a geometric representation of the Hahn–Banach theorem[8], a well-known mathematical result (see, e.g., Reed and Simon, 1972, Sec. III.3) that we will state here (without proof) in terms of the following theorem:

Theorem 16.9 (Geometric Hahn–Banach theorem)

Let A be a convex, compact set and let $b \notin A$. Then $\exists$ a hyperplane that separates b from the set A.

---

[8] *Hans Hahn* (1879–1934) was a prominent Austrian mathematician with a strong philosophical inclination. He installed a philosophical discussion circle, called the *"Kaffeehausphilosophen"* (coffee house philosophers), among them were the physicist *Philipp Frank* and the mathematician *Richard von Mises*. Hahn also became a driving force and prominent member of the so-called *"Wiener Kreis"* (Vienna Circle), a by now famous group of philosophers (*Moritz Schlick*, *Rudolf Carnap*, and *Ludwig Wittgenstein*, amongst others) and scientists (like, e.g., *Otto Neurath*), who regularly met from 1924 to 1936 for philosophical discussions at the Mathematical Seminar of the University of Vienna. Hahn's most famous student, *Kurt Gödel*, also participated in the Circle but not as a particularly active member since he could not reconcile his views with the positivist standpoint.

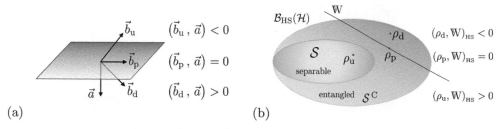

Fig. 16.3 Hyperplanes. In a Euclidean space a plane is determined by its normal vector $\vec{a}$ together with the values of the scalar products for the "upwards pointing" vectors $\vec{b}_{\mathrm{u}}$, the "in-plane" vectors $\vec{b}_{\mathrm{p}}$, and the "downwards pointing" vectors $\vec{b}_{\mathrm{d}}$. In the Hilbert–Schmidt space $\mathcal{B}_{\mathrm{HS}}(\mathcal{H})$, the relevant scalar product is the Hilbert–Schmidt inner product $(\cdot, \cdot)_{\mathrm{HS}}$ and, according to the Hahn–Banach theorem (Theorem 16.9) a hyperplane can be chosen such that the set $\mathcal{S}$ of separable states is not intersected.

Proof of Theorem 16.8 Since we are working in a Euclidean space, we can define a plane via its orthogonal vector $\vec{a}$. The plane separates "upwards pointing" vectors $\vec{b}_{\mathrm{u}}$, for which the (Euclidean) scalar product $(\vec{b}_{\mathrm{u}}, \vec{a}) = \vec{b}_{\mathrm{u}} \cdot \vec{a} < 0$ is negative, from "downwards pointing" vectors $\vec{b}_{\mathrm{d}}$ with positive scalar product $(\vec{b}_{\mathrm{d}}, \vec{a}) > 0$. The vectors lying in the plane $\vec{b}_{\mathrm{p}}$ have a vanishing scalar product with $\vec{a}$, that is, $(\vec{b}_{\mathrm{p}}, \vec{a}) = 0$, see Fig. 16.3.

In our case, the vector space in question is the Hilbert–Schmidt space $\mathcal{B}_{\mathrm{HS}}(\mathcal{H})$ where we have the scalar functional $(\rho, \mathrm{W})_{\mathrm{HS}} = 0$ that defines a hyperplane in the set of all states. The plane splits the "upwards lying" states ρ_{u} with $(\rho_{\mathrm{u}}, \mathrm{W})_{\mathrm{HS}} < 0$ from the "downwards lying" states ρ_{d} with $(\rho_{\mathrm{d}}, \mathrm{W})_{\mathrm{HS}} > 0$; the states ρ_{p} lie on the hyperplane, i.e., $(\rho_{\mathrm{p}}, \mathrm{W})_{\mathrm{HS}} = 0$, see Fig. 16.4. If we now choose this hyperplane such that all the separable states are "upwards lying", then we have an entanglement witness. Since the set of separable states is convex (Lemma 15.2) the Hahn–Banach theorem (Theorem 16.9) ensures that such a hyperplane exists. □

An entanglement witness $\mathrm{W}_{\mathrm{opt}}$ is called *optimal* if, apart from satisfying the inequalities of the entanglement-witness theorem (Theorem 16.8), there exists a separable state $\rho_0 \in \mathcal{S}$ such that

$$(\rho_0, \mathrm{W}_{\mathrm{opt}})_{\mathrm{HS}} = 0 \, . \tag{16.90}$$

It is optimal in the sense that $\mathrm{W}_{\mathrm{opt}}$ defines a tangent plane to set S of separable states, therefore the expression (16.90) is called *tangent functional* (Bertlmann *et al.*, 2002). Clearly, optimal witnesses represent an improvement on certain ("parallel") non-optimal witnesses, in the sense that they may detect more entangled states, but no (linear) entanglement witness can detect *all* entangled states, as illustrated in Fig. 16.4.

For further literature on entanglement witnesses, see for instance Lewenstein *et al.* (2000*b*), Bruß *et al.* (2002), Bruß (2002), Gühne *et al.* (2003), and Gühne and Tóth (2009).

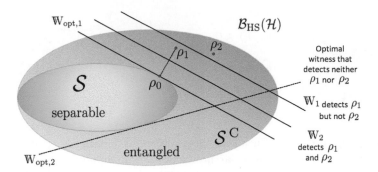

Fig. 16.4 Entanglement witnesses. Illustration of the convex set $\mathcal{S}$ of separable states and its complement $\mathcal{S}^C$ together with four hyperplanes, W_1 and W_2, as well as $W_{opt,1}$ and $W_{opt,2}$ (the straight lines) which are entanglement witnesses. Both $W_{opt,1}$ and $W_{opt,2}$ are optimal witnesses, but while there are some entangled states detected by both witnesses, there are also some (e.g., ρ_1 and ρ_2) that are detected by one ($W_{opt,1}$) but not by the other ($W_{opt,2}$). Nevertheless, the optimal witness $W_{opt,1}$ is an improvement over the witnesses W_1 and W_2, which detect fewer entangled states than $W_{opt,1}$. For instance, ρ_1 is detected by $W_{opt,1}$ and W_2, but not by W_1. The state ρ_0 is the separable state to the entangled state ρ_1.

16.3.2 Construction of an Entanglement Witness

The entanglement-witness theorem (Theorem 16.8) guarantees that any entangled state can be detected by an entanglement witness. Explicitly constructing such a witness, however, is not always an easy task. Fortunately there are some rather broad classes of examples for which there are known construction methods. Consider an NPT state, an entangled state ρ_{ent} with negative partial transposition, which means that the partially transposed operator $\rho_{ent}^{T_A} = (T_A \otimes \mathbb{1}_B)\, \rho_{ent}$ has at least one negative eigenvalue $\lambda_- < 0$ with the corresponding eigenvector $|\lambda_-\rangle$. Then, remembering the PPT criterion (Theorem 15.3), we can construct the following entanglement witness (Gühne *et al.*, 2002; Gühne *et al.*, 2003; Gühne and Tóth, 2009).

Lemma 16.1 (Entanglement witness for NPT state)

The operator

$$\mathbb{W} = |\lambda_-\rangle\langle\lambda_-|^{T_A} = (T_A \otimes \mathbb{1}_B)\,|\lambda_-\rangle\langle\lambda_-|$$

*is an **entanglement witness** for the entangled state $\rho_{ent} = |\lambda_-\rangle\langle\lambda_-|$, where $\lambda_- < 0$ is a negative eigenvalue of the partially transposed matrix*

$$\rho_{ent}^{T_A} = (T_A \otimes \mathbb{1}_B)\,\rho_{ent}$$

and $|\lambda_-\rangle$ is the corresponding eigenvector.

Equivalently, one may of course also choose the partial transposition T_B with respect to Bob's subsystem.

Proof To prove Lemma 16.1, we use the trace property that for any two matrices U, V: $\mathrm{Tr}(UV^{\mathsf{T}_A}) = \mathrm{Tr}(U^{\mathsf{T}_A}V)$, which can easily be seen by noting that the trace is invariant under (partial) transposition of its (entire) argument,

$$\mathrm{Tr}(UV^{\mathsf{T}_A}) = \mathrm{Tr}([U^{\mathsf{T}_A}V]^{\mathsf{T}_A}) = \mathrm{Tr}(U^{\mathsf{T}_A}V). \tag{16.91}$$

Then we have for the entangled state ρ_{ent}

$$(\rho_{\mathrm{ent}}, \mathrm{W})_{\mathrm{HS}} = \mathrm{Tr}(\mathrm{W}\,\rho_{\mathrm{ent}}) = \mathrm{Tr}\left(|\lambda_-\rangle\langle\lambda_-|^{\mathsf{T}_A}\rho_{\mathrm{ent}}\right) \tag{16.92}$$

$$= \mathrm{Tr}\left(|\lambda_-\rangle\langle\lambda_-|\,\rho_{\mathrm{ent}}^{\mathsf{T}_A}\right) = \lambda_- < 0\,,$$

and for all separable states $\rho_{\mathrm{sep}} \in \mathcal{S}$ we obtain positivity (non-negativity)

$$(\rho_{\mathrm{sep}}, \mathrm{W})_{\mathrm{HS}} = \mathrm{Tr}(\mathrm{W}\,\rho_{\mathrm{sep}}) = \mathrm{Tr}\left(|\lambda_-\rangle\langle\lambda_-|^{\mathsf{T}_A}\rho_{\mathrm{sep}}\right) \tag{16.93}$$

$$= \mathrm{Tr}\left(|\lambda_-\rangle\langle\lambda_-|\,\rho_{\mathrm{sep}}^{\mathsf{T}_A}\right) \geq 0\,,$$

since all separable states ρ_{sep} are PPT. These are precisely the properties of an entanglement witness as specified in Theorem 16.8. $\qquad\square$

Entanglement witness for Werner states: We now want to apply this method for the specific construction of the entanglement witness (Lemma 16.1) for the Werner states (15.68). Thus we study the following single-parameter family of states

$$\rho_{\mathrm{Werner}}(\alpha) = \alpha\,\rho + \frac{1-\alpha}{4}\,\mathbb{1}_4\,, \tag{16.94}$$

parameterized by α, where $\rho = |\psi\rangle\langle\psi|$, but we keep the state $|\psi\rangle$ more general than in Section 15.3.3, that is, $|\psi\rangle = a\,|01\rangle + b\,|10\rangle$ with the coefficients chosen real $a^* = a$, $b^* = b$, $a^2 + b^2 = 1$. For ρ and its partial transpose $\rho^{\mathsf{T}_B} = (\mathbb{1}_A \otimes \mathsf{T}_B)\rho$ we obtain the expressions

$$\rho = \begin{pmatrix} 0 & 0 & 0 & 0 \\ 0 & a^2 & ab & 0 \\ 0 & ab & b^2 & 0 \\ 0 & 0 & 0 & 0 \end{pmatrix}, \qquad \rho^{\mathsf{T}_B} = \begin{pmatrix} 0 & 0 & 0 & ab \\ 0 & a^2 & 0 & 0 \\ 0 & 0 & b^2 & 0 \\ ab & 0 & 0 & 0 \end{pmatrix}, \tag{16.95}$$

which leads to the following partial transposition of the Werner state

$$\rho^{T_B}_{\text{Werner}}(\alpha) = \alpha \rho^{T_A} + \frac{1-\alpha}{4}\mathbb{1}_4$$

$$= \frac{1}{4}\begin{pmatrix} 1-\alpha & 0 & 0 & 4ab\alpha \\ 0 & 1-\alpha(1-4a^2) & 0 & 0 \\ 0 & 0 & 1-\alpha(1-4b^2) & 0 \\ 4ab\alpha & 0 & 0 & 1-\alpha \end{pmatrix}, \qquad (16.96)$$

with eigenvalues

$$\left\{ \frac{1}{4}(1-\alpha(1-4a^2)), \frac{1}{4}(1-\alpha(1-4b^2)), \frac{1}{4}(1-\alpha(1-4ab)), \frac{1}{4}(1-\alpha(1+4ab)) \right\}. \qquad (16.97)$$

A negative eigenvalue is found for $(1-\alpha(1+4ab)) < 0$, i.e., when

$$\alpha > \frac{1}{1+4ab} \quad \text{if } b > 0 \quad \text{or} \quad \alpha > \frac{1}{1-4ab} \quad \text{if } b < 0, \qquad (16.98)$$

in which case $\rho_{\text{Werner}}(\alpha)$ is entangled. When ρ is indeed the Bell state ρ^-, that is, when $|\psi\rangle \to |\psi^-\rangle$, $a = -b = 1/\sqrt{2}$, then the condition for entanglement is $\alpha > 1/3$, as we have found before in Section 15.3.3.

For the explicit expression of the entanglement witness of Lemma 16.1 we have to calculate the eigenvector corresponding to the negative eigenvalue $\frac{1}{4}(1-\alpha(1+4ab))$ of the partial transposition in eqn (16.96). The result is quickly found; it is the Bell state

$$|\phi^-\rangle = \frac{1}{\sqrt{2}}\left(|00\rangle - |11\rangle\right) = \frac{1}{\sqrt{2}}\begin{pmatrix} 1 \\ 0 \\ 0 \\ -1 \end{pmatrix}. \qquad (16.99)$$

Therefore we find as entanglement witness for the family of Werner states (16.94)

$$W = |\phi^-\rangle\langle\phi^-|^{T_B} = (\mathbb{1}_A \otimes T_B)|\phi^-\rangle\langle\phi^-| = \frac{1}{2}\begin{pmatrix} 1 & 0 & 0 & 0 \\ 0 & 0 & -1 & 0 \\ 0 & -1 & 0 & 0 \\ 0 & 0 & 0 & 1 \end{pmatrix}$$

$$= \frac{1}{2}\mathbb{1} - |\psi^+\rangle\langle\psi^+|, \qquad (16.100)$$

which we have finally re-expressed in terms of the Bell state $|\psi^+\rangle$. This entanglement witness is optimal and independent of the parameter α, and it does not depend on the Schmidt coefficients a, b.

16.3.3 Bertlmann–Narnhofer–Thirring Theorem

In this section we present a method for constructing an entanglement witness, which relies on purely geometric considerations (Bertlmann *et al.*, 2002). To this end let us first recall the Hilbert–Schmidt distance (Definition 11.6) between two states given by the density matrices ρ_1 and ρ_2

$$D_{\mathrm{HS}}(\rho_1,\rho_2) = \|\rho_1 - \rho_2\|_{\mathrm{HS}} = \sqrt{(\rho_1 - \rho_2, \rho_1 - \rho_2)_{\mathrm{HS}}} = \sqrt{\mathrm{Tr}\big[(\rho_1 - \rho_2)^\dagger(\rho_1 - \rho_2)\big]}\,, \tag{16.101}$$

and the related Hilbert–Schmidt measure (16.79), the minimal distance of an entangled state ρ_{ent} to the set of separable states $\mathcal{S}$

$$\mathcal{D}_{\mathrm{HS}}(\rho_{\mathrm{ent}}) = \min_{\rho_{\mathrm{sep}} \in \mathcal{S}} D_{\mathrm{HS}}(\rho_{\mathrm{ent}}, \rho_{\mathrm{sep}}) = D_{\mathrm{HS}}(\rho_{\mathrm{ent}}, \rho_0)\,, \tag{16.102}$$

where ρ_0 denotes the closest separable state to ρ_{ent}. Although this quantity may not be a full entanglement measure itself (see the discussion following eqn (16.79)), it nevertheless plays a role in the construction of useful entanglement witnesses.

Starting from the Hilbert–Schmidt norm $\|A\|_{\mathrm{HS}}^2 = (A, A)_{\mathrm{HS}}$ we first rewrite it as $\|A\|_{\mathrm{HS}} = (A, \frac{A}{\|A\|_{\mathrm{HS}}})_{\mathrm{HS}}$ such that we can re-express the Hilbert–Schmidt distance as

$$D_{\mathrm{HS}}(\rho_1,\rho_2) = \|\rho_1 - \rho_2\|_{\mathrm{HS}} = (\rho_1 - \rho_2, \tfrac{\rho_1 - \rho_2}{\|\rho_1 - \rho_2\|_{\mathrm{HS}}})_{\mathrm{HS}} = (\rho_1 - \rho_2, \overline{A})_{\mathrm{HS}}\,, \tag{16.103}$$

with the normalized operator $\overline{A} := (\rho_1 - \rho_2)/\|\rho_1 - \rho_2\|_{\mathrm{HS}}$. Furthermore we may shift the operator $\overline{A}$ to $A = \overline{A} + a\,\mathbb{1}$ with $a \in \mathbb{C}$, which does not change the Hilbert–Schmidt distance

$$D_{\mathrm{HS}}(\rho_1,\rho_2) = (\rho_1 - \rho_2, \overline{A})_{\mathrm{HS}} = (\rho_1 - \rho_2, \overline{A})_{\mathrm{HS}} + (\rho_1 - \rho_2, a\,\mathbb{1})_{\mathrm{HS}} = (\rho_1 - \rho_2, A)_{\mathrm{HS}} \tag{16.104}$$

because $(\rho_1 - \rho_2, a\,\mathbb{1})_{\mathrm{HS}} = a\mathrm{Tr}(\rho_1) - a\mathrm{Tr}(\rho_2) = 0$. For any given pair of states ρ_1 and ρ_2 we can now fix the constant to $a = -(\rho_1, \rho_1 - \rho_2)_{\mathrm{HS}}/\|\rho_1 - \rho_2\|_{\mathrm{HS}}$, such that the operator A finally becomes

$$A = \frac{\rho_1 - \rho_2 - (\rho_1, \rho_1 - \rho_2)_{\mathrm{HS}}\,\mathbb{1}}{\|\rho_1 - \rho_2\|_{\mathrm{HS}}}\,. \tag{16.105}$$

Next we consider a hyperplane, as we did in Section 16.3.1, and we choose this hyperplane to include the state ρ_1 and is orthogonal to $\rho_1 - \rho_2$. This hyperplane, illustrated in Fig. 16.5, is thus defined by the set of states ρ_{p} satisfying

$$\frac{1}{\|\rho_1 - \rho_2\|_{\mathrm{HS}}} (\rho_{\mathrm{p}} - \rho_1, \rho_1 - \rho_2)_{\mathrm{HS}} = 0\,. \tag{16.106}$$

For all states "above" the hyperplane, the "upwards" states ρ_{u}, we have

$$\frac{1}{\|\rho_1 - \rho_2\|_{\mathrm{HS}}} (\rho_{\mathrm{u}} - \rho_1, \rho_1 - \rho_2)_{\mathrm{HS}} < 0\,, \tag{16.107}$$

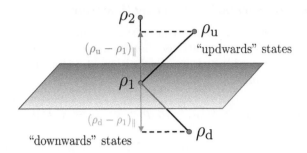

Fig. 16.5 Illustration of "upwards" and "downwards" states. For "upwards" states ρ_u from (16.107) above the hyperplane defined by (16.106) the Hilbert–Schmidt product $(\rho_\mathrm{u} - \rho_1, \rho_1 - \rho_2)_\mathrm{HS}$ is negative because the projection $(\rho_\mathrm{u} - \rho_1)_{\|}$ onto $\rho_1 - \rho_2$ "points" into the opposite direction as $\rho_1 - \rho_2$, whereas for "downwards" states ρ_d from (16.108) the Hilbert–Schmidt product $(\rho_\mathrm{d} - \rho_1, \rho_1 - \rho_2)_\mathrm{HS}$ is positive because $(\rho_\mathrm{d} - \rho_1)_{\|}$ "points" into the same direction as $\rho_1 - \rho_2$.

whereas for all states "below" the hyperplane, the "downwards" states ρ_d, we get

$$\frac{1}{\|\rho_1 - \rho_2\|_\mathrm{HS}} (\rho_\mathrm{d} - \rho_1, \rho_1 - \rho_2)_\mathrm{HS} > 0 \,. \tag{16.108}$$

In terms of the operator A from eqn (16.105), the conditions in (16.106), (16.107), and (16.108) can be re-expressed quite compactly by using the identity

$$(\rho, A)_\mathrm{HS} = (\rho, \tfrac{\rho_1 - \rho_2}{\|\rho_1 - \rho_2\|_\mathrm{HS}})_\mathrm{HS} - \tfrac{(\rho_1, \rho_1 - \rho_2)_\mathrm{HS}}{\|\rho_1 - \rho_2\|_\mathrm{HS}} (\rho, \mathbb{1})_\mathrm{HS} = \tfrac{1}{\|\rho_1 - \rho_2\|_\mathrm{HS}} (\rho - \rho_1, \rho_1 - \rho_2)_\mathrm{HS} \,. \tag{16.109}$$

Then the hyperplane is determined by

$$(\rho_\mathrm{p}, A)_\mathrm{HS} = 0 \,, \tag{16.110}$$

and the "upwards" and "downwards" states satisfy

$$(\rho_\mathrm{u}, A)_\mathrm{HS} < 0 \,, \qquad (\rho_\mathrm{d}, A)_\mathrm{HS} > 0 \,. \tag{16.111}$$

Now we are prepared to discuss the Bertlmann–Narnhofer–Thirring (BNT) theorem (Horodecki *et al.*, 1996*a*; Terhal, 2000; Bertlmann *et al.*, 2002). The theorem provides an interesting connection between the Hilbert–Schmidt measure (16.79)

$$(\rho, \mathrm{W})_\mathrm{HS} - (\rho_\mathrm{ent}, \mathrm{W})_\mathrm{HS} \geq 0 \qquad \forall \rho \in \mathcal{S} \,, \tag{16.112}$$

and define the maximal violation of the inequality in (16.112) as follows (note that the states ρ and the witnesses W are still freely at our disposal):

Definition 16.6 *The maximal violation of the entanglement-witness inequality is*

$$B(\rho_\mathrm{ent}) = \max_{\mathrm{W}, \|\mathrm{W} - a\mathbb{1}\|_\mathrm{HS} \leq 1} \left(\min_{\rho \in \mathcal{S}} (\rho, \mathrm{W})_\mathrm{HS} - (\rho_\mathrm{ent}, \mathrm{W})_\mathrm{HS} \right) .$$

The minimum is taken over all separable states and the maximum over all possible entanglement witnesses $\mathbb{W}$, suitably normalized, and $a \in \mathbb{C}$ is the coefficient of the identity term $\mathbb{1} = \mathbb{1}_A \otimes \mathbb{1}_B$. With this, the following theorem can be established.

Theorem 16.10 (BNT theorem)

For any entangled state ρ_{ent} the maximal violation of the witness inequality as captured by $B(\rho_{\text{ent}})$ is equal to the minimal Hilbert–Schmidt distance to the separable states, i.e., the Hilbert–Schmidt measure $\mathcal{D}_{\text{HS}}(\rho_{\text{ent}})$,

$$B(\rho_{\text{ent}}) = \mathcal{D}_{\text{HS}}(\rho_{\text{ent}}) \, .$$

The corresponding optimal witness is given by

$$\mathbb{W}_{\text{opt}} = \frac{\rho_0 - \rho_{\text{ent}} - (\rho_0, \rho_0 - \rho_{\text{ent}})_{\text{HS}} \, \mathbb{1}}{\|\rho_0 - \rho_{\text{ent}}\|_{\text{HS}}} \, ,$$

where ρ_0 is the separable state closest (in Hilbert–Schmidt distance) to ρ.

This is a remarkable result that we have illustrated in Fig. 16.6. Furthermore, the optimal entanglement witness $\mathbb{W}_{\text{opt}}$ is given explicitly in Theorem 16.10. However, we have to know the closest separable state ρ_0, which is easy to find in low dimensions, but not so in higher ones. Nevertheless, there exist approximation procedures to approach ρ_0 (see, e.g., Bertlmann and Krammer 2009 and Gühne and Tóth 2009).

Proof of Theorem 16.10 We start with the optimal entanglement witness. For an entangled state ρ_{ent} the minimum of the Hilbert–Schmidt distance is attained for some $\rho_0 \in \mathcal{S}$ since the norm is continuous and the set $\mathcal{S}$ is compact, see Fig. 16.6,

$$\exists \, \rho_0 \in \mathcal{S} : \quad \min_{\rho \in \mathcal{S}} \|\rho - \rho_{\text{ent}}\|_{\text{HS}} = \|\rho_0 - \rho_{\text{ent}}\|_{\text{HS}} \, . \tag{16.113}$$

Replacing the states ρ_1 and ρ_2 in eqns (16.104) and (16.105) with ρ_0 and ρ_{ent}, respectively, we obtain the Hilbert–Schmidt measure

$$\mathcal{D}_{\text{HS}}(\rho_{\text{ent}}) = D_{\text{HS}}(\rho_0, \rho_{\text{ent}}) = (\rho_0, A)_{\text{HS}} - (\rho_{\text{ent}}, A)_{\text{HS}} \, , \tag{16.114}$$

with A given explicitly by eqn (16.105). The operator A in eqn (16.114) is already an optimal entanglement witness since the state ρ_0 lies on the boundary of the separable states $\rho_0 \in \partial \mathcal{S}$ and the hyperplane defined by $(\rho_{\text{p}}, A)_{\text{HS}} = 0$ is orthogonal to $\rho_0 - \rho_{\text{ent}}$. Since ρ_0 is the closest separable state to ρ_{ent} the hyperplane must be a tangent to the set $\mathcal{S}$, as illustrated in Fig. 16.6. Then, eqns (16.110) and (16.111) imply the inequalities of the entanglement-witness theorem (Theorem 16.8), and the operator A hence has to be an optimal entanglement witness

$$A \longrightarrow \mathbb{W}_{\text{opt}} = \frac{\rho_0 - \rho_{\text{ent}} - (\rho_0, \rho_0 - \rho_{\text{ent}})_{\text{HS}} \, \mathbb{1}}{\|\rho_0 - \rho_{\text{ent}}\|_{\text{HS}}} \, . \tag{16.115}$$

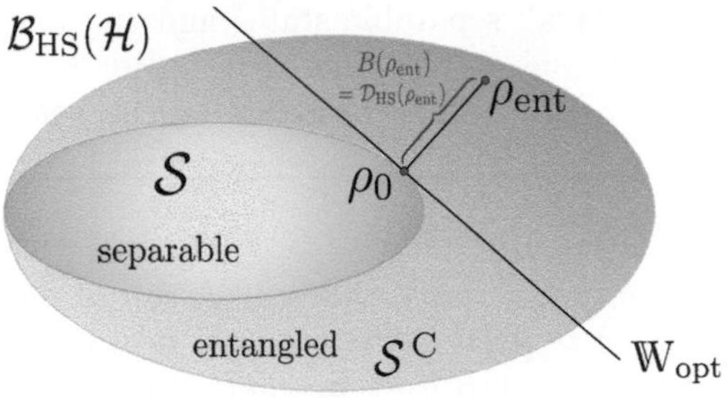

Fig. 16.6 Bertlmann–Narnhofer–Thirring theorem (Theorem 16.10). $\mathcal{D}_{\mathrm{HS}}(\rho_{\mathrm{ent}}) = B(\rho_{\mathrm{ent}})$. The minimal distance $\mathcal{D}_{\mathrm{HS}}(\rho_{\mathrm{ent}})$ of the entangled state ρ_{ent} to the set of separable states $\mathcal{S}$ in the Hilbert–Schmidt space is equal to the maximal violation $B(\rho_{\mathrm{ent}})$ of the entanglement-witness inequality. $\mathrm{W}_{\mathrm{opt}}$ represents the optimal entanglement witness.

With this, we now turn to proving the equality of the Hilbert–Schmidt measure and the maximal violation of the witness inequality. Choosing an optimal entanglement witness implies a vanishing distance to the set of separable states, $(\rho_0, \mathrm{W}_{\mathrm{opt}})_{\mathrm{HS}} = 0$, such that the maximum is attained for $\mathrm{W}_{\mathrm{opt}}$ in

$$\max_{\mathrm{W}, \|\mathrm{W}-a\mathbb{1}\|\leq 1} \left[- (\rho_{\mathrm{ent}}, \mathrm{W})_{\mathrm{HS}} \right] = - (\rho_{\mathrm{ent}}, \mathrm{W}_{\mathrm{opt}})_{\mathrm{HS}} . \tag{16.116}$$

When inserting eqn (16.116) into the Hilbert–Schmidt measure (16.114) we can rewrite it in the following way

$$\mathcal{D}_{\mathrm{HS}}(\rho_{\mathrm{ent}}) = (\rho_0, \mathrm{W}_{\mathrm{opt}})_{\mathrm{HS}} - (\rho_{\mathrm{ent}}, \mathrm{W}_{\mathrm{opt}})_{\mathrm{HS}} \tag{16.117}$$

$$= \max_{\mathrm{W}, \|\mathrm{W}-a\mathbb{1}\|\leq 1} \left[(\rho_0, \mathrm{W})_{\mathrm{HS}} - (\rho_{\mathrm{ent}}, \mathrm{W})_{\mathrm{HS}} \right]$$

$$= \max_{\mathrm{W}, \|\mathrm{W}-a\mathbb{1}\|\leq 1} \left[\min_{\rho\in\mathcal{S}} (\rho, \mathrm{W})_{\mathrm{HS}} - (\rho_{\mathrm{ent}}, \mathrm{W})_{\mathrm{HS}} \right] = B(\rho_{\mathrm{ent}}) ,$$

which completes the proof of the BNT theorem (Theorem 16.10). $\square$

16.3.4   Entanglement Witness for Werner States

It is instructive to explicitly calculate entanglement witnesses for several examples and to examine Theorems 16.8 and 16.10 discussed previously in this context. One of the simplest cases is that of the Werner states which we use in the Bloch decomposition form (16.80). We have already calculated the Hilbert–Schmidt measure for the Werner states, eqn (16.84), let us state the result again here,

$$\mathcal{D}_{\mathrm{HS}}[\rho_{\mathrm{Werner}}(\alpha > \tfrac{1}{3})] = \|\rho_0 - \rho_{\mathrm{Werner}}(\alpha > \tfrac{1}{3})\|_{\mathrm{HS}} = \tfrac{\sqrt{3}}{2}(\alpha - \tfrac{1}{3}), \tag{16.118}$$

for $\frac{1}{3} < \alpha \leq 1$. Using the result (16.82) for the difference of ρ_0 to an entangled state $\rho_{\mathrm{Werner}}(\alpha > \tfrac{1}{3})$, we further calculate

$$(\rho_0, \rho_0 - \rho_\alpha^{\text{ent}})_{\text{HS}} = \tfrac{1}{4}\tfrac{1}{4}(\alpha - \tfrac{1}{3})\,\text{Tr}\Big\{(\mathbb{1}\otimes\mathbb{1} - \tfrac{1}{3}\sigma_i\otimes\sigma_i)(\sigma_j\otimes\sigma_j)\Big\} = -\tfrac{1}{4}(\alpha - \tfrac{1}{3})\,,$$

(16.119)

and thus find the *optimal entanglement witness of Werner states* (Theorem 16.10)

$$\mathrm{W}_{\text{opt}} = \frac{\rho_0 - \rho_\alpha^{\text{ent}} - (\rho_0, \rho_0 - \rho_\alpha^{\text{ent}})_{\text{HS}}\,\mathbb{1}_4}{\|\rho_0 - \rho_\alpha^{\text{ent}}\|_{\text{HS}}}$$

(16.120)

$$= \big\{\tfrac{1}{4}(\alpha - \tfrac{1}{3})\sigma_i\otimes\sigma_i + \tfrac{1}{4}(\alpha - \tfrac{1}{3})\mathbb{1}\otimes\mathbb{1}\big\}\,\frac{2}{\sqrt{3}(\alpha - \tfrac{1}{3})} = \tfrac{1}{2\sqrt{3}}(\mathbb{1}\otimes\mathbb{1} + \sigma_i\otimes\sigma_i).$$

The condition for the tangent functional $(\rho_0, \mathrm{W}_{\text{opt}})_{\text{HS}} = 0$ of the witness as a hyperplane is clearly satisfied when inserting the expressions (16.81) and (16.120).

To demonstrate the entanglement-witness theorem we have to check the inequalities of Theorem 16.8. The first inequality for entangled states,

$$(\rho_{\text{Werner}}(\alpha > \tfrac{1}{3}), \mathrm{W}_{\text{opt}})_{\text{HS}} = \tfrac{1}{4}\tfrac{1}{2\sqrt{3}}\,\text{Tr}\Big\{(\mathbb{1}\otimes\mathbb{1} - \alpha\,\sigma_i\otimes\sigma_i)(\mathbb{1}\otimes\mathbb{1} + \sigma_j\otimes\sigma_j)\Big\}$$

$$= \tfrac{1}{2\sqrt{3}}(1 - 3\alpha) < 0$$

(16.121)

is indeed satisfied for $\tfrac{1}{3} < \alpha \le 1$.

For the second inequality of Theorem 16.8 we recall the Bloch expansion of a separable state (15.42), a product state, then we find that

$$(\rho_{\text{sep}}, \mathrm{W}_{\text{opt}})_{\text{HS}} = \tfrac{1}{4}\tfrac{1}{2\sqrt{3}}\,\text{Tr}\Big\{(\mathbb{1}\otimes\mathbb{1} + a_i\,\sigma_i\otimes\mathbb{1} + b_j\,\mathbb{1}\otimes\sigma_j + a_i b_j\,\sigma_i\otimes\sigma_j)$$

$$\times (\mathbb{1}\otimes\mathbb{1} + \sigma_k\otimes\sigma_k)\Big\}$$

$$= \tfrac{1}{2\sqrt{3}}\big[1 + \cos(\delta)\big] \ge 0\,,$$

(16.122)

is positive (non-negative) as it should be, and it remains positive for *any* separable state since it is the convex sum of product states. The angle δ is the angle between the Bloch vectors $\vec{a}$ and $\vec{b}$. For $\delta = 0$, i.e., $\vec{a}$ parallel to $\vec{b}$, we have $(\rho_{\text{sep}}, \mathrm{W}_{\text{opt}})_{\text{HS}} = 1/\sqrt{3}$ and for $\delta = \pi$, i.e., $\vec{a}$ antiparallel to $\vec{b}$, we get $(\rho_{\text{sep}}, \mathrm{W}_{\text{opt}})_{\text{HS}} = 0$, the condition for the tangent functional.

Finally we calculate the maximal violation of the entanglement-witness inequality (recall Definition 16.6) to illustrate the BNT theorem (Theorem 16.10)

$$B(\rho_\alpha^{\text{ent}}) = \max_{\mathrm{W},\,\|\mathrm{W}-a\mathbb{1}\|\le 1}\Big(\min_{\rho\in\mathcal{S}}(\rho, \mathrm{W})_{\text{HS}} - (\rho_\alpha^{\text{ent}}, \mathrm{W})_{\text{HS}}\Big)$$

$$= (\rho_0, \mathrm{W}_{\text{opt}})_{\text{HS}} - (\rho_\alpha^{\text{ent}}, \mathrm{W}_{\text{opt}})_{\text{HS}}\,.$$

(16.123)

Recalling that $(\rho_0, \mathrm{W}_{\text{opt}})_{\text{HS}} = 0$ we can use eqn (16.121) to obtain

$$B(\rho_\alpha^{\text{ent}}) = -(\rho_\alpha^{\text{ent}}, \mathrm{W}_{\text{opt}})_{\text{HS}} = \tfrac{\sqrt{3}}{2}(\alpha - \tfrac{1}{3}) \equiv \mathcal{D}_{\text{HS}}(\rho_\alpha^{\text{ent}})\,,$$

(16.124)

which clearly coincides with the Hilbert–Schmidt measure (16.118) as required by the BNT theorem (Theorem 16.10).

16.3.5 Entanglement Witness for Isotropic States

Next we want to calculate the entanglement witness for isotropic states $\rho_{\mathrm{iso}}(\alpha) \in \mathcal{B}_{\mathrm{HS}}(\mathcal{H}_A \otimes \mathcal{H}_B)$ in $d \times d$ dimensions, which we have introduced in Section 15.4.4.

As we shall elaborate on in Chapter 17, it is possible to choose the a matrix basis $\{\mathbb{1}_d, \Gamma_1, ..., \Gamma_{d^2-1}\}$ for both of the Hilbert–Schmidt subspaces $\mathcal{B}_{\mathrm{HS}}(\mathcal{H}_A)$ and $\mathcal{B}_{\mathrm{HS}}(\mathcal{H}_B)$, with the properties $\mathrm{Tr}(\Gamma_i) = 0$ and $\mathrm{Tr}(\Gamma_i \Gamma_j) = 2\,\delta_{ij}$. With this choice, we can generalize the Werner states in Bloch form, eqn (16.80), in the following way to obtain the *Bloch decomposition of the isotropic states*,

$$\rho_{\mathrm{iso}}(\alpha) = \frac{1}{d^2} \left(\mathbb{1}_d \otimes \mathbb{1}_d + \frac{d}{2}\, \alpha\, \Gamma \right), \qquad (16.125)$$

where $-\frac{1}{d^2-1} \leq \alpha \leq 1$ and where Γ is defined as

$$\Gamma := \sum_{i=1}^{d^2-1} c_i\, \Gamma_i \otimes \Gamma_i, \qquad c_i = \pm 1. \qquad (16.126)$$

The factor $\frac{d}{2}$ in eqn (16.125) is due to normalization and the sign $c_i = \pm 1$ in eqn (16.126) depends on the chosen basis. The expression (16.125) coincides with the definition of the isotropic states, Definition 15.12.

We have already established the splitting of $\rho_{\mathrm{iso}}(\alpha)$ (16.125) at $\alpha = 1/(d+1)$ into entangled and separable states in Section 15.4.4, more specifically in eqn (15.118).

For the Hilbert–Schmidt measure and the entanglement witness we first have to calculate the closest separable state ρ_0. Considering the geometry and symmetry of the isotropic states, ρ_0 obviously has to lie on the boundary of the separable states which are restricted by the parameter range $-\frac{1}{d^2-1} \leq \alpha \leq \frac{1}{d+1}$. Therefore we find

$$\rho_0 = \rho_{\mathrm{iso}}(\alpha = \tfrac{1}{d+1}) = \frac{1}{d^2} \left(\mathbb{1}_d \otimes \mathbb{1}_d + \frac{d}{2(d+1)}\, \Gamma \right). \qquad (16.127)$$

We further need the difference

$$\rho_0 - \rho_{\mathrm{iso}}^{\mathrm{ent}}(\alpha) = \frac{1}{2d} \left(\frac{1}{d+1} - \alpha \right) \Gamma, \qquad (16.128)$$

and we have to compute the norm

$$\|\Gamma\|_{\mathrm{HS}}^2 = \| \sum_{i=1}^{d^2-1} c_i\, \Gamma_i \otimes \Gamma_i \|_{\mathrm{HS}}^2 = 4 \sum_{i=1}^{d^2-1} c_i^2 = 4\,(d^2 - 1), \qquad (16.129)$$

which supplies us the *Hilbert–Schmidt measure for isotropic states*, explicitly

$$\mathcal{D}_{\mathrm{HS}}\big(\rho_{\mathrm{iso}}^{\mathrm{ent}}(\alpha)\big) = \|\rho_0 - \rho_{\mathrm{iso}}^{\mathrm{ent}}(\alpha)\|_{\mathrm{HS}} = \frac{1}{2d} \|\Gamma\|_{\mathrm{HS}} \left(\alpha - \frac{1}{d+1} \right) = \frac{\sqrt{d^2-1}}{d} \left(\alpha - \frac{1}{d+1} \right), \qquad (16.130)$$

with $1/(d+1) < \alpha \leq 1$. For the entanglement witness we further have to calculate

$$(\rho_0, \rho_0 - \rho_{\mathrm{iso}}^{\mathrm{ent}}(\alpha))_{\mathrm{HS}} = -\frac{1}{2d}\frac{1}{d^2}\left(\alpha - \frac{1}{d+1}\right)\mathrm{Tr}\left\{\left(\mathbb{1}_d \otimes \mathbb{1}_d + \frac{d}{2(d+1)}\Gamma\right)\Gamma\right\}$$

$$= -\frac{d-1}{d^2}\left(\alpha - \frac{1}{d+1}\right) \tag{16.131}$$

for $1/(d+1) < \alpha \leq 1$. Collecting the results (16.128), (16.129), and (16.131) we find the explicit expression for the *optimal entanglement witness of the isotropic states* (given by Theorem 16.10)

$$\mathbb{W}_{\mathrm{opt}} = \frac{\rho_0 - \rho_{\mathrm{iso}}^{\mathrm{ent}}(\alpha) - (\rho_0, \rho_0 - \rho_{\mathrm{iso}}^{\mathrm{ent}}(\alpha))_{\mathrm{HS}}\,\mathbb{1}_{d^2}}{\|\rho_0 - \rho_{\mathrm{iso}}^{\mathrm{ent}}(\alpha)\|_{\mathrm{HS}}} = \frac{d-1}{d\sqrt{d^2-1}}\left(\mathbb{1}_d \otimes \mathbb{1}_d - \frac{d}{2(d-1)}\Gamma\right). \tag{16.132}$$

The tensor Γ is given by eqn (16.126), and $(\rho_0, \mathbb{W}_{\mathrm{opt}})_{\mathrm{HS}} = 0$ is the tangent functional.

Next we want to exemplify the entanglement-witness theorem. For that we have to check the inequalities of Theorem 16.8. Considering the entangled states $\rho_{\mathrm{iso}}^{\mathrm{ent}}(\alpha)$ with the parameter range $\frac{1}{d+1} < \alpha \leq 1$ (recall eqn (15.118)), it is easy to see that the first inequality is indeed satisfied,

$$(\rho_{\mathrm{iso}}^{\mathrm{ent}}(\alpha), \mathbb{W}_{\mathrm{opt}})_{\mathrm{HS}} = \frac{1}{d^2}\frac{d-1}{d\sqrt{d^2-1}}\mathrm{Tr}\left\{\left(\mathbb{1}_d \otimes \mathbb{1}_d + \frac{d}{2}\alpha\Gamma\right)\left(\mathbb{1}_d \otimes \mathbb{1}_d - \frac{d}{2(d-1)}\Gamma\right)\right\}$$

$$= -\frac{\sqrt{d^2-1}}{d}\left(\alpha - \frac{1}{d+1}\right) < 0 \tag{16.133}$$

for $1/(d+1) < \alpha$. For the second inequality the separable states are involved, but there the parameter range changes $-1/(d^2-1) \leq \alpha \leq 1/(d+1)$, as we recall from eqn (15.118), and we obtain positivity as required

$$(\rho_{\mathrm{iso}}^{\mathrm{sep}}(\alpha), \mathbb{W}_{\mathrm{opt}})_{\mathrm{HS}} = -\frac{\sqrt{d^2-1}}{d}\left(\alpha - \frac{1}{d+1}\right) \geq 0, \tag{16.134}$$

for $\alpha \leq 1/(d+1)$. Positivity is maintained for *all* separable states since they form the convex hull of the product states that satisfy the positivity condition (the explicit expressions can be found in Section 17.3).

Let us finally calculate the maximal violation of the entanglement-witness inequality (recall Definition 16.6) to demonstrate the BNT theorem (Theorem 16.10),

$$B(\rho_{\mathrm{iso}}^{\mathrm{ent}}(\alpha)) = \max_{\mathbb{W},\,\|\mathbb{W}-a\mathbb{1}\|\leq 1}\left(\min_{\rho\in\mathcal{S}}(\rho, \mathbb{W})_{\mathrm{HS}} - (\rho_{\mathrm{iso}}^{\mathrm{ent}}(\alpha), \mathbb{W})_{\mathrm{HS}}\right)$$

$$= (\rho_0, \mathbb{W}_{\mathrm{opt}})_{\mathrm{HS}} - (\rho_{\mathrm{iso}}^{\mathrm{ent}}(\alpha), \mathbb{W}_{\mathrm{opt}})_{\mathrm{HS}}. \tag{16.135}$$

Noticing that $(\rho_0, \mathbb{W}_{\mathrm{opt}})_{\mathrm{HS}} = 0$, we obtain from eqn (16.133)

$$B(\rho_{\mathrm{iso}}^{\mathrm{ent}}(\alpha)) = -(\rho_{\mathrm{iso}}^{\mathrm{ent}}(\alpha), \mathbb{W}_{\mathrm{opt}})_{\mathrm{HS}} = \frac{\sqrt{d^2-1}}{d}\left(\alpha - \frac{1}{d+1}\right) \equiv D(\rho_{\mathrm{iso}}^{\mathrm{ent}}(\alpha)), \tag{16.136}$$

for $\frac{1}{d+1} < \alpha$ which clearly coincides with the Hilbert–Schmidt measure (16.130) as required by the BNT theorem (Theorem 16.10).

Résumé: An entanglement witness is a Hermitian operator—a measurable quantity—that enables us to decide whether a given quantum state is entangled or not. The entanglement-witness theorem (Theorem 16.8) provides a necessary and sufficient criterion for detecting entanglement. The theorem is represented by an inequality for the (Hilbert–Schmidt) inner product of the quantum state under consideration with the entanglement witness. For all separable states the inner product with the entanglement witness remains positive (non-negative) by construction, such that a negative inner product indicates an entangled state. To explicitly construct an entanglement witness is not always an easy task but for special cases, in particular for all NPT states, the construction is given by Lemma 16.1.

For general states the construction method relies on purely geometric considerations. In this context the BNT theorem is of importance. It states that the maximal violation of the entanglement-witness inequality is equal to the Hilbert–Schmidt measure. The entanglement witness in this case is given explicitly by the distance of the entangled state to the closest separable state. As examples, the entanglement witness and the BNT theorem have been examined explicitly for the Werner states and the isotropic states.

16.4   Entanglement and Separability—A Choice of Factorization

In the final section of this chapter, we want to draw attention to the fact that the question of entanglement versus separability is also always one about the choice of bipartition, that is, how one chooses to split a given system into two (or more) subsystems. Mathematically, this can be phrased as a choice of factorizing the operator algebra into tensor products. Indeed, depending on our choice of factorizing the algebra of the density matrix, the quantum state may appear either entangled or separable and one may switch between different factorizations—and thus between entanglement and separability—via (global) unitary transformations (see, e.g., the treatment in Thirring *et al.*, 2011). Therefore it only makes sense to talk about entanglement or separability with respect to a chosen factorization.

For a given quantum state, we are free to choose how to factorize the algebra to which a density matrix refers. Via global unitary transformations we can switch from one factorization to the other, where the quantum state appears entangled in one factorization but not in the other. Consequently, entanglement or separability of a quantum state depend on our choice of factorizing the algebra of the corresponding density matrix, and this choice is often suggested either by the setup of the experiment or left to the convenience of the theoretical discussion.

Equivalently to the algebra of the density matrix one may consider the tensor-product structure of quantum states, which closely relates the results of Thirring *et al.* (2011) and the work of *Paolo Zanardi* and collaborators (Zanardi *et al.*, 2004), who found the same *"democracy between the different tensor product structures, ... , without further physical assumptions no partition has an ontologically superior status with respect to any other"* (Zanardi, 2001).

For pure states the status is quite clear and this discussion may naively appear to be somewhat superfluous. Since any pure quantum state (in a fixed Hilbert space) can be reached from any other pure quantum state (in that Hilbert space) via global unitary transformations, any pure state can be factorized such that it appears separable or (maximally) entangled (Harshman and Ranade, 2011; Thirring *et al.*, 2011). For mixed states, however, the situation is much more complex (see, e.g., Lewenstein *et al.*, 2000*a*). The reason is that the maximally mixed state $\frac{1}{d_{AB}}\mathbb{1}_{AB}$ (with d_{AB} the dimension of the total Hilbert space considered), is separable for any factorization and therefore a sufficiently small neighbourhood of it also remains separable under all unitary transformations. Thus the question is: How mixed can a quantum state be in order to find a factorization for which it is entangled? Or, more quantitatively: How strongly entangled can a state with a given mixedness become for any factorization? For a generally mixed state the precise answer to these questions is not known. However, for some special cases clear answers are possible, as we shall demonstrate in the following.

16.4.1 Factorization Algebra

Let us consider quantum states represented by a density matrix ρ over the operator algebra $\mathcal{M}_{AB}$ on a Hilbert space $\mathcal{H}_{AB}$ with dimension $\dim(\mathcal{H}_{AB}) = d_{AB}$, and a factorization $\mathcal{M}_A^{d_A} \otimes \mathcal{M}_B^{d_B}$ of $\mathcal{M}_{AB}$ into algebras $\mathcal{M}_A^{d_A}$ and $\mathcal{M}_B^{d_B}$ on Hilbert spaces $\mathcal{H}_A$ and $\mathcal{H}_B$ with dimensions $\dim(\mathcal{H}_A) = d_A$ and $\dim(\mathcal{H}_B) = d_B$ such that $\mathcal{H}_{AB} = \mathcal{H}_A \otimes \mathcal{H}_B$ and $d_{AB} = d_A d_B$. Following Definition 15.5, the state ρ is called separable with respect to the factorization $\mathcal{M}_A^{d_A} \otimes \mathcal{M}_B^{d_B}$ if it can be written as $\rho = \sum_i p_i \rho_i^{(A)} \otimes \rho_i^{(B)}$, otherwise it is entangled. Choosing another factorization $U(\mathcal{M}_A^{d_A} \otimes \mathbb{1}_B) U^\dagger$ and $U(\mathbb{1}_A \otimes \mathcal{M}_B^{d_B}) U^\dagger$, where U represents a unitary transformation on the total space $\mathcal{H}_{AB}$, a formerly separable state might appear entangled and vice versa. For a separable ρ we can consider the effect of U on ρ, i.e., $\rho_U = U\rho U^\dagger$, and ρ_U might become entangled for $\mathcal{M}_A^{d_A} \otimes \mathcal{M}_B^{d_B}$.

Let us first concentrate on pure states, i.e. $\rho = |\psi\rangle\langle\psi|$, where the following theorem holds (Thirring *et al.*, 2011).

Theorem 16.11 (Factorization for pure states)

For any pure state ρ one can find a factorization $\mathcal{M}_{AB} = \mathcal{A}_A \otimes \mathcal{A}_B$ such that ρ is separable with respect to this factorization and another factorization $\mathcal{M}_{AB} = \mathcal{B}_A \otimes \mathcal{B}_B$ where ρ appears to be maximally entangled.

Proof For each pair of vectors of the same (here, unit) length there are unitary transformations which transform one vector into the other. The vector $|\psi\rangle$ defining the density matrix $\rho = |\psi\rangle\langle\psi|$ for a pure state can be transformed into any product vector $|\varphi\rangle_A \otimes |\phi\rangle_B$ by a unitary operator $U \in \mathcal{M}_{AB}$, i.e. $U|\psi\rangle = |\varphi\rangle_A \otimes |\phi\rangle_B$, or on the other hand we may choose U such that the state is maximally entangled $U|\psi\rangle = \frac{1}{\sqrt{d}} \sum_i |\varphi_i\rangle_A \otimes |\phi_i\rangle_B$, where $d = \min(d_A, d_B)$.

For the pure-state density matrix this means the following: Assuming the density matrix ρ_{ent} is entangled within the factorization algebra $\mathcal{M}_A^{d_A} \otimes \mathcal{M}_B^{d_B}$ then $\exists\, U :$ $U \rho_{\text{ent}} U^\dagger = \rho_{\text{sep}}$, i.e., after a unitary transformation the density matrix becomes separable within this factorization. However, when also transforming the factorization algebra $\mathcal{M}_{AB} = U(\mathcal{M}_A^{d_A} \otimes \mathcal{M}_B^{d_B}) U^\dagger$ we may consider ρ_{sep} within this unitarily transformed factorization, where ρ_{sep} is still entangled. Of course, we may choose either factorization $\mathcal{M}_{AB} = U(\mathcal{M}_A^{d_A} \otimes \mathcal{M}_B^{d_B}) U^\dagger$ or $\mathcal{M}_A^{d_A} \otimes \mathcal{M}_B^{d_B}$, since U preserves all algebraic relations used in the definitions. $\qquad\Box$

The extension of Theorem 16.11 to mixed states requires some restrictions, as we can see from the example of the maximally mixed state $\frac{1}{d_{AB}} \mathbb{1}_{AB}$ which is separable for any factorization.

Theorem 16.12 (Factorization for mixed states)

For any mixed state ρ one can find a factorization $\mathcal{M}_{AB} = \mathcal{A}_A \otimes \mathcal{A}_B$ such that ρ is separable with respect to this factorization. Another factorization $\mathcal{M}_{AB} = \mathcal{B}_A \otimes \mathcal{B}_B$ where ρ appears to be entangled only exists below a certain bound of mixedness.

Here we will only give the idea of the proof, which uses the fact that any mixed state can be unitarily transformed into a *Weyl state* (Thirring *et al.*, 2011) and for Weyl states a fairly good characterization of the regions of separable states, entangled, and bound-entangled states exists, see, e.g., the work by *Reinhold Bertlmann* and *Philipp Krammer* (2008a, 2008b, 2008c, 2009), *Bernhard Baumgartner, Beatrix Hiesmayr*, and *Heide Narnhofer* (2006, 2007, 2008), as well as of *Paweł, Michał*, and *Ryszard Horodecki* (1999).

We now want to make the statement of Theorem 16.12 more precise by formulating the following lemma (Thirring *et al.*, 2011):

Lemma 16.2 (Bound for split states)

For any state that can be split into a convex sum of a maximally entangled state $|\Psi_{\text{ME}}\rangle$, represented by a projector $P_{\text{ME}} = |\Psi_{\text{ME}}\rangle\langle\Psi_{\text{ME}}|$, and an orthogonal (but not necessarily pure) state σ,

$$\rho = \beta\, P_{\text{ME}} + (1 - \beta)\, \sigma \quad \text{with} \quad (\sigma, P_{\text{ME}})_{\text{HS}} = 0$$

and $0 \le \beta \le 1$, we have: If $\beta > \frac{1}{d}$ the state ρ is entangled.

Proof of Lemma 16.2 We first construct the following optimal entanglement witness (as explained in Section 16.3.1)

$$\mathbb{W} = \mathbb{1}_{d^2} - d\, P_{\mathrm{ME}} \tag{16.137}$$

to find a bound of separability or entanglement. The inner product of a witness with all separable states must remain positive semidefinite

$$(\rho_{\mathrm{sep}}, \mathbb{W})_{\mathrm{HS}} = \mathrm{Tr}(\rho_{\mathrm{sep}} \mathbb{W}) \geq 0. \tag{16.138}$$

To show that this is the case we first note that the maximally entangled state can be written in Schmidt decomposition (Theorem 15.1) as $|\Psi_{\mathrm{ME}}\rangle = \frac{1}{\sqrt{d}} \sum_{n=1}^{d} |\chi_n\rangle_A |\eta_n\rangle_B$, i.e., all Schmidt coefficients are equal to $\frac{1}{\sqrt{d}}$ and all relative phases are absorbed into the choice of the Schmidt basis. For notational convenience, we now write $|\chi_n\rangle_A \equiv |n\rangle_A$ and $|\eta_n\rangle_B \equiv |n\rangle_B$, and we drop the subscripts for the subsystems. For all pure product states $|\varphi\rangle \otimes |\psi\rangle$ with $|\varphi\rangle = \sum_m \varphi_m |m\rangle$ and $|\psi\rangle = \sum_n \psi_n |n\rangle$ we then have

$$\langle\varphi| \otimes \langle\psi| \left(\mathbb{1}_{d^2} - d\, P_{\mathrm{ME}}\right) |\varphi\rangle \otimes |\psi\rangle = 1 - d\, \langle\varphi| \otimes \langle\psi| P_{\mathrm{ME}} |\varphi\rangle \otimes |\psi\rangle$$

$$= 1 - d\, \langle\varphi| \otimes \langle\psi| \frac{1}{d} \sum_{m,n} |m\rangle\langle n| \otimes |m\rangle\langle n| |\varphi\rangle \otimes |\psi\rangle = 1 - d \sum_{m,n=1}^{d} \frac{1}{d}\, \varphi_m^* \psi_m^* \varphi_n \psi_n$$

$$= 1 - \langle\varphi^*|\psi\rangle \langle\psi|\varphi^*\rangle = 1 - |\langle\varphi^*|\psi\rangle|^2 \geq 0\,, \tag{16.139}$$

since $|\langle\varphi^*|\psi\rangle| \leq 1$, and $\langle\varphi| \otimes \langle\psi| \left(\mathbb{1}_{d^2} - d\, P_{\mathrm{ME}}\right) |\varphi\rangle \otimes |\psi\rangle = 0$ iff $|\psi\rangle = |\varphi^*\rangle$, which makes the witness optimal.

Now we apply the entanglement witness (16.137) to the state ρ of Lemma 16.2,

$$(\beta\, P_{\mathrm{ME}} + [1-\beta]\, \sigma, \mathbb{1}_{d^2} - d\, P_{\mathrm{ME}})_{\mathrm{HS}} \tag{16.140}$$

$$= \beta\, (P_{\mathrm{ME}}, \mathbb{1}_{d^2})_{\mathrm{HS}} - \beta\, d\, (P_{\mathrm{ME}}, P_{\mathrm{ME}})_{\mathrm{HS}} + (1-\beta)\, (\sigma, \mathbb{1}_{d^2})_{\mathrm{HS}} = 1 - \beta\, d\,, \tag{16.141}$$

where we have used $(\sigma, P_{\mathrm{ME}})_{\mathrm{HS}} = 0$. We thus see that the condition for entanglement, $(\rho, \mathbb{W})_{\mathrm{HS}} < 0$ is equivalent to $\beta > \frac{1}{d}$. $\qquad\square$

A particular family of states that is of the form specified in Lemma 16.2 is that of the generalized Werner state in $d \times d$ dimensions (we choose $d_A = d_B = d$), given by

$$\rho = \alpha\, P_{\mathrm{ME}} + \frac{1-\alpha}{d^2}\, \mathbb{1}_{d^2}\,, \tag{16.142}$$

where P_{ME} is a projector ($P_{\mathrm{ME}}^2 = P_{\mathrm{ME}}$) onto a maximally entangled state that we can leave unspecified, and we restrict the range of α to $0 \leq \alpha \leq 1$ here. Writing such a state in matrix form with respect to a basis that contains the maximally entangled state onto which P_{ME} projects as its first basis vector yields

$$\rho = \begin{pmatrix} \alpha + \frac{1-\alpha}{d^2} & 0 & \cdots & 0 \\ 0 & \frac{1-\alpha}{d^2} & & \\ \vdots & & \ddots & \\ 0 & & & \frac{1-\alpha}{d^2} \end{pmatrix}. \tag{16.143}$$

The maximal eigenvalue of this state is $\alpha + \frac{1-\alpha}{d^2}$, such that Lemma 16.2 implies that the state is entangled if $\alpha > \frac{1}{d+1}$, which in turn yields the well-known bound $\alpha > \frac{1}{3}$

for the (usual) Werner states in dimension $d = 2$, see Section 15.3.3. The bound is thus optimal for Werner states, and also for the states of the Gisin line, the Gisin states for the parameter $\theta = \frac{\pi}{4}$ (see Section 15.3.5), which are also of the form specified in Lemma 16.2. However, if not all eigenvalues of σ are equal then the witness $\mathbb{W} = \mathbb{1}_{d^2} - d\,P_{\mathrm{ME}}$ is no longer optimal with respect to the state ρ of Lemma 16.2. In this case the decomposition no longer represents a Werner state.

To conclude, for each state with maximal eigenvalue $\lambda_{\max} > \frac{1}{d}$ there exists a factorization like in Lemma 16.2 for which it is entangled. In other words, one can find a global unitary that maps the eigenstate corresponding to the eigenvalue $\lambda_{\max}$ to a maximally entangled state. However, the bound $\lambda_{\max} > \frac{1}{d}$ is generally not tight, i.e., the factorization of Lemma 16.2 is generally not optimal. But under certain constraints we can find a factorization that is indeed optimal, as we shall phrase in the following theorem (Thirring *et al.*, 2011).

Theorem 16.13 (Factorization under constraints)

For any state whose largest eigenvalue $\lambda_{\max}$ is bounded from below by $\frac{3}{d^2+2}$ there exists a choice of factorization such that the state is entangled.

Proof Let ρ be any mixed state with spectrum $\{\lambda_1 \geq \lambda_2 \geq ... \geq \lambda_{d^2-2} \geq \lambda_{d^2-1} \geq \lambda_{d^2}\}$, which we can assume to be ordered in decreasing order without loss of generality since there is always a global unitary that can rearrange the eigenvalues. Thus we have the decomposition (again we consider the case $d_A = d_B = d$)

$$\rho = \lambda_1 P_1 + \lambda_2 P_2 + ... + \lambda_{d^2} P_{d^2}\,, \tag{16.144}$$

where the P_i are the projectors onto the eigenstates corresponding to eigenvalues λ_i. Then, let us select four of these eigenstates, which span a two-qubit subspace of the entire d^2-dimensional Hilbert space, and choose a factorization (which we may do because of Theorem 16.11) such that two of these states are maximally entangled and two are separable (product states), i.e.,

$$P_1 = \tfrac{\lambda_1}{2}\left(|\,00\,\rangle + |\,11\,\rangle\right)\left(\langle\,00\,| + \langle\,11\,|\right),$$

$$P_{d^2-2} = \lambda_{d^2-2}\,|\,01\,\rangle\langle\,01\,|,$$

$$P_{d^2-1} = \tfrac{\lambda_{d^2-1}}{2}\left(|\,00\,\rangle - |\,11\,\rangle\right)\left(\langle\,00\,| - \langle\,11\,|\right),$$

$$P_{d^2} = \lambda_{d^2}\,|\,10\,\rangle\langle\,10\,|. \tag{16.145}$$

To detect entanglement of ρ with respect to this factorization we consider the partial transpose of the density matrix ρ in eqn (16.144) with the choice (16.145). The partial transpose is block diagonal, and the 2×2 block corresponding to the subspace spanned by the vectors $|\,01\,\rangle$ and $|\,10\,\rangle$ is of the form

$$\begin{pmatrix} \langle\,01\,|\rho^{T_B}|\,01\,\rangle & \langle\,01\,|\rho^{T_B}|\,10\,\rangle \\ \langle\,10\,|\rho^{T_B}|\,01\,\rangle & \langle\,10\,|\rho^{T_B}|\,10\,\rangle \end{pmatrix} = \begin{pmatrix} \lambda_{d^2-2} & \tfrac{1}{2}(\lambda_1 - \lambda_{d^2-1}) \\ \tfrac{1}{2}(\lambda_1 - \lambda_{d^2-1}) & \lambda_{d^2} \end{pmatrix}. \tag{16.146}$$

We now make use of the PPT criterion, Theorem 15.3, which in this case detects entanglement if the 2×2 block has a negative eigenvalue. This is the case if $\lambda_{d^2-2}\lambda_{d^2} - \frac{1}{4}(\lambda_1 - \lambda_{d^2-1})^2 < 0$, or in alternative form, if

$$\sqrt{\lambda_{d^2-2}\lambda_{d^2}} \;<\; \frac{1}{2}(\lambda_1 - \lambda_{d^2-1}). \qquad (16.147)$$

Now we recognize the left-hand side as a geometric mean, which is bounded from above by the corresponding arithmetic mean $\sqrt{\lambda_{d^2-2}\lambda_{d^2}} \leq (\lambda_{d^2-2} + \lambda_{d^2})/2$, and we can thus relax the previous condition to

$$\frac{\lambda_{d^2-2} + \lambda_{d^2}}{2} \;<\; \frac{1}{2}(\lambda_1 - \lambda_{d^2-1}) \quad\Rightarrow\quad \lambda_{d^2-2} + \lambda_{d^2-1} + \lambda_{d^2} < \lambda_1. \qquad (16.148)$$

At the same time we can bound the sum of the three smallest eigenvalues from above. Since normalization dictates $\sum_i \lambda_i = 1$, fixing the largest eigenvalue λ_1 implies $\sum_{i\geq 2}^{d^2} \lambda_i = 1 - \lambda_1$. The eigenvalues are ordered decreasingly such that the largest value that the sum of the smallest three eigenvalues can have is obtained when all remaining $d^2 - 1$ eigenvalues are the same, $\lambda_{i\geq 2} = (1 - \lambda_1)/(d^2 - 1)$, in which case

$$\lambda_{d^2-2} + \lambda_{d^2-1} + \lambda_{d^2} \;<\; 3\frac{1-\lambda_1}{d^2-1}. \qquad (16.149)$$

Combining this with the inequality on the right-hand side of (16.148) we see that the state must be entangled if

$$3\frac{1-\lambda_1}{d^2-1} \;<\; \lambda_1 \qquad \text{or} \qquad \lambda_1 \;>\; \frac{3}{d^2+2}. \qquad (16.150)$$

$\square$

For $d = 2$ we just have the bound $\lambda_{\max} > \frac{1}{2}$ as in Lemma 16.2, which we know is tight since we know from (15.144) that for any two-qubit state a fidelity with any maximally entangled state (a singlet fraction) above $\frac{1}{2}$ is a necessary and sufficient criterion for entanglement. For $d > 2$ we note that the bound $\lambda_{\max} > \frac{3}{d^2+2}$ improves upon the bound $\lambda_{\max} > \frac{1}{d}$ from Lemma 16.2, since $\frac{3}{d^2+2} < \frac{1}{d}$ in that case.

However, it also important to note that the criterion of Theorem 16.13 just tells us that a state with a maximal eigenvalue above the threshold can *become* entangled by choosing a suitable factorization (applying a suitable global unitary), not that it necessarily already *is* entangled. For instance, isotropic states from Section 15.4.4 with a parameter α below $\frac{1}{d}$ are separable, as we have shown, and a maximal eigenvalue larger than $\frac{3}{d^2+2}$ hence does not imply entanglement necessarily.

Thus, the message is, under certain constraints (on the largest eigenvalue) a mixed state is separable with respect to some factorization and entangled with respect to another.

Absolutely separable states: Since the maximally mixed state $\rho_{\text{mix}} = \frac{1}{d^2} \mathbb{1}_{d^2}$ is separable for any factorization and therefore a sufficiently small neighbourhood of it is separable as well, it is interesting now to look for those states which are separable with respect to all possible factorizations of the composite system into subsystems $\mathcal{M}_{AB} = \mathcal{A}_A \otimes \mathcal{A}_B$. This is the case if $\rho_U = U\rho U^\dagger$ remains separable for any unitary transformation U. Such states are called *absolutely separable states* (Kuś and Życzkowski, 2001; Życzkowski and Bengtsson, 2006; Bengtsson and Życzkowski, 2006; Thirring *et al.*, 2011). In this connection the *maximal ball* of states around the maximally mixed state with a radius $r = 1/(d\sqrt{d^2-1})$ of constant mixedness is considered, which can be inscribed into the separable states (Kuś and Życzkowski, 2001; Gurvits and Barnum, 2002). This radius is given in terms of the Hilbert–Schmidt distance

$$D_{\text{HS}}(\rho, \mathbb{1}_{d^2}) = \|\rho - \tfrac{1}{d^2}\mathbb{1}_{d^2}\|_{\text{HS}} = \sqrt{\text{Tr}\big[\big(\rho - \tfrac{1}{d^2}\mathbb{1}_{d^2}\big)^2\big]}. \qquad (16.151)$$

Theorem 16.14 (Absolute separability of the Kuś–Życzkowski ball)

All states within the maximal ball that can be inscribed into the set of bipartite separable states are not only separable but also absolutely separable.

In 2×2 dimensions we have illustrated the Kuś-Życzkowski ball in Fig. 16.7, it touches the double pyramid of separable states at the central points of its eight faces. However, note that geometrically, in terms of Theorem 16.13, the set of absolutely separable states is not as symmetric as one might be tempted to assume based on Theorem 16.14. The set of absolutely separable state is strictly larger than the maximal ball and, if graphically illustrated, would corresponds to what in Vienna would colloquially be described as a "Laberl"[9] rather than a ball.

Furthermore, we already know that in the two-qubit case the set of absolutely separable states is larger than the maximal ball devised by *Marek Kuś* and *Karol Życzkowski*. As conjectured by *Satoshi Ishizaka* and *Tohya Hiroshima* (2000) and proven by *Frank Verstraete, Koenraad Audenaert*, and *Bart De Moor* (2001b) the set of absolutely separable states contains any mixed state with certain constraints on the spectrum given by the following lemma.

Lemma 16.3 (Absolute separability in 2×2 dimensions)

Let ρ be any mixed state in 2×2 dimensions with (ordered) spectrum $\{\lambda_1 \geq \lambda_2 \geq \lambda_3 \geq \lambda_4\}$. If the spectrum is constrained by the inequality

$$\lambda_1 - \lambda_3 - 2\sqrt{\lambda_2\lambda_4} \leq 0\,,$$

then ρ is absolutely separable.

---

[9]In Viennese dialect, "Laberl" refers to a rounded but not perfectly spherical shape, often in the context of food, e.g., a burger patty, and "Fetzenlaberl" (a "Laberl" made of rags, or "Fetzen") is the Viennese expression for a self-made football out of shreds of fabric, which in the old times the Viennese boys liked to play with on the streets.

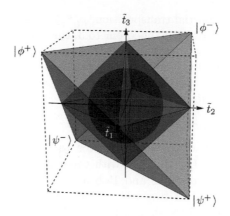

Fig. 16.7 Tetrahedron of Weyl states with absolutely separable states. The tetrahedron of all two-qubit Weyl states (compare with Fig. 15.3), the states with maximally mixed reduced states, is shown (green). The four Bell states $|\psi^{\pm}\rangle$, $|\phi^{\pm}\rangle$ are located at its corners. The singular values $(\tilde{t}_1, \tilde{t}_2, \tilde{t}_3)$ of the correlation tensor serve as coordinates. The set $\mathcal{S}$ of separable states forms a double pyramid (blue), of which the set of absolutely separable states is a subset. The ball of maximal radius $r = 1/(2\sqrt{3})$ that can be inscribed into set $\mathcal{S}$ is the Kuś-Życzkowski ball (purple). All states contained within this ball are absolutely separable but not all absolutely states lie within.

As an example let us simply quote the state with spectrum $\{0.47, 0.30, 0.13, 0.10\}$ (Kuś and Życzkowski, 2001) that does not belong to the maximal ball but satisfies the constraints of Lemma 16.3 and is, for this reason, absolutely separable.

Example—Narnhofer state: To illustrate the Kuś-Życzkowski ball from Theorem 16.14 we choose a separable state outside of the maximal ball that does not satisfy Lemma 16.3, say, the two-qubit Weyl state at the upper corner of the double pyramid in see Fig. 16.7. This state, called *Narnhofer state* (Thirring *et al.*, 2011), can be still transformed into an entangled state and is given by

$$\rho_{\mathrm{N}} = |\psi_{\mathrm{N}}\rangle\langle\psi_{\mathrm{N}}| = \tfrac{1}{2}\left(\rho^+ + \omega^+\right) = \tfrac{1}{4}\left(\mathbb{1} \otimes \mathbb{1} + \sigma_x \otimes \sigma_x\right). \tag{16.152}$$

This separable state has the smallest possible mixedness or largest purity $P(\rho) = \mathrm{Tr}(\rho^2)$. Now we apply the unitary transformation

$$U = \tfrac{1}{4}\left[(2+\sqrt{2})\,\mathbb{1} \otimes \mathbb{1} + i\sqrt{2}\left(\sigma_x \otimes \sigma_y + \sigma_y \otimes \sigma_x\right) - (2-\sqrt{2})\,\sigma_z \otimes \sigma_z\right], \tag{16.153}$$

which transforms the state ρ_{N} to

$$\rho_U = U\rho_{\mathrm{N}}\,U^\dagger = \tfrac{1}{4}\left[\mathbb{1} \otimes \mathbb{1} + \tfrac{1}{2}(\sigma_z \otimes \mathbb{1} + \mathbb{1} \otimes \sigma_z) + \tfrac{1}{2}(\sigma_x \otimes \sigma_x + \sigma_y \otimes \sigma_y)\right]. \tag{16.154}$$

Due to the occurrence of the term $(\sigma_z \otimes \mathbb{1} + \mathbb{1} \otimes \sigma_z)$ the transformation U (16.153) leads to a quantum state outside of the set of Weyl states. This new state ρ_U (16.154) is no longer positive under partial transposition, $\rho_U^{\mathsf{T}_B} \not\geq 0$, where $\rho_U^{\mathsf{T}_B} = (\mathbb{1} \otimes \mathsf{T}_B) \rho_U$. Therefore, due to the PPT criterion, Theorem 15.3, the state ρ_U (16.154) is entangled with the concurrence $C[\rho_U] = \frac{1}{2}$. The transformation U (16.153) is already optimal, i.e., it entangles ρ_N maximally. Thus ρ_U belongs to the so-called MEMS class, the class of *maximally entangled mixed states* for a given value of purity (Ishizaka and Hiroshima, 2000; Munro *et al.*, 2001).

16.4.2 Alice and Bob Factorizations

It is quite instructive to illustrate the free choice of the algebra factorization for a density matrix by examples. Let us begin with the example of two qubits. Here the dimensions of the subspaces of Alice and Bob are $d_A = d_B = 2$ and we span the two factors $\mathcal{M}^2$ with the aid of two sets of Pauli matrices $\vec{\sigma}_A$ and $\vec{\sigma}_B$, the subalgebras of Alice and Bob. The standard product basis is the computational basis $\{|00\rangle, |01\rangle, |10\rangle, |11\rangle\}$ and and the four maximally entangled Bell vectors are states are given by $|\psi^\pm\rangle = \frac{1}{\sqrt{2}}(|01\rangle \pm |10\rangle)$ and $|\phi^\pm\rangle = \frac{1}{\sqrt{2}}(|00\rangle \pm |11\rangle)$. In matrix notation we have

$$
\rho_{00} = \begin{pmatrix} 1 & 0 & 0 & 0 \\ 0 & 0 & 0 & 0 \\ 0 & 0 & 0 & 0 \\ 0 & 0 & 0 & 0 \end{pmatrix}, \quad
\rho_{01} = \begin{pmatrix} 0 & 0 & 0 & 0 \\ 0 & 1 & 0 & 0 \\ 0 & 0 & 0 & 0 \\ 0 & 0 & 0 & 0 \end{pmatrix}, \quad
\rho_{10} = \begin{pmatrix} 0 & 0 & 0 & 0 \\ 0 & 0 & 0 & 0 \\ 0 & 0 & 1 & 0 \\ 0 & 0 & 0 & 0 \end{pmatrix}, \quad
\rho_{11} = \begin{pmatrix} 0 & 0 & 0 & 0 \\ 0 & 0 & 0 & 0 \\ 0 & 0 & 0 & 0 \\ 0 & 0 & 0 & 1 \end{pmatrix},
$$

$$(16.155)$$

and for the entangled Bell states

$$
\rho^\pm = |\psi^\pm\rangle\langle\psi^\pm| = \frac{1}{2}\begin{pmatrix} 0 & 0 & 0 & 0 \\ 0 & 1 & \pm 1 & 0 \\ 0 & \pm 1 & 1 & 0 \\ 0 & 0 & 0 & 0 \end{pmatrix}, \quad
\omega^\pm = |\phi^\pm\rangle\langle\phi^\pm| = \frac{1}{2}\begin{pmatrix} 1 & 0 & 0 & \pm 1 \\ 0 & 0 & 0 & 0 \\ 0 & 0 & 0 & 0 \\ \pm 1 & 0 & 0 & 1 \end{pmatrix}.
$$

$$(16.156)$$

The unitary matrix U which transforms the entangled basis into the separable one is

$$
U = \frac{1}{\sqrt{2}}\left(\mathbb{1} \otimes \mathbb{1} + i\,\sigma_x \otimes \sigma_y\right) = \frac{1}{\sqrt{2}}\begin{pmatrix} 1 & 0 & 0 & 1 \\ 0 & 1 & -1 & 0 \\ 0 & 1 & 1 & 0 \\ -1 & 0 & 0 & 1 \end{pmatrix}. \tag{16.157}
$$

However, it is more illustrative to work with the Bloch decompositions of the states to show the subalgebras explicitly. In particular we have

$$
\rho^- = \tfrac{1}{4}\left(\mathbb{1} \otimes \mathbb{1} - \vec{\sigma} \otimes \vec{\sigma}\right), \tag{16.158a}
$$

$$
\rho_{01} = \tfrac{1}{4}\left(\mathbb{1} \otimes \mathbb{1} + \sigma_z \otimes \mathbb{1} - \mathbb{1} \otimes \sigma_z - \sigma_z \otimes \sigma_z\right). \tag{16.158b}
$$

Then U transforms the two sets of algebras of Alice and Bob as follows:

$$\sigma_x \otimes \mathbb{1} \xrightarrow{U} \sigma_x \otimes \mathbb{1} \qquad\qquad \mathbb{1} \otimes \sigma_x \xrightarrow{U} \sigma_x \otimes \sigma_z$$

$$\sigma_y \otimes \mathbb{1} \to -\sigma_z \otimes \sigma_y \qquad\qquad \mathbb{1} \otimes \sigma_y \to \mathbb{1} \otimes \sigma_y \qquad (16.159)$$

$$\sigma_z \otimes \mathbb{1} \to \sigma_y \otimes \sigma_y \qquad\qquad \mathbb{1} \otimes \sigma_z \to -\sigma_x \otimes \sigma_x\,,$$

which implies a change in Alice's and Bob's tensor products according to

$$\sigma_x \otimes \sigma_x \xrightarrow{U} \mathbb{1} \otimes \sigma_z$$

$$\sigma_y \otimes \sigma_y \to -\sigma_z \otimes \mathbb{1} \qquad (16.160)$$

$$\sigma_z \otimes \sigma_z \to \sigma_z \otimes \sigma_z\,.$$

Now we are prepared to demonstrate Theorems 16.11 and 16.12. Let us first focus on entanglement. It can be detected by an entanglement witness W that satisfies the *entanglement-witness inequality* (Theorem 16.8)

$$(\rho_{\mathrm{ent}}, \mathrm{W})_{\mathrm{HS}} = \mathrm{Tr}(\mathrm{W}\,\rho_{\mathrm{ent}}) < 0\,, \qquad \rho_{\mathrm{ent}} \in \mathcal{S}^{\mathrm{C}} \qquad (16.161\mathrm{a})$$

$$(\rho, \mathrm{W})_{\mathrm{HS}} = \mathrm{Tr}(\mathrm{W}\,\rho) \geq 0\,, \qquad \forall \rho \in \mathcal{S}\,. \qquad (16.161\mathrm{b})$$

with $\mathrm{W}_{\mathrm{opt}}$ being optimal $\langle \rho_0 \,|\, \mathrm{W}_{\mathrm{opt}} \rangle = 0$ for some separable state $\rho_0 \in \mathcal{S}$. The optimal entanglement witness can be constructed via the BNT Theorem 16.10

$$\mathrm{W}_{\mathrm{opt}} = \frac{\rho_0 - \rho_{\mathrm{ent}} - (\rho_0, \rho_0 - \rho_{\mathrm{ent}})_{\mathrm{HS}}\,\mathbb{1}}{\|\rho_0 - \rho_{\mathrm{ent}}\|_{\mathrm{HS}}}\,, \qquad (16.162)$$

where ρ_0 represents the closest separable state to the entangled state ρ_{ent}.

In particular, for the optimal entanglement witness for the Bell state ρ^- we get

$$\mathrm{W}^{\rho^-}_{\mathrm{opt}} = \frac{1}{2\sqrt{3}}\left(\mathbb{1} \otimes \mathbb{1} + \sum_i \sigma_i \otimes \sigma_i\right), \qquad (16.163)$$

which leads to the entanglement-witness inequality

$$(\rho^-, \mathrm{W}^{\rho^-}_{\mathrm{opt}})_{\mathrm{HS}} = \mathrm{Tr}(\rho^- \mathrm{W}^{\rho^-}_{\mathrm{opt}}) = -\frac{1}{\sqrt{3}} < 0, \qquad (16.164)$$

$$(\rho_{\mathrm{sep}}, \mathrm{W}^{\rho^-}_{\mathrm{opt}})_{\mathrm{HS}} = \mathrm{Tr}(\rho_{\mathrm{sep}} \mathrm{W}^{\rho^-}_{\mathrm{opt}}) = \frac{1}{2\sqrt{3}}[1 + \cos(\delta)] \geq 0 \qquad \forall \rho \in \mathcal{S}\,,$$

where δ represents the angle between the unit vectors $\vec{a}$ and $\vec{b}$ of a general separable state like in eqn (15.42).

Now transforming the entangled Bell state ρ^- according to eqn (16.157) we obtain

$$U \rho^- U^\dagger = \tfrac{1}{4}(\mathbb{1} \otimes \mathbb{1} + \sigma_z \otimes \mathbb{1} - \mathbb{1} \otimes \sigma_z - \sigma_z \otimes \sigma_z) \equiv \rho_{01}\,, \qquad (16.165)$$

$$(U \rho^- U^\dagger, \mathrm{W}^{\rho^-}_{\mathrm{opt}})_{\mathrm{HS}} = \mathrm{Tr}(U \rho^- U^\dagger\, \mathrm{W}^{\rho^-}_{\mathrm{opt}}) = 0\,, \qquad (16.166)$$

i.e., separability with respect to the algebra $\{\sigma_i \otimes \sigma_j\}_{i,j}$.

Thus the transformed state $U \rho^- U^\dagger$ represents a separable pure state as claimed in Theorem 16.11 and geometrically it has the Hilbert–Schmidt distance

$$D_{\mathrm{HS}}(U \rho^- U^\dagger, \rho^-) = \|U \rho^- U^\dagger - \rho^-\|_{\mathrm{HS}} = 1, \qquad (16.167)$$

to the state ρ^-. This distance represents the Hilbert–Schmidt measure (16.79)

$$\mathcal{D}_{\mathrm{HS}}(\rho^-) = \min_{\rho \in \mathcal{S}} \|\rho - \rho^-\|_{\mathrm{HS}} = \|\rho_0 - \rho^-\|_{\mathrm{HS}}, \qquad (16.168)$$

where ρ_0 denotes the closest separable state, for which the Hilbert–Schmidt distance is minimal. In our case of the maximally entangled Bell state ρ^- the closest separable state is mixed and is given by the Werner state (16.81)

$$\rho_0 = \rho_{\alpha=1/3} = \tfrac{1}{4} \left(\mathbb{1} \otimes \mathbb{1} - \tfrac{1}{3} \sigma_i \otimes \sigma_i \right). \qquad (16.169)$$

However, if we also transform the entanglement witness, i.e., if we choose a different algebra,

$$U \, \mathrm{W}^{\rho^-}_{\mathrm{opt}} \, U^\dagger = \tfrac{1}{4} \left(\mathbb{1} \otimes \mathbb{1} - \sigma_z \otimes \mathbb{1} + \mathbb{1} \otimes \sigma_z + \sigma_z \otimes \sigma_z \right), \qquad (16.170)$$

we then get

$$(U \rho^- U^\dagger, U \mathrm{W}^{\rho^-}_{\mathrm{opt}} U^\dagger)_{\mathrm{HS}} = (\rho^-, \mathrm{W}^{\rho^-}_{\mathrm{opt}})_{\mathrm{HS}} = -\tfrac{1}{\sqrt{3}} < 0, \qquad (16.171)$$

and the transformed state is again entangled with respect to the other algebra factorization $\{\sigma_i \otimes \mathbb{1}, \mathbb{1} \otimes \sigma_j, \sigma_i \otimes \sigma_j\}_{i,j}$. This nicely demonstrates Theorem 16.11.

Example—Werner states: As a typical example for mixed states we consider the Werner states in Bloch form (16.80) $\rho_{\mathrm{Werner}} = \tfrac{1}{4} \left(\mathbb{1} \otimes \mathbb{1} - \alpha \vec{\sigma} \otimes \vec{\sigma} \right)$. They are separable within the bound of mixedness, $\alpha \leq 1/3$. Transforming the state ρ_{Werner} according to eqn (16.157) we obtain

$$U \rho_{\mathrm{Werner}} U^\dagger = \tfrac{1}{4} \left(\mathbb{1} \otimes \mathbb{1} + \alpha \left[\sigma_z \otimes \mathbb{1} - \mathbb{1} \otimes \sigma_z \right] - \sigma_z \otimes \sigma_z \right), \qquad (16.172)$$

which is separable with respect to the algebra $\{\sigma_i \otimes \sigma_j\}_{i,j}$ for all values of $\alpha \in [0, 1]$ since the entanglement-witness inequality provides positivity (16.161b) (recall the entanglement witness $\mathrm{W}^{\rho_{\mathrm{Werner}}}_{\mathrm{opt}} \equiv \mathrm{W}^{\rho^-}_{\mathrm{opt}}$)

$$(U \rho_{\mathrm{Werner}} U^\dagger, \mathrm{W}^{\rho^-}_{\mathrm{opt}})_{\mathrm{HS}} = \mathrm{Tr}\left(U \rho_{\mathrm{Werner}} U^\dagger \mathrm{W}^{\rho^-}_{\mathrm{opt}} \right) = \tfrac{1}{2\sqrt{3}} (1 - \alpha) \geq 0. \qquad (16.173)$$

This is not entirely unexpected since the transformation U applied to the maximally entangled part $\rho_{\mathrm{Werner}}(\alpha = 1) = \rho^-$ is already separable. However, transforming also the entanglement witness $U \mathrm{W}^{\rho^-}_{\mathrm{opt}} U^\dagger$ in eqn (16.170), i.e., choosing a different factorization, we then get negativity

$$(U \rho_{\mathrm{Werner}} U^\dagger, U \mathrm{W}^{\rho^-}_{\mathrm{opt}} U^\dagger)_{\mathrm{HS}} = (\rho_{\mathrm{Werner}}, \mathrm{W}^{\rho^-}_{\mathrm{opt}})_{\mathrm{HS}} = \tfrac{1}{2\sqrt{3}} (1 - 3\alpha) < 0, \qquad (16.174)$$

for $\alpha > 1/3$. This means that the transformed Werner state is entangled again with respect to the other algebra factorization $\{\sigma_i \otimes \mathbb{1}, \mathbb{1} \otimes \sigma_j, \sigma_i \otimes \sigma_j\}$. But as claimed in Theorem 16.12 the entanglement occurs only beyond a certain bound of mixedness, here for the Werner state the bound is $\alpha > 1/3$.

Résumé: It is an interesting fact that the characteristics of a quantum state, entanglement or separability, depend on the choice of factorizing the algebra of the corresponding density matrix. We can switch between both features via a global unitary transformation. Therefore, only with respect to a chosen factorization—which we are free to choose from a mathematical perspective, no partition has an ontologically superior status with respect to any other—does it make sense to talk about entanglement or separability. In particular, any pure state can be factorized such that the state appears separable in one factorization and maximally entangled in another (Theorem 16.11). However, for general mixed states the situation is much more complex. For any mixed state one can find a factorization such that the state is separable but in another it appears entangled only if the state is below a certain bound of mixedness (Theorem 16.12). This bound can be specified more precisely for split states (Lemma 16.2) or for mixed states with constraints on the largest eigenvalue (Theorem 16.13). In addition there are states which remain separable with respect to all possible factorizations of the composite system into the subsystems. These are called absolutely separable states, for example, the Kuś-Życzkowski ball inside the double pyramid of separable states (Theorem 16.14) or certain states with constrained spectra (Lemma 16.3).

17
High-Dimensional Quantum Systems

In the previous chapters we have discussed pertinent concepts from entanglement theory, including techniques and criteria for detecting and quantifying entanglement. Many of these techniques apply in general, but are often sufficient for deciding between entanglement and separability only for special cases, most notably the case of two qubits. For quantum systems of higher local dimensions, the task of determining if a given quantum state is entangled, and if so how strongly or in what way, becomes notoriously difficult. At the same time we cannot ignore the fact that most quantum systems inherently have more than two dimensions. In this chapter, we therefore want to briefly present some of the techniques that are available to aid us when dealing with high-dimensional quantum systems. As we have discussed already in Section 11.5 the density matrix of a quantum system can be decomposed in many different ways. For higher dimensions the matrices become large, and identifying simple ways to express density matrices is thus of great interest. In this chapter we present three different matrix bases which turn out to be quite practical. Each decomposition can be identified with a vector, the *Bloch vector*, which we calculate for several examples. For the detection of entanglement in high-dimensional systems we further construct entanglement witnesses that allow us to study the geometric structure of the Hilbert–Schmidt space, the geometry of entanglement, and also detect bound entanglement. Finally, we discuss some methods for the quantification of entanglement beyond two qubits.

17.1 Bases for Density Matrices

The state of a d-dimensional quantum system, a *qudit*, is described by a $d \times d$ density matrix. Since the space of matrices is a vector space, there exist bases of matrices which can be used to decompose any matrix. For qubits such a basis is, for instance, that of the three Pauli matrices (plus the identity operator $\mathbb{1}_2$). Accordingly, a density matrix can be expressed in terms of a three-component vector, the *Bloch vector*, and any such vector has to lie within the so-called *Bloch ball* (Bloch, 1946), as we have discussed in Section 11.5.2. Unique for qubits is the fact that any point on the sphere, the *Bloch sphere*, and inside the ball corresponds to a physical state, i.e., to a density matrix. The pure states lie on the sphere and the mixed states lie inside.

In higher dimensions, one may similarly identify bases of $d^2 - 1$ operators (again, plus the identity operator $\mathbb{1}_d$) that allow one to write down a generalized Bloch decomposition. However, there the induced map—density matrix $\mapsto$ Bloch vector—is not bijective; not every point on the Bloch sphere in dimensions $d^2 - 1$ corresponds to a physical state. In spite of this the vectors are typically still referred to as Bloch vectors (see in this context Kimura, 2003; Kimura and Kossakowski, 2005; Kryszewski and Zachciał, 2006; Jakóbczyk and Siennicki, 2006; Mendaš, 2006; Bertlmann and Krammer, 2008*a*). In our presentation we will closely follow Krammer (2009, Sec. 2.3), which is based on Bertlmann and Krammer (2008*a*).

Any matrix in a Hilbert–Schmidt space of dimension d can be expanded into a matrix basis $\{\Gamma_i\}_i$. Here, we consider the properties of a "practical" matrix basis to be:

(i) The basis includes the identity matrix $\mathbb{1}_d$ and $d^2 - 1$ matrices Γ_i which are traceless $\mathrm{Tr}(\Gamma_i) = 0$.

(ii) The matrices of any basis $\{\Gamma_i\}_i$ are orthogonal, i.e., $\mathrm{Tr}(\Gamma_i^\dagger \Gamma_j) = N_\Gamma \delta_{ij}$ with N_Γ some normalization constant specific to the chosen operator basis.

Generalized Bloch decomposition of a density matrix: Since any matrix in the Hilbert–Schmidt space of dimension d can be expanded into a matrix basis $\{\Gamma_i\}_i$, we can, of course, expand a qudit density matrix as well and we thus get the *generalized Bloch decomposition* of the density matrix

$$\rho = \tfrac{1}{d}\left(\mathbb{1} + \sqrt{\tfrac{d(d-1)}{N_\Gamma}} \sum_{i=1}^{d^2-1} b_i\,\Gamma_i\right), \tag{17.1}$$

where the b_i are the real components of a vector $\vec{b} \in \mathbb{R}^{d^2-1}$ given by

$$b_i = \sqrt{\tfrac{d}{(d-1)N_\Gamma}}\,\langle\Gamma_i\rangle = \sqrt{\tfrac{d}{(d-1)N_\Gamma}}\,\mathrm{Tr}(\rho\Gamma_i). \tag{17.2}$$

The term $\frac{1}{d}\mathbb{1}_d$ is fixed because of the normalization condition $\mathrm{Tr}(\rho) = 1$. The prefactors in front of $\vec{b}\cdot\vec{\Gamma}$ and b_i may appear somewhat complicated, but are chosen such that $|\vec{b}| \leq 1$ for all density operators (Krammer, 2009, Sec. 2.3) and such that the special case in $d = 2$ where one chooses the Pauli matrices, $\Gamma_i = \sigma_i$, for which $N_\sigma = 2$, reduces to the familiar expressions from Section 11.5.2.

By construction, the expressions in eqns (17.1) and (17.2) thus yield a Hermitian ($\rho^\dagger = \rho$) and normalized ($\mathrm{Tr}(\rho) = 1$) density operator for any choice of $\vec{b}$, and $\vec{b}$ coincides with the usual Bloch vector for qubits ($d = 2$). However, in higher dimensions the condition $|\vec{b}| \leq 1$ alone does *not* guarantee that the corresponding operator ρ is positive semidefinite ($\rho \geq 0$), as we will illustrate with an example for $d = 3$ in Section 17.1.1. An example we give in Section 17.1.1. Consequently, we will refer to any $\vec{b}$ (with $|\vec{b}| \leq 1$) as (generalized) *Bloch vector* only when $\rho \geq 0$ indeed holds. In other words, each different matrix basis induces a different Bloch vector lying within a Bloch hypersphere of unit radius, but not every point of the hyperball corresponds to a physical state (with $\rho \geq 0$); these points are excluded, the state space is a hyperball with "holes".

All different Bloch hyperballs are isomorphic since they correspond to the same density matrix ρ. The interesting question is which Bloch hyperball—which matrix basis—is optimal for a specific purpose. In higher dimensions we have to decide which features of the operator basis are most desirable when generalizing the Pauli matrices. While the latter are both *Hermitian and unitary*, $\sigma_i^{\dagger} = \sigma_i = \sigma_i^{-1}$, it is not possible to maintain both of these properties for $d > 2$. In this context we will now present three different bases that are frequently used in the literature.

17.1.1 Generalized Gell–Mann Matrix Basis

As the first example for an operator basis in dimension d let us consider the *generalized Gell–Mann matrices* (GGMs). These can be considered to be generalizations of the Pauli matrices (for qubits, $d = 2$) and of the Gell–Mann matrices (for qutrits, $d = 3$) to higher dimensions in the sense that they are the standard generators of the special unitary group $\mathrm{SU}(d)$ in dimension d, i.e., every complex, unitary $d \times d$ matrix U with determinant $\det(U) = 1$ can be written as $U = \exp\left(-i\vec{\varphi} \cdot \vec{\Lambda}\right)$ for some $\vec{\varphi} \in \mathbb{R}^{d^2-1}$, where the components of $\vec{\Lambda}$ are the GGMs. The latter are Hermitian $d \times d$ matrices, each of which is one of the following three types:

(i) $\frac{d(d-1)}{2}$ symmetric GGMs given by

$$\Lambda_{\mathrm{s}}^{jk} = |j\rangle\langle k| + |k\rangle\langle j|, \quad 1 \le j < k \le d\,, \tag{17.3}$$

(ii) $\frac{d(d-1)}{2}$ antisymmetric GGMs given by

$$\Lambda_{\mathrm{a}}^{jk} = -i\,|j\rangle\langle k| + i\,|k\rangle\langle j|, \quad 1 \le j < k \le d\,, \tag{17.4}$$

(iii) $(d-1)$ diagonal GGMs given by

$$\Lambda^{l} = \sqrt{\frac{2}{l(l+1)}} \left(\sum_{j=1}^{l} |j\rangle\langle j| - l\,|l+1\rangle\langle l+1| \right), \quad 1 \le l \le d-1\,. \tag{17.5}$$

In total we thus have $d^2 - 1$ GGMs. It follows from the definitions that all GGMs are Hermitian and traceless. They are orthogonal and form a basis, the *generalized Gell–Mann basis* (GGMB).

Example I: Let us begin with dimension $d = 3$, the familiar eight *Gell–Mann matrices*:

(i) Three symmetric Gell–Mann matrices

$$\lambda_{\mathrm{s}}^{12} = \begin{pmatrix} 0 & 1 & 0 \\ 1 & 0 & 0 \\ 0 & 0 & 0 \end{pmatrix}, \quad \lambda_{\mathrm{s}}^{13} = \begin{pmatrix} 0 & 0 & 1 \\ 0 & 0 & 0 \\ 1 & 0 & 0 \end{pmatrix}, \quad \lambda_{\mathrm{s}}^{23} = \begin{pmatrix} 0 & 0 & 0 \\ 0 & 0 & 1 \\ 0 & 1 & 0 \end{pmatrix}, \tag{17.6}$$

(ii) Three antisymmetric Gell–Mann matrices

$$\lambda_{\mathrm{a}}^{12} = \begin{pmatrix} 0 & -i & 0 \\ i & 0 & 0 \\ 0 & 0 & 0 \end{pmatrix}, \quad \lambda_{\mathrm{a}}^{13} = \begin{pmatrix} 0 & 0 & -i \\ 0 & 0 & 0 \\ i & 0 & 0 \end{pmatrix}, \quad \lambda_{\mathrm{a}}^{23} = \begin{pmatrix} 0 & 0 & 0 \\ 0 & 0 & -i \\ 0 & i & 0 \end{pmatrix}, \quad (17.7)$$

(iii) Two diagonal Gell–Mann matrices

$$\lambda^{1} = \begin{pmatrix} 1 & 0 & 0 \\ 0 & -1 & 0 \\ 0 & 0 & 0 \end{pmatrix}, \quad \lambda^{2} = \frac{1}{\sqrt{3}} \begin{pmatrix} 1 & 0 & 0 \\ 0 & 1 & 0 \\ 0 & 0 & -2 \end{pmatrix}. \quad (17.8)$$

Example II: To see how the GGMs generalize for higher dimensions let us also explicitly state the matrices for qudits in dimension $d = 4$. There, we have

(i) Six symmetric GGMs

$$\Lambda_{\mathrm{s}}^{12} = \begin{pmatrix} 0 & 1 & 0 & 0 \\ 1 & 0 & 0 & 0 \\ 0 & 0 & 0 & 0 \\ 0 & 0 & 0 & 0 \end{pmatrix}, \quad \Lambda_{\mathrm{s}}^{13} = \begin{pmatrix} 0 & 0 & 1 & 0 \\ 0 & 0 & 0 & 0 \\ 1 & 0 & 0 & 0 \\ 0 & 0 & 0 & 0 \end{pmatrix}, \quad \Lambda_{\mathrm{s}}^{14} = \begin{pmatrix} 0 & 0 & 0 & 1 \\ 0 & 0 & 0 & 0 \\ 0 & 0 & 0 & 0 \\ 1 & 0 & 0 & 0 \end{pmatrix},$$

$$\Lambda_{\mathrm{s}}^{23} = \begin{pmatrix} 0 & 0 & 0 & 0 \\ 0 & 0 & 1 & 0 \\ 0 & 1 & 0 & 0 \\ 0 & 0 & 0 & 0 \end{pmatrix}, \quad \Lambda_{\mathrm{s}}^{24} = \begin{pmatrix} 0 & 0 & 0 & 0 \\ 0 & 0 & 0 & 1 \\ 0 & 0 & 0 & 0 \\ 0 & 1 & 0 & 0 \end{pmatrix}, \quad \Lambda_{\mathrm{s}}^{34} = \begin{pmatrix} 0 & 0 & 0 & 0 \\ 0 & 0 & 0 & 0 \\ 0 & 0 & 0 & 1 \\ 0 & 0 & 1 & 0 \end{pmatrix}, \quad (17.9)$$

(ii) Six antisymmetric GGMs

$$\Lambda_{\mathrm{a}}^{12} = \begin{pmatrix} 0 & -i & 0 & 0 \\ i & 0 & 0 & 0 \\ 0 & 0 & 0 & 0 \\ 0 & 0 & 0 & 0 \end{pmatrix}, \quad \Lambda_{\mathrm{a}}^{13} = \begin{pmatrix} 0 & 0 & -i & 0 \\ 0 & 0 & 0 & 0 \\ i & 0 & 0 & 0 \\ 0 & 0 & 0 & 0 \end{pmatrix}, \quad \Lambda_{\mathrm{a}}^{14} = \begin{pmatrix} 0 & 0 & 0 & -i \\ 0 & 0 & 0 & 0 \\ 0 & 0 & 0 & 0 \\ i & 0 & 0 & 0 \end{pmatrix},$$

$$\Lambda_{\mathrm{a}}^{23} = \begin{pmatrix} 0 & 0 & 0 & 0 \\ 0 & 0 & -i & 0 \\ 0 & i & 0 & 0 \\ 0 & 0 & 0 & 0 \end{pmatrix}, \quad \Lambda_{\mathrm{a}}^{24} = \begin{pmatrix} 0 & 0 & 0 & 0 \\ 0 & 0 & 0 & -i \\ 0 & 0 & 0 & 0 \\ 0 & i & 0 & 0 \end{pmatrix}, \quad \Lambda_{\mathrm{a}}^{34} = \begin{pmatrix} 0 & 0 & 0 & 0 \\ 0 & 0 & 0 & 0 \\ 0 & 0 & 0 & -i \\ 0 & 0 & i & 0 \end{pmatrix},$$

$$(17.10)$$

(iii) Three diagonal GGMs

$$
\Lambda^1 = \begin{pmatrix} 1 & 0 & 0 & 0 \\ 0 & -1 & 0 & 0 \\ 0 & 0 & 0 & 0 \\ 0 & 0 & 0 & 0 \end{pmatrix} , \quad
\Lambda^2 = \frac{1}{\sqrt{3}} \begin{pmatrix} 1 & 0 & 0 & 0 \\ 0 & 1 & 0 & 0 \\ 0 & 0 & -2 & 0 \\ 0 & 0 & 0 & 0 \end{pmatrix} , \quad
\Lambda^3 = \frac{1}{\sqrt{6}} \begin{pmatrix} 1 & 0 & 0 & 0 \\ 0 & 1 & 0 & 0 \\ 0 & 0 & 1 & 0 \\ 0 & 0 & 0 & -3 \end{pmatrix} .
$$
$$(17.11)$$

Using the GGMB we obtain, in general, the following Bloch-vector expansion of a density matrix:

$$
\rho = \tfrac{1}{d} \left(\mathbb{1} + \sqrt{\tfrac{d(d-1)}{2}} \ \vec{b} \cdot \vec{\Lambda} \right) ,
\tag{17.12}
$$

with $N_\Lambda = 2$ and a Bloch vector $\vec{b} = (\{b_s^{jk}\}, \{b_a^{jk}\}, \{b^l\})$ whose components are labelled by indices obeying the restrictions $1 \leq j < k \leq d$ and $1 \leq l \leq d-1$. Explicitly, the components are given by $b_s^{jk} = \sqrt{\tfrac{d}{2(d-1)}} \operatorname{Tr}(\Lambda_s^{jk} \rho)$, $b_a^{jk} = \sqrt{\tfrac{d}{2(d-1)}} \operatorname{Tr}(\Lambda_a^{jk} \rho)$ and $b^l = \sqrt{\tfrac{d}{2(d-1)}} \operatorname{Tr}(\Lambda^l \rho)$. Formally, the Bloch vector $\vec{b}$ is restricted to lie within the $(d^2 - 1)$-dimensional Bloch hyperball defined by $|\vec{b}| \leq 1$, but, as we shall see next, not all points within this hyperball correspond to quantum states with positive semidefinite density operators.

Example for unphysical Bloch vector: Let us now consider an example for qutrits, $d = 3$, where we choose the Gell–Mann matrices from eqns (17.6), (17.7), and (17.8) as operator basis, and the Bloch vector is hence $\vec{b} = (b_s^{12}, b_s^{13}, b_s^{23}, b_a^{12}, b_a^{13}, b_a^{23}, b^1, b^2)$. An example for a matrix that is of the form of (17.12) with $|\vec{b}| \leq 1$, actually exactly $|\vec{b}| = 1$, but which does not correspond to a density matrix is obtained for $\vec{b} = (0, 0, 0, 0, 0, 0, b_1, b_2)$, with $b_1 = \frac{\sqrt{7}}{4}$, and $b_2 = \frac{3}{4}$, while all other components of $\vec{b}$ are set to 0. For this choice we obtain the matrix

$$
\rho = \tfrac{1}{3} \left(\mathbb{1}_3 + \sqrt{3} \, \vec{b} \cdot \vec{\Lambda} \right) = \tfrac{1}{3} \left(\mathbb{1}_3 + \sqrt{3} \left[\tfrac{\sqrt{7}}{4} \lambda^1 + \tfrac{3}{4} \lambda^2 \right] \right)
\tag{17.13}
$$

$$
= \tfrac{1}{3} \begin{pmatrix} 1 & 0 & 0 \\ 0 & 1 & 0 \\ 0 & 0 & 1 \end{pmatrix}
+ \tfrac{\sqrt{3}}{3} \tfrac{\sqrt{7}}{4} \begin{pmatrix} 1 & 0 & 0 \\ 0 & -1 & 0 \\ 0 & 0 & 0 \end{pmatrix}
+ \tfrac{\sqrt{3}}{3} \tfrac{3}{4} \tfrac{1}{\sqrt{3}} \begin{pmatrix} 1 & 0 & 0 \\ 0 & 1 & 0 \\ 0 & 0 & -2 \end{pmatrix}
$$

$$
= \tfrac{1}{3} \begin{pmatrix} \tfrac{7+\sqrt{21}}{4} & 0 & 0 \\ 0 & \tfrac{7-\sqrt{21}}{4} & 0 \\ 0 & 0 & -\tfrac{1}{2} \end{pmatrix} , \quad \cdot
$$

which obviously has a negative eigenvalue $-\tfrac{1}{6}$, thus one has $\rho \not\geq 0$, the matrix is not a physical density operator any more.

Expansion of standard matrices in GGMB: Before we move on to the next example for a useful operator basis, we note that the GGMs have a simple relation to the standard matrices, which are simply the $d \times d$ matrices that have only one entry 1 and the other entries 0,

$$| j \rangle\langle k |, \qquad \text{with} \quad j, k = 1, \dots, d, \tag{17.14}$$

which trivially also form an orthonormal basis in the Hilbert–Schmidt space. Any matrix can easily be decomposed into a "vector" via a certain linear combination of the matrices (17.14). Knowing the expansion of matrices (17.14) in the GGMB we can therefore find the decomposition of any matrix in terms of the GGMB. The expansion of the standard matrices (17.14) in terms of the GGMs is (Bertlmann and Krammer, 2008a)

$$| j \rangle\langle k | = \begin{cases} \frac{1}{2} \left(\Lambda_s^{jk} + i\Lambda_a^{jk} \right) & \text{for } j < k \\ \frac{1}{2} \left(\Lambda_s^{kj} - i\Lambda_a^{kj} \right) & \text{for } j > k \\ -\sqrt{\frac{j-1}{2j}} \, \Lambda^{j-1} + \sum_{n=0}^{d-j-1} \frac{1}{\sqrt{2(j+n)(j+n+1)}} \, \Lambda^{j+n} + \frac{1}{d} \mathbb{1}_d & \text{for } j = k. \end{cases} \tag{17.15}$$

17.1.2 Polarization-Operator Basis

As a second example for a commonly used operator basis we want to present the *polarization operators*. They are operators in the Hilbert–Schmidt space of dimension d and defined as the following $d \times d$ matrices (Kryszewski and Zachciał, 2006; Varshalovich *et al.*, 1988):

$$T_{LM} = \sqrt{\frac{2L+1}{2s+1}} \sum_{k,l=1}^{d} C_{s,m_l;L,M}^{(s,m_k)} | k \rangle\langle l |. \tag{17.16}$$

The indices obey the relations

$$s = \frac{d-1}{2},$$
$$L = 0, 1, \dots, 2s,$$
$$M = -L, -L+1, \dots, L-1, L,$$
$$m_1 = s, \ m_2 = s-1, \dots, m_d = -s. \tag{17.17}$$

The coefficients $C_{s,m_l;L,M}^{(s,m_k)}$ are identified with the usual Clebsch–Gordan coefficients for the addition of angular momenta, see, e.g., the discussion in Section 8.3.2.

For $L = M = 0$ the polarization operator is proportional to the identity matrix

$$T_{00} = \frac{1}{\sqrt{d}} \mathbb{1}_d. \tag{17.18}$$

All polarization operators (except T_{00}) are traceless, $\text{Tr}(T_{LM}) = 0$, but in general *not* Hermitian, and their orthogonality relation is

$$\text{Tr}\left(T_{L_1 M_1}^\dagger T_{L_2 M_2} \right) = \delta_{L_1 L_2} \delta_{M_1 M_2}. \tag{17.19}$$

Therefore the d^2 polarization operators (17.16) form an orthonormal matrix basis—the *polarization-operator basis* (POB)—of the Hilbert–Schmidt space of dimension d.

Examples: The simplest example is of dimension $d = 2$, qubits. For a qubit the POB is given by the following matrices ($s = \frac{1}{2}$; $L = 0, 1$; $M = -1, 0, 1$)

$$T_{00} = \tfrac{1}{\sqrt{2}} \begin{pmatrix} 1 & 0 \\ 0 & 1 \end{pmatrix}, \quad T_{11} = - \begin{pmatrix} 0 & 1 \\ 0 & 0 \end{pmatrix},$$

$$T_{10} = \tfrac{1}{\sqrt{2}} \begin{pmatrix} 1 & 0 \\ 0 & -1 \end{pmatrix}, \quad T_{1-1} = \begin{pmatrix} 0 & 0 \\ 1 & 0 \end{pmatrix}. \tag{17.20}$$

For the next higher dimension $d = 3$ ($s = 1$), the case of qutrits, we get nine polarization operators T_{LM} with $L = 0, 1, 2$ and $M = -L, ..., L$ and we have

$$T_{11} = -\tfrac{1}{\sqrt{2}} \begin{pmatrix} 0 & 1 & 0 \\ 0 & 0 & 1 \\ 0 & 0 & 0 \end{pmatrix}, \quad T_{10} = \tfrac{1}{\sqrt{2}} \begin{pmatrix} 1 & 0 & 0 \\ 0 & 0 & 0 \\ 0 & 0 & -1 \end{pmatrix}, \quad T_{1-1} = \tfrac{1}{\sqrt{2}} \begin{pmatrix} 0 & 0 & 0 \\ 1 & 0 & 0 \\ 0 & 1 & 0 \end{pmatrix},$$

$$T_{22} = \begin{pmatrix} 0 & 0 & 1 \\ 0 & 0 & 0 \\ 0 & 0 & 0 \end{pmatrix}, \quad T_{21} = \tfrac{1}{\sqrt{2}} \begin{pmatrix} 0 & -1 & 0 \\ 0 & 0 & 1 \\ 0 & 0 & 0 \end{pmatrix}, \quad T_{20} = \tfrac{1}{\sqrt{6}} \begin{pmatrix} 1 & 0 & 0 \\ 0 & -2 & 0 \\ 0 & 0 & 1 \end{pmatrix},$$

$$T_{2-1} = \tfrac{1}{\sqrt{2}} \begin{pmatrix} 0 & 0 & 0 \\ 1 & 0 & 0 \\ 0 & -1 & 0 \end{pmatrix}, \quad T_{2-2} = \begin{pmatrix} 0 & 0 & 0 \\ 0 & 0 & 0 \\ 1 & 0 & 0 \end{pmatrix}. \tag{17.21}$$

Since $N_T = 1$, the generalized Bloch decomposition of any density matrix using the POB has the form

$$\rho = \tfrac{1}{d} \left(\mathbb{1}_d + \sqrt{d(d-1)} \sum_{L=1}^{2s} \sum_{M=-L}^{L} b_{LM} T_{LM} \right) = \tfrac{1}{d} \left(\mathbb{1}_d + \sqrt{d(d-1)} \, \vec{b} \cdot \vec{T} \right), \tag{17.22}$$

with the Bloch vector $\vec{b} = (b_{1-1}, b_{10}, b_{11}, b_{2-2}, b_{2-1}, b_{20}, ..., b_{LM})$, where the components are given by $b_{LM} = \sqrt{\frac{d}{(d-1)}} \operatorname{Tr}(T_{LM}^{\dagger} \rho)$. In general the components b_{LM} are complex-valued since the polarization operators T_{LM} are not Hermitian. Again, all Bloch vectors lie within a hypersphere of radius $|\vec{b}| \leq 1$ but not every point within necessarily corresponds to a physical state.

In two dimensions the Bloch vector $\vec{b} = (b_{1-1}, b_{10}, b_{11})$ forms a spheroid (Kryszewski and Zachciał, 2006), the pure states occupy the surface and the mixed ones lie in the volume. This decomposition is fully equivalent to the standard description of Bloch vectors with Pauli matrices.

In higher dimensions, however, the structure of the allowed range of $\vec{b}$ (due to the positivity requirement $\rho \geq 0$) is quite complicated, as can be seen already for $d = 3$. Nevertheless, pure states are on the surface, mixed ones lie within the volume and the maximally mixed state corresponds to $|\vec{b}| = 0$, thus $|\vec{b}|$ can be regarded as an indicator for the mixedness of the quantum state.

Expansion of standard matrices in POB: The standard matrices (17.14) can of course also be expanded in terms of the POB (Varshalovich *et al.*, 1988), which takes the form (Krammer, 2009, Sec. 2.5.2)

$$| i \rangle\langle j | = \sum_L \sum_M \sqrt{\frac{2L+1}{2s+1}} \, C^{(s,m_i)}_{s,m_j;L,M} \, T_{LM} \,, \tag{17.23}$$

where one should note that the sum over M is actually already fixed by the condition $m_j + M = m_i$.

17.1.3   Weyl-Operator Basis

Finally, we want to discuss a basis of the Hilbert–Schmidt space of dimension d that is frequently used in quantum information and which consists of the following d^2 operators that are called *Weyl operators*:

$$U_{nm} = \sum_{k=0}^{d-1} e^{\frac{2\pi i}{d} nk} | k \rangle\langle (k+m)_{\mathrm{mod}(d)} | \qquad n, m = 0, 1, \ldots, d-1 \,, \tag{17.24}$$

where we have expanded into the standard basis of the Hilbert space. The d^2 operators in (17.24) are unitary and form an orthonormal basis

$$\mathrm{Tr}\big(U^{\dagger}_{nm} U_{lj}\big) = d \, \delta_{nl} \, \delta_{mj} \,, \tag{17.25}$$

with $N_U = d$, called the *Weyl-operator basis* (WOB) of the Hilbert–Schmidt space, which will be convenient when we discuss the construction of maximally entangled qudit states, see, e.g., Sections 17.3 and 18.4.2, which might already be evident from comparison with the states we have encountered in eqn (14.11). Clearly the operator U_{00} represents the identity $U_{00} = \mathbb{1}$.

Example: An interesting example is that of dimension $d = 3$, the qutrit case. There the Weyl operators (17.24) have the following matrix form

$$U_{00} = \begin{pmatrix} 1 & 0 & 0 \\ 0 & 1 & 0 \\ 0 & 0 & 1 \end{pmatrix}, \qquad U_{01} = \begin{pmatrix} 0 & 1 & 0 \\ 0 & 0 & 1 \\ 1 & 0 & 0 \end{pmatrix}, \qquad U_{02} = \begin{pmatrix} 0 & 0 & 1 \\ 1 & 0 & 0 \\ 0 & 1 & 0 \end{pmatrix}, \qquad (17.26)$$

$$U_{10} = \begin{pmatrix} 1 & 0 & 0 \\ 0 & e^{\frac{2\pi i}{3}} & 0 \\ 0 & 0 & e^{-\frac{2\pi i}{3}} \end{pmatrix}, \quad U_{11} = \begin{pmatrix} 0 & 1 & 0 \\ 0 & 0 & e^{\frac{2\pi i}{3}} \\ e^{-\frac{2\pi i}{3}} & 0 & 0 \end{pmatrix}, \quad U_{12} = \begin{pmatrix} 0 & 0 & 1 \\ e^{\frac{2\pi i}{3}} & 0 & 0 \\ 0 & e^{-\frac{2\pi i}{3}} & 0 \end{pmatrix},$$

$$U_{20} = \begin{pmatrix} 1 & 0 & 0 \\ 0 & e^{-\frac{2\pi i}{3}} & 0 \\ 0 & 0 & e^{\frac{2\pi i}{3}} \end{pmatrix}, \quad U_{21} = \begin{pmatrix} 0 & 1 & 0 \\ 0 & 0 & e^{-\frac{2\pi i}{3}} \\ e^{\frac{2\pi i}{3}} & 0 & 0 \end{pmatrix}, \quad U_{22} = \begin{pmatrix} 0 & 0 & 1 \\ e^{-\frac{2\pi i}{3}} & 0 & 0 \\ 0 & e^{\frac{2\pi i}{3}} & 0 \end{pmatrix}.$$

Using the WOB we can generally decompose any density matrix into a generalized Bloch decomposition of the form

$$\rho = \tfrac{1}{d}\left(\mathbb{1}_d + \sqrt{d-1} \sum_{\substack{n,m=0 \\ m+n>0}}^{d-1} b_{nm} U_{nm}\right) = \tfrac{1}{d}\left(\mathbb{1}_d + \sqrt{d-1}\, \vec{b}\cdot\vec{U}\right), \qquad (17.27)$$

with the components of the Bloch vector $\vec{b} = (b_{nm})$ for $m,n = 0,1,\ldots,d-1$ with $m+n>0$ given by $b_{nm} = \frac{1}{\sqrt{d-1}}\operatorname{Tr}(U_{nm}\rho)$. For example, for qutrits the Bloch vector is $\vec{b} = (b_{01}, b_{02}, b_{10}, b_{11}, b_{12}, b_{20}, b_{21}, b_{22})^{\top}$. In general the components b_{nm} are complex since the Weyl operators are not Hermitian and the complex conjugates fulfil the relation $b_{nm}^{*} = \exp\left(-\frac{2\pi i}{d} nm\right) b_{(d-n)(d-m)}$, which follows from eqn (17.24) when considering that ρ is Hermitian. As for the other operator bases considered, the Bloch vectors lie within a hypersphere of radius $|\vec{b}| \le 1$ but in three or more dimensions the allowed values for of the Bloch vector are quite restricted within the hypersphere and the detailed structure is not known.

In dimension $d = 2$ the WOB as well as the GGMB coincide with the basis of Pauli operators and the POB represents a rotated Pauli basis (where $\sigma_{\pm} = \frac{1}{2}(\sigma_1 \pm i\sigma_2)$). Specifically, we have

$$\{U_{00}, U_{01}, U_{10}, U_{11}\} = \{\mathbb{1}, \sigma_1, \sigma_3, i\sigma_2\}, \qquad (17.28a)$$

$$\{\mathbb{1}, \Lambda_{\mathrm{s}}^{12}, \Lambda_{\mathrm{a}}^{12}, \Lambda^{1}\} = \{\mathbb{1}, \sigma_1, \sigma_2, \sigma_3\}, \qquad (17.28b)$$

$$\{T_{00}, T_{11}, T_{10}, T_{1-1}\} = \{\tfrac{1}{\sqrt{2}}\mathbb{1}, -\sigma_+, \tfrac{1}{\sqrt{2}}\sigma_3, \sigma_-\}. \qquad (17.28c)$$

Expansion of standard matrices in WOB: The standard matrices (17.14) can be expressed in terms of the WOB as (Bertlmann and Krammer, 2008a)

$$|j\rangle\langle k| = \tfrac{1}{d} \sum_{l=0}^{d-1} e^{-\frac{2\pi i}{d} lj} U_{l\,(k-j)_{\mathrm{mod}(d)}}. \qquad (17.29)$$

17.2   Applications of Operator Bases

In this section we want to present two applications of the operator bases that we have introduced in the previous section. First, we expand the isotropic two-qudit state into all three bases: GGMB, POB, and WOB. Then we show the convenience of GGMs for expressing an entanglement witness for qutrits into measurable quantities.

17.2.1   Generalized Bloch Decomposition for Two Qudits

Quite generally, a two-qudit density matrix ρ in the Hilbert–Schmidt space $\mathcal{B}_{\mathrm{HS}}(\mathcal{H}_A \otimes \mathcal{H}_B)$ with $\mathcal{H}_A = \mathcal{H}_B = \mathcal{H}_d$ and $\dim(\mathcal{H}_d) = d$ can be decomposed as

$$
\rho = \tfrac{1}{d^2} \left(\mathbb{1}_d \otimes \mathbb{1}_d + f_A \sum_{i=1}^{d^2-1} a_i \, \Gamma_i^{(A)} \otimes \mathbb{1}_d + f_B \sum_{i=1}^{d^2-1} b_i \, \mathbb{1}_d \otimes \Gamma_i^{(B)} \right.
$$

$$
\left. + f_A f_B \sum_{i,j=1}^{d^2-1} t_{ij} \, \Gamma_i^{(A)} \otimes \Gamma_j^{(B)} \right), \tag{17.30}
$$

where $a_i, b_i, t_{ij} \in \mathbb{C}$, $f_A = \sqrt{d(d-1)/N_{\Gamma^{(A)}}}$, $f_B = \sqrt{d(d-1)/N_{\Gamma^{(B)}}}$, and $\{\Gamma_i^{(A)}\}$ and $\{\Gamma_i^{(B)}\}$ represent some (in principle different) operator bases in the subspaces $\mathcal{B}_{\mathrm{HS}}(\mathcal{H}_A)$ and $\mathcal{B}_{\mathrm{HS}}(\mathcal{H}_B)$, respectively. Equation (17.30) is a generalization of the two-qubit Bloch decomposition of (15.39) in Section 15.2.3. Here, a_i and b_i correspond to the generalized Bloch-vector components of subsystems A and B, respectively, provided that ρ is a valid (positive semidefinite) density operator to begin with.

The matrix $T = (t_{ij})$ is the generalized correlation tensor $T = (t_{ij})$. In general, one can always choose unitary $(d^2 - 1) \times (d^2 - 1)$ transformations $U^{(A)} = (U_{ij}^{(A)})$ and $U^{(A)} = (U_{ij}^{(A)})$ for both subsystems such that the correlation tensor is diagonal with respect to the resulting operator bases $\widetilde{\Gamma}_i^{(A)} = \sum_j U_{ij}^{(A)} \Gamma_j^{(A)}$ and $\widetilde{\Gamma}_i^{(B)} = \sum_j U_{ij}^{(B)} \Gamma_j^{(B)}$, mirroring the decomposition of the Weyl states discussed in Section 15.5. The existence of such unitary transformations on the $(d^2 - 1)$-dimensional spaces of the respective operator bases follows directly from the singular-value decomposition of $T = (t_{ij})$, see, e.g,. the discussion following eqn (15.132).

Here it is crucial to note that the transformations are in general unitary, rather than orthogonal, since the correlation tensor itself can be complex-valued if the chosen operator bases are not Hermitian. However, in contrast to the two-qubit case, these transformations diagonalizing the correlation tensor may not always be realizable via local *unitary* transformations on the d-dimensional Hilbert spaces for local dimensions larger than $d = 2$. This can intuitively be understood by observing that the number of parameters necessary to fully specify a unitary in dimension d is not enough to fully specify a unitary in dimension $d^2 - 1$. For $d = 2$ it just so happens that the groups SU(2) and SO(3) are both three dimensional.

17.2.2 Isotropic Two-Qudit States

As an example for the generalized Bloch decomposition, let us now consider the family of isotropic two-qudit states from Section 15.4.4, given by

$$\rho_{\text{iso}}(\alpha) := \alpha \, |\phi_{00}\rangle\langle\phi_{00}| + \frac{1-\alpha}{d^2} \, \mathbb{1}_d \otimes \mathbb{1}_d \,, \tag{17.31}$$

for $-1/(d^2 - 1) \leq \alpha \leq 1$. The state $|\phi_{00}\rangle$, given by

$$|\phi_{00}\rangle = \frac{1}{\sqrt{d}} \sum_{k=0}^{d-1} |k\rangle \otimes |k\rangle \tag{17.32}$$

is maximally entangled, i.e., its reduced states ρ_A and ρ_B are maximally mixed, $\rho_A = \text{Tr}_B\big(|\phi_{00}\rangle\langle\phi_{00}|\big) = \frac{1}{d}\mathbb{1}_d = \rho_B$, which implies that the Bloch vectors $\vec{a}$ and $\vec{b}$ with respect to any chosen operator bases must vanish, $\vec{a} = \vec{b} = 0$, because the respective operators are always traceless. This leaves only the correlation tensor $T = (t_{ij})$ to be considered. Since we can always choose an operator basis such that T is diagonal, only the terms $t_{ii}\,\Gamma_i \otimes \Gamma_i$ are non-zero, the density matrix of an isotropic two-qudit state can be described by only $d^2 - 1$ independent parameters, which provides the dimension of the corresponding Bloch vector. Thus the isotropic two-qudit Bloch vector lies within a subspace of the same dimension as that of the Bloch vector of a general single-qudit state, which is a comfortable simplification. In the following we will see that, for all three examples of operator bases that we consider here, the GGMB, the POB, and the WOB, the correlation tensor is already conveniently diagonal (or can be brought to this form by a simple relabelling of the operators for one subsystem) for two-qudit isotropic states.

Expansion into GGMB: We begin with calculating the Bloch vector for the maximally entangled state $|\phi_{00}\rangle\langle\phi_{00}|$ in the GGMB. It is convenient to split the state into two parts

$$|\phi_{00}\rangle\langle\phi_{00}| = \frac{1}{d} \sum_{j,k=1}^{d} |j\rangle\langle k| \otimes |j\rangle\langle k| = D + O \,, \tag{17.33}$$

where D and O are operators representing the diagonal and off-diagonal elements (with respect to the computational basis) defined by

$$D := \frac{1}{d} \sum_j |j\rangle\langle j| \otimes |j\rangle\langle j| \,, \tag{17.34}$$

$$O := \frac{1}{d} \sum_{\substack{j,k \\ j \neq k}} |j\rangle\langle k| \otimes |j\rangle\langle k| \,, \tag{17.35}$$

which we will now calculate separately.

For the term D we need the case $j = k$ in the standard-matrix expansion (17.15), which provides

$$D = \frac{1}{2d} \sum_{m=1}^{d-1} \Lambda^m \otimes \Lambda^m + \frac{1}{d^2} \mathbb{1} \otimes \mathbb{1} . \tag{17.36}$$

For the off-diagonal term O we first split the sum over all j and k such that $j \neq k$ into a sum for $j < k$ and one for $j > k$, before inserting from the standard-matrix expansion of eqn (17.15), which yields

$$O = \frac{1}{4d} \left[\sum_{j<k} \left(\Lambda_{\mathrm{s}}^{jk} + i\Lambda_{\mathrm{a}}^{jk} \right) \otimes \left(\Lambda_{\mathrm{s}}^{jk} + i\Lambda_{\mathrm{a}}^{jk} \right) + \sum_{j>k} \left(\Lambda_{\mathrm{s}}^{kj} - i\Lambda_{\mathrm{a}}^{kj} \right) \otimes \left(\Lambda_{\mathrm{s}}^{kj} - i\Lambda_{\mathrm{a}}^{kj} \right) \right]$$

$$= \frac{1}{2d} \sum_{j<k} \left(\Lambda_{\mathrm{s}}^{jk} \otimes \Lambda_{\mathrm{s}}^{jk} - \Lambda_{\mathrm{a}}^{jk} \otimes \Lambda_{\mathrm{a}}^{jk} \right) . \tag{17.37}$$

Combining the terms for D and O we thus obtain the following GGMB expansion for the maximally entangled state (17.33):

$$| \phi_{00} \rangle\langle \phi_{00} | = \frac{1}{d^2} \mathbb{1} \otimes \mathbb{1} + \frac{1}{2d} \Lambda , \tag{17.38}$$

where we defined

$$\Lambda := \sum_{j<k} \Lambda_{\mathrm{s}}^{jk} \otimes \Lambda_{\mathrm{s}}^{jk} - \sum_{j<k} \Lambda_{\mathrm{a}}^{jk} \otimes \Lambda_{\mathrm{a}}^{jk} + \sum_{m=1}^{d-1} \Lambda^m \otimes \Lambda^m . \tag{17.39}$$

Finally for the isotropic two-qudit state (17.31) we have

$$\rho_{\mathrm{iso}}(\alpha) = \frac{1}{d^2} \mathbb{1} \otimes \mathbb{1} + \frac{\alpha}{2d} \Lambda . \tag{17.40}$$

Expansion into POB: Next we expand the maximally entangled state $| \phi_{00} \rangle\langle \phi_{00} |$ in the POB. Using (17.23) and the sum rule for the Clebsch–Gordan coefficients (Varshalovich *et al.*, 1988),

$$\sum_{\alpha,\gamma} C_{a\alpha,b\beta}^{c\gamma} C_{a\alpha,b'\beta'}^{c\gamma} = \frac{2c+1}{2b+1} \delta_{bb'} \delta_{\beta\beta'} , \tag{17.41}$$

we get the expression

$$| \phi_{00} \rangle\langle \phi_{00} | = \frac{1}{d} \sum_{i,j=1}^{d} | i \rangle\langle j | \otimes | i \rangle\langle j |$$

$$= \frac{1}{d} \sum_{L,L'} \frac{\sqrt{(2L+1)(2L'+1)}}{2s+1} \left(\sum_{i,j} C_{sm_j,LM}^{sm_i} C_{sm_j,L'M}^{sm_i} \right) T_{LM} \otimes T_{L'M}$$

$$= \frac{1}{d} \sum_{L} T_{LM} \otimes T_{LM} . \tag{17.42}$$

Extracting the identity (recall eqn (17.18)) and defining

$$T := \sum_{L,M \neq 0,0} T_{LM} \otimes T_{LM} \, , \tag{17.43}$$

we obtain the following expressions for the POB expansion of the maximally entangled state and, in turn, for the isotropic two-qudit state (17.31):

$$|\phi_{00}\rangle\langle\phi_{00}| = \frac{1}{d^2} \mathbb{1} \otimes \mathbb{1} + \frac{1}{d} T \, , \tag{17.44}$$

$$\rho_{\text{iso}}(\alpha) = \frac{1}{d^2} \mathbb{1} \otimes \mathbb{1} + \frac{\alpha}{d} T \, . \tag{17.45}$$

Expansion into WOB: Finally, we consider the expansion of the maximally entangled state and the isotropic two-qudit state into the WOB, which is straightforward (see, e.g., Krammer, 2009, p. 30), and the result is

$$|\phi_{00}\rangle\langle\phi_{00}| = \frac{1}{d^2} \mathbb{1} \otimes \mathbb{1} + \frac{1}{d^2} U \, , \tag{17.46}$$

$$\rho_{\text{iso}}(\alpha) = \frac{1}{d^2} \mathbb{1} \otimes \mathbb{1} + \frac{\alpha}{d^2} U \, , \tag{17.47}$$

where we have defined the operator

$$U := \sum_{l,m=0}^{d-1} U_{lm} \otimes U_{(d-l)m} \, , \qquad (l,m) \neq (0,0) \, . \tag{17.48}$$

Résumé: We have explicitly worked out the generalized Bloch decomposition for two-qudit isotropic states with respect to the generalized Gell–Mann basis (GGMB), the polarization-operator basis (POB), and the Weyl-operator basis WOB. Of the three, working out the corresponding Bloch vector (17.40) with expression (17.39) is most complicated for the GGMB. In the POB the Bloch vector (17.45) with expression (17.43) can be easily derived from the knowledge of the Clebsch-Gordon coefficient sum rule (17.41), and for the WOB the Bloch vector (17.47) with definition (17.48) is most easily constructed. For all three bases, the correlation tensor turns out to be diagonal for isotropic two-qudit states, at least up to a simple relabelling $(d - l \leftrightarrow l)$ of the operators in the WOB for Bob's subsystem.

17.2.3 Entanglement Witness in Terms of Spin-1 Operators

As we know from Section 16.3, entanglement witnesses are Hermitian operators and therefore observables. They should thus be measurable in principle in a given experimental setup and can thus provide direct experimental verification of entanglement. The quantity to be measured is the expectation value

$$\langle \mathbb{W} \rangle = \text{Tr}(\mathbb{W}\rho) \tag{17.49}$$

of an entanglement witness $\mathbb{W}$ for some state ρ. If $\langle \mathbb{W} \rangle < 0$ then the state ρ is entangled. But which measurements have to be performed?

Obviously it is possible to express entanglement witnesses (in finite-dimensional quantum systems) in terms of the generalized Gell–Mann matrices GGM (17.3), (17.4), and (17.5), since they are Hermitian. For $d = 3$—the case of qutrits—the Gell–Mann matrices (17.6), (17.7), and (17.8) can in turn be expressed in terms of eight physical operators, the *Spin-1 observables*,

$$S_x, \ S_y, \ S_z, \ S_x^2, \ S_y^2, \ \{S_x, S_y\}, \ \{S_y, S_z\}, \ \{S_z, S_x\}, \tag{17.50}$$

where $\vec{S} = (S_x, S_y, S_z)^\top$ is the vector of spin operators and

$$\{S_i, S_j\} = S_i S_j + S_j S_i \quad \text{with} \quad i, j = x, y, z \tag{17.51}$$

denote the corresponding anti-commutators.

The decomposition of the Gell–Mann matrices into spin-1 operators is as follows (Bertlmann and Krammer, 2008a):

$$\lambda_s^{12} = \frac{1}{\sqrt{2}\hbar^2}\left(\hbar S_x + \{S_z, S_x\}\right), \qquad \lambda_s^{13} = \frac{1}{\hbar^2}\left(S_x^2 - S_y^2\right), \tag{17.52}$$

$$\lambda_s^{23} = \frac{1}{\sqrt{2}\hbar^2}\left(\hbar S_x - \{S_z, S_x\}\right), \qquad \lambda_a^{12} = \frac{1}{\sqrt{2}\hbar^2}\left(\hbar S_y + \{S_y, S_z\}\right),$$

$$\lambda_a^{13} = \frac{1}{\hbar^2}\{S_x, S_y\}, \qquad \lambda_a^{23} = \frac{1}{\sqrt{2}\hbar^2}\left(\hbar S_y - \{S_y, S_z\}\right),$$

$$\lambda^1 = 2\mathbb{1} + \frac{1}{2\hbar^2}\left(\hbar S_z - 3S_x^2 - 3S_y^2\right), \qquad \lambda^2 = \frac{1}{\sqrt{3}}\left(-2\mathbb{1} + \frac{3}{2\hbar^2}\left(\hbar S_z + S_x^2 + S_y^2\right)\right).$$

Therefore all eight spin-1 observables can be represented by the following matrices:

$$S_x = \frac{\hbar}{\sqrt{2}}\begin{pmatrix} 0 & 1 & 0 \\ 1 & 0 & 1 \\ 0 & 1 & 0 \end{pmatrix}, \quad S_y = \frac{\hbar}{\sqrt{2}}\begin{pmatrix} 0 & -i & 0 \\ i & 0 & -i \\ 0 & i & 0 \end{pmatrix}, \quad S_z = \hbar\begin{pmatrix} 1 & 0 & 0 \\ 0 & 0 & 0 \\ 0 & 0 & -1 \end{pmatrix},$$

$$S_x^2 = \frac{\hbar^2}{2}\begin{pmatrix} 1 & 0 & 1 \\ 0 & 2 & 0 \\ 1 & 0 & 1 \end{pmatrix}, \quad S_y^2 = \frac{\hbar^2}{2}\begin{pmatrix} 1 & 0 & -1 \\ 0 & 2 & 0 \\ -1 & 0 & 1 \end{pmatrix}, \tag{17.53}$$

$$\{S_x, S_y\} = \hbar^2\begin{pmatrix} 0 & 0 & -i \\ 0 & 0 & 0 \\ i & 0 & 0 \end{pmatrix}, \quad \{S_y, S_z\} = \frac{\hbar^2}{\sqrt{2}}\begin{pmatrix} 0 & -i & 0 \\ i & 0 & i \\ 0 & -i & 0 \end{pmatrix}, \quad \{S_z, S_x\} = \frac{\hbar^2}{\sqrt{2}}\begin{pmatrix} 0 & 1 & 0 \\ 1 & 0 & -1 \\ 0 & -1 & 0 \end{pmatrix}.$$

In fact, any observable on an n-qutrit Hilbert space—a composite system of n particles of which each has a three-dimensional state space—can be expressed in terms of the spin-1 operators in (17.53).

We now want to study the *entanglement witness for an isotropic two-qutrit state*, i.e., the state (17.31) for $d = 3$. In this case the optimal entanglement witness is given by

$$W_{\text{opt}}^{\text{iso}} = \frac{1}{3\sqrt{2}} \left(\mathbb{1} \otimes \mathbb{1} - \frac{3}{4} \Lambda \right) , \tag{17.54}$$

where the operator Λ is defined in eqn (17.39) (compare with eqn (16.132) of Section 16.3.5 for $d = 3$).

Expressing the Gell–Mann matrices into which Λ in (17.39) is decomposed in terms of the spin-operator decomposition of eqn (17.52) we obtain the expectation value of the entanglement witness $W_{\text{opt}}^{\text{iso}}$ as

$$\langle W_{\text{opt}}^{\text{iso}} \rangle = \frac{1}{3\sqrt{2}} \langle \mathbb{1} \otimes \mathbb{1} \rangle - \frac{1}{4\sqrt{2}} \langle \Lambda \rangle , \tag{17.55}$$

where the expectation value $\langle \Lambda \rangle$ is given by the following observable quantities

$$\langle \Lambda \rangle = \frac{1}{\hbar^2} \Big(\langle S_x \otimes S_x \rangle - \langle S_y \otimes S_y \rangle + \langle S_z \otimes S_z \rangle \Big) + \frac{16}{3} \langle \mathbb{1} \otimes \mathbb{1} \rangle$$

$$- \frac{4}{\hbar^2} \Big(\langle \mathbb{1} \otimes S_x^2 \rangle + \langle \mathbb{1} \otimes S_y^2 \rangle + \langle S_x^2 \otimes \mathbb{1} \rangle + \langle S_y^2 \otimes \mathbb{1} \rangle \Big) \tag{17.56}$$

$$+ \frac{4}{\hbar^4} \Big(\langle S_x^2 \otimes S_x^2 \rangle + \langle S_y^2 \otimes S_y^2 \rangle \Big) + \frac{2}{\hbar^4} \Big(\langle S_x^2 \otimes S_y^2 \rangle + \langle S_y^2 \otimes S_x^2 \rangle \Big)$$

$$+ \frac{1}{\hbar^4} \Big(\langle \{S_z, S_x\} \otimes \{S_z, S_x\} \rangle - \langle \{S_y, S_z\} \otimes \{S_y, S_z\} \rangle - \langle \{S_x, S_y\} \otimes \{S_x, S_y\} \rangle \Big).$$

The decomposition in eqn (17.56) can then be determined experimentally by measuring the corresponding expectation values of the spin operators of Alice and Bob locally.

Résumé: Entanglement witnesses are observables and provide a method for experimentally detecting entanglement. The quantity to be measured is the expectation value of an entanglement witness W for some state ρ. If $\langle W \rangle < 0$ then the state ρ is entangled. The advantage of this entanglement-witness procedure is that for an experimental outcome $\langle W_{\text{opt}} \rangle < 0$ (with sufficiently small confidence intervals) it can be concluded that the considered quantum state is entangled without requiring full knowledge of the quantum state in question. At the same time, such witnesses can be found for any entangled state, not just for those for which one would also be able to violate a (suitable) Bell inequality. As we discussed already in Chapter 15, the entanglement-witness procedure enables us to detect more entangled states than could be detected purely based on Bell inequalities. However, the number of different measurement settings and measurements for each setting necessary to determine an entanglement witness is comparable to the respective numbers required for the Bell-inequality procedure.

17.3   Entanglement of Qutrits, Qudits ...

For a two-qubit system the geometric structure of the state space in terms of entangled and separable states is very well known. There, the PPT criterion, Theorem 15.3, provides a necessary and sufficient condition for separability, the positivity (non-negativity) of the partial transpose. The case of two qubits, however, is quite special due to its high symmetry and restricted local dimension and can therefore be somewhat misleading when one attempts to make naive conclusions about higher-dimensional systems.

In higher dimensions, the geometric structure within the state space is much more complicated and new phenomena emerge, for instance, *bound entanglement*, referring to entangled states that are not distillable (Horodecki *et al.*, 1998), see the discussion in Section 16.1.3. Although necessary and sufficient conditions à la Peres–Horodecki are not known in this case, we can construct an operator—an entanglement witness—which provides a criterion for the entanglement of the state via the violation of an inequality for the expectation value of the operator.

With the help of such an entanglement witness we can explore the geometric structure of two-qutrit and two-qudit states in the Hilbert–Schmidt space. Two qutrits are states in (3×3)-dimensional Hilbert spaces of bipartite quantum systems. In analogy to the familiar two-qubit case, which we will begin our discussion with in Section 17.3.1 for didactic reasons, we will introduce a two- and a three-parameter family of two-qutrit states in Section 17.3.2 which are part of the so-called *magic simplex* of states, and we determine their geometric properties. For the two-parameter family we will examine the entanglement via the Hilbert–Schmidt measure, recall Section 16.2.3, and for the three-parameter family we will discover bound-entangled states in addition to free-entangled (distillable) ones. Our presentation closely follows the presentation in (Bertlmann and Krammer, 2009).

17.3.1   Two-Parameter Entangled States—Qubits

Let us start with the familiar case of a two-qubit system to gain some intuition of the geometry of the states in the Hilbert–Schmidt space. We will determine the entanglement witness and the Hilbert–Schmidt measure (the minimal Hilbert–Schmidt distance to the set of separable states) for the following two-qubit states, which are particular mixtures of the Bell states $|\psi^-\rangle$, $|\psi^+\rangle$, $|\phi^-\rangle$, and $|\phi^+\rangle$ and are given by

$$\rho_{\alpha,\beta} = \frac{1-\alpha-\beta}{4}\, \mathbb{1} + \alpha\,|\phi^+\rangle\langle\phi^+| + \frac{\beta}{2}\left(|\psi^+\rangle\langle\psi^+| + |\psi^-\rangle\langle\psi^-|\right). \tag{17.57}$$

The states (17.57) are characterized by two parameters, α and β, and we will here refer to these states as the *two-parameter states*. Of course, the positivity requirement $\rho_{\alpha,\beta} \geq 0$ constrains the possible values of α and β, namely via the inequalities

$$\alpha \leq -\beta + 1, \quad \alpha \geq \frac{1}{3}\beta - \frac{1}{3}, \quad \alpha \leq \beta + 1, \tag{17.58}$$

which geometrically corresponds to a triangle, see Fig. 17.1.

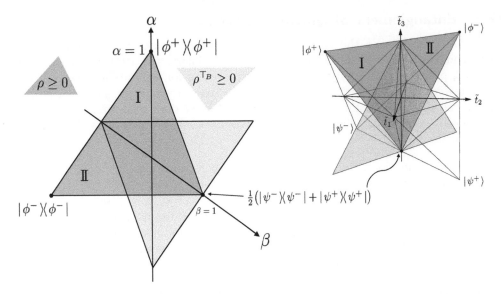

Fig. 17.1 Illustration of two-qubit two-parameter states. The family of states described by the density matrices $\rho = \rho_{\alpha,\beta}$ from eqn (17.57) (blue) and their respective partial transpositions ρ^{T_B} (green) are illustrated as regions in the Hilbert–Schmidt space. Since all $\rho = \rho_{\alpha,\beta}$ have maximally mixed reduced states, they are Weyl states (see Section 15.5.1) and can be described as lying in a two-dimensional plane within a three-dimensional parameter space with coordinates $\tilde{t}_1$, $\tilde{t}_2$, and $\tilde{t}_3$, where the set of physical states lies within a tetrahedron (right-hand side, compare with Fig. 15.3).

In the illustration of Fig. 17.1 the orthogonal lines indeed correspond to families of orthogonal states in Hilbert–Schmidt space. Therefore the coordinate axes for the parameters α and β are necessarily non-orthogonal. In particular, the α axis has to be orthogonal to the boundary line $\alpha = \frac{1}{3}(\beta - 1)$, and the β axis has to be orthogonal to $\alpha = \beta + 1$. The two-parameter states $\rho_{\alpha,\beta}$ define a plane in the Hilbert–Schmidt space, which can be located within the three-dimensional simplex, the tetrahedron of Weyl states (see Section 15.5.1), whose corners are the four two-qubit Bell states.

To calculate the Hilbert–Schmidt measure

$$\mathcal{D}_{\mathrm{HS}}(\rho_{\mathrm{ent}}) = \min_{\rho \in \mathcal{S}} \|\rho - \rho_{\mathrm{ent}}\|_{\mathrm{HS}} = \|\rho_0 - \rho_{\mathrm{ent}}\|_{\mathrm{HS}}, \qquad (17.59)$$

with ρ_0 the closest separable state relative to ρ_{ent}, for the two-parameter qubit family (17.57) we express the states in terms of the usual Pauli-matrix basis as

$$\rho_{\alpha,\beta} = \frac{1}{4}\left[\mathbb{1} + \alpha\left(\sigma_1 \otimes \sigma_1 - \sigma_2 \otimes \sigma_2\right) + (\alpha - \beta)\,\sigma_3 \otimes \sigma_3\right], \qquad (17.60)$$

where we have used the Pauli-matrix decomposition of the Bell states, eqn (15.134) of Section 15.5. In order to calculate the Hilbert–Schmidt measure for the entangled two-parameter states $\rho_{\alpha,\beta}^{\mathrm{ent}}$ we have to find the closest separable states ρ_0, which is generally a difficult task. Fortunately, there is a lemma that we can use to check whether a particular separable state is indeed the closest separable state to a given entangled state (Bertlmann and Krammer, 2009).

Lemma 17.1 (Criterion for closest separable state)

A state $\tilde{\rho}$ is equal to the closest separable state ρ_0 if and only if the operator

$$C = \frac{\tilde{\rho} - \rho_{\text{ent}} - (\tilde{\rho}, \tilde{\rho} - \rho_{\text{ent}})_{\text{HS}} \mathbb{1}}{\|\tilde{\rho} - \rho_{\text{ent}}\|_{\text{HS}}}$$

is an entanglement witness.

For the proof we refer to Bertlmann and Krammer (2009). Lemma 17.1 allows us to check if a guess $\tilde{\rho}$ is indeed correct for the closest separable state ρ_0, in which case the operator C represents the optimal entanglement witness W_{opt}. In the following we will make such a guess based on the geometric intuition obtained from Fig. 17.1. However, by itself, the geometric picture of Fig. 17.1 is not enough to conclude that ρ_0 has been reached, since Fig. 17.1 does not represent the full state space, e.g., states with terms like $a_i \sigma^i \otimes \mathbb{1}$ or $b_i \mathbb{1} \otimes \sigma^i$ for $i = 1, 2, 3$ are not contained. Therefore, we will use Lemma 17.1 in order to verify our guess for the closest separable state. For calculating the Hilbert–Schmidt measure of entanglement for the family of states (17.57) we consider the two regions I and II of Fig. 17.1 separately.

Region I: The entangled states of region I in Fig. 17.1 include the Bell state $|\phi^+\rangle$ and are characterized by (α, β) with the constraints $\alpha \leq \beta + 1$, $\alpha \leq -\beta + 1$, $\alpha > \frac{\beta + 1}{3}$. In the separable region of Fig. 17.1 the point that is closest (in the Euclidean sense) to the point (α, β) is given by $(\frac{1 + \beta}{3}, \beta)$, which corresponds to the state

$$\tilde{\rho}_\beta = \frac{1}{4} \left(\mathbb{1} + \frac{1 + \beta}{3} \left[\sigma_1 \otimes \sigma_1 - \sigma_2 \otimes \sigma_2 \right] + \frac{1 - 2\beta}{3} \sigma_3 \otimes \sigma_3 \right) . \qquad (17.61)$$

Thus the difference between a given entangled state and the state $\tilde{\rho}_\beta$ evaluates to

$$\tilde{\rho}_\beta - \rho_{\alpha,\beta}^{\text{ent}} = \frac{1}{4} \left(\frac{1 + \beta}{3} - \alpha \right) \Sigma , \qquad (17.62)$$

where Σ is defined as

$$\Sigma := \sigma_1 \otimes \sigma_1 - \sigma_2 \otimes \sigma_2 + \sigma_3 \otimes \sigma_3 . \qquad (17.63)$$

Using the Hilbert–Schmidt norm $\|\Sigma\|_{\text{HS}} = \sqrt{\text{Tr}(\Sigma^\dagger \Sigma)} = 2\sqrt{3}$ and identifying (as a guess) the state $\tilde{\rho}_\beta$ (17.61) with the closest separable state $\tilde{\rho}_\beta \equiv \rho_{0,\beta}$ of the total Hilbert–Schmidt space, we can express the Hilbert–Schmidt measure as

$$\mathcal{D}_{\text{HS}}^{\text{I}}(\rho_{\alpha,\beta}^{\text{ent}}) = \|\rho_{0,\beta} - \rho_{\alpha,\beta}^{\text{ent}}\|_{\text{HS}} = \frac{\sqrt{3}}{2} \left(\alpha - \frac{1 + \beta}{3} \right) . \qquad (17.64)$$

Of course we still have to check whether our guess indeed coincides with the closest separable state, i.e., whether the operator C of Lemma 17.1 represents an entanglement witness. The condition

$$(\rho_{\alpha,\beta}^{\mathrm{ent}}, C)_{\mathrm{HS}} = \mathrm{Tr}\left(C \, \rho_{\alpha,\beta}^{\mathrm{ent}}\right) = -\frac{\sqrt{3}}{2}\left(\alpha - \frac{1+\beta}{3}\right) < 0 \,, \qquad (17.65)$$

$$\text{with} \quad C = \frac{1}{2\sqrt{3}}\left(\mathbb{1} - \Sigma\right) \,, \qquad\qquad (17.66)$$

is satisfied by construction. The crucial test thus is that of the second inequality of the entanglement-witness theorem (Theorem 16.8 of Section 16.3.1), the positivity of the expectation value of C for all separable states. For this we will employ the following lemma proven in Bertlmann and Krammer (2009).

Lemma 17.2 (Positivity of expectation value in 2×2 dimensions)

For any Hermitian operator C on a Hilbert space of dimension 2×2 that is of the form

$$C = a \left(\mathbb{1} + \sum_{i=1}^{3} c_i \, \sigma_i \otimes \sigma_i \right) \quad a \in \mathbb{R}^+, \; c_i \in \mathbb{R}$$

the expectation value for all separable states is positive

$$(\rho, C)_{\mathrm{HS}} \geq 0 \quad \forall \rho \in \mathcal{S}, \quad \text{if } |c_i| \leq 1 \; \forall i \,.$$

Since the operator C (17.66) is of the form given in Lemma 17.2 its expectation value is positive (non-negative) for all separable states

$$(\rho, C)_{\mathrm{HS}} \geq 0 \qquad \forall \rho \in \mathcal{S} \,. \qquad\qquad (17.67)$$

Consequently the operator C (17.66) represents an entanglement witness $\mathrm{W}_{\mathrm{opt}}$ and due to Lemma 17.1 the state $\tilde{\rho}_\beta$ (17.61) is already the closest separable state $\rho_{0,\beta}$.

Region II: Next we determine the Hilbert–Schmidt measure for the entangled states $\rho_{\alpha,\beta}^{\mathrm{ent}}$ located in the triangle region that includes the Bell state $|\phi^-\rangle$, i.e., region II in Fig. 17.1. Here the parameters are constrained by $\alpha \leq \beta+1$, $\alpha \geq \frac{\beta-1}{3}$, $\alpha < -\beta-1$. The states in the separable region that are closest to an entangled state represented by the point (α, β) in the parameter space are denoted by $\tilde{\rho}_{\alpha,\beta}$. They are characterized by the pair of parameters $\left(\frac{1}{3}(-1 + 2\alpha - \beta), \frac{1}{3}(-2 - 2\alpha + \beta)\right)$. With this, we calculate the relevant quantities for the operator C of Lemma 17.1:

$$\tilde{\rho}_{\alpha,\beta} - \rho_{\alpha,\beta}^{\text{ent}} = -\frac{1}{12}\left(\alpha + \beta + 1\right)\Xi, \tag{17.68a}$$

$$\|\tilde{\rho}_{\alpha,\beta} - \rho_{\alpha,\beta}^{\text{ent}}\|_{\text{HS}} = -\frac{1}{2\sqrt{3}}\left(\alpha + \beta + 1\right), \tag{17.68b}$$

$$(\tilde{\rho}_{\alpha,\beta}, \tilde{\rho}_{\alpha,\beta} - \rho_{\alpha,\beta}^{\text{ent}})_{\text{HS}} = -\frac{1}{12}\left(\alpha + \beta + 1\right), \tag{17.68c}$$

with Ξ defined by

$$\Xi := \sigma_1 \otimes \sigma_1 - \sigma_2 \otimes \sigma_2 - \sigma_3 \otimes \sigma_3, \tag{17.69}$$

so that the explicit form of C is

$$C = \frac{1}{2\sqrt{3}}\left(\mathbb{1} + \Xi\right). \tag{17.70}$$

To test the operator C (17.70) for being an entanglement witness we have to check the first inequality of the entanglement-witness theorem (Theorem 16.8), which is

$$(\rho_{\alpha,\beta}^{\text{ent}}, C)_{\text{HS}} = \frac{1}{2\sqrt{3}}\left(\alpha + \beta + 1\right) < 0, \tag{17.71}$$

as expected, due to the third constraint for α. Since the operator C (17.70) is of the form given in Lemma 17.2 we obtain for the separable states

$$(\rho, C)_{\text{HS}} \geq 0 \qquad \forall \rho \in \mathcal{S}. \tag{17.72}$$

In region II the operator C (17.70) is therefore indeed an entanglement witness and $\tilde{\rho}_{\alpha,\beta}$ is the closest separable state $\tilde{\rho}_{\alpha,\beta} \equiv \rho_{0,\alpha\beta}$ for the entangled states $\rho_{\alpha,\beta}^{\text{ent}}$. Finally, we find the expression

$$\mathcal{D}_{\text{HS}}^{\text{II}}(\rho_{\alpha,\beta}^{\text{ent}}) = \|\rho_{0,\alpha,\beta} - \rho_{\alpha,\beta}^{\text{ent}}\|_{\text{HS}} = -\frac{1}{2\sqrt{3}}\left(\alpha + \beta + 1\right) \tag{17.73}$$

for the Hilbert–Schmidt measure of the entangled states

17.3.2 Two-Parameter Entangled States—Qutrits

The procedure of determining the geometry of separable and entangled states discussed in the previous Section 17.3.1 can be generalized to higher dimensions, and we will discuss how this works here for the case of two qutrits.

To this end, let us first examine how one may generalize the concept of the maximally entangled Bell basis to higher dimensions. A basis of maximally entangled two-qudit states can be obtained by starting with the maximally entangled qudit state $|\phi_{00}\rangle$, eqn (17.32), and constructing the other $d^2 - 1$ states in the following way:

$$|\phi_i\rangle = \tilde{U}_i \otimes \mathbb{1} |\phi_{00}\rangle \qquad i = 1, 2, \ldots, d^2 - 1, \tag{17.74}$$

where $\{\tilde{U}_i\}$ represents a suitable operator basis consisting of unitary matrices that are orthogonal in the sense that $\text{Tr}(\tilde{U}_i^\dagger \tilde{U}_j) = N_{\tilde{U}}\delta_{ij}$ and $\tilde{U}_0$ denotes the identity $\mathbb{1}$.

A practical choice of the basis of unitary matrices is the Weyl-operator basis (WOB) that we have discussed in Section 17.1.3. For this choice, we have

$$U_{nm} \otimes \mathbb{1} \, | \, \phi_{00} \, \rangle = \frac{1}{\sqrt{d}} \sum_{k,k'=0}^{d-1} e^{\frac{2\pi i}{d} nk} \, | \, k \, \rangle \langle \, (k+m)_{\text{mod}(d)} \, | \otimes \mathbb{1} \, | \, k' \, \rangle \otimes | \, k' \, \rangle$$

$$= \frac{1}{\sqrt{d}} \sum_{k=0}^{d-1} e^{\frac{2\pi i}{d} nk} \, | \, k \, \rangle \otimes | \, (k+m)_{\text{mod}(d)} \, \rangle = | \, \phi_{mn} \, \rangle \, , \qquad (17.75)$$

which matches exactly the maximally entangled states $| \, \phi_{mn} \, \rangle$ from eqn (14.11). As a remark that will be of relevance in later chapters, let us note here that the relationship between the operators U_{nm} on the one hand, which map states from a d-dimensional Hilbert space to states on another d-dimensional Hilbert space, and bipartite states $| \, \phi_{mn} \, \rangle = U_{nm} \otimes \mathbb{1} \, | \, \phi_{00} \, \rangle$ on a $d \times d$-dimensional Hilbert space is an expression of a more general duality called the *Choi–Jamiołkowski isomorphism*, which we will discuss in detail in Section 21.4.1. In this context, the d^2 projectors

$$P_{nm} := (U_{nm} \otimes \mathbb{1}) \, | \, \phi_{00} \, \rangle \langle \, \phi_{00} \, | \, (U_{nm}^\dagger \otimes \mathbb{1}) \qquad (17.76)$$

are (up to a merely conventional exchange of the two subsystems) the so-called *Choi states* of the maps U_{nm} for $n, m = 0, 1, \ldots, d-1$.

In the case of qutrits ($d = 3$) mixtures of the corresponding nine projectors of eqn (17.76)—for simplicity we will refer to them as "Bell projectors" from now on—form an eight-dimensional simplex which is the higher-dimensional analogue of the three-dimensional simplex, the tetrahedron for qubits, see Fig. 15.3. This eight-dimensional simplex has a very interesting geometry concerning the splitting between separability and entanglement (see, e.g., the works of *Bernhard Baumgartner*, *Beatrix Hiesmayr*, and *Heide Narnhofer*, 2006, 2007, 2008). Due to the high internal symmetry of this simplex—which is also the reason that it has been dubbed the *magic simplex*—it is enough to consider certain mixtures of maximally entangled states which form equivalence classes concerning their geometry.

We can express the Bell projectors as generalized Bloch vectors by using the Bloch-vector form from eqn (17.46) for $P_{00} := | \, \phi_{00} \, \rangle \langle \, \phi_{00} \, |$ and the relations (Narnhofer, 2006)

$$U_{nm}^\dagger = e^{\frac{2\pi i}{d} nm} U_{-n\,-m} \, , \qquad (17.77)$$

$$U_{nm} U_{lk} = e^{\frac{2\pi i}{d} ml} U_{n+l\,m+k} \, , \qquad (17.78)$$

where indices are understood to be mod(d). For the Bell projectors this provides the Bloch form

$$P_{nm} = \frac{1}{d^2} \sum_{k,l=0}^{d-1} e^{\frac{2\pi i}{d} (ml-nk)} U_{lk} \otimes U_{-lk} \, . \qquad (17.79)$$

Two-parameter two-qutrit states. Now we are interested in the following two-parameter family of states of two qutrits as a generalization of the two-qubit case from eqn (17.57),

$$\rho_{\alpha,\beta} = \frac{1-\alpha-\beta}{9}\,\mathbb{1} + \alpha\,P_{00} + \frac{\beta}{2}\,(P_{10}+P_{20})\ . \tag{17.80}$$

Following (Baumgartner *et al.*, 2006), the states $|\phi_{mn}\rangle$ represented by the projectors P_{nm} can be thought of as points in a discrete phase space. The indices n and m can be interpreted as "quantized" position and momentum coordinates, respectively. In this phase-space picture, the states P_{00}, P_{10}, P_{20} lie on a line with fixed "momentum" coordinate $m = 0$, but the geometric considerations are not altered if we were to consider a different line in this parameter space of maximally entangled states. As in the two-qubit case before (see Section 17.3.1), the two-parameter family of states $\rho_{\alpha,\beta}$ is thus a *linear* combination of maximally entangled states on a line and the maximally mixed state. Note that it is not necessarily a *convex* linear combination of the mentioned states, since the positivity constraints allow for non-positive values of α and β, as we shall see soon.

Inserting the Bloch-vector form of P_{00}, P_{10}, P_{20} from eqn (17.79) the Bloch-vector expansion of the two-parameter family of states (17.80) can be written as (Bertlmann and Krammer, 2008*a*)

$$\rho_{\alpha,\beta} = \frac{1}{9}\left(\mathbb{1} + \frac{2\alpha-\beta}{2}U_1 + (\alpha+\beta)\,U_2\right)\ , \tag{17.81}$$

where we have defined

$$U_1 := U_{01}\otimes U_{01} + U_{02}\otimes U_{02} + U_{11}\otimes U_{-11} + U_{12}\otimes U_{-12}$$
$$+\ U_{21}\otimes U_{-21} + U_{22}\otimes U_{-22}\,, \tag{17.82}$$

and $U_2 := U_2^{\mathrm{I}} + U_2^{\mathrm{II}}$ where

$$U_2^{\mathrm{I}} := U_{10}\otimes U_{-10}\,,\quad\text{and}\quad U_2^{\mathrm{II}} := U_{20}\otimes U_{-20}\ . \tag{17.83}$$

The constraints for the positivity requirement ($\rho_{\alpha,\beta}\geq 0$) are

$$\alpha \leq \frac{7}{2}\beta+1\,,\quad \alpha \leq -\beta+1\,,\quad \alpha \geq \frac{\beta-1}{8}\,, \tag{17.84}$$

while the positivity of the partial transpose requires

$$\alpha \leq -\beta-\frac{1}{2}\,,\quad \alpha \geq \frac{5}{4}\beta-\frac{1}{2}\,,\quad \alpha \leq \frac{\beta+2}{8}\ . \tag{17.85}$$

Taking these constraints into account, we can illustrate the region of the state space that represents the considered two-parameter family of states (17.80) in a picture of Euclidean geometry, which is presented in Fig. 17.2. As shown in Baumgartner *et al.* (2006), the states $\rho_{\alpha,\beta}$ with positive partial transpose are all separable states, thus

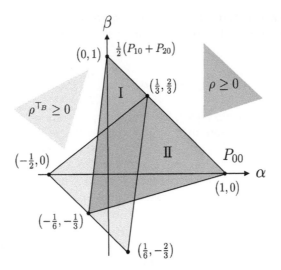

Fig. 17.2 Illustration of two-qutrit states. The regions in parameter space with coordinates (α, β) corresponding to the density matrices $\rho = \rho_{\alpha,\beta}$ from (17.80) (blue triangle) and their partial transpositions ρ^{T_B} (green triangle) are shown. The states in the regions I and II are entangled with negative partial transposition, while the states in the overlap region of both triangles are PPT and separable. The parameter axes are here shown as orthogonal, but when displaying them as non-orthogonal to each other, but instead orthogonal to the boundary lines of the positivity region, $\alpha = \frac{\beta-1}{8}$ and $\alpha = \frac{7}{2}\beta + 1$, on may reproduce the symmetry of the magic simplex, compare, e.g., Bertlmann and Krammer (2009).

there are no PPT entangled states of the form of (17.80). In order to detect PPT bound entanglement, we will therefore have to consider families of states that are described by more than two parameters in the chosen operator-basis expansion, and we will separately consider this case in Section 17.3.3.

To find the Hilbert–Schmidt measure for the entangled two-parameter two-qutrit states we apply the same procedure as in Section 17.3.1. First we determine the states that are the closest separable ones in the Euclidean sense of Fig. 17.2, where we use Lemma 17.1, i.e., we establish the required entanglement witness. Then we calculate the Hilbert–Schmidt measure to quantify the distance to the set of separable states.

Region I: Let us consider first region I in Fig. 17.2, i.e., the triangle region of entangled states around the α axis, constrained by the parameter values $\alpha \leq \frac{7}{2}\beta + 1$, $\alpha \leq -\beta+1$, $\alpha > \frac{\beta+2}{8}$. In the Euclidean picture the point that is closest to the point (α, β) in this region is given by $\left(\frac{2+\beta}{8}, \beta\right)$, which corresponds to the separable two-qutrit state

$$\tilde{\rho}_\beta = \frac{1}{9}\left(\mathbb{1} + \frac{2-3\beta}{8}U_1 + \frac{2+9\beta}{8}U_2\right), \tag{17.86}$$

with U_1 and U_2 as defined in eqn (17.82).

Next we calculate the difference between an entangled state $\rho_{\alpha,\beta}^{\text{ent}}$ and the corresponding closest separable state $\tilde{\rho}_\beta$, the Hilbert–Schmidt norm of this operator, and the Hilbert–Schmidt inner product of this difference with the closes separable state in order to construct the operator $\tilde{W}$, that is

$$\tilde{\rho}_\beta - \rho_{\alpha,\beta}^{\text{ent}} = \frac{1}{9}\left(\frac{2+\beta}{8} - \alpha\right)U\,, \tag{17.87}$$

$$\|\tilde{\rho}_\beta - \rho_{\alpha,\beta}^{\text{ent}}\|_{\text{HS}} = \frac{2\sqrt{2}}{3}\left(\alpha - \frac{2+\beta}{8}\right)\,, \tag{17.88}$$

$$(\tilde{\rho}_\beta, \tilde{\rho}_\beta - \rho_{\alpha,\beta}^{\text{ent}})_{\text{HS}} = \text{Tr}\big(\tilde{\rho}_\beta[\tilde{\rho}_\beta - \rho_{\alpha,\beta}^{\text{ent}}]\big) = -\frac{2}{9}\left(\alpha - \frac{2+\beta}{8}\right)\,, \tag{17.89}$$

where $U = U_1 + U_2$ as defined in eqn (17.48) with the Hilbert–Schmidt norm $\|U\|_{\text{HS}} = 3\sqrt{8} = 6\sqrt{2}$. With this, we can calculate the operator

$$\tilde{W} = \frac{\tilde{\rho}_\beta - \rho_{\alpha,\beta}^{\text{ent}} - (\tilde{\rho}_\beta, \tilde{\rho}_\beta - \rho_{\alpha,\beta}^{\text{ent}})_{\text{HS}}\,\mathbb{1}}{\|\tilde{\rho}_\beta - \rho_{\alpha,\beta}^{\text{ent}}\|_{\text{HS}}} = \frac{1}{6\sqrt{2}}\left(2\,\mathbb{1} - U\right)\,, \tag{17.90}$$

and test whether it represents an entanglement witness, i.e., whether $\tilde{W}$ (17.90) satisfies the inequalities of the entanglement-witness theorem (Theorem 16.8). As expected we find

$$(\rho_{\alpha,\beta}^{\text{ent}}, \tilde{W})_{\text{HS}} = -\frac{2\sqrt{2}}{3}\left(\alpha - \frac{2+\beta}{8}\right) < 0\,. \tag{17.91}$$

To check the second inequality of the entanglement-witness theorem (Theorem 16.8) we need the following lemma, similar to Lemma 17.2 (Bertlmann and Krammer, 2008*a*)

Lemma 17.3 (Positivity of expectation value in $d \times d$ dimensions)

For any Hermitian operator W *on a Hilbert space of dimension* $d \times d$ *that is of the form*

$$W = a\left((d-1)\,\mathbb{1}_{d^2} + \sum_{n,m=0}^{d-1} c_{nm}\,U_{nm} \otimes U_{-nm}\right),$$

for $a \in \mathbb{R}^+$ *and* $c_{nm} \in \mathbb{C}$ *the expectation value for all separable states is non-negative*

$$(\rho, W)_{\text{HS}} = \text{Tr}\big(W\rho\big) \geq 0 \qquad \forall \rho \in \mathcal{S}\,, \quad \text{if} \quad |c_{nm}| \leq 1 \quad \forall\, n, m\,.$$

Since the operator $\tilde{W}$ (17.90) is of the specified form we can use Lemma 17.3 to verify

$$(\rho, \tilde{W})_{\text{HS}} = \text{Tr}(\tilde{W}\rho) \geq 0 \qquad \forall \rho \in \mathcal{S}\,. \tag{17.92}$$

Therefore $\tilde{W}$ (17.90) is indeed an entanglement witness and $\tilde{\rho}_\beta$ is the closest separable state $\tilde{\rho}_\beta = \rho_{0;\beta}$ for the entangled states $\rho_{\alpha,\beta}^{\text{ent}}$ in Region I. For the Hilbert–Schmidt measure of the entangled two-parameter two-qutrit states (17.80) we find

$$\mathcal{D}^{\mathrm{I}}_{\mathrm{HS}}(\rho^{\mathrm{ent}}_{\alpha,\beta}) \;=\; \|\rho_{0;\,\beta} \,-\, \rho^{\mathrm{ent}}_{\alpha,\beta}\| \;=\; \frac{2\sqrt{2}}{3}\left(\alpha - \frac{2+\beta}{8}\right). \qquad (17.93)$$

Region II: In Region II of Fig. 17.2 the entangled two-parameter two-qutrit states are constrained by the parameter values $\alpha < \frac{5}{4}\beta - \frac{1}{2}$, $\alpha \le -\beta + 1$, $\alpha \ge \frac{\beta-1}{8}$. The points that have minimal Euclidean distance to the points (α, β) located in this region are $\left(\frac{1}{24}(-2 + 20\alpha + 5\beta), \frac{1}{6}(2 + 4\alpha + \beta)\right)$ and correspond to the states $\tilde{\rho}_{\alpha,\beta}$. The quantities needed for calculating $\tilde{\mathrm{W}}$ are

$$\tilde{\rho}_{\alpha,\beta} - \rho^{\mathrm{ent}}_{\alpha,\beta} \;=\; -\frac{1}{72}\,(2 + 4\alpha - 5\beta)\,(U_1 - U_2)\,, \qquad (17.94)$$

$$\|\tilde{\rho}_{\alpha,\beta} - \rho^{\mathrm{ent}}_{\alpha,\beta}\|_{\mathrm{HS}} \;=\; \frac{1}{6\sqrt{2}}\,(-2 - 4\alpha + 5\beta)\,, \qquad (17.95)$$

$$(\tilde{\rho}_{\alpha,\beta}, \tilde{\rho}_{\alpha,\beta} - \rho^{\mathrm{ent}}_{\alpha,\beta})_{\mathrm{HS}} \;=\; \frac{1}{36}\,(2 + 4\alpha - 5\beta)\,, \qquad (17.96)$$

so that the operator $\tilde{\mathrm{W}}$ can be expressed as

$$\tilde{\mathrm{W}} \;=\; \frac{1}{6\sqrt{2}}\,(2\,\mathbb{1} + U_1 - U_2)\,, \qquad (17.97)$$

with U_1 and U_2 as defined in eqn (17.82).

The check of the first inequality of the entanglement-witness theorem (Theorem 16.8) gives, unsurprisingly,

$$(\rho^{\mathrm{ent}}_{\alpha,\beta}, \tilde{\mathrm{W}})_{\mathrm{HS}} \;=\; \mathrm{Tr}\big(\tilde{\mathrm{W}}\,\rho^{\mathrm{ent}}_{\alpha,\beta}\big) \;=\; \frac{1}{6\sqrt{2}}\,(2 + 4\alpha - 5\beta) \;<\; 0\,, \qquad (17.98)$$

because of the constraint $2 + 4\alpha < 5\beta$. To test the second inequality of the theorem, which involves the separable states, we use the fact that the operator $\tilde{\mathrm{W}}$ from eqn (17.97) is of the form specified in Lemma 17.3. According to Lemma 17.3 we thus obtain

$$(\rho, \tilde{\mathrm{W}})_{\mathrm{HS}} \;=\; \mathrm{Tr}\big(\tilde{\mathrm{W}}\,\rho\big) \;\ge\; 0 \qquad \forall \rho \in \mathcal{S}\,. \qquad (17.99)$$

We have thus demonstrated that $\tilde{\mathrm{W}}$ (17.97) is indeed an entanglement witness and the states $\tilde{\rho}_{\alpha,\beta}$ are the closest separable ones with respect the entangled two-parameter states of eqn (17.80) of Region II, i.e., $\tilde{\rho}_{\alpha,\beta} = \rho_{0;\,\alpha,\beta}$. Finally, the Hilbert–Schmidt measure of these states evaluates to

$$\mathcal{D}^{\mathrm{II}}_{\mathrm{HS}}(\rho^{\mathrm{ent}}_{\alpha,\beta}) \;=\; \|\rho_{0;\,\alpha,\beta} - \rho^{\mathrm{ent}}_{\alpha,\beta}\| \;=\; \frac{1}{6\sqrt{2}}\,(-2 - 4\alpha + 5\beta)\,. \qquad (17.100)$$

17.3.3 Three-Parameter States and Bound Entanglement—Qutrits

As we have already discussed in detail, the PPT criterion, Theorem 15.3, is a necessary criterion for separability, and sufficient for Hilbert spaces of dimension 2×2 and 2×3. A separable state has to stay positive semidefinite under partial transposition.

Thus if a density matrix becomes indefinite under partial transposition, i.e., one or more eigenvalues of the partial transpose are negative, then the original state has to be entangled and is called NPT entangled. But there exist entangled states that remain positive semidefinite—PPT entangled states—these are bound-entangled states, i.e., they cannot be distilled to obtain pure, maximally entangled states, as we recall from our discussion in Section 15.3.

Quite generally, in a Hilbert space of dimension $d_A \times d_B$ with $d_{AB} = d_A d_B$ the set of all PPT states is convex, compact and contains the set $\mathcal{S}$ of separable states. Thus, the closest separable state ρ_0 appearing in the Hilbert–Schmidt measure in eqn (17.59) can be replaced by the closest PPT state τ_0, which provides the minimal distance of a state ρ to the set of PPT states

$$\mathcal{D}_{\mathrm{HS}}^{\mathrm{PPT}}(\rho) = \min_{\tau \ \mathrm{PPT}} \|\tau - \rho\|_{\mathrm{HS}} = \|\tau_0 - \rho\|_{\mathrm{HS}} . \tag{17.101}$$

If ρ is an NPT-entangled state ρ_{NPT} and τ_0 the closest PPT state, then the operator

$$A_{\mathrm{PPT}} := \tau_0 - \rho_{\mathrm{NPT}} - (\tau_0, \tau_0 - \rho_{\mathrm{NPT}})_{\mathrm{HS}} \, \mathbb{1}_{d_{AB}} \tag{17.102}$$

defines a tangent hyperplane to the set of PPT states for the same geometric reasons that make the operator (16.162) a tangent hyperplane to the set of separable states. Moreover, A_{PPT} has to be an entanglement witness since the set of all PPT stats is a subset of the set $\mathcal{S}$ of separable states (for convenience we do not normalize the operator (17.102) since it does not affect the following calculations). In principle, the entanglement of ρ_{NPT} can be measured in experiments that are able to verify that $(\rho_{\mathrm{NPT}}, A_{\mathrm{PPT}})_{\mathrm{HS}} = \mathrm{Tr}(A_{\mathrm{PPT}} \, \rho_{\mathrm{NPT}}) < 0$. If the state τ_0 is separable, it has to be the closest separable state ρ_0 since the operator (17.102) defines a tangent hyperplane to the set of separable states. In this case A_{PPT} is an optimal entanglement witness, $A_{\mathrm{PPT}} = \mathrm{W}_{\mathrm{opt}}$, and the Hilbert–Schmidt measure can be readily obtained. If τ_0 is not separable, i.e., PPT and entangled, then it has to be a bound-entangled state.

Unfortunately it is not trivial to check whether or not the state τ_0 is separable. In general, it is hard to find evidence of separability but it might be easier to reveal bound entanglement, not only for the state τ_0 but for a whole family of states. One method for detecting bound-entangled states has been proposed by Bertlmann and Krammer (2009), which we will present now. For other procedures see, e.g., the work of Baumgartner, Hiesmayr, and Narnhofer (2006, 2007, 2008).

Bertlmann–Krammer method: Let us now consider any PPT state ρ_{PPT} and the family of states ρ_λ that lie on the line between ρ_{PPT} and the maximally mixed (and of course separable) state $\frac{1}{d_{AB}} \mathbb{1}_{d_{AB}}$

$$\rho_\lambda := \lambda \rho_{\mathrm{PPT}} + \frac{(1 - \lambda)}{d_{AB}} \mathbb{1}_{d_{AB}} . \tag{17.103}$$

Then we can construct an operator W_λ as

$$\mathrm{W}_\lambda = \rho_\lambda - \rho_{\mathrm{PPT}} - (\rho_\lambda, \rho_\lambda - \rho_{\mathrm{PPT}})_{\mathrm{HS}} \, \mathbb{1}_{d_{AB}} . \tag{17.104}$$

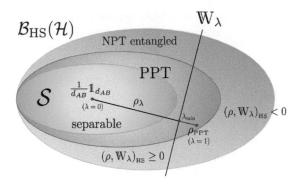

Fig. 17.3 Method to detect bound entanglement with entanglement witness W_λ.
An operator W_λ serves as entanglement witness since all separable states in $\mathcal{S}$ (blue) satisfy $(\rho, W_\lambda)_{\mathrm{HS}} \geq 0$. As convex combinations of the maximally mixed (and thus separable) state $\frac{1}{d_{AB}}\mathbb{1}_{d_{AB}}$ and a PPT state ρ_{PPT}, all states in the family ρ_λ from eqn (17.103) are PPT states by construction. For some $\lambda_{\min}$, all states ρ_λ with $\lambda_{\min} \leq \lambda \leq 1$ (dotted line) are detected as entangled by W_λ, that is, $(\rho_{\lambda \geq \lambda_{\min}}, W_\lambda)_{\mathrm{HS}} < 0$, and must therefore be PPT entangled (green region), rather than NPT entangled (orange region). The detected PPT entangled states are bound-entangled states.

If we can show that for some $\lambda_{\min} < 1$ we have $\mathrm{Tr}(W_{\lambda_{\min}}\rho) \geq 0, \quad \forall \rho \in \mathcal{S}$, then $W_{\lambda_{\min}}$ is an entanglement witness (due to the construction of W_λ the condition $\mathrm{Tr}(W_\lambda \rho_\lambda) < 0$ is already satisfied) and therefore ρ_{PPT} and all states ρ_λ with $\lambda_{\min} < \lambda \leq 1$ are bound entangled, see Fig. 17.3.

With this method in mind, let us now introduce the following family of three-parameter two-qutrit states:

$$\rho_{\alpha,\beta,\gamma} := \tfrac{1-\alpha-\beta-\gamma}{9}\,\mathbb{1} + \alpha P_{00} + \tfrac{\beta}{2}\left(P_{10} + P_{20}\right) + \tfrac{\gamma}{3}\left(P_{01} + P_{11} + P_{21}\right), \quad (17.105)$$

where the projectors P_{nk} onto the maximally entangled states are defined in eqn (17.79) and the parameters are constrained by the positivity requirement $\rho_{\alpha,\beta,\gamma} \geq 0$, specifically this means

$$\alpha \leq \frac{7}{2}\beta + 1 - \gamma, \quad \alpha \leq -\beta + 1 - \gamma,$$

$$\alpha \leq -\beta + 1 + 2\gamma, \quad \alpha \geq \frac{\beta}{8} - \frac{1}{8} + \frac{1}{8}\gamma. \quad (17.106)$$

The states of eqn (17.105) again lie in a "magic" simplex and for $\gamma = 0$ we simply recover the two-parameter states of eqn (17.80) that we have considered before. However, for $\gamma \neq 0$ it is now non-trivial to find the closest separable states since the PPT states no longer coincide with the separable states. But we can use the geometric entanglement witness (17.104) to detect bound entanglement.

To this end we make use of a one-parameter family of two-qutrit states introduced by *Paweł, Michał,* and *Ryszard Horodecki* (Horodecki *et al.*, 1999), these *Horodecki states*

$$\rho_{\mathrm{H}}(b) = \tfrac{2}{7}P_{00} + \tfrac{b}{7}P_{+} + \tfrac{5-b}{7}P_{-}\,, \tag{17.107}$$

where $0 \leq b \leq 5$ and the projectors $P_{\pm}$ are given by

$$P_{+} := \tfrac{1}{3}\left(|01\rangle\langle01| + |12\rangle\langle12| + |20\rangle\langle20|\right), \tag{17.108}$$

$$P_{-} := \tfrac{1}{3}\left(|10\rangle\langle10| + |21\rangle\langle21| + |02\rangle\langle02|\right). \tag{17.109}$$

The Horodecki states of eqn (17.107) are part of the three-parameter family of states from eqn (17.105), specifically

$$\rho_{\mathrm{H}}(b) \equiv \rho_{\alpha,\beta,\gamma} \quad \text{with} \quad \alpha = \frac{6-b}{21},\ \ \beta = -\frac{2b}{21},\ \ \gamma = \frac{5-2b}{7}\,, \tag{17.110}$$

and thus lie in the magic simplex. Testing the partial transposition the range for the parameter b of the Horodecki states (17.107) can be split into three regions

$$\text{for} \quad 0 \leq b < 1 \quad \rho_{\mathrm{H}}(b)\ \text{are NPT states}, \tag{17.111}$$

$$\text{for} \quad 1 \leq b \leq 4 \quad \rho_{\mathrm{H}}(b)\ \text{are PPT states}, \tag{17.112}$$

$$\text{for} \quad 4 < b \leq 5 \quad \rho_{\mathrm{H}}(b)\ \text{are NPT states}. \tag{17.113}$$

Moreover, it was shown in (Horodecki *et al.*, 1999) that the Horodecki states are separable for $2 \leq b \leq 3$, and are thus bound entangled for $3 < b \leq 4$. In what follows it is more convenient to use γ as the parameter of the Horodecki states. Using eqn (17.110) we therefore express b in terms of γ and obtain

$$\rho_{\mathrm{H}}(b) \equiv \rho_{\alpha,\beta,\gamma} \quad \text{with} \quad \alpha = \frac{1+\gamma}{6},\ \ \beta = \frac{-5+7\gamma}{21},\ \ \gamma\,. \tag{17.114}$$

The geometry of the three-parameter family of states $\rho_{\alpha,\beta,\gamma}$ as part of the magic simplex is illustrated in Fig. 17.4. In particular, the figure shows the states $\rho_{\alpha,\beta,\gamma}$ subject to the positivity requirement of (17.106) as well as the PPT states which are constrained by

$$\alpha \leq -\beta - \frac{1}{2} + \frac{1}{2}\gamma\,, \tag{17.115}$$

$$\left(16\alpha - 11\beta + \gamma + 2\right)^{2} \leq 4 + 9\beta^{2} + 4\gamma - 7\gamma^{2} - 6\beta(2+\gamma).$$

The positivity constraints (17.106) imply that the states $\rho_{\alpha,\beta,\gamma}$ form a pyramid with triangular base, whereas the constraints (17.115) restrict the PPT states to an elliptical cone that is cut off on one end by the plane $\alpha \leq -\beta + \frac{1}{2}(\gamma-1)$. The cone and pyramid overlap with each other in the way shown in Fig. 17.4, with the bound-entangled and separable states confined to the region where the pyramid and the cone intersect.

Application of the method: Now we want to sketch how the method we have described can be applied to detect bound entanglement within the three-parameter family of two-qutrit states of eqn (17.105). The idea is to choose PPT starting points on the boundary plane $\alpha = \frac{7}{2}\beta + 1 - \gamma$ of the positivity pyramid, more specifically, on

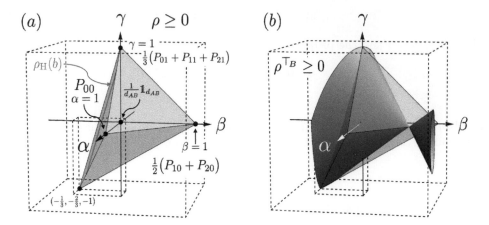

Fig. 17.4 Pyramid of three-parameter two-qutrit states. (a) Taking the parameters α, β, and γ as coordinates in $\mathbb{R}^3$, the states $\rho_{\alpha,\beta,\gamma}$ form a pyramid with a triangular base defined by the positivity constraints in (17.106), whose corners are located at $(\alpha, \beta, \gamma) = (1, 0, 0)$, $(0, 1, 0)$, $(0, 0, 1)$, and $(-1/3, -2/3, -1)$. The maximally mixed state $\frac{1}{d_{AB}}\mathbb{1}_{d_{AB}}$ is located at the origin, $(0, 0, 0)$, while the one-parameter family of Horodecki states $\rho_H(b)$ from eqn (17.107) forms a line (orange) on one of the four surfaces of the pyramid. (b) The positivity of the partial transpose translates to the constraints (17.115), which describe an elliptical (double) cone which overlaps with the pyramid. The intersection region of the pyramid and the cone contains all separable and all PPT entangled states, the latter are bound entangled.

the line of Horodecki states and in a region close to this line, and shift the operators W_λ along the parameterized lines that connect the chosen starting points with the maximally mixed state. If we can show that W_λ is an entanglement witness until a certain $\lambda_{\min}$, all states ρ_λ (17.103) with $1 \leq \lambda < \lambda_{\min}$ are PPT entangled, and thus bound entangled. The "starting states" on the boundary plane are parameterized as

$$\rho_{\text{plane}} \equiv \rho_{\alpha,\beta,\gamma} \quad \text{with} \quad \left(\alpha = \frac{1+\gamma+\epsilon}{6}, \ \beta = \frac{-5+7\gamma+\epsilon}{21}, \ \gamma\right), \quad \epsilon \in \mathbb{R}, \quad (17.116)$$

where we have introduced an additional parameter ϵ to account for the deviation from the line within the boundary plane. In terms of the parameters γ and ϵ the operator $W_{\gamma,\epsilon,\lambda}$, corresponding to the operator in eqn (17.104), can be written as

$$W_{\gamma,\epsilon,\lambda} = \rho_\lambda - \rho_{\text{plane}} - (\rho_\lambda, \rho_\lambda - \rho_{\text{plane}})_{\text{HS}} \, \mathbb{1} = \mu \left(2\,\mathbb{1} + c_1\, U_1 + c_2\, U_2^{\text{I}} + c_2^*\, U_2^{\text{II}}\right), \tag{17.117}$$

where we have defined the function $\mu(\gamma, \epsilon, \lambda) := \frac{\nu(\gamma,\epsilon)}{36}\, \lambda(1 - \lambda)$, as well as

$$c_1 = -\frac{4}{7\nu(\gamma,\epsilon)\lambda}\,(2+\epsilon), \quad c_2 = \frac{2}{7\nu(\gamma,\epsilon)\lambda}\,(1 - 7\sqrt{3}\gamma\, i - 3\epsilon), \tag{17.118}$$

which in turn depend on the function $\nu(\gamma, \epsilon) := 1 + 3\gamma^2 + \frac{3}{7}\epsilon(2+\epsilon)$. The operators U_1, U_2^{I}, and U_2^{II} are defined by eqn (17.82) and the family of states ρ_λ by

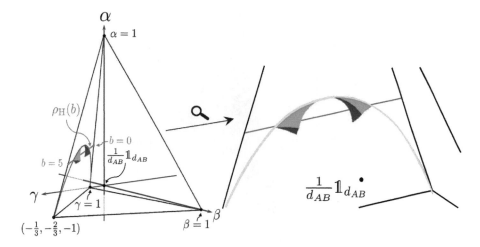

Fig. 17.5 Regions of detected bound-entangled states within the pyramid of states $\rho_{\alpha,\beta,\gamma}$. The coordinate axes for α, β, and γ have been rotated with respect to Fig. 17.4 for better visibility of the volume of PPT entangled states detected by the Bertlmann–Krammer method, and the axes are further being displayed as non-orthogonal (see, Bertlmann and Krammer 2008b) to show the symmetry of the magic simplex. The dot represents the maximally mixed state $\frac{1}{d_{AB}}\mathbb{1}_{d_{AB}}$, the Horodecki states $\rho_{\mathrm{H}}(b)$ are represented by the orange line through the boundary plane from which the regions of bound entanglement emerge. On the right-hand side a magnified picture is shown. Figure adapted from Fig. 7 in (Bertlmann and Krammer, 2009).

$$\rho_\lambda = \lambda\,\rho_{\mathrm{plane}} + \frac{1-\lambda}{9}\,\mathbb{1}\,. \tag{17.119}$$

Now one can attempt to find the minimal λ, denoted by λ_{min}, minimized over the parameters γ and ϵ, such that all states on the line (17.119) are bound entangled for $\lambda_{\mathrm{min}} < \lambda \le 1$. The result for the total minimum $\lambda_{\mathrm{min}}^{\mathrm{tot}}$ finally is (Bertlmann and Krammer, 2009)

$$\lambda_{\mathrm{min}}^{\mathrm{tot}} = \frac{1}{8}\left(3 + \sqrt{13}\right) \simeq 0.826\,, \tag{17.120}$$

which is significantly below 1 and the resulting volume of bound-entangled states is remarkably large. The total minimum (17.120) is attained at $|\gamma| \simeq 0.35$ and $\epsilon = \left(7\sqrt{13} - 25\right)/2 \simeq 0.12$. The entire line of states ρ_λ (17.119) in the interval $\lambda_{\mathrm{min}}^{\mathrm{tot}} < \lambda \le 1$ is found to be bound entangled. The volume of detected bound-entangled states is visualized in Fig. 17.5. For $\gamma = 0$, the point where the two bound-entangled regions touch, one is back at eqn (17.80), there is no bound entanglement.

Finally, let us take a closer look at the boundary plane defined by $\alpha = \frac{7}{2}\beta + 1 - \gamma$, specifically, the triangular region between the points $(\alpha, \beta, \gamma) = (1, 0, 0)$, $(0, 0, 1)$, and $\left(-\frac{1}{3}, -\frac{2}{3}, -1\right)$, as shown in Fig. 17.6. As mentioned, this triangle contains the line of Horodecki states, parameterized by $\gamma \in [-1, 1]$ (or $b \in [0, 5]$) as in eqn (17.114), which can be seen to be separable for $|\gamma| \le 1/7$ (corresponding to $2 \le b \le 3$), PPT (bound)

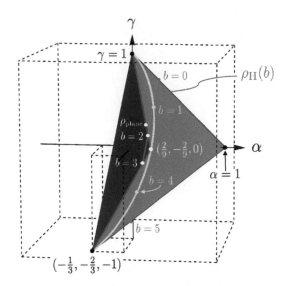

Fig. 17.6 Entanglement properties of $\rho_{\alpha,\beta,\gamma}$ on the boundary plane of the pyramid.
Within the triangular boundary region of the state space of the three-parameter two-qutrit
states that connects the points $(\alpha, \beta, \gamma) = (1,0,0)$, $(0,0,1)$, and $(-\frac{1}{3}, -\frac{2}{3}, -1)$, the separable
states are situated in a triangular region (red) between $(0,0,1)$, $(-\frac{1}{3}, -\frac{2}{3}, -1)$, and $(\frac{2}{9}, -\frac{2}{9}, 0)$,
and the PPT entangled (bound-entangled) states lie in between the triangular separable
region and the curve defining the intersection of the boundary plane with the PPT cone.
The remaining states are NPT entangled. The witness $\mathbb{W}$ used to establish the separation
between separability and entanglement is constructed from the starting state ρ_{plane}, and the
plane defined by the witness intersects the boundary plane along the line (red) connecting
the (separable) points $(0,0,1)$ and $(\frac{2}{9}, -\frac{2}{9}, 0)$. The Horodecki line (orange) runs through all
entanglement characteristics.

entangled for $1/7 < |\gamma| \le 3/7$ (corresponding to $1 \le b < 2$ and $3 < b \le 4$), and
NPT entangled otherwise (Horodecki *et al.*, 1999; Bertlmann and Krammer, 2009).
Meanwhile, the boundary plane is intersected by the PPT cone along a curve defined
by $\beta = \frac{1}{9}(3\gamma - 4 + \sqrt{4 - 3\gamma^2})$, which is shown in light green in Fig. 17.6.

By choosing a starting state ρ_{plane} on the boundary plane with $\gamma = 1/4$ and
$\epsilon = -1/4$ one can use Lemma 17.3 to find (Bertlmann and Krammer, 2009) that
the corresponding operator $\mathbb{W} := \frac{1}{\lambda(1-\lambda)}\mathbb{W}_{\gamma,\epsilon,\lambda}$, with $\mathbb{W}_{\gamma,\epsilon,\lambda}$ as in eqn (17.117), becomes an entanglement witness for $\lambda \to 1$. This witness represents a plane in the
parameter space that is defined by the condition $\text{Tr}(\mathbb{W}\,\rho_{\text{plane}}) = 0$, which translates to
the constraint $\alpha = \frac{2}{5}(1 + 2\beta - \gamma)$. The plane corresponding to the witness intersects
the boundary plane along the line (in red in Fig. 17.6) given by $\beta = \frac{2}{9}(\gamma - 1)$, and
this line in turn intersects with the PPT cone in the points $(0,0,1)$ and $(\frac{2}{9}, -\frac{2}{9}, 0)$. The
latter point is part of the two-parameter family of two-qutrit states discussed in
the previous section, for which all PPT states are separable. The former point corresponds to an equally weighted mixture of the maximally entangled states P_{01}, P_{11},

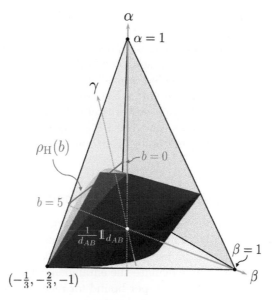

Fig. 17.7 Pyramid of physical states. Within the pyramid of $\rho_{\alpha,\beta,\gamma}$ (17.105) the PPT cone is plotted (purple), but only inside the pyramid of positivity. The region of bound-entangled states is pictured in translucent yellow. The coordinate axes for α, β, and γ are being displayed as non-orthogonal (see, Bertlmann and Krammer 2008b) to show the symmetry of the magic simplex. Figure adapted from Fig. 8 in (Bertlmann and Krammer, 2009).

and P_{21}, which is also separable (Baumgartner *et al.*, 2006). Since the set of separable states is convex all states on the line between these two points must be separable as well. And because the operator W is an entanglement witness no further separable states can lie in the triangle in the region between the line and the intersection with the PPT cone—all these states must be PPT entangled and thus bound entangled. Using symmetry arguments for the magic simplex, this reasoning can be extended to identify the line connecting $(\frac{2}{9}, -\frac{2}{9}, 0)$ with the point $(-\frac{1}{3}, -\frac{2}{3}, -1)$. Within the boundary plane, the separable states are thus confined to a smaller triangle (shown in red in Fig. 17.6, while all states between the triangle and the PPT cone are bound entangled.

As illustrated in Fig. 17.7, this method can be extended to investigate the whole three-parameter space (α, β, γ), which indeed allows one to detect *all* PPT (bound) entangled states that occur there (Bertlmann and Krammer, 2009). Furthermore, the presented method of "shifting" operators (geometric entanglement witnesses) along parameterized lines enables one to establish the shape of the set of separable states of the three-parameter family (17.105), so that the entanglement properties of this family of states can be fully characterized. Within the pyramid of physical states represented by the three-parameter family $\rho_{\alpha,\beta,\gamma}$ (17.105) the familiar PPT cone is plotted within the pyramid of positivity. The region of bound-entangled states, which lie on top of the separable states, is pictured in translucent yellow. The Horodecki states represented by the orange line run through all entanglement characteristics.

Résumé: The PPT criterion is a necessary criterion for separability, and sufficient for Hilbert spaces of dimensions 2×2 and 2×3. A separable state has to stay positive semidefinite under partial transposition. If a density matrix becomes indefinite under partial transposition, it describes a so-called NPT-entangled state. But in higher dimensions there exist further entangled states—bound-entangled states—including (but not necessarily limited to) those states that remain positive semidefinite under partial transposition. Bound entangled states occur, for instance, within the family of three-parameter two-qutrit states, and the PPT (bound) entangled states in this family can be identified using the method of Bertlmann–Krammer method.

17.4 Detecting and Quantifying High-Dimensional Entanglement

For systems of two qubits, both the questions of detecting entanglement as well as of quantifying it are essentially clear. Positive-map criteria, in particular, the PPT criterion provide necessary and sufficient conditions for separability (see Section 15.4), and, via the concurrence one may calculate the entanglement of formation exactly for two qubits, which in turn provides an upper bound for measures such as the entanglement cost, squashed entanglement, and distillable entanglement (see Section 16.2.2). And even in the absence of full information on the density operator in question, entanglement may be detected via the fidelity to maximally entangled states, the singlet fraction, see (15.144), which is an experimentally easily accessible quantity.

For systems of higher dimensions, two qudits of dimension $d \geq 3$, the situation is significantly more involved. Although the detection of some entangled states is still possible using the PPT criterion, this is no longer the case for all entangled states, and at the same time the experimental work required to reconstruct the density operator in order to evaluate the PPT criterion in the first place becomes an issue of scalability. Similarly, calculating any entanglement measures exactly is no longer a computationally viable option in higher dimensions. Consequently, a number of tools have been developed to detect entanglement and quantify it via suitable bounds on various entanglement measures. In this section we briefly discuss some of the promising approaches to the detection (Section 17.4.1) and quantification (Section 17.4.2) of entanglement between two d-dimensional quantum systems, and we refer the interested reader to the reviews (Gühne and Tóth, 2009; Horodecki *et al.*, 2009; Friis *et al.*, 2019) for alternative routes.

17.4.1 Detecting Entanglement in High Dimensions

Entanglement detection using linear contractions. One way to go beyond the possibilities of opened up by positive-map criteria is to stay with linear maps, but to consider another class of such maps, specifically, *linear contractions*, linear maps Λ^{LC} that do not increase the (trace) norm of product states, i.e., such that for all $\rho_A \otimes \rho_B$ we have

$$\|\Lambda^{\mathrm{LC}}[\rho_A \otimes \rho_B]\|_{\mathrm{tr}} \leq 1, \tag{17.121}$$

where $\|\rho\|_{\mathrm{tr}} = \mathrm{Tr}\sqrt{\rho^\dagger \rho}$ is the trace norm, which we will discuss in more detail in Section 23.4. As shown in (Horodecki *et al.*, 2006) such maps cannot increase the trace norm of any separable state and hence finding a value $\|\Lambda^{\mathrm{LC}}[\rho_{AB}]\|_{\mathrm{tr}} > 1$ for any bipartite state ρ_{AB} detects entanglement.

The most well-known example of such a map applied for entanglement detection is that of the *realignment map* Λ^{R},

$$\Lambda^{\mathrm{R}} : \ \rho_{AB} = \sum_{i,j,k,l} \rho_{ijkl} \,|\,i\,\rangle\!\langle\,j\,|_A \otimes |\,k\,\rangle\!\langle\,l\,|_B \longmapsto \sum_{i,j,k,l} \rho_{\pi(ijkl)} \,|\,i\,\rangle\!\langle\,j\,|_A \otimes |\,k\,\rangle\!\langle\,l\,|_B\,,$$

$$(17.122)$$

where the indices $(ijkl)$ of the coefficients ρ_{ijkl} of the density matrix are subject to the permutation $\pi(ijkl) = (ikjl)$. As was found independently by *Oliver Rudolph* (2001, 2003, 2005) and by *Kai Chen* and *Ling-An Wu* (2003), all separable states ρ_{AB} satisfy $\|\Lambda^{\mathrm{R}}[\rho_{AB}]\|_{\mathrm{tr}} \le 1$, and the resulting *cross-norm criterion* (Rudolph, 2005) or *realignment criterion* (Chen and Wu, 2003) hence detects entanglement if the realignment map results in a trace norm above 1. What is noticeable here is that the partial transposition is very similar in that one arrives at the PPT criterion when choosing the permutation $\pi(ijkl) = (ijlk)$, compare with Definition 15.9. However, the *computable cross-norm or realignment* (CCNR) criterion considered here is not equivalent to the PPT criterion, but complementary to it. While there are some entangled states that are detected by the PPT criterion but not by the CCNR criterion, there are also some states for which the converse is the case. Yet, neither of them can detect *all* entangled states.

The CCNR criterion thus extends our zoo of methods for entanglement detection in a non-trivial way, but the methods faces the same obstacle as the PPT criterion when it comes to experimental accessibility. In order to check these criteria, the full density operator must be reconstructed (estimated) from empirical data. Let us therefore now review a method that directly draws from measurement data in a few local bases of Alice and Bob.

Entanglement detection using mutually unbiased bases. From the point of view of experimental implementation, a practical way to detect entanglement that also respects the "distant-lab paradigm" of spatially separated observers is that of Alice and Bob each performing measurements in a set of different measurement bases. If the measurement data is incompatible with any separable state, then one can conclude that the state must be entangled. Indeed, the violation of a Bell inequality is exactly such an approach, with the difference that a Bell inequality does not make specific assumptions about the measurements that have to be performed, Bell inequality violation is a *device-independent* way of detecting entanglement, a flourishing research field of its own (see, e.g., the article collection in Pironio *et al.* 2016). Here, however, we do not aim to be this strict and will allow for certain assumptions to be made about the particular measurements performed to detect entanglement. A choice of measurement bases that has proven to be very useful for this task is that of so-called *mutually unbiased bases* (MUBs).

Definition 17.1 *A set of M bases $\{|\, i^{(1)}\,\rangle\}_i\,,\ \ldots,\ \{|\, i^{(M)}\,\rangle\}_i$, of a d-dimensional Hilbert space are **mutually unbiased** if $\forall\, i,j = 0,1,\ldots,d-1$ and for all $k,k' \in \{1,2,\ldots,M\}$ with $k \neq k'$ we have*

$$|\langle\, i^{(k)}\,|\, j^{(k')}\,\rangle|^2 \;=\; \tfrac{1}{d}\,.$$

The notion of mutual unbiasedness mathematically captures the mutual unpredictability of the measurement outcomes corresponding to the vectors of different bases. If a quantum system is prepared in any of the states $|\, i^{(1)}\,\rangle$ of the first basis, then by Born's rule, the different outcomes for the measurement in any of the mutually unbiased bases are all equally likely, each has the probability $\frac{1}{d}$ of occurring. An example for such a (as we shall see, complete) set of MUBs in dimension $d = 2$ is the set of eigenbases of the three Pauli matrices σ_x, σ_y, and σ_z. If an eigenstate $|\, 0\,\rangle$ or $|\, 1\,\rangle$ of σ_z is prepared, the probability of obtaining the result $|\, +\,\rangle$ or $|\, -\,\rangle$ is equally likely with probability $\frac{1}{2}$ each for the eigenstates $|\, \pm\,\rangle = (|\, 0\,\rangle \pm |\, 1\,\rangle)/\sqrt{2}$ of σ_x. In general, it is known (see, e.g., Bengtsson 2007) that the maximum number $M_{\max}(d)$ of MUBs in dimension d with prime factorization $d = p_1^{n_1} p_1^{n_1} \ldots p_k^{n_k}$, where the p_j are primes and $n_j \in \mathbb{N}$ with $p_1^{n_1} < p_1^{n_1} < \ldots < p_k^{n_k}$, is bounded by

$$p_1^{n_1} + 1 \;\leq\; M_{\max}(d) \;\leq\; d+1. \tag{17.123}$$

For any prime-power dimension $d = p^n$ the upper and lower bound coincide and there exist known constructions (see, for instance, Wootters and Fields 1989; Bandyopadhyay *et al.* 2002; Klappenecker and Rötteler 2004; Archer 2005; Kibler and 2006) for the corresponding complete sets of MUBs. A particularly compact construction for prime dimensions $d = p$ is that presented by *William Wootters* and *Brian Fields* (1989): Starting from a chosen "standard" basis (which can be taken to be the computational basis) $\{|\, m\,\rangle\}_m$, the basis vectors of d additional associated MUBs $\{|\, \tilde{j}_k\,\rangle\}_j$ for $k = 0,\ldots,d-1$ can be constructed as

$$|\, \tilde{j}_k\,\rangle \;=\; \frac{1}{\sqrt{d}} \sum_{m=0}^{d-1} e^{\frac{2\pi i}{d}(jm + km^2)} \,|\, m\,\rangle\,. \tag{17.124}$$

We see that $|\langle\, m\,|\, \tilde{j}_k\,\rangle|^2 = \frac{1}{d}$ for all m,j,k and d (even if d is not prime), while the relation $|\langle\, \tilde{l}_{k'}\,|\, \tilde{j}_k\,\rangle|^2 = \frac{1}{d}$ from Definition 17.1 can be shown to hold for all j,l (for prime d) among the bases labelled by $k \neq k'$ by using (slightly more complicated but) known results on quadratic Gauss sums (Herrera Valencia *et al.*, 2020, Appendix A.III.2). For non-prime-power dimensions the exact value of $M_{\max}(d)$ is generally unknown, and the example of $d = 6$ is (in)famously unsolved (see, e.g., Grassl 2004).

Now let us take a look at how MUBs can be useful for detecting entanglement. Following the approach of (Spengler *et al.*, 2012), let us consider the so-called *mutual predictability* $C_{a,b}$ for two measurement bases $\{|\, m_a\,\rangle_A\}_m$ and $\{|\, n_b\,\rangle_B\}_n$, where a and b

label the chosen bases for Alice and Bob, respectively. Then for an arbitrary bipartite state ρ_{AB} we define

$$C_{a,b}(\rho_{AB}) := \sum_{m=0}^{d-1} \langle m_a m_b | \rho_{AB} | m_a m_b \rangle . \tag{17.125}$$

If the quantity $C_{a,b}$ takes its maximal value $C_{a,b} = 1$, then Alice may predict the measurement outcome of Bob with certainty, provided that both measure in the bases $\{| m_a \rangle_A \}_m$ and $\{| n_b \rangle_B \}_n$, respectively. However, observing a value $C_{a,b} = 1$ in itself does not reveal anything about entanglement. Such a value can be obtained from a maximally entangled state $| \phi_{00} \rangle = \frac{1}{\sqrt{d}} \sum_m | m_a m_b \rangle$ but may as well be the result of a product state $| m_a m_b \rangle$. Now one may observe that the former is an isotropic state invariant under local unitaries $U \otimes U^*$ (see the discussion in Section 17.2.2). This implies that, if one were to choose a second pair of measurement bases $\{| j_{a'} \rangle_A \}_j$ and $\{| j_{b'} \rangle_B \}_j$ constructed according to eqn (17.124) such that $| j_{a'} \rangle_A = | \tilde{j}_{k=0} \rangle_A$ and $| j_{b'} \rangle_B = | \tilde{j}_{k=0}^* \rangle_B$, i.e., mutually unbiased with respect to the initial bases and (since both pairs are orthogonal bases) related to the former via unitaries U and U^*, respectively, the resulting mutual predictability would turn out the same as before, $C_{a',b'}(| \phi_{00} \rangle) = 1$. For the product state, meanwhile, the mutual unbiasedness would lead to a sum of d terms each with the value $\frac{1}{d^2}$, such that $C_{a',b'}(| m_a m_b \rangle) = \frac{1}{d}$. This observation can be made more formal in terms of the following lemma (Spengler *et al.*, 2012).

Lemma 17.4 (Entanglement detection using MUBs)

For any separable state ρ_{sep} in $d \times d$ dimensions, the sum of the mutual predictabilities $C_{a_i b_i}$ for $M \geq 2$ pairs of local MUBs labelled by a_i and b_i with $i = 1, 2, \ldots, M$ is bounded by

$$\sum_{i=1}^{M} C_{a_i b_i}(\rho_{\text{sep}}) \leq 1 + \frac{M-1}{d} .$$

Proof For the proof let us begin with an arbitrary pure product state $| \psi \rangle_A \otimes | \phi \rangle_B$, for which we have

$$\sum_{i=1}^{M} C_{a_i b_i}(| \psi \rangle_A \otimes | \phi \rangle_B) = \sum_{i=1}^{M} \sum_{m=0}^{d-1} |\langle m_{a_i} | \psi \rangle|^2 \, |\langle m_{b_i} | \phi \rangle|^2 . \tag{17.126}$$

Then the inequality $\sqrt{xy} \leq (x+y)/2$ between geometric and arithmetic means of two non-negative numbers x and y suggests

$$\sum_{i=1}^{M} C_{a_i b_i}(| \psi \rangle_A \otimes | \phi \rangle_B) \leq \frac{1}{2} \sum_{i=1}^{M} \sum_{m=0}^{d-1} \left(|\langle m_{a_i} | \psi \rangle|^4 + |\langle m_{b_i} | \phi \rangle|^4 \right). \tag{17.127}$$

At this stage one can leverage a result on mutually unbiased bases, (Wu *et al.*, 2009, Theorem 1), which suggest that for any single-qudit state ρ we have

$$\sum_{i=1}^{M} \sum_{m=0}^{d-1} \left(\langle m_{a_i} | \rho | m_{a_i} \rangle \right)^2 \leq \mathrm{Tr}(\rho^2) + \frac{M-1}{d}. \tag{17.128}$$

Here we have $\rho = |\psi\rangle\langle\psi|$ or $\rho = |\phi\rangle\langle\phi|$ with $\mathrm{Tr}(\rho^2) = 1$ and so the statement of Lemma 17.4 follows for all pure product states. However, all separable states are convex combinations of pure product states, and the mutual predictabilities are linear functions of the density operator, such that no separable state may surpass the bound either. $\qquad\square$

By performing measurements in two or more MUBs one may thus detect entanglement in any dimension in an (in principle) experimentally simple procedure. In the following, we shall see how one can take the usefulness of MUBs for entanglement certification even further by taking a look how one can also quantify (lower-bound) entanglement from such measurements.

17.4.2 Bounds on Entanglement Measures

Entanglement detection using entropic uncertainty relation. A first approach towards bounding entanglement in arbitrary dimensions comes from so-called *entropic uncertainty relations*, a detailed review of which can be found in (Coles *et al.*, 2017). There the idea is to formulate uncertainty relations not through confidence intervals (standard deviations and variances), but rather in terms of entropies, which feature more prominently in quantum information theory. The question whether such a description of uncertainty relations in terms of entropies is indeed possible was first brought up by *Hugh Everett* (1957) and addressed by *Isidore Isaac Hirschman Jr.* (1957). Subsequently, ideas in this direction were made more precise and extended by *William Beckner* (1975), *Iwo Białynicki-Birula* and *Jerzy Mycielski* (1975), as well as *David Deutsch* (1983), to name but a few, but the most commonly encountered version today is due to *Hans Maassen* and *Jos B. M. Uffink* (1988): For two observables $\mathbb{X}$ and $\mathbb{Y}$ with eigenvectors $|\mathbb{X}^x\rangle$ and $|\mathbb{Y}^y\rangle$ with associated eigenvalues x and y, the measurement outcomes can be treated as random variables X and Y, distributed according to probability distributions $\{p_X(x)\}_x$ and $\{p_Y(y)\}_y$. One may then show that (Maassen and Uffink, 1988)

$$H(X) + H(Y) \geq -\log\left(\max_{x,y} |\langle \mathbb{X}^x | \mathbb{Y}^y \rangle|^2\right), \tag{17.129}$$

where $H(X)$ denotes the Shannon entropy

$$H(X) = -\sum_x p_X(x) \log\left(p_X(x)\right), \tag{17.130}$$

which we will discuss in much detail in Section 19.2.1. Here, let it suffice to say that $H(X)$ quantifies our the lack of predictability, our lack of knowledge about the random variable X, or in the language of physics, about the measurement outcomes. the right-hand side of the entropic uncertainty relation (17.129) is strictly non-negative and bounded from above by $\log(d)$, where the upper bound is attained if the eigenbases of $\mathbb{X}$ and $\mathbb{Y}$ are MUB (see, e.g., the discussion around eqn (34) in Coles *et al.* 2017). What is particularly appealing about this approach is that, in contrast to the Robertson–Schrödinger relation of Theorem 2.4, the quantity on the right-hand side is independent of the particular quantum state and only depends on the observables considered.

Such (or similar) entropic uncertainty relations have found applications as separability criteria, e.g., in (Giovannetti, 2004; Gühne and Lewenstein, 2004; Huang, 2010), but here we will focus on a result by (Berta *et al.*, 2010). There, entropic uncertainty relations become relevant for entanglement detection in bipartite quantum states ρ_{AB} via the *quantum conditional entropy*

$$S(A|B) = S(\rho_{AB}) - S(\rho_B),\qquad(17.131)$$

which we will discuss in detail in Section 20.4.1. As we will see there, specifically in Theorem 20.4, is that a negative value of $S(A|B)$ implies entanglement. What is more, *Igor Devetak* and *Andreas Winter* were able to show (2005) that $-S(A|B)$, a quantity referred to as the *coherent information*, provides a lower bound to the entanglement of distillation E_D,

$$E_\mathrm{D}(\rho) \geq -S(A|B).\qquad(17.132)$$

At the same time, the quantum conditional entropy (the coherent information) appears in an entropic uncertainty relation for Alice's observables $\mathbb{X}$ and $\mathbb{Y}$ in the presence of a quantum memory represented by Bob. Derived in (Berta *et al.*, 2010), this relation takes the form

$$S(X|B) + S(Y|B) \geq S(A|B) - \log\big(\max_{x,y}|\langle \mathbb{X}^x|\mathbb{Y}^y\rangle|^2\big)\qquad(17.133)$$

where $S(X|B)$ is the quantum conditional entropy evaluated on the state

$$\rho_{XB} = \sum_x |\mathbb{X}^x\rangle\langle\mathbb{X}^x|\,(\langle\mathbb{X}^x|\otimes \mathbb{1}_B)\,\rho_{AB}\,(|\mathbb{X}^x\rangle\otimes\mathbb{1}_B),\qquad(17.134)$$

which represents the situation where Alice has performed a measurement of $\mathbb{X}$ (but the outcome x is not known), and analogously $S(Y|B)$ is defined with respect to a state ρ_{YB}. This entropic uncertainty relation can now be rearranged into a lower bound for the coherent information $-S(A|B)$, but the bound still contains the expressions $S(X|B)$ and $S(Y|B)$ which depend on the state ρ_{AB} in addition to Alice's observables $\mathbb{X}$ and $\mathbb{Y}$. However, by choosing (in principle arbitrary) observables $\mathbb{X}_B$ and $\mathbb{Y}_B$ for Bob (which do not have to have any particular relation to Alice's measurements), $S(X|B)$ and $S(Y|B)$ can be replaced by the corresponding classical conditional entropies $H(X_A|X_B)$ and $H(Y_A|Y_B)$, where $H(X_A|X_B) = H(X_A, X_B) - H(X_B)$ and $H(X_A, X_B) = -\sum_{x_A, x_B} p(x_A, x_B)\log\big(p(x_A, x_B)\big)$ is the Shannon entropy of the joint distribution $p(x_A, x_B)$, as will be analysed in detail in Section 19.3.2. Combined with the result from (Devetak and Winter, 2005) we thus have

$$E_{\mathrm{D}}(\rho) \geq -S(A|B) \geq -\log\big(\max_{x,y} |\langle \mathbb{X}^x | \mathbb{Y}^y \rangle|^2\big) - H(X_A|X_B) - H(Y_A|Y_B).$$

$$(17.135)$$

Entanglement can hence be detected and also quantified (via the distillable entanglement) by Alice and Bob both measuring pairs of complementary observables. In particular, when Alice's measurement bases are mutually unbiased, we have $|\langle \mathbb{X}^x | \mathbb{Y}^y \rangle|^2 = \frac{1}{d}$ for all x and y, and thus

$$E_{\mathrm{D}}(\rho) \geq \log(d) - H(X_A|X_B) - H(Y_A|Y_B).$$

$$(17.136)$$

Lower bounds on the entanglement of formation. Having seen how measurements in two complementary bases can be used to detect and quantify entanglement (via the distillable entanglement) by estimating conditional entropies, let us now take a closer look at an approach uses estimates of density-matrix elements to bound the entanglement of formation. Let us begin by stating a main technical ingredient for this method in terms of the following theorem.

Theorem 17.1 (Lower bound on EOF)

For any chosen pair of local bases $\{| m \rangle_A\}_m$ and $\{| n \rangle_B\}_n$, the entanglement of formation E_{F} of a general two-qudit state ρ can be bounded from below according to

$$E_{\mathrm{F}}(\rho) \geq -\log\big(1 - I^2(\rho)\big),$$

provided that the function $I(\rho)$, given by,

$$I(\rho) = \frac{1}{\sqrt{d(d-1)}} \sum_{m \neq n} \Big(|\langle mm | \rho | nn \rangle| - \sqrt{\langle mn | \rho | mn \rangle \langle nm | \rho | nm \rangle} \Big).$$

is non-negative, that is, if $I(\rho) \geq 0$.

The proof of Theorem 17.1 proceeds in two steps, first, we formulate and prove a lemma from (Huber and de Vicente, 2013; Huber *et al.*, 2013) that relates the function $I(\rho)$ for arbitrary mixed states to the convex-roof extension of the square root of the pure-state linear entropy of entanglement $\mathcal{E}_{\mathrm{L}}(| \psi \rangle) = S_{\mathrm{L}}(\rho_A)$, where $S_{\mathrm{L}}(\rho) = 1 - \mathrm{Tr}(\rho^2)$ is the linear entropy. In the next step, we will then examine this bound for pure states $| \psi \rangle$, and leverage known relations between the members of a family of entropy measures called *Rényi entropies* (which we will discuss in more detail in Section 19.5). Finally, we will extend the pure-state relation to mixed states via convexity/concavity arguments. The lemma that forms the starting point, first proven in (Huber and de Vicente, 2013; Huber *et al.*, 2013), can be formulated as follows:

Lemma 17.5 (Convex roof for square root of linear entropy)

For any chosen pair of local bases $\{|\, m\, \rangle_A\}_m$ and $\{|\, n\, \rangle_B\}_n$, the convex roof of the square root of the pure-state linear entropy of entanglement $\mathcal{E}_L(|\, \psi\, \rangle) = S_L(\rho_A)$ is bounded by

$$\inf_{\mathcal{D}(\rho)} \sum_i p_i \sqrt{\mathcal{E}_L(|\, \psi_i\, \rangle)} \geq I(\rho).$$

Proof of Lemma 17.5 First we note that the quantity $\inf_{\mathcal{D}(\rho)} \sum_i p_i \sqrt{\mathcal{E}_L(|\, \psi_i\, \rangle)}$, where $\mathcal{D}(\rho)$ is the set of all decomposition of ρ (i.e., the set of all sets $\{p_i, \rho^i_{\text{pure}}\}_i$ such that $\rho = \sum_i p_i\, |\, \psi_i\, \rangle\langle\, \psi_i\, |$ with $\sum_i p_i = 1$ and $0 \leq p_i \leq 1$), is similar but not identical to the tangle $\tau(\rho)$, which we recall from eqn (16.65) is the convex roof (up to a factor of 2) of the linear entropy itself. Nevertheless, when discussing the tangle, we encountered a useful formula for the linear entropy of entanglement for a pure state $|\, \psi\, \rangle = \sum_{m,n} c_{mn}\, |\, mn\, \rangle$ written with respect to an arbitrary pair of local bases $\{|\, m\, \rangle_A\}_m$ and $\{|\, n\, \rangle_B\}_n$: From eqn (16.63) we have

$$\mathcal{E}_L(|\, \psi\, \rangle) = S_L(\rho_A) = 1 - \text{Tr}(\rho_A^2) = \tfrac{1}{2} \sum_{\substack{m,n \\ m',n'}} |c_{mn}c_{m'n'} - c_{mn'}c_{m'n}|^2. \tag{17.137}$$

For this relation, which was derived in (Wu *et al.*, 2012), we can then obtain a simple bound by splitting the sum on the right-hand side into three contributions, first, the terms for which $m = n$ and $m' = n'$ but $m, n \neq m', n'$, second, those terms for which $m = n'$ and $m' = n$ but $m, n' \neq m', n$, and third, all remaining terms. The first two types of terms give exactly the same result, due to the absolute value they are identical up to a relabelling of the indices. The third term, meanwhile, is a sum of non-negative contributions, and we can drop it to obtain the inequality

$$\mathcal{E}_L(|\, \psi\, \rangle) \geq \sum_{m \neq n} |c_{mm}c_{nn} - c_{mn}c_{nm}|^2. \tag{17.138}$$

Then we apply the inequality $|a - b|^2 \geq \left(|a| - |b| \right)^2$ to further write

$$\mathcal{E}_L(|\, \psi\, \rangle) \geq \sum_{m \neq n} \left(|c_{mm}c_{nn}| - |c_{mn}c_{nm}| \right)^2. \tag{17.139}$$

Now we can make use of another inequality from (Wu *et al.*, 2012) that allows us to move the sum into the square, that is, let $\mathcal{I}$ be an index set with cardinality $|\mathcal{I}| = N$, then one has

$$|\mathcal{I}| \sum_{i \in \mathcal{I}} |a_i|^2 \geq \left| \sum_{i \in \mathcal{I}} a_i \right|^2. \tag{17.140}$$

This follows from the Cauchy–Schwarz inequality $|\vec{x}| \cdot |\vec{y}| \geq |\vec{x} \cdot \vec{y}|$ by considering two vectors $\vec{x}$ and $\vec{y}$ with N^2 components each, given by

$$\vec{x} = (a_1, \ldots, a_1, a_2, \ldots, a_2, \ldots, a_N, \ldots, a_N)^\top, \tag{17.141a}$$

$$\vec{y} = (a_1^*, a_2^*, \ldots, a_N^*, a_1^*, a_2^*, \ldots, a_N^*, \ldots, a_1^*, a_2^*, \ldots, a_N^*)^\top. \tag{17.141b}$$

In that case we have $|\vec{x}|^2 = |\vec{y}|^2 = N \sum_i |a_i|^2 = |\vec{x}| \cdot |\vec{y}|$, while $|\vec{x} \cdot \vec{y}| = |\sum_i a_i|^2$, such that the relation (17.140) follows. Here, we can set $a_i = |c_{mm}c_{nn}| - |c_{mn}c_{nm}|$ by interpreting $i = i(m,n)$ as a multi-index, and we thus arrive at

$$\mathcal{E}_{\rm L}(|\,\psi\,\rangle) \geq \frac{1}{N} \left| \sum_{m \neq n} |c_{mm}c_{nn}| - \sum_{m \neq n} |c_{mn}c_{nm}| \right|^2, \tag{17.142}$$

where the index set of all m and n such that $m \neq n$ has cardinality $N = d^2 - d = d(d-1)$. For any fixed decomposition of ρ into pure states $|\,\psi_i\,\rangle = \sum_{m,n} c_{mn}^{(i)} |\,mn\,\rangle$, we can thus write

$$\sum_i p_i \sqrt{\mathcal{E}_{\rm L}(|\,\psi_i\,\rangle)} \geq \sum_i p_i \frac{1}{\sqrt{d(d-1)}} \left(\sum_{m \neq n} |c_{mm}^{(i)} c_{nn}^{(i)}| - \sum_{m \neq n} |c_{mn}^{(i)} c_{nm}^{(i)}| \right) \tag{17.143}$$

$$= \frac{1}{\sqrt{d(d-1)}} \left(\sum_{m \neq n} \sum_i p_i |c_{mm}^{(i)} c_{nn}^{(i)}| - \sum_{m \neq n} \sum_i p_i \sum_{m \neq n} |c_{mn}^{(i)} c_{nm}^{(i)}| \right),$$

where we have dropped the modulus after taking the square root in the first step since the upper bound still trivially holds if the argument of the modulus is negative. Now let us examine the two terms in the last parenthesis separately. For the first term we can use $|c_{mm}^{(i)} c_{nn}^{(i)}| = |c_{mm}^{(i)} c_{nn}^{(i)*}|$, before we calculate

$$\sum_i p_i |c_{mm}^{(i)} c_{nn}^{(i)*}| \geq \left| \sum_i p_i c_{mm}^{(i)} c_{nn}^{(i)*} \right| = \left| \sum_i p_i \langle\,mm\,|\,\psi_i\,\rangle\langle\,\psi_i\,|\,nn\,\rangle \right| = |\langle\,mm\,|\,\rho\,|\,nn\,\rangle|. \tag{17.144}$$

For the second term we again use the Cauchy–Schwarz inequality, this time by treating $\vec{x}$ and $\vec{y}$ as vectors with components $x_i = \sqrt{p_i}|c_{mn}^{(i)}|$ and $y_i = \sqrt{p_i}|c_{nm}^{(i)}|$, such that

$$\sum_i p_i |c_{mn}^{(i)} c_{nm}^{(i)}| = |\vec{x} \cdot \vec{y}| \leq |\vec{x}| \cdot |\vec{y}| = \sqrt{\sum_i p_i |c_{mn}^{(i)}|^2} \sqrt{\sum_j p_j |c_{nm}^{(j)}|^2}$$

$$= \sqrt{\langle\,mn\,|\,\rho\,|\,mn\,\rangle \langle\,nm\,|\,\rho\,|\,nm\,\rangle}. \tag{17.145}$$

Taking into account the negative sign in front of the second term in the last line of (17.143), and including the infimum over all decompositions, we can finally write

$$\inf_{\mathcal{D}(\rho)} \sum_i p_i \sqrt{\mathcal{E}_{\rm L}(|\,\psi_i\,\rangle)} \geq \frac{1}{\sqrt{d(d-1)}} \sum_{m \neq n} \left(|\langle\,mm\,|\,\rho\,|\,nn\,\rangle| - \sqrt{\langle\,mn\,|\,\rho\,|\,mn\,\rangle\langle\,nm\,|\,\rho\,|\,nm\,\rangle} \right),$$

but we have dropped the infimum on the right-hand side since none of the terms there depend on the specific decomposition of ρ any more. We further recognize the right-hand side as the quantity $I(\rho)$ from Theorem 17.1, which concludes the proof of Lemma 17.5. $\square$

The lemma thus proven can now do some heavy lifting for us in the proof of Theorem 17.1, but before we continue, let us make some brief observations about this bound. First, we note that $I(\rho)$ need not be positive. For instance, for a mixed state that is diagonal in the basis $\{|\, mn\,\rangle\}_{m,n}$ the off-diagonals vanish, $|\langle\, mm\,|\,\rho\,|\, nn\,\rangle| = 0$, while the diagonal elements $\langle\, mn\,|\,\rho\,|\, mn\,\rangle$ and $\langle\, nm\,|\,\rho\,|\, nm\,\rangle$ are strictly non-negative. So we can quite trivially replace the right-hand side of the inequality in Lemma 17.5 by $\max\{0, I(\rho)\}$. Second, the bound is tight in the sense that there exist pure states, specifically the maximally entangled state $|\,\psi\,\rangle = |\,\phi_{00}\,\rangle = \frac{1}{\sqrt{d}}\sum_m |\, mm\,\rangle$, for which the inequality becomes an equality.

Proof of Theorem 17.1 To see how $I(\rho)$ is useful for bounding the entanglement of formation, let us go through the arguments laid out in (Erker *et al.*, 2017) and (Bavaresco *et al.*, 2018, Supplemental Material Sec. S.IV.). For pure states $|\,\psi\,\rangle$ we simply have

$$I(|\,\psi\,\rangle) \leq \sqrt{\mathcal{E}_{\mathrm{L}}(|\,\psi\,\rangle)} = \sqrt{1 - \mathrm{Tr}(\rho_A^2)}. \tag{17.146}$$

Whenever $I(|\,\psi\,\rangle) \geq 0$ we can therefore bound the subsystem purity $\mathrm{Tr}(\rho_A^2)$ by $\mathrm{Tr}(\rho_A^2) \leq 1 - I^2(|\,\psi\,\rangle)$, and, because $\log(x)$ is a monotonically increasing function, we can further write

$$-\log\big(\mathrm{Tr}(\rho_A^2)\big) \geq -\log\Big(1 - I^2(|\,\psi\,\rangle)\Big). \tag{17.147}$$

The introduction of the additional negative sign is convenient here because it lets us recognize the left-hand side as the so-called Rényi-2 (or collision) entropy. That is, we have the case $\alpha = 2$ of the family of Rényi entropies that we will discuss in more detail in Section 19.5, given by

$$S_\alpha(\rho) := \frac{1}{1-\alpha}\log\mathrm{Tr}(\rho^\alpha). \tag{17.148}$$

What is of interest for us here besides the identification of S_2 is that the limit $\alpha \to 1$ gives the familiar von Neumann entropy, $S_1(\rho) = \lim_{\alpha\to 1} S_\alpha(\rho) = S(\rho)$, as well as the hierarchy $S_\alpha(\rho) \geq S_\beta(\rho)$ that all Rényi entropies satisfy for $\alpha \leq \beta$ for all $\alpha, \beta \in \mathbb{N}$ and for all ρ (see Theorem 19.5 and the related discussion in Section 20.2). Here, this means we can rewrite (17.147) as

$$S(\rho_A) \geq -\log\Big(1 - I^2(|\,\psi\,\rangle)\Big). \tag{17.149}$$

For pure states this is already the desired bound since the von Neumann entropy of the subsystem just gives the entanglement of formation. For mixed states we now only have to note that $-\log(1 - x^2/2)$ is a convex function, and so is the function $I(\rho)$, because $|\langle\, mm\,|\,\rho\,|\, nn\,\rangle|$ is convex, while $\sqrt{\langle\, mn\,|\,\rho\,|\, mn\,\rangle\langle\, nm\,|\,\rho\,|\, nm\,\rangle}$ is concave. From Jensen's inequality (Jensen, 1906) we can then conclude that for all states ρ, for which $I(\rho) \geq 0$ we have

$$E_{\mathrm{F}}(\rho) \geq -\log\Big(1 - I^2(\rho)\Big). \tag{17.150}$$

$\square$

Again we find that the bound is tight, for the maximally entangled state $|\phi_{00}\rangle = \frac{1}{\sqrt{d}}\sum_m |mm\rangle$ both sides evaluate to $\log(d)$. More generally, the bound puts us in a position that lets us bound an entanglement measure from a few matrix elements of the density operator, $d(d-1)$ off-diagonal elements and $d(d-1)$ diagonal elements, whereas the full density operator is a $d^2 \times d^2$ matrix. Although this is computational simplification if these matrix elements are known, the question remains how one may easily determine the off-diagonal elements $\langle mm | \rho | nn \rangle$ in an experimental setup, and we shall look into this problem next.

Lower bounds on the entanglement of formation from MUBs. To realize how easily accessible the bound of Theorem 17.1 is in experiments, let us take a closer look at the fidelity $\mathcal{F}(\rho, |\phi_{00}\rangle)$ of the state ρ with the maximally entangled state $|\phi_{00}\rangle = \frac{1}{\sqrt{d}}\sum_m |mm\rangle$, which is given by

$$\mathcal{F}(\rho, |\phi_{00}\rangle) = \langle \phi_{00} | \rho | \phi_{00} \rangle = \frac{1}{d}\sum_{m,n} \langle mm | \rho | nn \rangle \qquad (17.151)$$

$$= \frac{1}{d}\sum_{m \neq n} \langle mm | \rho | nn \rangle + \frac{1}{d}\sum_m \langle mm | \rho | mm \rangle \,.$$

The first term in the last line is exactly the desired sum of off-diagonal elements appearing in $I(\rho)$, up to a modulus and a constant pre-factor, which means we can write

$$\sum_{m \neq n} |\langle mm | \rho | nn \rangle| \geq \sum_{m \neq n} \langle mm | \rho | nn \rangle = d\mathcal{F}(\rho, |\phi_{00}\rangle) - \sum_m \langle mm | \rho | mm \rangle \,.$$

$$(17.152)$$

This translates to a bound for $I(\rho)$ as (Bavaresco *et al.*, 2018)

$$I(\rho) \geq \frac{1}{\sqrt{d(d-1)}}\left(d\mathcal{F}(\rho, |\phi_{00}\rangle) - \sum_m \langle mm | \rho | mm \rangle - \sum_{m \neq n} \sqrt{\langle mn | \rho | mn \rangle \langle nm | \rho | nm \rangle}\right).$$

$$(17.153)$$

Except for the fidelity $\mathcal{F}(\rho, |\phi_{00}\rangle)$ all terms on the right-hand side can be estimated from the outcome statistics of measurements in the local bases $\{|m\rangle_A\}_m$ and $\{|n\rangle_B\}_n$.

The question we are thus facing is how to experimentally estimate the fidelity. The answer can be given in the form of a theorem from (Bavaresco *et al.*, 2018), but before we formulate this theorem, it makes sense to motivate it. The main observation can be made already for two qubits: There, one observes that the term $\sum_{m \neq n} \langle mm | \rho | nn \rangle$ also features when we add diagonal matrix elements with respect to the local Pauli-x bases of Alice and Bob, $|\pm\rangle = \frac{1}{\sqrt{2}}(|0\rangle \pm |1\rangle)$, i.e.,

$$2\sum_{\tilde{j}=\pm} \langle \tilde{j}\tilde{j} | \rho | \tilde{j}\tilde{j} \rangle = \sum_{m,n} \langle mn | \rho | mn \rangle + \sum_{m \neq n} \langle mm | \rho | nn \rangle + \sum_{m \neq n} \langle mn | \rho | nm \rangle \,.$$

On the right-hand side we have three terms, the first is just diagonal elements with respect to the original basis. The second is the term we were looking for, and the third can be bounded using the Cauchy–Schwarz inequality as in (17.145). For higher dimensions we can generalize this idea by going from the Pauli-z and Pauli-x bases to any two *mutually unbiased bases* (MUBs) by just including a complex conjugation (with respect to the computational basis) for Bob's vectors $|\tilde{j}^*\rangle$ relative to Alice's vectors $|\tilde{j}\rangle$. Moreover, since the terms we are looking for will appear for any MUB, we can include a sum (an average) over matrix elements in not just one but M MUBs, so that we can define the quantity

$$\Sigma^{(M)} = \frac{1}{M} \sum_{k=0}^{M-1} \sum_{j=0}^{d-1} \langle \tilde{j}_k \tilde{j}_k^* | \rho | \tilde{j}_k \tilde{j}_k^* \rangle , \qquad (17.154)$$

which, as we shall see, can improve the fidelity bounds we are about to construct. Finally, it will be useful later to notice already at this point that we can bound the fidelity not just for the maximally entangled state, but for *any* pure two-qudit state $|\phi\rangle$ with Schmidt decomposition $|\phi\rangle = \sum_m \lambda_m |mm\rangle$ by including the corresponding Schmidt coefficients λ_m and choosing vectors of the form

$$|\tilde{j}_k\rangle = \frac{1}{\sqrt{\sum_n \lambda_n}} \sum_{m=0}^{d-1} e^{\frac{2\pi i}{d}(jm + km^2)} \sqrt{\lambda_m} |m\rangle . \qquad (17.155)$$

With this, we can formulate the following theorem:

Theorem 17.2 (Lower bound on pure-state fidelity)

For any pure state with Schmidt decomposition $|\phi\rangle = \sum_m \lambda_m |mm\rangle$ the fidelity to an arbitrary mixed state ρ can be bounded by

$$\mathcal{F}(\rho, |\phi\rangle) \geq \tilde{\mathcal{F}}^{(M)}(\rho, |\phi\rangle) = \mathcal{F}_1 + \tilde{\mathcal{F}}_2^{(M)},$$

where the first term is

$$\mathcal{F}_1 = \sum_m \lambda_m^2 \langle mm | \rho | mm \rangle$$

and the second term is

$$\tilde{\mathcal{F}}_2^{(M)} := \frac{\left(\sum_m \lambda_m\right)^2}{d} \Sigma^{(M)} - \sum_{m,n=0}^{d-1} \lambda_m \lambda_n \langle mn | \rho | mn \rangle$$

$$- \sum_{\substack{m \neq m', m \neq n \\ n \neq n', n' \neq m'}} \tilde{\gamma}_{mm'nn'}^{(M)} \sqrt{\langle m'n' | \rho | m'n' \rangle \langle mn | \rho | mn \rangle}.$$

while the quantity $\tilde{\gamma}_{mm'nn'}^{(M)}$ is given by

$$\tilde{\gamma}_{mm'nn'}^{(M)} = \frac{1}{M} \left| \sum_{k=0}^{M-1} \omega^{k(m^2 - m'^2 - n^2 + n'^2)} \right| \times \begin{cases} 0 & \text{if } (m - m' - n + n')_{\text{mod}(d)} \neq 0 \\ \sqrt{\lambda_m \lambda_{m'} \lambda_n \lambda_{n'}} & \text{otherwise} \end{cases}.$$

Proof   Starting with the state $|\phi\rangle = \sum_m \lambda_m |mm\rangle$ in Schmidt decomposition, such that the λ_m are real, non-negative and ordered decreasingly, we can write the fidelity of this state with an arbitrary mixed state ρ as

$$\mathcal{F}(\rho, |\phi\rangle) = \langle\phi|\rho|\phi\rangle = \sum_{m,n} \lambda_m \lambda_n \langle mm|\rho|nn\rangle = \mathcal{F}_1 + \mathcal{F}_2, \qquad (17.156)$$

with $\mathcal{F}_1$ as defined in Theorem 17.2 and

$$\mathcal{F}_2(\rho, |\phi\rangle) = \sum_{m \neq n} \lambda_m \lambda_n \langle mm|\rho|nn\rangle. \qquad (17.157)$$

Our task will hence be to identify a suitable bound for this second term in the fidelity.

To do so, we examine the set of states $\{|\tilde{j}_k\rangle\}_{j=0,...,d-1}$ from eqn (17.155) for any fixed k. Following (Bavaresco *et al.*, 2018) we will refer to these sets as "*tilted bases*", where the word "tilted" alludes to the fact that vectors for different j are generally not orthogonal and, in fact, do not generally form a complete basis of the considered d-dimensional Hilbert space. In particular, when $|\phi\rangle$ is a product state, in which case only one of the λ_m is non-zero (equal to 1 in this case) while all other λ_m vanish, all vectors $|\tilde{j}_k\rangle$ coincide up to global phases. In contrast, when all λ_m are non-vanishing, the set $\{|\tilde{j}_k\rangle\}_{j=0,...,d-1}$ forms a (still generally non-orthogonal but) complete basis, and in the special case where $|\phi\rangle$ is maximally entangled, $|\phi\rangle = |\phi_{00}\rangle = \frac{1}{\sqrt{d}}\sum_m |mm\rangle$ such that $\lambda_m = \frac{1}{\sqrt{d}}$ for all m we recover the family of MUBs constructed according to Wootters and Fields (1989) in eqn (17.124). To proceed, we will first consider the special case of the sum $\Sigma^{(M)}$ from eqn (17.154) for $M = 1$, which means that only the tilted basis with $k = 0$ is included, and after going through the proof for this case, we will see how the more general case for general $M \geq 1$ follows. In this case we have

$$\Sigma^{(1)} = \sum_{j=0}^{d-1} \langle\tilde{j}_0\tilde{j}_0^*|\rho|\tilde{j}_0\tilde{j}_0^*\rangle = N \sum_{\substack{m,m'\\n,n'}} \sqrt{\lambda_m\lambda_n\lambda_{m'}\lambda_{n'}} \sum_{j=0}^{d-1} \omega^{j(m-m'-n+n')} \langle m'n'|\rho|mn\rangle, \qquad (17.158)$$

where we have defined $N^{-1} = \sum_l \lambda_l)^2$ and $\omega = \frac{2\pi i}{d}$ for a more compact notation. The sum over the labels m, m', n, and n' can then be separated into several qualitatively different contributions: In the case where $m = m'$ and $n = n'$ we have d equal contributions, and we hence have

$$\Sigma_1 := Nd \sum_{m,n} \lambda_m \lambda_n \langle mn|\rho|mn\rangle. \qquad (17.159)$$

When $m = m'$ but $n \neq n'$ or when $m \neq m'$ but $n = n'$ the relevant terms contain expressions of the form

$$\sum_{j=0}^{d-1} \omega^{j(n'-n)} = \delta_{nn'}, \qquad (17.160)$$

which vanish identically because $n \neq n'$. All the remaining contributions to the sum $\Sigma^{(1)}$ are such that $m \neq m'$ and $n \neq n'$, and these terms themselves fall into one of three categories. For the first we have $m = n$ and $m' = n'$, and one collect these contributions into the quantity

$$\Sigma_2 := N d \sum_{m \neq n} \lambda_m \lambda_n \langle mm | \rho | nn \rangle, \tag{17.161}$$

which is proportional to the sought-after term $\mathcal{F}_2$ in eqn (17.157). The second category collects terms with $m = n$ but $m' \neq n'$ or $m \neq n$ but $m' = n'$, which vanish because of eqn (17.160). This leaves us with the third category, terms of the form

$$\Sigma_3 := N \sum_{\substack{m \neq m', m \neq n \\ n \neq n', n' \neq m'}} \sqrt{\lambda_m \lambda_n \lambda_{m'} \lambda_{n'}} \sum_{j=0}^{d-1} \omega^{j(m-m'-n+n')} \langle m'n' | \rho | mn \rangle \tag{17.162}$$

$$= N \sum_{\substack{m \neq m', m \neq n \\ n \neq n', n' \neq m'}} \sqrt{\lambda_m \lambda_n \lambda_{m'} \lambda_{n'}} \ \mathrm{Re}\Big(\sum_{j=0}^{d-1} \omega^{j(m-m'-n+n')} \langle m'n' | \rho | mn \rangle \Big),$$

where we have introduced the real part in the last line because for each combination of m, m', n, and n' the sum over these indices contains another term for which the values of m and n are exchanged with those of m' and n', which yields the complex conjugate of the original term. Using the shorthand $c_{mnm'n'} := \sum_j \omega^{j(m-m'-n+n')}$, the real part can then be bounded via

$$\mathrm{Re}\Big(c_{mnm'n'} \langle m'n' | \rho | mn \rangle \Big) \leq |c_{mnm'n'} \langle m'n' | \rho | mn \rangle | = |c_{mnm'n'}| \cdot |\langle m'n' | \rho | mn \rangle|. \tag{17.163}$$

For the modulus of the matrix elements on the right-hand side we insert the spectral decomposition of the density operator, $\rho = \sum_i p_i | \psi_i \rangle \langle \psi_i |$, and denote the components of the eigenvectors with respect to the computational basis as $c_{mn}^{(i)} = \langle mn | \psi_i \rangle$ such that we can apply the Cauchy–Schwarz inequality $|\vec{x} \cdot \vec{y}| \leq |\vec{x}| \cdot |\vec{y}|$ for vectors $\vec{x}$ and $\vec{y}$ with components $x_i = \sqrt{p_i} c_{m'n'}^{(i)}$ and $y_i = \sqrt{p_i} c_{mn}^{(i)*}$, respectively, to obtain

$$|\langle m'n' | \rho | mn \rangle| = |\sum_i \sqrt{p_i} c_{m'n'}^{(i)} \sqrt{p_i} c_{mn}^{(i)*}|$$

$$\leq \sqrt{\sum_i p_i |c_{m'n'}^{(i)}|^2} \sqrt{\sum_i p_i |c_{mn}^{(i)}|^2} = \sqrt{\langle m'n' | \rho | m'n' \rangle \langle mn | \rho | mn \rangle}. \tag{17.164}$$

Meanwhile, we note that the factor $|c_{mnm'n'}|$ on the right-hand side of eqn (17.163) vanishes whenever $(m - m' - n + n')_{\mathrm{mod}(d)} \neq 0$, in which case one just has a complete sum of roots of unity, whereas the sum in $c_{mnm'n'}$ evaluates to d when $(m - m' - n + n')_{\mathrm{mod}(d)} = 0$. Extracting the factor d, we then collect $|c_{mnm'n'}|/d$ and the pre-factor

$\sqrt{\lambda_m \lambda_n \lambda_{m'} \lambda_{n'}}$ to obtain the quantity $\tilde{\gamma}^{(1)}_{mm'nn'}$ from Theorem 17.2. We thus arrive at the bound

$$\Sigma_3 \leq N d \sum_{\substack{m \neq m', \, m \neq n \\ n \neq n', \, n' \neq m' \\ m-m'-n+n'=0}} \tilde{\gamma}^{(1)}_{mm'nn'} \sqrt{\langle m'n' \,|\, \rho \,|\, m'n' \rangle \langle mn \,|\, \rho \,|\, mn \rangle}. \tag{17.165}$$

In turn, we can collect the terms Σ_1, Σ_2, and the bound on Σ_3 to arrive at the inequality

$$\Sigma^{(1)} = \Sigma_1 + \Sigma_2 + \Sigma_3 = \sum_{j=0}^{d-1} \langle \tilde{j}_0 \tilde{j}_0^* \,|\, \rho \,|\, \tilde{j}_0 \tilde{j}_0^* \rangle \leq N d \left(\sum_{m,n} \lambda_m \lambda_n \langle mn \,|\, \rho \,|\, mn \rangle \right.$$

$$\left. + \sum_{m \neq n} \lambda_m \lambda_n \langle mm \,|\, \rho \,|\, nn \rangle + \sum_{\substack{m \neq m', \, m \neq n \\ n \neq n', \, n' \neq m' \\ m-m'-n+n'=0}} \tilde{\gamma}^{(1)}_{mm'nn'} \sqrt{\langle m'n' \,|\, \rho \,|\, m'n' \rangle \langle mn \,|\, \rho \,|\, mn \rangle} \right). $$

$$\tag{17.166}$$

Finally, we can reformulate this as a bound for Σ_2, which appears in $\mathcal{F}_2$, which means that we can bound the latter by

$$\mathcal{F}_2 = \sum_{m \neq n} \lambda_m \lambda_n \langle mm \,|\, \rho \,|\, nn \rangle \geq \frac{(\sum_k \lambda_k)^2}{d} \sum_{j=0}^{d-1} \langle \tilde{j}_0 \tilde{j}_0^* \,|\, \rho \,|\, \tilde{j}_0 \tilde{j}_0^* \rangle \tag{17.167}$$

$$- \sum_{m,n} \lambda_m \lambda_n \langle mn \,|\, \rho \,|\, mn \rangle - \sum_{\substack{m \neq m', \, m \neq n \\ n \neq n', \, n' \neq m' \\ m-m'-n+n'=0}} \tilde{\gamma}^{(1)}_{mm'nn'} \sqrt{\langle m'n' \,|\, \rho \,|\, m'n' \rangle \langle mn \,|\, \rho \,|\, mn \rangle},$$

where we recognize the special case $M = 1$ of the quantity $\tilde{\mathcal{F}}_2^{(M)}$ from Theorem 17.2.

So far, we have only considered tilted-basis matrix elements with respect to the vectors $|\, \tilde{j}_k \rangle$ labelled by $k = 0$, but we now observe that the expressions for Σ_1 and Σ_2 in eqns (17.159) and (17.161) are unaffected by the choice of k, and so is the identity in eqn (17.160), which has eliminated some of the other terms. The only remaining term is Σ_3, where the additional phase factor that would arise from a value $k \neq 1$ would be absorbed into the definition of $c_{mnm'n'}$, and would disappear once (17.163) is used. In other words, the proof holds for any k. However, the bound obtained for any such k can be improved by replacing $\Sigma^{(1)}$ for a fixed k by an average (more generally speaking, one might even consider a weighted sum, but we will not do this here) over such matrix-element sums for more than one tilted basis, i.e., replacing $\Sigma^{(1)}$ by

$$\Sigma^{(M)} = \frac{1}{M} \sum_{k=0}^{M-1} \sum_{j=0}^{d-1} \langle \tilde{j}_k \tilde{j}_k^* \,|\, \rho \,|\, \tilde{j}_k \tilde{j}_k^* \rangle, \tag{17.168}$$

which means that the previous bound is modified by replacing $\tilde{\gamma}^{(1)}_{mm'nn'}$ by $\tilde{\gamma}^{(M)}_{mm'nn'}$ as defined in Theorem 17.2, which concludes the proof. $\qquad\square$

With the proof of Theorem 17.2, we see that measurements in two or more tilted bases are enough to give a lower bound on the fidelity to an arbitrary pure state, and this lower bound becomes tight when $M = d$ tilted bases are used, in which case the sum $\sum_{k=0}^{d-1} \omega^{k(m^2-m'^2-n^2+n'^2)}$ is a complete sum of the roots of unity, and hence vanishes, for any non-zero value of $(m^2 - m'^2 - n^2 + n'^2)$.

Although the problem of estimating the fidelity to arbitrary pure states $|\phi\rangle$ will be useful later on, see Section 17.4.3, the particular case where the target state is the maximally entangled state $|\phi\rangle = |\phi_{00}\rangle$ is of specific interest. In this case we have $\lambda_m = \frac{1}{\sqrt{d}}$ for all m, which simplifies many of the expressions, and which means that all tilted bases become unbiased with respect to the computational basis. When the dimension is a non-even prime, then all the corresponding tilted bases are MUBs. But in any case the resulting fidelity bound allows one to lower-bound the fidelity $\mathcal{F}(\rho, |\phi_{00}\rangle)$, which in turn appears in the lower bound for $I(\rho)$ in eqn (17.153),

$$I(\rho) \geq \frac{1}{\sqrt{d(d-1)}} \left(d\tilde{\mathcal{F}}^{(M)}(\rho, |\phi_{00}\rangle) - \sum_m \langle mm| \rho |mm\rangle - \sum_{m\neq n} \sqrt{\langle mn| \rho |mn\rangle\langle nm| \rho |nm\rangle} \right).$$

(17.169)

Consequently, measurements in as few as two (and up to $d + 1$) mutually unbiased bases can already provide a lower bound to the entanglement of formation by combining Theorems 17.1 and 17.2 (Bavaresco *et al.*, 2018). And, indeed, this entanglement-certification technique has been successfully used in experiments (Bavaresco *et al.*, 2018; Ecker *et al.*, 2019; Herrera Valencia *et al.*, 2020) with current record values for the certified entanglement of formation (without subtraction of accidental coincidences, and thus without assumptions on the state ρ) of, for instance, 4.0 ± 0.1 ebits (Herrera Valencia, Srivastav, Pivoluska, Huber, Friis, McCutcheon and Malik, 2020) for a local Hilbert-space dimension of $d = 31$, which corresponds to four maximally entangled two-qubit pairs and thus an effective dimension of $d_{\text{eff}} = 2^{4.0} = 16$, with $d_{\text{eff}}/d = 0.52$, or 2.2 ± 0.1 ebits in dimension $d = 5$, in which case $d_{\text{eff}}/d = 0.92$.

As we shall see next, there is also another prominent way of quantifying the effective dimension of a high-dimensionally entangled state, the so-called *Schmidt number*.

17.4.3 The Schmidt Number

While the entanglement of formation describes the resources (in terms of pairs of maximally entangled two-qubit states) to generate a given quantum state via LOCC, and one may derive a notion of the corresponding effective dimension of entanglement from this approach, this does not necessarily fully capture the idea of a high-dimensionally entangled states. In particular, we may think of a pure state of the form $|\phi\rangle = \sum_{m=0}^{d-1} \lambda_m |mm\rangle$ for which $\lambda_0 = 1 - \epsilon$ while $\lambda_1, \ldots, \lambda_{d-1} = \epsilon/(d-1)$. For a small but non-zero ϵ the state will have a small entanglement of formation (entropy of entanglement), but it is arguably an entangled state in dimension d that cannot be fully represented in lower effective dimension: The reduced state $\rho_A = \text{Tr}_B(|\phi\rangle\langle\phi|)$ is not maximally mixed but still has full (Schmidt) rank. Exactly this notion of entanglement dimensionality is captured by the generalization of the *Schmidt number* from pure to mixed states, which was first introduced by *Barbara Terhal* and *Paweł Horodecki* (2000):

Definition 17.2   *The Schmidt number $N_S(\rho)$ of a bipartite mixed state is defined as*

$$N_S(\rho) := \inf_{\mathcal{D}(\rho)} \left\{ \max_{|\psi_i\rangle} \left\{ \text{rank} \left[\text{Tr}_B \left(|\psi_i\rangle\langle\psi_i| \right) \right] \right\} \right\}.$$

Here, the infimum is taken over all pure-state decompositions of ρ, i.e., $\mathcal{D}(\rho)$ is the set of all sets $\{p_i, |\psi_i\rangle\}_i$ with $0 \le p_i \le 1$ and $\sum_i p_i = 1$ such that $\sum_i p_i |\psi_i\rangle\langle\psi_i| = \rho$, and for each given decomposition the maximum is taken over all pure states in the decomposition. The Schmidt number thus captures the minimal local dimension d that is needed to support the entanglement of the state, i.e., so that entanglement in a $d \times d$-dimensional Hilbert space is sufficient to describe the state. For this reason, the Schmidt number is sometimes referred to as *"entanglement dimensionality"* (see, e.g., Friis *et al.* 2019).

Similar to other convex-roof constructions, calculating the infimum in the definition of the Schmidt number exactly is practically impossible in many, if not most, cases. However, Theorem 17.2, which we have lengthily proven in the previous section, comes to the rescue, which we will see from combining the previous theorem with the following Schmidt-number witness (Fickler *et al.*, 2014, Supplementary Material, Sec. C):

Theorem 17.3 (Schmidt-number bound)

For any state ρ with Schmidt number $N_S = k \le d$, the fidelity $\mathcal{F}(\rho, |\phi\rangle)$ to any pure state $|\phi\rangle = \sum_{m=0}^{d-1} \lambda_m |mm\rangle$ with Schmidt number d is bounded by

$$\mathcal{F}(\rho, |\phi\rangle) \le B_k(|\phi\rangle) := \sum_{m=0}^{k-1} \lambda_m^2,$$

where the sum is over the squares of the k largest Schmidt coefficients of $|\phi\rangle$. Therefore, whenever $\mathcal{F}(\rho, |\phi\rangle) > B_k(|\phi\rangle)$ it follows that the Schmidt number of ρ is at least $k + 1$.

Proof   Following the proof in (Fickler *et al.*, 2014, Supplementary Material, Sec. C), we start by considering a pure state $\rho = |\psi(k)\rangle\langle\psi(k)|$ with a Schmidt number k for which we are interested in the overlap $|\langle\psi(k)|\phi\rangle|^2$ with another fixed pure state $|\phi\rangle = \sum_{m=0}^{d-1} \lambda_m |mm\rangle$ with Schmidt number d and Schmidt coefficients λ_m, ordered decreasingly with increasing m, $\lambda_0 \ge \lambda_1 \ge \ldots \lambda_{d-1}$. We can write the state $|\psi(k)\rangle$ with respect to the Schmidt bases of $|\phi\rangle$ as $|\psi(k)\rangle = \sum_{m,n} c_{mn} |m, n\rangle$. For fixed $|\phi\rangle$, we then maximize the overlap over all states $|\psi(k)\rangle$ with Schmidt number at most k:

$$B_k := \max_{|\psi(k)\rangle} \left| \langle \psi(k) | \phi \rangle \right|^2 = \max_{|\psi(k)\rangle} \left| \sum_{m,n,i} c^*_{mn} \lambda_i \langle m | i \rangle \langle n | i \rangle \right|^2 \tag{17.170}$$

$$= \max_{|\psi(k)\rangle} \left| \sum_{m,n,i} c^*_{mn} \lambda_i \langle i | m \rangle \langle n | i \rangle \right|^2 = \max_{|\psi(k)\rangle} \left| \sum_{m,n,i,j} c^*_{mn} \lambda_i \langle j | m \rangle \langle n | i \rangle \langle i | j \rangle \right|^2$$

$$= \max_{|\psi(k)\rangle} \left| \mathrm{Tr} \left[\left(\sum_{m,n} c^*_{mn} | m \rangle\langle n | \right) \left(\sum_i \lambda_i | i \rangle\langle i | \right) \right] \right|^2 = \max_{|\psi(k)\rangle} \left| \mathrm{Tr} \left[A^\dagger \sum_i \lambda_i | i \rangle\langle i | \right] \right|^2,$$

where we have defined $A := \sum_{m,n} c^*_{mn} | m \rangle\langle n |$. Since $|\phi\rangle$ has Schmidt rank k (or less), there exists a rank-k projector P_k with $P_k^2 = P_k$ such that $P_k A^\dagger = A^\dagger$, and we can therefore further write

$$B_k = \max_{|\psi(k)\rangle} \left| \mathrm{Tr} \left(P_k A^\dagger \sum_i \lambda_i | i \rangle\langle i | \right) \right|^2 = \max_{|\psi(k)\rangle} \left| \mathrm{Tr} \left(A^\dagger \sum_i \lambda_i | i \rangle\langle i | P_k \right) \right|^2. \tag{17.171}$$

In the last line we can identify the Hilbert–Schmidt inner product $(A, B)_{\mathrm{HS}} = \mathrm{Tr}(A^\dagger B)$ (see Definition 11.4) by defining an operator $B := \sum_i \lambda_i | i \rangle\langle i | P_k$. Since this inner product satisfies the Cauchy–Schwarz inequality $|(A, B)_{\mathrm{HS}}|^2 \leq (A, A)_{\mathrm{HS}} (B, B)_{\mathrm{HS}}$, we have

$$B_k = \max_{|\psi(k)\rangle} \left| (A, B)_{\mathrm{HS}} \right|^2 \leq \max_{|\psi(k)\rangle} \mathrm{Tr}(A^\dagger A) \, \mathrm{Tr} \left(P_k \sum_i \lambda_i | i \rangle\langle i | P_k \right)$$

$$= \max_{|\psi(k)\rangle} \mathrm{Tr} \left(P_k \sum_i \lambda_i^2 | i \rangle\langle i | P_k \right), \tag{17.172}$$

where we have used

$$\mathrm{Tr}(A^\dagger A) = \mathrm{Tr} \left(\sum_{\substack{m,n \\ m',n'}} c_{mn} c^*_{m'n'} | n \rangle\langle m | \, | m' \rangle\langle n' | \right) = \sum_{\substack{m,n \\ m',n'}} c_{mn} c^*_{m'n'} \langle n' | n \rangle \langle m | m' \rangle$$

$$= \sum_{m,n} c_{mn} c^*_{mn} = 1. \tag{17.173}$$

The maximum on the right-hand side of (17.172) is then obtained if P_k is chosen to be the projector onto the subspace corresponding to the k largest Schmidt coefficients, such that we arrive at

$$B_k \leq \sum_{i=0}^{k-1} \lambda_i^2. \tag{17.174}$$

For any mixed state ρ with Schmidt number of no more than k, the Definition of the mixed-state Schmidt number implies that it must be a convex combination of pure states with at most Schmidt number k, for which the bound in (17.174) holds individually, $\rho = \sum_i p_i | \psi_i(k) \rangle\langle \psi_i(k) |$ and thus

$$\mathcal{F}(\rho, |\phi\rangle) = \langle \phi | \rho | \phi \rangle = \sum_j p_j |\langle \psi_i(k) |) \langle \phi \|^2 \leq \sum_j p_j B_k = B_k. \tag{17.175}$$

$\square$

It now finally becomes obvious how Theorem 17.3 can be combined with Theorem 17.2. Whenever measurements in two or more tilted bases provide an estimate of the lower bound $\tilde{\mathcal{F}}^{(M)}(\rho, |\phi\rangle)$ on the fidelity from Theorem 17.2 that exceeds B_k from Theorem 17.3, the state ρ is certified to have a Schmidt number of (at least) $k + 1$ (Bavaresco *et al.*, 2018). In an experiment one has the choice of choosing a tilted basis, and thus the coefficients λ_m, which in turn also determine B_k, the bound to be surpassed. This is a non-trivial optimization problem. Nevertheless, even seemingly "simple" choices such as measurements in two MUBs (meaning, $\lambda_m = \frac{1}{\sqrt{d}}$ and $M = 1$) can lead to remarkable results: For photon pairs entangled in their spatial degrees of freedom, Schmidt numbers of $N_S = d$ have been certified for dimensions $d = 3, 5, 7, 11, 13$ and 17 (Herrera Valencia *et al.*, 2020). In higher dimensions, the certified Schmidt number has not reached the full subsystem dimension d, but the largest individual Schmidt number certified to date without assumptions on the state is $N_S = 55$ in dimension $d = 97$ (Herrera Valencia *et al.*, 2020).

18
Multipartite Entanglement

In this last section of Part II, we want to explore the structure of entanglement and separability for systems containing more than two particles. There the classification of entanglement is much richer than in the bipartite case presented so far. It turns out that several inequivalent classes of entanglement exist in general, and even the question of how to best meaningfully define in what sense entanglement is "multipartite" has several inequivalent answers. This of course complicates all of the aspects that we have previously discussed for the bipartite case: The distinction between separability, entanglement, and non-locality, the detection and quantification of entanglement, the transformation between different entangled states via LOCC, as well as the challenges and opportunities arising from considering higher-dimensional systems. Covering all of these aspects would in itself fill a book, so we shall restrict ourselves here to discussing a few of the fascinating features and facets of multipartite entanglement and refer the interested reader to the extensive reviews (Gühne and Tóth, 2009; Horodecki *et al.*, 2009; Friis *et al.*, 2019).

Here, we begin by discussing separability and entanglement for three qubits in Section 18.1, in particular, for the case of the well-known *Greenberger–Horne–Zeilinger* (GHZ) states and W states, and we explain their different features important for quantum information. The *GHZ theorem*—a generalization of Bell's theorem to three qubits—is presented in Section 18.2 in its spin version á la *David Mermin*. Then, in Section 18.3, we generalize our discussion to *n*-partite states, where we define the notion of *genuine multipartite entanglement* (GME), before we turn to specific examples of such states in Section 18.3.2, including so-called Dicke states and multi-qubit singlet states. Finally, in Section 18.4 we then present a procedure called *entangled entanglement*, which describes how to entangle the Bell states for two qubits with a third qubit in order to obtain all eight independent GHZ states, and we discuss a generalization to higher dimensions.

18.1 Tripartite Systems

18.1.1 Tripartite Pure States

The simplest case of a multipartite system is that of pure three-qubit states defined in a Hilbert space $\mathcal{H} = \mathcal{H}_A \otimes \mathcal{H}_B \otimes \mathcal{H}_C$ of three parties, Alice, Bob, and Charlie. When we attempt to classify separability and entanglement in such a system we observe that there are two qualitatively different types of separability:

(i) **Fully separable states**

The state is a pure product state of all three parties:

$$|\phi\rangle_{A|B|C} = |\alpha\rangle_A \otimes |\beta\rangle_B \otimes |\gamma\rangle_C \in \mathcal{H},$$

with $|\alpha\rangle_A \in \mathcal{H}_A$, $|\beta\rangle_B \in \mathcal{H}_B$, and $|\gamma\rangle_C \in \mathcal{H}_C$.

(ii) **Biseparable states**

Two of the three qubits form an entangled subsystem and are considered as one:

$$|\phi\rangle_{A|BC} = |\alpha\rangle_A \otimes |\kappa\rangle_{BC}, \; |\phi\rangle_{B|AC} = |\beta\rangle_B \otimes |\kappa\rangle_{AC}, \; |\phi\rangle_{C|AB} = |\gamma\rangle_C \otimes |\kappa\rangle_{AB},$$

where, e.g., $|\kappa\rangle_{AB} \in \mathcal{H}_A \otimes \mathcal{H}_B$ denotes a bipartite, potentially entangled state.

Although the two types of separability are different, this notion of biseparability does not yet contain anything that we have not previously already described in terms of bipartite entanglement, except that there is a third system which is separable from the other two. However, this does not cover all possible three-qubit pure states, and so we now turn to the definition of multipartite entanglement.

Definition 18.1 *Any pure tripartite state* $|\psi\rangle_{ABC}$ *that is neither fully separable nor biseparable is called **genuinely tripartite entangled**.*

Typical examples for genuinely tripartite entangled states are the GHZ state and the W state

$$|\mathrm{GHZ}\rangle = \frac{1}{\sqrt{2}}\left(|000\rangle + |111\rangle\right) \qquad \textbf{GHZ state,} \qquad (18.1\mathrm{a})$$

$$|\mathrm{W}\rangle = \frac{1}{\sqrt{3}}\left(|100\rangle + |010\rangle + |001\rangle\right) \qquad \textbf{W state.} \qquad (18.1\mathrm{b})$$

To produce a genuinely tripartite entangled state all three subsystems have to interact, while biseparable states can arise from bipartite interactions, and fully separable states do not require any interactions at all. Another important difference of the tripartite case with respect to the bipartite case was discovered by *Wolfgang Dür, Guifré Vidal*, and *Ignacio Cirac* (2000): Genuinely tripartite entangled three-qubit states are separated into two different equivalence classes, where an equivalence class is defined in the following way:

Definition 18.2 *Two tripartite pure states* $|\phi\rangle$ *and* $|\psi\rangle$ *in* $\mathcal{H}_A \otimes \mathcal{H}_B \otimes \mathcal{H}_C$ *are equivalent under SLOCC iff there exist invertible local operators* A, B, *and* C *in* $\mathcal{B}_{\mathrm{HS}}(\mathcal{H}_A)$, $\mathcal{B}_{\mathrm{HS}}(\mathcal{H}_B)$, *and* $\mathcal{B}_{\mathrm{HS}}(\mathcal{H}_C)$, *respectively, such that*

$$|\phi\rangle = A \otimes B \otimes C \, |\psi\rangle.$$

Since the operators are invertible, Definition 18.2 defines an equivalence relation. Further note that it would be sufficient to demand equivalence up to a constant pre-factor, i.e., that $|\phi\rangle$ is proportional to $A \otimes B \otimes C \, |\psi\rangle$, but if corresponding operators A, B, and C exist, then we can obviously multiply them by a suitable normalization factor. The generalization of Definition 18.2 to N-qubit states for $N > 3$ is straightforward and can be found in Dür *et al.* (2000). Now we are in a position to state the main result of Dür, Vidal, and Cirac (2000), and we shall summarize it in the following theorem.

Theorem 18.1 (Dür–Vidal–Cirac theorem)

There are two different equivalence classes of genuinely tripartite entangled states which cannot be transformed into each other via SLOCC:

  *(i) the **GHZ class** with $|\,\mathrm{GHZ}\,\rangle$ from eqn (18.1a) as a representative,*

  *(ii) and the **W class** with $|\,\mathrm{W}\,\rangle$ from eqn (18.1b) as a representative.*

For the proof, we refer to Dür *et al.* (2000). Hence, if a state $|\phi\rangle \in$ *GHZ class* and $|\psi\rangle \in$ *W class* then $|\phi\rangle$ can be converted into $|\,\mathrm{GHZ}\,\rangle$ from eqn (18.1a) and $|\psi\rangle$ into $|\,\mathrm{W}\,\rangle$ from eqn (18.1b) via SLOCC. However, $|\phi\rangle$ cannot be transformed into $|\psi\rangle$ and vice versa, via SLOCC.

However, the set of pure GHZ-class states is in some sense larger than the set of states forming the W class. This can be made more precise in terms of the following theorem proven by Acín *et al.* (2000), which is a generalization of Theorem 15.1.

Theorem 18.2 (Generalized Schmidt-decomposition theorem)

Any pure three-qubit state $|\phi\rangle \in \mathcal{H}_A \otimes \mathcal{H}_B \otimes \mathcal{H}_C$ can be brought to the form

$$|\phi\rangle = c_0 \,|\,000\,\rangle + c_1 e^{i\theta} \,|\,100\,\rangle + c_2 \,|\,101\,\rangle + c_3 \,|\,110\,\rangle + c_4 \,|\,111\,\rangle$$

by local unitary operations, where $c_i \geq 0$ are real, non-negative, and satisfy $\sum_i c_i^2 = 1$, and $\theta \in [0, \pi]$.

To characterize the properties of a general pure three-qubit state (which, as it turns out, is of GHZ type) we need six real parameters. For a state of the W class, however, fewer parameters are sufficient: Two parameters can be set to zero, $\theta = c_4 = 0$. The physical difference between the GHZ class and the W class can be understood from noting that the GHZ state of eqn (18.1a) is maximally entangled in any bipartition of one qubit versus the other two. As we will discuss for mixed states in Section 18.1.2, tracing out any single qubit leaves the other two in a separable state, and all single-qubit reduced states are maximally mixed. In this sense the GHZ state is thus a generalization of a Bell state for two qubits. The W state of eqn (18.1b), on

the other hand, is more robust against particle loss, the single-qubit reduced states
are not maximally mixed and the two-qubit reduced states are still entangled.

18.1.2   Tripartite Mixed States

For mixed states we approach the classification of separability and entanglement of
multipartite systems in full analogy to the bipartite case, but we take into account
the distinction between fully separable and biseparable states, and we note that, for
mixed states, there is now also a third type of separability. We can thus identify three
classes of separability for mixed tripartite states represented by density operators in
$\mathcal{B}_{\mathrm{HS}}(\mathcal{H}_A \otimes \mathcal{H}_B \otimes \mathcal{H}_C)$:

(i) **Fully separable states**

A density matrix $\rho_{A|B|C}$ is called *fully separable* if it can be expressed as a
convex sum of fully separable states

$$\rho_{A|B|C} = \sum_i p_i \, |\psi_i\rangle\langle\psi_i|_A \otimes |\phi_i\rangle\langle\phi_i|_B \otimes |\chi_i\rangle\langle\chi_i|_B$$

with $\sum_i p_i = 1$ and $0 \le p_i \le 1$.

(ii) **Partition-separable states**

A density matrix is called *partition-separable* if it is separable with respect to
(at least one) of the bipartitions $AB|C$, $A|BC$, or $AC|B$, i.e., if it can be
written as a convex sum of pure states that are separable with respect to a
fixed partition.

(iii) **Biseparable states**

The density matrix $\rho_{2-\mathrm{sep}}$ is called *biseparable* if it can be written as a convex
sum of biseparable pure states $|\psi_i^{2-\mathrm{sep}}\rangle$,

$$\rho_{2-\mathrm{sep}} = \sum_i p_i \, |\psi_i^{2-\mathrm{sep}}\rangle\langle\psi_i^{2-\mathrm{sep}}|,$$

where each $|\psi_i^{2-\mathrm{sep}}\rangle$ can be separable with respect to a different bipartition.

The notions of partition-separability and biseparability coincide for pure states, but
for mixed tripartite states the set of biseparable states is strictly larger. A partition-
separable state is separable with respect to $AB|C$, $A|BC$, or $AC|B$, and the intersec-
tion of these three sets of partition-separable states contains the set of fully separable
states, see Fig. 18.1. Yet, even separability with respect to all three bipartitions does
not imply full separability $A|B|C$. The intersection of $AB|C$, $A|BC$, or $AC|B$ contains
states that are not fully separable, these are the *multipartite bound-entangled states*
(Acín *et al.*, 2001; Horodecki *et al.*, 2009). The set of biseparable states represents

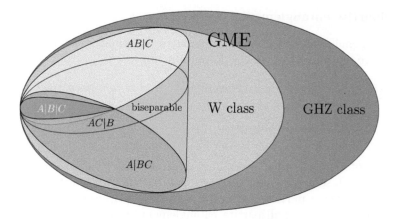

Fig. 18.1 Schematic structure of the set of three-qubit states. The state space of three qubits has a nested structure of convex sets. At the centre, containing the maximally mixed state, lies the set of fully separable (or three-separable) states $A|B|C$ (blue). The set of fully separable states is fully contained in each of the three partition-separable sets of states, i.e., states which are separable with respect to one of the bipartitions, $AB|C$ (yellow), $A|BC$ (green), and $AC|B$ (only shown schematically as a dashed line). The intersection of these three sets contains $A|B|C$, but is strictly larger, i.e., there are states that are separable with respect to all bipartitions, yet are not fully separable (light green). The convex hull of the sets of all partition-separable states forms the set of biseparable states. The region of biseparable but not partition-separable states is indicated in orange. All states that are not biseparable are genuinely multipartite (in this case, genuinely tripartite) entangled (GME), but the set of GME states is split into two classes, the W class and the GHZ class, with the union of the set of all W states and biseparable states forming a convex subset of the state space.

the convex hull of the partition-separable sets, and we note that biseparable states in general need not be separable with respect to any given bipartition, as we will also see for an example shortly. For any partition-separable state, separability of any of its subsystems with respect to the same bipartition is implied, e.g., a state in $AB|C$ remains separable with respect to $A|C$ if subsystem B is traced out. However, the converse is not true in general for biseparable states, as we shall see for the example in eqn (18.5): Separability of the reduced states with respect to, say, $A|C$ and $B|C$ does not imply separability with respect to $AB|C$, the global state of A, B, and C may still be entangled for this and/or other bipartitions, and may be biseparable or even genuinely multipartite entangled (which we will define shortly for mixed states).

The classes of separability thus introduced, the fully separable states, the three individual sets of partition-separable states (each containing the fully separable states as a subset), and their convex hull, the set of biseparable states are all convex and compact sets with the inclusion relation

$$A|B|C \ \subset \ A|BC \cap AB|C \cap AC|B \ \subset \ A|BC, AB|C, AC|B \ \subset \ 2-\text{sep}. \quad (18.2)$$

Genuine tripartite entanglement for mixed states:

We now define genuine multipartite (here, tripartite) entanglement for mixed states in full analogy to the pure-state case of Definition 18.1.

Definition 18.3 *Any mixed tripartite state ρ_{ABC} that is not biseparable (and thus also not partition-separable or fully separable) is called **genuinely tripartite entangled**.*

The description of multipartite entanglement as *genuine* serves the purpose of emphasizing that, although some biseparable states may contain multipartite entanglement in the sense of not being partition-separable, creating this biseparable entanglement does not require genuinely multipartite entanglement resources since it can be obtained via LOCC by mixing together states that only have bipartite entanglement for different bipartitions. For pure states this distinction was not required, but in this sense multipartite entanglement in pure states is always genuine. A feature that the mixed-state case shares with the pure-state case is that, again, we have two classes, the W states and the GHZ states.

W class: A genuinely tripartite entangled (and thus not biseparable) mixed state belongs to the W class if it is expressible as a convex combination of pure W-type states,

$$\rho_{\mathrm{W}} = \sum_i p_i \, |\phi_i^{\mathrm{W}} \rangle\langle \phi_i^{\mathrm{W}}| \qquad \text{with} \sum_i p_i = 1, \ \ 0 \le p_i \le 1 \,. \tag{18.3}$$

GHZ class: Conversely, a genuinely tripartite entangled mixed state belongs to the class of GHZ states if it *cannot* be expressed as a convex combination of pure W-type states.

Although neither of these classes on their own represents a convex set since $\mathrm{W} \cup \mathrm{GHZ} = \emptyset$, and $\mathrm{W} \cup 2-\mathrm{sep} = \emptyset$, the union of the W class with the biseparable states, $\overline{\mathrm{W}} := \mathrm{W} \cap 2-\mathrm{sep}$ is convex and compact (Acín *et al.*, 2001), and so is the union of the GHZ class with the former set, $\overline{\mathrm{GHZ}} := \mathrm{GHZ} \cap \overline{\mathrm{W}}$ such that we have

$$A|B|C \subset A|BC, \, AB|C, \, AC|B \subset 2-\mathrm{sep} \subset \overline{\mathrm{W}} \subset \overline{\mathrm{GHZ}}, \tag{18.4}$$

as illustrated in Fig. 18.1. However, deciding which class a state belongs to is not always easy to decide and it is worth bearing mind that the sketch of Fig. 18.1 is only schematic, since a two-dimensional representation can obviously only do limited justice to the structures and relations encountered in $2 \times 2 \times 2$ (or higher) dimensions.

Example I—mixed three-qubit state: As an example for a mixed three-qubit state we construct the following state

$$\rho_{ABC} = \tfrac{1}{3} \big(\, |\phi^+ \rangle\langle \phi^+|_{AB} \otimes |0\rangle\langle 0|_C + |\phi^+ \rangle\langle \phi^+|_{AC} \otimes |0\rangle\langle 0|_B$$
$$+ |\phi^+ \rangle\langle \phi^+|_{BC} \otimes |0\rangle\langle 0|_A \big) \,, \tag{18.5}$$

where $|\phi^+\rangle$ denotes the Bell state $|\phi^+\rangle = \frac{1}{\sqrt{2}}(|00\rangle + |11\rangle)$. The tripartite state in eqn (18.5) is a mixture of three partition-separable states and hence *biseparable* and thus *not* genuinely tripartite entangled. However, the state ρ_{ABC} is entangled with respect to each bipartition. To see this, we can just apply a partial transposition for the bipartition $AB|C$ by transposing subsystem C,

$$
\begin{aligned}
\rho_{ABC}^{T_C} = \tfrac{1}{6} \big(& 3|000\rangle\langle 000| + |011\rangle\langle 011| + |101\rangle\langle 101| + |110\rangle\langle 110| \\
& + |000\rangle\langle 110| + |110\rangle\langle 000| + |001\rangle\langle 100| + |100\rangle\langle 001| \\
& + |001\rangle\langle 010| + |010\rangle\langle 001| \big) ,
\end{aligned}
\tag{18.6}
$$

which can be seen to be block diagonal. The first six matrix elements form a 4×4 block with positive eigenvalues $\frac{1}{6}$ (twice degenerate) and $\frac{2 \pm \sqrt{2}}{6}$, while the latter four matrix elements form a 3×3 block with eigenvalues 0 and $\pm\frac{1}{3\sqrt{2}}$, i.e., we have a negative eigenvalue. According to the PPT criterion, the state ρ_{ABC} is thus entangled with respect to the bipartition $AB|C$, and for symmetry reasons the same is true for the bipartitions $A|BC$ and $AC|B$. Nevertheless, all three of the two-qubit reduced states are separable. Again, we can verify this by examining one of the three cases,

$$
\rho_{AB} = \mathrm{Tr}_C(\rho_{ABC}) = \tfrac{1}{3} \big(|\phi^+\rangle\langle\phi^+|_{AB} + \tfrac{1}{2}\mathbb{1}_A \otimes |0\rangle\langle 0|_B + |0\rangle\langle 0|_A \otimes \tfrac{1}{2}\mathbb{1}_B \big),
\tag{18.7}
$$

whose partial transpose $\rho_{AB}^{T_B}$ has eigenvalues $\frac{1}{2}$, $\frac{1}{3}$, $\frac{1}{6}$, and 0. Since the PPT criterion is necessary and sufficient for separability for two qubits, the reduced state ρ_{AB} is hence separable, and for symmetry reasons the reduced states ρ_{AC} and ρ_{BC} are as well.

Example II—W state: Next, let us consider the density matrix for the W state from eqn (18.1b), given by

$$
\begin{aligned}
\rho_{\mathrm{W}} = |W\rangle\langle W| = \tfrac{1}{3} \big(& |100\rangle\langle 100| + |010\rangle\langle 100| + |001\rangle\langle 100| \\
& + |100\rangle\langle 010| + |010\rangle\langle 010| + |001\rangle\langle 010| \\
& + |100\rangle\langle 001| + |010\rangle\langle 001| + |001\rangle\langle 001| \big),
\end{aligned}
\tag{18.8}
$$

and take the trace over one subspace, say over subsystem C. Then we obtain

$$
\begin{aligned}
\mathrm{Tr}_C(\rho_{\mathrm{W}}) &= \tfrac{1}{3} \big(|00\rangle\langle 00| + |10\rangle\langle 10| + |01\rangle\langle 10| + |10\rangle\langle 01| + |01\rangle\langle 01| \big) \\
&= \tfrac{1}{3} |00\rangle\langle 00| + \tfrac{2}{3} |\psi^+\rangle\langle\psi^+| ,
\end{aligned}
\tag{18.9}
$$

with the Bell state $|\psi^+\rangle = \frac{1}{\sqrt{2}}(|01\rangle + |10\rangle)$. First, we observe that the reduced state is mixed, it has eigenvalues $\frac{1}{3}$ and $\frac{2}{3}$, and since the W state is pure (and symmetric with respect to the exchange of the subsystems), this means $|W\rangle$ is entangled across every bipartition. Note that entanglement for all bipartitions is not in general sufficient to conclude that a mixed state is genuinely multipartite entangled, as our previous example has shown, but it is sufficient for a pure state such as the state $|W\rangle$ we consider here. What is more, in addition to the entanglement across all bipartitions, the W state also retains its entanglement when tracing out one subsystem (loosing, removing, or neglecting one particle). This can be seen by taking the partial transpose of the two-qubit reduced state in eqn (18.9), for which one finds the eigenvalues $\frac{1}{3}$ (twice degenerate) and $\frac{1}{6}(1 \pm \sqrt{5})$, where the negative eigenvalue again detects entanglement via the PPT criterion.

Example III—GHZ state: Next we illustrate, in contrast, the density matrix for the GHZ state of eqn (18.1a),

$$\rho_{\mathrm{GHZ}} = |\,\mathrm{GHZ}\,\rangle\langle\,\mathrm{GHZ}\,| = \tfrac{1}{2}\big(\,|\,000\,\rangle\langle\,000\,| + |\,111\,\rangle\langle\,000\,| \\ + |\,000\,\rangle\langle\,111\,| + |\,111\,\rangle\langle\,111\,|\,\big), \tag{18.10}$$

and we take the trace over one subsystem. For the reduced states we get mixed *separable* states, for instance, tracing out subsystem C we have

$$\mathrm{Tr}_C(\rho_{\mathrm{GHZ}}) = \tfrac{1}{2}\big(\,|\,00\,\rangle\langle\,00\,| + |\,11\,\rangle\langle\,11\,|\,\big), \tag{18.11}$$

which is a convex combination of product states and hence separable by definition. This means that none of the two-qubit reduced states retain any entanglement, the GHZ state is not robust against particle loss in this sense. Yet, since $|\,\mathrm{GHZ}\,\rangle$ is pure and its reduced states are all mixed, the state is entangled across every bipartition. Together with its purity, this implies genuine multipartite (here, tripartite) entanglement.

18.2   GHZ Theorem à la Mermin

The quantum correlations in the two-particle EPR–Bell experiments (see Sections 13.2 and 13.3) seem to imply some sort of "spooky action at a distance", at least in the context of the world view held by Einstein, a conclusion that the latter deemed unacceptable (recall Section 13.1).

In their celebrated 1989 article *Daniel Greenberger*, *Michael Horne*, and *Anton Zeilinger* (Greenberger *et al.*, 1989; Greenberger *et al.*, 1990) discovered an amazing extension of these two-particle experiments by involving three or more particles. Such GHZ experiments make the contradiction with the local-realist view even more apparent, and thus the "spookiness" within the local-realist mindset even more "spooky".

It was *David Mermin* (Mermin, 1990b; Mermin, 1990c), fascinated by this issue of non-local quantum correlations, who turned the original GHZ discussion into a gedanken experiment for a system consisting of three spin-$\frac{1}{2}$ particles. In his famous Physics Today article Mermin 1990c gave an extremely clear and most comprehensible presentation of the GHZ argument, which still makes it appealing to the physics community nowadays. Also John Bell, who had received a copy from Mermin, replied "I am full of admiration for your 3-spin trick" (mentioned in Mermin, 2016, p. 49). Here, we are going to present this *3-spin trick* invented by Mermin.

First we recall the basics of a hidden-variable theory (HVT, see Section 12.1) that assigns a set of values $v(A), v(B), v(C), \ldots$ to each individual system, which correspond to the outcomes for potential measurements of the observables $A, B, C, \ldots$. The essence is: If a functional relation $f(A, B, C, \ldots) = 0$ of commuting operators is satisfied then the same relation must also hold for the values $f\big(v(A), v(B), v(C), \ldots\big) = 0$ in the individual systems.

Keeping this in mind we consider a system of three spin-$\frac{1}{2}$ particles prepared in the GHZ-class state

$$|\,\text{GHZ}^-\,\rangle = \tfrac{1}{\sqrt{2}}\left(|\,000\,\rangle - |\,111\,\rangle\right), \qquad (18.12)$$

which is orthogonal to the state of eqn (18.1a). Altogether there are eight linearly independent pure GHZ-class states in this $(2 \times 2 \times 2)$-dimensional Hilbert space. The state in eqn (18.12) is an eigenstate of the following four commuting spin operators,

$$X_1 Y_2 Y_3, \quad Y_1 X_2 Y_3, \quad Y_1 Y_2 X_3, \quad X_1 X_2 X_3, \qquad (18.13)$$

with eigenvalues (± 1) since the square of each operator gives the identity. The individual spin operators $X_1, Y_2, Y_3 \ldots$ are as defined by eqn (12.45) in Section 12.5.3, for example, $X_1 = \sigma_x \otimes \mathbb{1} \otimes \mathbb{1}$, $Y_2 = \mathbb{1} \otimes \sigma_y \otimes \mathbb{1}$, $Y_3 = \mathbb{1} \otimes \mathbb{1} \otimes \sigma_y$, etc. The four operators in (18.13) constitute the horizontal line in Mermin's star, Fig. 12.5. Moreover, the operators satisfy the relation

$$X_1 X_2 X_3 = -Y_1 Y_2 X_3 \cdot Y_1 X_2 Y_3 \cdot X_1 Y_2 Y_3. \qquad (18.14)$$

The argumentation in favour of an HVT is then as follows. Let us carry out simultaneous spin measurements on the three particles, which are prepared in the GHZ state $|\,\text{GHZ}^-\,\rangle$ of eqn (18.12) and are assumed to be mutually well separated. Since the three particles are jointly in an eigenstate of the three spin operators $X_1 Y_2 Y_3$, $Y_1 X_2 Y_3$, and $Y_1 Y_2 X_3$ the product of the results of the three spin measurements is +1, regardless of which particle is selected for the spin measurement along the x direction. Therefore we can predict the result for the x component of the spin of one particle with certainty by measuring the y components of the other two, no matter how far apart they are. If the results of the y components are the same then the result for the x component must turn out to be +1. If the results of the y components differ, then the x component will yield the value -1. According to EPR's reality criterion and their demand of "no action at a distance" (recall Section 13.1) there *must* exist elements of reality corresponding to these spin-measurement results. Thus an HVT is required as in the case of Bell's inequality (Section 13.2), which assigns definite values to the individual systems.

Now we can construct a contradiction that finally leads to a *no-go theorem*, the GHZ theorem. Consider the GHZ state $|\,\text{GHZ}^-\,\rangle$ of eqn (18.12). It is an eigenstate of the spin operators $X_1 Y_2 Y_3$, $Y_1 X_2 Y_3$, and $Y_1 Y_2 X_3$, with eigenvalues $(+1, +1, +1)$ as well as of $X_1 X_2 X_3$ with eigenvalue -1. Therefore the products of the values must satisfy the relation

$$v(X_1)\,v(Y_2)\,v(Y_3) \;=\; +1$$
$$v(Y_1)\,v(X_2)\,v(Y_3) \;=\; +1$$
$$v(Y_1)\,v(Y_2)\,v(X_3) \;=\; +1 \qquad\qquad (18.15)$$
$$v(X_1)\,v(X_2)\,v(X_3) \;=\; -1 \;.$$

This, however, is a *contradiction*! Why? The product of the four operators on the left-hand sides of (18.15) is necessarily $+1$ since each observable occurs twice. Therefore we have to conclude:

Local realism has to be abandoned!

Furthermore, the assumption of elements of reality in an HVT is refuted in the GHZ case by a *single* measurement of the spin components, whereas in the case of Bell's argumentation, where only two particles are involved, it is required to collect statistics from *several* measurements. We summarize our analysis in the form of the following theorem:

Theorem 18.3 (GHZ theorem—of Bell-type)

For a three spin-$\frac{1}{2}$ system there exists a physical situation where all local realistic theories are inconsistent and incompatible with quantum mechanics.

Since the four spin operators (18.13) lie on the horizontal line of Mermin's pentagram, the star in Fig. 12.5, we can also argue in the spirit of *Kochen* and *Specker* to reformulate the GHZ argumentation, arriving at:

Theorem 18.4 (GHZ theorem—of Kochen–Specker-type)

For a three spin-$\frac{1}{2}$ system there is no way assigning definite values to each of the Mermin spin observables

$$X_1Y_2Y_3, \quad Y_1X_2Y_3, \quad Y_1Y_2X_3, \quad X_1X_2X_3$$

that is consistent with the functional identities.

The GHZ theorem in Bell form refutes the locality of realistic theories for mutually well-separated spin-$\frac{1}{2}$ particles by involving the properties of the quantum state. It implies a *non-locality* for the system if a realistic view is taken that assigns measurement outcomes independently of a measurement taking place.

In contrast, the GHZ theorem of the Kochen–Specker-type just focuses on certain operators and on the assignment of definite values for these operators. This leads to a contradiction giving rise to a *contextuality* for the system. This can be understood in the sense that GHZ in Bell form can be established regardless of which one of the eight simultaneous GHZ eigenstates we are considering.

Finally, let us note that the GHZ theorem has been confirmed by experiments which have been carried out by Zeilinger's group (Pan, 1999; Pan *et al.*, 2000). In those experiments all four Mermin spin observables (18.13) have been measured with locally well-separated photons, with the result that the data is in perfect agreement with quantum mechanics and in strong conflict with local realistic theories.

Résumé: The classification of entanglement for systems of more than two parties is much richer than in the bipartite case. For tripartite systems we have fully separable, partition-separable, and biseparable states. If a state is not in either of these categories, the state is called genuinely tripartite entangled. These states split into two different equivalence classes (with the GHZ and W states as representatives) which cannot be transformed into each other by SLOCC (Theorem 18.1), but the GHZ class is larger than the W class in the sense that states in the latter class can be obtained from convex combinations of the former, but not vice-versa. The schematic structure of the set of tripartite states is shown in Fig. 18.1. The GHZ theorem, an extension to Bell's theorem for two particles (Theorem 13.1), concludes that for tripartite quantum systems local realism has to be abandoned (Theorems 18.3 and 18.4).

18.3 Multipartite Systems

With what we have learned from tripartite systems, let us now generalize our discussion to more than three parties (and in principle arbitrary higher dimensions for each party). That is, we consider N-partite states defined in the Hilbert space $\mathcal{H} = \bigotimes_{i=1}^{N} \mathcal{H}_i = \mathcal{H}_1 \otimes \mathcal{H}_2 \otimes \ldots \otimes \mathcal{H}_N$. There, we approach the classification of separability and entanglement in the same way as before, first defining notions of separability for pure and mixed states before basing our definition of multipartite entanglement on these notions.

18.3.1 k-Separability and Genuine Multipartite Entanglement

Separability of pure states: As before, there are different types of separability that we distinguish for pure states, but instead of having only the distinction between full separability and biseparability, for N parties there is the possibility of separability with respect to partitions into k subsystems for any $k = 2, \ldots, N$. We hence define:

Definition 18.4 *An N-partite pure state $| \phi^{k\text{-sep}} \rangle \in \mathcal{H}$ is called k-**separable** for $1 < k \leq N$ if it can be written as a tensor product*

$$| \phi^{k\text{-sep}} \rangle = \bigotimes_{i=1}^{k} | \phi \rangle_{\mathcal{A}_i} \in \mathcal{H} \quad with \quad | \phi \rangle_{\mathcal{A}_i} \in \mathcal{H}_{\mathcal{A}_i},$$

with respect to a partition $\mathcal{A}_1 | \mathcal{A}_2 | \ldots | \mathcal{A}_k$ of the N parties into k subsets, i.e., such that $\mathcal{A}_i \subset \{1, 2, \ldots, N\}$ with $\bigotimes_{i=1}^{k} \mathcal{H}_{\mathcal{A}_i} = \mathcal{H}$.

This definition includes the case of *full separability* for $k = N$, whereas the case $k = 2$ is essentially the definition of biseparability (for pure states).

Separability of mixed states: As in the tripartite case we have discussed before, we again have different options of classifying separability in the Hilbert–Schmidt space $\mathcal{B}_{\mathrm{HS}}(\mathcal{H}) = \mathcal{B}_{\mathrm{HS}}(\bigotimes_{i=1}^{N} \mathcal{H}_i)$ of mixed states. As in the tripartite case, we have to distinguish between partition-separability and the extension of the pure-state separability definition to convex mixtures of states that are k-separable with respect to different partitions. For mixed states we can therefore now define:

(i) **Partition-separable states**

A density matrix of an N-partite system is called *partition-separable* if it is separable with respect to (at least one) partition $\mathcal{A}_1|\mathcal{A}_2|\ldots|\mathcal{A}_k$ for some k, i.e., if it can be written as a convex sum of pure states that are separable with respect to a fixed partition.

(ii) **k-separable states**

A density matrix $\rho_{k-\mathrm{sep}}$ is called *k-separable* if it can be written as a convex sum of (at least) k-separable pure states $|\psi_i^{k-\mathrm{sep}}\rangle$,

$$\rho_{k-\mathrm{sep}} = \sum_i p_i \, |\psi_i^{k-\mathrm{sep}}\rangle\langle\psi_i^{k-\mathrm{sep}}|,$$

where each $|\psi_i^{k-\mathrm{sep}}\rangle$ can be separable with respect to a different partition.

Once more the case $k = N$ refers to *full separability* and for $k = 2$ we have the definition of biseparability. In addition, we now have a hierarchy of k-separability in between full separability and biseparability (see Fig. 18.2): For every k with $1 < k < N$, the set of k-separable states contains the set of $(k+1)$-separable states as a convex subset by construction, while itself being a convex subset of the set of $(k-1)$-separable states,

$$N-\mathrm{sep} \subset (N-1)-\mathrm{sep} \subset \ldots \subset 3-\mathrm{sep} \subset 2-\mathrm{sep}. \tag{18.16}$$

While k-separability thus implies that there exists a decomposition such that each term is a tensor product of k subsystems, it does not mean each term has to be a tensor product of exactly k subsystems, as long it is a tensor product of *at least k* subsystems. And again the partitions into k subsystems can be different for each term.

Genuine N-partite entanglement:

Definition 18.5 *Any mixed N-partite state that is not k-separable for any $k \geq 2$ is called **genuinely N-partite entangled**.*

For instance, a biseparable ($2-\mathrm{sep}$) mixed state in $\mathcal{B}_{\mathrm{HS}}(\mathcal{H}_A \otimes \mathcal{H}_B \otimes \mathcal{H}_C \otimes \mathcal{H}_D)$ can be a convex mixture of terms that are partition-separable with respect to the bipartitions $A|BCD$, $B|ACD$, $C|ABD$, or $ABC|D$, but also terms separable with respect to bipartitions $AB|CD$, $AC|BD$, or $AD|BC$, and as long as there is always at least

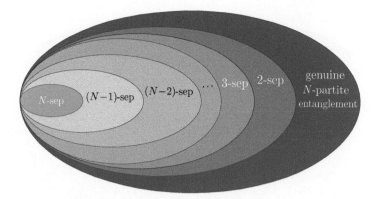

Fig. 18.2 Nested structure of k-separability. The state space of an N-partite system can be separated into a nested structure of convex sets of k-separable states for $k = 2, 3, \ldots, N$, where k'−sep $\subset k$−sep for all $k' > k$. The smallest k-separable set is the set of fully separable states ($k = N$), while the largest k-separable set is the set of biseparable states ($k = 2$). All states that are not contained in the biseparable states are genuinely N-partite entangled.

one term that is separable "only" with respect to these bipartitions, there can also be terms separable with respect to tripartitions $AB|C|D$, $AC|B|D$, $AD|B|C$, $BC|A|D$, $BD|A|C$, and $A|B|CD$, or even fully separable terms $A|B|C|D$, yet, the overall state would still be biseparable. Moreover, a k-separable state may still contain (genuine multipartite) entanglement between up to $N − k + 1$ parties.

Entanglement depth: As we have seen, the definition of genuine multipartite entanglement (Definition 18.5), captures some aspects of multipartite entanglement, but does not provide a complete characterization. For instance, let us compare a convex combination of partition-separable states $\rho_{AB|CD}$ and $\rho_{AC|BD}$ with a convex combination of partition-separable states $\rho_{A|BCD}$ and $\rho_{ABC|D}$. Both yield biseparable states, but the former strictly only requires bipartite entanglement, while the latter consists of two systems with tripartite entanglement. This difference can be captured by the notion of *entanglement depth*, introduced by *Anders Sørensen* and *Klaus Mølmer* in 2001, which is the contrapositive of *k-producibility* formalized by *Otfried Gühne*, *Géza Tóth*, and *Hans J. Briegel* (2005). That is, in analogy to genuine multipartite entanglement, entanglement depth is defined by what it is not:

Definition 18.6 *A density matrix $\rho_{k-\mathrm{prod}}$ is called k-**producible** if it can be written as a convex sum of product states for which each tensor factor is a state of at most k parties,*

$$\rho_{k-\mathrm{prod}} = \sum_i p_i \bigotimes_{m=1}^{M(i)} \rho_{\mathcal{A}_m^{(i)}} \,,$$

i.e., for all i and m, the set $\mathcal{A}_m^{(i)} \subset \{1, 2, \ldots, N\}$ with $|\mathcal{A}_m^{(i)}| \leq k$.

Definition 18.7 *A state that is not k-producible has an **entanglement depth** of at least $k + 1$.*

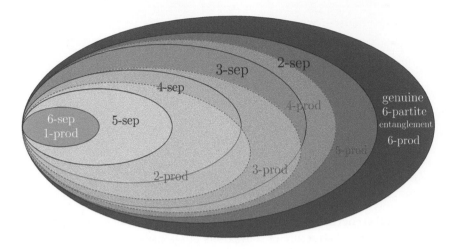

Fig. 18.3 Structure of k-producibility and k-separability. The nested structure of k-producibility and k-separability is illustrated for $N = 6$. There, the sets of fully separable (6-sep) and 1-producible states, as well as the genuinely 6-partite entangled and 6-producible sets coincide. The set 2-prod contains the set 5-sep but only partially overlaps with 4-sep and 3-sep, and 2-prod is entirely contained within 3-sep. Similarly, 3-prod contains 4-sep, and is a proper subset of 2-sep, while intersecting only partially with 3-sep. Meanwhile, 4-prod entirely contains 3-sep, but is itself contained in 2-sep, with the remainder of 2-sep being taken up by 5-prod.

The distinction between k-producibility and entanglement is thus rather different from the distinction between k-separability and genuine multipartite entanglement, but they coincide for the extremal cases. An N-separable (fully separable) state is 1-producible, while a genuinely N-partite entangled state is N-producible but not $(N-1)$-producible and hence has an entanglement depth of N. For values of k in between 1 and N, the relationship of k-producibility and k-separability is complicated. In general, the sets of k-producible states for different k form a nested convex structure, much in the same way as the sets of k-separable states illustrated in Fig. 18.2, but the sets of k-producible and k-separable sets intersect non-trivially, as shown in Fig. 18.3 for $N = 6$.

In general one can see from a simple counting argument that the set of k-separable but not $(k + 1)$-separable states of N parties contains (some of the) states that are k'-producible for values of k' from $\lceil N/k \rceil$ to $N-k+1$, where $\lceil x \rceil$ is the ceiling function giving the closest larger integer to x. Conversely, the set of k-producible but not $(k-1)$-producible states contains (some of the) states that are k'-separable for values of k' from $\lceil N/k \rceil$ to $N - k + 1$ as well. Despite this complicated relationship, the notion of entanglement depth is particularly useful for systems with many particles, specifically, when approaching the thermodynamic limit, since the resulting hierachy is somewhat independent from N, the total particle number. Entanglement depth is therefore often used for ensembles of cold atoms/atomic gases (see, e.g., Pezzè *et al.*, 2018). Here, however, we will focus on genuine multipartite entanglement in the remainder of this chapter, for which we now discuss some prominent examples.

18.3.2 Specific Multipartite Entangled States

Let us now discuss some families of multipartite entangled pure states that have interesting applications in quantum information.

Generalized GHZ states: The generalized GHZ state for N qubits is defined as

$$| \text{GHZ}_N \rangle = \frac{1}{\sqrt{2}} \left(|0\rangle^{\otimes N} + |1\rangle^{\otimes N} \right), \tag{18.17}$$

with the notation $|0\rangle^{\otimes N} = \bigotimes_{i=1}^{N} |0\rangle$. The generalized GHZ states have numerous applications in quantum computation, quantum cryptography, quantum teleportation, and quantum metrology (see Chapter 24). They represent maximally entangled multi-qubit states (in the sense that each qubit is maximally entangled with the rest) and lead to a maximal violation of local realism. For multi-qubit states local realism is restricted by inequalities that are multi-qubit extensions of the Bell inequalities for two qubits. Examples are the *Mermin inequalities* (Mermin, 1990a) or the *Svetlichny inequality* due to *George Svetlichny* (1987). For more details and reviews of progress in this field we refer to Gühne and Tóth (2009) and Brunner *et al.* (2014).

Dicke states: In (1954) *Robert H. Dicke* predicted the phenomenon of *superradiance*, that is, a sharp enhancement of the spontaneous radiation rate of an ensemble of N independent emitters—modelled as two-level atoms—compared to the radiation rate of a single emitter. This phenomenon can occur when the N emitters that are coupled by their own radiation field are spatially separated by distances much smaller than the wavelength of the emitted light. In this case, the emitters interact with the light in a collective and coherent way causing the group of atoms to emit light as a high-intensity pulse with a rate proportional to N^2 instead of the proportionality N for single emitters. In general, the Dicke states are simultaneous eigenstates of the *collective angular-momentum operators*

$$J_z = \frac{1}{2} \sum_i \sigma_z^{(i)} \quad \text{and} \quad \vec{J}^2. \tag{18.18}$$

In quantum optics, the *symmetric Dicke states* are of particular importance. These are eigenstates of the operator $\vec{J}^2$ with the maximal eigenvalue

$$\vec{J}^2 \longrightarrow \frac{N}{2} \left(\frac{N}{2} + 1 \right), \tag{18.19}$$

and are characterized uniquely by the eigenvalue of J_z,

$$J_z \longrightarrow m = \frac{N}{2}, \ldots, -\frac{N}{2}. \tag{18.20}$$

It is common to label the Dicke states by the integer

$$k = \frac{N}{2} - m, \tag{18.21}$$

which we call the "excitation number" for reasons that will become apparent shortly. With this we are ready to define the symmetric Dicke states:

Definition 18.8   *The **symmetric k-excitation Dicke states** of N qubits are defined as*

$$| \mathrm{D}_N^{(k)} \rangle := \binom{N}{k}^{-1/2} \sum_j \pi_j \left\{ | 1 \rangle^{\otimes k} \otimes | 0 \rangle^{\otimes (N-k)} \right\} ,$$

where the sum is over all possible permutations π_j of qubits and $\binom{N}{k} = \frac{n!}{k!(n-k)!}$ is the binomial coefficient.

Thus the symmetric Dicke states of N qubits represent equal superpositions of different states with k excitations each. They are genuinely multipartite entangled.

Examples of Dicke states:

$$| \mathrm{D}_3^{(1)} \rangle = \tfrac{1}{\sqrt{3}} \left(| 100 \rangle + | 010 \rangle + | 001 \rangle \right) \equiv | \mathrm{W} \rangle , \tag{18.22a}$$

$$| \mathrm{D}_4^{(2)} \rangle = \tfrac{1}{\sqrt{6}} \left(| 1100 \rangle + | 1010 \rangle + | 0101 \rangle + | 1001 \rangle + | 0110 \rangle + | 0011 \rangle \right), \tag{18.22b}$$

$$| \mathrm{D}_5^{(2)} \rangle = \tfrac{1}{\sqrt{10}} \left(| 11000 \rangle + | 10100 \rangle + | 10010 \rangle + | 10001 \rangle + | 01100 \rangle \right.$$
$$\left. + | 01010 \rangle + | 01001 \rangle + | 00110 \rangle + | 00101 \rangle + | 00011 \rangle \right). \tag{18.22c}$$

For the state $| \mathrm{D}_N^{(N/2)} \rangle$ with $k = N/2$ excitations the phenomenon of superradiance is strongest. The states $| \mathrm{D}_4^{(2)} \rangle$ and $| \mathrm{D}_N^{(N/2)} \rangle$ are relatively robust against decoherence, i.e., against particle loss (Gühne *et al.*, 2008; Gühne and Tóth, 2009). Experiments aiming at the verification of Dicke states and W have been carried out by several groups (Roos *et al.*, 2004; Eibl *et al.*, 2004; Kiesel *et al.*, 2007; Wieczorek *et al.*, 2008).

Generalized W states: The generalized W state for N qubits is defined as

$$| \mathrm{W}_N \rangle = \tfrac{1}{\sqrt{N}} \sum_j \pi_j \left\{ | 1 \rangle \otimes | 0 \rangle^{\otimes (N-1)} \right\} \equiv | \mathrm{D}_N^{(1)} \rangle . \tag{18.23}$$

For example, the W state for four qubits is of the form

$$| \mathrm{W}_4 \rangle = \tfrac{1}{\sqrt{4}} \left(| 1000 \rangle + | 0100 \rangle + | 0010 \rangle + | 0001 \rangle \right) \equiv | \mathrm{D}_4^{(1)} \rangle . \tag{18.24}$$

The entanglement of the generalized W states is very robust against particle loss, i.e., the state $| \mathrm{W}_N \rangle$ remains entangled even if $N-2$ parties are traced out. This is evident when we consider the reduced density matrix

$$\rho_{\mathrm{W}}^{AB} = \mathrm{Tr}_{[AB]CD\dots} | \mathrm{W}_N \rangle\langle \mathrm{W}_N | = \tfrac{2}{N} | \psi^+ \rangle\langle \psi^+ | + \tfrac{N-2}{N} | 00 \rangle\langle 00 | , \tag{18.25}$$

where we have traced out all systems except the first and second one as indicated by the square brackets $[AB]$. Because of the symmetry with respect to the exchange of the parties all reduced density matrices turn out to be identical and they remain entangled. The amount of entanglement can be calculated, for instance, using the *concurrence* (recall Section 16.2.2). The result is $C[\rho_{\mathrm{W}}^{AB}] = \tfrac{2}{N}$.

Multi-qubit singlet states: Finally, a further family of states which is important in quantum information is the family of *multi-qubit singlet states*. A singlet state is defined by the invariance under a simultaneous unitary operation U on all qubits

$$U^{\otimes N} | \psi \rangle = e^{i\vartheta} | \psi \rangle . \tag{18.26}$$

These states exist only for even N. For two qubits the only singlet state is the Bell state $| \psi^- \rangle = \frac{1}{\sqrt{2}} (| 01 \rangle - | 10 \rangle)$. For four qubits there exist two singlet states

$$| \psi_1 \rangle = | \psi^- \rangle \otimes | \psi^- \rangle , \tag{18.27}$$

$$| \psi_2 \rangle = \tfrac{1}{\sqrt{3}} \left(| 0011 \rangle + | 1100 \rangle - | \psi^+ \rangle \otimes | \psi^+ \rangle \right) , \tag{18.28}$$

where $| \psi^+ \rangle = \frac{1}{\sqrt{2}} (| 01 \rangle + | 10 \rangle)$ is the familiar symmetric Bell state. In general there are $d_N = \frac{N!}{(N/2)!(N/2+1)!}$ linearly independent N-qubit singlet states (Cabello, 2003; Cabello, 2007).

18.3.3   Detecting Genuine Multipartite Entanglement

One of the lessons we have learned from studying the detection and quantification of entanglement throughout the previous chapters is that these are generically very difficult tasks that have no one-fits-all solutions, or indeed, solutions in general. Having seen that, in addition to the challenges faced for bipartite entanglement, multipartite systems also present much more complicated structures of entanglement with respect to different partitions. One can therefore expect a difficult task ahead.

k-partitions: As a first step towards the characterization of general multipartite states, one can attempt to discern whether or not a given state is genuinely N-partite entangled, or more generally, whether or not it is k-separable for some k. However, already here one encounters the first bottleneck. In order to check for k-separability even for pure states, it is generally necessary to check separability with respect to a large number of k-partitions. For N parties the number of possible k-partitions is given by the *Sterling number* of the second kind, denoted by $\left\{ \begin{smallmatrix} N \\ k \end{smallmatrix} \right\}$ and expressible as

$$\left\{ \begin{matrix} N \\ k \end{matrix} \right\} = \tfrac{1}{k!} \sum_{j=0}^{k} (-1)^{k-j} j^N \binom{k}{j} . \tag{18.29}$$

The sum over all Sterling numbers (of the second kind) for fixed N is called the Nth Bell number, named after mathematician *Eric Temple Bell*, the total number of ways to partition a system of N parties. For any fixed k the Sterling number scales as $k^N / (k!)$ as $N \to \infty$, so the number of partitions to check when testing for k-separability quickly increases with the number of parties.

GME witnesses: As we have seen from our previous examples, detecting entanglement across various partitions is in itself not sufficient to detect GME since biseparable states can be mixtures of states that are separable with respect to different partitions. Nevertheless, we have seen that the set of biseparable states is convex, and one can therefore exploit this convex structure for the construction of *GME witnesses* in analogy to the witnesses for bipartite entanglement that we have discussed in Section 16.3. For any bipartition $\mathcal{A}|\overline{\mathcal{A}}$, one can start with some entangled state $|\psi\rangle$ whose maximal overlap with any pure state $|\phi\rangle$ that is separable with respect to $\mathcal{A}|\overline{\mathcal{A}}$ is $\alpha := \max_{|\phi\rangle \in \mathcal{A}|\overline{\mathcal{A}}} |\langle\phi|\psi\rangle|^2$ and write down an entanglement witness

$$\mathbb{W} = \alpha\mathbb{1} - |\psi\rangle\langle\psi| \,. \tag{18.30}$$

By construction the quantity $\mathrm{Tr}(\rho\,\mathbb{W})$ is non-negative for all separable states, but $|\psi\rangle$ and other entangled states that are sufficiently close to $|\psi\rangle$ are detected as entangled since $\alpha < 1$. We have already seen this construction being put to use in eqn (15.144). In full analogy we can construct witnesses for GME by defining (see, e.g., Acín *et al.*, 2001; Bourennane *et al.*, 2004)

$$\mathbb{W}_{\mathrm{GME}} = \alpha\mathbb{1} - |\psi\rangle\langle\psi| \quad \text{with} \quad \alpha := \max_{|\phi\rangle \in 2\text{–sep}} |\langle\phi|\psi\rangle|^2 \,. \tag{18.31}$$

As a prominent example of such a witness, take, for instance, the generalized N-qubit GHZ state from eqn (24.78), $|\psi\rangle = |\,\mathrm{GHZ}_N\,\rangle$. For any bipartition $\mathcal{A}|\overline{\mathcal{A}}$ of the N qubits into k and $N-k$ qubits we can write the state as a Bell state $|\psi\rangle = |\,\mathrm{GHZ}_N\,\rangle = \frac{1}{\sqrt{2}}\left(|\tilde{0}\rangle_{\mathcal{A}}|\tilde{0}\rangle_{\overline{\mathcal{A}}} + |\tilde{1}\rangle_{\mathcal{A}}|\tilde{1}\rangle_{\overline{\mathcal{A}}}\right)$ with $|\tilde{n}\rangle_{\mathcal{A}} = |n\rangle^{\otimes k}$ and $|\tilde{n}\rangle_{\mathcal{A}} = |n\rangle^{\otimes(N-k)}$ for $n = 0, 1$. From our discussion of the maximal overlap of any separable state with a Bell state, in particular from (15.144) we then see that the maximal overlap of the generalized GHZ state with any biseparable state is $\frac{1}{2}$. For any N-qubit state ρ we thus have the GME criterion:

$$\mathcal{F}(\rho, |\,\mathrm{GHZ}_N\,\rangle) > \tfrac{1}{2} \quad \Longrightarrow \quad \rho \text{ genuinely } N\text{-partite entangled}\,. \tag{18.32}$$

Because of the simplicity of this criterion it has been employed by a number of experimental groups for the verification of GME, and some of the records for the number of qubits found in a single genuinely N-partite entangled state are: $N = 12$ (Gong *et al.*, 2019), $N = 18$ (Song *et al.*, 2019), and $N = 27$ (Mooney *et al.*, 2021) superconducting qubits, $N = 14$ (Monz *et al.*, 2011), $N = 24$ (Pogorelov *et al.*, 2021), and $N = 32$ (Moses *et al.*, 2023) trapped-ion qubits, $N = 20$ Rydberg atoms (Omran *et al.*, 2019), and $N = 18$ qubits encoded in photonic degrees of freedom (Wang *et al.*, 2018). But apart from the example of the GHZ state, the construction of GME witnesses according to eqn (18.31) works quite generically for detecting specific (families of) states, and these and other more sophisticated methods have been employed for stabilizer states (see, e.g., Tóth and Gühne, 2005; Audenaert and Plenio, 2005; Smith and Leung, 2006), graph states (see, for instance, Hein *et al.*, 2005), Dicke states (Bergmann and Gühne, 2013), and states with more general symmetries (Tóth and Gühne, 2009).

GME witnesses from average Bell fidelities: The idea of using the fidelity to Bell states as an entanglement witness can also be generalized to the multipartite setting in a different way: Instead of considering the fidelity to a GME state, one can examine the average fidelities of the two-qubit reduced states of an N-qubit state with the respective closest Bell states. That is, for each of the two-qubit reduced states $\rho_{ij} = \mathrm{Tr}_{\neg i, j}(\rho)$

for $i \neq j \in \{1, 2, \ldots, N\}$ of an N-qubit state ρ, we can consider the fidelity to the Bell states from eqn (15.142). Taking into account the $b_N = \binom{N}{2} = \frac{1}{2} \frac{N!}{(N-2)!}$ possibilities of choosing a pair of qubits labelled by i and j, we can then write the average fidelity with the corresponding Bell states as (Friis *et al.*, 2018)

$$\frac{1}{b_N} \sum_{\substack{i,j=1 \\ i<j}}^{N} \mathcal{F}(\rho_{ij}, \rho_{\text{Bell},ij}) \leq \overline{\mathcal{F}}_{\text{Bell}}^{(N)} := \frac{1}{4b_N} \left(b_N + \sum_{\substack{i,j=1 \\ i<j}}^{N} \sum_{O=X,Y,Z} |\langle O_i O_j \rangle| \right), \qquad (18.33)$$

where X_i, Y_i, and Z_i denote the Pauli operators for the ith qubit. As shown in (Friis *et al.*, 2018, Appendix A.II.3), this *symmetric average Bell fidelity* can be used to detect GME. For every biseparable N-qubit state ρ the quantity $\overline{\mathcal{F}}_{\text{Bell}}^{(N)}$ is bounded from above by $\overline{\mathcal{F}}_{\text{Bell}}^{(N)2\text{-sep}}$:

$$\overline{\mathcal{F}}_{\text{Bell}}^{(N)} \leq \overline{\mathcal{F}}_{\text{Bell}}^{(N)2\text{-sep}} := \begin{cases} \frac{1}{12}(3 + \sqrt{15}) & \text{for } N = 3 \\ \frac{1}{4}(1 + \sqrt{3}) - \frac{1}{2N}(\sqrt{3} - 1) & \text{for } N \geq 4 \end{cases}. \qquad (18.34)$$

Conversely, whenever $\overline{\mathcal{F}}_{\text{Bell}}^{(N)}$ exceeds $\overline{\mathcal{F}}_{\text{Bell}}^{(N)2\text{-sep}}$ the state must be genuinely N-partite entangled. Conceptually, this means that, although entanglement between any two qubits alone is not sufficient to conclude that the state is GME, biseparability is not compatible with an arbitrary amount of (quantum) correlations between the qubits in the system. An advantage of such a criterion is that it is experimentally easily accessible. In (Friis *et al.*, 2018) it was used to detect simultaneous genuine tripartite entanglement between all groups of three neighbouring qubits realized in a linear chain of twenty trapped-ion qubits. An interesting feature of this technique is that it does not require *all* pairwise fidelities to be above the two-qubit separability threshold of $\frac{1}{2}$, as long as some of the pairs are sufficiently strongly entangled, some other pairs may only be classically correlated. As a downside, it is not known if these particular GME witnesses can detect any GME states beyond four qubits: For three and four qubits there provably exist states that can be detected in this way, e.g., the four-qubit two-excitation Dicke state $|\text{D}_4^{(2)}\rangle$ from eqn (18.22b) gives a value of $\overline{\mathcal{F}}_{\text{Bell}}^{(4)} = \frac{2}{3}$ compared to a bound $\frac{1}{4}(1 + \sqrt{3}) - \frac{1}{8}(\sqrt{3} - 1) \approx 0.591506$, but it is not clear if states with $\overline{\mathcal{F}}_{\text{Bell}}^{(N)} > \overline{\mathcal{F}}_{\text{Bell}}^{(N)2\text{-sep}}$ exist for $N \geq 5$.

Lifted bipartite witnesses and PPT mixers: In general, many of the techniques we have encountered for the detection of bipartite entanglement, e.g., separability criteria based on positive maps (see Section 15.4), are not directly applicable in the multipartite case. In spite of this, some of these criteria can be "lifted" from the bipartite to the multipartite setting (Huber and Sengupta, 2014; Clivaz *et al.*, 2017). For instance, it was shown in (Lancien *et al.*, 2015) that for any GME state ρ_{GME} one may find positive but not completely positive maps $\Lambda_{\mathcal{A}}$ for all bipartitions $\mathcal{A}|\overline{\mathcal{A}}$, together with states $|\psi_{\mathcal{A}}\rangle$ such that $\langle \psi_{\mathcal{A}}| (\Lambda_{\mathcal{A}} \otimes \mathbb{1}_{\overline{\mathcal{A}}})[\rho_{\text{GME}}] < 0$, from which one can construct GME witnesses of the form

$$\mathbb{W}_{\text{GME}} = \sum_{\mathcal{A}} (\Lambda_{\mathcal{A}}^* \otimes \mathbb{1}_{\overline{\mathcal{A}}}) \big[|\psi_{\mathcal{A}}\rangle\langle\psi_{\mathcal{A}}| \big] + M(\{\Lambda_{\mathcal{A}}, |\psi_{\mathcal{A}}\rangle\}_{\mathcal{A}}), \qquad (18.35)$$

where $\Lambda_{\mathcal{A}}^{*}$ denotes the dual of the map $\Lambda_{\mathcal{A}}$, and $M\big(\{\Lambda_{\mathcal{A}},|\psi_{\mathcal{A}}\rangle\}_{\mathcal{A}}\big)$ is a positive operator that depends on the set of positive maps $\Lambda_{\mathcal{A}}$ and associated states $|\psi_{\mathcal{A}}\rangle$.

Alternatively, one may observe that biseparable states being convex combinations of states that are separable with respect to some bipartitions implies the decomposability into PPT states, which allows for the construction of remarkably effective tools for numerical approaches to GME detection based on semidefinite programmes known as "PPT mixers" (Jungnitsch *et al.*, 2011*b*; Jungnitsch *et al.*, 2011*a*).

Non-linear GME witnesses: As a more general approach towards the detection of GME one can of course also consider non-linear functions of density-matrix elements. Spearheading efforts into this area were made by *Otfried Gühne* and *Michael Seevinck* (2010) as well as by *Marcus Huber, Florian Mintert, Andreas Gabriel*, and *Beatrix Hiesmayr* (2010). Here, let us focus on the latter technique, where one makes use of a seemingly mundane property of k-separable pure states: permutational symmetry in the exchange of corresponding subsystems of two copies of the state. To see how this works, let us start with two copies, $|\psi\rangle_{A_1 A_2}$ and $|\psi\rangle_{B_1 B_2}$, of a bi-separable pure state $|\psi\rangle_{1,2} = |\phi\rangle_1 |\chi\rangle_2$, where A_i and B_i $(i = 1,2)$ label the ith subsystem of two otherwise identical copies A and B, respectively. The subsystems A_2 and B_2 of the first and second copy may be freely exchanged,

$$\pi_{(A_1|A_2)_2} |\psi\rangle_{A_1 A_2} \otimes |\psi\rangle_{B_1 B_2} = |\psi\rangle_{A_1 B_2} \otimes |\psi\rangle_{B_1 A_2} , \tag{18.36}$$

where we have defined the permutation operator $\pi_{(A_1|A_2)_2}$ that exchanges the second subsystem of the two copies with respect to the partition $A_1|A_2$. This statement trivially extends to multipartite entanglement: A pure state that is N-separable with respect to an N-partition $\mathbb{P}(N) = A_1|A_2|\cdots|A_N$ is invariant under permutations $\pi_{\mathbb{P}(N)_i}$, i.e., exchanges of the ith subsystems A_i and B_i of the two copies. Using this statement the following theorem was formulated in (Gabriel *et al.*, 2010) as a generalization of results in (Huber *et al.*, 2010):

Theorem 18.5 (GME witnesses (Gabriel, Hiesmayr, Huber))

Every N-partite, k-separable state ρ satisfies

$$\left(\langle \Phi | \rho^{\otimes 2} \prod_i \pi_{\mathbb{P}(k)_i} | \Phi \rangle \right)^{1/2} \leq \sum_{\mathbb{P}(k)} \left(\prod_{i=1}^{k} \langle \Phi | \pi_{\mathbb{P}(k)_i}^{\dagger} \, \rho^{\otimes 2} \, \pi_{\mathbb{P}(k)_i} | \Phi \rangle \right)^{1/2k} ,$$

for every fully separable $(2N)$-partite pure state $|\Phi\rangle$.

Proof Let us quickly sketch the proof for this inequality. If ρ is a pure, N-partite, k-separable state, then it must be k-separable with respect to one of the k-partitions $\mathbb{P}(k)$ in the sum on the right-hand side. The corresponding term of the sum over all $\mathbb{P}(k)$ then cancels with the left-hand side since the fully separable state $|\Phi\rangle$ can trivially be written as a tensor product of two N-partite states, i.e., $|\Phi\rangle = |\Phi_A\rangle |\Phi_B\rangle$, and $\rho = |\psi\rangle\langle\psi|$ is pure. The remaining terms on the right-hand side are products of

$(2k)$th roots of diagonal entries ρ_{mm} of the density operator ρ, and therefore strictly positive. Thus the inequality is trivially satisfied. To extend the proof to mixed states one simply notes that the left-hand side is the modulus—a convex function—of a density-matrix element ρ_{mn}, while the $(2k)$th roots on the right-hand side are concave functions of the matrix elements, i.e.,

$$\left| \sum_i p_i\, \rho_{mn}^i \right| \leq \sum_i p_i\, |\rho_{mn}^i|\,, \tag{18.37a}$$

$$\left(\sum_i p_i\, \rho_{mm}^i \right)^{\frac{1}{2k}} \geq \sum_i p_i\, \left(\rho_{mm}^i \right)^{\frac{1}{2k}}\,, \tag{18.37b}$$

which concludes the proof. □

18.3.4 Characterizing Genuine Multipartite Entanglement

With some powerful techniques for the detection of GME now in our repertoire, let us take a closer look at the characterization and potential quantification of GME, which, as we will see, is a difficult task indeed.

Entanglement dimensionality for multipartite systems: Leaving aside the generalized "Schmidt-like" decomposition for three qubits (Theorem 18.2), there is a lack of a generalization of the Schmidt-decomposition theorem from bipartite systems to multipartite systems, i.e., not every N-partite pure state can be brought to the form $|\Psi\rangle = \sum_n \lambda_n\, |n\rangle^{\otimes N}$, which means that the spectra and ranks of reduced states of different subsystems do not match even for a fixed k-partition for $k > 2$. However, there are two popular ways of generalizing at least the concept of the Schmidt rank to multipartite systems: The *tensor rank* and the *Schmidt rank vector*.

The *tensor rank* r_{T}, a well-studied quantity in mathematics (see, e.g., Kruskal 1977) is defined as the minimum number of coefficients λ_n such that the state can be written as

$$|\Psi\rangle = \sum_{n=1}^{r_{\mathrm{T}}} \lambda_n \bigotimes_{\alpha=1}^{N} |\phi_n^{(\alpha)}\rangle\,. \tag{18.38}$$

Similar to the bipartite case, a value of $r_{\mathrm{T}} = 1$ implies full separability. Unfortunately, even for pure states calculating the tensor rank is NP-hard (Håstad, 1990) and its value is known only for some examples that exhibit specific symmetries (Chen *et al.*, 2010), and it is not even additive under tensor products (Christandl, Jensen and Zuiddam, 2018). But at least it can be bounded from below by simply taking the maximum over the Schmidt ranks of all bipartitions $\mathcal{A}|\overline{\mathcal{A}}$, such that $r_{\mathrm{T}} \geq \max_{\mathcal{A}} \mathrm{rank}\big(\mathrm{Tr}_{\overline{\mathcal{A}}}[\,|\Psi\rangle\langle\Psi|\,]\big)$.

Another, perhaps more straightforward option at generalizing the Schmidt rank is to collect the ranks of all reduced states for all bipartitions into a vector—the *Schmidt-rank vector* $\vec{r}_{\mathrm{S}}$—as suggested by *Marcus Huber* and *Julio de Vicente* (2013). The number of components of this Schmidt-rank vector grows exponentially with the number of parties, specifically as $2^{N-1}-1$, which is the number of possible bipartitions.

Keeping this difficulty in mind, the condition for full separability is that the rank of each reduced state is 1, which means that the state is fully separable if and only if $|\vec{r}_{\rm S}|^2 = 2^{N-1} - 1$. In general, the components, reflecting the respective Schmidt ranks for different bipartitions can be different, but not arbitrary, and they are restricted by non-trivial inequalities (see, e.g., Cadney *et al.*, 2014).

GME classes and maximal entanglement: The previously discussed generalizations of the Schmidt rank do provide a glimpse of the complicated structure of multipartite entanglement, but of course we also already know that even for the simple case of three qubits there are two different equivalence classes of GME, the GHZ class and the W class, which cannot be converted into each other via SLOCC (see Section 18.1). The general inequivalence and, more problematically, impossibility of even one-way conversion between multipartite entangled states under SLOCC severely complicates attempts at the operational quantification of GME via resource costs based on asymptotic formation or distillation costs from the outset. Put simply, for multipartite states there is no unique maximally entangled state from which all other states can be reached or which could be distilled from a sufficiently broad set of states. For tripartite systems, this could be alleviated by expanding to a set of two such states, one representative from the GHZ class and one from the W class, respectively.

More generally, one aims to identify *maximally entangled sets* (MES), i.e., minimal sets of N-partite states from which all other N-partite states can be obtained via LOCC while no state in the set can be reached via LOCC from outside the set, an idea first systematically investigated by *Julio de Vicente, Cornelia Spee*, and *Barbara Kraus* (2013) and later extended to the more realistic scenarios where only finitely many rounds of classical communication are permitted (Spee *et al.*, 2017; de Vicente *et al.*, 2017). However, already for four qubits the MES contains infinitely many states, but only a subset of measure zero is relevant since all other states of the MES correspond to isolated "islands" that are not convertible via LOCC. This, unfortunately, is a generic feature and it turns out that non-trivial LOCC transformations between generic, multipartite-entangled pure states are almost never possible (Gour *et al.*, 2017; Sauerwein *et al.*, 2018), resulting in an MES of infinite cardinality. In the sense just described there is no unique notion of "maximal entanglement" in the multipartite case from which all other forms of multipartite entanglement can be obtained via LOCC. It is hence operationally meaningless to quantify GME in terms of a function that assigns a non-negative number to each state that specifies the "strength" of the state's GME in units of the entanglement contained in any fixed, maximally entangled state.

Nevertheless, we can in principle talk about multipartite entanglement as being maximal if a pure state contains the maximum amount of bipartite entanglement across every possible bipartition, in which case it is called *absolutely maximally entangled* (AME). Such states are in principle useful for tasks such as quantum error correction (Scott, 2004) and quantum secret sharing (Helwig *et al.*, 2012), and, indeed, there exists a local dimension d for every N such that there is an AME state in $\mathcal{H} = \bigotimes_{n=1}^{N} \mathcal{H}_d$ with $\dim(\mathcal{H}_d) = d$. At the same time, AME states do not exist for every pair of N and d, in particular, for N qubits ($d = 2$), AME states exist only for $N = 2, 3, 5$, and 6 parties (Huber *et al.*, 2017).

Quantification of GME: For the quantification of multipartite entanglement it is hence necessary to think in terms of different categories than in the bipartite case, for instance, by dropping the idea that an entanglement measure quantifies the usefulness of a given state in absolute terms but rather by using relative usefulness. One such approach is to quantify multipartite entanglement via the *LOCC target volume*, the volume of states in the Hilbert–Schmidt space that can be reached from a given state via LOCC, and the *LOCC source volume*, the volume from which a given state can be reached via LOCC (Schwaiger *et al.*, 2015). Although such a volume-based approach does not give precise information about the usefulness of the states contained in the respective volumes, it gives some general insight into the utility of resource states for state transformations. In particular, states whose LOCC source volume is zero while their target volume is non-zero are extremal resource states. Other attempts to circumvent the restrictions of LOCC on the classification and quantification of resource states include considering sets of operations beyond LOCC which cannot create entanglement (Contreras-Tejada *et al.*, 2019), or considering probabilistic rather than deterministic operations, i.e., SLOCC (see, e.g., Verstraete *et al.*, 2002*b*; Verstraete *et al.*, 2003; Ritz *et al.*, 2019).

Alternatively one can of course forgo the operational aspects of potential entanglement measures and define multipartite equivalents of entanglement monotones such as the concurrence (see Section 16.2.2) or the negativity (see Section 16.2.4). For the generalization of the concurrence, one starts from the relation between the pure-state tangle and the pure-state concurrence in eqn (16.60), $\tau(|\psi\rangle) = C^2(|\psi\rangle)$. Since the pure-state tangle can further be related to the mixedness of the subsystem via eqn (16.64), $\tau(|\psi\rangle) = 2\big(1 - \mathrm{Tr}(\rho_A^2)\big)$, one can straightforwardly define the pure-state concurrence for arbitrary dimensions by taking the square root to obtain a monotone that vanishes for all bipartite separable pure states. To extend this to biseparability, we simply take the minimum over all bipartitions $\mathcal{A}|\overline{\mathcal{A}}$, defining the pure-state *GME concurrence* (Ma *et al.*, 2011)

$$C_{\mathrm{GME}}(|\psi\rangle) = \min_{\mathcal{A}|\overline{\mathcal{A}}} \sqrt{2\big(1 - \mathrm{Tr}(\rho_{\mathcal{A}}^2)\big)}\,. \tag{18.39}$$

The generalization to mixed states can then proceed in the familiar way via a convex-roof construction, i.e.,

$$C_{\mathrm{GME}}(|\rho\rangle) = \inf_{\mathcal{D}(\rho)} \sum_i p_i\, C(|\psi_i\rangle)\,. \tag{18.40}$$

Although the convex-roof represents (as usual) a significant obstacle for computing the GME concurrence, lower bounds can be obtained from the witnesses in Theorem 18.5, and for some special classes of states, e.g., N-partite states in so-called X-form, i.e., whose density matrices have non-vanishing entries only on the main diagonal and main anti-diagonal with respect to a chosen product basis, one may even obtain exact results for $C_{\mathrm{GME}}(|\rho\rangle)$ (Hashemi Rafsanjani *et al.*, 2012).

In complete analogy to the GME concurrence, one may define the *genuine multi-partite negativity* (Jungnitsch *et al.*, 2011*b*; Hofmann *et al.*, 2014) as a convex-roof construction over the bipartite negativity, after minimizing the latter over all bipartitions. The resulting quantity can be calculated numerically using semidefinite programming, which generally turns out to be a fruitful technique for evaluating entanglement measures (or monotones) based on convex-roof constructions (Tóth *et al.*, 2015). As we have discussed, the utility of these multipartite entanglement monotones is not so much to quantify GME in an operationally meaningful way, but to provide methods for the detection of GME that can be evaluated numerically.

Activation of genuine multipartite entanglement: Finally, we come to a feature of the distinction between biseparability and GME that entirely departs from the picture we are used to from bipartite entanglement. In the latter case it is clear that two parties will not be able to obtain or generate entanglement by sharing only separable states. In particular, if $\rho_{\mathcal{A}_1|\mathcal{A}_2}$ is separable with respect to the bipartition $\mathcal{A}_1|\mathcal{A}_2$, then obviously $\rho_{\mathcal{A}_1|\mathcal{A}_2} \otimes \rho_{\mathcal{B}_1|\mathcal{B}_2}$ is separable with respect to the bipartition $\mathcal{A}_1\mathcal{B}_1|\mathcal{A}_2\mathcal{B}_2$, including the case where $\rho_{\mathcal{A}_1|\mathcal{A}_2}$ and $\rho_{\mathcal{B}_1|\mathcal{B}_2}$ are two copies of the same state. However, for the multipartite setting the fact that biseparable states can be convex combinations of states that are separable with respect to different bipartitions means that it is not so clear a priori if an analogue statement holds for the separation between biseparability and GME. And indeed, as was first realized in Huber and Plesch (2011), there are biseparable mixed states for which two copies are GME. This was more systematically investigated in Yamasaki *et al.* (2022). There the authors considered a family of N-qubit *"isotropic"*[1] *GHZ states*, given by

$$\rho(p) = p \, |\,\mathrm{GHZ}_N \rangle\langle \mathrm{GHZ}_N| + (1-p)\tfrac{1}{2^N}\mathbb{1}_{2^N} \,. \tag{18.41}$$

Denoting the N subsystems of the first copy of such a state by $\mathcal{A}_1, \mathcal{A}_2, \ldots, \mathcal{A}_N$, the subsystems of the second copy by $\mathcal{B}_1, \mathcal{B}_2, \ldots, \mathcal{B}_N$, and so on, one can then examine biseparability and GME with respect to an N-partition $\mathcal{A}_1\mathcal{B}_1 \ldots |\mathcal{A}_2\mathcal{B}_2 \ldots | \ldots |\mathcal{A}_N\mathcal{B}_N \ldots$ for any number of copies based on the GME concurrence. That is, the states $\rho(p)$ are of the X-form required to employ the exact formula for $C_{\mathrm{GME}}(|\,\rho\,\rangle)$ from Hashemi Rafsanjani *et al.* (2012), and one may use this result for k copies by realizing that there is an SLOCC protocol (Lami and Huber, 2016) that maps a state $\rho \otimes \sigma \in \mathcal{B}_{\mathrm{HS}}(\mathcal{H} \otimes \mathcal{H})$ to the Hadamard product (or Schur product) $\rho \circ \sigma \in \mathcal{B}_{\mathrm{HS}}(\mathcal{H})$, i.e., the component-wise product of the two matrices, which preserves the X-form. As a result one obtains thresholds $p_{\mathrm{GME}}^{(k)}$ such that $p_{\mathrm{GME}}^{(k)} < p_{\mathrm{GME}}^{(k')}$ for $k' > k$, and for which $\rho(p)^{\otimes k}$ is GME for $p > p_{\mathrm{GME}}^{(k)}$, see Fig. 18.4. In particular, this means that states $\rho(p)$ with $p > p_{\mathrm{GME}}^{(1)}$ are GME and states with $p \in [p_{\mathrm{GME}}^{(2)}, p_{\mathrm{GME}}^{(1)}]$ are biseparable but two copies $\rho(p)^{\otimes 2}$ are GME. For states with $p < p_{\mathrm{GME}}^{(k)}$ at most k copies are required for $\rho(p)^{\otimes k}$ to be GME.

Moreover, it was demonstrated in Yamasaki *et al.* (2022) that two copies are generically not enough to activate GME, i.e., there are biseparable states for which two

---

[1]The $U \otimes U^*$ symmetry of the two-qudit isotropic states from Section 17.2.2 does not generalize to the multipartite setting, but viewing the GHZ state as a generalization of the Bell state $|\,\phi^+\rangle$ motivates this terminology.

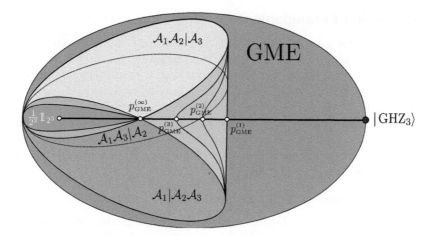

Fig. 18.4 GME activation. Illustration of the phenomenon of GME activation for tripartite systems ($N = 3$). The three-qubit "isotropic" GHZ states form a one-parameter family of states in the space of three-qubit states that connects the fully separable maximally mixed state $\mathbb{1}_{2^3}/2^3$ for $p = 0$ with the pure, genuinely tripartite entangled GHZ state for $p = 1$. For parameters p smaller than the threshold values $p_{\mathrm{GME}}^{(k)}$ but larger than $p_{\mathrm{GME}}^{(k+1)}$ at most k copies are required for $\rho^{\otimes k}$ to be GME. The region of biseparable but not partition-separable states is shown in orange and is subdivided schematically into the sets of k-copy activatable states for $k = 2$, $k = 3$, and $k \geq 4$.

copies are definitely still biseparable but three copies are GME. What is more, an example was provided for a biseparable state that has no distillable entanglement across any bipartition but still allows for GME activation. Based on these observations, two conjectures were put forward in Yamasaki *et al.* (2022):

(i) There exists a hierarchy of states with k-copy activatable GME, i.e., for all $k \geq 2$ there exists a biseparable but not partition-separable state ρ such that $\rho^{\otimes k-1}$ is biseparable, but $\rho^{\otimes k}$ is GME.

(ii) GME may be activated for any biseparable but not partition-separable state of any number of parties as $k \to \infty$.

Not long after, both conjectures were proven by *Carlos Palazuelos* and *Julio de Vicente* (2022), showing that (at least in the finite-dimensional case) all biseparable but not partition-separable states become GME for sufficiently large numbers of copies. As the number of copies goes to infinity, one thus arrives at a qualitative distinction only between partition-separability and GME, collapsing the hierarchy of k-separability asymptotically, and with it the comfortable notion of convex nested sets. Quite a departure from the bipartite scenario indeed. If this extends to infinite-dimensional Hilbert spaces is still under investigation.

18.4 Entangled Entanglement

In this last section we want to discuss a curious feature of multipartite systems that is called *entangled entanglement*. This term was coined by *Günther Krenn* and *Anton Zeilinger* (1996) to characterize the phenomenon that the entanglement of two qubits, represented by the Bell states, can further be entangled with a third qubit, resulting in a particular GHZ state. Interestingly, as discovered by *Gabriele Uchida, Reinhold Bertlmann*, and *Beatrix Hiesmayr* (2015), this idea of entangling entanglement can be generalized to a systematic construction of states that exhibit this entanglement feature for any higher dimension and number of particles. In our presentation we follow closely the work of Uchida *et al.* (2015).

18.4.1 Physical Aspects and Mathematical Structure

Let us begin to discuss the physics connected to the phenomenon of entangled entanglement. To this end, we express the GHZ state in terms of right-handed and left-handed circularly polarized photon states from eqn (3.12), $|R\rangle$ and $|L\rangle$, respectively, as

$$|\,\mathrm{GHZ}_1^-\,\rangle_{123} = \tfrac{1}{\sqrt{2}} \left(|R\rangle_1 \otimes |R\rangle_2 \otimes |R\rangle_3 + |L\rangle_1 \otimes |L\rangle_2 \otimes |L\rangle_3 \right), \qquad (18.42)$$

where we have kept the tensor-product notation "$\otimes$" explicitly (and we shall continue to do so in this section) to make the whole construction procedure more evident. Interestingly, the expression in (18.42) can be re-expressed by decomposing it into linearly polarized photon states $|H\rangle, |V\rangle$ and Bell states, i.e.,

$$|\,\mathrm{GHZ}_1^-\,\rangle_{123} = \tfrac{1}{\sqrt{2}} \left(|H\rangle_1 \otimes |\phi^-\rangle_{23} - |V\rangle_1 \otimes |\psi^+\rangle_{23} \right), \qquad (18.43)$$

where

$$|\phi^\pm\rangle = \tfrac{1}{\sqrt{2}} \left(|H\rangle \otimes |H\rangle \pm |V\rangle \otimes |V\rangle \right), \qquad (18.44a)$$

$$|\psi^\pm\rangle = \tfrac{1}{\sqrt{2}} \left(|H\rangle \otimes |V\rangle \pm |V\rangle \otimes |H\rangle \right) \qquad (18.44b)$$

denote the familiar maximally entangled Bell states, and we recall that the linearly polarized states $|H\rangle$ and $|V\rangle$ are related to the circularly polarized ones via $|R/L\rangle = \tfrac{1}{\sqrt{2}}(|H\rangle \pm i|V\rangle)$.

The GHZ state as expressed in eqn (18.43) represents entangled entanglement in the sense that there are quantum correlations between the linear polarization state of the first subsystem and the type of (maximally) entangled state of the second and third subsystem. This feature has been experimentally verified by Zeilinger's group (Walther *et al.*, 2006), who performed a Bell-type experiment on three particles. One party, Alice on line 1, projects onto the horizontally $|H\rangle_1$ or vertically $|V\rangle_1$ polarized state and the other party, Bob on lines 2 and 3, projects onto the maximally entangled states $|\phi^-\rangle_{23}$ or $|\psi^+\rangle_{23}$ via a Bell-state measurement that is based on a polarizing beam splitter (Pan and Zeilinger, 1998). Then (Walther *et al.*, 2006) test the CHSH inequality (recall Proposition 13.2 of Section 13.2) established between Alice and Bob and find a strong violation of the inequality (specifically, the Bell–CHSH parameter

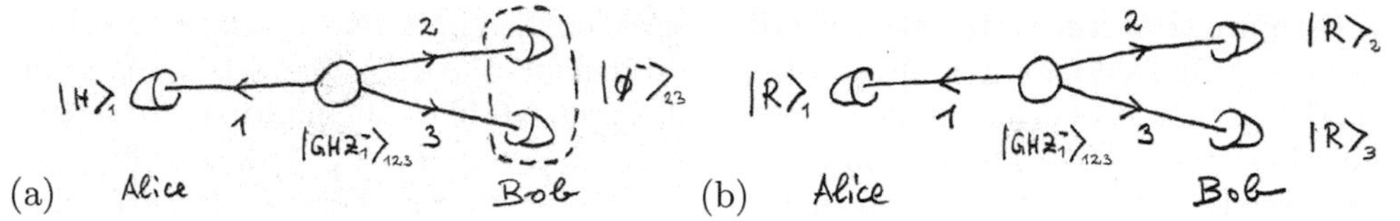

Fig. 18.5 (a) **Bob's photons.** In an entangled state. (b) **Bob's photons.** In a separable state. Figure adapted from Fig. 1 in Uchida *et al.* (2015).

S_{CHSH} exceeds the local-realistic bound of 2) by more than five standard deviations. Thus the entangled states of the two photons on Bob's side are found to be entangled with the single photon on Alice's side.

Physical aspects: But what is the physical significance of this phenomenon, in particular, in light of reasoning along the lines of EPR? As far as we can see, an EPR-like argumentation would go as follows (recall Section 13.1): If Alice is measuring the linearly polarized state $|H\rangle_1$ then Bob will find the Bell state $|\phi^-\rangle_{23}$ for his two photons, see Fig. 18.5 (a). If she measures the state $|V\rangle_1$ then Bob will get the Bell state $|\psi^+\rangle_{23}$. This perfect correlation between the polarization state of one photon on Alice's side and the entangled state of the two photons on Bob's side implies, under the EPR premises of *realism* and *no action at a distance*, that the entangled state of the two photons must represent an element of reality, whereas the individual photons of the latter state, which have no well-defined polarization property, do not correspond to elements of reality. For a realist we imagine this might perhaps be a surprising feature.

If, on the other hand, Alice is measuring a right-handed circularly polarized photon state $|R\rangle_1$ then Bob will find his two photons in a separable state $|R\rangle_2 \otimes |R\rangle_3$, see Fig. 18.5 (b), or if Alice measures $|L\rangle_1$ Bob will get the product state $|L\rangle_2 \otimes |L\rangle_3$. Then the polarizations of the two photons of Bob individually correspond to elements of reality, which might be satisfactory for a realist.

By the specific measurement, projecting onto linearly or circularly polarized photons, Alice is able to switch the properties of the two photons on Bob's side between entanglement and separability, and thereby, according to EPR, the reality of their associated polarization properties!

Mathematical structure: How can we understand this switching phenomenon between entanglement and separability? We have encountered it already in Section 16.4. It depends very much on how we factorize the algebra of a density matrix, accordingly the quantum state may appear either entangled or separable and via a global unitary transformation we can switch between both features. From a theorist's point of view this switch can thus be traced back to two different factorizations of the tensor product of three algebras $\mathcal{A}_1 \otimes \mathcal{A}_2 \otimes \mathcal{A}_3$, where $\mathcal{A}_1$ is associated to Alice and $\mathcal{A}_2 \otimes \mathcal{A}_3$ to Bob. Among the different factorizations none is singled out (Thirring *et al.*, 2011; Zanardi, 2001), no partition has an ontologically superior status over any of the other ones (if no specific physical realization is taken into account). For an experimentalist, however, a particular factorization is preferred and it is clearly fixed by the setup.

For pure tripartite states like the GHZ states, defined on a tensor product of three algebras, or indeed for any multipartite pure states for N parties defined on the product of N algebras, one may of course extend Theorem 16.11 to the multipartite setting (Thirring *et al.*, 2011): For any pure state $\rho = |\psi\rangle\langle\psi|$ on $\mathcal{H} = \bigotimes_{i=1}^{N} \mathcal{H}_i$ one can find a factorization $\mathcal{A}_1 \otimes \ldots \otimes \mathcal{A}_N$ such that ρ is k-separable for any chosen k. In particular, one may find a factorization $\mathcal{A}'_1 \otimes \ldots \otimes \mathcal{A}'_N$ for which ρ appears to be genuinely multipartite entangled. As in the bipartite case, however, such a unitary switching between separability and entanglement is possible for mixed states only beyond a certain bound of mixedness (Thirring *et al.*, 2011).

Complete GHZ system: Having discussed the structure of entangled entanglement for a particular GHZ state, we can now construct all the others in a similar way, arriving at a complete orthonormal system. Geometrically it is quite obvious how to proceed. We just have to entangle the first subsystem with all combinations of pairs of Bell states that correspond to opposite corners of the tetrahedron of Weyl states from Section 15.5.1. In this way we immediately find the *complete orthonormal basis* of eight GHZ states

$$|\,\mathrm{GHZ}_1^{\pm}\,\rangle_{123} = \tfrac{1}{\sqrt{2}}\left(|\,H\,\rangle_1 \otimes |\,\phi^-\,\rangle_{23} \pm |\,V\,\rangle_1 \otimes |\,\psi^+\,\rangle_{23}\right), \tag{18.45a}$$

$$|\,\mathrm{GHZ}_2^{\pm}\,\rangle_{123} = \tfrac{1}{\sqrt{2}}\left(|\,H\,\rangle_1 \otimes |\,\phi^+\,\rangle_{23} \pm |\,V\,\rangle_1 \otimes |\,\psi^-\,\rangle_{23}\right), \tag{18.45b}$$

$$|\,\mathrm{GHZ}_3^{\pm}\,\rangle_{123} = \tfrac{1}{\sqrt{2}}\left(|\,V\,\rangle_1 \otimes |\,\phi^-\,\rangle_{23} \pm |\,H\,\rangle_1 \otimes |\,\psi^+\,\rangle_{23}\right), \tag{18.45c}$$

$$|\,\mathrm{GHZ}_4^{\pm}\,\rangle_{123} = \tfrac{1}{\sqrt{2}}\left(|\,V\,\rangle_1 \otimes |\,\phi^+\,\rangle_{23} \pm |\,H\,\rangle_1 \otimes |\,\psi^-\,\rangle_{23}\right). \tag{18.45d}$$

These eight states form the vertices of a simplex —the *magic simplex of entangled entanglement* (MSEE)—in the corresponding $2 \otimes 2 \otimes 2$-dimensional Hilbert space in analogy to the tetrahedron of Weyl states in $2 \otimes 2$ dimensions. The magic simplex of entangled entanglement is formed by convex combinations of the eight states $|\,\mathrm{GHZ}_i^{\pm}\,\rangle$, i.e., a general state in the MSEE is of the form

$$\rho_{\mathrm{MSEE}} = \sum_{\substack{i=1,2,3,4 \\ k=\pm}} \lambda_i^k \,|\,\mathrm{GHZ}_i^k\,\rangle\langle\,\mathrm{GHZ}_i^k\,|. \tag{18.46}$$

The MSEE contains the maximally mixed state $\tfrac{1}{8}\mathbb{1}_8$ at its centre, and the coefficients λ_i^k with $\sum_{i,k}\lambda_i^k = 1$ are the (non-negative) eigenvalues of the corresponding state (Uchida, 2013).

All states of entangled entanglement (18.45) can certainly be re-expressed in terms of tensor products of right-handed and left-handed circularly polarized photon states, $|\,R\,\rangle$ and $|\,L\,\rangle$, respectively (Uchida, 2013)

$$| \text{GHZ}_1^+ \rangle_{123} = \frac{1}{\sqrt{2}} \left(| R \rangle_1 \otimes | L \rangle_2 \otimes | L \rangle_3 + | L \rangle_1 \otimes | R \rangle_2 \otimes | R \rangle_3 \right), \quad (18.47\text{a})$$

$$| \text{GHZ}_1^- \rangle_{123} = \frac{1}{\sqrt{2}} \left(| R \rangle_1 \otimes | R \rangle_2 \otimes | R \rangle_3 + | L \rangle_1 \otimes | L \rangle_2 \otimes | L \rangle_3 \right), \quad (18.47\text{b})$$

$$| \text{GHZ}_2^+ \rangle_{123} = \frac{1}{\sqrt{2}} \left(| R \rangle_1 \otimes | R \rangle_2 \otimes | L \rangle_3 + | L \rangle_1 \otimes | L \rangle_2 \otimes | R \rangle_3 \right), \quad (18.47\text{c})$$

$$| \text{GHZ}_2^- \rangle_{123} = \frac{1}{\sqrt{2}} \left(| R \rangle_1 \otimes | L \rangle_2 \otimes | R \rangle_3 + | L \rangle_1 \otimes | R \rangle_2 \otimes | L \rangle_3 \right), \quad (18.47\text{d})$$

$$| \text{GHZ}_3^+ \rangle_{123} = -\frac{i}{\sqrt{2}} \left(| R \rangle_1 \otimes | R \rangle_2 \otimes | R \rangle_3 - | L \rangle_1 \otimes | L \rangle_2 \otimes | L \rangle_3 \right), \quad (18.47\text{e})$$

$$| \text{GHZ}_3^- \rangle_{123} = -\frac{i}{\sqrt{2}} \left(| R \rangle_1 \otimes | L \rangle_2 \otimes | L \rangle_3 - | L \rangle_1 \otimes | L \rangle_2 \otimes | R \rangle_3 \right), \quad (18.47\text{f})$$

$$| \text{GHZ}_4^+ \rangle_{123} = -\frac{i}{\sqrt{2}} \left(| R \rangle_1 \otimes | L \rangle_2 \otimes | R \rangle_3 - | L \rangle_1 \otimes | R \rangle_2 \otimes | L \rangle_3 \right), \quad (18.47\text{g})$$

$$| \text{GHZ}_4^- \rangle_{123} = -\frac{i}{\sqrt{2}} \left(| R \rangle_1 \otimes | R \rangle_2 \otimes | L \rangle_3 - | L \rangle_1 \otimes | L \rangle_2 \otimes | R \rangle_3 \right). \quad (18.47\text{h})$$

Using local unitaries these states (18.47a) can of course be transformed into the corresponding states containing only linearly polarized states $| H \rangle$ and $| V \rangle$.

The appeal and importance of the construction of the entangled entanglement is that this procedure can be easily generalized to construct the corresponding states for higher dimensions d and arbitrary number N of particles. Entangling the GHZ states again with $| H \rangle$ and $| V \rangle$ we get the corresponding simplex in the $2 \otimes 2 \otimes 2 \otimes 2$ tensor space and so on. In this way we can construct all higher-dimensional simplices of entangled-entanglement states in a straightforward way just by entangling again the vertices of the simplex with $| H \rangle$ and $| V \rangle$. Thus we find the magic simplex for any particle number. The extension to higher-dimensional systems is obtained by the generalization of the Pauli matrices to the unitary Weyl operators. Next, we will present the procedure that constructs the simplices in a systematic and straightforward way, where the Weyl operators simplify the method.

18.4.2   Construction of Entangled Entanglement

We now briefly discuss the procedure for constructing a set of complete orthonormal basis states for any finite dimension and number of particles as established by *Gabriele Uchida, Reinhold Bertlmann,* and *Beatrix Hiesmayr* (2015). The point is that these basis states already exhibit the structure of entangled entanglement. We proceed in two steps. Firstly, we construct one entangled-entanglement state, and secondly, we find the remaining entangled-entanglement states of the simplex basis by applying Weyl operators to one of the subsystems.

Let us begin with qubits and, without loss of generality, we choose the initial state

$$|\Phi_1\rangle := |0\rangle, \quad (18.48)$$

where we have used the notation $|0\rangle$ and $|1\rangle$ instead of the notation $|H\rangle$ and $|V\rangle$ specific to the implementation using photon polarization, since the former is more appropriate for the generalization to higher dimensions. Inspired by the construction of the states in eqn (18.45) we carry out the following strategy, where we apply Weyl operators U_{nm} defined in eqn (17.24) to the subsystems. Then we consider a two-particle state $|+\rangle \otimes |\Phi_1\rangle$ with $|+\rangle = \frac{1}{\sqrt{2}}(|0\rangle + |1\rangle)$ from which we obtain the state $|\Phi_2\rangle$ by applying the Weyl operator $U_{1,1}$ on the second subsystem conditioned on the state of the first subsystem, i.e., we apply a controlled operation $|0\rangle\langle 0| \otimes \mathbb{1} + |1\rangle\langle 1| \otimes U_{1,1}$,

$$|\Phi_2\rangle = \tfrac{1}{\sqrt{2}}\big(|0\rangle \otimes |\Phi_1\rangle + |1\rangle \otimes U_{1,1}|\Phi_1\rangle\big) = \tfrac{1}{\sqrt{2}}\big(|0\rangle \otimes |0\rangle + |1\rangle \otimes U_{1,1}|0\rangle\big)$$

$$= \tfrac{1}{\sqrt{2}}\big(|0\rangle \otimes |0\rangle - |1\rangle \otimes |1\rangle\big) = |\phi^-\rangle . \tag{18.49}$$

We then iterate this step: We append a new "first" qubit in the state $|+\rangle$ and apply the Weyl operator $U_{1,1}$ to the last (now, third) subsystem, conditioned on the state of the first

$$|\Phi_3\rangle = \tfrac{1}{\sqrt{2}}\big(|0\rangle \otimes |\Phi_2\rangle + |1\rangle \otimes (\mathbb{1} \otimes U_{1,1})|\Phi_2\rangle\big)$$

$$= \tfrac{1}{\sqrt{2}}\big(|0\rangle \otimes |\phi^-\rangle - |1\rangle \otimes |\psi^+\rangle\big) = |\,\mathrm{GHZ}_1^-\rangle_{123} , \tag{18.50}$$

We thus get a GHZ state, specifically $|\,\mathrm{GHZ}_1^-\rangle_{123}$ of eqn (18.45). By iterating we generally find an N-qubit state in the entangled-entanglement form

$$|\Phi_N\rangle = \tfrac{1}{\sqrt{2}}\big(|0\rangle \otimes |\Phi_{N-1}\rangle + |1\rangle \otimes (\underbrace{\mathbb{1} \otimes \cdots \otimes \mathbb{1}}_{N-2} \otimes W_{1,1})|\Phi_{N-1}\rangle\big)$$

$$= \tfrac{1}{\sqrt{2}} \sum_{i=0}^{1} \big(\mathbb{1}^{\otimes(N-1)} \otimes U_{i,i}\big) |i\rangle \otimes |\Phi_{N-1}\rangle . \tag{18.51}$$

Finally, we generalize the states to higher dimensions d, which we can do in a similar manner

$$|\Phi_N^d\rangle = \tfrac{1}{\sqrt{d}}\big(|0\rangle \otimes |\Phi_{N-1}\rangle + |1\rangle \otimes (\underbrace{\mathbb{1} \otimes \cdots \otimes \mathbb{1}}_{N-2} \otimes W_{1,1})|\Phi_{N-1}\rangle$$

$$\dots + |d-1\rangle \otimes (\underbrace{\mathbb{1} \otimes \cdots \otimes \mathbb{1}}_{N-2} \otimes U_{d-1,d-1})|\Phi_{N-1}\rangle\big)$$

$$= \tfrac{1}{\sqrt{d}} \sum_{i=0}^{d-1} \big(\mathbb{1}^{\otimes(N-1)} \otimes U_{d-1,d-1}\big) |i\rangle \otimes |\Phi_{N-1}\rangle . \tag{18.52}$$

Having obtained an entangled-entanglement state for any finite dimension d and number N of particles, eqn (18.52), we go on to construct the (d^N-1) remaining entangled-entanglement states of the simplex by acting in one of the subsystems with all d^2 Weyl operators and in $(N-2)$ subsystems with Weyl operators $U_{s,0}$ that change the phase, i.e., we get

$$|\Phi_N^d(s_1, s_2, \dots, s_{N-2}, k, l)\rangle = \mathbb{1} \otimes U_{s_1,0} \otimes \dots U_{s_{N-2},0} \otimes U_{k,l} |\Phi_N^d\rangle . \tag{18.53}$$

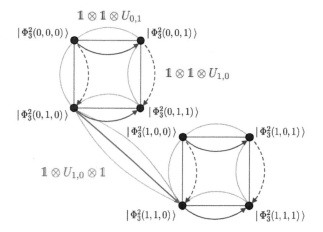

Fig. 18.6 Geometry of the basis states. The eight GHZ states $|\Phi_3^2(s,k,l)\rangle$ from eqn (18.45) form the magic simplex for three qubits. To change a certain GHZ state to another one within the square one needs to apply either a flip or phase operation in the last subsystem, whereas the phase operation $U_{1,0}$ applied to the second subsystem moves a certain GHZ state to a GHZ state in the other square. Figure adapted from Fig. 3 in (Uchida *et al.*, 2015).

Remark: A certain selection of all possible locally unitaries $U_{a_1,b_1} \otimes U_{a_2,b_2} \otimes \cdots \otimes U_{a_N,b_N}$ does the job as well for defining a complete orthogonal basis. Such a selection with its cyclic properties reminds us of the song *"Schön ist so ein Ringelspiel …"*[2] that describes the pleasure of a "Merry Go Round" in the *Viennese Prater*[3]. The illustration is given in Fig. 18.6 for three qubits, the original GHZ case, and in Fig. 18.7 for the extension to three qutrits.

Entangled entanglement simplex: Finally, the *entangled-entanglement simplex* $\mathcal{W}_N^d$ for N particles with dimension d is formed by all convex combinations of the constructed pure states, that is, every state in $\mathcal{W}_N^d$ can be written as

$$\mathcal{W}_N^d := \sum_{\substack{s_1,\ldots,s_{n-2},\\ k,l=0}}^{d-1} c_{s_1,\ldots,s_{N-2},k,l} \, |\Phi_N^d(s_1,s_2,\ldots,s_{N-2},k,l)\rangle\langle\Phi_N^d(s_1,s_2,\ldots,s_{N-2},k,l)|,$$

(18.54)

with $c_{s_1,\ldots,s_{N-2},k,l} \geq 0$ and $\sum c_{s_1,\ldots,s_{N-2},k,l} = 1$. The states $|\Phi_N^d(s_1,s_2,\ldots,s_{N-2},k,l)\rangle$ given in eqn (18.53) form an orthonormal system for all dimensions d and number of parties N. Based on the previous construction the states can be explicitly written as

---

[2] *"Schön ist so ein Ringelspiel …"*, famous Viennese song by Hermann Leopoldi about the pleasure of using the Carousel in the Viennese Prater.

[3] The *Viennese Prater* is a very popular amusement park in Vienna.

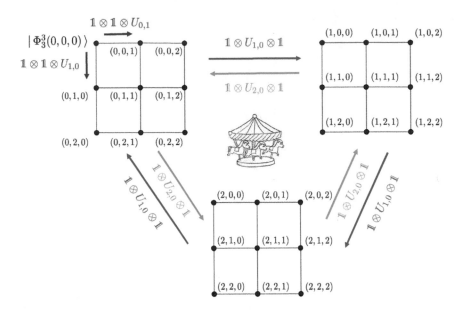

Fig. 18.7 *"Schön ist so ein Ringelspiel ..."* or the Merry Go Round *of the GHZ states.* In the construction procedure of Uchida, Bertlmann, and Hiesmayr the GHZ states possess a cyclic property that allows to move from one square to the next like in the carousel of the famous *Viennese Prater* amusement park. The geometry for three qutrits is a generalization of the one for three qubits, illustrated in Fig. 18.6. To change a certain GHZ state to another one within the square one needs to apply either a flip or phase operation in the third subsystem, whereas the phase operation $U_{1,0}$ or $U_{2,0}$ applied to the second subsystem moves a certain GHZ state from one square to a GHZ state in another square, in this way yielding the cyclic property in the construction of the GHZ states. This can be extended straightforwardly to even higher dimensions. Figure adapted from Fig. 4 in Uchida *et al.* (2015).

$$| \Phi_N^d(s_1, s_2, \ldots, s_{N-2}, k, l) \rangle = \frac{1}{\sqrt{d^{N-1}}} \Big(\gamma_{0,\ldots,0,0} |0 \ldots 0 0 z_{0,\ldots,0,0} \rangle \qquad (18.55)$$

$$+ \gamma_{0,\ldots,0,1} |0 \ldots 0 1 z_{0,\ldots,0,1} \rangle + \ldots + \gamma_{0,\ldots,0,d-1} |0 \ldots 0 (d-1) z_{0,\ldots,0,d-1} \rangle$$
$$+ \ldots + \gamma_{d-1,\ldots,d-1,0} |(d-1) \ldots (d-1) 0 z_{d-1,\ldots,d-1,0} \rangle$$
$$+ \ldots + \gamma_{d-1,\ldots,d-1,d-1} | \underbrace{(d-1),\ldots,(d-1)}_{(N-1)-times} z_{d-1,\ldots,d-1,d-1} \rangle \Big) \,,$$

where the coefficients $\gamma_{i_1,\ldots,i_{n-1}}$ are obtained from the different Weyl transformations, meaning that they are powers of the factor $e^{2\pi i/d}$. The state entries $z_{i_1,\ldots,i_{n-1}}$ are the results of the Weyl transformations $U_{k,l}$ for the last subsystem. The coefficients $z_{i_1,\ldots,i_{n-1}}$ take different values, but all in all there are equally many results giving the "digits" $0, 1, \ldots, (d-1)$.

Résumé: The GHZ-type entangled states provide particularly fertile grounds for investigations of the geometry of state space. They exhibit the interesting feature of entangled entanglement, where an entangled state of two qubits is further entangled with a third qubit. The procedure of Uchida *et al.* (2015) provides a generalization to systematically construct the entangled entanglement states for any higher dimension *and* number of parties. These states form the vertices of a magic simplex and the construction procedure reveals a remarkable cyclic geometric structure, akin to a "Merry Go Round" between the GHZ states among the simplex, as depicted in Fig. 18.6 for qubits and in Fig. 18.7 for qutrits.

The physical significance of entangled entanglement is that the naive concept of reality, that these GHZ-type entangled states always have well-defined local properties, fails completely. When Alice is measuring one particle the reality content of the remaining photons on Bob's side would have to switch accordingly (see Fig. 18.5), which would be quite puzzling for a realist. Mathematically, however, this switching phenomenon between entanglement and separability is understood; it can be traced back to the different allowed factorizations of the algebras for the quantum states of the involved particles.

Part III

Advanced Topics in Modern Quantum Physics

In Part III we have selected some basic topics in modern quantum physics, particularly in view of recent development of quantum-information technology. The topics and material presented in this part are more advanced, ranging from topics such as quantum entropies, quantum channels, open quantum systems and decoherence, atom-field coupling, quantum measurements and metrology, to quantum states of light. Finally, we highlight concepts of quantum information theory, like Bell inequalities, entanglement, and decoherence, within particle physics. Although the individual chapters of this part are mostly self-contained, knowledge of the previous parts of this book is helpful.

We start by investigating entropies which serve as information measures that quantify the amount of information that can be encoded in the state of a system. We review different notions of entropy in classical systems before we turn to quantum entropies, the von Neumann and quantum Rényi entropy, the quantum relative and joint entropy, and the mutual information. We also discuss the conditional-amplitude operator that generates the quantum conditional entropy. We discuss how the latter can even take on negative values for certain entangled states. These quantum states with negative conditional entropy are of particular interest in quantum information and are studied in detail.

Then we focus on quantum channels and quantum operations, where we explain that maps which represent unitary dynamics on a larger Hilbert space admit a so-called Kraus decomposition, which turn out to be completely positive and trace-preserving maps. We further study these maps in the context of open quantum systems. There, one studies the coupling of a system to its environment in terms of dynamical maps that can describe decoherence. Under Markovian conditions we derive a master equation, the Gorini–Kossakowski–Lindblad–Sudarshan equation, that separates the Hamiltonian dynamics of the system from decoherence and dissipation. We use this framework to describe the emission and absorption of photons and investigate the interaction of a two-level atom with the electromagnetic field in terms of the Jaynes–Cummings model.

Next we examine quantum measurements, in particular von Neumann measurements, as well as the more general positive operator-valued measures, non-ideal projective measurements, and discuss how to distinguish reliably between different quantum states, which is an important task in quantum information.

Measurement results provide estimates for the measured quantities. However, many physical quantities, such as phases, interaction strengths, unknown states, etc., are not directly observable—their values have to be estimated. We study these estimation procedures in a chapter dedicated to quantum metrology.

For quantum optics an accurate description of the quantum states of light is of utmost importance. We give a detailed discussion of the quantization of the electromagnetic field and present the theory of coherent states. These are superpositions of states with different photon numbers in the Fock space. Next we turn to a representation of quantum states that can be regarded as a fully quantum-mechanical analogue of a phase-space representation (in position and momentum), the Wigner function. With the knowledge of coherent states and the Wigner function at hand, we turn to the class of Gaussian states and operations.

At the end of this part we want to show how the quantum features discussed throughout the book, for which examples and demonstrations often involve experiments carried out with photons, appear in the context of particle physics. The application of tools and concepts from quantum information theory in this area, and the study of related phenomena, can provide new conceptual insights and lead to new tests of quantum mechanics. In contrast to photons, the considered particles are (very) massive, depend on several quantum numbers, and decay. We consider (mainly) the strange K mesons and the beauty B mesons, produced in accelerators, and discuss the entanglement of these particles and how their inevitable decay affects quantum features such as the violation of Bell inequalities. Finally we discuss a decoherence model for entangled particle systems.

19
Entropy of Classical Systems

Entropy is a fundamental concept in thermodynamics and statistical physics but also in classical and quantum information theory. In order to do justice to the multiple aspects of entropy we will first discuss its meaning in thermodynamics and statistical physics, where the notion of entropy was originally introduced, and include some simple examples to get an intuition for this basic quantity. Then we turn to the information-theoretical approach. In particular we study Shannon's perspective on entropy in classical information theory: There entropy represents the information content of a message on the one hand, and the uncertainty about the value of a random variable on the other. We further introduce the classical relative entropy, the joint entropy and the conditional entropy, together with their basic properties. We then elaborate on the concept of mutual information between two parties before we finally turn to some generalized entropic functions such as the Rényi entropy.

19.1   Entropy in Thermodynamics and Statistical Physics

The concept of entropy in thermodynamics and statistical physics is a topic certainly broadly covered in available literature, and we can only give a brief overview of this topic here to provide context and some major definitions. For further reading we refer to major books and reviews such as Wehrl (1978), Thirring (1980), Römer and Filk (1994), Alicki and Fannes (2001), Grimus (2010), or Swendsen (2012). Here we want to concentrate on entropy features which we find important for our intuitive understanding.

19.1.1   Thermodynamics

Work and heat: A conceptually fundamental distinction that is made in thermodynamics is that between *heat* and *work*. The first law of thermodynamics stipulates that the change in internal energy can be assigned to changes in either one of these two contributions. Whereas heat classically corresponds to changes of energy on the microscopic level, which are not practically accessible to us from the point of view of classical thermodynamics, work is associated to macroscopic changes in energy. In other words, heat is energy attributed to unordered thermal motion of the microscopic constituents of a system, whereas work performed by a system is energy that is transferred from one system to another via the exertion of macroscopic mechanical forces that are measurable and can in turn be stored for later use as potential energy, e.g., by lifting a weight against the gravitational pull. In this sense, work is "ordered" energy.

Heat and work can in principle be converted into each other, but the laws of thermodynamics enshrine the observation that there is a preferred direction to this conversion, i.e., work can in principle be turned into heat at unit efficiency, but the conversion of heat into work is limited in its efficiency. As argued in Swendsen (2012, p. 117), the first law of thermodynamics immediately provides the bound that a heat engine which uses an input $\delta Q \geq 0$ of heat and attempts to turn it into work δW must not violate the relation $\delta W \leq \delta Q$. Meanwhile, a heat engine that harnesses the flow of heat from a hot bath at temperature T_{H} to a cold bath at temperature T_{C} in order to perform work on a working substance can never achieve a conversion of heat into work at an efficiency beyond the Carnot efficiency $\eta_{\max} = 1 - T_{\mathrm{C}}/T_{\mathrm{H}}$. This implies a notion of irreversibility in the conversion of work to heat.

Entropy and the second law of thermodynamics: A different way of expressing this irreversibility is via the second law of thermodynamics phrased as a statement about *entropy S*, a quantity that was introduced by *Rudolf Clausius* in 1865 when considering the amount of heat that is transferred in a thermodynamic process. There, entropy represents a measure for the convertibility of heat, for the energy exchange by heat. For an idealized closed system that is initially at thermal equilibrium with its environment at temperature T and which may exchange no matter but only energy with the latter, infinitesimal changes dS of the entropy due to an irreversible process are bounded from below by the ratio of δQ, the infinitesimal increment[1] of heat transferred to the system, and the environment temperature T, that is

$$dS \geq \frac{\delta Q}{T}\,, \tag{19.1}$$

with equality when the process is reversible. Since $\delta Q \geq 0$, an implication of our discussion so far is the observation that the entropy of any *closed system*[2] may never decrease:

$$dS \geq 0\,, \tag{19.2}$$

and the entropy reaches its maximum (at fixed average energy) when the system is at thermal equilibrium. Thus heat flows from the warmer to the colder system. This statement is nowadays often chosen as the standard formulation of the *second law of thermodynamics*. In this sense, irreversibility is tied to an increase in entropy (or, from an information-theoretic point of view: loss of information), which is in turn associated (albeit not identified) with the transfer of heat. Conversely, a process in a closed system for which $dS = 0$ also satisfies $\delta Q = 0$ (for any finite environment temperature), and all changes in internal energy can be attributed to work in such a case. Heat and work can thus both be considered to be thermodynamical resources, but there is a clear hierarchy between them: Work is strictly more useful than heat.

---

[1] Note that the difference in notation between infinitesimal increments dS of entropy and δQ of heat highlights that the entropy is a function of the system state only, whereas heat is not.

[2] We will return to the discussion of the dynamics of open quantum systems in Chapter 22. In this context, we will also briefly re-examine the statement of the second law for closed quantum systems.

Returning to entropy, it was *Ludwig Boltzmann* who in 1877 defined entropy in terms of the probability of a macroscopic state. According to Boltzmann the second law of thermodynamics asserts that an isolated system develops from a less probable to a more probable macroscopic state and the thermodynamic entropy maximizes for the equilibrium state (see also the discussion in Uffink, 2022).

Boltzmann's concept of entropy was first formalized by *Max Planck* in 1900, when he calculated the entropy of n resonators (Planck, 1900*a*; Planck, 1900*b*). This resulted in the famous Boltzmann formula[3] that we will here phrase in terms of the following proposition:

Proposition 19.1 (Boltzmann's thermodynamic entropy)

The entropy S_{B} of an isolated system is proportional to the natural logarithm of the probability W of the macroscopic state of the system,

$$ S_{\mathrm{B}} = k_{\mathrm{B}} \log W . $$

The proportionality constant k_{B} is Boltzmann's constant: $k_{\mathrm{B}} = 1.38 \times 10^{-23}\,\mathrm{J\,K^{-1}}$.

In the formulation of this statement there is still some imprecision as to how and under which assumptions one may assign a probability to the macrostate of the system. The required context is given in terms of the framework of statistical mechanics, which we will briefly describe next.

19.1.2 Statistical Mechanics

Historically, thermodynamics is concerned with physical systems that have many degrees of freedom. For instance, a characteristic quantity indicative of the system sizes we encounter in our daily experience is the Loschmidt- or Avogadro-number $N_{\mathrm{A}} = 6.022 \times 10^{23}\,mol^{-1}$ particles per mol. It is not practical, or indeed, possible, to keep track of so many particles, and one hence typically has direct access only to a few macroscopically observable parameters, such as energy, temperature, or pressure. Classical thermodynamics provides a quantitative description of these parameters via the laws of thermodynamics. Meanwhile, statistical mechanics underpins this framework by providing microscopic explanations for the macroscopically observed phenomena. At the heart of these explanations lies the information that is available about a given system. In other words, the starting point of statistical mechanics is the assignment of probabilities to the microscopic states, or "microstates" for short, of a system via statistical ensembles, given information about the system's energy, composition, and interaction with its environment (e.g., whether the system is in thermal equilibrium

---

[3]The inscription "$S = k \log W$" was even added to Boltzmann's tombstone at the "Wiener Zentralfriedhof", the Vienna Central Cemetery. The use of the letter W for the probability goes back to the German word *"Wahrscheinlichkeit"*, which means probability.

with a heat bath or a non-equilibrium system), as discussed, for instance, by *E. T. Jaynes* (1957*a*). While information about a few macroscopic parameters of a quantum system—the *macrostate*—may thus be available, this typically means that there also is an enormous lack of information about the specific microstate of the system.

More precisely, microstates are energy eigenstates of the system Hamiltonian, $H\psi_n = E_n\psi_n$. In an isolated system the energy eigenvalues E_n lie within a fixed interval $E_n \in [U - \Delta U, U]$.

A macroscopic system is in equilibrium if the state of equilibrium is stationary, i.e., the corresponding density matrix can be written as

$$\rho = \sum_n \rho_n \, |\psi_n\rangle\langle\psi_n| \quad \text{with} \quad \dot\rho = 0 \,. \tag{19.3}$$

In addition, the postulate expressed in the following proposition is assumed to hold:

Proposition 19.2 (Fundamental postulate of statistical mechanics)

For an isolated system in equilibrium all microstates that are compatible with the macrostate have the same probability, i.e., the equilibrium can be described by the density matrix

$$\rho_{\text{mce}} = \frac{1}{\Omega} \sum_n |\psi_n\rangle\langle\psi_n|$$

where Ω is the number of microstates.

The density matrix ρ_{mce} represents the so-called *microcanonical ensemble*, which is applicable for system whose energy is fixed exactly. We will briefly comment on other ensembles at the end of this section.

While thermodynamics is the description of macroscopic states, statistical mechanics tells us how to obtain the macroscopic laws of thermodynamics with the help of the fundamental postulate (Proposition 19.2), by considering microstates.

Entropy: How do we now arrive at the concept of entropy as put forward in Proposition 19.1? Let us consider an isolated system with a number Ω of microstates that depends on an external parameter, like energy. Let the system $\mathcal{S}$ be split into two parts, $\mathcal{S}_1$ and $\mathcal{S}_2$, with energies U_1 and U_2, as well as particle numbers N_1 and N_2, respectively, such that the total system $\mathcal{S} = \mathcal{S}_1 \cup \mathcal{S}_2$ has energy $U = U_1 + U_2$ and a particle number $N = N_1 + N_2$, and we allow an energy exchange in terms of heat between $\mathcal{S}_1$ and $\mathcal{S}_2$, see Fig. 19.1.

Let us assume that, sometime after the preparation of the system, equilibrium has been reached and the microcanonical ensemble is described by the density matrix

$$\rho_{\text{mce}}(\mathcal{S}_1 \cup \mathcal{S}_2) = \frac{1}{\Omega} \sum_{n,n'} |\psi_n^{(1)}\rangle\langle\psi_n^{(1)}| \otimes |\psi_{n'}^{(2)}\rangle\langle\psi_{n'}^{(2)}| \,, \tag{19.4}$$

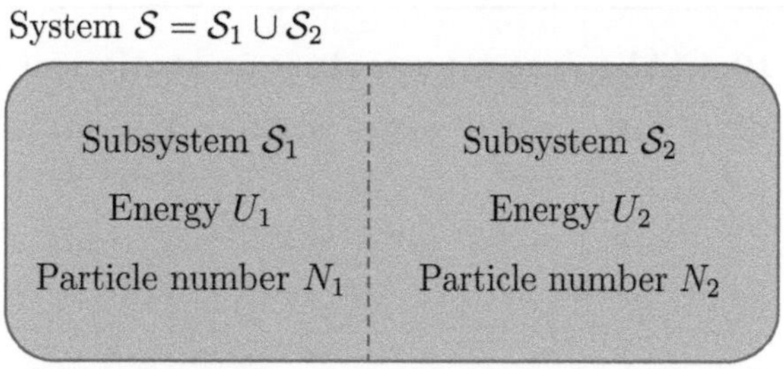

System $\mathcal{S} = \mathcal{S}_1 \cup \mathcal{S}_2$

Subsystem $\mathcal{S}_1$

Energy U_1

Particle number N_1

Subsystem $\mathcal{S}_2$

Energy U_2

Particle number N_2

Fig. 19.1 Thermodynamic systems. A total thermodynamic system $\mathcal{S} = \mathcal{S}_1 \cup \mathcal{S}_2$ consists of two subsystems $\mathcal{S}_1$ and $\mathcal{S}_2$ with energies U_1 and U_2, and particle numbers N_1 and N_2, respectively. We can imagine the subsystems $\mathcal{S}_1$ and $\mathcal{S}_2$ to be separated by a barrier that is permeable to heat.

and the number of states in the energy interval $[U - \Delta U, U]$ is given by the cardinality of the set

$$\left\{ (n, n') \,\middle|\, U - \Delta U \le E_n^{(1)} + E_{n'}^{(2)} \le U \right\}. \tag{19.5}$$

Phrasing the number of microstates as a function of the energy, we can write

$$\Omega(U) = \sum_{U_1} \Omega_1(U_1) \, \Omega_2(U_2 = U - U_1). \tag{19.6}$$

The most relevant contribution in $\rho_{\mathrm{mce}}(\mathcal{S}_1 \cup \mathcal{S}_2)$ comes from the maximum of $\Omega_1(U_1)\Omega_2(U - U_1)$, which is achieved at

$$\left. \frac{\partial \Omega}{\partial U_1} \right|_{\bar{U}_1} = 0. \tag{19.7}$$

This, in turn, implies the relations

$$\left. \frac{1}{\Omega_1} \frac{\partial \Omega_1}{\partial U_1} \right|_{\bar{U}_1} = \left. \frac{1}{\Omega_2} \frac{\partial \Omega_2}{\partial U_2} \right|_{\bar{U}_2}, \tag{19.8}$$

$$\left. \frac{\partial}{\partial U_1} \log \Omega(U_1) \right|_{\bar{U}_1} = \left. \frac{\partial}{\partial U_2} \log \Omega(U_2) \right|_{\bar{U}_2}. \tag{19.9}$$

where $\bar{U}_2 = U - \bar{U}_1$. Equation (19.9) suggests that equilibrium imposes conditions on the following two quantities:

Proposition 19.3 (Entropy in statistical mechanics)

The entropy of a macroscopic system is given by

$$S(U) = k_{\mathrm{B}} \log \Omega(U)$$

with Ω the number of microstates leading to the same macrostate.

Proposition 19.4 (Absolute temperature in statistical mechanics)

The variation of the entropy with respect to the energy provides the inverse absolute temperature T

$$\frac{1}{T} = \frac{\partial S(U)}{\partial U} .$$

Important implications of eqn (19.9), Proposition 19.3, and Proposition 19.4 are:

(i) The entropy of a closed system is non-decreasing: $dS = \frac{\partial S(U)}{\partial U} dU \geq 0$.

(ii) Entropy is maximal at equilibrium.

(iii) Two macroscopic systems in thermal equilibrium have the same temperature.

(iv) The entropy of two independent systems is additive: $S_{\mathcal{S}_1 \cup \mathcal{S}_2} = S_{\mathcal{S}_1} + S_{\mathcal{S}_2}$.

Example I—entropy of a spin system: Let us consider a spin system with $5N$ particles, and for the sake of the argument, let us say that $3N$ thereof have spin up ↑ and $2N$ have spin down ↓. How large is the entropy of such a system? For the formula in Proposition 19.3 we need Ω, the number of microstates, sometimes called the phase-space volume. For $N = 1$ we find ten possibilities, e.g., ↑↑↑↓↓, ↑↑↓↑↓, ↑↓↑↑↓, ... etc., which gives $\Omega = 10$. When selecting $2N$ elements out of $5N$, where the order of the elements does not matter, there are

$$\Omega = \binom{5N}{2N} = \frac{(5N)!}{(2N)!(3N)!} \tag{19.10}$$

possibilities. In general we thus have to use combinatorics, i.e., the binomial formula $\binom{n}{k} = \frac{n!}{k!(n-k)!}$. If we now consider N to be large, $N \to \infty$ so that we can apply *Stirling's formula*

$$N! \approx \sqrt{2\pi N} \left(\frac{N}{e}\right)^N, \qquad \text{for} \quad N \to \infty, \tag{19.11}$$

and keep only the leading term, then we get

$$\Omega \approx \frac{(5N)^{5N}}{(2N)^{2N}(3N)^{3N}} = e^{5N \log(5N)} \cdot e^{-2N \log(2N)} \cdot e^{-3N \log(3N)} \tag{19.12}$$

$$= e^{-5N[(-\log(5) + \frac{2}{5}\log(2) + \frac{3}{5}\log(3)]},$$

where we have made use of the rule $\log(ab) = \log(a) + \log(b)$ for the logarithm. Next, we take the logarithm of eqn (19.12) to obtain

$$\log(\Omega) \approx -5N\left(-\log(5) + \frac{2}{5}\log(2) + \frac{3}{5}\log(3)\right) = -5N\left(\frac{2}{5}\log\left(\frac{2}{5}\right) + \frac{3}{5}\log\left(\frac{3}{5}\right)\right),$$
$$(19.13)$$

which already has the form of the *binary entropy function* $H_{\text{bin}}\left(\frac{2}{5}\right)$ that we have encountered in Theorem 16.5 and which we will discuss in more detail later on, see eqn (19.37). For reasons of normalization we choose the logarithm to base 2 for the spin system, i.e., such that $\log_2(x) = \frac{\log(x)}{\log(2)}$, and we hence have to insert $\log_2\left(\frac{2}{5}\right) = -1.322$ and $\log_2\left(\frac{3}{5}\right) = -0.737$ so that we finally find the entropy of the considered spin system to be

$$S = 4.85\, k_{\text{B}}\, N.\qquad(19.14)$$

We note, the entropy is proportional to the particle number, $S \sim N$, i.e., the entropy is an *extensive* quantity.

Example II—entropy change of a gas: Next we want to calculate the entropy change of a gas when its volume is expanded. As a toy example, let us consider a gas consisting of thirty indistinguishable atoms in Brownian motion, in a cube of volume V_1, which we will coarse grain so that the atoms are distributed over 10^6 possible positions. Now let the walls of the cube become permeable such that the atoms spread into another cube with larger volume $V_2 > V_1$, where they can occupy 10^9 possible positions. *How large is the associated entropy increase?*

The number of microstates or the "phase-space volume" before and after the volume expansion are

$$\Omega_{\text{before}} = \frac{(10^6)^{30}}{30!} \quad\text{and}\quad \Omega_{\text{after}} = \frac{(10^9)^{30}}{30!},\qquad(19.15)$$

respectively. Calculating the entropy S, Proposition 19.3, before expansion, we find

$$S_{\text{before}} = k_{\text{B}}\log\left(\frac{(10^6)^{30}}{30!}\right) = \left(6\cdot 30\cdot\log(10) - \log(30!)\right)k_{\text{B}} = 339.81\,k_{\text{B}},\quad(19.16)$$

whereas the result after the expansion is

$$S_{\text{after}} = k_{\text{B}}\log\left(\frac{(10^9)^{30}}{30!}\right) = \left(9\cdot 30\cdot\log(10) - \log(30!)\right)k_{\text{B}} = (339.81 + 207.23)\,k_{\text{B}},$$
$$(19.17)$$

where we have computed $\log(30!)$ via Stirling's formula from eqn (19.11). Thus we find an increase in entropy of

$$\Delta S = S_{\text{after}} - S_{\text{before}} = 207.23\,k_{\text{B}} > 0,\qquad(19.18)$$

in agreement with our previous considerations and also in accordance with the second law of thermodynamics.

Entropy in the canonical ensemble: In the *microcanonical ensemble*, the overall system is closed, the overall energy is fixed at U and it is reasonable to assume that all microstates compatible with this energy have the same probability, i.e., all microstates contribute to the density operator in the same way. However, in many cases the assumption that the system is completely isolated from the environment is not justified. The basis for the *canonical ensemble* thus is the assumption that a system is in thermal equilibrium with respect to an external heat bath at fixed temperature and only the average energy is fixed. The question then is: Which probabilities do we assign to the different microstates? In particular, there can be many different probability distributions that give rise to the same average energy. Here, *Jaynes' maximum entropy principle* (Jaynes, 1957*a* and 1957*b*) gives a satisfying answer:

Proposition 19.5 (Jaynes' maximum entropy principle)

The macrostate that best represents our knowledge of a system is the state of ***maximal entropy*** *that is compatible with the available information about the system.*

For the canonical ensemble the average energy $\overline{E}$ is fixed, which means that the macrostate has contributions from microstates $|\psi_n\rangle$ with different energies E_n, i.e., we have a state of the form

$$\rho_{ce} = \sum_n p_n \,|\psi_n\rangle\langle\psi_n|\,, \qquad (19.19)$$

with the constraints $\overline{E} = \sum_n p_n E_n$ and $\sum_n p_n = 1$. A priori, the states $|\psi_n\rangle$ need not be energy eigenstates and thus ρ_{ce} need not be diagonal in the eigenbasis of the Hamiltonian. However, at fixed average energy it follows from the concavity of the entropy (see Section 20.1) that the entropy of a mixture of two states that are superpositions of energy eigenstates is always smaller than (or equal to) the entropy of a mixture of the two eigenstates. We can thus assume that ρ_{ce} is diagonal in the energy eigenbasis. To express the entropy in terms of the probabilities p_n, we note that for the microcanonical ensemble we have $p_n = \frac{1}{\Omega} \,\forall\, n = 1, 2, \ldots, \Omega$, and we can write the logarithm appearing in Proposition 19.3 as

$$\log\Omega = \sum_{n=1}^{\Omega} \tfrac{1}{\Omega}\log\Omega = \sum_n p_n \,\log\!\left(\tfrac{1}{p_n}\right) = -\sum_n p_n \,\log(p_n)\,. \qquad (19.20)$$

Adopting the expression on the right-hand side as our new definition of entropy and dropping the constant pre-factor k_{B}, we can then maximize the quantity

$$S = -\sum_n p_n \,\log(p_n) \qquad (19.21)$$

under the mentioned constraints using the method of *Lagrange multipliers*:

$$\frac{\partial}{\partial p_m}\left(-\sum_n p_n \log(p_n) - \lambda \left[\sum_n p_n E_n - \overline{E}\right] - \mu \left[\sum_n p_n - 1\right]\right)$$

$$= -\log(p_m) - 1 - \lambda E_m - \mu = 0. \tag{19.22}$$

Rearranging the last line, we can isolate the logarithm

$$\log(p_m) = -\lambda E_m - (\mu + 1), \tag{19.23}$$

and exponentiating this expression we obtain

$$p_m = e^{-\lambda E_m} e^{-(\mu+1)}. \tag{19.24}$$

The normalization condition $\sum_m p_m = e^{-(\mu+1)} \sum_m e^{-\lambda E_m} = 1$ then implies $e^{(\mu+1)} = \sum_m e^{-\lambda E_m}$ such that we arrive at

$$p_n = \frac{e^{-\lambda E_n}}{\sum_m e^{-\lambda E_m}}. \tag{19.25}$$

This is exactly the expression for the probabilities of the *thermal state* for a system in thermal equilibrium with a heat bath at temperature $T = 1/(k_\mathrm{B}\beta) = 1/(k_\mathrm{B}\lambda)$ from Section 11.4. We thus see that the (canonical) thermal state, also called the Gibbs state, is the state that maximizes the entropy at fixed average energy. Moreover, we see that the formula we have used for the entropy matches the expression for the von Neumann entropy that we have already defined (Definition 16.1) in the context of quantum information theory, and which we will discuss in more detail in Chapter 20.

19.2 Shannon Entropy in Classical Information Theory

The concept of entropy in information theory has been widely discussed in major books like Nielsen and Chuang (2000), Schumacher and Westmoreland (2010), Timpson (2013), and Wilde (2017). Here, we only give a brief overview and focus on those entropy concepts which we need for our further discussion of quantum-information processing.

19.2.1 Shannon Entropy

One of the basic quantities in classical information theory is Shannon's entropy. It quantifies how much information we gain from receiving a message, i.e., it is a measure for the amount of information contained in a message. To obtain such a quantity let us first make a statement in the spirit of *Claude Elwood Shannon* (1948), widely considered to be the father of information theory:

> *The amount of information gained from a message depends on how likely it is!*

Therefore, the less likely a message is—loosely, the more surprised one is by the message—the more information is gained upon receiving it. Let us formulate this intuition more mathematically. Let us consider a message that is exchanged between

Alice and Bob, the sender and receiver, respectively. The message consists of "letters" drawn from an alphabet with k symbols that have different probabilities of appearing in the message. This situation can be described by a random variable X that maps the letters from a sample space, here, the alphabet, to elements of a (measurable) set $\mathcal{X} = \{x_i, x_2, \ldots, x_k\}$ and associates a probability distribution $P(X) := \{p_i\}_i$. That is, X takes values x_i in $\mathcal{X}$, with probabilities $p_i = p(x_i) \geq 0$ satisfying $\sum_i p_i = 1$.

For the information content of the message à la Shannon the meaning of the symbols in the alphabet is of no consequence, and it can thus be characterized in terms of the probability distribution $P(X)$ alone. But before we consider a particular measure of information, let us formulate its desired properties, i.e., the conditions a reasonable measure of information should fulfil.

Proposition 19.6 (Measure of information)

A measure of information for a set X of messages is a real-valued function

$$H(X): \quad P(X) \quad \longrightarrow \quad \mathbb{R}$$

that satisfies

- *Continuity: $H(\{p_1, p_2, \ldots, p_k\})$ is continuous.*
- *Additivity: $H(\{p_1 q_1, p_2 q_2, \ldots, p_k q_k\}) = H(P) + H(Q)$, for probability distributions $P = \{p_i\}_i$ and $Q = \{q_j\}_j$.*
- *Monotonicity: for uniform distributions the information associated with each message increases with the number of possible symbols, i.e., $\forall\, k, k'$: $H(X') > H(X)$ with $P(X') = \{\frac{1}{k'}, \ldots, \frac{1}{k'}\}, P(X) = \{\frac{1}{k}, \ldots, \frac{1}{k}\}$.*
- *Bit normalization: the average information gained from two equally likely messages is one bit $H(\{\frac{1}{2}, \frac{1}{2}\}) = 1$.*

Now we are prepared to formulate Shannon's entropy.

Definition 19.1 *The information-theoretic entropy of a set of messages represented by a random variable X with associated probability distribution $P(X)$ is given by the **Shannon entropy***

$$H(X) = -\sum_{i=1}^{k} p_i \log_2(p_i).$$

Although we shall not prove this here, the entropy function $H(X)$ in Definition 19.1 satisfies the criteria for a measure of information as laid out in Proposition 19.6, and, recalling that $p_j \log_2(p_j) = 0$ for $p_j = 0$ one can see that the Shannon entropy has the following additional properties.

(i) $H(X)$ *is maximal for the uniform distribution* $p_1 = p_2 = \ldots = p_k = \frac{1}{k}$.

(ii) $H(X) = 0$ *if* $p_i = 1$ *for some* i *while* $p_j = 0$ $\forall j \neq i$.

Information content of a message: Now we are ready to motivate that Shannon's entropy indeed is a suitable measure for the information content of a message. We consider a message that consists of an N-letter word, where each "letter" $m_1, m_2, \ldots, m_N$ can take one of k values x_i (with $i = 1, 2, \ldots, k$) from a set $\mathcal{X} = \{x_i, x_2, \ldots, x_k\}$. This means the set of possible messages is of size k^N. In order to quantify the information content of the message, we need to make a statement about how likely it is that a specific message is received, and thus how surprised we are to receive it. To this end we assign probabilities p_i to the symbols x_i, and treat the letters of the message as a random variable X that can take values in $\mathcal{X}$. We denote the probability that $x = x_i$ as $p_i = p(x = x_i)$.

As an example, consider the case where the letters of the message are chosen from the set $\{a, b, c\}$, so $k = 3$, and the message is composed of four letters, $N = 4$. Among the possible messages one would then have instances such as $(m_1, m_2, m_3, m_4) = (a, b, a, c)$, (a, a, c, b), or (a, a, a, c). However, while the meaning of the message might change with the particular ordering of the letters, the information content should not be influenced by the ordering, so by all means we should attribute the same information content to, for instance, (a, b, a, c) and (a, a, c, b), and this information content will depend on how likely it is that a four-letter sequence with two instances of the letter a and one each for b and c is received. In general, a typical message will contain $N_i = N p_i$ instances of the letter associated to the value x_i, with $\sum_{i=1}^{k} N_i = N$. This brings us to the following question:

How many different N-letter sequences with N_i instances of the value x_i are there?

So what is the number of possibilities to distribute the value x_i exactly N_i times among the N possible letters of the message? There are $N!$ possibilities to arrange the N letters $m_1, \ldots, m_N$, but we have to divide by the number of possibilities to exchange different instances of x_i that occur multiple (N_i) times, of which there are $N_i!$ for each x_i. Denoting the number of N-letter sequences with N_i entries x_i as Z_N, the answer to the above question thus is

$$ Z_N = \frac{N!}{N_1!\, N_2!\, \ldots\, N_k!} \, . \tag{19.26} $$

This number can now be seen as an indicator for the information content of a particular N-letter message. If Z_N is large, which means there are many sequences of the same type, then receiving a particular one provides more information. However, the quantity is not additive. If we receive two independent messages of lengths N and N' out of the Z_N and $Z_{N'}$ different possible such messages, respectively, then there are $Z_{N,N'} = Z_N Z_{N'}$ different pairs of messages. This can be remedied by taking the logarithm, in which case we have $\log_2(Z_{N,N'}) = \log_2(Z_N) + \log_2(Z_{N'})$, and we take the logarithm to base 2 so that our information gain is measured in bits.

Next we consider the limit of large N, that is, $N \to \infty$, such that for all non-zero p_i we also have $N_i \to \infty$, i.e., we move from rates to probabilities, which allows us to make use of Stirling's formula (19.11)

$$\log(N!) = N \log(N) - N + \mathcal{O}(\log(N)) , \tag{19.27}$$

where $\mathcal{O}(x)$ denotes terms such that $\mathcal{O}(x)/x$ is finite as $x \to 0$. Taking the logarithm and inserting $N_i = N\, p_i$ we obtain

$$\log_2(Z_N) = N \log_2(N) - N - \sum_{i=1}^{k} \left[N_i \log_2(N_i) - N_i \right] , \tag{19.28}$$

$$= N \log_2(N) - N \sum_{i=1}^{k} p_i \log_2(p_i) - N \sum_{i=1}^{k} p_i \log_2(N) = -N \sum_{i=1}^{k} p_i \log_2(p_i) .$$

Thus, by normalizing (dividing by the length N of the message) we arrive at the following expression for the logarithm of Z_N,

$$H(\{p_i\}) = \lim_{N \to \infty} \frac{1}{N} \log_2(Z_N) = - \sum_{i=1}^{k} p_i \log_2(p_i) , \tag{19.29}$$

which is precisely Shannon's entropy, Definition 19.1.

Remark: Shannon's entropy can be interpreted in two ways. On one hand, it represents the *information gain* upon receiving a message, on the other, it is a measure for the *amount of uncertainty* about the value of a random variable. In fact, Shannon wanted to call it "uncertainty" but it was von Neumann who convinced him to name it "entropy"[4].

There is a link between information-theoretic entropy and thermodynamic entropy. If W is the number of microstates that correspond to a given macrostate, and each microstate occurs with the same probability $p_i = \frac{1}{W}\; \forall i$, then Shannon's entropy H yields precisely the entropy of Boltzmann S_B from Proposition 19.1,

$$H = - \sum_{i=1}^{k} p_i \log_2(p_i) = - \sum_{i=1}^{k} \tfrac{1}{W} \log_2\!\left(\tfrac{1}{W}\right) = \log_2(W) \equiv S_\mathrm{B} , \tag{19.30}$$

where we have disregard the discrepancy in the choice of units, i.e., the pre-factor k_B and the choice of the base of the logarithm.

19.2.2   Shannon Entropy—Message Compression

We now briefly return to the situation of the N-letter message. We recall that the letters of the message are chosen from an alphabet with k symbols, represented by the

---

[4]Legend has it that von Neumann told Shannon in a conversation that he should use the name "entropy" for his "uncertainty function" for two reasons. First, the function is already used in statistical mechanics under that name, and second, and more importantly, nobody knows what entropy really means, so in a discussion you will always have an advantage.

elements of the set $\mathcal{X} = \{x_1, \ldots, x_k\}$, and each x_i occurs with probability p_i. Thus, a typical message will contain $N_i = N p_i$ instances of the symbol corresponding to x_i. At the same time we have found that, in the limit of large N, the number of such sequences is Z_N, and according to eqn (19.28) we have

$$\log_2(Z_N) = N H(X) \qquad \text{or} \qquad Z_N = 2^{NH(X)} . \tag{19.31}$$

So we have $2^{NH(X)}$ typical messages with N letters, which means that we can encode them by using only $N H(X)$ bits. On the level of the different symbols in the alphabet, we may ask how many bits are needed to encode a message of N letters using a k-letter alphabet?

Writing the number of symbols in the alphabet as an exponential of some number x, we have $k = 2^x$, and conversely that we need $x = \log_2(k)$ bits to encode each symbol, or $N \log_2(k)$ bits to encode an N-letter message. And this is indeed the case if each symbol in the alphabet is equally likely to appear in the message. But if we take into account that each x_i appears with probability p_i, then we only need $N H(X)$ bits to encode a typical message. And, indeed, the maximal value of $H(X)$ is obtained exactly when the probability distribution is uniform, for $p_i = \frac{1}{k}$ for all $i = 1, 2, \ldots, k$ we have

$$H(\{\tfrac{1}{k}, \ldots, \tfrac{1}{k}\}) = -\sum_{i=1}^{k} p_i \log_2(p_i) = -k \tfrac{1}{k} \log_2\left(\tfrac{1}{k}\right) = \log_2(k) . \tag{19.32}$$

In this case $N \log_2(k) = N H(X)$ but in general $N \log_2(k) \geq N H(X)$.

Conclusion: *The Shannon entropy $H(X)$, Definition 19.1, represents the maximum amount that a message can be compressed with negligible risk of losing information!*

This forms the basis for what is today known as *Shannon's source-coding theorem* (see, e.g., Shannon, 1948 or MacKay, 2003, p. 81), which we will not discuss here in more detail but we refer the interested reader to MacKay (2003, Chap. I).

Example: Let us again consider a set of $k = 4$ symbols $\{a, b, c, d\}$ which we can encode into two bits, $a = 00$, $b = 01$, $c = 10$, $d = 11$. Then we need $\log_2(4) = 2$ bits per letter and $2N$ bits to encode a message with N letters. But this is without compression.

Suppose now that the letters a, b, c, and d occur with the following probabilities in a typical N-letter message

$$p_a = \tfrac{1}{2}, \quad p_b = \tfrac{1}{4}, \quad p_c = p_d = \tfrac{1}{8} . \tag{19.33}$$

Then the number of bits that is required to encode the message is

$$N H(X) = -N \left(\tfrac{1}{2} \log_2\left(\tfrac{1}{2}\right) + \tfrac{1}{4} \log_2\left(\tfrac{1}{4}\right) + \tfrac{1}{8} \log_2\left(\tfrac{1}{8}\right) + \tfrac{1}{8} \log_2\left(\tfrac{1}{8}\right) \right) = \tfrac{7}{4} N < 2N . \tag{19.34}$$

Therefore, instead of demanding $2N$ bits to encode the N-letter message, we manage with less, namely with $\tfrac{7}{4} N$ bits.

19.2.3 Shannon Entropy—Measure of Uncertainty

Another perspective that we can take on the Shannon entropy is as an expected value of the information content of a particular value x_i, a realization of the random variable X. Let $P = \{p_1, \dots, p_k\}$ again be a probability distribution over a set of values $\{x_1, \dots, x_k\}$ of a random variable X. Then the **expected value** $E[X]$ of the random variable X is

$$E[X] = \sum_{i=1}^{k} p_i \, x_i \,. \tag{19.35}$$

We then define the following quantity:

Definition 19.2 *The **information content** for a particular realization x_i of a random variable X is* $\quad I(x_i) := -\log_2(p_i) \,.$

This implies the larger the probability $p_i = p(x_i)$ is, the more certain it is to obtain the value x_i, the less information can be gained by learning that the random variable has taken on this value.

Properties of $I(x_i)$:

- $I(x_i) \geq 0$ *non-negative.*

- *If* $p_i \to 1 \Rightarrow I(x_i) \to 0$.

- $I(x_i)$ *increases as* $p_i \to 0$.

- *For* $x_1 \neq x_2$ *less probable events* $p_1 < p_2$ *contain more information* $I(x_1) > I(x_2)$.

- *For statistically independent joint events* $p(x_1, x_2) = p_1 \cdot p_2$ *the information is additive* $I(x_1, x_2) = I(x_1) + I(x_2)$.

The expected value of the *information content* is precisely Shannon's entropy, Definition 19.1

$$E[I(X)] = -\sum_{i=1}^{k} p_i \log_2(p_i) \equiv H(X) \,. \tag{19.36}$$

Thus Shannon's entropy tells us what information gain we can expect when measuring the random variable X.

19.2.4 Binary Entropy Function

A special case of the Shannon entropy that is of particular practical relevance in many situations, for instance, to a random coin toss, is that of the binary entropy, $H_{\text{bin}}(p)$, which is obtained for random variables that can only take one of two values, so $k = 2$. There the random variable X can take values in $\mathcal{X} = \{0, 1\}$ and the two possible outcomes have probabilities $p(0) = p$ and $p(1) = 1 - p$. Then the *binary entropy function* has the form

$$H_{\text{bin}}(p) = -p \log_2(p) - (1 - p) \log_2(1 - p), \tag{19.37}$$

where the logarithm is again taken to base 2. The binary entropy quantifies the information we obtain from tossing a coin, measured in bits. The maximum of the binary entropy is achieved when the coin is fair, that is, when both outcomes are equally likely, $p = \frac{1}{2}$, in which case the information gain per coin toss is exactly one bit, $H_{\text{bin}}(p = \frac{1}{2}) = 1$, see Fig. 19.2.

The minima occur at the values $p = 0$ and $p = 1$, where the outcome of the coin toss is deterministic and we do not gain any information $H_{\text{bin}}(p = 0) = H_{\text{bin}}(p = 1) = 0$. Therefore, if the outcomes "heads" and "tails" occur with the same frequency the binary entropy is maximal and the coin toss is fair. However, if the binary entropy is less than one bit $H_{\text{bin}} < 1$ the coin toss is biased.

Up to an overall sign, the derivative of the binary entropy coincides with the logit function

$$\frac{d}{dp} H_{\text{bin}}(p) = -\log_2\left(\frac{p}{1-p}\right) = -\operatorname{logit}(p), \tag{19.38}$$

which is frequently used in statistics and is the inverse of the standard logistic function $p(x) = 1/(1 + e^{-x})$.

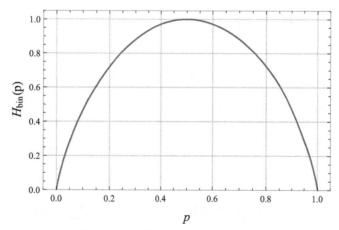

Fig. 19.2 Binary entropy function. The binary entropy function $H_{\text{bin}}(p)$ from eqn (19.37) is plotted as a function of the probability p. The maximum is achieved when the probabilities for the two outcomes are the same, $p(0) = p(1) = \frac{1}{2}$ and the minima occur when either of the outcomes are certain, $p = 0$ or $p = 1$.

19.3 Relative Entropy, Joint and Conditional Entropy

19.3.1 Classical Relative Entropy

Another conceptually important entropic quantity is the *classical relative entropy*, also called *Kullback–Leibler divergence* in reference to *Solomon Kullback* and *Richard Leibler* who introduced it in 1951.

Definition 19.3 *For probability distributions $P(X) = \{p(x)\}_x$ and $Q(X) = \{q(x)\}_x$ for the same random variable X with values $x \in \mathcal{X}$, the classical **relative entropy** or **Kullback–Leibler divergence** is given by*

$$H\big(P(X)\,\|\,Q(X)\big) := \sum_x p(x)\log_2\Big(\tfrac{p(x)}{q(x)}\Big)$$

$$= -H(X) - \sum_x p(x)\log_2\big(q(x)\big)\,.$$

The relative entropy is an entropy-like measure for the closeness of two probability distributions $P(X)$ and $Q(X)$. According to its definition the relative entropy is *not* symmetric under exchange of P and Q, nor does it satisfy the triangle inequality. Therefore it is *not* a geometric distance measure in a strict sense, which is alluded to by the use of the word "divergence" rather than "distance". Nevertheless, it is non-negative, a statement that follows from Gibbs' inequality:

Theorem 19.1 (Gibbs' inequality)

The classical relative entropy is non-negative,

$$H\big(P(X)\,\|\,Q(X)\big) \geq 0\,,$$

with equality if and only if $p(x) = q(x)\ \forall x \in \mathcal{X}$.

Proof Let us start with the inequality

$$\log_2(x)\cdot\ln(2) = \ln(x) \leq x - 1 \quad\Rightarrow\quad -\log_2(x) \geq \frac{1-x}{\ln(2)}\,. \tag{19.39}$$

Using inequality (19.39) and $\sum_x p(x) = \sum_x q(x) = 1$ we then obtain

$$H\big(P(X)\,\|\,Q(X)\big) = -\sum_x p(x)\log_2\big(\tfrac{q(x)}{p(x)}\big) \geq \tfrac{1}{\ln(2)}\sum_x p(x)\Big(1 - \tfrac{q(x)}{p(x)}\Big)$$

$$= \tfrac{1}{\ln(2)}\sum_x \big(p(x) - q(x)\big) = 0\,, \tag{19.40}$$

and we get equality in the second step if $\frac{p(x)}{q(x)} = 1\ \forall x \in \mathcal{X}$. $\qquad\square$

   The relative entropy is also useful to prove relations among entropic quantities. For instance, let us consider the case where X is a random variable over an alphabet with d symbols with (arbitrary) probability distribution $P(X) = \{p(x)\}_x$ and let $Q(X)$ be the uniform distribution, $q(x) = \frac{1}{d} \forall x$. Then the (non-negative) relative entropy satisfies the relation

$$0 \leq H(P(X) \| Q(X)) = H(\{p(x)\} \| \{\tfrac{1}{d}\}) = -H(\{p(x)\}) - \sum_x p(x) \log_2\left(\tfrac{1}{d}\right)$$

$$= \log_2(d) - H(\{p(x)\}) . \qquad (19.41)$$

This directly leads us to the following theorem:

Theorem 19.2 (Upper bound on the Shannon entropy)

Let X be a random variable that can take d values, then the Shannon entropy is bounded from above by

$$H(X) \leq \log_2(d) ,$$

with equality iff $P(X)$ is the uniform distribution over the d outcomes.

19.3.2 Joint and Conditional Entropy

Joint and conditional probabilities: Let us now consider situations with two random variables, X and Y, with probability distributions $P(X) = \{p_X(x)\}_x$ and $P(Y) = \{p_Y(y)\}_y$ for the possible values $x \in \mathcal{X}$ and $y \in \mathcal{Y}$, respectively. For instance, we can think of two parties, Alice and Bob, aiming to describe their respective measurement outcomes. In addition, we can now make statements about probabilities for events that involve both of the random variables taking specific values. In particular, the *joint probability distribution* $P(X \cap Y) \equiv P(X, Y) = \{p_{X,Y}(x,y)\}_{x,y}$ is the distribution of probabilities

$$p_{X,Y}(x,y) = P(X = x \text{ and } Y = y), \qquad (19.42)$$

so the probability that X takes on the value x, while Y takes on the value y. The individual distributions, $P(X)$ and $P(Y)$, are called the *marginal distributions* of the joint distribution $P(X \cap Y)$, and are obtained by summing over the values taken by the respective other random variable,

$$p_X(x) = \sum_y p_{X,Y}(x,y) \quad \text{and} \quad p_Y(y) = \sum_x p_{X,Y}(x,y) . \qquad (19.43)$$

Clearly, the joint probability is symmetric, that is,

$$P(X = x \text{ and } Y = y) = P(Y = y \text{ and } X = x) . \qquad (19.44)$$

If two events are *statistically independent* we have

$$p_{X,Y}(x,y) = p_X(x) \cdot p_Y(y) \,. \tag{19.45}$$

In the case that $p_{X,Y}(x,y) \neq p_X(x) \cdot p_Y(y)$ the events are not statistically independent and are called

$$\text{correlated} \quad \text{if} \quad p_{X,Y}(x,y) > p_X(x) \cdot p_Y(y)\,, \tag{19.46a}$$

$$\text{anticorrelated} \quad \text{if} \quad p_{X,Y}(x,y) < p_X(x) \cdot p_Y(y)\,. \tag{19.46b}$$

Alternatively we can also consider the probabilities for either of two events occurring,

$$P(X = x \text{ or } Y = y) = P(X = x) + P(Y = y) - P(X = x \text{ and } Y = y), \tag{19.47}$$

where the joint probability is subtracted since the probability for both $X = x$ and $Y = y$ occurring are included in the marginal distributions. Or, in more compact notation, this can be expressed as

$$P(X \cup Y) = P(X) + P(Y) - P(X \cap Y), \tag{19.48}$$

indicating that eqn (19.47) holds for all $x \in \mathcal{X}$ and $y \in \mathcal{Y}$. A simple consequence of this relation is the subadditivity of both $P(X \cap Y)$ and $P(X \cup Y)$,

$$P(X \cap Y) \leq P(X) + P(Y)\,, \tag{19.49a}$$

$$P(X \cup Y) \leq P(X) + P(Y)\,. \tag{19.49b}$$

Next we consider the *conditional probability distribution* $P(X|Y) = \{p(x|y)\}_{x,y}$, which gives the probabilities $p_{X,Y}(x|y) = P(X = x|Y = y)$ for the occurrence of results $X = x$ given the fact that $Y = y$ is already known. The joint probability is related to the conditional probability via

$$p_{X,Y}(x,y) = p_{X,Y}(x|y)\, p_Y(y), \tag{19.50}$$

which means that for all y for which $p_Y(y) > 0$ is non-zero, we can define the conditional probabilities

$$p_{X,Y}(x|y) = \frac{p_{X,Y}(x,y)}{p_Y(y)}\,. \tag{19.51}$$

If the two events are statistically independent then we have

$$p_{X,Y}(x|y) = p_X(x)\,, \tag{19.52}$$

i.e., the probability of the outcomes of X is independent of the results of Y.

From the symmetry of the joint probability distribution we can deduce

$$p_{X,Y}(y|x) = p_{X,Y}(x|y) \frac{p_Y(y)}{p_X(x)}\,, \tag{19.53}$$

which is known as *Bayes' law* in probability theory.

Joint and conditional entropy: With these preliminaries on notation and terminology established, we now turn to entropies.

Definition 19.4 *The **joint entropy** $H(X,Y)$ is the Shannon entropy associated to the joint probability $p_{X,Y}(x,y)$ of two random variables X and Y,*

$$H(X,Y) = -\sum_{x,y} p_{X,Y}(x,y) \log_2\big(p_{X,Y}(x,y)\big) \ .$$

The joint entropy describes the mean uncertainty for the occurrence of a pair of values $x \in \mathcal{X}$ and $y \in \mathcal{Y}$. From the symmetry of the joint probability, eqn (19.44), it follows that the joint entropy is symmetric as well, which we denote as

$$H(X,Y) = H(Y,X) \ . \tag{19.54}$$

Lemma 19.1 (Subadditivity of joint entropy)

The joint entropy is subadditive

$$H(X,Y) \leq H(X) + H(Y)$$

with equality if and only if the random variables X and Y are statistically independent.

Proof We start by considering the (non-negative) relative entropy between an arbitrary joint distribution $P(X \cup Y) = \{p_{X,Y}(x,y)\}_{x,y}$ and an uncorrelated joint distribution with the same marginal probabilities $p_X(x)$ and $p_Y(y)$,

$$0 \leq H\big(\{p_{X,Y}(x,y)\} \,\|\, p_X(x)\,p_Y(y)\big)$$

$$= -H(\{p_{X,Y}(x,y)\}) - \sum_{x,y} p_{X,Y}(x,y)\big[\log_2\big(p_X(x)\big) + \log_2\big(p_Y(y)\big)\big]$$

$$= -H(\{p_{X,Y}(x,y)\}) - \sum_{x} p_X(x) \log_2\big(p_X(x)\big) - \sum_{y} p_Y(y) \log_2\big(p_Y(y)\big)$$

$$= -H(\{p_{X,Y}(x,y)\}) + H(\{p_X(x)\}) + H(\{p_Y(y)\}) \ , \tag{19.55}$$

since $\sum_y p_{X,Y}(x,y) = p_X(x)$ and $\sum_x p_{X,Y}(x,y) = p_Y(y)$. Gibbs' inequality (Theorem 19.1) then ensures that the three terms add to zero if and only if the random variables are statistically independent, $p_{X,Y}(x,y) = p_X(x)\,p_Y(y) \ \forall x \in \mathcal{X}$ and $y \in \mathcal{Y}$. $\square$

Now let us focus on the interesting case of two random variables X and Y that are correlated, i.e., which are *not* statistically independent. Associating these random variables with two observers, Alice and Bob, this means that Bob can get some information about Alice's random variable X from the values taken by his random variable Y. How much information can be obtained by Bob in this way can be phrased in terms of the information content (see Definition 19.2) of the conditional probability. That is, Bob's information gain upon obtaining the value $X = x$ given that Alice obtained $Y = y$ is

$$I(x|y) = -\log_2\big(p_{X,Y}(x|y)\big)\,. \tag{19.56}$$

Since the value of $I(x|y)$ can be different for each pair of x and y, we can take the average, weighted with the probability for the occurrence of both x and y, i.e., the joint probability $p_{X,Y}(x,y)$, to obtain the expected information content.

Definition 19.5   *The **classical conditional entropy** $H(X|Y)$ of two random variables X and Y is given by the expected information content $I(x|y)$, i.e.,*

$$H(X|Y) = E\big[I(X|Y)\big] = -\sum_{x,y} p_{X,Y}(x,y)\log_2\big[p_{X,Y}(x|y)\big]\,.$$

Thus, the conditional entropy represents the mean uncertainty about the value of x of X if we already know the value y of Y. In that sense it quantifies the remaining rest-uncertainty.

Lemma 19.2 (Relation of conditional and joint entropy)

The conditional entropy is related to the joint entropy by

$$H(X|Y) = H(X,Y) - H(Y)\,.$$

Proof   From the conditional probability (19.51) we have

$$\log_2\big(p(x|y)\big) = \log_2\big(p_{X,Y}(x,y)\big) - \log_2\big(p_Y(y)\big)\,. \tag{19.57}$$

Inserting eqn (19.57) into the conditional entropy, Definition 19.5, we obtain

$$
\begin{aligned}
H(X|Y) &= -\sum_{x,y} p_{X,Y}(x,y)\,\log_2\big(p_{X,Y}(x|y)\big)\\
&= -\sum_{x,y} p_{X,Y}(x,y)\,\log_2\big(p_{X,Y}(x,y)\big) + \sum_{y} p_Y(y)\,\log_2\big(p_Y(y)\big)\\
&= H(X,Y) - H(Y)\,, \tag{19.58}
\end{aligned}
$$

where we used $\sum_{x} p_{X,Y}(x,y) = p_Y(y)$. $\square$

The classical conditional entropy is non-negative,

$$H(X|Y) \geq 0 \,, \tag{19.59}$$

which follows from Definition 19.5, but it is generally *not* symmetric, since

$$H(Y|X) = H(X|Y) - H(X) + H(Y) \,. \tag{19.60}$$

The form of eqn (19.60) is reminiscent of Bayes' law from eqn (19.53), and can in this sense be seen as an extension of Bayes' law to entropies.

Both the joint and conditional entropy are bounded by the Shannon entropy of a subsystem.

Theorem 19.3 (Bounds on the joint and conditional entropy)

The joint entropy $H(X,Y)$ of a pair of random variables X and Y is greater or equal than the Shannon entropy of X,

$$H(X) \leq H(X,Y).$$

The conditional entropy $H(X|Y)$ is smaller or equal than the Shannon entropy of X,

$$H(X|Y) \leq H(X).$$

Thus conditioning does *not* increase the entropy!

Proof From the non-negativity of the conditional entropy, eqn (19.59), the symmetry of the joint entropy, eqn (19.54), and relabelling $X \leftrightarrow Y$, Lemma 19.2 allows us to obtain

$$0 \leq H(X|Y) = H(X,Y) - H(Y)$$
$$\Rightarrow \quad H(Y) \leq H(X,Y)$$
$$\Rightarrow \quad H(X) \leq H(X,Y) \,. \tag{19.61}$$

Again using Lemmas 19.2 and 19.1 we further have

$$H(X|Y) - H(X) = H(X,Y) - H(X) - H(Y) \tag{19.62}$$
$$\leq H(X) + H(Y) - H(X) - H(Y) = 0$$
$$\Rightarrow \quad H(X|Y) \leq H(X) \,.$$

$\square$

19.4   Mutual Information

Next we introduce an entropic measure for the *mutual information* that two parties Alice and Bob can share. Suppose, as usual, Alice describes the outcomes of some process by a random variable X and Bob those of another process by another random variable Y. What then is the information content shared between them? If we add to the information content of X, which is $H(X)$, the information content of Y, which is $H(Y)$, then the information content that is common to X and Y is counted twice. Therefore we have to subtract the joint information of the pair (X, Y), the joint entropy $H(X, Y)$, to arrive at the correct mutual information.

Definition 19.6   *The **classical mutual information** $I(X:Y)$ of two random variables X and Y is given by*

$$I(X:Y) = H(X) + H(Y) - H(X, Y) .$$

Physically, the mutual information specifies how much the uncertainty $H(X, Y)$ of the pair (X, Y) is smaller than the sum of the individual uncertainties $H(X) + H(Y)$. It represents a meaningful measure for the correlation between the values of X and Y. If the mutual information vanishes the joint entropy is just equal to the entropies of the marginals, i.e., the events of X and Y are statistically independent.

We can connect the mutual information with the conditional entropy via Lemma 19.2,

$$I(X:Y) = H(X) - H(X|Y) . \tag{19.63}$$

In this sense, the mutual information evaluates how much the uncertainty of the random variable X is reduced by knowing the outcome of the other variable Y. If Bob knows the outcome of his variable Y he has an uncertainty $H(X|Y)$ about Alice's variable X. Thus the information gain $H(X|Y)$ about X reduces the uncertainty $H(X)$ about X, which Bob would have if he did not have access to additional information.

Clearly, the mutual information is symmetric

$$I(X:Y) = I(Y:X) . \tag{19.64}$$

And it immediately follows from Theorem 19.3 that the mutual information is non-negative.

Corollary 19.1 (Non-negativity of classical mutual information)

The classical mutual information of any two random variables X and Y is non-negative

$$I(X:Y) \geq 0 .$$

Furthermore, the classical mutual information is bounded from above by the Shannon entropies of both subsystems as expressed in the following theorem.

Theorem 19.4 (Upper bound on classical mutual information)

The classical mutual information of any two random variables X and Y is bounded from above by the minimum of the entropies of the subsystems

$$I(X\!:\!Y) \ \leq \ \min\{H(X), H(Y)\}\,.$$

Proof Starting with the mutual information from Definition 19.6 and recalling Theorem 19.3, which assures us that $H(X) \leq H(X,Y)$ and $H(Y) \leq H(X,Y)$ we obtain

$$
\begin{aligned}
I(X\!:\!Y) \ &= \ H(X) + H(Y) - H(X,Y) \\
&\leq \ H(X,Y) + H(Y) - H(X,Y) = H(Y)\,, \qquad (19.65a) \\
\text{or} \quad &\leq \ H(X) + H(X,Y) - H(X,Y) = H(X)\,. \qquad (19.65b)
\end{aligned}
$$

Thus we have to choose the minimum, $I(X\!:\!Y) \leq \min\{H(X), H(Y)\}$. $\qquad\square$

However, when we extend these entropies to the quantum case, which we will do in Section 20.4, we shall notice that the classical bound can be surpassed for the quantum mutual information.

Alternative forms of the mutual information: Finally, let us note that—in the classical case—there are different equivalent ways of defining the mutual information. One such way is to express the mutual information as the expected value of the ratio of the joint probability distribution $p_{X,Y}(x,y)$ and the product of the marginal distributions, $p_X(x) \cdot p_Y(y)$ for x and y such that $p_X(x)$ and $p_Y(y)$ are non-vanishing, that is, we first define the quantity

$$p_{X,Y}(x\!:\!y) \ := \ \frac{p_{X,Y}(x,y)}{p_X(x) \cdot p_Y(y)} \ = \ \frac{p_{X,Y}(x|y)}{p_X(x)}\,, \qquad (19.66)$$

where $p_{X,Y}(x|y)$ is the conditional probability (19.51). Since $p_{X,Y}(x,y) \geq 0$ and we assume $p_X(x), p_Y(y) > 0$, their ratio is positive, but since the joint probability exceeds the product of the marginal probabilities when the respective values are correlated, the ratio $p_{X,Y}(x\!:\!y)$ can be larger than 1 and is hence not a probability. Nevertheless we can use it to define $P(X\!:\!Y) := \{p_{X,Y}(x\!:\!y)\}_{x,y}$, such that

$$I(X\!:\!Y) \ = \ E\big[P(X\!:\!Y)\big] \ = \ \sum_{x,y} p_{X,Y}(x,y)\log_2\big(p_{X,Y}(x\!:\!y)\big)\,. \qquad (19.67)$$

We can then easily check that this matches our previous Definition 19.6,

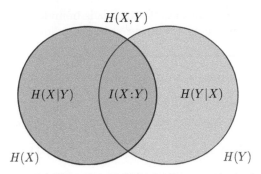

Fig. 19.3 Information diagram for classical entropies. The relation between different entropies is illustrated. The marginal entropies $H(X)$ (purple) and $H(Y)$ (blue) that quantify the average information content of the random variables X and Y respectively, are illustrated as disks. The joint entropy $H(X,Y) \equiv H(X \cup Y)$ is illustrated as the union of the sets representing the marginal entropies. Following Lemma 19.2, the conditional entropies are given by $H(X|Y) = H(X,Y) - H(Y)$ and $H(Y|X) = H(X,Y) - H(X)$ and are thus represented here as the sets of elements that are "in" $H(X,Y)$ but not "in" $H(Y)$ and $H(X)$, respectively, while the mutual information from Definition 19.6 is $I(X:Y) = H(X) + H(Y) - H(X,Y)$, and is hence illustrated as the intersection of the two disks.

$$
\begin{aligned}
I(X:Y) &= \sum_{x,y} p_{X,Y}(x,y) \log_2\Big(\tfrac{p_{X,Y}(x,y)}{p_X(x) \cdot p_Y(y)}\Big) \\
&= \sum_{x,y} p_{X,Y}(x,y) \log_2\big(p_{X,Y}(x,y)\big) - \sum_{x,y} p_{X,Y}(x,y)\big[\log_2\big(p_X(x)\big) + \log_2\big(p_Y(y)\big)\big] \\
&= -H(X,Y) + H(X) + H(Y).
\end{aligned}
\tag{19.68}
$$

We directly observe what we noticed already in Definition 19.6, namely that the mutual information $I(X:Y)$ of two random variables X and Y vanishes if the events are statistically independent, i.e., $p_{X,Y}(x,y) = p(x) \cdot p(y)$ or $p(x:y) = 1$. Conversely we can say, only in case of correlations, i.e., statistically dependent events, do we obtain a non-vanishing mutual information. In fact, we can even write the classical mutual information as the (Kullback–Leibler) divergence—the classical relative entropy—between the joint probability distribution $p_{X,Y}(x,y)$ and the product of the marginals, and indeed, we have already done this in the proof of Lemma 19.1. We can thus give a third equivalent form for the classical mutual information:

$$
I(X:Y) = H\big(P(X,Y) \,\|\, P(X)\,P(Y)\big).
\tag{19.69}
$$

The entropies we have discussed so far can be nicely illustrated in a so-called information diagram that is reminiscent of a Venn diagram, as long as one keeps in mind that the illustrated quantities are not in fact sets, see Fig. 19.3.

Example—Bob watching Alice's socks: Let us apply our entropy-information discussion to a simple example. Imagine that Alice describes the colour of her two socks by two random variables, L for the colour of her left sock, and R for the colour

Fig. 19.4 Bob watching Alice's socks. The colours of Alice's left and right socks can be described as random variables L and R.

of her right sock. Alice only has red and green socks, but she always wears socks of different colours. Bob is aware of this but he also noticed that Alice prefers to choose red on her left foot. The probabilities for red and green on Alice's feet are $p_L(\text{red}) = \frac{3}{4}$, $p_R(\text{red}) = \frac{1}{4}$. When Bob observes Alice, as illustrated in Fig. 19.4, the joint probabilities for seeing the colours on Alice's feet are

$$p(\text{red}, \text{green}) = \tfrac{3}{4}, \qquad p(\text{red}, \text{red}) = 0 , \tag{19.70a}$$

$$p(\text{green}, \text{red}) = \tfrac{1}{4}, \qquad p(\text{green}, \text{green}) = 0 .$$

Thus the mean uncertainty for the pair of socks, the joint entropy of L and R, is

$$H(L, R) = H(L) = H(R) = - \sum_{\text{red,green}} p(L, R) \log_2\big(p(L, R)\big) \tag{19.71}$$

$$= - \tfrac{3}{4} \log_2\big(\tfrac{3}{4}\big) - \tfrac{1}{4} \log_2\big(\tfrac{1}{4}\big) \approx 0.81 .$$

Since the joint entropy matches the marginal entropies, we get the same result for the mutual information, $I(L{:}R) = 0.81$. The remaining rest-uncertainty given by the conditional entropy vanishes $H(L|R) = H(L, R) - H(R) = 0$. So even though Bob can only predict with some certainty which colour of sock will appear on which foot, he can be sure that the colours on the two feet will not match. Of course, the information would be biggest $I(L : R) = 1$ for equal colour probabilities, if $p_{L, R}(\text{red}) = \frac{1}{2} = p_{L, R}(\text{green})$.

19.5 Rényi Entropy

The Hungarian mathematician *Alfréd Rényi* (1961) constructed an important extension to Shannon's entropy. In fact, he established a whole family of entropies labelled by a parameter α, which provide a range of options to quantify uncertainty and randomness in (quantum) information theory.

Definition 19.7 *Let X be a random variable that can take values $x_i \in \mathcal{X}$ with probabilities $p_i = p_X(x_i)$. Then the **Rényi entropy** of order $\alpha \geq 0$ is given by*

$$H_\alpha(X) = \tfrac{1}{1-\alpha} \log_2\Big(\sum_{i=1}^{n} p_i^\alpha \Big) .$$

If the probability distribution is uniform, that is, if all probabilities are equal, $p_i = \frac{1}{n} \ \forall i = 1, \ldots, n$, then the Rényi entropies of all orders coincide and evaluate to

$$H_\alpha(X) = \frac{1}{1-\alpha} \log_2\left(n \left(\tfrac{1}{n}\right)^\alpha\right) = \frac{1}{1-\alpha} \left[\log_2(n) + \alpha \log_2(\tfrac{1}{n})\right] = \log_2(n). \quad (19.72)$$

For any fixed discrete random variable X the Rényi entropy $H_\alpha(X)$, interpreted as a function of the order parameter α, is a non-increasing function in α, which we will formulate in terms of the following theorem.

Theorem 19.5 (Hierarchy of Rényi entropies)

For any $\alpha, \beta \geq 0$ and for any random variable X the Rényi entropies $H_\alpha(X)$ and $H_\beta(X)$ satisfy

$$H_\alpha(X) \leq H_\beta(X) \quad \forall \quad \alpha \geq \beta \geq 0 \,.$$

Proof To see this, we start by assuming $\alpha \neq 1$ and consider the derivative

$$\frac{\partial}{\partial \alpha} H_\alpha = \frac{1}{(1-\alpha)^2} \log_2\left(\sum_{i=1}^n p_i^\alpha\right) + \frac{1}{1-\alpha} \left(\sum_{j=1}^n p_j^\alpha\right)^{-1} \sum_{i=1}^n p_i^\alpha \log_2(p_i)$$

$$= \frac{1}{(1-\alpha)^2} \left[\log_2\left(\sum_{i=1}^n p_i^\alpha\right) + (1-\alpha)\left(\sum_{j=1}^n p_j^\alpha\right)^{-1} \sum_{i=1}^n p_i^\alpha \log_2(p_i)\right]$$

$$= \frac{1}{(1-\alpha)^2} \left[\log_2\left(\sum_{i=1}^n p_i^\alpha\right) - \frac{\sum_{i=1}^n p_i^\alpha \, \alpha \log_2(p_i)}{\sum_{j=1}^n p_j^\alpha} + \frac{\sum_{i=1}^n p_i^\alpha \log_2(p_i)}{\sum_{j=1}^n p_j^\alpha}\right]. \quad (19.73)$$

If we multiply the first term in the square brackets by $1 = \left(\sum_j p_j^\alpha\right)/\left(\sum_{j'} p_{j'}^\alpha\right)$ and in the second term we rewrite $\alpha \log_2(p_i) = \log_2(p_i^\alpha)$, then (with some renaming of summation indices) the first two terms can be collected into

$$\frac{\sum_i p_i^\alpha}{\sum_{j'} p_{j'}^\alpha} \log_2\left(\sum_{j=1}^n p_j^\alpha\right) - \frac{\sum_i p_i^\alpha \log_2(p_i^\alpha)}{\sum_j p_j^\alpha} = -\sum_{i=1}^n \frac{p_i^\alpha}{\sum_{j'} p_{j'}^\alpha} \log_2\left(\frac{p_i^\alpha}{\sum_j p_j^\alpha}\right). \quad (19.74)$$

The quantities $q_i := p_i^\alpha / \left(\sum_j p_j^\alpha\right) \geq 0$ sum to one, and can thus be identified with the probabilities in a distribution $Q(X) = \{q_i\}_i$. We thus see that, up to a non-positive pre-factor $-(1-\alpha)^{-2}$, the derivative of H_α is proportional to the relative entropy (see Definition 19.3) of the distributions $Q(X)$ and $P(X)$,

$$\frac{\partial}{\partial \alpha} H_\alpha = -\frac{1}{(1-\alpha)^2} H\big(Q(X)\,\|\,P(X)\big) \leq 0. \quad (19.75)$$

From Gibbs' inequality (Theorem 19.1) it then follows that the derivative is non-positive. The special case $\alpha \to 1$ is included in this consideration, since the limits from the left and right coincide, $\lim_{\alpha \to 1_-} H_\alpha(X) = \lim_{\alpha \to 1_+} H_\alpha(X) = \lim_{\alpha \to 1} H_\alpha(X)$ and evaluate to

$$\lim_{\alpha \to 1} H_\alpha(X) = \lim_{\alpha \to 1} \frac{\frac{\partial}{\partial \alpha} \log_2 \left(\sum_i p_i^\alpha \right)}{\frac{\partial}{\partial \alpha} (1 - \alpha)} = - \lim_{\alpha \to 1} \Big(\sum_{j=1}^n p_j^\alpha \Big)^{-1} \sum_{i=1}^n p_i^\alpha \log_2(p_i)$$

$$= - \sum_{i=1}^n p_i \log_2(p_i) = H(X), \tag{19.76}$$

where we have used de l'Hôpital's rule, and we in fact find that the limit yields the Shannon entropy $H(X)$. In conclusion, we thus see that the Rényi entropy $H_\alpha(X)$ must hence be decreasing with increasing α for any fixed random variable. $\qquad\square$

Some of the expressions for the Rényi entropies for particular α values that are commonly used in information theory are:

- **Hartley entropy**: $\alpha = 0$, $H_0(X) = \log_2(n)$.

- **Shannon entropy**: $\alpha \to 1$, $H_1(X) = - \sum_{i=1}^n p_i \log_2(p_i)$.

- **Collision entropy** or **Rényi-2 entropy**: $\alpha = 2$, $H_2(X) = - \log_2 \Big(\sum_{i=1}^n (p_i^2) \Big)$.

- **Min-entropy**: $\alpha \to \infty$, $H_\alpha \to H_\infty(X) = \min_i \big(- \log_2(p_i) \big) = - \log_2(\max_i p_i)$.

The **min-entropy** $H_\infty(X)$ represents the smallest entropy measure in the α spectrum of the Rényi entropies according to Theorem 19.5. In this sense, it is the strongest method to measure the information content of a random variable X. In particular, the min-entropy is always smaller or equal than the Shannon entropy.

To illustrate these quantities, we can consider their expressions for the case of a binary random variable X with outcomes 0 and 1 and probabilities $P(X = 0) = p$ and $P(X = 1) = 1 - p$. The explicit formulas obtained for $\alpha = 0$, $\alpha \to 1$, $\alpha = 2$, and for $\alpha \to \infty$, plotted in Fig. 19.5 as functions of p, are:

- Hartly entropy: $\alpha = 0$, $H_0(\{p, 1 - p\}) = \log_2(2)$.

- Shannon entropy: $\alpha \to 1$, $H_1(\{p, 1 - p\}) = - p \log_2(p) - (1 - p) \log_2(1 - p)$.

- Collision/Rényi-2 entropy: $\alpha = 2$, $H_2(\{p, 1 - p\}) = - \log_2 \big(p^2 + (1 - p)^2 \big)$.

- Min-entropy: $\alpha \to \infty$, $H_\infty(\{p, 1 - p\}) = - \log_2 \big(\max\{p, 1 - p\} \big)$.

Rényi divergence: Another important quantity is the *Rényi divergence* of order α, which generalizes the Kullback–Leibler divergence, i.e., the relative entropy from Definition 19.3.

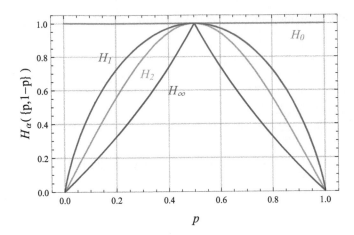

Fig. 19.5 Rényi entropies. The quantities H_0, H_1, H_2, H_∞, i.e., the Rényi entropies for $\alpha = 0, 1, 2$ and $\alpha \to \infty$ are plotted for a binary random variable with probability distribution $\{p, 1 - p\}$ as a function of $p \in [0, 1]$.

Definition 19.8   *For probability distributions $P(X) = \{p_i\}_i$ and $Q(X) = \{q_i\}_i$ for the same random variable X with values $x_i \in \mathcal{X}$, the **Rényi divergence** of order α with $0 < \alpha < \infty$, $\alpha \neq 1$ is given by*

$$D_\alpha(P \| Q) = \frac{1}{\alpha - 1} \log_2 \left(\sum_{i=1}^{n} \frac{p_i^\alpha}{q_i^{\alpha - 1}} \right).$$

The Rényi divergence is a non-decreasing function of its order α (see, e.g., Theorem 3 in van Erven and Harremoës, 2007) and being a divergence, a kind of "distance" between probability distributions, it is positive (non-negative).

Theorem 19.6 (Hierarchy of Rényi divergences)

For any $0 < \alpha, \beta < \infty$ with $\alpha, \beta \neq 1$, and any probability distributions $P(X)$ and $Q(X)$ over a discrete random variable X the Rényi divergences $D_\alpha(P \| Q)$ satisfy

$$D_\alpha(P \| Q) \leq D_\beta(P \| Q) \quad \forall \quad \alpha \leq \beta.$$

Proof   For the proof we follow the argument in van Erven and Harremoës (2007, Theorem 3) and first note that for $\alpha < \beta$ and $x \geq 0$ the function $f(x) = x^{(\alpha-1)/(\beta-1)}$ is convex if $\alpha < 1$ and concave if $\alpha > 1$, which we can see from the sign of the second derivative,

$$\frac{\partial^2}{\partial x^2} x^{\frac{\alpha-1}{\beta-1}} = \left(\frac{\alpha-1}{\beta-1} \right)\left(\frac{\alpha-1}{\beta-1} - 1 \right) x^{\frac{\alpha-1}{\beta-1}-2} = \frac{(\alpha-1)(\alpha-\beta)}{(\beta-1)^2} x^{\frac{\alpha-1}{\beta-1}-2}, \quad (19.77)$$

since $x^\gamma \geq 0$ for $x \geq 0$. We then write the Rényi divergence of order α as

$$D_\alpha(P \,\|\, Q) = \frac{1}{\alpha - 1} \log_2 \Big(\sum_{i=1}^n \frac{p_i^\alpha}{q_i^{\alpha - 1}} \Big) = \frac{1}{\alpha - 1} \log_2 \Big(\sum_{i=1}^n p_i \big(\tfrac{q_i}{p_i} \big)^{(1-\beta)\frac{\alpha-1}{\beta-1}} \Big), \quad (19.78)$$

and use Jensen's inequality (Jensen, 1906) for $x_i = \big(\tfrac{q_i}{p_i} \big)^{(1-\beta)}$. When $\alpha < 1$ and $f(x)$ is convex, we have $\sum_i p_i f(x_i) \geq f(\sum_i p_i x_i)$ but also $\alpha - 1 < 0$, whereas for $\alpha > 1$ the function $f(x)$ is concave and the inequality is reversed, but also $\alpha - 1 > 0$. In either case we have

$$\frac{1}{\alpha-1} \log_2 \Big(\sum_{i=1}^n p_i \big(\tfrac{q_i}{p_i} \big)^{(1-\beta)\frac{\alpha-1}{\beta-1}} \Big) \leq \frac{1}{\alpha-1} \log_2 \Big(\Big[\sum_{i=1}^n p_i \big(\tfrac{q_i}{p_i} \big)^{(1-\beta)} \Big]^{\frac{\alpha-1}{\beta-1}} \Big) \qquad (19.79)$$

$$= \frac{1}{\alpha-1} \frac{\alpha-1}{\beta-1} \log_2 \Big(\sum_{i=1}^n p_i \big(\tfrac{q_i}{p_i} \big)^{(1-\beta)} \Big) = D_\beta(P \,\|\, Q).$$

$\square$

The orders α in $]0, 1[\cup]1, \infty[$ are called "simple orders". For the so-called extended orders $\alpha = 0, 1$ and $\alpha \to \infty$, Definition 19.8 cannot be used directly, but the Rényi divergences for these orders are instead defined via the continuity in α as (van Erven and Harremoës, 2007)

$$D_{\alpha=0}(P \,\|\, Q) := \inf_{0 < \alpha < 1} D_\alpha(P \,\|\, Q) = - \log_2 \Big(\sum_{i : p_i \neq 0} q_i \Big), \qquad (19.80a)$$

$$D_{\alpha=1}(P \,\|\, Q) := H\big(P(X) \,\|\, Q(X)\big), \qquad (19.80b)$$

$$D_{\alpha \to \infty}(P \,\|\, Q) := \sup_{\alpha > 1} D_\alpha(P \,\|\, Q) = \log_2 \big(\sup_i \tfrac{p_i}{q_i} \big), \qquad (19.80c)$$

where $H\big(P(X) \,\|\, Q(X)\big)$ is the relative entropy (the Kullback–Leibler divergence) from Definition 19.3. And the hierarchy of Theorem 19.6 thus holds for all orders, not just for the simple ones. In addition, all orders give non-negative divergences.

Theorem 19.7 (Non-negativity of Rényi divergence)

The classical Rényi divergence is non-negative

$$D_\alpha(P \,\|\, Q) \geq 0,$$

with equality if and only if $P = Q$ for $\alpha > 0$.

Proof To prove that $D_\alpha(P \,\|\, Q) \geq 0$ we can just observe that for $\alpha = 0$ the argument of the logarithm in the Rényi divergence is in the interval $]0, 1]$, and thus $D_{\alpha=0}(P \,\|\, Q) \geq 0$. Since the Rényi divergences are non-decreasing (Theorem 19.6), this means we must have $D_\alpha(P \,\|\, Q) \geq 0$ also for all $\alpha > 0$. For the proof of the conditions for equality we refer to (van Erven and Harremoës, 2007, Theorem 8). $\square$

Finally, we note that the Rényi entropy from Definition 19.7 can be expressed in terms of the Rényi divergence, Definition 19.8. To this end we consider the Rényi divergence for a probability distribution P from the uniform distribution $P_{\mathrm{U}} = \{\frac{1}{n}, \ldots, \frac{1}{n}\}$, in which case we have (van Erven and Harremoës, 2007)

$$H_\alpha(P) = H_\alpha(P_{\mathrm{U}}) - D_\alpha(P \| P_{\mathrm{U}}) = \log_2(n) - D_\alpha(P \| P_{\mathrm{U}}) . \qquad (19.81)$$

Résumé: Entropy is a concept that historically originates from thermodynamics and statistical physics, originally associated with the number of microstates leading to the same macrostate. Its original interpretation derives from the fact that the variation with respect to the energy provides the absolute temperature.

Shannon later introduced the concept of entropy into classical information theory, where it represents the gain of information upon receiving a message. But, at the same time, entropy is also a measure for the uncertainty of a random variable. Based on this notion, different types of entropic measures can be defined, like the relative entropy, the joint and conditional entropy. These entropic measures can be used to express the mutual information shared by two parties. The mutual information specifies how much smaller the uncertainty of a pair of random variables is compared to the sum of the uncertainties of the individual random variables. In this sense, it is a meaningful measure for the correlation between the values of the random variables.

Finally, the notion of entropy can be generalized to the family of Rényi entropies, which provide useful tools for describing additional structure encoded in the probability distributions of two random variables, and we will employ these entropies in the study of quantum correlations in the next chapter of our book.

20

Quantum Entropy and Correlations in Quantum Information

In this chapter we discuss the extension of the concept of entropy from classical to quantum information theory. In particular, this means that we are interested in information measures that quantify the amount of information in a quantum system. In this regard, the most fundamental quantity is the von Neumann entropy, the quantum analogue of Shannon's entropy. It measures the degree of uncertainty that we have about a quantum state. Here, we will study the properties of the von Neumann entropy and the important inequalities it obeys in detail. We then discuss the quantum relative entropy and joint entropy, the quantum conditional entropy and mutual information, and finally, the quantum Rényi entropy. We then turn to the quantum mutual information, where the Shannon entropies are replaced by the von Neumann entropies. But this "simple" replacement has profound consequences: The quantum version exceeds its classical bound, which turns out to capture an essential feature of quantum information. Also the quantum conditional entropy deviates radically from its classical counterpart. It can even become *negative* for certain quantum states. This negativity of the conditional entropy—a typical quantum feature—is of high significance and we therefore investigate its relation to non-locality and entanglement in a separate section.

Although many aspects of quantum entropies are thus covered here, there is much more to know about this interesting topic than can be conveyed here alone, and we therefore direct the reader to further books and reviews in this field (e.g., Wehrl, 1978; Thirring, 1980; Nielsen and Chuang, 2000; Alicki and Fannes, 2001; Vedral, 2002; Horodecki *et al.*, 2009; Schumacher and Westmoreland, 2010; Wilde, 2017).

20.1 Von Neumann Entropy

A quantum state is fully described by a density matrix that expresses all inherent uncertainties. Therefore, a quantum measure of uncertainty should be related to the density matrix similarly as the classical measure of uncertainty is associated to the probability distribution of a random variable. And indeed, we obtain *von Neumann's entropy* for quantum systems by identifying the probabilities in Shannon's classical formula (Definition 19.1) with the eigenvalues of density matrices:

Definition 20.1 *The **von Neumann entropy** of a density matrix ρ is defined as*

$$S(\rho) := -\operatorname{Tr}\big(\rho \log(\rho)\big).$$

Unless otherwise specified, the logarithm in Definition 20.1 is understood to be the natural logarithm, $\log(x) \equiv \ln(x)$. But in many cases in quantum information, in particular, when considering qubits, it is customary to instead consider the logarithm to base 2, i.e., $\log(x) \rightarrow \log_2(x) = \frac{\ln(x)}{\ln(2)}$.

The trace of an operator is defined via its eigenvalues, so we first write the density operator ρ on a d-dimensional Hilbert space in *spectral decomposition*

$$\rho = \sum_{i=0}^{d-1} p_i \, |i\rangle\langle i| \,, \quad \text{with} \quad p_i \geq 0, \, \sum_{i=0}^{d-1} p_i = 1 \,, \tag{20.1}$$

where $|i\rangle$ are the (pure) eigenstates of ρ. Let us calculate the von Neumann entropy of the state in (20.1),

$$S(\rho) = -\operatorname{Tr}\big(\rho \log(\rho)\big) = -\operatorname{Tr}\Big[\sum_{i=0}^{d-1} p_i \, |i\rangle\langle i| \, \log\Big(\sum_{j=0}^{d-1} p_j \, |j\rangle\langle j|\Big)\Big] \tag{20.2}$$

$$= -\sum_{i=0}^{d-1} p_i \sum_{n=0}^{d-1} \langle n|i\rangle \langle i| \, \log\Big(\sum_{j=0}^{d-1} p_j \, |j\rangle\langle j|\Big) \, |n\rangle = -\sum_{i=0}^{d-1} p_i \log(p_i) = H(\{p_i\}) \,.$$

We find that the von Neumann entropy arises[1] as the Shannon entropy for a random variable X whose associated probability distribution $P(X) = \{p_i\}_i$ corresponds to the spectrum of the density operator.

For a maximally mixed state $\rho_{\text{mix}} = \frac{1}{d}\mathbb{1}_d$ we get $S(\rho_{\text{mix}}) = \log(d)$. Meanwhile, for any pure quantum state $|\psi\rangle$ the density matrix is a rank-one projector $|\psi\rangle\langle\psi|$, with one eigenvalue $p_0 = 1$ while all other eigenvalues vanish. For the eigenvalue $p_0 = 1$ we just have $p_0 \log(p_0) = 1 \log(1) = 0$. For the eigenvalues $p_{i>0} = 0$ we apply the rule of de l'Hôpital's rule,

$$\lim_{p \to 0} p \log(p) = \lim_{p \to 0} \frac{\log(p)}{p^{-1}} = \lim_{p \to 0} \frac{\frac{\partial}{\partial p} \log(p)}{\frac{\partial}{\partial p} p^{-1}} = \lim_{p \to 0} \frac{p^{-1}}{-p^{-2}} = = -\lim_{p \to 0} p = 0 \,. \tag{20.3}$$

For the von Neumann entropy we thus have

$$S\big(|\psi\rangle\langle\psi|\big) = -\sum_{i=0}^{d-1} p_i \log(p_i) = -1 \log(1) - (d-1)\lim_{p \to 0} p \log(p) = 0 \,, \tag{20.4}$$

the entropy of a pure state vanishes, while the entropy of all other states takes values in between those of pure and maximally mixed states,

---

[1]Since quantum information theory arose long after classical information theory, one might think that the von Neumann entropy was defined after Shannon considered entropy of classical systems. But in fact, von Neumann established his quantity in the context of statistical physics long before Shannon established his information-theoretic entropy.

$$0 \leq S(\rho) \leq \log(d) . \qquad (20.5)$$

By choosing the base of the logarithm to be the system dimension, we can normalize the entropy, that is, for $\log(x) \rightarrow \log_d(x) = \frac{\ln(x)}{\ln(d)}$ we have the entropy range

$$0 \leq S(\rho) \leq 1 . \qquad (20.6)$$

Physical interpretation: Let us consider the von Neumann entropy in the context of a quantum communication scenario between two observers, Alice and Bob. Suppose Alice prepares a quantum system in a state that she randomly selects from an ensemble $\{|i\rangle\}_i$, and the probability that $|i\rangle$ is chosen is p_i. Alice then sends this to Bob, who knows the ensemble and the associated probabilities but does not yet know which specific state the system sent by Alice is in. Then the density matrix ρ that Bob assigns to the system he receives is given by eqn (20.1). The von Neumann entropy $S(\rho)$ quantifies Bob's uncertainty about the state Alice is sending. Alternatively we can view $S(\rho)$ as representing the average information gain, measured in qubits, after having received the system. Quite generally we can state:

Proposition 20.1 (Von Neumann entropy—degree of uncertainty)

The von Neumann entropy quantifies the degree of uncertainty—the lack of knowledge—about a quantum state.

Or, to phrase it in the spirit of *Walter Thirring* (e.g., as discussed in Thirring 1980):

Proposition 20.2 (Entropy statement à la Thirring)

Mixed states provide only partial information about a quantum system. The von Neumann entropy expresses how much of the maximal information is missing.

Properties of the von Neumann entropy:

(i) **Unitary invariance:** $S(\rho)$ is invariant under unitary transformations U

$$S(U\rho U^\dagger) = S(\rho) \quad \text{for} \quad UU^\dagger = U^\dagger U = \mathbb{1} . \qquad (20.7)$$

(ii) **Concavity:** $S(\rho)$ is a concave functional on the Hilbert–Schmidt space

$$S\left(\sum_i p_i \rho_i\right) \geq \sum_i p_i S(\rho_i) \quad \forall\, p_i \geq 0, \ \sum_i p_i = 1, \ \text{and} \ \forall \ \text{sets} \ \{\rho_i\}_i . \qquad (20.8)$$

Equality holds if and only if $\rho_i = \rho \ \forall i$.

In words, the uncertainty about the state ρ is greater (or equal) than the average uncertainty of the states ρ_i in the ensemble.

(iii) **Subadditivity**: For a density matrix ρ of a bipartite composite quantum system on the Hilbert–Schmidt space $\mathcal{B}_{HS}(\mathcal{H}_A \otimes \mathcal{H}_B)$ with subsystems A and B, and corresponding reduced density matrices $\rho_A = \mathrm{Tr}_B(\rho)$ and $\rho_B = \mathrm{Tr}_A(\rho)$, the following inequalities hold:

$$S(\rho) \leq S(\rho_A) + S(\rho_B) \qquad \textbf{subadditivity,} \tag{20.9}$$

$$S(\rho) = S(\rho_A) + S(\rho_B) \qquad \text{iff } \rho = \rho_A \otimes \rho_B \text{ is a } \textbf{product state,} \tag{20.10}$$

$$S(\rho) \geq |S(\rho_A) - S(\rho_B)| \qquad \textbf{Araki–Lieb inequality.} \tag{20.11}$$

Physically, eqn (20.9) means that the uncertainty about the composite system is smaller (or equal) than the combined uncertainties of the separate subsystems. Clearly, by tracing over one subsystems, A or B, information about the correlations between A and B is lost, the entropy increases.

(iv) **Vanishing for pure states**: The von Neumann entropy vanishes if and only if ρ is a pure quantum state $\rho = |\psi\rangle\langle\psi|$,

$$S(\rho) = 0 \quad \text{iff} \quad \rho \text{ is pure.} \tag{20.12}$$

(v) **Matching subsystem entropies for pure bipartite states**: For any bipartite system in a *pure* state $|\psi\rangle_{AB}$, the reduced states ρ_A and ρ_B have the same eigenvalues λ_i^2, where the $\lambda_i \geq 0$ are the Schmidt coefficients of $|\psi\rangle_{AB}$. Consequently the entropies of the subsystems are equal

$$S(\rho_A) = S(\rho_B) = -\sum_i \lambda_i^2 \log_2(\lambda_i^2) \geq 0, \tag{20.13}$$

and are strictly positive $S(\rho_{A/B}) > 0$ iff $|\psi\rangle_{AB}$ is entangled.

The proof of the unitary invariance, property (i), is quite straightforward. Since unitaries leave the eigenvalues of ρ invariant and $S(\rho)$ is a function of only the eigenvalues, the entropy must be unitarily invariant. We have already proven that the entropy vanishes for pure states in eqn (20.4), and it is clear that any probability distribution $\{p_i\}_i$ in which more than one value p_i is different from zero must have an entropy larger than zero, since the entropy is a sum of non-negative terms. Thus both directions of property (iv) are proven. Similarly, property (v) follows directly from Lemma 15.1.

For the remaining properties, (ii) and (iii), including the Araki–Lieb inequality, we require a little more work. We will therefore first formulate and prove two useful lemmas, which we will then employ, in conjunction with the subadditivity property, to prove the concavity property (ii) at the end of this section on page 664. To prove the subadditivity property (iii) itself, we will make use of the quantum relative entropy and its non-negativity, a relation known as Klein's inequality, as we will discuss in Section 20.3.1, where we also prove the Araki–Lieb inequality.

Let us therefore now continue with the following lemmas:

Lemma 20.1 (Entropy for mixed quantum states)

Let $\{p_i \,|\, p_i \geq 0, \sum_i p_i = 1\}$ be a probability distribution and let ρ_i be quantum states with support on orthogonal subspaces, then the von Neumann entropy satisfies the relation

$$S\Big(\rho = \sum_i p_i \, \rho_i \Big) = H(\{p_i\}) + \sum_i p_i \, S(\rho_i),$$

where $H(\{p_i\})$ is the Shannon entropy of the probability distribution.

If the ρ_i are pure quantum states then $S(\rho_i) = 0$ and Lemma 20.1 coincides with eqn (20.2).

Proof Let us denote the eigenvalues and eigenvectors of ρ_i as $p_j^{(i)}$ and $|\,\psi_j^{(i)}\,\rangle$, respectively, such that $\rho_i = \sum_j p_j^{(i)} |\,\psi_j^{(i)}\,\rangle\langle\,\psi_j^{(i)}\,|$. Since the ρ_i are assumed to have orthogonal support, the eigenvalues of $\rho = \sum_i p_i \, \rho_i$ are $\tilde{p}_{i,j} = p_i \, p_j^{(i)}$, and thus

$$S(\rho) = -\sum_{i,j} p_i \, p_j^{(i)} \, \log(p_i \, p_j^{(i)}) = -\sum_{i,j} p_i p_j^{(i)} \big(\log(p_i) + \log(p_j^{(i)})\big) \tag{20.14}$$

$$= -\sum_i p_i \, \log(p_i) \Big(\sum_j p_j^{(i)}\Big) - \sum_i p_i \sum_j p_j^{(i)} \, \log(p_j^{(i)})$$

$$= H(\{p_i\}) + \sum_i p_i \, S(\rho_i). \qquad\qquad \square$$

Now we turn to a generalization of the previous lemma to certain quantum states of bipartite composite quantum systems.

Lemma 20.2 (Entropy of bipartite systems)

Let $\{p_i \,|\, p_i \geq 0, \sum_i p_i = 1\}$ be a probability distribution and let ρ_i be quantum states on a Hilbert space $\mathcal{H}_A$, and $\{|\,i\,\rangle\}_i$ a set of orthonormal quantum states on Hilbert space $\mathcal{H}_B$, then the entropy of the quantum state

$$\rho_{AB} = \sum_i p_i \, \rho_i \otimes |\,i\,\rangle\langle\,i\,|$$

satisfies the relation

$$S(\rho_{AB}) = H(\{p_i\}) + \sum_i p_i \, S(\rho_i).$$

Proof For the proof we just note that appending the (orthogonal) pure states $|i\rangle$ to the (arbitrary) density operators ρ_i means that the density operators $\rho_i \otimes |i\rangle\langle i|$ have support on orthogonal subspaces, independently of the choice of the ρ_i. Moreover, the non-zero eigenvalues of $\rho_i \otimes |i\rangle\langle i|$ are given by the at most $d_A = \dim(\mathcal{H}_A)$ non-zero eigenvalues of ρ_i. Therefore $S(\rho_i \otimes |i\rangle\langle i|) = S(\rho_i)$ and from Lemma 20.1 it follows that

$$S(\rho_{AB}) = H(\{p_i\}) + \sum_i p_i\, S(\rho_i)\,. \tag{20.15}$$

$\square$

Proof of the concavity of the von Neumann entropy (property (ii)) With Lemmas 20.1 and 20.2 at hand, we are now in a position to prove the concavity property of the von Neumann entropy. Let us consider a composite system as in Lemma 20.2, described by the density matrix $\rho_{AB} = \sum_i p_i\, \rho_i \otimes |i\rangle\langle i|$, where ρ_i is any density matrix on the subsystem A and $\{|i\rangle\}_i$ a set of orthonormal quantum states on the auxiliary subsystem B corresponding to the index i of ρ_i. Then we know from Lemma 20.2 that the entropy of the composite system can be expressed by $S(\rho_{AB}) = H(\{p_i\}) + \sum_i p_i\, S(\rho_i)$. Meanwhile, the entropies of the subsystems are given by

$$S(\rho_A) = S\Big(\sum_i p_i\, \rho_i\Big)\,, \tag{20.16a}$$

$$S(\rho_B) = S\Big(\sum_i p_i\, |i\rangle\langle i|\Big) = -\sum_i p_i\, \log(p_i) = H(\{p_i\})\,. \tag{20.16b}$$

Now we make use of the subadditivity, property (iii) from eqn (20.9), for the composite system, we insert the corresponding entropies and find

$$S(\rho_{AB}) \leq S(\rho_A) + S(\rho_B)$$

$$H(\{p_i\}) + \sum_i p_i\, S(\rho_i) \leq S\Big(\sum_i p_i \rho_i\Big) + H(\{p_i\}) \tag{20.17}$$

$$\sum_i p_i\, S(\rho_i) \leq S\Big(\sum_i p_i \rho_i\Big)\,. \qquad \square$$

20.2 Quantum Rényi Entropy

Next, we introduce the *quantum Rényi entropy* in analogy to the classical case, Section 19.5, generalizes the von Neumann entropy and quantifies the uncertainty and randomness of a quantum system with different weights depending on the parameter α. A detailed discussion of this quantity can be found in Müller-Lennert *et al.* (2013).

Definition 20.2 *For a d-dimensional quantum system described by a density matrix ρ with eigenvalues p_i the **Rényi entropy** of order $\alpha \geq 0$ is given by*

$$S_\alpha(\rho) = \tfrac{1}{1-\alpha}\log \mathrm{Tr}\big(\rho^\alpha\big) = \tfrac{1}{1-\alpha}\log\Big(\sum_{i=1}^{d} p_i^\alpha\Big)\,.$$

As in classical information theory, some of the α values give rise to quantities of particular relevance in quantum information:

- **Hartley entropy**: $\alpha = 0$, $S_0(\rho) = \log(d)$.

- **Von Neumann entropy**: $\alpha \to 1$, $S_1(\rho) = -\operatorname{Tr}(\rho \log(\rho))$.

- **Quantum Rényi-2 entropy**: $\alpha = 2$, $S_2(\rho) = -\log(\operatorname{Tr}[\rho^2]) = -\log\left(\sum_{i=1}^{d} p_i^2\right)$.

- **Min-entropy**: $\alpha \to \infty$, $S_{\min}(\rho) = S_{\alpha \to \infty}(\rho) = -\log \|\rho\|$,

  where $\|\rho\| = \sup\{\frac{\|\rho|\psi\rangle\|}{\||\psi\rangle\|} \,|\, |\psi\rangle \in \mathcal{H}, |\psi\rangle \neq 0\}$ denotes the operator norm of ρ.

As we recall from Section 19.2.1 entropy quite generally quantifies how much information is gained from receiving a message. The less likely a particular message is, the higher the level of surprise upon receiving, the more information is gained. With regard to the Rényi entropies, the larger the parameter α is the more weight is put on higher likelihood. As in classical information theory, the quantum Rényi entropy is a decreasing (non-increasing) function in the parameter α,

$$S_\alpha(\rho) \leq S_\beta(\rho) \quad \forall \quad \alpha \geq \beta \geq 0, \tag{20.18}$$

with the min-entropy $S_{\min}(\rho)$ representing the smallest entropy measure in the α-spectrum of Rényi entropies, i.e., it is the strongest method to measure the information content of a message. This follows directly from Theorem 19.5, since the quantum Rényi entropies for any fixed state ρ are just the classical Rényi entropies $H_\alpha(X)$ for a random variable X distributed according to the probability distribution given by the eigenvalues of ρ,

$$S_\alpha(\rho) = H_\alpha(\{p_i\}). \tag{20.19}$$

The hierarchy of the quantum Rényi entropies from (20.18) can be useful, for example, for the detection of entanglement, i.e., for $\alpha \geq 2$, entropy measures based on Rényi entropies can provide stronger entanglement criteria than corresponding measures based on the von Neumann's entropy ($\alpha = 1$), as we will discuss in Section 20.7.

Finally, let us also again mention the **quantum linear entropy**,

$$S_{\mathrm{L}}(\rho) = \operatorname{Tr}(\rho - \rho^2) = 1 - \operatorname{Tr}(\rho^2). \tag{20.20}$$

Although it is not within the family of Rényi entropies, the linear entropy is a measure for the mixedness of a quantum state that we have already discussed before, recall Section 11.2, which finds application in many problems in quantum information theory (e.g., recall Lemma 17.5).

20.3 Quantum Relative Entropy and Quantum Joint Entropy

Analogously to the classical case of Section 19.3 we now want to discuss the quantum relative entropy and the quantum joint entropy in Sections 20.3.1 and 20.3.2, respectively.

20.3.1   Quantum Relative Entropy

The *quantum relative entropy* is defined analogously to the classical relative entropy (or Kullback–Leibler divergence) but replacing the probability distributions by density matrices.

Definition 20.3 *For density matrices ρ and σ in $\mathcal{B}_{\mathrm{HS}}(\mathcal{H})$, the **quantum relative entropy** or **quantum Kullback–Leibler divergence** is given by*

$$S(\rho \| \sigma) \; := \; \mathrm{Tr}\big(\rho \big[\log(\rho) - \log(\sigma)\big]\big) \, .$$

Similar to the classical case the quantum relative entropy is an entropic measure for the closeness of two density matrices ρ and σ. However, in a strict sense it is not a geometric distance (but a divergence) since it is not symmetric under the exchange of ρ and σ, and it does not satisfy the triangle inequality. Nevertheless, the quantum relative entropy is often used as a substitute for a distance measure, e.g., to quantify the (one-sided) "distance" of an entangled state to the set of separable states, in this connection recall eqn (16.72) in Section 16.2.3.

Like the classical relative entropy the quantum relative entropy is not bounded from above.

Lemma 20.1 (Divergence of relative entropy)

The quantum relative entropy diverges if the kernel $\ker(\sigma)$ of σ has a non-trivial intersection with the support $\mathrm{supp}(\rho)$ of ρ,

$$S(\rho \| \sigma) \longrightarrow \infty \quad \textit{if} \;\; \ker(\sigma) \cap \mathrm{supp}(\rho) \neq \emptyset$$
$$S(\rho \| \sigma) \; < \; \infty \quad \textit{if} \;\; \ker(\sigma) \cap \mathrm{supp}(\rho) = \emptyset \, .$$

Proof   Recall, the kernel $\ker(\sigma)$ represents the vector space $\mathcal{H}$ spanned by the eigenvectors of σ with eigenvalue zero, thus $\ker(\sigma) = \{|\psi\rangle \in \mathcal{H} \,|\, \sigma \,|\psi\rangle = 0\}$, and the support $\mathrm{supp}(\rho)$ is the vector space spanned by the eigenvectors of ρ with non-zero eigenvalues, thus $\mathrm{supp}(\rho) = \mathrm{span}(\{|\phi\rangle \in \mathcal{H} \,|\, \rho \,|\phi\rangle = p\,|\phi\rangle \,, p > 0\})$. In particular, the kernel is the complement of the support and $\ker(\sigma) \cap \mathrm{supp}(\rho) \neq \emptyset$ hence indicates that the states ρ and σ are (partially) supported on orthogonal subspaces.

There are thus contributions to these states that are perfectly distinguishable (see, e.g., the discussion in Section 23.4), and the two states are in this sense (loosely speaking) "infinitely different". Formally, the statement $\ker(\sigma) \cap \mathrm{supp}(\rho) \neq \emptyset$ implies that the term $-\mathrm{Tr}\big(\rho \log(\sigma)\big)$ in $S(\rho \| \sigma)$ has at least one contribution of the form $-p \log(q)$ for $p > 0$ and $q = 0$, compare, e.g., to the expressions in eqns (20.22) and (20.23). Technically this is not well-defined, but we can think of such expressions as limits $q \to 0$, for which the relative entropy diverges. Meanwhile, the condition $\ker(\sigma) \cap \mathrm{supp}(\rho) = \emptyset$ means that all terms in $S(\rho \| \sigma)$ for which the argument of the logarithm vanishes also have a vanishing pre-factor, in which case the product vanishes, see, e.g., eqn (20.3). □

Here, one should be careful to note that a divergence of the relative entropy does not necessarily correspond to two states that are very different according to other measures. In particular, let $\sigma = \sum_i p_i |i\rangle\langle i|$ be state for which $p_i = 0$ for $i = 0$ and $p_i > 0$ for $i > 0$, and let ρ be given by

$$\rho = \sigma + \varepsilon \, |0\rangle\langle 0| - \varepsilon \, |1\rangle\langle 1| \,, \tag{20.21}$$

such that $p_1 - \varepsilon > 0$. Then $|0\rangle$ is in the support of ρ and in the kernel of σ, and the relative entropy diverges, despite the fact that ε, and thus any reasonable measure of distance between ρ and σ can be made arbitrarily small.

Nevertheless, the quantum relative entropy is non-negative like its classical analogue, a statement known as *Klein's inequality* (Klein, 1931).

Theorem 20.1 (Klein's inequality)

The quantum relative entropy is non-negative

$$S(\rho \| \sigma) \geq 0 \,,$$

with equality if and only if $\rho = \sigma$.

Proof Let $\rho = \sum_i p_i |i\rangle\langle i|$ and $\sigma = \sum_j q_j |\tilde{j}\rangle\langle\tilde{j}|$ be quantum states with eigenbases $\{|i\rangle\}_i$ and $\{|\tilde{j}\rangle\}_j$, respectively. Then the definition of the relative entropy, Definition 20.3, gives

$$S(\rho \| \sigma) = \sum_i p_i \log(p_i) - \sum_i \langle i | \rho \log(\sigma) | i \rangle \,. \tag{20.22}$$

We know that $\langle i | \rho = p_i \langle i |$ and are left with

$$\langle i | \log(\sigma) | i \rangle = \langle i | \sum_j \log(q_j) |\tilde{j}\rangle\langle\tilde{j}| i \rangle = \sum_j P_{ij} \log(q_j) \,, \tag{20.23}$$

where $P_{ij} = \langle i | \tilde{j}\rangle\langle\tilde{j}| i \rangle = |\langle i | \tilde{j}\rangle|^2$ with $\sum_j P_{ij} = \sum_i P_{ij} = 1$. Since the logarithm is a concave function we can use Jensen's inequality (Jensen, 1906) which gives the inequality

$$\sum_j P_{ij} \log(q_j) \leq \log\Big(\sum_j P_{ij} \, q_j\Big) =: \log(Q_i) \,, \tag{20.24}$$

with the probability distribution $Q = \{Q_i = \sum_j P_{ij}\, q_j\}_i$ with $\sum_i Q_i = 1$. We thus have a lower bound on the quantum relative entropy,

$$S(\rho\,\|\,\sigma) \geq \sum_i p_i \log(p_i) - \sum_i p_i \log(Q_i) = H(P\,\|\,Q) \geq 0\,, \qquad (20.25)$$

which can be seen to be non-negative by virtue of Gibbs' inequality, Theorem 19.1, for the classical relative entropy for the probability distributions $P = \{p_i\}_i$ and $Q = \{Q_i\}_i$. $\qquad\square$

Next we consider a general bipartite state ρ_{AB} in a Hilbert–Schmidt space $\mathcal{B}_{\mathrm{HS}}(\mathcal{H}_A \otimes \mathcal{H}_B)$ and use the relative entropy to compare it to an uncorrelated state, more specifically, a tensor product $\rho_A \otimes \rho_B$ of its marginals $\rho_{A/B} = \mathrm{Tr}_{B/A}(\rho)$, then the following lemma is valid.

Lemma 20.2 (Relative entropy to product state of marginals)

For a density matrix $\rho_{AB} \in \mathcal{B}_{\mathrm{HS}}(\mathcal{H}_A \otimes \mathcal{H}_B)$ of a bipartite system with reduced states ρ_A and ρ_B the relative entropy between ρ and the product state $\rho_A \otimes \rho_B$ is given by

$$S(\rho_{AB}\,\|\,\rho_A \otimes \rho_B) = S(\rho_A) + S(\rho_B) - S(\rho_{AB})\,.$$

Proof   Let us begin by writing the reduced states in spectral decomposition as $\rho_A = \sum_i p_i^{(A)}\,|i\rangle\langle i|$ and $\rho_B = \sum_j p_j^{(B)}\,|j\rangle\langle j|$, respectively, then we have

$$\log(\rho_A \otimes \rho_B) = \log\Big(\sum_{i,j} p_i^{(A)}\, p_j^{(B)}\,|i\rangle\langle i| \otimes |j\rangle\langle j|\Big) = \sum_{i,j}\log(p_i^{(A)} p_j^{(B)})\,|i\rangle\langle i| \otimes |j\rangle\langle j|$$

$$= \sum_{i,j}(\log(p_i^{(A)}) + \log(p_j^{(B)}))\,|i\rangle\langle i| \otimes |j\rangle\langle j|$$

$$= \sum_i \log(p_i^{(A)})\,|i\rangle\langle i| \otimes \sum_j |j\rangle\langle j| + \sum_i |i\rangle\langle i| \otimes \sum_j \log(p_j^{(B)})\,|j\rangle\langle j|$$

$$= \log(\rho_A) \otimes \mathbb{1}_B + \mathbb{1}_A \otimes \log(\rho_B)\,. \qquad (20.26)$$

With this, we can easily calculate

$$S(\rho_{AB}\,\|\,\rho_A \otimes \rho_B) = \mathrm{Tr}\big(\rho\,[\log(\rho_{AB}) - \log(\rho_A \otimes \rho_B)]\big)$$

$$= \mathrm{Tr}(\rho_{AB}\,\log(\rho_{AB})) - \mathrm{Tr}_A(\rho_A\,\log(\rho_A)) - \mathrm{Tr}_B(\rho_B\,\log(\rho_B))$$

$$= S(\rho_A) + S(\rho_B) - S(\rho_{AB})\,. \qquad (20.27)$$

$\qquad\square$

Proof of the subadditivity of the von Neumann entropy (property (iii)) The subadditivity from eqn (20.9) now follows directly by combining Lemma 20.2 with Klein's inequality (Theorem 20.1), since

$$S(\rho_{AB} \| \rho_A \otimes \rho_B) = S(\rho_A) + S(\rho_B) - S(\rho_{AB}) \geq 0 \tag{20.28}$$

implies
$$S(\rho_{AB}) \leq S(\rho_A) + S(\rho_B). \tag{20.29}$$

□

Proof of the Araki–Lieb inequality (property (iii)) From the subadditivity it is in turn easy to derive the Araki–Lieb inequality, first proven by *Huzihiro Araki* and *Elliott Lieb* in 1970. For the proof let us assume that $\rho_{ABC} = |\psi\rangle\langle\psi|_{ABC}$ is a purification of ρ_{AB} such that $\text{Tr}_C(\rho_{ABC}) = \rho_{AB}$, see Theorem 15.2. But we can of course also take the partial trace over subsystem C to get $\rho_{AC} = \text{Tr}_B(\rho_{ABC})$, for which subadditivity implies

$$S(\rho_{AC}) \leq S(\rho_A) + S(\rho_C). \tag{20.30}$$

But since ρ_{ABC} is pure, the Schmidt-decomposition theorem (see Theorem 15.1) implies that the reduced states ρ_{AB} and ρ_C have the same eigenvalues and hence the same entropy. Thus we can insert $S(\rho_{AC}) = S(\rho_B)$ and similarly $S(\rho_C) = S(\rho_{AB})$ in (20.30) to obtain

$$S(\rho_B) \leq S(\rho_A) + S(\rho_{AB}) \qquad \Rightarrow \qquad S(\rho_{AB}) \geq S(\rho_B) - S(\rho_A). \tag{20.31}$$

Since one can make the same argument with the roles of A and B exchanged, we arrive at the Araki–Lieb inequality

$$S(\rho_{AB}) \geq |S(\rho_A) - S(\rho_B)|. \tag{20.32}$$

□

Additional properties of the quantum relative entropy:

(i) **Unitary Invariance**: $S(\rho\|\sigma)$ is invariant unitary transformations U,

$$S(U\rho U^\dagger \| U\sigma U^\dagger) = S(\rho\|\sigma) \quad \text{for } U: \quad UU^\dagger = U^\dagger U = \mathbb{1}. \tag{20.33}$$

(ii) **Joint convexity**: Let $\rho = \sum_i p_i \rho_i$ and $\sigma = \sum_i p_i \sigma_i$ with $0 \leq p_i \leq 1$ and $\sum_i p_i = 1$ be convex combinations of density operators ρ_i and σ_i, respectively, then the relative entropy is jointly convex, i.e.,

$$S(\rho\|\sigma) = S\left(\sum_i p_i \rho_i \,\Big\|\, \sum_i p_i \sigma_i\right) \leq \sum_i p_i\, S(\rho_i \| \sigma_i). \tag{20.34}$$

(iii) **Additivity**: The quantum relative entropy is additive on tensor products, i.e., for any pair of product states $\rho_A \otimes \rho_B$ and $\sigma_A \otimes \sigma_B$,

$$S(\rho_A \otimes \rho_B \| \sigma_A \otimes \sigma_B) = S(\rho_A \| \sigma_A) + S(\rho_B \| \sigma_B). \tag{20.35}$$

(iv) **Monotonicity**: Let ρ_{AB} and σ_{AB} be density matrices in $\mathcal{B}_{\mathrm{HS}}(\mathcal{H}_A \otimes \mathcal{H}_B)$ with reduced states $\rho_A = \mathrm{Tr}_B(\rho_{AB})$ and $\sigma_A = \mathrm{Tr}_B(\sigma_{AB})$, respectively, then

$$S(\rho_A \| \sigma_A) \leq S(\rho_{AB} \| \sigma_{AB}), \qquad (20.36)$$

i.e., tracing out a subsystem reduces the relative entropy. This is a desirable property for a quantifier of distance between states since ignoring part of the system makes it more difficult to distinguish the two states—the reduced states are more "similar" to each other than the joint states.

Proof of the unitary invariance of the quantum relative entropy (property (i))
Let ρ and σ be density operators with spectral decompositions $\rho = \sum_i p_i |i\rangle\langle i|$ and $\sigma = \sum_j q_j |\tilde{j}\rangle\langle\tilde{j}|$, respectively. Then any unitary U applied to either of these operators will leave the respective eigenvalues p_i and q_j invariant, but transform the corresponding eigenvectors to $U|i\rangle$ and $U|\tilde{j}\rangle$, such that $\log(U\rho U^\dagger) = \sum_i \log(p_i) U|i\rangle\langle i|U^\dagger$ and $\log(U\sigma U^\dagger) = \sum_j \log(q_j) U|\tilde{j}\rangle\langle\tilde{j}|U^\dagger$. It is then obvious that

$$S(U\rho U^\dagger \| U\sigma U^\dagger) = \mathrm{Tr}\left(U\rho U^\dagger\left[\sum_i \log(p_i) U|i\rangle\langle i|U^\dagger - \sum_j \log(q_j) U|\tilde{j}\rangle\langle\tilde{j}|U^\dagger\right]\right)$$

$$= \mathrm{Tr}\left(U\rho U^\dagger U\left[\log(\rho) - \log(\sigma)\right]U^\dagger\right) = S(\rho\|\sigma), \qquad (20.37)$$

where we have used the invariance of the trace under cyclic permutations of its arguments and $UU^\dagger = U^\dagger U = \mathbb{1}$.   □

The proof of joint convexity (property (ii)) can be carried out quite compactly based on a theorem by Elliott Lieb (1973) regarding trace inequalities for positive matrices, a problem closely tied to the proof of the so-called *strong subadditivity* of the von Neumann entropy (which we will discuss in Section 20.4.3), which we will here use in the following form:

Theorem 20.2 (Lieb's concavity theorem)

Let K be an $m \times n$ matrix, and let A and B be positive definite $m \times m$ and $n \times n$ matrices, respectively, then the function

$$f(A, B) := \mathrm{Tr}\left(K^\dagger A^p K B^q\right)$$

is jointly concave *in A and B, i.e., for matrices $A = \sum_i p_i A_i$ and $B = \sum_i p_i B_i$ with $0 \leq p_i \leq 1$ and $\sum_i p_i = 1$:*

$$f\left(\sum_i p_i A_i, \sum_i p_i B_i\right) \geq \sum_i p_i f(A_i, B_i).$$

Here, we are not going to prove Theorem 20.2, which was originally proven by Lieb (1973), with alternative proofs presented later by other authors (e.g., by Andô, 1979 and Simon, 2005, see also Nielsen and Chuang, 2000, Appendix 6). However, using Lieb's concavity theorem, it is now easy to see that the relative entropy is jointly convex, as was first pointed out by *Göran Lindblad* (1974).

Proof of the joint convexity of the quantum relative entropy (property (ii)) Let ρ and σ denote two density operators in a Hilbert–Schmidt space $\mathcal{B}_{HS}(\mathcal{H})$, such that $\text{supp}(\rho) = \text{supp}(\rho)$. Then we can replace both ρ and σ by their respective restrictions to their support, where they are positive definite operators. Then according to Lieb's concavity result (Theorem 20.2) the map $(\rho, \sigma) \mapsto \text{Tr}(\rho^p \sigma^{1-p})$ must be jointly concave for all p with $0 \leq p \leq 1$, which means that the map

$$(\rho, \sigma) \longmapsto \frac{1}{p-1} \left[\text{Tr}(\rho^p \sigma^{1-p}) - \text{Tr}(\rho) \right] \tag{20.38}$$

is jointly convex for all $0 \leq p \leq 1$, since $\text{Tr}(\rho)$ is trivially convex, $\text{Tr}(\rho) = \text{Tr}(\lambda \rho_1 + (1-\lambda)\rho_2)$ for all normalized density operators. Now we can consider the limit $p \to 1$, for which both the numerator and denominator go to 0, so we apply de l'Hôpital's rule by noting that $\frac{\partial}{\partial p} \rho^p = \rho^p \log(\rho)$ and $\frac{\partial}{\partial p} \sigma^{1-p} = -\sigma^{1-p} \log(\sigma)$, such that

$$\lim_{p \to 1} \frac{\text{Tr}(\rho^p \sigma^{1-p}) - \text{Tr}(\rho)}{p-1} = \lim_{p \to 1} \frac{\text{Tr}\left[\rho^p \log(\rho) \sigma^{1-p} - \rho^p \sigma^{1-p} \log(\sigma) \right]}{1}$$

$$= \text{Tr}(\rho \log(\rho) - \rho \log(\sigma)) = S(\rho \| \sigma). \tag{20.39}$$

So we see that relative entropy must be jointly convex. In the case that the supports of the operators do not match but we have $\ker(\sigma) \cap \text{supp}(\rho) = \emptyset$, we can just replace σ with its restriction to the support of ρ and the proof goes through as before. Finally, in case that $\ker(\sigma) \cap \text{supp}(\rho) \neq \emptyset$, the left-hand side of the joint-convexity relation diverges, but so does the right-hand side, since this implies that for at least one of the indices i, we have $\ker(\sigma_i) \cap \text{supp}(\rho_i) \neq \emptyset$, so joint convexity still holds trivially. $\square$

Proof of the additivity of the quantum relative entropy (property (iii)) Let us compare two pairs of uncorrelated quantum states, $\rho_A \otimes \rho_B$ and $\sigma_A \otimes \sigma_B$, in the same Hilbert–Schmidt space $\mathcal{B}_{HS}(\mathcal{H}_A \otimes \mathcal{H}_B)$ by calculating their relative entropy,

$$S(\rho_A \otimes \rho_B \| \sigma_A \otimes \sigma_B) = \text{Tr}(\rho_A \otimes \rho_B \left[\log(\rho_A \otimes \rho_B) - \log(\sigma_A \otimes \sigma_B) \right])$$

$$= \text{Tr}(\rho_A \otimes \rho_B \left[\log(\rho_A) \otimes \mathbb{1}_B + \mathbb{1}_A \otimes \log(\rho_B) \right]) \tag{20.40}$$

$$- \text{Tr}(\rho_A \otimes \rho_B \left[\log(\sigma_A) \otimes \mathbb{1}_B + \mathbb{1}_A \otimes \log(\sigma_B) \right]),$$

where we have inserted for the logarithm of the product state from eqn (20.26). Taking the partial traces over the subsystems for which we only have identity operators in the square brackets and pairing up the terms pertaining to subsystems A and B we thus have

$$S(\rho_A \otimes \rho_B \,\|\, \sigma_A \otimes \sigma_B) = \mathrm{Tr}_A\big(\rho_A \log(\rho_A)\big) - \mathrm{Tr}_A\big(\rho_A \log(\sigma_A)\big)$$

$$+ \mathrm{Tr}_B\big(\rho_B \log(\rho_B)\big) - \mathrm{Tr}_B\big(\rho_B \log(\sigma_B)\big)$$

$$= S(\rho_A \,\|\, \sigma_A) + S(\rho_B \,\|\, \sigma_B)\,. \qquad (20.41)$$

We thus see that the relative entropy is *additive*.   □

Proof of the monotonicity of the quantum relative entropy (property (iv))
Here, we follow a similar line of argument for the monotonicity of the relative entropy
as presented originally by *Armin Uhlmann* (1973) and Göran Lindblad (1974). The
monotonicity of the relative entropy under the partial-trace operation follows from the
properties we have already proven by making use of some results on quantum channels
that we will derive in Chapter 21. In particular, we note that the so-called *complete
depolarizing channel* obtained for $p = 0$ in eqn (21.12) is a map $\Lambda[\rho]$ that takes the
input quantum state ρ on a d-dimensional Hilbert space to the maximally mixed
state, $\Lambda[\rho] = \mathrm{Tr}(\rho)\frac{1}{d}\mathbb{1}_d$. At the same time we know from eqn (21.55) that this channel
can be expressed as a weighted sum of unitaries U_i applied to ρ with probability p_i,
$\Lambda[\rho] = \sum_i p_i U_i \rho U_i^\dagger$, where both the U_i and the p_i are independent from ρ. If we now
apply this map to subsystem B of a bipartite system in a state ρ_{AB}, the map traces
out B and replaces it with the maximally mixed state, such that

$$\big(\Lambda \otimes \mathbb{1}_B\big)[\rho_{AB}] = \rho_A \otimes \tfrac{1}{d}\mathbb{1}_d = \sum_i p_i\,(\mathbb{1} \otimes U_i)\,\rho_{AB}\,(\mathbb{1} \otimes U_i^\dagger)\,. \qquad (20.42)$$

On the one hand, the additivity of the relative entropy, property (iii) now lets us
calculate

$$S\big(\rho_A \otimes \tfrac{1}{d}\mathbb{1}_d \,\|\, \sigma_A \otimes \tfrac{1}{d}\mathbb{1}_d\big) = S(\rho_A \,\|\, \sigma_A) + S\big(\tfrac{1}{d}\mathbb{1}_d \,\|\, \tfrac{1}{d}\mathbb{1}_d\big) = S(\rho_A \,\|\, \sigma_A)\,, \qquad (20.43)$$

since $S(\rho \,\|\, \rho) = 0$ for all ρ. On the other hand, using the shorthand $\tilde{U}_i := \mathbb{1} \otimes U_i$ the
joint convexity (property (ii)) and unitary invariance (property (i)) lead us to

$$S\Big(\sum_i p_i \tilde{U}_i \rho_{AB} \tilde{U}_i^\dagger \,\|\, \sum_i p_i \tilde{U}_i \sigma_{AB} \tilde{U}_i^\dagger\Big) \le \sum_i p_i\, S\big(\tilde{U}_i \rho_{AB} \tilde{U}_i^\dagger \,\|\, \tilde{U}_i \sigma_{AB} \tilde{U}_i^\dagger\big) \qquad (20.44)$$

$$= \sum_i p_i\, S\big(\rho_{AB} \,\|\, \sigma_{AB}\big) = S\big(\rho_{AB} \,\|\, \sigma_{AB}\big)\,.$$

We thus arrive at the conclusion that the quantum relative entropy decreases under
the partial-trace operation, $S\big(\rho_{AB} \,\|\, \sigma_{AB}\big) \ge S\big(\rho_A \,\|\, \sigma_A\big)$.   □

We have shown that the quantum relative entropy is decreasing under the partial
trace, which can be understood as an important example of what is called a completely
positive and trace-preserving (CPTP) map. As we will discuss in detail in Chapter 21,
such maps represent the most general kind of process that takes density operators as

input and returns density operators. Any CPTP map $\Lambda : \rho \mapsto \Lambda[\rho]$ can be written (see Theorem 21.2) as a unitary transformation on a larger Hilbert space,

$$\Lambda[\rho] = \mathrm{Tr}_B \left(U \rho \otimes |\omega\rangle\langle\omega| U^\dagger \right). \qquad (20.45)$$

If we apply this map to both arguments of the relative entropy, and use the monotonicity under the partial trace, property (iv) followed by the unitary invariance (property (i)), we have

$$
\begin{aligned}
S\big(\Lambda[\rho] \,\|\, \Lambda[\sigma]\big) &= S\Big(\mathrm{Tr}_B\big(U\rho\otimes|\omega\rangle\langle\omega|\,U^\dagger\big) \,\big\|\, \mathrm{Tr}_B\big(U\sigma\otimes|\omega\rangle\langle\omega|\,U^\dagger\big)\Big) \\
&\leq S\big(U\rho\otimes|\omega\rangle\langle\omega|\,U^\dagger \,\|\, U\sigma\otimes|\omega\rangle\langle\omega|\,U^\dagger\big) \\
&= S\big(\rho\otimes|\omega\rangle\langle\omega| \,\|\, \sigma\otimes|\omega\rangle\langle\omega|\big) = S\big(\rho\|\sigma\big),
\end{aligned} \qquad (20.46)
$$

where we made use of the additivity (property (iii)) and $S(\rho\|\rho) = 0 \ \forall \rho$ in the last step. We thus arrive at an important corollary:

Corollary 20.1 (Monotonicity under CPTP maps)

The quantum relative entropy is decreasing under CPTP maps Λ:

$$S\big(\Lambda[\rho]\,\|\,\Lambda[\sigma]\big) \leq S\big(\rho\|\sigma\big).$$

20.3.2 Quantum Joint Entropy

As for the classical Shannon entropy, we now next examine the joint entropy. To this end we consider a bipartite system of two parties, Alice and Bob, described by a density matrix of ρ_{AB} in a Hilbert–Schmidt space $\mathcal{B}_{\mathrm{HS}}(\mathcal{H}_A \otimes \mathcal{H}_B)$. The subsystems of Alice and Bob are determined by the reduced density matrices $\rho_A = \mathrm{Tr}_B(\rho_{AB}) \in \mathcal{B}_{\mathrm{HS}}(\mathcal{H}_A)$ and $\rho_B = \mathrm{Tr}_A(\rho_{AB}) \in \mathcal{B}_{\mathrm{HS}}(\mathcal{H}_B)$, respectively. Then the *quantum joint entropy* describing the uncertainty—the lack of knowledge—about the joint quantum system, and any measurements performed thereon, is given by the von Neumann entropy of ρ_{AB}.

Definition 20.4 *For a bipartite system with density matrix $\rho_{AB} \in \mathcal{B}_{\mathrm{HS}}(\mathcal{H}_A \otimes \mathcal{H}_B)$ the **quantum joint entropy** is given by*

$$S(\rho_{AB}) = -\mathrm{Tr}\big(\rho_{AB}\,\log(\rho_{AB})\big).$$

In particular, the joint entropy encodes information about the correlations between the subsystems, and thus about the correlations between measurement outcomes of Alice and Bob. Of course, the quantum joint entropy also satisfies all of the properties of the von Neumann entropy listed in the previous Section 20.1, i.e., properties (i)–(v) from eqns (20.7)–(20.13).

However, the quantum joint entropy deviates from the classical joint entropy in terms of some properties that are essential for quantum information. In classical information theory the joint entropy $H(X, Y)$ of a pair of random variables X and Y is always larger than (or equal to) the entropies of either of the individual random variables X and Y, $H(X, Y) \geq H(X)$, as we recall from Theorem 19.3. This is intuitively clear since we cannot be more uncertain about a single random variable X than we are about the pair (X, Y) of random variables of the joint state. However, this intuition fails for a quantum state!

Consider, for example, the familiar Bell state ρ^- from eqn (15.17), which is maximally entangled

$$\rho_{AB} \equiv \rho^- = |\psi^-\rangle\langle\psi^-| = \frac{1}{2} \begin{pmatrix} 0 & 0 & 0 & 0 \\ 0 & 1 & -1 & 0 \\ 0 & -1 & 1 & 0 \\ 0 & 0 & 0 & 0 \end{pmatrix}, \qquad (20.47)$$

for which the quantum joint entropy vanishes

$$S(\rho_{AB}) = S(\rho^-) = 0, \qquad (20.48)$$

since ρ^- is a pure state. Yet, the entropies of the reduced states of the subsystems of Alice and Bob are maximal,

$$S(\rho_A) = S(\rho_B) = 1, \qquad (20.49)$$

since $\rho_A = \rho_B = \frac{1}{2}\mathbb{1}_2$ are maximally mixed.

Thus, for quantum systems it is possible to satisfy the inequality $S(\rho_{AB}) < S(\rho_A)$ when the joint state is pure and the subsystems A and B are entangled. The less entangled the pure state of the total system ρ_{AB} is, the less mixed the subsystems are, approaching pure states when the joint state approaches a separable state. Then the entropies of Alice's and Bob's subsystems vanish, $S(\rho_A) = S(\rho_B) = 0$, in accordance with property (v) of the von Neumann entropy in eqn (20.13), which we will re-state here in terms of the following lemma, which follows directly from Corollary 15.2:

Lemma 20.3 (Equality of subsystem entropies)

For a pure bipartite state $\rho_{AB} = |\psi\rangle\langle\psi|$ the joint entropy vanishes

$$S(\rho_{AB}) = 0,$$

whereas the entropies $S(\rho_A)$ and $S(\rho_B)$ of subsystems A and B are equal and non-negative,

$$S(\rho_A) = S(\rho_B) \geq 0,$$

with strict positivity > 0 if and only if ρ_{AB} is entangled.

20.4 Quantum Conditional Entropy and Mutual Information

In this section we investigate the quantum conditional entropy and the quantum mutual information in analogy to the classical cases discussed previously in Sections 19.3.2 and 19.4. We discuss the similarities between the quantities of the classical and the quantum cases but also stress the important differences.

20.4.1 Quantum Conditional Entropy

Inspired by the classical relation, Lemma 19.2, the *quantum conditional entropy* is defined as:

Definition 20.5   *The **quantum conditional entropy** of a bipartite system described by a density matrix $\rho_{AB} \in \mathcal{B}_{\mathrm{HS}}(\mathcal{H}_A \otimes \mathcal{H}_B)$ conditioned on the state ρ_B of subsystem B is given by the difference between the joint entropy $S(\rho_{AB})$ and the subsystem entropy $S(\rho_B)$,*

$$S(A|B) \; := \; S(\rho_{AB}) \; - \; S(\rho_B) \,.$$

The quantity $S(A|B)$ represents the entropy of Alice's subsystem when she already knows the state of Bob's subsystem, and all related predictions for the outcomes of measurements Bob might make. The quantum conditional entropy thus corresponds to the remaining rest-uncertainty. The expression for the quantum conditional entropy satisfies the equivalent relation to the upper bound on the classical conditional entropy in Theorem 19.3, $S(A|B) \leq S(\rho_A)$, which follows from the subadditivity of the von Neumann entropy (property (iii), eqn (20.9)), $S(\rho_{AB}) \leq S(\rho_A) + S(\rho_B)$, and equality is achieved for product states $\rho_{AB} = \rho_A \otimes \rho_B$.

However, as we already emphasized in Lemma 20.3, for entangled pure states like the Bell state $\rho_{AB} = \rho^-$ the quantum entropy vanishes $S(\rho_{AB}) = S(\rho^-) = 0$, while the subsystem entropies are maximal, $S(\rho_A) = S(\rho_B) = \log(2)$. This is in stark contrast to the relation $H(X,Y) \geq H(X)$ from Theorem 19.3 for the classical conditional entropy for a pair of two random variables X and Y. For the quantum conditional entropy of a Bell state we find

$$S_{\mathrm{Bell}}(A|B) \; = \; S(\rho^-) - S(\rho_B) \; = \; -\log(2) \; < \; 0 \,, \tag{20.50}$$

the quantum conditional entropy is *negative*! This is a typical quantum feature, not possible in the classical case. We thus find the following bounds on the quantum conditional entropy:

Theorem 20.3 (Bounds on quantum conditional entropy)

The quantum conditional entropy $S(A|B)$ of a bipartite system is bounded from above by the subsystem entropy $S(\rho_A)$ and from below by the negative subsystem entropy $S(\rho_B)$,

$$-S(\rho_B) \leq S(A|B) \leq S(\rho_A).$$

Of course, the negativity of the quantum conditional entropy is not a special feature of the Bell state but holds more generally. There is a theorem that connects the negativity of the quantum conditional entropy and the entanglement of a *general* bipartite system.

Theorem 20.4 (Conditional entropy—entanglement)

Let $\rho_{AB} \in \mathcal{B}_{\mathrm{HS}}(\mathcal{H}_A \otimes \mathcal{H}_B)$ be the density matrix describing a bipartite system. If the quantum conditional entropy is negative then the system is entangled,

$$\text{if} \quad S(A|B) < 0 \quad \Longrightarrow \quad \rho_{AB} \quad \text{entangled.}$$

However, the reverse is not true $\neq$!

For the proof of Theorem 20.4 it is useful to first formulate the following lemma.

Lemma 20.3 (Concavity of conditional entropy)

The quantum conditional entropy is a concave functional in ρ_{AB}.

Proof of Lemma 20.3   Let us calculate the quantum relative entropy between the bipartite state ρ_{AB} under consideration and a product state $\frac{1}{d}\mathbb{1}_d \otimes \rho_B$, where $\rho_B = \mathrm{Tr}_A(\rho_{AB})$ is the reduced state of subsystem B and subsystem A, assumed to have dimension d, has been replaced with a maximally mixed state. Then from Definition 20.3 we have

$$S(\rho_{AB} \| \tfrac{1}{d}\mathbb{1}_d \otimes \rho_B) = \mathrm{Tr}\big(\rho_{AB}\big[\log(\rho_{AB}) - \log(\tfrac{1}{d}\mathbb{1}_d \otimes \rho_B)\big]\big). \tag{20.51}$$

We can then (up to a sign flip) identify the first term on the right-hand side as the quantum joint entropy, $-S(\rho_{AB})$, and for the second term we insert from eqn (20.26),

$$
\begin{aligned}
S(\rho_{AB} \| \tfrac{1}{d}\mathbb{1}_d \otimes \rho_B) &= -S(\rho_{AB}) - \mathrm{Tr}\big(\rho_{AB}\big[\log(\tfrac{1}{d}\mathbb{1}_d) \otimes \mathbb{1}_B + \mathbb{1}_A \otimes \log(\rho_B)\big]\big) \\
&= -S(\rho_{AB}) + \log(d)\,\mathrm{Tr}\big(\rho_{AB}\,\mathbb{1}_A \otimes \mathbb{1}_B\big) - \mathrm{Tr}_B\big(\rho_B\,\log(\rho_B)\big) \\
&= -S(A|B) + \log(d). \tag{20.52}
\end{aligned}
$$

Now we see that the conditional entropy can be written as

$$S(A|B) = -S(\rho_{AB} \| \tfrac{1}{d}\mathbb{1}_d \otimes \rho_B) + \log(d), \tag{20.53}$$

and the concavity of the conditional entropy in ρ_{AB} is a consequence of the joint convexity of the relative entropy, see property (ii).   □

With this, we can return to Theorem 20.4.

Proof of Theorem 20.4 We know that the quantum conditional entropy vanishes for any pure product state $|\psi\rangle_A \otimes |\phi\rangle_B$. As we recall from the definition of entanglement and separability for mixed states, Definition 15.5, any mixed separable state can be written as a convex combination of product states, $\rho_{AB} = \sum_i p_i \rho_i^{(A)} \otimes \rho_i^{(B)}$, and since any of the mixed stated $\rho_i^{(A)}$ and $\rho_i^{(B)}$ can themselves be expanded into convex combinations of pure states, we can generally write any separable state as

$$\rho_{AB} = \sum_i p_i \, |\psi_i\rangle\langle\psi_i|_A \otimes |\phi_i\rangle\langle\phi_i|_A \, . \tag{20.54}$$

Now it is easy to see from Lemma 20.3 that the relative entropy calculated from a state ρ_{AB} that is a convex sum of pure product states is larger than (or equal to) the convex sum of relative entropies calculated from the respective pure product states $|\psi_i\rangle_A \otimes |\phi_i\rangle_B$. But since the latter relative entropies all vanish, the relative entropy of the original separable state must be bounded from below by 0. Conversely, this means that any negative value of the relative entropy can only occur if the state ρ_{AB} is entangled.

As we will see by examining some counterexamples in Section 20.6, the reverse direction is not true, i.e., non-negative conditional entropy does not imply separability.   □

We will further investigate the detailed geometry of quantum states, particularly the connection between separability, entanglement, non-locality, and conditional entropy in Section 20.6.

How can we conceptually understand the fact that the conditional entropy can become negative, which is impossible in the classical world? Mathematically, this is clear: In the quantum case the analogue to Theorem 19.3 does not hold, the quantum joint entropy *can* become smaller than the subsystem entropy.

Already *Erwin Schrödinger* realized this fundamental difference between the classical world and the quantum domain in his seminal work (Schrödinger, 1935*a*, 1935*c*, and 1936) and emphasized: "I would not call that one but rather the characteristic trait of quantum mechanics, the one that enforces its entire departure from classical lines of thought. By the interaction the two representatives have become entangled ... the best possible knowledge of a whole does not necessarily include the best possible knowledge of all its parts ..." Recall our discussion of Schrödinger's reply to the EPR paradox, Section 13.1.

Within quantum information negative values of the conditional entropy can be given an operational meaning. *Michał Horodecki, Jonathan Oppenheim*, and *Andreas Winter* (2005) described a process called *quantum state merging*, which is the optimal transfer of a quantum state between Alice and Bob, when Bob already has access to part of the state. The information sent is given by the quantum conditional entropy. In this context one finds that positive values of the *quantum mutual information*—which we discuss in Section 20.4.2—quantify the partial information in qubits that needs to be sent from A to B, whereas a negative conditional entropy indicates that, in addition to successfully running the protocol, a surplus of qubits remains for potential future quantum communication, such as quantum teleportation or dense coding. Thus in case of a negative conditional entropy Alice and Bob gain entanglement, rather than use it[2].

Moreover, a classical analogue of negative partial information was presented in Oppenheim *et al.* (2005). Other physical interpretations of negative conditional entropy arise in quantum thermodynamics (del Rio *et al.*, 2011), and come up when considering measurements of quantum systems, where the negative conditional entropy quantifies the amount of information in the post-selected ensembles (Salek *et al.*, 2014). These interesting interpretations motivate considering *"entropic Bell inequalities"* set up by (Cerf and Adami, 1997*a*), whose violation implies a negative conditional entropy.

Remark: We would like to mention that the negative conditional entropy in the quantum case differs from the concept of *negative entropy* introduced by Schrödinger in his book *What is life?* (Schrödinger, 1944). Later on, *Léon Brillouin* (Brillouin, 1953) shortened the phrase to *negentropy* to express that living systems import negentropy and store it. In information theory, the *negentropy* is used as a measure of distance between a given distribution and a normal (or Gaussian) distribution which has the highest entropy.

20.4.2 Quantum Mutual Information

In full analogy to the classical relation from Definition 19.6, we now define the *quantum mutual information* shared between two parties Alice and Bob as the information content of Alice and Bob individually, but subtracting the joint information which would be counted twice otherwise.

Definition 20.6 *The **quantum mutual information** of a bipartite system described by a density matrix $\rho_{AB} \in \mathcal{B}_{\mathrm{HS}}(\mathcal{H}_A \otimes \mathcal{H}_B)$ is given by the difference of the sum of subsystem entropies $S(\rho_A)$ and $S(\rho_B)$ and the joint entropy $S(\rho_{AB})$,*

$$S(A\!:\!B) \ := \ S(\rho_A) + S(\rho_B) - S(\rho_{AB}).$$

---

[2]Referring to the paper of Horodecki, Oppenheim, and Winter, the newspaper *Bristol Evening Post* featured an article (5 August 2005) about Andreas Winter with the amusing title "Scientist knows less than nothing". Of course, it sounds rather meaningless but nevertheless it reflects somehow the fact that we can know less about a part of the quantum system than we do about the whole.

In this chapter we will use the notation $S(A:B)$ for the quantum mutual information, in this context sometimes called *mutual entropy*, but note that it is often customary to use the alternative notation $\mathcal{I}(\rho_{AB}) \equiv S(A:B)$ (e.g., we have done so in eqn (16.73) in Chapter 16) that more closely mimics the classical case where one typically uses the letter I.

The quantum mutual information provides a way to quantify correlations between the subsystems A and B. But, owing to its definition, $S(A:B)$ does not distinguish classical correlations from quantum correlations (entanglement). It is considered as a general measure of correlations in information theory for pure *and* mixed states. It thus can be used to characterize not just the quantum correlations between Alice and Bob, but rather both classical correlations *and* quantum entanglement.

The quantum mutual information $S(A:B)$ is obviously symmetric, $S(A:B) = S(B:A)$, and also has a close relation to the conditional entropy and the subsystem entropies $S(\rho_A)$ and $S(\rho_B)$,

$$S(A:B) = S(\rho_A) - S(A|B) = S(\rho_B) - S(B|A) = S(B:A). \qquad (20.55)$$

However, in contrast to the conditional entropy the quantum mutual information remains non-negative, like its classical counterpart (recall Corollary 19.1), which trivially follows from the subadditivity of the von Neumann entropy (eqn (20.9) in property (iii)).

Theorem 20.5 (Non-negativity of quantum mutual information)

The quantum mutual information of any bipartite system is non-negative

$$S(A:B) \geq 0.$$

At the same time, a feature of the quantum conditional entropy that the quantum mutual information shares is that it can exceed its classical bound, Theorem 19.4, by a factor of 2.

Theorem 20.6 (Bound of quantum mutual information)

The quantum mutual information of any bipartite system is bounded from above by twice the value of the smaller of its subsystem entropies,

$$S(A:B) \leq 2 \min\{S(\rho_A), S(\rho_B)\}.$$

This inequality can be interpreted as allowing for channels with a capacity for the transmission of (quantum) information that can reach twice the classical upper bound, as is the case, for example, in dense coding (Bennett and Wiesner, 1992, see the discussion in Section 14.3.5).

Proof   For the proof of Theorem 20.6, we first assume that $S(\rho_B) \leq S(\rho_A)$ and rewrite the Araki–Lieb inequality (20.11) for a bipartite system ρ_{AB},

$$\left| S(\rho_A) - S(\rho_B) \right| \leq S(\rho_{AB}) \quad \Longrightarrow \quad -S(\rho_{AB}) \leq -S(\rho_A) + S(\rho_B). \qquad (20.56)$$

We then insert the result from eqn (20.56) into Definition 20.6 and obtain

$$S(A:B) = S(\rho_A) + S(\rho_B) - S(\rho_{AB}) \qquad (20.57)$$
$$\leq S(\rho_A) + S(\rho_B) - S(\rho_A) + S(\rho_B) = 2\,S(\rho_B).$$

Analogously we obtain $S(A:B) \leq S(\rho_A)$ for the case $S(\rho_A) \leq S(\rho_B)$. Thus, we finally have

$$S(A:B) \leq 2 \cdot \min[S(\rho_A), S(\rho_B)]. \qquad (20.58)$$

$\square$

From Lemma 20.2 we also see that the quantum mutual information can be expressed as the relative entropy of a given state ρ_{AB} to the product state $\rho_A \otimes \rho_B$ of its marginals ρ_A and ρ_B,

$$S(A:B) = S(\rho_{AB} \| \rho_A \otimes \rho_B). \qquad (20.59)$$

Diagram representation: To aid our intuition on quantum entropies, their relationship can be illustrated in a "Venn-like" information diagram, shown in Fig. 20.1, in analogy to the diagrammatic representation of classical entropies in Fig. 19.3, see, e.g., Cerf and Adami (1996*b* and 1997*b*). There, the subsystem entropies $S(\rho_A)$ and $S(\rho_B)$ of Alice and Bob, respectively, can be represented as two sets whose union corresponds to the quantum joint entropy $S(\rho_{AB})$, while their intersection—the information shared by Alice and Bob—can be interpreted as the quantum mutual information $S(A:B)$. The complements of each of the subsystem-entropy sets corresponding to $S(\rho_A)$ and $S(\rho_B)$ can be identified with the conditional entropies $S(B|A)$ and $S(A|B)$, respectively.

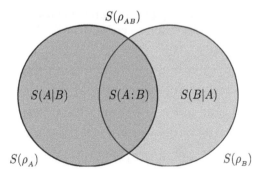

Fig. 20.1 Information diagram for quantum entropies. The union of the two sets representing the subsystem entropies $S(\rho_A)$ and $S(\rho_B)$ can be associated to the quantum joint entropy, symbolically we write $S(\rho_{AB}) = S(\rho_A) \cup S(\rho_B)$, while their intersection can be identified with the quantum mutual information $S(A:B) = S(\rho_A) \cap S(\rho_B)$, and the respective complements with the quantum conditional entropies $S(A|B)$ and $S(B|A)$.

Examples: To illustrate the quantum joint entropy, the quantum conditional entropy and the quantum mutual information, let us consider a few examples for two-qubit systems, and we focus on the special case where the subsystem entropies are both maximal (Weyl states, as we recall from Section 15.5.1), that is, $S(\rho_A) = S(\rho_B) = 1$, where we have chosen the logarithm to base 2. We discuss three possible cases and illustrate them in Fig. 20.2.

(a) **Two qubits in a product state:** $\rho_{AB} = \rho_A \otimes \rho_B$

Joint entropy: $S(\rho_{AB}) = S(\rho_A \otimes \rho_B) = S(\rho_A) + S(\rho_B) = 1 + 1 = 2$.

Quantum conditional entropy: $S(A|B) = S(\rho_{AB}) - S(\rho_B) = 2 - 1 = 1$,

$\Rightarrow$ the upper bound (Theorem 20.3) is achieved.

Quantum mutual information: $S(A:B) = S(\rho_A) + S(\rho_B) - S(\rho_{AB}) = 1 + 1 - 2 = 0$,

$\Rightarrow$ the lower bound (Theorem 20.5) is achieved.

(b) **Two classically correlated qubits:** $\rho_{\mathrm{mix}} = \frac{1}{2}\big(|01\rangle\langle 01| + |10\rangle\langle 10|\big)$

Joint entropy: $S(\rho_{AB}) = S(\rho_{\mathrm{mix}}) = \frac{1}{2} + \frac{1}{2} = 1$.

Quantum conditional entropy: $S(A|B) = S(\rho_{AB}) - S(\rho_B) = 1 - 1 = 0$,

$\Rightarrow$ the classical lower bound of eqn (19.59) is achieved.

Quantum mutual information: $S(A:B) = S(\rho_A) + S(\rho_B) - S(\rho_{AB}) = 1 + 1 - 1 = 1$,

$\Rightarrow$ the classical upper bound (Theorem 19.4) is achieved.

(c) **Two qubits in a Bell state:** $\rho_{AB} = \rho^- = |\psi^-\rangle\langle\psi^-|$

Joint entropy: $S(\rho_{AB}) = S(\rho^-) = 0$, since the Bell state is a pure state.

Quantum conditional entropy: $S(A|B) = S(\rho_{AB}) - S(\rho_B) = 0 - 1 = -1$,

$\Rightarrow$ the negative lower bound (Theorem 20.3) is achieved.

Quantum mutual information: $S(A:B) = S(\rho_A) + S(\rho_B) - S(\rho_{AB}) = 1 + 1 - 0 = 2$,

$\Rightarrow$ the quantum upper bound (Theorem 20.6) is achieved.

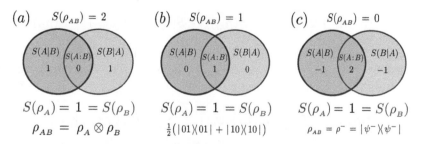

Fig. 20.2 Information diagrams for quantum entropies of two-qubit examples. The joint entropy, conditional entropies, and mutual information (with logarithms to base 2) are illustrated in information diagrams for (a) an (uncorrelated) product state, (b) a classically correlated state, and (c) a maximally entangled state.

20.4.3 Entropies of Multipartite Systems

Of course one can also calculate entropies of quantum systems that consist of more than two subsystems. In the following, we focus on the case of three subsystems, A, B, and C, which already allows us to make some interesting observations.

Strong subadditivity: We now come to an important property of quantum entropy, *strong subadditivity*, which arises as an extension of the subadditivity property of the von Neumann entropy when considering three subsystems, A, B, and C. There, the three-party density operator ρ_{ABC} has six different reduced states, three where one of the parties is traced out, $\rho_{AB} = \mathrm{Tr}_C(\rho_{ABC})$, $\rho_{AC} = \mathrm{Tr}_B(\rho_{ABC})$, and $\rho_{BC} = \mathrm{Tr}_A(\rho_{ABC})$, and three for which only one subsystem is left, $\rho_A = \mathrm{Tr}_{B,C}(\rho_{ABC})$, $\rho_B = \mathrm{Tr}_{A,C}(\rho_{ABC})$, and $\rho_C = \mathrm{Tr}_{A,B}(\rho_{ABC})$. Strong subadditivity then provides non-trivial constraints on the relations between the entropies of these quantities.

Theorem 20.7 (Strong subadditivity of von Neumann entropy)

Let $\rho_{ABC} \in \mathcal{B}_{\mathrm{HS}}(\mathcal{H}_A \otimes \mathcal{H}_B \otimes \mathcal{H}_C)$ be the density operator of a tripartite quantum system with reduced states ρ_{AB}, ρ_{AC}, ρ_{BC}, ρ_A, ρ_B, and ρ_C. Then their entropies satisfy

$$S(\rho_{ABC}) + S(\rho_B) \leq S(\rho_{AB}) + S(\rho_{BC}).$$

Based on analogous (but easier to prove) constraints on the classical Shannon entropy, it was conjectured that strong subadditivity should hold also for the von Neumman entropy already in 1967 (see Robinson and Ruelle, 1967 and Lanford and Robinson, 1968), but the proof, based on earlier breakthroughs by Lieb (1973), was finally provided by *Elliott Lieb* and *Mary Beth Ruskai* (1973). Here, we follow the arguments in Nielsen (1999, p. 521) for the proof.

Proof Let us begin by defining a functional f on ρ_{ABC} given by

$$f[\rho_{ABC}] := S(C|A) + S(C|B) = S(\rho_{AC}) - S(\rho_A) + S(\rho_{BC}) - S(\rho_B). \quad (20.60)$$

From Lemma 20.3 we know that the conditional entropies in this expression are concave in ρ_{ABC}, therefore $f[\rho_{ABC}]$ is concave in ρ_{ABC}. In particular, this means that when writing the state ρ_{ABC} in spectral decomposition, $\rho_{ABC} = \sum_i p_i |\psi_i\rangle\langle\psi_i|$ we have

$$f[\rho_{ABC}] = f\left[\sum_i p_i |\psi_i\rangle\langle\psi_i|\right] \geq \sum_i p_i f\left[|\psi_i\rangle\langle\psi_i|\right]. \quad (20.61)$$

At the same time, for pure states $|\psi_i\rangle_{ABC}$ we know from the Schmidt-decomposition theorem (see Theorem 15.1) that the reduced states ρ_{AC} and ρ_B have the same eigenvalues and hence the same entropy, $S(\rho_{AC}) = S(\rho_B)$, and similarly $S(\rho_{BC}) = S(\rho_A)$. We thus have $f\left[|\psi_i\rangle\langle\psi_i|\right] = 0$ and consequently $f[\rho_{ABC}] \geq 0$, which we rewrite as

$$S(\rho_A) + S(\rho_B) \leq S(\rho_{AC}) + S(\rho_{BC}). \tag{20.62}$$

This last inequality, also first proven in (Lieb and Ruskai, 1973), is known as the *weak monotonicity* of the von Neumann entropy (see, e.g., Pippenger, 2003), since it can be read as a statement about the decrease of (sums of) entropies when tracing out a subsystem (C in this case), similar to the monotonicity of the relative entropy (property (iv), eqn (20.36)). However, this monotonicity is *weak* in that it clearly does not hold when just considering the transition from AC to subsystem A, since a pure state ρ_{AC} will have vanishing entropy, $S(\rho_{AC}) = 0$ while $S(\rho_A) \geq 0$. Therefore, one needs to jointly consider the pair AC and BC to see the decrease when going to the subsystem entropies of A and B. In our pure-state example $\rho_{AC} = |\phi\rangle\langle\phi|_{AC}$, this would mean that joint state of A, B, C must be of the form $\rho_{AC} \otimes \rho_B$, and via the additivity of the von Neumann entropy for product states, eqn (20.10), we have $S(\rho_{BC}) = S(\rho_B) + S(\rho_C) = S(\rho_B) + S(\rho_A)$, such that equality is achieved for the weak-monotonicity inequality.

From the weak monotonicity, strong subadditivity follows by purification (see Section 15.1.4), that is, suppose we add a fourth subsystem D and consider a pure state ρ_{ABCD} such that $\mathrm{Tr}_D(\rho_{ABCD}) = \rho_{ABC}$, then from (20.62) we have (exchanging A and D)

$$S(\rho_D) + S(\rho_B) \leq S(\rho_{CD}) + S(\rho_{BC}). \tag{20.63}$$

But since ρ_{ABCD} is assumed to be pure, we have $S(\rho_D) = S(\rho_{ABC})$ and $S(\rho_{CD}) = S(\rho_{AB})$, so the strong subadditivity follows. Moreover, the converse also holds, starting from strong subadditivity and considering a purification ρ_{ABCD}, we have

$$S(\rho_A) + S(\rho_B) = S(\rho_{BCD}) + S(\rho_B) \leq S(\rho_{BD}) + S(\rho_{BC}) = S(\rho_{AC}) + S(\rho_{BC}), \tag{20.64}$$

where we have used the purity of ρ_{ABCD} in the first and last step. We have thus not only proven strong subadditivity, Theorem 20.7, but also that it is equivalent to the weak-monotonicity property in (20.62). □

The strong subadditivity (or, equivalently) weak monotonicity thus provide non-trivial constraints on the subsystem entropies for tripartite systems, and an interesting open questions relates to the existence and form of potential non-trivial entropy inequalities for four or more parties (see, e.g., Ibinson, 2007). At the same time, it is of interest to understand if (and if so which) inequalities constrain the subsystem entropies when other entropies are used, in particular, from the family of Rényi entropies S_α from Section 20.2. For the so-called "simple orders" α in $]0,1[\cup]1,\infty[$ the Rényi entropies generally only satisfy the trivial inequality of non-negativity and no other (linear) constraints apply (Linden *et al.*, 2013). The case $\alpha \to 1$ (i.e., the von Neumann entropy) hence takes a special role among the Rényi entropies, together with the case $\alpha \to 0$. That is, there are also non-trivial linear inequalities on the ranks of the reduced states (Cadney *et al.*, 2014), which can be seen as equivalent to constraints on the extended-order Rényi entropy for $\alpha \to 0$. These results can be complemented by dimension-dependent non-trivial constraints on the linear entropies of the subsystems (Appel *et al.*, 2020; Morelli *et al.*, 2020).

Next, we turn to extensions of the quantum conditional entropy and the quantum mutual information, which have been investigated, e.g., by *Nicolas Cerf* and *Christoph Adami* in a series of works (Cerf and Adami, 1996*b*; Cerf and Adami, 1997*b*).

Generalized quantum conditional entropy: For a tripartite system with subsystems A, B, and C we can define the generalized quantum conditional entropies in the following way

$$S(A|B,C) := S(A,B,C) - S(B,C), \tag{20.65a}$$

$$S(B|A,C) := S(A,B,C) - S(A,C), \tag{20.65b}$$

$$S(C|A,B) := S(A,B,C) - S(A,B), \tag{20.65c}$$

where we have switched to the simplified notation $S(A,B,C) = S(\rho_{ABC})$, and similarly $S(A,B) = S(\rho_{AB})$, $S(B,C) = S(\rho_{BC})$, and $S(A,C) = S(\rho_{AC})$ for the respective subsystems.

   Let us consider an example. For a three-qubit GHZ state with density matrix $\rho_{\mathrm{GHZ}} = |\,\mathrm{GHZ}\,\rangle\langle\,\mathrm{GHZ}\,|$ from eqn (18.10), where $|\,\mathrm{GHZ}\,\rangle = \frac{1}{\sqrt{2}}\left(|\,000\,\rangle + |\,111\,\rangle\right)$ as in eqn (18.1a), the joint entropy for the tripartite system vanishes, $S(A,B,C) = S(\rho_{\mathrm{GHZ}}) = 0$, while tracing out one subsystem leads to a mixed (but not maximally mixed) two-qubit state with eigenvalues 0 and $\frac{1}{2}$, each twice degenerate, see eqn (18.11). We thus find the entropies $S(A,B) = S(B,C) = S(A,C) = \log(2)$, and the generalized conditional entropies thus give $S(A|B,C) = S(B|A,C) = S(C|A,B) = -\log(2) < 0$.

Conditional mutual information: Likewise we can generalize the mutual information in the spirit of eqn (20.55) and define the conditional mutual information (sometimes called the conditional mutual entropy) as the quantum mutual information between two subsystems when the third one is known

$$S(A{:}B|C) := S(A|C) - S(A|B,C), \tag{20.66a}$$

$$S(B{:}C|A) := S(B|A) - S(B|C,A), \tag{20.66b}$$

$$S(C{:}A|B) := S(C|B) - S(C|A,B). \tag{20.66c}$$

Of course, the mutual information conditioned on the state of a certain subsystem, say B, is symmetric, e.g., $S(C : A|B) = S(A : C|B)$. To see this, we simply insert the definitions of the quantum conditional entropy (Definition 20.5) and generalized conditional entropy, eqn (20.65),

$$S(C{:}A|B) := \quad S(C|B) - S(C|A,B) = S(C,B) - S(B) - S(A,B,C) + S(A,B), \tag{20.67}$$

from which the symmetry with respect to the exchange of A and C becomes obvious.

As for the unconditional case (see Theorem 20.5), the quantum mutual information remains positive also in case of conditioning.

Theorem 20.8 (Non-negativity of conditional mutual information)

The conditional mutual information of a tripartite system is non-negative

$$S(A\!:\!B|C) \geq 0.$$

Proof Inserting $S(A|C) = S(A,C) - S(C)$ and $S(A|B,C) = S(A,B,C) - S(B,C)$ for the conditional entropies we can rewrite eqn (20.66a) as

$$S(A\!:\!B|C) = S(A,C) - S(C) - S(A,B,C) + S(B,C). \qquad (20.68)$$

Meanwhile, employing the subadditivity of the von Neumann entropy (property (iii), eqn (20.9)) twice we have

$$-S(A,B,C) \geq -S(A) - S(BC) \geq -S(A) - S(B) - S(C), \qquad (20.69)$$

such that we find

$$S(A\!:\!B|C) \geq S(A,C) + S(B,C) \geq 0. \qquad (20.70)$$

$\square$

This means the conditional mutual information is non-negative as a consequence of the subadditivity (20.9) of the von Neumann entropy.

As a direct extension to the mutual information of a bipartite system Cerf and Adami (1998, 1996*b*) define the *ternary mutual information*, also called *ternary mutual entropy* for a tripartite system:

Definition 20.7 *The **quantum ternary mutual information** of a tripartite system with subsystems A, B, and C is given by*

$$S(A\!:\!B\!:\!C) = S(A\!:\!B) - S(A\!:\!B|C).$$

Definition 20.7 is a direct extension of eqn (20.55). Inserting the mutual information, Definition 20.6, and the conditional mutual information (20.66a) we find explicitly

$$S(A\!:\!B\!:\!C) = S(A) + S(B) + S(C) - S(AB) - S(A,C) - S(B,C) + S(A,B,C). \qquad (20.71)$$

From this formula it is obvious that the ternary mutual information is symmetric with respect to the exchange of any pair of the subsystem. And for a tripartite system in

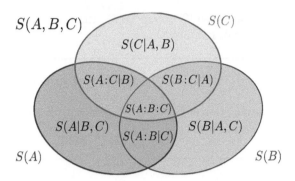

Fig. 20.3 Information diagram for tripartite system. When representing the subsystem entropies $S(A)$, $S(B)$, and $S(C)$ as sets illustrated in a Venn diagram, the joint entropy $S(A, B, C)$, conditional entropies (such as $S(A|B, C)$), as well as the values of the conditional mutual information (e.g., $S(A : B|C)$), and finally the ternary mutual information $S(A : B : C)$ can be identified with the union (in case of the joint entropy) and intersections of the sets corresponding to the subsystem entropies. For instance, the conditional entropy $S(A|B, C)$ is represented as the complement of the set identified with $S(B) \cup S(C)$, the conditional mutual information $S(A : B|C)$ is represented as the intersection $S(A : B|C) = S(A) \cap S(B) \setminus S(A : B : C)$, and the ternary mutual information $S(A : B : C)$ is the intersection of all three subsystem entropies $S(A : B : C) = S(A) \cap S(B) \cap S(C)$.

a pure state we have $S(A, B, C) = 0$ and the complementary subsystem entropies are equal, i.e., $S(AB) = S(C)$, $S(A, C) = S(B)$, and $S(B, C) = S(A)$, such that the ternary mutual information vanishes $S(A : B : C) = 0$.

The relationship between the different entropic quantities for tripartite systems is illustrated for general states in the information diagram in Fig. 20.3. There, the subsystem entropies $S(A)$, $S(B)$, $S(C)$ are interpreted as sets, whose union can be identified with the joint entropy $S(A, B, C)$, while the intersections of different combinations of entropy sets yield the conditional and mutual entropies. For example, the conditional entropy $S(A|B, C)$ can be viewed as the complement to $S(B) \cup S(C)$, the conditional mutual information $S(A : B|C)$ as excluding the set $S(A : B : C)$ from the intersection $S(B) \cap S(C)$, that is $S(A : B|C) = S(B) \cap S(C) \setminus S(A : B : C)$. In this picture, the ternary mutual information $S(A : B : C)$ is the intersection of all three subsystem entropies, which is denoted by $S(A : B : C) = S(A) \cap S(B) \cap S(C)$.

To illustrate our general discussion of tripartite states by a specific example, let us consider the GHZ state $\rho_{\mathrm{GHZ}} = |\,\mathrm{GHZ}\,\rangle\langle\,\mathrm{GHZ}\,|$ from eqn (18.10) with $|\,\mathrm{GHZ}\,\rangle = \frac{1}{\sqrt{2}}\left(|\,000\,\rangle + |\,111\,\rangle\right)$ as in eqn (18.1a). As the GHZ state is a pure state the joint entropy vanishes $S(A, B, C) = 0$. The entropies $S(A) = S(B) = S(C) = S(A, B) = S(A, C) = S(B, C) = 1$ (again considering the logarithm to base 2 for simplicity) are all equal and maximal, therefore all generalized quantum conditional entropies are negative $S(A|B, C) = S(B|A, C) = S(C|A, B) = -1$. This negativity reveals that the GHZ state, a pure quantum state, deviates drastically from classical physics. The ternary mutual information vanishes $S(A : B : C) = 0$ as can be seen from eqn (20.71),

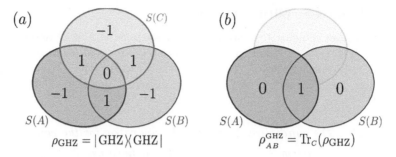

Fig. 20.4 Ternary entropy Venn diagrams. (a) For the GHZ state ρ_{GHZ}, where the
ternary mutual information vanishes $S(A:B:C) = 0$, the conditional mutual information
for different subsystems evaluates to $S(A:B|C) = S(B:C|A) = S(A:C|B) = 1$, and the
conditional entropies are negative $S(A|B,C) = S(B|C,A) = S(C|A,B) = -1$. (b) For the
reduced density matrix $\rho_{AB}^{\mathrm{GHZ}} = \mathrm{Tr}_C(\rho_{\mathrm{GHZ}})$ for the subspace AB, which is a mixed state and
separable, i.e., the remaining two qubits are classically correlated, the mutual information is
maximal $S(A:B) = 1$ and the conditional entropies vanish $S(A|B) = S(B|A) = 0$.

which is generic for any tripartite system being in a pure quantum state. The different
values of the entropic measures in the tripartite system are illustrated in Fig. 20.4 (a).

Tracing out one of the subsystems, say C, from the density matrix of the GHZ
state, we obtain the state $\rho_{AB}^{\mathrm{GHZ}} = \mathrm{Tr}_C(\rho_{\mathrm{GHZ}})$ from eqn (18.11), which is mixed and
separable. The remaining two qubits are classically correlated and their joint entropy is
maximal $S(A,B) = 1$, whereas their conditional entropies vanish $S(A|B) = S(B|A) = 0$, as illustrated in Fig. 20.4 (b). The joint entropy of Alice and Bob $S(A,B) = 1$ is
balanced by the "ignored" entropy $S(C|A,B) = -1$ of Charly conditioned on Alice
and Bob.

20.5 Conditional and Mutual Amplitude Operators

As we had noticed in Chapter 19, the classical joint entropy $H(X,Y)$, classical con-
ditional entropy $H(X|Y)$, and classical mutual information $I(X:Y)$ can each be
represented as the expected values of other (probability) distributions, $p_{X,Y}(x,y)$,
$p_{X,Y}(x|y)$, and $(p_{X,Y}(x:y)$, see Definitions 19.4 and 19.5, and eqn (19.67), respectively. The quantum joint entropy of $S(A,B) = S(\rho_{AB})$ of a bipartite system can, as
we know, formally be obtained by switching from a probability distribution $p_{X,Y}(x,y)$
to a density operator ρ_{AB} and replacing the sum with a trace. In this sense, the (joint)
quantum entropy can just be regarded as the expected value of an operator $-\log(\rho_{AB})$
for the quantum state ρ_{AB},

$$S(A,B) = S(\rho_{AB}) = \langle -\log(\rho_{AB}) \rangle_{\rho_{AB}} . \tag{20.72}$$

In a series of papers Nicolas Cerf and Christoph Adami (1996a, 1996b, 1997b, 1998,
1999) constructed operators that similarly allow us to generate the quantum conditional entropy and the quantum mutual information as expected values of operators,
the *conditional amplitude operator* and the *mutual amplitude operator*, respectively.

20.5.1   Conditional Amplitude Operator

Let us again consider a bipartite system with subsystems A and B described by the joint density operator $\rho_{AB} \in \mathcal{B}_{\mathrm{HS}}(\mathcal{H}_A \otimes \mathcal{H}_B)$. The subsystems of Alice and Bob are described by the reduced matrices $\rho_A = \mathrm{Tr}_B(\rho_{AB}) \in \mathcal{B}_{\mathrm{HS}}(\mathcal{H}_A)$ and $\rho_B = \mathrm{Tr}_A(\rho_{AB}) \in \mathcal{B}_{\mathrm{HS}}(\mathcal{H}_B)$. Then one may define the following operator following Cerf and Adami (1997b, 1999).

Definition 20.8   *The **conditional amplitude operator** for the conditioning of subsystem A on subsystem B is defined as*

$$\rho_{A|B} := \exp\left[\log(\rho_{AB}) - \log\left(\mathbb{1}_A \otimes \rho_B \right) \right].$$

The conditional amplitude operator $\rho_{A|B}$ is a positive semidefinite Hermitian operator defined on the support of ρ_{AB}. It is referred to as an *amplitude operator* to emphasize that it retains information about the phases of the quantum state, in contrast to the conditional probabilities $p_{X,Y}(x|y)$ from eqn (19.51). Furthermore, the conditional amplitude operator $\rho_{A|B}$ can be re-expressed with the help of the *Lie–Trotter formula* for non-commuting matrices, well-known in the mathematical literature (for a proof, we refer to Reed and Simon, 1972, Theorem VIII.29).

Lemma 20.4 (Lie–Trotter formula)

For two (finite-dimensional) matrices U and V that do not commute, $[U, V] \neq 0$, *we have*

$$e^{U+V} = \lim_{n \to \infty} \left[e^{\frac{U}{n}} e^{\frac{V}{n}} \right]^n,$$

where e^U denotes the matrix exponential.

The Lie–Trotter formula can be extended to unbounded operators U and V if they are self-adjoint and if $U + V$ is self-adjoint on the union of the domains of U and V (see Reed and Simon, 1972, Theorem VIII.30). Then the expression of $\rho_{A|B}$ in Definition 20.8 can be reformulated by the Lie–Trotter limiting procedure.

Theorem 20.9 (Conditional amplitude operator as Lie–Trotter limit)

Defining the matrix

$$\sigma_{AB} := \mathbb{1}_A \otimes \log(\rho_B) - \log(\rho_{AB}),$$

the conditional amplitude operator can be obtained as the Lie–Trotter limit

$$\rho_{A|B} = e^{-\sigma_{AB}} = \lim_{n \to \infty} \left[(\rho_{AB})^{\frac{1}{n}} \left(\mathbb{1}_A \otimes \rho_B \right)^{-\frac{1}{n}} \right]^n.$$

Proof Let us define

$$e^U := \rho_{AB} \quad \text{and} \quad e^V := (\mathbb{1}_A \otimes \rho_B)^{-1} \tag{20.73a}$$

$$\implies \quad U = \log \rho_{AB} \quad \text{and} \quad V = -\log(\mathbb{1}_A \otimes \rho_B) = -\mathbb{1}_A \otimes \log(\rho_B). \tag{20.73b}$$

Then using the Lie–Trotter formula, Lemma 20.4, we have

$$\lim_{n \to \infty} \left[(\rho_{AB})^{\frac{1}{n}} (\mathbb{1}_A \otimes \rho_B)^{-\frac{1}{n}} \right]^n = \exp\left[\log(\rho_{AB}) - \mathbb{1}_A \otimes \log(\rho_B) \right] = e^{-\sigma_{AB}} = \rho_{A|B} . \tag{20.74}$$

$\square$

Remark: The Lie–Trotter limit of the operator $\rho_{A|B}$ in Theorem 20.9 can be regarded as the quantum generalization of the classical conditional probability $p(a|b) = p(a,b)/p(b)$, eqn (19.51). This means $\rho_{A|B}$ reduces to $p(a|b)$ in the classical limit. This is quite obvious in the case when ρ_{AB} and $\mathbb{1}_A \otimes \rho_B$ commute (e.g., when ρ_{AB} is diagonal). Then the Lie–Trotter limit for the conditional amplitude operator $\rho_{A|B}$, Theorem 20.9, reduces to

$$\rho_{A|B} = \rho_{AB} (\mathbb{1}_A \otimes \rho_B)^{-1} . \tag{20.75}$$

With the help of the operator $\rho_{A|B}$ Cerf and Adami (1997*b*, 1999) then proceeded to construct the quantum extension to the classical conditional entropy from Definition 19.5:

Theorem 20.10 (Quantum conditional entropy—operator form)

The quantum conditional entropy for a bipartite system is given by the expectation value of the operator $-\log(\rho_{A|B})$, *where* $\rho_{A|B}$ *is the conditional amplitude operator, in the joint state with density matrix* ρ_{AB},

$$S(A|B) = \left\langle -\log(\rho_{A|B}) \right\rangle_{\rho_{AB}} = -\text{Tr}\left(\rho_{AB} \log(\rho_{A|B}) \right) .$$

With this construction we can reproduce the *conditional entropy* from Definition 20.5.

Proof Recalling Definition 20.8 and noticing that

$$\text{Tr}\left[\rho_{AB} \log(\mathbb{1}_A \otimes \rho_B) \right] = \text{Tr}\left[\rho_{AB} \mathbb{1}_A \otimes \log(\rho_B) \right] \tag{20.76}$$

$$= \text{Tr}_A(\rho_A \mathbb{1}_A) \text{Tr}_B(\rho_B \log(\rho_B)) = -S(\rho_B) ,$$

we find

$$S(A|B) = -\text{Tr}\left(\rho_{AB} \log(\rho_{A|B}) \right) = -\text{Tr}\left(\rho_{AB} \log(\rho_{AB}) \right) + \text{Tr}\left[\rho_{AB} \log(\mathbb{1}_A \otimes \rho_B) \right]$$

$$= S(\rho_{AB}) - S(\rho_B) . \tag{20.77}$$

$\square$

20.5.2 Mutual Amplitude Operator

In analogy to Definition 20.8, Cerf and Adami (1996*b*, 1997*b*, 1998) further defined an operator whose negative logarithm gives rise to the *quantum mutual information*.

Definition 20.9 *The **mutual amplitude operator** for a bipartite system with subsystems A and B is defined as*

$$\rho_{A:B} := \exp\left[\log\left(\rho_A \otimes \rho_B\right) - \log(\rho_{AB})\right].$$

As before, the expression of $\rho_{A:B}$ in Definition 20.9 can be reformulated in terms of a Lie–Trotter limiting procedure (see Lemma 20.4).

Theorem 20.11 (Mutual amplitude operator as Lie–Trotter limit)

Defining the matrix

$$\tau_{AB} := \log(\rho_{AB}) - \log\left(\rho_A \otimes \rho_B\right),$$

the mutual amplitude operator is given by the Lie–Trotter limit

$$\rho_{A:B} = e^{-\tau_{AB}} = \lim_{n\to\infty}\left[\left(\rho_A \otimes \rho_B\right)^{\frac{1}{n}}\left(\rho_{AB}\right)^{-\frac{1}{n}}\right]^n.$$

In the case that the matrices $\rho_A \otimes \rho_B$ and ρ_{AB} commute the mutual amplitude operator reduces to

$$\rho_{A:B} = \left(\rho_A \otimes \rho_B\right)\left(\rho_{AB}\right)^{-1}, \tag{20.78}$$

which makes the quantum extension to the classical case with its mutual probability $p(x{:}y) = p_{X,Y}(x,y)/[p(x)p(y)]$, eqn (19.66), more obvious.

With the help of $\rho_{A:B}$ we can now construct the quantum mutual information.

Theorem 20.12 (Mutual information—operator form)

The quantum mutual information for a bipartite system is given by the expectation value of the operator $-\log(\rho_{A:B})$, where $\rho_{A:B}$ is the mutual amplitude operator, in the joint state with density matrix ρ_{AB},

$$S(A{:}B) = \left\langle -\log(\rho_{A:B})\right\rangle_{\rho_{AB}} = -\mathrm{Tr}\left(\rho_{AB}\log(\rho_{A:B})\right).$$

Note the sign change in our expressions for the classical mutual information in eqn 19.67 and the quantum version in Theorem 20.12. Nevertheless, the latter construction agrees with our definition of the quantum mutual information, Definition 20.6.

Proof

$$
\begin{aligned}
S(A\!:\!B) &= -\operatorname{Tr}\!\big(\rho_{AB}\left[\log(\rho_A \otimes \rho_B) - \log(\rho_{AB})\right]\big) \\
&= -\operatorname{Tr}\!\left[\rho_{AB}\log(\rho_A \otimes \mathbb{1}_B)\right] - \operatorname{Tr}\!\left[\rho_{AB}\log(\mathbb{1}_A \otimes \rho_B)\right] + \operatorname{Tr}\!\left[\rho_{AB}\log(\rho_{AB})\right] \\
&= -\operatorname{Tr}_A\!\left[\rho_A\log(\rho_A)\right] - \operatorname{Tr}_B\!\left[\rho_B\log(\rho_B)\right] + \operatorname{Tr}\!\left[\rho_{AB}\log(\rho_{AB})\right] \\
&= S(\rho_A) + S(\rho_B) - S(\rho_{AB})\,.
\end{aligned}
\tag{20.79}
$$

$\square$

With the operator forms of the quantum mutual information thus also established, we now return to the conditional amplitude operator and its properties, in particular, with a view to its usefulness for entanglement detection.

20.5.3 Properties of the Conditional Amplitude Operator

The conditional amplitude operators discussed so far have interesting properties, in particular, in relation to the separability or entanglement of the quantum states. First, it is crucial to note that, although the conditional amplitude operator itself is not invariant under local unitary transformations, its spectrum is, since an operation of the form $U_A \otimes U_B$ for some unitaries U_A and U_B on $\mathcal{H}_A$ and $\mathcal{H}_B$, respectively, leave the spectra of both ρ_{AB} and $\mathbb{1}_A \otimes \rho_B$ invariant.

Thus, the quantum conditional entropy $S(A|B)$ is invariant under local unitary transformations, as is also obvious from Definition 20.5. Since local unitary transformations do not change the entanglement of a quantum state, this already suggests that the spectrum of $\rho_{A|B}$ is related to the separability of the corresponding bipartite state ρ_{AB}. We will focus on this aspect now.

Theorem 20.13 (Cerf–Adami theorem I)

If the bipartite state characterized by the density matrix ρ_{AB} is separable, the operator

$$
\sigma_{AB} := -\log(\rho_{A|B}) = \mathbb{1}_A \otimes \log(\rho_B) - \log(\rho_{AB}) \geq 0
$$

is positive semidefinite.

Proof We consider a separable bipartite system characterized by the density matrix

$$
\rho_{AB}^{\text{sep}} = \sum_i p_i\, \rho_i^{(A)} \otimes \rho_i^{(B)} \quad \text{with} \quad 0 \leq p_i \leq 1,\ \sum_i p_i = 1\,,
\tag{20.80}
$$

where $\rho_i^{(A)} \in \mathcal{B}_{\text{HS}}(\mathcal{H}_A)$ and $\rho_i^{(B)} \in \mathcal{B}_{\text{HS}}(\mathcal{H}_B)$ describe states in the subsystems of Alice and Bob. Next we define the operator

$$\omega_{AB} := \mathbb{1}_A \otimes \rho_B - \rho_{AB} \tag{20.81}$$

and check that ω_{AB} is positive semidefinite if ρ_{AB} is separable. Inserting the separability, eqn (20.80), we get

$$\omega_{AB} = \sum_i p_i \left(\mathbb{1}_A \otimes \rho_i^{(B)} - \rho_i^{(A)} \otimes \rho_i^{(B)}\right) \geq 0 , \tag{20.82}$$

since a sum of positive operators is a positive operator again. Note that this is the basis for the reduction criterion that we have discussed in Section 15.4.3.

To complete the proof, we make use of a well-known theorem in linear algebra for monotone functions of matrices, which was first proven by *Karl Löwner* (1934). Here, we state a simplified version for the logarithmic function.

Theorem 20.14 (Löwner's theorem)

Let X and Y be two Hermitian matrices, then

$$if \ \ X \geq Y \geq 0 \quad \Longrightarrow \quad \log(X) \geq \log(Y).$$

Choosing $X = \mathbb{1}_A \otimes \rho_B$ and $Y = \rho_{AB}$, we see that positivity of ω_{AB} implies $X \geq Y$, and thus Theorem 20.14 assures that $\log(\mathbb{1}_A \otimes \rho_B) \geq \log(\rho_{AB})$, or $\sigma_{AB} \geq 0$.   □

The properties of the operator σ_{AB} obviously determine those of $\rho_{A|B}$. Since $\sigma_{AB} \geq 0$ for separable states, Theorem 20.13 implies the following two corollaries.

Corollary 20.2 (Conditional amplitude operator of separable states)

For any bipartite separable state the conditional amplitude operator is bounded from above by

$$\rho_{A|B} = e^{-\sigma_{AB}} \leq 1.$$

Corollary 20.3 (Conditional entropy of separable states)

For any bipartite separable state the quantum conditional entropy is non-negative

$$S(A|B) \geq 0.$$

Corollary 20.3 can be obtained directly from Theorem 20.10 by observing $S(A|B) = S(\rho_{AB}) - S(\rho_B) = \mathrm{Tr}(\rho_{AB}\,\sigma_{AB}) \geq 0$. Indeed, recall that we had already seen that Corollary 20.3 holds when proving Theorem 20.4.

Here we again see that the situation changes for entangled quantum states. There exist entangled states such that $\sigma_{AB} \ngeq 0$ or $\rho_{A|B} \nleq 1$. In such cases, the conditional amplitude operator $\rho_{A|B}$ does not represent a valid (normalized) density operator. We will present a detailed geometric investigation of such states in Section 20.6.

Let us finally consider Theorem 20.4 again. If the conditional entropy is negative $S(A|B) < 0$ then the joint state ρ_{AB} of Alice and Bob is entangled. The reverse, however, is not true. A corresponding statement can also be made for the conditional amplitude operator.

Theorem 20.15 (Cerf–Adami theorem II)

If the conditional entropy of a bipartite quantum state is negative, $S(A|B) < 0$, then the conditional amplitude operator exceeds the separability bound of one, $\rho_{A|B} \nleq 1$, i.e., the corresponding state is entangled. However, the reverse is not true!

$$S(A|B) < 0 \quad \overset{\Longrightarrow}{\nLeftarrow} \quad \rho_{A|B} \nleq 1 \quad or \quad \sigma_{AB} \ngeq 0 \, .$$

The conditional amplitude operator can thus serve as an entanglement witness, and provides a strictly stronger criterion ($\rho_{A|B} > 1$) for detecting entanglement than the negativity of the conditional entropy. The interesting question now is: Can $\rho_{A|B}$ be used to detect *all* entangled states?

The answer is *no*, nevertheless, for a large class of familiar quantum states the conditional amplitude nevertheless provides a useful entanglement-detection criterion, as we will discuss in detail in Section 20.6. Here, let us just state this observation in terms of the following theorem (which we will prove later on).

Theorem 20.16 (Cerf–Adami theorem III)

There exist inseparable bipartite quantum states ρ_{AB} such that the operator σ_{AB} is positive semidefinite. Consequently, $\sigma_{AB} \geq 0$ or $\rho_{A|B} \leq 1$ are not sufficient conditions for the separability of a general quantum state ρ_{AB}.

Together, we can refer to Theorems 20.13–20.16 as the *conditional amplitude operator criterion* or *CAO criterion* for short.

Theorem 20.17 (CAO criterion)

Let ρ_{AB} be the density operator of a bipartite system with conditional amplitude operator $\rho_{A|B}$ and $\sigma_{AB} = -\log(\rho_{A|B})$, then

$$\rho_{A|B} \not\leq 1 \quad or \quad \sigma_{AB} \not\geq 0 \quad \overset{\Longrightarrow}{\nLeftarrow} \quad \rho_{AB} \text{ is entangled.}$$

20.6 Negative Conditional Entropy and Geometry of Quantum States

In both classical and quantum information theory entropies play a crucial role, but also properties like separability, entanglement, and non-locality are of fundamental importance. And indeed, we have already discussed in detail how entropic measures can be used for the quantification of entanglement in Chapter 16. In this context, it is therefore instructive to study the relationship between entanglement-detection criteria based on the conditional entropy (Theorem 20.4), the conditional amplitude operator (the CAO criterion from Theorem 20.17) and previously discussed techniques such as the PPT criterion (Theorem 15.3). In particular, we are interested in comparing these criteria for the family of Weyl states from Section 15.5.1, which allows us to examine these quantities in a simple geometric picture. In doing so, we closely follow the presentation in (Friis *et al.*, 2017*a*).

Before we begin, let us briefly recall that the set $\mathcal{W}$ of Weyl states is defined as the set of two-qubit states whose reduced states are maximally mixed. Up to local unitaries every Weyl state can thus be represented in two-qubit Bloch decomposition with diagonal correlation tensor,

$$\mathcal{W} = \left\{ \rho \,\middle|\, \rho = \frac{1}{4}\left(\mathbb{1}_A \otimes \mathbb{1}_B + \sum_{i=1}^{3} \tilde{t}_i\, \sigma_i^A \otimes \sigma_i^B \right) \right\}. \tag{20.83}$$

The components of the correlation tensor $\tilde{t}_i$ serve as coordinates and are constrained by $-1 \leq \tilde{t}_i \leq +1$. The four Bell states $\rho^{\mp} = |\psi^{\mp}\rangle\langle\psi^{\mp}|$ and $\omega^{\mp} = |\phi^{\mp}\rangle\langle\phi^{\mp}|$ lie at the corners $(\tilde{t}_1, \tilde{t}_2, \tilde{t}_3) = (1,-1,1)$ ($|\phi^{+}\rangle$), $(-1,1,1)$ ($|\phi^{-}\rangle$), $(1,1,-1)$ ($|\psi^{+}\rangle$), and $(-1,-1,-1)$ ($|\psi^{-}\rangle$) of a tetrahedron, which contains all Weyl states, as shown in Fig. 15.3 of Section 15.5, and at whose centre $(0,0,0)$ lies the maximally mixed state $\rho_{\text{mix}} = \frac{1}{4}\mathbb{1}_4$.

Within the tetrahedron of Weyl states, the set $\mathcal{S}$ of separable states forms a double pyramid, the maximal Kuś–Życzkowski ball of absolutely separable states lies within the blue double pyramid and touches the faces of the pyramids at the points where $|\tilde{t}_1| = |\tilde{t}_2| = |\tilde{t}_3| = \frac{1}{3}$, see Fig. 16.7, and the entangled states are located outside the double pyramid in the remaining regions towards the tips of the tetrahedron, see

Fig. 15.3. On the other hand, all local states, the states that satisfy the CHSH–Bell inequality, Proposition 13.2, lie within a region with curved surface that includes but is strictly larger than the separable double pyramid, thus including also some entangled states. Consequently, there are regions of non-locality in the tips of the tetrahedron.

20.6.1 Geometry of the Conditional Entropy and Cerf–Adami Operator

Now we incorporate the conditional entropy and the conditional amplitude operator into the geometric picture of the Weyl-state tetrahedron. To this end let us first concentrate on the single-parameter family of Werner states $\rho_{\text{Werner}}(\alpha)$ with $\alpha \in [-\frac{1}{3}, 1]$, given by eqns (15.68) and (15.69), which form a line in the tetrahedron of Weyl states, connecting the maximally entangled state $|\Psi^-\rangle$, obtained for $\alpha = 1$, with the maximally mixed state at the origin (for $\alpha = 0$), and reaching all the way to the opposite side of the double pyramid of separable states (for $\alpha = -\frac{1}{3}$), as illustrated in Fig. 15.3. Let us further recall from Section 15.3.3 that the Werner states are entangled for $\alpha > \frac{1}{3}$ and violate the CHSH inequality for $\alpha > \frac{1}{\sqrt{2}}$.

Next we compute the conditional entropy for the Werner states. We note that $\rho_{\text{Werner}}(\alpha)$ has a thrice degenerate eigenvalue $(1-\alpha)/4$ and a single eigenvalue $(1 + 3\alpha)/4$, based on which we want to determine the boundary between negative and non-negative conditional entropy. Formally, this boundary is represented by the state $\rho_{\text{Werner}}(\alpha_0)$, where α_0 is the solution of the transcendental equation

$$3(1-\alpha)\log(1-\alpha) + (1+3\alpha)\log(1+3\alpha) = 4\log 2, \qquad (20.84)$$

which can easily be numerically determined to be

$$\alpha_0 \doteq 0.7476 > \tfrac{1}{\sqrt{2}} = 0.7071. \qquad (20.85)$$

We thus see that the condition $S(A|B) < 0$ is a strictly stronger condition than non-locality (here defined as the violation of the CHSH inequality using projective measurements, as in Proposition 13.2) for the family of Werner states, as illustrated in Fig. 20.5. Indeed, a numerical analysis shows that this is the case for all Weyl states, i.e., the curved surfaces beyond which the conditional entropy becomes negative lie strictly outside of the local region that is confined by the dark-yellow parachutes in the tetrahedron of Weyl states in Fig. 15.3.

On the other hand, examining the conditional amplitude operator from Definition 20.8 for the Werner states we find

$$\rho_{A|B} = 2\,\rho_{\text{Werner}}, \qquad (20.86)$$

since $\rho_B = \text{Tr}_A(\rho_{\text{Werner}}) = \frac{1}{2}\mathbb{1}$ such that $\log(\mathbb{1}_A \otimes \rho_B)$ commutes with $\log(\rho_{\text{Werner}})$. Therefore, the condition $\rho_{A|B} \leq \mathbb{1}$ is met as long as $\alpha \leq \frac{1}{3}$, i.e., as long as ρ_{Werner} is separable, whereas $\rho_{A|B}$ has an eigenvalue larger than 1 for $\alpha > \frac{1}{3}$. For the Werner states the condition $\rho_{A|B} \not\leq \mathbb{1}$ of the CAO criterion (Theorem 20.17) is thus equivalent to the PPT criterion, Theorem 15.3, a fact already noticed by Cerf and Adami (1999).

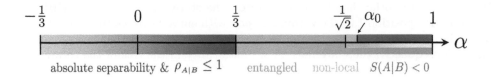

$$
-\tfrac{1}{3} \qquad 0 \qquad \tfrac{1}{3} \qquad \tfrac{1}{\sqrt{2}} \quad \alpha_0 \qquad 1
$$

absolute separability & $\rho_{A|B} \leq 1$ entangled   non-local $S(A|B) < 0$

Fig. 20.5 Parameter range of Werner states. The line illustrates the regions of the parameter α for which the Werner states are (absolutely) separable (lilac/purple), entangled (green), violate the CHSH inequality (orange), and have negative conditional entropy (red), respectively.

Indeed, as observed in Friis *et al.* (2017*a*) the statement just made for the Werner states also holds for all other Weyl states in addition to this one-parameter subfamily.

Theorem 20.18 (CAO criterion for Weyl states)

For every Weyl state $\rho_{AB} \in \mathcal{W}$ the CAO criterion $\rho_{A|B} \leq 1$ for the conditional amplitude operator $\rho_{A|B}$, given by Definition 20.8, is equivalent to the PPT criterion, Theorem 15.3, i.e.,

$$\rho_{A|B} \leq 1 \quad \textit{if and only if} \quad \rho_{AB} \in \mathcal{W} \quad \textit{is separable} ,$$

$$\rho_{A|B} \not\leq 1 \quad \textit{if and only if} \quad \rho_{AB} \in \mathcal{W} \quad \textit{is entangled} .$$

Proof For the proof of Theorem 20.18 we recall Wootter's concurrence formula for a state ρ. displayed in Theorem 16.6,

$$C(\rho) = \max\{0, \lambda_1 - \lambda_2 - \lambda_3 - \lambda_4\} , \qquad (20.87)$$

where the λ_i's are the square roots of the eigenvalues of the matrix $\rho\tilde{\rho}$ in decreasing order. The spin-flipped state is given by $\tilde{\rho} = (\sigma_y \otimes \sigma_y)\rho^*(\sigma_y \otimes \sigma_y)$ and ρ^* denotes the complex conjugate of ρ. For the Weyl states we immediately notice that $\tilde{\rho} = \rho$. The square roots of the eigenvalues of $\rho\tilde{\rho}$ needed for the concurrence are hence just the eigenvalues p_n with $n = 1, 2, 3, 4$ of ρ, which satisfy $\sum_n p_n = 1$. Consequently, the concurrence of all Weyl states can be written as

$$C(\rho) = \max\{0, p_1 - p_2 - p_3 - p_4\} = \max\{0, 2p_1 - 1\} , \qquad (20.88)$$

where the largest eigenvalue p_1 must exceed the value of $\tfrac{1}{2}$ for ρ to be entangled.

Next, we recall that for all Weyl states ρ_{AB} we have $\rho_A = \rho_B = \tfrac{1}{2}\mathbb{1}_2$ implying that the conditional amplitude operator is twice the Weyl state itself,

$$\rho_{A|B} = \exp\big(\log(\rho_{AB}) - \log \mathbb{1}_A \otimes \rho_B\big) = 2\,\rho_{AB} . \qquad (20.89)$$

Thus we have $\rho_{A|B} \not\leq 1$ when the largest eigenvalue of $\rho_{A|B} = 2\rho_{AB}$ exceeds 1, i.e., when the largest eigenvalue of ρ_{AB} exceeds $\tfrac{1}{2}$. By virtue of eqn (20.88) this means that the state is entangled. Conversely, since all entangled Weyl states (in fact, all entangled two-qubit states) are detected as entangled by a non-zero concurrence (or non-zero entanglement of formation), see, e.g., eqn (16.89), they must have an eigenvalue larger than $\frac{1}{2}$ such that $\rho_{A|B} \not\leq \mathbb{1}$. The fact that all entangled two-qubit states have a non-positive partial transposition concludes the proof. $\square$

While the CAO criterion is thus equivalent to the PPT criterion for the two-qubit Weyl states, and while the CAO criterion is a strictly stronger entanglement criterion than the negativity of the conditional entropy, a numerical evaluation of the conditional entropy of all states of the set $\mathcal{W}$, eqn (20.83), shows that the negativity of $S(A|B)$ in turn is a strictly stronger condition than non-locality for these states. As we shall see in Fig. 20.7, the red curved surfaces in Fig. 15.3 that indicate where the conditional entropy changes sign lie outside the dark-yellow "parachutes" that mark the boundary of non-locality for all Weyl states.

However, as we shall see shortly, the CAO and PPT criteria are generally inequivalent and there is generally also no hierarchy between non-locality and negative conditional (von Neumann) entropy.

20.6.2 Inequivalence of the CAO and PPT Criteria

Having established the significance of the conditional amplitude operator and the relationship of entanglement, negative conditional entropy, and non-locality for the Weyl states, we now want to investigate whether the observations made so far also hold for other states. But, as discussed in Friis *et al.* (2017*a*), this is not the case. To see this, let us examine some examples.

Narnhofer states: Starting from a specific Weyl state, we consider its unitary orbit which takes us outside of this set. As a Weyl state to start with we choose the Narnhofer state

$$\rho_{\mathrm{N}} = \tfrac{1}{4}\left(\mathbb{1} \otimes \mathbb{1} + \sigma_x \otimes \sigma_x\right), \tag{20.90}$$

introduced in Section 16.4.1, which is represented by the point at the upper corner of the double pyramid of separable states in the tetrahedron of Fig. 15.3, located half-way on the line connecting $|\Psi^+\rangle$ and $|\Phi^+\rangle$. Next we apply the unitary given by

$$U = \tfrac{1}{4}\left[(2+\sqrt{2})\,\mathbb{1} \otimes \mathbb{1} + i\sqrt{2}\left(\sigma_x \otimes \sigma_y + \sigma_y \otimes \sigma_x\right) - (2-\sqrt{2})\,\sigma_z \otimes \sigma_z\right], \tag{20.91}$$

to transform the state ρ_{N} into

$$\rho_U = U\rho_{\mathrm{N}}U^\dagger = \tfrac{1}{4}\left[\mathbb{1} \otimes \mathbb{1} + \tfrac{1}{2}(\sigma_z \otimes \mathbb{1} + \mathbb{1} \otimes \sigma_z) + \tfrac{1}{2}(\sigma_x \otimes \sigma_x + \sigma_y \otimes \sigma_y)\right]. \tag{20.92}$$

Because of the term $\frac{1}{2}(\sigma_z \otimes \mathbb{1} + \mathbb{1} \otimes \sigma_z)$ the state (20.92) lies outside of the set $\mathcal{W}$. The purity $\mathrm{Tr}(\rho_{\mathrm{N}}^2) = \mathrm{Tr}(\rho_U^2) = \frac{1}{2}$ is left unchanged by the unitary transformation but the final state ρ_U is entangled. In fact, the concurrence takes the maximally possible value at fixed purity, $C(\rho_U) = \frac{1}{2}$, i.e., the state ρ_U belongs to the MEMS class of states. In other words, no global unitary may entangle this state any further.

Keeping this in mind, we now consider a family of states in the Hilbert–Schmidt space along the line from ρ_U to ρ_{top}, i.e., we define the *Narnhofer line* as

$$\rho_U(\beta) := \beta\,\rho_U + (1-\beta)\,\rho_{\text{top}}, \tag{20.93}$$

with $0 \leq \beta \leq 1$. The state ρ_{top} represents the top corner of the separable double pyramid in Fig. 15.3, and can be expressed as

$$\rho_{\text{top}} = \tfrac{1}{2}\left(|00\rangle\langle 00| + |11\rangle\langle 11|\right) = \tfrac{1}{4}\left(\mathbb{1}\otimes\mathbb{1} + \sigma_z\otimes\sigma_z\right). \tag{20.94}$$

The eigenvalues of the partial transpose $\rho_U^{T_B}(\beta) = (\mathbb{1}\otimes T)[\rho_U(\beta)]$ are $\tfrac{\beta}{4}$ (twice degenerate), as well as $\tfrac{1}{4}\left[2 - \beta(1\pm\sqrt{2})\right]$. According to the PPT criterion the states along the Narnhofer line are hence entangled if

$$\beta > \beta_{\text{PPT}} = 2(\sqrt{2}-1) = 0.8284. \tag{20.95}$$

Now, if we consider the CAO criterion, we first compute the reduced state

$$\rho_U^B(\beta) = \text{Tr}_A\big(\rho_U(\beta)\big) = \tfrac{1}{4}\begin{pmatrix} 2+\beta & 0 \\ 0 & 2-\beta \end{pmatrix}, \tag{20.96}$$

and then the spectrum of $\rho_U(\beta)$, which yields

$$\text{spectr}\big[\rho_U(\beta)\big] = \{\tfrac{1}{2}, 0, \tfrac{1-\beta}{2}, \tfrac{\beta}{2}\}. \tag{20.97}$$

In order to finally compute the spectrum of $\rho_{A|B}$, we note that $\rho_U(\beta)$ has (at least) one vanishing eigenvalue (see eqn (20.97)), which is problematic when evaluating $\log \rho_U(\beta)$. However, one can simply replace the vanishing eigenvalue by $\epsilon > 0$ throughout the computation and take the limit $\epsilon \to 0$ at the end. With this procedure the eigenvalues of the conditional amplitude operator $\rho_{A|B}$ along the Narnhofer line evaluate to

$$\left\{0, \frac{2}{2+\beta}, \frac{2-2\beta}{2-\beta}, \frac{2\beta}{\sqrt{4-\beta^2}}\right\}. \tag{20.98}$$

The first three eigenvalues are always smaller than 1, but the last eigenvalue becomes larger than 1 when

$$\beta > \beta_{\text{CAO}} = \frac{2}{\sqrt{5}} = 0.8944. \tag{20.99}$$

Consequently, what we find is

$$\beta_{\text{CAO}} > \beta_{\text{PPT}}, \tag{20.100}$$

i.e., the CAO criterion provides as weaker condition for detecting entanglement among the states of the Narnhofer line outside the set of Weyl states.

Summarizing, the CAO criterion and the PPT criterion are inequivalent in general. In addition, one may easily check that the conditional entropy of the state $\rho_U(\beta)$ remains non-negative for all values of β, and none of these states allows for a violation of the CHSH inequality either.f

Gisin states: For this next example we also incorporate the (negative) conditional entropy and non-locality into the picture. Hence we turn to the Gisin states, discussed already in Section 15.3.5,

$$\rho_{\text{Gisin}}(\lambda, \theta) = \lambda \rho_\theta + (1 - \lambda) \rho_{\text{top}} \tag{20.101}$$

$$= \frac{1}{4} \left(\mathbb{1} \otimes \mathbb{1} - \lambda \cos(2\theta) \left(\sigma_z \otimes \mathbb{1} - \mathbb{1} \otimes \sigma_z \right) \right.$$

$$\left. + \lambda \sin(2\theta) \left(\sigma_x \otimes \sigma_x + \sigma_y \otimes \sigma_y \right) + (1 - 2\lambda) \sigma_z \otimes \sigma_z \right),$$

with the parameter range $0 \leq \lambda \leq 1$. The spectrum of the density operator $\rho_{\text{Gisin}}(\lambda, \theta)$ is computed to be

$$\text{spectr}\left[\rho_{\text{Gisin}}(\lambda, \theta) \right] = \left\{ 0, \tfrac{1-\lambda}{2}, \tfrac{1-\lambda}{2}, \lambda \right\}, \tag{20.102}$$

while the reduced states ρ_A and ρ_B are diagonal with eigenvalues $\frac{1}{2}\left(1 \pm \lambda \cos(2\theta)\right)$. The graphical analysis of the parameter region for λ and θ for which the conditional entropy is negative reveals an interesting feature, see Fig. 20.6.

In summary, while some Gisin states are both non-local and have negative conditional entropy, some only have one of these properties, but not the other. This means, contrary to what was found for the Weyl states, in general not all states for which $S(A|B) < 0$ are non-local. And, as before, not all non-local states have negative conditional entropy.

When we examine the condition $\rho_{A|B} \leq \mathbb{1}$ for the Gisin states we observe, as in the example for the Narnhofer line, there exist entangled states for which $\rho_{A|B} \leq \mathbb{1}$ indeed holds. But then, due to Theorem 20.18, these must lie outside the set $\mathcal{W}$ of Weyl states.

When computing the spectrum of the conditional amplitude operator (20.102) for the Gisin states, we again adopt the procedure of replacing the vanishing eigenvalue by $\epsilon > 0$ in the computation and taking the limit $\epsilon \to 0$ at the end. With this method the four eigenvalues κ_i of $\rho_{A|B}$ are found to be

$$\kappa_1 = 0, \quad \kappa_2 = \frac{1 - \lambda}{1 - \lambda \cos(2\theta)}, \quad \kappa_3 = \frac{1 - \lambda}{1 + \lambda \cos(2\theta)},$$

$$\kappa_4 = \frac{2\lambda \exp\left(-\cos(2\theta)\, \text{artanh}[\lambda \cos(2\theta)]\right)}{\sqrt{1 - \lambda^2 \cos^2(2\theta)}}. \tag{20.103}$$

While κ_2 and κ_3 are smaller than 1 for all values of λ and θ, the fourth eigenvalue κ_4 can become larger than 1. The corresponding region, delimited by the purple lines in Fig. 20.6 (a), is contained within the region of entangled states, but there is a region of entanglement where $\kappa_4 < 1$ and hence $\rho_{A|B} \leq \mathbb{1}$. This clearly demonstrates that the condition $\rho_{A|B} \leq \mathbb{1}$ for the conditional amplitude operator provides a necessary but in general not sufficient condition for separability.

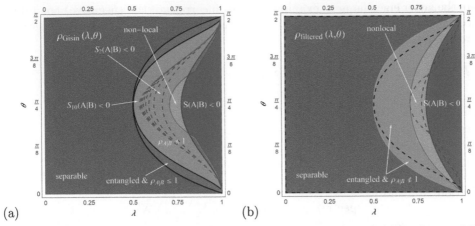

Fig. 20.6 Conditional entropy of Gisin states. The parameter regions of separability (lilac), entanglement (green), non-locality (orange), and negative conditional entropy (red) are shown for the family of unfiltered Gisin states $\rho_{\text{Gisin}}(\lambda, \theta)$ from eqn (20.101) in (a), and for the filtered Gisin states $\rho_{\text{filtered}}(\lambda, \theta)$ from eqn (20.104) in (b). In (a), the region of entangled states whose conditional amplitude operator is bounded by the identity is delimited in purple, showing that the condition $\rho_{A|B} \not\leq \mathbb{1}$ is strictly weaker than the PPT criterion. For comparison the boundaries of the conditional Rényi entropies $S_\alpha(A|B)$, Definition 20.10, for $\alpha = 2, 3, \ldots, 10$ are plotted as dashed lines. In addition, it can be clearly seen in (a) that there is no clear hierarchy between the regions of non-locality and negative conditional entropy. That is, there exist local states with negative conditional entropy, as well as non-local states with positive conditional entropy $S(A|B) \geq 0$. However, the hierarchy is recovered when considering conditional Rényi entropies $S_\alpha(A|B)$ for $\alpha \geq 2$. In (b), the boundaries for $S(A|B) < 0$, non-locality, and $\rho_{A|B} \not\leq \mathbb{1}$ for the unfiltered states are indicated by the corresponding dashed lines. Modified from Figs. 2 and 3, respectively, in *Geometry of two-qubit states with negative conditional entropy*, Friis *et al.*, J. Phys. A: Math. Theor. **50**, 125301 (2017) © IOP Publishing. Reproduced with permission. All rights reserved.

Finally, let us also consider the filtered Gisin states, introduced in Section 15.3.5,

$$\rho_{\text{filtered}}(\lambda, \theta) = \frac{1}{N(\lambda, \theta)} \left(\lambda \sin(2\theta) \rho^+ + (1 - \lambda) \rho_{\text{top}} \right) \tag{20.104}$$

$$= \frac{1}{4} \left(\mathbb{1} \otimes \mathbb{1} + \frac{\lambda \sin(2\theta)}{N(\lambda, \theta)} \left(\sigma_x \otimes \sigma_x + \sigma_y \otimes \sigma_y \right) + \frac{1 - \lambda - \lambda \sin(2\theta)}{N(\lambda, \theta)} \sigma_z \otimes \sigma_z \right),$$

with the normalization $N(\lambda, \theta) = \lambda \sin(2\theta) + 1 - \lambda$. The filtered Gisin states fully lie within the set of Weyl states. Therefore Theorem 20.18 applies and the boundary between $\rho_{A|B} \leq \mathbb{1}$ and $\rho_{A|B} \not\leq \mathbb{1}$ coincides with the boundary between separability and entanglement. To determine the conditional entropy of $\rho_{\text{filtered}}(\lambda, \theta)$ we note that the non-zero eigenvalues of the filtered Gisin states are

$$\frac{\lambda \sin(2\theta)}{1 - \lambda + \lambda \sin(2\theta)} \quad \text{and} \quad \frac{1 - \lambda}{2 \left(1 - \lambda + \lambda \sin(2\theta) \right)}, \tag{20.105}$$

where the latter eigenvalue is twice degenerate. With these eigenvalues, we can evaluate the conditional entropy and find that the region where it is negative is contained within the region of non-locality, see Fig. 20.6 (b).

20.7   Conditional Rényi Entropy and Non-Locality

In our investigation of the Gisin states we observed, see Fig. 20.6, that for the conditional entropy based on the von Neumann entropy $S(\rho)$, no clear hierarchy with respect to non-locality can be established. Some states may be non-local and satisfy $S(A|B) \geq 0$, while other states may have negative values of $S(A|B)$, while being local. However, the hierarchy is regained when considering generalized entropy measures, i.e., the conditional Rényi entropies. The quantum Rényi entropy we introduced already in Section 20.2 and we recall its definition

$$S_\alpha(\rho) \;=\; \frac{1}{1-\alpha}\,\log\big(\mathrm{Tr}(\rho^\alpha)\big) \;=\; \frac{1}{1-\alpha}\,\log\Big(\sum_{i=1}^{n} p_i^\alpha\Big)\,, \tag{20.106}$$

where the p_i denote the eigenvalues of ρ and $\alpha > 0$. For $\alpha \to 1$ the von Neumann entropy $S_1(\rho) = -\mathrm{Tr}\big(\rho\,\log(\rho)\big) = S(\rho)$ is recovered. For $\alpha \geq 2$ the Rényi family of entropies provide stronger entanglement criteria than the von Neumann entropy. Now, let us define the *quantum conditional Rényi entropy*.

Definition 20.10 *The quantum **conditional Rényi entropy** for a bipartite system described by the density operator ρ_{AB} with joint and subsystem Rényi entropies $S_\alpha(\rho_{AB}) = S_\alpha(A,B)$ and $S_\alpha(\rho_B) = S_\alpha(B)$, respectively, is defined as*
$$S_\alpha(A|B) \;:=\; S_\alpha(A,B) - S_\alpha(B).$$

These generalized conditional entropies can be shown (Horodecki and Horodecki, 1996) to be non-negative for all separable states, such that $S_\alpha(A|B) < 0$ implies that the quantum state is entangled. Moreover, the negativity of the conditional Rényi-2 entropy, $S_2(A|B) < 0$, is already a necessary condition for non-locality, as proven in Horodecki *et al.* (1996*b*).

In other words, the positivity of $S_2(A|B)$ implies that the CHSH inequality cannot be violated. This means, choosing the angles of the measurement orientations such that the expectation value of the CHSH operator for a general state ρ takes its maximal value $\overline{\mathcal{B}}_{\mathrm{CHSH}}^{\mathrm{max}}(\rho)$, we have

$$S_2(A|B) \geq 0 \quad \Rightarrow \quad \overline{\mathcal{B}}_{\mathrm{CHSH}}^{\mathrm{max}} \leq 2\,\sqrt{1 - \big|\,|\vec{a}|^2 - |\vec{b}|^2\,\big|}\,, \tag{20.107}$$

where $\vec{a}$ and $\vec{b}$ are the Bloch vectors of the state ρ (15.39). The conditional Rényi-2 entropy hence provides a strictly stronger condition than non-locality for two qubits, but not in general for higher dimensions (Horodecki *et al.*, 1996*b*). We illustrate these properties for the Weyl states in one sector of the tetrahedron in Fig. 20.7.

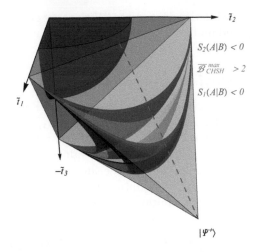

Fig. 20.7 Conditional Rényi entropies and non-locality. The sector of the tetrahedron of Weyl states defined by $\tilde{t}_1, \tilde{t}_2 \geq 0$ and $\tilde{t}_3 \leq 0$ is shown, including the corresponding regions of the Kuś–Życzkowski ball of absolute separability within the (sector of the) separable double pyramid. Within the region of entanglement (green), the boundaries between states with positive and negative conditional von Neumann and Rényi-2 entropies from Definitions 20.5 and 20.10 are shown, in addition to the surface (yellow) that separates the states violating the CHSH inequality from the corresponding local states. From Fig. 6 in *Geometry of two-qubit states with negative conditional entropy*, Friis *et al.*, J. Phys. A: Math. Theor. **50**, 125301 (2017) © IOP Publishing. Reproduced with permission. All rights reserved.

The boundary between states with positive and negative conditional Rényi-2 entropy, defined in Definition 20.10, is illustrated. As can be seen, all local states lie within the set of states for which $S_2(A|B) \geq 0$, whereas all non-local states have negative conditional Rényi-2 entropy. The dashed red line indicates the Gisin states at $\theta = \pi/4$, where filtered and unfiltered Gisin states coincide, corresponding to (parts of) the dashed red lines in Fig. 15.3 and Fig. 15.2.

Moreover, let us formulate (without proof) the following theorem connecting the separability of Weyl states with the positivity of the quantum conditional Rényi entropies for all α (Horodecki and Horodecki, 1996).

Theorem 20.19 (Conditional Rényi entropy—separability)

If all conditional Rényi α-entropies of a Weyl state $\rho_{AB} \in \mathcal{W}$ are positive semidefinite then the state is separable and vice versa:

$$S_\alpha(A|B) \geq 0 \;\; \forall \alpha \quad \Longleftrightarrow \quad \rho_{AB} \in \mathcal{W} \quad separable\,.$$

Hence, the positivity of the entire family of conditional Rényi α-entropies provides an entanglement criterion equivalent to the PPT criterion for Weyl states.

For other two-qubit states the situation is less clear. For instance, for the unfiltered Gisin state $\rho_{\text{Gisin}}(\lambda, \theta)$ from eqn (20.101), the conditions $S_\alpha(A|B) \geq 0$ are shown in Fig. 20.6 (a) for $\alpha = 2, 3, \ldots, 10$, which are clearly stronger than non-locality, but weaker than the PPT criterion or the conditional amplitude operator condition $\rho_{A|B} \leq \mathbb{1}$ for detecting entanglement. Nonetheless, conditional entropies provide straightforward entanglement witnesses that can in principle be employed for arbitrary systems. In general, however, the exact relationship between entanglement, non-locality, and the negativity of conditional entropies remains complicated.

Résumé: The entropy concept of classical thermodynamics and information theory can be extended to quantum systems, where the entropy serves as an information measure to quantify the amount of information inherent in a quantum system. The most fundamental measure of a quantum system's entropy is the von Neumann entropy, which measures the degree of uncertainty that we have about a quantum system. The quantum Rényi entropy generalizes this concept, providing more flexibility and in some cases stronger methods for characterizing the information content and the correlations of quantum systems.

In this context, other important quantities are the quantum relative entropy that gives information about the closeness of two density matrices, the quantum joint entropy that describes the uncertainty for the occurrence of measurement outcomes of measurements carried out by two parties, Alice and Bob, and the quantum mutual information that characterizes both the classical *and* the quantum correlations between Alice and Bob.

The quantum conditional entropy represents the entropy of Alice when Bob's quantum state is given, thus it represents the remaining rest-uncertainty. It has the interesting property to become negative for (certain) entangled states and, together with the conditional amplitude operator, it is an interesting tool for investigating the geometry of two-qubit systems.

When investigating the geometry of entanglement for two-qubit systems we observe that, despite its simplicity, this bipartite system already reveals many of the intricacies in the relationship of the numerous criteria for detecting entanglement and separability, and hence serves as an important guiding example. In particular, the roles of negative conditional entropy and of the conditional amplitude operator criterion as entanglement-detection methods are quite interesting. For the locally maximally mixed Weyl states, a clear hierarchy emerges, in which the set of CHSH-non-local states fully contains the set of states with negative conditional (von Neumann) entropy, while it is itself fully contained within the set of states with negative conditional Rényi 2-entropy. Furthermore, one finds that the CAO criterion is equivalent to the PPT criterion for all Weyl states, but not in general, as we have seen for selected examples.

21

Quantum Channels and Quantum Operations

In our previous treatment, we have encountered pure and mixed quantum states (see Chapter 11), and we have quantified the entropy of classical (see Chapter 19) and quantum systems (see Chapter 20). We have seen that mixedness and entropy in general can be attributed to a lack of information about a system that can arise from correlations—specifically, entanglement—with another system. At the same time, all dynamics that we have considered this far has been unitary. This fits the picture: Global unitary dynamics conserve the total entropy but can entangle or disentangle subsystems, leading to changes in local entropies. Moreover, unitary dynamics is physically well motivated, because it can be understood as being generated by a Hamiltonian and it preserves probability. But, given that subsystems do not evolve unitarily for general global unitary dynamics, how do they evolve? And what is the most general evolution of a quantum system in general? In this chapter, we will attempt to address these questions in a mathematically precise way.

21.1 Purification of Quantum States Revisited

Let us start by returning to the notion of a mixed state represented by a density operator ρ, as introduced in Chapter 11. As we have argued before, recall, e.g., Proposition 20.2, we can think of the mixedness (or entropy) of ρ as a consequence of our lack of information about the state of the system. But is this lack of knowledge fundamental or the result of a potentially incomplete description of our system? In a sense, both of these options are true. Indeed, as we have already shown in Section 15.1.4, any mixed state of any quantum system can be understood as the description of a subsystem of a larger quantum system that is actually in a pure entangled state.

Although we have mentioned purification before in Section 15.1.4, this observation is an important element of a larger picture that is worth discussing. As we have seen previously, this can be seen directly from the Schmidt decomposition (Theorem 15.1), that is, the eigenvalues of any mixed-state density operator can be considered to be the squared Schmidt coefficients of the corresponding purification, that is, any state ρ can be written as

$$\rho = \sum_n p_n \, |\psi_n \rangle\langle \psi_n| \,, \tag{21.1}$$

with respect to an orthonormal basis $\{|\psi_n\rangle_A\}_n$ of the corresponding Hilbert space $\mathcal{H}_A$, and a purification $|\psi\rangle_{AB} \in \mathcal{H}_{AB} = \mathcal{H}_A \otimes \mathcal{H}_B$ is given by

$$| \psi \rangle_{AB} = \sum_n \sqrt{p_n} \, | \psi_n \rangle_A \otimes | \psi_n \rangle_B \, . \tag{21.2}$$

Here, $\mathcal{H}_B$ can be the same Hilbert space, or even a different Hilbert space of the same (or larger dimension) than $\mathcal{H}_A$. Indeed, for any given state ρ, even a space with smaller dimension than $\mathcal{H}_A$ is sufficient, as long as $\dim(\mathcal{H}_B) \geq \mathrm{rank}(\rho)$, i.e., as long as the dimension of $\mathcal{H}_B$ is at least as large as the number of non-zero eigenvalues of ρ. For the interested reader, let us note that, conceptually, the notion of a state's purification can be understood as part of a more general mathematical framework called the Gelfand–Neumark–Segal (GNS) construction, named after *Israel Gelfand, Mark Naimark* (1943), and *Irving Segal* (1947).

An interesting aspect of purification is that the mixedness (or entropy) of any quantum system can in this way be seen as arising from locally inaccessible information. In other words, one can take the point of view that ρ is mixed *because* $| \psi \rangle_{AB}$ is entangled. From this perspective, the lack of information about ρ is fundamental. At the same time, Theorem 15.2 assures us that pure and mixed states are part of the same quantum-mechanical framework. Indeed, we shall see in the remainder of this chapter that the simple quantum-mechanical rules for pure states, unitary dynamics, and projective measurements discussed in Part I of this book are sufficient to describe the most general quantum-mechanical systems, evolution, and measurements. The habit of thinking about mixed states and (irreversible) dynamics in this broader context is sometimes referred to as *the church of the larger Hilbert space*, a phrase attributed to *John Smolin* (see, for instance Bennett, 2007).

21.2 Quantum Operations and Quantum Channels

21.2.1 CPTP Maps

Having discussed the general description of quantum states, we can of course ask about the most general evolution of quantum states permitted by nature. Here, we want to be as inclusive as possible. In addition to free time evolution, we would also like to cover procedures such as state preparation, potentially noisy quantum channels, loss, and all conceivable combinations, and refer to them as *quantum operations*. If we do not make any specific assumptions about the particular dynamics of a quantum system, then the only thing that we can stipulate is that an arbitrary quantum operation must map quantum states to quantum states. We may even allow for states in one Hilbert space to be mapped to states in another Hilbert space, e.g., consider a situation where one measures a two-dimensional quantum system (a qubit) and, depending on the outcome, prepares different states of three-dimensional quantum system, or vice versa. Quantum operations such as state preparation, quantum channels, and measurements on a Hilbert space $\mathcal{H}_A$ can be represented as maps

$$\Lambda : \mathcal{B}_{\mathrm{HS}}(\mathcal{H}_A) \rightarrow \mathcal{B}_{\mathrm{HS}}(\mathcal{H}_B) \tag{21.3}$$

$$\rho \mapsto \Lambda[\rho]$$

from the Hilbert–Schmidt space $\mathcal{B}_{\mathrm{HS}}(\mathcal{H}_A)$ of (bounded, finite-norm) linear operators over the Hilbert space $\mathcal{H}_A$ (see Definition 11.5) onto $\mathcal{B}_{\mathrm{HS}}(\mathcal{H}_B)$, the Hilbert–Schmidt space of (bounded, finite-norm) linear operators over the Hilbert space $\mathcal{H}_B$. Since such maps should be defined for all states, they must be linear, that is

$$\textbf{linear} \qquad \Lambda\Big[\sum_n p_n\,\rho_n\Big] = \sum_n p_n\,\Lambda[\rho_n]\ \forall\, \rho_n \in \mathcal{B}_{\mathrm{HS}}(\mathcal{H}_A)\ \&\ \forall\, p_n \in \mathbb{R}\,. \qquad (21.4)$$

Moreover, we have to demand that the obtained objects are valid density operators. In particular, $\Lambda[\rho]$ must be

$$\textbf{positive semidefinite} \qquad \Lambda[\rho] \geq 0\ \forall\, \rho \geq 0\,, \qquad (21.5)$$

and hence Hermitian. Moreover, the resulting states should be normalized, i.e., the map Λ should be

$$\textbf{trace preserving} \qquad \mathrm{Tr}\big(\Lambda[\rho]\big) = \mathrm{Tr}\big(\rho\big) = 1\,. \qquad (21.6)$$

However, there is one more requirement. Interestingly, to guarantee that physical states are mapped to physical states, it is not enough to consider positive maps. Recall, e.g., the PPT criterion from Theorem 15.3 and our discussion of positive maps in Section 15.4. The transposition is an example of a positive map that can result in states with negative eigenvalues, when applied to a subsystem (of an entangled state). To account for the possibility that the state we use as input for the map Λ describes a subsystem of a larger system with Hilbert space $\mathcal{H}_A \otimes \mathcal{H}_C$, we have to demand

$$\textbf{complete positivity} \qquad \big(\Lambda \otimes \mathbb{1}_C\big)[\rho_{AC}] \geq 0\ \forall\, \rho_{AC} \geq 0\ \&\ \forall\, \mathcal{H}_C\,, \qquad (21.7)$$

where $\rho_{AC} \in \mathcal{B}_{\mathrm{HS}}(\mathcal{H}_A \otimes \mathcal{H}_C)$. In summary, any physical evolution map[1] for any quantum system must be a *completely positive*[2] and *trace-preserving* (CPTP) map. While this may seem complicated at first glance, it turns out that concepts such as mixed states, generalized measurements (so-called POVMs, see Chapter 23), and quantum operations can all be thought of as unitary dynamics of pure states and projective measurements on a larger Hilbert space. In principle, most of these statements can be collected in an even broader mathematical context first developed by *William Forrest Stinespring* (1955). However, for pedagogical reasons and conforming with common usage in the field of quantum information (see, e.g. Nielsen and Chuang, 2000 or Paris, 2012), we will split this more general formulation into several individual technical statements, and reserve the name *Stinespring dilation* for the specific statement in Theorem 21.2.

21.2.2 Dephasing and Depolarizing Channel

To begin, let us consider some simple examples. Let $\mathcal{H}_A = \mathbb{C}^2$ be the Hilbert space of a single qubit and let us define a map Λ from $\mathcal{B}_{\mathrm{HS}}(\mathcal{H}_A)$ to itself. To specify this map, we make use of a second qubit, $\mathcal{H}_B = \mathbb{C}^2$, such that

$$\Lambda[\rho] = \mathrm{Tr}_B\big(U_{\mathrm{CNOT}}\,\rho \otimes |0\rangle\langle 0|_B\,U_{\mathrm{CNOT}}^\dagger\big), \qquad (21.8)$$

where U_{CNOT} is the familiar CNOT transformation from eqn (14.47) that flips the second qubit conditional on the state of the first qubit. Inserting into eqn (21.8) and

---

[1] Here, let us note that there is a distinction between evolution maps and inference maps, as we will discuss in Section 21.4.

[2] Recall from our discussion in Section 15.4.1 that not all positive maps are completely positive.

writing the initial state with respect to the chosen basis as $\rho_A = \sum_{i,j=0,1} \rho_{ij} \, |i\rangle\langle j|$, we can take the trace to find

$$
\Lambda[\rho] = \sum_{i,j=0,1} \rho_{ij} \mathrm{Tr}_B \Big(\langle 0|i\rangle\langle j|0\rangle \, |0\rangle\langle 0| \otimes |0\rangle\langle 0| + \langle 1|i\rangle\langle j|1\rangle \, |1\rangle\langle 1| \otimes |1\rangle\langle 1| \Big)
$$

$$
= \rho_{00} \, |0\rangle\langle 0| + \rho_{11} \, |1\rangle\langle 1| = \sum_{i=0,1} \rho_{ii} \, |i\rangle\langle i|. \tag{21.9}
$$

That is, this particular map leaves the diagonal elements ρ_{ii} with respect to the chosen basis invariant, but all off-diagonal elements are removed. Such a map is thus referred to as a *completely dephasing channel* and it is common to use the symbol $ instead of Λ to denote it. Furthermore, the complete dephasing map for qubits is a special case of a dephasing channel given by

$$
\$_\lambda[\rho] = \lambda \rho + (1-\lambda) \sigma_z \rho \sigma_z. \tag{21.10}
$$

For $\frac{1}{2} \leq \lambda \leq 1$, the channel reduces the off-diagonal elements of the state ρ by a factor $(2\lambda - 1)$. Geometrically, the dephasing map can be understood on the Bloch sphere (see Chapter 11) as a rescaling of the x and y components of the Bloch vector, while the z component remains invariant, see Fig. 21.1 (a).

For arbitrary system dimension d the *complete dephasing map* can be defined with respect to a basis $\{|i\rangle\}_{0,\ldots,d-1}$ as

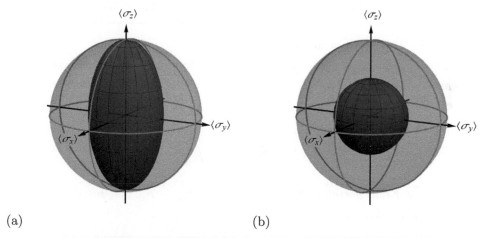

(a)   (b)

Fig. 21.1 Quantum channels for qubits. The action of the dephasing and depolarization channels on the Bloch sphere of a single qubit are shown in (a) and (b), respectively. The dephasing channel illustrated in (a) for $\lambda = \frac{3}{4}$ maps the sphere of pure single-qubit states to an ellipsoid by reducing the components $a_x = \mathrm{Tr}(\rho \sigma_x)$ and $a_y = \mathrm{Tr}(\rho \sigma_y)$ by a factor of $(2\lambda-1)$, while $a_z = \mathrm{Tr}(\rho \sigma_z)$ is unaffected. In the limit of $\lambda = \frac{1}{2}$, one obtains a complete dephasing channel, corresponding to a projection onto the z axis. (b) The action of the depolarizing channel from eqn (21.12) (for $d = 2$) effectively shrinks the Bloch sphere by a factor p, here chosen as $p = \frac{1}{2}$, but preserves the spherical symmetry.

$$\$[\rho] = \sum_{i=0}^{d-1} \langle i | \rho | i \rangle \, | i \rangle\langle i | \, . \tag{21.11}$$

Similarly, we can consider the *depolarizing channel* for arbitrary d-dimensional systems, given by

$$\Lambda[\rho] = p\,\rho + \frac{1-p}{d}\,\mathrm{Tr}(\rho)\,\mathbb{1}_d. \tag{21.12}$$

For values[3] $0 \le p \le 1$, this channel can be thought of as white noise (the maximally mixed state $\mathbb{1}_d/d$) that is added to the initial state ρ with some probability p. For qubits, the action of this channel can be illustrated on the Bloch sphere as a compression, all points (states) on the Bloch sphere are moved radially inwards towards the origin, as illustrated in Fig. 21.1 (b).

21.3   Kraus Decomposition

21.3.1   Unitary Dynamics on Larger Hilbert Spaces

For both examples, the dephasing channel in eqn (21.8) and the depolarizing map in eqn (21.12), it is straightforward to see that both maps are linear[4], and positive. But are they completely positive? And if so, under which conditions can CPTP maps be written as unitaries on acting on a larger system? Or, conversely, we can ask if all unitaries on larger Hilbert spaces induce CPTP maps. An extremely useful tool in addressing these questions is the *Kraus decomposition* named after *Karl Kraus (1983)*.

Lemma 21.1 (Kraus decomposition)

*Every map $\Lambda : \mathcal{B}_{\mathrm{HS}}(\mathcal{H}_A) \to \mathcal{B}_{\mathrm{HS}}(\mathcal{H}_B)$ that can be represented as unitary dynamics on a larger Hilbert space with factorizing boundary condition, i.e., which can be written as $\Lambda[\rho_A] = \mathrm{Tr}_{AC}\big(U\,\rho_A \otimes | \omega \rangle\langle \omega |_{BC}\, U^\dagger\big)$, admits a **Kraus decomposition***

$$\Lambda[\rho_A] = \sum_n K_n\,\rho_A K_n^\dagger \, ,$$

where the Kraus operators K_n are maps from $\mathcal{H}_A$ to $\mathcal{H}_B$ that satisfy $\sum_n K_n^\dagger K_n = \mathbb{1}_A$.

Proof   To prove Lemma 21.1, note that we can write $\rho_A \otimes | \omega \rangle\langle \omega |_{BC}$ as

$$\rho_A \otimes | \omega \rangle\langle \omega |_{BC} = (\rho_A \otimes \mathbb{1}_{BC})(\mathbb{1}_A \otimes | \omega \rangle_{BC})(\mathbb{1}_A \otimes \langle \omega |_{BC})$$

$$= (\mathbb{1}_A \otimes | \omega \rangle_{BC})(\rho_A \otimes \mathbb{1}_{BC})(\mathbb{1}_A \otimes \langle \omega |_{BC}). \tag{21.13}$$

---

[3]One can also obtain valid quantum channels for some negative values of p, as we shall see.

[4]In particular, note that the factor of $\mathrm{Tr}(\rho)$ was included explicitly for this reason on the right-hand side of eqn (21.12).

Inserting into the expression from Lemma 21.1, we have

$$\Lambda[\rho_A] = \mathrm{Tr}_{AC}\big(U\rho_A \otimes |\omega\rangle\langle\omega|_{BC}\, U^\dagger\big)$$

$$= \sum_n \big(\langle n|_{AC} \otimes \mathbb{1}_B\big)U\big(\mathbb{1}_A \otimes |\omega\rangle_{BC}\big)\big(\rho_A \otimes \mathbb{1}_{BC}\big)\big(\mathbb{1}_A \otimes \langle\omega|_{BC}\big)U^\dagger\big(|n\rangle_{AC} \otimes \mathbb{1}_B\big)$$

$$= \sum_n \langle n|U|\omega\rangle\, \rho_A \,\langle\omega|U^\dagger|n\rangle = \sum_n K_n\, \rho_A\, K_n^\dagger, \qquad (21.14)$$

where we have omitted the subsystem labels and identity operators in the last line for ease of notation, and we have defined the *Kraus operators*

$$K_n := \big(\langle n|_{AC} \otimes \mathbb{1}_B\big)U\big(\mathbb{1}_A \otimes |\omega\rangle_{BC}\big) = \langle n|U|\omega\rangle. \qquad (21.15)$$

As stated, these operators are linear maps from $\mathcal{H}_A$ to $\mathcal{H}_B$, and we can check that they form a complete set. That is, we calculate

$$\sum_n K_n^\dagger K_n = \big(\mathbb{1}_A \otimes \langle\omega|_{BC}\big)U^\dagger \sum_n \big(|n\rangle\langle n|_{AC} \otimes \mathbb{1}_B\big)U\big(\mathbb{1}_A \otimes |\omega\rangle_{BC}\big) \qquad (21.16)$$

$$= \big(\mathbb{1}_A \otimes \langle\omega|_{BC}\big)U^\dagger \mathbb{1}_{ABC} U\big(\mathbb{1}_A \otimes |\omega\rangle_{BC}\big)$$

$$= \mathbb{1}_A \otimes \langle\omega|\omega\rangle_{BC} = \mathbb{1}_A. \qquad \square$$

The Kraus decomposition is a very useful tool for practical calculations. As an example, let us return to the complete dephasing channel from eqn (21.8). There, $|\omega\rangle = |0\rangle$, and we obtain the Kraus operators

$$K_0 = \big(\langle 0| \otimes \mathbb{1}\big)U_{\mathrm{CNOT}}\big(\mathbb{1} \otimes |0\rangle\big) \qquad (21.17a)$$

$$= \big(\langle 0| \otimes \mathbb{1}\big)\big(|0\rangle\langle 0| \otimes \mathbb{1} + |1\rangle\langle 1| \otimes \sigma_x\big)\big(\mathbb{1} \otimes |0\rangle\big)$$

$$= \big(\langle 0| \otimes \mathbb{1}\big)\big(\mathbb{1} \otimes |0\rangle\big) = \langle 0| \otimes |0\rangle = {}_B|0\rangle\langle 0|_A,$$

$$K_1 = \big(\langle 1| \otimes \mathbb{1}\big)U_{\mathrm{CNOT}}\big(\mathbb{1} \otimes |0\rangle\big) \qquad (21.17b)$$

$$= \big(\langle 1| \otimes \mathbb{1}\big)\big(|0\rangle\langle 0| \otimes \mathbb{1} + |1\rangle\langle 1| \otimes \sigma_x\big)\big(\mathbb{1} \otimes |0\rangle\big)$$

$$= \langle 1| \otimes \big(|0\rangle\langle 1| + |1\rangle\langle 0|\big)\big(\mathbb{1} \otimes |0\rangle\big) = \langle 1| \otimes |1\rangle = {}_B|1\rangle\langle 1|_A.$$

In other words, this means that we can write the single-qubit complete dephasing channel as

$$\$[\rho] = \sum_{n=0,1} |n\rangle\langle n|\, \rho\, |n\rangle\langle n| = \sum_{n=0,1} \langle n|\rho|n\rangle\, |n\rangle\langle n| = \sum_{n=0,1} \rho_{nn}\, |n\rangle\langle n|,$$

$$(21.18)$$

which is indeed (and as expected) the exact same expression as in eqn (21.9).

21.3.2 Kraus Decomposition for the Amplitude-Damping Channel

Although the Kraus operators in the previous example are a complete set of projectors, this is not always the case, Kraus operators need not be projectors, need not be Hermitian, need not be diagonalizable, and need not even be square matrices. The last property is rather clear from the fact that we have already shown that Kraus operators exist also for maps between different Hilbert spaces $\mathcal{H}_A$ and $\mathcal{H}_B$, in particular, one can have $\dim(\mathcal{H}_A) \neq \dim(\mathcal{H}_B)$.

Before we continue, let us therefore now also consider an example with Kraus operators that are not just projectors. Consider the single-qubit channel defined as $\Lambda[\rho] = \mathrm{Tr}_B\left(U\,\rho \otimes |0\rangle\langle 0|_B\,U^\dagger\right)$, with U given by

$$U = |00\rangle\langle 00| + |11\rangle\langle 11| + \cos(\theta)\left(|01\rangle\langle 01| + |10\rangle\langle 10|\right)$$

$$+ \sin(\theta)\left(|01\rangle\langle 10| - |10\rangle\langle 01|\right). \tag{21.19}$$

This unitary transformation can be considered as a rotation (similar to a spatial rotation around the y axis) between the singlet states $|\Psi^\pm\rangle = \frac{1}{\sqrt{2}}(|01\rangle \pm |10\rangle)$. A similar calculation as in eqns (21.17a) and (21.17b) then gives the result

$$K_0 = |0\rangle\langle 0| + \cos(\theta)\,|1\rangle\langle 1|, \tag{21.20a}$$

$$K_1 = \sin(\theta)\,|0\rangle\langle 1|. \tag{21.20b}$$

These operators still satisfy

$$K_0^\dagger K_0 + K_1^\dagger K_1 = |0\rangle\langle 0| + \cos^2(\theta)\,|1\rangle\langle 1| + \sin^2(\theta)\,|1\rangle\langle 1| = \mathbb{1}. \tag{21.21}$$

Using Lemma 21.1, a short calculation reveals the particular effect of this channel on an input state $\rho = \sum_{i,j=0,1} \rho_{ij}\,|i\rangle\langle j|$ to be

$$\Lambda[\rho] = K_0\,\rho\,K_0^\dagger + K_1\,\rho\,K_1^\dagger = \left[\rho_{00} + \rho_{11}\sin^2(\theta))\right]|0\rangle\langle 0| + \rho_{11}\cos^2(\theta)\,|1\rangle\langle 1|$$

$$+ \cos(\theta)\left(\rho_{01}\,|0\rangle\langle 1| + \rho_{10}\,|1\rangle\langle 0|\right). \tag{21.22}$$

As we see, for $\theta \in\,]0, \frac{\pi}{2}]$, the (absolute) values of the off-diagonal elements are reduced and population is shifted from $|1\rangle\langle 1|$ to $|0\rangle\langle 0|$. That is, the map in question is an *amplitude-damping channel*, illustrated in Fig. 21.2.

21.3.3 Kraus Decomposition and CPTP Maps

The Kraus decomposition further allows us to prove an important lemma connecting the unitary dynamics on a larger Hilbert space to CPTP maps:

Lemma 21.2 *Every map* $\Lambda : \mathcal{B}_{\mathrm{HS}}(\mathcal{H}_A) \to \mathcal{B}_{\mathrm{HS}}(\mathcal{H}_B)$ *that admits a Kraus decomposition is completely positive and trace-preserving (CPTP).*

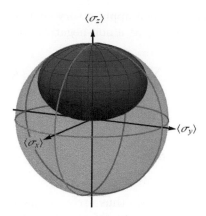

Fig. 21.2 Amplitude-damping channel for qubits. The action of the amplitude–damping channel from eqn (21.22) on the Bloch sphere of a single qubit is shown for $\theta = \frac{\pi}{4}$. Qualitatively, the amplitude-damping channel causes the Bloch sphere to shrink and compress towards the north pole (the state $|0\rangle$). For a state with Bloch vector $\vec{a} = (a_x, a_y, a_z)^\top$, the Bloch-vector components $\tilde{a}_i$ of $\Lambda[\rho]$ are $\tilde{a}_x = \cos(\theta)\, a_x$, $\tilde{a}_y = \cos(\theta)\, a_y$, and $\tilde{a}_z = \frac{1+a_z}{2} - \frac{1-a_z}{2}\cos(2\theta)$.

Proof To prove Lemma 21.2, we have to verify that maps with a Kraus decomposition satisfy the properties stated in eqns (21.4) to (21.7). Starting with linearity, we quite simply have

$$\Lambda\Big[\sum_n p_n \rho_n\Big] = \sum_m K_m \sum_n p_n \rho_n \, K_m^\dagger = \sum_n p_n \sum_m K_m \, \rho_n \, K_m^\dagger = \sum_n p_n \Lambda[\rho_n].$$

$$(21.23)$$

Then, we show that such maps are trace preserving, that is,

$$\mathrm{Tr}\big(\Lambda[\rho]\big) = \mathrm{Tr}\Big(\sum_n K_n \rho K_n^\dagger\Big) = \sum_n \mathrm{Tr}\big(\rho\, K_n^\dagger K_n\big) = \mathrm{Tr}\Big(\rho \sum_n K_n^\dagger K_n\Big) = \mathrm{Tr}(\rho) = 1,$$

$$(21.24)$$

where we have used the linearity and cyclicity of the trace and the normalization of the Kraus operators. Next, let us examine the positivity condition. For any positive operator $\rho \in \mathcal{B}_{\mathrm{HS}}(\mathcal{H}_A)$, that is, for every operator such that $\langle \phi | \rho | \phi \rangle \geq 0$ for all $|\phi\rangle \in \mathcal{H}_A$, we must have $\Lambda[\rho] \geq 0$. We thus simply check that for any vector $|\psi\rangle \in \mathcal{H}_B$, we have

$$\langle \psi | \Lambda[\rho] | \psi \rangle = \sum_n \langle \psi | K_n \rho \, K_n^\dagger | \psi \rangle = \sum_n \langle \phi_n | \rho | \phi_n \rangle \geq 0, \qquad (21.25)$$

which is non-negative, since $\rho \geq 0$ and $|\phi_n\rangle := K_n^\dagger |\psi\rangle \in \mathcal{H}_A$. Note, the vectors $|\phi_n\rangle$ are generally not normalized here, but nonetheless, the expectation value for any positive operator satisfies $\langle \phi_n | \rho | \phi_n \rangle \geq 0$, and the sum of non-negative terms is likewise non-negative. Finally, we can apply a very similar method to show complete positivity. Let $\rho_{AC} \in \mathcal{B}_{HS}(\mathcal{H}_A \otimes \mathcal{H}_C)$ be a non-negative operator, $\rho_{AC} \geq 0$, then for all $|\tilde{\psi}\rangle \in \mathcal{H}_B \otimes \mathcal{H}_C$, we have

$$\langle \tilde{\psi} | (\Lambda \otimes \mathbb{1}_C)[\rho_{AC}] | \tilde{\psi} \rangle = \sum_n \langle \tilde{\psi} | (K_n \otimes \mathbb{1}_C) \rho_{AC} (K_n^\dagger \otimes \mathbb{1}_C) | \tilde{\psi} \rangle \qquad (21.26)$$

$$= \sum_n \langle \tilde{\phi}_n | \rho_{AC} | \tilde{\phi}_n \rangle \geq 0,$$

where we have defined the (unnormalized) vectors $| \tilde{\phi}_n \rangle := (K_n^\dagger \otimes \mathbb{1}_C) | \tilde{\psi} \rangle \in \mathcal{H}_A \otimes \mathcal{H}_C$.
□

Combining Lemmas 21.1 and 21.2, we thus arrive at the corollary that indeed every map that can be realized by a unitary transformation on a larger Hilbert space (with factoring initial condition) is a CPTP map. In the next section, we want to prove that the converse statement is also true, i.e., that every CPTP map admits a Kraus decomposition and can be represented as unitary dynamics on a larger Hilbert space.

21.4 The Church of the Larger Hilbert Space

21.4.1 The Choi–Jamiołkowski Isomorphism

In the last section, we have seen that any map that can be represented as unitary dynamics on a larger Hilbert space (with factoring initial condition) is indeed a CPTP map. However, the class of CPTP maps (which we have argued is the most general form of evolution map for any quantum system) could still contain dynamics that deviate from the picture of a unitary on a larger space. To show that this is not the case, let us consider a CPTP map $\Lambda : \mathcal{B}_{HS}(\mathcal{H}_A) \to \mathcal{B}_{HS}(\mathcal{H}_B)$ with $\dim(\mathcal{H}_A) = d_A$ and $\dim(\mathcal{H}_B) = d_B$. We then define the operator

$$\rho_\Lambda := (\Lambda \otimes \mathbb{1}) \big[| \Phi_{00} \rangle\langle \Phi_{00} | \big], \qquad (21.27)$$

where $| \Phi_{00} \rangle \in \mathcal{H}_A \otimes \mathcal{H}_{A'}$ with $\mathcal{H}_A = \mathcal{H}_{A'}$ is a pure state of two copies of the initial Hilbert space $\mathcal{H}_A$ given by

$$| \Phi_{00} \rangle = \frac{1}{\sqrt{d_A}} \sum_{k=0}^{d_A-1} | k \rangle \otimes | k \rangle, \qquad (21.28)$$

i.e., a particular maximally entangled state $| \Phi_{mn} \rangle$ from eqn (14.11) with $m = n = 0$. Note that, in defining the state $| \Phi_{00} \rangle$, we have chosen the same bases $\{| k \rangle\}_{k=0,\dots,d_A-1}$ for both copies of $\mathcal{H}_A$. Given this choice of basis, the map Λ fully determines the operator ρ_Λ, and we can write it as

$$\rho_\Lambda = \frac{1}{d_A} \sum_{k,k'} \Lambda\big[| k \rangle\langle k' | \big] \otimes | k \rangle\langle k' | \ \in \mathcal{B}_{HS}(\mathcal{H}_B \otimes \mathcal{H}_A). \qquad (21.29)$$

But what is the meaning of this operator? First, observe that $\rho_\Lambda \geq 0$ is positive semidefinite because Λ is completely positive, eqn (21.7). Then, let us consider the trace of ρ_Λ, i.e.,

$$\mathrm{Tr}(\rho_\Lambda) = \sum_{m,n} \langle m| \otimes \langle n| \, \rho_\Lambda \, |m\rangle \otimes |n\rangle \tag{21.30}$$

$$= \sum_{m,n,k,k'} \tfrac{1}{d_A} \langle m|\Lambda\big[|k\rangle\langle k'|\big]|m\rangle \langle n|k\rangle \langle k'|n\rangle$$

$$= \tfrac{1}{d_A} \sum_{m,k} \langle m|\Lambda\big[|k\rangle\langle k|\big]|m\rangle = \tfrac{1}{d_A}\sum_k \mathrm{Tr}\big(\Lambda\big[|k\rangle\langle k|\big]\big) = \tfrac{1}{d_A}\sum_k 1 = 1,$$

where we have used the assumption that Λ is trace preserving, eqn (21.6). We thus observe that ρ_Λ is a positive semidefinite (and hence Hermitian) and normalized operator. In other words, ρ_Λ is a density operator on the Hilbert space $\mathcal{H}_B \otimes \mathcal{H}_A$. This correspondence between CPTP maps from $\mathcal{H}_A$ to $\mathcal{H}_B$ and states on $\mathcal{H}_B \otimes \mathcal{H}_A$ is called the *Choi isomorphism*, named after *Man-Duen Choi* (1975) and ρ_Λ is called the *Choi state* corresponding to the map Λ. The intuition behind this isomorphism is indeed rather simple: $|\psi\rangle\langle\phi|$ is a linear operator on a Hilbert space $\mathcal{H}$ but flipping the "bra" $\langle\phi|$ to a "ket" $|\phi\rangle$ one obtains a state $|\psi\rangle \otimes |\phi\rangle$.

For the reverse direction of the Choi isomorphism we again have to make use of the state $|\Phi_{00}\rangle$. That is, using the Choi state ρ_Λ, we can obtain the action of the map Λ on a state ρ via

$$d_A^2 \big(\,_{AA'}\langle\Phi_{00}| \otimes \mathbb{1}_B\big) \rho \otimes \rho_\Lambda \big(|\Phi_{00}\rangle_{AA'} \otimes \mathbb{1}_B\big) = \tag{21.31}$$

$$= \sum_{m,n,k,k'} \big(\langle m| \otimes \mathbb{1}_B \otimes \langle m|\big) \big(\rho \otimes \Lambda\big[|k\rangle\langle k'|\big] \otimes |k\rangle\langle k'|\big) \big(|n\rangle \otimes \mathbb{1}_B \otimes \langle n|\big)$$

$$= \sum_{k,k'} \langle k|\rho|k'\rangle \, \Lambda\big[|k\rangle\langle k'|\big] = \Lambda\Big[\sum_{k,k'}\langle k|\rho|k'\rangle|k\rangle\langle k'|\Big] = \Lambda[\rho],$$

which can be understood as a teleportation protocol (see Chapter 14) from the input Hilbert space $\mathcal{H}_A$ to the output Hilbert space $\mathcal{H}_B$, see Fig. 21.3.

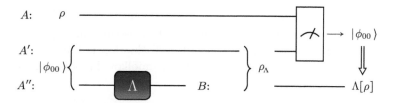

Fig. 21.3 Choi isomorphism and teleportation. The "reverse direction" of the Choi isomorphism in eqn (21.31) can be understood in terms of quantum teleportation (see Chapter 14). More specifically, when Λ is the identity channel, the state ρ can be teleported from the input Hilbert space $\mathcal{H}_A$ to the output Hilbert space $\mathcal{H}_B$ by preparing the entangled resource state $|\Phi_{00}\rangle_{A'B}$ and post-selecting on the measurement outcome $|\Phi_{00}\rangle_{AA'}$ for the Bell measurement. For an arbitrary channel Λ from $\mathcal{H}_{A''}$ to $\mathcal{H}_B$, the resource state for the teleportation protocol becomes the Choi state ρ_Λ instead of $|\Phi_{00}\rangle$ and the teleported state is $\Lambda[\rho]$.

One can thus use the Choi isomorphism to map channels Λ to states ρ_Λ via eqn (21.27), and bipartite states ρ_Λ to channels Λ via eqn (21.31), which establishes a channel-state duality. However, this particular way of expressing this duality has the drawback of being basis-dependent. In order to obtain the correct map Λ from ρ_Λ we require the specific basis $\{|\,k\,\rangle\}_k$ to construct the correct $|\,\Phi_{00}\,\rangle$.

This basis dependence can be avoided by switching to *Andrzej Jamiołkowski's* version of the channel-state duality (Jamiołkowski, 1972). There, one defines the operator corresponding to a CPTP map Λ as

$$\tilde\rho_\Lambda := (\Lambda \otimes \mathbb{1})\big[|\,\Phi_{00}\,\rangle\langle\,\Phi_{00}\,|^{T_{A'}}\big] = \frac{1}{d_A}\sum_{k,k'}\Lambda\big[|\,k\,\rangle\langle\,k'\,|\big]\otimes|\,k'\,\rangle\langle\,k\,|, \qquad (21.32)$$

where the superscript $T_{A'}$ denotes partial transposition of the part of the operator pertaining to $\mathcal{H}_{A'}$ (the second copy of $\mathcal{H}_A$), resulting in an exchange of the indices k and k' in the last tensor product factor in the last line of eqn (21.32) compared to eqn (21.29). The partial transposition preserves the trace, and consequently $\mathrm{Tr}(\tilde\rho_\Lambda) = 1$. However, since transposition is a positive but not completely positive map (see, e.g., the discussion in Section 15.3), $\tilde\rho_\Lambda$ is not necessarily positive semidefinite and $\tilde\rho_\Lambda$ can hence generally not be considered to be a density operator. While this version of the isomorphism hence loosens the stricter equivalence between channels and states of the Choi isomorphism, one gains the advantage that the mapping is now basis-independent. That is, given $\tilde\rho_\Lambda$ we can obtain the action of Λ on any input state ρ_A by just calculating

$$d_A\,\mathrm{Tr}_A\big(\tilde\rho_\Lambda\,(\mathbb{1}_B \otimes \rho_A)\big) = \sum_n (\mathbb{1}\otimes\langle\,n\,|)\sum_{k,k'}\Lambda\big[|\,k\,\rangle\langle\,k'\,|\big]\otimes|\,k'\,\rangle\langle\,k\,|\,(\mathbb{1}\otimes\rho_A)(\mathbb{1}\otimes|\,n\,\rangle)$$

$$= \sum_{n,k,k'}\Lambda\big[|\,k\,\rangle\langle\,k'\,|\big]\,\langle\,n\,|\,k'\,\rangle\,\langle\,k\,|\,\rho_A\,|\,n\,\rangle = \Lambda\Big[\sum_{k,k'}\langle\,k\,|\,\rho_A\,|\,k'\,\rangle\,|\,k\,\rangle\langle\,k'\,|\Big] = \Lambda[\rho_A].$$

$$(21.33)$$

Apart from their technical details, both versions of the isomorphism provide a mapping between channels and operators, and often the distinction is blurred by collectively referring to such duality relations as the Choi–Jamiołkowski isomorphism (see, e.g., Holevo, 2011*a*).

Even though we are primarily interested in physical channels represented by completely positive maps in this chapter, it is also worth noting that these isomorphisms can also be applied for other maps, in particular, in the case when the linear map in question is positive but *not* completely positive, an example that is of particular relevance in the context of entanglement detection, see Section 15.4. To make this distinction more obvious, let us consider a positive (but not necessarily completely positive) map Λ from $\mathcal{B}_{\mathrm{HS}}(\mathcal{H}_A)$ to $\mathcal{B}_{\mathrm{HS}}(\mathcal{H}_B)$, for which we have the following lemma (Horodecki *et al.*, 1996*a*):

Lemma 21.3 (Choi states for positive maps)

The Choi state ρ_Λ of a linear map $\Lambda : \mathcal{B}_{\mathrm{HS}}(\mathcal{H}_A) \to \mathcal{B}_{\mathrm{HS}}(\mathcal{H}_B)$ is Hermitian, $\rho_\Lambda^\dagger = \rho_\Lambda$, and satisfies $\mathrm{Tr}(\rho_\Lambda P \otimes Q) \geq 0$ for all projectors P and Q on $\mathcal{H}_B$ and $\mathcal{H}_A$, respectively, if and only if the map is positive, $\Lambda \geq 0$.

Proof We split the proof into two parts. For the first part, we assume that $\Lambda \geq 0$ is a positive map, which implies that the map preserves Hermiticity. That is, if we apply the map to an arbitrary Hermitian operator $H^\dagger = H$, then $\Lambda[H]^\dagger = \Lambda[H]$. This can be seen by noting that H must have real eigenvalues $h_i \in \mathbb{R}$, that can be separated into those that are negative, and those that are non-negative, meaning that we can rewrite the spectral decomposition (see Theorem 3.1) of H as $H = H_+ - H_-$, such that both H_+ and H_- are positive operators, $H_\pm \geq 0$. The positivity of Λ then suggests that for every vector $|\psi\rangle$ we have $\langle\psi|\Lambda[H_\pm]|\psi\rangle \geq 0$ and hence

$$\langle\psi|\Lambda[H]|\psi\rangle = \langle\psi|\Lambda[H_+]|\psi\rangle - \langle\psi|\Lambda[H_-]|\psi\rangle \in \mathbb{R}, \tag{21.34}$$

the eigenvalues of $\Lambda[H]$ are real-valued, and $\Lambda[H]$ must hence be Hermitian when H is Hermitian. This will be useful for us here since it also means that for any bounded linear operator A we have $\Lambda[A]^\dagger = \Lambda[A^\dagger]$. Specifically, one may see this by considering Hermitian operators $B = A + A^\dagger$ and $C = iA - iA^\dagger$. Applying Λ to B as well as C and using the linearity of Λ we have

$$\Lambda[B] = \Lambda[B]^\dagger \implies \Lambda[A + A^\dagger] = \Lambda[A + A^\dagger]^\dagger$$

$$\implies \Lambda[A] + \Lambda[A^\dagger] = \Lambda[A]^\dagger + \Lambda[A^\dagger]^\dagger, \tag{21.35a}$$

$$\Lambda[C] = \Lambda[C]^\dagger \implies i\Lambda[A - A^\dagger] = \left(i\Lambda[A - A^\dagger]\right)^\dagger$$

$$\implies i\Lambda[A] - i\Lambda[A^\dagger] = -i\Lambda[A]^\dagger + i\Lambda[A^\dagger]^\dagger. \tag{21.35b}$$

Multiplying eqn (21.35b) by i and adding it to eqn (21.35a), we wee that $\Lambda[A]^\dagger = \Lambda[A^\dagger]$ follows. We then apply this result when considering the Hermitian conjugate of the Choi state ρ_Λ in the form of eqn (21.29),

$$\rho_\Lambda^\dagger = \left(\frac{1}{d_A}\sum_{k,k'}\Lambda[|k\rangle\langle k'|] \otimes |k\rangle\langle k'|\right)^\dagger = \frac{1}{d_A}\sum_{k,k'}\left(\Lambda[|k\rangle\langle k'|]\right)^\dagger \otimes |k'\rangle\langle k|$$

$$= \frac{1}{d_A}\sum_{k,k'}\Lambda[|k'\rangle\langle k|] \otimes |k'\rangle\langle k| = \rho_\Lambda, \tag{21.36}$$

where we have swapped the labels of the summation indices in the last step. We have thus shown that $\Lambda \geq 0$ implies $\rho_\Lambda^\dagger = \rho_\Lambda$.

The reverse statement is generally not true, and a simple counterexample is the map $\Lambda[A] = -A$, which is clearly linear but does not preserve positivity, $\Lambda \not\geq 0$. At the same time, it does maps all Hermitian operators to Hermitian operators, and it is easy to see that the Choi "state" for this map is Hermitian but not positive, $\rho_\Lambda = -|\Phi_{00}\rangle\langle\Phi_{00}| = \rho_\Lambda^\dagger$. So a Hermitian Choi state alone is not sufficient for the associated map to be positive.

However, we can also show that the Choi state of any positive map has an additional property: We consider a pair of projection operators P and Q, and using the shorthand notation $P_{00} = |\Phi_{00}\rangle\langle\Phi_{00}|$ we calculate

$$\mathrm{Tr}\big(\rho_\Lambda\,P\otimes Q\big) = \mathrm{Tr}\Big((\Lambda\otimes\mathbb{1})[P_{00}]\,P\otimes Q\Big) = \big((\Lambda\otimes\mathbb{1})[P_{00}]^\dagger, P\otimes Q\big)_{\mathrm{HS}}$$

$$= \big((\Lambda\otimes\mathbb{1})[P_{00}], P\otimes Q\big)_{\mathrm{HS}} = \big(P_{00}, (\Lambda^\dagger\otimes\mathbb{1})[P\otimes Q]\big)_{\mathrm{HS}}$$

$$= \big(P_{00}, \Lambda^\dagger[P]\otimes Q\big)_{\mathrm{HS}} \tag{21.37}$$

which we have rewritten using the Hilbert–Schmidt inner product $(A, B)_{\mathrm{HS}} := \mathrm{Tr}(A^\dagger B)$ from Definition 11.4 with $A = \rho_\Lambda$ and $B = P\otimes Q$, and we have used $\rho_\Lambda^\dagger = \rho_\Lambda$ in going from the first to the second line, before switching from the map $\Lambda\otimes\mathbb{1}$ to its dual (with respect to the Hilbert–Schmidt inner product) denoted here as $\Lambda^\dagger\otimes\mathbb{1}$. Since Λ preserves positivity, so does the dual map $\Lambda^\dagger$, which we can see by considering projectors onto two arbitrary states $|\psi\rangle\in\mathcal{H}_A$ and $|\phi\rangle\in\mathcal{H}_B$, for which the dual map is defined via

$$\big(\Lambda[|\psi\rangle\langle\psi|], |\phi\rangle\langle\phi|\big)_{\mathrm{HS}} = \big(|\psi\rangle\langle\psi|, \Lambda^\dagger[|\phi\rangle\langle\phi|]\big)_{\mathrm{HS}}. \tag{21.38}$$

The left-hand side can be rewritten as

$$\big(\Lambda[|\psi\rangle\langle\psi|], |\phi\rangle\langle\phi|\big)_{\mathrm{HS}} = \mathrm{Tr}(\Lambda[|\psi\rangle\langle\psi|]^\dagger\,|\phi\rangle\langle\phi|) = \mathrm{Tr}(\Lambda[|\psi\rangle\langle\psi|]\,|\phi\rangle\langle\phi|)$$

$$= \langle\phi|\,\Lambda[|\psi\rangle\langle\psi|]\,|\phi\rangle \geq 0 \quad \forall\,\psi,\phi, \tag{21.39}$$

where we have again used the fact that $\Lambda[A]^\dagger = \Lambda[A^\dagger]$ for positive maps. The right-hand side of eqn (21.38) can similarly be written as

$$\big(|\psi\rangle\langle\psi|, \Lambda^\dagger[|\phi\rangle\langle\phi|]\big)_{\mathrm{HS}} = \mathrm{Tr}(|\psi\rangle\langle\psi|^\dagger\,\Lambda^\dagger[|\phi\rangle\langle\phi|]) = \mathrm{Tr}(|\psi\rangle\langle\psi|\,\Lambda^\dagger[|\phi\rangle\langle\phi|])$$

$$= \langle\psi|\,\Lambda^\dagger[|\phi\rangle\langle\phi|]\,|\psi\rangle, \tag{21.40}$$

Since this must be non-negative for all ψ and ϕ, the dual map must be positive, $\Lambda^\dagger \geq 0$. Returning to eqn (21.37), we see that $\Lambda^\dagger[P] \geq 0$ since the projector $P \geq 0$. We thus arrive at the trace of the product of two positive operators, a projector $P_{00} \geq 0$ and $\Lambda^\dagger[P]\otimes Q \geq 0$, such that

$$\mathrm{Tr}\big(\rho_\Lambda\,P\otimes Q\big) = \mathrm{Tr}\big(P_{00}\,\Lambda^\dagger[P]\otimes Q\big) \geq 0. \tag{21.41}$$

Now we want to show the reverse direction, i.e., that the combination of a Hermitian Choi state ρ_Λ and $\mathrm{Tr}\big(\rho_\Lambda\,P\otimes Q\big) \geq 0$ for all projectors P and Q implies $\rho_\Lambda \geq 0$. To see this we go back to eqn (21.37) and note that positivity of this expression for all P and Q implies (since $Q \geq 0$ anyways), that $\Lambda^\dagger[P] \geq 0$ for all P, which satisfy $P \geq 0$ since P is a projector as well. Via the argument supporting the positivity of the dual map, this means that $\Lambda^\dagger \geq 0$, which implies $\Lambda \geq 0$. $\qquad\square$

21.4.2 Kraus-Decomposition Theorem

With the help of the Choi isomorphism we can now show the Kraus decomposition theorem (Kraus, 1983).

Theorem 21.1 (Kraus decomposition theorem)

Every CPTP map $\Lambda : \mathcal{B}_{\mathrm{HS}}(\mathcal{H}_A) \to \mathcal{B}_{\mathrm{HS}}(\mathcal{H}_B)$ *admits a Kraus decomposition*

$$\Lambda[\rho] = \sum_n K_n \, \rho \, K_n^\dagger \,,$$

where the Kraus operators $K_n : \mathcal{H}_A \to \mathcal{H}_B$ *satisfy* $\sum_n K_n^\dagger K_n = \mathbb{1}_A$.

Proof Let us consider the Choi state of the CPTP map as given by eqn (21.27). It is a density operator that we can diagonalize, i.e., there exists a set of non-negative weights $\{p_n\}_n$ with $\sum_n p_n = 1$ and normalized states $|\,\omega_n\,\rangle \in \mathcal{H}_B \otimes \mathcal{H}_A$ such that

$$\rho_\Lambda = \sum_n p_n \, |\,\omega_n\,\rangle\langle\,\omega_n\,| \,. \tag{21.42}$$

Inserting the spectral decomposition of the Choi state into eqn (21.31), we have

$$\Lambda[\rho_A] = d_A^2 \big(\,_{AA'}\langle\,\Phi_{00}\,| \otimes \mathbb{1}_B\big)\, \rho_A \otimes \rho_\Lambda \,\big(|\,\Phi_{00}\,\rangle_{AA'} \otimes \mathbb{1}_B\big) \tag{21.43}$$

$$= \sum_n d_A p_n \sum_{k,k'} \Big(\,_A\langle\, k\,| \otimes \mathbb{1}_B \otimes \,_A\langle\, k\,|\Big)\, \rho_A \otimes |\,\omega_n\,\rangle\langle\,\omega_n\,|_{BA'} \,\Big(|\,k'\,\rangle_A \otimes \mathbb{1}_B \otimes |\,k'\,\rangle_{A'}\Big)$$

$$= \sum_n \Big(\sqrt{d_A p_n} \sum_k \,_{AA'}\langle\, kk\,|\,\omega_n\,\rangle_{BA'}\Big)\, \rho_A \,\Big(\sqrt{d_A p_n} \sum_{k'} \,_{BA'}\langle\,\omega_n\,|\, k'k'\,\rangle_{AA'}\Big)$$

$$= \sum_n K_n \, \rho_A \, K_n^\dagger,$$

where we have used

$$\rho_A \otimes |\,\omega_n\,\rangle\langle\,\omega_n\,|_{BA'} = \big(\rho_A \otimes \mathbb{1}_{BA'}\big)\big(\mathbb{1}_A \otimes |\,\omega_n\,\rangle\langle\,\omega_n\,|_{BA'}\big) \tag{21.44}$$

$$= \big(\mathbb{1}_A \otimes |\,\omega_n\,\rangle_{BA'}\big)\big(\rho_A \otimes \mathbb{1}_{BA'}\big)\big(\mathbb{1}_A \otimes \langle\,\omega_n\,|_{BA'}\big),$$

and we have identified the Kraus operators

$$K_n = \sqrt{d_A p_n} \sum_k \,_{AA'}\langle\, kk\,|\,\omega_n\,\rangle_{BA'} \,. \tag{21.45}$$

If the eigenstates $|\,\omega_n\,\rangle$ of the Choi state are specified with respect to the basis $\{|\, k\,\rangle\}_k$ used for the Choi isomorphism, that is, given the decomposition $|\,\omega_n\,\rangle = \sum_{i,j} c_{ij}^{(n)} |\, ij\,\rangle$, one can write the Kraus operators explicitly in terms of the complex coefficients $c_{ij}^{(n)}$ as

$$K_n = \sqrt{d_A p_n} \sum_k \sum_{i,j} {}_{AA'}\langle kk | ij \rangle_{BA'} = \sqrt{d_A p_n} \sum_{i,j} c_{ij}^{(n)} {}_B| i \rangle\langle j |_A . \tag{21.46}$$

To confirm that these operators satisfy the correct normalization condition, we just have to note that Λ is trace preserving to obtain

$$\mathrm{Tr}(\Lambda[\rho_A]) = \mathrm{Tr}\Big(\sum_n K_n \rho_A K_n^\dagger \Big) = \mathrm{Tr}\Big(\rho_A \sum_n K_n^\dagger K_n \Big) = 1 \; \forall \rho_A, \tag{21.47}$$

which can only be satisfied for all ρ_A if $\sum_n K_n^\dagger K_n = \mathbb{1}_A$.   □

21.4.3 Kraus Decomposition for the Depolarizing Channel

With Theorem 21.1 at hand, we can now return to the example of the depolarizing channel from eqn (21.12) and determine its Kraus operators. To do this, we first require the Choi representation of the depolarizing map $\Lambda[\rho] = p\rho + (1-p)\mathrm{Tr}(\rho)\mathbb{1}/d$. From eqn (21.29) we see that we first need to understand the action of the map on an arbitrary matrix element, which is

$$\Lambda\big[| k \rangle\langle k' |\big] = p | k \rangle\langle k' | + (1-p)\mathrm{Tr}\big(| k \rangle\langle k' |\big) \tfrac{1}{d}\mathbb{1} = p | k \rangle\langle k' | + (1-p)\delta_{kk'} \tfrac{1}{d}\mathbb{1} . \tag{21.48}$$

Inserting into eqn (21.29), we find

$$
\begin{aligned}
\rho_\Lambda &= \tfrac{1}{d} \sum_{k,k'} (p | k \rangle\langle k' | + (1-p)\delta_{kk'} \tfrac{1}{d}\mathbb{1}) \otimes | k \rangle\langle k' | \\
&= p\tfrac{1}{d} \sum_{k,k'} | k \rangle\langle k' | \otimes | k \rangle\langle k' | + (1-p)\tfrac{1}{d}\mathbb{1} \otimes \tfrac{1}{d} \sum_k | k \rangle\langle k | \\
&= p | \Phi_{00} \rangle\langle \Phi_{00} | + (1-p)\tfrac{1}{d}\mathbb{1} \otimes \tfrac{1}{d}\mathbb{1} .
\end{aligned} \tag{21.49}
$$

To determine the Kraus operators, we now have to diagonalize the Choi state. The eigenvalues p_n are easily found by noting that the second term in eqn (21.49) is proportional to the identity, which can be expanded in any orthonormal basis. In particular, we can choose a basis that contains $| \Phi_{00} \rangle$ as a basis vector. We can then just read off the eigenvalues to be $(p+(1-p)/d^2)$ and $(1-p)/d^2$, where the last value is (d^2-1)-fold degenerate. However, to continue we also require the specific eigenvectors. A convenient basis of $\mathcal{H}_A \otimes \mathcal{H}_A$ that contains $| \Phi_{00} \rangle$ is given by $\{| \phi_{mn} \rangle\}_{m,n=0,\dots,d-1}$, with (see also eqn (14.11))

$$| \phi_{mn} \rangle = \frac{1}{\sqrt{d}} \sum_{k=0}^{d-1} e^{\frac{2\pi i}{d}nk} | k \rangle \otimes | (k+m)_{\mathrm{mod}(d)} \rangle . \tag{21.50}$$

Before making further use of this basis, let us briefly confirm that the states in eqn (21.50) indeed form an orthonormal basis by calculating

$$\langle \phi_{m'n'} | \phi_{mn} \rangle = \frac{1}{d} \sum_{k,k'=0}^{d-1} e^{\frac{2\pi i}{d}(nk-n'k')} \langle k' | k \rangle \langle (k'+m')_{\mathrm{mod}(d)} | (k+m)_{\mathrm{mod}(d)} \rangle$$

$$= \frac{1}{d} \sum_{k=0}^{d-1} e^{\frac{2\pi i}{d}k(n-n')} \langle (k+m')_{\mathrm{mod}(d)} | (k+m)_{\mathrm{mod}(d)} \rangle$$

$$= \delta_{mm'} \frac{1}{d} \sum_{k=0}^{d-1} e^{\frac{2\pi i}{d}k(n-n')} = \delta_{mm'} \delta_{nn'} , \tag{21.51}$$

where we have used the fact that for $n \neq n'$ the dth roots of unity sum to zero, while for $n = n'$ we have $\sum_k 1 = d$. The states $| \phi_{mn} \rangle$ hence form an orthonormal basis with elements labelled by the multi-index (m, n). For $(m, n) = (0, 0)$ we have $| \phi_{00} \rangle = | \Phi_{00} \rangle$, the maximally entangled[5] state appearing in eqn (21.49). We can thus rewrite ρ_Λ in its spectral decomposition

$$\rho_\Lambda = \left(p + \tfrac{1-p}{d^2} \right) | \phi_{00} \rangle\langle \phi_{00} | + \tfrac{1-p}{d^2} \sum_{\substack{m,n=0 \\ (m,n)\neq(0,0)}}^{d-1} | \phi_{mn} \rangle\langle \phi_{mn} | = \sum_{m,n=0}^{d-1} p_{mn} | \phi_{mn} \rangle\langle \phi_{mn} | , \tag{21.52}$$

where the eigenvalues p_{mn} are given by

$$p_{mn} = \begin{cases} p + \tfrac{1-p}{d^2} & \text{for } m = n = 0 \\ \tfrac{1-p}{d^2} & \text{otherwise} \end{cases} . \tag{21.53}$$

From eqn (21.50) we can then read off the coefficients $c_{jk}^{(m, n)} = \frac{1}{\sqrt{d}} e^{2\pi inj/d} \delta_{k(j+m)_{\mathrm{mod}(d)}}$ such that $| \phi_{mn} \rangle = \sum_{j,k} c_{jk}^{(m, n)} | j \rangle | k \rangle$. Inserting into eqn (21.46) we thus find the Kraus operators for the depolarizing channel can be written as

$$K_{mn} = \sqrt{d \, p_{mn}} \sum_{j,k} c_{jk}^{(m, n)} | j \rangle\langle k | = \sqrt{p_{mn}} \sum_{j,k} e^{\frac{2\pi inj}{d}} \delta_{k(j+m)_{\mathrm{mod}(d)}} | j \rangle\langle k |$$

$$= \sqrt{p_{mn}} \sum_{j} e^{\frac{2\pi inj}{d}} | j \rangle\langle (j+m)_{\mathrm{mod}(d)} | . \tag{21.54}$$

It is interesting to note here that, up to the pre-factors $\sqrt{p_{mn}}$, the Kraus operators are given by unitary operators that form a complete orthonormal operator basis, the *Weyl operators* U_{nm} with $U_{nm}U_{nm}^\dagger = U_{nm}^\dagger U_{nm} = \mathbb{1}$ we have discussed in Section 17.1.3. We can thus understand the depolarizing channel as a convex sum of unitaries applied to the state ρ with probabilities p_{mn},

$$\Lambda[\rho] = \sum_{m,n} p_{mn} U_{nm} \rho U_{nm}^\dagger . \tag{21.55}$$

---

[5]Interestingly, all basis states $| \phi_{mn} \rangle$ are maximally entangled, i.e., one can easily check that the reduced states for both subsystems are maximally mixed. This observation is the premise for allowing teleportation in arbitrary dimensions, see Chapter 14 in Part II, and Bennett *et al.* (1993).

21.4.4 Stinespring Dilation

Finally, we can use our insights about the Kraus decomposition to show that every CPTP map Λ can indeed be represented as unitary dynamics on a larger Hilbert space. Here, we will refer to this statement as the *Stinespring dilation theorem* (Stinespring, 1955):

Theorem 21.2 (Stinespring dilation)

For every CPTP map $\Lambda : \mathcal{B}_{\text{HS}}(\mathcal{H}_A) \to \mathcal{B}_{\text{HS}}(\mathcal{H}_B)$, there exists a Hilbert space $\mathcal{H}_C$ with a state $|\omega\rangle \in \mathcal{H}_C$ and a unitary U on $\mathcal{H}_A \otimes \mathcal{H}_C$ such that

$$\Lambda[\rho] = \text{Tr}_C \left(U \rho \otimes |\omega\rangle\langle\omega| \, U^\dagger \right).$$

Proof To show Theorem 21.2, we start with a generic state $|\omega\rangle \in \mathcal{H}_C$ and for this fixed state we formally define the operator U via its action $U|\varphi\rangle \otimes |\omega\rangle = \sum_k K_k |\varphi\rangle \otimes |k\rangle$ for arbitrary states $|\varphi\rangle \in \mathcal{H}_A$. Since we assume that Λ is CPTP, Theorem 21.1 tells us that a Kraus decomposition exists, and we define U via these Kraus operators K_k. We then show that U is unitary in the subspace spanned by $|\omega\rangle$ and any basis of $\mathcal{H}_A$, i.e., we calculate

$$\langle\varphi'| \otimes \langle\omega| U^\dagger U |\varphi\rangle \otimes |\omega\rangle = \sum_{k,l} ((\langle\varphi'|K_k^\dagger) \otimes \langle k| (K_l|\varphi\rangle) \otimes |l\rangle) \tag{21.56}$$

$$= \sum_{k,l} \langle\varphi'|K_k^\dagger K_l|\varphi\rangle \langle k|l\rangle = \langle\varphi'| \sum_k K_k^\dagger K_k |\varphi\rangle = \langle\varphi'|\varphi\rangle,$$

to find that U leaves the scalar product of the subspace spanned by $\{|\varphi\rangle \otimes |\omega\rangle\}_\varphi$ invariant. We then again make use of the fact that map Λ has a Kraus decomposition and write the input state in its spectral decomposition $\rho = \sum_n p_n |n\rangle\langle n|$ to obtain

$$\Lambda[\rho] = \sum_k K_k \rho K_k^\dagger = \sum_{n,k} p_n K_k |n\rangle\langle n| K_k^\dagger \tag{21.57}$$

$$= \sum_n p_n \text{Tr}_C \left[\sum_{k,k'} K_k |n\rangle\langle n| K_{k'}^\dagger \otimes |k\rangle\langle k'|_C \right]$$

$$= \sum_n p_n \text{Tr}_C \left[\left(\sum_k K_k |n\rangle \otimes |k\rangle \right) \left(\sum_{k'} \langle n| K_{k'}^\dagger \otimes \langle k'| \right) \right]$$

$$= \sum_n p_n \text{Tr}_C \left[U |n\rangle \otimes |\omega\rangle \langle n| \otimes \langle\omega| U^\dagger \right]$$

$$= \text{Tr}_C \left[U \sum_n p_n |n\rangle\langle n| \otimes |\omega\rangle\langle\omega| U^\dagger \right] = \text{Tr}_C \left[U \rho \otimes |\omega\rangle\langle\omega| U^\dagger \right].$$

$\square$

21.5   Impossible Operations—No Cloning

In this chapter we have so far discussed types of operations and transformations that *are* allowed by quantum mechanics, ranging from unitary operations to CPTP maps. In Chapter 23 we will further inspect possible operations that are collected under the umbrella of measurements, projective measurements to POVM measurements, and we have seen in previous chapters that this range of different operations allows us to execute sophisticated protocols such as the violation of a Bell inequality, quantum teleportation, and quantum key distribution to name only a few examples. But there are of course also a number of things that *cannot* be done within quantum mechanics. Examples include perfectly distinguishing non-orthogonal quantum states (see Section 23.4.2), perfect "cloning" of quantum states (which we will discuss here), and related statements such as the no-deleting and no-broadcasting theorems.

These kind of statements are of interest conceptually, because the corresponding tasks are often possible in classical information theory, so it is somewhat surprising initially that quantum mechanics does not allow the respective generalizations. At the same time, the fact that, for instance, perfectly cloning a quantum state is not possible is what guarantees the usefulness of certain more complicated protocols.

21.5.1   The No-Cloning Theorem

The no-cloning theorem, although mathematically relatively simple, is a conceptually remarkable and practically crucial feature of quantum information theory, which was first formulated by *James Park* in (Park, 1970), but was then (re)discovered[6] and formally proven simultaneously by (and is often associated with) *William Wootters* and *Wojciech H. Zurek* (1982) and *Dennis Dieks* (1982). Loosely speaking, the no-cloning theorem says that it is impossible to make a (perfect) copy of an unknown quantum state. Mathematically, it can be formulated in the following way:

Theorem 21.3 (No-cloning theorem)

There exists **no** *unitary U such that for all states $|\psi\rangle$ and for some fixed state $|0\rangle$:*

$$U\,|\psi\rangle\,|0\rangle = |\psi\rangle\,|\psi\rangle\,.$$

Proof   To prove this statement, let us assume that it was true for two arbitrary pure states $|\psi\rangle$ and $|\phi\rangle$, and let us attach labels A and B to the two Hilbert spaces of the original state and its copy, respectively, i.e.,

$$U\,|\psi\rangle_A\,|0\rangle_B = |\psi\rangle_A\,|\psi\rangle_B\,, \tag{21.58a}$$

$$U\,|\phi\rangle_A\,|0\rangle_B = |\phi\rangle_A\,|\phi\rangle_B\,. \tag{21.58b}$$

---

[6]For a historical perspective on this (re)discovery see the discussion in Ortigoso (2018).

We can then take the inner product between the left-hand sides of eqn (21.58),

$$_A\langle\psi|\ _B\langle 0|\ U^\dagger U\ |\phi\rangle_A\ |0\rangle_B = \ _A\langle\psi|\phi\rangle_A\ _B\langle 0|0\rangle_B = \ _A\langle\psi|\phi\rangle_A\ , \qquad (21.59)$$

and compare it to the inner product of the right-hand sides of eqn (21.58), which just gives $_A\langle\psi|\phi\rangle_A\ _B\langle\psi|\phi\rangle_B$. If perfect cloning is to be possible for the two states $|\psi\rangle$ and $|\phi\rangle$, then we thus need to have

$$\langle\psi|\phi\rangle = \langle\psi|\phi\rangle^2\ , \qquad (21.60)$$

but this is only possible if $\langle\psi|\phi\rangle = 0$ or if $\langle\psi|\phi\rangle = 1$, the two states either need to be the same, or orthogonal, but the statement certainly cannot hold for arbitrary pairs of states, and therefore cannot hold in general.   □

21.5.2   Cloning with More General Operations

As a first observation, let us note that the formulation of the no-cloning theorem that we have chosen is rather simple, involving pure states and unitary maps, and one can of course ask if the restriction imposed by the theorem might not simply be an artefact of this idealization. In other words, maybe one could clone by using more general maps than unitaries, or maybe mixed states can be cloned perfectly? The answer to both questions is: *No*. Suppose we had a CPTP map Λ of the form:

$$\Lambda[|\psi\rangle\langle\psi| \otimes |0\rangle\langle 0|] = |\psi\rangle\langle\psi| \otimes |\psi\rangle\langle\psi|\ . \qquad (21.61)$$

By applying the Stinespring dilation (Theorem 21.2), we could represent any such CPTP map as a unitary on a larger space, that is, for some $|\omega\rangle \in \mathcal{H}(C)$ we could write

$$\Lambda[|\psi\rangle\langle\psi| \otimes |0\rangle\langle 0|] = \mathrm{Tr}_C\Big(U\ |\psi\rangle\langle\psi| \otimes |0\rangle\langle 0| \otimes |\omega\rangle\langle\omega|\ U^\dagger\Big) = |\psi\rangle\langle\psi| \otimes |\psi\rangle\langle\psi|\ . \qquad (21.62)$$

However, the requirement of returning a pure state $|\psi\rangle|\psi\rangle$ as output means that subsystem C (which is traced out in the last step), must also be in a pure state (this follows from the Schmidt decomposition, Theorem 15.1), so we must have

$$U\ |\psi\rangle \otimes |0\rangle \otimes |\omega\rangle = |\psi\rangle \otimes |\psi\rangle \otimes |\tilde\omega\rangle \qquad (21.63)$$

for some state $|\tilde\omega\rangle$, which may depend on the input state, $|\tilde\omega\rangle = |\tilde\omega(\psi)\rangle$. If we take this as the premise of the proof of the no-cloning theorem (Theorem 21.3), we obtain the condition

$$\langle\psi|\phi\rangle = \langle\psi|\phi\rangle^2\ \langle\tilde\omega(\psi)|\tilde\omega(\phi)\rangle\ . \qquad (21.64)$$

This does not much improve things, indeed, once more the statement is true if $\langle\psi|\phi\rangle = 0$ or if $\langle\psi|\phi\rangle = 1$, but it cannot be true in general, since the absolute values of all involved inner products are smaller or equal than 1. We thus conclude, perfectly cloning arbitrary quantum states is also not possible using general CPTP maps.

However, we can ask which states could potentially be cloned. That is, the no-cloning theorem prevents us from finding a unitary that can successfully clone *all* input states. But, certainly, there are unitaries that can clone some input states, in particular (as we have seen), orthogonal input states. A trivial example are the states in the computational basis: Suppose we consider $|\psi\rangle = |0\rangle$ and $|\phi\rangle = |1\rangle$ such that we are looking for a unitary that achieves

$$U\,|0\rangle|0\rangle = |0\rangle|0\rangle\,, \qquad \text{and} \qquad U\,|1\rangle|0\rangle = |1\rangle|1\rangle\,. \tag{21.65}$$

This is exactly the transformation realized by the CNOT transformation in eqn (14.47) that flips the second qubit conditional on the state of the first qubit.

The no-cloning problem can also be extended to mixed states, that is, one may ask if it is possible to find a unitary (or, again, any CPTP map) that provides two copies of any arbitrary mixed single-copy input state. Such a process (referred to as "broadcasting") is also not possible (except if one restricts to a set of mutually commuting states) and this statement is known as the *no-broadcasting theorem* (Barnum *et al.*, 1996).

21.5.3 Cloning Classical Information

We have seen that states (density operators) that can be cloned perfectly by the same process must commute (which includes orthogonal pure states). This implies that the states in question can be written as density operators that are diagonal with respect to the same basis. In other words:

Classical information can be perfectly cloned!

To make the previous statement more precise, simply note that "classical information" in this context just means that we have fixed a basis and that any coherence with respect to this basis is lost or is irrelevant. In a d-dimensional Hilbert space we can then simply nominate basis states $|0\rangle, |1\rangle, |2\rangle, \ldots, |d-1\rangle$ and write the unitary

$$\begin{aligned}
U &= |0\rangle\langle 0| \otimes \mathbb{1} + |1\rangle\langle 1| \otimes \big(|0\rangle\langle 1| + |1\rangle\langle 0| + |2\rangle\langle 2| + \ldots + |d-1\rangle\langle d-1|\big) \\
&\quad + |2\rangle\langle 2| \otimes \big(|1\rangle\langle 1| + |0\rangle\langle 2| + |2\rangle\langle 0| + |3\rangle\langle 3| + \ldots + |d-1\rangle\langle d-1|\big) \\
&\quad + \ldots + |d-1\rangle\langle d-1| \otimes \big(|1\rangle\langle 1| + |2\rangle\langle 2| + \ldots + |0\rangle\langle d-1| + |d-1\rangle\langle 0|\big) \\
&= |0\rangle\langle 0| \otimes \mathbb{1} + \sum_{n=1}^{d-1} |n\rangle\langle n| \otimes \big(\mathbb{1} - |0\rangle\langle 0| - |n\rangle\langle n| + |0\rangle\langle n| + |n\rangle\langle 0|\big), \tag{21.66}
\end{aligned}$$

such that $U\,|n\rangle|0\rangle = |n\rangle|n\rangle$ for all $n = 0, 1, \ldots, d-1$. We thus have an explicit unitary that allows for perfectly copying classical information, whereas we cannot generally copy quantum information perfectly.

An obvious exception is the case where we already know the information to be copied. In other words, it is no problem in principle to prepare an arbitrary number of systems in any pre-agreed state. But if we think about this carefully, then this just means that we have classical information about the state (i.e., we know which state it is).

This is an interesting observation in relation to the fact that we cannot measure a system without disturbing it. If we could, then we could repeatedly obtain classical information about the state (until we know exactly which state it is) and then prepare arbitrarily many copies, thus effectively cloning it.

21.5.4   Imperfect Cloning

Although it is not possible to perfectly clone quantum states, one can approximately clone quantum states to some extent. Suppose we are interested in a process that can make copies of two selected states that are not orthogonal, e.g., $|0\rangle$ and $|+\rangle = \frac{1}{\sqrt{2}}(|0\rangle + |1\rangle)$. That is, we are looking for a unitary U such that $U|00\rangle$ is close to $|00\rangle$, and $U|+0\rangle$ is close to $|++\rangle$. Mathematically, we can formulate this requirement in the following way, we can try to find a unitary that maximizes the average fidelity

$$\bar{\mathcal{F}} = \tfrac{1}{2}|\langle 00|U|00\rangle|^2 + \tfrac{1}{2}|\langle ++|U|+0\rangle|^2. \tag{21.67}$$

To gauge how well we can do in this task, we can write the matrix element $\langle ++|U|+0\rangle$ in the computational basis

$$\langle ++|U|+0\rangle = \tfrac{1}{2\sqrt{2}}\big(\langle 00|U|00\rangle + \langle 00|U|10\rangle + \langle 01|U|00\rangle + \langle 01|U|10\rangle$$
$$+ \langle 10|U|00\rangle + \langle 10|U|10\rangle + \langle 11|U|00\rangle + \langle 11|U|10\rangle\big). \tag{21.68}$$

We then see, even for a trivial solution (doing nothing, $U = \mathbb{1}$), we have

$$\bar{\mathcal{F}} = \tfrac{1}{2}|\langle 00|U|00\rangle|^2 + \tfrac{1}{2}|\langle ++|U|+0\rangle|^2 \tag{21.69}$$
$$= \tfrac{1}{2}|\langle 00|00\rangle|^2 + \tfrac{1}{2}|\tfrac{1}{2\sqrt{2}}\langle 00|00\rangle + \tfrac{1}{2\sqrt{2}}\langle 10|10\rangle|^2 = \tfrac{1}{2} + \tfrac{1}{4} = \tfrac{3}{4}.$$

But we can do even better. By making the simple ansatz

$$U = \begin{pmatrix} \cos(\theta) & 0 & 0 & -\sin(\theta) \\ 0 & \cos(\varphi) & -\sin(\varphi) & 0 \\ 0 & \sin(\varphi) & \cos(\varphi) & 0 \\ \sin(\theta) & 0 & 0 & \cos(\theta) \end{pmatrix}, \tag{21.70}$$

we find the matrix elements

$$\tfrac{1}{2}|\langle 00|U|00\rangle|^2 = \tfrac{1}{2}\cos^2(\theta), \tag{21.71a}$$

$$\tfrac{1}{2}|\langle ++|U|+0\rangle|^2 = \tfrac{1}{16}|\cos(\theta) + \sin(\theta) + \cos(\varphi) + \sin(\varphi)|^2. \tag{21.71b}$$

Setting $\varphi = \tfrac{\pi}{4}$ and performing numerical optimization over the other angle θ using some suitable software, one finds $\bar{\mathcal{F}} = 0.90188$ for $\theta = 0.07893\pi$. We thus see that approximate cloning can do quite well, at least for two particular non-orthogonal states.

Other no-go results: The no-cloning theorem is a, if not *the*, prime example of a no-go theorem in quantum mechanics, and there are a number of other statements of things that cannot be achieved using quantum operations despite the fact that corresponding (restricted) tasks can be achieved in classical information processing. For instance, the *no-deleting theorem* (Pati and Braunstein, 2000) provides a statement dual to the no-cloning theorem, stipulating that given two (or more) copies of a quantum state, one cannot "delete" one of them and turn it into a generic state independent of the initial state.

Another instance of a no-go result in quantum-information processing is the impossibility of turning arbitrary unitaries U into "controlled" versions $|0\rangle\langle0|_A \otimes \mathbb{1}_B + |1\rangle\langle1|_A \otimes U_B$ conditioned on the state of a control qubit (here, the first qubit A), i.e., there are no unitaries V and W independent of U on the larger system such that (Araújo *et al.*, 2014)

$$V_{AB}\, \mathbb{1}_A \otimes U_B\, W_{AB} = |0\rangle\langle0|_A \otimes \mathbb{1}_B + |1\rangle\langle1|_A \otimes U_B\,. \tag{21.72}$$

In this particular case it is interesting to note that, although there of course is no way around the mathematical statement of this *no-adding-control* theorem, there are practical ways of mimicking the procedure of "adding control" by exploiting pre-existing conditioning of the physical operations that are used to implement certain (unitary) operations (as was demonstrated in Zhou *et al.*, 2011; Friis *et al.*, 2014; Friis *et al.*, 2015a; Rambo *et al.*, 2016; Thompson *et al.*, 2018).

22

Open Quantum Systems, Decoherence, Atom-Field Coupling

The time evolution of a *closed* quantum system is unitary and at least in the non-relativistic limit governed by the Schrödinger equation, or in case of mixed states the von Neumann equation, as we have discussed in Chapter 2 and Section 11.3, respectively. Indeed, for all intents and purposes we may raise obeying unitary dynamics to the definition of a quantum system being *closed* in the first place. An open quantum system, in contrast, cannot generally be described by unitary time evolution. Yet, in all practical physical situations we must assume that the system of interest interacts with its environment in some way and that losses, noise, dissipation, and decoherence have to be taken into account. In general, describing the time evolution of an open quantum system relies on the availability of some information about the environment and its interaction with the system. Nevertheless, under some reasonable assumptions on the strength of the interaction and on the involved timescales—in the Born–Markov approximation—we can formulate a suitable equation of motion for the time evolution of a density matrix describing an open system, a so-called *Markovian master equation*. Based on this description, we then discuss spontaneous and stimulated emission, as well as the absorption of photons by a two-level atom, and present the important example of the *Jaynes–Cummings model* for the interaction of an atom with the electromagnetic field in a cavity.

22.1 Interaction of System and Environment

How can we describe the dynamics of a system interacting with its environment, in particular, if we do not have much knowledge about the environment? As we shall see, the CPTP maps we introduced in Chapter 21 will play an essential role.

22.1.1 Open Quantum Systems

An open quantum system is a system S that is coupled to its environment E. Practically, this is always the case, the idea of an isolated system is an idealization, albeit a reasonable approximation in many cases. Here, we are primarily considering situations where the environment is a much larger system with infinitely many degrees of freedom. We call it a *reservoir*, or a *heat bath* if the reservoir is at thermal equilibrium.

The total system $S + E$ is described by a density matrix $\rho_{SE} \in \mathcal{B}_{\mathrm{HS}}(\mathcal{H}) = \mathcal{B}_{\mathrm{HS}}(\mathcal{H}_S \otimes \mathcal{H}_E)$, an element of the total Hilbert–Schmidt space $\mathcal{B}_{\mathrm{HS}}(\mathcal{H})$ over the Hilbert

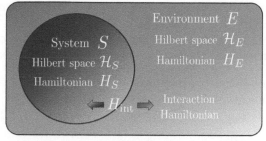

Joint system $S+E$ $\mathcal{H}_S \otimes \mathcal{H}_E$ $H = H_S + H_E + H_{\text{int}}$

Fig. 22.1 Open quantum system. The system S interacts with its environment E.

space $\mathcal{H}$, which is a tensor product of the spaces of the system $\mathcal{H}_S$ and the environment $\mathcal{H}_E$. The state of the system S is given by the reduced density operator $\rho_S = \text{Tr}_E(\rho_{SE})$, i.e., it is obtained by tracing out the environment from ρ_{SE}.

We consider a situation where all observables of interest, denoted as A, refer to the open system S and are of the form

$$A \otimes \mathbb{1}_E \quad \text{with} \quad A \in \mathcal{B}_{\text{HS}}(\mathcal{H}_S) \,. \tag{22.1}$$

Then the expectation value of the observable A is given by

$$\langle\, A \,\rangle \;=\; \text{Tr}_S(\rho_S\, A) \,. \tag{22.2}$$

Thus the knowledge of the reduced density matrix ρ_S is of central interest to describe the open system S.

When there is no interaction between the system and the environment, then their time evolutions are governed by unitary dynamics determined by the system Hamiltonian H_S and the environment Hamiltonian H_E, respectively. However, for an open quantum system there is also an interaction Hamiltonian H_{int} that couples S and E. Therefore the dynamics of the subsystem, regarded as an open quantum system, is determined by the Hamiltonian H of the total system $S + E$,

$$H \;=\; H_S + H_E + H_{\text{int}} \,, \tag{22.3}$$

as we have sketched in Fig. 22.1. All parts of the Hamiltonian may in principle be time-dependent, and from the outset the situation we are facing is reminiscent of the setting in which we have applied time-dependent perturbation theory in Section 10.5. As we recall from this discussion, the total (closed) system $S + E$ follows a unitary time evolution which is, in the case of a time-dependent Hamiltonian $H(t)$, represented by the unitary time-evolution operator

$$U(t, t_o) \;=\; \mathcal{T} \left\{ \exp\left(-\frac{i}{\hbar} \int_{t_o}^{t} dt'\, H(t') \right) \right\} \,, \tag{22.4}$$

also called *propagator*, with the property $U(t_o, t_o) = \mathbb{1}$. The symbol $\mathcal{T}$ is the *time-ordering* operator, which gives rise to a time-ordered product in which products of time-dependent operators are rearranged such that their time arguments increase from right to left[1], see eqn (10.146). If the Hamiltonian of a system is time-independent, $H(t) = H = \text{const.}$, we call the system *isolated*, its energy remains constant over time, and the propagator simplifies to the familiar expression $U(t, t_o) = \exp\left(-iH(t-t_o)/\hbar\right)$.

For a general (time-dependent) Hamiltonian the total system is determined by the von Neumann equation (recall Theorem 11.3 of Section 11.3)

$$\frac{\partial}{\partial t}\, \rho_{SE}(t) = -\frac{i}{\hbar}\, [\, H(t)\, ,\, \rho_{SE}(t)\,]\ , \tag{22.5}$$

and the subsystem S is obtained by tracing out the environment E in eqn (22.5),

$$\frac{\partial}{\partial t}\, \rho_S(t) = -\frac{i}{\hbar}\, \text{Tr}_E\{[\, H(t)\, ,\, \rho_{SE}(t)\,]\}\ . \tag{22.6}$$

In principle, we can now make a power-series ansatz similar to what we have done for the propagator in eqn (10.145),

$$\rho_S(t) = \rho_S(t_o) - \frac{i}{\hbar} \int_{t_o}^{t} dt_1\, \text{Tr}_E\{[\, H(t_1)\, ,\, \rho_{SE}(t_o)\,]\} \tag{22.7}$$

$$+ \left(\frac{-i}{\hbar}\right)^2 \int_{t_o}^{t} dt_1 \int_{t_o}^{t_1} dt_2\, \text{Tr}_E\{[\, H(t_1)\, ,\, [\, H(t_2)\, ,\, \rho_{SE}(t_o)\,]\,]\} + \cdots\ ,$$

As already mentioned in Section 11.3 the equation (22.5) is the quantum-mechanical analogue of a Liouville-type equation in statistical mechanics, and for this reason we can write eqn (22.5) in the form

$$\frac{\partial}{\partial t}\, \rho_{SE}(t) = \mathcal{L}(t)\big[\rho_{SE}(t)\big]\ , \tag{22.8}$$

where $\mathcal{L}(t)$ is to be understood as a super-operator acting on quantum states, which leads to the formal solution

$$\rho_{SE}(t) = \mathcal{T}\left\{\exp\left(\int_{t_o}^{t} dt'\, \mathcal{L}(t')\right)\big[\rho(t_o)\big]\right\}\ . \tag{22.9}$$

In case of a time-independent Hamiltonian we obviously get

$$\rho_{SE}(t) = \exp\left(\mathcal{L}(t - t_o)\right)\big[\rho_{SE}(t_o)\big]\ , \tag{22.10}$$

and in both eqns (22.9) and (22.10) the system state is obtained by tracing out E. However, the solutions (22.9) and (22.10) are just formal since in practice it is quite difficult to model the environment such that it both corresponds to a relevant physical situation and is mathematically solvable. Therefore a simpler and more practicable approach has been developed, which focuses, under certain approximations, on the reduced quantum system and its observables.

---

[1]A mnemonic device that makes it easy to keep this in mind is that the time-ordering symbol arranges factors that appear *Later* to the *Left*.

22.1.2   Dynamics of Open Quantum Systems—Dynamical Maps

The dynamics of an open quantum system can—under some assumptions—be described using so-called *dynamical maps*. Here, we give a brief overview of their definition and refer to dedicated literature such as, e.g., Breuer and Petruccione (2002) and Dominy *et al.* (2016), for a more comprehensive treatment.

Let us suppose that we can prepare the total system $S + E$ such that at the initial time $t = t_o$ the joint state is an uncorrelated product state of the system $\rho_S \in \mathcal{B}_{\mathrm{HS}}(\mathcal{H}_S)$ and the environment $\rho_E \in \mathcal{B}_{\mathrm{HS}}(\mathcal{H}_E)$

$$\rho_{SE}(t_o) = \rho_S(t_o) \otimes \rho_E(t_o) . \tag{22.11}$$

The time evolution of the total system is determined by the unitary operator (22.4)

$$\rho_{SE}(t_o) \xrightarrow{\text{unitary evolution}} \rho_{SE}(t) = U(t,t_o) \left[\rho_S(t_o) \otimes \rho_E(t_o) \right] U^\dagger(t,t_o) . \tag{22.12}$$

The change of the reduced system S from the initial time $t = t_o$ to some fixed $t > 0$ can be obtained by tracing out the environment E in eqn (22.12),

$$\rho_S(t_o) \xrightarrow{\text{dynamical map}} \rho_S(t) = \mathrm{Tr}_E \left\{ U(t,t_o) \left[\rho_S(t_o) \otimes \rho_E(t_o) \right] U^\dagger(t,t_o) \right\}$$

$$=: \Lambda(t,t_o) \left[\rho_S(t_o) \right] , \tag{22.13}$$

which defines a *dynamical map* $\Lambda(t,t_o)$ on the Hilbert–Schmidt space $\mathcal{B}_{\mathrm{HS}}(\mathcal{H}_S)$, see Fig. 22.2.

If we assume that the environment is initially in a pure state, $\rho_E(t_o) = |\phi_0\rangle\langle\phi_0|$, then we immediately see from the Stinespring dilation, Theorem 21.2, that the dynami-

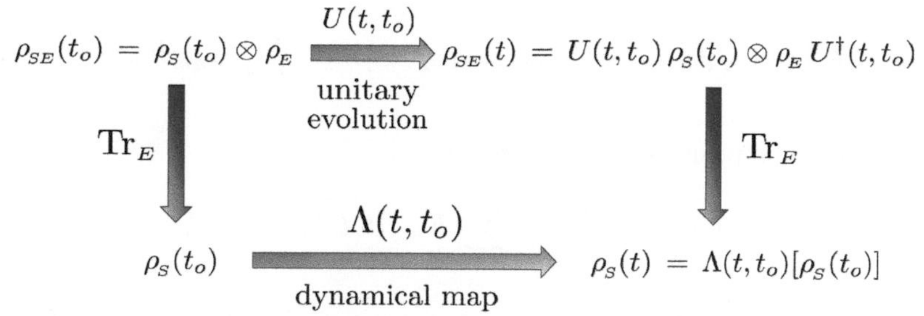

$$\rho_{SE}(t_o) = \rho_S(t_o) \otimes \rho_E \xrightarrow{U(t,t_o)} \rho_{SE}(t) = U(t,t_o)\,\rho_S(t_o) \otimes \rho_E\, U^\dagger(t,t_o)$$

$$\text{unitary evolution}$$

$$\mathrm{Tr}_E \Big\downarrow \qquad\qquad \mathrm{Tr}_E \Big\downarrow$$

$$\rho_S(t_o) \xrightarrow[\text{dynamical map}]{\Lambda(t,t_o)} \rho_S(t) = \Lambda(t,t_o)[\rho_S(t_o)]$$

Fig. 22.2 Unitary evolution versus dynamical map. Whereas the states ρ_{SE} of the total system $S+E$ undergo unitary time evolution governed by the propagator $U(t,t_o)$, the reduced density operator $\rho_S(t)$ of the subsystem S at any time t is obtained from ρ_{SE} by tracing out the environment. If system and environment are uncorrelated at t_o, $\rho_{SE}(t_o) = \rho_S(t_o)\otimes\rho_E(t_o)$, then the environment state $\rho_S(t)$ at any later time t can be obtained from $\rho_S(t_o)$ by a dynamical map $\Lambda(t,t_o)$ that is completely positive and trace-preserving (CPTP). In the Born–Markov approximation, one assumes $\rho_{SE}(t) = \rho_S(t)\otimes\rho_E$ holds for all times, with $\rho_E(t) \equiv \rho_E$ constant.

cal map $\Lambda(t, t_o)$ is a completely positive and trace-preserving (CPTP) map. In particular, starting from the state $|\phi_0\rangle$ we can construct an orthonormal basis $\{|\phi_n\rangle \in \mathcal{H}_E\}_n$ of the environment, and take the partial trace over E with respect to this basis, such that we get the expression

$$\rho_S(t) = \Lambda(t, t_o)\big[\rho_S(t_o)\big] = \sum_n \langle \phi_n | U(t, t_o)\big[\rho_S(t_o) \otimes |\phi_0\rangle\langle\phi_0|\big]U^\dagger(t, t_o)|\phi_n\rangle .$$

(22.14)

Defining the *Kraus operators* as

$$K_n(t, t_o) := \langle \phi_n | U(t, t_o)|\phi_0\rangle \quad \text{with} \quad \sum_n K_n^\dagger K_n = \mathbb{1}_S ,$$

(22.15)

we find the *Kraus decomposition* for the CPTP map $\Lambda(t, t_o)$,

$$\rho_S(t) = \Lambda(t, t_o)\big[\rho_S(t_o)\big] = \sum_n K_n(t, t_o)\, \rho_S(t_o)\, K_n^\dagger(t, t_o) ,$$

(22.16)

in accordance with Lemma 21.1.

If $\rho_E(t_o)$ is not a pure state, then we can simply extend the environment to include the purification of $\rho_E(t_o)$, see Section 21.1, and thus still have a CPTP map. To express this on the level of the Kraus operators we first write the initial environment state in its *spectral decomposition*,

$$\rho_E(t_o) = \sum_n p_n |\phi_n\rangle\langle\phi_n| .$$

(22.17)

Then, taking the trace over E with respect to the eigenvectors of $\rho_E(t_o)$, the Kraus operators can be defined as

$$K_{nm}(t, t_o) := \sqrt{p_m}\, \langle \phi_n | U(t, t_o)|\phi_m\rangle \quad \text{with} \quad \sum_{n,m} K_{nm}^\dagger K_{nm} = \mathbb{1}_S , \quad (22.18)$$

and the Kraus decomposition is of the form

$$\rho_S(t) = \Lambda(t, t_o)\big[\rho_S(t_o)\big] = \sum_{n,m} K_{nm}(t, t_o)\, \rho_S(t_o)\, K_{nm}^\dagger(t, t_o) .$$

(22.19)

The map $\Lambda(t, t_o)$ is a CPTP map that can be completely characterized by operators acting only on the space of the system. These are the Kraus operators introduced in Section 21.3. As such, $\Lambda(t, t_o)$ has the following properties (see Lemma 21.2 and its proof).

Properties of the dynamical map $\Lambda(t, t_o)$:

$$\textbf{Linearity:} \quad \Lambda(t, t_o)\Big[\sum_n p_n\, \rho_n\Big] = \sum_n p_n\, \Lambda(t, t_o)\big[\rho_n\big] .$$

(22.20)

(ii) **Trace-preserving:**

$$\mathrm{Tr}\big(\Lambda(t,t_o)\big[\rho_S(t_o)\big]\big) = \mathrm{Tr}\Big(\sum_n K_n(t,t_o)\,\rho_S(t_o)\,K_n^\dagger(t,t_o)\Big)$$

$$= \mathrm{Tr}\Big(\rho_S(t_o)\sum_n K_n^\dagger K_n\Big) = \mathrm{Tr}\big(\rho_S(t_o)\big) = 1\,. \quad (22.21)$$

(iii) **Completely positive:** Let ρ_{SA} be a density operating in $\mathcal{B}_{\mathrm{HS}}(\mathcal{H}_S\otimes\mathcal{H}_A)$, where $\mathcal{H}_A$ is an auxiliary Hilbert space of arbitrary dimension, then also the extended map $\Lambda(t,t_o)\otimes\mathbb{1}_A$, with $\mathbb{1}_A$ the identity map on $\mathcal{H}_A$, is a positive map,

$$\big(\Lambda(t,t_o)\otimes\mathbb{1}_A\big)[\rho_{SA}] \geq 0\,, \quad \forall\rho_{SA}\geq 0,\ \forall\mathcal{H}_A\,. \quad (22.22)$$

Why do we need complete positivity and not just positivity?

While the dynamical map allows us to describe the system's time evolution without specific reference to the environment or its specific state, let us note that the Kraus operators do implicitly depend on the initial state of E. Moreover, the dependence of $\rho_E(t_o)$ and $U(t,t_o)$ on t_o means that also the map $\Lambda(t,t_o)$ and its Kraus operators $K_{nm}(t,t_o)$ depend on the specific starting time t_o, rather than just the evolution time $\Delta t = t - t_o$. In particular, that we can even write down such Kraus operators, which is tied to the fact that the map is CPTP (see Theorem 21.1), is a result of our assumption that the initial state at $t = t_o$ is a product state. This means, without further assumptions, the dynamical map $\Lambda(t,t_o)$ allows us to map the system state $\rho_S(t_o)$ to the reduced state $\rho_S(t > t_o)$ for any fixed later time, but since the joint state $\rho_{SE}(t)$ will in general not be a product state, we cannot simply apply the map (or in fact any CPTP map) again to $\rho_{SE}(t)$ in order to obtain a later state $\rho_{SE}(t' > t)$.

In other words, complete positivity ensures that we can apply a map to a system that potentially is a subsystem of a larger joint system without the danger of obtaining unphysical negative eigenvalues as a result. When we apply the unitary time evolution to a joint system $S + E$ in a product state, we always get a CPTP map on the system (see Lemma 21.1 in combination with Theorem 21.1). However, when we start from a correlated (potentially even entangled) state $\rho_{SE}(t)$, apply the unitary operator $U(t',t)$ and trace out E, the induced map from $\rho_S(t) = \mathrm{Tr}_E\big(\rho_{SE}(t)\big)$ to $\rho_S(t') = \mathrm{Tr}_E\big(U(t',t)\rho_{SE}(t)U^\dagger(t',t)\big)$ is not completely positive. This means, although $\rho_S(t')$ will still be a positive semidefinite and normalized state, we cannot apply this map to any other state of S, e.g., at some other time, and still expect to generally obtain a well-defined density operator. Moreover, such an evolution map would not admit a Kraus decomposition.

From a purely practical point of view, it would thus be desirable for the description of open quantum systems to be able to use dynamical maps that are CPTP for all starting times. As we shall see next, this is exactly what is achieved by the Born–Markov approximation.

22.2   Markovian Dynamics and Master Equations

Up to now we have considered the dynamical map $\Lambda(t, t_o)$ for a fixed initial time t_o, which takes the form of a CPTP for any chosen $t \geq t_o$, provided that the initial state at t_o factorizes into a product state $\rho_S(t_o) \otimes \rho_E(t_o)$. But ultimately, we are interested in obtaining a dynamical map that can be applied independently of the starting time and only depends on the evolution time t. To obtain such a *one-parameter family* $\{\Lambda(t) | t \geq 0\}$ with $\Lambda(t = 0) = \mathbb{1}_S$. This family of maps describes the time evolution of open quantum system interacting with much larger, memoryless environments.

22.2.1   The Born–Markov Approximation

A key assumption that drastically simplifies the mathematical description of open quantum systems is the so-called *Born–Markov approximation*. The origin of this terminology lies in approximations used by *Max Born* (1926) for weakly interacting systems in the context of scattering theory, and the description of *memoryless* stochastic processes by *Andrey Markov*. Conceptually, the *Born approximation* corresponds to the following assumptions:

> (i) **Weak interaction** between S and E.

> (ii) The **environment** is **much larger** than the system.

Physically, these assumptions can be phrased as statements in reference to the characteristic timescales τ_S and τ_E of the system and the environment, respectively. For instance, for a system consisting of two energy levels with energies E_0 and E_1 with $E_1 > E_0$ the characteristic time is inversely proportional to the energy gap, $\tau_S \sim \frac{1}{E_1 - E_0}$. The assumption of weak interaction can then be interpreted as demanding that correlations between the system and the environment are built up only slowly compared to τ_S. In this sense, information about the system "leaks" into the environment slowly. Meanwhile, the assumption that the environment is much larger than the system then corresponds, on the one hand, to saying that the information about S is quickly "dissipated" into the environment. That is, we assume that the characteristic timescale τ_E of the environment, i.e., the timescale for the decay of the correlation functions of the environment, which are proportional to $\propto \exp(-\frac{t}{\tau_E})$, is much smaller than the characteristic time of the system τ_S, or

$$\tau_E \ll \tau_S . \tag{22.23}$$

In combination, this can be translated to the assumption that any correlations between S and E are built up much more slowly (weak interaction) than it takes for them to be "diluted" in the environment, such that we can assume that

$$\rho_{SE}(t) = \rho_S(t) \otimes \rho_E(t) \tag{22.24}$$

for all times t, not just for the initial time $t = t_o$. Moreover, if we assume the interaction between S and E to be weak and E to be "large", we can take this to mean that the interaction with S has no lasting effect on the state of E, so that we can assume that the environment state is independent of time, $\rho_E(t) = \rho_E(t_o) \equiv \rho_E$. With this, we can now see that the dynamical map becomes *composable*. That is, for every t and t' with $t' > t$ and $t > t_o$ we have

$$
\begin{aligned}
\rho_S(t') &= \Lambda(t', t)\big[\rho_S(t)\big] = \text{Tr}_E \left\{ U(t', t) \big[\rho_S(t) \otimes \rho_E\big] U^\dagger(t', t) \right\} \\
&= \text{Tr}_E \left\{ U(t', t)\, U(t, t_o)\big[\rho_S(t_o) \otimes \rho_E\big] U^\dagger(t, t_o)\, U^\dagger(t', t) \right\} \\
&= \text{Tr}_E \left\{ U(t', t_o)\big[\rho_S(t_o) \otimes \rho_E\big] U^\dagger(t', t_o) \right\} = \Lambda(t', t_o)\big[\rho_S(t_o)\big]\,,
\end{aligned}
\tag{22.25}
$$

where we have used, according to our previous assumption,

$$
\rho_S(t) \otimes \rho_E = \rho_{SE}(t) = U(t, t_o)\, \rho_{SE}(t_o)\, U^\dagger(t, t_o),
\tag{22.26}
$$

and the fact that $U(t', t)\, U(t, t_o) = U(t', t_o)$, see eqn (10.149). But since we can write $\rho_S(t) = \Lambda(t, t_o)\big[\rho_S(t_o)\big]$ in the first line, we have

$$
\Lambda(t', t)\Big[\Lambda(t, t_o)\big[\rho_S(t_o)\big]\Big] = \Lambda(t', t) \circ \Lambda(t, t_o)\big[\rho_S(t_o)\big] = \Lambda(t', t_o)\big[\rho_S(t_o)\big]\,.
\tag{22.27}
$$

Formally, we thus have the composition rule

$$
\Lambda(t_3, t_2) \circ \Lambda(t_2, t_1) = \Lambda(t_3, t_1) \qquad \text{for all } t_3 \geq t_2 \geq t_1\,.
\tag{22.28}
$$

However, we see that there still is a dependence on the initial time t_o, the dynamical map still needs two parameters to be described.

This is where the *Markov approximation* comes in, which represents the idea that it is in many cases legitimate to neglect memory effects in the dynamics of the system, i.e., the future of the system is only determined by the present state and not influenced by the past. In other words, it is the assumption that what happens to the system at time $t + \delta t$ does not depend on what happened to the system at time t, the environment has no "memory". This is very much in keeping with the idea of a "large" environment that quickly carries away information about the system. Open-system evolution that obeys this principle is called *Markovian dynamics* in statistical physics. Here, in the quantum analogy, this feature is expressed as the *semigroup property* of the dynamical maps for Markovian processes, i.e.,

$$
\Lambda(t_1) \circ \Lambda(t_2) = \Lambda(t_1 + t_2) \qquad \text{for all } t_1, t_2 \geq 0\,.
\tag{22.29}
$$

Given a quantum dynamical semigroup $\Lambda(t)$ then there exists a linear operator $\mathcal{L}$, the generator of the semigroup, such that (see, e.g., Alicki and Lendi, 1987; Alicki, 2002)

$$
\Lambda(t) = \exp(\mathcal{L} \cdot t)\,.
\tag{22.30}
$$

The application to the density matrix at some initial time $t_o = 0$, i.e.,

$$
\rho_S(t) = \Lambda(t)\big[\rho_S(0)\big] = \exp(\mathcal{L} \cdot t)\big[\rho_S(0)\big]\,,
\tag{22.31}
$$

which leads to a differential equation for the open quantum system

$$
\frac{\partial}{\partial t}\, \rho_S(t) = \mathcal{L}\big[\rho_S(t)\big]\,.
\tag{22.32}
$$

Equation (22.32) already represents the general form of a *Markovian quantum master equation*.

The technical assumptions that are made in arriving at this form, especially with regards to Markovianity, may seem somewhat *ad hoc*, and that is because they are! In some sense these approximations arise from the wish to model the problem in such a way that it can be more easily treated, but they derive their legitimacy from the fact that the predictions made using this approximation match empirical observations in many (but not all!) practical situations. Nevertheless, the assumption of Markovian dynamics is a subject of discussion and we refer to Rivas *et al.* (2010) for a critical reflection on this matter.

22.2.2 Markovian Master Equations—GKLS Equation

For a finite-dimensional Hilbert space one can construct a general form of a Markovian master equation of the type of (22.32) that explicitly separates the (unitary) Hamiltonian dynamics of the system from the dissipate dynamics. Equivalent representations of such master equations were independently formulated by Swedish physicist *Göran Lindblad* (1976) and by a team at the University of Texas at Austin consisting of *Vittorio Gorini, Andrzej Kossakowski*, and *Ennackal Chandy George Sudarshan* (1976).

Let us begin with Lindblad's expression, a form which is frequently used in the literature and which we will also further use in this book. Lindblad showed that the von Neumann equation, Theorem 11.3, can be extended to

$$\frac{\partial}{\partial t}\, \rho_S(t) \;=\; -\frac{i}{\hbar}\,[\,H_S\,,\,\rho_S(t)\,] \;-\; \mathcal{D}[\rho_S(t)]\,, \tag{22.33}$$

where H_S is the Hamilton of the system and the additional term $\mathcal{D}[\rho_S(t)]$ is called the *dissipator*. Lindblad proved that the complete positivity of the dynamical map $\Lambda(t): \rho_S(0) \to \rho_S(t)$ completely determines the general structure of $\mathcal{D}[\rho_S(t)]$

$$\mathcal{D}[\rho_S] \;=\; \frac{1}{2}\sum_k \lambda_k \left(A_k^\dagger A_k \rho_S + \rho_S A_k^\dagger A_k - 2 A_k \rho_S A_k^\dagger \right)\,, \tag{22.34a}$$

$$= \frac{1}{2}\sum_k \lambda_k \left([A_k^\dagger, A_k \rho_S] + [\rho_S A_k^\dagger, A_k] \right)\,, \tag{22.34b}$$

where the operators A_k, called *Lindblad operators*, act on the Hilbert space $\mathcal{H}_S$. The Lindblad operators have to be determined for the specific physical situation one considers and the $\lambda_k \geq 0$ are positive constants. Equation (22.33) together with expression (22.34a) or (22.34b) is usually called *Lindblad equation* or Markovian master equation in the Lindblad form. We will provide a derivation in Section 22.2.3. (For further literature, see, e.g., Gorini and Kossakowski, 1976; Gorini *et al.*, 1978; Alicki and Lendi, 1987; Holevo, 2001.)

However, the decompositions (22.34a) and (22.34b) are not unique. We can shift the operators A_k such that a part of $\mathcal{D}[\rho_S]$ is absorbed by the Hamiltonian, then we arrive at the expression of Gorini, Kossakowski, and Sudarshan (1976).

Following the treatment of Bertlmann and Grimus (2002) we can shift the Lindblad operator A_k by

$$A_k = B_k + s_k \mathbb{1} \quad \text{with} \quad \text{Tr}(B_k) = 0 , \tag{22.35}$$

then the dissipative term, (22.34a) or (22.34b), can be reformulated as (setting $\lambda_k = 1, \hbar = 1$ for convenience)

$$\mathcal{D}[\rho_S] = -i[\Delta H, \rho_S] + D'[\rho_S] \quad \text{with} \quad \Delta H = \frac{i}{2} \sum_k \left(s_k B_k^\dagger - s_k^* B_k \right) , \tag{22.36}$$

where $D'[\rho_S]$ is obtained from $\mathcal{D}[\rho_S]$ by the replacement $A_k \to B_k$. This reformulation has the effect that a part of $\mathcal{D}[\rho_S]$ is shifted into the quantum-mechanical term of the time evolution (22.33), such that a new Hamiltonian $H' = H + \Delta H$ appears. Note that for Hermitian operators A_j we have $\Delta H = 0$. In the space of traceless operators on the Hilbert space $\mathcal{H}_S$ we can choose a basis $\{F_k \,|\, k = 1, \ldots, d^2 - 1\}$ with the property

$$\text{Tr}\left(F_k^\dagger F_l\right) = \delta_{kl} , \tag{22.37}$$

and expand the operators B_k as

$$B_k = \sum_{l=1}^{d^2-1} C_{lk} F_l , \tag{22.38}$$

with $C_{lk} \in \mathbb{C}$. Then we arrive at the *Gorini–Kossakowski–Sudarshan dissipator*

$$\mathcal{D}'[\rho_S] = -\frac{1}{2} \sum_k \left([B_k, \rho_S B_k^\dagger] + [B_k \rho_S, B_k^\dagger] \right) , \tag{22.39a}$$

$$= -\frac{1}{2} \sum_{k,l=1}^{d^2-1} c_{kl} \left([F_k, \rho_S F_l^\dagger] + [F_k \rho_S, F_l^\dagger] \right) , \tag{22.39b}$$

where the matrix (c_{kl}) is positive and its components are defined by

$$c_{kl} = \sum_j C_{kj} C_{lj}^* . \tag{22.40}$$

The forms (22.39a) and (22.39b) of the dissipative term have been derived by Gorini, Kossakowski, and Sudarshan (1976). They are equivalent to the Lindblad forms (22.34a) and (22.34b). Thus the time evolution with H and dissipator $\mathcal{D}[\rho_S]$ is equivalent to the one with H' and $\mathcal{D}'[\rho_S]$. Owing to this equivalence and the, at this point, unspecified nature of the (Lindblad) operators, both forms can be treated as conceptually on an equal footing. It is therefore common (see, e.g., Chruściński and Pascazio) to refer to a Markovian master equation of either of the discussed forms as the *Gorini–Kossakowski–Lindblad–Sudarshan* or simply *GKLS equation*, with its namesakes in alphabetical order. Other equivalent forms of the dissipative term $\mathcal{D}[\rho_S]$ can also be found, see, for instance, the investigation in Bertlmann and Grimus (2002).

Remark: In the master equation (22.33) the dissipator $\mathcal{D}[\rho_S]$ describes two phenomena occurring in an open quantum system, namely decoherence and dissipation (see, e.g., Breuer and Petruccione, 2002). When the system S interacts with the environment E the initially uncorrelated product state evolves into an entangled state of S and E over time (Joos, 1996; Kübler and Zeh, 1973). This leads to mixed states of the system S—which means decoherence—and to an energy exchange between S and E—which is called dissipation.

Decoherence destroys the occurrence of long-range quantum correlations by suppressing the off-diagonal elements of the system density matrix in a given basis and leads to an information transfer from the system S to the environment E, but does not necessarily lead to a change in the diagonal elements of ρ_S. Dissipation, on the other hand, affects the diagonal elements.

In general, both effects are present, but decoherence typically acts on a (much) shorter timescale than dissipation (Joos, 1996; Kübler and Zeh, 1973; Joos and Zeh, 1985; Zurek, 1991; Alicki, 2002) and therefore often is the more important effect for the purpose of quantum-information processing. Accordingly, we call the constants λ_k in the dissipator $\mathcal{D}[\rho_S]$ in (22.34a) or (22.34b) *decoherence parameters* that denote the strength of the decoherence.

22.2.3 Derivation of the GKLS Equation

To derive the specific GKLS form of a Markovian master equation for an open quantum system we start from its Kraus decomposition (22.16). For the time evolution, for its dynamics, we make the following assumption in view of the Born–Markov condition (22.23)

$$\tau_E \ll \delta t \ll \tau_S. \tag{22.41}$$

That means we consider time evolution of the system for a time δt that is short from the point of view of the system, but long compared to the timescale of the environment. At the same time, we consider the environment to be memoryless, which we have seen leads to CPTP maps $\Lambda(t)$ that form a dynamical semigroup, see eqn (22.29), and whose Kraus operators hence only depend on the evolution time t but not the starting time t_o, which we here choose as $t_o = 0$ for convenience. With this we can approximate the density matrix of the system in the following way

$$\rho_S(\delta t) = \sum_k K_k(\delta t)\,\rho_S(0)\,K_k^\dagger(\delta t) = \rho_S(0) + \mathcal{O}(\delta t) + \mathcal{O}(\delta t^2) + \cdots, \tag{22.42}$$

where $\mathcal{O}(\delta t)$ denotes a quantity for which $\mathcal{O}(\delta t)/\delta t$ is finite as $\delta t \to 0$. We then only keep the terms to order $\mathcal{O}(\delta t)$, and neglect the higher orders, $\mathcal{O}(\delta t^2)$. Making a power-series ansatz in δt for the Kraus operators, we see that the only terms that can contribute at linear order in δt must arise from Kraus operators whose leading order in the expansion is proportional to $\sqrt{\delta t}$, or which (again, neglecting higher orders) are of the form $\mathbb{1}_S + \mathcal{O}(\delta t)$. Without loss of generality, we can thus make the ansatz

$$K_0 = \mathbb{1}_S + \left(F - \frac{i}{\hbar}H\right)\delta t\,, \tag{22.43a}$$

$$K_k = A_k\sqrt{\delta t}\,, \quad k \geq 1\,, \tag{22.43b}$$

where F, H are Hermitian operators and the A_k are in general neither Hermitian nor unitary, but bounded. In particular, we note that any linear operator can be written as the sum of a Hermitian operator H and an anti-Hermitian operator $-iF/\hbar$ with $F^\dagger = F$. From the normalization condition (22.15) we then have

$$\mathbb{1}_S = \mathbb{1}_S + \left(2F + \sum_k A_k^\dagger A_k\right)\delta t + \mathcal{O}(\delta t^2)\,, \tag{22.44}$$

and since the pre-factors of any non-zero power of δt must vanish identically, we find

$$F = -\frac{1}{2}\sum_k A_k^\dagger A_k\,. \tag{22.45}$$

With the ansatz from eqns (22.43a) and (22.43b) we calculate the time evolution of the system for the time δt

$$\rho_S(\delta t) = K_0(\delta t)\,\rho_S(0)\,K_0^\dagger(\delta t) + \sum_{k\geq 1} K_k(\delta t)\,\rho_S(0)\,K_k^\dagger(\delta t) \tag{22.46}$$

$$= \left(\mathbb{1}_S + \delta t\left(F - \frac{i}{\hbar}H\right)\right)\rho_S(0)\left(\mathbb{1}_S + \delta t\left(F + \frac{i}{\hbar}H\right)\right) + \delta t\sum_{k\geq 1} A_k\,\rho_S(0)\,A_k^\dagger\,.$$

Subtracting the initial state we further have

$$\rho_S(\delta t) - \rho_S(0) = \delta t\left(-\frac{i}{\hbar}[H,\rho_S] - \sum_{k\geq 1}\left(\frac{1}{2}\{\rho_S, A_k^\dagger A_k\} - A_k\rho_S A_k^\dagger\right)\right)\,, \tag{22.47}$$

where $[\,\dot{}\,,\,]$ and $\{\,\dot{}\,,\,\}$ denote the commutator and anti-commutator, respectively, as usual. Rescaling $A_k \longrightarrow \sqrt{\lambda_k}A_k$ with $\lambda \geq 0$ and taking the limit $\delta t \to 0$ we thus obtain

$$\lim_{\delta t\to 0}\frac{\rho_S(\delta t) - \rho_S(0)}{\delta t} = \left.\frac{\partial}{\partial t}\rho_S(t)\right|_{t=0}\,. \tag{22.48}$$

Since we assumed that our Kraus operators are independent of the starting time, we can drop the restriction to $t = 0$ and we thus arrive at the GKLS master equation

$$\frac{\partial}{\partial t}\rho_S(t) = -\frac{i}{\hbar}\,[H_S\,,\rho_S(t)] - \frac{1}{2}\sum_{k\geq 1}\lambda_k\left(A_k^\dagger A_k\rho_S + \rho_S A_k^\dagger A_k - 2A_k\rho_S A_k^\dagger\right)\,, \tag{22.49}$$

where we have identified the operator H with the system Hamiltonian H_S and A_k are the Lindblad operators.

As already mentioned, the Hamiltonian is not unique. The master equation is invariant under the transformations (again $\lambda_k = 1, \hbar = 1$ for convenience)

$$A_k \longrightarrow A_k + a_k \mathbb{1}_S \,, \tag{22.50a}$$

$$H_S \longrightarrow H_S + \frac{1}{2i} \sum_k \left(a_k^* A_k - a_k A_k^\dagger \right) + b \mathbb{1}_S \,. \tag{22.50b}$$

Finally, also note that unitary transformations U_{kj} of the Lindblad operators, $A_k \longrightarrow \sum_j U_{kj} A_j$, do not change the form of the master equation.

Résumé: An isolated system is an idealization. Practically, there is always an interaction of the system with its environment—whatever it may be—either small and negligible or large and substantial, and this interaction must generally be taken into account, it is of physical interest. The influence of the environment, known or unknown, on the quantum system can be described by treating the latter as an open quantum system. Its dynamics are determined by the dynamical map which is completely characterized by the Kraus operators—the Kraus decomposition. These operators act only on the space of the quantum system and can in principle be determined for any given physical situation.

Assuming the Born–Markov approximation holds—in particular, that the characteristic time of the environment is much shorter than that of the system, that the interaction is weak and the environment memoryless—we can derive a Markovian master equation of Lindblad-type or equivalently of Gorini–Kossakowski–Sudarshan-type for the density matrix of the system. It modifies the von Neumann equation by an additional term, the dissipator, whose structure is given in terms of the Lindblad operators which can be calculated for specific physical settings. The dissipator describes decoherence and dissipation of the open quantum system, where decoherence is typically the more important effect in the context of quantum-information processing.

22.3   Emission and Absorption of Photons

The emission and absorption of photons from or by atoms are processes that we have already described from various perspectives throughout this book, in the context of Planck's law in Section 1.1.3 or in terms of time-dependent perturbation theory in Section 10.5.3. Here, we want to briefly discuss how the dynamics of the atom can be described by a GKLS master equation, and we refer to further literature (e.g., Walls and Milburn, 1994; Breuer and Petruccione, 2002) for a more in-depth discussion. We consider the interaction of photons and atoms in the *two-level atom approximation*, which is sufficient for our purpose. This approximation works well when the frequency of the photon coincides with one of the optical transitions of the atom. That is, we consider a resonance phenomenon that occurs, for instance, when the atom is placed within a cavity and the transition of the atom coincides with one of the eigenfrequencies of the resonator.

Another example is the interaction of an atom with laser light at the atom's transition frequency. There, the light beam induces dipole oscillations in the atom, which re-radiate at the same frequency. If the photon frequency corresponds to an atom transition, we speak of an *on-resonance phenomenon*, the magnitude of the dipole oscillations is large and the photon-atom interaction is strong. If the photon frequency does not correspond to an atom transition, i.e., if it is *off-resonance*, then the magnitude of the driven oscillations is small and the photon-atom interaction is small and negligible.

22.3.1 Spontaneous Emission

Here, the system under consideration is a two-level atom, the environment is the photon field. The system Hamiltonian is given by

$$H = -\frac{\hbar\omega}{2}\sigma_z , \tag{22.51}$$

where σ_z is the usual Pauli matrix, and we drop the subscripts designating the system and environment that we have used in the previous sections since we now exclusively deal with quantities pertaining to the two-level atom. The two eigenstates H and their corresponding energies, illustrated in Fig. 22.3, are

$$|\,1\,\rangle = \begin{pmatrix} 0 \\ 1 \end{pmatrix} \quad \text{excited state:} \quad E_1 = \frac{\hbar\omega}{2} , \tag{22.52a}$$

$$|\,0\,\rangle = \begin{pmatrix} 1 \\ 0 \end{pmatrix} \quad \text{ground state:} \quad E_0 = -\frac{\hbar\omega}{2} . \tag{22.52b}$$

The transition operators that switch between the two eigenstates, $|\,1\,\rangle \longleftrightarrow |\,0\,\rangle$, are

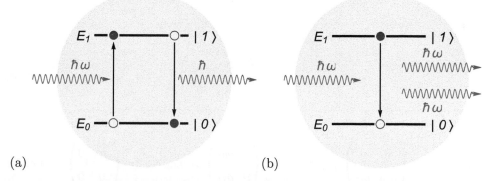

(a)   (b)

Fig. 22.3 Absorption and emission of a photon. The interaction of an atom with photons can be modelled as a two-level system (a qubit, representing the atom) with ground state $|\,0\,\rangle$ of energy E_0 and excited states $|\,1\,\rangle$ of energy E_1, with $E_1 - E_0 = \hbar\omega$ that can (a) absorb or spontaneously emit a photon of frequency $\hbar\omega$ when it is initially in the ground or excited state, respectively. (b) When the photon field is populated, the presence of photons at frequency $\hbar\omega$ in the field leads to stimulated emission at a rate proportional to the average photon number, see Section 10.5.3.

$$\sigma_+ = |0\rangle\langle 1| = \begin{pmatrix} 0 & 1 \\ 0 & 0 \end{pmatrix}, \quad \sigma_- = |1\rangle\langle 0| = \begin{pmatrix} 0 & 0 \\ 1 & 0 \end{pmatrix}, \qquad (22.53)$$

with $\sigma_\pm = \frac{1}{2}(\sigma_x \pm i\sigma_y)$ and the commutator

$$[H, \sigma_\pm] = \mp \hbar\omega\,\sigma_\pm . \qquad (22.54)$$

Note that in this convention the subscripts $+$ and $-$ do not refer to the raising and lowering of the energy of the atom; $\sigma_+|1\rangle = |0\rangle$ lowers the energy, and $\sigma_-|0\rangle = |1\rangle$ increases the energy of the two-level system, respectively, but by applying σ_+ a photon with energy $E = \hbar\omega$ is emitted and by σ_- it is absorbed. In this sense σ_+ and σ_- raise and lower the energy of the field, respectively.

As Lindblad operator (we just need one) we choose

$$A_1 = \sqrt{\Gamma}\,\sigma_+, \quad A_1^\dagger = \sqrt{\Gamma}\,\sigma_- , \qquad (22.55)$$

where Γ is the *rate for spontaneous emission*, such that the general GKLS form in eqn (22.49) becomes the *master equation for spontaneous emission*,

$$\frac{\partial}{\partial t}\rho(t) = -\frac{i}{\hbar}[H, \rho(t)] - \frac{\Gamma}{2}(\sigma_-\sigma_+\rho + \rho\sigma_-\sigma_+ - 2\sigma_+\rho\sigma_-) . \qquad (22.56)$$

To solve the equation for ρ, the density matrix of the atom, we write the latter in components with respect to eigenbasis of H as

$$\rho = \begin{pmatrix} \rho_{00} & \rho_{01} \\ \rho_{10} & \rho_{11} \end{pmatrix} . \qquad (22.57)$$

We then need the commutator

$$[\sigma_z, \rho] = 2\begin{pmatrix} 0 & \rho_{01} \\ -\rho_{10} & 0 \end{pmatrix} , \qquad (22.58)$$

and the terms

$$\sigma_-\sigma_+\rho = \begin{pmatrix} 0 & 0 \\ \rho_{10} & \rho_{11} \end{pmatrix}, \quad \rho\sigma_-\sigma_+ = \begin{pmatrix} 0 & \rho_{01} \\ 0 & \rho_{11} \end{pmatrix}, \quad \sigma_+\rho\sigma_- = \begin{pmatrix} \rho_{11} & 0 \\ 0 & 0 \end{pmatrix} \qquad (22.59)$$

of the dissipator, so that we finally obtain the master equation for the components of ρ:

$$\frac{\partial}{\partial t}\begin{pmatrix} \rho_{00} & \rho_{01} \\ \rho_{10} & \rho_{11} \end{pmatrix} = \frac{i\omega}{\hbar}\begin{pmatrix} 0 & \rho_{01} \\ -\rho_{10} & 0 \end{pmatrix} + \Gamma\begin{pmatrix} \rho_{11} & -\frac{1}{2}\rho_{01} \\ -\frac{1}{2}\rho_{10} & -\rho_{11} \end{pmatrix} . \qquad (22.60)$$

To solve eqn (22.60), we first consider the diagonal component ρ_{11},

$$\dot{\rho}_{11} = -\Gamma\,\rho_{11} \quad\Longrightarrow\quad \rho_{11}(t) = \rho_{11}(0)\,e^{-\Gamma t} \,, \tag{22.61}$$

which we can insert into the differential equation for ρ_{00} to obtain

$$\dot{\rho}_{00} = \Gamma\,\rho_{11}(0)\,e^{-\Gamma t} \quad\Longrightarrow\quad \rho_{00}(t) = \rho_{00}(0) + \rho_{11}(0)\left(1 - e^{-\Gamma t}\right)\,. \tag{22.62}$$

For the off-diagonal elements we simply have

$$\rho_{01}(t) = \rho_{01}(0)\,e^{\frac{i\omega t}{\hbar} - \frac{\Gamma t}{2}}\,, \tag{22.63a}$$

$$\rho_{10}(t) = \rho_{10}(0)\,e^{-\frac{i\omega t}{\hbar} - \frac{\Gamma t}{2}}\,. \tag{22.63b}$$

What we observe is that the population in the excited state $|1\rangle$ decays exponentially with a *mean decay time*

$$\tau_1 = \frac{1}{\Gamma}\,. \tag{22.64}$$

But the off-diagonal elements, the coherences, have a twice longer mean decay time

$$\tau_2 = \frac{2}{\Gamma}\,, \tag{22.65}$$

i.e., $\tau_2 = 2\,\tau_1$. In reference to their different roles in terms of dissipation and decoherence the times τ_1 and τ_2 are usually referred to as *relaxation time* or *amplitude-damping timescale* τ_1 and the *dephasing time* or *coherence time* τ_2, respectively. In the simple model we have considered here, the dephasing time τ_2 is twice as long as the relaxation time τ_1, but in more realistic situations and other physical systems this is often reversed and dephasing is often the main problem for quantum-information processing, e.g., in trapped-ion quantum computers (Schindler *et al.*, 2013).

22.3.2 Emission Process—Amplitude-Damping Channel

The process of emission of a photon from a two-level atom, described by the master equation (22.56), corresponds to the amplitude-damping channel that we have already discussed in Section 21.3.2. To make this more explicit, we first transform the density matrix into the interaction picture (see Section 10.5.2) by defining

$$\tilde{\rho} = e^{\frac{i}{\hbar}Ht}\rho(t)\,e^{-\frac{i}{\hbar}Ht}\,. \tag{22.66}$$

Then the master equation (22.56) takes the form

$$\frac{\partial}{\partial t}\left(e^{-\frac{i}{\hbar}Ht}\,\tilde{\rho}(t)\,e^{\frac{i}{\hbar}Ht}\right) = -\frac{i}{\hbar}\,e^{-\frac{i}{\hbar}Ht}\big(H\,\tilde{\rho} - \tilde{\rho}\,H\big)e^{\frac{i}{\hbar}Ht} \tag{22.67}$$
$$-\frac{\Gamma}{2}\left(\sigma_-\sigma_+\,e^{-\frac{i}{\hbar}Ht}\,\tilde{\rho}\,e^{\frac{i}{\hbar}Ht} + e^{-\frac{i}{\hbar}Ht}\,\tilde{\rho}\,e^{\frac{i}{\hbar}Ht}\,\sigma_-\sigma_+ - 2\sigma_+\,e^{-\frac{i}{\hbar}Ht}\tilde{\rho}\,e^{\frac{i}{\hbar}Ht}\,\sigma_-\right).$$

Differentiating the left-hand side the first term (pertaining to the unitary evolution of the system governed by the Hamiltonian H) disappears. Also transforming the operators $\sigma_\pm$ to the interaction picture, i.e., defining

$$\tilde{\sigma}_\pm = e^{\frac{i}{\hbar}Ht}\, \sigma_\pm\, e^{-\frac{i}{\hbar}Ht}\,, \tag{22.68}$$

provides the master equation for spontaneous emission in the interaction picture,

$$\frac{\partial}{\partial t}\,\tilde{\rho}(t) = -\frac{\Gamma}{2}\,(\tilde{\sigma}_-\tilde{\sigma}_+\,\tilde{\rho} + \tilde{\rho}\,\tilde{\sigma}_-\tilde{\sigma}_+ - 2\tilde{\sigma}_+\tilde{\rho}\,\tilde{\sigma}_-)\,. \tag{22.69}$$

For the Hamiltonian of eqn (22.51) of the considered two-level atom the master equation can be simplified further. In the next step we then use the familiar *Hadamard lemma* of the *Baker–Campbell–Hausdorff*, eqn (2.39) of Section 2.3.3, here written in the form

$$e^A\, B\, e^{-A} = \sum_{n=0}^\infty \frac{1}{n!}\,[A,B]_n \tag{22.70}$$

with $[A,B]_n = [A,[A,B]]_{n-1}$ and $[A,B]_0 = B$. Here, we have $A = \frac{i}{\hbar}Ht$ and $B = \sigma_\pm$, for which we calculate the commutator $[A,B]_1$ using eqn (22.54),

$$[A,B] = [\tfrac{i}{\hbar}Ht,\sigma_\pm] = \tfrac{it}{\hbar}\,[H,\sigma_\pm] = \mp i\omega t\,\sigma_\pm\,. \tag{22.71}$$

With this we can further calculate $[A,B]_n = (\mp i\omega t)^n\,\sigma_\pm$ such that we obtain $\tilde{\sigma}_\pm$ from eqn (22.68) as

$$\tilde{\sigma}_\pm = \sum_{n=0}^\infty \frac{1}{n!}\,(\mp i\omega t)^n\,\sigma_\pm = e^{\mp i\omega t}\,\sigma_\pm\,, \tag{22.72}$$

and we hence immediately find $\tilde{\sigma}_-\tilde{\sigma}_+ = \sigma_-\sigma_+$. Finally, we arrive at the following master equation for the spontaneous emission of a photon from a two-level atom

$$\frac{\partial}{\partial t}\,\tilde{\rho}(t) = -\frac{\Gamma}{2}\,(\sigma_-\sigma_+\,\tilde{\rho} + \tilde{\rho}\,\sigma_-\sigma_+ - 2\sigma_+\tilde{\rho}\,\sigma_-)\,. \tag{22.73}$$

Solving the differential equation (22.73) for the components of the density matrix with respect to the eigenbasis of H (recall eqns (22.61), (22.62), (22.63a), and (22.63b)) we have

$$\tilde{\rho}_{00}(t) = \tilde{\rho}_{00}(0) + \tilde{\rho}_{11}(0)\,(1 - e^{-\Gamma t})\,, \tag{22.74a}$$

$$\tilde{\rho}_{11}(t) = \tilde{\rho}_{11}(0)\,e^{-\Gamma t}\,, \tag{22.74b}$$

$$\tilde{\rho}_{01}(t) = \tilde{\rho}_{01}(0)\,e^{-\frac{\Gamma}{2}t}\,, \tag{22.74c}$$

$$\tilde{\rho}_{10}(t) = \tilde{\rho}_{10}(0)\,e^{-\frac{\Gamma}{2}t}\,. \tag{22.74d}$$

Choosing as initial condition

$$\tilde{\rho}(0) = |1\rangle\langle 1| = \rho_1 = \begin{pmatrix} 0 & 0 \\ 0 & 1 \end{pmatrix}\,, \tag{22.75}$$

i.e., an initially excited state with $\tilde{\rho}_{00}(0) = \tilde{\rho}_{01}(0) = \tilde{\rho}_{10}(0) = 0$, $\tilde{\rho}_{11}(0) = 1$, the solutions reduce to

$$\tilde{\rho}_{00}(t) = 1 - e^{-\Gamma t} , \tag{22.76a}$$

$$\tilde{\rho}_{11}(t) = e^{-\Gamma t} . \tag{22.76b}$$

Similarly writing the ground-state density operator with respect to the energy eigenbasis,

$$\rho_0 = |0\rangle\langle 0| = \begin{pmatrix} 1 & 0 \\ 0 & 0 \end{pmatrix} , \tag{22.77}$$

we can express the time evolution of the density matrix $\tilde{\rho}(t)$ as

$$\tilde{\rho}(t) = \left(1 - e^{-\Gamma t}\right) \rho_0 + e^{-\Gamma t} \rho_1 = \begin{pmatrix} 1 - e^{-\Gamma t} & 0 \\ 0 & e^{-\Gamma t} \end{pmatrix} . \tag{22.78}$$

In eqn (22.78) we rediscover the Kraus decomposition for the amplitude-damping channel of Section 21.3.2. That means we can rewrite eqn (22.78) in the form

$$\tilde{\rho}(t) = K_0 \rho_1 K_0^\dagger + K_1 \rho_1 K_1^\dagger = \sum_{n=0}^{1} K_n \tilde{\rho}(0) K_n^\dagger = \Lambda(t) \tilde{\rho}(0) , \tag{22.79}$$

with the Kraus operators

$$K_0 = \begin{pmatrix} 1 & 0 \\ 0 & \sqrt{1-p} \end{pmatrix} \quad \text{and} \quad K_1 = \begin{pmatrix} 0 & \sqrt{p} \\ 0 & 0 \end{pmatrix} \quad \text{with} \quad p = 1 - e^{-\Gamma t} , \tag{22.80}$$

where we note that $K_0^\dagger K_0 + K_1^\dagger K_1 = \mathbb{1}$, as it should be. The expression in eqn (22.79) is precisely the Kraus decomposition for the amplitude-damping channel and the Kraus operators (22.80) coincide with the operators of eqns (21.20a) and (21.20b) when we make the identification $p = \sin^2(\theta)$, i.e.

$$K_0 = \begin{pmatrix} 1 & 0 \\ 0 & \cos(\theta) \end{pmatrix} = |0\rangle\langle 0| + \cos(\theta) |1\rangle\langle 1| , \tag{22.81a}$$

$$K_1 = \begin{pmatrix} 0 & \sin(\theta) \\ 0 & 0 \end{pmatrix} = \sin(\theta) |0\rangle\langle 1| . \tag{22.81b}$$

22.3.3   Emission-Absorption Process

The model for describing the effects of spontaneous emission on the two-level system operates under the (so far implicit) assumption that the photon field is in its ground state; it is not populated initially. Now, we want to consider the case where the photon field is populated such that the two-level atom can emit *and* absorb photons. More specifically, we will assume that the photon field is in thermal equilibrium at temperature T. In this sense, the environment of the atom represents a heat reservoir. We have indeed already discussed such a situation before in the derivation of Planck's law in Section 1.1.3. Now we apply the master equation technique developed so far to this emission-absorption process. Recall that in this case the dynamics are split into the absorption process on the one hand, and on the other hand the spontaneous emission along with stimulated emission caused by the photonic background, see Fig. 22.3. The stimulated emission and absorption of photons is proportional to the average photon number (see, e.g., the discussion in Section 10.5.3) given by eqn (11.50) in Section 1.1.3,

$$\overline{n} = \frac{1}{e^{\hbar\omega/k_{\mathrm{B}}T} - 1} \, , \tag{22.82}$$

whereas the spontaneous emission does not depend on $\overline{n}$.

For this reason we can use our result (22.73) for the spontaneous emission and include another identical term with a pre-factor $\overline{n}$ to describe stimulated emission, while the inclusion of the adjoint of this term represents absorption. In this way we find the *master equation for spontaneous and stimulated emission and absorption* of photons of a two-level atom interacting (resonantly) with a single mode of the electromagnetic field (photons at the transition frequency of the atom) in a thermal state,

$$\frac{\partial}{\partial t} \, \tilde{\rho}(t) = - \frac{\Gamma}{2} \, (1 + \overline{n}) \, (\sigma_- \sigma_+ \, \tilde{\rho} + \tilde{\rho} \, \sigma_- \sigma_+ - 2\sigma_+ \tilde{\rho} \, \sigma_-)$$
$$- \frac{\Gamma}{2} \, \overline{n} \, (\sigma_+ \sigma_- \, \tilde{\rho} + \tilde{\rho} \, \sigma_+ \sigma_- - 2\sigma_- \tilde{\rho} \, \sigma_+) \, , \tag{22.83}$$

where $\tilde{\rho}$ is the system density operator in the interaction picture as defined in eqn (22.66). For $\overline{n} = 0$ there is only spontaneous emission in which case we return to the model discussed in Section 22.3.1, whereas for $\overline{n} \neq 0$ stimulated emission and absorption of photons occurs in addition.

To solve eqn (22.83), we proceed as before, and write the interaction-picture density operator in components with respect to the eigenbasis of the system Hamiltonian H, in which case the master equation becomes

$$\frac{\partial}{\partial t} \begin{pmatrix} \tilde{\rho}_{00} & \tilde{\rho}_{01} \\ \tilde{\rho}_{10} & \tilde{\rho}_{11} \end{pmatrix} = \Gamma \begin{pmatrix} (1 + \overline{n}) \, \tilde{\rho}_{11} - \overline{n} \, \tilde{\rho}_{00} & -\frac{1}{2}(2\overline{n} + 1) \, \tilde{\rho}_{01} \\ -\frac{1}{2}(2\overline{n} + 1) \, \tilde{\rho}_{10} & -(1 + \overline{n}) \, \tilde{\rho}_{11} + \overline{n} \, \tilde{\rho}_{00} \end{pmatrix} . \tag{22.84}$$

We can again solve this differential equation separately for the different components. For the off-diagonal elements it is easy to see that the solutions are just

$$\tilde{\rho}_{01}(t) = \tilde{\rho}_{01}(0)\, e^{-\frac{\Gamma}{2}(2\bar{n}+1)\,t}, \tag{22.85a}$$

$$\tilde{\rho}_{10}(t) = \tilde{\rho}_{10}(0)\, e^{-\frac{\Gamma}{2}(2\bar{n}+1)\,t}. \tag{22.85b}$$

As expected, they decay exponentially with a decay time, the dephasing time, proportional to $2/\Gamma$, and we note that the dephasing also happens faster the larger the mean photon number $\bar{n}$ is. For the diagonal elements, we start with $\tilde{\rho}_{00}$, and use the normalization of the density operator to write $\tilde{\rho}_{11} = 1 - \tilde{\rho}_{00}$, such that the relevant differential equation becomes

$$\frac{\partial}{\partial t}\tilde{\rho}_{00}(t) = \Gamma\,(1+\bar{n})\,(1-\tilde{\rho}_{00}) - \Gamma\,\bar{n}\,\tilde{\rho}_{00} = -(2\,\bar{n}+1)\,\tilde{\rho}_{00} + \Gamma\,(\bar{n}+1). \tag{22.86}$$

As can be easily seen, this inhomogeneous differential equation is solved by

$$\tilde{\rho}_{00}(t) = \left(\tilde{\rho}_{00}(0) - \frac{\bar{n}+1}{2\bar{n}+1}\right) e^{-\Gamma\,(2\bar{n}+1)\,t} + \frac{\bar{n}+1}{2\bar{n}+1}. \tag{22.87}$$

Inspecting the factor $(\bar{n}+1)/(2\bar{n}+1)$ more closely, we see that we can rewrite it using eqn (22.82) to find

$$\frac{\bar{n}+1}{2\bar{n}+1} = \frac{1+1/\bar{n}}{2+1/\bar{n}} = \frac{1+e^{\hbar\omega/k_{\mathrm{B}}T}-1}{2+e^{\hbar\omega/k_{\mathrm{B}}T}-1} = \frac{1}{1+e^{-\hbar\omega/k_{\mathrm{B}}T}}, \tag{22.88}$$

which is exactly the ground-state population of a qubit with energy gap $\hbar\omega$ in a thermal state (see eqn (11.28) in Section 11.4) at temperature T, i.e.,

$$\tau_{00} = \langle 0\,|\,\tau(\beta)\,|\,0\rangle = \langle 0\,|\,\frac{e^{-\beta H}}{\mathrm{Tr}(e^{-\beta H})}\,|\,0\rangle = \frac{1}{1+e^{-\hbar\omega/k_{\mathrm{B}}T}}, \tag{22.89}$$

for $\beta = 1/(k_{\mathrm{B}}T)$. And from the normalization condition $\tilde{\rho}_{11}(t) = 1 - \tilde{\rho}_{00}(t)$ we immediately get the time evolution of $\tilde{\rho}_{11}$. We thus see that, irrespective of the initial state, the limit of large times, $t \to \infty$ always results in the complete decay of the off-diagonal elements, while the diagonal of the density operator approaches that of a thermal state at the temperature T of the environment (the photon field in this case),

$$\lim_{t\to\infty} \tilde{\rho}(t) = \frac{1}{1+e^{-\hbar\omega/k_{\mathrm{B}}T}} \begin{pmatrix} 1 & 0 \\ 0 & e^{-\hbar\omega/k_{\mathrm{B}}T} \end{pmatrix} = \tau(\beta). \tag{22.90}$$

The system thermalizes with the heat bath that it interacts with.

Indeed, this is what we expect to happen thermodynamically. Yet, it is not so clear that this is a generic feature of open-system dynamics. Indeed, for higher-dimensional systems with more complicated Hamiltonians there a no guarantees that a simple ansatz for the Lindblad operators will lead to thermalization in the long-time limit. Indeed, the particular circumstances under which a system in contact with its environment reaches a thermal, or even a more general equilibrium state are hard to specify, if not to say generically unknown (see Gogolin *et al.*, 2011 or the extensive review on this topic in Gogolin and Eisert, 2016).

An interesting approach to the question of equilibration is that of so-called closed-system equilibration, where one considers the unitary time evolution of a large system under some assumptions on the degeneracies of the overall Hamiltonian, but without having to resort to the Born–Markov approximation. If one then focuses on a small subsystem, one can describe equilibration *on average*, that is, how close the state of the subsystem is to its long-time average and how likely large deviations from this time average are. In particular, it was shown by *Noah Linden, Sandu Popescu, Anthony Short*, and *Andreas Winter* (2009) that the deviation from the time average are small on average for subsystems that are (in a sense that can be made technically precise) small compared to the overall system. Although we could go on here and elaborate on this both mathematically and physically very satisfying approach, no summary or review we could present here would be able to compete with the already exceptionally clear and enjoyable presentation in the original paper (Linden *et al.*, 2009), which we therefore strongly encourage the interested reader of this book to read for themselves.

22.4   The Jaynes–Cummings Model

In the previous sections of this chapter we have described the emission and absorption of photons from a two-level atom, where the photons were not explicitly modelled, but were assumed to be thermally distributed, part of a (perhaps even classical) heat reservoir. Now we want to study the interaction of a two-level atom with the electromagnetic field more generally, where the electromagnetic field is also quantized, but we will limit our discussion to a single mode here and refer to a more general discussion of the quantization of the electromagnetic field to Section 22.4.2. This simplification was proposed by *Edwin Thompson Jaynes* and *Frederick W. Cummings* (1963) and became a very successful and popular model—the *Jaynes–Cummings model*—in quantum optics. It describes, for example, the experimental situation of Rydberg atoms interacting with a field in a cavity. A *Rydberg atom* is an excited atom with an electron in a state corresponding to a very high principal quantum number (e.g., in rubidium), thus becoming very sensitive to microwaves and well adapted to study atom-field interactions. (For reviews and textbooks in this field, see, e.g., Knight and Milonni, 1980; Shore and Knight, 1993; Scully and Zubairy, 1997; Cohen-Tannoudji *et al.*, 2004; Schleich, 2004; Haroche and Raimond, 2013.)

22.4.1   Two-Level Atom

The Jaynes–Cummings model is based on the following assumptions.

(i) **Atom**: The atom is represented by two energy levels, all others are neglected.

(ii) **Electromagnetic field**: The electromagnetic field is represented by a single quantized mode.

Then the Hamiltonian H_{atom} of the atom has two levels, the ground state $|\, g \,\rangle$ with energy $E_g = 0$ scaled to zero, and the excited state $|\, e \,\rangle$ with energy $E_e = \hbar\omega$,

$$H_{\text{atom}} \, | \, g \, \rangle = 0 \, , \tag{22.91a}$$

$$H_{\text{atom}} \, | \, e \, \rangle = \hbar \omega \, , \tag{22.91b}$$

which is exactly the situation we have assumed when discussing the process of spontaneous emission, see Fig. 22.3, except that the energy levels have been shifted by $+\frac{\hbar\omega}{2}$ so that the ground-state energy is set to 0. In analogy[2] to the transition operators $\sigma_\pm$ from eqn (22.53) we now define the *excitation operator* S^+ and the *relaxation operator* S^-,

$$S^+ = |\, e \, \rangle\langle \, g \,| \qquad S^- = |\, g \, \rangle\langle \, e \,| \, , \tag{22.92a}$$

$$S^+ \, | \, g \, \rangle = |\, e \, \rangle \qquad S^- \, | \, e \, \rangle = |\, g \, \rangle \, , \tag{22.92b}$$

$$S^+ \, | \, e \, \rangle = 0 \qquad S^- \, | \, g \, \rangle = 0 \, , \tag{22.92c}$$

such that we can express the Hamiltonian of the two-level atom as

$$H_{\text{atom}} = \hbar \omega \, |\, e \, \rangle\langle \, e \,| = \hbar \omega \, S^+ S^- \, . \tag{22.93}$$

22.4.2 Quantization of Electric and Magnetic Field

As we will discuss in more detail in Section 22.4.2, the electromagnetic field is quantized, that is, each mode function—the linearly independent solutions of the respective equation of motion, here Maxwell's equations—is treated as a quantum-mechanical harmonic oscillator as discussed in Chapter 5. For a free field there can nevertheless be a continuum of such solutions. Here, we therefore consider the electric and magnetic field confined within a cylindrical resonator cavity, see Fig. 22.4. Much like in our discussion of the time-independent Schrödinger equation in Chapter 4, this leads to quantized mode functions with well-defined (and well-separated) frequencies.

For quantizing the electric and magnetic field we consider two mirrors with surface areas A that are placed at a distance L, which defines a volume of $V = A \cdot L$. The mode functions in the z direction of such a cavity are given by

$$Q_j(t) = q_j(t) \sin(k_j z) \quad \text{with} \quad k_j = j\frac{\pi}{L} \, , \quad j \in \mathbb{N} \, , \tag{22.94}$$

where $q_j(t)$ are time dependent amplitudes. We can expand the linearly polarized electric field in x direction into such mode functions of the cavity,

$$E_x(z,t) = \sum_j A_j \, Q_j(t) = \sum_j A_j \, q_j(t) \sin(k_j z) \, , \tag{22.95}$$

where we choose the constants

$$A_j = \left(\frac{2m_j \omega_j^2}{V} \right)^{\frac{1}{2}} \tag{22.96}$$

where the m_j are constants that we have named such that, later on, we find a close analogy with the harmonic oscillator of mass m_j. The frequencies $\omega_j = c\, k_j$ denote

---

[2]We have chosen the intuitive notation $S_\pm$ here instead of $\sigma_\pm$ to indicate that S_+ raises the energy of the atom, while S_- lowers it.

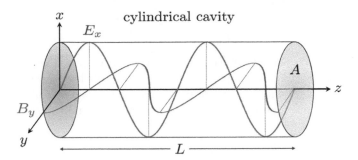

Fig. 22.4 Cylindrical resonator cavity. A cylindrical resonator cavity is sketched. Two mirrors with surface areas A are placed at a distance L from each other, confining the electromagnetic field within, which leads to quantized mode functions for the electromagnetic field inside, as illustrated for the x component E_x of the electromagnetic field, and the y component B_y of the magnetic field.

the discrete resonator frequencies of the cavity.

According to Maxwell's equations, the induced magnetic field along the y direction (we set the permittivity and permeability $\varepsilon_0 = \mu_0 = 1$ here for convenience) is then expressed as

$$B_y(z,t) = \sum_j \frac{A_j}{k_j}\, q_j(t)\, \cos(k_j z) \ . \tag{22.97}$$

Now we calculate the electric and magnetic energy in the cavity

$$E_{\text{electric}} = \frac{1}{2} \int_V \vec{E}^{\,2} dV \ , \quad E_{\text{magnetic}} = \frac{1}{2} \int_V \vec{B}^{\,2} dV \ , \tag{22.98}$$

and after integration over the volume V we find the total energy

$$E_{\text{total}} = E_{\text{electric}} + E_{\text{magnetic}} = \tfrac{1}{2} \sum_j \left(m_j \omega_j^2 q_j^2 + m_j \dot{q}_j^2 \right) \ . \tag{22.99}$$

We see that the total energy (22.99) of both fields inside the cavity is the sum of the energies of independent harmonic oscillators labelled by j with masses m_j and frequencies ω_j.

We already know from Chapter 5 how to quantize a harmonic oscillator. Thus the Hamiltonian of the electromagnetic field is just the sum of the Hamiltonians corresponding to the individual quantized oscillators

$$H_{\text{field}} = \hbar \sum_j \omega_j \left(a_j^\dagger a_j + \tfrac{1}{2} \right) \ , \tag{22.100}$$

where the respective creation-, annihilation-, and occupation-number operators are given by

$$a_j^\dagger = \frac{1}{\sqrt{2m_j\omega_j\hbar}}(m_j\omega_j q_j + ip_j)\,, \tag{22.101a}$$

$$a_j = \frac{1}{\sqrt{2m_j\omega_j\hbar}}(m_j\omega_j q_j - ip_j)\,, \tag{22.101b}$$

$$N_j = a_j^\dagger a_j\,. \tag{22.101c}$$

These operators act on the occupation-number states $|n\rangle$ as (suppressing the index j, see Section 25.1.3 for a more detailed treatment)

$$a^\dagger|n\rangle = \sqrt{n+1}\,|n+1\rangle\,, \tag{22.102a}$$

$$a|n\rangle = \sqrt{n}\,|n-1\rangle\,, \tag{22.102b}$$

$$N|n\rangle = n|n\rangle \qquad n = 0,1,2,\dots\,. \tag{22.102c}$$

Inserting eqns (22.101a), (22.101b) into (22.95) and (22.97) we finally get the electric and the magnetic field in their quantized form

$$E_x(z,t) = \sum_j \sqrt{\frac{\hbar\omega_j}{V}}\left(a_j^\dagger(t) + a_j(t)\right)\sin(k_j z)\,, \tag{22.103a}$$

$$B_y(z,t) = \sum_j \sqrt{\frac{\hbar\omega_j}{V}}\,i\!\left(a_j^\dagger(t) - a_j(t)\right)\cos(k_j z)\,. \tag{22.103b}$$

22.4.3 Uncoupled Atom and Field

To begin our analysis, let us consider the atom and the field as uncoupled. We also make the following approximation. We neglect the field modes with frequencies strongly different from the atom transition ω, following Fermi's golden rule (see Section 10.5.3) their contributions are minor. Thus only one mode, the *resonance mode* ω_R, is involved in the coupling and the field Hamiltonian reduces to

$$H_{\text{field}} = \hbar\omega_R\left(a_R^\dagger a_R + \tfrac{1}{2}\right)\,, \tag{22.104}$$

so that the *total Hamiltonian of the atom and field* becomes

$$H = H_{\text{atom}} + H_{\text{field}} = \hbar\omega\,S^+S^- + \hbar\omega_R\left(a_R^\dagger a_R + \tfrac{1}{2}\right)\,. \tag{22.105}$$

The state space is spanned by the orthonormal basis $\{|g\rangle|n\rangle, |e\rangle|n\rangle\}_n$, i.e., the energy eigenstates satisfying

$$H|g\rangle|n\rangle = E_{g,n}|g\rangle|n\rangle \quad \text{with}\quad E_{g,n} = \hbar\omega_R\left(n + \tfrac{1}{2}\right)\,, \tag{22.106a}$$

$$H|e\rangle|n\rangle = E_{e,n}|e\rangle|n\rangle \quad \text{with}\quad E_{e,n} = \hbar\omega + \hbar\omega_R\left(n + \tfrac{1}{2}\right)\,. \tag{22.106b}$$

Now we take into account that the driving field and the atom are generally slightly *detuned*

$$\delta := \omega_R - \omega \neq 0, \quad \text{with}\quad \delta \ll \omega_R,\ \omega_R > \omega \text{ assumed}\,, \tag{22.107}$$

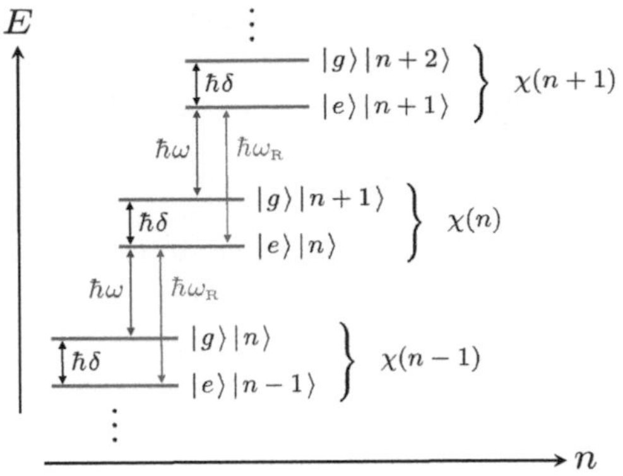

Fig. 22.5 Energy levels of uncoupled atom-field system. The states $|g\rangle|n+1\rangle$ and $|e\rangle|n\rangle$ are grouped into families $\chi(n)$ and split by the detuning energy $\hbar\delta$, which provides a small spacing, whereas the states $|e\rangle|n\rangle$ and $|g\rangle|n\rangle$ are separated by a larger energy gap $\hbar\omega$.

then the degeneracy of the states is slightly lifted. The energy levels $E_{g,n}$ and $E_{e,n}$ are separated by

$$E_{e,n} - E_{g,n} = \hbar\omega, \qquad (22.108)$$

which gives a large spacing, however, the states $|g\rangle|n\rangle$ and $|e\rangle|n-1\rangle$ are separated by

$$E_{g,n} - E_{e,n-1} = \hbar\delta, \qquad (22.109)$$

which is much smaller than the energy gap between $|e\rangle|n\rangle$ and $|e\rangle|n-1\rangle$

$$E_{e,n} - E_{e,n-1} = \hbar\omega_{\mathrm{R}}. \qquad (22.110)$$

Altogether we thus get the family of pairs of states $\chi(n) = \{|g\rangle|n+1\rangle, |e\rangle|n\rangle\}$, a ladder of doublets, see Fig. 22.5. Thus the nth excited doublet $\chi(n)$ spans the state space with $(n+1)$ elementary excitations, either in the form of $(n+1)$ field quanta (photons) and an atomic ground state $|g\rangle|n+1\rangle$, or as n field quanta and the atomic excited state $|e\rangle|n\rangle$.

22.4.4 Interacting Atom and Field

Now we extend our analysis to include the interaction of the atom with the quantized field. Ignoring the spatial variation of the electromagnetic field on the characteristic length of the atom, we can apply the dipole approximation, such that the (classical) expression for the energy is given by the interaction of the electric dipole $\vec{d}$ with the electric radiation field $\vec{E}$ (compare, e.g., with Section 10.4)

$$E_{\mathrm{class}} = -\vec{E} \cdot \vec{d}, \qquad (22.111)$$

where $\vec{d} = q\vec{x}$ denotes the classical dipole moment and $q = -q_{\mathrm{e}}$ is the electron charge (with the elementary charge $q_{\mathrm{e}} \approx 1.602 \times 10^{-19}$ C).

To obtain the corresponding quantity in the quantized picture, we replace the position vector $\vec{x}$ with the corresponding operator $\vec{X}$. We now need to determine the form of this operator in the subspace spanned by the two considered energy levels $|g\rangle$ and $|e\rangle$. Although we have not specified the specific type of atom or which pair of energy levels is chosen, we may assume that for any such atom we are considering bound states of a symmetric potential, in which case Lemma 4.3 suggests that we can choose the bound states, including the wave functions $\psi_g(\vec{x})$ and $\psi_e(\vec{x})$ to be either even or odd functions. In either case, $|\psi_g(\vec{x})|^2$ and $|\psi_e(\vec{x})|^2$ are even functions, and we hence have

$$\langle g\,|\,\vec{X}\,|\,g\rangle = \langle e\,|\,\vec{X}\,|\,e\rangle = 0\,. \tag{22.112}$$

Thus, the electric dipole moment $\vec{\mu}_e$ of any state in the subspace spanned by $|g\rangle$ and $|e\rangle$ must arise from the off-diagonal elements of $\vec{X}$ in this basis,

$$\langle g\,|\,\vec{X}\,|\,e\rangle = \langle e\,|\,\vec{X}\,|\,g\rangle = \vec{\mu}_e\,. \tag{22.113}$$

Since the position operator is Hermitian, the electric dipole moment $\vec{\mu}_e$, has real components. This means that superpositions of the two eigenstates of H_{atom} exhibit a permanent dipole moment, whereas the atom eigenstates do not.

With this we can rewrite the dipole operator in terms of the excitation and relaxation operator (22.92a)

$$\vec{d} = \vec{\mu}_e\left(S^+ + S^-\right), \tag{22.114}$$

and insert eqn (22.114) and the electric field operator (22.103a) into the interaction energy (22.111) to obtain the quantized version, the *interaction Hamiltonian* of the atom and the field

$$
\begin{aligned}
H_{\text{int}} &= -\mu_e\left[S^+(t) + S^-(t)\right]\sqrt{\frac{\hbar\omega_{\text{R}}}{V}}\left[a_{\text{R}}^\dagger(t) + a_{\text{R}}(t)\right]\sin(k_{\text{R}}z)\\
&= c\left(S^+ a^\dagger + S^+ a + S^- a^\dagger + S^- a\right),
\end{aligned}
\tag{22.115}
$$

where we have introduced the coupling constant

$$c := -\mu_e\sqrt{\frac{\hbar\omega_{\text{R}}}{V}}\,\sin(k_{\text{R}}z) \tag{22.116}$$

with $\mu_e = |\vec{\mu}_e|$ and we have dropped the subscripts R for the operators a and $a^\dagger$, which we will do from now on to make the notation less cumbersome.

The operators in H_{int} are time dependent, and explicitly given by (recall Section 2.6.2)

$$a^\dagger(t) = a^\dagger(0)\,e^{i\omega_{\text{R}}t}\,, \qquad S^+(t) = S^+(0)\,e^{i\omega t}\,, \tag{22.117a}$$

$$a(t) = a(0)\,e^{i\omega_{\text{R}}t}\,, \qquad S^-(t) = S^-(0)\,e^{i\omega t}\,. \tag{22.117b}$$

Therefore the operators in the interaction Hamiltonian (22.115) have the following time dependence

$$S^+ a^\dagger \sim e^{i(\omega + \omega_R)t} \quad \text{non-resonant}, \tag{22.118a}$$

$$S^+ a \sim e^{i(\omega - \omega_R)t} \quad \text{absorption}, \tag{22.118b}$$

$$S^- a^\dagger \sim e^{-i(\omega - \omega_R)t} \quad \text{stimulated emission}, \tag{22.118c}$$

$$S^- a \sim e^{-i(\omega + \omega_R)t} \quad \text{non-resonant}. \tag{22.118d}$$

The terms $S^+ a^\dagger$ and $S^- a$ represent non-energy preserving processes. The term $S^+ a^\dagger$ corresponds to a simultaneous excitation of the atom and the emission of a photon, whereas the term $S^- a$ describes the relaxation of the atom and the simultaneous absorption of a photon. We have assumed the system to be close to resonance, eqn (22.107), and we can therefore neglect the terms $S^+ a^\dagger$ and $S^- a$ since they oscillate much faster (with the sum of the frequencies ω and ω_R) than the other two terms, and their contributions thus average out over time. This approximation is called the *rotating-wave approximation*. In the interaction picture one may think of the interaction Hamiltonian to be transformed to a co-rotating frame, and only the approximately co-rotating terms are kept, while the counter-rotating terms are neglected.

Therefore we are left with the very simple *interaction Hamiltonian*

$$H_{\text{int}} = c\left(S^+ a + S^- a^\dagger\right). \tag{22.119}$$

The only non-vanishing matrix elements in the state space spanned by $|g\rangle |n\rangle$ and $|e\rangle |n\rangle$ are (for clarity we show once again the tensor-product structure of the components explicitly)

$$\langle e| \otimes \langle n-1| H_{\text{int}} |g\rangle \otimes |n\rangle = = c\,\langle e| \otimes \langle n-1| S^+ \otimes a + S^- \otimes a^\dagger |g\rangle \otimes |n\rangle$$

$$= c\,\langle e|S^+|g\rangle\langle n-1|a|n\rangle = c\sqrt{n}, \tag{22.120}$$

and its Hermitian conjugate. Thus, the interaction Hamiltonian only leads to transitions between states within the same family $\chi(n)$. Keeping this in mind and collecting all terms we thus arrive at the *Jaynes–Cummings Hamiltonian*

$$H_{\text{JC}} = H_{\text{atom}} + H_{\text{field}} + H_{\text{int}} = \hbar\omega\, S^+ S^- + \hbar\omega_R\left(a_R^\dagger a_R + \frac{1}{2}\right) + c\left(S^+ a + S^- a^\dagger\right). \tag{22.121}$$

22.4.5 Eigenvalues and Eigenstates of the Jaynes–Cummings Hamiltonian

Next we want to study the eigenvalues and eigenstates of the Jaynes–Cummings Hamiltonian H_{JC} (22.121). The Hamiltonian of the atom contains just diagonal elements

$$\langle f| \langle n| H_{\text{atom}} |f\rangle |n\rangle = \begin{cases} \hbar\omega & \text{for} \quad f = e \\ 0 & \text{for} \quad f = g. \end{cases}$$

As we can see, every second diagonal element vanishes. Also the field Hamiltonian contains only diagonal elements

$$\langle f | \langle n | H_{\text{field}} | f \rangle | n \rangle = \hbar \omega_{\text{R}} \left(n + \tfrac{1}{2} \right). \tag{22.122}$$

But the interaction Hamiltonian contains (only) off-diagonal elements

$$\langle e | \langle n - 1 | H_{\text{int}} | g \rangle | n \rangle = \langle g | \langle n | H_{\text{int}} | e \rangle | n - 1 \rangle = c \sqrt{n}. \tag{22.123}$$

Considering all matrix elements together we obtain a block-diagonal matrix, where the n block matrices are formed by the Hamiltonians of the states in the nth family $\chi(n)$,

$$H_{\text{JC}} = \begin{pmatrix} H_{\text{JC}}(1) & 0 & 0 & 0 & \cdots \\ 0 & H_{\text{JC}}(2) & 0 & 0 & \cdots \\ 0 & 0 & \ddots & \vdots & \\ 0 & 0 & \cdots & H_{\text{JC}}(n) & \cdots \\ \vdots & \vdots & \vdots & & \ddots \end{pmatrix}. \tag{22.124}$$

The individual blocks are just 2×2 matrices

$$\begin{aligned} H_{\text{JC}}(n) &= \begin{pmatrix} \langle e | \langle n - 1 | H_{\text{JC}} | e \rangle | n - 1 \rangle & \langle e | \langle n - 1 | H_{\text{JC}} | g \rangle | n \rangle \\ \langle g | \langle n | H_{\text{JC}} | e \rangle | n - 1 \rangle & \langle g | \langle n | H_{\text{JC}} | g \rangle | n \rangle \end{pmatrix} \\ &= \begin{pmatrix} E_{e,n-1} & c\sqrt{n} \\ c\sqrt{n} & E_{g,n} \end{pmatrix} \end{aligned} \tag{22.125}$$

with the ground-state energy $E_{g,n}$ and excited-state energy $E_{e,n-1}$ of the atom from eqns (22.106a) and (22.106b), respectively.

Energy eigenvalues: The matrix (22.125) can be easily diagonalized, from which one obtains the two eigenvalues

$$E_{1,n} = \tfrac{1}{2} \left(E_{g,n} + E_{e,n-1} + \hbar \Omega_n \right), \tag{22.126a}$$

$$E_{2,n} = \tfrac{1}{2} \left(E_{g,n} + E_{e,n-1} - \hbar \Omega_n \right). \tag{22.126b}$$

Here, we have introduced a new quantity, the so-called *splitting quantity*

$$\Omega_n = \sqrt{\delta^2 + \omega_n^2} \quad \text{with} \quad \omega_n = \frac{2}{\hbar} c \sqrt{n}. \tag{22.127}$$

So the splitting is determined by the detuning parameter δ (22.107) and a frequency ω_n, which is the nth *Rabi frequency*. What we find is that *Rabi oscillations* occur in the system we consider: a two-level atom in the presence of an oscillatory driving field. The atom cyclically absorbs a quantum of the field and re-emits it by stimulated emission. The cycles are called *Rabi cycles* and the frequency is given by the inverse of its duration.

Eigenstates—dressed states: Accordingly, we obtain a new set of eigenstates for the individual nth Jaynes–Cummings Hamiltonian (22.125)

$$| 1, n \rangle = \sin(\theta_n) | g \rangle | n \rangle + \cos(\theta_n) | e \rangle | n - 1 \rangle , \qquad (22.128a)$$

$$| 2, n \rangle = \cos(\theta_n) | g \rangle | n \rangle - \sin(\theta_n) | e \rangle | n - 1 \rangle , \qquad (22.128b)$$

with the angles θ_n defined by

$$\tan(2\theta_n) := -\frac{\Omega_n}{\delta} . \qquad (22.129)$$

These states form orthonormal bases $\{| 1, n \rangle, | 2, n \rangle\}$ for the respective two-dimensional subspaces $\chi(n)$ and are called *dressed states* in contrast to the uncoupled states $\{| g \rangle | n \rangle, | e \rangle | n \rangle\}$, the *bare states*.

Note that at resonance $\delta = 0$, such that $\theta_n = \frac{\pi}{4}$, which in turn implies $\sin(\theta_n) = \cos(\theta_n) = 1/\sqrt{2}$, in which case the splitting quantity $\Omega_n = \omega_n$ equals the Rabi frequency.

Rewriting the energy eigenvalues by introducing the quantity

$$\Delta_n = \frac{\hbar}{2} (\Omega_n - \delta) , \qquad (22.130)$$

the amount of the energy splitting due to "dressing" can be expressed by

$$E_{1,n} = E_{g,n} + \Delta_n , \qquad (22.131a)$$

$$E_{2,n} = E_{e,n-1} - \Delta_n , \qquad (22.131b)$$

which we have illustrated in Fig. 22.6. The whole ladder of the energies of the dressed states is called the *Jaynes–Cummings ladder*. The energy gaps of the pairs of dressed states on this ladder scale like $\sqrt{n}$, and thus non-linearly with n.

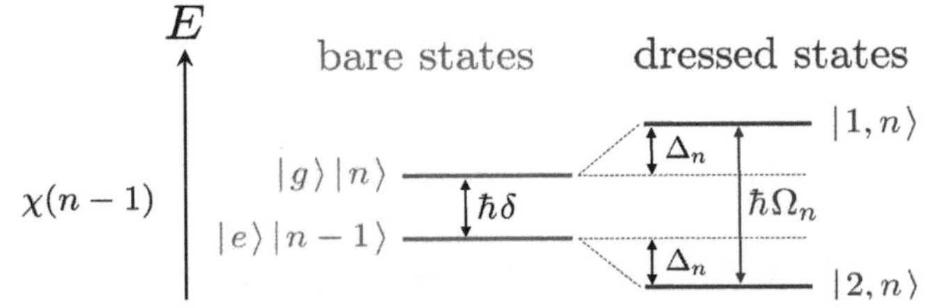

Fig. 22.6 Jaynes–Cummings states. The dressed states $| 1, n \rangle$ and $| 2, n \rangle$ with energy eigenvalues $E_{1,n}$ and $E_{2,n}$, respectively, are compared with the eigenvalues $E_{g,n}$ and $E_{e,n}$ of the bare states $| g \rangle | n \rangle$ and $| e \rangle | n \rangle$.

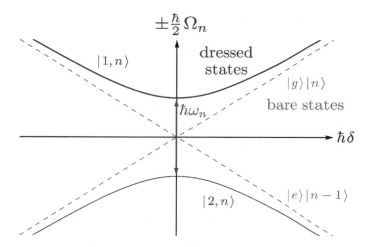

Fig. 22.7 Anti-crossing. The variation of the energy levels in dependence of the driving field frequency ω_R is sketched. The parabolas represent the dressed states; they show an anti-crossing at ω, in contrast to the bare states which cross.

Properties of dressed states: Next we study the behaviour of the dressed states in relation to the bare states, when the detuning δ is varied. The energies of the dressed states $|1,n\rangle$ and $|2,n\rangle$ form parabolas in dependence of δ, where the asymptotes are given by the bare states $|g\rangle|n\rangle$ and $|e\rangle|n\rangle$, see Fig. 22.7.

When the detuning is large, i.e., when $\omega_R \gg \omega$, which implies $\Omega_n \to \delta$ and $\Delta_n \to 0$, the dressed energies approach the bare ones. If the detuning is small, say $\delta = 0$, such that $\omega_R = \omega$, the driving field and the atom are in resonance. In this case the energy levels of the bare states are degenerate within one family $\chi(n)$ and we get a *crossing* of the levels. However, due to the coupling of the atom with the quantized field the dressed states converge to the minimum coupling energy $\hbar\Omega_n = \hbar\omega_n$, determined by the Rabi frequency, and "repel" each other. This leads to an *anti-crossing* of the energy levels, see Fig. 22.7, indeed, this is exactly the phenomenon of an *avoided crossing* that we have already discussed in Section 10.1.4.

Result: What we observe in the Jaynes–Cummings model is a typical quantum feature. For a fixed initial excitation number n, the system consisting of the atom and the quantized field oscillates between the states of a family $\chi(n)$: $|g\rangle|n\rangle \longleftrightarrow |e\rangle|n-1\rangle$ with Rabi frequency $\omega_n = \frac{2}{\hbar}c\sqrt{n}$. Thus we have continued *stimulated emission* and absorption of a photon quanta.

In particular, for $n = 1$ we see that Rabi oscillations can also occur without photons initially present in the cavity if the atom is initially in the excited state: $|e\rangle|0\rangle \longleftrightarrow |g\rangle|1\rangle$. In this case the process corresponds to the continuing transition between *spontaneous emission* and absorption. This is a fascinating quantum feature called *vacuum Rabi oscillations*, which cannot happen classically!

22.4.6   Atom-Level Probability

Finally, we are also interested in the probabilities to find the atom in the cavity in its excited state or ground state, respectively, specifically when the atom and the driving field are in resonance $\delta = 0$. The result can be phrased in terms of the following lemma.

Lemma 22.1 (Atom level probability)

In the system comprising the two-level atom and the quantized field the probabilities to find the atom in the excited state $|\,e\,\rangle$ or in the ground state $|\,g\,\rangle$, respectively, are given by

$$P_e(t) = \sin^2\left(\tfrac{\omega_n}{2}t\right) \qquad for \qquad |\,e\,\rangle,$$
$$P_g(t) = \cos^2\left(\tfrac{\omega_n}{2}t\right) \qquad for \qquad |\,g\,\rangle,$$

if the atom is in its ground state $|\,g\,\rangle$ at $t = 0$.

Note, the probability oscillates with half of the Rabi frequency.

Proof   We start from the set of dressed states at resonance $\delta = 0$ such that $\theta_n = \tfrac{\pi}{4}$ and thus $\sin(\theta_n) = \cos(\theta_n) = 1/\sqrt{2}$

$$|\,1,n\,\rangle = \tfrac{1}{\sqrt{2}}\left(|\,g\,\rangle|\,n\,\rangle + |\,e\,\rangle|\,n-1\,\rangle\right), \qquad (22.132\text{a})$$

$$|\,2,n\,\rangle = \tfrac{1}{\sqrt{2}}\left(|\,g\,\rangle|\,n\,\rangle - |\,e\,\rangle|\,n-1\,\rangle\right). \qquad (22.132\text{b})$$

The state of the total system at any time t is a superposition of the states $|\,g\,\rangle|\,n\,\rangle$ and $|\,e\,\rangle|\,n-1\,\rangle$ from the family $\chi(n)$, i.e.,

$$|\,\psi(t)\,\rangle = C_g(t)\,|\,g\,\rangle|\,n\,\rangle + C_e(t)\,|\,e\,\rangle|\,n-1\,\rangle, \qquad (22.133)$$

and, in the interaction picture, has to satisfy the time-dependent Schrödinger equation

$$i\hbar\frac{\partial}{\partial t}\,|\,\psi(t)\,\rangle = H_{\text{int}}\,|\,\psi(t)\,\rangle, \qquad (22.134)$$

with the interaction Hamiltonian H_{int} and the off-diagonal elements (22.123). Inserting the state $|\,\psi(t)\,\rangle$ from eqn (22.133) into the Schrödinger equation (22.134) and applying the covectors $\langle\,e\,|\langle\,n-1\,|$ and $\langle\,g\,|\langle\,n\,|$ from the left, respectively, yields differential equations for the time-dependent weights $C_e(t)$ and $C_g(t)$,

$$\dot{C}_e(t) = -\frac{i}{\hbar}\,c\sqrt{n}\,C_g(t), \qquad (22.135\text{a})$$

$$\dot{C}_g(t) = -\frac{i}{\hbar}\,c\sqrt{n}\,C_e(t). \qquad (22.135\text{b})$$

Decoupling the set of differential equations by repeated differentiation gives

$$\ddot{C}_e(t) = \left(-\frac{i}{\hbar}c\sqrt{n}\right)^2 C_e(t) \, , \tag{22.136a}$$

$$\ddot{C}_g(t) = \left(-\frac{i}{\hbar}c\sqrt{n}\right)^2 C_g(t) \, . \tag{22.136b}$$

Assuming as initial condition that the atom is in its ground state $|g\rangle$ at $t = 0$, i.e., $C_g(0) = 1$ and $C_e(0) = 0$, the solutions to the equations (22.136a) and (22.136b) are

$$C_e(t) = \sin\left(\frac{c\sqrt{n}}{\hbar}t\right) \, , \tag{22.137a}$$

$$C_g(t) = \cos\left(\frac{c\sqrt{n}}{\hbar}t\right) \, . \tag{22.137b}$$

Recalling the nth Rabi frequency $\frac{\omega_n}{2} = \frac{c\sqrt{n}}{\hbar}$ we finally obtain the probabilities when squaring the amplitudes,

$$P_e(t) = |C_e(t)|^2 = \sin^2\left(\frac{\omega_n}{2}t\right) \, , \tag{22.138a}$$

$$P_g(t) = |C_g(t)|^2 = \cos^2\left(\frac{\omega_n}{2}t\right) \, . \tag{22.138b}$$

This proves Lemma 22.1. □

When the atom is initially in the excited state, $|\psi(t = 0)\rangle = |e\rangle|n - 1\rangle$ then one similarly finds the probabilities

$$P_e(t) = \cos^2\left(\frac{\omega_n}{2}t\right) \, , \tag{22.139a}$$

$$P_g(t) = \sin^2\left(\frac{\omega_n}{2}t\right) \, . \tag{22.139b}$$

This oscillatory behaviour is in stark contrast to the decay that we have seen to arise as a prediction from the Markovian model for open-system dynamics in Section 22.3, but this can be taken as a consequence of the fact that a single mode of the electromagnetic field is far from a Markovian heat bath. However, if we only consider timescales much shorter than the inverse Rabi frequency, then we similarly see a decline in the excited-state population, but if we wait long enough, then a revival[3] of the excited-state population can be observed.

---

[3]The decay and revival is more drastic when one considers a coherent state (with fixed average photon number, as we will discuss in Section 25.2) instead of a state with fixed photon number, in which case interference effects lead to a pronounced exponential decay for short times, followed by a revival of probability at long times (see, e.g., the discussion in Garrison and Chiao, 2008, Sec. 12.1).

23
Quantum Measurements

At the heart of every physical theory lie the predictions that the theory makes for the outcomes of measurements. In quantum theory, the collection of the available information about a physical system—including predictions for the outcomes of any measurement of the system—is represented by a quantum state, a density operator. The previous focus of this book has hence been on the description of quantum states. In this chapter, we shall turn our attention more explicitly on the description of the possible measurements that can be performed on a given quantum system. In particular, we will discuss the mathematical description of general measurements in quantum theory, physical restrictions on their implementation, and applications for distinguishing quantum states.

23.1 Von Neumann Measurements

The measurements that we have so far considered (both implicitly and explicitly) in this book are so-called *projective* or *von Neumann measurements* (von Neumann, 2018). Mathematically, such measurements can be described by a set of *measurement operators* $\{M_n\}_{n=1,\ldots,k}$ that is a complete set of orthogonal projectors (recall Section 3.2.1 in Part I), i.e., satisfying

$$M_m M_n = \delta_{mn} M_n \quad \text{and} \quad \sum_n M_n^\dagger M_n = \sum_{n=1}^k E_n = \mathbb{1}. \tag{23.1}$$

Here, $k \leq \dim(\mathcal{H})$ is the number of distinguishable outcomes labelled by $n = 1, 2, \ldots, k$. Because the measurement operators that we consider here are projection operators, we have $M_n = M_n^\dagger$ and $E_n = M_n^\dagger M_n = M_n^2 = M_n$, but it is nonetheless useful to note the distinction between E_n and M_n for comparison with more general measurements in Section 23.2. A common situation is that $k = \dim(\mathcal{H})$, in which case we can further write $M_n = |n\rangle\langle n|$ and the states $|n\rangle$ form an orthonormal basis of the Hilbert space $\mathcal{H}$. Given a quantum state $\rho \in L(\mathcal{H})$, we can calculate the probability $p(n)$ to obtain outcome n as

$$p(n) = \text{Tr}(E_n \rho) = \text{Tr}(M_n \rho) = \text{Tr}(|n\rangle\langle n|\rho) = \langle n|\rho|n\rangle. \tag{23.2}$$

Physically, such a measurement corresponds to measuring an observable O, i.e., a measurement in the eigenbasis $\{|n\rangle\}_n$ of the observable $O = \sum_n \lambda_n |n\rangle\langle n|$. Upon observing a measurement outcome n one then concludes that the measured value of the observable is the corresponding eigenvalue λ_n. But what can we say about the post-measurement system? From a Bayesian point of view, the post-measurement system state should reflect one's knowledge about the system, given that one has obtained a particular measurement outcome n and hence corresponds to a simple update rule called *Lüders rule* after *Gerhart Lüders* (1951) resulting in the conditional state

$$\rho(n) = \frac{M_n \, \rho \, M_n^\dagger}{p(n)}, \tag{23.3}$$

where the factor $1/p(n)$ ensures normalization since $\text{Tr}(M_n\rho M_n^\dagger) = \text{Tr}(M_n^\dagger M_n \rho) = \text{Tr}(E_n\rho) = p(n)$. For $M_n = |n\rangle\langle n|$, the update rule in eqn (23.3) becomes $\rho(n) = |n\rangle\langle n|$, a statement which is usually raised to the status of a postulate of quantum mechanics.

Proposition 23.1 (Projection postulate)

A projective measurement with measurement operators $M_i = |i\rangle\langle i|$ for $i = 1, \ldots, \dim(\mathcal{H})$ resulting in outcome n leaves the system in a post-measurement state that is the eigenstate $|n\rangle$ of the corresponding observable $O = \sum_i \lambda_i |i\rangle\langle i|$, i.e.,

$$\rho(n) = |n\rangle\langle n|.$$

A prominent pedagogical example of such a projective measurement is a spin measurement using a Stern–Gerlach apparatus, as discussed in Section 8.2.1 in Part I. There, after obtaining the outcome labelled "↑" ($m_s = \frac{1}{2}$), one considers the particle to be in the state $|\uparrow\rangle = |s = \frac{1}{2}, m_s = \frac{1}{2}\rangle$. More generally, a particular measurement result n may correspond to more than one orthogonal quantum state, in which case $k < \dim(\mathcal{H})$ and the corresponding projector has a rank greater than 1, i.e., we can write $M_n = \sum_i |\psi_n^{(i)}\rangle\langle\psi_n^{(i)}|$. In this case, the measurement preserves the coherences (the off-diagonal elements) with respect to the degenerate states $|\psi_n^{(i)}\rangle$. This corresponds to a situation where two (or more) commuting observables can be measured. For instance, consider a measurement of the total spin $\vec{S}^2$ of a system of two spin-$\frac{1}{2}$ particles where $\vec{S} = \vec{S}_1 + \vec{S}_2$. The possible measurement results can be labelled by the spin quantum numbers $s = 0$ and $s = 1$. If the result is $s = 0$, then one may assume that the system has been projected into the singlet state $|s = 0, m_s = 0\rangle$, whereas a result $s = 1$ means that the state has been projected into the three-dimensional subspace spanned by the triplet $|s = 1, m_s = -1\rangle$, $|s = 1, m_s = 0\rangle$, and $|s = 1, m_s = +1\rangle$, see Section 8.3.3. The corresponding measurement operators are thus the rank-one projector $M_0 = |s = 0, m_s = 0\rangle\langle s = 0, m_s = 0|$ and the rank-three projector $M_1 = \sum_{m_s=-1}^{1} |s = 1, m_s\rangle\langle s = 1, m_s|$. However, projective measurements can also be understood as a mathematically more general way of describing measurements, as we will see in the next section.

23.2 Positive Operator-Valued Measures (POVMs)

23.2.1 Mathematical Description of POVMs

Starting from the point of view of the measurement operators M_n, we see that the basic requirement for these operators to provide reasonable predictions in terms of outcome probabilities $p(n)$ and conditional states $\rho(n)$ allows for operators other than projectors. That is, the requirement of non-negative probabilities $p(n) \geq 0$ with $\sum_n p(n) = 1$ is satisfied independently of the initial state ρ for any set of positive operators E_n that sum to the identity. Such a set

$$\{E_n | E_n \geq 0, \sum_n E_n = \mathbb{1}\}_n \tag{23.4}$$

is called a *positive operator-valued measure* (POVM). The probability to obtain outcome n is then simply

$$p(n) = \mathrm{Tr}(E_n \rho). \tag{23.5}$$

For any POVM, we can then find a set of measurement operators $\{M_n\}_n$ such that $E_n = M_n^\dagger M_n$, which determine the conditional state $\rho(n)$ given outcome n according to eqn (23.3). To see this, simply note that, since $E_n \geq 0$, the operator E_n must be Hermitian, and diagonalizable. That is, there exists a unitary U_n such that $E_n = U_n D_n U_n^\dagger$ where D_n is a diagonal matrix. This means we can define

$$M_n := \sqrt{E_n} = U_n \sqrt{D_n} U_n^\dagger. \tag{23.6}$$

However, note that this choice is not unique. Indeed, for any set of unitaries $\{V_n\}_n$ (on the Hilbert space in question) such that $V_n^\dagger V_n = \mathbb{1}$ the operators $M_n := V_n \sqrt{E_n}$ satisfy $M_n^\dagger M_n = \sqrt{E_n} V_n^\dagger V_n \sqrt{E_n} = E_n$. While the probabilities $p(n)$ for any particular outcome n are thus the same for any realization of a given POVM, the post-measurement states given by eqn (23.3) can differ depending on the specific choice of M_n.

Let us consider an example for a POVM in a two-dimensional Hilbert space with orthonormal basis $\{|\, n \,\rangle\}_{n=0,1}$. For some p with $0 < p < 1$ we define the measurement operators

$$M_0 = |\,0\,\rangle\langle\,0\,| + \sqrt{p}\,|\,1\,\rangle\langle\,1\,|, \quad M_1 = \sqrt{1-p}\,|\,1\,\rangle\langle\,1\,|. \tag{23.7}$$

Although the measurement operators are Hermitian, they are not projectors since $M_n^\dagger M_n \neq M_n$ and they are not orthogonal, $M_0 M_1 \neq 0$. The POVM we consider here has two possible outcomes, 0 and 1 with probabilities

$$p(0) = \mathrm{Tr}(E_0 \rho) = \langle\,0\,|\rho|\,0\,\rangle + p\,\langle\,1\,|\rho|\,1\,\rangle, \tag{23.8a}$$

$$p(1) = \mathrm{Tr}(E_1 \rho) = (1-p)\,\langle\,1\,|\rho|\,1\,\rangle. \tag{23.8b}$$

We see that the first outcome can always be obtained in principle for any state ρ, but the probability depends on the initial state. The second outcome, on the other hand, can only be obtained if $\langle 1 | \rho | 1 \rangle$ is non-zero, i.e., if $\rho \neq |0\rangle\langle0|$. Denoting the matrix elements of the initial state by $\rho_{ij} = \langle i | \rho | j \rangle$, the resulting conditional post-measurement states can be written as

$$\rho(0) = \tfrac{1}{p(0)} M_0 \, \rho \, M_0^\dagger = \tfrac{1}{p(0)} \begin{pmatrix} \rho_{00} & \sqrt{p}\,\rho_{01} \\ \sqrt{p}\,\rho_{10} & p\rho_{11} \end{pmatrix}, \tag{23.9a}$$

$$\rho(1) = \tfrac{1}{p(1)} M_1 \, \rho \, M_1^\dagger = |1\rangle\langle1|. \tag{23.9b}$$

We see that the first outcome leads to a state whose component in the direction of $|0\rangle$ is amplified, while the second outcome always results in the pure state $|1\rangle$. Post-selecting on the outcome 0 thus makes the measurement procedure described by the POVM $\{M_0, M_1\}$ into a filtering or attenuating channel. On the other hand, when we consider the unconditional state, averaged over all outcomes, then we find

$$\rho \mapsto p(0)\rho(0) + p(1)\rho(1) = \begin{pmatrix} \rho_{00} & \sqrt{p}\,\rho_{01} \\ \sqrt{p}\,\rho_{10} & \rho_{11} \end{pmatrix}, \tag{23.10}$$

corresponding to a dephasing channel from eqn (21.10). Indeed, the relation between POVMs and channels is no coincidence, simply note that for any initial state and any POVM, the unconditional (i.e., when disregarding or not knowing the measurement outcome) post-measurement state is given by $\sum_n p(n)\rho(n) = \sum_n M_n\rho M_n^\dagger$ for operators satisfying $\sum_n M_n^\dagger M_n = \mathbb{1}$. That is, the measurement operators of any POVM always constitute a set of Kraus operators, see Section 21.3. At the same time, the map from any initial state ρ to any conditional post-measurement state, i.e., post-selecting on one (or several) outcomes, is always completely positive and trace non-increasing since $M_n^\dagger M_n \leq \mathbb{1}$, and can be extended to be trace-preserving (CPTP) by adding a POVM element $\mathbb{1} - M_n^\dagger M_n$. However, in that case the corresponding measurement operators and Kraus operators generally do not coincide, as we have seen for the example discussed.

23.2.2 Symmetric Informationally Complete POVMs

An interesting class of POVMs are so-called informationally complete POVMs which span the space $L(\mathcal{H})$ for a given Hilbert space $\mathcal{H}$. A set of at least d^2 operators E_i with $\sum_i E_i = \mathbb{1}$ can be informationally complete, and is called minimal if it is informationally complete and consists of exactly d^2 operators. A special case of particular importance are those minimal informationally complete POVMs that can be constructed from d^2 rank-one projectors Π_i ($i = 1, 2, \ldots, d^2$) that satisfy

$$\mathrm{Tr}\big(\Pi_i \Pi_j\big) = \frac{d\delta_{ij} + 1}{d + 1} \, \forall i, j. \tag{23.11}$$

Since the Hilbert–Schmidt inner product (see Section 15.5) between these projectors is symmetric with respect to the two indices i and j, the corresponding POVMs are called *symmetric, informationally complete* (SIC) POVMs, given by (Prugovečki, 1977; Busch, 1991; Renes *et al.*, 2004)

$$\left\{ E_i = \tfrac{1}{d}\Pi_i \,\middle|\, \Pi_i^2 = \Pi_i, \ \mathrm{Tr}(\Pi_i\Pi_j) = \tfrac{d\delta_{ij}+1}{d+1} \right\}_{i=1}^{d^2}. \tag{23.12}$$

These SIC-POVMs have important applications, for instance, for the statistical reconstruction of quantum states that is called quantum state tomography (Caves *et al.*, 2002), where minimally informationally complete POVMs are used in generalizing the de Finetti theorem from statements about classical random variables to the domain of random quantum states. Although minimal informationally complete POVMs are sometimes already enough for practical applications, the more specific SIC-POVMs have beautiful mathematical properties whose description goes beyond the scope of this chapter, but we direct the interested reader to Zauner (2011) and Wootters (2004). Moreover, it is to date not known whether or not SIC-POVMs exist for all Hilbert space dimensions. While a general answer is, thus far, not available, a conjecture due to *Gerhard Zauner* (2011) provides a construction for SIC-POVMs that we present here in a version due to Renes *et al.* (2004):

Conjecture 23.1 (Zauner's conjecture)

For any Hilbert space $\mathcal{H} = \mathbb{C}^d$ with dimension $d \in \mathbb{N}$ and orthonormal basis $\{\,|\,k\,\rangle\}_{k=0}^{d-1}$, let

$$D_{jk} = \sum_{m=0}^{d-1} e^{\frac{2\pi i j}{d}(m+\frac{k}{2})} \,|\,(m+k)_{\mathrm{mod}(d)}\,\rangle\langle\, m\,|\,.$$

Then there exists a vector $|\,\Phi\,\rangle \in \mathbb{C}^d$ such that a SIC-POVM is given by

$$\left\{ \tfrac{1}{d} D_{jk} \,|\,\Phi\,\rangle\langle\,\Phi\,|\, D_{jk}^\dagger \right\}_{j,k=1,\dots,d}.$$

Although Zauner's conjecture has not yet been proven (or disproven), exact examples in all dimensions from two to sixteen and numerically calculated SIC-POVMs for all dimensions up to $d = 151$, along with a number of isolated additional (exact and numerical) results, seem to support the conjecture (Fuchs *et al.*, 2017; Appleby *et al.*, 2018), with new methods for finding additional SIC-POVMs being developed (Appleby *et al.*, 2018). Here, let us illustrate the conjecture by using it to construct a SIC-POVM for dimension $d = 2$. In this case, we find that the operators D_{jk} correspond to the Pauli operators, more specifically, up to irrelevant global signs, the expression in Conjecture 23.1 yields

$$D_{11} = \sigma_y, \quad D_{12} = \sigma_z, \quad D_{21} = \sigma_x, \quad D_{22} = \mathbb{1}. \tag{23.13}$$

We then use a multi-label to write $P_{jk} = D_{jk}\,|\,\Phi\,\rangle\langle\,\Phi\,|\,D_{jk}$ for the projectors from eqn (23.11). Since all D_{jk} satisfy $D_{jk}^\dagger = D_{jk}$ and $D_{jk}^2 = \mathbb{1}$, it is then easy to see that for any normalized state $|\,\Phi\,\rangle$ we have

$$\mathrm{Tr}(\Pi_{jk}\Pi_{jk}) = \mathrm{Tr}\big(D_{jk}\,|\,\Phi\,\rangle\langle\,\Phi\,|\,D_{jk}^2\,|\,\Phi\,\rangle\langle\,\Phi\,|\,D_{jk}\big) = \mathrm{Tr}(|\,\Phi\,\rangle\langle\,\Phi\,|) = 1. \tag{23.14}$$

For the Hilbert–Schmidt inner products of two different projectors, we have

$$\mathrm{Tr}\big(\Pi_{jk}\Pi_{j'k'}\big) = |\langle\,\Phi\,|\,D_{jk}D_{j'k'}\,|\,\Phi\,\rangle|^2\,,\tag{23.15}$$

which leads to the condition in the condition

$$|\langle\,\Phi\,|\,\sigma_x\,|\,\Phi\,\rangle|^2 = |\langle\,\Phi\,|\,\sigma_y\,|\,\Phi\,\rangle|^2 = |\langle\,\Phi\,|\,\sigma_z\,|\,\Phi\,\rangle|^2 = \tfrac{1}{d+1} = \tfrac{1}{3}\,.\tag{23.16}$$

Using the ansatz $|\,\Phi\,\rangle = \alpha\,|\,0\,\rangle + \beta\,|\,1\,\rangle$ with $\alpha,\beta\in\mathbb{C}$ and $|\alpha|^2+|\beta|^2=1$, we obtain

$$|\langle\,\Phi\,|\,\sigma_x\,|\,\Phi\,\rangle|^2 = 4\,\mathrm{Re}^2(\alpha\beta^*),\tag{23.17a}$$

$$|\langle\,\Phi\,|\,\sigma_y\,|\,\Phi\,\rangle|^2 = 4\,\mathrm{Im}^2(\alpha\beta^*),\tag{23.17b}$$

$$|\langle\,\Phi\,|\,\sigma_z\,|\,\Phi\,\rangle|^2 = \big(|\alpha|^2-|\beta|^2\big)^2 = \big(2|\alpha|^2-1\big)^2.\tag{23.17c}$$

Starting from the last line, one can then see that $\alpha = \sqrt{\frac{\sqrt{3}+1}{2\sqrt{3}}}$ and $\beta = e^{i\frac{\pi}{4}}\sqrt{\frac{\sqrt{3}-1}{2\sqrt{3}}}$ provides a solution to eqn (23.16). A SIC-POVM in $d=2$ can hence be realized by

$$\{E_i\}_{i=1}^4 = \big\{\tfrac{1}{2}\,|\,\Phi\,\rangle\!\langle\,\Phi\,|\,,\,\tfrac{1}{2}\sigma_x\,|\,\Phi\,\rangle\!\langle\,\Phi\,|\,\sigma_x,\,\tfrac{1}{2}\sigma_y\,|\,\Phi\,\rangle\!\langle\,\Phi\,|\,\sigma_y,\,\tfrac{1}{2}\sigma_z\,|\,\Phi\,\rangle\!\langle\,\Phi\,|\,\sigma_z,\,\big\}\tag{23.18}$$

with the state

$$|\,\Phi\,\rangle = \sqrt{\tfrac{\sqrt{3}+1}{2\sqrt{3}}}\,|\,0\,\rangle + e^{i\frac{\pi}{4}}\sqrt{\tfrac{\sqrt{3}-1}{2\sqrt{3}}}\,|\,1\,\rangle\,.\tag{23.19}$$

23.2.3 Naimark Dilation

Interestingly, there is another connection between quantum channels and quantum measurements. Loosely speaking, POVMs are to projective measurements what CPTP maps are to unitary maps. In a similar way that all CPTP maps can be represented as unitary dynamics on a larger Hilbert space, POVMs can be represented as projective measurements on larger Hilbert spaces, a fact that can be understood as part of Stinespring's theorem (Stinespring, 1955), but is usually attributed to *Mark Naimark* (1940), see Paulsen (2003).

Theorem 23.1 (Naimark dilation)

For any POVM on a Hilbert space $\mathcal{H}_A$ with elements $\{E_n\}_n$ there exists a Hilbert space $\mathcal{H}_B$ with a state $|\,\omega\,\rangle \in \mathcal{H}_B$ and a projective measurement with measurement operators $\{\tilde{P}_n\}_n$ on $\mathcal{H}_A\otimes\mathcal{H}_B$ such that $\forall\rho\in\mathcal{B}_{\mathrm{HS}}(\mathcal{H}_A)$:

$$p(n) = \mathrm{Tr}\big(E_n\,\rho\big) = \mathrm{Tr}\big(\tilde{P}_n\,\rho\otimes|\,\omega\,\rangle\!\langle\,\omega\,|\big)\quad\forall n.$$

Proof   For a given POVM $\{E_n\}_n$ on the Hilbert space $\mathcal{H}_A$ with measurement opera-tors M_n for $n = 1, \ldots, k$, let us define a unitary operator U on a larger Hilbert space $\mathcal{H}_A \otimes \mathcal{H}_B$ via

$$U \, |\psi\rangle \, |\omega\rangle = \sum_n M_n \, |\psi\rangle \, |n\rangle \,, \tag{23.20}$$

where $|\psi\rangle$ is an arbitrary state in $\mathcal{H}_A$, $|\omega\rangle$ is a fixed state in $\mathcal{H}_B$, and $\{|n\rangle\}_n$ is an orthonormal basis of $\mathcal{H}_B$. In complete analogy to eqn (21.56) in the proof of the Stinespring theorem (see Theorem 21.2), we can confirm that $\langle\psi'| \otimes \langle\omega| U^\dagger U |\psi\rangle \otimes |\omega\rangle = \langle\psi'|\psi\rangle$ for all $|\psi\rangle, |\psi'\rangle \in \mathcal{H}_A$, i.e., U is unitary on subspace of fixed $|\omega\rangle$ and can be extended to be unitary everywhere on $\mathcal{H}_A \otimes \mathcal{H}_B$. We then define the projection operators $\tilde{P}_n$ for $n = 1, \ldots, k$ via

$$\tilde{P}_n = U^\dagger \, \mathbb{1} \otimes |n\rangle\langle n| \, U \,. \tag{23.21}$$

For any state $\rho \otimes |\omega\rangle\langle\omega|$ on the Hilbert space $\mathcal{H}_A \otimes \mathcal{H}_B$ we can then write $\rho = \sum_i p_i \, |\psi_i\rangle\langle\psi_i|$ and calculate the probability to obtain outcome n in the projective measurement defined by the operators $\{\tilde{P}_n\}_n$ as

$$
\begin{aligned}
p(n) &= \mathrm{Tr}\big(\rho \otimes |\omega\rangle\langle\omega| \, \tilde{P}_n\big) \ = \ \mathrm{Tr}\big(\rho \otimes |\omega\rangle\langle\omega| U^\dagger \, \mathbb{1} \otimes |n\rangle\langle n| \, U\big) \\
&= \sum_i p_i \, \langle\psi_i| \langle\omega| U^\dagger \, \mathbb{1} \otimes |n\rangle\langle n| \, U |\psi_i\rangle |\omega\rangle \\
&= \sum_i p_i \sum_{j,k} \big(\langle\psi_i| M_j^\dagger\big) \langle j| \, \mathbb{1} \otimes |n\rangle\langle n| \, \big(M_k |\psi_i\rangle\big) |k\rangle \\
&= \sum_i p_i \sum_{j,k} \langle\psi_i| M_j^\dagger M_k |\psi_i\rangle \, \langle j|n\rangle \, \langle n|k\rangle \\
&= \sum_i p_i \, \langle\psi_i| M_n^\dagger M_n |\psi_i\rangle \ = \ \mathrm{Tr}\big(M_n^\dagger M_n \rho\big) \ = \ \mathrm{Tr}\big(E_n\rho\big). \tag{23.22}
\end{aligned}
$$

$\square$

Let us illustrate Naimark's dilation theorem by returning to our previous example from eqn (23.7). We select the state $|\omega\rangle = |0\rangle$ and use the measurement operators M_0 and M_1 to determine the action of the unitary U from eqn (23.20), i.e.,

$$U|0\rangle|0\rangle = M_0|0\rangle|0\rangle + M_1|0\rangle|1\rangle = |0\rangle|0\rangle \,, \tag{23.23a}$$

$$U|1\rangle|0\rangle = M_0|1\rangle|0\rangle + M_1|1\rangle|1\rangle = \sqrt{p}\,|1\rangle|0\rangle + \sqrt{1-p}\,|1\rangle|1\rangle \,. \tag{23.23b}$$

A unitary that is defined on the entire two-qubit Hilbert space and is compatible with these requirements is given by

$$U = |0\rangle\langle 0| \otimes \mathbb{1} + |1\rangle\langle 1| \otimes U_p \,, \tag{23.24a}$$

$$U_p = \sqrt{p}\big(|0\rangle\langle 0| + |1\rangle\langle 1|\big) + \sqrt{1-p}\big(|1\rangle\langle 0| - |0\rangle\langle 1|\big) \,. \tag{23.24b}$$

A projective measurement on the two-qubit space corresponding to the POVM with

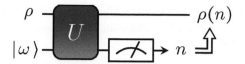

Fig. 23.1 POVM as interaction with measurement device. Every POVM $\{E_n\}_n$ can be understood as an interaction of the measured system with a measurement device—a "pointer"—that is originally in a pure state, followed by a projective measurement of the pointer. After obtaining the outcome n, one concludes that the system is left in the state $\rho(n)$ from eqn (23.3).

measurement operators M_0 and M_1 is thus represented by the projectors $\tilde{P}_0$ and $\tilde{P}_1$ with

$$\tilde{P}_n = |0\rangle\langle0| \otimes |n\rangle\langle n| + |1\rangle\langle1| \otimes U_p^\dagger |n\rangle\langle n| U_p, \tag{23.25}$$

satisfying $\tilde{P}_m \tilde{P}_n = \delta_{mn} \tilde{P}_n$.

Upon closer inspection of Theorem 23.1 in combination with eqn (23.21) we further see that Naimark's theorem can also be interpreted in a very physical way: For every POVM $\{E_n\}_n$ on a system $\mathcal{H}_A$ we can write the probability for any outcome n as

$$p(n) = \mathrm{Tr}(E_n\,\rho) = \mathrm{Tr}(U^\dagger \mathbb{1} \otimes |n\rangle\langle n|\, U\,\rho \otimes |\omega\rangle\langle\omega|)$$

$$= \mathrm{Tr}(\mathbb{1} \otimes |n\rangle\langle n|\, U\,\rho \otimes |\omega\rangle\langle\omega|\, U^\dagger). \tag{23.26}$$

In other words, the POVM on $\mathcal{H}_A$ can be thought of as projective measurement (with rank-one projectors $|n\rangle\langle n|$) on $\mathcal{H}_B$ after a unitary interaction U between the system initially in state ρ, and another system initially in the pure state $|\omega\rangle$. As illustrated in Fig. 23.1, the system $\mathcal{H}_B$ acts as a measurement device. Finding this measurement apparatus (the "pointer") in a particular state $|n\rangle$, which happens with probability $p(n)$ one records the measurement result n and assumes that the measured system is left in the post-measurement state $\rho(n) = M_n \rho M_n^\dagger / p(n)$. The question remains: Is the converse statement true? Can every measurement procedure be represented as a unitary interaction between two quantum systems followed by projective measurement on one of them be described purely as a POVM on just the measured system?

Indeed, the answer is: yes. Suppose we start with the situation pictured in Fig. 23.1, we have a system S in the state ρ_S that is to be measured, and the pointer P is prepared in the state $|\omega\rangle_P$ and interacts with our system via a unitary U, before the pointer is measured in the basis $|n\rangle_P$. We can then simply define the operators

$$E_n = \mathrm{Tr}_P(\mathbb{1}_S \otimes |\omega\rangle\langle\omega|_P\, U^\dagger\, \mathbb{1}_S \otimes |n\rangle\langle n|_P\, U). \tag{23.27}$$

Taking the trace $\mathrm{Tr}_S(\rho\, E_n)$ one immediately finds

$$\mathrm{Tr}_S(\rho_S\, E_n) = \mathrm{Tr}_S\Big(\rho_S\, \mathrm{Tr}_P(\mathbb{1}_S \otimes |\omega\rangle\langle\omega|_P\, U^\dagger\, \mathbb{1}_S \otimes |n\rangle\langle n|_P\, U)\Big)$$

$$= \mathrm{Tr}\Big(\rho_S \otimes |\omega\rangle\langle\omega|_P\, U^\dagger\, \mathbb{1}_S \otimes |n\rangle\langle n|_P\, U\Big) = p(n), \tag{23.28}$$

where we have used the cyclic property of the trace in the last step. Since $p(n) \geq 0$ for all $\rho_S \geq 0$, we have $E_n \geq 0$ and by choosing the measurement on the pointer to be complete in the sense that $\sum_n |n\rangle\langle n| = \mathbb{1}_P$, we have

$$\sum_n E_n = \mathrm{Tr}_P\big(\mathbb{1}_S \otimes |\omega\rangle\langle\omega|_P\, U^\dagger\, \mathbb{1}_S \otimes \sum_n |n\rangle\langle n|\, U\big) = \mathrm{Tr}_P\big(\mathbb{1}_S \otimes |\omega\rangle\langle\omega|_P\big) = \mathbb{1}_S.$$
(23.29)

The corresponding measurement operators M_n that determine the (conditional) post-measurement states can then be defined via their action on arbitrary pure states $|\phi\rangle$ on the Hilbert space of the measured system as

$$M_n\, |\phi\rangle_S := \mathbb{1}_S \otimes \langle n|_P\, U\, |\phi\rangle_S \otimes |\omega\rangle_P,$$
(23.30)

such that we have

$$M_n\, |\phi\rangle\langle\phi|\, M_n^\dagger = \mathbb{1}_S \otimes \langle n|_P\, U\, |\phi\rangle\langle\phi|_S \otimes |\omega\rangle\langle\omega|_P\, U^\dagger\, \mathbb{1}_S \otimes |n\rangle_P$$
$$= \mathrm{Tr}_P\big(\mathbb{1}_S \otimes |n\rangle\langle n|\, U\, |\phi\rangle\langle\phi|_S \otimes |\omega\rangle\langle\omega|_P\, U^\dagger\big),$$
(23.31)

which is the product of the conditional state $\rho(n)$ and the corresponding probability $p(n)$ by definition.

The keen observer might notice a subtlety here that distinguishes Naimark's theorem (as we have stated it in Theorem 23.1) and its "reverse". For Theorem 23.1, we can choose the dimension of the second Hilbert space (containing $|\omega\rangle_P$) such that it matches exactly the number of POVM elements. For every E_n there is then a $|n\rangle_P$ on the pointer space such that $\sum_n |n\rangle\langle n|_P = \mathbb{1}_P$. In the reverse statement, we can choose the projective measurement on the pointer along with its dimension. In particular, the projectors need not be rank-one projectors $|n\rangle\langle n|$, but we might select a set of orthogonal projectors $\Pi_n = |\psi_i^{(n)}\rangle\langle\psi_i^{(n)}|$ of any rank such that $\Pi_n \Pi_m = \delta_{mn}\Pi_n$ and $\sum_n \Pi_n = \mathbb{1}_P$. Physically, this can be understood as a *coarse-graining*. That is, although there are in principle distinguishable outcomes for all rank-one projectors $|\psi_i^{(n)}\rangle\langle\psi_i^{(n)}|$ labelled by n and i, some of them are not distinguished, either because it is practically not possible or simply because one chooses not to. The corresponding POVM elements and measurement operators can hence be defined just as in eqn (23.27) and eqn (23.30), respectively, by simply replacing the label n by the two labels n and i. The coarse-graining is then achieved by taking the POVM elements $E_{n,i}$ and defining a new (smaller) set of POVM elements via $E_n = \sum_i E_{n,i}$ with $p(n) = \mathrm{Tr}(E_n \rho)$. Similarly, the conditional state after obtaining the coarse-grained outcome n is

$$\rho(n) = \frac{1}{p(n)} \sum_i M_{n,i}\, \rho\, M_{n,i}^\dagger.$$
(23.32)

Having presented the mathematical background of generalized measurements along with some examples of classes of POVMs, let us discuss some conceptually relevant questions about these measurements. First, let us summarize what we can say about practically implementing such generalized measurements. In this regard, Naimark's theorem gives us useful insights both from a conceptual and from a practical point of view. It informs us that we can regard projective measurements as a fundamental conceptual ingredient of quantum theory. At the same time, it provides a way of implementing a given POVM by implementing a corresponding projective measurement on a larger system. However, while Naimark's theorem tells us that we *can* find a suitable projective measurement on an extended Hilbert space, the required control over a potentially much larger Hilbert space and its interaction with the measured system can bring about new technological challenges. An interesting intermediate path is the restriction to such POVMs that can be simulated by projective measurements in combination with classical randomness and classical post-processing (Oszmaniec *et al.*, 2017), although not all POVMs are projectively simulable in this way. While we cannot go into much more detail on this approach here, we encourage the interested reader to consult Oszmaniec *et al.* (2017) for more information.

However, a question that we will investigate here more closely concerns the implementation of projective measurements themselves. As we will discuss in Section 23.3, projective measurements are an idealization and are never realized exactly, but they can be approximated to arbitrary precision and are therefore a mathematically convenient description. Finally, one may wonder, if POVMs can be represented as projective measurements on larger spaces, why should one be interested in using a POVM in the first place? To give at least a partial answer, we will examine a particular task, distinguishing quantum states, where POVMs can be more useful than simple projective measurements in Section 23.4.

23.3   Non-Ideal Projective Measurements

23.3.1   A Model for Ideal Projective Measurements

The dilation theorems that we have presented here and in Chapter 21 provide a unified picture: Theorem 15.2 asserts that general states are described by density operators ρ but can be thought of as pure states on larger Hilbert spaces. Theorem 21.2 ensures us that general processes are described by CPTP maps Λ but can be thought of as unitaries on larger Hilbert spaces. And Theorem 23.1 affirms that general measurements are described by POVMs but can be thought of as projective measurements on larger Hilbert spaces. At the same time, pure states, unitary maps, and projective measurements can be seen as the idealized versions of the more general concepts which more accurately describe what can practically be realized in a laboratory. In practice, imprecisions, losses, inaccurate timing, unavoidable coupling to the environment, and other effects always mean that quantum states are mixed and only approximately pure, and operations on quantum systems are only approximately unitary.

For quantum states this fundamental observation is acknowledged quite prominently in the *third law of thermodynamics* in the form of the *unattainability principle* (see, e.g., the discussion in Levy *et al.*, 2012 and Masanes and Oppenheim, 2017). That is, cooling any system to the ground state requires infinite time (infinitely many steps), infinite energy, or operations of infinite complexity (see Taranto *et al.*, 2023 for a discussion). Since any pure state necessarily has a higher energy than the ground state and can be reached deterministically from the latter via a unitary, the principle extends to all pure states. However, this appears to be in contrast to the projection postulate, Proposition 23.1, which suggests that one can prepare a pure state by performing a projective measurement. As we shall see in this section, there is no conflict between these statements because perfect projective measurements require preparing the measurement apparatus in a pure state to begin with (Guryanova *et al.*, 2020).

To better understand this, let us employ a similar model for projective measurements that we have previously used for POVMs in Fig. 23.1. That is, we model a measurement as an interaction between a system (that is to be measured) and a measurement apparatus, a "pointer". For simplicity, let us consider an example first. Suppose that the system to be measured is a single qubit in an unknown state ρ_S and the pointer is modelled by another qubit that is initialized in the state $|0\rangle_P$. We wish to carry out a measurement with respect to the basis $\{|0\rangle_S, |1\rangle_S\}$ and we can associate the pointer basis states $|0\rangle_P$ and $|1\rangle_P$ with the corresponding outcomes. The measurement procedure can then be viewed as an interaction between the system and pointer via a unitary U,

$$\rho_S \otimes |0\rangle\langle 0|_P \mapsto U \rho_S \otimes |0\rangle\langle 0|_P U^\dagger = \tilde{\rho}_{SP}, \tag{23.33}$$

which results in a correlated final state $\tilde{\rho}_{SP}$. More specifically, we can here consider the unitary given by

$$U = |0\rangle\langle 0|_S \otimes \mathbb{1}_P + |1\rangle\langle 1|_S \otimes (|0\rangle\langle 1|_P + |1\rangle\langle 0|_P), \tag{23.34}$$

which happens to be the controlled NOT operation that we have encountered already in eqn (14.47). Denoting the matrix elements of ρ_S as $\rho_{ij} := \langle i | \rho_S | j \rangle$, we then have

$$\tilde{\rho}_{SP} = \rho_{00} |00\rangle\langle 00|_{SP} + \rho_{11} |11\rangle\langle 11|_{SP} + (\rho_{01} |00\rangle\langle 11|_{SP} + \text{H.c.}). \tag{23.35}$$

This joint state can be interpreted in the following way: First, the probability to find the post-interaction pointer state $\tilde{\rho}_P = \text{Tr}_S(\tilde{\rho}_{SP})$ in the state $|i\rangle_P$ is the same as the probability according to the initial system state, i.e., $\langle i | \tilde{\rho}_P | i \rangle = \langle i | \rho_S | i \rangle = \rho_{ii}$ for all i. Second, whenever the post-interaction pointer state is found to be $|i\rangle_P$, then the system is also left in the state $|i\rangle_S$ after the interaction. This second property corresponds to the projection postulate of Proposition 23.1. Finally, we note that the system is also disturbed by the interaction, such that $\tilde{\rho}_S = \text{Tr}_P(\tilde{\rho}_{SP}) \neq \rho_S$, but the diagonal (with respect to the measurement basis) is left invariant, $\langle i | \tilde{\rho}_S | i \rangle = \langle i | \rho_S | i \rangle \, \forall i$. This third property can be seen to arise from combining the first two properties. If we know that a measurement has been performed but we don't know the outcome, then we know the system is left in the state $|i\rangle_S$ with probability $p_i = \rho_{ii}$, such that we assign the state $\tilde{\rho}_S = \sum_i p_i |i\rangle\langle i|$ to the system, which coincides with $\text{Tr}_P(\tilde{\rho}_{SP}) \neq \rho_S$.

Now, let us reconsider the unattainability principle. The practical impossibility of exactly preparing pure states means that the assumption of preparing the pointer in the state $|0\rangle_P$ is an idealization and we generally have to consider the pointer to be in a mixed state ρ_P initially. More specifically, this mixed state will be of full rank,

Fig. 23.2 Non-ideal measurement. When finite resources (energy, time, complexity of operations) are used, measurements are non-ideal because the pointer can only be prepared in a mixed state, e.g., a thermal state τ_P. For such non-ideal measurements (Guryanova *et al.*, 2020), the system state $\rho(n)$ after obtaining the pointer outcome n is not a pure state either. Nonetheless, the measurement can still be unbiased such that the probability to obtain the outcome n is still $p(n) = \langle n \,|\, \rho_S \,|\, n \rangle$ for any measured state ρ_S.

i.e., the probabilities for most outcomes might be very small, but not exactly zero. We are hence forced to make the subtle transition from the situation pictured in Fig. 23.1 to that shown in Fig. 23.2. So what happens when we replace the initial pointer state with a mixed state? For instance, suppose that one invests energy to cool the pointer as much as possible, reaching a thermal state

$$\tau_P(\beta) = \tfrac{1}{\mathcal{Z}_P} \exp(-\beta\, H_P) = \tfrac{1}{\mathcal{Z}_P} \sum_n \exp(-\beta E_n^{(\mathrm{P})}) \, |E_n^{(\mathrm{P})}\rangle\langle E_n^{(\mathrm{P})}| \,, \tag{23.36}$$

where $H_P = \sum_n E_n^{(\mathrm{P})} |E_n^{(\mathrm{P})}\rangle\langle E_n^{(\mathrm{P})}|$ is the pointer Hamiltonian, $\mathcal{Z}_P = \mathrm{Tr}(H_P e^{-\beta H_P})$ is the partition function, and $\beta = (k_{\mathrm{B}} T)$ for some finite temperature T. For $T \to 0$, the thermal state is just the ground state, but for any non-zero temperature we have a mixed state of full rank. Applying the unitary from eqn (23.34) to $\rho_S \otimes \tau_P$ we obtain a final state $\tilde{\rho}_{SP}$ for which $\langle i \,|\, \tilde{\rho}_P \,|\, i \rangle = \rho_{ii}(2p-1) + 1 - p$ for $p = \big(1 + \exp[-\beta(E_1^P - E_0^P)]\big)^{-1}$. The "measurement" is now biased, one of the outcomes is now more likely to occur than one would assume based on ρ_S, the other is less likely. Moreover, upon finding the pointer in a given state, say $|0\rangle_P$, the probability for finding the post-measurement system in the corresponding state is not one. Clearly, this could be an artefact of the particular unitary that we have used to correlate system and pointer. The question is, can this be fixed?

Although we have considered a quite specific example, we can observe that the discussed features can be formulated more generally. To this end, we can consider a system with arbitrary Hilbert space $\mathcal{H}_S$ with dimension $d_S = \dim(\mathcal{H}_S)$. For the pointer one may similarly consider a Hilbert space $\mathcal{H}_P$ with dimension $d_P = \dim(\mathcal{H}_P)$, but since we assume that the measurement can in principle distinguish between all d_S states of the measurement basis, let $d_P \geq d_S$.

One may then simply assign orthogonal projectors Π_i on $\mathcal{H}_P$ to the different outcomes, such that $\Pi_i \Pi_j = \delta_{ij} \Pi_j$ and $\sum_i \Pi_i = \mathbb{1}_P$. With this, we can reformulate the three properties, which we take to be the defining features of ideal measurements. That is, according to *Yelena Guryanova, Nicolai Friis*, and *Marcus Huber* (2020), any ideal projective measurement is:

(i) **Unbiased**: The post-interaction pointer reproduces the measurement statistics of the (pre-interaction) system,

$$\mathrm{Tr}\big(\mathbb{1}_S \otimes \Pi_j \tilde{\rho}_{SP}\big) = \mathrm{Tr}\big(|j\rangle\langle j|_S\, \rho_S\big)\ \forall j\ \forall \rho_S\,. \tag{23.37}$$

(ii) **Faithful**: The post-interaction pointer and the post-interaction system are perfectly correlated with respect to the measurement basis (and the corresponding projectors),

$$C(\tilde{\rho}_{SP}) = \sum_j \mathrm{Tr}\big(|j\rangle\langle j|_S \otimes \Pi_j\, \tilde{\rho}_{SP}\big) = 1\,. \tag{23.38}$$

That is, given a specific measurement outcome i, the probability for the system to be left in the state $|i\rangle_S$ is 1.

(iii) **Non-invasive**: The diagonal entries of the system state with respect to the measurement basis are not disturbed by the measurement procedure, i.e.,

$$\mathrm{Tr}\big(|j\rangle\langle j|_S\, \rho_S\big) = \mathrm{Tr}\big(|j\rangle\langle j|_S \otimes \mathbb{1}_P\, \tilde{\rho}_{SP}\big)\ \forall j\ \forall \rho_S\,. \tag{23.39}$$

As discussed in Guryanova *et al.* (2020), the unattainability principle, which forbids exactly preparing pure states, implies that property (ii) does not hold. Therefore, the correlation function $C(\tilde{\rho}_{SP})$ can take values arbitrarily close to 1, but ultimately $C(\tilde{\rho}_{SP}) < 1$. In practice, this means that even if one is very confident that the system is left in a particular pure state after observing a particular outcome, e.g., by reading off the display of a measurement device, this does not mean there is no room for error. Confidence is not certainty. Consequently, we have to accept ideal projective measurements as a convenient idealization.

23.3.2   Unbiased Non-Ideal Measurements

Following Guryanova *et al.* (2020), we call measurements *non-ideal* if they fail to satisfy all of the properties (i)–(iii). One can then ask how closely non-ideal measurements can approximate ideal measurements. Since we know that measurements can in practice not be faithful, one could hope that at least the other two properties hold and that non-ideal measurement can still be unbiased and non-invasive, but alas, this cannot be: Although any single property does not mean the other two have to hold, any two of the three properties can be shown to imply the third (Guryanova *et al.*, 2020). When faced with a choice between properties (i) and (iii), one thus resorts to *unbiased measurements*, because one can at least trust the measurement statistics. That is, if the measurement procedure is unbiased, one can collect data from many such measurements and correctly estimate mean values of observables even if one is not sure which particular state the system is left in after each measurement.

Mathematically, this leaves us with the task of determining the structure of such unbiased measurements, e.g., the most general form of CPTP maps that realize unbiased non-ideal measurements, or more restricted cases such as unbiased non-ideal measurements reaching maximal correlations $C(\tilde{\rho}_{SP})$ while using minimal energy (Guryanova *et al.*, 2020) or while disturbing the measured system only minimally (Debarba *et al.*, 2019). Here, let us briefly illustrate this approach by concentrating on such measurement procedures that can be realized by acting unitarily on the system and pointer, as shown in Fig. 23.2. As shown in Guryanova *et al.* (2020), the pointer Hilbert space must in this case satisfy (or be truncated to satisfy) $d_P = \lambda d_S$ with $\lambda \in \mathbb{N}$. Then, any unitary U realizing an unbiased measurement for all system states ρ_S and for initially mixed pointer states ρ_P is of the form $U = V\tilde{U}$, where V and $\tilde{U}$ are given by

$$V = \sum_{m,n=0}^{d_S-1} \sum_{j=1}^{\lambda} | m \rangle\langle n |_S \otimes | \psi_j^{(n)} \rangle\langle \psi_j^{(m)} |_P, \tag{23.40a}$$

$$\tilde{U} = \sum_{n=0}^{d_S-1} | n \rangle\langle n |_S \otimes \tilde{U}^{(n)}, \tag{23.40b}$$

and the operators $\tilde{U}^{(n)}$ are arbitrary unitaries on the pointer Hilbert space.

Although the initial state of the pointer is not pure, we can nonetheless think of such a non-ideal measurement as a POVM on the system space $\mathcal{H}_S$. Indeed, recalling Theorem 15.2, we can simply consider an extension of the pointer space, i.e., its purification. That is, we can diagonalize the initial pointer state, $\rho_P = \sum_i q_i | \omega_i \rangle\langle \omega_i |$ and use the pure state $| \omega \rangle = \sum_i \sqrt{q_i} | \omega_i \rangle_{P_1} \otimes | \omega_i \rangle_{P_2} \in \mathcal{H}_P^{\otimes 2}$ as the pure state in Fig. 23.1. At the same time, we do not want to increase the number of outcomes that are distinguished by the measurement by adding the second copy of the pointer space. This means we have to coarse-grain as discussed in Section 23.2.3 and formally introduce a complete set of rank-one projectors on $\mathcal{H}_P^{\otimes 2}$. For simplicity, let us begin by assuming that $d_{P_1} = d_P = d_S$ such that $\Pi_n = | n \rangle\langle n |_{P_1}$. Then we can select the projectors $\{| n \rangle\langle n |_{P_1} \otimes | \omega_i \rangle\langle \omega_i |_{P_2}\}_{n,i}$ and coarse-grain over the outcomes i for the second copy. Before the coarse-graining, the measurement operators and POVM elements are thus defined as

$$M_{n,i} | \phi \rangle_S = \mathbb{1}_S \otimes \langle n |_{P_1} \otimes \langle \omega_i |_{P_2} \left(U_{SP_1} \otimes \mathbb{1}_{P_2} \right) | \phi \rangle_S \otimes | \omega \rangle_{P_1 P_2}. \tag{23.41}$$

Then, as in the coarse-graining discussed before, the post-measurement system state $\rho(n)$, given that the measurement has resulted in the outcome n is as in eqn (23.32), and the probability $p(n) = \text{Tr}(E_n \rho_S)$ for obtaining this outcome, can be calculated from the coarse-grained POVM elements $E_n = \sum_i E_{n,i}$, where

$$E_{n,i} = \text{Tr}_{P_1 P_2} \left(\mathbb{1}_S \otimes | \omega \rangle\langle \omega |_{P_1 P_2} U_{SP_1}^\dagger \otimes \mathbb{1}_{P_2} \mathbb{1}_S \otimes | n \rangle\langle n |_{P_1} \otimes | \omega_i \rangle\langle \omega_i |_{P_2} U_{SP_1} \otimes \mathbb{1}_{P_2} \right). \tag{23.42}$$

However, in general, the pointer in a non-ideal measurement can have a higher dimension than the measured system, such that further coarse-graining is in order. That is, one has $\Pi_n = | \psi_i^{(n)} \rangle\langle \psi_i^{(n)} |$, and consequently we have to introduce a third index for the POVM elements

$$E_{n,i,j} = \text{Tr}_{P_1 P_2} \left(\mathbb{1}_S \otimes | \omega \rangle\langle \omega |_{P_1 P_2} U_{SP_1}^\dagger \mathbb{1}_S \otimes | \psi_i^{(n)} \rangle\langle \psi_i^{(n)} |_{P_1} \otimes | \omega_j \rangle\langle \omega_j |_{P_2} U_{SP_1} \right), \tag{23.43}$$

where we have dropped the identity operators on the unitaries for ease of notation. Using the completeness relation $\sum_j |\omega_j\rangle\langle\omega_j|_{P_2} = \mathbb{1}_{P_2}$, the coarse-grained POVM elements then become

$$E_n = \sum_{i,j} E_{n,i,j} = \mathrm{Tr}_{P_1 P_2}\left(\mathbb{1}_S \otimes |\omega\rangle\langle\omega|_{P_1 P_2}\, U^\dagger_{SP_1}\, \mathbb{1}_S \otimes \Pi_n \otimes \mathbb{1}_{P_2}\, U_{SP_1}\right)$$

$$= \mathrm{Tr}_{P_1}\left(\mathbb{1}_S \otimes \rho_{P_1}\, U^\dagger_{SP_1}\, \mathbb{1}_S \otimes \Pi_n\, U_{SP_1}\right). \tag{23.44}$$

We then assume that the measurement is unbiased and split the unitary $U = V\tilde{U}$ into $\tilde{U}$ and V according to eqn (23.40) to obtain

$$E_n = \mathrm{Tr}_{P_1}\left(\mathbb{1}_S \otimes \rho_{P_1}\, \tilde{U}^\dagger_{SP_1}\, V^\dagger_{SP_1}\, \mathbb{1}_S \otimes \Pi_n\, V_{SP_1}\, \tilde{U}_{SP_1}\right) \tag{23.45}$$

$$= \mathrm{Tr}_{P_1}\left(\mathbb{1}_S \otimes \rho_{P_1}\, \tilde{U}^\dagger_{SP_1} \sum_{i,k} |n\rangle\langle n|_S \otimes |\psi_k^{(i)}\rangle\langle\psi_k^{(i)}|_{P_1}\, \tilde{U}_{SP_1}\right)$$

$$= \mathrm{Tr}_{P_1}\left(\mathbb{1}_S \otimes \rho_{P_1}\, \tilde{U}^\dagger_{SP_1}\, |n\rangle\langle n|_S \otimes \mathbb{1}_{P_1}\, \tilde{U}_{SP_1}\right), = \mathrm{Tr}_{P_1}\left(|n\rangle\langle n|_S \otimes \rho_{P_1}\right) = |n\rangle\langle n|_S.$$

So, as desired for unbiased measurements, the probability for obtaining the pointer outcome n is just determined by the initial system state, $p(n) = \mathrm{Tr}(E_n \rho_S) = \langle n|\rho_S|n\rangle$. The measurement operators for this POVM before coarse-graining now also feature three indices and are given by

$$M_{n,i,j}|\phi\rangle_S = \mathbb{1}_S \otimes \langle\psi_i^{(n)}|_{P_1} \otimes \langle\omega_j|_{P_2}\, U_{SP_1}\, |\phi\rangle_S \otimes |\omega\rangle_{P_1 P_2}. \tag{23.46}$$

The corresponding post-measurement state after obtaining outcome n (but without having or disregarding information about i and j) is then

$$\rho(n) = \frac{1}{p(n)} \sum_{i,j} M_{n,i,j}\, \rho_S\, M^\dagger_{n,i,j} \tag{23.47}$$

$$= \frac{1}{p(n)} \sum_{i,j} \mathbb{1}_S \otimes \langle\psi_i^{(n)}|_{P_1} \otimes \langle\omega_j|_{P_2}\, U_{SP_1}\, \rho_S \otimes |\omega\rangle\langle\omega|_{P_1 P_2}\, U^\dagger_{SP_1}\, \mathbb{1}_S \otimes |\psi_i^{(n)}\rangle_{P_1} \otimes |\omega_j\rangle_{P_2}.$$

Since we have $_{P_2}\langle\omega_j|\omega\rangle_{P_1 P_2} = \sqrt{q_j}\,|\omega_j\rangle_{P_1}$ and $\rho_{P_1} = \sum_j q_j |\omega_j\rangle\langle\omega_j|$, we can further simplify this to

$$\rho(n) = \frac{1}{p(n)} \sum_i \mathbb{1}_S \otimes \langle\psi_i^{(n)}|_{P_1}\, U_{SP_1}\, \rho_S \otimes \rho_{P_1}\, U^\dagger_{SP_1}\, \mathbb{1}_S \otimes |\psi_i^{(n)}\rangle_{P_1} \tag{23.48}$$

$$= \frac{1}{p(n)} \sum_i \mathbb{1}_S \otimes \langle\psi_i^{(n)}|_{P_1}\, V_{SP_1}\tilde{U}_{SP_1}\, \rho_S \otimes \rho_{P_1}\, \tilde{U}^\dagger_{SP_1} V^\dagger_{SP_1}\, \mathbb{1}_S \otimes |\psi_i^{(n)}\rangle_{P_1}$$

$$= \frac{1}{p(n)} \sum_{i,j,k} |j\rangle\langle n|_S \otimes \left(\langle\psi_i^{(j)}|_{P_1}\, \tilde{U}_{P_1}^{(n)}\right) \rho_S \otimes \rho_{P_1}\, |n\rangle\langle k|_S \otimes \left(\tilde{U}_{P_1}^{(n)\dagger}|\psi_i^{(k)}\rangle_{P_1}\right)$$

$$= \frac{1}{p(n)} \sum_{j,k} |j\rangle\langle k|_S\, \langle n|\rho_S|n\rangle \sum_i \langle\psi_i^{(j)}|_{P_1}\, \tilde{U}_{P_1}^{(n)}\, \rho_{P_1}\, \tilde{U}_{P_1}^{(n)\dagger}|\psi_i^{(k)}\rangle_{P_1}$$

$$= \sum_{j,k} |j\rangle\langle k|_S \sum_i \langle\psi_i^{(j)}|_{P_1}\, \tilde{U}_{P_1}^{(n)}\, \rho_{P_1}\, \tilde{U}_{P_1}^{(n)\dagger}|\psi_i^{(k)}\rangle_{P_1} = \sum_{j,k} \rho_{jk}(n)\, |j\rangle\langle k|_S,$$

where we have inserted $\tilde{U}$ and V as in eqn (23.40), we have used $p(n) = \langle n | \rho_S | n \rangle$, and we have defined the matrix elements

$$\rho_{jk}(n) = \langle j | \rho(n) | k \rangle = \sum_i \langle \psi_i^{(j)} |_{P_1} \tilde{U}_{P_1}^{(n)} \rho_{P_1} \tilde{U}_{P_1}^{(n)\dagger} | \psi_i^{(k)} \rangle_{P_1} . \qquad (23.49)$$

It is interesting to note here that the $\rho_{jk}(n)$, and hence also the post-measurement states $\rho(n)$, are independent of the initial system state ρ_S. Indeed, this is always the case for ideal projective measurements: Given an outcome n a perfect projective measurement always leaves the system in the state $| n \rangle$, independently of the state the system was in before. As we see, this initial-state independence is also recovered for non-ideal measurements, if they are unbiased.

We have thus learned that, although every POVM can be viewed abstractly as a perfect projective measurement on a larger Hilbert space, in practice, even projective measurements are typically not perfect and should hence be thought of as idealized versions of POVMs, even though ideal projective measurements may be approximated arbitrarily well by investing resources into suitably preparing the measurement apparatus.

23.4 Distinguishing Quantum States

23.4.1 Distinguishing Orthogonal States

Previously, we have encountered POVMs as mathematically more general versions of projective measurements, but also as descriptions of the in practice non-ideal realizations of theoretically ideal projective measurements. Now, we turn to a particular problem—distinguishing quantum states—that illustrates that performing a measurement represented by a (specific) POVM can even be advantageous as compared to (ideal) projective measurements. For many tasks in quantum information it is of importance to reliably distinguish between different quantum states.

Suppose Alice prepares a state $| \psi_i \rangle$ from a previously agreed upon set $\{| \psi_j \rangle\}_{j=1,\dots,n}$ of states from a Hilbert space $\mathcal{H}$, with $\dim \mathcal{H} = d > n$ and sends it to Bob. The goal of Bob is then to identify the state Alice sent *with certainty* using only a single measurement. That is, Bob performs a POVM with elements E_j and outcomes j, and applies a rule $f(.)$ to guess the index i if the outcome j was obtained, i.e., such that $f(j) = i$. The states $\{| \psi_i \rangle\}$ can be perfectly distinguished if the probability to obtain the outcome j such that $f(j) = i$ is 1 when the state $| \psi_i \rangle$ has been prepared.

First, let us find a POVM with elements $\{E_i\}$ that perfectly distinguishes between *orthogonal* states $| \psi_i \rangle$, i.e., where $\langle \psi_i | \psi_j \rangle = 0$ for $i \neq j$. This is indeed rather simple. We begin by assigning the POVM elements

$$E_i = | \psi_i \rangle\langle \psi_i | \quad \text{for} \quad i = 1, \dots, n . \qquad (23.50)$$

Since $n < d$, we have to add some POVM elements, e.g., we can pick

$$E_0 = \mathbb{1} - \sum_{i=1}^{n} E_i, \tag{23.51}$$

such that $\sum_{i=0,1,\dots,n} E_i = \mathbb{1}$. Then, clearly, if one obtains one of the outcomes $i = 1,\dots,n$, one concludes—with certainty—that the system was in the state $|\,\psi_i\,\rangle$ before the measurement. Technically speaking, if promised that one of these n states has been prepared, the outcome corresponding to E_0 is never obtained, but even if it was obtained in this scenario, one would conclude, again, with certainty, that the promise was broken.

23.4.2   Distinguishing Non-Orthogonal States

Now let us see what happens if the states $|\,\psi_j\,\rangle$ are *non-orthogonal*. In particular, we focus on the case of two non-orthogonal states $|\,\psi_1\,\rangle$ and $|\,\psi_2\,\rangle$, with $\langle\,\psi_1\,|\,\psi_2\,\rangle \neq 0$. As we will show now using a proof by contradiction no POVM exists that *perfectly* distinguishes between two non-orthogonal states. Suppose that it was possible to perform a measurement that distinguishes between $|\,\psi_1\,\rangle$ and $|\,\psi_2\,\rangle$ perfectly, such that when $|\,\psi_1\,\rangle$ $(|\,\psi_2\,\rangle)$ is prepared, the probability of measuring j such that $f(j) = 1$ $(f(j) = 2)$ is one. For such a case we can define the POVM elements E_i such that

$$E_i = \sum_{j \text{ s.t. } f(j)=i} M_j^\dagger M_j. \tag{23.52}$$

If the measurement is to distinguish perfectly, we also have

$$\langle\,\psi_1\,|\,E_1\,|\,\psi_1\,\rangle = 1 = \langle\,\psi_2\,|\,E_2\,|\,\psi_2\,\rangle. \tag{23.53}$$

Now, since $\sum_i E_i = \mathbb{1}$, we have $\sum_i \langle\,\psi_1\,|\,E_i\,|\,\psi_1\,\rangle = 1$, and since we further already know that $\langle\,\psi_1\,|\,E_1\,|\,\psi_1\,\rangle = 1$ this implies that $\langle\,\psi_1\,|\,E_2\,|\,\psi_1\,\rangle = 0$. The POVM element E_2 is a positive operator by definition, so its square root is well defined, and because the inner product is non-degenerate, we must have $\sqrt{E_2}\,|\,\psi_1\,\rangle = 0$. Now, because the states in question are non-orthogonal, $\langle\,\psi_1\,|\,\psi_2\,\rangle \neq 0$, $|\,\psi_2\,\rangle$ must have a non-vanishing component in the direction of $|\,\psi_1\,\rangle$, and we may hence write $|\,\psi_2\,\rangle = \alpha\,|\,\psi_1\,\rangle + \beta\,|\,\varphi\,\rangle$, where $|\,\varphi\,\rangle$ is some state that is orthogonal to $|\,\psi_1\,\rangle$, $\langle\,\varphi\,|\,\psi_1\,\rangle = 0$, and where $\alpha, \beta \in \mathbb{C}$ with $|\alpha|^2 + |\beta|^2 = 1$, and more specifically, $|\beta| < 1$. However, we can now calculate

$$\sqrt{E_2}\,|\,\psi_2\,\rangle = \sqrt{E_2}(\alpha\,|\,\psi_1\,\rangle + \beta\,|\,\varphi\,\rangle) = \beta\,\sqrt{E_2}\,|\,\varphi\,\rangle. \tag{23.54}$$

Consequently, we can further compute

$$\langle\,\psi_2\,|\,E_2\,|\,\psi_2\,\rangle = |\beta|^2\,\langle\,\varphi\,|\,E_2\,|\,\varphi\,\rangle \leq \sum_i |\beta|^2\,\langle\,\varphi\,|\,E_i\,|\,\varphi\,\rangle = |\beta|^2, \tag{23.55}$$

where we have used the fact that the POVM elements are positive operators such that $\langle\,\varphi\,|\,E_i\,|\,\varphi\,\rangle \geq 0\,\forall i$. Since we now have $\langle\,\psi_2\,|\,E_2\,|\,\psi_2\,\rangle \leq |\beta|^2 < 1$, we arrive at a contradiction to the assumption in eqn (23.53). In other words, it is not possible to find a POVM that perfectly distinguishes two non-orthogonal states.

In general, one can think of a scenario where one is provided a quantum system that is promised to be in one of two states, ρ_1 or ρ_2, with equal probability. Then, the optimal probability p_{opt} for guessing which of the two quantum states was provided can be related to the *trace distance* $D_{\text{tr}}(\rho_1, \rho_2)$ via (Barnett, 2009, Chap. 4)

$$p_{\text{opt}} = \tfrac{1}{2}\big(1 + D_{\text{tr}}(\rho_1, \rho_2)\big), \tag{23.56}$$

where the trace distance between two density operators is

$$D_{\text{tr}}(\rho_1, \rho_2) = \tfrac{1}{2}\|\rho_1 - \rho_2\|_{\text{tr}} = \tfrac{1}{2}\text{Tr}\sqrt{(\rho_1 - \rho_2)^2} = \tfrac{1}{2}\sum_i |\lambda_i|. \tag{23.57}$$

Here, the λ_i are the eigenvalues of the Hermitian operator $\rho_1 - \rho_2$, and $\|\cdot\|_{\text{tr}}$ denotes the *trace norm*, which is defined for arbitrary operators A as $\|A\|_{\text{tr}} = \text{Tr}\sqrt{AA^\dagger}$. One way to evaluate the trace norm for general (non-Hermitian) operators is via the singular values of A. That is, there always exist orthonormal bases $\{|u_i\rangle\}_i$ and $\{|v_j\rangle\}_j$, such that $A = \sum_i s_i |u_i\rangle\langle v_i|$, where the real, non-negative s_i are the singular values of A. It is then easy to see that the trace norm is nothing but the sum of the singular values,

$$\|A\|_{\text{tr}} = \text{Tr}\sqrt{AA^\dagger} = \text{Tr}\sqrt{\sum_{i,j} s_i s_j |u_i\rangle\langle v_i| v_j\rangle\langle u_j|}$$

$$= \text{Tr}\sqrt{\sum_i s_i^2 |u_i\rangle\langle u_i|} = \text{Tr}\Big(\sum_i |s_i| |u_i\rangle\langle u_i|\Big) = \sum_i s_i. \tag{23.58}$$

Here we can note that the trace norm is the special case $p = 1$ of the family of Schatten p norms from eqn (11.95), and the trace distance induced by the trace norm is a metric on the Hilbert–Schmidt space that satisfies the properties stated for the Hilbert–Schmidt distance in (11.97).

Returning to the problem of distinguishing two quantum states, let us consider the simple example of a single qubit with Hilbert space $\mathcal{H} = \mathbb{C}^2$ and computational basis $\{|0\rangle, |1\rangle\}$. Then, let $|\psi_1\rangle = |0\rangle$ and $|\psi_2\rangle = |+\rangle$, where $|\pm\rangle = \frac{1}{\sqrt{2}}(|0\rangle \pm |1\rangle)$. The eigenvalues of $\rho_1 - \rho_2 = |0\rangle\langle 0| - |+\rangle\langle +|$ evaluate to $\pm\frac{1}{\sqrt{2}}$, such that $p_{\text{opt}} = \tfrac{1}{2}\big(1 + \frac{1}{\sqrt{2}}\big) \approx 0.854$. However, this still leaves a probability of around 15% to misidentify the quantum state.

Alternatively, we have the option of constructing a POVM that distinguishes the two states in such a way that there are measurement outcomes that unambiguously identify one of the non-orthogonal states, while other outcomes simply give no information at all. Although such a POVM may not always succeed (i.e., it only succeeds probabilistically), it *never misidentifies* the states in question. To see how this works, let us reconsider the previous example where $|\psi_1\rangle = |0\rangle$ and $|\psi_2\rangle = |+\rangle$. We then construct the POVM $\{E_i\}_{i=1,2,3}$ with elements

$$E_1 = a|-\rangle\langle -|, \quad E_2 = b|1\rangle\langle 1|, \quad E_3 = \mathbb{1} - a|-\rangle\langle -| - b|1\rangle\langle 1|. \tag{23.59}$$

Clearly, if the outcome of the measurement is 1, then the state in question must be $|\psi_1\rangle$, because the probability for obtaining outcome 1 for state $|\psi_2\rangle$ is zero. Similarly, when the outcome 2 is observed, the state must have been $|\psi_2\rangle$, since it could not have been $|\psi_1\rangle$. However, if the third outcome is obtained, then no conclusive statement can be made; outcome 3 is compatible with either of the two initial states. The only remaining question here is: What are the probabilities for obtaining the three outcomes? That is, we still have to determine the allowed range of a and b so that E_3 yields a positive operator. Obviously, we must have $a \geq 0$ and $b \geq 0$ so that $E_1 \geq 0$ and $E_2 \geq 0$, respectively. Then, let us consider the case where we do not wish to bias the measurement to work better (or worse) for one of the two initial states as compared to the other, so we choose $a = b$. The third POVM element then takes the form

$$E_3 = \left(1 - \tfrac{a}{2}\right)|0\rangle\langle 0| + \left(1 - \tfrac{3a}{2}\right)|1\rangle\langle 1| + \tfrac{a}{2}\left(|0\rangle\langle 1| + |1\rangle\langle 0|\right). \tag{23.60}$$

When we diagonalize this matrix, we find the eigenvalues $\lambda_{\pm} = 1 - a\left(1 \pm \tfrac{1}{\sqrt{2}}\right)$. For both of these values to be positive, we have $0 \leq a \leq \tfrac{\sqrt{2}}{1+\sqrt{2}}$. Therefore, if one believes that either initial state is equally likely to begin with, the optimal success probability for distinguishing the states by means of this POVM is given by $\tfrac{\sqrt{2}}{1+\sqrt{2}} \approx 0.586$.

24
Quantum Metrology

24.1   Quantum Parameter Estimation

As we have learned already early on in this book (in Section 2.4 of Part I), many physical quantities can be measured in a straightforward way, at least theoretically. In particular, measurements of observable quantities—or, to be more precise, directly observable quantities—can be represented as projective measurements (see Chapter 23). The results of these measurements provide data that we process with statistical methods to obtain estimates for the measured quantities, along with (estimates of) the corresponding confidence intervals. Meanwhile, many physical quantities, such as phases, interaction strengths, unknown states, or quantum channels and other parameters, are not directly observable but have to be estimated based on the measurement of other, themselves directly measurable quantities. While this may seem cumbersome initially, it also opens up the opportunity to determine the best way of estimating a given parameter (or set thereof). The field of study that (among other interesting problems such as quantum state discrimination, see, e.g., Audenaert *et al.*, 2007; Audenaert *et al.*, 2008) investigates such estimation procedures is *quantum metrology*. From early seminal papers such as Wineland *et al.* (1992) and Huelga *et al.* (1997), and more recent developments (Giovannetti *et al.*, 2006), the field of metrology comprises a vast body of applications and corresponding literature (see, for instance, the reviews Paris; Giovannetti *et al.*, 2009, 2011; Tóth and Apellaniz, 2014; Demkowicz-Dobrzański *et al.*, 2015; Sidhu and Kok 2020). In this chapter, we give a brief overview of some elementary concepts and approaches to parameter estimation and the potential advantages one may obtain from using genuine quantum features.

24.1.1   Measurement Statistics

Before we discuss the estimation of parameters that cannot be measured directly, let us recall how *directly measurable quantities* are estimated—how observables are measured. Observables are represented as Hermitian operators $X = \sum_i x_i \, | \, i \, \rangle\langle \, i \, |$ and measuring these quantities mathematically corresponds to projective measurements with respect to the observable's eigenbasis $\{| \, i \, \rangle\}_i$. The eigenvalues $x_i \in \mathbb{R}$ represent the possible measurement outcomes. For a given quantum state ρ, we can predict the probability of obtaining the outcome labelled "i" to be $p_i = \mathrm{Tr}(\rho \, | \, i \, \rangle\langle \, i \, |)$ and the predicted mean value is the expectation (or expected) value $\langle \, X \, \rangle_\rho = \mathrm{Tr}(X\rho)$. When an actual measurement is performed on a quantum system, a particular result (an outcome) is obtained. In the language of probability theory, this can be described by a random variable with possible values x_i (which may form a discrete or continuous set) and a corresponding probability distribution $\{p_i\}_i$. When repeating such a measurement procedure a total of N times, we obtain outcomes $x_{i_1}, x_{i_2}, \ldots, x_{i_N}$, where we have used a double-subscript since any of the possible values x_i may in principle be obtained multiple times (or not at all). For instance, for a measurement of the spin along the z axis of a spin-$\frac{1}{2}$ particle, we have $X \equiv S_z = \frac{\hbar}{2}\sigma_z$, with two possible outcomes, $x_1 = +\frac{\hbar}{2}$ and $x_2 = -\frac{\hbar}{2}$, see Chapter 8. Repeating this measurement for identically prepared systems might, for instance, result in outcomes $+\frac{\hbar}{2}, +\frac{\hbar}{2}, -\frac{\hbar}{2}$, and so on, such that $i_1 = 1$, $i_2 = 1$, $i_3 = 2$, etc. The outcomes $x_{i_1}, x_{i_2}, \ldots, x_{i_N}$ can then be collected into a *sample mean*

$$\bar{X}^{(N)} = \tfrac{1}{N} \sum_{j=1}^{N} x_{i_j}, \tag{24.1}$$

where we have indicated the size N of the sample that results in this particular mean value for clarity. The sample mean is generally not identical to the expected value $\langle X \rangle$ that we wish to estimate. However, provided that the preparation of ρ and the measurements are carried out in such a way that the corresponding N individual random variables are independent and identically distributed (i.i.d.), the sample mean is an *unbiased estimator* of the expected value. That is, the average of the sample mean over all possible samples of size N matches $\langle X \rangle$,

$$\langle \bar{X}^{(N)} \rangle = \langle \tfrac{1}{N} \sum_{j=1}^{N} x_{i_j} \rangle = \tfrac{1}{N} \sum_{j=1}^{N} \langle x_{i_j} \rangle = \tfrac{1}{N} \sum_{j=1}^{N} \langle X \rangle = \langle X \rangle. \tag{24.2}$$

Moreover, the law of large numbers further suggests that the sample mean $\bar{X}^{(N)}$ converges to the expected value $\langle X \rangle_\rho$ as $N \to \infty$ (see Ross, 2009, Chap. 8 for more details). Although equality only holds asymptotically, i.e., $\lim_{N \to \infty} \bar{X}^{(N)} = \langle X \rangle$, for finite N, *Hoeffding's inequality* informs us that the probability $P(|\bar{X}^{(N)} - \langle X \rangle| \geq \varepsilon)$ for a deviation of $\bar{X}^{(N)}$ from $\langle X \rangle$ by some amount ε is exponentially suppressed in both ϵ and N (Hoeffding, 1963), that is,

$$P(|\bar{X}^{(N)} - \langle X \rangle| \geq \varepsilon) \leq 2e^{-2N\varepsilon^2}. \tag{24.3}$$

Given some desired precision (specified by ε) and level of confidence $1 - C$ (where C is the probability for a mistake), Hoeffding's inequality can be used to determine a lower bound on the number of required measurements. By setting $C = P(\bar{X}^{(N)} \notin [\langle X \rangle - \varepsilon, \langle X \rangle + \varepsilon]) \leq 2e^{-2N\varepsilon^2}$, we find that, for given ε and C, the number N cannot exceed $\ln(2/C)/(2\varepsilon^2)$. Or, in other words, to obtain a precision of at least ε with confidence of at least $1 - C$, one requires

$$N \geq \frac{\ln(2/C)}{2\varepsilon^2}. \tag{24.4}$$

In practice, one of course bases confidence on the actually obtained data. This means, one is interested in being able to quantify how likely it is that the mean of a sample of size N is within some confidence interval of the "true" value. Loosely speaking, one wishes to be able to make a quantitative statement about the likely range of mean values if one were to repeat the same experiment (with N measurements each). The first step towards such a statement is the estimation of the variance of X. That is, the expected value of the variance—the *uncertainty* of an observable—is

$$\mathrm{Var}(X) = (\Delta X)^2 = \left\langle \left(X - \langle X \rangle\right)^2 \right\rangle = \langle X^2 \rangle - \langle X \rangle^2. \tag{24.5}$$

To estimate the variance, one can calculate the *mean square error* (MSE), that is, the arithmetic mean of the square deviations $(x_{i_j} - \bar{X}^{(N)})^2$ from the sample mean $\bar{X}^{(N)}$. Similarly to the sample mean, one finds that the sample variance converges to the (expected) variance as $N \to \infty$, $\lim_{N \to \infty} \frac{1}{N} \sum_{j=1}^{N} (x_{i_j} - \bar{X}^{(N)})^2 = \mathrm{Var}(X)$. However, for any finite N, the sample variance is a biased estimator of the variance if the true population mean is unknown (see Montgomery and Runger, 1994, p. 193 and Ross, 2009, Chap. 7). That is, the population variance disagrees with the expected value of the sample variance by a factor $N/(N-1)$ (known as Bessel's correction), and one therefore defines the *unbiased sample variance* as

$$(\Delta X^{(N)})^2 = \frac{1}{N-1} \sum_{j=1}^{N} (x_{i_j} - \bar{X}^{(N)})^2. \tag{24.6}$$

Bessel's correction can be understood as accounting for the fact that there are only $N-1$ independent deviations $(x_{i_j} - \bar{X}^{(N)})$, because their sum is constrained by $\sum_j (x_{i_j} - \bar{X}^{(N)}) = N(\bar{X}^{(N)} - \bar{X}^{(N)}) = 0$, whereas there are N independent values of x_{i_j}. The variance characterizes the spread of the individual values x_{i_j}, and estimating the variance hence allows us to make predictions on the likely deviation from the mean (depending on the underlying probability distribution) for the outcomes of individual future measurements.

However, to quantify the precision of the mean value itself we further need to recall the *central limit theorem*, which we shall state here only in its essence. For more details see, for instance, Montgomery and Runger (1994, p. 240). The central limit theorem assures us that, for a random (in particular, i.i.d. but otherwise arbitrarily distributed) sample of size N taken from a finite or infinite population, the distribution of mean values $\bar{X}^{(N)}$ is Gaussian with variance $\mathrm{Var}(X)/N$, centred around $\langle X \rangle$. Most importantly, this means that the distribution of sample means becomes ever more narrow with increasing sample size N. More specifically, since the variance of the mean decreases linearly with N, one can obtain estimates $\bar{X}^{(N)}$ and $(\Delta X^{(N)})^2$ (of $\langle X \rangle$ and $\mathrm{Var}(X)$, respectively) with any desired precision by making N sufficiently large. Usually, the final result is then specified in terms of a confidence interval (see, e.g., Montgomery and Runger, 1994, Chap. 8), i.e., in terms of the desired level of confidence and the *standard deviation* of the mean $\sigma = \Delta X^{(N)}/\sqrt{N}$.

For the measurement of an observable, there is not much freedom to improve upon this procedure. If we want to estimate, say, the energy of a quantum system in the state ρ, then the observable is the system Hamiltonian. For each of the N measurements, the system must be prepared in the state ρ and the measurement must be a projective measurement in the eigenbasis of H. However, as we shall see in Section 24.1.2, the estimation of a parameter such as a phase leaves us some options, in particular, to select the best probe states and measurements for a given problem.

24.1.2 Local Parameter Estimation

Let us now consider a typical parameter estimation protocol of the variety usually referred to as the "local" or "frequentist" approach, see, e.g., Demkowicz-Dobrzański *et al.* (2015, p. 30), as illustrated in Fig. 24.1. The name "local" will become apparent in due course (see page 785). The problem is the following: We wish to estimate a parameter θ that is not directly measurable but which determines a transformation acting on a suitable (quantum) system—the probe—that we can control. The parameter θ is assumed to have a "true" value that is unknown a priori but which could be stated to any desired precision if it was known. Although we will here focus on the case where θ is a single real parameter with a continuous range of possible values, this need not be the case in general. In particular, θ might be a collection of such parameters (multi-parameter estimation, see, e.g., Sidhu and Kok, 2020, p. 16) that one wishes to estimate simultaneously. For instance, the parameters in question could be the Euler angles of a three-dimensional rotation. Estimating these parameters can be understood as aligning two reference frames that differ by a spatial rotation. When the axis of rotation is fixed but the angle is unknown, one recovers the single-parameter estimation problem known as phase estimation. In order to estimate the parameter (or collection of parameters), one prepares the probe on which the transformation then encodes information about the parameter. Subsequently, a measurement is performed on the system. Repeating this procedure a number of times one collects the individual outcomes, which are processed using statistical tools to obtain an estimate of the true value of θ.

Let us now take a closer look at the individual steps of the estimation procedure. First, the probe system is prepared in a chosen (quantum) state ρ. The choice of this probe is usually assumed to be a degree of freedom that can be optimized over, but restrictions might apply that limit this choice, e.g., the average energy of the state. The probe system is then subjected to a transformation that can, in general, be represented as a completely positive and trace-preserving (CPTP) map Λ_θ (see Chapter 21) that takes ρ to $\rho(\theta) = \Lambda_\theta[\rho]$. Generally speaking, the encoding transformation of course need not be unitary, take, for example, the problem of estimating the loss in a channel. However, here, we shall focus on what is referred to as *Hamiltonian parameter estimation*. That is, the transformation Λ_θ is a unitary map,

$$\Lambda_\theta[\rho] = U_\theta \, \rho \, U_\theta^\dagger = e^{-i\tilde{H}\theta} \, \rho \, e^{i\tilde{H}\theta} =: \rho(\theta), \tag{24.7}$$

where the generating "Hamiltonian"[1] $\tilde{H}$ is known exactly but the (dimensionless) coupling parameter θ is to be estimated. The transformation (in particular, the Hamiltonian $\tilde{H}$ and the value of θ) is assumed to be the same in every round.

---

[1] Here, we consider a dimensionless self-adjoint operator $\tilde{H}$ rather than an actual Hamiltonian H with dimension energy, and we have added the tilde as a reminder.

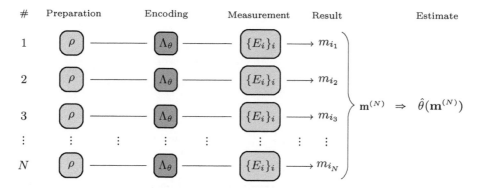

#	Preparation	Encoding	Measurement	Result	Estimate

Fig. 24.1 Parameter estimation. In the standard paradigm of (local/frequentist) parameter estimation, N independent rounds of preparation and measurement are performed. In each round, a probe state ρ is prepared, on which the parameter θ is imprinted by way of a (fixed) transformation Λ_θ (generally a CPTP map but often chosen as a unitary map U_θ), before the encoded state $\rho(\theta) = \Lambda_\theta[\rho]$ is measured (via a POVM with elements E_i and possible outcomes m_i). The individual results $m_{i_j} \in \{m_i\}_i$ for $j = 1, \ldots, N$ are finally collected into a total result $\mathbf{m}^{(N)} = (m_{i_1}, m_{i_2}, \ldots, m_{i_N})$, which is processed using an estimator function $\hat{\theta}$ to obtain an estimate $\hat{\theta}(\mathbf{m}^{(N)})$ of θ.

After the parameter has been encoded in the state $\rho(\theta)$, one performs a series of measurements, each represented by a POVM with elements $\{E_i\}_i$ and corresponding possible outcomes m_i (see Chapter 23). After N measurements one can gather the individual outcomes $m_{i_1}, m_{i_2}, \ldots, m_{i_N} \in \{m_i\}$ in a list $\mathbf{m}^{(N)} := (m_{i_1}, \ldots, m_{i_N})$ that we can call "the outcome" or the "measurement data" of the entire measurement procedure. This outcome is finally processed using a suitable estimator function $\hat{\theta}(m)$ to provide an estimate of the true value of θ, where we write m instead of $\mathbf{m}^{(N)}$ to simplify the notation for the following discussion.

As in Section 24.1.1, we are interested in obtaining estimates without *bias*. The estimator $\hat{\theta}(m)$ is called *unbiased* if it assigns the value θ on average,

$$\langle \hat{\theta}(m) \rangle = \sum_m p(m|\theta)\, \hat{\theta}(m) = \theta, \tag{24.8}$$

i.e., if the expected value of the estimator yields the true parameter. Here, the conditional probability that the sample (of size N) results in the outcome m, given the parameter value θ, is

$$p(m|\theta) = \mathrm{Tr}\big(E_m \rho(\theta)^{\otimes N}\big) = \mathrm{Tr}\big(E_{m_{i_1}} \otimes \ldots \otimes E_{m_{i_N}} \rho(\theta)^{\otimes N}\big) = \prod_{j=1}^{N} \mathrm{Tr}\big(E_{m_{i_j}} \rho(\theta)\big). \tag{24.9}$$

The unbiasedness of the estimator ensures the *accuracy* of the measurement procedure in the sense that the estimates $\hat{\theta}(m)$ will be distributed around the true value θ, but

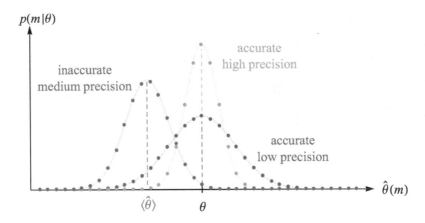

Fig. 24.2 Accuracy versus precision. Illustration of the accuracy and precision of estimators. The dots at positions $(\hat{\theta}(m), p(m|\theta))$ represent pairs of estimates $\hat{\theta}(m)$ (horizontal axis) and probabilities $p(m|\theta)$ (vertical axis) for three different estimators and different outcomes m. Two of the estimators are unbiased and hence accurate, but provide different precision (widths of the distributions of estimates). The third estimator achieves a precision in between the performance of the other two, but is biased, i.e., inaccurate in the sense that the average estimate is off from the true value θ.

the individual estimates might still be far away from the mean. Loosely speaking, distributions of different "width" can still have the same mean: The *precision* of the measurement is independent of its accuracy, as illustrated in Fig. 24.2. This notion of precision can be captured by the variance $V[\hat{\theta}]$ of the estimator. For instance, in many situations the mean square error (MSE) variance, given by

$$V[\hat{\theta}] = \sum_m p(m|\theta) \left(\hat{\theta}(m) - \theta \right)^2, \tag{24.10}$$

or the associated standard deviation $\sigma = \sqrt{V[\hat{\theta}]}$, are useful quantifiers of the "width" of a distribution.

In a parameter estimation procedure it is hence desirable to increase the precision as much as possible, or, equivalently, to obtain a variance that is as small as possible given the available resources, without introducing a bias. To achieve this, we have three ingredients that we can optimize over:

    (i) the probe state $\rho^{\otimes N}$,
    (ii) the measurement $\{E_m\}_m$,
    (iii) the (unbiased) estimator $\hat{\theta}(m)$.

Although determining the best estimation strategy in terms of these three variables may appear to be a daunting optimization problem, it can be addressed systematically, as we will see in the following.

24.1.3 The Cramér–Rao Bound

To optimize the estimation procedure, one may start by attempting to find the best way of processing the data (the measurement results) from a setup with fixed probe state ρ and fixed measurements represented by a POVM $\{E_m\}_m$. That is, one can try to determine the unbiased estimator that gives the minimal possible variance for such a choice. As it turns out, this is in generally a difficult problem with some subtleties since such a choice depends on the true value of the parameter and even on the number N of repetitions. However, one may (at least temporarily) sidestep the particular question of choosing an estimator by way of the *Cramér–Rao* bound on the variance of the estimated parameter θ. This bound, which we shall state as a theorem below, was independently found in the 1940s by *Maurice Fréchet* (1943), *Georges Darmois* (1945), *Calyampudi Radhakrishna Rao* (1992), and *Harald Cramér* (1946), but we will follow the typical (see, e.g., Demkowicz-Dobrzański *et al.*, 2015) convention and refer to it as the *Cramér–Rao* bound for the sake of brevity.

Theorem 24.1 (Cramér–Rao Inequality)

Let $p(m|\theta)$ be the conditional probability for obtaining the measurement outcome m given the true parameter θ. For any unbiased estimator $\hat{\theta}(m)$, the corresponding variance $V[\hat{\theta}]$ is then bounded from below by

$$V[\hat{\theta}] \geq \frac{1}{I\big(p(m|\theta)\big)}$$

where $I\big(p(m|\theta)\big)$ is the Fisher information *given by*

$$I\big(p(m|\theta)\big) = \sum_m p(m|\theta) \left(\frac{\partial}{\partial \theta} \log p(m|\theta) \right)^2 .$$

Proof For the proof of the Cramér–Rao bound from Theorem 24.1, we will follow the treatment by *Harry van Trees* (1968) and *Roy Frieden* (2004). We begin by rewriting the unbiasedness condition from eqn (24.8) for an estimator $\hat{\theta}(m)$. Using the normalization $\sum_m p(m|\theta) = 1$ of the conditional probability we write

$$\sum_m p(m|\theta)\,\hat{\theta}(m) - \theta \sum_m p(m|\theta) = \sum_m p(m|\theta)\,\big(\hat{\theta}(m) - \theta\big) = 0\,. \qquad (24.11)$$

Then, we differentiate the condition of eqn (24.11) with respect to the parameter θ, which gives

$$\sum_m \frac{\partial p(m|\theta)}{\partial \theta}\,\big(\hat{\theta}(m) - \theta\big) - \sum_m p(m|\theta) = 0\,. \qquad (24.12)$$

We can further rewrite the last condition as

$$\sum_{m} \left(\hat{\theta}(m) - \theta\right) p(m|\theta) \frac{\partial}{\partial\theta} \log p(m|\theta) = \sum_{m} \xi_m \chi_m = 1 \,, \qquad (24.13)$$

where we have defined the quantities

$$\xi_m := \sqrt{p(m|\theta)} \left(\hat{\theta}(m) - \theta\right), \qquad (24.14a)$$

$$\chi_m := \sqrt{p(m|\theta)} \frac{\partial}{\partial\theta} \log p(m|\theta). \qquad (24.14b)$$

Finally, we may use the Cauchy–Schwarz inequality in the form

$$\left| \sum_{m} \xi_m \chi_m \right|^2 \leq \sum_{m} |\xi_m|^2 \sum_{n} |\chi_n|^2 \qquad (24.15)$$

to arrive at the inequality

$$\left[\sum_{m} p(m|\theta) \left(\hat{\theta}(m) - \theta\right)^2 \right] \left[\sum_{n} p(n|\theta) \left(\frac{\partial}{\partial\theta} \log p(n|\theta)\right)^2 \right] \geq 1. \qquad (24.16)$$

The quantity in the first bracket on the left-hand side of eqn (24.16) is just the variance $V[\hat{\theta}]$ as in eqn (24.10), while the second factor corresponds to the Fisher information $I\big(p(m|\theta)\big)$ as stated in Theorem 24.1. Dividing by the Fisher information thus yields the *Cramér–Rao* inequality

$$V[\hat{\theta}] \geq \frac{1}{I\big(p(m|\theta)\big)}. \qquad (24.17)$$

$$\square$$

For given probe state and measurement, the (inverse) Fisher information hence allows us to determine how well this measurement strategy could potentially perform, provided one finds a suitable estimator. In particular, if a given unbiased estimator $\hat{\theta}$ provides an expected variance identical to $1/I$, one can conclude that it is optimal. In practice, however, one faces two problems:

First, it is not always possible to find an estimator that is unbiased for all possible values of the true parameter. At the same time, this requirement might be too limiting in the first place. That is, one can relax it by allowing an estimator that is generally biased, but which is unbiased *locally*[2] in the neighbourhood of a particular parameter value $\theta = \theta_o$. One may then regard the estimation procedure as a way of determining the precise value of θ in a sufficiently small vicinity of θ_o. Technically, this would mean that one has to know the parameter value in advance (at least approximately) before even estimating it. Practically, however, one may use a certain number of initial measurements to determine a rough estimate, which can in turn be used to nominate an estimator that is unbiased in the vicinity of the initial approximate estimate. If the overall number of measurements (the sample size N) becomes very large (technically,

---

[2]Note that the word "local" here has nothing to do with spatial locality restrictions relevant in entanglement theory, as discussed in Chapter 16.

infinite) such a *local parameter estimation* procedure is well justified and any comparably small number of initial measurements to gauge θ_o can be ignored for the purpose of evaluating the performance of the estimation strategy.

Second, even when allowing estimators that are only locally unbiased instead of globally, the *Cramér–Rao* bound is generally not tight (see, e.g., Kay, 1993). That is, one may not be able to identify an unbiased estimator (not even locally unbiased) such that eqn (24.17) is satisfied for any conditional probability distribution $p(m|\theta)$. Nonetheless, the *Cramér–Rao* bound is tight asymptotically when $N \to \infty$. In particular, this can be shown (Kay, 1993) via the *maximum-likelihood estimator*

$$\hat{\theta}_{\mathrm{ML}}(m) = \arg \max_{\theta} p(m|\theta), \tag{24.18}$$

with $m = \mathbf{m}^{(N)}$. For fixed outcome $m = \mathbf{m}^{(N)}$, the estimator $\hat{\theta}_{\mathrm{ML}}(m)$ assigns the value $\theta = \theta_{\max}$ for which the probability of obtaining the outcome m is maximal, i.e., $\max_{\theta} p(m|\theta) = p(m|\theta_{\max})$. For finite N, the maximum-likelihood estimator is generally not unbiased, but for $N \to \infty$, it becomes unbiased and saturates the Cramér–Rao inequality. So, once more, we see that local parameter estimation is well justified for large sample sizes, in particular, in the limit $N \to \infty$.

With these restrictions on the applicability in mind, we can make some more concrete statements about the dependence of the Fisher information $I\big(p(m|\theta)\big)$ on N. To this end, note that, although we write $I\big(p(m|\theta)\big)$ for brevity, for any given θ the Fisher information depends on the entire distribution $\{p(m|\theta)\}_m$ of conditional probabilities $p(m|\theta)$, not just on the value for a particular outcome m. At the same time, the Fisher information depends only on this (classical) distribution, regardless of the type of particular probe states or measurements giving rise to the distribution. When we consider a parallel measurement strategy, i.e., N independent probes, the probability distribution factorizes,

$$p\big(\mathbf{m}^{(N)}|\theta\big) = \prod_{j=1}^{N} p(m_{i_j}|\theta). \tag{24.19}$$

This is the case in the strategy that we have previously described, see, e.g., Fig. 24.1 and eqn (24.9), where the probe states and measurements are the same for all of the N probes. This factorization condition would also hold true if the probes and measurements were different, as long as they are chosen independently of the other (previous) measurement outcomes. For a probability distribution $\{p(\mathbf{m}^{(N)}|\theta)\}_{\mathbf{m}}$ that factorizes into N independent (and potentially different) distributions $\{p(m_{i_1}|\theta)\}_{i_1}$, $\{p(m_{i_2}|\theta)\}_{i_2}$, to $\{p(m_{i_N}|\theta)\}_{i_N}$, the derivative of the logarithm (called the *score*) evaluates to

$$\tfrac{\partial}{\partial\theta} \log p(\mathbf{m}^{(N)}|\theta) = \tfrac{\partial}{\partial\theta} \log \prod_{j=1}^{N} p(m_{i_j}|\theta) = \tfrac{\partial}{\partial\theta} \sum_{j=1}^{N} \log p(m_{i_j}|\theta) = \sum_{j=1}^{N} \tfrac{\partial}{\partial\theta} \log p(m_{i_j}|\theta).$$

$$\tag{24.20}$$

With this, we can evaluate the square of the score appearing in the Fisher information to calculate

$$I\big(p(\mathbf{m}^{(N)}|\theta)\big) = \sum_{i_1,\ldots,i_N} \prod_{j=1}^{N} p(m_{i_j}|\theta)\Big(\sum_{k=1}^{N} \tfrac{\partial}{\partial\theta}\log p(m_{i_k}|\theta)\Big)^2 \qquad (24.21)$$

$$= \sum_{k=1}^{N}\sum_{i_k} p(m_{i_k}|\theta)\big(\tfrac{\partial}{\partial\theta}\log p(m_{i_k}|\theta)\big)^2 \prod_{j\neq k}\sum_{i_j} p(m_{i_j}|\theta)$$

$$+ \sum_{\substack{k,k'=1\\k\neq k'}}\sum_{i_k,i_{k'}} p(m_{i_k}|\theta)p(m_{i_{k'}}|\theta)\big(\tfrac{\partial}{\partial\theta}\log p(m_{i_k}|\theta)\big)\big(\tfrac{\partial}{\partial\theta}\log p(m_{i_{k'}}|\theta)\big) \prod_{j\neq k,k'}\sum_{i_j} p(m_{i_j}|\theta).$$

This expression can be simplified by noting that $\sum_{i_j} p(m_{i_j}|\theta) = 1$ for all j, and $\tfrac{\partial}{\partial\theta}\log p(m_{i_j}|\theta) = \big(\tfrac{\partial}{\partial\theta}p(m_{i_j}|\theta)\big)/p(m_{i_j}|\theta)$, such that we obtain

$$I\big(p(\mathbf{m}^{(N)}|\theta)\big) = \sum_{k=1}^{N}\sum_{i_k} p(m_{i_k}|\theta)\big(\tfrac{\partial}{\partial\theta}\log p(m_{i_k}|\theta)\big)^2$$

$$+ \sum_{\substack{k,k'=1\\k\neq k'}}^{N}\sum_{i_k,i_{k'}} \big(\tfrac{\partial}{\partial\theta}p(m_{i_k}|\theta)\big)\big(\tfrac{\partial}{\partial\theta}p(m_{i_{k'}}|\theta)\big) \qquad (24.22)$$

$$= \sum_{k=1}^{N} I\big(p(m_{i_k}|\theta)\big) + \sum_{\substack{k,k'=1\\k\neq k'}}^{N} \Big(\tfrac{\partial}{\partial\theta}\sum_{i_k} p(m_{i_k}|\theta)\Big)\Big(\tfrac{\partial}{\partial\theta}\sum_{i_{k'}} p(m_{i_{k'}}|\theta)\Big) = \sum_{k=1}^{N} I\big(p(m_{i_k}|\theta)\big),$$

where we have used that $\sum_{i_j} p(m_{i_j}|\theta) = 1$ is a constant whose derivative vanishes. In particular, this means that, when the probe states and measurements chosen in the N rounds are identical, as in Fig. 24.1, we have $p(m_{i_j}|\theta) = p(m_{i_k}|\theta) \ \forall \ j,k$, and

$$I\big(p(\mathbf{m}^{(N)}|\theta)\big) = N\, I\big(p(m_{i_1}|\theta)\big). \qquad (24.23)$$

This linear scaling of the Fisher information reproduces exactly what we expect from the central limit theorem for N i.i.d. samples. That is, if we insert eqn (24.23) into the Cramér–Rao bound from Theorem 24.1, we find that the (minimal) variance of the mean is obtained by dividing the sample variance by N. In other words, suppose we consider any single run of the estimation procedure, e.g., the first run, corresponding to the conditional probability distribution $p(m_{i_1}|\theta)$. Then for any (unbiased) estimator, the resulting variance is bounded from below by $1/I\big(p(m_{i_1}|\theta)\big)$. If we repeat the same preparation and measurement N times, then *at best* we can achieve a variance that is N times smaller.

This kind of scaling behaviour is typically referred to as *standard quantum limit* (see Demkowicz-Dobrzański *et al.*, 2015) or *shot-noise limit*[3] (see Sidhu and Kok, 2020),

---

[3]This terminology arises in the context of interferometric setups for phase estimation, which we will comment on in Chapter 25.

even if it arises from non-factoring conditional probability distributions for sample size N. To illustrate the methods discussed here, let us apply them to a particular example in the next section.

24.1.4 Phase Estimation with Individual Qubits

Let us now illustrate the concepts discussed in the previous section for a simple example: phase estimation with single-qubit probe states. This means the Hilbert space we consider is $\mathcal{H} = \mathbb{C}^2$, and the encoding transformation is given by $U_\theta = e^{-i\theta\tilde{H}}$ with

$$\tilde{H} = \tfrac{1}{2}\sigma_z = \tfrac{1}{2}\big(|0\rangle\langle0| - |1\rangle\langle1|\big). \tag{24.24}$$

In other words, the transformation is a rotation around the z axis of the Bloch sphere (see Chapter 11) with unknown angle θ. As a probe state in this Hilbert space, let us select $|+\rangle = \frac{1}{\sqrt{2}}\big(|0\rangle + |1\rangle\big)$. The intuition behind this, which we shall make more mathematically precise in Section 24.2, is the following. Suppose we were to select either of the states $|0\rangle$ or $|1\rangle$, corresponding to the north and south poles of the Bloch sphere, respectively. Since these are the eigenstates of $\tilde{H}$, the transformation U_θ would imprint a non-measurable global phase on the state. However, if we choose a superposition of $|0\rangle$ and $|1\rangle$, the transformation generates a relative phase. Among these superpositions, the equally weighted ones, corresponding to vectors on the equator of the Bloch sphere (see Fig. 24.3) are most sensitive to the value of θ, i.e., they change "a lot" under the influence of U_θ. With this choice, the encoded state is

$$\rho(\theta) = U_\theta|+\rangle\langle+|U_\theta^\dagger, \tag{24.25}$$

with $U_\theta|+\rangle = \frac{e^{-i\theta/2}}{\sqrt{2}}\big(|0\rangle + e^{i\theta}|1\rangle\big)$, i.e., the Bloch vector of the initial state rotates along the equator by an angle θ.

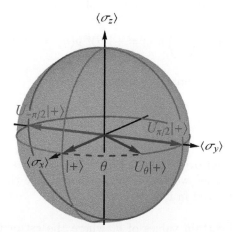

Fig. 24.3 Optimal single-qubit phase estimation. Illustration of the probe states $|+\rangle$, encoding transformation U_θ, and (optimal) measurement basis $\{U_{\pi/2}|\pm\rangle\}$ for single-qubit phase estimation.

Suppose now that we perform a measurement in the basis $\{U_{\pi/2}|\pm\rangle\} = \{\frac{1}{\sqrt{2}}(|0\rangle \pm i|1\rangle)\}$ for each single-qubit probe. The possible measurement outcomes are $+1$ and -1, and the corresponding conditional probabilities for obtaining an outcome $m \in \{\pm 1\}$ for the jth qubit given the true parameter value θ are

$$p(m = \pm 1|\theta) = |\tfrac{1}{2}(\langle 0| \mp i\langle 1|)(|0\rangle + e^{i\theta}|1\rangle)|^2 = \tfrac{1}{2}(1 \pm \sin(\theta)). \qquad (24.26)$$

This measurement is nothing but a measurement of the observable σ_y, with the expected value

$$\langle \sigma_y \rangle_{\rho(\theta)} = \langle +|U_\theta^\dagger \sigma_y U_\theta|+\rangle = (+1)p(+1|\theta) + (-1)p(-1|\theta) = \sin(\theta). \qquad (24.27)$$

Already at this point we see that this kind of measurement strategy only allows determining θ within an interval of width π. Let us therefore assume for now that $\theta \in \left[-\frac{\pi}{2}, \frac{\pi}{2}\right]$. The fact that the sample mean $\bar{m}^{(N)}$ is an unbiased estimator of the expected value (see Section 24.1.1) then motivates to invert eqn (24.27) to nominate the estimator

$$\hat{\theta}(m) = \arcsin(m). \qquad (24.28)$$

To see when this estimator is (or can be) unbiased, consider the possible values of the sample mean $\bar{m}^{(N)}$ for doing N measurements on individual qubits. Each outcome can be $+1$ or -1, but since the probe-state preparation and measurements are identical for all qubits, only the number k of $+1$ outcomes is relevant, not their order. We can thus parameterize the possible outcomes for the sample mean by k,

$$\bar{m}^{(N)}(k) = \frac{2k - N}{N}, \qquad (24.29)$$

and we write the corresponding conditional probabilities as

$$p(k|\theta) = \binom{N}{k} p(m = +1|\theta)^k\, p(m = -1|\theta)^{N-k}. \qquad (24.30)$$

The expected value of the estimator $\hat{\theta}$ is then

$$\langle \hat{\theta} \rangle_k = \sum_{k=0}^{N} p(k|\theta)\, \arcsin\!\big(\bar{m}^{(N)}(k)\big), \qquad (24.31)$$

where the average is taken over all values of k. While we have argued that this estimator is a reasonable choice, we also need to acknowledge that it is not unbiased for all θ and all N, as illustrated in Fig. 24.4.

However, there are certain values of θ, where the estimator *is* unbiased. When $\theta = 0$, the conditional probability $p(k|\theta = 0) = \frac{1}{2^N}\binom{N}{k}$ is symmetric with respect to reflections around $k = \frac{N}{2}$, whereas $\bar{m}^{(N)}(k)$ and $\hat{\theta}\big(\bar{m}^{(N)}(k)\big)$ are antisymmetric around this argument. Consequently, we have $\langle \hat{\theta} \rangle_k = 0$ for $\theta = 0$. Similarly, we

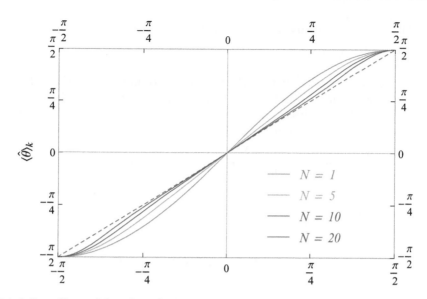

Fig. 24.4 Locally unbiased estimator. The average of the estimator over all outcomes (solid curves, shown for different sample sizes N) generally does not match the true value θ (dashed straight line), meaning the estimator is generally biased, except for some isolated points ($\theta = 0, \pm\frac{\pi}{2}$).

note that for $\theta = \pm\frac{\pi}{2}$, only a single one of the conditional probabilities is non-zero, $p(k = N|\theta = \frac{\pi}{2}) = 1$ and $p(k = 0|\theta = -\frac{\pi}{2}) = 1$, respectively, and since $\arcsin\big(\bar{m}^{(N)}(k = N)\big) = \arcsin(1) = \frac{\pi}{2}$ and $\arcsin\big(\bar{m}^{(N)}(k = 0)\big) = \arcsin(-1) = -\frac{\pi}{2}$, we find that the estimator is unbiased when $\theta = 0, \pm\frac{\pi}{2}$, independently of N. Therefore, a local estimation procedure of fluctuations around (for instance) $\theta = 0$ is well justified.

Moreover, we observe in Fig. 24.4 that the curve for $\langle\hat{\theta}\rangle_k$ more closely resembles the straight line θ in an ever larger neighbourhood around $\theta = 0$ as N increases. This suggests that for reasonably large N, we do not have to worry about the unbiasedness of the estimator even if the true parameter does not exactly have the value $\theta = 0$. Indeed, this is no coincidence. The estimator $\hat{\theta}\big(\bar{m}^{(N)}(k)\big) = \arcsin\big(\frac{2k-N}{N}\big)$ we have used is in fact the *maximum-likelihood estimator* from eqn (24.18) for this problem. To see this, let us calculate $\arg\max_\theta p(k|\theta)$, and focus first on the case where $k \neq 0, N$. Since the binomial factor is independent of θ, we have to evaluate

$$\frac{\partial}{\partial\theta}p(m = +1|\theta)^k\, p(m = -1|\theta)^{N-k} = \tag{24.32}$$

$$= \frac{1}{2^N}\cos(\theta)\big(1 + \sin(\theta)\big)^{k-1}\big(1 - \sin(\theta)\big)^{N-k-1}\big[(2k - N) - N\sin(\theta)\big].$$

For $\theta \in]-\frac{\pi}{2}, \frac{\pi}{2}[$, the pre-factor $\cos(\theta)\big(1 + \sin(\theta)\big)^{k-1}\big(1 - \sin(\theta)\big)^{N-k-1}$ is non-zero, and we find the condition $\sin(\theta) = \frac{2k-N}{N}$ for the maximum. Consequently, the maximum is obtained for the argument $\arcsin\big(\frac{2k-N}{N}\big)$. When $k = 0$, we thus find that

$p(0|\theta) = \left(1 - \sin(\theta)\right)^N$ has its maximum at $\theta = \frac{\pi}{2}$, and for $k = N$, the maximum of $p(N|\theta) = \left(1 + \sin(\theta)\right)^N$ is at $\theta = -\frac{\pi}{2}$.

Nevertheless, we are still left with the task of determining the variance for this estimator. Since our estimate $\hat{\theta}$ depends only on a single "outcome"—the sample mean $\bar{m}^{(N)}(k)$, eqn (24.10) provides no meaningful way for determining an estimate $(\Delta\theta)^2$ (corresponding to a sample variance for θ) of the variance $V[\hat{\theta}]$. Instead, we can evaluate the (unbiased) sample variance $\left(\Delta m^{(N)}\right)^2$ associated to the mean value $\bar{m}^{(N)}$ and propagate the error to θ. More specifically, we can evaluate

$$\left(\Delta m^{(N)}\right)^2 = \tfrac{1}{N-1} \sum_{i=1}^{N} \left(m_{i_j} - \bar{m}^{(N)}\right)^2 \tag{24.33}$$

$$= \tfrac{1}{N-1}\left[k\left(1 - \tfrac{2k-N}{N}\right)^2 + (N-k)\left(1 + \tfrac{2k-N}{N}\right)^2\right] = \frac{4k(N-k)}{N(N-1)},$$

where k is number of individual +1 outcomes as before. The corresponding variance of the mean is then obtained by dividing by N. To propagate this error to a variance of the estimate of θ, we calculate the (square of the) derivative of the estimator with respect to m and evaluate it at $m = \bar{m}^{(N)}(k) = \frac{2k-N}{N}$, that is,

$$\left(\frac{\partial\hat{\theta}(m)}{\partial m}\right)^2\bigg|_{m=\bar{m}^{(N)}} = \left(\frac{\partial}{\partial m}\arcsin(m)\right)^2\bigg|_{m=\bar{m}^{(N)}} = \frac{1}{1-m^2}\bigg|_{m=\bar{m}^{(N)}} = \frac{N^2}{4k(N-k)}. \tag{24.34}$$

As long as the variance of the mean $\left(\Delta m^{(N)}\right)^2/N$ is sufficiently small (which corresponds, again, to sufficiently large N), the (estimated) variance $(\Delta\theta)^2$ is well approximated by

$$(\Delta\theta)^2 = \left(\frac{\partial\hat{\theta}(m)}{\partial m}\right)^2\bigg|_{m=\bar{m}^{(N)}} \frac{\left(\Delta m^{(N)}\right)^2}{N} = \frac{N^2}{4k(N-k)}\frac{4k(N-k)}{N^2(N-1)} = \frac{1}{N-1}. \tag{24.35}$$

The variance we obtain does not depend on the specific outcome k (or $m^{(N)}(k)$), and $(\Delta\theta)^2$ decreases with increasing N. After N individual single-qubit probes, of which k result in the outcome "+1", the result of the estimation procedure for θ can hence be stated as

$$\theta = \hat{\theta}\big(m^{(N)}(k)\big) \pm \Delta\theta = \arcsin\left(\tfrac{2k-N}{N}\right) \pm \tfrac{1}{\sqrt{N-1}}, \tag{24.36}$$

i.e., in terms of the estimator evaluated at the mean $m^{(N)}(k)$, and the associated standard deviation.

We may nonetheless wonder whether we could have chosen a better estimator. We therefore calculate the Fisher information for the chosen measurement. To this end, let us first rewrite the expression from Theorem 24.1, i.e.,

$$I\big(p(m|\theta)\big) = \sum_m p(m|\theta)\left(\frac{\partial}{\partial\theta}\log p(m|\theta)\right)^2 = \sum_m p(m|\theta)\left(\frac{\dot{p}(m|\theta)}{p(m|\theta)}\right)^2 = \sum_m \frac{\big(\dot{p}(m|\theta)\big)^2}{p(m|\theta)}, \tag{24.37}$$

where $\dot{p} = \frac{\partial}{\partial\theta}p$. For the single-qubit measurements we consider, we have $m_{i_j} \in \{\pm1\}$ for $i = 1,\dots,N$, and $p(\pm1|\theta) = \frac{1}{2}\big(1 \pm \sin(\theta)\big)$ and $\dot{p}(\pm1|\theta) = \pm\frac{1}{2}\cos(\theta)$, such that

$$I\big(p(m_{i_j}|\theta)\big) = \frac{1}{4}\cos^2(\theta)\left(\frac{1}{\frac{1}{2}\big(1+\sin(\theta)\big)} + \frac{1}{\frac{1}{2}\big(1-\sin(\theta)\big)}\right) = 1. \tag{24.38}$$

Following eqn (24.23) this implies that we have to calculate the Fisher information for the individual single-qubit measurements and multiply the result by N to obtain the overall Fisher information. The Cramér–Rao bound for the measurement strategy we consider is hence

$$V[\hat{\theta}] \geq \frac{1}{N I\big(p(m_{i_j}|\theta)\big)} = \frac{1}{N}. \tag{24.39}$$

We thus see that for any finite N, the estimator we have chosen does not saturate the Cramér–Rao bound. The ratio of the (right-hand side of the) bound and the estimated variance $(\Delta\theta)^2$ is $N/(N-1) > 1$, but as $N \to \infty$, we can approach the bound from above, $\lim_{N\to\infty}\frac{N}{N-1} = 1$.

Although our freedom to optimize the estimator in the given estimation problem is thus practically exhausted, we might have chosen different probe states and different measurements. We shall explore such possibilities in the next section, where we will see that—as far as single-qubit probes go—the chosen strategy was already optimal. However, there is still room for making use of the Hilbert space of the N-qubit probe state more efficiently, as we will see in Section 24.2.4.

24.2 The Quantum Cramér–Rao Bound and Heisenberg Scaling

24.2.1 The Quantum Cramér–Rao Bound

As we have seen in Theorem 24.1, the Cramér–Rao bound allows us to (temporarily) eliminate the choice of the estimator from the problem of determining the best estimation strategy. For any given probe state and measurement, the (inverse of the) Fisher information $I\big(p(m|\theta)\big)$ then provides a lower bound on the achievable variance. In a second step, one may then further simplify the optimization problem by maximizing the Fisher information over all possible POVMs that can be performed on the system.

Theorem 24.2 (Quantum Cramér–Rao Inequality)

Let $\rho(\theta)$ be a density operator describing the state of a quantum system, and let θ be a parameter that is to be estimated. For any POVM that is performed on $\rho(\theta)$ and any unbiased estimator, the variance $V[\hat{\theta}]$ is bounded from below by

$$V[\hat{\theta}] \geq \frac{1}{\mathcal{I}\big(\rho(\theta)\big)}$$

where $\mathcal{I}\big(\rho(\theta)\big)$ is the quantum Fisher information *(QFI) given by*

$$\mathcal{I}\big(\rho(\theta)\big) = 2\,\mathrm{Tr}\big(\hat{S}_\theta\,\dot{\rho}(\theta)\big)\,.$$

Here, $\dot{\rho}(\theta) = \frac{\partial}{\partial\theta}\rho(\theta)$, and the operator $\hat{S}_\theta \equiv \hat{S}[\rho(\theta)]$, called the symmetric logarithmic derivative *(SLD), is implicitly given by*

$$\hat{S}_\theta\rho(\theta) + \rho(\theta)\hat{S}_\theta = \dot{\rho}(\theta)\,.$$

This form of the Cramér–Rao bound is a special case of a more general bound for multi-parameter estimation originally derived by *Carl Helstrom* (1969, 1976). For multi-parameter problems, the corresponding bound is not tight (not even asymptotically), but one may derive another bound due to *Alexander Holevo* (see, e.g., Holevo, 2011*b* or Tsang, 2019) that is tight and attainable asymptotically (Gill and Guţă, 2013; Yamagata *et al.*, 2013; Yang *et al.*, 2019). Here, we do not prove Theorem 24.2, but we refer the interested reader to Braunstein and Caves (1994) for a compact proof. For the estimation of a single parameter (as discussed here), any issues with tightness of the quantum Cramér–Rao bound arise from the previously (see Section 24.1.3) mentioned issues of the attainability of the usual Cramér–Rao bound from Theorem 24.1.

In particular, there always exists a POVM for which the Fisher information and the QFI coincide. This optimal POVM is a projective measurement in the eigenbasis $\{|\lambda_m\rangle\}_m$ of the SLD. To see this, let us first note that for any fixed POVM with elements E_m, the Fisher information takes the form

$$I\big(p(m|\theta)\big) = \sum_m \frac{\big(\mathrm{Tr}(E_m\dot{\rho})\big)^2}{\mathrm{Tr}(E_m\rho)}\,. \tag{24.40}$$

Then, we write $\hat{S}_\theta = \sum_m \lambda_m\,|\lambda_m\rangle\langle\lambda_m|$, such that the relation defining the SLD becomes

$$\dot{\rho}(\theta) = \hat{S}_\theta\rho(\theta) + \rho(\theta)\hat{S}_\theta = \sum_m \lambda_m\,\big(|\lambda_m\rangle\langle\lambda_m|\,\rho(\theta) + \rho(\theta)\,|\lambda_m\rangle\langle\lambda_m|\big). \tag{24.41}$$

Inserting $E_m = |\lambda_m\rangle\langle\lambda_m|$ for the projective measurement in the SLD's eigenbasis into the Fisher information and making use of eqn (24.41), we have

$$I\big(p(m|\theta)\big) = \sum_m \frac{\big(\mathrm{Tr}(|\lambda_m\rangle\langle\lambda_m|\dot\rho)\big)^2}{\mathrm{Tr}(|\lambda_m\rangle\langle\lambda_m|\rho)} = \sum_m \frac{\big(\langle\lambda_m|\dot\rho|\lambda_m\rangle\big)^2}{\langle\lambda_m|\rho|\lambda_m\rangle} \tag{24.42}$$

$$= \sum_{m,n} \lambda_n^2 \frac{\big(\langle\lambda_m|(|\lambda_n\rangle\langle\lambda_n|\rho(\theta) + \rho(\theta)|\lambda_n\rangle\langle\lambda_n|)|\lambda_m\rangle\big)^2}{\langle\lambda_m|\rho|\lambda_m\rangle}$$

$$= 4\sum_m \lambda_m^2 \frac{\big(\langle\lambda_m|\rho|\lambda_m\rangle\big)^2}{\langle\lambda_m|\rho|\lambda_m\rangle} = 4\sum_m \lambda_m^2 \langle\lambda_m|\rho|\lambda_m\rangle \ .$$

Meanwhile, the QFI gives the result

$$\mathcal{I}\big(\rho(\theta)\big) = 2\mathrm{Tr}\big(\hat{S}_\theta\,\dot\rho(\theta)\big) = 2\sum_m \lambda_m \mathrm{Tr}\big(|\lambda_m\rangle\langle\lambda_m|\,\dot\rho(\theta)\big) \tag{24.43}$$

$$= 2\sum_m \lambda_m \langle\lambda_m|\big(\hat{S}_\theta\rho(\theta) + \rho(\theta)\hat{S}_\theta\big)|\lambda_m\rangle$$

$$= 2\sum_{m,n} \lambda_m\lambda_n \langle\lambda_m|\big(|\lambda_n\rangle\langle\lambda_n|\rho(\theta) + \rho(\theta)|\lambda_n\rangle\langle\lambda_n|\big)|\lambda_m\rangle$$

$$= 4\sum_m \lambda_m^2 \langle\lambda_m|\rho|\lambda_m\rangle = I\big(p(m|\theta)\big).$$

The quantum Cramér–Rao bound is hence asymptotically attainable.

The QFI hence gives us a figure of merit to evaluate the potential usefulness of any given state ρ in a fixed local parameter estimation problem. In simple terms, the QFI can be understood to capture how sensitive a given state is to small changes in the parameter. This can be expressed via the *Uhlmann fidelity* $\mathcal{F}$ (Uhlmann, 1976). Because the fidelity is itself important in many areas of modern quantum theory we shall discuss it in a little more detail in the next section, before returning to the discussion of the QFI in Section 24.2.3.

24.2.2 The Uhlmann Fidelity

The Uhlmann fidelity $\mathcal{F}$ (Uhlmann, 1976) is one way[4] of establishing a notion of distance between quantum states. It can be defined for arbitrary mixed states ρ_1 and ρ_2 as

$$\mathcal{F}(\rho_1,\rho_2) = \Big(\mathrm{Tr}\sqrt{\sqrt{\rho_1}\rho_2\sqrt{\rho_1}}\Big)^2. \tag{24.44}$$

Alternatively, the fidelity[5] can be expressed as

$$\mathcal{F}(\rho_1,\rho_2) = \|\sqrt{\rho_1}\sqrt{\rho_2}\|_{\mathrm{tr}}^2, \tag{24.45}$$

---

[4] Another way, for instance, is the trace distance, see Section 23.4.

[5] Note that some authors (e.g., Nielsen and Chuang, 2000), use different conventions and refer to $\sqrt{\mathcal{F}}$ as fidelity.

where $\| \cdot \|_{\text{tr}}$ is the trace norm from eqn (23.58). To verify this form, let us briefly calculate

$$\text{Tr}\sqrt{\sqrt{\rho_1}\rho_2\sqrt{\rho_1}} = \text{Tr}\sqrt{\sqrt{\rho_1}\sqrt{\rho_2}\sqrt{\rho_2}\sqrt{\rho_1}} = \text{Tr}\sqrt{\sqrt{\rho_1}\sqrt{\rho_2}\left(\sqrt{\rho_1}\sqrt{\rho_2}\right)^{\dagger}} = \|\sqrt{\rho_1}\sqrt{\rho_2}\|_{\text{tr}}\,.$$
$$(24.46)$$

From the formula in eqn (24.45), we can then immediately see that the fidelity is symmetric with respect to the exchange of its two arguments, $\mathcal{F}(\rho_1, \rho_2) = \mathcal{F}(\rho_2, \rho_1)$.

The fidelity can be understood as the overlap of two quantum states ρ_1 and ρ_2. In particular, when one of the states is pure, we have $\mathcal{F}(\rho, |\psi\rangle) = \langle\psi|\rho|\psi\rangle$, and, consequently, when both states are pure $\mathcal{F}(|\phi\rangle, |\psi\rangle) = |\langle\phi|\psi\rangle|^2$. Moreover, for arbitrary mixed states, one can express the fidelity via the following theorem due to *Armin Uhlmann* (1976).

Theorem 24.3 (Uhlmann's theorem)

Let $\rho_i \in L(\mathcal{H}_A)$ for $i = 1, 2$ be two arbitrary density operators on the same Hilbert space $\mathcal{H}_A$, then

$$\mathcal{F}(\rho_1, \rho_2) = \max_{\psi, \phi} |\langle\psi|\phi\rangle|^2\,,$$

where the maximization is carried out over all purifications $|\psi\rangle$ and $|\phi\rangle$ of ρ_1 and ρ_2, respectively, i.e., such that $|\psi\rangle, |\phi\rangle \in \mathcal{H}_A \otimes \mathcal{H}_B$ satisfy $\text{Tr}_B(|\psi\rangle\langle\psi|) = \rho_1$ and $\text{Tr}_B(|\phi\rangle\langle\phi|) = \rho_2$.

Proof To prove Uhlmann's theorem, let us first write the two density operators in spectral decomposition,

$$\rho_1 = \sum_i \lambda_i |\psi_i\rangle\langle\psi_i|\,, \qquad \rho_2 = \sum_j \mu_j |\phi_j\rangle\langle\phi_j|\,. \qquad (24.47)$$

From eqn (21.2), we can then write any two corresponding purifications $|\psi\rangle$ and $|\phi\rangle$ as

$$|\psi\rangle = \sum_i \sqrt{\lambda_i}|\psi_i\rangle_A|\chi_i\rangle_B\,, \qquad |\phi\rangle = \sum_j \sqrt{\mu_j}|\phi_j\rangle_A|\xi_j\rangle_B\,, \qquad (24.48)$$

where any phase factors that might appear in addition to the square roots $\sqrt{\lambda_i}$ and $\sqrt{\mu_j}$ can be absorbed into the choice of the orthonormal bases $\{|\chi_i\rangle\}_i$ and $\{|\xi_j\rangle\}_j$. Next, we write the scalar product of the two purifications and insert $\delta_{ik} = \text{Tr}(|\psi_i\rangle\langle\psi_k|)$ and $\delta_{jl} = \langle\phi_j|\phi_l\rangle$ along with sums over the additional indices k and l, that is,

$$\langle \psi | \phi \rangle = \sum_{i,j} \sqrt{\lambda_i \mu_j} \, \langle \psi_i | \phi_j \rangle \, \langle \chi_i | \xi_j \rangle \tag{24.49}$$

$$= \sum_{i,j,k} \sqrt{\lambda_i \mu_j} \, \langle \psi_i | \phi_j \rangle \, \langle \chi_k | \xi_j \rangle \, \mathrm{Tr}(| \psi_i \rangle\langle \psi_k |)$$

$$= \mathrm{Tr}\Big(\sum_{i,j,k,l} \sqrt{\lambda_i \mu_j} \, \langle \psi_i | \phi_j \rangle \, \langle \chi_k | \xi_l \rangle | \psi_i \rangle\langle \phi_j | \phi_l \rangle\langle \psi_k | \Big)$$

$$= \mathrm{Tr}\Big[\Big(\sum_{i,j} \sqrt{\lambda_i \mu_j} \, \langle \psi_i | \phi_j \rangle \, | \psi_i \rangle\langle \phi_j | \Big) \Big(\sum_{k,l} \langle \chi_k | \xi_l \rangle | \phi_l \rangle\langle \psi_k | \Big) \Big]$$

$$= \mathrm{Tr}\Big[\Big(\sum_{i,j} \sqrt{\lambda_i \mu_j} \, | \psi_i \rangle\langle \psi_i | \, | \phi_j \rangle\langle \phi_j | \Big) U \Big] = \mathrm{Tr}\big[\sqrt{\rho_1} \sqrt{\rho_2} U \big],$$

$$\tag{24.50}$$

where we have further defined the unitary U via

$$U = \sum_{k,l} \langle \chi_k | \xi_l \rangle | \phi_l \rangle\langle \psi_k |, \tag{24.51}$$

for which one may easily check that $U^\dagger U = U U^\dagger = \mathbb{1}$. We can then continue by noting that for any operator A and unitary U, we have

$$|\mathrm{Tr}(AU)| \le \|A\|_{\mathrm{tr}}. \tag{24.52}$$

To see that this is the case, let us write A in its singular-value decomposition, $A = \sum_i s_i | u_i \rangle\langle v_i |$, where $\{| u_i \rangle\}_i$ and $\{| v_i \rangle\}_i$ are orthonormal bases. Since U is unitary, also the vectors $| \tilde{v}_i \rangle = U | v_i \rangle$ form an orthonormal basis, which allows us to write

$$|\mathrm{Tr}(AU)| = | \sum_k \langle \tilde{v}_k | AU | \tilde{v}_k \rangle | = | \sum_{i,k} \langle \tilde{v}_k | s_i | u_i \rangle\langle v_i | U | \tilde{v}_k \rangle | \tag{24.53}$$

$$= | \sum_i s_i \langle \tilde{v}_i | u_i \rangle | \le \sum_i | s_i \langle \tilde{v}_i | u_i \rangle |,$$

where we have used the triangle inequality in the last step. Moreover, since the singular values $s_i \ge 0$ and $|\langle \tilde{v}_i | u_i \rangle| \le 1$, we have $|\mathrm{Tr}(AU)| \le \sum_i s_i$, which is equal to the trace norm of A, see eqn (23.58), which proves eqn (24.52). Moreover, this bound is tight, i.e., equality can be achieved in eqn (24.52), for $U = U_{\mathrm{max}} \equiv \sum_k | v_k \rangle\langle u_k |$ we have

$$|\mathrm{Tr}(AU_{\mathrm{max}})| = |\mathrm{Tr}(\sum_{i,k} s_i | u_i \rangle\langle v_i | v_k \rangle\langle u_k |)| = |\mathrm{Tr}(\sum_i s_i | u_i \rangle\langle u_i |)| = \sum_i s_i = \|A\|_{\mathrm{tr}}.$$

$$\tag{24.54}$$

Now, since such a unitary exists for any A, it also exists for $A = \sqrt{\rho_1} \sqrt{\rho_2}$. In other words, the maximum of $|\langle \psi | \phi \rangle|$ over all purifications $| \psi \rangle$ and $| \phi \rangle$ is $\|\sqrt{\rho_1} \sqrt{\rho_2}\|_{\mathrm{tr}}$, and it can be achieved for $| \chi_k \rangle = \sum_l \langle \psi_k | U_{\mathrm{max}}^\dagger | \phi_l \rangle | \xi_l \rangle$. From the definition of the fidelity $\mathcal{F}(\rho_1, \rho_2)$ from eqn (24.45) via the trace norm, we hence recover the statement of Uhlmann's theorem. $\qquad\square$

The fidelity can further be related to the *Bures distance* D_{B} (Bures, 1969) via

$$\left(D_{\mathrm{B}}(\rho_1, \rho_2)\right)^2 = 2\left(1 - \sqrt{\mathcal{F}(\rho_1, \rho_2)}\right). \tag{24.55}$$

The Bures distance can in turn be used to define a metric (the Bures metric, also called Helstrom metric, see Helstrom (1967) on the space of density operators, which can be understood as a mixed state generalization of the Fisher-information metric (see Facchi *et al.*, 2010). In this sense, the Uhlmann fidelity and the Bures distance have clear operational and geometric interpretations in terms of distances in Hilbert space, which also translate to the QFI, as we will see next.

24.2.3   The Quantum Fisher Information

For arbitrary mixed probe states, one may express the QFI as

$$\mathcal{I}\bigl(\rho(\theta)\bigr) = \lim_{d\theta \to 0} 8 \frac{1 - \sqrt{\mathcal{F}\bigl(\rho(\theta), \rho(\theta + d\theta)\bigr)}}{d\theta^2} = \lim_{d\theta \to 0} 4 \frac{\left[D_{\mathrm{B}}\bigl(\rho(\theta), \rho(\theta + d\theta)\bigr)\right]^2}{d\theta^2}, \tag{24.56}$$

where $\mathcal{F}$ and D_{B} are the *Uhlmann fidelity* (Uhlmann, 1976) and the *Bures distance* (Bures, 1969) discussed in the previous section, respectively. The QFI thus captures how "far away" a given probe state "moves" in Hilbert space, when the estimated parameter changes infinitesimally, i.e., how sensitive a given probe is to small variations of the parameter. Besides this nice interpretation, expressing the QFI via the fidelity has some practical advantages in cases where the fidelity takes a particularly compact form. For instance, when restricting to the family of Gaussian continuous-variable states (see Section 25.4), the fidelity (Scutaru, 1998; Marian and Marian, 2012; Banchi *et al.*, 2015) and the QFI (Gaiba and Paris, 2009; Pinel *et al.*, 2013; Monràs, 2013; Jiang, 2014; Šafránek *et al.*, 2015; Šafránek and Fuentes, 2016; Rigovacca *et al.*, 2017; Šafránek, 2019) can be evaluated analytically in some generality, and for general probe states one may still obtain compact expressions when the encoding operations are Gaussian[6] as well (Friis *et al.*, 2015b).

Let us now discuss some interesting properties of the QFI. The QFI is convex (Cohen, 1968):

$$\mathcal{I}\Bigl(\sum_i p_i\, \rho_i\Bigr) \leq \sum_i p_i\, \mathcal{I}(\rho_i). \tag{24.57}$$

To see this, we follow a compact proof given in Hyllus *et al.* (2010). Let us first show that convexity is also true already for the Fisher information $I\bigl(p(m|\theta)\bigr)$. That is, let $\rho(\theta) = \sum_i p_i\rho_i(\theta)$, then for any fixed POVM $\{E_m\}_m$ we have the conditional probability to obtain outcome m,

$$p(m|\theta) = \mathrm{Tr}\bigl(E_m\, \rho(\theta)\bigr) = \sum_i p_i\, \mathrm{Tr}\bigl(E_m\, \rho_i(\theta)\bigr) = \sum_i p_i\, p_i(m|\theta). \tag{24.58}$$

---

[6]For a definition and discussion of these operations see Section 25.4.

We then have

$$\frac{(\dot{p}(m|\theta))^2}{p(m|\theta)} = \frac{\left(\sum\limits_i p_i\,\dot{p}_i(m|\theta)\right)^2}{\sum\limits_j p_j\,p_j(m|\theta)} \le \sum_i p_i\,\frac{(\dot{p}_i(m|\theta))^2}{p_i(m|\theta)}, \tag{24.59}$$

where we have used the Cauchy–Schwarz inequality by writing

$$\left(\sum_i p_i\,\dot{p}_i(m|\theta)\right)^2 = \left(\sum_i \sqrt{p_i\,p_i(m|\theta)}\,\frac{\sqrt{p_i}\,\dot{p}_i(m|\theta)}{\sqrt{p_i(m|\theta)}}\right)^2 \tag{24.60}$$

$$\le \left(\sum_j p_j\,p_j(m|\theta)\right)\left(\sum_i p_i\,\frac{(\dot{p}_i(m|\theta))^2}{p_i(m|\theta)}\right).$$

Following eqn (24.37), we obtain the Fisher information by summing the left-hand side of eqn (24.59) over the possible measurement outcomes labelled by m, such that

$$I\big(p(m|\theta)\big) \le \sum_i p_i \sum_m \frac{(\dot{p}_i(m|\theta))^2}{p_i(m|\theta)} = \sum_i p_i\,I\big(p_i(m|\theta)\big). \tag{24.61}$$

The Fisher information is thus convex. In particular, this is true also for the POVM that is optimal for the state $\rho(\theta)$, in which case the left-hand side of eqn (24.61) is equal to the QFI $\mathcal{I}(\rho(\theta))$. At the same time, this POVM is not necessarily optimal for each of the ρ_i in the decomposition of ρ, such that $I\big(p_i(m|\theta)\big) \le \mathcal{I}\big(\rho_i(\theta)\big)$, and we hence arrive at eqn (24.57). Incoherently mixing two states always makes the resulting probe less sensitive for parameter estimation.

As a consequence, the optimum of the QFI amongst all quantum states is always attained within the set of pure probe states. This is very convenient, since the QFI for pure states $\rho = |\psi(\theta)\rangle\langle\psi(\theta)|$ takes on the particularly simple form

$$\mathcal{I}(|\psi\rangle) = 4\big(\langle\dot{\psi}(\theta)|\dot{\psi}(\theta)\rangle - |\langle\psi(\theta)|\dot{\psi}(\theta)\rangle|\big). \tag{24.62}$$

To see this, let us first write the QFI from Theorem 24.2 for $\rho(\theta) = |\psi(\theta)\rangle\langle\psi(\theta)|$

$$\mathcal{I}(|\psi(\theta)\rangle) = 2\,\mathrm{Tr}\big(\hat{S}_\theta\,\dot{\rho}(\theta)\big) = 2\,\mathrm{Tr}\big(\hat{S}_\theta\,[\,|\dot{\psi}(\theta)\rangle\langle\psi(\theta)| + |\psi(\theta)\rangle\langle\dot{\psi}(\theta)|\,]\big) \tag{24.63}$$

$$= 2\left(\langle\psi(\theta)|\hat{S}_\theta|\dot{\psi}(\theta)\rangle + \langle\dot{\psi}(\theta)|\hat{S}_\theta|\psi(\theta)\rangle\right),$$

where we have used the fact that the trace is invariant under cyclic permutations of its arguments. Meanwhile, the relation that determines the SLD (see Theorem 24.2) takes the form

$$\dot{\rho}(\theta) = |\dot{\psi}(\theta)\rangle\langle\psi(\theta)| + |\psi(\theta)\rangle\langle\dot{\psi}(\theta)| = \hat{S}_\theta|\psi(\theta)\rangle\langle\psi(\theta)| + |\psi(\theta)\rangle\langle\psi(\theta)|\hat{S}_\theta. \tag{24.64}$$

Applying $\langle\psi(\theta)|$ and $|\psi(\theta)\rangle$, or $\langle\dot{\psi}(\theta)|$ and $|\psi(\theta)\rangle$, or $\langle\psi(\theta)|$ and $|\dot{\psi}(\theta)\rangle$ to eqn (24.64) from the left and right, respectively, we obtain

$$\langle\psi|\dot{\rho}|\psi\rangle = \langle\psi|\dot{\psi}\rangle + \langle\dot{\psi}|\psi\rangle = 2\langle\psi|\hat{S}_\theta|\psi\rangle, \tag{24.65a}$$

$$\langle\dot{\psi}|\dot{\rho}|\psi\rangle = \langle\dot{\psi}|\dot{\psi}\rangle + \langle\dot{\psi}|\psi\rangle^2 = \langle\dot{\psi}|S_\theta|\psi\rangle + \langle\dot{\psi}|\psi\rangle\langle\psi|S_\theta|\psi\rangle, \tag{24.65b}$$

$$\langle\psi|\dot{\rho}|\dot{\psi}\rangle = \langle\psi|\dot{\psi}\rangle^2 + \langle\dot{\psi}|\dot{\psi}\rangle = \langle\psi|S_\theta|\psi\rangle\langle\psi|\dot{\psi}\rangle + \langle\psi|S_\theta|\dot{\psi}\rangle, \tag{24.65c}$$

where we have not explicitly stated the dependence on θ for ease of notation. One can then sum insert $\langle\psi|\hat{S}_\theta|\psi\rangle$ from eqn (24.65a) into eqns (24.65b) and (24.65c) to express the sum of $\langle\psi|\hat{S}_\theta|\dot{\psi}\rangle$ and $\langle\dot{\psi}|\hat{S}_\theta|\psi\rangle$ as

$$\langle\psi|\hat{S}_\theta|\dot{\psi}\rangle + \langle\dot{\psi}|\hat{S}_\theta|\psi\rangle = 2\langle\dot{\psi}|\dot{\psi}\rangle + \langle\dot{\psi}|\psi\rangle^2 + \langle\psi|\dot{\psi}\rangle^2 - \tfrac{1}{2}\big(\langle\psi|\dot{\psi}\rangle + \langle\dot{\psi}|\psi\rangle\big)^2$$

$$= 2\big(\langle\dot{\psi}|\dot{\psi}\rangle - \mathrm{Im}^2(\langle\psi|\dot{\psi}\rangle)\big). \tag{24.66}$$

Finally, note that since

$$0 = \tfrac{\partial}{\partial\theta}1 = \tfrac{\partial}{\partial\theta}\langle\psi|\psi\rangle = \big(\tfrac{\partial}{\partial\theta}\langle\psi|\big)|\psi\rangle + \langle\psi|\tfrac{\partial}{\partial\theta}|\psi\rangle, \tag{24.67}$$

the operator $\tfrac{\partial}{\partial\theta}$ is anti-Hermitian, which implies that $i\langle\psi|\tfrac{\partial}{\partial\theta}|\psi\rangle = i\langle\psi|\dot{\psi}\rangle \in \mathbb{R}$, and $\mathrm{Im}^2(\langle\psi|\dot{\psi}\rangle) = |\langle\psi|\dot{\psi}\rangle|^2$. Combining eqns (24.63) and (24.66), we thus arrive at the following expression for the quantum Fisher information for pure states

$$\mathcal{I}(|\psi(\theta)\rangle) = 4\big(\langle\dot{\psi}|\dot{\psi}\rangle - |\langle\psi|\dot{\psi}\rangle|^2\big). \tag{24.68}$$

At least formally, one may also write the QFI for arbitrary mixed states in a similar form by formulating it in terms of a minimum over all purifications (Fujiwara and Imai, 2008; Escher *et al.*, 2012), i.e.,

$$\mathcal{I}(\rho(\theta)) = \min_{\psi(\theta)} \mathcal{I}(|\psi(\theta)\rangle), \tag{24.69}$$

where the minimum is taken over all $|\psi(\theta)\rangle \in \mathcal{H}_A \otimes \mathcal{H}_B$ such that $\mathrm{Tr}_B\big(|\psi(\theta)\rangle\langle\psi(\theta)|\big) = \rho(\theta) \in L(\mathcal{H}_A)$. Note the formal analogy to Uhlmann's theorem for the fidelity of two mixed states, except that the maximum in Theorem 24.3 is replaced by a minimum.

Another simplification of the QFI is achieved when restricting to Hamiltonian parameter estimation, where the parameter encoding is unitary, $|\psi(\theta)\rangle = U_\theta|\psi(0)\rangle$ with $U_\theta = e^{-i\theta\tilde{H}}$ as in eqn (24.7). In this case, we have $|\dot{\psi}(\theta)\rangle = -i\tilde{H}U_\theta|\psi(0)\rangle$, and, consequently, the QFI from eqn (24.68) becomes

$$\mathcal{I}(|\psi\rangle) = 4\big(\langle\psi|\tilde{H}^2|\psi\rangle - |\langle\psi|\tilde{H}|\psi\rangle|^2\big) = 4\big(\Delta\tilde{H}_\psi\big)^2. \tag{24.70}$$

The QFI just turns out to be proportional to the variance of the encoding Hamiltonian $\tilde{H}$. Moreover, here we have $|\psi\rangle \equiv |\psi(0)\rangle$, and the QFI hence does not depend on the particular true value of the parameter θ.

This independence of the QFI from θ in the case of unitary encoding also holds for arbitrary mixed probe states, that is,

$$\mathcal{I}(\rho(\theta)) = \mathcal{I}(\rho(\theta')) \quad \forall\, \theta, \theta'. \tag{24.71}$$

This can be seen by first noting that the encoded states are in this case related by $\rho(\theta') = U_{\theta'-\theta}\rho(\theta)U_{\theta'-\theta}^\dagger$. Moreover, for unitary encoding, the derivative of $\rho(\theta)$

is $\dot{\rho}(\theta) = i\left[\rho(\theta), H\right]$. Consequently, we have $\dot{\rho}(\theta') = U_{\theta'-\theta}\dot{\rho}(\theta)U_{\theta'-\theta}^{\dagger}$. This means that the SLD (see Theorem 24.2) transforms in the same way as the encoded state, $\hat{S}_{\theta'} = U_{\theta'-\theta}\hat{S}_{\theta}U_{\theta'-\theta}^{\dagger}$. Inserting these relations in $\mathcal{I}(\rho(\theta')) = 2\,\mathrm{Tr}(\hat{S}_{\theta'}\,\dot{\rho}(\theta'))$ and cyclically permuting the unitaries in the trace then results in $\mathcal{I}(\rho(\theta)) = \mathcal{I}(\rho(\theta'))$.

Also, once more we can formally find a similar expression for general mixed probe states (as long as the encoding transformation is unitary) in terms of a convex-roof construction[7]. That is, as shown in Tóth and Petz (2013) and Yu (2013), one can write

$$\mathcal{I}(\rho) = 4 \min_{\mathcal{D}(\rho)} \sum_i p_i \left(\Delta \tilde{H}_{\psi_i}\right)^2, \tag{24.72}$$

where the minimum is taken over all pure-state decompositions of ρ, i.e., $\mathcal{D}(\rho)$ is the set of all sets $\{p_i, |\psi_i\rangle\}_i$ with $0 \leq p_i \leq 1$ and $\sum_i p_i = 1$ such that $\sum_i p_i |\psi_i\rangle\langle\psi_i| = \rho$.

With these different expressions at hand, let us now (re)turn to some practical examples.

24.2.4 Phase Estimation with N-Qubit Probe States

In this section, we want to revisit our previous problem of phase estimation with qubits from Section 24.1.4 to determine, whether we could have chosen a better estimation procedure. To begin, let us re-examine the strategy we considered for preparing and measuring individual qubits. First, we know that the optimal probe state is pure (see Section 24.2.3). Second, phase estimation is a Hamiltonian estimation problem, such that optimal pure probe states should maximize the variance of $\tilde{H}$, see eqn (24.70). Therefore, the probe state $|+\rangle$ was indeed optimal. At the same time, the QFI for this probe state and $\tilde{H}$ as in eqn (24.24) evaluates to

$$\mathcal{I}(|+\rangle) = 4\left(\Delta \tilde{H}_{\psi}\right)^2 = \langle+|\sigma_z^2|+\rangle - \langle+|\sigma_z|+\rangle^2 = \langle+|+\rangle = 1. \tag{24.73}$$

This matches the expression we have obtained in eqn (24.38) for the Fisher information. Consequently, we can conclude that, as far as single-qubit measurements are concerned, we had already chosen optimally. Nonetheless, there is yet another possibility, that we have not considered so far. In principle, we may have performed joint measurements on any number (say N) copies of the encoded states, $\rho(\theta)^{\otimes N}$. For instance, one might naively expect that one might come up with a measurement scheme where information obtained from the first outcome could be used to select a better measurement for the second copy, and so on. However, as we shall see next, this is not possible.

> **Lemma 24.1 (Additivity of the QFI)** *The QFI is additive on product states, that is, for any two states ρ_1 and ρ_2,*
>
> $$\mathcal{I}(\rho_1 \otimes \rho_2) = \mathcal{I}(\rho_1) + \mathcal{I}(\rho_2).$$

---

[7]This is the same type of construction that is used, for instance, to define the entanglement of formation, see Section 16.2.2.

Proof Let us consider a probe state that is a mixed product state, $\rho(\theta) = \rho_1(\theta) \otimes \rho_2(\theta)$, and let $\hat{S}_1$ and $\hat{S}_2$ be the SLD operators for ρ_1 and ρ_2, respectively, such that

$$\dot{\rho}_i = \hat{S}_i \rho_i + \rho_i \hat{S}_i, \tag{24.74}$$

and $\mathcal{I}(\rho_i) = 2\text{Tr}(\hat{S}_i \rho_i)$. For the joint product state ρ, we can write

$$\dot{\rho} = \dot{\rho}_1 \otimes \rho_2 + \rho_1 \otimes \dot{\rho}_2 = (\hat{S}_1 \rho_1 + \rho_1 \hat{S}_1) \otimes \rho_2 + \rho_1 \otimes (\hat{S}_2 \rho_2 + \rho_2 \hat{S}_2) \tag{24.75}$$

$$= (\hat{S}_1 \otimes \mathbb{1}_2 + \mathbb{1}_1 \otimes \hat{S}_2) \rho_1 \otimes \rho_2 + \rho_1 \otimes \rho_2 (\hat{S}_1 \otimes \mathbb{1}_2 + \mathbb{1}_1 \otimes \hat{S}_2) = \hat{S} \rho + \rho \hat{S}.$$

In other words, the SLD $\hat{S}$ for the product state takes the form $\hat{S} = (\hat{S}_1 \otimes \mathbb{1}_2 + \mathbb{1}_1 \otimes \hat{S}_2)$. Consequently, the QFI evaluates to

$$\mathcal{I}(\rho) = 2\text{Tr}[\hat{S}\dot{\rho}] = 2\text{Tr}[(\hat{S}_1 \otimes \mathbb{1}_2 + \mathbb{1}_1 \otimes \hat{S}_2)(\dot{\rho}_1 \otimes \rho_2 + \rho_1 \otimes \dot{\rho}_2)] \tag{24.76}$$

$$= 2\text{Tr}[\hat{S}_1 \dot{\rho}_1] + 2\text{Tr}[\hat{S}_2 \dot{\rho}_2] = \mathcal{I}(\rho_1) + \mathcal{I}(\rho_2),$$

where we have used $\text{Tr}[\hat{S}_1 \dot{\rho}_1 \otimes \rho_2] = \text{Tr}[\hat{S}_1 \dot{\rho}_1] \text{Tr}[\rho_2]$ and $\text{Tr}(\dot{\rho}_i) = \frac{\partial}{\partial \theta} \text{Tr}(\rho_i) = \frac{\partial}{\partial \theta} 1 = 0$. This concludes the proof of the additivity of the QFI. $\quad\square$

A consequence of Lemma 24.1 is that there is no gain in performing potentially complicated measurements jointly on an ensemble of N probes in a product state. In particular, it does not matter whether the preparation and measurement of the individual probes is performed one after the other or if all of them are executed in parallel, the variance is limited by the *standard quantum limit*, i.e., the (sample) variance decreases (at best) as $\frac{1}{N}$. However, as shall see next, there can be an advantage in preparing a joint probe state of N qubits.

Suppose that, instead of preparing the states of all qubits individually, we prepare groups of N qubits in a joint probe state, which is subjected to the unitary encoding channel $U_\theta^{\otimes N}$ with U_θ as in eqn (24.25). As illustrated in Fig. 24.5, we can (for now) consider a joint measurement of this N-qubit probe, and we repeat the procedure with groups of N qubits ν times. To optimize the precision of the local phase estimation procedure, we are interested in maximizing the corresponding QFI. As explained on page 797, we therefore choose a pure probe state $|\psi\rangle$, and following eqn (24.70), we see that this state should maximize the variance of the corresponding Hamiltonian $\tilde{H}$. For N qubits, we have

$$\tilde{H} = \sum_{i=1}^{N} \tilde{H}_i = \sum_{i=1}^{N} \mathbb{1}_1 \otimes \mathbb{1}_2 \otimes \ldots \otimes \mathbb{1}_{i-1} \otimes \tfrac{1}{2}\sigma_z \otimes \mathbb{1}_{i+1} \otimes \ldots \otimes \mathbb{1}_N. \tag{24.77}$$

The variance of $\tilde{H}$ becomes maximal for an equally weighted superposition of the two eigenstates of $\tilde{H}$ corresponding to its minimal and maximal eigenvalues, i.e., the optimal probe state for N qubits is the generalized GHZ state for N qubits from eqn (18.17), i.e.,

$$|\psi\rangle = |\text{GHZ}_N\rangle = \tfrac{1}{\sqrt{2}}(|0\rangle^{\otimes N} + |1\rangle^{\otimes N}). \tag{24.78}$$

Let us now evaluate the QFI for this generalized N-qubit GHZ state (compare with eqn (18.1a)). Inspecting the Hamiltonian $\tilde{H}$ in eqn (24.77) and the state in eqn (24.78),

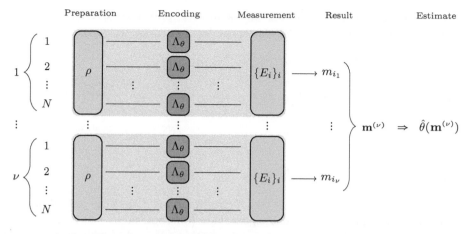

Fig. 24.5 Parameter estimation with N-Qubit probe states. In quantum parameter estimation, joint probe states ρ can be prepared for groups of N qubits in each of ν rounds of preparation, encoding Λ_θ and measurement. The measurements, represented by POVMs with elements E_i can also in principle be performed jointly on any number of qubits. However, the additivity of the QFI implies that measurements can performed independently on independent probe states without loss in precision. As usually, the individual results $m_{i_j} \in \{m_i\}_i$ for $j = 1, \ldots, \nu$ are finally collected into a total result $\mathbf{m}^{(\nu)} = (m_{i_1}, m_{i_2}, \ldots, m_{i_\nu})$, which is processed using an estimator function $\hat{\theta}$ to obtain an estimate $\hat{\theta}(\mathbf{m}^{(\nu)})$ of θ.

we note $\tilde{H} \,|\, 0 \,\rangle^{\otimes N} = \sum_{i=1}^{N}(-\tfrac{1}{2}) \,|\, 0 \,\rangle^{\otimes N} = -\tfrac{N}{2} \,|\, 0 \,\rangle^{\otimes N}$, and similarly, $\tilde{H} \,|\, 1 \,\rangle^{\otimes N} = \tfrac{N}{2} \,|\, 1 \,\rangle^{\otimes N}$, such that $\tilde{H} \,|\, \psi \,\rangle = \tfrac{N}{2} \,|\, \psi^\perp \,\rangle$ with $\langle\, \psi \,|\, \psi^\perp \,\rangle = 0$ and $\langle\, \psi^\perp \,|\, \psi^\perp \,\rangle = 1$. Inserting into eqn (24.70), we therefore obtain

$$\mathcal{I}\big(|\,\psi\,\rangle\big) = 4\big(\Delta\tilde{H}_\psi\big)^2 = N^2\big(\langle\, \psi^\perp \,|\, \psi^\perp \,\rangle - \langle\, \psi \,|\, \psi^\perp \,\rangle^2\big) = N^2. \tag{24.79}$$

We thus find that the QFI increases quadratically with N, as opposed to a linear increase obtained for N individual probes. This kind of scaling advantage with respect to the best "classical" (here, meaning product state probe) estimation strategy is called *Heisenberg scaling*.

One might expect that the measurements to obtain such a scaling advantage could in principle be complicated N-qubit measurements, but, as we shall see, simple single-qubit projective measurements are here sufficient. Consider, for instance[8], a measurement where each of the N qubits of the encoded probe state is measured with respect to the basis $|\pm_\varphi \,\rangle := U_\varphi \,|\, \pm \,\rangle$, where $|\, \pm \,\rangle$ are the eigenstates of σ_x and $U_\varphi = \exp(-i\varphi\tfrac{\sigma_z}{2})$. In principle, this corresponds to a projective measurement with 2^N outcomes, each a string of N individual "+" and "-" outcomes corresponding to the pure states $|\pm_\varphi \,\rangle$. However, the permutational symmetry of the probe state implies that the specific

[8]The particular measurement direction along the Bloch-sphere equator is not important, but we choose the particular direction here for simpler comparison with the discussion in Section 24.1.4.

distribution of "+" and "-" results does not matter, only their overall number. The probability for obtaining m "+" outcomes is then just

$$\binom{N}{m} \operatorname{Tr}\left[\left(|+_\varphi\rangle\langle+_\varphi|\right)^{\otimes m} \otimes \left(|-_\varphi\rangle\langle-_\varphi|\right)^{\otimes N-m} |\psi(\theta)\rangle\langle\psi(\theta)|\right] \tag{24.80}$$

$$= \binom{N}{m} \left|\langle+_\varphi|^{\otimes m} \otimes \langle-_\varphi|^{\otimes N-m} |\psi(\theta)\rangle\right|^2 = \binom{N}{m} \left|\langle+|^{\otimes m} \otimes \langle-|^{\otimes N-m} U_\varphi^{\dagger \otimes N} |\psi(\theta)\rangle\right|^2$$

$$= \binom{N}{m} \left|\langle+|^{\otimes m} \otimes \langle-|^{\otimes N-m} |\psi(\theta - \varphi)\rangle\right|^2 = \frac{1}{2^{N+1}} \binom{N}{m} \left|1 + (-1)^{N-m} e^{i(\theta-\varphi)N}\right|^2$$

$$= \frac{1}{2^N} \binom{N}{m} \left(1 + (-1)^{N-m} \sin(N\theta)\right),$$

where we have set $\varphi = \frac{\pi}{2N}$ in the last step. Now further note that the overall projective measurement only has two different eigenvalues (outcomes), $n = +1$ and $n = -1$, depending on whether there is an even or odd number $(N-m)$ of "-" results. Denoting the corresponding projectors of rank 2^{N-1} by Π_+ and Π_-, respectively, we calculate the probability of the even/odd outcome as

$$p(n = \pm 1|\theta) = \operatorname{Tr}(\Pi_\pm \rho(\theta)) = \sum_{\substack{N-m \\ \text{even/odd}}} \frac{1}{2^N} \binom{N}{m} (1 \pm \sin(N\theta)) = \tfrac{1}{2}(1 \pm \sin(N\theta)).$$

$$\tag{24.81}$$

We can now note the similarity with eqn (24.26) for individual qubits, with the notable difference that the argument θ is now replaced by $N\theta$. We hence find the expected mean value for this measurement to be

$$p(n = +1|\theta) - p(n = -1|\theta) = \sin(N\theta), \tag{24.82}$$

and therefore we nominate the estimator

$$\hat{\theta}(n) = \tfrac{1}{N} \arcsin(n). \tag{24.83}$$

Note that this implies that we can now only determine θ up to integer multiples of π/N. If this N-qubit measurement is repeated ν times with outcomes $n_{i_1}, n_{i_2}, \ldots, n_{i_\nu}$ and we obtain k outcomes of $n = +1$ (an even number of individual $m = +1$ outcomes), we obtain the sample mean

$$\bar{n}^{(\nu)}(k) = \frac{2k - \nu}{\nu}. \tag{24.84}$$

The expected value of the estimator consequently is

$$\langle \hat{\theta} \rangle_k = \sum_{k=0}^{\nu} \binom{\nu}{k} p(n = +1|\theta)^k \, p(n = -1|\theta)^{\nu-k} \tfrac{1}{N} \arcsin\left(\tfrac{2k-\nu}{\nu}\right). \tag{24.85}$$

As illustrated in Fig. 24.6, the estimator again corresponds to the maximum-likelihood estimator and is unbiased only for $\theta = 0$ (as before in the single-qubit case) and

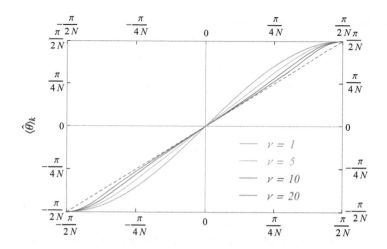

Fig. 24.6 Locally unbiased estimator for N-qubit phase estimation. The expected value of the estimator (solid curves, shown for different sample sizes ν) from eqn (24.85) generally does not match the true value θ (dashed straight line), meaning the estimator is generally biased, except for some isolated points ($\theta = 0, \pm\frac{\pi}{2N}$).

$\theta = \pm\frac{\pi}{2N}$, unless the sample size ν approaches infinity, in which case the estimator is unbiased in the full range.

In analogy to the estimation problem for individual qubits, the sample variance for the outcomes n_{i_j} is

$$\left(\Delta n^{(\nu)}\right)^2 = \tfrac{1}{\nu-1} \sum_{i=1}^{\nu} \left(n_{i_j} - \bar{n}^{(\nu)}\right)^2 = \frac{4k(\nu-k)}{\nu(\nu-1)}, \tag{24.86}$$

and the variance of the mean is obtained by dividing by ν. Once more, we propagate the error to obtain the variance of the estimate of θ. However, now, the (square of the) derivative of the estimator with respect to n, evaluated at $n = \bar{n}^{(\nu)}(k)$, is,

$$\left(\frac{\partial\hat{\theta}(n)}{\partial n}\right)^2 \Bigg|_{n=\bar{n}^{(\nu)}} = \left(\frac{\partial}{\partial n} \tfrac{1}{N} \arcsin(n)\right)^2 \Bigg|_{n=\bar{n}^{(\nu)}} = \frac{1}{N^2} \frac{\nu^2}{4k(\nu-k)}. \tag{24.87}$$

That is, we get an additional factor of $1/N^2$, such that

$$(\Delta\theta)^2 = \left(\frac{\partial\hat{\theta}(n)}{\partial n}\right)^2 \Bigg|_{n=\bar{n}^{(\nu)}} \frac{\left(\Delta n^{(\nu)}\right)^2}{\nu} = \frac{1}{N^2} \frac{1}{\nu-1}. \tag{24.88}$$

As long as the variance of the mean $\left(\Delta n^{(\nu)}\right)^2/\nu$ is sufficiently small (which here means sufficiently large ν), the (estimated) variance $(\Delta\theta)^2$ thus decreases quadratically with N, we obtain Heisenberg scaling.

Let us make some interesting observations about this result. First, one cannot simply make one single measurement with a total of $N\nu$ qubits, i.e., one still requires sufficiently many individual data points to get around the issues of unbiasedness and to formally make sense of the error-propagation formula. In particular, this estimation strategy can only determine (or rather, estimate) θ within an interval of width π/N. Therefore, θ must already be known sufficiently well before this strategy can be meaningfully applied in the first place.

Second, although there is an advantage in preparing probe states jointly for groups of N qubits, one achieves Heisenberg scaling in this example using single-qubit measurements (and classical post-processing).

Finally, let us remark that the scaling advantage of the kind discussed here is very sensitive to noise and is typically lost asymptotically, i.e., in the presence of noise one typically expects only a constant advantage (with respect to the best "classical" estimation strategy) with increasing N as N becomes very large (Knysh *et al.*, 2011; Escher *et al.*, 2012; Demkowicz-Dobrzański *et al.*, 2012; Knysh *et al.*, 2014). Nevertheless, there exist strategies to recover scaling advantages for certain types of noise provided that one can very precisely control the measuring apparatus (Dür *et al.*, 2014; Sekatski *et al.*, 2016; Sekatski *et al.*, 2017), but discussing these approaches in more detail goes beyond the scope of this chapter.

However, next we want to give a brief overview over another approach to quantum parameter estimation that is well defined even for as little as a single measurement but follows a different philosophy: Bayesian estimation.

24.3 Bayesian Parameter Estimation

The "local" (or "frequentist") paradigm for parameter estimation that we have discussed so far in this chapter is certainly not the only way of quantifying one's statistical confidence. A prominent alternative approach that we want to discuss now is called Bayesian estimation. Although well known in statistics, the Bayesian approach to parameter estimation is less widely used in (quantum) physics, but nevertheless has its place in quantum metrology (Personick, 1971; Demkowicz-Dobrzański *et al.*, 2015). The main philosophy of Bayesian estimation is the idea that knowledge is updated in the light of new observations (via Bayes' law, see eqn (19.53)), as we shall explain in the following.

24.3.1 The Bayesian-Estimation Paradigm

The central tenet of Bayesian estimation lies in the idea that, even before any measurements are performed, one has some prior knowledge (or belief) about the parameter that is to be estimated. And, moreover, that this knowledge can be captured by a suitable probability distribution—the *prior*. Depending on the parameter that is to be estimated, this probability distribution can be over a discrete or continuous, finite or infinite range of values. Some values or ranges of values may be excluded. The prior may represent very precise information obtained from previous observations, or may be based entirely upon one's prior belief. Data that is obtained from measurements is then used to update this knowledge: The prior is replaced with a *posterior* probability distribution, conditioned on the observed outcomes. The posterior thus becomes the new prior, which is itself updated by further measurements, and so on.

The most important conceptual difference to the frequentist paradigm thus lies in the central role (and choice) of the prior, whereas statements within frequentist estimation are based solely on the data. However, as we have seen in the previous sections, many techniques in frequentist estimation rely on identifying suitably narrow "local" parameter regions, which could be understood as taking into account certain prior information. This leads to a formal correspondence, a relation up to a constant factor π (Jarzyna and Demkowicz-Dobrzański, 2015; Górecki *et al.*, 2020) between the variances obtained from Bayesian and frequentist estimation in the limit of infinitely large available data sets. At the same time, the reliance on the prior means that Bayesian estimation is also well defined when very little data is available, and can be applied even for individual probes and measurements.

To see how this works, let us again consider the estimation of a parameter θ that has been encoded in a quantum state $\rho(\theta) = \Lambda_\theta[\rho]$ by some process (a CPTP map, see Chapter 21) Λ_θ that depends on θ. A measurement of the encoded probe state is represented by a POVM $\{E_m\}$ (see Chapter 23) with elements E_m and outcomes labelled by m. As before in eqn (24.9), the conditional probability to obtain outcome m given the parameter value θ is

$$p(m|\theta) = \mathrm{Tr}\big(E_m \rho(\theta)\big). \tag{24.89}$$

At this point, the local estimation scenario only requires that θ is close to a (known) value for which an unbiased estimator is available. This can, at least approximately, be satisfied when m is a collection of (sufficiently) many outcomes. For Bayesian estimation, no such assumption is required. Instead, this is where Bayesian estimation introduces the prior, i.e., a probability distribution $p(\theta) \geq 0$, satisfying $\int d\theta\, p(\theta) = 1$. Then, the probability for any single measurement outcome m is simply

$$p(m) = \int d\theta\, p(\theta)\, p(m|\theta) = \mathrm{Tr}\big(E_m \Gamma\big), \tag{24.90}$$

where we have used the notation of Personick (1971) to define the operator

$$\Gamma = \int d\theta\, p(\theta)\, \rho(\theta). \tag{24.91}$$

Further note that although we have written integrals here for simplicity, these should be replaced with sums in case that the parameter takes on only discrete values.

With the distributions $p(\theta)$, $p(m)$, and $p(m|\theta)$ at hand, we can use *Bayes' law* from eqn (19.53) to obtain the conditional probability distribution for θ, given that the measurement outcome m is observed, or the *posterior* for short, which is given by

$$p(\theta|m) = \frac{p(m|\theta)\, p(\theta)}{p(m)}. \tag{24.92}$$

The posterior captures the state of one's entire knowledge (or belief) given the observed data. However, in practice, stating the entire probability distribution to state one's belief about a parameter can be rather unwieldy. Instead, we can study certain key characteristics of the distribution, such as its mean or variance. However, note that one needs to take some care in choosing suitable quantifiers for mean and variance, depending on the specific type of parameter and prior.

For instance, let us first consider a parameter that can in principle take on arbitrary real values, $\theta \in \mathbb{R}$, and thus assume that the prior has support over all of $\mathbb{R}$. A single-value estimate for θ comparable to the estimator $\hat{\theta}(m)$ used in local estimation can then be obtained from the mean of the posterior, i.e., we can nominate

$$\hat{\theta}(m) = \int d\theta \, p(\theta|m) \, \theta = \frac{1}{p(m)} \int d\theta \, p(m|\theta) \, p(\theta) \, \theta = \frac{\mathrm{Tr}(E_m \eta)}{\mathrm{Tr}(E_m \Gamma)}, \qquad (24.93)$$

where we have used eqns (24.92) and (24.90), and we have defined the operator

$$\eta = \int d\theta \, p(\theta) \, \rho(\theta) \, \theta. \qquad (24.94)$$

This assignment of an estimator is not tied to the notion of unbiasedness in the same sense as the estimator in local estimation. However, a similar property arises by construction directly from the prior, i.e., when averaging over all possible outcomes m, the estimate is *the same* as the mean value provided by the prior,

$$\bar{\theta} = \sum_m p(m) \, \hat{\theta}(m) = \sum_m \mathrm{Tr}(E_m \eta) = \mathrm{Tr}(\eta) = \int d\theta \, p(\theta) \, \theta. \qquad (24.95)$$

In a similar fashion, the precision of the estimate, or rather, the precision of our updated knowledge of the parameter given the outcome m, can be quantified using the MSE variance $V_{\mathrm{post}}^{(m)}$ of the *posterior* $p(\theta|m)$, i.e.,

$$V_{\mathrm{post}}^{(m)} = V[p(\theta|m)] = \int d\theta \, p(\theta|m) \, (\theta - \hat{\theta}(m))^2 \qquad (24.96)$$

$$= \frac{1}{p(m)} \left[\mathrm{Tr}\left(E_m \int d\theta \, p(\theta) \, \rho(\theta) \, \theta^2 \right) - \frac{(\mathrm{Tr}(E_m \eta))^2}{\mathrm{Tr}(E_m \Gamma)} \right],$$

where we have again made use of Bayes' law from eqn (24.92) and the definition of the estimator in eqn (24.93). Here it is interesting to compare with the variance $V[\hat{\theta}]$ of the estimator from eqn (24.10) that we have considered for local estimation. In local estimation, the variance depends (among other things) on the chosen estimator and the estimate of the variance derives from the distribution $p(m|\theta)$ of the outcomes m, given some value of θ. Conversely, the quantity $V_{\mathrm{post}}^{(m)}$ concerns statements about one's belief about the distribution $p(\theta|m)$ of the possible values of θ given a specific outcome m. Moreover, the width of the posterior may not always be smaller than that of the prior; certain measurement outcomes can lead to a broadening of the distribution. To similarly describe the increase in knowledge that one expects to achieve in a given Bayesian estimation problem, we therefore average over different measurement outcomes, that is, we define

$$\overline{V}_{\text{post}} = \sum_m p(m) \, V_{\text{post}}^{(m)} = \int d\theta \, p(\theta) \, \theta^2 - \sum_m \frac{\left(\text{Tr}(E_m \eta)\right)^2}{\text{Tr}(E_m \Gamma)}. \qquad (24.97)$$

Before discussing situations where other quantifiers for the mean and width of the posterior may be more appropriate, let us examine a simple example. Consider a parameter $\theta \in \mathbb{R}$, and a Gaussian prior of width $\sigma > 0$ centred at $\theta = \theta_o$, that is,

$$p(\theta) = \frac{1}{\sqrt{2\pi}\sigma} \, e^{-\frac{(\theta - \theta_o)^2}{2\sigma^2}}. \qquad (24.98)$$

On the one hand, using the integral formulas from eqns (2.64) and (2.67), the mean and variance of the prior evaluate to

$$\bar{\theta} = \int_{-\infty}^{\infty} d\theta \, p(\theta) \, \theta = \theta_o, \quad V[p(\theta)] = \int_{-\infty}^{\infty} d\theta \, p(\theta) \, (\theta - \theta_o)^2 = \sigma^2. \qquad (24.99)$$

For the posterior, on the other hand, we can inspect the right-hand side of eqn (24.97), where the first term evaluate to $\int d\theta \, p(\theta) \, \theta^2 = \sigma^2 + \theta_o^2$. Together with the condition $0 \leq \overline{V}_{\text{post}} \leq V[p(\theta)]$, this implies that the remaining term in eqn (24.97), which determines the average decrease in width of the posterior with respect to the prior, is bounded by

$$\theta_o^2 \leq \sum_m \frac{\left(\text{Tr}(E_m \eta)\right)^2}{\text{Tr}(E_m \Gamma)} < \sigma^2 + \theta_o^2. \qquad (24.100)$$

Any further specific details of the decrease in width of the posterior depend on the encoding, probe state, and measurements.

As we have mentioned, for parameters and priors with support on all of $\mathbb{R}$, the approach described thus far is certainly justified. However, for parameters subject to boundary conditions, other quantifiers may be more appropriate. For example, when θ is a phase, such that θ is identified with $\theta + 2\pi k$ for all integers k, the MSE variance would assign large penalties to estimates that differ by $2\pi k$, even though they correspond to the same phase. Nevertheless, the MSE variance can still be useful also in phase estimation when priors are very narrow (see, e.g., Sec. A.2.5 in Friis *et al.*, 2017*b*). In addition, the MSE variance allows to establish lower bounds on the variance of the posterior for arbitrary priors, as we shall explain in Section 24.3.3. Before we do so, let us consider a more detailed example for phase estimation in the next section.

24.3.2   Bayesian Phase Estimation with Single Qubits

We now want to consider the Bayesian phase estimation problem analogous to the local phase estimation problem in Section 24.1.4, that is, for single qubits with Hilbert space $\mathcal{H} = \mathbb{C}^2$. In particular, we again use the encoding transformation is $U_\theta = e^{-i\tilde{H}\theta}$ with

$\tilde{H} = \frac{1}{2}\sigma_z$, and we will use the probe state $|+\rangle$, such that the encoded state is $U_\theta |+\rangle = \frac{e^{-i\theta/2}}{\sqrt{2}}(|0\rangle + e^{i\theta}|1\rangle)$. Let us then perform a projective measurement corresponding to state vectors on the equator of the Bloch sphere, which we parameterize using an angle μ with respect to the direction perpendicular to the probe state, i.e., a measurement in the basis $\{U_{\mu+\pi/2}|\pm\rangle\}$, where the parameter μ represents our freedom to adjust the measurement direction along the equator. The probability for the outcomes $m = \pm 1$ is then found from a calculation analogous to eqn (24.26), which results in

$$p(m = \pm 1|\theta) == \tfrac{1}{2}\big(1 \pm \sin(\theta - \mu)\big). \tag{24.101}$$

In order to obtain the unconditional probabilities for the measurement outcomes, we now have to take into account the prior. Here, we will consider a generalization of the Gaussian distribution of eqn (24.98) that is suitable for quantities with cyclic boundary conditions. That is, we will consider the prior to be a *wrapped Gaussian* of the form

$$p(\theta) = \frac{1}{\sqrt{2\pi}\sigma} \sum_{k=-\infty}^{\infty} \exp\Big(\frac{-(\theta - \theta_o + 2\pi k)^2}{2\sigma^2}\Big), \tag{24.102}$$

where the sums runs over all integers k. We use the same symbol $p(\theta)$, since it is clear from the context when it refers to wrapped or regular Gaussians. Using the substitution $\theta' = \theta + 2\pi k$, it is easy to see that the wrapped Gaussian is normalized, i.e.,

$$\int_0^{2\pi} d\theta\, p(\theta) = \frac{1}{\sqrt{2\pi}\sigma} \sum_{k=-\infty}^{\infty} \int_0^{2\pi} d\theta\, e^{-\frac{(\theta-\theta_o+2\pi k)^2}{2\sigma^2}} = \frac{1}{\sqrt{2\pi}\sigma} \sum_{k=-\infty}^{\infty} \int_{2\pi k}^{2\pi(k+1)} d\theta'\, e^{-\frac{(\theta'-\theta_o)^2}{2\sigma^2}}. \tag{24.103}$$

The sum over all k can then be converted to an integration from $-\infty$ to ∞, recovering the normalization of the regular Gaussian in eqn (24.98). With the prior and conditional probability at hand, we can proceed with the calculation of $p(m)$. Writing the *sin* function in terms of exponentials and performing a calculation similar to that in eqn (24.103), using $e^{2\pi i k} = 1$ for all integers k, we obtain

$$p(m) = \int_0^{2\pi} d\theta\, p(\theta)\, p(m|\theta) = \tfrac{1}{2} \pm \frac{1}{\sqrt{2\pi}\sigma} \sum_{k=-\infty}^{\infty} \int_0^{2\pi} d\theta\, \sin(\theta - \mu)\, \exp\Big(\frac{-(\theta - \theta_o + 2\pi k)^2}{2\sigma^2}\Big)$$

$$= \tfrac{1}{2}\big(1 \pm e^{-\sigma^2/2}\sin(\theta_o - \mu)\big). \tag{24.104}$$

Consequently, we obtain the posterior distributions for the outcomes $m = \pm 1$,

$$p(\theta|m) = \frac{1 \pm \sin(\theta - \mu)}{1 \pm e^{-\sigma^2/2}\sin(\theta_o - \mu)} \frac{1}{\sqrt{2\pi}\sigma} \sum_{k=-\infty}^{\infty} \exp\Big(\frac{-(\theta - \theta_o + 2\pi k)^2}{2\sigma^2}\Big). \tag{24.105}$$

Before proceeding with our calculations of means and variances of the prior and posteriors, let us pause to examine the properties of these distribution. First, note that an average over θ, weighted with $p(\theta)$ or $p(\theta|m)$, is problematic, since the integrand would not be invariant under the operation $\theta \mapsto \theta + 2\pi$. To meaningfully assign a mean value on the unit circle, one instead uses an average over $z := e^{i\theta}$ and assigns the mean value

$$\hat{\theta} = \arg(\langle z \rangle) = \arg\left(\int_0^{2\pi} d\theta\, e^{i\theta}\, p(\theta)\right). \tag{24.106}$$

For the example of the wrapped Gaussian prior, this results in

$$\langle z \rangle = \frac{1}{\sqrt{2\pi}\sigma} \int_{-\infty}^{\infty} d\theta'\, e^{i\theta'}\, e^{-\frac{(\theta'-\theta_o)^2}{2\sigma^2}} = e^{i\theta_o} e^{-\frac{\sigma^2}{2}}. \tag{24.107}$$

The argument of the mean value of z for the prior is thus $\hat{\theta} = \arg(\langle z \rangle) = \theta_o$. The width of the distribution, meanwhile, can be characterized by the magnitude of $\langle z \rangle$. More specifically, the *circular standard deviation* s evaluates to

$$s := \sqrt{\ln\left(|\langle z \rangle|^{-2}\right)} = \sigma. \tag{24.108}$$

This measure is particularly useful here, since it relates to the standard deviation of the underlying Gaussian distribution, but one could use other measures such as the *Holevo phase variance* $V_\phi := |\langle z \rangle|^{-2} - 1$ (see Holevo, 1984).

In any case, we can continue by evaluating $|\langle z \rangle|^{-2}$ for the posterior. Integrals of the same type as we have encountered before and some algebra (which we leave to the interested reader as an exercise) then provide the expression

$$|\langle z \rangle_{p(\theta|m)}|^{-2} = \frac{e^{\sigma^2}\left(1 \pm e^{-\sigma^2/2}\sin(\theta_o - \mu)\right)^2}{1 + e^{-\sigma^2}\sinh^2(\sigma^2) + e^{-\sigma^2}\sin^2(\theta_o - \mu)}. \tag{24.109}$$

We observe that the variances of the posterior depend on the chosen measurement direction (parameterized by μ). In order to select the optimal value of μ, let us examine the average Holevo phase variance, which here takes the form

$$\bar{V}_\phi := \sum_m p(m)|\langle z \rangle_{p(\theta|m)}|^{-2} - 1 = \frac{e^{\sigma^2}\left(1 + 3e^{-\sigma^2}\sin^2(\theta_o - \mu)\right)}{1 + e^{-\sigma^2}\sinh^2(\sigma^2) + e^{-\sigma^2}\sin^2(\theta_o - \mu)} - 1. \tag{24.110}$$

From this expression it is quite easy to see that the minimum average variance is obtained when $\theta_o = \mu$. Simply note that $\bar{V}_\phi$ only depends on μ via $x \equiv \sin^2(\theta_o - \mu)$, which takes values between 0 and 1. For non-negative constants $a = e^{-\sigma^2}$ and $b = e^{-\sigma^2}\sinh^2(\sigma^2)$, one then examines the function $f(x) = (1 + 3ax)/(1 + ax + b)$, which is strictly monotonically increasing for $a > 0$ and $b > 0$. The minimum is hence at $x = 0$, which translates to $\theta_o = \mu$.

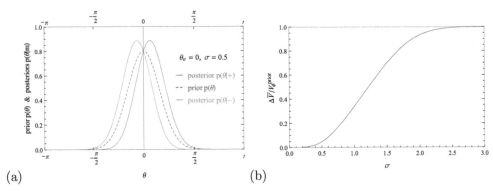

Fig. 24.7 Optimal Bayesian single-qubit phase estimation. (a) Illustration of a wrapped Gaussian prior with mean $\theta_o = 0$ and circular standard deviation $s = \sigma = 0.5$ (dashed, blue curve) and the posteriors resulting from the optimal single-qubit measurements for the results $m = +1$ (orange) and $m = -1$ (green). (b) Relative average decrease of the Holevo phase variance for the optimal single-qubit measurement as a function of the circular standard deviation of the prior.

For this optimal measurement direction we thus have $p(m) = \frac{1}{2}$, the outcomes are both equally likely, and the measurement is thus maximally informative. The posterior distributions are simply obtained from eqn (24.105) and are illustrated in Fig. 24.7 (a). The corresponding means $\langle z \rangle_{p(\theta|m)}$ then take the simple form

$$\langle z \rangle_{p(\theta|m)} = e^{i\theta_o}e^{-\sigma^2/2}\big(1 \pm ie^{-\sigma^2/2}\sinh(\sigma^2)\big). \qquad (24.111)$$

While $\langle z \rangle_{p(\theta|m)}$, and hence also the estimate $\hat{\theta}$, depend on the measurement outcome, the width of the posterior as captured by $|\langle z \rangle_{p(\theta|m)}|^{-2}$ is independent of m. Therefore, the average Holevo phase variance (or the average circular standard deviation, if desired) can be easily calculated, which provides

$$\overline{V}_\phi^{\text{post}} = \frac{e^{\sigma^2}}{1 + e^{-\sigma^2}\sinh^2(\sigma^2)} - 1. \qquad (24.112)$$

Here, it is interesting to make a direct comparison with the phase variance $V_\phi^{\text{prior}} = e^{\sigma^2} - 1$ of the prior (no average is required). The average relative decrease in phase variance, illustrated in Fig. 24.7, is given by

$$\frac{\Delta \overline{V}_\phi}{V_\phi^{\text{prior}}} = \frac{V_\phi^{\text{prior}} - \overline{V}_\phi^{\text{post}}}{V_\phi^{\text{prior}}} = \frac{e^{-\sigma^2}\sinh^2(\sigma^2)}{1 + e^{-\sigma^2}\sinh^2(\sigma^2)}. \qquad (24.113)$$

One observes that the relative decrease in variance, or increase in knowledge about θ, varies quite drastically, depending on the amount of initial information. If the initial prior is very narrow (for small σ close to 0), a single-qubit measurement only provides a very small relative decrease. At the same time, when there is hardly any initial information, e.g., when the prior is flat, $p(\theta) = \frac{1}{2\pi}$ in the limit $\sigma \to \infty$, the relative decrease in variance approaches 1. The question that we will turn to in the next section is then how quickly the variance can be decreased on average when one performs additional measurements.

24.3.3 A Bayesian Cramér–Rao Bound

In the last section, we have presented an example of the updating procedure in Bayesian phase estimation for a single measurement. In general, one is of course interested in being able to make several measurements and updates. In particular, it is desirable to identify useful strategies for sequences of probe states and measurements that will most increase our knowledge of the estimated parameter. Because of the dependence on the specific choice of the initial prior, and the generally complicated shape of the posterior after updates have been performed[9] the formulation of optimal strategies for Bayesian estimation is typically very challenging (if at all possible). However, it is nonetheless possible to give lower bounds on the average MSE variance from eqn (24.97). A very prominent such bound is the *van Trees inequality*, first derived by Harry L. van Trees (1968), which is usually considered as a Bayesian version of the Cramér–Rao bound (Theorem 24.1), see Gill and Levit (1995).

Theorem 24.4 (Van Trees Inequality)

The average MSE variance of the posterior can be bounded from below by

$$\overline{V}_{\text{post}} \geq \frac{1}{I\big(p(\theta)\big) + \bar{I}\big(p(m|\theta)\big)} \, ,$$

where $I\big(p(\theta)\big)$ is the Fisher information of the prior, given by

$$I\big(p(\theta)\big) = \int d\theta \, p(\theta) \left(\frac{\partial}{\partial \theta} \log p(\theta) \right)^2 ,$$

and $\bar{I}\big(p(m|\theta)\big)$ is the average Fisher information associated to the conditional probability distribution $p(m|\theta)$

$$\bar{I}\big(p(m|\theta)\big) = \int d\theta \, p(\theta) \, I\big(p(m|\theta)\big) = \int d\theta \, p(\theta) \sum_m p(m|\theta) \left(\frac{\partial}{\partial \theta} \log p(m|\theta) \right)^2 .$$

Before we present an explicit proof for the van Trees inequality, let us briefly remark on the involved quantities and range of validity. Since the variance $\overline{V}_{\text{post}}$ in question is the average MSE variance, one would typically expect that the parameter in question can take any values in $\mathbb{R}$ but that the prior satisfies $p(\theta \to \pm\infty) = 0$. However, as we will see, the van Trees inequality also holds if we assume some periodic boundary conditions, e.g., that θ is to be understood as $\theta \mod 2\pi$ and $p(\theta) = p(\theta \mod 2\pi)$. In the latter case, the MSE variance might not be a useful quantifier when the prior

---

[9]For instance, note that, even after a single update, the posterior arising from a (wrapped) Gaussian prior is usually no longer itself a (wrapped) Gaussian, see, e.g., the posterior in eqn (24.105).

has a large width, e.g., a wrapped Gaussian with large σ. However, as we have seen in the last section, already very few measurements can drastically reduce the width of the distribution, and for sufficiently narrow priors, an approach based on the MSE variance is a very good approximation (see, e.g., Sec. A.2.5 in Friis *et al.*, 2017b). With this in mind, let us now turn to the formal proof of Theorem 24.4.

Proof   To begin, we consider the integration limits $(a, b) = (-\infty, +\infty)$ (or $(a, b) = (0, 2\pi)$ for periodic priors) for the integral

$$\int_a^b d\theta \, \hat{\theta}(m) \, \frac{\partial}{\partial \theta} (p(\theta) \, p(m|\theta)) = \hat{\theta}(m) \, [p(\theta) \, p(m|\theta)]_a^b = 0 \,, \qquad (24.114)$$

where we have integrated by parts and used the fact that $\hat{\theta}$ is itself obtained from averaging over θ and hence does not depend on θ, while $p(a)p(m|a) = p(b)p(m|b)$. Another similar calculation yields the identity

$$\int_a^b d\theta \, \theta \, \frac{\partial}{\partial \theta} (p(\theta) \, p(m|\theta)) = - \int_a^b d\theta \, p(\theta) \, p(m|\theta) \,. \qquad (24.115)$$

We can then combine the two expressions of eqns (24.114) and (24.115) and make use of Bayes' law in eqn (24.92) to obtain

$$\int d\theta \, (\hat{\theta}(m) - \theta) \frac{\partial}{\partial \theta} (p(\theta) \, p(m|\theta)) = \int d\theta \, p(\theta) \, p(m|\theta) = p(m) \int d\theta \, p(\theta|m) = p(m) \,. \qquad (24.116)$$

Since $p(m)$ is normalized, summing over all measurement outcomes yields

$$\sum_m \int d\theta \, (\hat{\theta}(m) - \theta) \frac{\partial}{\partial \theta} (p(\theta) \, p(m|\theta)) = \sum_m p(m) = 1 \,. \qquad (24.117)$$

Alternatively, we may also rewrite the integrand using a natural logarithm, i.e.,

$$\frac{\partial}{\partial \theta} (p(\theta) \, p(m|\theta)) = p(\theta) \, p(m|\theta) \frac{\partial}{\partial \theta} \log(p(\theta) \, p(m|\theta)) \,. \qquad (24.118)$$

We then apply essentially the same technique as in the proof of Theorem 24.1, except that there is now a dependence also on θ. That is, we define

$$\xi_m(\theta) := \sqrt{p(\theta)p(m|\theta)}(\hat{\theta}(m) - \theta), \qquad (24.119a)$$

$$\chi_m(\theta) := \sqrt{p(\theta)p(m|\theta)} \, \frac{\partial}{\partial \theta} \log p(m|\theta), \qquad (24.119b)$$

and then use the Cauchy–Schwarz inequality such that

$$1 = \left| \sum_m \int d\theta \, \xi_m(\theta) \chi_m(\theta) \right|^2 \leq \sum_m \int d\theta \, |, \xi_m(\theta)|^2 \sum_n \int d\theta' \, |\chi_n(\theta')|^2 \qquad (24.120)$$

$$\leq \sum_m \int d\theta \, p(\theta) \, p(m|\theta) \, (\hat{\theta}(m) - \theta)^2 \sum_n \int d\theta' \, p(\theta') \, p(n|\theta') \left(\frac{\partial}{\partial \theta'} \log [p(\theta') \, p(n|\theta')] \right)^2 \,.$$

On the right-hand side of eqn (24.120), the first factor is identified as $\overline{V}_{\text{post}}$ as in eqn (24.97). The remaining second factor needs to be examined in more detail. Using the product rule, we have

$$\frac{\partial}{\partial \theta} \log [p(\theta)\, p(m|\theta)] = \frac{\partial}{\partial \theta} \log p(\theta) + \frac{\partial}{\partial \theta} \log p(m|\theta)\,, \qquad (24.121)$$

which results in three terms when squared. In the first of these terms, containing $\left(\frac{\partial}{\partial \theta} \log p(\theta)\right)^2$, we can sum over the conditional probability $p(n|\theta')$ to obtain the Fisher information of the prior, i.e.,

$$\int d\theta\, p(\theta) \sum_n p(n|\theta') \left(\frac{\partial}{\partial \theta} \log p(\theta)\right)^2 = \int d\theta\, p(\theta) \left(\frac{\partial}{\partial \theta} \log p(\theta)\right)^2 = I\big(p(\theta)\big)\,. \quad (24.122)$$

The second term, containing the square of $\frac{\partial}{\partial \theta} \log p(m|\theta)$, quite simply results in the average Fisher information $\bar{I}\big(p(m|\theta)\big)$ appearing in the van Trees inequality in Theorem 24.4. Finally, the cross term takes of the form

$$\sum_m \int d\theta\, p(\theta)\, p(m|\theta) \left(\frac{\partial}{\partial \theta} \log p(\theta)\right) \left(\frac{\partial}{\partial \theta} \log p(m|\theta)\right) \qquad (24.123)$$

$$= \sum_m \int d\theta \left(\frac{\partial}{\partial \theta} p(\theta)\right) \left(\frac{\partial}{\partial \theta} p(m|\theta)\right) = \int d\theta \left(\frac{\partial}{\partial \theta} p(\theta)\right) \left(\frac{\partial}{\partial \theta} \sum_m p(m|\theta)\right).$$

However, since $\sum_m p(m|\theta) = 1$, its derivative vanishes, and so does the cross term. From the inequality in (24.120) we thus arrive at the van Trees inequality by dividing both sides by the sum of $I\big(p(\theta)\big)$ and $\bar{I}\big(p(m|\theta)\big)$, i.e.,

$$\overline{V}_{\text{post}} \geq \frac{1}{I\big(p(\theta)\big) + \bar{I}\big(p(m|\theta)\big)}\,, \qquad (24.124)$$

which concludes the proof. $\qquad\qquad\qquad\qquad\qquad\qquad\qquad\qquad\qquad\qquad\qquad\qquad\qquad\square$

Although there are some similarities to the Cramér–Rao bound (Theorem 24.1), there are also important conceptual differences. First, the van Trees inequality is generally not tight. Given some prior $p(\theta)$, along with an encoded probe state $\rho(\theta)$ and POVM elements E_m such that $p(m|\theta) = \text{Tr}\big(E_m \rho(\theta)\big)$, the average posterior variance is generally strictly larger than the right-hand side of eqn (24.124). At the same time, if we have the necessary information to work out the bound, we could also calculate $\overline{V}_{\text{post}}$ exactly.

Nevertheless, the van Trees inequality can provide some important insights into the scaling behaviour of the average variance with repeated measurements. In particular, let us examine the situation where N consecutive probes are used to update our knowledge. What kind of scaling can we expect in such a case? If we examine the average Fisher information $\bar{I}\big(p(m|\theta)\big)$ appearing in eqn (24.124), we note that the Fisher information $I\big(p(m|\theta)\big)$ for every fixed value of θ can be bounded from above by the corresponding QFI, $I\big(p(m|\theta)\big) \leq \mathcal{I}\big(\rho(\theta)\big)$, since the QFI arises from the optimization over all measurements. In the special case of unitary encoding, where the QFI is

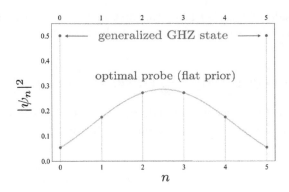

Fig. 24.8 Probe-state probabilities for Bayesian and local estimation. Illustration of the probability weights $|\psi_n|^2$ for the optimal probe states for N-qubit Bayesian phase estimation for $N = 5$ with a flat prior (blue) and local N-qubit phase estimation (red), corresponding to the state in eqn (24.127) with weights as in eqn (24.128), and the generalized GHZ probe of eqn (24.78).

independent of θ, see eqn (24.71), we can then easily write $\bar{I}\big(p(m|\theta)\big) \leq \mathcal{I}(\rho)$, which allows to formulate the following modified van Trees inequality

$$\overline{V}_{\text{post}} \geq \frac{1}{I\big(p(\theta)\big) + \mathcal{I}(\rho)} \, . \tag{24.125}$$

If we then make use of Lemma 24.1, we further see that for a probe state of N consecutive probes, $\rho = \rho_1 \otimes \rho_2 \otimes \ldots \otimes \rho_N$, we can write $\mathcal{I}(\rho) \leq N\mathcal{I}(\rho_{\text{opt}})$, where ρ_{opt} is the state optimizing the QFI for a single probe. As N increases, the Fisher information of the prior remains constant. We thus see that any "classical" strategy of individual probes interspersed with updates of the prior, probe states, and measurement directions can at most lead to a decrease of the average variance with $1/N$, i.e.,

$$\overline{V}_{\text{post}}^{\text{classical}} \geq \frac{1}{I\big(p(\theta)\big) + \mathcal{I}(\rho)} \geq \frac{1}{I\big(p(\theta)\big) + N\mathcal{I}(\rho_{\text{opt}})} \propto \frac{1}{N} \quad \text{as } N \to \infty. \tag{24.126}$$

However, as in the local estimation paradigm, there exist strategies that go beyond this *standard quantum limit* and achieve Heisenberg scaling, as we will discuss in the next section.

24.3.4 Bayesian Phase Estimation with N Qubits

In this section, we want to study a particular example for Bayesian phase estimation that serves as a counterpart to the example in Section 24.2.4. The probe system again consists of N qubits, but instead of the (generalized) GHZ state from eqn (24.78), we will use a more general probe state of the form

$$| \psi \rangle = \sum_{n=0}^{N} \psi_n \, | n \rangle_{\text{un}} = \sum_{n=0}^{N} \psi_n \, | 1 \rangle^{\otimes n} | 0 \rangle^{\otimes (N-n)} \, . \tag{24.127}$$

Here, $|n\rangle_{\text{un}}$ represents a *unary encoding* of the number n. The crucial observation that leads to the selection of this probe state is that the superposition contains exactly one vector contribution for each of the different eigenvalues of the total encoding Hamiltonian $\tilde{H}$ from eqn (24.77). The $N+1$ different eigenvalues are degenerate, since the total Hilbert space has dimension 2^N, but the subspace of each fixed eigenvalue is left invariant by the unitary generated by $\tilde{H}$. Therefore, there is no benefit in considering superpositions of more than one vector from each subspace. We also see that the choice $\psi_0 = \psi_N = \frac{1}{\sqrt{2}}$ results in the generalized GHZ state we have previously used. However, for Bayesian phase estimation, it is in general better to use a probabilities $|\psi_n|^2$ that are non-zero for all n. In particular, for a flat prior[10] $p(\theta) = \frac{1}{2\pi}$, which we will consider here, the optimal choice is given by (Berry and Wiseman, 2000)

$$\psi_n = \sqrt{\frac{2}{N+2}} \sin\left(\frac{(n+1)\pi}{N+2}\right). \tag{24.128}$$

As illustrated in Fig. 24.8, the probability distribution of these weights is not flat, despite the prior being flat. The corresponding optimal measurement (Berry and Wiseman, 2000) is realized by a projective measurement in the "basis" $\{|e_k\rangle\}_{k=0,1,\dots,N}$ representing the discrete Fourier transform of the vectors $|n\rangle_{\text{un}}$, given by

$$|e_k\rangle = \frac{1}{\sqrt{N+1}} \sum_{n=0}^{N} e^{in\frac{2\pi k}{N+1}} |n\rangle_{\text{un}}. \tag{24.129}$$

Technically speaking, the set $\{|e_k\rangle\}_{k=0,1,\dots,N}$ does not span the entire Hilbert space of N qubits, but it does form a basis of the subspace spanned by the $|n\rangle_{\text{un}}$. Since this subspace is in turn left invariant by the encoding transformation, considering the measurement outcomes corresponding to these vectors is hence sufficient in our case. In particular, the conditional probability for obtaining outcome k given θ is

$$p(k|\theta) = |\langle e_k | U_\theta | \psi \rangle|^2 = \frac{1}{N+1} \left| \sum_{n=0}^{N} \psi_n e^{in(\theta - \frac{2\pi k}{N+1})} \right|^2. \tag{24.130}$$

By inserting ψ_n from eqn (24.128) and noting that the Kronecker delta can be written as $\delta_{mn} = \frac{1}{2\pi} \int d\theta\, e^{-i\theta(m-n)}$, we can then easily evaluate the unconditional outcome probabilities, i.e.,

$$p(k) = \int_0^{2\pi} d\theta\, p(\theta)\, p(k|\theta) = \frac{1}{2\pi} \int_0^{2\pi} d\theta\, p(k|\theta) = \frac{2}{(N+1)(N+2)} \sum_{n=0}^{N} \sin^2\left(\frac{(n+1)\pi}{N+2}\right) = \frac{1}{N+1}. \tag{24.131}$$

As in Section 24.3.2, we thus see that the optimal measurement is such that all possible measurement outcomes are equally likely. Finally, the posterior can be obtained directly from eqn (24.130) as

---

[10]Note that this can be considered to be a limiting case of the wrapped Gaussian from eqn (24.102) for $\sigma \to \infty$, and corresponds to having no initial knowledge about θ.

$$p(\theta|k) = \tfrac{N+1}{2\pi} \, p(k|\theta) \,. \tag{24.132}$$

Next, we want to evaluate the average $\langle\, e^{i\theta}\,\rangle$ for the posterior, i.e.,

$$\langle\, e^{i\theta}\,\rangle_{p(\theta|k)} = \tfrac{N+1}{2\pi} \int_{0}^{2\pi} d\theta \, e^{i\theta} \, p(k|\theta) \tag{24.133}$$

$$= \frac{1}{\pi(N+2)} \sum_{m,n=0}^{N} e^{i(m-n)\frac{2\pi k}{N+1}} \sin\left(\tfrac{(m+1)\pi}{N+2}\right) \sin\left(\tfrac{(n+1)\pi}{N+2}\right) \int_{0}^{2\pi} d\theta \, e^{-i\theta(m-n-1)} \,.$$

The integral gives $2\pi\delta_{m,n+1}$, and we can use the identity

$$\frac{2}{N+2} \sum_{n=0}^{N} \sin\left(\frac{(n+1)\pi}{N+2}\right) \sin\left(\frac{(n+2)\pi}{N+2}\right) = \cos\left(\frac{\pi}{N+2}\right) \tag{24.134}$$

to arrive at the simple result

$$\langle\, z\,\rangle_{p(\theta|k)} = \langle\, e^{i\theta}\,\rangle_{p(\theta|k)} = e^{i\frac{2\pi k}{N+1}} \cos\left(\frac{\pi}{N+2}\right) \,. \tag{24.135}$$

We thus see that, given the outcome k, the estimate for θ is just $\frac{2\pi k}{N+1}$, whereas the phase variance is

$$V_{\phi}^{(k)} = \left|\langle\, z\,\rangle_{p(\theta|k)}\right|^{-2} - 1 = \frac{1}{\cos^2\left(\frac{\pi}{N+2}\right)} - 1 = \tan^2\left(\frac{\pi}{N+2}\right) \,. \tag{24.136}$$

Since this value is independent of k, it matches the average phase variance, $\overline{V}_{\phi}^{\,\text{post}} = V_{\phi}^{(k)}$ for all k. Moreover, for $N \gg 1$, we can expand the tangent for small values of $(N+2)^{-1}$, such that

$$\overline{V}_{\phi}^{\,\text{post}} \propto \frac{\pi^2}{N^2} \quad \text{as } N \to \infty \,. \tag{24.137}$$

We thus also find Heisenberg scaling for Bayesian phase estimation, but note that the pre-factor π^2 appears in the asymptotics of the variance with respect to the variance of the local estimation problem. For further discussion of this scaling behaviour, see, e.g., Jarzyna and Demkowicz-Dobrzański (2015) and Górecki *et al.* (2020). Finally, let us remark that, although the probe states and optimal measurements for Bayesian and local phase estimation with N qubits can be quite different, both can be efficiently prepared within a unified estimation setting based on measurement-based computation (see Friis *et al.*, 2017*b*).

25

Quantum States of Light

In this chapter, we return to the very beginning of this book, the quantum-mechanical description of light, and thus sketch out the connection between the origins of quantum mechanics—in particular, the contributions by *Max Planck* and *Albert Einstein* as laid out in Chapter 1—and modern quantum optics, i.e., the quantum-mechanical description of the electromagnetic field and its interaction with matter. Providing a full overview of this large field of study is certainly beyond the scope of this chapter, and, indeed, of this book—for an in-depth study, we refer to the textbooks by Chistopher C. Gerry (2005), Vogel and Welsch (2006), or Garrison and Chiao (2008)—but here we aim to give an introduction to some of the most fundamental concepts. To this end, we will outline the idea of second quantization for the electromagnetic field in Section 25.1. We then explore the family of *coherent states* in Section 25.2, which are of central importance in quantum optics, since they form a conceptual bridge connecting classical and quantum harmonic oscillators. Moreover, coherent states admit a compact and intuitive description in phase space. As we will see in Section 25.3, such a phase-space representation can be chosen for any quantum state, for instance, in terms of the famous Wigner function, and hence serves as a corner stone of any further study of quantum optics. With this basis established, we can then turn to the larger class of so-called Gaussian states in Section 25.4. This class of states includes the aforementioned coherent states, and permits an elegant mathematical description in phase space, but is also extremely relevant from the point of view of experimental accessibility (see, e.g., Klauder and Skagerstam, 1985; Andersen *et al.*, 2016).

25.1 Quantization of the Electromagnetic Field

25.1.1 Modes of the Classical Electromagnetic Field

If we turn our attention back to the origins of quantum mechanics, specifically to the derivation of Planck's law of black-body radiation in Section 1.1.3, we recall that we considered the electromagnetic field confined to a cavity with which it is at thermal equilibrium. Let us now re-examine some of the assumptions that went into this derivation in more technical detail—and with the benefit of hindsight.

The classical electromagnetic field can be represented by the gauge potential $\vec{A}$ and the electric potential ϕ, which we can collect into a four-vector $A^\mu = (\phi, \vec{A})^\top$, using Gaussian units and abstract index notation. That is, here A^μ denotes both the four-vector itself and its μth component, where Greek indices run from 0 (denoting the temporal component) to 3, $\mu = 0, 1, 2, 3$, with spatial components $i = 1, 2, 3$ indexed by Latin letters. The electric and magnetic fields are given by $\vec{E} = -\vec{\nabla}\phi - \frac{1}{c}\frac{\partial \vec{A}}{\partial t}$ and $\vec{B} = \vec{\nabla} \times \vec{A}$. In terms of the electromagnetic field strength tensor

$$F^{\mu\nu} := \partial^\mu A^\nu - \partial^\nu A^\mu = \begin{pmatrix} 0 & -E_1/c & -E_2/c & -E_3/c \\ E_1/c & 0 & -B_3 & B_2 \\ E_2/c & B_3 & 0 & -B_1 \\ E_3/c & -B_2 & B_1 & 0 \end{pmatrix}, \tag{25.1}$$

Maxwell's equations can then be expressed in a Lorentz-covariant and gauge-invariant way as

$$\partial^\mu F^{\mu\nu} = \mu_0 J^\nu, \quad \partial^\mu \left(\tfrac{1}{2}\varepsilon^{\mu\nu\alpha\beta} F_{\alpha\beta}\right) = 0, \tag{25.2}$$

where $J^\nu = (c\rho, \vec{j})^\top$ is the electromagnetic current density. However, by choosing a gauge, we can write them even more compactly. In the Lorenz gauge, with gauge condition $\partial_\mu A^\mu = 0$, where the Einstein summation convention is implied, Maxwell's equations from eqn (25.2) simply become

$$\Box A^\mu = \partial_\nu \partial^\nu A^\mu = \frac{4\pi}{c} J^\mu. \tag{25.3}$$

In the absence of source terms, i.e., in vacuum where $J^\mu = 0$, we simply have $\Box A^\mu = 0$. To determine the solutions of the vacuum field equations, we can make a plane-wave ansatz $A^\mu = \epsilon^\mu(q^\nu) \exp(-iq^\alpha x_\alpha)$ for the gauge potential based on the (photon) four-momentum $q^\mu = \left(\omega/c, \vec{k}\right)^\top$. Here, ω is the photon's (angular) frequency and $\vec{k}$ is the photon wave vector satisfying $|\vec{k}| = \omega/c$ such that q^μ is a null vector, $q^\mu q_\mu = 0$. The four-vector ϵ^μ may depend on the four-momentum and is to be determined. Inserting into the vacuum Maxwell equations, we have

$$\partial_\nu \partial^\nu A^\mu = -\epsilon^\mu \exp(-iq^\alpha x_\alpha) q^\nu q_\nu = 0, \tag{25.4}$$

which is satisfied for any choice of q^μ since the latter is a null vector. The choice of $\epsilon^\mu(q^\nu) = \epsilon^\mu(\vec{k})$ is restricted by the Lorenz-gauge condition,

$$\partial_\mu A^\mu = -i q_\mu \epsilon^\mu = 0 \quad \rightarrow \quad q_\mu \epsilon^\mu = 0, \tag{25.5}$$

but the Lorenz-gauge condition has not uniquely fixed the gauge yet, because we can still perform the transformation $A_\mu \rightarrow A_\mu + \partial_\mu \Lambda$ for any scalar function Λ satisfying $\partial_\mu \partial^\mu \Lambda = 0$ without changing the physics. This means we can fix the gauge via the choice $\Lambda = i\kappa \exp(-iq^\alpha x_\alpha)$, which implies $\partial_\mu \Lambda = \kappa q_\mu \exp(-iq^\alpha x_\alpha)$. The remaining gauge freedom is then parameterized by κ and corresponds to transformations $\epsilon^\mu \rightarrow \epsilon^\mu + \kappa q^\mu$. In particular, we can select κ such that $\epsilon^0 = 0$, which reduces the original Lorenz-gauge condition to $\vec{q} \cdot \vec{\epsilon} = 0$. For any fixed photon three-momentum $\vec{q} = \vec{k}$, which fixes the photon frequency, we are thus left with two linearly independent solutions for $\vec{\epsilon}_s$ with $s = \pm$ in the plane perpendicular to $\vec{k}$, corresponding to the two photon polarizations. In the absence of boundary conditions, the non-trivial general solution of Maxwell's equations (in this gauge) can hence be written as

$$\vec{A} = \int d^3k \sum_{s=\pm} \left(a_s(\vec{k})\, \vec{\epsilon}_s(\vec{k})\, e^{-i(\omega t - \vec{k}\cdot\vec{x})} + \text{c.c.} \right). \tag{25.6}$$

For the classical electromagnetic field, the amplitudes $a_s(\vec{k})$ are complex coefficients (with appropriate units), and the vectors $\vec{\epsilon}_s(\vec{k})$ can be assumed to be normalized without loss of generality, while the $e^{-i(\omega t - \vec{k}\cdot\vec{x})}$ are the (time-dependent) plane-wave mode functions of the field.

In the derivation of the black-body spectrum (see Section 1.1.3) we then considered the electromagnetic field to be confined within a three-dimensional box. The resulting boundary conditions restrict the possible mode functions to a discrete set, much in the same way that the solutions of the Schrödinger equation are restricted in the presence of boundary conditions (see Section 4.2). In particular, in the absence of sources (assuming a vacuum within the box) and using the mentioned gauge, where $A^0 = 0$, one may solve the resulting wave equation $\Box \vec{A}(t, \vec{x}) = 0$ via a separation ansatz as we have used in Section 4 for the Schrödinger equation. That is, the gauge potential can be decomposed into linearly independent solutions $h_m(t)\, \vec{u}_m(\vec{x})$ that factor into time-dependent functions $h_m(t)$ and time-independent (here, vector-valued) mode functions $\vec{u}_m(\vec{x})$, both labelled by a discrete (multi-) index m, i.e.,

$$\vec{A}(t, \vec{x}) = \sum_m h_m(t)\, \vec{u}_m(\vec{x}). \tag{25.7}$$

The mode functions $\vec{u}(\vec{x})$ are then defined by the eigenvalue equation

$$\Delta \vec{u}_m(\vec{x}) = -k_m^2\, \vec{u}_m(\vec{x}), \tag{25.8}$$

and additional constraints that depend on the specific boundary conditions. For instance, for perfect-conductor boundary conditions, which require the tangential components of $\vec{E}$ and the orthogonal components of $\vec{B}$ to vanish at any conducting surface with unit normal $\vec{n}$, these constraints read $\vec{\nabla} \cdot \vec{u}_m(\vec{x}) = 0$ and $\vec{n} \times \vec{u}_m(\vec{x}) = 0$ (see, e.g., Ballentine, 2014, p. 526). The resulting mode functions are normalized, such that

$$\int d^3x\ \vec{u}_{m'}(\vec{x}) \cdot \vec{u}_m(\vec{x}) = \delta_{m'm}, \tag{25.9}$$

but since we will not further consider their specific form here, we refer to (Friis *et al.*, 2013*b*, Sections IV and V) for more detailed expressions of the mode functions (for both polarizations). With the ansatz of eqn (25.7) at hand, the wave equation then becomes

$$\Box \vec{A}(t, \vec{x}) = -\frac{1}{c^2}\frac{\partial^2 \vec{A}}{\partial t^2} + \Delta \vec{A} = -\sum_m \frac{1}{c^2}\frac{\partial^2}{\partial t^2} h_m(t)\, \vec{u}_m(\vec{x}) + \sum_m h_m(t)\, \Delta \vec{u}_m(\vec{x})$$

$$= -\sum_m \left(\frac{1}{c^2}\frac{\partial^2}{\partial t^2} + k_m^2 \right) h_m(t)\, \vec{u}_m(\vec{x}) = 0, \tag{25.10}$$

where we have inserted from eqn (25.8) in the third step. Due to the linear independence of the $\vec{u}$, we see that each of the functions h_m satisfies the differential equation of a harmonic oscillator with frequency $\omega_m = ck_m$. In other words, the classical electromagnetic field in the box can be decomposed into a discrete (but infinite) set of independent harmonic oscillators. In particular, the electric and magnetic fields can be decomposed as

$$\vec{E}(t, \vec{x}) = \sum_m f_m(t)\, \vec{u}_m(\vec{x}), \tag{25.11}$$

$$\vec{B}(t, \vec{x}) = \sum_m h_m(t)\, \vec{\nabla} \times \vec{u}_m(\vec{x}), \tag{25.12}$$

where the functions f_m are related to the h_m via

$$f_m(t) = -\tfrac{1}{c}\tfrac{\partial h_m(t)}{\partial t}, \tag{25.13}$$

and it can easily be shown (Ballentine, 2014, p. 528) that the functions f_m also correspond to harmonic oscillators with frequencies $\omega_m = ck_m$. In particular, we can now obtain the Hamiltonian of the classical electromagnetic field by integrating over the energy-density $(|\vec{E}|^2 + |\vec{B}|^2)/(8\pi)$. Using the perfect-conductor boundary conditions, it can then be shown straightforwardly (e.g., in Ballentine, 2014, p. 528) that the classical electromagnetic field Hamiltonian takes the form

$$H_{\mathrm{EM}} = \tfrac{1}{8\pi} \int d^3r \left(|\vec{E}(\vec{x}, t)|^2 + |\vec{B}(\vec{x}, t)|^2 \right) = \tfrac{1}{8\pi} \sum_m (f_m^2 + k_m^2 h_m^2). \tag{25.14}$$

We can then define the canonical position and momentum variables $Q_m := \frac{f_m}{2\omega_m\sqrt{\pi}}$ and

$$P_m := \frac{\partial Q_m}{\partial t} = \frac{1}{2\omega_m\sqrt{\pi}} \frac{\partial f_m}{\partial t} = \frac{1}{2\omega_m\sqrt{\pi}} \left(-\tfrac{1}{c}\frac{\partial^2 h_m}{\partial t^2} \right) = \frac{c\,k_m^2\,h_m}{2\omega_m\sqrt{\pi}} = \frac{\omega_m}{2c\sqrt{\pi}}\,h_m, \tag{25.15}$$

where we have inserted from eqns (25.13) and (25.10). In terms of Q_m and P_m, the Hamiltonian takes the form

$$H_{\mathrm{EM}} = \tfrac{1}{2} \sum_m (\omega_m^2\, Q_m^2 + P_m^2). \tag{25.16}$$

25.1.2   The Quantized Electromagnetic Field

Comparing the Hamiltonian (function) in eqn (25.16) to the expression for the Hamiltonian (operator) quantum harmonic oscillator in eqn (5.3), the path to the quantization of the electromagnetic field is clearly laid out. Each mode of frequency ω_m can be associated with a quantum harmonic oscillator (with unit mass) via the replacement

$$Q_m \longrightarrow \hat{X}_m, \qquad P_m \longrightarrow \hat{P}_m, \tag{25.17}$$

where the linear operators $\hat{X}_m$ and $\hat{P}_m$ satisfy the canonical commutation relations

$$[\hat{X}_m, \hat{P}_n] = i\hbar\delta_{mn}, \quad [\hat{X}_m, \hat{X}_n] = [\hat{P}_m, \hat{P}_n] = 0 \tag{25.18}$$

that we recall from Theorem 2.2. The subscripts now label the (infinitely many) modes of the electromagnetic field (instead of the three spatial directions as in Chapter 2). Since we are dealing with independent harmonic oscillators, we may define corresponding annihilation and creation operators a_m and $a_m^\dagger$ for each mode, i.e.,

$$a_m := \frac{1}{\sqrt{2\hbar\omega_m}}(\omega_m\hat{X}_m + i\hat{P}_m), \tag{25.19a}$$

$$a_m^\dagger := \frac{1}{\sqrt{2\hbar\omega_m}}(\omega_m\hat{X}_m - i\hat{P}_m), \tag{25.19b}$$

in full analogy to Definition 5.1 in Chapter 5. These mode operators then satisfy the commutation relations

$$[a_m, a_n] = [a_m^\dagger, a_n^\dagger] = 0, \quad [a_m, a_n^\dagger] = \delta_{mn}. \tag{25.20}$$

Mirroring the expression in eqn (5.9), the Hamiltonian of the classical electromagnetic field from eqn (25.16) can then be written in its quantized form

$$H = \sum_m \frac{1}{2}\left(\omega_m\hat{X}_m^2 + \hat{P}_m^2\right) = \sum_m \hbar\omega_m\left(a_m^\dagger a_m + \frac{1}{2}\right). \tag{25.21}$$

Here, two remarks are in order. First, it is of course not just the Hamiltonian that is quantized by this procedure, turning a function into a self-adjoint operator. Indeed, the field itself is quantized. For a real choice of mode functions $\vec{u}_m(\vec{x})$, the coefficient functions $h_m(t)$ of the expansion in eqn (25.7), and hence the gauge field $\vec{A}(t,\vec{x})$ become self-adjoint operators. That is, from eqns (25.15) and (25.19) we have $h_m = \frac{2c\sqrt{\pi}}{\omega_m}\hat{P}_m$ and $\hat{P}_m = -i\sqrt{\frac{\hbar\omega_m}{2}}\left(a_m - a_m^\dagger\right)$ and thus we can write

$$\vec{A}(t,\vec{x}) = -i\sum_m \sqrt{\frac{2\hbar c^2\pi}{\omega_m}}\,\vec{u}_m(\vec{x})\left(a_m - a_m^\dagger\right). \tag{25.22}$$

Although we should note that it is often customary (and more convenient) to expand the field in terms of complex-valued mode functions in which case the coefficients of the annihilation and creation operators in the expansion are the mode functions and their complex conjugates, respectively (see, e.g., Ballentine, 2014, p. 532, for a discussion). Regardless of the particular choice of mode functions, the resulting electromagnetic field is operator-valued, a *quantum field*, and must be treated within the framework of quantum field theory.

Second, let us briefly discuss the ground state of the Hamiltonian in eqn (25.21). In analogy to Definition 5.2, we can define $N_m := a_m^\dagger a_m$ as the occupation-number operator of the mode labelled by m, such that $H = \sum_m \hbar\omega_m\left(N_m + \frac{1}{2}\right)$. Since the N_m are positive (semidefinite) operators, the ground state of the field is the joint ground state of all N_m. For simplicity, let us denote this state as $|0\rangle$, such that $N_m|0\rangle = 0$. Each individual harmonic oscillator (each mode labelled by m) is associated to a non-zero but finite zero-point energy $\frac{\hbar\omega_m}{2}$. However, for the ground state of the quantum field we then have $H|0\rangle = \sum_m \frac{\hbar\omega_m}{2}$, which generally diverges unless a cutoff frequency is introduced. In particular, infinite zero-point energies are obtained in principle for the free field (see, e.g., p. 22 and the discussion on p. 790 in the textbook by Peskin and Schroeder, 1995) and the field within a three-dimensional box with perfect-conductor boundary conditions (Ballentine, 2014, Sec. 19.3).

On the one hand, it is often argued that only differences in energy are measurable. The additional (infinite) constant term is therefore typically dropped for simplicity to work with the more compact Hamiltonian $H = \sum_m \hbar\omega_m N_m$ (see, for instance, Peskin and Schroeder, 1995, p. 22). On the other hand, the difference between two (infinite) vacuum energies, for example, on two sides of a metal plate inserted into a conducting cavity, can be finite, as argued by *Hendrik Casimir* (1948) based on earlier work with *Dirk Polder* (Casimir and Polder, 1948). This creates a force—the *Casimir force*—that can be attributed solely to the differences in the vacuum energy densities, as outlined concisely by *Leslie Ballentine* (2014, p. 535).

The *Casimir effect*, which has been validated both qualitatively (Sparnaay, 1958) and quantitatively (Lamoreaux, 1997), thus indicates that entirely dismissing the reality of the vacuum energy may not be called for. Indeed, this question is both unresolved and may point to a resolution outside the current understanding of quantum field theory (see, e.g., the discussion in Peskin and Schroeder, 1995, p. 790). For the remainder of this book, however, we shall not touch upon this question again and will hence work with the simpler Hamiltonian $H = \sum_m \hbar\omega_m N_m$.

In particular, we can employ this Hamiltonian for the calculation of the average photon number of the electromagnetic field in thermal equilibrium with a surrounding cavity, as it appears, for instance, in the derivation of the black-body spectrum in Section 1.1.3. For the purpose of such a calculation, the electromagnetic field within the cavity can be treated as a closed quantum system. Any such system in thermal equilibrium with a heat bath at temperature T can be described by a thermal state[1] $\tau(\beta)$ with $\beta = 1/(k_\mathrm{B}T)$, given by

$$\tau(\beta) = \frac{e^{-\beta H}}{Z}, \tag{25.23}$$

where $Z = \mathrm{Tr}(e^{-\beta H}) = \sum_i \beta E_i$ and the E_i are the eigenvalues of H. From the commutation relations in eqn (25.20), it follows that $[N_m, N_n] = 0$. Consequently, the individual number operators have joint eigenstates and we can therefore calculate the expected value of the average photon number for each mode individually, i.e.,

---

[1] Also called a *Gibbs state*. When the system is assumed to be of fixed composition but may be in states of varying energy, this corresponds to the canonical ensemble, which we will consider here unless stated otherwise.

$$\bar{N}_m = \text{Tr}\big[N_m \tau(\beta)\big] = \text{Tr}\Big(N_m \frac{\exp\big(-\beta \hbar \omega_m \, N_m\big)}{\text{Tr}\big[\exp\big(-\beta \hbar \omega_m \, N_m\big)\big]}\Big) = \frac{\sum\limits_n n e^{-\beta \hbar \omega_m n}}{\sum\limits_{n'} e^{-\beta \hbar \omega_m n'}} = \frac{1}{e^{-\beta \hbar \omega_m} - 1}.$$

$$(25.24)$$

Here we have made use of the previously established fact (see eqn (5.13)) that the eigenvalues of the number operator (for any mode) are non-negative integers $n = 0, 1, 2, 3, \ldots$ and we leave the evaluation of the infinite sums as an exercise. Inserting $\beta = 1/(k_{\text{B}}T)$ then provides the result that we have used in eqn (1.9). However, a noteworthy detail in the last calculation that we have not spelled out explicitly is the specific construction of the joint eigenstates of all number operators. We shall therefore take a closer look at these states next.

25.1.3   The Fock Space

Here, we want to examine the joint eigenstates of the number operators N_m for different modes, and the Hilbert space spanned by these states. For a single mode, we can simply drop the mode label and write an eigenvalue equation

$$N \, | \, n \, \rangle = n \, | \, n \, \rangle \qquad \text{for} \ \ n = 0, 1, 2, \ldots \tag{25.25}$$

as in eqn (5.13). Indeed, a single mode just corresponds to a single harmonic oscillator, discussed in detail in Section 5, for which the annihilation and creation operators act on the eigenstates of the number operator according to

$$a \, | \, n \, \rangle = \sqrt{n} \, | \, n - 1 \, \rangle, \tag{25.26a}$$

$$a^\dagger \, | \, n \, \rangle = \sqrt{n+1} \, | \, n+1 \, \rangle, \tag{25.26b}$$

and $N = a^\dagger a$. The eigenstates $| \, n \, \rangle$ of N are called *Fock states* (of a single mode). They correspond to states with fixed photon number n and their respective energies are equally spaced, i.e., two states with photon numbers n and $n + 1$ differ in their energy by one unit of $\hbar \omega$ for a mode with frequency ω. In other words, transitions between two such neighbouring eigenstates are associated with the absorption or emission of one photon of frequency ω.

Now, as we have discussed, the electromagnetic field consists of more than one mode—in general infinitely many—and different modes may correspond to different frequencies or different polarizations, but may in general also relate to other degrees of freedom, such as, e.g., orbital angular momentum of photons (see, e.g., Allen *et al.*, 1992; Leach *et al.*, 2002; Barnett *et al.*, 2017). In particular, the spectrum may be continuous, e.g., for a free field, or (in approximation) in a cavity that is sufficiently large. For simplicity, we will here use a discrete mode label $m = 0, 1, 2, \ldots$, as in the previous section, and we will assume that the set of modes can be finite or infinite. The Fock states for such a set of modes are then the joint eigenstates of all N_m, and can be compactly written as $| \, n_0, n_1, n_2, \ldots \, \rangle$, such that

$$N_m \, | \, n_0, n_1, \ldots, n_m, \ldots \, \rangle = n_m \, | \, n_0, n_1, \ldots, n_m, \ldots \, \rangle. \tag{25.27}$$

And, in full analogy to eqn (25.26), we then have the action of the ladder operators

$$a_m \,|\, n_0, n_1, \ldots, n_m, \ldots \rangle = \sqrt{n_m} \,|\, n_0, n_1, \ldots, n_m - 1, \ldots \rangle, \qquad (25.28\text{a})$$

$$a_m^\dagger \,|\, n_0, n_1, \ldots, n_m, \ldots \rangle = \sqrt{n_m + 1} \,|\, n_0, n_1, \ldots, n_m + 1, \ldots \rangle. \qquad (25.28\text{b})$$

Again, each of these states has a definite photon number, given by the eigenvalue of $N_{\text{tot}} = \sum_m N_m$. However, since the different modes may have different frequencies (except, e.g., for modes with the same frequency and different polarizations, or free-field modes with different directions of propagation), different Fock states do not necessarily differ in their energies by integer multiples of $\hbar\omega$, except when the occupation numbers in all but one mode are the same.

The Hilbert space spanned by the $|\, n_0, n_1, n_2, \ldots \rangle$ is called the *Fock space*, and it is instructive to study its construction in some more detail, starting from the unique state $|\, 0 \rangle$ with zero occupation number, i.e., for which $N \,|\, 0 \rangle = 0$. This vacuum state is defined by the condition that it is annihilated by all annihilation operators, i.e.,

$$a_m \,|\, 0 \rangle = 0 \quad \forall\, m. \qquad (25.29)$$

Following the previously used notation we should technically write $|\, 0_0, 0_1, 0_2, \ldots \rangle$, but it is customary to simple write $|\, 0 \rangle$ or $|\, \Omega \rangle$, since there is no ambiguity. Similarly, we can omit writing the 0 entries for any unoccupied modes when we start populating the vacuum with particles (photons). For a single photon, we can simply write

$$a_m^\dagger \,|\, 0 \rangle = |\, 1_m \rangle = |\, \psi_m \rangle, \qquad (25.30)$$

where we have indicated in the last step that such a state can be interpreted as a single particle in a state with wave function ψ_m corresponding to the respective solution of the equation of motion (here, Maxwell's equations). In other words, we can regard the set of states $\{|\, \psi_m \rangle\}_m$ as a basis of the single-particle Hilbert space $\mathcal{H}_{1-\text{p}}$, whereas the vacuum state $|\, 0 \rangle$ is the (sole) representative for the 0-particle sector that we will denote as $\mathcal{H}_{0-\text{p}}$.

When constructing states of two particles we have to remind ourselves that photons are indistinguishable, and considerations such as those discussed in Section 8.3.3 play a role. More specifically, since photons are bosons[2], the state of two photons is symmetric under the exchange of the two particles according to Definition 8.3. When a second particle is added, we have to take this symmetrization into account on the level of the single-particle wave functions. When the particle is added to the same mode, we have

---

[2]As a remark for experts, photons are massless spin-1 particles. In particular, for freely propagating plane-wave photons, the projection of the spin angular momentum onto their direction of momentum—their helicity—is $\pm\hbar$, where the two signs correspond to left- and right-circular polarization. This identification can be understood from a more fundamental point of view to arise from the classification of the unitary irreducible representations of the (proper, orthochronous) Lorentz group due to *Eugene Wigner* (1939), and we refer the interested reader to the textbook by *Roman Sexl* and *Helmuth Urbantke* (2001) for an in-detail treatment of this topic and to Friis (2010, Chap. 3) for a shorter overview.

$$a_m^\dagger a_m^\dagger \,|\,0\,\rangle = a_m^\dagger \,|\,1_m\,\rangle = \sqrt{2}\,|\,2_m\,\rangle = \sqrt{2}\,|\,\psi_m\,\rangle_1 \otimes |\,\psi_m\,\rangle_2\,, \qquad (25.31)$$

where we have added the subscripts indicating the first and second photon, respectively, and we have included the tensor-product symbol explicitly to emphasize the structure of the Hilbert space, i.e., a tensor product of two (identical) single-particle Hilbert spaces. Here it is evident that the exchange of the two particles leaves the state invariant. Using the commutation relations from (25.20), it is also easy to confirm that $|\,2_m\,\rangle$ is properly normalized, i.e., $\langle\,2_m|\,2_m\,\rangle = 1$. But what happens when the two particles are added in *different* modes? Then we have

$$a_m^\dagger a_n^\dagger \,|\,0\,\rangle = a_m^\dagger \,|\,1_n\,\rangle =: |\,1_m, 1_n\,\rangle\,, \qquad (25.32)$$

where we can raise the last equality to the status of a definition for the notation $|\,1_m, 1_n\,\rangle$. Here, we first note that since the creation operators commute, see eqn (25.20), the ordering of the modes in this notation makes no difference, i.e., we could have also written[3] $a_m^\dagger a_n^\dagger \,|\,0\,\rangle = |\,1_n, 1_m\,\rangle$. On the level of the normalization, we have to take this indistinguishability into account by noting that while $\langle\,1_m|\,1_n\,\rangle = \delta_{mn}$, we now have

$$
\begin{aligned}
\langle\,1_m, 1_n|\,1_k, 1_l\,\rangle &= \langle\,0\,|\,a_m a_n a_k^\dagger a_l^\dagger\,|\,0\,\rangle = \langle\,0\,|\,a_m (a_k^\dagger a_n + \delta_{nk}) a_l^\dagger\,|\,0\,\rangle \\
&= \langle\,0\,|\,a_m a_k^\dagger a_n a_l^\dagger + \delta_{nk} a_m a_l^\dagger\,|\,0\,\rangle \\
&= \langle\,0\,|\,(a_k^\dagger a_m + \delta_{mk})(a_l^\dagger a_n + \delta_{nl}) + \delta_{nk} a_l^\dagger a_m + \delta_{nk}\delta_{ml}\,|\,0\,\rangle \\
&= \langle\,0\,|\,\delta_{mk}\delta_{nl} + \delta_{nk}\delta_{ml}\,|\,0\,\rangle = \delta_{mk}\delta_{nl} + \delta_{nk}\delta_{ml}, \qquad (25.33)
\end{aligned}
$$

where we have used $a_i\,|\,0\,\rangle = 0\,\forall\,i$ and $\langle\,0|\,0\,\rangle = 1$ in the last step. Since the single-particle wave functions satisfy $\langle\,\psi_m|\,\psi_k\,\rangle = \delta_{mk}$, we thus see that the state $|\,1_m, 1_n\,\rangle$ is not simply given by $|\,\psi_m\,\rangle_1 \otimes |\,\psi_n\,\rangle_2$, but we must have

$$|\,1_m, 1_n\,\rangle = \tfrac{1}{\sqrt{2}}\left(|\,\psi_m\,\rangle_1 \otimes |\,\psi_n\,\rangle_2 + |\,\psi_n\,\rangle_1 \otimes |\,\psi_m\,\rangle_2\right). \qquad (25.34)$$

That is, we cannot say that particle 1 is in the state $|\,\psi_m\,\rangle_1$ and particle 2 is in the state $|\,\psi_n\,\rangle_2$, but instead we can say that $|\,1_m, 1_n\,\rangle$ is a state of two particles, one of which is in the state $|\,\psi_m\,\rangle$, while the other is in the state $|\,\psi_n\,\rangle$. The two-photon (or, more generally, two-boson) states are thus elements of the *symmetrized* tensor-product space of two single-particle Hilbert spaces $\mathcal{H}_{1-\mathrm{p}}$,

$$\mathcal{H}_{2-\mathrm{p}} = S\left(\mathcal{H}_{1-\mathrm{p}} \otimes \mathcal{H}_{1-\mathrm{p}}\right), \qquad (25.35)$$

that is, the subspace of $\mathcal{H}_{1-\mathrm{p}} \otimes \mathcal{H}_{1-\mathrm{p}}$ that is symmetric with respect to the exchange of the particles. Similarly, states with higher particle content need to be symmetrized as well, e.g., the k-particle sector of the Fock space is $S\left(\mathcal{H}_{1-\mathrm{p}}^{\otimes k}\right)$. Overall, the *bosonic*

---

[3]As a remark, the same is not true if one considers fermions, since the corresponding creation and annihilation operators satisfy anti-commutation relations instead of commutation relations, which leads to some peculiar features in the description of the reduced states when only subsets of modes are considered. We direct the curious reader to Friis *et al.* (2013a) and Friis (2016) for a discussion of these issues.

Fock space $\mathbb{F}$ is then obtained as the direct sum over all sectors with fixed particle numbers, i.e.,

$$\mathbb{F}(\mathcal{H}_{1-\mathrm{p}}) = \bigoplus_{k=0}^{\infty} S\left(\mathcal{H}_{1-\mathrm{p}}^{\otimes k}\right) = \mathcal{H}_{0-\mathrm{p}} \oplus \mathcal{H}_{1-\mathrm{p}} \oplus S\left(\mathcal{H}_{1-\mathrm{p}} \otimes \mathcal{H}_{1-\mathrm{p}}\right) \oplus \dots . \qquad (25.36)$$

This general state space can thus contain states with fixed photon numbers, where the photons may be distributed across the available modes in any fashion, in particular, one may have a superposition of these photons being found in different modes. Importantly, however, the Fock space also contains coherent superpositions of *different* photon numbers, which are generally of the form

$$|\,\Phi\,\rangle = \theta_0\,|\,0\,\rangle + \sum_{i\neq 0}\theta_i\,|\,1_i\,\rangle + \sum_{j,k}\theta_{jk}\,|\,1_j, 1_k\,\rangle + \dots , \qquad (25.37)$$

where θ_0, θ_i, θ_{jk}, etc. are complex-valued coefficients whose squared moduli sum to one. The possibilities for interesting quantum states in this state space are thus almost endless. Nevertheless, there are a number of particularly important states—and indeed, families of states—in the Fock space that we will now inspect more closely in the remainder of this chapter.

25.2   Coherent States

Coherent states are an important family of states in the Fock space which feature superpositions of different photon numbers, but no superpositions of photons being distributed across different modes. Therefore, it will be sufficient to focus the discussion in this section on a single mode, i.e., a single harmonic oscillator, and we hence drop the mode labels for now. Today, coherent states play an important role in quantum optics, especially in laser physics, where they provide a fully quantum description of the oscillations of the electromagnetic field. Much of the initial ground-breaking work in this field was carried out by *Roy J. Glauber* (1963) who was awarded the 2005 Nobel Prize in Physics for his contribution to the quantum theory of optical coherence. Glauber's work on formalizing the theory of optical coherence can certainly be considered as a milestone that propelled and popularized research in this area, but contributions by *Ennackal Chandy George Sudarshan* (1963) were also most crucial, and the concept of coherent states can even be traced back to *John Klauder* (1960) and even *Julian Schwinger* (1953*b*).

Here, we aim to provide a detailed overview of the theory of coherent states, which can be briefly characterized as quantum analogues of classical harmonic oscillators in the sense that they describe systems with two oscillating conjugate observables—the *quadratures* $\hat{X}$ and $\hat{P}$ that are associated, e.g., with the electric and magnetic field strength—that have

(i) an indefinite number of photons (a genuine quantum feature),
(ii) but an unambiguously defined phase and amplitude.

This is in contrast to states with a fixed particle number, for a single harmonic oscillator we now consider, the eigenstates $|\,n\,\rangle$ of N, for which the amplitude has a precisely defined value n, but where the phase is completely random. In particular, from eqns (5.44) and (5.48) we have[4] to $\langle\,n\,|\,\hat{X}\,|\,n\,\rangle = \langle\,n\,|\,\hat{P}\,|\,n\,\rangle = 0$. For the coherent states, on the other hand, we will find the sinusoidal behaviour of these quantities as we would expect from a classical oscillator.

Amplitude-phase uncertainty relation: However, since a quantum system is considered after all, an uncertainty relation (which we will state but not prove here) places constraints on the amplitude and phase. Although the idea (or perhaps, the desire) of obtaining such an uncertainty relation is intuitively clear, formalizing this idea is made difficult by the fact that the phase itself *cannot* be represented as a Hermitian operator. Ways around this issue had already been discussed by *Paul Dirac* (1927), but significant advances were made by *Leonard Susskind* and *Jonathan Glogower* (1964) as well as *Jean-Marc Lévy-Leblond* (1976). In the formalism of (Susskind and Glogower, 1964), the amplitude is represented by the occupation number $\langle\,N\,\rangle$ and the phase is represented by a pair of operators "sin Φ" and "cos Φ", which are not to be understood as the functions $\sin(\cdot)$ and $\cos(\cdot)$ whose argument is a Hermitian operator, but rather as Hermitian operators in their own right, defined via their decomposition into matrix elements with respect to the Fock-state basis (see, e.g., Thun-Hohenstein, 2010, Sec. 3.1.3, for details and a pedagogical introduction). For these operators, Susskind and Glogower (1964) derived the inequality

$$\Delta N\,\Delta(\sin\Phi) \geq \frac{1}{2}\,|\,\langle\,\cos\Phi\,\rangle\,|\,, \tag{25.38}$$

which, for small phases, reduces to

$$\Delta N\,\Delta\Phi \geq \frac{1}{2}\,, \tag{25.39}$$

where the left-hand side of the inequality is minimized for coherent states.

25.2.1 Definition and Properties of Coherent States

As we have seen in Chapter 5, the eigenstates $|\,n\,\rangle$ of the quantum harmonic oscillator have a well-defined, in fact, a sharp amplitude $\langle\,N\,\rangle = n$, since they are eigenstates of the number operator, $N\,|\,n\,\rangle = n\,|\,n\,\rangle$. However, they are also stationary states and hence do not oscillate. In particular, they have no well-defined phase. For a quantum analogue of a classical oscillator, one wishes to have both a fixed amplitude and frequency, and hence a time-dependent phase. However, fixing the amplitude $\langle\,N\,\rangle = \langle\,\psi\,|\,a^{\dagger}a\,|\,\psi\,\rangle$ can also be achieved by states that are not eigenstates of N, but which are eigenstates of the (non-Hermitian) annihilation operator a. We will define coherent states as follows:

---

[4]Note that we are using the $\hat{\ }$ symbol for the quadrature operators $\hat{X}$ and $\hat{P}$ in this chapter to better distinguish from the canonical classical variables Q_m and P_m used in Section 25.1, whereas in Chapter 5 we have denoted the position and momentum operators by X and P, respectively.

828   *Quantum States of Light*

Definition 25.1   *A **coherent state** or **Glauber state** $|\alpha\rangle$ is defined as an eigenstate of the annihilation operator a, with eigenvalues $\alpha \in \mathbb{C}$*

$$a\,|\alpha\rangle = \alpha\,|\alpha\rangle.$$

Since a is not a Hermitian operator, the eigenvalue $\alpha = |\alpha|\,e^{i\varphi} \in \mathbb{C}$ is a complex number which, as we shall see, can be interpreted as a complex wave amplitude as one would encounter in classical optics. We may therefore think of the annihilation operator as an *amplitude operator*. To provide a more rigorous analysis of this parameter, let us now study the properties of coherent states as defined above. First, we note that the vacuum $|0\rangle$ is a special case of a coherent state with $\alpha = 0$, but is of course also an eigenstate of the Hamiltonian with eigenvalue $E_0 = \hbar\omega/2$, see eqn (5.27). In general, coherent states are not eigenstates of H, but we can nonetheless easily calculate their average energy,

$$\langle H\rangle = \langle\alpha|\,H\,|\alpha\rangle = \hbar\omega\,\langle\alpha|\,a^\dagger a + \tfrac{1}{2}\,|\alpha\rangle = \hbar\omega\,(|\alpha|^2 + \tfrac{1}{2}). \tag{25.40}$$

Here, the first term on the right-hand side represents the classical wave intensity, while the second term reflects the vacuum energy.

Next we introduce the unitary *phase-shifting operator*

$$U(\theta) = e^{-i\theta N}, \tag{25.41}$$

where θ is a real phase. Then we have

$$U^\dagger(\theta)\,a\,U(\theta) = a\,e^{-i\theta} \tag{25.42}$$

that is, $U(\theta)$ results in a phase shift of θ for the amplitude operator. To see this, note that the left-hand side $\frac{d}{d\theta}U^\dagger(\theta)\,a\,U(\theta) = i\,U^\dagger(\theta)\,[N,a]\,U(\theta) = -i\,U^\dagger(\theta)\,a\,U(\theta)$ and the right-hand side $\frac{d}{d\theta}\,a\,e^{-i\theta} = -i\,a\,e^{-i\theta}$ obey the same differential equation. Since the phase-shifting operator satisfies $U^\dagger(\theta) = U(-\theta)$, we further have $U(\theta)\,a\,U^\dagger(\theta) = a\,e^{i\theta}$. Now, let us apply the phase-shifting operator $U(\theta)$ to the left-hand side of Definition 25.1, i.e.,

$$U(\theta)\,a\,|\alpha\rangle = U(\theta)\,a\,U^\dagger(\theta)\,U(\theta)\,|\alpha\rangle = a\,e^{-i\theta}\,U(\theta)\,|\alpha\rangle. \tag{25.43}$$

Comparing to the right-hand side of Definition 25.1, this must match $\alpha\,U(\theta)\,|\alpha\rangle$, and we hence see $U(\theta)\,|\alpha\rangle$ must be an eigenstate of the amplitude operator a with eigenvalue $e^{-i\theta}\,\alpha$,

$$U(\theta)\,|\alpha\rangle = \big|\,e^{-i\theta}\,\alpha\,\big\rangle. \tag{25.44}$$

We thus note an important difference to Fock states $|n\rangle$: While the application of the phase-shifting operator only results in a physically irrelevant global phase for $|n\rangle$, i.e., $U(\theta)\,|n\rangle = e^{-i\theta n}\,|n\rangle$, the same operation applied to a coherent state changes the argument α, and hence the state itself. To discover what this means in practice, we thus need to obtain an operational interpretation for the parameter α.

The displacement operator: In order to do so, we introduce the so-called *displacement operator* $D(\alpha)$, which can generate coherent states from the vacuum in a similar fashion to how the creation operator $a^\dagger$ creates Fock states $|n\rangle$, see eqn (5.26).

Definition 25.2 *The **displacement operator** $D(\alpha)$ is defined as*

$$D(\alpha) = e^{\alpha a^\dagger - \alpha^* a},$$

where $\alpha = |\alpha| e^{i\varphi} \in \mathbb{C}$ is a complex number, and $a^\dagger$ and a are the creation and annihilation operators, respectively.

We can first note that the displacement operator is a unitary operator, i.e., $D^\dagger D = \mathbb{1}$, since the exponent $\alpha a^\dagger - \alpha^* a$ is a Hermitian operator. We can further rewrite $D(\alpha)$ using eqn (2.38), but to do so, we have to check if the commutators $[[A, B], A]$ and $[[A, B], B]$ vanish, where $A = \alpha a^\dagger$ and $B = \alpha^* a$. We therefore start by calculating the commutator of A and B,

$$[A, B] = [\alpha a^\dagger, \alpha^* a] = \alpha \alpha^* \underbrace{[a^\dagger, a]}_{-1} = -|\alpha|^2. \tag{25.45}$$

Since the result is a real number it commutes with both A and B, and we can use eqn (2.38). Directly inserting the result of eqn (25.45), we obtain the displacement operator in the form

$$D(\alpha) = e^{-\frac{1}{2}|\alpha|^2} e^{\alpha a^\dagger} e^{-\alpha^* a}. \tag{25.46}$$

With this result at hand, let us now state a few important properties of the displacement operator.

Properties of the displacement operator:

(i) $D^\dagger(\alpha) = D^{-1}(\alpha) = D(-\alpha)$ unitarity, $\qquad\qquad$ (25.47)

(ii) $D^\dagger(\alpha) \, a \, D(\alpha) = a + \alpha,$ $\qquad\qquad\qquad\qquad$ (25.48)

(iii) $D^\dagger(\alpha) \, a^\dagger \, D(\alpha) = a^\dagger + \alpha^*,$ $\qquad\qquad\qquad$ (25.49)

(iv) $D(\alpha + \beta) = D(\alpha) \, D(\beta) \, e^{-i \, \mathrm{Im}(\alpha \beta^*)}.$ $\qquad\qquad$ (25.50)

Proof To see that property (i) holds, we straightforwardly calculate

$$D^\dagger(\alpha) = \left(e^{\alpha a^\dagger - \alpha^* a}\right)^\dagger = e^{\alpha^* a - \alpha a^\dagger} = e^{(-\alpha) a^\dagger - (-\alpha)^* a} = D(-\alpha). \tag{25.51}$$

To prove (ii) we then use the Hadamard lemma of the Baker–Campbell–Hausdorff formula, eqn (2.39), with $A = \alpha^* a - \alpha a^\dagger$ and $B = a$, such that

$$D^\dagger(\alpha)\,a\,D(\alpha) = e^{\alpha^*a - \alpha a^\dagger}\,a\,e^{\alpha a^\dagger - \alpha^*a} = a + \left[\alpha^*a - \alpha a^\dagger,\, a\right]$$
$$= a + \alpha^*\underbrace{\left[a,\,a\right]}_{0} - \alpha\underbrace{\left[a^\dagger,\,a\right]}_{-1} = a + \alpha, \qquad (25.52)$$

where the higher-order commutators vanish because the commutator of A and B is a complex number that commutes with the other operators. Property (iii) then follows by taking the adjoint of (ii), i.e.,

$$D^\dagger(\alpha)\,a^\dagger\,D(\alpha) = \left(D^\dagger(\alpha)\,a\,D(\alpha)\right)^\dagger = (a + \alpha)^\dagger = a^\dagger + \alpha^*. \qquad (25.53)$$

To prove the property in statement (iv), we again use eqn (2.38), with $A = \alpha\,a^\dagger - \alpha^*a$ and $B = \beta\,a^\dagger - \beta^*a$, which is justified because the condition that the higher-order commutators vanish is satisfied due to $[A,\,B] \in \mathbb{C}$. We thus have

$$D(\alpha + \beta) = e^{\alpha a^\dagger - \alpha^*a + \beta a^\dagger - \beta^*a} = e^{\alpha a^\dagger - \alpha^*a}\,e^{\beta a^\dagger - \beta^*a}\,e^{-\frac{1}{2}\left[\alpha a^\dagger - \alpha^*a,\,\beta a^\dagger - \beta^*a\right]}$$
$$= D(\alpha)\,D(\beta)\,e^{-\frac{1}{2}(\alpha\beta^* - \alpha^*\beta)} = D(\alpha)\,D(\beta)\,e^{-i\,\mathrm{Im}(\alpha\beta^*)}. \qquad (25.54)$$

The result of the last commutator can also be seen easily by noting that the creation and annihilation operators commute with themselves but not with each other, and commutator of the latter gives ± 1. $\qquad \square$

With this knowledge about the displacement operator we can now turn to the creation of coherent states from the vacuum, which we will cast in the form of the following theorem.

Theorem 25.1 (Vacuum displacement)

The coherent state $|\alpha\rangle$ is generated from the vacuum $|0\rangle$ by the displacement operator $D(\alpha)$, i.e.,

$$|\alpha\rangle = D(\alpha)\,|0\rangle.$$

Proof   Applying a negative displacement to $|\alpha\rangle$, followed by the application of the annihilation operator a, we can make use of properties (i) and (ii), eqns (25.47) and (25.48), respectively, such that

$$a\,D(-\alpha)\,|\alpha\rangle = D(-\alpha)\,\underbrace{D^\dagger(-\alpha)\,a\,D(-\alpha)}_{a - \alpha}\,|\alpha\rangle = D(-\alpha)(a - \alpha)\,|\alpha\rangle = 0, \quad (25.55)$$

which vanishes according to our definition of coherent states as eigenstates of a, see Definition 25.1. This implies that $D(-\alpha)\,|\alpha\rangle$ is an eigenstate of the annihilation operator with eigenvalue 0, in other words, the vacuum state $|0\rangle$, and hence $D(-\alpha)\,|\alpha\rangle = |0\rangle$.

Applying the displacement operator $D(\alpha)$ and using its unitarity from eqn (25.47), we then have

$$D(\alpha)D(-\alpha)\,|\,\alpha\,\rangle = D(\alpha)D^{\dagger}(\alpha)\,|\,\alpha\,\rangle = |\,\alpha\,\rangle = D(\alpha)\,|\,0\,\rangle. \qquad (25.56)$$

$\square$

On the other hand, we also could raise Theorem 25.1 to the level of a definition of coherent states, in which case our Definition 25.1 would follow as a theorem.

Proof

$$a\,|\,\alpha\,\rangle = \underbrace{D(\alpha)\,D^{\dagger}(\alpha)}_{1}\,a\,|\,\alpha\,\rangle = D(\alpha)\,\underbrace{D^{\dagger}(\alpha)\,a\,D(\alpha)}_{a+\alpha}\,|\,0\,\rangle = D(\alpha)\,\Big(\underbrace{a\,|\,0\,\rangle}_{0}+\alpha\,|\,0\,\rangle\Big)$$

$$= \alpha\,D(\alpha)\,|\,0\,\rangle = \alpha\,|\,\alpha\,\rangle. \qquad (25.57)$$

$\square$

A physical interpretation of Definition 25.1 becomes apparent when thinking of a coherent state as describing a coherent photon beam, e.g., a laser. There, it is to be expected that the state of the laser remains unchanged if one photon is annihilated, e.g., if it is detected.

Meanwhile, we can now combine Theorem 25.1 with property (iv) from eqn (25.50) to see that, up to a global phase, the displacement operator acts cumulatively, that is

$$D(\beta)\,|\,\alpha\,\rangle = D(\beta)\,D(\alpha)\,|\,0\,\rangle = e^{i\,\mathrm{Im}(\alpha^*\beta)}\,D(\alpha+\beta)\,|\,0\,\rangle = e^{i\,\mathrm{Im}(\alpha^*\beta)}\,|\,\alpha+\beta\,\rangle. \qquad (25.58)$$

Coherent states in Fock-basis decomposition: While the coherent states $|\,\alpha\,\rangle$ with $\alpha = |\alpha|\,e^{i\varphi}$ have a well-defined phase φ and amplitude $|\alpha|$, their photon number is indefinite, which will become evident from an expansion in terms of the Fock states $|\,n\,\rangle$, which form a complete orthonormal system. To arrive at such an expansion, let us start by inserting the completeness relation of the occupation-number states, $\mathbb{1} = \sum_n |\,n\,\rangle\langle\,n\,|$, i.e.,

$$|\,\alpha\,\rangle = \sum_n |\,n\,\rangle\langle\,n\,|\,\alpha\,\rangle. \qquad (25.59)$$

Then we calculate the transition amplitude $\langle\,n\,|\,\alpha\,\rangle$ by using Definition 25.1, where we multiply the whole eigenvalue equation with $\langle\,n\,|$ from the left

$$\langle\,n\,|\,a\,|\,\alpha\,\rangle = \langle\,n\,|\,\alpha\,|\,\alpha\,\rangle. \qquad (25.60)$$

Using the adjoint of the relation in eqn (25.26b), that is,

$$\langle\,n\,|\,a = \sqrt{n+1}\,\langle\,n+1\,|, \qquad (25.61)$$

we rewrite the left-hand side of eqn (25.60) as

$$\sqrt{n+1}\,\langle n+1\,|\,\alpha\rangle = \alpha\,\langle n\,|\,\alpha\rangle\,. \tag{25.62}$$

Since this must hold for any n, we can replace the occupation number n with $n-1$ to obtain

$$\langle n\,|\,\alpha\rangle = \frac{\alpha}{\sqrt{n}}\,\langle n-1\,|\,\alpha\rangle\,. \tag{25.63}$$

By iterating the last step, i.e., once more replacing n with $n-1$ in eqn (25.63) and reinserting the result into the right-hand side, we get

$$\langle n\,|\,\alpha\rangle = \frac{\alpha^2}{\sqrt{n(n-1)}}\,\langle n-2\,|\,\alpha\rangle = \cdots = \frac{\alpha^n}{\sqrt{n!}}\,\langle 0\,|\,\alpha\rangle\,. \tag{25.64}$$

Finally inserting into the expansion in eqn (25.59) results in

$$|\alpha\rangle = \langle 0\,|\,\alpha\rangle \sum_{n=0}^{\infty} \frac{\alpha^n}{\sqrt{n!}}\,|n\rangle\,. \tag{25.65}$$

The remaining transition amplitude $\langle 0\,|\,\alpha\rangle$ can be calculated in two different ways: first, using the normalization condition, and second, by employing the displacement operator. For the sake of variety we will use the second method here and leave the other one as an exercise,

$$\langle 0\,|\,\alpha\rangle = \langle 0|\,D(\alpha)\,|0\rangle = e^{-\frac{1}{2}|\alpha|^2}\,\langle 0|\,e^{\alpha a^\dagger}\,e^{\alpha^* a}\,|0\rangle\,, \tag{25.66}$$

where we have used the expression from eqn (25.46) for the displacement operator. We then expand the exponentials of operators in the scalar product into their Taylor series,

$$\langle 0|\,e^{\alpha a^\dagger}\,e^{\alpha^* a}\,|0\rangle = \langle 0|\,(1 + \alpha\,a^\dagger + \cdots)(1 + \alpha^*\,a + \cdots)\,|0\rangle = 1\,, \tag{25.67}$$

and we apply the operators inside the left-hand side and right-hand side parentheses to the left and to the right, respectively. In this way, we can use property $a\,|0\rangle = 0$ of the vacuum state to eliminate all terms of the expansion except the leading constant. Therefore the transition amplitude is given by

$$\langle 0\,|\,\alpha\rangle = e^{-\frac{1}{2}|\alpha|^2}\,, \tag{25.68}$$

and we can write down the coherent state in terms of an exact expansion in terms of the Fock basis as

$$|\alpha\rangle = e^{-\frac{1}{2}|\alpha|^2} \sum_{n=0}^{\infty} \frac{\alpha^n}{\sqrt{n!}}\,|n\rangle = e^{-\frac{1}{2}|\alpha|^2} \sum_{n=0}^{\infty} \frac{(\alpha\,a^\dagger)^n}{n!}\,|0\rangle\,. \tag{25.69}$$

Photon-number distribution of coherent states: We can now analyse the probability distribution of the different photon numbers featuring in a coherent state, i.e., the probability of detecting n photons in a coherent state $|\alpha\rangle$, which is given by

$$P(n) = |\langle n\,|\,\alpha\rangle|^2 = \frac{|\alpha|^{2n}\,e^{-|\alpha|^2}}{n!} \,. \qquad (25.70)$$

By noting, that the *mean photon number* is determined by the expectation value of the particle-number operator

$$\bar{n} = \langle\alpha|\,N\,|\alpha\rangle = \langle\alpha|\,a^\dagger a\,|\alpha\rangle = |\alpha|^2 \,, \qquad (25.71)$$

where we have used Definition 25.1, we can rewrite the probability distribution to get

$$P(n) = \frac{\bar{n}^n\,e^{-\bar{n}}}{n!} \,, \qquad (25.72)$$

which is a *Poissonian distribution*.

Remarks: Before moving on, let us make some remarks about what we have learned about coherent states so far. First, eqn (25.71) can be viewed as a connection between the mean photon number—the "particle view"—and the complex amplitude squared, the intensity of the wave—the "wave view". In other words, coherent states describe (amongst other examples) a laser beam consisting of individual photons but at the same time with a well-defined amplitude and phase like a classical wave. Second, classical particles obey the same statistical law, the Poisson formula (25.70), when their number is randomly sampled from an ensemble with $|\alpha|^2$ particles on average. Counting the photons in a coherent state, one therefore encounters a behaviour as for randomly distributed classical particles, which might not be too surprising since coherent states feature wave-like properties. Nevertheless, it's amusing that the photons in a coherent state behave—pictorially speaking—like the raisins in a "Gugelhupf"[5], which have been distributed from the baker randomly with $|\alpha|^2$ on average per unit volume. Then the probability of finding n raisins per unit volume in the Gugelhupf follows precisely the distribution in eqn (25.70).

Scalar product of two coherent states: As we have seen, the relationship between the Fock states $|n\rangle$ and coherent states is not trivial. In particular, since the occupation-number operator $N = a^\dagger a$ is Hermitian, it follows from Theorem 4.2 that its eigenvalues n are real and its eigenstates $|n\rangle$ are orthogonal. But coherent states are eigenstates of the annihilation operator a, which is surely not Hermitian and, as we already know, the eigenvalues α are complex numbers. Therefore we cannot automatically assume that coherent states are orthogonal, and, as we shall see, they indeed are not. To make this more explicit, we have to calculate their scalar product. Again using Theorem 25.1 and eqn (25.46) we find

---

[5]A traditional Austrian cake, which often contains raisins, but whose number and distribution, however it may be, is most certainly not as desired by the recipient.

$$\langle \beta \,|\, \alpha \rangle = \langle 0 |\, D^{\dagger}(\beta)\, D(\alpha)\, | 0 \rangle = \langle 0 |\, e^{-\beta\, a^{\dagger}}\, e^{\beta^{*}\, a}\, e^{\alpha\, a^{\dagger}}\, e^{-\alpha^{*}\, a}\, | 0 \rangle\, e^{-\frac{1}{2}(|\alpha|^2 + |\beta|^2)} . \tag{25.73}$$

We now use the same method as in eqn (25.67), i.e., we expand the operators in their Taylor series to see that the two "outer" operators $e^{-\beta\, a^{\dagger}} = (1 - \beta\, a^{\dagger} + \cdots)$ and $e^{-\alpha^{*}\, a} = (1 - \alpha^{*} a + \cdots)$, acting to the left and to the right, respectively, annihilate the vacuum state, save for the scalar-valued term "1" and we can thus ignore the operator-valued contributions to these terms. We then expand the remaining operators as well and consider their action on the vacuum, that is,

$$\langle 0 |\, \big(1 + \beta^{*} a + \frac{1}{2!}(\beta^{*}a)^2 + \cdots \big)\, \big(1 + \alpha\, a^{\dagger} + \frac{1}{2!}(\alpha\, a^{\dagger})^2 + \cdots \big)\, | 0 \rangle \tag{25.74}$$

$$= \Big(\cdots + \langle 2 |\, \frac{1}{2!}\, \sqrt{2!}\, (\beta^{*})^2 + \langle 1 |\, \beta^{*} + \langle 0 | \Big)\, \Big(| 0 \rangle + \alpha\, | 1 \rangle + \frac{1}{2!}\, \sqrt{2!}\, \alpha^2\, | 2 \rangle + \cdots \Big) .$$

Using the orthogonality of the particle number states, $\langle n \,|\, m \rangle = \delta_{nm}$, we get

$$\langle \beta \,|\, \alpha \rangle = e^{-\frac{1}{2}(|\alpha|^2 + |\beta|^2)} \big(1 + \alpha\beta^{*} + \frac{1}{2!}(\alpha\beta^{*})^2 + \cdots \big) = e^{-\frac{1}{2}(|\alpha|^2 + |\beta|^2) + \alpha\beta^{*}} . \tag{25.75}$$

We can then further simplify the exponent by noting that

$$|\alpha - \beta|^2 = (\alpha - \beta)(\alpha^{*} - \beta^{*}) = |\alpha|^2 + |\beta|^2 - \alpha\beta^{*} - \alpha^{*}\beta , \tag{25.76}$$

and, finally, we can write down the transition probability in a compact way as

$$|\langle \beta \,|\, \alpha \rangle|^2 = e^{-|\alpha - \beta|^2} . \tag{25.77}$$

This means that coherent states indeed are *not orthogonal* and their transition probability only vanishes in the limit of large differences $|\alpha - \beta| \gg 1$.

Completeness of coherent states: Although coherent states are not orthogonal, they nevertheless form a complete basis of the Hilbert space, and it is thus possible to expand any quantum state in terms of the complete set of coherent states. The *completeness relation* for the coherent states reads

$$\frac{1}{\pi} \int d^2\alpha\, | \alpha \rangle\langle \alpha | = \mathbb{1} . \tag{25.78}$$

In fact, the set of coherent states is "overcomplete", which means that, as a consequence of their non-orthogonality, any coherent state can be expanded in terms of all coherent states. So the coherent states $| \alpha \rangle$ for different α are not linearly independent,

$$| \beta \rangle = \frac{1}{\pi} \int d^2\alpha\, | \alpha \rangle\langle \alpha \,|\, \beta \rangle = \frac{1}{\pi} \int d^2\alpha\, | \alpha \rangle\, e^{-\frac{1}{2}(|\alpha|^2 + |\beta|^2) + \alpha\beta^{*}} . \tag{25.79}$$

Proof To prove the completeness relation in eqn (25.78), we first make use of the Fock-state expansion from eqn (25.69)

$$\frac{1}{\pi} \int d^2\alpha \, |\alpha\rangle\langle\alpha| = \frac{1}{\pi} \sum_{n,m} \frac{1}{\sqrt{n!\,m!}} \, |n\rangle\langle m| \int d^2\alpha \, e^{-|\alpha|^2} \alpha^n (\alpha^*)^m \,. \qquad (25.80)$$

The integral on the right-hand side of eqn (25.80) can be solved using polar coordinates $\alpha = |\alpha|e^{i\varphi} = r\,e^{i\varphi}$ in the complex plane, i.e.,

$$\int d^2\alpha \, e^{-|\alpha|^2} \alpha^n (\alpha^*)^m = \int_0^\infty r\,dr\,e^{-r^2} r^{n+m} \underbrace{\int_0^{2\pi} d\varphi \, e^{i(n-m)\varphi}}_{2\pi\,\delta_{nm}} = 2\pi \int_0^\infty r\,dr\,e^{-r^2} r^{2n} \,.$$

$$(25.81)$$

Substituting $r^2 = t$, $2r\,dr = dt$ to rewrite the integral

$$2 \int_0^\infty r\,dr\,e^{-r^2} r^{2n} = \int_0^\infty e^{-t} t^n \, dt = \Gamma(n+1) = n! \,, \qquad (25.82)$$

we recover the exact definition of the gamma function, compare, e.g., with eqn (7.29.) With this we can finally write down the completeness relation

$$\frac{1}{\pi} \int d^2\alpha \, |\alpha\rangle\langle\alpha| = \frac{1}{\pi} \sum_n \frac{1}{n!} \, |n\rangle\langle n| \, \pi\,n! = \sum_n |n\rangle\langle n| = \mathbb{1}. \qquad (25.83)$$

$\square$

25.2.2   Coordinate Representation of Coherent States

With all the insights we have gathered about coherent states, displacement operators and their relationship, we have nonetheless stayed on a rather abstract level of states and operators in a Hilbert space. We now want to connect what we have learned to predictions for certain measurable quantities. More specifically, we want to study the properties of coherent states related to the canonically conjugate position and momentum operators—the quadratures $\hat{X}$ and $\hat{P}$—in terms of which we have defined our annihilation and creation operators in the first place. These observables have continuous spectra, i.e., we can formally write eigenvalue equations

$$\hat{X} \, |x\rangle = x \, |x\rangle \,, \qquad (25.84\text{a})$$

$$\hat{P} \, |p\rangle = p \, |p\rangle \,, \qquad (25.84\text{b})$$

and their eigenvalues x and p can thus be thought of as coordinates in a phase space much like that of a single particle in classical mechanics. Indeed, if we recall our discussion in Chapter 5, this is the picture we had in mind there. However, as we have seen in this chapter, "position" and "momentum" associated to a harmonic oscillator need not correspond to spatial degrees of freedom of a single massive particle. The quadrature operators can, for instance, correspond to the electric and magnetic field strengths (as discussed in Section 25.1), but could also be, say, the (operators for the) flux $\boldsymbol{\Phi}$ through the inductor and the charge $\mathbf{Q}$ on the capacitor plates of a superconducting LC-circuit (see, e.g., Devoret *et al.*, 2004). Nevertheless, the quadrature operators are associated with the measurable quantities that are observed to be oscillating in a classical harmonic oscillator. We therefore now want to focus on the coordinate representation of the coherent states to gain some better understanding of these physically measurable properties of a coherent state and their time evolution.

To begin, let us reiterate here the expansion of the coherent states in terms of Fock states from eqn (25.69)

$$ | \alpha \rangle = e^{-\frac{1}{2}|\alpha|^2} \sum_{n=0}^{\infty} \frac{\alpha^n}{\sqrt{n!}} | n \rangle = e^{-\frac{1}{2}|\alpha|^2} \sum_{n=0}^{\infty} \frac{(\alpha \, a^\dagger)^n}{n!} | 0 \rangle \,, \tag{25.85} $$

as well as the coordinate representation of the harmonic-oscillator states from eqn (5.41),

$$ \psi_n(x) = \langle x \, | \, n \rangle = \psi_n(x) = \frac{1}{\sqrt{2^n (n!) \sqrt{\pi} x_0}} \, H_n(\xi) \, e^{-\frac{\xi^2}{2}} \,, \tag{25.86} $$

$$ \psi_0(x) = \langle x \, | \, 0 \rangle = \psi_0(x) = \frac{1}{\sqrt{\sqrt{\pi} x_0}} \, e^{-\frac{\xi^2}{2}} \,, \tag{25.87} $$

where $\xi = \frac{x}{x_0}$ and $x_0 = \sqrt{\frac{\hbar}{m\omega}}$. Combining these expressions, we can easily give the coordinate representation of the coherent states as

$$ \langle x \, | \, \alpha \rangle = \phi_\alpha(x) = e^{-\frac{1}{2}|\alpha|^2} \sum_{n=0}^{\infty} \frac{\alpha^n}{\sqrt{n!}} \, \psi_n(x). \tag{25.88} $$

Time evolution: With the expression in eqn (25.88) at hand, it is then easy to study the time evolution of coherent states, simply by taking a look at the time evolution of the harmonic oscillator. The eigenstates $| n \rangle$ of the harmonic oscillator with associated wave functions $\psi_n(x) = \langle x | n \rangle$ are stationary states, and hence satisfy

$$ \psi_n(t, x) = \psi_n(x) \, e^{-\frac{i}{\hbar} E_n t} = \psi_n(x) \, e^{-in\,\omega t} \, e^{-\frac{i\omega t}{2}} \,, \tag{25.89} $$

where we have inserted the harmonic-oscillator energy $E_n = \hbar\omega(n+\frac{1}{2})$ from eqn (5.27). We are then in the position to write down the time evolution of the coherent state as

$$ \phi_\alpha(t, x) = e^{-\frac{1}{2}|\alpha|^2} \, e^{-\frac{i\omega t}{2}} \sum_{n=0}^{\infty} \frac{(\alpha \, e^{-i\omega t})^n}{\sqrt{n!}} \, \psi_n(x) \,. \tag{25.90} $$

With a little notational trick, i.e., making the label α time-dependent, $\alpha(t) = \alpha \, e^{-i\omega t}$, we can bring eqn (25.90) into the form

$$\phi_\alpha(t, x) = \phi_{\alpha(t)}(x)\, e^{-\frac{i\omega t}{2}}, \tag{25.91}$$

that is, apart from a (physically irrelevant) global phase $e^{-\frac{i\omega t}{2}}$, the time evolution of coherent states can be understood purely as a rotation of the complex amplitude in the complex plane. Of course, none of the arguments we have used to arrive at this statement depend on the specific basis we choose to represent the state in (including the basis $\{\,|\,x\,\rangle\}_{x\in\mathbb{R}}$), and we hence also have

$$e^{-iHt/\hbar}\,|\,\alpha\,\rangle = e^{-i\hbar\omega(a^\dagger a + \frac{1}{2})}\,|\,\alpha\,\rangle = e^{-\frac{i\omega t}{2}}\,|\,\alpha(t)\,\rangle = e^{-\frac{i\omega t}{2}}\,|\,\alpha e^{-i\omega t}\,\rangle. \tag{25.92}$$

We are therefore now interested in providing a clear physical interpretation for the coherent-state amplitude α. As we shall see next, such a physically meaningful picture for the significance of α arises from the mean values of the quadrature operators, which we will calculate next.

Statistical moments of the quadrature operators: Before we begin with the calculation of the mean values of the quadratures, let us recall the corresponding mean values obtained for the harmonic-oscillator eigenstates $|\,n\,\rangle$ in eqns (5.44) and (5.48), which, in the notation of this chapter, take the form

$$\langle\,\hat{X}\,\rangle_n = \langle\,n\,|\,\hat{X}\,|\,n\,\rangle = 0 = \langle\,n\,|\,\hat{P}\,|\,n\,\rangle = \langle\,\hat{P}\,\rangle_n. \tag{25.93}$$

To reiterate, the quantum-mechanical harmonic-oscillator eigenstates $|\,n\,\rangle$ show no oscillating behaviour. Now, let us more closely inspect the corresponding relations obtained for coherent states. Recalling the expression of the creation and annihilation operator in terms of $\hat{X}$ and $\hat{P}$ in eqn (25.19), we have (again, for fixed mode label m, which we hence drop)

$$\hat{X} = \sqrt{\frac{\hbar}{2\omega}}\,(a + a^\dagger), \tag{25.94a}$$

$$\hat{P} = -i\sqrt{\frac{\hbar\omega}{2}}\,(a - a^\dagger), \tag{25.94b}$$

in analogy to eqn (5.5) for a unit-mass oscillator. With this, we can easily compute the expectation values of the position and momentum quadratures for the time-dependent coherent states $|\,\alpha(t)\,\rangle$,

$$\langle\,\hat{X}\,\rangle_{\alpha(t)} = \sqrt{\tfrac{\hbar}{2\omega}}\,\langle\,\alpha(t)\,|\,a + a^\dagger\,|\,\alpha(t)\,\rangle = \sqrt{\tfrac{\hbar}{2\omega}}\,(\alpha(t) + \alpha(t)^*) = \sqrt{\tfrac{2\hbar}{\omega}}\,\mathrm{Re}(\alpha(t))$$

$$= \sqrt{\tfrac{2\hbar}{\omega}}\,|\alpha|\,\cos(\varphi - \omega t), \tag{25.95}$$

$$\langle\,\hat{P}\,\rangle_{\alpha(t)} = -i\sqrt{\tfrac{\hbar\omega}{2}}\,\langle\,\alpha(t)\,|\,a - a^\dagger\,|\,\alpha(t)\,\rangle = -i\sqrt{\tfrac{\hbar\omega}{2}}\,(\alpha(t) - \alpha(t)^*)$$

$$= \sqrt{2\hbar\omega}\,\mathrm{Im}(\alpha(t)) = \sqrt{2\hbar\omega}\,|\alpha|\,\sin(\varphi - \omega t), \tag{25.96}$$

where we have applied the annihilation and creation operators to the right and left on $|\,\alpha(t)\,\rangle$ and on $\langle\,\alpha(t)\,|$, respectively, and we have used $\alpha(t) = \alpha\,e^{-i\omega t} = |\alpha|\,e^{-i(\omega t - \varphi)}$. To summarize the calculation, we conclude that the coherent state, unlike the quantum-mechanical harmonic oscillator, *does oscillate* with angular frequency ω. Moreover, we now see that α (or $|\alpha|$) can really be interpreted as the amplitude of this oscillation. The quadrature mean values of coherent states thus behave exactly like one would expect from a classical harmonic oscillator. However, since we are now considering a quantum system, there are still uncertainties associated to these mean values which have to obey an uncertainty relation as discussed in Section 2.5. Let us therefore also calculate the variances of the quadratures. To begin, we calculate

$$\langle \hat{X}^2 \rangle_{\alpha(t)} = \tfrac{\hbar}{2\omega} \langle \alpha(t) | (a + a^\dagger)^2 | \alpha(t) \rangle = \tfrac{\hbar}{2\omega} \langle \alpha(t) | a^2 + aa^\dagger + a^\dagger a + a^{\dagger 2} | \alpha(t) \rangle$$

$$= \tfrac{\hbar}{2\omega} \langle \alpha(t) | a^2 + 2a^\dagger a + 1 + a^{\dagger 2} | \alpha(t) \rangle = \tfrac{\hbar}{2\omega} (\alpha^2 + 2|\alpha|^2 + 1 + \alpha^{*2})$$

$$= \tfrac{\hbar}{2\omega} (\alpha + \alpha^*)^2 + \tfrac{\hbar}{2\omega} = \langle \hat{X} \rangle^2_{\alpha(t)} + \tfrac{\hbar}{2\omega}, \tag{25.97}$$

where we have used the commutator $aa^\dagger = a^\dagger a + 1$ in going from the first to the second line. The corresponding variance then simply becomes

$$\left(\Delta \hat{X}_{\alpha(t)} \right)^2 = \langle \hat{X}^2 \rangle_{\alpha(t)} - \langle \hat{X} \rangle^2_{\alpha(t)} = \frac{\hbar}{2\omega}. \tag{25.98}$$

Similarly, we find for the momentum quadrature,

$$\langle \hat{P}^2 \rangle_{\alpha(t)} = -\tfrac{\hbar\omega}{2} \langle \alpha(t) | (a - a^\dagger)^2 | \alpha(t) \rangle = -\tfrac{\hbar\omega}{2} \langle \alpha(t) | a^2 - aa^\dagger - a^\dagger a + a^{\dagger 2} | \alpha(t) \rangle$$

$$= -\tfrac{\hbar\omega}{2} \langle \alpha(t) | a^2 - 2a^\dagger a - 1 + a^{\dagger 2} | \alpha(t) \rangle = -\tfrac{\hbar\omega}{2} (\alpha^2 - 2|\alpha|^2 - 1 + \alpha^{*2})$$

$$= -\tfrac{\hbar\omega}{2} (\alpha - \alpha^*)^2 + \tfrac{\hbar\omega}{2} = \langle \hat{P} \rangle^2_{\alpha(t)} + \tfrac{\hbar\omega}{2}, \tag{25.99}$$

such that we obtain the variance

$$\left(\Delta \hat{P}_{\alpha(t)} \right)^2 = \langle \hat{P}^2 \rangle_{\alpha(t)} - \langle \hat{P} \rangle^2_{\alpha(t)} = \frac{\hbar\omega}{2}. \tag{25.100}$$

We thus find that coherent states minimize the Robertson–Schrödinger uncertainty relation. That is, the relation from Theorem 2.4 becomes

$$\Delta \hat{X}_{\alpha(t)} \, \Delta \hat{P}_{\alpha(t)} = \frac{\hbar}{2}. \tag{25.101}$$

In particular, it can be checked (which we leave as an exercise) that the covariance $\tfrac{1}{2} \langle \{ \hat{X}, \hat{P} \} \rangle_\alpha - \langle \hat{X} \rangle \langle P \rangle_\alpha = 0$ for coherent states, and since $[\hat{X}, \hat{P}] = i\hbar$, the minimal value for the product of the uncertainties is indeed $\tfrac{\hbar}{2}$. Moreover, we observe that the uncertainties of both quadratures are independent of the coherent-state amplitude α. From our discussion of the uncertainty of Gaussian wave packets in Section 2.5.3, we can therefore here already anticipate that all coherent-state wave functions will be of just such Gaussian shape, but to confirm this, let us calculate these wave functions explicitly.

The coherent-state wave function: To work out the explicit coherent-state wave function $\phi_\alpha(x) = \langle x \,|\, \alpha \rangle$, we again turn to the eigenvalue equation from Definition 25.1 and write it in its coordinate representation by applying $\langle x \,|$ from the left, i.e.,

$$\langle x \,|\, a \,|\, \alpha \rangle = \alpha \langle x \,|\, \alpha \rangle. \tag{25.102}$$

We then express the annihilation operator in terms of the quadratures as in eqn (25.19), which allows us to rewrite eqn (25.102) as

$$\tfrac{1}{\sqrt{2\hbar\omega}} \langle x \,|\, \omega \hat{X} + i\hat{P} \,|\, \alpha \rangle = \sqrt{\tfrac{\omega}{2\hbar}} \Big(x + \tfrac{\hbar}{\omega} \tfrac{d}{dx} \Big) \phi_\alpha(x) = \alpha\,\phi_\alpha(x). \tag{25.103}$$

As we will show in the following, this differential equation is solved by

$$\phi_\alpha(x) = \Big(\tfrac{\omega}{\hbar\pi} \Big)^{\frac{1}{4}} e^{i\,\mathrm{Im}(\alpha)\sqrt{\frac{\omega}{2\hbar}}\left(2x - \langle \hat{X} \rangle_\alpha\right)}\, \exp\Big(-\tfrac{(x - \langle \hat{X} \rangle_\alpha)^2}{2\hbar/\omega} \Big). \tag{25.104}$$

Proof To see this, let us first consider the derivatives of the two exponential terms

$$\tfrac{\hbar}{\omega}\tfrac{d}{dx} e^{i\,\mathrm{Im}(\alpha)\sqrt{\frac{\omega}{2\hbar}}\left(2x - \langle \hat{X} \rangle_\alpha\right)} = i\,\mathrm{Im}(\alpha)\sqrt{\tfrac{2\hbar}{\omega}}\, e^{i\,\mathrm{Im}(\alpha)\sqrt{\frac{\omega}{2\hbar}}\left(2x - \langle \hat{X} \rangle_\alpha\right)}, \tag{25.105a}$$

$$\tfrac{\hbar}{\omega}\tfrac{d}{dx} \exp\Big(-\tfrac{(x - \langle \hat{X} \rangle_\alpha)^2}{2\hbar/\omega} \Big) = -(x - \langle \hat{X} \rangle_\alpha) \exp\Big(-\tfrac{(x - \langle \hat{X} \rangle_\alpha)^2}{2\hbar/\omega} \Big). \tag{25.105b}$$

Combining these terms and inserting $\langle \hat{X} \rangle_\alpha = \sqrt{\tfrac{2\hbar}{\omega}}\,\mathrm{Re}(\alpha)$ from eqn (25.95), we thus see that

$$\tfrac{\hbar}{\omega}\tfrac{d}{dx}\phi_\alpha(x) = \Big[i\,\mathrm{Im}(\alpha)\sqrt{\tfrac{2\hbar}{\omega}} - (x - \langle \hat{X} \rangle_\alpha) \Big]\phi_\alpha(x) \tag{25.106}$$

$$= \Big[\sqrt{\tfrac{2\hbar}{\omega}}\,(\mathrm{Re}(\alpha) + i\,\mathrm{Im}(\alpha)) - x \Big]\phi_\alpha(x) = \Big[\sqrt{\tfrac{2\hbar}{\omega}}\,\alpha - x \Big]\phi_\alpha(x).$$

It is thus easy to see that $\phi_\alpha(x)$ from eqn (25.104) satisfies the differential equation in eqn (25.103), and we leave it as an exercise to check the normalization constant $\big(\tfrac{\omega}{\hbar\pi} \big)^{1/4}$. $\qquad\square$

The result in eqn (25.104) also immediately provides the time-dependent wave function via eqn (25.91), such that we obtain

$$\phi_\alpha(t, x) = \Big(\tfrac{\omega}{\hbar\pi} \Big)^{\frac{1}{4}} e^{-\frac{i\omega t}{2}}\, e^{i\,\mathrm{Im}\left(\alpha(t)\right)\sqrt{\frac{\omega}{2\hbar}}\left(2x - \langle \hat{X} \rangle_{\alpha(t)}\right)}\, \exp\Big(-\tfrac{(x - \langle \hat{X} \rangle_{\alpha(t)})^2}{2\hbar/\omega} \Big). \tag{25.107}$$

Moreover, by just considering the corresponding probability density, we can disregard the global phases and arrive at

$$|\phi_\alpha(t,x)|^2 = \sqrt{\frac{\omega}{\hbar\pi}}\,\exp\!\left(-\frac{(x-\langle\hat{X}\rangle_{\alpha(t)})^2}{\hbar/\omega}\right). \tag{25.108}$$

As we have suspected from the minimal product of uncertainties in eqn (25.101), we find that coherent states are Gaussian wave packets of constant width $\sigma = \Delta\hat{X} = \sqrt{\frac{\hbar}{2\omega}}$. However, unlike the free wave packets that we have considered in Section 2.6.4, coherent states are bound in a central oscillator potential, as illustrated in Fig. 25.1.

Consequently, the wave packets are not spreading in width during their time evolution, but their mean values oscillate sinusoidally. In this sense, coherent states represent the closest quantum-mechanical analogue to the free classical single-mode field. Indeed, the analogy to the case of classical oscillation represented in a (classical) phase space can be made even more precise. In the following, we therefore want to establish a clear mathematical picture for a quantum-mechanical phase-space representation, see Section 25.3, before we consider the broader class of so-called Gaussian states, which contains coherent states as special cases, but also comprises states with more exotic quantum behaviour, see Section 25.4.

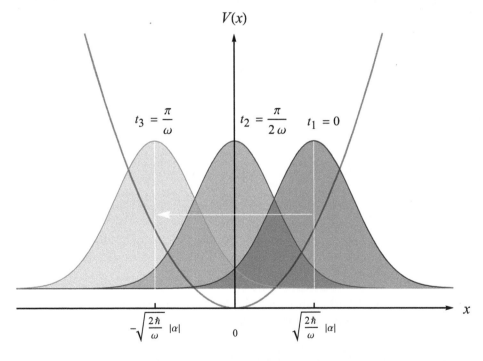

Fig. 25.1 Time evolution of coherent-state wave function. The probability density $|\phi_\alpha(t,x)|^2$ of a coherent state $|\alpha(t)\rangle$ is a Gaussian distribution, whose mean $\langle\hat{X}\rangle_{\alpha(t)} = \sqrt{\frac{2\hbar}{\omega}}\,|\alpha|\,\cos(\varphi-\omega t)$ oscillates sinusoidally in a harmonic-oscillator potential between its extremal values $\pm\sqrt{\frac{2\hbar}{\omega}}|\alpha|$, illustrated here for $\varphi=0$ such that $\alpha = \alpha(0) = |\alpha|$.

25.3 Phase Space in Quantum Mechanics—The Wigner Function

We have seen that coherent states can in many ways be seen as quantum-mechanical analogues of classical harmonic oscillators, in particular, the expectation values of the quadratures behave in this way. In terms of a classical phase-space picture, we can thus think of a harmonic oscillator as tracing out a circle around the origin, provided that we use suitable coordinates. As we shall see in this section, we can also provide a quantum-mechanical analogue of such a phase space. As a first step towards such a picture, we note that the conventions we have used so far do not treat the two quadratures, $\hat{X}$ and $\hat{P}$, on an equal footing. In particular, the expectation values $\langle \hat{X} \rangle_{\alpha(t)}$ and $\langle \hat{P} \rangle_{\alpha(t)}$ from eqn (25.95) have different units. In the following, it will thus be convenient to define dimensionless quadrature operators

$$\hat{x} := \sqrt{\tfrac{\omega}{\hbar}} \, \hat{X} \;=\; \tfrac{1}{\sqrt{2}} (a + a^\dagger), \tag{25.109a}$$

$$\hat{p} := \tfrac{1}{\sqrt{\hbar\omega}} \, \hat{P} \;=\; \tfrac{-i}{\sqrt{2}} (a - a^\dagger), \tag{25.109b}$$

such that $[\hat{x}, \hat{p}] = \tfrac{1}{\hbar} [\hat{X}, \hat{P}] = i$. Because we have just changed the scalar pre-factor of these operators, $\hat{x}$ and $\hat{p}$ have the same respective eigenstates as $\hat{X}$ and $\hat{P}$, but their eigenvalues $\tilde{x}$ and $\tilde{p}$ are related to the eigenvalues x and p of the original operators by the corresponding pre-factors, i.e.,

$$\hat{x} \, | x \rangle := \sqrt{\tfrac{\omega}{\hbar}} \, \hat{X} \, | x \rangle \;=\; \sqrt{\tfrac{\omega}{\hbar}} \, x \, | x \rangle \;=\; \tilde{x} \, | x \rangle, \tag{25.110a}$$

$$\hat{p} \, | p \rangle := \tfrac{1}{\sqrt{\hbar\omega}} \, \hat{P} \, | p \rangle \;=\; \tfrac{1}{\sqrt{\hbar\omega}} \, p \, | p \rangle \;=\; \tilde{p} \, | p \rangle. \tag{25.110b}$$

With this redefinition, we can consider a vector of mean values for the coherent state in a two-dimensional real vector space, that is,

$$\begin{pmatrix} \langle \hat{x} \rangle_{\alpha(t)} \\ \langle \hat{p} \rangle_{\alpha(t)} \end{pmatrix} = \sqrt{2} |\alpha| \begin{pmatrix} \cos(\varphi - \omega t) \\ \sin(\varphi - \omega t) \end{pmatrix}, \tag{25.111}$$

whose time evolution corresponds to a circle with radius $\sqrt{2}|\alpha|$ in $\mathbb{R}^2$. Moreover, in this picture, eqn (25.58) tells us that we can think of the displacement operator $D(\beta)$ as a (like the name suggests) *displacement* from position $\sqrt{2}(\mathrm{Re}(\alpha), \mathrm{Im}(\alpha))^\top$ to $\sqrt{2}(\mathrm{Re}(\alpha + \beta), \mathrm{Im}(\alpha + \beta))^\top$, as illustrated in Fig. 25.2.

The Wigner function: While the phase-space representation of coherent states via their mean values and uncertainties gives some intuition and is, as we shall see, already a complete description of coherent states, it is—at least at this point—not clear at all if, and if so in what sense, this description can be regarded as complete in fully representing the system. We shall therefore now turn to a representation of quantum states that can be regarded as a fully quantum-mechanical analogue of a phase-space representation: the *Wigner function*. The Wigner function (Wigner, 1932) is a representative from a family of so-called *quasi-probability distributions*, of which the most

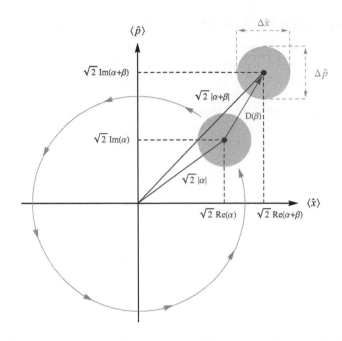

Fig. 25.2 Coherent states in phase space. The time evolution of the mean values $\langle \hat{x} \rangle$ and $\langle \hat{p} \rangle$ for a coherent state $|\alpha\rangle$ describes a circle in phase space, where the displacement operator $D(\beta)$ acts as a coordinate shift from position $\sqrt{2}(\text{Re}(\alpha), \text{Im}(\alpha))^{\top}$ to $\sqrt{2}(\text{Re}(\alpha + \beta), \text{Im}(\alpha + \beta))^{\top}$. For any fixed time t, a circular uncertainty region of diameter $\Delta \hat{x} = \Delta \hat{p} = \frac{1}{\sqrt{2}}$ can be associated to each mean value.

well-known members are the *Glauber–Sudarshan P-function* (Glauber, 1963; Sudarshan, 1963) and the *Husimi Q-function* (Husimi, 1940). All of these three functions have different advantages and disadvantages (see, e.g., the textbook Vogel and Welsch, 2006, or the discussion on experimental accessibility in Agudelo *et al.*, 2013) but can be seen to arise from the same one-parameter family of characteristic functions (Cahill and Glauber, 1969). Their most noticeable common feature is that they are not non-negative, i.e., may take negative values in some regions for some states. Here, we want to explore the features of the Wigner function as a phase-space description of quantum systems.

Before we take a closer look at the Wigner function itself, it is useful to first briefly recall the features of classical phase-space descriptions. For classical systems of N particles, each particle would be ascribed a set of two conjugate variables, x_n and p_n for $n = 1, 2, \ldots, N$, that can be interpreted as canonical positions and momenta, and which can be collected into joint position and momentum vectors, $x = (x_1, x_2, \ldots, x_N)^{\top} \in \mathbb{R}^N$ and $p = (p_1, p_2, \ldots, p_N)^{\top} \in \mathbb{R}^N$, respectively. It is then convenient to collect all variables into one vector, such that at any given time, the state of the N-particle system corresponds to a point $(x_1, p_1, x_2, p_2, \ldots, x_N, p_N)^{\top} \in \mathbb{R}^{2N}$. The time evolution of the system then traces out a curve in this $2N$-dimensional phase space. A statistical ensemble of such systems can then be represented by a phase-space density function $\varrho(x,p)$ with $\int dx\, dp\, \varrho(x,p) = 1$ and of course $\varrho(x,p) \geq 0$.

When we now turn to a quantum-mechanical analogue, we have to start by specifying the type of quantum system that such a description is to apply to. Clearly, the notion of "particle" may not be the best guiding principle when considering that coherent states have no fixed particle number. Instead, we can simply refer to a collection of N quantum systems for which canonically conjugate operators $\hat{x}_n$ and $\hat{p}_n$ can be defined for $n = 1, 2, \ldots, N$ such that $[\hat{x}_m, \hat{p}_n] = i\delta_{mn}$. The eigenvalue equations for these observables can be written compactly as

$$\hat{x}_n \,|\, x \,\rangle = x_n \,|\, x \,\rangle = x_n \,|\, x_1 \,\rangle \otimes |\, x_2 \,\rangle \otimes \ldots \otimes |\, x_N \,\rangle , \qquad (25.112a)$$

$$\hat{p}_n \,|\, p \,\rangle = p_n \,|\, p \,\rangle = p_n \,|\, p_1 \,\rangle \otimes |\, p_2 \,\rangle \otimes \ldots \otimes |\, p_N \,\rangle . \qquad (25.112b)$$

For instance, this can be done in the case of the Hilbert space $\mathcal{L}_2(\mathbb{R}^N, dx)$, i.e., the space of square-integrable (with respect to the Lebesgue measure dx) functions over $\mathbb{R}^N$. Since the eigenvalues x_n and p_n of $\hat{x}$ and $\hat{p}$, respectively, can take values in a continuous range, such systems are referred to as *continuous-variable* (CV) systems. In particular, we note that this includes (but is not limited to) a set of N harmonic oscillators, e.g., N modes of the electromagnetic field. For any such operators, we can also define ladder operators a_n and $a_n^\dagger$ by using the relations in eqn (25.109), that is

$$a_n = \tfrac{1}{\sqrt{2}}(\hat{x} + i\hat{p}), \qquad (25.113a)$$

$$a_n^\dagger = \tfrac{1}{\sqrt{2}}(\hat{x} - i\hat{p}), \qquad (25.113b)$$

with $[a_m, a_n^\dagger] = \delta_{mn}$ and $[a_m, a_n] = [a_m^\dagger, a_n^\dagger] = 0$. However, a CV system need not consist of harmonic oscillators in the sense that the system Hamiltonian may not be of the form of eqn (25.21), but since the existence of the N pairs of canonically conjugate quadrature operators allows one to associate mode operators a_n and $a_n^\dagger$, we shall refer to the system in question as an N-mode system. With these preliminaries and definitions at hand, we can now define the Wigner function $\mathcal{W}(x,p)$ associated to any given density operator ρ of the considered quantum system.

Definition 25.3   *The **Wigner function** $\mathcal{W}(x,p)$ of an N-mode quantum system described by the density operator ρ is given by*

$$\mathcal{W}(x,p) = \tfrac{1}{(2\pi\hbar)^N} \int dy\, e^{-ipy/\hbar} \,\langle\, x + \tfrac{y}{2} \,|\, \rho \,|\, x - \tfrac{y}{2} \,\rangle ,$$

where $x, y, p \in \mathbb{R}^N$.

Let us now inspect some of the properties of this function. We observe that for N modes, the Wigner function is a function of $2N$ real variables, collected in the vectors $x, p \in \mathbb{R}^N$. The Wigner function itself is also real-valued, which is easy to see since

$$W(x,p)^* = \frac{1}{(2\pi\hbar)^N} \int\limits_{-\infty}^{\infty} dy\, e^{ipy/\hbar} \langle\, x - \tfrac{y}{2}\, |\, \rho\, |\, x + \tfrac{y}{2}\, \rangle \qquad (25.114)$$

$$= \frac{1}{(2\pi\hbar)^N} \int\limits_{-\infty}^{\infty} dy'\, e^{-ipy'/\hbar} \langle\, x + \tfrac{y'}{2}\, |\, \rho\, |\, x - \tfrac{y'}{2}\, \rangle = W(x,p),$$

where we have simply substituted $y' = -y$. The Wigner function further has the nice property that integrating over either one of the two types of variables, x or p, results in the marginal probability distribution for the respective other variable, i.e., integrating over the momenta gives

$$\int dp\, W(x,p) = \frac{1}{(2\pi\hbar)^N} \int dy\, dp\, e^{-ipy/\hbar} \langle\, x + \tfrac{y}{2}\, |\, \rho\, |\, x - \tfrac{y}{2}\, \rangle \qquad (25.115)$$

$$= \frac{1}{(2\pi\hbar)^N} \int dy\, (2\pi\hbar)^N\, \delta^{(N)}(y)\, \langle\, x + \tfrac{y}{2}\, |\, \rho\, |\, x - \tfrac{y}{2}\, \rangle = \langle\, x\, |\, \rho\, |\, x\, \rangle,$$

the diagonal density matrix elements with respect to the position basis. And similarly, integrating over the positions results in

$$\int dx\, W(x,p) = \frac{1}{(2\pi\hbar)^N} \int dx\, dy\, e^{-ipy/\hbar} \langle\, x + \tfrac{y}{2}\, |\, \rho\, |\, x - \tfrac{y}{2}\, \rangle$$

$$= \frac{1}{(2\pi\hbar)^N} \int dx'\, dx''\, e^{-ip(x'-x'')/\hbar} \langle\, x'\, |\, \rho\, |\, x''\, \rangle \qquad (25.116)$$

$$= \int dx'\, dx''\, \langle\, p\, |\, x'\, \rangle\langle\, x'\, |\, \rho\, |\, x''\, \rangle\langle\, x''\, |\, p\, \rangle = \langle\, p\, |\, \rho\, |\, p\, \rangle,$$

where we have used $e^{-ipx'/\hbar} = (2\pi\hbar)^{N/2} \langle\, p\, |\, x'\, \rangle$ from eqn (3.45) and completeness of the position eigenstates from eqn (3.46). Both eqns (25.115) and (25.116) then quickly allow us to conclude that the Wigner function is normalized, i.e.,

$$\int dx\, dp\, W(x,p) = \int dp\, \langle\, p\, |\, \rho\, |\, p\, \rangle = \int dx\, \langle\, x\, |\, \rho\, |\, x\, \rangle = \mathrm{Tr}(\rho) = 1. \qquad (25.117)$$

From what we have seen so far, we can conclude that the Wigner function provides a representation of a given quantum state's properties in position and momentum space, respectively. However, so far it is not yet clear in what sense this description is more useful than, say, naively considering the pair of functions $f(x) := \langle\, x\, |\, \rho\, |\, x\, \rangle$ and $g(p) := \langle\, p\, |\, \rho\, |\, p\, \rangle$. The answer to this question lies in realizing that, on the one hand, the pair of function $f(x)$ and $g(p)$ does not provide a full description of a quantum state, i.e., quantum states are not uniquely identified by measurements in two bases[6].

---

[6] For a quantum system of dimension d, one requires measurements in $d + 1$ so-called mutually unbiased bases to fully determine a quantum state (for full state tomography), but in general a full set of $d + 1$ mutually unbiased bases can only be constructed if d is the power of a prime (Wootters and Fields, 1989).

The Wigner function, on the other hand, is in one-to-one correspondence with a density operator ρ. More specifically, while $\mathcal{W}$ is obtained from ρ via the *Wigner transform* $\mathcal{W}[\rho](x,p) := W(x,p)$ in Definition 25.3, the density operator ρ can be obtained from the Wigner function via the *Weyl transform* $\overline{\mathcal{W}}[\mathcal{W}] = \rho$. That is, for a given $\mathcal{W}(x,p)$, we can obtain the density-matrix elements $\langle x | \rho | y \rangle$ of ρ with respect to the position basis via

$$\langle x | \rho | y \rangle = \int dp \, e^{ip(x-y)/\hbar} \, \mathcal{W}(\tfrac{x+y}{2}, p). \tag{25.118}$$

Proof

$$\int dp \, e^{ip(x-y)/\hbar} \, \mathcal{W}(\tfrac{x+y}{2}, p) = \tfrac{1}{(2\pi\hbar)^N} \int dp \, dy' \, e^{ip(x-y-y')/\hbar} \, \langle \tfrac{x+y+y'}{2} | \rho | \tfrac{x+y-y'}{2} \rangle$$

$$= \tfrac{1}{(2\pi\hbar)^N} \int dy' \, (2\pi\hbar)^N \delta^{(N)}(x-y-y') \langle \tfrac{x+y+y'}{2} | \rho | \tfrac{x+y-y'}{2} \rangle$$

$$= \langle \tfrac{x+y+(x-y)}{2} | \rho | \tfrac{x+y-(x-y)}{2} \rangle = \langle x | \rho | y \rangle. \tag{25.119}$$

$\square$

Since we can write the density operator as $\rho = \int dx \, dy \, \langle x | \rho | y \rangle \, | x \rangle\langle y |$, this fully determines the state. This means that we can switch back and forth between the density-operator representations and the corresponding Wigner function, and use whichever representation of the state suits our needs best. In particular, we can also calculate expectation values of any observable in the Wigner representation, that is, we can write

$$\langle G \rangle_\rho = \mathrm{Tr}(\rho \, G) = (2\pi\hbar)^N \int dx \, dp \, \mathcal{W}[\rho](x,p) \, \mathcal{W}[G](x,p). \tag{25.120}$$

Proof

$$(2\pi\hbar)^N \int dx \, dp \, \mathcal{W}[\rho](x,p) \, \mathcal{W}[G](x,p)$$

$$= \tfrac{1}{(2\pi\hbar)^N} \int dx \, dp \, dy \, dy' \, e^{-ip(y+y')/\hbar} \, \langle x + \tfrac{y}{2} | \rho | x - \tfrac{y}{2} \rangle\langle x + \tfrac{y'}{2} | G | x - \tfrac{y'}{2} \rangle$$

$$= \int dx \, dy \, dy' \, \delta^{(N)}(y+y') \langle x + \tfrac{y}{2} | \rho | x - \tfrac{y}{2} \rangle\langle x + \tfrac{y'}{2} | G | x - \tfrac{y'}{2} \rangle$$

$$= \int dx \, dy \, \langle x + \tfrac{y}{2} | \rho | x - \tfrac{y}{2} \rangle\langle x - \tfrac{y}{2} | G | x + \tfrac{y}{2} \rangle. \tag{25.121}$$

We now insert resolutions of the identity in terms of the position basis, such that

$$\int dx \, dy \, \langle x + \tfrac{y}{2} | \rho | x - \tfrac{y}{2} \rangle\langle x - \tfrac{y}{2} | G | x + \tfrac{y}{2} \rangle = \tag{25.122}$$

$$= \int dx \, dy \, dx' \, dy' \, \langle x + \tfrac{y}{2} | x' \rangle\langle x' | \rho | y' \rangle\langle y' | x - \tfrac{y}{2} \rangle\langle x - \tfrac{y}{2} | G | x + \tfrac{y}{2} \rangle.$$

Evaluating the first inner product in the integrand, we then further have

$$
= \int dx\, dy\, dx'\, dy'\, \delta^{(N)}\big(x - (x' - \tfrac{y}{2})\big)\, \langle\, x'\,|\,\rho\,|\,y'\,\rangle\, \langle\, y'\,|\, x - \tfrac{y}{2}\,\rangle\, \langle\, x - \tfrac{y}{2}\,|\, G\,|\, x + \tfrac{y}{2}\,\rangle
$$

$$
= \int dy\, dx'\, dy'\, \langle\, x'\,|\,\rho\,|\,y'\,\rangle\, \langle\, y'\,|\, x' - y\,\rangle\, \langle\, x' - y\,|\, G\,|\, x'\,\rangle \tag{25.123}
$$

$$
= \int dy\, dx'\, dy'\, \langle\, x'\,|\,\rho\,|\,y'\,\rangle\, \delta^{(N)}\big(y - (x' - y')\big)\, \langle\, x' - y\,|\, G\,|\, x'\,\rangle
$$

$$
= \int dx'\, \langle\, x'\,|\,\rho\, \int dy'\, |\, y'\,\rangle\langle\, y'\,|\, G\,|\, x'\,\rangle = \int dx'\, \langle\, x'\,|\,\rho G\,|\, x'\,\rangle = \mathrm{Tr}(\rho G).
$$

□

The observation that we can calculate expressions such as $\mathrm{Tr}(\rho G)$ in the Wigner representation immediately implies that we can replace the observable G with a second density operator, i.e., for any two states ρ_1 and ρ_2 we have

$$
\mathrm{Tr}\big(\rho_1\, \rho_2\big) = (2\pi\hbar)^N \int dx\, dp\, \mathcal{W}[\rho_1](x,p)\, \mathcal{W}[\rho_2](x,p). \tag{25.124}
$$

Since we can find pairs of states (for instance, orthogonal pure states) for which $\mathrm{Tr}(\rho_1\, \rho_2) = 0$, we thus see from the right-hand side of eqn (25.124) that the Wigner function cannot generally be non-negative everywhere for all states. Or, conversely, there exist states for which the Wigner function is not positive semidefinite. We thus have to conclude that, although the Wigner function is a normalized distribution in phase space, it is not a proper probability distribution. Instead, it is an example of so-called *quasi-probability distribution.*

Let us now briefly consider a few examples for Wigner functions. For instance, if we consider the single-mode ($N = 1$) Fock states $|\, n\,\rangle$ such that $\rho = |\, n\,\rangle\langle\, n\,|$ and denote the corresponding Wigner function by $\mathcal{W}_n(x,p)$, we have

$$
\mathcal{W}_n(x,p) = \frac{1}{2\pi\hbar} \int dy\, e^{-ipy/\hbar}\, \langle\, x + \tfrac{y}{2}\,|\, n\,\rangle\langle\, n\,|\, x - \tfrac{y}{2}\,\rangle \tag{25.125}
$$

$$
= \frac{\omega^{1/2}}{2^{n+1}(n!)\,\pi^{3/2}\,\hbar^{3/2}} \int dy\, e^{-ipy/\hbar}\, H_n\!\left(\tfrac{x+y/2}{\sqrt{\hbar/\omega}}\right) H_n^*\!\left(\tfrac{x-y/2}{\sqrt{\hbar/\omega}}\right) \exp\!\left(-\tfrac{x^2+(y/2)^2}{\hbar/\omega}\right),
$$

where we have inserted the expressions $\psi_n(x) = \langle\, x\,|\, n\,\rangle$ for the harmonic-oscillator stationary states from eqn (5.41) for unit mass ($m = 1$, such that $\sigma_o = \sqrt{\hbar/\omega}$), and H_n are the Hermite polynomials. Evaluating the integral for the ground state and the first few excited states leads to

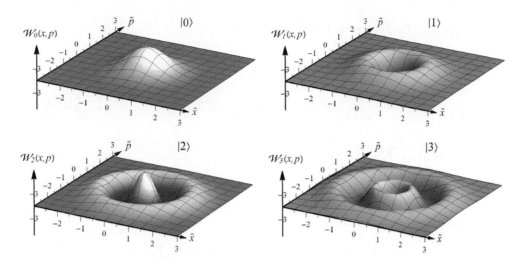

Fig. 25.3 Wigner functions of Fock states. The Wigner functions $\mathcal{W}_n(x,p)$ of the first four eigenstates $|n\rangle$ for $n = 0, 1, 2, 3$ of the harmonic oscillator are illustrated here. The axes show the dimensionless variables $\tilde{x} = \sqrt{\omega/\hbar}x$ and $\tilde{p} = p/\sqrt{\hbar\omega}$.

$$|0\rangle: \quad W_0(x,p) = \tfrac{1}{\pi} e^{-\tilde{x}^2 - \tilde{p}^2}, \tag{25.126a}$$

$$|1\rangle: \quad W_1(x,p) = \tfrac{1}{\pi} e^{-\tilde{x}^2 - \tilde{p}^2} \left(2\tilde{x}^2 + 2\tilde{p}^2 - 1\right), \tag{25.126b}$$

$$|2\rangle: \quad W_2(x,p) = \tfrac{1}{\pi} e^{-\tilde{x}^2 - \tilde{p}^2} \left(2[\tilde{x}^2 + \tilde{p}^2 - 1]^2 - 1\right), \tag{25.126c}$$

$$|3\rangle: \quad W_3(x,p) = \tfrac{1}{\pi} e^{-\tilde{x}^2 - \tilde{p}^2} \left(6[\tilde{x}^2 + \tilde{p}^2 - 1]^3 - 2[\tilde{x}^2 + \tilde{p}^2]^3 + 3\right), \tag{25.126d}$$

where we have again used $\tilde{x} = \sqrt{\omega/\hbar}x$ and $\tilde{p} = p/\sqrt{\hbar\omega}$ as in eqn (25.110). We observe that all of these functions, illustrated in Fig. 25.3, are symmetric with respect to the exchange of $\tilde{x}$ and $\tilde{p}$, indeed, they are even invariant under arbitrary rotations in the $\tilde{x}$–$\tilde{p}$ plane. This is not generally the case for Wigner functions of arbitrary states. In addition, we see that the ground state, which we recall from eqn (5.41) has a Gaussian probability distribution $|\psi_0(x)|^2$ in position (and thus also a Gaussian probability distribution in momentum space), also has a Gaussian Wigner function. For the excited states, on the other hand, the Wigner function is not Gaussian, and is not even positive semidefinite. Similar features, in particular, the Gaussian character of the respective phase-space functions, can also be observed for other quasi-probability distributions such as the Glauber–Sudarshan P-function (Glauber, 1963; Sudarshan, 1963) and the Husimi Q-function (Husimi, 1940).

We can also take a closer look at the Wigner function $\mathcal{W}_\alpha(x,p)$ of the coherent states $|\alpha\rangle$ of a single mode (meaning, again, $N = 1$). In fact, we can just insert $\rho = |\alpha\rangle\langle\alpha|$ into Definition 25.3 and insert the (time-independent, for now) coherent-state wave functions from eqn (25.104), that is

$$\mathcal{W}_\alpha(x,p) = \tfrac{1}{2\pi\hbar} \int dy\, e^{-ipy/\hbar} \left\langle x + \tfrac{y}{2} \,\middle|\, \alpha \right\rangle\!\!\left\langle \alpha \,\middle|\, x - \tfrac{y}{2} \right\rangle$$

$$= \tfrac{1}{2\pi\hbar} \int dy\, e^{-ipy/\hbar}\, \phi_\alpha\!\left(x + \tfrac{y}{2}\right) \phi_\alpha^*\!\left(x - \tfrac{y}{2}\right)$$

$$= \tfrac{1}{2\pi\hbar} \int dy\, e^{-ipy/\hbar}\, \sqrt{\tfrac{\omega}{\hbar\pi}}\, e^{i\,\mathrm{Im}(\alpha)\sqrt{\frac{\omega}{2\hbar}}\,2y} \exp\!\left(-\frac{(x+\frac{y}{2}-\langle \hat{X}\rangle_\alpha)^2}{2\hbar/\omega} - \frac{(x-\frac{y}{2}-\langle \hat{X}\rangle_\alpha)^2}{2\hbar/\omega}\right)$$

$$= \tfrac{1}{2\pi\hbar} \sqrt{\tfrac{\omega}{\hbar\pi}} \int dy\, e^{-i\left(p - \sqrt{2\hbar\omega}\,\mathrm{Im}(\alpha)\right)y/\hbar} \exp\!\left(-\frac{(x-\langle\hat{X}\rangle_\alpha)^2 + (y/2)^2}{\hbar/\omega}\right). \tag{25.127}$$

In the last line, we now recognize $\sqrt{2\hbar\omega}\,\mathrm{Im}(\alpha) = \langle \hat{P}\rangle_\alpha$ from eqn (25.96). Comparing the result to the Wigner function $\mathcal{W}_{n=0}(x,p)$ of the ground state $|\,0\,\rangle$ in eqn (25.125), we find the simple correspondence

$$\mathcal{W}_\alpha(x,p) = \mathcal{W}_0\!\left(x - \langle\,\hat{X}\,\rangle_\alpha,\, p - \langle\,\hat{P}\,\rangle_\alpha\right), \tag{25.128}$$

which holds also for all $\alpha = \alpha(t)$. That is, the Wigner functions of all coherent states are identical to the Wigner function of the ground state up to displacements by the vector of mean values. In particular, this means that all coherent states have Gaussian Wigner functions. Indeed, coherent states can be understood as a special case of the larger family of Gaussian states, which we will discuss in the next section.

25.4   Gaussian States and Operations

Gaussian states: In the previous section, we found that all coherent states have Gaussian Wigner functions, which are hence positive semidefinite in the entire range of x and p. Moreover, for coherent states, the uncertainties of the dimensionless variables $\tilde{x} = \sqrt{\omega/\hbar}x$ and $\tilde{p} = p/\sqrt{\hbar\omega}$ are identical, and their product minimizes the Robertson–Schrödinger uncertainty relation. However, attaining the minimum product of uncertainties does not imply that the uncertainties in $\tilde{x}$ and $\tilde{p}$ have to be the same: Indeed, stretching the Wigner function in one phase-space direction by a constant factor and squeezing it in the orthogonal direction by the inverse factor would again result in a bivariate Gaussian distribution minimizing the uncertainty relation. Meanwhile, stretching the Wigner function in both directions would result in a Gaussian distribution whose uncertainties are the same for all direction, but do not minimize the uncertainty relation. In all of these cases, we could nevertheless use the Weyl transform to map the respective Wigner functions to valid density operators. One can thus define an entire family of states via their Wigner functions (Schumaker, 1986).

Definition 25.4   *Any N-mode quantum state whose Wigner function $\mathcal{W}(x,p)$ is a multivariate Gaussian distribution is called a **Gaussian state**.*

Multivariate Gaussians can be fully described by their first and second statistical moments. Here, the first moments are collected in the *vector of first moments*, e.g., as in eqn (25.111), while the second statistical moments can be conveniently collected in the *covariance matrix*. To give a compact definition for this object, let us first establish the following notation for the quadrature operators of N modes,

$$\mathbb{X}_{2n-1} = \hat{x}_n = \frac{1}{\sqrt{2}}(a_n + a_n^\dagger), \tag{25.129}$$

$$\mathbb{X}_{2n} = \hat{p}_n = \frac{-i}{\sqrt{2}}(a_n - a_n^\dagger). \tag{25.130}$$

In this notation, we can define the vector of first moments, $\overline{\mathbb{X}} \in \mathbb{R}^{2N}$ as

$$\overline{\mathbb{X}} = \langle \mathbb{X} \rangle_\rho = \begin{pmatrix} \langle \mathbb{X}_1 \rangle_\rho \\ \langle \mathbb{X}_2 \rangle_\rho \\ \langle \mathbb{X}_3 \rangle_\rho \\ \langle \mathbb{X}_4 \rangle_\rho \\ \vdots \\ \langle \mathbb{X}_{2N-1} \rangle_\rho \\ \langle \mathbb{X}_{2N} \rangle_\rho \end{pmatrix} = \begin{pmatrix} \langle \hat{x}_1 \rangle_\rho \\ \langle \hat{p}_1 \rangle_\rho \\ \langle \hat{x}_2 \rangle_\rho \\ \langle \hat{p}_2 \rangle_\rho \\ \vdots \\ \langle \hat{x}_N \rangle_\rho \\ \langle \hat{p}_N \rangle_\rho \end{pmatrix}. \tag{25.131}$$

The commutation relations $[a_m, a_n^\dagger] = \delta_{mn}$, or, equivalently, the canonical commutators $[\hat{x}_m, \hat{p}_n] = i\delta_{mn}$, then imply the commutation relations

$$-i[\mathbb{X}_m, \mathbb{X}_n] = \delta_{m,n-1} - \delta_{n,m+1} = \Omega_{mn}, \tag{25.132}$$

which we use to define the components Ω_{mn} of a $2N \times 2N$-matrix Ω that is called the *symplectic form*, and which we can write as

$$\Omega = \bigoplus_n \Omega^{(n)} \quad \text{with} \quad \Omega^{(n)} = \begin{pmatrix} 0 & 1 \\ -1 & 0 \end{pmatrix} \quad \forall n. \tag{25.133}$$

The symplectic form encodes the structure of the classical phase space, i.e., the asymmetry between the canonical position and momentum variables that appears as a sign change, e.g., in Hamilton's equations. We will return to the symplectic form shortly. Now, let us further define a vector $\xi \in \mathbb{R}^{2N}$ of dimensionless phase-space coordinates, $\xi = (\tilde{x}_1, \tilde{p}_1, \tilde{x}_2, \tilde{p}_2, \ldots, \tilde{x}, \tilde{p}_N)^\top$, which allows us to make the following definition:

Definition 25.5 *The **covariance matrix** $\Gamma = (\Gamma_{ij})$ of any N-mode quantum state is a real, symmetric $2N \times 2N$ matrix with components*

$$\Gamma_{ij} = \langle \mathbb{X}_i \mathbb{X}_j + \mathbb{X}_j \mathbb{X}_i \rangle - 2\langle \mathbb{X}_i \rangle \langle \mathbb{X}_j \rangle.$$

That the entries Γ_{ij} of the covariance matrix are real and that the matrix is symmetric, $\Gamma_{ij} = \Gamma_{ji}$, is easy to see from the definition. The diagonal entries $\Gamma_{ii} = \mathrm{Var}(\mathbb{X}_i)$ provide the variances (up to a conventional factor of 2) of the quadrature operators $\mathbb{X}_i$, that is, $\Gamma_{ii} = 2(\Delta\hat{x}_n)^2$ for $i = 2n - 1$ and $\Gamma_{ii} = 2(\Delta\hat{p}_n)^2$ for $i = 2n$. Meanwhile, the off-diagonal elements $\Gamma_{ij} = 2\,\text{Cov}(\mathbb{X}_i, \mathbb{X}_j)$ for $i \neq j$ represent the covariances of operator pairs (again, up to a factor 2) that we have encountered in Theorem 2.4. In further interpreting these covariances, let us first consider the off-diagonals pertaining to the same mode n, and return to those pertaining to different modes later. For the same mode n, we have $\Gamma_{n,n+1} = 2\,\text{Cov}(\hat{x}_n, \hat{p}_n)$ and the covariance matrix for any single mode then takes the form

$$\Gamma^{(n)} = 2 \begin{pmatrix} (\Delta\hat{x}_n)^2 & \text{Cov}(\hat{x}_n, \hat{p}_n) \\ \text{Cov}(\hat{x}_n, \hat{p}_n) & (\Delta\hat{p}_n)^2 \end{pmatrix}. \tag{25.134}$$

Here we have made use of the fact that the equivalent operation to the partial trace for density operators is extremely simple when it comes to the description via the statistical moments. Ignoring or removing any modes simply corresponds to removing the corresponding rows (and columns) from the vector of first moments, covariance matrix, and symplectic form Ω. Using the latter, one can even formulate a bona fide uncertainty relation (Simon *et al.*, 1994)

$$\Gamma + i\Omega \geq 0. \tag{25.135}$$

Proof   Here, we will explicitly prove this relation only for a single mode, and for a more general proof for the N-mode case we refer to Simon *et al.* (1994). When considering a single mode labelled n, we calculate the eigenvalues $\lambda_\pm$ of the corresponding matrix $\Gamma^{(n)} + i\Omega^{(n)}$. Combining eqs (25.134) and (25.133), we obtain the expression

$$\lambda_\pm = (\Delta\hat{x}_n)^2 + (\Delta\hat{p}_n)^2 \tag{25.136}$$

$$\pm \sqrt{\left[(\Delta\hat{x}_n)^2 + (\Delta\hat{p}_n)^2\right]^2 - 4(\Delta\hat{x}_n)^2(\Delta\hat{p}_n)^2 + 4\,\text{Cov}^2(\hat{x}_n, \hat{p}_n) + 1}.$$

Now, since $\Gamma^{(n)} + i\Omega^{(n)}$ is Hermitian (recall that $\Gamma^{(n)}$ is real and symmetric, and $\Omega^{(n)}$ is antisymmetric), its eigenvalues are real, which we can also see by noting that the discriminant under the square-root is the sum of non-negative terms, i.e.,

$$\left[(\Delta\hat{x}_n)^2 + (\Delta\hat{p}_n)^2\right]^2 - 4(\Delta\hat{x}_n)^2(\Delta\hat{p}_n)^2 + 4\,\text{Cov}^2(\hat{x}_n, \hat{p}_n) + 1 \tag{25.137}$$

$$= \left[(\Delta\hat{x}_n)^2 - (\Delta\hat{p}_n)^2\right]^2 + 4\,\text{Cov}^2(\hat{x}_n, \hat{p}_n) + 1 \geq 0.$$

We thus see that the non-negativity of the eigenvalues, specifically of λ_-, reduces to the condition

$$4(\Delta\hat{x}_n)^2(\Delta\hat{p}_n)^2 - 4\,\text{Cov}^2(\hat{x}_n, \hat{p}_n) - 1 \geq 0, \tag{25.138}$$

which can be rewritten as

$$(\Delta\hat{x}_n)^2(\Delta\hat{p}_n)^2 \geq \text{Cov}^2(\hat{x}_n, \hat{p}_n) + \tfrac{1}{4}, \tag{25.139}$$

which is exactly the Robertson–Schrödinger uncertainty relation from Theorem 2.4 for the operators $\hat{x}_n$ and $\hat{p}_n$. $\square$

The relation in eqn (25.135) is satisfied by the covariance matrix of any (not necessarily Gaussian) quantum state, and, indeed, any real, symmetric $2N \times 2N$ matrix Γ that satisfies $\Gamma + i\Omega \geq 0$ is the covariance matrix of a valid N-mode CV quantum system (Simon *et al.*, 1994). Moreover, eqn (25.135) implies positive semidefiniteness of Γ, since $\Gamma + i\Omega \geq 0$ means that $(\Gamma + i\Omega)^\top \geq 0$ (the transposition is a positive map), and the sum of two positive operators must be positive, hence

$$\tfrac{1}{2}(\Gamma + i\Omega) + \tfrac{1}{2}(\Gamma + i\Omega)^\top = \tfrac{1}{2}(\Gamma + i\Omega) + \tfrac{1}{2}(\Gamma^\top + i\Omega^\top) \tag{25.140}$$

$$= \tfrac{1}{2}(\Gamma + i\Omega) + \tfrac{1}{2}(\Gamma - i\Omega) = \Gamma \geq 0.$$

For any Gaussian state, the Wigner function can then be written in terms of its covariance matrix Γ and vector of first moments $\overline{\mathbb{X}}$ as

$$\mathcal{W}(x, p) = \frac{1}{\pi^N \sqrt{\det(\Gamma)}} \exp\left[-(\xi - \overline{\mathbb{X}})^\top \Gamma^{-1} (\xi - \overline{\mathbb{X}})\right]. \tag{25.141}$$

In particular, this means that any Gaussian state is thus completely determined by $\overline{\mathbb{X}}$ and Γ. In turn, this means that for any covariance matrix satisfying eqn (25.135), there exists a corresponding Gaussian state (with arbitrary vector of first moments, we will come back to this point shortly) defined by eqn (25.141). For Gaussian states, the Wigner function is non-negative everywhere by construction, but, interestingly, the positivity of the Wigner function is also not just a necessary but also a sufficient criterion for any pure state to be Gaussian (Hudson, 1974). For mixed states, however, positivity of the Wigner function does not imply it is Gaussian.

Examples for single-mode Gaussian states: Let us now consider some important examples for single-mode Gaussian states. For coherent states, we have already obtained the vector of first moments in eqn (25.111), and the Wigner function is given by eqn (25.128). From eqns (25.98) and (25.100) we also immediately get $(\Delta \hat{x}_n)^2 = (\Delta \hat{p}_n)^2 = \tfrac{1}{2}$ and one also finds $\text{Cov}(\hat{x}_n, \hat{p}_n) = 0$ for coherent states. For the covariance matrix, with our convention including the additional factor 2 with respect to the covariances, i.e., $\Gamma_{ij} = 2\,\text{Cov}^2(\mathbb{X}_i \mathbb{X}_j)$, we thus simply have

$$\Gamma_\alpha = \begin{pmatrix} 1 & 0 \\ 0 & 1 \end{pmatrix} = \mathbb{1}_2\,, \tag{25.142}$$

and we note that covariance matrix does not depend on the displacement parameter α. The fact that the diagonal elements of this covariance matrix are the same expresses that the uncertainties in $\hat{x}$ and $\hat{p}$ have the same magnitude for coherent states, while their product is minimal. Of course we can also consider a covariance matrix where the two values are different from each other, while keeping their product constant, e.g.,

$$\Gamma_r = \begin{pmatrix} e^{-2r} & 0 \\ 0 & e^{2r} \end{pmatrix}. \tag{25.143}$$

Here, the variance of the first (position) quadrature is reduced (for $r > 0$) by a factor e^{-2r}, while the uncertainty in the second (momentum) quadrature is increased by the inverse factor e^{2r}. States of this type are called (single-mode) *squeezed states*. As we shall see, both of these covariance matrices can (and, indeed must) be associated with pure states, but of course we can also have mixed Gaussian states. For instance, the thermal state from eqn (25.23) is an example for a Gaussian state (which we will prove a little bit later in this chapter). For a system of non-interacting harmonic oscillators with Hamiltonian $H = \sum_m H_m$ with $H_m = \hbar\omega_n N_m$, the thermal state is a product state $\tau(\beta) = \bigotimes_m \tau_m(\beta)$, and the covariance matrix is hence of block-diagonal form $\Gamma_{\text{th}}(\beta) = \bigoplus_m \Gamma_{m,\text{th}}(\beta)$. We can therefore again focus on a single mode labelled m with thermal state $\tau_n(\beta)$. To calculate the corresponding statistical moments, it is convenient to write the thermal state in the Fock basis, where $N_m = \sum_n n |n\rangle\langle n|$, such that

$$\tau_m(\beta) = \frac{e^{-\beta H_m}}{Z_m} = \left(1 - e^{-\beta\hbar\omega_m}\right) \sum_n e^{-\beta\hbar\omega_m n} |n\rangle\langle n|. \tag{25.144}$$

With this, we can write the expectation value of an operator A in the (single-mode) thermal state as

$$\langle A \rangle_{\tau_m} = \text{Tr}\left(A\tau_m(\beta)\right) = \left(1 - e^{-\beta\hbar\omega_m}\right) \sum_n e^{-\beta\hbar\omega_m n} \langle n | A | n \rangle. \tag{25.145}$$

We can then insert from eqns (5.44)–(5.49), from which we obtain $\langle n | \hat{x}_m | n \rangle = \langle n | \hat{p}_m | n \rangle = 0$ and $\langle n | \hat{x}_m^2 | n \rangle = \langle n | \hat{p}_m^2 | n \rangle = n + \frac{1}{2}$, such that $\langle \hat{x}_m \rangle_{\tau_m} = \langle \hat{p}_m \rangle_{\tau_m} = 0$ and

$$(\Delta\hat{x}_m)^2_{\tau_m} = (\Delta\hat{p}_m)^2_{\tau_m} = \left(1 - e^{-\beta\hbar\omega_m}\right) \sum_n e^{-\beta\hbar\omega_m n} \left(n + \frac{1}{2}\right) = \frac{1}{2}\coth\left(\frac{\beta\hbar\omega_m}{2}\right). \tag{25.146}$$

To obtain the covariances, we quickly calculate

$$\langle n | \hat{x}_m\hat{p}_m | n \rangle = \tfrac{-i}{2}\langle n | (a_n + a_n^\dagger)(a_n - a_n^\dagger) | n \rangle = \tfrac{-i}{2}\langle n | a_n^2 + a_n^\dagger a_n - a_n a_n^\dagger - a_n^{\dagger 2} | n \rangle$$
$$= \tfrac{-i}{2}\langle n | a_n^\dagger a_n - (a_n^\dagger a_n + 1) | n \rangle = \tfrac{i}{2}, \tag{25.147}$$

$$\langle n | \hat{p}_m\hat{x}_m | n \rangle = \tfrac{-i}{2}\langle n | (a_n - a_n^\dagger)(a_n + a_n^\dagger) | n \rangle = \tfrac{-i}{2}\langle n | a_n^2 - a_n^\dagger a_n + a_n a_n^\dagger - a_n^{\dagger 2} | n \rangle$$
$$= \tfrac{i}{2}\langle n | a_n^\dagger a_n - (a_n^\dagger a_n + 1) | n \rangle = \tfrac{-i}{2}, \tag{25.148}$$

and we hence have $\text{Cov}(\hat{x}_n, \hat{p}_n) = 0$ once more. The covariance matrix of an N-mode thermal state can thus be compactly written as

$$\Gamma_{\text{th}}(\beta) = \bigoplus_m \Gamma_{m,\text{th}}(\beta) = \bigoplus_m \coth\left(\frac{\beta\hbar\omega_m}{2}\right) \mathbb{1}_2, \tag{25.149}$$

and the corresponding vector of first moments is just $\overline{\mathbb{X}}_{\text{th}} = 0$. A comparison of the resulting Wigner functions is shown in Fig. 25.4.

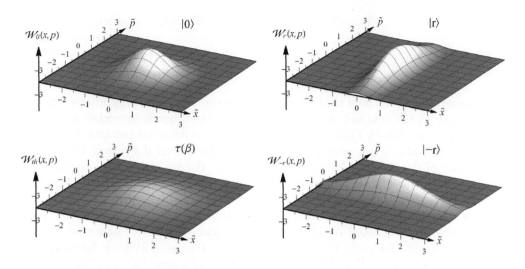

Fig. 25.4 Single-mode Gaussian states. The plots show the Wigner functions $\mathcal{W}_\alpha(x,p)$, $\mathcal{W}_r(x,p)$, $\mathcal{W}_{-r}(x,p)$, and $\mathcal{W}_{\mathrm{th}}(x,p)$ for a coherent state $|\alpha = 0\rangle$ (the vacuum), two single-mode squeezed states $|r\rangle$ and $|-r\rangle$ with $r = \ln(2) > 0$ for squeezing along different quadratures, and a single-mode thermal state $\tau(\beta)$ for $\beta = 1/(k_\mathrm{B}T) = 1$, respectively. The axes show the dimensionless variables $\tilde{x} = \sqrt{\omega/\hbar}x$ and $\tilde{p} = p/\sqrt{\hbar\omega}$.

Now that we have gathered some intuition on Gaussian states and Wigner functions, let us take a closer look at some properties of the respective quantum systems and how we can obtain these from the statistical moments. To begin, let us consider the average energy $E(\rho) = \mathrm{Tr}(H\rho)$ for the case of N non-interacting harmonic oscillators, as before, with $H = \sum_m H_m$ where $H_m = \hbar\omega_n N_m$. Since the operator H is a quadratic polynomial in the mode operators a_n and $a_n^\dagger$, we can indeed obtain its expectation value from the first and second statistical moments, even if the state is not Gaussian and hence not fully described by $\overline{\mathbb{X}}$ and Γ. More specifically, the average energy takes the form

$$E(\rho) = \sum_{n=1}^{N} \hbar\omega_n \left(\tfrac{1}{4}\big[\mathrm{Tr}(\Gamma^{(n)}) - 2\big] + \tfrac{1}{2}\|\overline{\mathbb{X}}^{(n)}\|^2 \right). \qquad (25.150)$$

Proof To see this, we first straightforwardly evaluate $\mathrm{Tr}(\Gamma^{(n)})$, that is

$$\mathrm{Tr}(\Gamma^{(n)}) = 2(\Delta\hat{x}_n)^2 + 2(\Delta\hat{p}_n)^2 = 2\big(\langle\hat{x}_n^2\rangle - \langle\hat{x}_n\rangle^2 + \langle\hat{p}_n^2\rangle - \langle\hat{p}_n\rangle^2\big)$$

$$= 2\langle\hat{x}_n^2 + \hat{p}_n^2\rangle - \langle\hat{x}_n\rangle^2 - \langle\hat{p}_n\rangle^2. \qquad (25.151)$$

Here, we can identify $\hat{x}_n^2 + \hat{p}_n^2 = 2(a_n^\dagger a_n + \tfrac{1}{2}) = 2(N_n + \tfrac{1}{2})$, while $\langle\hat{x}_n\rangle^2 + \langle\hat{p}_n\rangle^2 = \|\overline{\mathbb{X}}^{(n)}\|^2$ corresponds to the squared norm of the vector of first moments $\overline{\mathbb{X}}^{(n)} = (\langle\hat{x}_n\rangle, \langle\hat{p}_n\rangle)^\top$ of mode n. We thus have

$$\mathrm{Tr}(\Gamma^{(n)}) = 4\langle N_n \rangle + 2 - 2\|\overline{\mathbb{X}}^{(n)}\|^2. \tag{25.152}$$

Rearranging this expression and summing over all modes then results in the formula in eqn (25.150). □

It is interesting to note here that the formula for the average energy in eqn (25.150) depends only on the first and second statistical moments, but is nevertheless valid for all states, i.e., not just for Gaussian states. For operators that are not quadratic in the mode operators this is not necessarily the case. In fact, even if we want to calculate the variance of the energy, $(\Delta H)^2 = \langle H^2 \rangle - \langle H \rangle^2$, the presence of the operator H^2 means that we will generally need to consider higher statistical moments. However, since Gaussian states are fully determined by Γ and $\overline{\mathbb{X}}$, it is in principle possible to calculate any expectation value from these quantities. A direct way to do this lies in transforming the considered observables to their phase-space representation via the Wigner transform and using the expression for the Gaussian Wigner function from eqn (25.141) in terms of Γ and $\overline{\mathbb{X}}$. Inserting both expressions into eqn (25.120), one obtains the expectation value as a function of the first and second moments of the Gaussian states. For instance, for the squared single-mode number operator N_n^2, this procedure leads to the expression (Friis and Huber, 2018)

$$\langle N_n^2 \rangle = \left(\tfrac{1}{4}\left[\mathrm{Tr}(\Gamma^{(n)}) - 2\right] + \tfrac{1}{2}\|\overline{\mathbb{X}}^{(n)}\|^2 \right)^2 + \tfrac{1}{2}\overline{\mathbb{X}}^{(n)\top}\Gamma^{(n)}\,\overline{\mathbb{X}}^{(n)} + \tfrac{1}{8}\left[\mathrm{Tr}(\Gamma^{(n)2}) - 2\right]. \tag{25.153}$$

Since the first term on the right-hand side can be recognized as the squared average energy of a single mode, see eqn (25.150), one immediately obtains the variance of the single-mode number operator for any Gaussian state as

$$(\Delta N_n)^2 = \tfrac{1}{2}\overline{\mathbb{X}}^{(n)\top}\Gamma^{(n)}\,\overline{\mathbb{X}}^{(n)} + \tfrac{1}{8}\left[\mathrm{Tr}(\Gamma^{(n)2}) - 2\right]. \tag{25.154}$$

For Gaussian states one can of course also calculate other state properties (that are functions of the density matrix only) directly from the first and second moments. For instance, the *purity* $P(\rho) = \mathrm{Tr}(\rho^2)$ of any Gaussian state with covariance matrix Γ takes the simple form (Adesso *et al.*, 2004)

$$P(\rho) = \frac{1}{\sqrt{\det \Gamma}}. \tag{25.155}$$

That is, the purity is determined just by the covariance matrix, which is intuitively clear since all coherent states are pure states, but their first moments may take on arbitrary values (determined by α). Using this expression for the purity, it is easy to see that the covariance matrix Γ_r of the single-mode squeezed state in eqn (25.143) represents a pure state $|r\rangle$, since $\det(\Gamma_r) = 1$. For more complicated entropic functions, it is convenient to introduce the notion of the *symplectic spectrum*, which requires that we briefly discuss Gaussian operations first.

Gaussian operations: Since the family of Gaussian states includes pure states such as coherent state or squeezed states, it is evident that there must exist unitary operations that map Gaussian states to Gaussian states.

Definition 25.6 *A unitary transformation U that maps all Gaussian states to Gaussian states is called a **Gaussian unitary**.*

Any such unitary U must map any valid pair of first and second statistical moments $(\overline{\mathbb{X}}, \Gamma)$ to valid statistical moments. As discussed in Simon *et al.* (1994) this means that any Gaussian unitary is represented in phase space by an affine map

$$(S, \xi) : \mathbb{X} \longmapsto S\mathbb{X} + \xi. \tag{25.156}$$

Here, the real vector $\xi \in \mathbb{R}^{2N}$ represents displacements of all modes in phase space, and simply corresponds to the unitary Weyl operators $D(\xi) = \exp(i\mathbb{X}^{\mathsf{T}}\Omega\xi)$, i.e., acting upon mode n as the displacement operator $D(\alpha_n)$ from Definition 25.2 with $\xi_{2n-1} = \sqrt{2}\,\mathrm{Re}(\alpha_n)$ and $\xi_{2n} = \sqrt{2}\,\mathrm{Im}(\alpha_n)$.

The linear operators S transform the commutation relations that define the symplectic form Ω according to

$$\mathbb{X}\mathbb{X}^{\mathsf{T}} + (\mathbb{X}\mathbb{X}^{\mathsf{T}})^{\mathsf{T}} \mapsto S\mathbb{X}\mathbb{X}^{\mathsf{T}}S^{\mathsf{T}} + (S\mathbb{X}\mathbb{X}^{\mathsf{T}}S^{\mathsf{T}})^{\mathsf{T}} = S[\mathbb{X}\mathbb{X}^{\mathsf{T}} + (\mathbb{X}\mathbb{X}^{\mathsf{T}})^{\mathsf{T}}]S^{\mathsf{T}}, \tag{25.157}$$

which implies that the transformation S must leave the symplectic form invariant to respect the phase-space structure,

$$S\,\Omega\,S^{\mathsf{T}} = \Omega. \tag{25.158}$$

The $2N \times 2N$ matrices S that satisfy eqn (25.158) are called *symplectic transformations*. The symplectic transformations on n modes form the *real, symplectic group* $\mathrm{Sp}(2n, \mathbb{R})$. Any Gaussian unitary corresponds to a symplectic operation in phase space, and the Gaussian unitaries are generated by Hamiltonians that are (at most) quadratic in the mode (or, equivalently, the quadrature) operators.

An example for such transformations that we have already encountered are the displacement operators from Definition 25.2, for which $S = \mathbb{1}$, as we can see from eqns (25.48) and (25.49), and for which the generating Hamiltonian is a linear function of the mode operators. Symplectic operations as defined in eqn (25.156) transform the covariance matrix Γ according to

$$\Gamma \longmapsto S\Gamma S^{\mathsf{T}}. \tag{25.159}$$

From this result, it is easy to see that displacements have no effect on the covariance matrix.

Symplectic transformations for single modes: Some non-trivial examples for symplectic transformations on single modes include single-mode squeezing, that can be represented on the phase space and Hilbert space of a single mode n by the symplectic and unitary operators S_r and U_r, respectively, which are given by

$$S_r = \begin{pmatrix} e^{-r} & 0 \\ 0 & e^r \end{pmatrix}, \quad \text{and} \quad U_S(r) = \exp\left[\frac{r}{2}(a_n^2 - a_n^{\dagger 2})\right]. \tag{25.160}$$

Applying these operations to the vacuum, represented by $|0\rangle$ and $\Gamma_0 = \mathbb{1}$, we have $U_S(r)|0\rangle = |r\rangle$ and $\Gamma_r = S_r \Gamma_0 S_r^\mathsf{T}$. Another example for a single-mode Gaussian transformation is a *rotation in phase space*, which is generated by the number operator (and thus corresponds to free time evolution in a harmonic-oscillator potential), and represented by

$$S_\theta = \begin{pmatrix} \cos(\theta) & \sin(\theta) \\ -\sin(\theta) & \cos(\theta) \end{pmatrix}, \quad \text{and} \quad U_R(\theta) = \exp\left[-i\theta\, a_n^\dagger a_n\right]. \tag{25.161}$$

Together with the displacement operator, this already exhausts the list of qualitatively different Gaussian operations for individual modes. Indeed, any pure single-mode Gaussian state can be written as $D(\alpha)U_R(\theta)U_S(r)|0\rangle$ with covariance matrix

$$\Gamma = \begin{pmatrix} \cosh 2r - \cos 2\theta\, \sinh 2r & \sin\varphi\, \sinh 2r \\ \sin\varphi\, \sinh 2r & \cosh 2r + \cos 2\theta\, \sinh 2r \end{pmatrix}. \tag{25.162}$$

To obtain the covariance matrix of arbitrary mixed Gaussian single-mode states, all that is required is to add a pre-factor ν with $\nu \geq 1$ that accounts for the mixedness of the state and which, for a single mode, can be interpreted as a non-zero temperature by setting $\nu = \coth\left(\frac{\beta\hbar\omega_n}{2}\right)$.

Bogoljubov transformations: Another way to think about Gaussian unitary transformations is in terms of linear transformations of the mode operators known as *Bogoljubov transformations* (Bogoljubov, 1958; Valatin, 1958). That is, any Gaussian unitary transformation maps a set $\{a_n, a_n^\dagger\}_n$ of bosonic mode operators with commutation relations as in eqn (25.20) to another set $\{\tilde{a}_n, \tilde{a}_n^\dagger\}_n$ of such operators with the same commutation relations. Such a Gaussian unitary transformation can be written in the form

$$\tilde{a}_m = \sum_n \left(\alpha_{mn}^* a_n - \beta_{mn}^* a_n^\dagger\right) + \alpha_m, \tag{25.163}$$

where the parameters α_m correspond to the displacements of each mode, and the complex coefficients[7] α_{mn} and β_{mn} are known as the Bogoljubov coefficients, which

---

[7]Not to be confused with the displacement parameters. Unfortunately, both the notations for displacements and Bogoljubov coefficients is very well established and both use the Greek letter α.

were first introduced in the context of superconductivity (Bogoljubov, 1958; Valatin, 1958) but have been widely used in various areas of quantum optics and quantum field theory (see, e.g., Birrell and Davies, 1982, or Friis, 2013). Symplectic transformations can be written explicitly in terms of the Bogoljubov coefficients in a straightforward fashion (Friis and Fuentes, 2013),

$$
S = \begin{pmatrix} \mathcal{M}_{11} & \mathcal{M}_{12} & \mathcal{M}_{13} & \cdots \\ \mathcal{M}_{21} & \mathcal{M}_{22} & \mathcal{M}_{23} & \cdots \\ \mathcal{M}_{31} & \mathcal{M}_{32} & \mathcal{M}_{33} & \cdots \\ \vdots & \vdots & \vdots & \ddots \end{pmatrix},
\tag{25.164}
$$

where we have decomposed the transformation matrix into the 2×2 sub-blocks $\mathcal{M}_{mn}$ given by

$$
\mathcal{M}_{mn} = \begin{pmatrix} \mathrm{Re}(\alpha_{mn} - \beta_{mn}) & \mathrm{Im}(\alpha_{mn} + \beta_{mn}) \\ -\mathrm{Im}(\alpha_{mn} - \beta_{mn}) & \mathrm{Re}(\alpha_{mn} + \beta_{mn}) \end{pmatrix}.
\tag{25.165}
$$

The symplectic spectrum: A particularly noteworthy property of the symplectic formalism is captured by *Williamson's theorem* (Williamson, 1936):

Theorem 25.2 (Williamson theorem)

For any N-mode Gaussian state with covariance matrix Γ there exists a symplectic transformation S that brings Γ to its **Williamson normal form** Γ_ν,

$$
\Gamma_\nu = S\,\Gamma\,S^\top = \bigoplus_{n=1}^{N} \begin{pmatrix} \nu_n & 0 \\ 0 & \nu_n \end{pmatrix}.
$$

For a proof, we will refer the interested reader to (Williamson, 1936). The set of values $\{\nu_1, \nu_2, \ldots, \nu_N\}$ is called the *symplectic spectrum* of Γ, which provides a useful tool to describe properties of Gaussian states (Serafini *et al.*, 2004). The twice-degenerate *symplectic eigenvalues* ν_n can be calculated as the eigenvalues of $|i\Omega\Gamma|$. To see this, we first note that we can express the diagonal form of the covariance matrix as $\Gamma_\nu = S^{-1}\Gamma S^{\top-1}$. Then, we note that the inverse and transpose of the any symplectic matrix are also symplectic matrices satisfying eqn (25.158) and that $\det \Omega = \det S = 1$. We can then turn to the characteristic polynomial of the matrix $i\Omega\Gamma$ and write

$$\begin{aligned}
\det(i\Omega\Gamma - \lambda\mathbb{1}) &= \det(i\Omega S^{-1}\Gamma_\nu (S^\mathsf{T})^{-1} - \lambda\mathbb{1}) \\
&= \det((S^{-1})^\mathsf{T})\det(i\Omega S^{-1}\Gamma_\nu(S^\mathsf{T})^{-1} - \lambda\mathbb{1})\det(\Omega)\det(S^{-1}) \\
&= \det(i(S^{-1})^\mathsf{T}\Omega S^{-1}\Gamma_\nu(S^\mathsf{T})^{-1}\Omega S^{-1} - \lambda(S^{-1})^\mathsf{T}\Omega S^{-1}) \\
&= \det(i\Omega\Gamma_\nu\Omega - \lambda\Omega) = \det(i\Omega\Gamma_\nu - \lambda)\det(\Omega) \\
&= \det(i\Omega\Gamma_\nu - \lambda\mathbb{1}).
\end{aligned}$$

(25.166)

We thus see that the eigenvalues of $i\Omega\Gamma$ match the eigenvalues of $i\Omega\Gamma_\nu$, but the latter matrix is just

$$i\Omega\Gamma_\nu = \bigoplus_{n=1}^{N}\begin{pmatrix} 0 & i \\ -i & 0 \end{pmatrix}\begin{pmatrix} \nu_n & 0 \\ 0 & \nu_n \end{pmatrix} = \bigoplus_{n=1}^{N}\begin{pmatrix} 0 & i\nu_n \\ -i\nu_n & 0 \end{pmatrix},$$

(25.167)

such that it is easy to see that $\lambda = \pm\nu_n$. In terms of the symplectic eigenvalues, the uncertainty relation $\Gamma + i\Omega \geq 0$ from eqn (25.135) just becomes

$$\nu_n \geq 1.$$

(25.168)

The Williamson normal form of the covariance matrix corresponds to the normal-decomposition of a system of harmonic oscillators, and Γ_ν can be interpreted as a thermal product state of these normal modes, where the ν_n can be associated with the corresponding temperatures via the identification $\nu_n = \coth\left(\frac{\beta\hbar\omega_n}{2}\right)$, and the purity from eqn (25.155) simply becomes

$$P(\rho) = \frac{1}{\sqrt{\det\Gamma}} = \prod_{n=1}^{N}\frac{1}{\nu_n}.$$

(25.169)

Similarly, the von Neumann entropy of any Gaussian state ρ can be expressed as (Holevo *et al.*, 1999; Serafini *et al.*, 2004)

$$S_{\mathrm{vN}}(\rho) = \sum_{n=1}^{N}\mathfrak{h}(\nu_n),$$

(25.170)

where the entropic function $\mathfrak{h}$ is given by

$$\mathfrak{h}(\nu_n) = \tfrac{\nu_n+1}{2}\ln\left(\tfrac{\nu_n+1}{2}\right) - \tfrac{\nu_n-1}{2}\ln\left(\tfrac{\nu_n-1}{2}\right).$$

(25.171)

Here, let us simply note that the hierarchy of Gaussian mixed states induced by the purity $P(\rho)$, or, equivalently, by the mixedness $1 - P(\rho)$, is not the same as that induced by the von Neumann entropy beyond the case of single-mode Gaussian states (see Adesso *et al.*, 2004 for a discussion).

In summary, we can say that the symplectic representation of Gaussian states is in many respects mirroring aspects of the usual Hilbert-space representation, by noting the following correspondences for Gaussian states (ignoring the first moments):

	Hilbert space	**phase space**
composition	$\otimes$	$\oplus$
state description	$\rho = \rho^\dagger$	$\Gamma = \Gamma^\top$
condition	$\rho \geq 0$	$\Gamma + i\Omega \geq 0$
operations	$\rho \mapsto U\rho U^\dagger$	$\Gamma \mapsto S\Gamma S^\top$
condition	$UU^\dagger = \mathbb{1}$	$S\Omega S^\top = \Omega$
diagonalization	$U\rho U^\dagger = \text{diag}\{p_n\}$	$S\Gamma S^\top = \Gamma_\nu$
condition	$p_n \geq 0,\ \sum_n p_n = 1$	$\nu_n \geq 1$

Gaussian states of two modes: Finally, let us also consider some examples for Gaussian states of more than one mode. Since any symplectic transformation can be decomposed as the product of two-mode transformations (Huang and Agarwal, 1994), it will be most instructive to consider the case of two modes here. An important example for a Gaussian state of two modes is the *two-mode squeezed state* with $\overline{\mathbf{X}}_{\text{TMS}} = 0$ and covariance matrix

$$\Gamma_{\text{TMS}}(r) = \begin{pmatrix} \cosh(2r) & 0 & \sinh(2r) & 0 \\ 0 & \cosh(2r) & 0 & -\sinh(2r) \\ \sinh(2r) & 0 & \cosh(2r) & 0 \\ 0 & -\sinh(2r) & 0 & \cosh(2r) \end{pmatrix}, \tag{25.173}$$

where $r \in \mathbb{R}$ is the squeezing parameter. This covariance matrix can be seen to arise from applying the symplectic operation $S_{\text{TMS}}(r)$ to the two-mode vacuum $\Gamma_0 \oplus \Gamma_0$, where

$$S_{\text{TMS}}(r) = \begin{pmatrix} \cosh(r) & 0 & \sinh(r) & 0 \\ 0 & \cosh(r) & 0 & -\sinh(r) \\ \sinh(r) & 0 & \cosh(r) & 0 \\ 0 & -\sinh(r) & 0 & \cosh(r) \end{pmatrix}. \tag{25.174}$$

On the level of the Hilbert space of two modes n and n', this Gaussian state is obtained from the vacuum by applying the unitary operator

$$U_{\text{TMS}}(r) = e^{r(a_n^\dagger a_{n'}^\dagger - a_n a_{n'})}, \tag{25.175}$$

and the corresponding pure state can be written with respect to the Fock basis as

$$|\psi_{\text{TMS}}\rangle = U_{\text{TMS}}(r)\,|0\rangle = \frac{1}{\cosh(r)}\sum_{n=0}^{\infty}\tanh^{n}(r)\,|n,n\rangle. \qquad (25.176)$$

Obtaining this particular result from $U_{\text{TMS}}(r)$ to $|0\rangle$ can be achieved via some elegant operator-ordering theorems, derived in detail in (Barnett and Radmore, 1997, Sec. 3), but takes some dedication. In turn, it is much more straightforward to verify that the state in eqn (25.176) leads to the covariance matrix in eqn (25.173).

A simple observation that we can make right away is that $|\psi_{\text{TMS}}\rangle$ is a pure entangled state, and we can hence easily evaluate its entanglement, e.g., as quantified by the entropy of entanglement from Section 16.1.1. Since the reduced state of either mode is now just

$$\Gamma^{(n)} = \Gamma^{(n')} = \begin{pmatrix} \cosh(2r) & 0 \\ 0 & \cosh(2r) \end{pmatrix}, \qquad (25.177)$$

we can easily evaluate the von Neumann entropy of the reduced states by inserting $\nu_{n} = \cosh(2r)$ in eqn (25.170). We then see that the resulting function is monotonically increasing with increasing magnitude $|r|$ of the squeezing parameter. The transformation $U_{\text{TMS}}(r)$ can hence be regarded as a Gaussian entangling transformation. Moreover, we note that the reduced states have covariance matrices that are proportional to the identity, and are thus thermal states whose temperature rises with increasing $|r|$. In particular, these thermal states are also Gaussian states. In full analogy to the Stinespring dilation from Theorem 21.2, general Gaussian operations on a set of one or more modes can thus be realized via Gaussian unitaries on a larger set of modes.

In addition, we observe that the feature that sets apart the covariance matrix $\Gamma_{\text{TMS}}(r)$ from a purely thermal state are the off-diagonal 2×2 blocks, i.e., the covariances of operators pertaining to pairs of operators from different modes, e.g., $\text{Cov}(\hat{x}_{n}, \hat{x}_{n'})$, which capture correlations between the observables of different subsystems.

We can further observe that the symplectic operation in eqn (25.174) is not the only way to obtain the two-mode squeezed state from the vacuum. Alternatively, one may initially prepare the two modes in an antisymmetrically (i.e., $-r_{k} = r_{k'} = r$) single-mode squeezed state $\Gamma_{-r}^{(n)} \oplus \Gamma_{r}^{(n')}$ by applying local squeezing operations and combine the two modes on an ideal beam splitter, an operation represented by the symplectic matrix

$$S_{\text{BS}}(\tau) = \begin{pmatrix} \sqrt{\tau} & 0 & \sqrt{1-\tau} & 0 \\ 0 & \sqrt{\tau} & 0 & \sqrt{1-\tau} \\ \sqrt{1-\tau} & 0 & -\sqrt{\tau} & 0 \\ 0 & \sqrt{1-\tau} & 0 & -\sqrt{\tau} \end{pmatrix}. \qquad (25.178)$$

For a balanced beam splitter, $\tau = 1/2$, one then has

$$
\begin{aligned}
\Gamma_{\mathrm{TMS}}(r) &= S_{\mathrm{BS}}(\tfrac{1}{2})\, \Gamma_{-r}^{(n)} \oplus \Gamma_r^{(n')}\, S_{\mathrm{BS}}(\tfrac{1}{2})^\top \\
&= S_{\mathrm{BS}}(\tfrac{1}{2})\, S_{-r} \oplus S_r\, \Gamma_0^{(n)} \oplus \Gamma_0^{(n')}\, S_{-r}^\top \oplus S_r^\top\, S_{\mathrm{BS}}(\tfrac{1}{2})^\top
\end{aligned}
\tag{25.179}
$$

This decomposition of the symplectic operation $S_{\mathrm{TMS}}(r)$ is a special case of a more general result, the Bloch–Messiah decomposition, that we will discuss next.

Bloch–Messiah decomposition: It is convenient to characterize different types of symplectic transformations via their associated changes in the average photon number. Following the nomenclature in (Wolf *et al.*, 2003) we distinguish:

    (i) *Passive symplectic transformations* S_{P} are represented by orthogonal, symplectic matrices $S_{\mathrm{P}}^T S_{\mathrm{P}} = \mathbb{1}$ and they form a subgroup of the symplectic group $\mathrm{Sp}(2n, \mathbb{R})$. Practically, passive transformations can be realized, for instance, by passive/linear optical elements, such as (ideal) *beam splitters* or *phase-space rotations*.

    (ii) *Active symplectic transformations* S_{A} are represented by symmetric, symplectic matrices $S_{\mathrm{A}}^T = S_{\mathrm{A}}$. Active transformations, such as single- and two-mode squeezing, can be realized by active/non-linear optical elements and they change the energy and average particle number, as opposed to passive transformations.

The label "passive" simply alludes to the fact that orthogonal symplectic transformations leave the average photon number invariant. To quickly see this, we return to eqn (25.152) to express $\langle N_n \rangle$ via the first and second moments, and obtain

$$
\begin{aligned}
\langle N \rangle &= \sum_{n=1}^{N} \langle N_n \rangle = \sum_{n=1}^{N} \left(\tfrac{1}{4} \big[\mathrm{Tr}(\Gamma^{(n)}) - 2 \big] + \tfrac{1}{2} \| \overline{\mathbb{X}}^{(n)} \|^2 \right) \\
&= \tfrac{1}{4} \big[\mathrm{Tr}(\Gamma) - 2 \big] + \tfrac{1}{2} \| \overline{\mathbb{X}} \|^2,
\end{aligned}
\tag{25.180}
$$

where we have used $\mathrm{Tr}(\Gamma) = \sum_{n=1}^{N} \mathrm{Tr}(\Gamma^{(n)})$ and $\| \overline{\mathbb{X}} \|^2 = \sum_{n=1}^{N} \| \overline{\mathbb{X}}^{(n)} \|^2$. For any optically passive symplectic N-mode transformation S_{P} we then have

$$
\mathrm{Tr}(S_{\mathrm{P}} \Gamma S_{\mathrm{P}}^\top) = \mathrm{Tr}(S_{\mathrm{P}}^\top S_{\mathrm{P}} \Gamma) = \mathrm{Tr}(\Gamma),
\tag{25.181a}
$$

$$
\| S_{\mathrm{P}} \overline{\mathbb{X}} \|^2 = (\overline{\mathbb{X}})^\top S_{\mathrm{P}}^\top S_{\mathrm{P}} \overline{\mathbb{X}} = \| \overline{\mathbb{X}} \|^2.
\tag{25.181b}
$$

Every symplectic transformation can be decomposed into passive and active transformations, in particular we may decompose any symplectic matrix as (Arvind *et al.*, 1995)

$$
S = S_{\mathrm{P}}\, S_{\mathrm{A}}.
\tag{25.182}
$$

We have already encountered examples for optically passive transformations, which include the phase-space rotation S_θ from eqn (25.161) and the beam splitter S_{BS} from eqn (25.178), whereas the single-mode squeezing operation S_r in eqn (25.160) and the two-mode squeezing operation $S_{\mathrm{TMS}}(r)$ from eqn (25.174) are examples for optically active symplectic transformations. The single-mode squeezing transformation gains particular significance via the *Bloch–Messiah decomposition* (Braunstein, 2005).

Theorem 25.3 (Bloch–Messiah decomposition)

Every N-mode symplectic transformation S can be written in the Bloch–Messiah decomposition

$$S = S_{\mathrm{P}} \bigoplus_n S_{r_n}^{(n)} \, S_{\mathrm{P}}',$$

where $S_{r_n}^{(n)}$ is the symplectic representation of the single-mode squeezing of eqn (25.160) in mode n with squeezing parameter r_n, while S_{P} and S_{P}' are passive N-mode operations.

For a proof of Theorem 25.3 we refer the interested reader to Braunstein (2005).

The Bloch–Messiah decomposition is remarkable in that it isolates single-mode squeezing as an irreducible resource when it comes to Gaussian operations, and the usefulness of many Gaussian states for quantum-information processing tasks can be traced back to the presence and strength of single-mode squeezing via Theorem 25.3, including quantum advantages in parameter estimation (see, e.g., Friis *et al.*, 2015*b*), or measurement-based quantum computation with continuous-variable systems (see, e.g., Menicucci *et al.*, 2006; Zhang and Braunstein, 2006). On a fundamental level, the strength of the single-mode squeezing determines how entangled the individual modes can become when passive multi-mode operations are applied.

Entanglement of Gaussian states: Finally, let us briefly turn to the quantification of entanglement of Gaussian states. As mentioned, the entanglement between two modes in a pure Gaussian state is completely characterized by the two-mode squeezing parameter r. However, we wish to find a quantification that also relates to our previous treatment of non-Gaussian states in Chapters 15 and 16. As we will see, two-mode Gaussian states provide a qualitative analogy to the case of two qubits when it comes to the ability to detect and quantify entanglement. First, it was shown in Simon (2000) that the Peres–Horodecki criterion (Theorem 15.3) provides a necessary and sufficient condition for entanglement of two-mode Gaussian states. Here, the partial transposition is implemented on the phase space by a mirror operation—a sign flip—of the $\hat{p}$-quadrature of one of the modes. The "partially transposed" covariance matrix $\breve{\Gamma}$ of two modes n and n' is then simply $\breve{\Gamma} = \breve{T}_{n'} \Gamma \breve{T}_{n'}$, where $\breve{T}_{k'} = \mathbb{1} \oplus \mathrm{diag}\{1, -1\}$. In complete analogy to the usual partial transposition, the symplectic eigenvalues (see Theorem 25.2) of $\breve{\Gamma}$ do not necessarily correspond to a physical state any more.

The smallest eigenvalue $\breve{\nu}_- \geq 0$ of $|i\Omega\breve{\Gamma}|$, where $\mathrm{spectr}(i\Omega\breve{\Gamma}) = \{\pm\breve{\nu}_-, \pm\breve{\nu}_+\}$ with $\breve{\nu}_+ \geq \breve{\nu}_-$, can be smaller than 1 (Simon, 2000):

Theorem 25.4 (PPT criterion for Gaussian states)

Any two-mode Gaussian state represented by the covariance matrix Γ is entangled if and only if the smallest eigenvalue of $|i\Omega\breve{\Gamma}|$ is smaller than 1,

$$0 \leq \breve{\nu}_- < 1.$$

A proof of Theorem 25.4 can be found in (Simon, 2000). The smallest symplectic eigenvalue $\breve{\nu}_-$ (we will omit the suffix "of the partial transpose" from now on and rely on the distinction made by the "$\smile$" symbol) can then be used to construct the usual negativity measures.

Both the logarithmic negativity $E_{\mathcal{N}}$, see eqn (16.88), and the negativity $\mathcal{N}$ (see Definition 16.5) are monotonously decreasing functions of $\breve{\nu}_-$ (Adesso *et al.*, 2004) such that we can write the simple expressions

$$E_{\mathcal{N}}(\Gamma) = \max\{0, -\log_2(\breve{\nu}_-)\}, \tag{25.183a}$$

$$\mathcal{N}(\Gamma) = \max\{0, (1 - \breve{\nu}_-)/2\breve{\nu}_-\}. \tag{25.183b}$$

For symmetric two-mode states, i.e., for which the local reduced-state covariance matrices $\Gamma^{(n)}$ and $\Gamma^{(n')}$ for the two modes n and n', respectively, have the same mixedness such that $\det(\Gamma^{(n)}) = \Gamma^{(n')}$, it is even possible to compute the entanglement of formation E_{F} (see Section 16.2.2). The involved minimization procedure reveals that the corresponding state decomposition is realized within the set of two-mode Gaussian states (Giedke *et al.*, 2003) and the entanglement of formation can be expressed as

$$E_{\mathrm{F}} = \begin{cases} \mathfrak{h}(\breve{\nu}_-) & \text{if } 0 \leq \breve{\nu}_- < 1 \\ 0 & \text{if } \breve{\nu}_- \geq 1 \end{cases}, \tag{25.184}$$

where $\mathfrak{h}$ is the entropic function from eqn (25.171). Moreover, for symmetric states the smallest symplectic eigenvalue $\breve{\nu}_-$ provides a unique characterization of the entanglement, i.e., all (known) entanglement measures are monotonously decreasing functions of $\breve{\nu}_-$ and they provide the same ordering of entangled states (Adesso and Illuminati, 2005). Unfortunately, this is no longer true for non-symmetric two-mode Gaussian states—the answer to the question "Is one state more entangled than another?" generally depends on the chosen measure of entanglement (Adesso and Illuminati, 2005).

As an example, let us again consider the two-mode squeezed state of eqn (25.173). The smallest symplectic eigenvalue for this state is directly related to the two-mode squeezing parameter r via the relation $\breve{\nu}_-(r) = e^{-2|r|}$. It can be easily seen that maximal entanglement can only be achieved in the limit $r \to \infty$.

Generally, there is no known way of explicitly calculating the entanglement of formation when the Gaussian two-mode state is not symmetric, or when more than two modes are considered (or when the state is non-Gaussian), and it would thus be highly desirable to be able to provide useful (lower) bounds on the entanglement of formation for Gaussian states, and some bounds are already known (Wolf *et al.*, 2004; Tserkis and Ralph, 2017; Tserkis *et al.*, 2019).

As a concluding remark to this chapter, we can note that the study of Gaussian states and operations, and of non-Gaussian states beyond, has opened a vast field of research that we have only been able to cover here in the broadest of strokes, but we refer the interested reader to the large body of available literature on this topic: For more information on Gaussian operations and states in quantum optics, see, e.g., Olivares (2012), whereas reviews with a focus on the theory and processing of quantum information with Gaussian states can, for instance, be found in Ferraro *et al.* (2005), Braunstein and van Loock (2005), Adesso (2007), Weedbrook *et al.* (2012), Adesso *et al.* (2014). For more information on non-Gaussian states, we refer to Walschaers (2021) and literature cited therein.

26
Particle Physics—Bell Inequalities

In discussing entanglement as well as the non-local features of quantum mechanics discovered by John Bell, we have focused on photon experiments since most of the (early) experiments were carried out with photons. However, it is also of fundamental interest to investigate these quantum features in the field of particle physics. In contrast to photons the particles typically considered in this context are (very) massive, their characteristics depend on several quantum numbers, and they typically decay rapidly. These features restrict our options to employ previously discussed entanglement tests and Bell inequalities. This presents both a challenge and an opportunity to formulate novel Bell inequalities, which provide new tests of quantum mechanics.

In this chapter and in Chapter 27 we are going to discuss these foundational issues of quantum mechanics within particle physics. We begin by introducing the particles we consider: the *strange* K mesons and the *beauty* B mesons produced in the huge accelerators employed in experimental particle physics. Then we show how the particles are entangled, namely in their quantum numbers *strangeness* and *beauty*. For the entanglement of other particles, such as for instance neutrinos, we refer to the literature (e.g., Blasone *et al.*, 2009).

The particles we consider here are not stable but decay, decoherence, and dissipation occur, implying entanglement loss for the whole system. We will present a quantitative description for these phenomena and discuss several typical examples. There is already a lot of activity in this field (to mention a few contributions, see Eberhard, 1993*b*; Di Domenico, 1995; Bertlmann, 2006; Bertlmann and Hiesmayr, 2007; Amelino-Camelia *et al.*, 2010; Khan *et al.*, 2020; Qian *et al.* 2020), and we refer to references in these papers for further literature.

26.1   K Mesons

Let us start with the description of K mesons, which carry strangeness. We know that for each particle there is a corresponding anti-particle; both have equal masses, spins, and lifetimes. But all types of charges—electric, leptonic, baryonic, flavour, colour—have opposite sign. Kaons are pseudoscalar particles ($J^P = 0^-$), charged $K^+(\bar{s}u)$, $K^-(\bar{u}s)$ with mass $m = 493$ MeV, or neutral $K^0(\bar{s}d)$, $\bar{K}^0(\bar{d}s)$ with $m = 497$ MeV. They are bound states composed of the quarks u, d, and s and anti-quarks $\bar{u}$, $\bar{d}$, and $\bar{s}$. Kaons are subject to strong interactions that conserve strangeness S and leave charge conjugation-parity CP invariant, but they are also influenced by the weak interaction that violates both of these symmetries. Indeed, neutral K mesons are amazing quantum objects. They seem to be specially selected by nature to showcase typical quantum features. To quote *Lev Borisovich Okun* (as reported in Ho-Kim and Pham 1998, p. 377):

> *Even if they* [the neutral K mesons] *did not exist* (...) *we would have invented them in order to illustrate the fundamental principles of quantum physics.*

Four features characterize their "strange" behaviour as attributed by the work of *Abraham Pais* (Gell-Mann and Pais, 1955; Pais and Piccioni, 1955):

(i) *strangeness,*

(ii) *CP violation,*

(iii) *strangeness oscillation,*

(iv) *regeneration.*

Next, we will examine the properties of neutral kaons in more detail to provide a basis for the discussion that is to follow.

26.1.1 Strangeness and *CP*

Strangeness as quantum number: K mesons are characterized by their *strangeness* quantum number (and associated operator) S, which also refers to their quark content

$$S\,|\mathrm{K}^0\rangle \;=\; +|\mathrm{K}^0\rangle\,, \tag{26.1a}$$

$$S\,|\bar{\mathrm{K}}^0\rangle \;=\; -|\bar{\mathrm{K}}^0\rangle\,. \tag{26.1b}$$

As the K mesons are pseudoscalars they have negative parity P and charge conjugation C transforms particles K^0 and anti-particles $\bar{\mathrm{K}}^0$ into each other so that the combined transformation CP (with our choice of phases) gives

$$CP\,|\mathrm{K}^0\rangle \;=\; -|\bar{\mathrm{K}}^0\rangle\,, \tag{26.2a}$$

$$CP\,|\bar{\mathrm{K}}^0\rangle \;=\; -|\mathrm{K}^0\rangle\,. \tag{26.2b}$$

It then follows that the orthogonal linear combinations

$$|\mathrm{K}_1^0\rangle \;=\; \frac{1}{\sqrt{2}}\big\{|\mathrm{K}^0\rangle - |\bar{\mathrm{K}}^0\rangle\big\}\,, \tag{26.3a}$$

$$|\mathrm{K}_2^0\rangle \;=\; \frac{1}{\sqrt{2}}\big\{|\mathrm{K}^0\rangle + |\bar{\mathrm{K}}^0\rangle\big\}\,, \tag{26.3b}$$

are eigenstates of CP, a quantum number conserved in strong interactions

$$CP\,|\mathrm{K}_1^0\rangle \;=\; +|\mathrm{K}_1^0\rangle\,, \tag{26.4a}$$

$$CP\,|\mathrm{K}_2^0\rangle \;=\; -|\mathrm{K}_2^0\rangle\,. \tag{26.4b}$$

CP **violation:** Due to weak interactions, which do not conserve *strangeness* and in addition break *CP* invariance, the kaons are subject to decay processes. The physical states, which differ only slightly in mass, $\Delta m = m_\mathrm{L} - m_\mathrm{S} = 3.49 \times 10^{-6}$ eV, but differ immensely in their lifetimes and decay modes, are the short- and long-lived states

$$|\mathrm{K_S}\rangle \;=\; \frac{1}{N}\{p|\mathrm{K}^0\rangle - q|\bar{\mathrm{K}}^0\rangle\}\,, \tag{26.5a}$$

$$|\mathrm{K_L}\rangle \;=\; \frac{1}{N}\{p|\mathrm{K}^0\rangle + q|\bar{\mathrm{K}}^0\rangle\}\,. \tag{26.5b}$$

The weights $p = 1 + \varepsilon$, $q = 1 - \varepsilon$, with $N^2 = |p|^2 + |q|^2$ contain the complex *CP-violation parameter* ε with $|\varepsilon| \approx 10^{-3}$. *CPT invariance* is assumed; thus the short- and long-lived states contain the same *CP*-violation parameter $\varepsilon_S = \varepsilon_L = \varepsilon$. Then the *CPT* theorem (Schwinger, 1951; Schwinger, 1953a; Bell, 1955; Lüders, 1954; Pauli, 1955) implies that time reversal T is violated too.

The short-lived K meson decays dominantly into $\mathrm{K_S} \to 2\pi$ with a width or lifetime $\Gamma_\mathrm{S}^{-1} \sim \tau_\mathrm{S} = 0.89 \times 10^{-10}$ sec and the long-lived K meson decays dominantly into $\mathrm{K_L} \to 3\pi$ with $\Gamma_\mathrm{L}^{-1} \sim \tau_\mathrm{L} = 5.17 \times 10^{-8}$ sec. However, due to *CP* violation we observe a small amount of long-lived kaons decaying in the channel $\mathrm{K_L} \to 2\pi$ (Christenson *et al.*, 1964)[1]. To appreciate the importance of *CP* violation let us be reminded here that there is an enormous disproportion of matter to anti-matter in our universe, and so far *CP* violation is the only known non-trivial difference between matter and anti-matter that could serve as an explanation.

Strangeness oscillation: The kaon states $\mathrm{K_S}, \mathrm{K_L}$ are eigenstates of the non-Hermitian *"effective mass"* Hamiltonian

$$H \;=\; M - \frac{i}{2}\,\Gamma\,, \tag{26.6}$$

satisfying

$$H\,|\mathrm{K_{S,L}}\rangle \;=\; \lambda_\mathrm{S,L}\,|\mathrm{K_{S,L}}\rangle\,, \tag{26.7}$$

with

$$\lambda_\mathrm{S,L} \;=\; m_\mathrm{S,L} - \frac{i}{2}\,\Gamma_\mathrm{S,L}\,. \tag{26.8}$$

Both mesons K^0 and $\bar{\mathrm{K}}^0$ have transitions to common states (due to *CP* violation), therefore they mix, which means they *oscillate* between K^0 and $\bar{\mathrm{K}}^0$ before decaying. Since the decaying states evolve—according to the Wigner–Weisskopf approximation—exponentially in time

---

[1] In 1980 *James Watson Cronin* and his co-researcher *Val Logsdon Fitch* were awarded the 1980 Nobel Prize in Physics for this 1964 experiment that proved *CP* violation.

$$|K_{S,L}(t)\rangle = e^{-i\lambda_{S,L}t}|K_{S,L}\rangle, \tag{26.9}$$

the subsequent time evolution for K^0 and $\bar{K}^0$ is given by

$$|K^0(t)\rangle = g_+(t)|K^0\rangle + \frac{q}{p}g_-(t)|\bar{K}^0\rangle, \tag{26.10a}$$

$$|\bar{K}^0(t)\rangle = \frac{p}{q}g_-(t)|K^0\rangle + g_+(t)|\bar{K}^0\rangle, \tag{26.10b}$$

with

$$g_\pm(t) = \frac{1}{2}\left[\pm e^{-i\lambda_S t} + e^{-i\lambda_L t}\right]. \tag{26.11}$$

Let us suppose that a K^0 beam is produced at $t = 0$, e.g., by strong interactions, then the probabilities of finding K^0 or $\bar{K}^0$ in the beam are calculated to be

$$\left|\langle K^0|K^0(t)\rangle\right|^2 = \frac{1}{4}\left\{e^{-\Gamma_S t} + e^{-\Gamma_L t} + 2e^{-\Gamma t}\cos(\Delta m\, t)\right\}, \tag{26.12a}$$

$$\left|\langle \bar{K}^0|K^0(t)\rangle\right|^2 = \frac{1}{4}\frac{|q|^2}{|p|^2}\left\{e^{-\Gamma_S t} + e^{-\Gamma_L t} - 2e^{-\Gamma t}\cos(\Delta m\, t)\right\}, \tag{26.12b}$$

with $\Delta m = m_L - m_S$ and $\Gamma = \frac{1}{2}(\Gamma_L + \Gamma_S)$.

The K^0 beam oscillates with frequency $\Delta m/2\pi$, where $\Delta m\, \tau_S = 0.47$. The oscillation is clearly visible at times of the order of a few τ_S, before all K_S have died out leaving only the K_L in the beam. So in a beam, which contains only K^0 mesons at the time $t = 0$, the $\bar{K}^0$ will occur far from the production source through its presence in the K_L meson (with equal probability as the K^0 meson). A similar feature occurs when starting with a $\bar{K}^0$ beam.

Regeneration of K_S: In a K-meson beam, after a few centimetres, only the long-lived kaon state survives. But if one places a thin slab of matter into the remaining K_L beam then the short-lived state K_S is regenerated. The reason is that the K^0 and $\bar{K}^0$ components of the beam are scattered/absorbed differently in matter. Only $\bar{K}^0$ is absorbed in matter through the process $\bar{K}^0 + p \rightarrow \Lambda + \pi^+$ or $\bar{K}^0 + n \rightarrow \Lambda + \pi^0$, where p(uud) is the proton, n(udd) the neutron and Λ(uds) the lambda particle, $\pi^+(\bar{d}u)$ and $\pi^0(\bar{u}u - \bar{d}d)$ denote the pions. The analogous processes for K^0 are forbidden by strangeness conservation.

Denoting the scattering amplitudes of K^0 and $\bar{K}^0$ by f and $\bar{f}$, respectively, when either are scattered from the protons and neutrons in matter, then we have $f \neq \bar{f}$. The initial wave function entering the slab of matter is that of K_L (we neglect CP violation here for simplicity)

$$|\psi_i\rangle = |K_L\rangle_{\varepsilon=0} = |K_2^0\rangle = \frac{1}{\sqrt{2}}\left\{|K^0\rangle + |\bar{K}^0\rangle\right\}. \tag{26.13}$$

The interaction with matter causes the wave function to change so that the final wave function behind the slab is

$$|\psi_f\rangle = \frac{1}{\sqrt{2}}\{f|K^0\rangle + \bar{f}|\bar{K}^0\rangle\} = \frac{1}{\sqrt{2}}\{(f+\bar{f})|K_L\rangle + (f-\bar{f})|K_S\rangle\}$$

$$= \frac{f+\bar{f}}{\sqrt{2}}\{|K_L\rangle + r|K_S\rangle\}, \qquad (26.14)$$

where $r = (f - \bar{f})/(f + \bar{f}) \neq 0$, is the regeneration parameter, quantifying the amount of K_S regeneration. This phenomenon, predicted by *Abraham Pais* and *Oreste Piccioni* (1955), has been experimentally observed by Piccioni and his group (Muller *et al.*, 1960).

26.2 Analogies and Quasi-Spin

A good *optical analogy* to the phenomenon of strangeness oscillation is the following situation. Let us take a crystal that absorbs the different polarization states of a photon differently, say horizontally (H) polarized light strongly, but vertically (V) polarized light only weakly. Then if we shine right-circularly (R) polarized light through the crystal, after some distance there is a large probability for finding left-circularly (L) polarized light.

In comparison with spin-$\frac{1}{2}$ particles, or with photons having the polarization directions H and V, it is especially useful to work with the *quasi-spin* picture for kaons (see, e.g., Bertlmann and Hiesmayr, 2001), originally introduced by *Tsung Dao Lee* and *Chien-Shiung Wu* (1966) and *Harry J. Lipkin* (1968). The two states $|K^0\rangle$ and $|\bar{K}^0\rangle$ are regarded as the quasi-spin states up $|\uparrow\rangle$ and down $|\downarrow\rangle$ and the operators acting in this quasi-spin space are expressed in terms of the Pauli matrices. So the strangeness operator S can be identified with the Pauli matrix $\sigma_z = \sigma_3$, the CP operator with $-\sigma_x = -\sigma_1$ and CP violation is proportional to $\sigma_y = \sigma_2$. Then the Hamiltonian (26.6) can be rewritten as

$$H = a \cdot \mathbb{1} + \vec{b} \cdot \vec{\sigma}, \qquad (26.15)$$

with $b_1 = b\cos(\alpha)$, $b_2 = b\sin(\alpha)$, while $b_3 = 0$ (due to CPT invariance), and

$$a = \frac{1}{2}(\lambda_L + \lambda_S), \quad b = \frac{1}{2}(\lambda_L - \lambda_S), \qquad (26.16)$$

where the phase α is related to the CP parameter ε by

$$e^{i\alpha} = \frac{1-\varepsilon}{1+\varepsilon}. \qquad (26.17)$$

To summarize, we thus have the following analogy between K mesons, spin-$\frac{1}{2}$ particles, and photon polarization:

K meson	spin-$\frac{1}{2}$	photon
$|K^0\rangle$	$|\uparrow\rangle_z$	$|V\rangle$
$|\bar{K}^0\rangle$	$|\downarrow\rangle_z$	$|H\rangle$
$|K_S\rangle$	$|\rightarrow\rangle_y$	$|L\rangle = \frac{1}{\sqrt{2}}(|V\rangle - i|H\rangle)$
$|K_L\rangle$	$|\leftarrow\rangle_y$	$|R\rangle = \frac{1}{\sqrt{2}}(|V\rangle + i|H\rangle)$.

26.3   Entanglement of Strangeness

Now we are interested in entangled states of $K^0\bar{K}^0$ pairs, in analogy to the entangled spin-up and spin-down pairs, or photon pairs. Such states are produced by e^+e^- machines through the reaction $e^+e^- \to \Phi \to K^0\bar{K}^0$, in particular at DAΦNE in Frascati (Italy), or they are produced in $p\bar{p}$ collisions, e.g., at LEAR, CERN. There, a $K^0\bar{K}^0$ pair is created in a $J^{PC} = 1^{--}$ quantum state, which is thus antisymmetric under C and P. At time $t = 0$ we have the entangled state

$$|\psi^-(t=0)\rangle \;=\; \frac{1}{\sqrt{2}}\left\{|K^0\rangle_l \otimes |\bar{K}^0\rangle_r - |\bar{K}^0\rangle_l \otimes |K^0\rangle_r\right\}, \qquad (26.18)$$

of the particles travelling to the left (l) and right (r), respectively. In the basis consisting of K_S and K_L this state can be expressed as

$$|\psi^-(t=0)\rangle \;=\; \frac{N_{\mathrm{SL}}}{\sqrt{2}}\left\{|K_S\rangle_l \otimes |K_L\rangle_r - |K_L\rangle_l \otimes |K_S\rangle_r\right\}, \qquad (26.19)$$

with $N_{\mathrm{SL}} = \frac{N^2}{2pq}$. The neutral kaons fly apart and can be detected on the left-hand (l) and right-hand (r) side of the source, see Fig. 26.1. Of course, during their propagation there is an oscillation of the K^0 and $\bar{K}^0$ components, while the K_S and K_L components decay. This is an important difference to the case of spin-$\frac{1}{2}$ particles or photons which are quite stable.

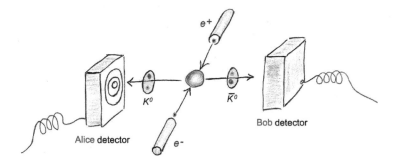

Fig. 26.1 Entangled kaons. Sketch of an experiment, where an entangled $K^0\bar{K}^0$ pair is created in an e^+e^- machine.

26.3.1   Time Evolution and Unitarity

To continue we first have to discuss the time evolution of kaon states more closely (Bell and Steinberger, 1965). At any time t the state $|K^0(t)\rangle$ decays to a specific final state $|f\rangle$ with a probability proportional to the modulus squared of the transition-matrix element. Because of the unitarity of the time evolution the norm of the total state must be conserved. This means that the decrease in the norm of the state $|K^0(t)\rangle$ must be compensated for by the increase in the norm of the final states.

Starting at $t = 0$ with a K^0 meson, the state we have to consider for the time evolution is given by

$$|K^0\rangle \longrightarrow a(t)|K^0\rangle + b(t)|\bar{K}^0\rangle + \sum_f c_f(t)|f\rangle \,, \qquad (26.20)$$

where $a(t) = g_+(t)$ and $b(t) = \frac{q}{p} g_-(t)$. The functions $g_\pm(t)$ are defined in eqn (26.11). Denoting the amplitudes of the decays of the K^0 and $\bar{K}^0$ components to a specific final state f by

$$\mathcal{A}(K^0 \longrightarrow f) \equiv \mathcal{A}_f \qquad \text{and} \qquad \mathcal{A}(\bar{K}^0 \longrightarrow f) \equiv \bar{\mathcal{A}}_f \,, \qquad (26.21)$$

respectively, then the time variation for the total amplitude squared is

$$\frac{d}{dt}|c_f(t)|^2 = |a(t)\mathcal{A}_f + b(t)\bar{\mathcal{A}}_f|^2 \,, \qquad (26.22)$$

and for the probability of the decay $K^0 \to f$ at a certain time τ we finally get

$$P_{K^0 \to f}(\tau) = \int_0^\tau \frac{d}{dt}|c_f(t)|^2 \, dt \,. \qquad (26.23)$$

Since the state $|K^0(t)\rangle$ evolves according to a Schrödinger equation with "effective mass" Hamiltonian (26.6) the decay amplitudes are related to the matrix Γ by

$$\Gamma_{11} = \sum_f |\mathcal{A}_f|^2, \quad \Gamma_{22} = \sum_f |\bar{\mathcal{A}}_f|^2, \quad \Gamma_{12} = \sum_f \mathcal{A}_f^* \bar{\mathcal{A}}_f \,. \qquad (26.24)$$

These are the *Bell–Steinberger unitarity relations* (Bell and Steinberger, 1965). They are a consequence of probability conservation and play an important role in kaon physics.

For our purpose the following formalism generalized to arbitrary quasi-spin states is quite convenient (Ghirardi *et al.*, 1991; Bertlmann and Hiesmayr, 2001). We describe a complete evolution of mass eigenstates by a unitary operator $U(t,0)$ whose effect can be written as

$$U(t,0)\,|K_{S,L}\rangle = e^{-i\lambda_{S,L}t}\,|K_{S,L}\rangle + |\Omega_{S,L}(t)\rangle \,, \qquad (26.25)$$

where $|\Omega_{S,L}(t)\rangle$ denotes the state of all decay products. For the transition amplitudes of the decay-product states we then have

$$\langle \Omega_S(t)|\Omega_S(t)\rangle = 1 - e^{-\Gamma_S t} \,, \qquad (26.26)$$

$$\langle \Omega_L(t)|\Omega_L(t)\rangle = 1 - e^{-\Gamma_L t} \,, \qquad (26.27)$$

$$\langle \Omega_L(t)|\Omega_S(t)\rangle = \langle K_L|K_S\rangle (1 - e^{i\,\Delta m\,t}\, e^{-\Gamma t}) \,, \qquad (26.28)$$

$$\langle K_{S,L}|\Omega_S(t)\rangle = \langle K_{S,L}|\Omega_L(t)\rangle = 0 \,. \qquad (26.29)$$

The mass eigenstates of eqns (26.5a) and (26.5b) are normalized but due to the CP violation they are not orthogonal,

$$\langle K_L | K_S \rangle = \frac{2\,\mathrm{Re}\{\varepsilon\}}{1 + |\varepsilon|^2} =: \delta \,. \tag{26.30}$$

Now we consider entangled states of kaon pairs, and we start at time $t = 0$ from the entangled state given in the basis consisting of K_S and K_L, eqn (26.19),

$$|\psi(t = 0)\rangle = \frac{N^2}{2\sqrt{2}pq}\big\{ |K_S\rangle_l \otimes |K_L\rangle_r - |K_L\rangle_l \otimes |K_S\rangle_r \big\} \,. \tag{26.31}$$

Then we get the state at time t from (26.31) by applying the unitary operator

$$U(t, 0) = U_l(t, 0) \otimes U_r(t, 0) \,, \tag{26.32}$$

where the operators $U_l(t, 0)$ and $U_r(t, 0)$ act on the subspaces of the left-wards travelling and of the right-wards travelling mesons, respectively, according to the time evolution (26.25).

What we finally need are the quantum-mechanical probabilities for detecting—or not detecting—a specific quasi-spin state on the left-hand side $|k_n\rangle_l$ and on the right-hand side $|k_n\rangle_r$ of the source. For that we need the projection operators $P_{l,r}(k_n)$ onto the left-hand, or right-hand side quasi-spin states $|k_n\rangle_{l,r}$ together with the projection operators $Q_{l,r}(k_n)$ that act onto the orthogonal states

$$P_l(k_n) = |k_n\rangle_l \langle k_n|_l \quad \text{and} \quad P_r(k_n) = |k_n\rangle_r \langle k_n|_r \,, \tag{26.33}$$

$$Q_l(k_n) = \mathbb{1} - P_l(k_n) \quad \text{and} \quad Q_r(k_n) = \mathbb{1} - P_r(k_n) \,. \tag{26.34}$$

So starting from the initial state (26.31) the unitary time evolution (26.32) determines the state at a time t_r

$$|\psi(t_r)\rangle = U(t_r, 0) |\psi(t = 0)\rangle = U_l(t_r, 0) \otimes U_r(t_r, 0) |\psi(t = 0)\rangle \,. \tag{26.35}$$

Measuring now a certain quasi-spin k_m at time t_r on the right-hand side means that we project k_m onto the state

$$|\tilde{\psi}(t_r)\rangle = P_r(k_m) |\psi(t_r)\rangle \,. \tag{26.36}$$

This state, which is now a single-particle state of the left-moving particle, evolves until the time t_l when we measure another quasi-spin k_n on the left-hand side. Then we get

$$|\tilde{\psi}(t_l, t_r)\rangle = P_l(k_n) U_l(t_l, t_r) P_r(k_m) |\psi(t_r)\rangle \,. \tag{26.37}$$

The probability of the joint measurement is given by the squared norm of the state (26.37). It coincides (due to unitarity, composition laws, and commutation properties of operators acting on the left-hand and right-hand side subsystems) with the state

$$|\psi(t_l, t_r)\rangle = P_l(k_n) P_r(k_m) U_l(t_l, 0) U_r(t_r, 0) |\psi(t = 0)\rangle \,, \tag{26.38}$$

which corresponds to a factorization of the time into a characteristic time t_l on the left-hand side and into a characteristic time t_r on the right-hand side.

Then we finally calculate the quantum-mechanical probability $P_{n,m}(Y, t_1; Y, t_r)$ for finding a quasi-spin k_n at t_1 on the left-hand side *and* a quasi-spin k_m at t_r on the right-hand side and the probability $P_{n,m}(N, t_1; N, t_1)$ for finding *no* such kaons by the following norms; and similarly the probability $P_{n,m}(Y, t_1; N, t_r)$ when a k_n at t_1 is detected on the left but *no* k_m at t_r on the right

$$P_{n,m}(Y, t_1; Y, t_r) = \| P_1(k_n) P_r(k_m) U_1(t_1, 0) U_r(t_r, 0) |\psi(t=0)\rangle \|^2 , \quad (26.39a)$$

$$P_{n,m}(N, t_1; N, t_r) = \| Q_1(k_n) Q_r(k_m) U_1(t_1, 0) U_r(t_r, 0) |\psi(t=0)\rangle \|^2 , \quad (26.39b)$$

$$P_{n,m}(Y, t_1; N, t_r) = \| P_1(k_n) Q_r(k_m) U_1(t_1, 0) U_r(t_r, 0) |\psi(t=0)\rangle \|^2 . \quad (26.39c)$$

26.4 Analogies and Differences for K Mesons

Before turning to Bell inequalities we want to point out the following analogies and differences with respect to the stable spin-$\frac{1}{2}$ particles or photons (Bertlmann and Hiesmayr, 2001).

Analogies: If we measure a K^0 meson at time t on the left-hand side (denoted by Y, yes), we will at the same time t find *no* K^0 on the right-hand side (denoted by N, no) with certainty. This is an EPR–Bell correlation analogously to the spin-$\frac{1}{2}$ or photon case (Bramon and Nowakowski, 1999; Bertlmann and Hiesmayr, 2001; Gisin and Go, 2001).

The analogy would be perfect if the kaons were stable ($\Gamma_S = \Gamma_L = 0$). Then the quantum probabilities yield the result

$$P(Y, t_1; Y, t_r) = P(N, t_1; N, t_r) = \frac{1}{4}\left\{1 - \cos\left[\Delta m(t_1 - t_r)\right]\right\} , \quad (26.40a)$$

$$P(Y, t_1; N, t_r) = P(N, t_1; Y, t_r) = \frac{1}{4}\left\{1 + \cos\left[\Delta m(t_1 - t_r)\right]\right\} . \quad (26.40b)$$

It coincides with the probability result of simultaneously finding two entangled spin-$\frac{1}{2}$ particles in spin directions $\uparrow \uparrow$ or $\uparrow \downarrow$ along two chosen directions $\vec{n}$ and $\vec{m}$

$$P(\vec{n}, \uparrow; \vec{m}, \uparrow) = P(\vec{n}, \downarrow; \vec{m}, \downarrow) = \frac{1}{4}\left\{1 - \cos(\theta)\right\} , \quad (26.41a)$$

$$P(\vec{n}, \uparrow; \vec{m}, \downarrow) = P(\vec{n}, \downarrow; \vec{m}, \uparrow) = \frac{1}{4}\left\{1 + \cos(\theta)\right\} . \quad (26.41b)$$

Analogy—times and angles: There is perfect analogy between times and angles.

- *The time differences $\Delta m(t_1 - t_r)$ in the kaon case plays the role of the angle differences θ in the spin-$\frac{1}{2}$ or photon case.*

    K propagation   **photon propagation**

left 1^{--} right   Alice   singlet   Bob

- $K^0 \bar{K}^0$ oscillation • K_S, K_L decay   • stable

- for $t_l = t_r$: EPR-like correlation

$$K^0 \longleftarrow\!\!\!\bigcirc\!\longrightarrow \bar{K}^0 \qquad \uparrow V \longleftarrow\!\!\!\bigcirc\!\longrightarrow \downarrow H$$

$$K^0 \qquad \text{NO} \qquad K^0 \qquad \uparrow V \qquad \text{NO} \qquad \uparrow V$$

- for $t_l \neq t_r$: EPR–Bell correlation

$$K^0 \longleftarrow\!\!\!\bigcirc\!\longrightarrow K^0 \bar{K}^0 \qquad \uparrow V \longleftarrow\!\!\!\bigcirc\!\longrightarrow \nearrow\!\!\swarrow V H$$

Differences: There are important physical differences.

(i) While in the spin-$\frac{1}{2}$ or photon case one can test whether a system is in an arbitrary spin state $\alpha | \uparrow\rangle + \beta | \downarrow\rangle$, but one cannot perform such a test for an arbitrary superposition $\alpha | K^0\rangle + \beta | \bar{K}^0\rangle$.

(ii) For entangled spin-$\frac{1}{2}$ particles or photons it is clearly sufficient to consider the direct product space $H_{\text{spin}}^l \otimes H_{\text{spin}}^r$ to account for all spin or polarization properties of the entangled system, however, this is not so for kaons. The unitary time evolution of a kaon state also involves the decay-product states (see Section 26.3.1), therefore one has to include the Hilbert space of the decay products $H_\Omega^l \otimes H_\Omega^r$ which is orthogonal to the space $H_{\text{kaon}}^l \otimes H_{\text{kaon}}^r$ of the surviving kaons.

Consequently, the appropriate dichotomic question for the system is: "Are you a K^0 or not?" It is clearly different from the question "Are you a K^0 or a $\bar{K}^0$?" since all decay products—additional characteristics of the quantum system—are ignored by the latter question.

26.5   Bell Inequalities for K Mesons

Now we are prepared to study Bell inequalities for our entangled K mesons in the quasi-spin picture. We start from the state $|\psi(t = 0)\rangle$ of entangled kaons, given by eqn (26.18) or eqn (26.19), and consider its time evolution $U(t, 0)|\psi(0)\rangle$.

Measuring a $\bar{K}^0$ (the anti-particle that is actually measured via strong interactions in matter) on the left-hand side we can predict with certainty to find at the same time *no* $\bar{K}^0$ at the right-hand side. In any local realistic theory this property of finding *no* $\bar{K}^0$ must be present on the right-hand side irrespective of having performed the measurement or not. In order to discriminate between quantum mechanics and a local realistic theory we set up a Bell inequality for the kaon system, where now the different times play the role of the different angles in the spin-$\frac{1}{2}$ or photon case. But, in addition, we may also use the freedom of choosing a particular quasi-spin state of the kaon, e.g., the strangeness eigenstate, the mass eigenstate, or the CP eigenstate. Thus an expectation value for the combined measurement $E(k_n, t_a; k_m, t_b)$ depends on a certain quasi-spin k_n measured on the left-hand side at a time t_a *and* on a (possibly different) quasi-spin k_m on the right-hand side at time t_b. Using similar arguments as those presented in Section 13.2 one can then derive the following *Bell–CHSH inequality* (Bertlmann and Hiesmayr, 2001)

$$|E(k_n, t_a; k_m, t_b) - E(k_n, t_a; k_{m'}, t_{b'})| \tag{26.42}$$

$$+ |E(k_{n'}, t_{a'}; k_{m'}, t_{b'}) + E(k_{n'}, t_{a'}; k_m, t_b)| \leq 2,$$

which expresses both the freedom of choice in quasi-spin *and* in time. If we identify $E(k_n, t_a; k_m, t_b) \equiv E(n, m)$ we are back at the inequality for the spin-$\frac{1}{2}$ case (Proposition 13.2).

The expectation value for a series of independent and identically distributed measurements can be expressed in terms of the probabilities which are defined by eqns (26.39a), (26.39b), and (26.39c). Then the expectation value for a joint measurement is given by the following probabilities

$$E(k_n, t_a; k_m, t_b) = P_{n,m}(Y, t_a; Y, t_b) + P_{n,m}(N, t_a; N, t_b)$$

$$- P_{n,m}(Y, t_a; N, t_b) - P_{n,m}(N, t_a; Y, t_b). \tag{26.43}$$

Since the sum of the probabilities for (Y, Y), (N, N), (Y, N), and (N, Y) just add up to one, we find

$$E(k_n, t_a; k_m, t_b) = -1 + 2\{P_{n,m}(Y, t_a; Y, t_b) + P_{n,m}(N, t_a; N, t_b)\}. \tag{26.44}$$

Note that the relation (26.43) between the expectation value and the probabilities is satisfied for both quantum mechanics and local realistic theories.

26.5.1 Bell Inequality for Time Variation

In Bell inequalities for meson systems we have two options:

> (i) fixing the quasi-spin—freedom in time, or

> (ii) fixing the time—freedom in quasi-spin.

Let us elaborate on the first option (i). We choose a definite quasi-spin, say strangeness $S = +1$, which means $k_n = k_m = k_{n'} = k_{m'} = \mathrm{K}^0$, and we neglect CP violation here (which does not play a role at the level of accuracy we are interested in here), then we obtain the following formula for the expectation value

$$E(t_l; t_r) = -\cos(\Delta m \Delta t) \cdot e^{-\Gamma(t_l + t_r)} + \frac{1}{2}(1 - e^{-\Gamma_L t_l})(1 - e^{-\Gamma_S t_r})$$

$$+ \frac{1}{2}(1 - e^{-\Gamma_S t_l})(1 - e^{-\Gamma_L t_r}). \tag{26.45}$$

Since expectation value (26.45) corresponds to a unitary time evolution, it contains terms originating from the decay-product states $|\Omega_{L,S}(t)\rangle$, in addition to the pure meson-state contribution.

However, in the kaon system we can neglect the width of the long-lived K meson as compared to the short-lived one, $\Gamma_L \ll \Gamma_S$, so that, in good approximation, we have

$$E^{\text{approx}}(t_l; t_r) = -\cos(\Delta m \Delta t) \cdot e^{-\Gamma(t_l + t_r)}, \qquad (26.46)$$

which coincides with an expectation value where all decay products are ignored (the probabilities, e.g., $P(K^0, t_a; \bar{K}^0, t_b)$, just contain the meson states). This is certainly not the case for other meson systems, like the $B^0 \bar{B}^0$, $D^0 \bar{D}^0$, or $B_s^0 \bar{B}_s^0$ systems.

Inserting the quantum-mechanical expectation value (26.46) into inequality (26.42) we arrive at the result of *Giancarlo Ghirardi*, *Roberto Grassi*, and *Tullio Weber* (1991)

$$|e^{-\frac{\Gamma_S}{2}(t_a + t_{a'})} \cos(\Delta m(t_a - t_{a'})) - e^{-\frac{\Gamma_S}{2}(t_a + t_{b'})} \cos(\Delta m(t_a - t_{b'}))| \qquad (26.47)$$

$$+|e^{-\frac{\Gamma_S}{2}(t_{a'} + t_b)} \cos(\Delta m(t_{a'} - t_b)) + e^{-\frac{\Gamma_S}{2}(t_b + t_{b'})} \cos(\Delta m(t_b - t_{b'}))| \leq 2.$$

Of course, we could have chosen $\bar{K}^0$ instead of K^0 without any change.

No violation for K mesons: Unfortunately, inequality (26.47) *cannot* be violated for any choice of the four (positive) times $t_a, t_b, t_{a'}, t_{b'}$ due to the interplay between the kaon decay and strangeness oscillations (Ghirardi *et al.*, 1991; Ghirardi *et al.*, 1992). As demonstrated by *Beatrix Hiesmayr* (1999) a possible violation depends very much on the ratio $x = \Delta m / \Gamma$. The numerically determined range for *no violation* is $0 < x < 2$ and the experimental value $x_K^{\text{exper}} = 0.95$ lies exactly in this region.

Remark on B-meson and D-meson systems: Instead of K mesons we can also consider entangled B mesons, produced via the resonance decay $\Upsilon(4S) \to B^0 \bar{B}^0$, e.g., at the KEK-B asymmetric e^+e^- collider in Japan. In such a system, the *beauty* quantum number $B = +, -$ is the analogue of the *strangeness* $S = +, -$ and instead of short- and long-lived states we have the heavy and light B-meson states $|B_H\rangle$ and $|B_L\rangle$, respectively, as eigenstates of the non-Hermitian "effective mass" Hamiltonian. Since for B mesons the decay widths are equal, $\Gamma_H = \Gamma_L = \Gamma_B$, the expectation value for a unitary time evolution yields

$$E(t_l; t_r) = -\cos(\Delta m_B \Delta t) \cdot e^{-\Gamma_B(t_l + t_r)} + (1 - e^{-\Gamma_B t_l})(1 - e^{-\Gamma_B t_r}), \qquad (26.48)$$

where $\Delta m_B = m_H - m_L$ is the mass difference of the heavy and light B meson. Here, the additional term from the decay products cannot be ignored.

Inserting expectation value (26.48) into inequality (26.42) for a fixed quasi-spin, say, for flavour $B = +1$, i.e., B^0, we find that the CHSH–Bell inequality *cannot* be violated in the x range $0 < x < 2.6$. Here too, the experimental value $x_B^{\text{exper}} = 0.77$ lies inside the corresponding range where no violation can occur.

In this connection we also want to mention the work of *Apollo Go* (2004), who analysed the entangled $B^0 \bar{B}^0$ meson pairs produced at the KEK-B collider and collected at the Belle detector in the context of a specific CHSH–Bell inequality. The considered inequality, however, is not a genuine Bell inequality since the unitary time evolution of the unstable quantum state—the decay property of the meson—is ignored. Therefore, it cannot discriminate between quantum mechanics and local realistic theories, for a more detailed criticism (see Bertlmann *et al.*, 2004; Bramon *et al.*, 2005). Nevertheless, the work by Go (2004) represents a notable test of quantum-mechanical correlations exhibited by $B^0\bar{B}^0$ entangled pairs.

The feature that a violation of the CHSH–Bell inequality (26.42) depends crucially on the value of x also occurs for other meson-antimeson systems, e.g., for the *charmed* system $D^0\bar{D}^0$.

The experimental x values for selected meson systems are as follows:

x	meson system
0.95	$K^0\bar{K}^0$
0.77	$B^0\bar{B}^0$
< 0.03	$D^0\bar{D}^0$
> 19.00	$B_s^0\bar{B}_s^0$

Therefore, *no violation* of the CHSH–Bell inequality occurs for the familiar meson-antimeson systems; only for the last system a violation is expected. However, the Bell inequality for decaying systems can be generalized for a restricted observable space (i.e., this *generalized Bell inequality* is valid for a restricted but nevertheless large class of local realistic theories) such that quantum mechanics produces a violation in a range which is feasible at accelerator facilities with current technology (Hiesmayr, 2001; Hiesmayr *et al.*, 2012).

Conclusion: In the familiar meson-antimeson systems, like those of the K, D, or B mesons, one cannot use the time-variation type of Bell inequalities to exclude local realistic theories. However, with *generalized Bell inequalities* a test for a large class of local realistic theories is feasible.

26.5.2 Bell Inequality for Quasi-Spin States—*CP* Violation

Let us next investigate the second option mentioned at the beginning of Section 26.5.1. We fix the time, say at $t = 0$, and vary the quasi-spin of the K meson. This corresponds to a rotation in quasi-spin space analogously to the spin-$\frac{1}{2}$ or photon case.

Analogy: Recall that the quasi-spin of kaons plays the role of spin or photon polarization. Hence, a rotation in the quasi-spin space can correspond to one in the polarization space,

$$|k\rangle = a\,|K^0\rangle + b\,|\bar{K}^0\rangle \quad \longleftrightarrow \quad |\vec{n}\rangle = \cos\left(\tfrac{\alpha}{2}\right)|{\uparrow}\rangle + \sin\left(\tfrac{\alpha}{2}\right)|{\downarrow}\rangle. \qquad (26.49)$$

For a Bell inequality we need three different "angles"—"quasi-spins"—and we may choose the H, S, and CP eigenstates

$$|k_n\rangle = |\mathrm{K_S}\rangle , \qquad |k_m\rangle = |\bar{\mathrm{K}}^0\rangle , \qquad |k_{n'}\rangle = |\mathrm{K}_1^0\rangle . \tag{26.50}$$

Denoting the probability of measuring the short-lived state $\mathrm{K_S}$ on the left-hand side and the antikaon $\bar{\mathrm{K}}^0$ on the right-hand side, at the time $t = 0$, by $P(\mathrm{K_S}, \bar{\mathrm{K}}^0)$, and analogously the probabilities $P(\mathrm{K_S}, \mathrm{K}_1^0)$ and $P(\mathrm{K}_1^0, \bar{\mathrm{K}}^0)$ we can derive the following Bell inequality under the hypothesis of Bell's locality (recall Proposition 13.4 of Section 13.2):

Proposition 26.1 (Uchiyama–Bell inequality)

$$P(\mathrm{K_S}, \bar{\mathrm{K}}^0) \;\leq\; P(\mathrm{K_S}, \mathrm{K}_1^0) + P(\mathrm{K}_1^0, \bar{\mathrm{K}}^0) .$$

The Uchiyama–Bell inequality first considered by *Fumiyo Uchiyama* (1997) can be converted into an inequality for the *CP* violation parameter ε: $\Re\{\varepsilon\} \leq |\varepsilon|^2$. This inequality is obviously violated by the experimental value of ε, having an absolute value of order 10^{-3} and a phase of about $45°$.

Here, however, we want to stay as general and loophole-free as possible and follow the approach of *Reinhold Bertlmann*, *Walter Grimus*, and *Beatrix Hiesmayr* (2001) to construct a Bell inequality for the *CP* violation parameter in the following way.

The Uchiyama–Bell inequality is rather formal because it involves the unphysical *CP*-even state $|\mathrm{K}_1^0\rangle$, but it implies an inequality on a *physical CP* violation parameter, which is experimentally testable. For the derivation, recall the H and *CP* eigenstates, eqns (26.5a), (26.5b), and (26.3a), (26.3b), then we have the following transition amplitudes

$$\langle \bar{\mathrm{K}}^0 | \mathrm{K_S}\rangle = -\frac{q}{N} , \quad \langle \bar{\mathrm{K}}^0 | \mathrm{K}_1^0 \rangle = -\frac{1}{\sqrt{2}} , \quad \langle \mathrm{K_S} | \mathrm{K}_1^0 \rangle = \frac{1}{\sqrt{2}N}(p^* + q^*) , \tag{26.51}$$

which we use to calculate the probabilities in the inequality of Proposition 26.1. Optimizing the inequality we find, independent of any phase conventions of the kaon states,

$$|p| \leq |q| , \tag{26.52}$$

which one can put forward as an experimentally testable proposal.

Proposition 26.2 (Kaon transition coefficients)

The inequality $|p| \leq |q|$ is experimentally testable!

Semileptonic decays: Let us consider the semileptonic decays of the K mesons. The strange quark s decays weakly as a constituent of $\bar{\mathrm{K}}^0$ as depicted in Fig. 26.2. Due

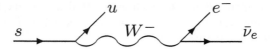

Fig. 26.2 Feynman diagram. The strange quark is decaying weakly into an up quark u and an electron e^- together with its anti-neutrino $\bar{\nu}_e$.

to their quark content the kaon $K^0(\bar{s}d)$ and the antikaon $\bar{K}^0(s\bar{d})$ have the following definite decays following the rule $\Delta S = \Delta Q$:

decay of strange particles				quark level		
$K^0(d\bar{s})$	$\longrightarrow$	$\pi^-(d\bar{u})$	$l^+\ \nu_l$	$\bar{s}$	$\longrightarrow$	$\bar{u}\ \ l^+\ \nu_l$
Q 0		-1		$\frac{1}{3}$		$-\frac{2}{3}$
S 1		0		1		0
$\bar{K}^0(\bar{d}s)$	$\longrightarrow$	$\pi^+(\bar{d}u)$	$l^-\ \bar{\nu}_l$	s	$\longrightarrow$	$u\ \ l^-\ \bar{\nu}_l$
Q 0		$+1$		$-\frac{1}{3}$		$\frac{2}{3}$
S -1		0		-1		0

In particular, we study the *leptonic charge asymmetry*

$$\delta = \frac{\Gamma(K_L \to \pi^- l^+ \nu_l) - \Gamma(K_L \to \pi^+ l^- \bar{\nu}_l)}{\Gamma(K_L \to \pi^- l^+ \nu_l) + \Gamma(K_L \to \pi^+ l^- \bar{\nu}_l)} \quad \text{with} \quad l = \mu, e\,, \qquad (26.53)$$

where l represents either a muon or an electron. The rule $\Delta S = \Delta Q$ for the decays of the strange particles implies that—due to their quark content—the kaon $K^0(\bar{s}d)$ and the antikaon $\bar{K}^0(s\bar{d})$ have definite decays. Thus, l^+ and l^- tag K^0 and $\bar{K}^0$, respectively, in the K_L state, and the leptonic asymmetry (26.53) is expressed by the probabilities $|p|^2$ and $|q|^2$ of finding a K^0 and a $\bar{K}^0$, respectively, in the K_L state

$$\delta = \frac{|p|^2 - |q|^2}{|p|^2 + |q|^2}\,. \qquad (26.54)$$

Then inequality (26.52) turns into the bound

$$\delta \leq 0 \qquad (26.55)$$

for the leptonic charge asymmetry which measures CP violation. The Uchiyama–Bell inequality (Proposition 26.1) is thus converted into the inequality (26.55) which is experimentally testable.

If CP were conserved, we would have $\delta = 0$. Experimentally, however, the asymmetry is non-vanishing[2], namely

$$\delta = (3.27 \pm 0.12) \cdot 10^{-3}\,, \qquad (26.56)$$

and is thus a clear sign of CP violation.

[2]It is the weighted average over electron and muon events (see Caso *et al.*, 1998).

The bound (26.55) dictated by the Uchiyama–Bell inequality is in contradiction to the experimental value (26.56) which is definitely positive. Thus we can state the interesting fact in the following proposition:

Proposition 26.3 (*CP* violation—Bell inequality)

The CP violation in a $K^0\bar{K}^0$ *system is related to the violation of a Bell inequality.*

On the other hand, we can replace $\bar{K}^0$ by K^0 in the Uchiyama–Bell inequality (Proposition 26.1) and along the same lines as discussed before we obtain the inequality

$$|p| \geq |q| \quad \text{or} \quad \delta \geq 0 \,. \tag{26.57}$$

Altogether inequalities (26.52), (26.55), and (26.57) imply the strict equality

$$|p| = |q| \quad \text{or} \quad \delta = 0 \,, \tag{26.58}$$

which is in contradiction to experiment.

Conclusions: The premises of local realistic theories are *only* compatible with strict *CP* conservation in $K^0\bar{K}^0$ mixing. Conversely, *CP* violation in $K^0\bar{K}^0$ mixing, no matter which sign the experimental asymmetry (26.53) actually has, always leads to a *violation* of a Bell inequality, either of inequalities (26.52), (26.55), or of (26.57). In this way, $\delta \neq 0$ is a manifestation of the entanglement of the considered state.

We also want to remark that in case of the Uchiyama–Bell inequality (Proposition 26.1), since it is considered at $t = 0$, it is rather *contextuality* than nonlocality which is tested. *Non-contextuality*—a main hypothesis in hidden-variable theories—means that the value of an observable does not depend on the experimental context. The measurement of the observable must yield the value independently of other simultaneous measurements, recall our discussion on hidden-variable theories in Section 12.5, in particular, the Kochen–Specker Theorem 12.4. Thus we can state:

Proposition 26.4 (Contextuality and $K^0\bar{K}^0$ entanglement)

The quantum feature of contextuality is verified in the entangled $K^0\bar{K}^0$ *case!*

Other types of Bell inequalities and non-local tests: In a series of papers the group of *Albert Bramon* (Bramon and Nowakowski, 1999; Ancochea *et al.*, 1999; Bramon and Garbarino, 2002*a*; Bramon and Garbarino, 2002*b*; Bramon *et al.*, 2006; Bramon *et al.*, 2007) constructed and examined various types of Bell inequalities, also including the regeneration of kaons. All inequalities are experimentally feasible. More details can be found in Tataroglu (2009).

Interestingly, note that when considering non-maximally entangled $K^0\bar{K}^0$ states it is in principle possible to violate a Bell–CHSH inequality by measuring the time (Hiesmayr, 2007). Furthermore, *tests of local realism and separability* of the wave function in the $K^0\bar{K}^0$ system have been proposed by a number of authors (Thörnqvist, 1981; Six, 1982; Six, 1990; Selleri, 1983; Privitera and Selleri, 1992; Eberhard, 1993*b*; Di Domenico, 1995; Selleri, 1997; Bertlmann *et al.*, 1999; Foadi and Selleri, 1999*a*; Foadi and Selleri, 1999*b*; Afriat and Selleri, 1999; Genovese *et al.*, 2001; Bramon and Garbarino, 2002*a*; Bramon and Garbarino, 2002*b*); the experimental data given by the CPLEAR collaboration (Apostolakis *et al.*, 1998) refute both.

There has also been a lot of activity in constructing and testing Bell inequalities for the η_c-decay and χ_c-decay specially designed for the charm factory BESC II and BES III in Beijing (see, e.g., Ding *et al.*, 2007; Li and Qiao, 2009; Li and Qiao, 2010; Khan *et al.*, 2020; Qian *et al.*, 2020; Shi and Yang, 2020).

In entangled $K^0\bar{K}^0$ systems one may also study phenomena that make it seem like "the future impacts the past" (Bernabeu and Di Domenico, 2022), and we also have "quantum marking and quantum erasure for neutral kaons" (Bramon *et al.*, 2004*a*; Bramon *et al.*, 2004*d*) and "duality and kaon interferometry" (Bramon *et al.*, 2004*b*; Bramon *et al.*, 2004*c*). Last but not least, "High energy quantum teleportation using neutral kaons" has been proposed by Shi (2006*a* and 2006*b*). Thus we encounter the fascinating features of optical quantum information also in massive particle-anti-particle systems of high-energy physics.

27

Particle Physics—Entanglement and Decoherence

As we have seen in the previous chapter, the experimental test of Bell inequalities in particle physics is not an easy task that comes along with loopholes for local realistic models. Although such Bell tests with massive particles are of fundamental interest, the examination of entanglement rather than non-locality for the created meson-antimeson pairs might be more feasible. To examine the possibilities and challenges for such tests, it is useful to study possible mechanisms for decoherence and dissipation that might influence the particle pairs under consideration. More specifically, in this chapter we will revisit the creation of an entangled kaon state at the Φ resonance, where the particles comprising the relevant pair propagate in to opposite directions, to the left and to the right, respectively, until the kaons are measured, recall Fig. 26.1. Possible decoherence will provide us with the information about the quality of the entangled state. This leads us to the question: How can we describe and measure the decoherence of an entangled state?

In the following we consider possible decoherence effects arising from an interaction of the quantum system with its environment. Sources for standard decoherence effects for the systems we consider here are the strong-interaction scatterings of kaons with nucleons, the weak-interaction decays, as well as the noise of the experimental setup. "Non-standard" decoherence effects that might result from fundamental modifications of quantum mechanics and which might be traced back to the influence of quantum gravity (Hawking, 1982; 't Hooft, 1999, 2000, and 2005)—quantum fluctuations in the space-time structure on the scale of the Planck mass—or to dynamical state-reduction models (Ghirardi *et al.*, 1986; Pearle, 1989; Penrose, 1996 and 1998), are not considered here and would anyways arise on a different energy scale, if at all. In the following we will concentrate on a specific model of decoherence developed by (Bertlmann and Grimus, 2001), where we can quantify the strength of such possible decoherence effects by the data of existing experiments.

27.1 Decoherence Model for Entangled Particle Systems

Before discussing the decoherence of the $K^0\bar{K}^0$ systems let us first introduce the decoherence model of Bertlmann and Grimus (2001). This model treats decoherence in a two-dimensional Hilbert space of states $\mathcal{H} = \mathbb{C}^2$. There we allow for a non-Hermitian Hamiltonian H, in order to include the possibility of incorporating particle decay in the Wigner–Weisskopf approximation (Weisskopf and Wigner, 1930a and 1930b). We denote the normalized energy eigenstates by $|e_j\rangle$ ($j = 1, 2$) and have, therefore,

$$H|e_j\rangle = \lambda_j|e_j\rangle \quad \text{with} \quad \lambda_j = m_j - \frac{i}{2}\Gamma_j\,, \tag{27.1}$$

where m_j and Γ_j are real and the latter quantities are positive in addition. Furthermore, we make the crucial assumption that

$$\langle e_1|e_2\rangle = 0\,, \tag{27.2}$$

despite H being non-Hermitian.

Now including decoherence, the time evolution of the density matrix ρ has the form

$$\frac{\partial \rho}{\partial t} = -i\,H\rho + i\,\rho H^\dagger - \mathcal{D}[\rho]\,. \tag{27.3}$$

The model of decoherence we now consider amounts to the assumption that the *dissipator* $\mathcal{D}[\rho]$ can be expressed in terms of the projection operators onto the eigenstates of the Hamiltonian,

$$\mathcal{D}[\rho] = \lambda\,(P_1\rho P_2 + P_2\rho P_1)\,, \quad \text{where} \quad P_j = |e_j\rangle\langle e_j|\,, \tag{27.4}$$

where the *decoherence parameter* λ is a positive constant. It can readily be checked that the decoherence term in eqn (27.4) is of Lindblad-type (recall Section 22.2.2)

$$\mathcal{D}[\rho] = \frac{1}{2}\left(\sum_j A_j^\dagger A_j\rho + \rho\sum_j A_j^\dagger A_j\right) - \sum_j A_j\rho A_j^\dagger\,, \tag{27.5}$$

if we make the identification $A_j = \sqrt{\lambda}P_j$. Thus, the term (27.4) generates a completely positive map; moreover, since $P_j^\dagger = P_j$ and $[P_j, H] = 0$, the decoherence term would increase the von Neumann entropy and conserve energy in the case of a Hermitian Hamiltonian.

However, what is more important in our discussion is the fact that with the choice (27.4) the equations for the components of ρ decouple. Indeed, with

$$\rho = \sum_{j,k=1}^{2} \rho_{jk}\,|e_j\rangle\langle e_k|\,, \tag{27.6}$$

where $\rho_{jk} = \rho_{kj}^*$, and with the time evolution (27.3), we obtain

$$\rho_{11}(t) = \rho_{11}(0)\,\exp(-\Gamma_1 t)\,, \tag{27.7a}$$

$$\rho_{22}(t) = \rho_{22}(0)\,\exp(-\Gamma_2 t)\,, \tag{27.7b}$$

$$\rho_{12}(t) = \rho_{12}(0)\,\exp\{-[i(m_1 - m_2) + (\Gamma_1 + \Gamma_2)/2 + \lambda]t\}\,. \tag{27.7c}$$

27.2 Measurement of Entangled Kaons

In this section we discuss what it means to measure entangled kaons.

27.2.1 Entangled Kaons

Let us first describe pairs of entangled neutral kaons. We use the abbreviations

$$|e_1\rangle = |K_S\rangle_l \otimes |K_L\rangle_r \quad \text{and} \quad |e_2\rangle = |K_L\rangle_l \otimes |K_S\rangle_r \,, \tag{27.8}$$

and regard—as usual—the total Hamiltonian as a tensor product of the single-particle Hilbert spaces: $H = H_l \otimes \mathbb{1}_r + \mathbb{1}_l \otimes H_r$, where l denotes the left-moving and r the right-moving particle. The initial Bell singlet state

$$|\psi^-\rangle = \frac{1}{\sqrt{2}}\big(|e_1\rangle - |e_2\rangle\big) \tag{27.9}$$

is expressed by the density matrix

$$\rho(0) = |\psi^-\rangle\langle\psi^-| = \frac{1}{2}\Big(|e_1\rangle\langle e_1| + |e_2\rangle\langle e_2| - |e_1\rangle\langle e_2| - |e_2\rangle\langle e_1|\Big). \tag{27.10}$$

Then the time evolution given by (27.3) with ansatz (27.4), where now the operators $P_j = |e_j\rangle\langle e_j|$ ($j = 1, 2$) project to the eigenstates of the two-particle Hamiltonian H, also decouples

$$\rho_{11}(t) = \rho_{11}(0)\, e^{-2\Gamma t} \,, \tag{27.11a}$$

$$\rho_{22}(t) = \rho_{22}(0)\, e^{-2\Gamma t} \,, \tag{27.11b}$$

$$\rho_{12}(t) = \rho_{12}(0)\, e^{-2\Gamma t - \lambda t} \,. \tag{27.11c}$$

Consequently, for the time-dependent density matrix we obtain

$$\rho(t) = \frac{1}{2}\, e^{-2\Gamma t}\Big(|e_1\rangle\langle e_1| + |e_2\rangle\langle e_2| - e^{-\lambda t}\big(|e_1\rangle\langle e_2| + |e_2\rangle\langle e_1|\big)\Big). \tag{27.12}$$

The decoherence arises through the factor $e^{-\lambda t}$ which only effects the off-diagonal elements. It means that for $t > 0$ and $\lambda \neq 0$ the density matrix $\rho(t)$ does not correspond to a pure state any more. Finally, for a proper density matrix for the kaon system, conditioned on having not decayed up to time t, we have to divide $\rho(t)$ in eqn (27.12) by the trace $\text{Tr}\big(\rho(t)\big)$.

27.2.2 Measurement

In the model by Bertlmann and Grimus (2001) the parameter λ quantifies the strength of possible decoherence of the whole entangled state. The range of values can be determined by experiment. Concerning the procedure of the measurement, one can then take the following point of view. The two-particle density matrix follows the time evolution given by eqns (27.3) and (27.5) with the Lindblad operators $A_j = \sqrt{\lambda}|e_j\rangle\langle e_j|$ and thereby undergoes some decoherence. We measure the strangeness content S of the right-moving particle at time t_r and of the left-moving particle at time t_l. For the sake of definiteness we choose $t_\mathrm{r} \leq t_\mathrm{l}$. Then for times $t_\mathrm{r} \leq t \leq t_\mathrm{l}$ we have a single-particle state which evolves exactly according to quantum mechanics, i.e., no further decoherence is picked up.

In theory we describe the measurement of the strangeness content, i.e., the right-moving particle being a K^0 or a $\bar{\mathrm{K}}^0$ at time t_r, by the following projection onto ρ

$$\mathrm{Tr}_\mathrm{r}\big(\mathbb{1}_\mathrm{l} \otimes |S'\rangle\langle S'|_\mathrm{r}\, \rho(t_\mathrm{r})\big) \;\equiv\; \rho^\mathrm{l}(t = t_\mathrm{r}; t_\mathrm{r})\,, \qquad (27.13)$$

where strangeness $S' = +,-$ and $|+\rangle = |\mathrm{K}^0\rangle$, $|-\rangle = |\bar{\mathrm{K}}^0\rangle$. Consequently, $\rho^\mathrm{l}(t; t_\mathrm{r})$ for times $t \geq t_\mathrm{r}$ is the single-particle density matrix for the left-moving particle and evolves as a single-particle state according to pure quantum mechanics. At $t = t_\mathrm{l}$ the strangeness content ($S = +,-$) of the second particle is measured and we finally calculate the probability

$$P_\lambda(S, t_\mathrm{l}; S', t_\mathrm{r}) \;=\; \mathrm{Tr}_\mathrm{l}\big(|S\rangle\langle S|_\mathrm{l}\, \rho^\mathrm{l}(t_\mathrm{l}; t_\mathrm{r})\big)\,. \qquad (27.14)$$

Explicitly, we find the following results for the probabilities of finding the strangeness to be correlated or anti-correlated, respectively,

$$P_\lambda(\mathrm{K}^0, t_\mathrm{l}; \mathrm{K}^0, t_\mathrm{r}) \;=\; P_\lambda(\bar{\mathrm{K}}^0, t_\mathrm{l}; \bar{\mathrm{K}}^0, t_\mathrm{r}) \qquad (27.15\mathrm{a})$$

$$= \frac{1}{8}\Big(e^{-\Gamma_\mathrm{S} t_\mathrm{l} - \Gamma_\mathrm{L} t_\mathrm{r}} + e^{-\Gamma_\mathrm{L} t_\mathrm{l} - \Gamma_\mathrm{S} t_\mathrm{r}} - e^{-\lambda t_\mathrm{r}}\, 2\cos(\Delta m \Delta t)\cdot e^{-\Gamma(t_\mathrm{l} + t_\mathrm{r})}\Big)\,,$$

$$P_\lambda(\mathrm{K}^0, t_\mathrm{l}; \bar{\mathrm{K}}^0, t_\mathrm{r}) \;=\; P_\lambda(\bar{\mathrm{K}}^0, t_\mathrm{l}; \mathrm{K}^0, t_\mathrm{r}) \qquad (27.15\mathrm{b})$$

$$= \frac{1}{8}\Big(e^{-\Gamma_\mathrm{S} t_\mathrm{l} - \Gamma_\mathrm{L} t_\mathrm{r}} + e^{-\Gamma_\mathrm{L} t_\mathrm{l} - \Gamma_\mathrm{S} t_\mathrm{r}} + e^{-\lambda t_\mathrm{r}}\, 2\cos(\Delta m \Delta t)\cdot e^{-\Gamma(t_\mathrm{l} + t_\mathrm{r})}\Big)\,,$$

with $\Delta t = t_\mathrm{l} - t_\mathrm{r}$. Note that for matching times $t_\mathrm{l} = t_\mathrm{r} = t$, the probabilities for matching strangeness do not vanish,

$$P_\lambda(\mathrm{K}^0, t; \mathrm{K}^0, t) \;=\; P_\lambda(\bar{\mathrm{K}}^0, t; \bar{\mathrm{K}}^0, t) \;=\; \frac{1}{4}\, e^{-2\Gamma t}\,(1 - e^{-\lambda t}) \qquad (27.16)$$

in contrast to the pure quantum-mechanical EPR-correlations.

Asymmetry of probabilities: The interesting quantity is the *asymmetry of probabilities*. It is directly sensitive to the interference term and can be measured experimentally. For pure quantum mechanics we have

$$A^\mathrm{QM}(\Delta t) = \frac{P(\mathrm{K}^0, t_\mathrm{l}; \bar{\mathrm{K}}^0, t_\mathrm{r}) + P(\bar{\mathrm{K}}^0, t_\mathrm{l}; \mathrm{K}^0, t_\mathrm{r}) - P(\mathrm{K}^0, t_\mathrm{l}; \mathrm{K}^0, t_\mathrm{r}) - P(\bar{\mathrm{K}}^0, t_\mathrm{l}; \bar{\mathrm{K}}^0, t_\mathrm{r})}{P(\mathrm{K}^0, t_\mathrm{l}; \bar{\mathrm{K}}^0, t_\mathrm{r}) + P(\bar{\mathrm{K}}^0, t_\mathrm{l}; \mathrm{K}^0, t_\mathrm{r}) + P(\mathrm{K}^0, t_\mathrm{l}; \mathrm{K}^0, t_\mathrm{r}) + P(\bar{\mathrm{K}}^0, t_\mathrm{l}; \bar{\mathrm{K}}^0, t_\mathrm{r})}$$

$$= \frac{\cos(\Delta m \Delta t)}{\cosh(\frac{1}{2}\Delta\Gamma \Delta t)}\,, \qquad (27.17)$$

with $\Delta\Gamma = \Gamma_\mathrm{L} - \Gamma_\mathrm{S}$. By inserting the probabilities (27.15a) and (27.15b) we find

$$A^\lambda(t_\mathrm{l}, t_\mathrm{r}) = \frac{\cos(\Delta m \Delta t)}{\cosh(\frac{1}{2}\Delta\Gamma \Delta t)}\, e^{-\lambda\, \min\{t_\mathrm{l}, t_\mathrm{r}\}} = A^\mathrm{QM}(\Delta t)\, e^{-\lambda\, \min\{t_\mathrm{l}, t_\mathrm{r}\}}\,. \qquad (27.18)$$

Thus the decoherence effect, which is simply given by the factor $e^{-\lambda \min\{t_1, t_r\}}$, depends only—according to the philosophy of the chosen decoherence model—on the time that the first kaon is measured, in our case: $\min\{t_1, t_r\} = t_r$.

27.2.3 CPLEAR Experiment

Now we compare the model by Bertlmann and Grimus (2001) with the results of an experiment carried out at CERN. It is the CPLEAR experiment (Apostolakis *et al.*: CPLEAR Coll., 1998), where $K^0 \bar{K}^0$ pairs are produced in the $p\bar{p}$ collider: $p\bar{p} \to K^0 \bar{K}^0$. These pairs are predominantly in an antisymmetric state with quantum numbers $J^{PC} = 1^{--}$ and the strangeness of the kaons is detected via strong interactions in surrounding absorbers (made of copper and carbon).

Examples:

$$S = +1: \quad K^0(d\bar{s}) + N \longrightarrow K^+(u\bar{s}) + X \,,$$
$$S = -1: \quad \bar{K}^0(\bar{d}s) + N \longrightarrow K^-(\bar{u}s) + X \,,$$
$$S = -1: \quad \bar{K}^0(\bar{d}s) + N \longrightarrow \Lambda(uds) + X \quad \text{and} \quad \Lambda \longrightarrow p\pi^- \,;$$

Correlated strangeness events: (K^-, Λ) ,

Anti-correlated strangeness events: (K^+, K^-), (K^+, Λ) .

The experimental setup has two configurations, see Fig. 27.1. In configuration $C(0)$ both kaons propagate 2 cm; they have nearly equal proper times ($t_r \approx t_1$) when they are measured by the absorbers. This fulfils the condition for an EPR-type experiment. In configuration $C(5)$ one kaon propagates 2 cm and the other kaon 7 cm, thus the flight-path difference is 5 cm on average, corresponding to a proper-time difference $|t_r - t_1| \approx 1.2\,\tau_S$.

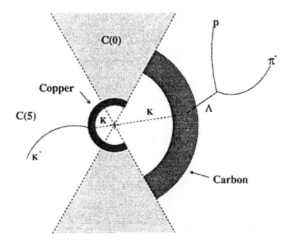

Fig. 27.1 Example of a CPLEAR event. An event with matching strangeness, (K^-, Λ). Figure from Fig. 3(a) in Apostolakis *et al.*: CPLEAR Coll. (1998).

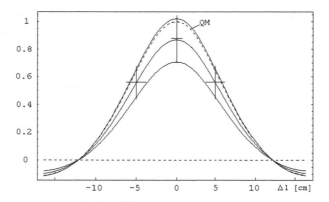

Fig. 27.2 Asymmetry of probabilities. The asymmetry (27.18) as a function of the flight-path difference of the kaons is plotted. The dashed curve corresponds to quantum mechanics, the solid curves represent the best fit (27.19) of the asymmetry to the CPLEAR data Apostolakis *et al.*: CPLEAR Coll. (1998) given by the crosses. The horizontal line indicates the Furry–Schrödinger hypothesis (Schrödinger, 1935*a*; Schrödinger, 1935*b*; Furry, 1936*a*; Furry, 1936*b*), as explained in Section 27.3. Reprinted (Fig. 1) with permission from Reinhold A. Bertlmann, Walter Grimus, and Beatrix C. Hiesmayr, Phys. Rev. D **60**, 114032 (1999). Copyright (1999) by the American Physical Society.

Now fitting the decoherence parameter λ by comparing the asymmetry (27.18) with the experimental data (Apostolakis *et al.*: CPLEAR Coll., 1998), see Fig.27.2, we find (Bertlmann, 2006) the following bounds when averaging over both configurations:

$$\bar{\lambda} = (1.84^{+2.50}_{-2.17}) \cdot 10^{-12} \text{ MeV} \quad \text{and} \quad \bar{\Lambda} = \frac{\bar{\lambda}}{\Gamma_S} = 0.25^{+0.34}_{-0.32}. \tag{27.19}$$

The results (27.19) are certainly compatible with quantum mechanics ($\lambda = 0$), nevertheless, the experimental data allows an upper bound $\bar{\lambda}_{\text{up}} = 4.34 \cdot 10^{-12}$ MeV for possible decoherence in the entangled $K^0\bar{K}^0$ system. For an introduction of decoherence including kaons and neutron systems, see Traxler (2009, 2011).

27.3 Connection to Phenomenological Model

At first, Bertlmann and Grimus (1997, 1998) proposed a simple procedure for introducing decoherence in a rather phenomenological way, which is very well suited to experimental measurements. Actually this was first noted for the $B^0\bar{B}^0$ system (see also Dass and Sarma, 1998). The model is more in the spirit of *Erwin Schrödinger* (1935*a*, 1935*b*) and *Wendell Furry* (1936*a*, 1936*b*) to describe the process of spontaneous factorization of the wave function. Quantum mechanics is modified in the sense that we multiply the interference term of the transition amplitude by the factor $(1-\zeta)$. The quantity ζ is called *effective decoherence parameter*. Starting again from the Bell singlet state $|\psi^-\rangle$, which is given by the mass eigenstate representation (27.9), we have for the correlated strangeness probability

$$P(K^0, t_l; K^0, t_r) = || \langle K^0|_l \otimes \langle K^0|_r |\psi^-(t_l, t_r)\rangle ||^2, \tag{27.20}$$

which is then modified to

$$P_\zeta(K^0, t_1; K^0, t_r) = \frac{1}{2} \Big(e^{-\Gamma_S t_1 - \Gamma_L t_r} |\langle K^0 | K_S \rangle_1|^2 \, |\langle K^0 | K_L \rangle_r|^2$$

$$+ \, e^{-\Gamma_L t_1 - \Gamma_S t_r} |\langle K^0 | K_L \rangle_1|^2 \, |\langle K^0 | K_S \rangle_r|^2 \; - \; 2 \underbrace{(1 - \zeta)}_{\text{modification}} \, e^{-\Gamma(t_1 + t_r)}$$

$$\times \, \text{Re} \left\{ \langle K^0 | K_S \rangle_1^* \langle K^0 | K_L \rangle_r^* \langle K^0 | K_L \rangle_1 \langle K^0 | K_S \rangle_r \, e^{-i \Delta m \Delta t} \right\} \Big)$$

$$= \frac{1}{8} \Big(e^{-\Gamma_S t_1 - \Gamma_L t_r} + e^{-\Gamma_L t_1 - \Gamma_S t_r} \; - \; 2 \underbrace{(1 - \zeta)}_{\text{modification}} e^{-\Gamma(t_1 + t_r)} \cos(\Delta m \Delta t) \Big), \tag{27.21}$$

and for the anti-correlated strangeness probability we just have to change the sign of the interference term.

Feature of the model: The value $\zeta = 0$ corresponds to pure quantum mechanics and $\zeta = 1$ to total decoherence or spontaneous factorization of the wave function, which is commonly known as the *Furry–Schrödinger hypothesis* (Schrödinger, 1935a; Schrödinger, 1935b; Furry, 1936a; Furry, 1936b). The decoherence parameter ζ, introduced by "by hand" by Bertlmann and Grimus, is quite effective to test possible decoherence of the created $K^0 \bar{K}^0$ pairs and also (as we shall see later in Section 27.5) of the produced $B^0 \bar{B}^0$ pairs at the $\Upsilon(4S)$ resonance (see Bertlmann and Grimus, 1997; Bertlmann and Grimus, 1998; Dass and Sarma, 1998; Bertlmann *et al.*, 1999). It interpolates continuously between the two limits: quantum mechanics on the one hand, and spontaneous factorization on the other. It represents a measure for the amount of decoherence, which results in a loss of entanglement of the total quantum state (we come back to this point in Section 27.4).

Correspondence: There exists a remarkable one-to-one correspondence between the decoherence model of Bertlmann and Grimus, eqn (27.3) with eqn (27.4), and their phenomenological ansatz (27.20). This implies a relation between $\lambda \longleftrightarrow \zeta$ (already noticed in Bertlmann and Grimus, 2001 and Bertlmann *et al.*, 2003). We can see it quickly in the following way. Calculating the asymmetry of strangeness events, as defined in eqn (27.17), with the probabilities (27.20) we obtain

$$A^\zeta(t_1, t_r) = A^{QM}(\Delta t) \left(1 - \zeta(t_1, t_r) \right). \tag{27.22}$$

When we now compare the two approaches, i.e., eqn (27.18) with eqn (27.22), we find the formula

$$\zeta(t_1, t_r) = 1 - e^{-\lambda \min\{t_1, t_r\}}. \tag{27.23}$$

Of course, when fitting the experimental data with the two models, the λ values (27.19) are in agreement with the corresponding ζ values, averaged over both experimental setups, as determined in (Bertlmann *et al.*, 1999)

$$\bar{\zeta} = 0.13^{+0.16}_{-0.15}. \tag{27.24}$$

Remark: Note that the deviations from quantum mechanics, the ζ parameter term and the ζ values clearly depend on the chosen basis for the initial state (Bertlmann and Grimus, 1997; Bertlmann and Grimus, 1998; Bertlmann and Grimus, 2001; Bertlmann *et al.*, 1999). We have chosen here the "best" basis, the H eigenstates or mass eigenstates, i.e., the basis $\{K_S, K_L\}$. Furthermore, the values of ζ (27.24) can be improved considerably by concentrating on special (*CP* suppressed) decay modes, e.g., $\zeta = 0.018 \pm 0.040_{\text{stat}} \pm 0.007_{\text{syst}}$ (Ambrosino *et al.*: KLOE Coll., 2006).

Conclusions: The phenomenological model demonstrates that the $K^0\bar{K}^0$ system is close to quantum mechanics, i.e., $\zeta = 0$, *and* (at the same time) far away from total decoherence, i.e., $\zeta = 1$. It confirms nicely and in a quantitative way the existence of entangled massive particles over macroscopic distances, namely over 9 cm.

We consider the decoherence parameter λ to be a fundamental constant of nature, whereas the value of the effective decoherence parameter ζ depends on the time when a measurement is performed. In the time evolution of the state $|\psi^-\rangle$, eqn (27.9), represented by the density matrix (27.12), enters the relation

$$\zeta(t) = 1 - e^{-\lambda t}, \tag{27.25}$$

which after measurement of the left- and right-moving particles at t_l and t_r turns into the formula (27.23), if decoherence occurs as described in Section 27.2.2.

The decoherence model by Bertlmann and Grimus (2001) has a very specific time evolution given by eqn (27.23). Measuring the strangeness content of the entangled kaons at definite times we have the possibility to distinguish experimentally, on the basis of time-dependent event rates, between the prediction of our model (27.23) and the results of other models (which would differ from eqn (27.23)). Indeed, it is of high interest to experimentally measure the asymmetry of the strangeness events at various times (or the corresponding quantum number for other particles like the D or B mesons), in order to confirm the very specific time dependence of the decoherence effect.

27.4 Decoherence and Entanglement Loss for Kaonic Systems

In the master equation (27.3) the dissipator $\mathcal{D}[\rho]$ describes two phenomena occurring in an open quantum system, *decoherence* and *dissipation* (see, e.g., Breuer and Petruccione, 2002). When the system S interacts with the environment E the initial product state evolves into entangled states of S and E in the course of time (Joos, 1996; Kübler and Zeh, 1973). It leads to mixed states in S, which means decoherence, and to an energy exchange between S and E, which is called dissipation.

Decoherence destroys the occurrence of long-range quantum correlations by suppressing the off-diagonal elements of the density matrix in a given basis and leads to an information transfer from the system S to the environment E. In general, both effects are present, however, decoherence typically acts on a much shorter timescale (Joos and Zeh, 1985; Zurek, 1991; Joos, 1996; Alicki, 2002) than dissipation and thus is the more important effect in quantum-information processing.

The model by Bertlmann and Grimus (2001) describes decoherence but *not* dissipation. The increase of decoherence of the initially totally entangled $K^0\bar{K}^0$ system—as time evolves—implies a decrease of entanglement in the system. This loss of entanglement can be measured explicitly (Hiesmayr, 2002; Bertlmann *et al.*, 2003).

In quantum-information theory, the entanglement of a state is quantified by introducing certain *entanglement measures*, recall Section 16. In this connection the entropy plays a fundamental role.

27.4.1   Von Neumann Entropy for $K^0\bar{K}^0$ Systems

Recalling our discussion of the entropy of a quantum system, in particular, the von Neumann entropy (Section 20.1), we know that "the entropy measures the degree of uncertainty—the lack of knowledge—of a quantum state" (Proposition 20.1). In general, a quantum state can be in a pure or mixed state. Whereas the pure state supplies us with maximal information about the system, a mixed state does not. In a mixed state "the von Neumann entropy measures how much of the maximal information is missing", recall Thirring's entropy statement (Proposition 20.2).

For mixed states the entanglement measures cannot be defined uniquely and each has its advantages and disadvantages. A common measure of entanglement is, e.g., the entanglement of formation E, while the concurrence C is an example for an entanglement monotone.

Let's start with von Neumann's entropy (Definition 20.1 in Section 20.1). Since we are only interested in the effect of decoherence, we normalize the entangled kaon state (27.12) properly in order to compensate for the decay property

$$\rho_N(t) = \frac{\rho(t)}{\mathrm{Tr}\big(\rho(t)\big)} \;. \tag{27.26}$$

Then the *von Neumann entropy* for the kaonic quantum state (27.26) is defined by

$$S\big(\rho_N(t)\big) = -\,\mathrm{Tr}\big[\rho_N(t)\log\rho_N(t)\big] = -\tfrac{1-e^{-\lambda t}}{2}\log\big(\tfrac{1-e^{-\lambda t}}{2}\big) - \tfrac{1+e^{-\lambda t}}{2}\log\big(\tfrac{1+e^{-\lambda t}}{2}\big)\,, \tag{27.27}$$

where we have chosen the logarithm to base 2, the dimension of the Hilbert space (the qubit space), such that S varies between $0 \le S \le 1$.

Entropy features:

(i) $S\big(\rho_N(t)\big) = 0$ for $t = 0$; the entropy is zero at time $t = 0$, there is no uncertainty in the system. The quantum state is pure and maximally entangled.

(ii) $S\big(\rho_N(t)\big) = 1$ for $t \to \infty$; the entropy increases for increasing t and approaches the value 1 at infinity. Hence the state becomes more and more mixed.

Reduced density matrices: Let us consider quite generally the familiar composite quantum system consisting of subsystems A (Alice) and B (Bob). Then the reduced density matrix of subsystem A is given by tracing the density matrix of the joint state over all states of B. In our case the subsystems are propagating kaons on the left-hand (l) and right-hand (r) side, thus we have the *reduced density matrices*

$$\rho_N^l(t) = \text{Tr}_r\big(\rho_N(t)\big) \quad \text{and} \quad \rho_N^r(t) = \text{Tr}_l\big(\rho_N(t)\big) . \tag{27.28}$$

The *uncertainty in the subsystem l* before the subsystem r is measured is given by the von Neumann entropy $S(\rho_N^l(t))$ of the corresponding reduced density matrix $\rho_N^l(t)$ (and alternatively we can replace l and r). In our case we find

$$S\big(\rho_N^l(t)\big) = S\big(\rho_N^r(t)\big) = 1 \qquad \forall\, t \geq 0 . \tag{27.29}$$

What we observe is:

The reduced-state entropies are independent of λ !

This means that the correlation stored in the composite system is, with increasing time, lost into the environment—what we would intuitively expect—and this information is *not* available in the subsystems, i.e., in the individual kaons.

For pure quantum states von Neumann's entropy function (27.27) is a good measure for entanglement—the entropy of entanglement—in which case the subsystems of A (Alice) and B (Bob) are most entangled when their reduced density matrices are maximally mixed. For mixed states, however, von Neumann's entropy and the reduced-state entropies are no longer a good measure for entanglement so that we have to take a different path to quantify entanglement, see Chapter 16.

27.4.2   Separability and Entanglement of Kaonic Systems

In the following we want to show that the initially entangled Bell singlet state—although subjected to decoherence and thus to entanglement loss in the course of time—remains entangled. It is convenient to work with the quasi-spin description for the $K^0\bar{K}^0$ system (recall Section 26.2). The projection operators of the mass eigenstates correspond to the spin projection operators "up" and "down"

$$P_S = |K_S\rangle\langle K_S| = \sigma_\uparrow = \frac{1}{2}\left(\mathbb{1} + \sigma_z\right) = \begin{pmatrix} 1 & 0 \\ 0 & 0 \end{pmatrix} , \tag{27.30a}$$

$$P_L = |K_L\rangle\langle K_L| = \sigma_\downarrow = \frac{1}{2}\left(\mathbb{1} - \sigma_z\right) = \begin{pmatrix} 0 & 0 \\ 0 & 1 \end{pmatrix} , \tag{27.30b}$$

and the transition operators are the "spin ladder" operators

$$P_{SL} = |K_S\rangle\langle K_L| = \sigma_+ = \frac{1}{2}\left(\sigma_x + i\,\sigma_y\right) = \begin{pmatrix} 0 & 1 \\ 0 & 0 \end{pmatrix} , \tag{27.31a}$$

$$P_{LS} = |K_L\rangle\langle K_S| = \sigma_- = \frac{1}{2}\left(\sigma_x - i\,\sigma_y\right) = \begin{pmatrix} 0 & 0 \\ 1 & 0 \end{pmatrix} . \tag{27.31b}$$

Then density matrix (27.26) with eqn (27.12) is expressed in terms of the Pauli spin matrices in the following way

$$\rho_N(t) = \frac{1}{4}\{\mathbb{1} - \sigma_z \otimes \sigma_z - e^{-\lambda t}[\sigma_x \otimes \sigma_x + \sigma_y \otimes \sigma_y]\}, \tag{27.32}$$

which at $t = 0$ coincides with the well-known expression for the pure spin-singlet state $\rho_N(t = 0) = \frac{1}{4}(\mathbb{1} - \vec{\sigma} \otimes \vec{\sigma})$ (see, e.g., Bertlmann *et al.*, 2002). The operator (27.32) can be written as 4×4 matrix

$$\rho_N(t) = \frac{1}{2}\begin{pmatrix} 0 & 0 & 0 & 0 \\ 0 & 1 & -e^{-\lambda t} & 0 \\ 0 & -e^{-\lambda t} & 1 & 0 \\ 0 & 0 & 0 & 0 \end{pmatrix}. \tag{27.33}$$

For yet another representation of the density matrix $\rho_N(t)$ we choose the so-called "Bell basis"

$$\rho^{\mp} = |\psi^{\mp}\rangle\langle\psi^{\mp}| \quad \text{and} \quad \omega^{\mp} = |\phi^{\mp}\rangle\langle\phi^{\mp}|, \tag{27.34}$$

with $|\psi^{-}\rangle$ given by eqn (27.9) and $|\psi^{+}\rangle$ by

$$|\psi^{+}\rangle = \frac{1}{\sqrt{2}}\left(|e_1\rangle + |e_2\rangle\right). \tag{27.35}$$

The states $|\phi^{\mp}\rangle = \frac{1}{\sqrt{2}}(|\uparrow\uparrow\rangle \mp |\downarrow\downarrow\rangle)$ (in spin notation) do not contribute here.

Recall now Section 15.2, the Definitions 15.5 and 15.6 of separable and entangled states. Then, in general, a state ρ is called *entangled* if it is *not separable*, i.e. $\rho \in \mathcal{S}^C$ where $\mathcal{S}^C$ is the complement of the set of separable states $\mathcal{S}$ given by Definition 15.7; and $\mathcal{S} \cup \mathcal{S}^C = \mathcal{H} = \mathcal{H}_A \otimes \mathcal{H}_B$.

The important question is now to judge whether a quantum state is entangled or conversely separable or not. Several *separability criteria* give an answer to that. A first one is the Peres–Horodecki *positive partial transpose* (PPT) *criterion*, Theorem 15.3, and a second one we consider here is the *reduction criterion*, Theorem 15.11.

However, Theorems 15.3 and 15.11 are necessary and sufficient separability conditions only for dimensions 2×2 and 2×3. A more general separability criterion, valid in any dimensions, does exist; it is formulated by the so-called *entanglement-witness theorem*, Theorem 16.8 of Section 16.3.1.

Now let us return to the question of entanglement and separability for our kaonic quantum state described by density matrix $\rho_N(t)$ (27.26) with (27.12) as it evolves in time. This question is clarified by the following proposition set up in Bertlmann *et al.* (2003).

Proposition 27.1 (Time evolution of $K^0\bar{K}^0$ state)

The state represented by the density matrix $\rho_N(t)$ (27.26) with eqn (27.12) becomes mixed for $0 < t < \infty$ but remains entangled. Separability is reached asymptotically $t \to \infty$ with the weight $e^{-\lambda t}$. Explicitly, $\rho_N(t)$ is the following mixture of the Bell states ρ^- and ρ^+:

$$\rho_N(t) = \frac{1}{2}\left(1 + e^{-\lambda t}\right)\rho^- + \frac{1}{2}\left(1 - e^{-\lambda t}\right)\rho^+ .$$

That means that the entangled $K^0\bar{K}^0$ state remains entangled before the kaons decay.

27.4.3 Entanglement of Formation and Concurrence of Kaonic Systems

For pure states the entropy of the reduced density matrices is sufficient, for mixed states, however, we need another measure, e.g., the *entanglement of formation* or the *concurrence*.

Entanglement of formation: Let us first recall the entanglement measure *entanglement of formation* from Chapter 16. We know that every density matrix ρ can be decomposed into an ensemble of pure states $\rho_i = |\psi_i\rangle\langle\psi_i|$ with probabilities p_i such that $\rho = \sum_i p_i \rho_i$. The entanglement of formation for a pure state is given by the entropy of either of the two subsystems. For a mixed state the *entanglement of formation* is defined as the average entanglement entropy of the pure states of the decomposition, minimized over the set $\mathcal{D}(\rho)$ of all decompositions of ρ, recall Section 16.2.2,

$$E_{\mathrm{F}}(\rho) = \inf_{\mathcal{D}(\rho)} \sum_i p_i\, S(\rho_i^1) . \tag{27.36}$$

It quantifies (gives an upper bound on) the resources needed to create a given entangled state.

In Bennett *et al.* (1996*c*) a remarkably simple formula for the *entanglement of formation* (recall Theorem 16.5 of Section 16.2.2) was found

$$E_{\mathrm{F}}(\rho) \geq h\big(f(\rho)\big) , \tag{27.37}$$

where the function $h\big(f(\rho)\big)$ is defined by

$$h(f) = H_{\mathrm{bin}}\Big(\frac{1}{2} + \sqrt{f(1-f)}\Big) \quad \text{for} \quad f \geq \frac{1}{2} , \tag{27.38}$$

and $h(f) = 0$ for $f < \frac{1}{2}$. The function H represents the familiar binary entropy function $H_{\mathrm{bin}}(x) = -x\log_2(x) - (1-x)\log_2(1-x)$. The quantity $f(\rho)$ is called the *fully entangled fraction* of ρ

$$f(\rho) = \max \langle e|\rho|e\rangle , \tag{27.39}$$

which is the maximum over all maximally entangled states $|e\rangle$.

For general mixed states ρ the function $h(f(\rho))$ is only a lower bound to the entropy $E_F(\rho)$. For pure states and mixtures of Bell states—the case we consider here—the bound is saturated, $E_F = h$, and we have the formula (27.38) for calculating the entanglement of formation.

Concurrence: Next we consider the entanglement monotone called *concurrence* from Section 16.2.2. Wootters and Hill (Hill and Wootters, 1997; Wootters, 1998; Wootters, 2001) discovered that the entanglement of formation for a general mixed state ρ of two qubits can be expressed by another quantity, the *concurrence* C, see Theorem 16.7 in Section 16.2.2

$$E_F(\rho) = E_F\big(C(\rho)\big) = H\left(\frac{1}{2} + \frac{1}{2}\sqrt{1 - C^2}\right) \quad \text{with} \quad 0 \leq C \leq 1. \quad (27.40)$$

Inserting the entropy function H_{bin} the entanglement of formation can be expressed as

$$E_F(C(\rho)) = -\tfrac{1}{2}(1 + \sqrt{1 - C^2}) \log\left(\tfrac{1}{2}(1 + \sqrt{1 - C^2})\right)$$
$$- \tfrac{1}{2}(1 - \sqrt{1 - C^2}) \log\left(\tfrac{1}{2}(1 - \sqrt{1 - C^2})\right), \quad (27.41)$$

where again the logarithm is taken to base 2. The function $E_F(\rho)$ is monotonically increasing from 0 to 1 as C runs from 0 to 1, and C itself is an entanglement monotone.

Defining the spin-flipped state $\tilde{\rho}$ of ρ as

$$\tilde{\rho} = (\sigma_y \otimes \sigma_y)\, \rho^*(\sigma_y \otimes \sigma_y), \quad (27.42)$$

where ρ^* is the complex conjugate with respect to the chosen standard basis, i.e., here the basis $\{|\uparrow\uparrow\rangle, |\downarrow\downarrow\rangle, |\uparrow\downarrow\rangle, |\downarrow\uparrow\rangle\}$, the *concurrence* C is given by the formula (Theorem 16.6)

$$C(\rho) = \max\{0, \lambda_1 - \lambda_2 - \lambda_3 - \lambda_4\}. \quad (27.43)$$

The λ_i are the square roots of the eigenvalues, in decreasing order, of the matrix $\rho\tilde{\rho}$.

Application to the decoherence model: For the density matrix $\rho_N(t)$ (27.26) with eqn (27.12) according to our chosen decoherence model, which is invariant under spin flips, i.e., $\tilde{\rho}_N = \rho_N$ and thus $\rho_N\tilde{\rho}_N = \rho_N^2$, the *concurrence* is

$$C\big(\rho_N(t)\big) = \max\big\{0, e^{-\lambda t}\big\} = e^{-\lambda t}, \quad (27.44)$$

and the *fully entangled fraction* of $\rho_N(t)$ is

$$f\big(\rho_N(t)\big) = \frac{1}{2}\big(1 + e^{-\lambda t}\big), \quad (27.45)$$

which is simply the largest eigenvalue of $\rho_N(t)$. The functions C and f are related by

$$C\big(\rho_N(t)\big) = 2\, f\big(\rho_N(t)\big) - 1. \quad (27.46)$$

Finally, we have for the *entanglement of formation* of the $K^0\bar{K}^0$ system

$$E\big(\rho_N(t)\big) = -\frac{1+\sqrt{1-e^{-2\lambda t}}}{2}\log\Big(\frac{1+\sqrt{1-e^{-2\lambda t}}}{2}\Big)$$
$$-\frac{1-\sqrt{1-e^{-2\lambda t}}}{2}\log\Big(\frac{1-\sqrt{1-e^{-2\lambda t}}}{2}\Big). \qquad (27.47)$$

Now using the relation (27.25) between the decoherence parameters λ and ζ we find a striking connection between the entanglement measure, defined by the entropy of the state, and the decoherence of the quantum system, which describes the amount of factorization into product states (Furry–Schrödinger hypothesis).

Loss of entanglement: Defining the *loss of entanglement* as the gap between the value of an entanglement for a given state and its maximum value for any state, we find (Bertlmann *et al.*, 2003)

$$1 - C\big(\rho_N(t)\big) = \zeta(t), \qquad (27.48\text{a})$$

$$1 - E_{\mathrm{F}}\big(\rho_N(t)\big) \doteq \frac{1}{\ln(2)}\,\zeta(t) \doteq \frac{\lambda}{\ln(2)}\,t, \qquad (27.48\text{b})$$

where in eqn (27.48b) we have expanded expression (27.47) for small values of the parameters λ or ζ. In this way we arrive at the proposition by *Reinhold Bertlmann, Katharina Durstberger*, and *Beatrix Hiesmayr* (2003):

Proposition 27.2 (Entanglement loss—decoherence)

The entanglement loss matches the decoherence!

It implies that, under the previous assumptions on the decoherence model, we are able to experimentally determine the degree of entanglement of the $K^0\bar{K}^0$ system, namely by considering the asymmetry (27.18) and fitting the parameter ζ or λ to the data.

Experimental results: In Fig. 27.3 we have plotted the loss of entanglement $1 - E_{\mathrm{F}}$, given by eqn (27.47), as compared to the loss of information, the von Neumann entropy function S, eqn (27.27), in dependence of the time t/τ_s of the propagating $K^0\bar{K}^0$ system. The loss of entanglement of formation increases slower with time and visualizes the resources needed to create a given entangled state. At $t = 0$ the pure Bell state ρ^- is created and becomes mixed for $t > 0$ by the other Bell state ρ^+. In the total state the amount of entanglement decreases until separability is reached (exponentially fast) for $t \to \infty$.

In case of the CPLEAR experiment, where one kaon propagates about 2 cm, corresponding to a propagation time $t_0/\tau_s \approx 0.55$, until it is measured by an absorber, the entanglement loss is about 18% for the mean value and maximal 38% for the upper bound of the decoherence parameter λ.

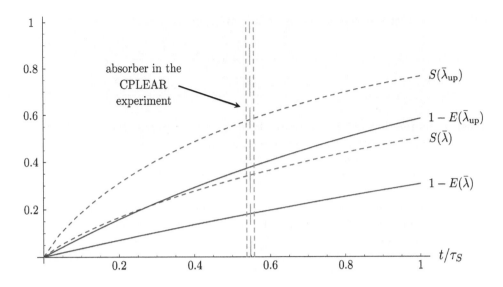

Fig. 27.3 The time-dependence of the entropy. The von Neumann entropy (dashed lines), eqn (27.27), and the loss of entanglement of formation $1 - E$ (solid lines), given by eqn (27.47), are plotted for the experimental mean value $\bar{\lambda} = 1.84 \cdot 10^{-12}$ MeV (lower curve) and the upper bound $\bar{\lambda}_{\text{up}} = 4.34 \cdot 10^{-12}$ MeV (upper curve), eqn (27.19), of the decoherence parameter λ. The time t is scaled versus the lifetime τ_s of the short-lived kaon K_S: $t \to t/\tau_s$. The vertical lines represent the propagation time $t_0/\tau_s \approx 0.55$ of one kaon, including the experimental error bars, until it is measured by the absorber in the CPLEAR experiment. Reprinted (Fig. 1) with permission from Reinhold A. Bertlmann, Katharina Durstberger, and Beatrix C. Hiesmayr, Phys. Rev. A **68**, 012111 (2003). Copyright (2003) by the American Physical Society.

Résumé: The model of Bertlmann and Grimus (2001) provides a quite simple and practicable procedure to quantitatively estimate the degree of possible decoherence of a quantum state due to some interaction of the state with its "environment". This is of special interest in the case of entangled states where a single parameter, the decoherence parameter λ or ζ, quantifies the strength of decoherence between the two subsystems or the amount of spontaneous wave-function factorization. The asymmetry of correlated and anti-correlated flavour events is directly sensitive to λ or ζ.

Furthermore, there exists a remarkable bridge to common entanglement measures and monotones, such as entanglement of formation and concurrence, which can be phrased in the proposition: "The entanglement loss matches the decoherence"! This provides a different view on making statements on the amount of entanglement in the state. Using the experimental data of the CPLEAR experiment provides us with upper bounds on the decoherence parameters and the entanglement loss of the $K^0\bar{K}^0$ state produced in $p\bar{p}$ collisions, which is of macroscopic extent (9 cm). Of course, these decoherence investigations can be performed with other entangled systems, like D-meson or B-meson systems (Section 27.5 and 27.6) or photonic systems, as well.

27.5   Entanglement of Beauty

In this section we consider the entanglement of B mesons. They are produced to a huge amount at the $\Upsilon(4S)$ resonance in the B-factories KEK-B in Tsukuba or PEP-II at SLAC.

27.5.1   B mesons

Let us first discuss the main properties of the B mesons, also called *bottom* mesons or *beauty* mesons, which we are interested in here. They contain the heavy *bottom* or *beauty* quark b ($m_b = 4.20 \pm 0.07$ GeV) and are characterized therefore by the quantum number B, named *bottom* or *beauty*. The B mesons are pseudoscalar particles ($J^P = 0^-$), charged $B^+(\bar{b}u)$, $B^-(\bar{u}b)$ and neutral $B^0(\bar{b}d)$, $\bar{B}^0(\bar{d}b)$ with mass $m = 5279$ MeV, and a tiny mass difference $m_{B^0} - m_{B^\pm} = 0.33 \pm 0.28$ MeV. They are bound states composed of the quarks u, d, and b, and anti-quarks $\bar{u}$, $\bar{d}$, and $\bar{b}$.

The quantum-mechanical formalism that describes them is analogous to that of the K mesons (see Section 26.1), so we just have to replace the strange quark s by the bottom quark b. Let us begin with the specific properties of the B mesons we are interested in.

Beauty oscillation $B^0 \leftrightarrow \bar{B}^0$: In case of B mesons the eigenstates $|B_H\rangle$ and $|B_L\rangle$ of the non-Hermitian effective mass Hamiltonian H (26.6) are called heavy and light states, respectively, satisfying

$$H\,|B_{H,L}\rangle = \lambda_{H,L}\,|B_{H,L}\rangle \quad \text{with} \quad \lambda_{H,L} = m_{H,L} - \frac{i}{2}\Gamma_{H,L}\,. \tag{27.49}$$

The heavy and light neutral B states are defined via

$$|B_H\rangle = p|B^0\rangle + q|\bar{B}^0\rangle \quad \text{and} \quad |B_L\rangle = p|B^0\rangle - q|\bar{B}^0\rangle\,, \tag{27.50}$$

where $p = 1 + \varepsilon$, $q = 1 - \varepsilon$ are the weights when CP violation with the parameter ε is included, analogously to eqns (26.5a) and (26.5b). Since both mesons B^0 and $\bar{B}^0$ have transitions to common states (due to CP violation) they mix; that means they *oscillate* between B^0 and $\bar{B}^0$ before decaying.

Beauty decay: The decaying states evolve, according to the Wigner–Weisskopf approximation, exponentially in time as

$$|B_{H,L}(t)\rangle = e^{-i\lambda_{H,L}t}\,|B_{H,L}\rangle\,, \tag{27.51}$$

implying the time evolution for B^0 and $\bar{B}^0$

$$|B^0(t)\rangle = h_+(t)|B^0\rangle + \tfrac{q}{p}\,h_-(t)|\bar{B}^0\rangle\,, \tag{27.52a}$$

$$|\bar{B}^0(t)\rangle = \tfrac{p}{q}\,h_-(t)|B^0\rangle + h_+(t)|\bar{B}^0\rangle\,, \tag{27.52b}$$

with

$$h_{\pm}(t) \;=\; \frac{1}{2}\left[\pm e^{-i\lambda_{\mathrm{H}}t} + e^{-i\lambda_{\mathrm{L}}t}\right] . \tag{27.53}$$

Let us now suppose that a B^0 beam is produced at $t = 0$, then the probability of finding a B^0 or $\bar{\mathrm{B}}^0$ in the beam is calculated to be

$$\left|\langle \mathrm{B}^0|\mathrm{B}^0(t)\rangle\right|^2 \;=\; \frac{1}{4}\left\{e^{-\Gamma_{\mathrm{H}}t} + e^{-\Gamma_{\mathrm{L}}t} + 2\,e^{-\Gamma t}\cos(\Delta m\, t)\right\} , \tag{27.54a}$$

$$\left|\langle \bar{\mathrm{B}}^0|\mathrm{B}^0(t)\rangle\right|^2 \;=\; \tfrac{1}{4}\tfrac{|q|^2}{|p|^2}\left\{e^{-\Gamma_{\mathrm{H}}t} + e^{-\Gamma_{\mathrm{L}}t} - 2\,e^{-\Gamma t}\cos(\Delta m\, t)\right\} , \tag{27.54b}$$

with $\Delta m = m_{\mathrm{H}} - m_{\mathrm{L}}$ and $\Gamma = \tfrac{1}{2}(\Gamma_{\mathrm{H}} + \Gamma_{\mathrm{L}})$.

In the following we assume CP conservation in B^0-$\bar{\mathrm{B}}^0$ mixing, which is a good approximation and corresponds to $|p/q| = 1$. In this case we have $\langle \mathrm{B}_{\mathrm{H}}|\mathrm{B}_{\mathrm{L}}\rangle = 0$, otherwise $\langle \mathrm{B}_{\mathrm{H}}|\mathrm{B}_{\mathrm{L}}\rangle \neq 0$. That means finally, we will set $p = q = 1/\sqrt{2}$.

The main difference of the B mesons in contrast to the K mesons is:

(i) $\Delta m = m_{\mathrm{H}} - m_{\mathrm{L}} = 3 \times 10^{-4}$ eV is large,

(ii) $\Delta\Gamma = \Gamma_{\mathrm{H}} - \Gamma_{\mathrm{L}} \approx 0$ is small and $\Gamma_{\mathrm{B}^0}^{-1} = \tau_{\mathrm{B}^0} = 1.5 \times 10^{-12}$ sec.

27.5.2 Production of *B* Mesons

Next we want to discuss briefly the production of B-meson pairs at the asymmetric $\mathrm{e}^+\mathrm{e}^-$ collider KEK-B (operating 1998–2010) in Tsukuba, Japan. The low-energy ring (LER) of the collider contained positrons e^+ at 3.5 GeV and the high-energy ring (HER) the electrons e^- at 8.0 GeV, see Fig. 27.4, giving 10.58 GeV centre-of-mass energy, which is equal to the mass of the $\Upsilon(4S)$ resonance. The transition $\mathrm{e}^+\mathrm{e}^- \to \gamma \to \Upsilon \to \mathrm{B}^0\bar{\mathrm{B}}^0$ is called *hadronization*, illustrated in Fig. 27.5. In the interaction region the particles were detected by the huge Belle detector (Fig. 27.6). The resonance decays at the primary vertex into the B mesons $\Upsilon \to \mathrm{B}^0\bar{\mathrm{B}}^0$ and the B mesons decay at the secondary vertices, see Fig. 27.7. The created $\mathrm{B}^0\bar{\mathrm{B}}^0$ system, which oscillates $\mathrm{B}^0\bar{\mathrm{B}}^0 \leftrightarrow \bar{\mathrm{B}}^0\mathrm{B}^0$, is entangled until the first B meson decays at a time t_1, while the second B meson, still oscillating further, decays at a time t_2, as illustrated in Fig. 27.7.

27.5.3 Entanglement of *B* Mesons

In the Belle detector of the KEK-B experiment is a B-meson pair produced at the primary vertex, see Fig. 27.7. It is created in the antisymmetric Bell state $|\psi^-\rangle$ which we can express in the basis consisting of B_{H} and B_{L} as

$$|\psi^-(t=0)\rangle \;=\; \frac{N_{\mathrm{HL}}}{\sqrt{2}}\left\{|\mathrm{B}_{\mathrm{H}}\rangle_{\mathrm{l}} \otimes |\mathrm{B}_{\mathrm{L}}\rangle_{\mathrm{r}} - |\mathrm{B}_{\mathrm{L}}\rangle_{\mathrm{l}} \otimes |\mathrm{B}_{\mathrm{H}}\rangle_{\mathrm{r}}\right\} , \tag{27.55}$$

with $N_{\mathrm{HL}} = \frac{N^2}{2pq}$, $N^2 = |p|^2 + |q|^2$ and p, q are the weights of the heavy and light mesons B_{H} and B_{L}, respectively, which are defined in eqn (27.50). Thus the B-meson formalism proceeds analogously to the kaon case we discussed before, we just have to replace K_{S} and K_{L} by B_{H} and B_{L}.

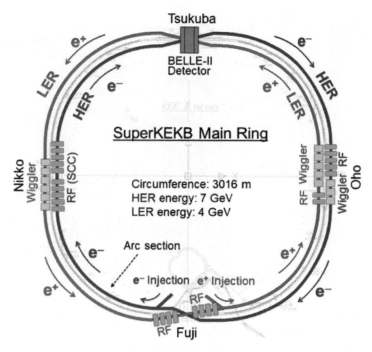

Fig. 27.4 Layout of the SuperKEK-B ring. The two storage rings for the high- (HER) and low-energy (LER) rings. The SuperKEK-B accelerator and Belle II detector shown in the illustration are upgraded versions of the KEK-B accelerator and Belle detector, but with similar layout, in particular, with the detector located at the beam crossing at the Tsukuba area. Figure from Suetsugu *et al.* 2019, Fig. 1).

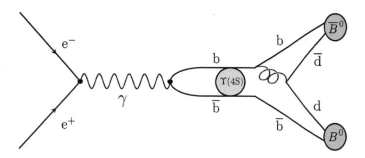

Fig. 27.5 Hadronisation of beauty. In the transition $e^+e^- \to \gamma \to \Upsilon(4S) \to B^0\bar{B}^0$ the beauty mesons are created at the $\Upsilon(4S)$ resonance by hadronization, i.e., via gluon exchange between the quarks.

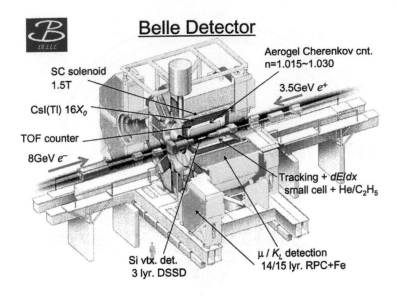

Fig. 27.6 Belle detector. Sketch of the huge Belle detector at the KEK-B ring, which tracks the B mesons via their decays. A figure of a typical human is shown for scale on the bottom left. Figure adapted from Widhalm (2005, Fig. 4).

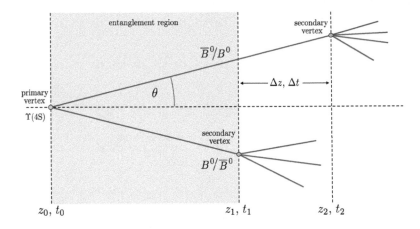

Fig. 27.7 $B^0\bar{B}^0$ **creation.** At the primary vertex the Υ resonance decays into an entangled $B^0\bar{B}^0$ pair at a time t_0. The entangled pair propagates by oscillating $B^0\bar{B}^0 \leftrightarrow \bar{B}^0B^0$ inside the entanglement region until the first B meson decays at a time t_1, while the second B meson, still oscillating, decays at a time t_2. The angle θ is quite small such that $\tan(\theta) \approx 0.1$.

27.6 Decoherence of Entangled Beauty

27.6.1 Decoherence Model

Next we consider the time evolution of the B mesons according to the Wigner–Weisskopf approximation

$$|B_{H,L}(t)\rangle = e^{-i\lambda_{H,L}t}|B_{H,L}\rangle, \quad \text{with} \quad \lambda_{H,L} = m_{H,L} - \frac{i}{2}\Gamma_{H,L}, \qquad (27.56)$$

and follow the model by Bertlmann and Grimus (2001) described in Section 27.1 for the measurement of the entangled B-meson pair. It describes the possible decoherence of the pair. Then we find the probabilities for correlated and anti-correlated beauty—analogously to the kaon case with eqns (27.15a), (27.15b)—the formulas

$$P_\lambda(B^0, t_1; B^0, t_r) = P_\lambda(\bar{B}^0, t_1; \bar{B}^0, t_r) \qquad (27.57a)$$

$$= \frac{1}{8}\left(e^{-\Gamma_H t_1 - \Gamma_L t_r} + e^{-\Gamma_L t_1 - \Gamma_H t_r} - e^{-\lambda t_r} 2\cos(\Delta m \Delta t) \cdot e^{-\Gamma(t_1 + t_r)} \right),$$

$$P_\lambda(B^0, t_1; \bar{B}^0, t_r) = P_\lambda(\bar{B}^0, t_1; B^0, t_r) \qquad (27.57b)$$

$$= \frac{1}{8}\left(e^{-\Gamma_H t_1 - \Gamma_L t_r} + e^{-\Gamma_L t_1 - \Gamma_H t_r} + e^{-\lambda t_r} 2\cos(\Delta m \Delta t) \cdot e^{-\Gamma(t_1 + t_r)} \right), \quad (27.57c)$$

with $\Delta t = t_1 - t_r$. The interesting quantity directly sensitive to the interference term is the asymmetry of probabilities

$$A^{QM}(\Delta t) = \frac{P(B^0, t_1; \bar{B}^0, t_r) + P(\bar{B}^0, t_1; B^0, t_r) - P(B^0, t_1; B^0, t_r) - P(\bar{B}^0, t_1; \bar{B}^0, t_r)}{P(B^0, t_1; \bar{B}^0, t_r) + P(\bar{B}^0, t_1; B^0, t_r) + P(B^0, t_1; B^0, t_r) + P(\bar{B}^0, t_1; \bar{B}^0, t_r)}$$

$$= \frac{\cos(\Delta m \Delta t)}{\cosh(\frac{1}{2}\Delta\Gamma\Delta t)}, \qquad (27.58)$$

with $\Delta\Gamma = \Gamma_L - \Gamma_H$. Thus we find by inserting the probabilities (27.57a) and (27.57b)

$$A^\lambda(t_1, t_r) = \frac{\cos(\Delta m \Delta t)}{\cosh(\frac{1}{2}\Delta\Gamma\Delta t)} e^{-\lambda \min\{t_1, t_r\}} = A^{QM}(\Delta t) e^{-\lambda \min\{t_1, t_r\}}. \qquad (27.59)$$

Replacing $e^{-\lambda \min\{t_1, t_r\}} = (1 - \zeta(t_1, t_r))$ according to eqn (27.23) we finally get

$$A^\zeta(t_1, t_r) = A^{QM}(\Delta t)\ (1 - \zeta(t_1, t_r)). \qquad (27.60)$$

Not that for B mesons the difference $\Delta\Gamma$ is practically zero so that in eqn (27.58) the denominator $\cosh(\frac{1}{2}\Delta\Gamma\Delta t) \approx 1$.

27.6.2 Experiment at KEK-B

Let us now discuss an analysis of B-meson data at KEK-B. Working with the previously described formalism, *Gerald Richter* (2008) analysed the B-meson data of the Belle detector at KEK-B. For the λ-parameter he obtained the values

$$\lambda_{\text{Richter}} = -0.044 \, {}^{+0.057}_{-0.048} \pm 0.103 \,, \tag{27.61}$$

in units of B_d^0 lifetimes, where the first error is of statistical, the second one of systematic origin. Translated to the dimensionless ζ-parameter via eqn (27.23) we find

$$\zeta_{\text{Richter}} = -0.045 \pm 0.155 \,. \tag{27.62}$$

Apollo Go from the Belle collaboration (2007) also analysed the data of the neutral B-meson pairs produced at $\Upsilon(4S)$ and deduced the value

$$\zeta_{\text{Go}} = 0.029 \pm 0.057 \,. \tag{27.63}$$

Finally the ζ values (27.62) and (27.63) can be compared with an analysis from a different experiment at DESY in the 1990s, which was derived by Bertlmann and Grimus (2001) from the $\Upsilon(4S)$ data of the ARGUS collaboration (1994) and CLEO collaboration (1993)

$$\zeta_{\text{BG}} = -0.06 \pm 0.10 \,. \tag{27.64}$$

We notice that all ζ values from experiments are consistent with quantum mechanics ($\zeta = 0$) and far from the total decoherence ($\zeta = 1$) of the local realistic theories.

Résumé: Analysing the neutral B-meson pairs produced at the $\Upsilon(4S)$ resonance we notice that they indeed exhibit the non-local correlations of the type EPR–Bell. The data are consistent with the predictions of entanglement by quantum mechanics ($\zeta = 0$) and far from the total decoherence ($\zeta = 1$) of the local realistic theories. Thus also these very massive particle systems (≈ 10 GeV) produced in high-energy accelerators are entangled at a macroscopic scale (≈ 1 mm). The validity of the energy/mass range (from very low to very high) of quantum mechanics is quite impressive.

27.7 Open Quantum System and Particle Decay

In this section we consider an open quantum system which contains unstable states like the K mesons or B mesons discussed in previous sections. Then the time evolution of the system can be described by an effective non-Hermitian Hamiltonian H_{eff}, in accordance with the Wigner–Weisskopf approximation, and an additional term of the Lindblad form, the so-called dissipator. After enlarging the original Hilbert space by states which represent the decay products of the unstable states, the non-Hermitian part of H_{eff}—the "particle decay"—can be incorporated into the dissipator of the enlarged space via a specific Lindblad operator. As a consequence the new formulation of the time evolution on the enlarged space has a Hermitian Hamiltonian and conserves probability. The equivalence of the new formulation with the original one demonstrates that the time evolution which is governed by a *non-Hermitian* Hamiltonian and a dissipator of the Lindblad form is nevertheless completely positive, just as systems with Hermitian Hamiltonians. In this section we closely follow Bertlmann *et al.* (2006).

27.7.1 Physical Setup—Open Quantum System

In reality a quantum system is not isolated but always interacts with its environment and has to be considered as an open quantum system (Alicki and Fannes, 2001; Breuer and Petruccione, 2002). This leads to a mixing of the states in the system—that is, decoherence—and to an energy exchange between system and environment—dissipation (Kübler and Zeh, 1973; Joos and Zeh, 1985; Zurek, 1991; Joos, 1996). Decoherence and dissipation weaken or destroy the typical quantum phenomena such as interference.

We recall the master equation (27.3) for such an open quantum system described by the density matrix $\rho(t)$

$$\frac{\partial \rho}{\partial t} = -iH_{\text{eff}}\, \rho + i\rho\, H_{\text{eff}}^{\dagger} - \mathcal{D}[\rho]\,, \qquad (27.65)$$

where the dissipator is of Linblad type (27.5)

$$\mathcal{D}[\rho] = \frac{1}{2} \sum_j \left(A_j^{\dagger} A_j\, \rho + \rho A_j^{\dagger} A_j - 2 A_j \rho A_j^{\dagger} \right)\,, \qquad (27.66)$$

and A_j are the Lindblad operators. The effective Hamiltonian of the system is given, according to the Wigner–Weisskopf approximation, by

$$H_{\text{eff}} = H - \frac{i}{2}\, \Gamma, \qquad (27.67)$$

where H and Γ are both Hermitian and, in addition, $\Gamma \geq 0$.

In terms of open quantum systems, the time evolution of the density matrix represents a dynamical map which transforms initial density matrices $\rho(0)$ on the Hilbert space of states, $\mathcal{H}_S$, to final density matrices $\rho_S(t) = \Lambda(t)\,\rho_S(0)$, while the system is interacting with an environment (recall Section 22.1.2). Such dynamical maps are:

    (i) linear,
    (ii) trace preserving, and
    (iii) completely positive.

Complete positivity (see eqn (22.22) of Section 22.1.2) is a rather strong and important property. It is defined by demanding that all extensions $\Lambda(t) \otimes \mathbb{1}_n$ on the Hilbert space $\mathcal{H}_S \otimes \mathbb{C}^n$ are positive, i.e.

$$\left(\Lambda(t) \otimes \mathbb{1}_n \right) \rho \geq 0\,, \quad \forall \rho \geq 0\,, \quad \forall n = 1, 2, \dots \qquad (27.68)$$

where ρ is a density matrix on $\mathcal{H}_S \otimes \mathbb{C}^n$. Physically, it is a reasonable condition since the extension $\Lambda(t) \otimes \mathbb{1}_n$ can be considered as an operator on the composite quantum system of two parties, Alice and Bob, which acts locally on Alice's system without influencing Bob. Furthermore, complete positivity ensures that tensor products of maps $\Lambda(t)$ remain positive, an important property especially when considering entangled states.

27.7.2   Extended Formalism

We will show that both approaches—each one having its own appeal—are indeed equivalent to each other for quite general quantum systems. To be more precise, we ask the following question.

Question: Can we work with a Hermitian Hamiltonian and incorporate the decay as Lindblad operator such that the time evolution of the system represents a completely positive map?

Moreover, does it describe the Wigner–Weisskopf approximation properly without effecting decoherence and/or dissipation?

Answer: *Yes, this is possible!*

But in order to do so we have to enlarge the Hilbert space and formally include the decay products of the unstable states. Thus we assume in the following that the total Hilbert space $\mathcal{H}_{\text{tot}}$ is the direct sum

$$\mathcal{H}_{\text{tot}} = \mathcal{H}_S \oplus \mathcal{H}_f , \tag{27.69}$$

where $\mathcal{H}_S$ contains the states of the system we are interested in, whereas $\mathcal{H}_f$ is the space of the "decay states" defined in the following way. First we mention that $d_s := \dim(\mathcal{H})_S < \infty$. Then the non-Hermitian part Γ of the effective Hamiltonian in eqn (27.67) can be expressed as

$$\Gamma = \sum_{j=1}^{r} \gamma_j \, |\varphi_j\rangle\langle\varphi_j| \quad \text{with} \quad \gamma_j > 0 \; \forall j , \tag{27.70}$$

with a basis $\{\varphi_j\}_j$ of $\mathcal{H}_S$ and $r = \dim(\mathcal{H}_S) - n_0$, where n_0 denotes the degeneracy of the eigenvalue zero of Γ. We also demand that $\dim(\mathcal{H}_f) =: d_f \geq r$, otherwise $\mathcal{H}_f$ is arbitrary.

In the following, ρ denotes a density matrix on $\mathcal{H}_{\text{tot}}$, denoted by

$$\rho = \begin{pmatrix} \rho_{ss} & \rho_{sf} \\ \rho_{fs} & \rho_{ff} \end{pmatrix} \quad \text{with} \quad \rho_{ss}^\dagger = \rho_{ss} , \quad \rho_{ff}^\dagger = \rho_{ff} , \quad \rho_{fs}^\dagger = \rho_{sf} . \tag{27.71}$$

Next we have to define the time evolution of ρ on $\mathcal{H}_{\text{tot}}$. The Hamiltonian H and the Lindblad operators A_j are easily extended to the total Hilbert space by the matrices

$$\mathcal{H} = \begin{pmatrix} H & 0 \\ 0 & 0 \end{pmatrix} \qquad \mathcal{A}_j = \begin{pmatrix} A_j & 0 \\ 0 & 0 \end{pmatrix} . \tag{27.72}$$

Note that for the definition of $\mathcal{H}$ we have used the *Hermitian* part H of H_{eff}.

Decay: Now we turn to the decay. We need a Lindblad operator $\mathcal{B}$ on the full space $\mathcal{H}_{\text{tot}}$ which describes the decay in the subspace $\mathcal{H}_S$. As we will see, the decay is described by

$$\mathcal{B} = \begin{pmatrix} 0 & 0 \\ B & 0 \end{pmatrix} \quad \text{with} \quad B : \mathcal{H}_S \to \mathcal{H}_f \, , \tag{27.73}$$

and

$$\Gamma = B^\dagger B \, . \tag{27.74}$$

Let $\{f_k\}$ be an orthonormal basis of $\mathcal{H}_f$. Then we can decompose B as

$$B = \sum_{k=1}^{d_f} \sum_{j=1}^{r} b_{kj} \, |f_k\rangle\langle\varphi_j| \, . \tag{27.75}$$

In order to satisfy eqn (27.74) we require

$$\sum_{k=1}^{d_f} b_{ki}^* b_{kj} = \delta_{ij} \, \gamma_j \, . \tag{27.76}$$

For $d_f \geq r$, such a $d_f \times r$ matrix (b_{kj}) always exists. The simplest case is $d_f = r$, where we can choose

$$B = \sum_{j=1}^{d_f} \sqrt{\gamma_j} \, |f_j\rangle\langle\varphi_j| \, . \tag{27.77}$$

In that case each unstable decaying state $|\varphi_j\rangle$ would decay into just one specific decay state $|f_j\rangle$.

27.7.3 Extended Master Equation

We can write the following extended master equation for $\rho \in \mathcal{H}_{\text{tot}}$:

$$\frac{\partial \rho}{\partial t} = -i \, [\mathcal{H}, \rho] - \mathcal{D}[\rho] \, , \tag{27.78}$$

with the dissipator

$$\mathcal{D}[\rho] = \frac{1}{2} \sum_j \left(A_j^\dagger A_j \, \rho + \rho \, A_j^\dagger A_j - 2 \, A_j \rho A_j^\dagger \right) + \frac{1}{2} \left(\mathcal{B}^\dagger \mathcal{B} \rho + \rho \mathcal{B}^\dagger \mathcal{B} - 2 \mathcal{B} \rho \mathcal{B}^\dagger \right) \, . \tag{27.79}$$

To prove that this extended master equation contains the original equation (27.65) we decompose it with respect to the components of ρ:

$$\dot{\rho}_{ss} = -i \, [H, \rho_{ss}] - \frac{1}{2} \{ B^\dagger B, \rho_{ss} \}$$
$$- \frac{1}{2} \sum_j \left(A_j^\dagger A_j \, \rho_{ss} + \rho_{ss} A_j^\dagger A_j - 2 \, A_j \rho_{ss} A_j^\dagger \right) \, , \tag{27.80a}$$

$$\dot{\rho}_{sf} = -i \, H \rho_{sf} - \frac{1}{2} B^\dagger B \, \rho_{sf} - \frac{1}{2} \sum_j A_j^\dagger A_j \, \rho_{sf} \, , \tag{27.80b}$$

$$\dot{\rho}_{ff} = B \, \rho_{ss} \, B^\dagger \, . \tag{27.80c}$$

Indeed, with eqn (27.74), we immediately see that eqn (27.80a) for ρ_{ss} reproduces the original equation (27.65). Furthermore, $\text{Tr}(\rho(t)) = 1 \quad \forall t \geq 0$.

Remark I: At this point, some general remarks are in order. By construction, the time evolution of ρ_{ss} is independent of ρ_{sf}, ρ_{fs}, and ρ_{ff}. Actually, the time evolution of ρ_{sf} or ρ_{fs} completely decouples from that of ρ_{ss} and ρ_{ff}. The time evolution of ρ_{ff}, the density matrix of the decay states, is determined solely by $\rho_{ss}(t)$ and is the characteristic for a decay process. With the initial condition $\rho_{ff}(0) = 0$ we have

$$\rho_{ff}(t) = B \int_0^t dt' \rho_{ss}(t') B^\dagger . \tag{27.81}$$

For the initial condition $\rho_{sf}(0) = 0$ elements ρ_{sf} and ρ_{fs} remain zero for all times.

Remark II: Generally, for decaying systems the following properties of the time evolution (27.78) are most important:

(i) $\rho_{ss}(t)$ and $\rho_{ff}(t)$ are positive $\forall t \geq 0$, and

(ii) the time evolution of ρ_{ss} is *completely positive*.

27.7.4 Case of Non-Singular Decay Γ

Finally, it is interesting to consider the special case of a non-singular decay matrix Γ, where all states of $\mathcal{H}_S$ decay. Then, $\dim(\mathcal{H}_f) \geq \dim(\mathcal{H}_S)$ and, in addition to what we discussed before, the following properties of the time evolution (27.78) hold:

(1) $\lim_{t \to \infty} \rho_{sf}(t) = 0$,

(2) $\lim_{t \to \infty} \rho_{ss}(t) = 0$,

(3) $\lim_{t \to \infty} \text{Tr}(\rho_{ff}(t)) = 1$.

Simplest possible example: We choose dimensions $d_s = d_f = 1$, the Hamiltonian

$$\mathcal{H} = \begin{pmatrix} m & 0 \\ 0 & 0 \end{pmatrix} , \tag{27.82}$$

and set $A_j = 0$, i.e., we neglect any decoherence part. Then, eqns (27.80a), (27.80c), and (27.81) simplify considerably, providing the result

$$\rho(t) = \begin{pmatrix} e^{-\Gamma t} & 0 \\ 0 & 1 - e^{-\Gamma t} \end{pmatrix} , \tag{27.83}$$

where we have chosen the initial conditions $\rho_{ss}(0) = 1$, $\rho_{ff}(0) = \rho_{sf}(0) = 0$. This example illustrates the general fact that, for non-singular Γ, the probability loss of the system is balanced by the probability increase of the decay states. Considering the mixedness of the quantum states, we use the measure

$$\delta(t) = \text{Tr}\left(\rho^2(t)\right) = 1 - 2\,e^{-\Gamma t} + 2\,e^{-2\Gamma t}. \tag{27.84}$$

We have $\delta(0) = 1$, $\lim_{t\to\infty} \delta(t) = 1$ and $\delta(t) < 1$ for $0 < t < \infty$, i.e., in the beginning ρ represents a pure state, then the state is mixed, whereas for large times it approaches a pure state again, when the system definitely has changed into the decay state.

Kraus operators: Finally, in the case of our example, we would like to consider the time evolution of the density matrix as a dynamical map $\Lambda(t)$. In our case, $\Lambda(t)$ can be represented by a sum of operators, the Kraus operators (recall Section 21.3)

$$\rho(0) \longrightarrow \Lambda[\rho(0)] = \sum_j K_j(t)\,\rho(0)K_j^\dagger(t) = \rho(t), \tag{27.85}$$

with the normalization $\sum_j K_j^\dagger(t)K_j(t) = \mathbb{1}$ for the Kraus operators $K_j(t)$.

The operator-sum representation (27.85) is a very useful approach in quantum information to describe the quantum operations or the specific quantum channels. In our example, the decay of a particle corresponds to the amplitude-damping channel of a quantum operation (Section 21.3.2), e.g., the spontaneous emission of a photon. We can calculate the Kraus operators needed:

$$M_0(t) = \begin{pmatrix} \sqrt{1 - p(t)} & 0 \\ 0 & 1 \end{pmatrix} \quad \text{and} \quad M_1(t) = \begin{pmatrix} 0 & 0 \\ \sqrt{p(t)} & 0 \end{pmatrix}, \tag{27.86}$$

with the probability $p(t) = 1 - e^{-\Gamma t}$.

Résumé: In this last section we have considered a time evolution given by eqn (27.65), with a non-Hermitian Hamiltonian H_{eff} and a dissipator of the Lindblad-type (27.66). Assuming that the non-Hermitian part of H_{eff} describes "particle decay", we have shown that such a time evolution is completely positive. The strategy—according to Bertlmann *et al.* (2006)—was to add the space of "decay states" $\mathcal{H}_f$ to the space of the "system" states $\mathcal{H}_S$ and to extend the time evolution in a straightforward way to the full space $\mathcal{H}_{\text{tot}} = \mathcal{H}_S \oplus \mathcal{H}_f$, such that this full time evolution conserves probability. With the initial condition that the density matrix is only non-zero on $\mathcal{H}_S$, the Lindblad operator $\mathcal{B}$ of eqn (27.73), which is responsible for the decay, shifts states from $\mathcal{H}_S$ to $\mathcal{H}_f$, whereas at the same time the $\mathcal{A}_j$ terms cause decoherence and dissipation on $\mathcal{H}_S$. Thus, particle decay and decoherence/dissipation are related phenomena, which can be described by the same formalism of a completely positive time evolution.

References

Acín, Antonio; Andrianov, Alexander; Costa, Laura; Jané, Enric; Latorre, José Ignacio; and Tarrach, Rolf (2000). Generalized Schmidt Decomposition and Classification of Three-Quantum-Bit States. *Phys. Rev. Lett.*, **85**, 1560–1563. Preprint https://arxiv.org/abs/quant-ph/0003050. `https://doi.org/10.1103/PhysRevLett.85.1560`.

Acín, Antonio; Bruß, Dagmar; Lewenstein, Maciej; and Sanpera, Anna (2001). Classification of mixed three-qubit states. *Phys. Rev. Lett.*, **87**, 040401–040404. Preprint https://arxiv.org/abs/quant-ph/0103025. `https://doi.org/10.1103/PhysRevLett.87.040401`.

Acín, Antonio; Gisin, Nicolas; and Toner, Benjamin (2006). Grothendieck's constant and local models for noisy entangled quantum states. *Phys. Rev. A*, **73**, 062105. Preprint https://arxiv.org/abs/quant-ph/0606138. `https://doi.org/10.1103/PhysRevA.73.062105`.

Adams, Robert A. and Fournier, John J. F. (2003). *Sobolev Spaces* (2nd edn). Academic Press, Amsterdam.

Adesso, Gerardo (2007). *Entanglement of Gaussian states*. Ph.D. thesis, University of Salerno. Preprint https://arxiv.org/abs/quant-ph/0702069.

Adesso, Gerardo and Illuminati, Fabrizio (2005). Gaussian measures of entanglement versus negativities: Ordering of two-mode Gaussian states. *Phys. Rev. A*, **72**, 032334. Preprint https://arxiv.org/abs/quant-ph/0506124. `https://doi.org/10.1103/PhysRevA.72.032334`.

Adesso, Gerardo; Ragy, Sammy; and Lee, Antony R. (2014). Continuous Variable Quantum Information: Gaussian States and Beyond. *Open Syst. Inf. Dyn.*, **21**(01n02), 1440001. Preprint https://arxiv.org/abs/1401.4679. `https://doi.org/10.1142/S1230161214400010`.

Adesso, Gerardo; Serafini, Alessio; and Illuminati, Fabrizio (2004). Extremal entanglement and mixedness in continuous variable systems. *Phys. Rev. A*, **70**, 022318. Preprint https://arxiv.org/abs/quant-ph/0402124. `https://doi.org/10.1103/PhysRevA.70.022318`.

Afriat, Alexander and Selleri, Franco (1999). *The Einstein, Podolsky, and Rosen Paradox in Atomic, Nuclear, and Particle Physics*. Plenum Press, New York. `https://doi.org/10.1007/978-1-4899-0254-2`.

Agudelo, Elizabeth; Sperling, Jan; and Vogel, Werner (2013). Quasiprobabilities for multipartite quantum correlations of light. *Phys. Rev. A*, **87**, 033811. Preprint https://arxiv.org/abs/1211.1585. `https://doi.org/10.1103/PhysRevA.87.033811`.

Aharonov, Yakir and Anandan, Jeeva S. (1987). Phase change during a cyclic quantum evolution. *Phys. Rev. Lett.*, **58**, 1593–1596. `https://doi.org/10.1103/PhysRevLett.58.1593`.

Aharonov, Yakir and Bohm, David (1959). Significance of electromagnetic potentials in the quantum theory. *Phys. Rev.*, **115**, 485–491. https://doi.org/10.1103/PhysRev.115.485.

Aharonov, Yakir and Bohm, David (1961). Further considerations on electromagnetic potentials in the quantum theory. *Phys. Rev.*, **123**, 1511–1524. https://doi.org/10.1103/PhysRev.123.1511.

Alberti, Peter M. and Uhlmann, Armin (1982). *Stochasticity and Partial Order: Doubly Stochastic Maps and Unitary Mixing*. Springer, Dordrecht, Netherlands.

Albrecht, H.; *et al.*: ARGUS Coll. (1994). A study of $B^0 \to D^{*+}$ and $B^0 B^0$ mixing using partial D^{*+} reconstruction. *Phys. Lett. B*, **324**, 249–254. https://doi.org/10.1016/0370-2693(94)90415-4.

Alicki, Robert (2002). Search for a border between classical and quantum worlds. *Phys. Rev. A*, **65**, 034104. Preprint https://arxiv.org/abs/quant-ph/0105089. https://doi.org/10.1103/PhysRevA.65.034104.

Alicki, Robert and Fannes, Mark (2001). *Quantum Dynamical Systems*. Oxford University Press, Oxford. https://doi.org/10.1093/acprof:oso/9780198504009.001.0001.

Alicki, Robert and Fannes, Mark (2003). Continuity of quantum mutual information. *J. Phys. A.*, **37**, L55–L57. Preprint https://arxiv.org/abs/quant-ph/0312081. https://doi.org/10.1088/0305-4470/37/5/L01.

Alicki, Robert and Lendi, Karl (1987). *Quantum Dynamical Semigroups and Applications, Lecture Notes in Physics 286*. Springer Verlag, Heidelberg. https://doi.org/10.1007/3-540-70861-8.

Allen, Les; Beijersbergen, Marco W.; Spreeuw, Robert J. C.; and Woerdman, Johannes Petrus (1992). Orbital angular momentum of light and the transformation of Laguerre-Gaussian laser modes. *Phys. Rev. A*, **45**, 8185. https://doi.org/10.1103/PhysRevA.45.8185.

Ambrosino, Fabio; *et al.*: KLOE Coll. (2006). First observation of quantum interference in the process $\phi \to K_S K_L \to \pi^+\pi^-\pi^+\pi^-$: A test of quantum mechanics and CPT symmetry. *Phys. Lett. B*, **642**, 315–321. Preprint https://arxiv.org/abs/hep-ex/0607027. https://doi.org/10.1016/j.physletb.2006.09.046.

Amelino-Camelia, Giovanni; Archilli, Flavio; and Babusci, Danilo; *et al.* (2010). Physics with the KLOE-2 experiment at the upgraded DAΦNE. *Eur. Phys. J. C*, **68**, 619–681. Preprint https://arxiv.org/abs/1003.3868. https://doi.org/10.1140/epjc/s10052-010-1351-1.

Anandan, Jeeva S. (1992). The geometric phase. *Nature*, **360**, 307–313. https://doi.org/10.1038/360307a0.

Anandan, Jeeva S.; Christian, Joy; and Wanelik, Kazimir (1997). Geometric phases in physics. *Am. J. Phys.*, **65**, 180–185. https://doi.org/10.1119/1.18570.

Ancochea, Bernat; Bramon, Albert; and Nowakowski, Marek (1999). Bell inequalities for $K^0 K^0$ pairs from Φ-resonance decays. *Phys. Rev. D*, **60**, 094008. Preprint https://arxiv.org/abs/hep-ph/9811404. https://doi.org/10.1103/PhysRevD.60.094008.

Andersen, Ulrik L.; Gehring, Tobias; Marquardt, Christoph; and Leuchs, Gerd (2016). 30 years of squeezed light generation. *Phys. Scr.*, **91**(5), 053001. Preprint https://arxiv.org/abs/1511.03250. `https://doi.org/10.1088/0031-8949/91/5/` `053001`.

Andô, Tsuyoshi (1979). Concavity of certain maps on positive definite matrices and applications to Hadamard products. *Linear Algebra Appl.*, **26**, 203–241. `https://doi.org/10.1016/0024-3795(79)90179-4`.

Andrae, Dirk (1997). Recursive evaluation of expectation values $\langle r^k \rangle$ for arbitrary states of the relativistic one-electron atom. *J. Phys. B: At. Mol. Opt. Phys.*, **30**, 4435–4451. `https://doi.org/10.1088/0953-4075/30/20/008`.

Apostolakis, A.; *et al.*: CPLEAR Coll. (1998). An EPR experiment testing the non-separability of the $K^0 \bar{K}^0$ wave function. *Phys. Lett. B*, **422**, 339–348. `https://doi.org/10.1016/S0370-2693(97)01545-1`.

Appel, Paul; Klöckl, Claude; and Huber, Marcus (2020). Monogamy of correlations and entropy inequalities in the Bloch picture. *J. Phys. Commun.*, **4**, 025009. Preprint https://arxiv.org/abs/1710.02473. `https://doi.org/10.1088/2399-6528/ab6fb4`.

Appleby, Marcus; Chien, Tuan-Yow; Flammia, Steven; and Waldron, Shayne (2018). Constructing exact symmetric informationally complete measurements from numerical solutions. *J. Phys. A: Math. Theor.*, **51**, 165302. Preprint https://arxiv.org/abs/1703.05981. `https://doi.org/10.1088/1751-8121/aab4cd`.

Araki, Huzihiro and Lieb, Elliott H. (1970). Entropy inequalities. *Commun. Math. Phys.*, **18**, 160–170. `https://doi.org/10.1007/BF01646092`.

Araújo, Mateus; Feix, Adrien; Costa, Fabio; and Brukner, Časlav (2014). Quantum circuits cannot control unknown operations. *New J. Phys.*, **16**, 093026. Preprint https://arxiv.org/abs/1309.7976. `https://doi.org/10.1088/1367-2630/16/9/093026`.

Archer, Claude (2005). There is no generalization of known formulas for mutually unbiased bases. *J. Math. Phys.*, **46**, 022106. Preprint https://arxiv.org/abs/quant-ph/0312204. `https://doi.org/10.1063/1.1829153`.

Armstrong Jr., Lloyd (1971). Group properties of hydrogenic radial functions. *Phys. Rev. A*, **3**, 1546–1550. `https://doi.org/10.1103/PhysRevA.3.1546`.

Arndt, Markus; Nairz, Olaf; Voss-Andreae, Julian; Keller, Claudia; van der Zouw, Gerbrand; and Zeilinger, Anton (1999). Wave-particle duality of C60 molecules. *Nature*, **401**, 680–682. `https://doi.org/10.1038/44348`.

Arvind; Dutta, B.; Mukunda, Narasimhaiengar; and Simon, Rajiah (1995). The real symplectic groups in quantum mechanics and optics. *Pramana*, **45**, 471. Preprint https://arxiv.org/abs/quant-ph/9509002. `https://doi.org/10.1007/BF02848172`.

Aspect, Alain (1976). Proposed experiment to test the nonseparability of quantum mechanics. *Phys. Rev. D*, **14**, 1944–1951. `https://doi.org/10.1103/PhysRevD.14.1944`.

Aspect, Alain; Dalibard, Jean; and Roger, Gérard (1982*a*). Experimental Test of Bell's Inequalities Using Time-Varying Analyzers. *Phys. Rev. Lett.*, **49**, 1804–1807. `https://doi.org/10.1103/PhysRevLett.49.1804`.

Aspect, Alain and Grangier, Philippe (1985). About resonant scattering and other hypothetical effects in the Orsay atomic-cascade experiment tests of Bell inequalities: A discussion and some new experimental data. *Lett. Nouvo Cimento*, **43**, 345–348. https://doi.org/10.1007/BF02746964.

Aspect, Alain; Grangier, Philippe; and Roger, Gérard (1981). Experimental Tests of Realistic Local Theories via Bell's Theorem. *Phys. Rev. Lett.*, **47**, 460–463. https://doi.org/10.1103/PhysRevLett.47.460.

Aspect, Alain; Grangier, Philippe; and Roger, Gérard (1982*b*). Experimental Realization of Einstein-Podolsky-Rosen-Bohm Gedankenexperiment: A New Violation of Bell's Inequalities. *Phys. Rev. Lett.*, **49**, 91–94. https://doi.org/10.1103/PhysRevLett.49.91.

Aubrun, Guillaume and Szarek, Stanisław J. (2015). Two proofs of Størmer's theorem. Preprint https://arxiv.org/abs/1512.03293.

Audenaert, Koenraad; Verstraete, Frank; and De Moor, Bart (2001). Variational characterizations of separability and entanglement of formation. *Phys. Rev. A*, **64**, 052304. Preprint https://arxiv.org/abs/quant-ph/0006128. https://doi.org/10.1103/PhysRevA.64.052304.

Audenaert, Koenraad M. R.; Calsamiglia, John; Masanes, Lluis; Muñoz-Tapia, Ramon; Acín, Antonio; Bagan, Emilio; and Verstraete, Frank (2007). Discriminating States: The Quantum Chernoff Bound. *Phys. Rev. Lett.*, **98**, 160501. Preprint https://arxiv.org/abs/quant-ph/0610027. https://doi.org/10.1103/PhysRevLett.98.160501.

Audenaert, Koenraad M. R.; Nussbaum, Michael; Szkoła, Arleta; and Verstraete, Frank (2008). Asymptotic Error Rates in Quantum Hypothesis Testing. *Commun. Math. Phys.*, **279**, 251. Preprint https://arxiv.org/abs/0708.4282. https://doi.org/10.1007/s00220-008-0417-5.

Audenaert, Koenraad M. R. and Plenio, Martin B. (2005). Entanglement on mixed stabilizer states: normal forms and reduction procedures. *New J. Phys.*, **7**(1), 170. Preprint https://arxiv.org/abs/quant-ph/0505036. https://doi.org/10.1088/1367-2630/7/1/170.

Badziąg, Piotr; Deuar, Piotr; Horodecki, Michał; Horodecki, Paweł; and Horodecki, Ryszard (2002). Concurrence in arbitrary dimensions. *J. Mod. Opt.*, **49**, 1289. Preprint https://arxiv.org/abs/quant-ph/0107147. https://doi.org/10.1080/09500340210121589.

Balasubramanian, S. (2000). A simple derivation of the recurrence relation for $\langle r^N \rangle$. *Am. J. Phys.*, **68**, 959. https://doi.org/10.1119/1.1287351.

Ballentine, Leslie E. (2014). *Quantum Mechanics* (2nd edn). World Scientific, Singapore. https://doi.org/10.1142/9038.

Banchi, Leonardo; Braunstein, Samuel L.; and Pirandola, Stefano (2015). Quantum Fidelity for Arbitrary Gaussian States. *Phys. Rev. Lett.*, **115**, 260501. Preprint https://arxiv.org/abs/1507.01941. https://doi.org/10.1103/PhysRevLett.115.260501.

Bandyopadhyay, Somshubhro; Boykin, P. Oscar; Roychowdhury, Vwani; and Vatan, Farrokh (2002). A New Proof for the Existence of Mutually Unbiased Bases. *Algorithmica*, **34**, 512–528. Preprint https://arxiv.org/abs/quant-ph/0103162. https://doi.org/10.1007/s00453-002-0980-7.

Barnett, Stephen M. (2009). *Quantum Information*. Oxford University Press, Oxford, U.K.

Barnett, Stephen M.; Babiker, Mohamed; and Padgett, Miles J. (2017). Optical orbital angular momentum. *Philos. Trans. R. Soc. A*, **375**(2087), 20150444. https://doi.org/10.1098/rsta.2015.0444.

Barnett, Stephen M. and Radmore, Paul M. (1997). *Methods in Theoretical Quantum Optics*. Clarendon Press, Oxford, U.K.

Barnum, Howard; Caves, Carlton M.; Fuchs, Christopher A.; Jozsa, Richard; and Schumacher, Benjamin (1996). Noncommuting Mixed States Cannot Be Broadcast. *Phys. Rev. Lett.*, **76**, 2818–2821. Preprint https://arxiv.org/abs/quant-ph/9511010. https://doi.org/10.1103/PhysRevLett.76.2818.

Barrett, Jonathan (2002). Nonsequential positive-operator-valued measurements on entangled mixed states do not always violate a Bell inequality. *Phys. Rev. A*, **65**, 042302. Preprint https://arxiv.org/abs/quant-ph/0107045. https://doi.org/10.1103/PhysRevA.65.042302.

Barrett, Murray D.; Chiaverini, John; Schaetz, Tobias; Britton, Joseph W.; Itano, Wayne M.; Jost, John D.; Knill, Emanuel; Langer, Christopher E.; Leibfried, Dietrich.; Ozeri, Roee; and Wineland, David J. (2004). Deterministic Quantum Teleportation of Atomic Qubits. *Nature*, **429**, 737–739. https://doi.org/10.1038/nature02608.

Bartelt, John E.; *et al.*: CLEO Coll. (1993). Two measurements of $B^0\bar{B}^0$ mixing. *Phys. Rev. Lett.*, **71**, 1680–1684. https://doi.org/10.1103/PhysRevLett.71.1680.

Bartosik, Hannes; Klepp, Jürgen; Schmitzer, Claus-Stefan; Sponar, Stephan; Cabello, Adán; Rauch, Helmut; and Hasegawa, Yuji (2009). Experimental Test of Quantum Contextuality in Neutron Interferometry. *Phys. Rev. Lett.*, **103**, 040403. https://doi.org/10.1103/PhysRevLett.103.040403.

Basler, Carl (2014). *Quantum phase control and interrogation: fundamental tools and experimental determination of the curvature of the projective Hilbert space*. Ph.D. thesis, University of Freiburg. https://freidok.uni-freiburg.de/data/9925.

Bassi, Angelo; Lochan, Kinjalk; Satin, Seema; Singh, Tejinder P.; and Ulbricht, Hendrik (2013). Models of wave-function collapse, underlying theories, and experimental tests. *Rev. Mod. Phys.*, **85**, 471–527. Preprint https://arxiv.org/abs/1204.4325. https://doi.org/10.1103/RevModPhys.85.471.

Baumann, Veronika (2021). *Agents in superposition*. Ph.D. thesis, University of Vienna. https://doi.org/10.25365/thesis.70420.

Baumann, Veronika and Brukner, Časlav (2020). Wigner's Friend as a Rational Agent. In *Quantum, Probability, Logic. Jerusalem Studies in Philosophy and History of Science* (eds. M. Hemmo and O. Shenker). Springer, Cham. Preprint https://arxiv.org/abs/1901.11274. https://doi.org/10.1007/978-3-030-34316-3_4.

Baumann, Veronika; Hansen, Arne; and Wolf, Stefan (2016). The measurement problem is the measurement problem is the measurement problem. Preprint https://arxiv.org/abs/1611.01111.

Baumann, Veronika and Wolf, Stefan (2018). On formalisms and interpretations.

Quantum, **2**, 99. https://doi.org/10.22331/q-2018-10-15-99.

Baumgartner, Bernhard; Hiesmayr, Beatrix C.; and Narnhofer, Heide (2006). State space for two qutrits has a phase space structure in its core. *Phys. Rev. A*, **74**, 032327. Preprint https://arxiv.org/abs/quant-ph/0606083. https://doi.org/10.1103/PhysRevA.74.032327.

Baumgartner, Bernhard; Hiesmayr, Beatrix C.; and Narnhofer, Heide (2007). A special simplex in the state space for entangled qudits. *J. Phys. A: Math. Theor.*, **40**, 7919–7938. Preprint https://arxiv.org/abs/quant-ph/0610100. https://doi.org/10.1088/1751-8113/40/28/S03.

Baumgartner, Bernhard; Hiesmayr, Beatrix C.; and Narnhofer, Heide (2008). The geometry of bipartite qutrits including bound entanglement. *Phys. Lett. A*, **372**, 2190–2195. Preprint https://arxiv.org/abs/0705.1403. https://doi.org/10.1016/j.physleta.2007.11.028.

Bavaresco, Jessica; Herrera Valencia, Natalia; Klöckl, Claude; Pivoluska, Matej; Erker, Paul; Friis, Nicolai; Malik, Mehul; and Huber, Marcus (2018). Measurements in two bases are sufficient for certifying high-dimensional entanglement. *Nat. Phys.*, **14**, 1032–1037. Preprint https://arxiv.org/abs/1709.07344. https://doi.org/10.1038/s41567-018-0203-z.

Becker, Adam (2018). *What is Real? The Unfinished Quest for the Meaning of Quantum Physics*. John Murray Publishing, London, U.K.

Beckner, William (1975). Inequalities in Fourier Analysis. *Ann. Math.*, **102**, 159–182. https://doi.org/10.2307/1970980.

Bell, John Stewart (1955). Time reversal in field theory. *Proceedings of the Royal Society of London A*, **231**, 479–495. https://doi.org/10.1098/rspa.1955.0189.

Bell, John Stewart (1964). On the Einstein-Podolsky-Rosen paradox. *Physics*, **1**, 195–200. https://doi.org/10.1103/PhysicsPhysiqueFizika.1.195.

Bell, John Stewart (1966). On the problem of hidden variables in quantum mechanics. *Rev. Mod. Phys.*, **38**, 447–452. https://doi.org/10.1103/RevModPhys.38.447.

Bell, John Stewart (1971). Introduction to the hidden-variable question. In *Foundations of Quantum Mechanics* (ed. B. D'Espagnat), New York, p. 171–181. Academic.

Bell, John Stewart (1976). The theory of local beables. *Epistemol. Lett.*, **9**, March 1976,11. https://doi.org/10.1007/BF02823296.

Bell, John Stewart (1980). Bertlmann's socks and the nature of reality. CERN preprint Ref.TH.2926-CERN. https://cds.cern.ch/record/142461/files/198009299.pdf.

Bell, John Stewart (1981). Bertlmann's socks and the nature of reality. *J. Phys. Colloques*, **42**, C2 41–61. https://doi.org/10.1051/jphyscol:1981202.

Bell, John Stewart (1990). Interview by Elisabeth Guggenberger, recorded as source material for the production "Die Raupe kann den Schmetterling nicht verstehen" by the Austrian national public broadcaster ORF on the occasion of the 100th anniversary of Erwin Schrödinger's birthday, published by Cinevision in 1990, available online at https://ubdata.univie.ac.at/AC03111215.

Bell, John Stewart (2004). The theory of local beables. In *Speakable and Unspeakable in Quantum Mechanics* (2nd edn) (ed. J. S. Bell), p. 52–62. Cambridge University Press, Cambridge, U.K. https://doi.org/10.1017/CBO9780511815676.009.

Bell, John Stewart and Jackiw, Roman (1969). A PCAC puzzle: $\pi \rightarrow \gamma\gamma$ in the σ−model. *Nuovo Cim. A*, **60**, 47–61. https://doi.org/10.1007/BF02823296.

Bell, John Stewart and Steinberger, Jack (1965). Weak Interactions of Kaons. In *Proceedings of the Oxford International Conference on Elementary Particles*, p. 195. https://cds.cern.ch/record/99127/files/Oxford-1965.pdf.

Benatti, Fabio and Narnhofer, Heide (2001). Additivity of the entanglement of formation. *Phys. Rev. A*, **63**, 042306. Preprint https://arxiv.org/abs/quant-ph/0005126. https://doi.org/10.1103/PhysRevA.63.042306.

Bengtsson, Ingemar (2007). Three Ways to Look at Mutually Unbiased Bases. *AIP Conf. Proc.*, **889**, 40–51. Preprint https://arxiv.org/abs/quant-ph/0610216. https://doi.org/10.1063/1.2713445.

Bengtsson, Ingemar and Życzkowski, Karol (2006). *Geometry of Quantum States*. Cambridge University Press, Cambridge, U.K. https://doi.org/10.1017/CBO9780511535048.

Bennett, Charles H. (2007). More about entanglement and cryptography. last accessed 1 August 2023. http://www.lancaster.ac.uk/users/esqn/windsor07/Lectures/Bennett2.pdf.

Bennett, Charles H.; Bernstein, Herbert J.; Popescu, Sandu; and Schumacher, Benjamin (1996a). Concentrating partial entanglement by local operations. *Phys. Rev. A*, **53**, 2046–2052. Preprint https://arxiv.org/abs/quant-ph/9511030. https://doi.org/10.1103/PhysRevA.53.2046.

Bennett, Charles H. and Brassard, Gilles (2014). Quantum cryptography: Public key distribution and coin tossing. *Theor. Comput. Sci.*, **560**, 7–11. Preprint https://arxiv.org/abs/2003.06557, originally published in Proceedings of IEEE International Conference on Computers, Systems and Signal Processing, pp. 175–179 (1984). https://doi.org/10.1016/j.tcs.2014.05.025.

Bennett, Charles H.; Brassard, Gilles; Crépeau, Claude; Jozsa, Richard; Peres, Asher; and Wootters, William K. (1993). Teleporting an Unknown Quantum State via Dual Classical and Einstein-Podolsky-Rosen Channels. *Phys. Rev. Lett.*, **70**, 1895–1899. https://doi.org/10.1103/PhysRevLett.70.1895.

Bennett, Charles H.; Brassard, Gilles; Popescu, Sandu; Schumacher, Benjamin; Smolin, John A.; and Wooters, William K. (1996b). Purification of Noisy Entanglement and Faithful Teleportation via Noisy Channels. *Phys. Rev. Lett.*, **76**, 722–725. Preprint https://arxiv.org/abs/quant-ph/9511027. https://doi.org/10.1103/physrevlett.76.722.

Bennett, Charles H.; Di Vincenzo, David P.; Smolin, John A.; and Wootters, William K. (1996c). Mixed-state entanglement and quantum error correction. *Phys. Rev. A*, **54**, 3824. Preprint https://arxiv.org/abs/quant-ph/9604024. https://doi.org/10.1103/PhysRevA.54.3824.

Bennett, Charles H. and Wiesner, Stephen J. (1992). Communication via one- and two-particle operators on Einstein-Podolsky-Rosen states. *Phys. Rev. Lett.*, **69**, 2281–2884. https://doi.org/10.1103/PhysRevLett.69.2881.

Bergmann, Marcel and Gühne, Otfried (2013). Entanglement criteria for Dicke states. *J. Phys. A: Math. Theor.*, **46**, 385304. Preprint https://arxiv.org/abs/1305.2818. https://doi.org/10.1088/1751-8113/46/38/385304.

Bernabeu, Jose and Di Domenico, Antonio (2022). Can future observation of the living partner post-tag the past decayed state in entangled neutral K mesons? *Phys. Rev. D*, **105**, 116004. https://doi.org/10.1103/PhysRevD.105.116004.

Berry, Dominic W. and Wiseman, Howard M. (2000). Optimal States and Almost Optimal Adaptive Measurements for Quantum Interferometry. *Phys. Rev. Lett.*, **85**, 5098–5101. Preprint https://arxiv.org/abs/quant-ph/0009117. https://doi.org/10.1103/PhysRevLett.85.5098.

Berry, Michael Victor (1984). Quantal phase factors accompanying adiabatic changes. *Proc. R. Soc. Lond. A*, **392**, 45–57. https://doi.org/10.1098/rspa.1984.0023.

Berry, Michael Victor (1987). Quantum phase corrections from adiabatic iteration. *Proc. R. Soc. Lond. A*, **414**, 31–46. https://doi.org/10.1098/rspa.1987.0131.

Berry, Michael Victor (1990). Anticipations of the geometric phase. *Physics Today*, **43**, 34–40. https://doi.org/10.1063/1.881219.

Berta, Mario; Christandl, Matthias; Colbeck, Roger; Renes, Joseph M.; and Renner, Renato (2010). The uncertainty principle in the presence of quantum memory. *Nat. Phys.*, **6**, 659–662. Preprint https://arxiv.org/abs/0909.0950. https://doi.org/10.1038/nphys1734.

Bertlmann, Reinhold A. (1988). Bell's Theorem and the Nature of Reality. Preprint University of Vienna.

Bertlmann, Reinhold A. (1990). Bell's Theorem and the Nature of Reality. *Found. Phys.*, **20**, 1191–1212. https://doi.org/10.1007/BF01889465.

Bertlmann, Reinhold A. (2000). *Anomalies in Quantum Field Theory* (2nd edn). Oxford University Press, Oxford. https://doi.org/10.1093/acprof:oso/9780198507628.001.0001.

Bertlmann, Reinhold A. (2002). Magic Moments: A Collaboration with John Bell. In *Quantum [Un]speakables* (eds. R. A. Bertlmann and A. Zeilinger), p. 29–47. Springer, Berlin, Heidelberg. https://doi.org/10.1007/978-3-662-05032-3_4.

Bertlmann, Reinhold A. (2006). Entanglement, Bell Inequalities and Decoherence in Particle Physics. In *Quantum Coherence, from Quarks to Solids* (eds. W. Pötz, J. Fabian, and U. Hohenester), p. 1–45. Springer, Berlin, Heidelberg. Preprint https://arxiv.org/abs/quant-ph/0410028. https://doi.org/10.1007/11398448_1.

Bertlmann, Reinhold A. (2014). John Bell and the Nature of the Quantum World. *J. Phys. A: Math. Theor.*, **47**, 424007. Preprint https://arxiv.org/abs/1411.5322. https://doi.org/10.1088/1751-8113/47/42/424007.

Bertlmann, Reinhold A. (2015). Magic Moments with John Bell. *Physics Today*, **68**, 7, 40–45. https://doi.org/10.1063/PT.3.2847.

Bertlmann, Reinhold A. (2017). Bell's Universe: A Personal Recollection. In *Quantum [Un]speakables II* (eds. R. A. Bertlmann and A. Zeilinger), p. 17–80. Springer, Berlin, Heidelberg. Preprint https://arxiv.org/abs/1605.08081. https://doi.org/10.1007/978-3-319-38987-5_3.

Bertlmann, Reinhold A. (2020). Real or not real that is the question... *Eur. Phys. J. H*, **45**, 205–236. https://doi.org/10.1140/epjh/e2020-10022-x.

Bertlmann, Reinhold A.; Bramon, Albert; Garbarino, Gianni; and Hiesmayr, Beatrix C. (2004). Violation of a Bell inequality in particle physics experimentally verified? *Phys. Lett. A*, **332**, 355–360. Preprint https://arxiv.org/abs/quant-ph/0409051. `https://doi.org/10.1016/j.physleta.2004.10.006`.

Bertlmann, Reinhold A.; Durstberger, Katharina; and Hiesmayr, Beatrix C. (2003). Decoherence of entangled kaons and its connection to entanglement measures. *Phys. Rev. A*, **68**, 012111. Preprint https://arxiv.org/abs/quant-ph/0209017. `https://doi.org/10.1103/PhysRevA.68.012111`.

Bertlmann, Reinhold A. and Grimus, Walter (1997). Quantum-Mechanical Interference over Macroscopic Distances in the $B^0\bar{B}^0$ System. *Phys. Lett. B*, **392**, 426–432. Preprint https://arxiv.org/abs/hep-ph/9610301. `https://doi.org/10.1016/S0370-2693(96)01558-4`.

Bertlmann, Reinhold A. and Grimus, Walter (1998). How devious are deviations from quantum mechanics: The case of the $B^0\bar{B}^0$ system. *Phys. Rev. D*, **58**, 034014. Preprint https://arxiv.org/abs/hep-ph/9710236. `https://doi.org/10.1103/PhysRevD.58.034014`.

Bertlmann, Reinhold A. and Grimus, Walter (2001). Model for decoherence of entangled beauty. *Phys. Rev. D*, **64**, 056004. Preprint https://arxiv.org/abs/hep-ph/0101160. `https://doi.org/10.1103/PhysRevD.64.056004`.

Bertlmann, Reinhold A. and Grimus, Walter (2002). Dissipation in a 2-dimensional Hilbert space: various forms of complete positivity. *Phys. Lett. A*, **300**, 107–114. Preprint https://arxiv.org/abs/quant-ph/0201142. `https://doi.org/10.1016/S0375-9601(02)00816-2`.

Bertlmann, Reinhold A.; Grimus, Walter; and Hiesmayr, Beatrix C. (1999). Quantum mechanics, Furry's hypothesis, and a measure of decoherence in the $K^0\bar{K}^0$ system. *Phys. Rev. D*, **60**, 114032. Preprint https://arxiv.org/abs/hep-ph/9902427. `https://doi.org/10.1103/PhysRevD.60.114032`.

Bertlmann, Reinhold A.; Grimus, Walter; and Hiesmayr, Beatrix C. (2001). Bell inequality and CP violation in the neutral kaon system. *Phys. Lett. A*, **289**, 21–26. Preprint https://arxiv.org/abs/quant-ph/0107022. `https://doi.org/10.1016/S0375-9601(01)00577-1`.

Bertlmann, Reinhold A.; Grimus, Walter; and Hiesmayr, Beatrix C. (2006). Open-quantum-system formulation of particle decay. *Phys. Rev. A*, **73**, 054101. Preprint https://arxiv.org/abs/quant-ph/0602116. `https://doi.org/10.1103/PhysRevA.73.054101`.

Bertlmann, Reinhold A. and Hiesmayr, Beatrix C. (2001). Bell inequalities for entangled kaons and their unitary time evolution. *Phys. Rev. A*, **63**, 062112. Preprint https://arxiv.org/abs/hep-ph/0101356. `https://doi.org/10.1103/PhysRevA.63.062112`.

Bertlmann, Reinhold A. and Hiesmayr, Beatrix C. (2007). Strangeness measurements of kaon pairs, CP violation and Bell inequalities. In *Handbook on neutral kaon interferometry at a Phi-factory* (ed. A. Di Domenico), Frascati, p. 197–216. Servizio Documentazione dei Laboratori Nazionale di Frascati. Preprint https://arxiv.org/abs/hep-ph/0609251. `http://www.lnf.infn.it/sis/frascatiseries/Volume43/volume43.pdf`.

Bertlmann, Reinhold A. and Krammer, Philipp (2008a). Bloch vectors for qudits. *J. Phys. A: Math. Theor.*, **41**, 235303. Preprint https://arxiv.org/abs/0806.1174.

https://doi.org/10.1088/1751-8113/41/23/235303.

Bertlmann, Reinhold A. and Krammer, Philipp (2008*b*). Bound entanglement in the set of Bell-state mixtures of two qutrits. *Phys. Rev. A*, **78**, 014303. Preprint https://arxiv.org/abs/0804.1525. https://doi.org/10.1103/PhysRevA.78.014303.

Bertlmann, Reinhold A. and Krammer, Philipp (2008*c*). Geometric entanglement witnesses and bound entanglement. *Phys. Rev. A*, **77**, 024303. Preprint https://arxiv.org/abs/0710.1184. https://doi.org/10.1103/PhysRevA.77.024303.

Bertlmann, Reinhold A. and Krammer, Philipp (2009). Entanglement witnesses and geometry of entanglement of two-qutrit states. *Ann. Phys.*, **324**, 1388–1407. Preprint https://arxiv.org/abs/0901.4729. https://doi.org/10.1016/j.aop.2009.01.008.

Bertlmann, Reinhold A.; Narnhofer, Heide; and Thirring, Walter (2002). A Geometric Picture of Entanglement and Bell Inequalities. *Phys. Rev. A*, **66**, 032319. Preprint https://arxiv.org/abs/quant-ph/0111116. https://doi.org/10.1103/PhysRevA.66.032319.

Bertlmann, Reinhold A.; Narnhofer, Heide; and Thirring, Walter (2013). Time-ordering Dependence of Measurements in Teleportation. *Eur. Phys. J. D*, **67**, 62. Preprint https://arxiv.org/abs/1210.5646. https://doi.org/10.1140/epjd/e2013-30647-y.

Bertlmann, Reinhold A. and Zeilinger, Anton (eds.) (2002). *Quantum [Un]speakables*. Springer, Berlin, Heidelberg. https://doi.org/10.1007/978-3-662-05032-3.

Bertlmann, Reinhold A. and Zeilinger, Anton (eds.) (2017). *Quantum [Un]speakables II*. Springer, Berlin, Heidelberg. https://doi.org/10.1007/978-3-319-38987-5.

Bethe, Hans A. (1947). The Electromagnetic Shift of Energy Levels. *Phys. Rev.*, **72**, 339–341. https://doi.org/10.1103/PhysRev.72.339.

Białynicki-Birula, Iwo and Mycielski, Jerzy (1975). Uncertainty Relations for Information Entropy in Wave Mechanics. *Commun. Math. Phys.*, **44**, 129. https://doi.org/10.1007/BF01608825.

Bierhorst, Peter (2015). A robust mathematical model for a loophole-free Clauser–Horne experiment. *J. Phys. A: Math. Theor.*, **48**, 195302. Preprint https://arxiv.org/abs/1312.2999. https://doi.org/10.1088/1751-8113/48/19/195302.

Birkhoff, Garrett (1946). Tres observaciones sobre el algebra lineal (Three observations on linear algebra). *Univ. Nac. Tucumán. Revista A.*, **5**, 147–151. in Spanish.

Birrell, Nicholas David and Davies, Paul C. W. (1982). *Quantum Fields in Curved Space*. Cambridge University Press, Cambridge, U.K. https://doi.org/10.1017/CBO9780511622632.

Blasone, Massimo; Dell'Anno, Fabio; De Siena, Silvio; and Illuminati, Fabrizio (2009). Entanglement in neutrino oscillations. *Eur. Phys. Lett.*, **85**, 50002. Preprint https://arxiv.org/abs/1210.5646. https://doi.org/10.1209/0295-5075/85/50002.

Bloch, Felix (1946). Nuclear Induction. *Phys. Rev.*, **70**, 460–474. https://doi.org/10.1103/PhysRev.70.460.

Bogoljubov, Nikolay N. (1958). On a new method in the theory of superconductivity. *Nuovo Cim.*, **7**, 794–805. https://doi.org/10.1007/BF02745585.

Bohm, David (1951). *Quantum Theory*. Prentice-Hall Inc., New York.

Bohm, David (1952). A suggested interpretation of the quantum theory in terms of "hidden" variables. *Phys. Rev.*, **85**, 166–179. https://doi.org/10.1103/PhysRev.85.166.

Bohm, David and Aharonov, Yakir (1957). Discussion of Experimental Proof for the Paradox of Einstein, Rosen, and Podolsky. *Phys. Rev.*, **108**, 1070–1076. https://doi.org/10.1103/PhysRev.108.1070.

Bohr, Niels (1913). On the constitution of atoms and molecules. *London, Edinburgh Dublin Philos. Mag. J. Sci.*, **26**(151), 1–25. https://doi.org/10.1080/14786441308634955.

Bohr, Niels (1935). Can Quantum-Mechanical Description of Physical Reality Be Considered Complete? *Phys. Rev.*, **48**, 696–702. https://doi.org/10.1103/PhysRev.48.696.

Bohr, Niels (1983). Diskussion mit Einstein über Erkenntnistheoretische Probleme in der Atomphysik. In *Albert Einstein als Philosoph und Naturforscher. Facetten der Physik* (ed. P. A. Schilpp), p. 84–119. Vieweg+Teubner Verlag, Wiesbaden. In German. https://doi.org/10.1007/978-3-322-88795-5_6.

Boltzmann, Ludwig (1877). Über die Beziehung zwischen dem zweiten Hauptsatze der mechanischen Wärmetheorie und der Wahrscheinlichkeitsrechnung, respektive den Sätzen über das Wäremegleichgewicht. *Wiener Berichte*, **76**, 373–435. in: Boltzmann, Ludwig. (1909). *Wissenschaftliche Abhandlungen* (ed. F. Hasenöhrl), Vol. II, paper 42. Leipzig: Barth; reissued New York: Chelsea, 1969. In German.

Bonneau, Guy; Faraut, Jacques; and Valent, Galliano (2001). Self-adjoint extensions of operators and the teaching of quantum mechanics. *Am. J. Phys.*, **69**, 322. https://doi.org/10.1119/1.1328351.

Born, Max (1926). Zur Quantenmechanik der Stoßvorgänge. *Z. Physik*, **37**, 863–867. https://doi.org/10.1007/BF01397477.

Born, Max and Fock, Vladimir Aleksandrovich (1928). Beweis des Adiabatensatzes. *Z. Phys.*, **51**, 165–180. https://doi.org/10.1007/BF0134319.

Boschi, D.; Branca, Sara; De Martini, Francesco; Hardy, Lucien; and Popescu, Sandu (1998). Experimental Realization of Teleporting an Unknown Pure Quantum State via Dual Classical and Einstein-Podolsky-Rosen Channels. *Phys. Rev. Lett.*, **80**, 1121–1125. Preprint https://arxiv.org/abs/quant-ph/9710013. https://doi.org/10.1103/PhysRevLett.80.1121.

Bourennane, Mohamed; Eibl, Manfred; Kurtsiefer, Christian; Gaertner, Sascha; Weinfurter, Harald; Gühne, Otfried; Hyllus, Philipp; Bruß, Dagmar; Lewenstein, Maciej; and Sanpera, Anna (2004). Experimental Detection of Multipartite Entanglement using Witness Operators. *Phys. Rev. Lett.*, **92**, 087902. Preprint https://arxiv.org/abs/quant-ph/0309043. https://doi.org/10.1103/PhysRevLett.92.087902.

Bouwmeester, Dik; Pan, Jian-Wei; Mattle, Klaus; Eibl, Manfred; Weinfurter, Harald; and Zeilinger, Anton (1997). Experimental Quantum Teleportation. *Nature*, **390**, 575–579. Preprint https://arxiv.org/abs/1901.11004. https://doi.org/10.1038/

37539.

Bragg, William Henry and Bragg, William Lawrence (1913). The reflection of X-rays by crystals. *Proc. R. Soc. Lond. A*, **88**, 428–438. https://doi.org/10.1098/rspa. 1913.0040.

Bramon, Albert; Escribano, Rafel; and Garbarino, Gianni (2005). Bell's inequality tests: from photons to *B*-mesons. *J. Mod. Optics*, **52**, 1681–1684. Preprint https://arxiv.org/abs/quant-ph/0410122. https://doi.org/10.1080/ 09500340500072893.

Bramon, Albert; Escribano, Rafel; and Garbarino, Gianni (2006). Bell's Inequality Tests with Meson-Antimeson Pairs. *Found. Phys.*, **36**, 563–584. Preprint https://arxiv.org/abs/quant-ph/0501069. https://doi.org/10.1007/ s10701-005-9030-z.

Bramon, Albert; Escribano, Rafel; and Garbarino, Gianni (2007). A review on Bell inequality tests with neutral kaons. In *Handbook on neutral kaon interferometry at a Phi-factory* (ed. A. Di Domenico), Frascati, p. 217–253. Servizio Documentazione dei Laboratori Nazionale di Frascati. http://www.lnf.infn.it/sis/ frascatiseries/Volume43/volume43.pdf.

Bramon, Albert and Garbarino, Gianni (2002*a*). Novel Bell's Inequalities for Entangled $K^0\bar{K}^0$ Pairs. *Phys. Rev. Lett.*, **88**, 040403. Preprint https://arxiv.org/abs/quant-ph/0108047. https://doi.org/10.1103/ PhysRevLett.88.040403.

Bramon, Albert and Garbarino, Gianni (2002*b*). Test of Local Realism with Entangled Kaon Pairs and without Inequalities. *Phys. Rev. Lett.*, **89**, 160401. Preprint https://arxiv.org/abs/quant-ph/0205112. https://doi.org/10.1103/ PhysRevLett.89.160401.

Bramon, Albert; Garbarino, Gianni; and Hiesmayr, Beatrix C. (2004*a*). Active and passive quantum erasers for neutral kaons. *Phys. Rev. A*, **69**, 062111. Preprint https://arxiv.org/abs/quant-ph/0402212. https://doi.org/10.1103/PhysRevA. 69.062111.

Bramon, Albert; Garbarino, Gianni; and Hiesmayr, Beatrix C. (2004*b*). Quantitative complementarity in two-path interferometry. *Phys. Rev. A*, **69**, 022112. Preprint https://arxiv.org/abs/quant-ph/0311179. https://doi.org/10.1103/PhysRevA. 69.022112.

Bramon, Albert; Garbarino, Gianni; and Hiesmayr, Beatrix C. (2004*c*). Quantitative duality and neutral kaon interferometry. *Eur. Phys. J. C*, **32**, 377–380. Preprint https://arxiv.org/abs/hep-ph/0307047. https://doi.org/10. 1140/epjc/s2003-01444-5.

Bramon, Albert; Garbarino, Gianni; and Hiesmayr, Beatrix C. (2004*d*). Quantum Marking and Quantum Erasure for Neutral Kaons. *Phys. Rev. Lett.*, **92**, 020405. Preprint https://arxiv.org/abs/quant-ph/0306114. https://doi.org/10. 1103/PhysRevLett.92.020405.

Bramon, Albert and Nowakowski, Marek (1999). Bell Inequalities for Entangled Pairs of Neutral Kaons. *Phys. Rev. Lett.*, **83**, 1–5. https://doi.org/10.1103/ PhysRevLett.83.1.

Brandão, Fernando G. S. L.; Christandl, Matthias; and Yard, Jon (2011).

Faithful Squashed Entanglement. *Commun. Math. Phys*, **306**, 805–830. Preprint https://arxiv.org/abs/1010.1750. `https://doi.org/10.1007/s00220-011-1302-1`.

Braunstein, Samuel L. (2005). Squeezing as an irreducible resource. *Phys. Rev. A*, **71**, 055801. Preprint https://arxiv.org/abs/quant-ph/9904002. `https://doi.org/10.1103/PhysRevA.71.055801`.

Braunstein, Samuel L. and Caves, Carlton M. (1994). Statistical distance and the geometry of quantum states. *Phys. Rev. Lett.*, **72**, 3439–3443. `https://doi.org/10.1103/PhysRevLett.72.3439`.

Braunstein, Samuel L. and Mann, Ady (1995). Measurement of the Bell operator and quantum teleportation. *Phys. Rev. A*, **51**, R1727–R1730. `https://doi.org/10.1103/PhysRevA.51.R1727`.

Braunstein, Samuel L. and van Loock, Peter (2005). Quantum information with continuous variables. *Rev. Mod. Phys.*, **77**, 513–577. Preprint https://arxiv.org/abs/quant-ph/0410100. `https://doi.org/10.1103/RevModPhys.77.513`.

Brendel, Jürgen; Mohler, Ernst; and Martienssen, Werner (1992). Experimental Test of Bell's Inequality for Energy and Time. *Europhys. Lett.*, **20**, 575–580. `https://doi.org/10.1209/0295-5075/20/7/001`.

Breuer, Heinz-Peter and Petruccione, Francesco (2002). *The Theory of Open Quantum Systems*. Oxford University Press, Oxford. `https://doi.org/10.1093/acprof:oso/9780199213900.001.0001`.

Briegel, Hans J.; Browne, Dan E.; Dür, Wolfgang; Raussendorf, Robert; and Van den Nest, Maarten (2009). Measurement-based quantum computation. *Nat. Phys.*, **5**, 19. Preprint https://arxiv.org/abs/0910.1116. `https://doi.org/10.1038/nphys1157`.

Briegel, Hans J.; Dür, Wolfgang; Cirac, Juan Ignacio; and Zoller, Peter (1998). Quantum Repeaters: The Role of Imperfect Local Operations in Quantum Communication. *Phys. Rev. Lett.*, **81**, 5932. Preprint https://arxiv.org/abs/quant-ph/9803056. `https://doi.org/10.1103/PhysRevLett.81.5932`.

Briegel, Hans J. and Raussendorf, Robert (2001). Persistent Entanglement in Arrays of Interacting Particles. *Phys. Rev. Lett.*, **86**, 910–913. Preprint https://arxiv.org/abs/quant-ph/0004051. `https://doi.org/10.1103/PhysRevLett.86.910`.

Brillouin, Léon (1953). Negentropy principle of information. *J. of Applied Physics*, **24**, 1152–1163. `https://doi.org/10.1063/1.1721463`.

Brukner, Časlav (2021). Qubits are not observers—a no-go theorem. Preprint https://arxiv.org/abs/2107.03513.

Brukner, Časlav (2016). On the quantum measurement problem. In *Quantum [Un]speakables II* (eds. R. A. Bertlmann and A. Zeilinger), p. 95–117. Springer, Berlin, Heidelberg. `https://doi.org/10.1007/978-3-319-38987-5_5`.

Brukner, Časlav and Zeilinger, Anton (1999). Operationally invariant information in quantum measurements. *Phys. Rev. Lett.*, **83**, 3354–3357. `https://doi.org/10.1103/PhysRevLett.83.3354`.

Brukner, Časlav and Zeilinger, Anton (2001). Conceptually indequacy of the Shannon information in quantum measurements. *Phys. Rev. A*, **63**, 3354–3357.

https://doi.org/10.1103/PhysRevA.63.022113.

Brukner, Časlav and Zeilinger, Anton (2003). Information and fundamental elements of the structure of quantum theory. In *Time, Quantum, Information* (eds. L. Castell and O. Ischebeck), p. 323–354. Springer, Berlin, Heidelberg. https://doi.org/10.1007/978-3-662-10557-3_21.

Brukner, Časlav; Żukowski, Marek; and Zeilinger, Anton (2001). The essence of entanglement. Preprint https://arxiv.org/abs/quant-ph/0106119.

Brunner, Nicolas; Cavalcanti, Daniel; Pironio, Stefano; Scarani, Valerio; and Wehner, Stephanie (2014). Bell nonlocality. *Rev. Mod. Phys.*, **86**, 419–478. Preprint https://arxiv.org/abs/1303.2849. https://doi.org/10.1103/RevModPhys.86.419.

Bruß, Dagmar (2002). Characterizing entanglement. *J. Math. Phys.*, **43**, 4237–4251. Preprint https://arxiv.org/abs/quant-ph/0110078. https://doi.org/10.1063/1.1494474.

Bruß, Dagmar; Cirac, Juan Ignacio; Horodecki, Paweł; Hulpke, Florian; Kraus, Barbara; Lewenstein, Maciej; and Sanpera, Anna (2002). Reflections upon separability and distillability. *J. Mod. Opt.*, **49**, 1399–1418. https://doi.org/10.1080/09500340110105975.

Bures, Donald (1969). An extension of Kakutani's theorem on infinite product measures to the tensor product of semifinite w^*-algebra. *T. Am. Math. Soc.*, **135**, 199–212. https://doi.org/10.1090/S0002-9947-1969-0236719-2.

Busch, Paul (1991). Informationally complete sets of physical quantities. *Int. J. Theor. Phys.*, **30**, 1217–1227. https://doi.org/10.1007/BF00671008.

Cañas, Gustavo; Arias, Mauricio; Etcheverry, Sebastián; Gómez, Esteban S.; Cabello, Adán; Xavier, Guilherme B.; and Lima, Gustavo (2014). Applying the Simplest Kochen-Specker Set for Quantum Information Processing. *Phys. Rev. Lett.*, **113**, 090404. Preprint https://arxiv.org/abs/1408.6857. https://doi.org/10.1103/PhysRevLett.113.090404.

Cabello, Adán (2003). Supersinglets. *J. Mod. Opt.*, **50**, 10049. Preprint https://arxiv.org/abs/quant-ph/0306074. https://doi.org/10.1080/09500340308234551.

Cabello, Adán (2007). Six-qubit permutation-based decoherence-free orthogonal basis. *Phys. Rev. A*, **75**, 020301. Preprint https://arxiv.org/abs/quant-ph/0702118. https://doi.org/10.1103/PhysRevA.75.020301.

Cabello, Adán (2016). Simple method for experimentally testing any form of quantum contextuality. *Phys. Rev. A*, **93**, 032102. Preprint https://arxiv.org/abs/1512.05370. https://doi.org/10.1103/PhysRevA.93.032102.

Cabello, Adán (2017). Interpretations of Quantum Theory: A Map of Madness. In *What is Quantum Information?* (eds. Olimpia Lombardi, Sebastian Fortin, Federico Holik, and Cristian López), Chapter 7, p. 138–144. Cambridge University Press, Cambridge, U.K. Preprint https://arxiv.org/abs/1509.04711. https://doi.org/10.1017/9781316494233.009.

Cadney, Josh; Huber, Marcus; Linden, Noah; and Winter, Andreas (2014). Inequalities for the ranks of multipartite quantum states. *Lin. Alg. Appl.*, **452**, 153–171. Preprint https://arxiv.org/abs/1308.0539. https://doi.org/10.1016/

j.laa.2014.03.035.

Cahill, Kevin Eric and Glauber, Roy J. (1969). Density Operators and Quasiprobability Distributions. *Phys. Rev.*, **177**, 1882–1902. https://doi.org/10.1103/PhysRev.177.1882.

Campbell, Mark L. (1991). The correct interpretation of Hund's rule as applied to "uncoupled states" orbital diagrams. *J. Chem. Educ.*, **68**, 134. https://doi.org/10.1021/ed068p134.

Casimir, Hendrik Brugt Gerhard (1948). On the Attraction Between Two Perfectly Conducting Plates. *Indag. Math.*, **10**, 261–263.

Casimir, Hendrik Brugt Gerhard and Polder, Dirk (1948). The Influence of retardation on the London-van der Waals forces. *Phys. Rev.*, **73**, 360–372. https://doi.org/10.1103/PhysRev.73.360.

Caso, Carlo; *et al.* (1998). Review of particle physics. Particle Data Group. *Eur. Phys. J. C*, **3**, 1–794. https://doi.org/10.1007/s10052-998-0104-x.

Caves, Carlton M.; Fuchs, Christopher A.; and Schack, Rüdiger (2002). Unknown quantum states: The quantum de Finetti representation. *J. Math. Phys.*, **43**, 4537. Preprint https://arxiv.org/abs/quant-ph/0104088. https://doi.org/10.1063/1.1494475.

Cerf, Nicolas J. and Adami, Christoph (1996*a*). Negative Entropy in Quantum Information Theory. In *New Developments on Fundamental Problems in Quantum Physics* (eds. M. Ferrero and A. van der Merwe), p. 77–84. Springer, Dordrecht. Preprint https://arxiv.org/abs/quant-ph/9610005. https://doi.org/10.1007/978-94-011-5886-2_11.

Cerf, Nicolas J. and Adami, Christoph (1996*b*). Quantum Mechanics of Measurement. Preprint https://arxiv.org/abs/quant-ph/9605002.

Cerf, Nicolas J. and Adami, Christoph (1997*a*). Entropic Bell inequalities. *Phys. Rev. A*, **55**, 3371–3374. Preprint https://arxiv.org/abs/quant-ph/9608047. https://doi.org/10.1103/PhysRevA.55.3371.

Cerf, Nicolas J. and Adami, Christoph (1997*b*). Negative entropy and information in quantum mechanics. *Phys. Rev. Lett.*, **79**, 5194–5197. Preprint https://arxiv.org/abs/quant-ph/9512022. https://doi.org/10.1103/PhysRevLett.79.5194.

Cerf, Nicolas J. and Adami, Christoph (1998). Information theory of quantum entanglement and measurement. *Phys. D: Nonlinear Phenom.*, **120**, 62–81. Preprint https://arxiv.org/abs/quant-ph/9605039. https://doi.org/10.1016/S0167-2789(98)00045-1.

Cerf, Nicolas J. and Adami, Christoph (1999). Quantum extension of conditional probability. *Phys. Rev. A*, **60**, 893–897. Preprint https://arxiv.org/abs/quant-ph/9710001. https://doi.org/10.1103/PhysRevA.60.893.

Cerf, Nicolas J.; Adami, Christoph; and Gingrich, Robert M. (1999). Reduction criterion for separability. *Phys. Rev. A*, **60**, 898–909. Preprint https://arxiv.org/abs/quant-ph/9710001. https://doi.org/10.1103/PhysRevA.60.898.

Chadan, Khosrow; Khuri, Nicola N.; Martin, Andre; and Wu, Tai Tsun (2003). Bound states in one and two spatial dimensions. *J. Math. Phys.*, **44**, 406. https://doi.

org/10.1063/1.1532538.

Chambers, Robert G. (1960). Shift of an Electron Interference Pattern by Enclosed Magnetic Flux. *Phys. Rev. Lett.*, **5**, 3–5. https://doi.org/10.1103/PhysRevLett.5.3.

Chen, Kai and Wu, Ling-An (2003). A matrix realignment method for recognizing entanglement. *Quant. Inf. Comput.*, **3**, 193–202. Preprint https://arxiv.org/abs/quant-ph/0205017. https://doi.org/10.26421/QIC3.3-1.

Chen, Lin; Chitambar, Eric; Duan, Runyao; Ji, Zhengfeng; and Winter, Andreas (2010). Tensor Rank and Stochastic Entanglement Catalysis for Multipartite Pure States. *Phys. Rev. Lett.*, **105**, 200501. Preprint https://arxiv.org/abs/1003.3059. https://doi.org/10.1103/PhysRevLett.105.200501.

Cheung, Albert C.; Rank, David M.; Townes, Charles H.; Thornton, Donald D.; and Welch, William J. (1968). Detection of NH3 Molecules in the Interstellar Medium by Their Microwave Emission. *Phys. Rev. Lett.*, **21**, 1701. https://doi.org/10.1103/PhysRevLett.21.1701.

Chiribella, Giulio; D'Ariano, Giacomo Mauro; and Perinotti, Paolo (2011). Informational derivation of quantum theory. *Phys. Rev. A*, **84**, 012311. https://doi.org/10.1103/PhysRevA.84.012311.

Chisolm, Eric D. (2001). Generalizing the Heisenberg uncertainty relation. *Am. J. Phys.*, **69**, 368–371. Preprint https://arxiv.org/abs/quant-ph/0011115. https://doi.org/10.1119/1.1317561.

Chistopher C. Gerry, Peter L. Knight (2005). *Introductory Quantum Optics*. Cambridge University Press, Cambridge, U.K.

Chitambar, Eric; Leung, Debbie; Mančinska, Laura; Ozols, Maris; and Winter, Andreas (2014). Everything You Always Wanted to Know About LOCC (But Were Afraid to Ask). *Commun. Math. Phys.*, **328**, 303–326. Preprint https://arxiv.org/abs/1210.4583. https://doi.org/10.1007/s00220-014-1953-9.

Choi, Man-Duen (1975). Completely Positive Linear Maps on Complex Matrices. *Linear Algebra Appl.*, **10**, 285–290. https://doi.org/10.1016/0024-3795(75)90075-0.

Christandl, Matthias (2006). *The Structure of Bipartite Quantum States - Insights from Group Theory and Cryptography*. Ph.D. thesis, University of Cambridge. Preprint https://arxiv.org/abs/quant-ph/0604183.

Christandl, Matthias; Jensen, Asger Kjærulff; and Zuiddam, Jeroen (2018). Tensor rank is not multiplicative under the tensor product. *Lin. Alg. Appl.*, **543**, 125–139. Preprint https://arxiv.org/abs/1705.09379. https://doi.org/10.1016/j.laa.2017.12.020.

Christandl, Matthias and Winter, Andreas (2004). "Squashed entanglement": An additive entanglement measure. *J. Math. Phys.*, **45**(3), 829–840. Preprint https://arxiv.org/abs/quant-ph/0308088. https://doi.org/10.1063/1.1643788.

Christenson, James H.; Cronin, James Watson; Fitch, Val Logsdon; and Turlay, René (1964). Evidence for the 2π Decay of the K_2^0 Meson. *Phys. Rev. Lett.*, **13**, 138–140. https://doi.org/10.1103/PhysRevLett.13.138.

Chruściński, Dariusz and Jamiołkowski, Andrzej (2004). *Geometric phases in classical and quantum mechanics*. Birkhäuser, Basel.

Chruściński, Dariusz and Pascazio, Saverio. A Brief History of the GKLS Equation. *Open Sys. Inf. Dyn.*, **24**, 1740001. Preprint https://arxiv.org/abs/1710.05993. https://doi.org/10.1142/S1230161217400017.

Clauser, John F. (1969). Proposed experiment to test local hidden-variable theories. *Bull. Am. Phys. Soc.*, **14**, 578. https://www.johnclauser.com/_files/ugd/36ef59_e9b993b0864c45148ca0c25c5ddaaf71.pdf.

Clauser, John F. (1976). Experimental Investigation of a Polarization Correlation Anomaly. *Phys. Rev. Lett.*, **36**, 1223–1226. https://doi.org/10.1103/PhysRevLett.36.1223.

Clauser, John F. (2002). Early history of Bell's theorem. In *Quantum [Un]speakables* (eds. R. A. Bertlmann and A. Zeilinger), p. 61–98. Springer, Berlin, Heidelberg. https://doi.org/10.1007/978-3-662-05032-3_6.

Clauser, John F. and Freedman, Stuart J. (1972). Experimental Test of Local Hidden-Variable Theories. *Phys. Rev. Lett.*, **28**, 938–941. https://doi.org/10.1103/PhysRevLett.28.938.

Clauser, John F. and Horne, Michael A. (1974). Experimental consequences of objective local theories. *Phys. Rev. D*, **10**, 526–535. https://doi.org/10.1103/PhysRevD.10.526.

Clauser, John F.; Horne, Michael A.; Shimony, Abner; and Holt, Richard A. (1969). Proposed Experiment to Test Local Hidden-Variable Theories. *Phys. Rev. Lett.*, **23**, 880–884. https://doi.org/10.1103/PhysRevLett.23.880.

Clausius, Rudolf (1865). Über verschiedene für die Anwendung bequeme Formen der Hauptgleichungen der mechanischen Wärmetheorie. *Annalen der Physik*, **201**(7), 353–400. in German. https://doi.org/10.1002/andp.18652010702.

Clifton, Rob (1993). Getting contextual and nonlocal elements-of-reality the easy way. *Am. J. Phys.*, **61**(5), 443–447. https://doi.org/10.1119/1.17239.

Clifton, Rob; Bub, Jeffrey; and Halvorson, Hans (2003). Characterizing quantum theory in terms of information-theoretic constraints. *Found. Phys.*, **33**, 1561–1591. https://doi.org/10.1023/A:1026056716397.

Clivaz, Fabien; Huber, Marcus; Lami, Ludovico; and Murta, Gláucia (2017). Genuine-multipartite entanglement criteria based on positive maps. *J. Math. Phys.*, **58**(8), 082201. Preprint https://arxiv.org/abs/1609.08126. https://doi.org/10.1063/1.4998433.

Coffman, Valerie; Kundu, Joydip; and Wootters, William K. (2000). Distributed entanglement. *Phys. Rev. A*, **61**, 052306. Preprint https://arxiv.org/abs/quant-ph/9907047. https://doi.org/10.1103/PhysRevA.61.052306.

Cohen, Martin L. (1968). The Fisher information and convexity. *IEEE T. Inform. Theory*, **14**, 591–592. https://doi.org/10.1109/TIT.1968.1054175.

Cohen-Tannoudji, Claude; Diu, Bernard; and Laloë, Franck (1991). *Quantum Mechanics* (1st edn). Volume 2. Wiley, New York.

Cohen-Tannoudji, Claude; Diu, Bernard; and Laloë, Franck (1999). *Quantum Mechanics* (2nd edn). Volume 1. Wiley, New York.

Cohen-Tannoudji, Claude; Dupont-Roc, Jacques; and Grynberg, Gilbert (2004). *Atom-Photon Interactions*. Wiley-VCH, Weinheim. https://doi.org/10.1002/

9783527617197.

Colbeck, Roger and Renner, Renato (2012). Is a System's Wave Function in One-to-One Correspondence with Its Elements of Reality? *Phys. Rev. Lett.*, **108**, 150402. Preprint https://arxiv.org/abs/111.6597. https://doi.org/10.1103/PhysRevLett.108.150402.

Colbeck, Roger and Renner, Renato (2017). A system's wave function is uniquely determined by its underlying physical state. *New J. Phys.*, **19**, 013016. Preprint https://arxiv.org/abs/1312.7353. https://doi.org/10.1088/1367-2630/aa515c.

Coles, Patrick J.; Berta, Mario; Tomamichel, Marco; and Wehner, Stephanie (2017). Entropic uncertainty relations and their applications. *Rev. Mod. Phys.*, **89**, 015002. Preprint https://arxiv.org/abs/1511.04857. https://doi.org/10.1103/RevModPhys.89.015002.

Compton, Arthur Holly (1923). A quantum theory of the scattering of X-rays by light elements. *Phys. Rev.*, **21**, 483–502. https://doi.org/10.1103/PhysRev.21.483.

Condon, Edward Uhler and Shortley, George H. (1970). *The Theory of Atomic Spectra.* Cambridge University Press, Cambridge, U.K.

Contreras-Tejada, Patricia; Palazuelos, Carlos; and de Vicente, Julio I. (2019). Resource Theory of Entanglement with a Unique Multipartite Maximally Entangled State. *Phys. Rev. Lett.*, **122**, 120503. Preprint https://arxiv.org/abs/1807.11395. https://doi.org/10.1103/PhysRevLett.122.120503.

Conway, John (1990). *A course in functional analysis.* Springer, New York.

Conway, John and Kochen, Simon B. (2002). The Geometry of the Quantum Paradoxes. In *Quantum [Un]speakables* (eds. R. A. Bertlmann and A. Zeilinger), p. 257–269. Springer, Berlin, Heidelberg. https://doi.org/10.1007/978-3-662-05032-3_18.

Cramér, Harald (1946). *Mathematical Methods of Statistics.* Princeton University Press, Princeton, NJ.

Cushing, James T. (1994). *Quantum Mechanics: Historical Contingency and the Copenhagen Hegemony.* The University of Chicago Press, Chicago, U.S.A.

Cushing, James T. and McMullin, Ernan (ed.) (1989). *Philosophical Consequences of Quantum Theory: Reflections on Bell's Theorem.* University of Notre Dame Press, Notre Dame, Indiana, U.S.A.

Dakić, Borivoje and Brukner, Časlav (2011). Quantum theory and beyond: Is entanglement special? In *Deep Beauty: Understanding the Quantum World through Mathematical Innovation* (ed. H. Halvorson), p. 365–392. Cambridge University Press, Cambridge, U.K.

Dakić, Borivoje and Brukner, Časlav (2016). The Classical Limit of a Physical Theory and the Dimensionality of Space. In *Quantum Theory: Informational Foundations and Foils* (eds. Giulio Chiribella and Robert Spekkens), p. 249–282. Springer, Dordrecht.

Darmois, Georges (1945). Sur les limites de la dispersion de certaines estimations. *Rev. Inst. Int. Statist.*, **13**(1/4), 9–15. https://doi.org/10.2307/1400974.

Darwin, Charles Galton (1928). The wave equations of the electron. *roc. R. Soc. Lond. A*, **118**, 654–680. https://doi.org/10.1098/rspa.1928.0076.

Dass, G. and Sarma, K. (1998). Two-particle correlations and $B^0\bar{B}^0$ mixing. *Eur.*

Phys. J. C, **5**, 283–286. `https://doi.org/10.1007/s100529800854`.

Davisson, Clinton J. and Germer, Lester H. (1928). Reflection of Electrons by a Crystal of Nickel. *Proc. Natl. Acad. Sci. U.S.A.*, **14**(4), 317–322. `https://doi.org/10.1073/pnas.14.4.317`.

de Broglie, Louis (1923). Radiations—ondes et quanta. *Compt. Rend.*, **177**, 507–510. in French, https://www.academie-sciences.fr/pdf/dossiers/Broglie/Broglie_pdf/CR1923_p507.pdf.

de Broglie, Louis (1924). A tentative theory of light quanta. *London, Edinburgh Dublin Philos. Mag. J. Sci.*, **47**(278), 446–458. `https://doi.org/10.1080/14786442408634378`.

de Broglie, Louis (1925). Recherches sur la théorie des quanta. *Ann. Phys.*, **10**, 22–128. in French. `https://doi.org/10.1051/anphys/192510030022`.

de Broglie, Louis (1927). La mécanique ondulatoire et la structure atomique de la matière et du rayonnement. *J. Phys. Radium*, **8**(5), 225–241. in French. `https://doi.org/10.1051/jphysrad:0192700805022500`.

de Vicente, Julio I.; Spee, Cornelia; and Kraus, Barbara (2013). Maximally Entangled Set of Multipartite Quantum States. *Phys. Rev. Lett.*, **111**, 110502. Preprint https://arxiv.org/abs/1305.7398. `https://doi.org/10.1103/PhysRevLett.111.110502`.

de Vicente, Julio I.; Spee, Cornelia; Sauerwein, David; and Kraus, Barbara (2017). Entanglement manipulation of multipartite pure states with finite rounds of classical communication. *Phys. Rev. A*, **95**, 012323. Preprint https://arxiv.org/abs/1607.05145. `https://doi.org/10.1103/PhysRevA.95.012323`.

Debarba, Tiago; Manzano, Gonzalo; Guryanova, Yelena; Huber, Marcus; and Friis, Nicolai (2019). Work estimation and work fluctuations in the presence of non-ideal measurements. *New J. Phys.*, **21**, 113002. Preprint https://arxiv.org/abs/1902.08568. `https://doi.org/10.1088/1367-2630/ab4d9d`.

DeBrota, John B.; Fuchs, Christopher A.; and Schack, Rüdiger (2020). Respecting One's Fellow: QBism's Analysis of Wigner's Friend. Preprint https://arxiv.org/abs/2008.03572.

del Rio, Lídia; Å berg, Johan; Renner, Renato; Dahlstein, Oscar; and Vedral, Vlatko (2011). The thermodynamic meaning of negative entropy. *Nature*, **474**, 61–63. Preprint https://arxiv.org/abs/1009.1630. `https://doi.org/10.1038/nature10123`.

Demkowicz-Dobrzański, Rafał; Jarzyna, Marcin; and Kołodyński, Janek (2015). Quantum limits in optical interferometry. *Prog. Optics*, **60**, 345. Preprint https://arxiv.org/abs/1405.7703. `https://doi.org/10.1016/bs.po.2015.02.003`.

Demkowicz-Dobrzański, Rafał; Kołodyński, Janek; and Guţă, Mădălin (2012). The elusive Heisenberg limit in quantum-enhanced metrology. *Nat. Commun.*, **3**, 1063. Preprint https://arxiv.org/abs/1201.3940. `https://doi.org/10.1038/ncomms2067`.

D'Espagnat, Bernard (1999). *Conceptual Foundations of Quantum Mechanics* (2nd edn). Perseus Books, Reading, Massachusetts.

Deutsch, David (1983). Uncertainty in Quantum Measurements. *Phys. Rev. Lett.*, **50**, 631–633. `https://doi.org/10.1103/PhysRevLett.50.631`.

Deutsch, David (1985). Quantum theory as a universal physical theory. *Int. J. Th. Phys.*, **24**, 1–41. `https://doi.org/10.1007/BF00670071`.

Devetak, Igor and Winter, Andreas (2005). Distillation of secret key and entanglement from quantum states. *Proc. R. Soc. Lond. A*, **461**, 207–235. Preprint https://arxiv.org/abs/quant-ph/0306078. `https://doi.org/10.1098/rspa.2004.1372`.

Devoret, Michel H.; Wallraff, Andreas; and Martinis, John M. (2004). Superconducting Qubits: A Short Review. Preprint https://arxiv.org/abs/cond-mat/0411174.

Di Biagio, Andrea and Rovelli, Carlo (2021). Stable Facts, Relative Facts. *Found. Phys.*, **51**, 30. Preprint https://arxiv.org/abs/2006.15543. `https://doi.org/10.1007/s10701-021-00429-w`.

Di Biagio, Andrea and Rovelli, Carlo (2022). Relational Quantum Mechanics is About Facts, Not States: A Reply to Pienaar and Brukner. *Found. Phys.*, **52**, 62. Preprint https://arxiv.org/abs/2110.03610. `https://doi.org/10.1007/s10701-022-00579-5`.

Di Domenico, Antonio (1995). Testing quantum mechanics in the neutral kaon system at a Φ-factory. *Nucl. Phys. B*, **450**, 293–324. `https://doi.org/10.1016/0550-3213(95)00283-X`.

Dicke, Robert Henry (1954). Coherence in spontaneous radiation processes. *Phys. Rev.*, **93**, 99. `https://doi.org/10.1103/PhysRev.93.99`.

Dieks, Dennis (1982). Communication by EPR devices. *Phys. Lett. A*, **92**, 271–272. `https://doi.org/10.1016/0375-9601(82)90084-6`.

Ding, Yi-Bing; Li, Junli; and Qiao, Cong-Feng (2007). Bell Inequalities in High Energy Physics. *Chinese Phys. C*, **31**, 1086–1097. Preprint https://arxiv.org/abs/hep-ph/0702271. `http://cpc.ihep.ac.cn/article/id/96af946d-0738-408e-b964-e188f79b6784`.

Dirac, Paul Adrien Maurice (1927). The quantum theory of the emission and absorption of radiation. *Proc. Roy. Soc. A Math. Phys.*, **114**(767), 243–265. `https://doi.org/10.1098/rspa.1927.0039`.

Dirac, Paul Adrien Maurice (1930). *The Principles of Quantum Mechanics* (4th edn). Oxford University Press, Oxford.

Dirac, Paul Adrien Maurice (1931). Quantised singularities in the electromagnetic field. *Proc. Roy. Soc. A*, **133**, 60–73. `https://doi.org/10.1098/rspa.1931.0130`.

Dominy, Jason M.; Shabani, Alireza; and Lidar, Daniel A. (2016). A general framework for complete positivity. *Quantum Inf. Process*, **15**, 465–494. Preprint https://arxiv.org/abs/1312.0908. `https://doi.org/10.1007/s11128-015-1148-0`.

Donald, Matthew J. and Horodecki, Michał (1999). Continuity of Relative Entropy of Entanglement. *Phys. Lett. A*, **264**, 257–260. Preprint https://arxiv.org/abs/quant-ph/9910002. `https://doi.org/10.1016/S0375-9601(99)00813-0`.

Donald, Matthew J.; Horodecki, Michał; and Rudolph, Oliver (2002). The uniqueness theorem for entanglement measures. *J. Math. Phys.*, **43**, 4252–4272. Preprint https://arxiv.org/abs/quant-ph/0105017. `https://doi.org/10.1063/1.1495917.`

Duan, Luming; Cirac, Juan Ignacio; and Zoller, Peter (2001). Geometric manipulation of trapped ions for quantum computation. *Science*, **292**, 1695–1697. `https://doi.org/10.1126/science.1058835.`

Dür, Wolfgang (2001). Multipartite Bound Entangled States that Violate Bell's Inequality. *Phys. Rev. Lett.*, **87**, 230402. `https://doi.org/10.1103/PhysRevLett.87.230402.`

Dür, Wolfgang; Skotiniotis, Michalis; Fröwis, Florian; and Kraus, Barbara (2014). Improved Quantum Metrology Using Quantum Error Correction. *Phys. Rev. Lett.*, **112**, 080801. Preprint https://arxiv.org/abs/1310.3750. `https://doi.org/10.1103/PhysRevLett.112.080801.`

Dür, Wolfgang; Vidal, Guifré; and Cirac, Juan Ignacio (2000). Three qubits can be entangled in two inequivalent ways. *Phys. Rev. A*, **62**, 062314. Preprint https://arxiv.org/abs/quant-ph/0005115. `https://doi.org/10.1103/PhysRevA.62.062314.`

Dürr, Detlef and Teufel, Stefan (2009). *Bohmian Mechanics: The Physics and Mathematics of Quantum Theory*. Springer, Berlin, Heidelberg. `https://doi.org/10.1007/b99978.`

Durstberger, Katharina (2003). Geometric phases in quantum theory. Diploma thesis, University of Vienna. `https://homepage.univie.ac.at/reinhold.bertlmann/pdfs/dipl_diss/Durstberger_Diplomarbeit.pdf.`

Durstberger, Katharina (2005). *Geometry of entanglement and decoherence in quantum systems*. Ph.D. thesis, University of Vienna.

Eberhard, Philippe H. (1993*a*). Background level and counter efficiencies required for a loophole-free Einstein-Podolsky-Rosen experiment. *Phys. Rev. A*, **47**, R747–R750. `https://doi.org/10.1103/PhysRevA.47.R747.`

Eberhard, Philippe H. (1993*b*). Testing the non-locality of quantum theory in two-kaon systems. *Nucl. Phys. B*, **398**, 155–183. `https://doi.org/10.1016/0550-3213(93)90631-X.`

Ecker, Sebastian; Bouchard, Frédéric; Bulla, Lukas; Brandt, Florian; Kohout, Oskar; Steinlechner, Fabian; Fickler, Robert; Malik, Mehul; Guryanova, Yelena; Ursin, Rupert; and Huber, Marcus (2019). Overcoming Noise in Entanglement Distribution. *Phys. Rev. X*, **9**, 041042. Preprint https://arxiv.org/abs/1904.01552. `https://doi.org/10.1103/PhysRevX.9.041042.`

Eguchi, T.; Gilkey, P. B.; and Hanson, A. J. (1980). Gravitation, gauge theories and differential geometry. *Phys. Rep.*, **66**, 213–393. `https://doi.org/10.1016/0370-1573(80)90130-1.`

Ehrenberg, Werner and Siday, Raymond Eldred (1949). The Refractive Index in Electron Optics and the Principles of Dynamics. *Proc. Phys. Soc. B*, **62**(1), 8–21. `https://doi.org/10.1088/0370-1301/62/1/303.`

Ehrenfest, Paul (1916). Adiabatische Invarianten und Quantentheorie. *Ann. Phys.*, **53**, 327. in German. `https://doi.org/10.1002/andp.19163561905.`

Ehrenfest, Paul (1927). Bemerkung über die angenäherte Gültigkeit der klassischen Mechanik innerhalb der Quantenmechanik. *Z. Physik*, **45**, 455–457. in German. `https://doi.org/10.1007/BF01329203.`

Eibl, Manfred; Kiesel, Nikolai; Bourennane, Mohamed; Kurtsiefer, Christian; and Weinfurter, Harald (2004). Experimental Realization of a Three-Qubit Entangled W State. *Phys. Rev. Lett.*, **92**, 077901–077904. https://doi.org/10.1103/PhysRevLett.92.077901.

Einstein, Albert (1905). Über einen die Erzeugung und Verwandlung des Lichtes betreffenden heuristischen Gesichtspunkt. *Ann. Phys. (Berl.)*, **322**(6), 132–148. in German. https://doi.org/10.1002/andp.19053220607.

Einstein, Albert (1916). Strahlungs-Emission und -Absorption nach der Quantentheorie. *Verhandlungen der Deutschen Physikalischen Gesellschaft*, **18**, 318–323. in German, English translation in: *The Collected Papers of Albert Einstein, Volume 6 (English): The Berlin Years: Writings, 1914-1917*, page 212. https://einsteinpapers.press.princeton.edu/vol6-trans/224.

Einstein, Albert (1926). Letter to Max Born, December 4th, 1926. In *Albert Einstein Max Born, Briefwechsel 1916 - 1955*. Rowohlt 1969. In German.

Einstein, Albert (1947). Letter to Max Born, March 3rd, 1947. In *Albert Einstein Max Born, Briefwechsel 1916 - 1955*. Rowohlt 1969. In German.

Einstein, Albert (1952). Letter to Max Born, May 12th, 1952. In *Albert Einstein Max Born, Briefwechsel 1916 - 1955*. Rowohlt 1969. In German.

Einstein, Albert; Podolsky, Boris; and Rosen, Nathan (1935). Can Quantum-Mechanical Description of Physical Reality Be Considered Complete? *Phys. Rev.*, **47**, 777–780. https://doi.org/10.1103/PhysRev.47.777.

Ekert, Artur K. (1991). Quantum cryptography based on Bell's theorem. *Phys. Rev. Lett.*, **67**, 661–663. https://doi.org/10.1103/PhysRevLett.67.661.

Eltschka, Christopher; Huber, Marcus; Morelli, Simon; and Siewert, Jens (2021). The shape of higher-dimensional state space: Bloch-ball analog for a qutrit. *Quantum*, **5**, 485. Preprint https://arxiv.org/abs/2012.00587. https://doi.org/10.22331/q-2021-06-29-485.

Eltschka, Christopher and Siewert, Jens (2014). Quantifying entanglement resources. *J. Phys. A: Math. Theor.*, **47**, 424005. Preprint https://arxiv.org/abs/1402.6710. https://doi.org/10.1088/1751-8113/47/42/424005.

Engel, Thomas and Reid, Philip (2006). *Physical Chemistry* (3rd edn). Pearson Benjamin-Cummings, San Francisco.

Erker, Paul; Krenn, Mario; and Huber, Marcus (2017). Quantifying high dimensional entanglement with two mutually unbiased bases. *Quantum*, **1**, 22. Preprint https://arxiv.org/abs/1512.05315. https://doi.org/10.22331/q-2017-07-28-22.

Esaki, Leo (1958). New Phenomenon in Narrow Germanium $p - n$ Junctions. *Phys. Rev.*, **109**, 603–604. https://doi.org/10.1103/PhysRev.109.603.

Escher, Bruno de Moura; Davidovich, Luiz; Zagury, Nicim; and de Matos Filho, Ruynet Lima (2012). Quantum Metrological Limits via a Variational Approach. *Phys. Rev. Lett.*, **109**, 190404. Preprint https://arxiv.org/abs/1207.3307. https://doi.org/10.1103/PhysRevLett.109.190404.

Even, Shimon; Goldreich, Oded; and Lempel, Abraham (1985). A Randomized Protocol for Signing Contracts. *Commun. ACM*, **28**(6), 637–647. https://doi.org/10.1145/3812.3818.

Everett, Hugh (1957). "Relative State" Formulation of Quantum Mechanics. *Rev. Mod. Phys.*, **29**, 454–462. https://doi.org/10.1103/RevModPhys.29.454.

Ewen, Harold I. and Purcell, Edward Mills (1951). Observation of a Line in the Galactic Radio Spectrum: Radiation from Galactic Hydrogen at 1,420 Mc./sec. *Nature*, **168**, 356. https://doi.org/10.1038/168356a0.

Facchi, Paolo; Kulkarni, Ravi; Man'ko, Vladimir Ivanovich; Marmo, Giuseppe; Sudarshan, Ennackal Chandy George; and Ventriglia, Franco (2010). Classical and Quantum Fisher Information in the Geometrical Formulation of Quantum Mechanics. *Phys. Lett. A*, **374**, 4801. Preprint https://arxiv.org/abs/1009.5219. https://doi.org/10.1016/j.physleta.2010.10.005.

Fermi, Enrico (1927). Un Metodo Statistico per la Determinazione di alcune Proprietà dell'Atomo. *Rend. Accad. Naz. Lincei.*, **6**, 602–607. In Italian.

Fermi, Enrico (1950). *Nuclear Physics*. University of Chicago Press, Chicago, U.S.A.

Fernandez, D. J.; Nietot, L. M.; del Olmo, M. A.; and Santander, M. (1992). Aharonov-Anandan geometric phase for spin-1/2 periodic Hamiltonians. *J. Phys. A: Math. Gen.*, **25**, 5151–5164. http://doi.org/10.1088/0305-4470/25/19/023.

Ferraro, Alessandro; Olivares, Stefano; and Paris, Matteo G. A. (2005). *Gaussian States in Quantum Information*. Napoli Series on Physics and Astrophysics. Bibliopolis. Preprint https://arxiv.org/abs/quant-ph/0503237.

Feynman, R. P.; Leighton, R. B.; and Sands, M. (1971). *The Feynman lectures on physics*. Volume 3. Addison-Wesley, Reading, Massachusetts. online version accessible at https://www.feynmanlectures.caltech.edu/III_toc.html, last accessed 21 January 2022.

Fickler, Robert; Lapkiewicz, Radek; Huber, Marcus; Lavery, Martin P. J.; Padgett, Miles J.; and Zeilinger, Anton (2014). Interface between path and orbital angular momentum entanglement for high-dimensional photonic quantum information. *Nat. Commun.*, **5**, 4502. Preprint https://arxiv.org/abs/1402.2423. https://doi.org/10.1038/ncomms5502.

Filipp, Stefan (2006). *New aspects of the quantum geometric phase*. Ph.D. thesis, Technical University of Vienna. http://hdl.handle.net/20.500.12708/12021.

Filipp, Stefan and Sjöqvist, Erik (2003). Off-diagonal geometric phase for mixed states. *Phys. Rev. Lett.*, **90**, 050403–050406. Preprint https://arxiv.org/abs/quant-ph/0209087. https://doi.org/10.1103/PhysRevLett.90.050403.

Foadi, R. and Selleri, Franco (1999*a*). Quantum mechanics versus local realism and a recent EPR experiment on $K^0 \bar{K}^0$ pairs. *Phys. Lett. B*, **461**, 123–130. https://doi.org/10.1016/S0370-2693(99)00799-6.

Foadi, R. and Selleri, Franco (1999*b*). Quantum mechanics versus local realism for neutral kaon pairs. *Phys. Rev. A*, **61**, 012106. https://doi.org/10.1103/PhysRevA.61.012106.

Frankel, Theodore (1997). *The Geometry of Physics*. Cambridge University Press, Cambridge, U.K. https://doi.org/10.1017/CBO9781139061377.

Frauchiger, Daniela and Renner, Renato (2018). Quantum theory cannot consistently describe the use of itself. *Nat. Commun.*, **9**, 3711. Preprint https://arxiv.org/abs/1604.07422. https://doi.org/10.1038/s41467-018-05739-8.

Fréchet, Maurice (1907). Sur les ensembles de fonctions et les opérations linéaires. *C. R. Acad. Sci. Paris*, **144**, 1414–1416. In French.

Fréchet, Maurice (1943). Sur l'extension de certaines evaluations statistiques au cas de petits echantillons. *Rev. Inst. Int. Statist.*, **11**(3/4), 182–205. in French. https://doi.org/10.2307/1401114.

Freedman, Stuart J. (1972*a*). *Lawrence Berkeley National Laboratory, LBL Rep. No. 191*.

Freedman, Stuart J. (1972*b*). *Experimental Test of Local Hidden-Variable Theories*. Ph.D. thesis, University of California, Berkeley. Lawrence Berkeley laboratory Report No. LBL-391.

Freire Jr., Olival (2015). *The Quantum Dissidents* (1st edn). Springer, Berlin, Heidelberg. https://doi.org/10.1007/978-3-662-44662-1.

Frieden, B. Roy (2004). *Science from Fisher Information: A Unification*. Cambridge University Press, Cambridge, U.K.

Friedrich, Bretislav (2004). ... hasn't it? A commentary on Eric Scerri's Paper "Has Quantum Mechanics Explained the Periodic Table?". *Found. Chem.*, **6**, 117–132. https://doi.org/10.1023/B:FOCH.0000020999.69326.89.

Friis, Nicolai (2010). Relativistic Effects in Quantum Entanglement. Diploma thesis, University of Vienna. https://doi.org/10.25365/thesis.8370.

Friis, Nicolai (2013). *Cavity mode entanglement in relativistic quantum information*. Ph.D. thesis, University of Nottingham. Preprint https://arxiv.org/abs/1311.3536.

Friis, Nicolai (2016). Reasonable fermionic quantum information theories require relativity. *New J. Phys.*, **18**, 033014. Preprint https://arxiv.org/abs/1502.04476. https://doi.org/10.1088/1367-2630/18/3/033014.

Friis, Nicolai; Bulusu, Sridhar; and Bertlmann, Reinhold A. (2017*a*). Geometry of two-qubit states with negative conditional entropy. *J. Phys. A: Math. Theor.*, **50**, 125301. Preprint https://arxiv.org/abs/1609.04144. https://doi.org/10.1088/1751-8121/aa5dfd.

Friis, Nicolai; Dunjko, Vedran; Dür, Wolfgang; and Briegel, Hans J. (2014). Implementing quantum control for unknown subroutines. *Phys. Rev. A*, **89**, 030303(R). Preprint https://arxiv.org/abs/1401.8128. https://doi.org/10.1103/PhysRevA.89.030303.

Friis, Nicolai and Fuentes, Ivette (2013). Entanglement generation in relativistic quantum fields. *J. Mod. Opt.*, **60**, 22. Preprint https://arxiv.org/abs/1204.0617. https://doi.org/10.1080/09500340.2012.712725.

Friis, Nicolai and Huber, Marcus (2018). Precision and Work Fluctuations in Gaussian Battery Charging. *Quantum*, **2**, 61. Preprint https://arxiv.org/abs/1708.00749. https://doi.org/10.22331/q-2018-04-23-61.

Friis, Nicolai; Lee, Antony R.; and Bruschi, David Edward (2013*a*). Fermionic mode entanglement in quantum information. *Phys. Rev. A*, **87**, 022338. Preprint https://arxiv.org/abs/1211.7217. https://doi.org/10.1103/PhysRevA.87.022338.

Friis, Nicolai; Lee, Antony R.; and Louko, Jorma (2013*b*). Scalar, spinor, and photon fields under relativistic cavity motion. *Phys. Rev. D*, **88**, 064028. Preprint https://arxiv.org/abs/1307.1631. https://doi.org/10.1103/PhysRevD.

88.064028.

Friis, Nicolai; Marty, Oliver; Maier, Christine; Hempel, Cornelius; Holzäpfel, Milan; Jurcevic, Petar; Plenio, Martin B.; Huber, Marcus; Roos, Christian; Blatt, Rainer; and Lanyon, Ben (2018). Observation of Entangled States of a Fully Controlled 20-Qubit System. *Phys. Rev. X*, **8**, 021012. Preprint https://arxiv.org/abs/1711.11092. `https://doi.org/10.1103/PhysRevX.8.021012`.

Friis, Nicolai; Melnikov, Alexey A.; Kirchmair, Gerhard; and Briegel, Hans J. (2015*a*). Coherent controlization using superconducting qubits. *Sci. Rep.*, **5**, 18036. Preprint https://arxiv.org/abs/1508.00447. `https://doi.org/10.1038/srep18036`.

Friis, Nicolai; Orsucci, Davide; Skotiniotis, Michalis; Sekatski, Pavel; Dunjko, Vedran; Briegel, Hans J.; and Dür, Wolfgang (2017*b*). Flexible resources for quantum metrology. *New J. Phys.*, **19**, 063044. Preprint https://arxiv.org/abs/1610.09999. `https://doi.org/10.1088/1367-2630/aa7144`.

Friis, Nicolai; Skotiniotis, Michalis; Fuentes, Ivette; and Dür, Wolfgang (2015*b*). Heisenberg scaling in gaussian quantum metrology. *Phys. Rev. A*, **92**, 022106. Preprint https://arxiv.org/abs/1502.07654. `https://doi.org/10.1103/PhysRevA.92.022106`.

Friis, Nicolai; Vitagliano, Giuseppe; Malik, Mehul; and Huber, Marcus (2019). Entanglement Certification From Theory to Experiment. *Nat. Rev. Phys.*, **1**, 72–87. Preprint https://arxiv.org/abs/1906.10929. `https://doi.org/10.1038/s42254-018-0003-5`.

Fritz, Tobias (2012). Tsirelson's problem and Kirchberg's conjecture. *Rev. Math. Phys.*, **24**, 1250012. Preprint https://arxiv.org/abs/1008.1168. `https://doi.org/10.1142/S0129055X12500122`.

Fröwis, Florian; Sekatski, Pavel; Dür, Wolfgang; Gisin, Nicolas; and Sangouard, Nicolas (2018). Macroscopic quantum states: Measures, fragility, and implementations. *Rev. Mod. Phys.*, **90**, 025004. Preprint https://arxiv.org/abs/1706.06173. `https://doi.org/10.1103/RevModPhys.90.025004`.

Fry, Edward S. and Thompson, Randall C. (1976). Experimental Test of Local Hidden-Variable Theories. *Phys. Rev. Lett.*, **37**, 465–468. `https://doi.org/10.1103/PhysRevLett.37.465`.

Fuchs, Christopher A.; Hoang, Michael C.; and Stacey, Blake C. (2017). The SIC Question: History and State of Play. *Axioms*, **6**, 21. Preprint https://arxiv.org/abs/0906.2187. `https://doi.org/10.3390/axioms6030021`.

Fuchs, Christopher A. and Schack, Rüdiger (2013). Quantum-Bayesian coherence. *Rev. Mod. Phys.*, **85**, 1693–1715. Preprint https://arxiv.org/abs/1703.07901. `https://doi.org/10.1103/RevModPhys.85.1693`.

Fuchs, Christopher A. and Schack, Rüdiger (2015). QBism and the Greeks: why a quantum state does not represent an element of physical reality. *Phys. Scr.*, **90**, 015104. Preprint https://arxiv.org/abs/1412.4211. `https://doi.org/10.1088/0031-8949/90/1/015104`.

Fujii, Kazuyuki (2001). Mathematical foundations of holonomic quantum computer. *Rep. Math. Phys.*, **48**, 75–82. Preprint https://arxiv.org/abs/quant-ph/0004102. `https://doi.org/10.1016/S0034-4877(01)80066-5`.

Fujiwara, Akio and Imai, Hiroshi (2008). A fibre bundle over manifolds of quantum channels and its application to quantum statistics. *J. Phys. A: Math. Theor.*, **41**, 255304. https://doi.org/10.1088/1751-8113/41/25/255304.

Fukuda, Motohisa and Wolf, Michael M. (2007). Simplifying additivity problems using direct sum constructions. *J. Math. Phys.*, **48**, 072101. Preprint https://arxiv.org/abs/0704.1092. https://doi.org/10.1063/1.2746128.

Furry, Wendell H. (1936*a*). Note on the Quantum-Mechanical Theory of Measurement. *Phys. Rev.*, **49**, 393–399. https://doi.org/10.1103/PhysRev.49.393.

Furry, Wendell H. (1936*b*). Remarks on Measurements in Quantum Theory. *Phys. Rev.*, **49**, 476–476. https://doi.org/10.1103/PhysRev.49.476.

Furusawa, Akira; Sørensen, Jens L.; Braunstein, Samuel L.; Fuchs, Christopher A.; Kimble, Harry Jeffrey; and Polzik, Eugene S. (1998). Unconditional Quantum Teleportation. *Science*, **282**, 706–709. https://doi.org/10.1126/science.282.5389.706.

Gabriel, Andreas; Hiesmayr, Beatrix C.; and Huber, Marcus (2010). Criterion for k-separability in mixed multipartite systems. *Quantum Inf. Comput.*, **10**, 0829–0836. Preprint https://arxiv.org/abs/1002.2953. https://doi.org/10.26421/QIC10.9-10-8.

Gaiba, Roberto and Paris, Matteo G. A. (2009). Squeezed vacuum as a universal quantum probe. *Phys. Lett. A*, **373**, 934. Preprint https://arxiv.org/abs/0802.1682. https://doi.org/10.1016/j.physleta.2009.01.026.

Gamow, George (1928). Zur Quantentheorie des Atomkernes. *Z. Physik*, **51**, 204–212. in German, English translation available at http://web.ihep.su/dbserv/compas/src/gamow28/eng.pdf, last accessed: 9 February 2022. https://doi.org/10.1007/BF01343196.

Garrison, John C. and Chiao, Raymond Y. (2008). *Quantum Optics*. Oxford University Press, Oxford, U.K.

Gelfand, Israel and Naimark, Mark Aronovich (1943). On the imbedding of normed ringsinto the ring of operators in Hilbert space. *Rec. Math. [Mat. Sbornik]*, **12(54)**, 197–217. https://mi.mathnet.ru/eng/msb6155.

Gell-Mann, Murray and Hartle, James B. (1993). Classical equations for quantum systems. *Phys. Rev. D*, **47**, 3345–3382. https://doi.org/10.1103/PhysRevD.47.3345.

Gell-Mann, Murray and Pais, Abraham (1955). Behavior of Neutral Particles under Charge Conjugation. *Phys. Rev.*, **97**, 1387–1389. https://doi.org/10.1103/PhysRev.97.1387.

Genovese, Marco; Novero, C.; and Predazzi, Enrico (2001). Can experimental tests of Bell inequalities performed with pseudoscalar mesons be definitive? *Phys. Lett. B*, **513**, 401–405. Preprint https://arxiv.org/abs/hep-ph/0103298. https://doi.org/10.1016/S0370-2693(01)00580-9.

Gerlach, Walther and Stern, Otto (1922*a*). Das magnetische Moment des Silberatoms. *Z. Phys.*, **9**, 353–355. In German. https://doi.org/10.1007/BF01326984.

Gerlach, Walther and Stern, Otto (1922*b*). Der experimentelle Nachweis der Richtungsquantelung im Magnetfeld. *Z. Phys.*, **9**, 349–352. In German. https://doi.org/10.1007/BF01326983.

Gerlach, Walther and Stern, Otto (1922*c*). Der experimentelle Nachweis des magnetischen Moments des Silberatoms. *Z. Phys.*, **8**, 110–111. In German. `https://doi.org/10.1007/BF01329580`.

Gharibian, Sevag (2010). Strong NP-hardness of the quantum separability problem. *Quantum Inf. Comput.*, **10**, 343–360. Preprint https://arxiv.org/abs/0810.4507. `https://doi.org/10.26421/QIC10.3-4-11`.

Ghirardi, Giancarlo; Grassi, Renata; and Ragazzon, Renzo (1992). Can one test quantum mechanics at the *Phi*-factory? In *The DAΦNE Physics Handbook* (eds. L. Maiani, G. Pancheri, and N. Paver), Frascati, p. 283. Servizio Documentazione dei Laboratori Nazionale di Frascati. `https://www.lnf.infn.it/sis/preprint/getfilepdf.php?filename=LNF-92-082(P).pdf`.

Ghirardi, Giancarlo; Grassi, Renata; and Weber, Tullio (1991). Quantum Mechanics Paradoxes at the Φ Factory. In *Physics and Detectors for DAΦNE, the Frascati Φ Factory* (ed. G. Pancheri), Frascati, p. 261. Servizio Documentazione dei Laboratori Nazionale di Frascati.

Ghirardi, Giancarlo; Rimini, Alberto; and Weber, Tullio (1986). Unified dynamics for microscopic and macroscopic systems. *Phys. Rev. D*, **34**, 470–491. `https://doi.org/10.1103/PhysRevD.34.470`.

Giaever, Ivar (1960). Energy Gap in Superconductors Measured by Electron Tunneling. *Phys. Rev. Lett.*, **5**, 147–148. `https://doi.org/10.1103/PhysRevLett.5.147`.

Giedke, Geza; Wolf, Michael M.; Küger, O.; Werner, Reinhard F.; and Cirac, Juan Ignacio (2003). Entanglement of Formation for Symmetric Gaussian States. *Phys. Rev. Lett.*, **91**, 107901. Preprint https://arxiv.org/abs/quant-ph/0304042. `https://doi.org/10.1103/PhysRevLett.91.107901`.

Gill, Richard D. and Guţă, Mădălin I. (2013). *On asymptotic quantum statistical inference*, Volume 9, Collections, p. 105–127. Institute of Mathematical Statistics, Beachwood, Ohio, USA. Preprint https://arxiv.org/abs/1112.2078. `https://doi.org/10.1214/12-IMSCOLL909`.

Gill, Richard D. and Levit, Boris Y. (1995). Applications of the van Trees Inequality: A Bayesian Cramér-Rao Bound. *Bernoulli*, **1**, 59–79. https://projecteuclid.org/euclid.bj/1186078362. `https://doi.org/10.2307/3318681`.

Giovannetti, Vittorio (2004). Separability conditions from entropic uncertainty relations. *Phys. Rev. A*, **70**, 012102. Preprint https://arxiv.org/abs/quant-ph/0307171. `https://doi.org/10.1103/PhysRevA.70.012102`.

Giovannetti, Vittorio; Lloyd, Seth; and Maccone, Lorenzo (2006). Quantum Metrology. *Phys. Rev. Lett.*, **96**, 010401. Preprint https://arxiv.org/abs/quant-ph/0509179. `https://doi.org/10.1103/PhysRevLett.96.010401`.

Giovannetti, Vittorio; Lloyd, Seth; and Maccone, Lorenzo (2011). Advances in quantum metrology. *Nat. Photon.*, **5**, 222. Preprint https://arxiv.org/abs/1102.2318. `https://doi.org/10.1038/nphoton.2011.35`.

Gisin, Nicolas (1996). Hidden quantum nonlocality revealed by local filters. *Phys. Lett. A*, **210**, 151–156. `https://doi.org/10.1016/S0375-9601(96)80001-6`.

Gisin, Nicolas and Go, Apollo (2001). EPR test with photons and kaons: Analogies. *Am. J. Phys.*, **69**, 264. Preprint https://arxiv.org/abs/quant-ph/0004063. `https://doi.org/10.1119/1.1326080`.

Gisin, Nicolas; Ribordy, Grégoire; Tittel, Wolfgang; and Zbinden, Hugo (2002). Quantum cryptography. *Rev. Mod. Phys.*, **74**, 145–195. Preprint https://arxiv.org/abs/quant-ph/0101098. `https://doi.org/10.1103/RevModPhys.74.145`.

Giustina, Marissa; Mech, A.; Ramelow, S.; Wittmann, B.; Kofler, Johannes; Beyer, J.; Lita, Adriana E.; Calkins, B.; Gerrits, T.; Nam, S. W.; Ursin, Rupert; and Zeilinger, Anton (2013). Bell violation using entangled photons without the fair-sampling assumption. *Nature*, **497**, 227–230. `https://doi.org/10.1038/nature12012`.

Giustina, Marissa; Versteegh, Marijn A. M.; Wengerowsky, Sören; Handsteiner, Johannes; Hochrainer, Armin; Phelan, Kevin; Steinlechner, Fabian; Kofler, Johannes; Larsson, Jan-Åke; Abellán, Carlos; Amaya, Waldimir; Pruneri, Valerio; Mitchell, Morgan W.; Beyer, Jörn; Gerrits, Thomas; Lita, Adriana E.; Shalm, Lynden K.; Nam, Sae Woo; Scheidl, Thomas; Ursin, Rupert; Wittmann, Bernhard; and Zeilinger, Anton (2015). Significant-Loophole-Free Test of Bell's Theorem with Entangled Photons. *Phys. Rev. Lett.*, **115**, 250401. Preprint https://arxiv.org/abs/1511.03190. `https://doi.org/10.1103/PhysRevLett.115.250401`.

Glaser, Vladimir; Martin, André; Grosse, Harald; and Thirring, Walter (1976). A Family of Optimal Conditions for the Absence of Bound States in a Potential. In *Studies in Mathematical Physics* (ed. E. H. Lieb), p. 169–194. Princeton University Press, Princeton, New Jersey. `https://doi.org/10.1515/9781400868940-009`.

Glauber, Roy J. (1963). Coherent and Incoherent States of the Radiation Field. *Phys. Rev.*, **131**, 2766–2788. `https://doi.org/10.1103/PhysRev.131.2766`.

Gleason, Andrew M. (1957). Measures on Closed Subspaces of a Hilbert Space. *J. Math. Mech.*, **6**, 885–893. https://www.jstor.org/stable/24900629. `https://doi.org/10.1007/978-94-010-1795-4_7`.

Go, Apollo (2004). Observation of Bell inequality violation in B mesons. *J. Mod. Optics*, **51**, 991–998. `https://doi.org/10.1080/09500340408233614`.

Go, Apollo; *et al.*: BELLE Coll. (2007). Measurement of Einstein-Podolsky-Rosen-Type Flavor Entanglement in $\Upsilon(4S) \to B^0\overline{B}^0$ Decays. *Phys. Rev. Lett.*, **99**, 131802. `https://doi.org/10.1103/PhysRevLett.99.131802`.

Gogolin, Christian and Eisert, Jens (2016). Equilibration, thermalisation, and the emergence of statistical mechanics in closed quantum systems. *Rep. Prog. Phys.*, **79**(5), 056001. Preprint https://arxiv.org/abs/1503.07538. `https://doi.org/10.1088/0034-4885/79/5/056001`.

Gogolin, Christian; Müller, Markus P.; and Eisert, Jens (2011). Absence of Thermalization in Nonintegrable Systems. *Phys. Rev. Lett.*, **106**(4), 040401. Preprint https://arxiv.org/abs/1009.2493. `https://doi.org/10.1103/PhysRevLett.106.040401`.

Gong, Ming; Chen, Ming-Cheng; Zheng, Yarui; Wang, Shiyu; Zha, Chen; Deng, Hui; Yan, Zhiguang; Rong, Hao; Wu, Yulin; Li, Shaowei; Chen, Fusheng; Zhao, Youwei; Liang, Futian; Lin, Jin; Xu, Yu; Guo, Cheng; Sun, Lihua; Castellano, Anthony D.; Wang, Haohua; Peng, Chengzhi; Lu, Chao-Yang; Zhu, Xiaobo; and Pan, Jian-Wei (2019). Genuine 12-Qubit Entanglement on a Superconducting Quantum Processor. *Phys. Rev. Lett.*, **122**, 110501. Preprint https://arxiv.org/abs/1811.02292. `https:`

//doi.org/10.1103/PhysRevLett.122.110501.

Górecki, Wojciech; Demkowicz-Dobrzański, Rafał; Wiseman, Howard M.; and Berry, Dominic W. (2020). π-corrected heisenberg limit. *Phys. Rev. Lett.*, **124**, 030501. Preprint https://arxiv.org/abs/1907.05428. https://doi.org/10.1103/PhysRevLett.124.030501.

Gorini, Vittorio; Frigerio, Alberto; Verri, Maurizio; Kossakowski, Andrzej; and Sudarshan, Ennackal Chandy George (1978). Properties of quantum Markovian master equations. *Rep. Math. Phys.*, **13**, 149–173. https://doi.org/10.1016/0034-4877(78)90050-2.

Gorini, Vittorio and Kossakowski, Andrzej (1976). N-level system in contact with a singular reservoir. *J. Math. Phys.*, **17**, 1298. https://doi.org/10.1063/1.523057.

Gorini, Vittorio; Kossakowski, Andrzej; and Sudarshan, Ennackal Chandy George (1976). Completely positive dynamical semigroups of N-level systems. *J. Math. Phys.*, **17**, 821–825. https://doi.org/10.1063/1.522979.

Goudsmit, Samuel Abraham and Richards, Paul Irving (1964). The Order of Electron Shells in Ionized Atoms. *Proc. Natl. Acad. Sci.*, **51**, 664–671. https://doi.org/10.1073/pnas.51.4.664.

Gour, Gilad; Kraus, Barbara; and Wallach, Nolan R. (2017). Almost all multipartite qubit quantum states have trivial stabilizer. *J. Math. Phys.*, **58**(9), 092204. Preprint https://arxiv.org/abs/1609.01327. https://doi.org/10.1063/1.5003015.

Grassl, Markus (2004). On SIC-POVMs and MUBs in Dimension 6. In *Proc. ERATO Conference on Quantum Information Science* (ed. J. Gruska), EQUIS. Preprint https://arxiv.org/abs/quant-ph/0406175.

Greenberger, Daniel M.; Horne, Michael A.; Shimony, Abner; and Zeilinger, Anton (1990). Bell's theorem without inequalities. *Am. J. Phys.*, **58**, 1131–1143. https://doi.org/10.1119/1.16243.

Greenberger, Daniel M.; Horne, Michael A.; and Zeilinger, Anton (1989). Going beyond Bell's theorem. In *Bell's Theorem, Quantum Theory, and Conceptions of the Universe* (ed. M. Kafalatos), p. 73–76. Kluwer Academics, Dordrecht, The Netherlands.

Greiner, Walter (2000). *Relativistic Quantum Mechanics. Wave Equations* (3rd edn). Springer, Berlin Heidelberg. https://doi.org/10.1007/978-3-662-04275-5.

Griffiths, David J. (1982). Hyperfine splitting in the ground state of hydrogen. *Am. J. Phys.*, **50**(8), 698–703. https://doi.org/10.1119/1.12733.

Griffiths, David J. (1995). *Introduction to Quantum Mechanics* (2nd edn). Prentice Hall, New Jersey.

Griffiths, Robert B. (1996). Consistent histories and quantum reasoning. *Phys. Rev. A*, **54**, 2759–2774. https://doi.org/10.1103/PhysRevA.54.2759.

Griffiths, Robert B. (1998). Choice of consistent family, and quantum incompatibility. *Phys. Rev. A*, **57**, 1604–1618. https://doi.org/10.1103/PhysRevA.57.1604.

Grimus, Walter (2010). *Einführung in die Statistische Physik und Thermodynamik.* Oldenbourg Verlag, München. In German.

Grosse, Harald and Martin, André (1997). *Particle Physics and the Schrödinger Equation.* Cambridge University Press, Cambridge, U.K.

Guérin, Philippe Allard; Baumann, Veronika; Del Santo, Flavio; and Brukner, Časlav

(2021). A no-go theorem for the persistent reality of Wigner's friend's perception. *Commun. Phys.*, **4**, 93. Preprint https://arxiv.org/abs/2009.09499. https://doi.org/10.1038/s42005-021-00589-1.

Gühne, Otfried; Bodoky, Fabian; and Blaauboer, Miriam (2008). Multiparticle entanglement under the influence of decoherence. *Phys. Rev. A*, **78**, 060301–060304. Preprint https://arxiv.org/abs/0805.2873. https://doi.org/10.1103/PhysRevA.78.060301.

Gühne, Otfried; Budroni, Costantino; Cabello, Adán; Kleinmann, Matthias; and Larsson, Jan-Åke (2014). Bounding the quantum dimension with contextuality. *Phys. Rev. A*, **89**, 062107. Preprint https://arxiv.org/abs/1302.2266. https://doi.org/10.1103/PhysRevA.89.062107.

Gühne, Otfried; Hyllus, Philipp; Bruß, Dagmar; Ekert, Artur K.; Lewenstein, Maciej; Macchiavello, Chiara; and Sanpera, Anna (2002). Detection of entanglement with few local measurements. *Phys. Rev. A*, **66**, 062305. Preprint https://arxiv.org/abs/quant-ph/0205089. https://doi.org/10.1103/PhysRevA.66.062305.

Gühne, Otfried; Hyllus, Philipp; Bruß, Dagmar; Ekert, Artur K.; Lewenstein, Maciej; Macchiavello, Chiara; and Sanpera, Anna (2003). Experimental detection of entanglement via witness operators and local measurements. *J. Mod. Opt.*, **50**, 1079–1102. Preprint https://arxiv.org/abs/quant-ph/0210134. https://doi.org/10.1080/09500340308234554.

Gühne, Otfried and Lewenstein, Maciej (2004). Entropic uncertainty relations and entanglement. *Phys. Rev. A*, **70**, 022316. Preprint https://arxiv.org/abs/quant-ph/0403219. https://doi.org/10.1103/PhysRevA.70.022316.

Gühne, Otfried and Seevinck, Michael (2010). Separability criteria for genuine multiparticle entanglement. *New J. Phys.*, **12**(5), 053002. Preprint https://arxiv.org/abs/0905.1349. https://doi.org/10.1088/1367-2630/12/5/053002.

Gühne, Otfried and Tóth, Géza (2009). Entanglement detection. *Phys. Rep.*, **474**, 1–75. Preprint https://arxiv.org/abs/0811.2803. https://doi.org/10.1016/j.physrep.2009.02.004.

Gühne, Otfried; Tóth, Géza; and Briegel, Hans J. (2005). Multipartite entanglement in spin chains. *New J. Phys.*, **7**, 229. Preprint https://arxiv.org/abs/quant-ph/0502160. https://doi.org/10.1088/1367-2630/7/1/229.

Gurvits, Leonid (2003). Classical Deterministic Complexity of Edmonds' Problem and Quantum Entanglement. In *Proceedings of the Thirty-fifth Annual ACM Symposium on Theory of Computing*, STOC '03, New York, NY, USA, p. 10–19. ACM. Preprint https://arxiv.org/abs/quant-ph/0303055. https://doi.org/10.1145/780542.780545.

Gurvits, Leonid (2004). Classical complexity and quantum entanglement. *J. Comput. Syst. Sci.*, **69**(3), 448–484. Special Issue on STOC 2003. https://doi.org/10.1016/j.jcss.2004.06.003.

Gurvits, Leonid and Barnum, Howard (2002). Largest separable balls around the maximally mixed bipartite quantum state. *Phys. Rev. A*, **66**, 062311. Preprint https://arxiv.org/abs/quant-ph/0204159. https://doi.org/10.1103/PhysRevA.

66.062311.

Guryanova, Yelena; Friis, Nicolai; and Huber, Marcus (2020). Ideal projective measurements have infinite resource costs. *Quantum*, **4**, 222. Preprint https://arxiv.org/abs/1805.11899. https://doi.org/10.22331/q-2020-01-13-222.

Hagiwara, Kaoru; *et al.*: Particle Data Group (2002). Review of Particle Properties. *Phys. Rev. D*, **66**, 010001. https://doi.org/10.1103/PhysRevD.66.010001.

Hall, Brian C. (2013). *Quantum Theory for Mathematicians*. Springer, Berlin.

Hannay, John H. (1985). Angle variable holonomy in adiabatic excursion of an integrable Hamiltonian. *J. Phys. A: Math. Gen.*, **18**, 221–230. https://doi.org/10.1088/0305-4470/18/2/011.

Hardy, Lucien (2001). Quantum theory from five reasonable axioms. Preprint https://arxiv.org/abs/quant-ph/0101012.

Hardy, Lucien and Spekkens, Robert (2010). Why Physics Needs Quantum Foundations. *Physics in Canada*, **66**, 73–76. Preprint https://arxiv.org/abs/1003.5008.

Haroche, Serge and Raimond, Jean-Michel (2013). *Exploring the Quantum* (Paperback edn). Oxford University Press, Oxford. https://doi.org/10.1093/acprof:oso/9780198509141.001.0001.

Harshman, Nathan L. and Ranade, Kedar S. (2011). Observables can be tailored to change the entanglement of any pure state. *Phys. Rev. A*, **84**, 012303. Preprint https://arxiv.org/abs/1102.0955. https://doi.org/10.1103/PhysRevA.84.012303.

Hasegawa, Yuji; Durstberger-Rennhofer, Katharina; Sponar, Stephan; and Rauch, Helmut (2011). Kochen-Specker theorem studied with neutron interferometer. *Nucl. Instrum. Methods Phys. Res.*, **634**, S21–S24. Preprint https://arxiv.org/abs/1004.2836. https://doi.org/10.1016/j.nima.2010.06.234.

Hasegawa, Yuji; Loidl, Rudolf; Badurek, Gerald; Baron, Matthias; and Rauch, Helmut (2003). Violation of a Bell-like inequality in single-neutron interferometry. *Nature*, **425**, 45–48. https://doi.org/10.1038/nature01881.

Hashemi Rafsanjani, Seyed Mohammad; Huber, Marcus; Broadbent, Curtis J.; and Eberly, Joseph H. (2012). Genuinely multipartite concurrence of N-qubit X matrices. *Phys. Rev. A*, **86**, 062303. Preprint https://arxiv.org/abs/1208.2706. https://doi.org/10.1103/PhysRevA.86.062303.

Håstad, Johan (1990). Tensor rank is NP-complete. *J. Algorithm*, **11**(4), 644–654. https://doi.org/10.1016/0196-6774(90)90014-6.

Hastings, Matthew B. (2009). Superadditivity of communication capacity using entangled inputs. *Nat. Phys.*, **5**, 255–257. Preprint https://arxiv.org/abs/0809.3972. https://doi.org/10.1038/nphys1224.

Hawking, Stephen W. (1982). The unpredictability of quantum gravity. *Commun. Math. Phys.*, **87**, 395–415. https://doi.org/10.1007/BF01206031.

Hayden, Patrick; Jozsa, Richard; Petz, Denes; and Winter, Andreas (2004). Structure of states which satisfy strong subadditivity of quantum entropy with equality. *Commun. Math. Phys.*, **246**, 359–374. Preprint https://arxiv.org/abs/quant-ph/0304007. https://doi.org/10.1007/s00220-004-1049-z.

Hayden, Patrick and Winter, Andreas (2008). Counterexamples to the maximal *p*-norm multiplicativity conjecture for all $p > 1$. *Comm. Math. Phys.*, **284**, 263–280. Preprint https://arxiv.org/abs/0807.4753. `https://doi.org/10.1007/s00220-008-0624-0`.

Hayden, Patrick M.; Horodecki, Michał; and Terhal, Barbara M. (2001). The asymptotic entanglement cost of preparing a quantum state. *J. Phys. A: Math. Gen.*, **34**(35), 6891–6898. Preprint https://arxiv.org/abs/quant-ph/0008134. `https://doi.org/10.1088/0305-4470/34/35/314`.

Hein, Marc; Dür, Wolfgang; Eisert, Jens; Raussendorf, Robert; Van den Nest, Maarten; and Briegel, Hans. J. (2005). Entanglement in graph states and its applications. *Proceedings of the International School of Physics "Enrico Fermi"*, **162**, 115–218. Preprint https://arxiv.org/abs/quant-ph/0602096. `https://doi.org/10.3254/978-1-61499-018-5-115`.

Heisenberg, Werner (1927). Über den anschaulichen Inhalt der quantentheoretischen Kinematik und Mechanik. *Z. Physik*, **43**, 172–198. in German. `https://doi.org/10.1007/BF01397280`.

Heisenberg, Werner and Jordan, Pascual (1926). Anwendung der Quantenmechanik auf das Problem der anomalen Zeemaneffekte. *Z. Physik*, **37**, 263–277. in German. `https://doi.org/10.1007/BF01397100`.

Hellwig, Helmut; Vessot, Robert F. C.; Levine, Martin W.; Zitzewitz, Paul W.; Allan, David W.; and Glaze, David J. (1970). Measurement of the Unperturbed Hydrogen Hyperfine Transition Frequency. *IEEE Trans. Instrum. Meas.*, **19**(4), 200–209. `https://doi.org/10.1109/TIM.1970.4313902`.

Helstrom, Carl W. (1967). Minimum mean-squared error of estimates in quantum statistics. *Phys. Lett. A*, **25**, 101–102. `https://doi.org/10.1016/0375-9601(67)90366-0`.

Helstrom, Carl W. (1969). Quantum detection and estimation theory. *J. Stat. Phys.*, **1**, 231–252. `https://doi.org/10.1007/BF01007479`.

Helstrom, Carl W. (1976). *Quantum Detection and Estimation Theory*. Academic Press, New York.

Helwig, Wolfram; Cui, Wei; Latorre, José Ignacio; Riera, Arnau; and Lo, Hoi-Kwong (2012). Absolute maximal entanglement and quantum secret sharing. *Phys. Rev. A*, **86**, 052335. Preprint https://arxiv.org/abs/1204.2289. `https://doi.org/10.1103/PhysRevA.86.052335`.

Hensen, B.; Bernien, H.; Dréau, A. E.; Reiserer, A.; Kalb, N.; Blok, M. S.; Ruitenberg, J.; Vermeulen, R. F. L.; Schouten, R. N.; Abellán, C.; Amaya, W.; Pruneri, V.; Mitchell, Morgan W.; Markham, M.; Twitchen, D. J.; Elkouss, D.; Wehner, S.; Taminiau, T. H.; and Hanson, R. (2015). Loophole-free Bell inequality violation using electron spins separated by 1.3 kilometres. *Nature*, **526**, 682–686. Preprint https://arxiv.org/abs/1508.05949. `https://doi.org/10.1038/nature15759`.

Hermann, Grete (1935*a*). Die naturphilosophischen Grundlagen der Quantenmechanik. *Abhandlungen der Fries'schen Schule*, **6**, 69–152. In German.

Hermann, Grete (1935*b*). Die naturphilosophischen Grundlagen der Quantenmechanik. *Naturwissenschaften*, **23**, 718–721. In German. `https://doi.org/10.1007/BF01491142`.

Herrera Valencia, Natalia; Srivastav, Vatshal; Pivoluska, Matej; Huber, Marcus; Friis, Nicolai; McCutcheon, Will; and Malik, Mehul (2020). High-Dimensional Pixel Entanglement: Efficient Generation and Certification. *Quantum*, **4**, 376. Preprint https://arxiv.org/abs/2004.04994. `https://doi.org/10.22331/q-2020-12-24-376`.

Hertz, Heinrich (1887). Ueber einen Einfluss des ultravioletten Lichtes auf die electrische Entladung. *Ann. Phys. (Berl.)*, **267**, 983–1000. in German. `https://doi.org/10.1002/andp.18872670827`.

Hiesmayr, Beatrix C. (1999). The puzzling story of the $K^0\bar{K}^0$—system or about quantum mechanical interference and Bell inequalities in particle physics. Diploma thesis, University of Vienna.

Hiesmayr, Beatrix C. (2001). A Generalized Bell Inequality and Decoherence for the K0 anti-K0 system. *Found. of Phys. Lett.*, **14**, 231–245. Preprint https://arxiv.org/abs/hep-ph/0010108. `https://doi.org/10.1023/A:1013457210230`.

Hiesmayr, Beatrix C. (2002). *The puzzling story of the neutral kaon system or what we can learn of entanglement*. Ph.D. thesis, University of Vienna.

Hiesmayr, Beatrix C. (2007). Nonlocality and entanglement in a strange system. *Eur. Phys. J. C*, **50**, 73–79. Preprint https://arxiv.org/abs/quant-ph/0607210. `https://doi.org/10.1140/epjc/s10052-006-0199-x`.

Hiesmayr, Beatrix C.; Di Domenico, Antonio; Curceanu, Catalina; Gabriel, Andreas; Huber, Marcus; Larsson, Jan-Åke; and Moskal, Pawel (2012). Revealing Bell's nonlocality for unstable systems in high energy physics. *Eur. Phys. J. C*, **72**, 1856. Preprint https://arxiv.org/abs/1111.4797. `https://doi.org/10.1140/epjc/s10052-012-1856-x`.

Hill, Scott and Wootters, William K. (1997). Entanglement of a Pair of Quantum Bits. *Phys. Rev. Lett.*, **78**, 5022–5025. Preprint https://arxiv.org/abs/quant-ph/9703041. `https://doi.org/10.1103/PhysRevLett.78.5022`.

Hiroshima, Tohya; Adesso, Gerardo; and Illuminati, Fabrizio (2007). Monogamy Inequality for Distributed Gaussian Entanglement. *Phys. Rev. Lett.*, **98**, 050503. Preprint https://arxiv.org/abs/quant-ph/0605021. `https://doi.org/10.1103/PhysRevLett.98.050503`.

Hirsch, Flavien; Quintino, Marco Túlio; Bowles, Joseph; Vértesi, Tamás; and Brunner, Nicolas (2016). Entanglement without hidden nonlocality. *New J. Phys.*, **18**, 113019. Preprint https://arxiv.org/abs/1606.02215. `https://doi.org/10.1088/1367-2630/18/11/113019`.

Hirsch, Flavien; Quintino, Marco Túlio; Vértesi, Tamás; Navascués, Miguel; and Brunner, Nicolas (2017). Better local hidden variable models for two-qubit Werner states and an upper bound on the Grothendieck constant $K_G(3)$. *Quantum*, **1**, 3. Preprint https://arxiv.org/abs/1609.06114. `https://doi.org/10.22331/q-2017-04-25-3`.

Hirschman Jr., Isidore Isaac (1957). A note on entropy. *Am. J. Math.*, **79**, 152–156. `https://doi.org/10.2307/2372390`.

Hiskett, Philip A.; Rosenberg, Danna; Peterson, Charles G.; Hughes, Richard J.; Nam, Sae Woo; Lita, Adriana E.; Miller, Aaron J.; and Nordholt, Jane E. (2006).

Long-distance quantum key distribution in optical fibre. *New J. Phys.*, **8**(9), 193–193. Preprint https://arxiv.org/abs/quant-ph/0607177. https://doi.org/10.1088/1367-2630/8/9/193.

Ho-Kim, Quang and Pham, Xuan-Yem (1998). *Elementary Particles and Their Interactions* (1st edn). Springer, Berlin, Heidelberg. https://doi.org/10.1007/978-3-662-03712-6.

Hoeffding, Wassily (1963). Probability Inequalities for Sums of Bounded Random Variables. *J. Am. Stat. Assoc.*, **58**(301), 13–30. https://doi.org/10.1080/01621459.1963.10500830.

Hoffman, Kenneth and Kunze, Ray (1971). *Linear Algebra* (2nd edn). Prentice-Hall, Englewood Cliffs, New Jersey.

Hofmann, Martin; Moroder, Tobias; and Gühne, Otfried (2014). Analytical characterization of the genuine multiparticle negativity. *J. Phys. A: Math. Theor.*, **47**, 155301. Preprint https://arxiv.org/abs/1401.2424. https://doi.org/10.1088/1751-8113/47/15/155301.

Holevo, Alexander S. (1984). Covariant measurements and imprimitivity systems. *Lect. Notes Math.*, **1055**, 153. https://doi.org/10.1007/BFb0071720.

Holevo, Alexander S. (2001). *Statistical Structure of Quantum Theory, Lecture Notes in Physics M67*. Springer Verlag, Heidelberg.

Holevo, Alexander S. (2011*a*). On the Choi-Jamiolkowski Correspondence in Infinite Dimensions. *J. Math. Phys.*, **52**, 042202. Preprint https://arxiv.org/abs/1004.0196. https://doi.org/10.1063/1.3581879.

Holevo, Alexander S. (2011*b*). *Probabilistic and Statistical Aspects of Quantum Theory* (2nd edn). Edizioni della Normale, Pisa, Italy.

Holevo, Alexander S.; Sohma, M.; and Hirota, O. (1999). Capacity of quantum Gaussian channels. *Phys. Rev. A*, **59**, 1820–1828. https://doi.org/10.1103/PhysRevA.59.1820.

Holland, Peter R. (1993). *The Quantum Theory of Motion: An Account of the de Broglie-Bohm Causal Interpretation of Quantum Mechanics*. Cambridge University Press, Cambridge, U.K. https://doi.org/10.1017/CBO9780511622687.

Holt, Richard A. and Pipkin, Francis M. (1974). Quantum Mechanics versus Hidden Variables. Preprint, Havard University.

Horn, Alfred (1954). Doubly stochastic matrices and the diagonal of a rotation matrix. *Am. J. Math.*, **76**, 620–630. https://doi.org/10.2307/2372705.

Horodecki, Michał and Horodecki, Paweł (1999). Reduction criterion of separability and limits for a class of distillation protocols. *Phys. Rev. A*, **59**, 4206–4216. Preprint https://arxiv.org/abs/quant-ph/9708015. https://doi.org/10.1103/PhysRevA.59.4206.

Horodecki, Michał; Horodecki, Paweł; and Horodecki, Ryszard (1996*a*). Separability of mixed states: necessary and sufficient conditions. *Phys. Lett. A*, **223**, 25. Preprint https://arxiv.org/abs/quant-ph/9605038. https://doi.org/10.1016/S0375-9601(96)00706-2.

Horodecki, Michał; Horodecki, Paweł; and Horodecki, Ryszard (1997). Inseparable two spin-$\frac{1}{2}$ density matrices can be distilled to a singlet form. *Phys. Rev. Lett.*, **78**, 574–577. Preprint https://arxiv.org/abs/quant-ph/9607009. https://doi.org/

10.1103/PhysRevLett.78.574.

Horodecki, Michał; Horodecki, Paweł; and Horodecki, Ryszard (1998). Mixed-state entanglement and distillation: Is there a "bound" entanglement in nature? *Phys. Rev. Lett.*, **80**, 5239. Preprint https://arxiv.org/abs/quant-ph/9801069. https://doi.org/10.1103/PhysRevLett.80.5239.

Horodecki, Michał; Horodecki, Paweł; and Horodecki, Ryszard (2006). Separability of Mixed Quantum States: Linear Contractions and Permutation Criteria. *Open Syst. Inf. Dyn.*, **13**, 103. Preprint https://arxiv.org/abs/quant-ph/0206008. https://doi.org/10.1007/s11080-006-7271-8.

Horodecki, Michał; Oppenheim, Jonathan; and Winter, Andreas (2005). Partial quantum information. *Nature*, **436**, 673–676. Preprint https://arxiv.org/abs/quant-ph/0505062. https://doi.org/10.1038/nature03909.

Horodecki, Paweł; Horodecki, Michał; and Horodecki, Ryszard (1999). Bound Entanglement Can Be Activated. *Phys. Rev. Lett.*, **82**, 1056–1059. https://doi.org/10.1103/PhysRevLett.82.1056.

Horodecki, Ryszard and Horodecki, Michał (1996). Information-theoretic aspects of inseparability of mixed states. *Phys. Rev. A*, **54**, 1838–1857. https://doi.org/10.1103/PhysRevA.54.1838.

Horodecki, Ryszard; Horodecki, Paweł; and Horodecki, Michał (1995). Violating Bell inequality by mixed spin-1/2 states: necessary and sufficient condition. *Phys. Lett. A*, **200**, 340–344. https://doi.org/10.1016/0375-9601(95)00214-N.

Horodecki, Ryszard; Horodecki, Paweł; and Horodecki, Michał (1996*b*). Quantum α-entropy inequalities: independent condition for local realism? *Phys. Lett. A*, **210**, 377–381. https://doi.org/10.1016/0375-9601(95)00930-2.

Horodecki, Ryszard; Horodecki, Paweł; Horodecki, Michał; and Horodecki, Karol (2009). Quantum entanglement. *Rev. Mod. Phys.*, **81**, 865–942. Preprint https://arxiv.org/abs/quant-ph/0702225. https://doi.org/10.1103/RevModPhys.81.865.

Horvath, Ferdinand (2013). Information theoretical reconstructions of quantum theory. Bachelor thesis, University of Vienna. https://homepage.univie.ac.at/reinhold.bertlmann/pdfs/dipl_diss/Horvath_BA_InformationTheoreticalReconstructionsOfQuantumTheory.pdf.

Hou, Bo-Yu and Hou, Bo-Yuan (1997). *Differential Geometry for Physicists, Advanced Series on Theoretical Physical Sciences, Vol. 6.* World Scientific, Singapore.

Huang, H. and Agarwal, Girish S. (1994). General linear transformations and entangled states. *Phys. Rev. A*, **49**, 52–60. https://doi.org/10.1103/PhysRevA.49.52.

Huang, Yichen (2010). Entanglement criteria via concave-function uncertainty relations. *Phys. Rev. A*, **82**, 012335. https://doi.org/10.1103/PhysRevA.82.012335.

Huang, Yichen (2014). Computing quantum discord is NP-complete. *New J. Phys.*, **16**, 033027. Preprint https://arxiv.org/abs/1305.5941. https://doi.org/10.1088/1367-2630/16/3/033027.

Huber, Felix; Gühne, Otfried; and Siewert, Jens (2017). Absolutely Maximally Entangled States of Seven Qubits Do Not Exist. *Phys. Rev. Lett.*, **118**, 200502. Preprint https://arxiv.org/abs/1608.06228. https://doi.org/10.1103/PhysRevLett.118.200502.

Huber, Marcus and de Vicente, Julio I. (2013). Structure of multidimensional entanglement in multipartite systems. *Phys. Rev. Lett.*, **110**, 030501. Preprint https://arxiv.org/abs/1210.6876. `https://doi.org/10.1103/PhysRevLett.110.030501`.

Huber, Marcus; Mintert, Florian; Gabriel, Andreas; and Hiesmayr, Beatrix C. (2010). Detection of High-Dimensional Genuine Multipartite Entanglement of Mixed States. *Phys. Rev. Lett.*, **104**, 210501. Preprint https://arxiv.org/abs/0912.1870. `https://doi.org/10.1103/PhysRevLett.104.210501`.

Huber, Marcus; Perarnau-Llobet, Martí; and de Vicente, Julio I. (2013). The entropy vector formalism and the structure of multidimensional entanglement in multipartite systems. *Phys. Rev. A*, **88**, 042328. Preprint https://arxiv.org/abs/1307.3541. `https://doi.org/10.1103/PhysRevA.88.042328`.

Huber, Marcus and Plesch, Martin (2011). Purification of genuine multipartite entanglement. *Phys. Rev. A*, **83**, 062321. Preprint https://arxiv.org/abs/1103.4294. `https://doi.org/10.1103/PhysRevA.83.062321`.

Huber, Marcus and Sengupta, Ritabrata (2014). Witnessing Genuine Multipartite Entanglement with Positive Maps. *Phys. Rev. Lett.*, **113**, 100501. Preprint https://arxiv.org/abs/1404.7449. `https://doi.org/10.1103/PhysRevLett.113.100501`.

Hudson, Robin Lyth (1974). When is the Wigner quasi-probability density non-negative? *Rep. Math. Phys.*, **6**(2), 249–252. `https://doi.org/10.1016/0034-4877(74)90007-X`.

Huelga, Susana F.; Macchiavello, Chiara; Pellizzari, Thomas; Ekert, Artur K.; Plenio, Martin B.; and Cirac, Juan Ignacio (1997). On the Improvement of Frequency Stardards with Quantum Entanglement. *Phys. Rev. Lett.*, **79**, 3865. Preprint https://arxiv.org/abs/quant-ph/9707014. `https://doi.org/10.1103/PhysRevLett.79.3865`.

Hund, Friedrich (1927). *Linienspektren und Periodisches System der Elemente*. Springer, Berlin. In German.

Hungerford, Thomas W. (1974). *Algebra* (Graduate Texts in Mathematics edn). Springer, New York.

Husimi, Kôdi (1940). Some Formal Properties of the Density Matrix. *Proceedings of the Physico-Mathematical Society of Japan. 3rd Series*, **22**(4), 264–314. `https://doi.org/10.11429/ppmsj1919.22.4_264`.

Hyllus, Philipp; Pezzé, Luca; and Smerzi, Augusto (2010). Entanglement and Sensitivity in Precision Measurements with States of a Fluctuating Number of Particles. *Phys. Rev. Lett.*, **105**, 120501. Preprint https://arxiv.org/abs/1003.0649. `https://doi.org/10.1103/PhysRevLett.105.120501`.

Ibinson, Ben (2007). *Quantum Information and Entropy*. Ph.D. thesis, University of Bristol. `https://ethos.bl.uk/OrderDetails.do?uin=uk.bl.ethos.492602`.

Ishizaka, Satoshi and Hiroshima, Tohya (2000). Maximally entangled mixed states under nonlocal unitary operations in two qubits. *Phys. Rev. A*, **62**, 022310. Preprint https://arxiv.org/abs/quant-ph/0003023. `https://doi.org/10.1103/PhysRevA.62.022310`.

Jackiw, Roman (1995). *Diverse Topics in Theoretical and Mathematical Physics*.

World Scientific, Singapore.

Jackson, John David (1999). *Classical Electrodynamics* (3rd edn). Wiley, New York.

Jakóbczyk, L. and Siennicki, M. (2006). Geometry of Bloch vectors in two-qubit system. *Phys. Lett. A*, **286**, 383–390. https://doi.org/10.1016/S0375-9601(01) 00455-8.

Jamiołkowski, Andrzej (1972). Linear transformations which preserve trace and positive semidefiniteness of operators. *Rep. Math. Phys.*, **3**, 275–278. https://doi.org/10.1016/0034-4877(72)90011-0.

Janet, Charles (1929). The helicoidal classification of the elements. *Chemical News*, **138**, 372–374, 388–393.

Jarzyna, Marcin and Demkowicz-Dobrzański, Rafał (2015). True precision limits in quantum metrology. *New J. Phys.*, **17**, 013010. Preprint https://arxiv.org/abs/1407.4805. https://doi.org/10.1088/1367-2630/17/1/013010.

Jaynes, Edwin Thompson (1957a). Information Theory and Statistical Mechanics. *Phys. Rev.*, **106**, 620–630. https://doi.org/10.1103/PhysRev.106.620.

Jaynes, Edwin Thompson (1957b). Information Theory and Statistical Mechanics. II. *Phys. Rev.*, **108**, 171–190. https://doi.org/10.1103/PhysRev.108.171.

Jaynes, Edwin Thompson and Cummings, Frederick W. (1963). Comparison of quantum and semiclassical radiation theories with application to the beam maser. *Proc. IEEE.*, **51**, 89–109. https://doi.org/10.1109/PROC.1963.1664.

Jensen, Johan Ludwig William Valdemar (1906). Sur les fonctions convexes et les inégalités entre les valeurs moyennes. *Acta Math.*, **30**, 175. in French. https://doi.org/10.1007/BF02418571.

Ji, Zhengfeng; Natarajan, Anand; Vidick, Thomas; Wright, John; and Yuen, Henry (2020). MIP*=RE. Preprint https://arxiv.org/abs/2001.04383.

Jiang, Zhang (2014). Quantum Fisher information for states in exponential form. *Phys. Rev. A*, **89**, 032128. Preprint https://arxiv.org/abs/1310.2687. https://doi.org/10.1103/PhysRevA.89.032128.

Jonathan, Daniel and Plenio, Martin B. (1999). Minimal conditions for local pure-state entanglement manipulation. *Phys. Rev. Lett.*, **83**, 1455–1458. Preprint https://arxiv.org/abs/quant-ph/9903054. https://doi.org/10.1103/PhysRevLett.83.1455.

Jones, Jonathan A.; Vedral, Vlatko; Ekert, Artur K.; and Castagnoli, Giuseppe (2000). Geometric quantum computation using nuclear magnetic resonance. *Nature*, **403**, 869–871. https://doi.org/10.1038/35002528.

Jönsson, Claus (1961). Elektroneninterferenzen an mehreren künstlich hergestellten Feinspalten. *Z. Phys.*, **161**, 454–474. In German. https://doi.org/10.1007/BF01342460.

Joos, Erich (1996). Decoherence through interaction with the environment. In *Decoherence and the apperance of a classical world in quantum theory* (eds. D. Giulini *et al.*), p. 35. Springer Verlag, Heidelberg.

Joos, Erich and Zeh, H. Dieter (1985). The emergence of classical properties through interaction with the environment. *Z. Phys. B*, **59**, 223–243. https://doi.org/10.1007/BF01725541.

Josephson, Brian David (1962). Possible new effects in superconductive tunnelling. *Phys. Lett.*, **1**(7), 251–253. https://doi.org/10.1016/0031-9163(62)91369-0.

Juffmann, Thomas; Milic, Adriana; Müllneritsch, Michael; Asenbaum, Peter; Tsukernik, Alexander; Tüxen, Jens; Mayor, Marcel; Cheshnovsky, Ori; and Arndt, Markus (2012). Real-time single-molecule imaging of quantum interference. *Nat. Nanotechnol.*, **7**, 297–300. Preprint https://arxiv.org/abs/1402.1867. https://doi.org/10.1038/nnano.2012.34.

Jungnitsch, Bastian; Moroder, Tobias; and Gühne, Otfried (2011*a*). Entanglement witnesses for graph states: General theory and examples. *Phys. Rev. A*, **84**, 032310. Preprint https://arxiv.org/abs/1106.1114. https://doi.org/10.1103/PhysRevA.84.032310.

Jungnitsch, Bastian; Moroder, Tobias; and Gühne, Otfried (2011*b*). Taming Multiparticle Entanglement. *Phys. Rev. Lett.*, **106**, 190502. Preprint https://arxiv.org/abs/1010.6049. https://doi.org/10.1103/PhysRevLett.106.190502.

Karol, Paul J.; Barber, Robert C.; Sherrill, Bradley M.; Vardaci, Emanuele; and Yamazaki, Toshimitsu (2016*a*). Discovery of the element with atomic number $Z = 118$ completing the 7th row of the periodic table (IUPAC Technical Report). *Pure Appl. Chem.*, **88**, 155–160. https://doi.org/10.1515/pac-2015-0501.

Karol, Paul J.; Barber, Robert C.; Sherrill, Bradley M.; Vardaci, Emanuele; and Yamazaki, Toshimitsu (2016*b*). Discovery of the elements with atomic numbers $Z = 113$, 115 and 117 (IUPAC Technical Report). *Pure Appl. Chem.*, **88**, 139–153. https://doi.org/10.1515/pac-2015-0502.

Kato, Tosio (1950). On the Adiabatic Theorem of Quantum Mechanics. *J. Phys. Soc. Jpn.*, **5**, 435–439. https://doi.org/10.1143/JPSJ.5.435.

Kay, Steven M. (1993). *Fundamentals of Statistical Signal Processing: Estimation Theory*. Prentice Hall, Upper Saddle River, NJ.

Kennard, Earle Hesse (1927). Zur Quantenmechanik einfacher Bewegungstypen. *Z. Physik*, **44**, 326–352. in German. https://doi.org/10.1007/BF01391200.

Kerber, Gabriele; Dick, Auguste; and Kerber, Wolfgang (1987). *Dokumente, Materialien und Bilder zur 100. Wiederkehr des Geburtstages von Erwin Schrödinger*. Fassbaender, Vienna, Austria. in German.

Khalfin, Leonid Aleksandrovich and Tsirelson, Boris Semyonovich (1985). Quantum and quasi-classical analogs of Bell inequalities. In *Symposium on the Foundations of Modern Physics* (eds. P. Lahti and P. Mittelstaedt), p. 441–460. World Sci. Publ. http://www.math.tau.ac.il/~tsirel/download/khts85.html.

Khan, Abdul Sattar; Li, Jun-Li; and Qiao, Cong-Feng (2020). Test of nonlocal hidden variable theory by the Leggett inequality in high energy physics. *Phys. Rev. D*, **101**, 096016. Preprint https://arxiv.org/abs/2003.04669. https://doi.org/10.1103/PhysRevD.101.096016.

Kibler, Maurice and , Michel (2006). A SU(2) recipe for mutually unbiased bases. *Int. J. Mod. Phys. B*, **20**, 1802. Preprint https://arxiv.org/abs/quant-ph/0601092. https://doi.org/10.1142/S0217979206034303.

Kiesel, Nikolai; Schmid, Christian; Tóth, Géza; Solano, Enrique; and Weinfurter, Harald (2007). Experimental Observation of Four-Photon Entangled Dicke State with High Fidelity. *Phys. Rev. Lett.*, **98**, 063604–063607. Preprint https://arxiv.org/abs/quant-ph/0606234. `https://doi.org/10.1103/PhysRevLett.98.063604`.

Kim, Jeong San; Das, Anirban; and Sanders, Barry C. (2009). Entanglement monogamy of multipartite higher-dimensional quantum systems using convex-roof extended negativity. *Phys. Rev. A*, **79**, 012329. Preprint https://arxiv.org/abs/0811.2047. `https://doi.org/10.1103/PhysRevA.79.012329`.

Kim, Yoon-Ho; Kulik, Sergei P.; and Shih, Yanhua (2001). Quantum Teleportation of a Polarization State with a complete Bell State Measurement. *Phys. Rev. Lett.*, **86**, 1370–1373. Preprint https://arxiv.org/abs/quant-ph/0010046. `https://doi.org/10.1103/PhysRevLett.86.1370`.

Kimura, Gen (2003). The Bloch vector for N-level systems. *Phys. Lett. A*, **314**, 339–349. Preprint https://arxiv.org/abs/quant-ph/0301152. `https://doi.org/10.1016/S0375-9601(03)00941-1`.

Kimura, Gen and Kossakowski, Andrzej (2005). The Bloch-Vector Space for N-Level Systems: the Spherical-Coordinate Point of View. *Open Syst. Info. Dynamics*, **12**, 207–229. Preprint https://arxiv.org/abs/quant-ph/0408014. `https://doi.org/10.1007/s11080-005-0919-y`.

Klappenecker, Andreas and Rötteler, Martin (2004). Constructions of Mutually Unbiased Bases. In *Finite Fields and Applications. Fq 2003. Lecture Notes in Computer Science* (eds. Gary L. Mullen, Alain Poli, and Henning Stichtenoth), Volume 2948, p. 137–144. Springer, Berlin, Heidelberg. Preprint https://arxiv.org/abs/quant-ph/0309120. `https://doi.org/10.1007/978-3-540-24633-6_10`.

Klauder, John R. (1960). The action option and a Feynman quantization of spinor fields in terms of ordinary c-numbers. *Ann. Phys. (N. Y.)*, **11**(2), 123–168. `https://doi.org/10.1016/0003-4916(60)90131-7`.

Klauder, John R. and Skagerstam, Bo-Sture K. (1985). *Coherent States*. World Scientific Publishing, Singapore. `https://doi.org/10.1142/0096`.

Klechkowski, Vsevolod Mavrikievich (1962). Justification of the Rule for Successive Filling of $(n+l)$ Groups. *J. Exp. Theor. Phys.*, **14**, 334. [Translated version, Russian original: Zh. Exsperim. i Teor. Fiz. **41**, 465 (1962)].

Klein, Oskar (1931). Zur quantenmechanischen Begründung des zweiten Hauptsatzes der Wärmelehre. *Z. Physik*, **72**, 767–775. `https://doi.org/10.1007/BF01341997`.

Knight, Peter L. and Milonni, Peter W. (1980). The Rabi frequency in optical spectra. *Phys. Rep.*, **66**, 21–107. `https://doi.org/10.1016/0370-1573(80)90119-2`.

Knysh, Sergey; Chen, Edward H.; and Durkin, Gabriel A. (2014). True Limits to Precision via Unique Quantum Probe. Preprint https://arxiv.org/abs/1402.0495.

Knysh, Sergey; Smelyanskiy, Vadim N.; and Durkin, Gabriel A. (2011). Scaling laws for precision in quantum interferometry and the bifurcation landscape of the optimal state. *Phys. Rev. A*, **83**, 021804. Preprint https://arxiv.org/abs/1006.1645. `https://doi.org/10.1103/PhysRevA.83.021804`.

Koashi, Masato and Winter, Andreas (2004). Monogamy of quantum entanglement and other correlations. *Phys. Rev. A*, **69**, 022309. Preprint https://arxiv.org/abs/quant-ph/0310037. `https://doi.org/10.1103/PhysRevA.`

69.022309.

Kochen, Simon B. (2000). Proof of the Kochen-Specker Theorem by Conway and Kochen with a set of 31 directions. private communication to Reinhold Bertlmann.

Kochen, Simon B. and Specker, Ernst (1967). The problem of hidden variables in quantum mechanics. *J. Math. Mech.*, **17**, 59–87. https://doi.org/10.1512/iumj.1968.17.17004.

Kofler, Johannes; Giustina, Marissa; Larsson, Jan-Åke; and Mitchell, Morgan W. (2016). Requirements for a loophole-free photonic Bell test using imperfect setting generators. *Phys. Rev. A*, **93**, 032115. Preprint https://arxiv.org/abs/1411.4787. https://doi.org/10.1103/PhysRevA.93.032115.

König, Dénes (1936). *Theorie der endlichen und unendlichen Graphen.* Akademische Verlags Gesellschaft, Leipzig.

Kramers, Hendrik Anthony (1938). *Die Grundlagen der Quantentheorie: Quantentheorie des Elektrons und der Strahlung, Vol. 2.* Akademische Verlagsgesellschaft, Leipzig.

Krammer, Philipp (2009). *Entanglement beyond two qubits—geometry and entanglement witnesses.* Ph.D. thesis, University of Vienna. https://doi.org/10.25365/thesis.6968.

Kraus, Karl (1983). *States, Effects and Operations.* Springer, Berlin.

Krenn, Günther and Zeilinger, Anton (1996). Entangled entanglement. *Phys. Rev. A*, **54**, 1793. https://doi.org/10.1103/PhysRevA.54.1793.

Kruskal, Joseph B. (1977). Three-way arrays: rank and uniqueness of trilinear decompositions, with application to arithmetic complexity and statistics. *Lin. Alg. Appl.*, **18**, 95–138. https://doi.org/10.1016/0024-3795(77)90069-6.

Kryszewski, Stanislaw and Zachciał, Mateusz (2006). Alternative representation of $N \times N$ density matrix. Preprint https://arxiv.org/abs/quant-ph/0602065.

Kübler, Olaf and Zeh, H. Dieter (1973). Dynamics of quantum correlations. *Ann. Phys. (N.Y.)*, **76**, 405–418. https://doi.org/10.1016/0003-4916(73)90040-7.

Kuś, Marek and Życzkowski, Karol (2001). Geometry of entangled states. *Phys. Rev. A*, **63**, 032307. Preprint https://arxiv.org/abs/quant-ph/0006068. https://doi.org/10.1103/PhysRevA.63.032307.

Kullback, Solomon and Leibler, Richard A. (1951). On Information and Sufficiency. *Ann. Math. Statist.*, **22**(1), 79–86. https://doi.org/10.1214/aoms/1177729694.

Kurz, Heinrich and Rauch, Helmut (1969). Beugung thermischer Neutronen an einem Strichgitter. *Z. Phys. A*, **220**, 419–426. In German. https://doi.org/10.1007/BF01394786.

Lamb, Willis E. and Retherford, Robert C. (1947). Fine Structure of the Hydrogen Atom by a Microwave Method. *Phys. Rev.*, **72**, 241–243. https://doi.org/10.1103/PhysRev.72.241.

Lami, Ludovico and Huber, Marcus (2016). Bipartite depolarizing channels. *J. Math. Phys.*, **57**, 092201. Preprint https://arxiv.org/abs/1603.02158. https://doi.org/10.1063/1.4962339.

Lamoreaux, Steven K. (1997). Demonstration of the Casimir Force in the 0.6 to $6\mu m$ Range. *Phys. Rev. Lett.*, **78**, 5–8. https://doi.org/10.1103/PhysRevLett.78.5.

Lancien, Cécilia; Gühne, Otfried; Sengupta, Ritabrata; and Huber, Marcus (2015).

Relaxations of separability in multipartite systems: Semidefinite programs, witnesses and volumes. *J. Phys. A: Math. Theor.*, **48**(50), 505302. Preprint https://arxiv.org/abs/1504.01029. https://doi.org/10.1088/1751-8113/48/50/505302.

Landau, Lev D. and Lifshitz, Evgeny M. (1971). *The Classical Theory of Fields. Course of Theoretical Physics* (3rd revised English edn). Volume 2. Pergamon Press, New York. https://doi.org/10.1016/C2009-0-14608-1.

Landau, Lev D. and Lifshitz, Evgeny M. (1977). *Quantum Mechanics Non-Relativistic Theory* (3rd edn). Pergamon Press, New York. https://doi.org/10.1016/C2013-0-02793-4.

Landau, Lev D. and Lifshitz, Evgeny M. (1980). *Statistical Physics* (3rd edn). Butterworth-Heinemann, Oxford. https://doi.org/10.1016/C2009-0-24487-4.

Landé, Alfred (1921). Über den anomalen Zeemaneffekt (Teil I). *Z. Physik*, **5**, 231–241. in German. https://doi.org/10.1007/BF01335014.

Lanford, Oscar E. and Robinson, Derek W. (1968). Mean Entropy of States in Quantum-Statistical Mechanics. *J. Math. Phys.*, **9**, 1120–1125. https://doi.org/10.1063/1.1664685.

Leach, Jonathan; Padgett, Miles J.; Barnett, Stephen M.; Franke-Arnold, Sonja; and Courtial, Johannes (2002). Measuring the Orbital Angular Momentum of a Single Photon. *Phys. Rev. Lett.*, **88**, 257901. https://doi.org/10.1103/PhysRevLett.88.257901.

Lee, Tsung Dao and Wu, Chien-Shiung (1966). Decays of neutral K mesons. *Ann. Rev. Nucl. Sci.*, **76**, 511. https://doi.org/10.1146/annurev.ns.16.120166.002455.

Leibfried, Dietrich; Blatt, Rainer; Monroe, Christopher; and Wineland, David (2003). Quantum dynamics of single trapped ions. *Rev. Mod. Phys.*, **75**, 281–324. https://doi.org/10.1103/RevModPhys.75.281.

Levine, Ira N. (2000). *Quantum Chemistry* (7th edn). Prentice Hall, New Jersey.

Levy, Amikam; Alicki, Robert; and Kosloff, Ronnie (2012). Quantum refrigerators and the third law of thermodynamics. *Phys. Rev. E*, **85**, 061126. Preprint https://arxiv.org/abs/1205.1347. https://doi.org/10.1103/PhysRevE.85.061126.

Lévy-Leblond, Jean-Marc (1967). Nonrelativistic particles and wave equations. *Comm. Math. Phys.*, **6**, 286–311. https://doi.org/10.1007/BF01646020.

Lévy-Leblond, Jean-Marc (1976). Who is afraid of nonhermitian operators? a quantum description of angle and phase. *Ann. Phys.*, **101**(1), 319–341. https://doi.org/10.1016/0003-4916(76)90283-9.

Lewenstein, Maciej; Bruß, Dagmar; Cirac, Juan Ignacio; Kraus, Barbara; Kuś, Marek; Samsonowicz, Jan; Sanpera, Anna; and Tarrach, Rolf (2000*a*). Separability and distillability in composite quantum systems - a primer. *J. Mod. Opt.*, **47**, 2841–2499. Preprint https://arxiv.org/abs/quant-ph/0006064. https://doi.org/10.1080/09500340008232176.

Lewenstein, Maciej; Kraus, Barbara; Cirac, Juan Ignacio; and Horodecki, Paweł (2000*b*). Optimization of entanglement witnesses. *Phys. Rev. A*, **62**, 052310. https://doi.org/10.1103/PhysRevA.62.052310.

Li, Junli and Qiao, Cong-Feng (2009). New possibilities for testing local realism in high energy physics. *Phys. Lett. A*, **47**, 4311–4314. Preprint https://arxiv.org/abs/0812.0869. `https://doi.org/10.1016/j.physleta.2009.09.057`.

Li, Junli and Qiao, Cong-Feng (2010). Testing local realism in $P \to VV$ decays. *Sci. China Phys. Mech. Astron.*, **53**, 870–875. Preprint https://arxiv.org/abs/0903.1246. `https://doi.org/10.1007/s11433-010-0202-2`.

Lide, David R. (ed.) (2005). Electron Work Function of the Elements. In *CRC Handbook of Chemistry and Physics, Internet Version*, Chapter 12, p. 124. CRC Press, Boca Raton, FL. last accessed 1 August 2023. `http://webdelprofesor.ula.ve/ciencias/isolda/libros/handbook.pdf`.

Lieb, Elliott H. (1973). Convex trace functions and the Wigner-Yanase-Dyson conjecture. *Adv. Math.*, **11**, 267–288. `https://doi.org/10.1016/0001-8708(73)90011-x`.

Lieb, Elliott H. and Ruskai, Mary Beth (1973). Proof of the strong subadditivity of quantum-mechanical entropy. *J. Math. Phys.*, **14**(12), 1938–1941. `https://doi.org/10.1063/1.1666274`.

Lindblad, Göran (1974). Expectations and entropy inequalities for finite quantum systems. *Commun. Math. Phys.*, **39**, 111–119. `https://doi.org/10.1007/bf01608390`.

Lindblad, Göran (1976). On the Generators of Quantum Dynamical Semigroups. *Commun. Math. Phys.*, **48**, 119–130. `https://doi.org/10.1007/BF01608499`.

Linden, Noah; Mosonyi, Milán; and Winter, Andreas (2013). The structure of Rényi entropic inequalities. *P. Roy. Soc. A Math. Phy.*, **469**(2158), 20120737. Preprint https://arxiv.org/abs/1212.0248. `https://doi.org/10.1098/rspa.2012.0737`.

Linden, Noah; Popescu, Sandu; Short, Anthony J.; and Winter, Andreas (2009). Quantum mechanical evolution towards thermal equilibrium. *Phys. Rev. E*, **79**, 061103. Preprint https://arxiv.org/abs/0812.2385. `https://doi.org/10.1103/PhysRevE.79.061103`.

Lipkin, Harry J. (1968). CP Violation and Coherent Decays of Kaon Pairs. *Phys. Rev.*, **176**, 1715–1718. `https://doi.org/10.1103/PhysRev.176.1715`.

Llewellyn, Daniel; Ding, Yunhong; Faruque, Imad I.; Paesani, Stefano; Bacco, Davide; Santagati, Raffaele; Qian, Yan-Jun; Li, Yan; Xiao, Yun-Feng; Huber, Marcus; Malik, Mehul; Sinclair, Gary F.; Zhou, Xiaoqi; Rottwitt, Karsten; O'Brien, Jeremy L.; Rarity, John G.; Gong, Qihuang; Oxenlowe, Leif K.; Wang, Jianwei; and Thompson, Mark G. (2020). Chip-to-chip quantum teleportation and multi-photon entanglement in silicon. *Nat. Phys.*, **16**, 148–153. Preprint https://arxiv.org/abs/1911.07839. `https://doi.org/10.1038/s41567-019-0727-x`.

Lo, Hoi-Kwong and Popescu, Sandu (1999). Concentrating entanglement by local actions—beyond mean values. Preprint https://arxiv.org/abs/quant-ph/9707038.

López-Incera, Andrea; Sekatski, Pavel; and Dür, Wolfgang (2019). All macroscopic quantum states are fragile and hard to prepare. *Quantum*, **3**, 118. Preprint https://arxiv.org/abs/1805.09868. `https://doi.org/10.22331/q-2019-01-25-118`.

Lostaglio, Matteo and Bowles, Joseph (2021). The original Wigner's friend paradox within a realist toy model. *Proc. R. Soc. A*, **477**, 20210273. Preprint https://arxiv.org/abs/2101.11032. https://doi.org/10.1098/rspa.2021.0273.

Löwner, Karl (1934). Über monotone Matrixfunktionen. *Math. Z.*, **38**, 177–216. In German. https://doi.org/10.1007/BF01170633.

Lüders, Gerhart (1951). Über die Zustandsänderung durch den Meßprozeß. *Ann. Phys. (Leipzig)*, **8**, 322–328. available online at http://www.physik.uni-augsburg.de/annalen/history/historic-papers/1951_443_322-328.pdf, English translation by K. A. Kirkpatrick: Concerning the state-change due to the measurement process, Ann. Phys. (Leipzig) **15**, 663–670 (2006), Preprint https://arxiv.org/abs/quant-ph/0403007. https://doi.org/10.1002/andp.200610207.

Lüders, Gerhart (1954). On the equivalence of invariance under time reversal and under particle-antiparticle conjugation for relativistic field theories. *Dan. Mat. Fys. Medd.*, **28**, 1–17. https://cds.cern.ch/record/1071765/files/mfm-28-5.pdf.

Ma, Xiao-Song; Herbst, Thomas; Scheidl, Thomas; Wang, Daqing; Kropatschek, Sebastian; Naylor, William; Mech, Alexandra; Wittmann, Bernhard; Kofler, Johannes; Anisimova, Elena; Makarov, Vadim; Jennewein, Thomas; Ursin, Rupert; and Zeilinger, Anton (2012a). Quantum Teleportation over 143 Kilometers Using Active Feed-Forward. *Nature*, **489**, 269–273. Preprint https://arxiv.org/abs/1205.3909. https://doi.org/10.1038/nature11472.

Ma, Xiao-Song; Kofler, Johannes; and Zeilinger, Anton (2016). Delayed-choice gedanken experiments and their realizations. *Rev. Mod. Phys.*, **88**, 015005. Preprint https://arxiv.org/abs/1407.2930. https://doi.org/10.1103/RevModPhys.88.015005.

Ma, Xiao-Song; Zotter, Stefan; Kofler, Johannes; Ursin, Rupert; Jennewein, Thomas; Brukner, Časlav; and Zeilinger, Anton (2012b). Experimental delayed-choice entanglement swapping. *Nat. Phys.*, **8**, 480–485. Preprint https://arxiv.org/abs/1203.4834. https://doi.org/10.1038/nphys2294.

Ma, Zhi-Hao; Chen, Zhi-Hua; Chen, Jing-Ling; Spengler, Christoph; Gabriel, Andreas; and Huber, Marcus (2011). Measure of genuine multipartite entanglement with computable lower bounds. *Phys. Rev. A*, **83**, 062325. Preprint https://arxiv.org/abs/1101.2001. https://doi.org/10.1103/PhysRevA.83.062325.

Maassen, Hans and Uffink, Jos B. M. (1988). Generalized entropic uncertainty relations. *Phys. Rev. Lett.*, **60**, 1103–1106. https://doi.org/10.1103/PhysRevLett.60.1103.

MacKay, David J. C. (2003). *Information Theory, Inference, and Learning Algorithms*. Cambridge University Press, Cambridge, U.K.

Madelung, Erwin (1936). *Mathematische Hilfsmittel des Physikers* (3rd edn). Springer, Berlin. In German.

Malin, Simon (2006). What are quantum states? *Quantum Information Processing*, **5**, 233–237. https://doi.org/10.1007/s11128-006-0016-3.

Manini, Nicola and Pistolesi, Fabio (2000). Off-Diagonal Geometric Phases. *Phys. Rev. Lett.*, **85**, 3067–3070. Preprint https://arxiv.org/abs/quant-ph/9911083.

https://doi.org/10.1103/PhysRevLett.85.3067.

Marcikic, Ivan; de Riedmatten, Hugues; Tittel, Wolfgang; Zbinden, Hugo; and Gisin, Nicolas (2003). Long-Distance Teleportation of Qubits at Telecommunication Wavelengths. *Nature*, **421**, 509–513. Preprint https://arxiv.org/abs/quant-ph/0301178. https://doi.org/10.1038/nature01376.

Marian, Paulina and Marian, Tudor A. (2012). Uhlmann fidelity between two-mode Gaussian states. *Phys. Rev. A*, **86**, 022340. Preprint https://arxiv.org/abs/1111.7067. https://doi.org/10.1103/PhysRevA.86.022340.

Marshall, Albert W.; Olkin, Ingram; and Arnold, Barry C. (2011). *Inequalities: Theory of Majorization and its Applications* (2nd edn). Springer, New York, NY. https://doi.org/10.1007/978-0-387-68276-1.

Masanes, Lluis and Müller, Markus P. (2011). A derivation of quantum theory from physical requirements. *New J. Phys.*, **13**, 063001. Preprint https://arxiv.org/abs/1004.1483. https://doi.org/10.1088/1367-2630/13/6/063001.

Masanes, Lluis and Oppenheim, Jonathan (2017). A general derivation and quantification of the third law of thermodynamics. *Nat. Commun.*, **8**, 14538. Preprint https://arxiv.org/abs/1412.3828. https://doi.org/10.1038/ncomms14538.

Matsukevich, Dzmitry N.; Maunz, Peter; Moehring, David L.; Olmschenk, Steven; and Monroe, Christopher (2008). Bell Inequality Violation with Two Remote Atomic Qubits. *Phys. Rev. Lett.*, **100**, 150404. Preprint https://arxiv.org/abs/0801.2184. https://doi.org/10.1103/PhysRevLett.100.150404.

Meek, Terry L. and Allen, Leland C. (2002). Configuration irregularities: deviations from the Madelung rule and inversion of orbital energy levels. *Chem. Phys. Lett.*, **362**, 362–64. https://doi.org/10.1016/S0009-2614(02)00919-3.

Meissner, Walther and Ochsenfeld, Robert (1933). Ein neuer Effekt bei Eintritt der Supraleitfähigkeit. *Naturwissenschaften*, **21**(44), 787–788. in German. https://doi.org/10.1007/BF01504252.

Mendaš, Istok P. (2006). The classification of three-parameter density matrices for a qutrit. *J. Phys. A: Math. Gen.*, **39**, 11313. https://doi.org/10.1088/0305-4470/39/36/012.

Menicucci, Nicolas C.; van Loock, Peter; Gu, Mile; Weedbrook, Christian; Ralph, Timothy C.; and Nielsen, Michael A. (2006). Universal Quantum Computation with Continuous-Variable Cluster States. *Phys. Rev. Lett.*, **97**, 110501. Preprint https://arxiv.org/abs/quant-ph/0605198. https://doi.org/10.1103/PhysRevLett.97.110501.

Mermin, N. David (1990*a*). Extreme quantum entanglement in a superposition of macroscopically distinct states. *Phys. Rev. Lett.*, **65**, 1838–1840. https://doi.org/10.1103/PhysRevLett.65.1838.

Mermin, N. David (1990*b*). Simple unified form for the major no-hidden-variables theorems. *Phys. Rev. Lett.*, **65**, 3373–3376. https://doi.org/10.1103/PhysRevLett.65.3373.

Mermin, N. David (1990*c*). What's wrong with these elements of reality? *Physics Today*, **43**, 9. https://doi.org/10.1063/1.2810588.

Mermin, N. David (1993). Hidden variables and two theorems of John Bell. *Rev. Mod. Phys.*, **65**, 803–815. Preprint https://arxiv.org/abs/1802.10119. https://doi.org/10.1103/RevModPhys.65.803.

Mermin, N. David (2014). Physics: QBism puts the scientist back into science. *Nature*, **507**, 421–423. https://doi.org/10.1038/507421a.

Mermin, N. David (2016). *Why quark rhymes with pork—and other scientific diversions*. Cambridge University Press, Cambridge.

Mermin, N. David (2017). Why QBism Is Not the Copenhagen Interpretation and What John Bell Might Have Thought of It. In *Quantum [Un]speakables II* (eds. R. A. Bertlmann and A. Zeilinger), p. 83–93. Springer, Berlin, Heidelberg. Preprint https://arxiv.org/abs/1409.2454. https://doi.org/10.1007/978-3-319-38987-5_4.

Merzbacher, Eugen (1998). *Quantum Mechanics* (3rd edn). Wiley, New York.

Messiah, Albert (1969). *Quantum Mechanics Vol. II*. North Holland Publishing Company, Amsterdam.

Meyenn, Karl von and Brown, Laurie M. (1996). Wolfgang Pauli: Scientific Correspondence with Bohr, Einstein, Heisenberg a.o.; Volume IV, Part I: 1950-1952. Springer, Berlin, Heidelberg. https://doi.org/10.1007/978-3-540-78803-4.

Michelson, Albert A. and Morley, Edward W. (1887). On a method of making the wavelength of sodium light the actual and practical standard of length. *Am. J. Sci.*, **s3-34**(204), 427–430. https://doi.org/10.2475/ajs.s3-34.204.427.

Millikan, Robert Andrews (1914). A Direct Determination of "h". *Phys. Rev.*, **4**, 73–75. https://doi.org/10.1103/PhysRev.4.73.2.

Mironova, Polina V. (2010). *Berry phase in atom optics*. Ph.D. thesis, University of Ulm. http://dx.doi.org/10.18725/OPARU-1899.

Mironova, Polina V.; Efremov, Maxim A.; and Schleich, Wolfgang P. (2013). Berry phase in atom optics. *Phys. Rev. A*, **87**, 013627–013638. Preprint https://arxiv.org/abs/1212.4399. https://doi.org/10.1103/PhysRevA.87.013627.

Möllenstedt, Gottfried and Jönsson, Claus (1959). Elektronen-Mehrfachinterferenzen an regelmäßig hergestellten Feinspalten. *Z. Phys.*, **155**, 472–474. In German. https://doi.org/10.1007/BF01333127.

Monràs, Alex (2013). Phase space formalism for quantum estimation of Gaussian states. Preprint https://arxiv.org/abs/1303.3682.

Montgomery, Douglas C. and Runger, George C. (1994). *Applied statistics and probability for engineers*. John Wiley & Sons, New York.

Monz, Thomas; Schindler, Philipp; Barreiro, Julio T.; Chwalla, Michael; Nigg, Daniel; Coish, William A.; Harlander, Maximilian; Hänsel, Wolfgang; Hennrich, Markus; and Blatt, Rainer (2011). 14-Qubit Entanglement: Creation and Coherence. *Phys. Rev. Lett.*, **106**, 130506. Preprint https://arxiv.org/abs/1009.6126. https://doi.org/10.1103/PhysRevLett.106.130506.

Mooney, Gary J.; White, Gregory A. L.; Hill, Charles D.; and Hollenberg, Lloyd C. L. (2021). Whole-Device Entanglement in a 65-Qubit Superconducting Quantum Computer. *Adv. Quantum Technol.*, **4**, 2100061. Preprint https://arxiv.org/abs/2102.11521. https://doi.org/10.1002/qute.202100061.

Moore, Walter J. (1989). Discovery of wave mechanics. In *Schrödinger: Life and Thought*, Chapter 6, p. 191–229. Cambridge University Press, Cambridge, U.K. https://doi.org/10.1017/CBO9780511600012.009.

Morelli, Simon; Klöckl, Claude; Eltschka, Christopher; Siewert, Jens; and Huber, Marcus (2020). Dimensionally sharp inequalities for the linear entropy. *Linear Algebra Its Appl.*, **584**, 294–325. Preprint https://arxiv.org/abs/1903.11887. https://doi.org/10.1016/j.laa.2019.09.008.

Moroder, Tobias; Gittsovich, Oleg; Huber, Marcus; and Gühne, Otfried (2014). Steering Bound Entangled States: A Counterexample to the Stronger Peres Conjecture. *Phys. Rev. Lett.*, **113**, 050404. Preprint https://arxiv.org/abs/1405.0262. https://doi.org/10.1103/PhysRevLett.113.050404.

Moses, Steven A. and Baldwin, Charles H.; *et al.* (2023). A Race Track Trapped-Ion Quantum Processor. Preprint https://arxiv.org/abs/2305.03828.

Muller, Francis; Birge, Robert W.; Fowler, William B.; Good, Robert H.; Hirsch, Warner; Matsen, Robert P.; Oswald, Larry; Powell, Wilson M.; White, Howard S.; and Piccioni, Oreste (1960). Regeneration and Mass Difference of Neutral K Mesons. *Phys. Rev. Lett.*, **4**, 539–539. https://doi.org/10.1103/PhysRevLett.4.539.

Müller-Lennert, Martin; Dupuis, Frédéric; Szehr, Oleg; Fehr, Serge; and Tomamichel, Marco (2013). On quantum Rényi entropies: a new generalization and some properties. *J. Math. Phys.*, **54**, 122203. Preprint https://arxiv.org/abs/1306.3142. https://doi.org/10.1063/1.4838856.

Munro, William J.; James, Daniel F. V.; White, Andrew G.; and Kwiat, Paul G. (2001). Maximizing the entanglement of two mixed qubits. *Phys. Rev. A*, **64**, 030302. https://doi.org/10.1103/PhysRevA.64.030302.

Naimark, Mark Aronovich (1940). Spectral functions of a symmetric operator. *Izv. Akad. Nauk SSSR Ser. Mat.*, **4**, 277–318. https://mi.mathnet.ru/eng/izv/v4/i3/p277.

Nakahara, Mikio (1990). *Geometry, topology and physics*. Adam Hilger, Bristol.

Narnhofer, Heide (2006). Entanglement reflected in Wigner functions. *J. Phys. A: Math. Gen.*, **39**, 7051–7064. https://doi.org/10.1088/0305-4470/39/22/017.

Navascués, Miguel; Pironio, Stefano; and Acín, Antonio (2007). Bounding the Set of Quantum Correlations. *Phys. Rev. Lett.*, **98**, 010401. Preprint https://arxiv.org/abs/quant-ph/0607119. https://doi.org/10.1103/PhysRevLett.98.010401.

Navascués, Miguel; Pironio, Stefano; and Acín, Antonio (2008). A convergent hierarchy of semidefinite programs characterizing the set of quantum correlations. *New J. Phys.*, **10**, 073013. Preprint https://arxiv.org/abs/0803.4290. https://doi.org/10.1088/1367-2630/10/7/073013.

Nielsen, Michael A. (1999). Conditions for a Class of Entanglement Transformations. *Phys. Rev. Lett.*, **83**, 436. Preprint https://arxiv.org/abs/quant-ph/9811053. https://doi.org/10.1103/PhysRevLett.83.436.

Nielsen, Michael A. and Chuang, Isaac L. (2000). *Quantum Computation and Quantum Information*. Cambridge University Press, Cambridge, U.K.

Nielsen, Michael A.; Knill, Emanuel; and Laflamme, Raymond (1998). Complete quantum teleportation using nuclear magnetic resonance. *Nature*, **396**, 52–55.

Preprint https://arxiv.org/abs/quant-ph/9811020. `https://doi.org/10.1038/` `23891.`

Nielsen, Michael A. and Vidal, Guifré (2001). Majorization and the interconversion of bipartite states. *Quantum Inf. Comput.*, **1**(1), 76–93. `https://doi.org/10.` `26421/QIC1.1-5.`

NobelPrize.org. The Nobel Prize in Physics 1932. *Nobel Prize Outreach AB*. https://www.nobelprize.org/prizes/physics/1932/summary/ last accessed 20 January 2022.

NobelPrize.org. The Nobel Prize in Physics 1994. *Nobel Prize Outreach AB*. https://www.nobelprize.org/prizes/physics/1994/summary/ last accessed 21 January 2022.

Nölleke, Christian; Neuzner, Andreas; Reiserer, Andreas; Hahn, Carolin; Rempe, Gerhard; and Ritter, Stephan (2013). Efficient Teleportation between Remote Single-Atom Quantum Memories. *Phys. Rev. Lett.*, **110**, 140403. Preprint https://arxiv.org/abs/1212.3127. `https://doi.org/10.1103/PhysRevLett.110.` `140403.`

Ohanian, Hans C. (1988). *Classical Electrodynamics*. Allyn and Bacon, Newton, Massachusetts.

Ohanian, Hans C. (1990). *Principles of Quantum Mechanics*. Prentice Hall, Englewood Cliffs, New Jersey.

Olivares, Stefano (2012). Quantum optics in the phase space - A tutorial on Gaussian states. *Eur. Phys. J. Spec. Top.*, **203**, 3–24. Preprint https://arxiv.org/abs/1111.0786. `https://doi.org/10.1140/epjst/` `e2012-01532-4.`

Olmschenk, Steven; Matsukevich, Dzmitry N.; Maunz, Peter; Hayes, D.; Duan, L. M.; and Monroe, Christopher (2009). Quantum Teleportation between Distant Matter Qubits. *Science*, **323**, 486–489. `https://doi.org/10.1126/science.1167209.`

Omran, Ahmed; Levine, Harry; Keesling, Alexander; Semeghini, Giulia; Wang, Tout T.; Ebadi, Sepehr; Bernien, Hannes; Zibrov, Alexander S.; Pichler, Hannes; Choi, Soonwon; Cui, Jian; Rossignolo, Marco; Rembold, Phila; Montangero, Simone; Calarco, Tommaso; Endres, Manuel; Greiner, Markus; Vuletić, Vladan; and Lukin, Mikhail D. (2019). Generation and manipulation of Schrödinger cat states in Rydberg atom arrays. *Science*, **365**, 570–574. Preprint https://arxiv.org/abs/1905.05721. `https://doi.org/10.1126/science.aax9743.`

Oppenheim, Jonathan; Spekkens, Robert W.; and Winter, Andreas (2005). A classical analogue of negative information. Preprint https://arxiv.org/abs/quant-ph/0511247.

Ortigoso, Juan (2018). Twelve years before the quantum no-cloning theorem. *Am. J. Phys.*, **86**, 201–205. Preprint https://arxiv.org/abs/1707.06910. `https://doi.` `org/10.1119/1.5021356.`

Osakabe, Nobuyuki; Matsuda, Tsuyoshi; Kawasaki, Takeshi; Endo, Junji; Tonomura, Akira; Yano, Shinichiro; and Yamada, Hiroji (1986). Experimental confirmation of Aharonov-Bohm effect using a toroidal magnetic field confined by a superconductor. *Phys. Rev. A*, **34**, 815–822. `https://doi.org/10.1103/PhysRevA.34.815.`

Osborne, Tobias J. (2005). Entanglement measure for rank-2 mixed states. *Phys.*

Rev. A, **72**, 022309. Preprint https://arxiv.org/abs/quant-ph/0203087. `https://doi.org/10.1103/PhysRevA.72.022309`.

Osborne, Tobias J. and Verstraete, Frank (2006). General Monogamy Inequality for Bipartite Qubit Entanglement. *Phys. Rev. Lett.*, **96**, 220503. Preprint https://arxiv.org/abs/quant-ph/0502176. `https://doi.org/10.1103/PhysRevLett.96.220503`.

Oszmaniec, Michał; Guerini, Leonardo; Wittek, Peter; and Acín, Antonio (2017). Simulating Positive-Operator-Valued Measures with Projective Measurements. *Phys. Rev. Lett.*, **119**, 190501. Preprint https://arxiv.org/abs/1609.06139. `https://doi.org/10.1103/PhysRevLett.119.190501`.

Ou, Yong-Cheng (2007). Violation of monogamy inequality for higher-dimensional objects. *Phys. Rev. A*, **75**, 034305. Preprint https://arxiv.org/abs/quant-ph/0612127. `https://doi.org/10.1103/PhysRevA.75.034305`.

Ozawa, Masanao (2000). Entanglement measures and the Hilbert-Schmidt distance. *Phys. Lett. A*, **268**, 158. Preprint https://arxiv.org/abs/quant-ph/0002036. `https://doi.org/10.1016/S0375-9601(00)00171-7`.

Pais, Abraham and Piccioni, Oreste (1955). Note on the Decay and Absorption of the θ^0. *Phys. Rev.*, **100**, 1487–1489. `https://doi.org/10.1103/PhysRev.100.1487`.

Palazuelos, Carlos and de Vicente, Julio I. (2022). Genuine multipartite entanglement of quantum states in the multiple-copy scenario. *Quantum*, **6**, 735. Preprint https://arxiv.org/abs/2201.08694. `https://doi.org/10.22331/q-2022-06-13-735`.

Pan, Jian-Wei (1999). *Quantum teleportation and multi-photon entanglement*. Ph.D. thesis, University of Vienna.

Pan, Jian-Wei; Bouwmeester, Dik; Daniell, Matthew; Weinfurter, Harald; and Zeilinger, Anton (2000). Experimental test of quantum nonlocality in three-photon Greenberger-Horne-Zeilinger entanglement. *Nature*, **403**, 515–519. `https://doi.org/10.1038/35000514`.

Pan, Jian-Wei; Bouwmeester, Dik; Weinfurter, Harald; and Zeilinger, Anton (1998). Experimental entanglement swapping: Entangling photons that never interacted. *Phys. Rev. Lett.*, **80**, 3891–3894. `https://doi.org/10.1103/PhysRevLett.80.3891`.

Pan, Jian-Wei; Daniell, Matthew; Gasparoni, Sara; Weihs, Gregor; and Zeilinger, Anton (2001). Experimental Demonstration of Four-Photon Entanglement and High-Fidelity Teleportation. *Phys. Rev. Lett.*, **86**, 4435–4438. Preprint https://arxiv.org/abs/quant-ph/0104047. `https://doi.org/10.1103/PhysRevLett.86.4435`.

Pan, Jian-Wei and Zeilinger, Anton (1998). Greenberger-Horne-Zeilinger-state analyzer. *Phys. Rev. A*, **57**, 2208–2211. `https://doi.org/10.1103/PhysRevA.57.2208`.

Pancharatnam, Shivaramakrishnan (1956). Generalized theory of interference and its applications. *Proc. Indian Acad. Sci. A*, **44**, 247–262. `https://doi.org/10.1007/BF03046050`.

Pankowski, Łukasz; Piani, Marco; Horodecki, Michał; and Horodecki, Paweł (2010). A Few Steps More Towards NPT Bound Entanglement. *IEEE Trans. Inf. The-*

ory, **56**(8), 4085–4100. Preprint https://arxiv.org/abs/0711.2613. `https://doi.org/10.1109/TIT.2010.2050810`.

Paris, Matteo G. A. (2009). Quantum estimation for quantum technology. *Int. J. Quantum Inf.*, **7**, 125. Preprint https://arxiv.org/abs/0804.2981. `https://doi.org/10.1142/S0219749909004839`.

Paris, Matteo G. A. (2012). The modern tools of quantum mechanics (A tutorial on quantum states, measurements, and operations). *Eur. Phys. J. S. T.*, **203**, 61. Preprint https://arxiv.org/abs/1110.6815. `https://doi.org/10.1140/epjst/e2012-01535-1`.

Park, James L. (1970). The concept of transition in quantum mechanics. *Found. Phys.*, **1**, 23–33. `https://doi.org/10.1007/BF00708652`.

Paschen, Friedrich and Back, Ernst E. A. (1921). Liniengruppen magnetisch vervollständigt. *Physica*, **1**, 261–273. in German, available at https://www.lorentz.leidenuniv.nl/history/proefschriften/Physica/Physica_1_1921_05391.pdf.

Pasternack, Simon (1937). On the Mean Value of r^s for Keplerian Systems. *Proc. Natl. Acad. Sci. U.S.A.*, **23**(2), 91–94. `https://doi.org/10.1073/pnas.23.2.91`.

Pati, Arun Kumar and Braunstein, Samuel L. (2000). Impossibility of deleting an unknown quantum state. *Nature*, **404**, 164–165. Preprint https://arxiv.org/abs/quant-ph/9911090. `https://doi.org/10.1038/404130b0`.

Pauli, Wolfgang (1927). Zur Quantenmechanik des magnetischen Elektrons. *Z. Phys.*, **43**, 601–623. In German. `https://doi.org/10.1007/BF01397326`.

Pauli, Wolfgang (1940). The Connection Between Spin and Statistics. *Phys. Rev.*, **58**, 716. `https://doi.org/10.1103/PhysRev.58.716`.

Pauli, Wolfgang (1953). Remarques sur le problème des paramètres cachés dans la mècanique quantique et sur la thèorie de l'onde pilote. In *Louis de Broglie. Physicien et penseur* (ed. A. George), p. 33–42. Albin Michel, Paris.

Pauli, Wolfgang (1955). Exclusion principle, Lorentz group and reflection of space-time and charge. In *Niels Bohr and the Development of Physics* (ed. W. Pauli), p. 30–51. McGraw-Hill, New York.

Pauli, Wolfgang (1999). Letter of W. Pauli to M. Born, March 31st, 1954. In *Wolfgang Pauli, scientific correspondence with Bohr, Einstein, Heisenberg, Vol. IV, Part II, 1953-1954* (ed. K. von Meyenn), p. 545–548. Springer, Berlin, Heidelberg. `https://doi.org/10.1007/978-3-540-78804-1`.

Paulsen, Vern (2003). *Completely Bounded Maps and Operator Algebras*. Cambridge University Press, Cambridge, U.K.

Pearle, Philip (1989). Combining stochastic dynamical state-vector reduction with spontaneous localization. *Phys. Rev. A*, **39**, 2277–2289. `https://doi.org/10.1103/PhysRevA.39.2277`.

Penrose, Roger (1996). On Gravity's role in Quantum State Reduction. *Gen. Relat. Gravit.*, **28**, 581–600. `https://doi.org/10.1007/BF02105068`.

Penrose, Roger (1998). Quantum computation, entanglement and state reduction. *Phil. Trans. Roy. Soc. Lond. A*, **356**, 1927–1938. `https://doi.org/10.1098/rsta.1998.0256`.

Peres, Asher (1978). Unperformed experiments have no results. *Am. J. Phys.*, **46**,

745. https://doi.org/10.1119/1.11393.

Peres, Asher (1990). Incompatible results of quantum measurements. *Phys. Lett. A*, **151**, 107–108. https://doi.org/10.1016/0375-9601(90)90172-K.

Peres, Asher (1991). Two simple proofs of the Kochen-Specker theorem. *J. Phys. A: Math. Gen.*, **24**, L175. https://doi.org/10.1088/0305-4470/24/4/003.

Peres, Asher (1996). Separability Criterion for Density Matrices. *Phys. Rev. Lett.*, **77**, 1413–1415. https://doi.org/10.1103/PhysRevLett.77.1413.

Peres, Asher (2000). Delayed Choice for Entanglement Swapping. *J. Mod. Opt.*, **47**, 139–143. https://doi.org/10.1080/09500340008244032.

Personick, Stewart D. (1971). Application of quantum estimation theory to analog communication over quantum channels. *IEEE Trans. Inf. Theory*, **17**, 240. https://doi.org/10.1109/TIT.1971.1054643.

Peskin, Michael E. and Schroeder, Daniel V. (1995). *An Introduction to Quantum Field Theory*. Westview Press, Reading, Massachusetts. https://doi.org/10.1201/9780429503559.

Petersen, Aage (1963). The Philosophy of Niels Bohr. *Bulletin of the Atomic Scientists*, **19**, 8–14. https://doi.org/10.1080/00963402.1963.11454520.

Pezzè, Luca; Smerzi, Augusto; Oberthaler, Markus K.; Schmied, Roman; and Treutlein, Philipp (2018). Quantum metrology with nonclassical states of atomic ensembles. *Rev. Mod. Phys.*, **90**, 035005. Preprint https://arxiv.org/abs/1609.01609. https://doi.org/10.1103/RevModPhys.90.035005.

Pfaff, Wolfgang; Hensen, Bas; Bernien, Hannes; van Dam, Suzanne B.; Blok, Machiel S.; Taminiau, Tim H.; Tiggelman, Marijn J.; Schouten, Raymond N.; Markham, Matthew; Twitchen, Daniel J.; and Hanson, Ronald (2014). Unconditional quantum teleportation between distant solid-state qubits. *Science*, **345**, 532–535. Preprint https://arxiv.org/abs/1404.4369. https://doi.org/10.1126/science.1253512.

Pienaar, Jacques L. (2021). A Quintet of Quandaries: Five No-Go Theorems for Relational Quantum Mechanics. *Found. Phys.*, **51**, 97. Preprint https://arxiv.org/abs/2107.00670. https://doi.org/10.1007/s10701-021-00500-6.

Pinel, Olivier; Jia, Pu; Fabre, Claude; Treps, Nicolas; and Braun, Daniel (2013). Quantum parameter estimation using general single-mode Gaussian states. *Phys. Rev. A*, **88**, 040102(R). Preprint https://arxiv.org/abs/1307.5318. https://doi.org/10.1103/PhysRevA.88.040102.

Pippenger, Nicholas (2003). The inequalities of quantum information theory. *IEEE Trans. Inf. Theory*, **49**(4), 773–789. https://doi.org/10.1109/TIT.2003.809569.

Pirandola, Stefano; Andersen, Ulrik L.; Banchi, Leonardo; Berta, Mario; Bunandar, Darius; Colbeck, Roger; Englund, Dirk; Gehring, Tobias; Lupo, Cosmo; Ottaviani, Carlo; Pereira, Jason L.; Razavi, Mohsen; Shamsul Shaari, Jesni; Tomamichel, Marco; Usenko, Vladyslav C.; Vallone, Giuseppe; Villoresi, Paolo; and Wallden, Petros (2020). Advances in quantum cryptography. *Adv. Opt. Photon.*, **12**(4), 1012–1236. Preprint https://arxiv.org/abs/1906.01645. https://doi.org/10.1364/AOP.361502.

Pironio, Stefano; Scarani, Valerio; and Vidick, Thomas (2016). Focus on device independent quantum information. *New J. Phys.*, **18**(10), 100202. https://doi.org/10.1088/1367-2630/18/10/100202.

Planck, Max (1900*a*). Entropie und Temperatur strahlender Wärme. *Annalen d. Physik*, **306**, 719–737. In German. https://doi.org/10.1002/andp.19003060410.

Planck, Max (1900*b*). Zur Theorie des Gesetzes der Energieverteilung im Normalspectrum. *Verhandlungen der Deutschen Physikalischen Gesellschaft*, **2**, 237–245. In German.

Planck, Max (1901). Ueber das Gesetz der Energieverteilung im Normalspectrum. *Annalen der Physik*, **309**(3), 553–563. https://doi.org/10.1002/andp.19013090310.

Plenio, Martin B. (2005). Logarithmic Negativity: A Full Entanglement Monotone That is not Convex. *Phys. Rev. Lett.*, **95**, 090503. Preprint https://arxiv.org/abs/quant-ph/0505071. https://doi.org/10.1103/PhysRevLett.95.090503.

Plenio, Martin B. and Virmani, Shashank (2007). An introduction to entanglement measures. *Quant. Inf. Comput.*, **7**, 1–51. Preprint https://arxiv.org/abs/quant-ph/0504163. https://doi.org/10.26421/QIC7.1-2-1.

Pogorelov, Ivan; Feldker, Thomas; Marciniak, Christian D.; Jacob, Georg; Podlesnic, Verena; Meth, Michael; Negnevitsky, Vlad; Stadler, Martin; Lakhmanskiy, Kirill; Blatt, Rainer; Schindler, Philipp; and Monz, Thomas (2021). Compact Ion-Trap Quantum Computing Demonstrator. *PRX Quantum*, **2**, 020343. Preprint https://arxiv.org/abs/2101.11390. https://doi.org/10.1103/PRXQuantum.2.020343.

Poincaré, Henri (1885). Sur les courbes définies par les équations différentielles. *J. Math. Pures Appl.*, **4**, 167–244. in French. http://portail.mathdoc.fr/JMPA/afficher_notice.php?id=JMPA_1885_4_1_A6_0.

Popescu, Sandu (1995). Bell's Inequalities and Density Matrices: Revealing "Hidden" Nonlocality. *Phys. Rev. Lett.*, **74**, 2619–2622. https://doi.org/10.1103/PhysRevLett.74.2619.

Popescu, Sandu and Rohrlich, Daniel (1994). Quantum nonlocality as an axiom. *Found. Phys.*, **24**, 379–385. https://doi.org/10.1007/BF02058098.

Preston, Thomas (1899). Radiation Phenomena in the Magnetic Field. *Nature*, **59**, 224–229. https://doi.org/10.1038/059224c0.

Privitera, Paolo and Selleri, Franco (1992). Quantum mechanics versus local realism for neutral kaon pairs. *Phys. Lett. B*, **296**, 261–272. https://doi.org/10.1016/0370-2693(92)90838-U.

Prugovečki, Eduard (1977). Information-theoretical aspects of quantum measurement. *Int. J. Theor. Phys.*, **16**, 321–331. https://doi.org/10.1007/BF01807146.

Pusey, Matthew F.; Barrett, Jonathan; and Rudolph, Terry (2012). On the reality of the quantum state. *Nat. Phys.*, **8**, 475. Preprint https://arxiv.org/abs/1111.3328. https://doi.org/10.1038/nphys2309.

Qian, Chen; Li, Jun-Li; Khan, Abdul Sattar; and Qiao, Cong-Feng (2020). Nonlocal correlation of spin in high energy physics. *Phys. Rev. D*, **101**, 116004. Preprint https://arxiv.org/abs/2002.04283. https://doi.org/10.1103/PhysRevD.101.116004.

Rabin, Michael O. (1981). How to exchange secrets with oblivious transfer. Technical report, Aiken Computation Laboratory, Harvard University. Cryptology ePrint Archive, Report 2005/187 https://ia.cr/2005/187.

Rambo, Timothy; Altepeter, Joseph; D'Ariano, Giacomo Mauro; and Kumar, Prem (2016). Functional Quantum Computing: An Optical Approach. *Phys. Rev. A*, **93**, 052321. Preprint https://arxiv.org/abs/1211.1257. https://doi.org/10.1103/PhysRevA.93.052321.

Ramsauer, Carl (1921). Über den Wirkungsquerschnitt der Gasmoleküle gegenüber langsamen Elektronen. *Ann. Phys.*, **369**, 513–540. In German. https://doi.org/10.1002/andp.19213690603.

Rana, Swapan (2013). Negative eigenvalues of partial transposition of arbitrary bipartite states. *Phys. Rev. A*, **87**, 054301. Preprint https://arxiv.org/abs/1304.6775. https://doi.org/10.1103/PhysRevA.87.054301.

Rao, Calyampudi Radhakrishna (1992). Information and the Accuracy Attainable in the Estimation of Statistical Parameters. In *Breakthroughs in Statistics: Foundations and Basic Theory* (eds. Samuel Kotz and Norman L. Johnson), p. 235–247. Springer, New York, NY. https://doi.org/10.1007/978-1-4612-0919-5_16.

Rasel, Ernst M.; Oberthaler, Markus K.; Batelaan, Herman; Schmiedmayer, Jörg; and Zeilinger, Anton (1995). Atom Wave Interferometry with Diffraction Gratings of Light. *Phys. Rev. Lett.*, **75**, 2633–2637. https://doi.org/10.1103/PhysRevLett.75.2633.

Rastall, Peter (1985). Locality, Bell's theorem, and quantum mechanics. *Found. Phys.*, **15**, 963–972. https://doi.org/10.1007/BF00739036.

Rauch, Dominik; Handsteiner, Johannes; Hochrainer, Armin; Gallicchio, Jason; Friedman, Andrew S.; Leung, Calvin; Liu, Bo; Bulla, Lukas; Ecker, Sebastian; Steinlechner, Fabian; Ursin, Rupert; Hu, Beili; Leon, David; Benn, Chris; Ghedina, Adriano; Cecconi, Massimo; Guth, Alan H.; Kaiser, David I.; Scheidl, Thomas; and Zeilinger, Anton (2018). Cosmic Bell Test Using Random Measurement Settings from High-Redshift Quasars. *Phys. Rev. Lett.*, **121**, 080403. Preprint https://arxiv.org/abs/1808.05966. https://doi.org/10.1103/PhysRevLett.121.080403.

Raussendorf, Robert and Briegel, Hans J. (2001). A One-Way Quantum Computer. *Phys. Rev. Lett.*, **86**, 5188–5191. Preprint https://arxiv.org/abs/quant-ph/0010033. https://doi.org/10.1103/PhysRevLett.86.5188.

Redhead, Michael (1987). *Incompleteness, Nonlocality, and Realism. A Prolegomenon to the Philosophy of Quantum Mechanics* (2nd edn). Clarendon Press, Oxford.

Reed, Michael and Simon, Barry (1972). *Methods of Modern Mathematical Physics I: Functional Analysis.* Academic Press, New York and London.

Regner, Christoph (2015). Information as foundation principle for quantum mechanics. Bachelor thesis, University of Vienna. https://homepage.univie.ac.at/reinhold.bertlmann/pdfs/dipl_diss/Regner_Bachelorarbeit.pdf.

Ren, Ji-Gang; Xu, Ping; Yong, Hai-Lin; Zhang, Liang; Liao, Sheng-Kai; Yin, Juan; Liu, Wei-Yue; Cai, Wen-Qi; Yang, Meng; Li, Li; Yang, Kui-Xing; Han, Xuan; Yao, Yong-Qiang; Li, Ji; Wu, Hai-Yan; Wan, Song; Liu, Lei; Liu, Ding-Quan; Kuang, Yao-Wu; He, Zhi-Ping; Shang, Peng; Guo, Cheng; Zheng, Ru-Hua; Tian, Kai; Zhu,

Zhen-Cai; Liu, Nai-Le; Lu, Chao-Yang; Shu, Rong; Chen, Yu-Ao; Peng, Cheng-Zhi; Wang, Jian-Yu; and Pan, Jian-Wei (2017). Ground-to-satellite quantum teleportation. *Nature*, **549**(7670), 70–73. Preprint https://arxiv.org/abs/1707.00934. `https://doi.org/10.1038/nature23675`.

Renes, Joseph M. (2021). Consistency in the description of quantum measurement: Quantum theory can consistently describe the use of itself. Preprint https://arxiv.org/abs/2107.02193.

Renes, Joseph M.; Blume-Kohout, Robin; Scott, Andrew J.; and Caves, Carlton M. (2004). Symmetric Informationally Complete Quantum Measurements. *J. Math. Phys.*, **45**, 2171. Preprint https://arxiv.org/abs/quant-ph/0310075. `https://doi.org/10.1063/1.1737053`.

Rényi, Alfréd (1961). On measures of entropy and information. *Proc. Fourth Berkeley Symp. Math. Stat. Prob., 1960*, **1**, 547–561. `https://digitalassets.lib.berkeley.edu/math/ucb/text/math_s4_v1_article-27.pdf`.

Resch, Kevin J.; Lindenthal, Michael; Blauensteiner, Bibiane; Böhm, Hannes R.; Fedrizzi, Alessandro; Kurtsiefer, Christian; Poppe, Andreas; Schmitt-Manderbach, Tobias; Taraba, Michael; Ursin, Rupert; Walther, Philip; Weier, Henning; Weinfurter, Harald; and Zeilinger, Anton (2005). Distributing entanglement and single photons through an intra-city, free-space quantum channel. *Opt. Express*, **13**, 202–209. Preprint https://arxiv.org/abs/quant-ph/0501008. `https://doi.org/10.1364/OPEX.13.000202`.

Richeson, David S. (2008). *Euler's Gem: The Polyhedron Formula and the Birth of Topology*. Princeton University Press, Princeton and Oxford.

Richter, Gerald (2008). *Stability of nonlocal quantum correlations in neutral B-meson systems*. Ph.D. thesis, TU Wien. `https://resolver.obvsg.at/urn:nbn:at:at-ubtuw:1-21946`.

Riebe, Mark; Häffner, Hartmut; Roos, Christian F.; Hänsel, Wolfgang; Benhelm, Jan; Lancaster, Gavin P. T.; Körber, Timo W.; Becher, Christoph; Schmidt-Kaler, Ferdinand; James, Daniel F. V.; and Blatt, Rainer (2004). Deterministic Quantum Teleportation with Atoms. *Nature*, **429**, 734–737. `https://doi.org/10.1038/nature02570`.

Riesz, Frigyes (1907). Sur une espèce de géométrie analytique des systèmes de fonctions sommables. *C. R. Acad. Sci. Paris*, **144**, 1409–1411. In French.

Rigovacca, Luca; Farace, Alessandro; Souza, Leonardo A. M.; De Pasquale, Antonella; Giovannetti, Vittorio; and Adesso, Gerardo (2017). Versatile Gaussian probes for squeezing estimation. *Phys. Rev. A*, **95**, 052331. Preprint https://arxiv.org/abs/1703.05554. `https://doi.org/10.1103/PhysRevA.95.052331`.

Ritz, Christina; Spee, Cornelia; and Gühne, Otfried (2019). Characterizing multipartite entanglement classes via higher-dimensional embeddings. *J. Phys. A: Math. Theor.*, **52**, 335302. Preprint https://arxiv.org/abs/1901.08847. `https://doi.org/10.1088/1751-8121/ab2f54`.

Rivas, Ángel; Plato, A. Douglas K.; Huelga, Susana F.; and Plenio, Martin B. (2010). Markovian master equations: a critical study. *New J. Phys.*, **12**, 113032. Preprint https://arxiv.org/abs/1006.4666. `https://doi.org/10.1088/1367-2630/12/11/`

113032.

Robertson, Howard Percy (1929). The Uncertainty Principle. *Phys. Rev.*, **34**, 163–164. https://doi.org/10.1103/PhysRev.34.163.

Robinson, Derek W. and Ruelle, David (1967). Mean entropy of states in classical statistical mechanics. *Commun. Math. Phys.*, **5**, 288–300. https://doi.org/10.1007/bf01646480.

Rogers, Daniel J. (2010). Broadband Quantum Cryptography. *Synthesis Lectures on Quantum Computing*, **2**(1), 1–97. https://doi.org/10.2200/S00265ED1V01Y201004QMC003.

Römer, Hartmann and Filk, Thomas (1994). *Statistische Mechanik*. VCH, Weinheim. In German. https://doi.org/10.1002/cite.330680129.

Roos, Christian F.; Riebe, Mark; Häffner, Hartmut; Hänsel, Wolfgang; Benhelm, Jan; Lancaster, Gavin P. T.; Becher, Christoph; Schmidt-Kaler, Ferdinand; and Blatt, Rainer (2004). Control and measurement of three-qubit entangled states. *Science*, **304**, 1478–1480. https://doi.org/10.1126/science.1097522.

Ross, Sheldon (2009). *A first course in probability* (8th edn). Prentice Hall press, Upper Saddle River, NJ.

Rovelli, Carlo (1996). Relational quantum mechanics. *Int. J. Theor. Phys.*, **35**, 1637–1678. Preprint https://arxiv.org/abs/quant-ph/9609002. https://doi.org/10.1007/BF02302261.

Rowe, Mary A.; Kielpinsky, Dave; Meyer, V.; Sackett, Cass A.; Itano, Wayne M.; and Wineland, David J. (2001). Experimental violation of a Bell's inequality with efficient detection. *Nature*, **409**, 791–794. https://doi.org/10.1038/35057215.

Rudin, Walter (1991). *Functional Analysis*. McGraw-Hill, New York.

Rudolph, Oliver (2001). A new class of entanglement measures. *J. Math. Phys.*, **42**, 5306–5314. Preprint https://arxiv.org/abs/math-ph/0005011. https://doi.org/10.1063/1.1398062.

Rudolph, Oliver (2003). Some properties of the computable cross norm criterion for separability. *Phys. Rev. A*, **67**, 032312. Preprint https://arxiv.org/abs/quant-ph/0212047. https://doi.org/10.1103/PhysRevA.67.032312.

Rudolph, Oliver (2005). Further results on the cross norm criterion for separability. *Quantum Inf. Process.*, **4**, 219–239. Preprint https://arxiv.org/abs/quant-ph/0202121. https://doi.org/10.1007/s11128-005-5664-1.

Rungta, Pranaw; Bužek, Vladimir; Caves, Carlton M.; Hillery, Mark; and Milburn, Gerard J. (2001). Universal state inversion and concurrence in arbitrary dimensions. *Phys. Rev. A*, **64**, 042315. Preprint https://arxiv.org/abs/quant-ph/0102040. https://doi.org/10.1103/PhysRevA.64.042315.

Rutherford, Ernest (1911). The Scattering of α and β Particles by Matter and the Structure of the Atom. *London, Edinburgh Dublin Philos. Mag. J. Sci.*, **21**(125), 669–688. https://doi.org/10.1080/14786440508637080.

Rydberg, Johannes R. (1889). Researches sur la constitution des spectres d'émission des éléments chimiques. *Kongliga Svenska Vetenskaps-Akademiens Handlingar [Proceedings of the Royal Swedish Academy of Science]*, **23**, 1–177. in French, https://portal.research.lu.se/en/publications/recherches-sur-la-constitution-des-spectres-d%C3%A9mission-des-%C3%A9l%C3%A9ment.

Ryder, Lewis H. (1996). *Quantum Field Theory* (2nd edn). Cambridge University Press, Cambridge, U.K.

Šafránek, Dominik (2019). Estimation of Gaussian quantum states. *J. Phys. A: Math. Theor.*, **52**, 035304. Preprint https://arxiv.org/abs/1801.00299. https://doi.org/10.1088/1751-8121/aaf068.

Šafránek, Dominik and Fuentes, Ivette (2016). Optimal probe states for the estimation of Gaussian unitary channels. *Phys. Rev. A*, **94**, 062313. Preprint https://arxiv.org/abs/1603.05545. https://doi.org/10.1103/PhysRevA.94.062313.

Šafránek, Dominik; Lee, Antony R.; and Fuentes, Ivette (2015). Quantum parameter estimation using multi-mode Gaussian states. *New J. Phys.*, **17**, 073016. Preprint https://arxiv.org/abs/1502.07924. https://doi.org/10.1088/1367-2630/17/7/073016.

Sakurai, Jun John and Napolitano, Jim (2017). *Modern Quantum Mechanics* (2nd edn). Cambridge University Press, Cambridge, U.K. https://doi.org/10.1017/9781108499996.

Salek, Sina; Schubert, Roman; and Wiesner, Karoline (2014). Negative conditional entropy of postselected states. *Phys. Rev. A*, **90**, 022116–022119. Preprint https://arxiv.org/abs/1305.0932. https://doi.org/10.1103/PhysRevA.90.022116.

Samuel, Joseph and Bhandari, Rajendra (1988). General setting for Berry's phase. *Phys. Rev. Lett.*, **60**, 2339–2342. https://doi.org/10.1103/PhysRevLett.60.2339.

Sanpera, Anna; Tarrach, Rolf; and Vidal, Guifré (1998). Local description of quantum inseparability. *Phys. Rev. A*, **58**, 826–830. Preprint https://arxiv.org/abs/quant-ph/9801024. https://doi.org/10.1103/PhysRevA.58.826.

Sauerwein, David; Wallach, Nolan R.; Gour, Gilad; and Kraus, Barbara (2018). Transformations among Pure Multipartite Entangled States via Local Operations are Almost Never Possible. *Phys. Rev. X*, **8**, 031020. Preprint https://arxiv.org/abs/1711.11056. https://doi.org/10.1103/PhysRevX.8.031020.

Saunders, Simon; Barrett, Jonathan; Kent, Adrian; and Wallace, David (ed.) (2010). *Many Worlds?: Everett, Quantum Theory, and Reality*. Oxford University Press, Oxford, U.K. https://doi.org/10.1093/acprof:oso/9780199560561.001.0001.

Scarani, Valerio; Bechmann-Pasquinucci, Helle; Cerf, Nicolas J.; Dušek, Miloslav; Lütkenhaus, Norbert; and Peev, Momtchil (2009). The security of practical quantum key distribution. *Rev. Mod. Phys.*, **81**, 1301–1350. Preprint https://arxiv.org/abs/0802.4155. https://doi.org/10.1103/RevModPhys.81.1301.

Scerri, Eric R. (1998). How Good Is the Quantum Mechanical Explanation of the Periodic System? *J. Chem. Ed.*, **75**, 1384. https://doi.org/10.1021/ed075p1384.

Scerri, Eric R. (2004). Just how ab initio is ab initio quantum chemistry? *Found. Chem.*, **6**, 93–116. https://doi.org/10.1023/B:FOCH.0000020998.31689.16.

Schatten, Robert (1960). *Norm Ideals of Completely Continuous Operators* (1st edn). Springer, Berlin, Heidelberg. https://doi.org/10.1007/978-3-642-87652-3.

Scheidl, Thomas; Ursin, Rupert; Kofler, Johannes; Ramelow, Sven; Ma, Xiao-Song; Herbst, Thomas; Ratschbacher, Lothar; Fedrizzi, Alessandro; Langford, Nathan K.; Jennewein, Thomas; and Zeilinger, Anton (2010). Violation of local realism with freedom of choice. *Proc. Natl. Acad. Sci. U.S.A.*, **107**, 19708–19713. Preprint https://arxiv.org/abs/0811.3129. `https://doi.org/10.1073/pnas.1002780107`.

Schindler, Philipp; Nigg, Daniel; Monz, Thomas; Barreiro, Julio T.; Martinez, Esteban; Wang, Shannon X.; Quint, Stephan; Brandl, Matthias F.; Nebendahl, Volckmar; Roos, Christian F.; Chwalla, Michael; Hennrich, Markus; and Blatt, Rainer (2013). A quantum information processor with trapped ions. *New J. Phys.*, **15**(12), 123012. Preprint https://arxiv.org/abs/1308.3096. `https://doi.org/10.1088/1367-2630/15/12/123012`.

Schleich, Wolfgang P. (2004). *Quantum Optics in Phase Space*. Wiley-VCH, Weinheim. `https://doi.org/10.1002/3527602976`.

Schmidt, Erhard (1907). Zur Theorie der linearen und nichtlinearen Integralgleichungen. *Math. Ann.*, **63**, 433–476. in German. `https://doi.org/10.1007/BF01449770`.

Schmitt-Manderbach, Tobias; Weier, Henning; Fürst, Martin; Ursin, Rupert; Tiefenbacher, Felix; Scheidl, Thomas; Perdigues, Josep; Sodnik, Zoran; Kurtsiefer, Christian; Rarity, John G.; Zeilinger, Anton; and Weinfurter, Harald (2007). Experimental demonstration of free-space decoy-state quantum key distribution over 144 km. *Phys. Rev. Lett.*, **98**, 010504. `https://doi.org/10.1103/PhysRevLett.98.010504`.

Schrödinger, Erwin (1926*a*). An Undulatory Theory of the Mechanics of Atoms and Molecules. *Phys. Rev.*, **28**, 1049–1070. `https://doi.org/10.1103/PhysRev.28.1049`.

Schrödinger, Erwin (1926*b*). Quantisierung als Eigenwertproblem. *Ann. Phys. (Berl.)*, **384**(4), 361–376. in German. `https://doi.org/10.1002/andp.19263840404`.

Schrödinger, Erwin (1926*c*). Quantisierung als Eigenwertproblem. *Ann. Phys. (Leipzig)*, **385**(13), 437–490. in German. `https://doi.org/10.1002/andp.19263851302`.

Schrödinger, Erwin (1930). Zum Heisenbergschen Unschärfeprinzip. *Sitzungsberichte der Preußischen Akademie der Wissenschaften. Physikalisch-mathematische Klasse*, **14**, 296–303. in German.

Schrödinger, Erwin (1935*a*). Die gegenwärtige Situation in der Quantenmechanik. *Naturwissenschaften*, **23**, 844–849. In German. `https://doi.org/10.1007/BF01491987`.

Schrödinger, Erwin (1935*b*). Die gegenwärtige Situation in der Quantenmechanik. *Naturwissenschaften*, **23**, 807–812. In German. `https://doi.org/10.1007/BF01491891`.

Schrödinger, Erwin (1935*c*). Discussion of probability relations between separated systems. *Math. Proc. Cambridge Phil. Soc.*, **31**, 555–563. `https://doi.org/10.1017/S0305004100013554`.

Schrödinger, Erwin (1936). Probability relations between separated systems. *Math. Proc. Cambridge Phil. Soc.*, **32**, 446–452. `https://doi.org/10.1017/`

S0305004100019137.

Schrödinger, Erwin (1944). *What is life? The physical aspect of the living cell"*. Cambridge University Press, Cambridge, U.K.

Schumacher, Benjamin and Westmoreland, Michael (2010). *Quantum Processes, Systems, and Information*. Cambridge University Press, Cambridge, U.K. `https://doi.org/10.1017/CBO9780511814006`.

Schumaker, Bonny L. (1986). Quantum mechanical pure states with gaussian wave functions. *Phys. Rep.*, **135**, 317–408. `https://doi.org/10.1016/0370-1573(86)90179-1`.

Schur, Issai (1923). Über eine Klasse von Mittelbildungen mit Anwendungen auf die Determinantentheorie. *Sitzungsber. Berl. Math. Ges.*, **22**, 9–20. in German.

Schwaiger, Katharina; Sauerwein, David; Cuquet, Martí; de Vicente, Julio I.; and Kraus, Barbara (2015). Operational Multipartite Entanglement Measures. *Phys. Rev. Lett.*, **115**, 150502. Preprint https://arxiv.org/abs/1503.00615. `https://doi.org/10.1103/PhysRevLett.115.150502`.

Schwinger, Julian Seymour (1951). The Theory of Quantized Fields. I. *Phys. Rev.*, **82**, 914–927. `https://doi.org/10.1103/PhysRev.82.914`.

Schwinger, Julian Seymour (1953*a*). The Theory of Quantized Fields. II. *Phys. Rev.*, **91**, 713–728. `https://doi.org/10.1103/PhysRev.91.713`.

Schwinger, Julian Seymour (1953*b*). The Theory of Quantized Fields. III. *Phys. Rev.*, **91**, 728–740. `https://doi.org/10.1103/PhysRev.91.728`.

Scott, Andrew J. (2004). Multipartite entanglement, quantum-error-correcting codes, and entangling power of quantum evolutions. *Phys. Rev. A*, **69**, 052330. Preprint https://arxiv.org/abs/quant-ph/0310137. `https://doi.org/10.1103/PhysRevA.69.052330`.

Scully, Marlan O. and Zubairy, M. Suhail (1997). *Quantum Optics*. Cambridge University Press, Cambridge, U.K. `https://doi.org/10.1017/CBO9780511813993`.

Scutaru, Horia (1998). Fidelity for displaced squeezed states and the oscillator semigroup. *J. Phys. A: Math. Gen.*, **31**, 3659–3663. Preprint https://arxiv.org/abs/quant-ph/9708013. `https://doi.org/10.1088/0305-4470/31/15/025`.

Segal, Irving E. (1947). Irreducible representations of operator algebras. *Bull. Am. Math. Soc.*, **53**, 73–88. `https://doi.org/10.1090/s0002-9904-1947-08742-5`.

Sekatski, Pavel; Skotiniotis, Michalis; and Dür, Wolfgang (2016). Dynamical decoupling leads to improved scaling in noisy quantum metrology. *New J. Phys.*, **18**, 073034. Preprint https://arxiv.org/abs/1512.07476. `https://doi.org/10.1088/1367-2630/18/7/073034`.

Sekatski, Pavel; Skotiniotis, Michalis; Kołodyński, Janek; and Dür, Wolfgang (2017). Quantum metrology with full and fast quantum control. *Quantum*, **1**, 27. Preprint https://arxiv.org/abs/1603.08944. `https://doi.org/10.22331/q-2017-09-06-27`.

Selleri, Franco (1983). Einstein locality and the $K^0 \bar{K}^0$ system. *Lett. Nuovo Cim.*, **36**, 521–526. `https://doi.org/10.1007/BF02725928`.

Selleri, Franco (1997). Incompatibility between local realism and quantum mechanics for pairs of neutral kaons. *Phys. Rev. A*, **56**, 3493–3506. `https://doi.org/10.`

1103/PhysRevA.56.3493.

Serafini, Alessio; Illuminati, Fabrizio; and De Siena, Silvio (2004). Symplectic invariants, entropic measures and correlations of Gaussian states. *J. Phys. B: At. Mol. Opt. Phys.*, **37**, L21. Preprint https://arxiv.org/abs/quant-ph/0307073. https://doi.org/10.1088/0953-4075/37/2/L02.

Sexl, Roman U. and Urbantke, Helmuth K. (2001). *Relativity, Groups, Particles.* Springer, Wien, New York.

Shalm, L. K.; Meyer-Scott, E.; Christensen, B. G.; Bierhorst, P.; Wayne, M. A.; Stevens, M. J.; Gerrits, T.; Glancy, S.; Hamel, D. R.; Allman, M. S.; Coakley, K. J.; Dyer, S. D.; Hodge, C.; Lita, A. E.; Verma, V. B.; Lambrocco, C.; Tortorici, E.; Migdall, A. L.; Zhang, Y.; Kumor, D. R.; Farr, W. H.; Marsili, F.; Shaw, M. D.; Stern, J. A.; Abellán, C.; Amaya, W.; Pruneri, V.; Jennewein, Thomas; Mitchell, Morgan W.; Kwiat, Paul G.; Bienfang, J. C.; Mirin, R. P.; Knill, Emanuel; and Nam, S. W. (2015). Strong Loophole-Free Test of Local Realism. *Phys. Rev. Lett.*, **115**, 250402. Preprint https://arxiv.org/abs/1511.03189. https://doi.org/10.1103/PhysRevLett.115.250402.

Shankar, Ramamurti (1994). *Principles of Quantum Mechanics* (2nd edn). Springer, New York, NY. https://doi.org/10.1007/978-1-4757-0576-8.

Shannon, Claude Elwood (1948). A mathematical theory of communication. *Bell System Technical Journal*, **27**, 379–423, 623–656.

Shapere, Alfred and Wilczek, Frank (ed.) (1989). *Geometric phases in physics.* World Scientific, Singapore. https://doi.org/10.1142/0613.

Shi, Yu (2006*a*). Erratum to: High energy quantum teleportation using neutral kaons. *Phys. Lett. B*, **641**, 492. https://doi.org/10.1016/j.physletb.2006.09.007.

Shi, Yu (2006*b*). High energy quantum teleportation using neutral kaons. *Phys. Lett. B*, **641**, 75–80. Preprint https://arxiv.org/abs/quant-ph/0605070. https://doi.org/10.1016/j.physletb.2006.08.015.

Shi, Yu and Yang, Ji-Chong (2020). Entangled baryons: violation of inequalities based on local realism assuming dependence of decays on hidden variables. *Eur. Phys. J. C*, **80**, 116. Preprint https://arxiv.org/abs/1912.04111. https://doi.org/10.1140/epjc/s10052-020-7684-5.

Shor, Peter W. (2004). Equivalence of additivity questions in quantum information theory. *Commun. Math. Phys.*, **246**(3), 473–473. Preprint https://arxiv.org/abs/quant-ph/0305035. https://doi.org/10.1007/s00220-004-1071-1.

Shor, Peter W.; Smolin, John A.; and Terhal, Barbara M. (2001). Non-additivity of Bipartite Distillable Entanglement Follows from a Conjecture on Bound Entangled Werner States. *Phys. Rev. Lett.*, **86**, 2681–2684. Preprint https://arxiv.org/abs/quant-ph/0010054. https://doi.org/10.1103/PhysRevLett.86.2681.

Shore, Bruce W. and Knight, Peter L. (1993). The Jaynes-Cummings Model. *J. Mod. Opt.*, **40**, 1195–1238. https://doi.org/10.1080/09500349314551321.

Shull, Clifford Glenwood (1969). Single-Slit Diffraction of Neutrons. *Phys. Rev.*, **179**, 752–754. https://doi.org/10.1103/PhysRev.179.752.

Sidhu, Jasminder S. and Kok, Pieter (2020). A Geometric Perspective on

Quantum Parameter Estimation. *AVS Quantum Sci.*, **2**, 014701. Preprint https://arxiv.org/abs/1907.06628. `https://doi.org/10.1116/1.5119961`.

Simon, Barry (1983). Holonomy, the Quantum Adiabatic Theorem, and Berry's Phase. *Phys. Rev. Lett.*, **51**, 2167–2170. `https://doi.org/10.1103/PhysRevLett.51.2167`.

Simon, Barry (2005). *Trace Ideals and their Applications* (2nd edn). Amer. Math. Soc., Providence, RI. originally published in 1979 by Cambridge University Press. `https://doi.org/10.1090/surv/120`.

Simon, Rajiah (2000). Peres-Horodecki Separability Criterion for Continuous Variable Systems. *Phys. Rev. Lett.*, **84**, 2726–2729. Preprint https://arxiv.org/abs/quant-ph/9909044. `https://doi.org/10.1103/PhysRevLett.84.2726`.

Simon, Rajiah; Mukunda, Narasimhaiengar; and Dutta, Biswadeb (1994). Quantum-noise matrix for multimode systems: U(n) invariance, squeezing, and normal forms. *Phys. Rev. A*, **49**, 1567–1583. `https://doi.org/10.1103/PhysRevA.49.1567`.

Six, Jules (1982). Test of the non-separability of the $K^0 \bar{K}^0$ system. *Phys. Lett. B*, **114**, 200–202. `https://doi.org/10.1016/0370-2693(82)90146-0`.

Six, Jules (1990). Local models with hidden variables for a $K^0 \bar{K}^0$ system. *Phys. Lett. A*, **150**, 243–202. `https://doi.org/10.1016/0375-9601(90)90088-6`.

Sjöqvist, Erik; Pati, Arun K.; Ekert, Artur; Anandan, Jeeva S.; Ericsson, Marie; Oi, Daniel K. L.; and Vedral, Vlatko (2000). Geometric Phases for Mixed States in Interferometry. *Phys. Rev. Lett.*, **85**, 2845–2849. Preprint https://arxiv.org/abs/quant-ph/0005072. `https://doi.org/10.1103/PhysRevLett.85.2845`.

Skotiniotis, Michalis; Dür, Wolfgang; and Sekatski, Pavel (2017). Macroscopic superpositions require tremendous measurement devices. *Quantum*, **1**, 34. Preprint https://arxiv.org/abs/1705.07053. `https://doi.org/10.22331/q-2017-11-21-34`.

Smith, Graeme and Leung, Debbie (2006). Typical entanglement of stabilizer states. *Phys. Rev. A*, **74**, 062314. Preprint https://arxiv.org/abs/quant-ph/0510232. `https://doi.org/10.1103/PhysRevA.74.062314`.

Sommerfeld, Arnold (1940). Zur Feinstruktur der Wasserstofflinien. Geschichte und gegenwärtiger Stand der Theorie. *Naturwissenschaften*, **28**, 417–423. in German. `https://doi.org/10.1007/BF01490583`.

Song, Chao; Xu, Kai; Li, Hekang; Zhang, Yuran; Zhang, Xu; Liu, Wuxin; Guo, Qiujiang; Wang, Zhen; Ren, Wenhui; Hao, Jie; Feng, Hui; Fan, Heng; Zheng, Dongning; Wang, Dawei; Wang, H.; and Zhu, Shiyao (2019). Observation of multi-component atomic Schrödinger cat states of up to 20 qubits. *Science*, **365**, 574–577. Preprint https://arxiv.org/abs/1905.00320. `https://doi.org/10.1126/science.aay0600`.

Sontz, Stephen Bruce (2015). *Principal Bundles, The Classical Case*. Springer, Cham. `https://doi.org/10.1007/978-3-319-14765-9`.

Sørensen, Anders and Mølmer, Klaus (2001). Entanglement and Extreme Spin Squeezing. *Phys. Rev. Lett.*, **86**, 4431—4434. Preprint https://arxiv.org/abs/quant-ph/0011035. `https://doi.org/10.1103/PhysRevLett.86.4431`.

Sparnaay, Marcus Johannes (1958). Measurements of attractive forces between flat plates. *Physica*, **24**, 751–764. `https://doi.org/10.1016/S0031-8914(58)`

80090-7.

Spee, Cornelia; de Vicente, Julio I.; Sauerwein, David; and Kraus, Barbara (2017). Entangled Pure State Transformations via Local Operations Assisted by Finitely Many Rounds of Classical Communication. *Phys. Rev. Lett.*, **118**, 040503. Preprint https://arxiv.org/abs/1606.04418. https://doi.org/10.1103/PhysRevLett.118.040503.

Spengler, Christoph; Huber, Marcus; Brierley, Stephen; Adaktylos, Theodor; and Hiesmayr, Beatrix C. (2012). Entanglement detection via mutually unbiased bases. *Phys. Rev. A*, **86**, 022311. Preprint https://arxiv.org/abs/1202.5058. https://doi.org/10.1103/PhysRevA.86.022311.

Stewart, Philip J. (2010). Charles Janet: unrecognized genius of the Periodic System. *Found. Chem.*, **12**, 5–15. https://doi.org/10.1007/s10698-008-9062-5.

Stinespring, William Forrest (1955). Positive functions on C^*-algebras. *Proc. Amer. Math. Soc.*, **6**, 211–216. https://doi.org/10.1090/S0002-9939-1955-0069403-4.

Størmer, Erling (1963). Positive linear maps of operator algebras. *Acta Math.*, **110**, 233–278. preprint available at https://projecteuclid.org/euclid.acta/1485889359. https://doi.org/10.1007/BF02391860.

Strutt (Lord Rayleigh), John William (1906). On electrical vibrations and the constitution of the atom. *The London, Edinburgh, and Dublin Philosophical Magazine and Journal of Science*, **11**(61), 117–123. https://doi.org/10.1080/14786440609463428.

Strutt (Lord Rayleigh), John William (2011). *The Theory of Sound.* Volume 1, Cambridge Library Collection - Physical Sciences. Cambridge University Press, Cambridge, U.K. first published in 1877. https://doi.org/10.1017/CBO9781139058087.

Sudarshan, Ennackal Chandy George (1963). Equivalence of Semiclassical and Quantum Mechanical Descriptions of Statistical Light Beams. *Phys. Rev. Lett.*, **10**, 277–279. https://doi.org/10.1103/PhysRevLett.10.277.

Suetsugu, Yusuke; *et al.* (2019). Mitigating the electron cloud effect in the SuperKEKB positron ring. *Phys. Rev. Accel. Beams*, **22**, 023201. https://doi.org/10.1103/PhysRevAccelBeams.22.023201.

Suslov, Sergei K. (2010). Relativistic Kramers-Pasternack recurrence relations. *J. Phys. B: At. Mol. Opt. Phys.*, **43**, 074006. https://doi.org/10.1088/0953-4075/43/7/074006.

Susskind, Leonard and Glogower, Jonathan (1964). Quantum mechanical phase and time operator. *Phys. Phys. Fiz.*, **1**, 49–61. https://doi.org/10.1103/PhysicsPhysiqueFizika.1.49.

Svetlichny, George (1987). Distinguishing three-body from two-body nonseparability by a Bell-type inequality. *Phys. Rev. D*, **35**, 3066–3069. https://doi.org/10.1103/PhysRevD.35.3066.

Swendsen, Robert H. (2012). *An Introduction to Statistical Mechanics and Thermodynamics.* Oxford University Press, Oxford.

Szymanski, Tomasz and Freericks, James K. (2021). Algebraic derivation of Kramers–Pasternack relations based on the Schrödinger factorization method. *Eur.*

J. Phys., **42**(2), 025409. Preprint https://arxiv.org/abs/2007.11158. `https://doi.org/10.1088/1361-6404/abd228`.

't Hooft, Gerard (1999). Quantum gravity as a dissipative deterministic system. *Class. Quant. Grav.*, **16**, 3263. Preprint https://arxiv.org/abs/gr-qc/9903084. `https://doi.org/10.1088/0264-9381/16/10/316`.

't Hooft, Gerard (2000). Determinism and Dissipation in Quantum Gravity, Erice lecture. Preprint https://arxiv.org/abs/hep-th/0003005.

't Hooft, Gerard (2005). Determinism Beneath Quantum Mechanics. In *Quo Vadis Quantum Mechanics? The Frontiers Collection* (ed. A. C. Elitzur, S. Dolev, and N. Kolenda), p. 99. Springer, Berlin, Heidelberg.

Tapster, P. R.; Rarity, John G.; and Owens, P. C. M. (1994). Violation of Bell's Inequality over 4 km of Optical Fiber. *Phys. Rev. Lett.*, **73**, 1923–1926. `https://doi.org/10.1103/PhysRevLett.73.1923`.

Taranto, Philip; Bakhshinezhad, Faraj; Bluhm, Andreas; Silva, Ralph; Friis, Nicolai; Lock, Maximilian P. E.; Vitagliano, Giuseppe; Binder, Felix C.; Debarba, Tiago; Schwarzhans, Emanuel; Clivaz, Fabien; and Huber, Marcus (2023). Landauer Versus Nernst: What is the True Cost of Cooling a Quantum System? *PRX Quantum*, **4**, 010332. Preprint https://arxiv.org/abs/2106.05151. `https://doi.org/10.1103/PRXQuantum.4.010332`.

Tataroglu, Hatice (2009). Nichtlokale Korrelation in Kaonischen Systemen. Diploma thesis, University of Vienna. `https://doi.org/10.25365/thesis.8315`.

Terhal, Barbara M. (2000). Bell inequalities and the separability criterion. *Phys. Lett. A*, **271**, 319–326. `https://doi.org/10.1016/S0375-9601(00)00401-1`.

Terhal, Barbara M. and Horodecki, Paweł (2000). Schmidt number for density matrices. *Phys. Rev. A*, **61**, 040301(R). Preprint https://arxiv.org/abs/quant-ph/9911117. `https://doi.org/10.1103/PhysRevA.61.040301`.

Thirring, Walter (1980). *Lehrbuch der Mathematischen Physik 4: Quantenmechanik großer Systeme*. Springer-Verlag, Wien. In German.

Thirring, Walter; Bertlmann, Reinhold A.; Köhler, Philipp; and Narnhofer, Heide (2011). Entanglement or Separability: The Choice of How to Factorize the Algebra of a Density Matrix. *Eur. Phys. J. D*, **64**, 181–196. Preprint https://arxiv.org/abs/1106.3047. `https://doi.org/10.1140/epjd/e2011-20452-1`.

Thomas, Llewellyn Hilleth (1926). The Motion of the Spinning Electron. *Nature*, **117**, 514. `https://doi.org/10.1038/117514a0`.

Thomas, Llewellyn Hilleth (1927). The calculation of atomic fields. *Proc. Cambridge Phil. Soc.*, **23**, 542–548. `https://doi.org/10.1017/S0305004100011683`.

Thompson, Jayne; Modi, Kavan; Vedral, Vlatko; and Gu, Mile (2018). Quantum plug n' play: modular computation in the quantum regime. *New J. Phys.*, **20**, 013004. Preprint https://arxiv.org/abs/1310.2927. `https://doi.org/10.1088/1367-2630/aa99b3`.

Thomson, George Paget (1927). The Diffraction of Cathode Rays by Thin Films of Platinum. *Nature*, **120**, 802. `https://doi.org/10.1038/120802a0`.

Thomson, George Paget and Reid, Alexander (1927). Diffraction of Cathode Rays by a Thin Film. *Nature*, **119**, 890. `https://doi.org/10.1038/119890a0`.

Thörnqvist, Nils A. (1981). Suggestion for Einstein-Podolsky-Rosen experiments using reactions like $e^+e^- \to \Lambda\bar{\Lambda} \to \pi^- p\pi^+\bar{p}$. *Found. Phys.*, **11**, 171–177. https://doi.org/10.1007/BF00715204.

Thun-Hohenstein, Philipp (2010). Quantum phase and uncertainty. Diploma thesis, University of Vienna. https://doi.org/10.25365/thesis.8540.

Timpson, Christopher G. (2013). *Quantum Information Theory and the Foundations of Quantum Mechanics*. Oxford University Press, Oxford. Preprint https://arxiv.org/abs/quant-ph/0412063.

Tipler, Paul A. and Llewellyn, Ralph A. (2008). *Modern Physics* (5th edn). W. H. Freeman and Company, New York.

Tittel, Wolfgang; Brendel, Jürgen; Zbinden, Hugo; and Gisin, Nicolas (1998). Violation of Bell Inequalities by Photons More Than 10 km Apart. *Phys. Rev. Lett.*, **81**, 3563–3566. https://doi.org/10.1103/PhysRevLett.81.3563.

Tonomura, Akira; Osakabe, Nobuyuki; Matsuda, Tsuyoshi; Kawasaki, Takeshi; Endo, Junji; Yano, Shinichiro; and Yamada, Hiroji (1986). Evidence for Aharonov-Bohm effect with magnetic field completely shielded from electron wave. *Phys. Rev. Lett.*, **56**, 792–795. https://doi.org/10.1103/PhysRevLett.56.792.

Tóth, Géza and Apellaniz, Iagoba (2014). Quantum metrology from a quantum information science perspective. *J. Phys. A: Math. Theor.*, **47**, 424006. Preprint https://arxiv.org/abs/1405.4878. https://doi.org/10.1088/1751-8113/47/42/424006.

Tóth, Géza and Gühne, Otfried (2005). Entanglement detection in the stabilizer formalism. *Phys. Rev. A*, **72**, 022340. Preprint https://arxiv.org/abs/quant-ph/0501020. https://doi.org/10.1103/PhysRevA.72.022340.

Tóth, Géza and Gühne, Otfried (2009). Entanglement and Permutational Symmetry. *Phys. Rev. Lett.*, **102**, 170503. Preprint https://arxiv.org/abs/0812.4453. https://doi.org/10.1103/PhysRevLett.102.170503.

Tóth, Géza; Moroder, Tobias; and Gühne, Otfried (2015). Evaluating Convex Roof Entanglement Measures. *Phys. Rev. Lett.*, **114**, 160501. Preprint https://arxiv.org/abs/1409.3806. https://doi.org/10.1103/PhysRevLett.114.160501.

Tóth, Géza and Petz, Dénes (2013). Extremal properties of the variance and the quantum fisher information. *Phys. Rev. A*, **87**, 032324. Preprint https://arxiv.org/abs/1109.2831. https://doi.org/10.1103/PhysRevA.87.032324.

Townsend, John Sealy and Bailey, Victor Albert (1921). The motion of electrons in gases. *Phil. Mag.*, **42**, 873–891. https://doi.org/10.1080/14786442108633831.

Traxler, Tanja (2009). Decoherence and the Physics of Open Quantum Systems. Lecture notes accompanying Reinhold Bertlmann's lectures, University of Vienna. https://homepage.univie.ac.at/reinhold.bertlmann/pdfs/decoscript_v2.pdf.

Traxler, Tanja (2011). Decoherence and entanglement of two qubit systems. Diploma thesis, University of Vienna. https://doi.org/10.25365/thesis.15239.

Tsang, Mankei (2019). The Holevo Cramér-Rao bound is at most thrice the Helstrom version. Preprint https://arxiv.org/abs/1911.08359.

Tserkis, Spyros; Onoe, Sho; and Ralph, Timothy C. (2019). Quantifying entanglement of formation for two-mode Gaussian states: Analytical expressions for upper and lower bounds and numerical estimation of its exact value. *Phys. Rev. A*, **99**, 052337. Preprint https://arxiv.org/abs/1903.09961. https://doi.org/10.1103/PhysRevA.99.052337.

Tserkis, Spyros and Ralph, Timothy C. (2017). Quantifying entanglement in two-mode Gaussian states. *Phys. Rev. A*, **96**, 062338. Preprint https://arxiv.org/abs/1705.03612. https://doi.org/10.1103/PhysRevA.96.062338.

Tsirelson, Boris Semyonovich (1980). Quantum generalizations of Bell's inequality. *Lett. Math. Phys.*, **4**, 93–100. https://doi.org/10.1007/BF00417500.

Tsirelson, Boris Semyonovich (1987). Quantum analogues of the Bell inequalities. The case of two spatially separated domains. *J. Math. Sci.*, **36**, 557–570. https://doi.org/10.1007/BF01663472.

Tucci, Robert R. (1999). Quantum Entanglement and Conditional Information Transmission. Preprint https://arxiv.org/abs/quant-ph/9909041.

Tucci, Robert R. (2002). Entanglement of Distillation and Conditional Mutual Information. Preprint https://arxiv.org/abs/quant-ph/0202144.

Tung, Wu-Ki (1985). *Group Theory in Physics*. World Scientific, Singapore.

Uchida, Gabriele (2013). Geometry of GHZ-type quantum states. Diploma thesis, University of Vienna. https://doi.org/10.25365/thesis.27159.

Uchida, Gabriele; Bertlmann, Reinhold A.; and Hiesmayr, Beatrix C. (2015). Entangled entanglement: A construction procedure. *Phys. Lett. A*, **379**, 2698–2703. Preprint https://arxiv.org/abs/1410.7145. https://doi.org/10.1016/j.physleta.2015.07.045.

Uchiyama, Fumiyo (1997). Generalized Bell inequality in two neutral kaon systems. *Phys. Lett. A*, **231**, 295–298. https://doi.org/10.1016/S0375-9601(97)00347-2.

Uffink, Jos (2022). Boltzmann's Work in Statistical Physics. In *The Stanford Encyclopedia of Philosophy* (Summer 2022 edn) (ed. E. N. Zalta). Metaphysics Research Lab, Stanford University. https://plato.stanford.edu/archives/sum2022/entries/statphys-Boltzmann/.

Uhlmann, Armin (1971). Sätze über Dichtematrizen. *Wiss. Z. Karl-Marx-Univ. Leipzig, Math.-Nat. R.*, **20**, 633–637. in German.

Uhlmann, Armin (1972). Endlich-dimensionale Dichtematrizen I. *Wiss. Z. Karl-Marx-Univ. Leipzig, Math.-Nat. R.*, **21**, 421–452. in German.

Uhlmann, Armin (1973). Endlich-dimensionale Dichtematrizen II. *Wiss. Z. Karl-Marx-Univ. Leipzig, Math.-Nat. R.*, **22**, 139–177. in German.

Uhlmann, Armin (1976). The "transition probability" in the state space of a *-algebra. *Rep. Math. Phys.*, **9**, 273–279. https://doi.org/10.1016/0034-4877(76)90060-4.

Uhlmann, Armin (1986). Parallel transport and quantum holonomy along density operators. *Rep. Math. Phys.*, **24**, 229–240. https://doi.org/10.1016/0034-4877(86)90055-8.

Uhlmann, Armin (1991). Parallel transport of phases. In *Differential Geometry,*

Group Representations, and Quantization. Lecture Notes In Physics (eds. J. D. Hennig, W. Lücke, and J. Tolar), Volume 379, p. 55–72. Springer, Berlin, Heidelberg. https://doi.org/10.1007/3-540-53941-7_4.

Uhlmann, Armin (1998). Entropy and Optimal Decompositions of States Relative to a Maximal Commutative Subalgebra. *Open Syst. Inf. Dyn.*, **5**, 209–228. Preprint https://arxiv.org/abs/quant-ph/9701014. https://doi.org/10.1023/A:1009664331611.

Uhlmann, Armin (2003). Antilinearity in Bipartite Quantum Sytems and Imperfect Quantum Teleportation. In *Quantum Probability and Infinite Dimensional Analysis* (ed. W. Freudenberg), Volume 15, p. 255–268. World Scientific, Singapore. Preprint https://arxiv.org/abs/quant-ph/0407244.

Ursin, Rupert; Jennewein, Thomas; Aspelmeyer, Markus; Kaltenbaek, Rainer; Lindenthal, Michael; Walther, Philip; and Zeilinger, Anton (2004). Quantum Teleportation Across the Danube. *Nature*, **430**, 849–849. https://doi.org/10.1038/430849a.

Ursin, Rupert; Tiefenbacher, Felix; Schmitt-Manderbach, Tobias; Weier, Henning; Scheidl, Thomas; Lindenthal, Michael; Blauensteiner, Bibiane; Jennewein, Thomas; Perdigues, Josep; Trojek, Pavel; Ömer, Bernhard; Fürst, Martin; Meyenburg, Michael; Rarity, John G.; Sodnik, Zoran; Barbieri, Cesare; Weinfurter, Harald; and Zeilinger, Anton (2007). Entanglement-based quantum communication over 144 km. *Nat. Phys.*, **3**, 481–486. Preprint https://arxiv.org/abs/quant-ph/0607182. https://doi.org/10.1038/nphys629.

Vaidman, Lev (2012). Role of potentials in the Aharonov-Bohm effect. *Phys. Rev. A*, **86**, 040101. Preprint https://arxiv.org/abs/1110.6169. https://doi.org/10.1103/PhysRevA.86.040101.

Valatin, John George (1958). Comments on the theory of superconductivity. *Nuovo Cim.*, **7**, 843–857. https://doi.org/10.1007/BF02745589.

van Erven, Tim and Harremoës, Peter (2007). Rényi divergence and Kullback-Leibler divergence. *Journ. Latex Class Files*, **6**, 1–24. Preprint https://arxiv.org/abs/1206.2459. https://doi.org/10.1109/TIT.2014.2320500.

van Trees, Harry L. (1968). *Detection, Estimation, and Modulation Theory, Part I.* Wiley, New York.

Varshalovich, Dmitry Alexandrovich; Moskalev, A. N.; and Khersonskii, V. K. (1988). *Quantum Theory of Angular Momentum.* World Scientific, Singapore. https://doi.org/10.1142/0270.

Vedral, Vlatko (2002). The role of relative entropy in quantum information theory. *Rev. Mod. Phys.*, **74**, 197–234. Preprint https://arxiv.org/abs/quant-ph/0102094. https://doi.org/10.1103/RevModPhys.74.197.

Vedral, Vlatko and Plenio, Martin B. (1998). Entanglement measures and purification procedures. *Phys. Rev. A*, **57**, 1619–1633. Preprint https://arxiv.org/abs/quant-ph/9707035. https://doi.org/10.1103/PhysRevA.57.1619.

Vedral, Vlatko; Plenio, Martin B.; Jacobs, Kurt; and Knight, Peter L. (1997*a*). Statistical inference, distinguishability of quantum states, and quantum entanglement. *Phys. Rev. A*, **56**, 4452–4455. Preprint https://arxiv.org/abs/quant-ph/9703025. https://doi.org/10.1103/PhysRevA.56.4452.

Vedral, Vlatko; Plenio, Martin B.; Rippin, M. A.; and Knight, Peter L. (1997*b*). Quantifying Entanglement. *Phys. Rev. Lett.*, **78**, 2275–2279. Preprint https://arxiv.org/abs/quant-ph/9702027. `https://doi.org/10.1103/PhysRevLett.78.2275`.

Verstraete, Frank; Audenaert, Koenraad; Dehaene, Jeroen; and De Moor, Bart (2001*a*). A comparison of the entanglement measures negativity and concurrence. *J. Phys. A*, **34**, 10327. Preprint https://arxiv.org/abs/quant-ph/0108021. `https://doi.org/10.1088/0305-4470/34/47/329`.

Verstraete, Frank; Audenaert, Koenraad M. R.; and De Moor, Bart (2001*b*). Maximally entangled mixed states of two qubits. *Phys. Rev. A*, **64**, 012316. Preprint https://arxiv.org/abs/quant-ph/0011110. `https://doi.org/10.1103/PhysRevA.64.012316`.

Verstraete, Frank; Dehaene, Jeroen; and De Moor, Bart (2002*a*). On the geometry of entangled states. *J. Mod. Opt.*, **49**, 1277. Preprint https://arxiv.org/abs/quant-ph/0107155. `https://doi.org/10.1080/09500340110115488`.

Verstraete, Frank; Dehaene, Jeroen; and De Moor, Bart (2003). Normal forms and entanglement measures for multipartite quantum states. *Phys. Rev. A*, **68**, 012103. Preprint https://arxiv.org/abs/quant-ph/0105090. `https://doi.org/10.1103/PhysRevA.68.012103`.

Verstraete, Frank; Dehaene, Jeroen; De Moor, Bart; and Verschelde, Henri (2002*b*). Four qubits can be entangled in nine different ways. *Phys. Rev. A*, **65**, 052112. Preprint https://arxiv.org/abs/quant-ph/0109033. `https://doi.org/10.1103/PhysRevA.65.052112`.

Verstraete, Frank and Verschelde, Henri (2002). Fidelity of mixed states of two qubits. *Phys. Rev. A*, **66**, 022307. Preprint https://arxiv.org/abs/quant-ph/0203073. `https://doi.org/10.1103/PhysRevA.66.022307`.

Vértesi, Tamás (2008). More efficient Bell inequalities for Werner states. *Phys. Rev. A*, **78**, 032112. Preprint https://arxiv.org/abs/0806.0096. `https://doi.org/10.1103/PhysRevA.78.032112`.

Vértesi, Tamás and Brunner, Nicolas (2012). Quantum Nonlocality Does Not Imply Entanglement Distillability. *Phys. Rev. Lett.*, **108**, 030403. Preprint https://arxiv.org/abs/1106.4850. `https://doi.org/10.1103/PhysRevLett.108.030403`.

Vértesi, Tamás and Brunner, Nicolas (2014). Disproving the Peres conjecture by showing Bell nonlocality from bound entanglement. *Nature Commun.*, **5**, 5297. Preprint https://arxiv.org/abs/1405.4502. `https://doi.org/10.1038/ncomms6297`.

Vidal, Guifré (1999). Entanglement of pure states for a single copy. *Phys. Rev. Lett.*, **83**, 1046–1049. Preprint https://arxiv.org/abs/quant-ph/9902033. `https://doi.org/10.1103/PhysRevLett.83.1046`.

Vidal, Guifré; Dür, Wolfgang; and Cirac, Juan Ignacio (2002). Entanglement Cost of Bipartite Mixed States. *Phys. Rev. Lett.*, **89**, 027901. Preprint https://arxiv.org/abs/quant-ph/0112131. `https://doi.org/10.1103/PhysRevLett.89.027901`.

Vidal, Guifré; Jonathan, Daniel; and Nielsen, Michael A. (2000). Approximate transformations and robust manipulation of bipartite pure-state entanglement. *Phys. Rev. A*, **62**, 012304. Preprint https://arxiv.org/abs/quant-ph/9910099. `https://doi.org/10.1103/PhysRevA.62.012304`.

Vidal, Guifré and Werner, Reinhard F. (2002). Computable measure of entanglement. *Phys. Rev. A*, **65**, 032314. Preprint https://arxiv.org/abs/quant-ph/0102117. `https://doi.org/10.1103/PhysRevA.65.032314`.

Vogel, Werner and Welsch, Dirk-Gunnar (2006). *Quantum Optics* (3rd edn). WILEY-VCH Verlag, Weinheim, Germany.

Vollbrecht, Karl G. H. and Werner, Reinhard F. (2001). Entanglement measures under symmetry. *Phys. Rev. A*, **64**, 062307. Preprint https://arxiv.org/abs/quant-ph/0010095. `https://doi.org/10.1103/PhysRevA.64.062307`.

von Neumann, John (1929). Allgemeine Eigenwerttheorie Hermitischer Funktional-operatoren. *Math. Ann.*, **102**, 49–131. `https://doi.org/10.1007/BF01782338`.

von Neumann, John (1932). *Mathematische Grundlagen der Quantenmechanik*. Springer, Berlin. English translation: Princeton University Pess 1955.

von Neumann, John (1953). A certain zero-sum two-person game equivalent to an optimal assignment problem. *Ann. Math. Studies*, **28**, 5–12.

von Neumann, John (2018). *Mathematical Foundations of Quantum Mechanics: New Edition*. Princeton University Press, Princeton, New Jersey. Translated by Robert T. Beyer.

von Neumann, John and Wigner, Eugene P. (1993). Über merkwürdige diskrete Eigenwerte. In *The Collected Works of Eugene Paul Wigner vol A1* (ed. A. S. Wightman), p. 291–293. Springer, Berlin, Heidelberg. in German. `https://doi.org/10.1007/978-3-662-02781-3_19`.

von Weizsäcker, C. F. (1985). *Aufbau der Physik*. Carl Hanser Verlag, München, Wien.

Walls, Daniel Frank and Milburn, Gerard J. (1994). *Quantum Optics*. Springer, Berlin.

Walschaers, Mattia (2021). Non-Gaussian Quantum States and Where to Find Them. Preprint https://arxiv.org/abs/2104.12596.

Walther, Philip; Resch, Kevin J.; Brukner, Časlav; and Zeilinger, Anton (2006). Experimental entangled entanglement. *Phys. Rev. Lett.*, **97**, 020501–020504. `https://doi.org/10.1103/PhysRevLett.97.020501`.

Wang, Shun-Jin (1990). Nonadiabatic Berry's phase for a spin particle in a rotating magnetic field. *Phys. Rev. A*, **42**, 5107–5110. `http://doi.org/10.1103/PhysRevA.42.5107`.

Wang, Xi-Lin; Luo, Yi-Han; Huang, He-Liang; Chen, Ming-Cheng; Su, Zu-En; Liu, Chang; Chen, Chao; Li, Wei; Fang, Yu-Qiang; Jiang, Xiao; Zhang, Jun; Li, Li; Liu, Nai-Le; Lu, Chao-Yang; and Pan, Jian-Wei (2018). 18-Qubit Entanglement with Six Photons' Three Degrees of Freedom. *Phys. Rev. Lett.*, **120**, 260502. Preprint https://arxiv.org/abs/1801.04043. `https://doi.org/10.1103/PhysRevLett.120.260502`.

Warner, Frank W. (1983). *Foundations of Differentiable Manifolds and Lie Groups* (2nd edn). Springer, Berlin.

Watrous, John (2004). Many Copies May Be Required for Entanglement Distillation.

Phys. Rev. Lett., **93**, 010502. Preprint https://arxiv.org/abs/quant-ph/0312123. `https://doi.org/10.1103/PhysRevLett.93.010502`.

Weedbrook, Christian; Pirandola, Stefano; García-Patrón, Raúl; Cerf, Nicolas J.; Ralph, Timothy C.; Shapiro, Jeffrey H.; and Lloyd, Seth (2012). Gaussian quantum information. *Rev. Mod. Phys.*, **84**, 621–669. Preprint https://arxiv.org/abs/1110.3234. `https://doi.org/10.1103/RevModPhys.84.621`.

Wehrl, Alfred (1978). General properties of entropy. *Rev. Mod. Phys.*, **50**, 221–260. `https://doi.org/10.1103/RevModPhys.50.221`.

Weihs, Gregor (1998). *Ein Experiment zum Test der Bellschen Ungleichung unter Einstein Lokalität*. Ph.D. thesis, University of Innsbruck.

Weihs, Gregor; Jennewein, Thomas; Simon, Christoph; Weinfurter, Harald; and Zeilinger, Anton (1998). Violation of Bell's Inequality under Strict Einstein Locality Conditions. *Phys. Rev. Lett.*, **81**, 5039–5043. `https://doi.org/10.1103/PhysRevLett.81.5039`.

Weinberger, Peter (2006). Revisiting Louis de Broglie's famous 1924 paper in the Philosophical Magazine. *Phil. Mag. Lett.*, **86**(7), 405–410. `https://doi.org/10.1080/09500830600876565`.

Weinfurter, Harald (1994). Experimental Bell-State Analysis. *Europhys. Lett.*, **25**, 559–564. `https://doi.org/10.1209/0295-5075/25/8/001`.

Weisskopf, Victor F. and Wigner, Eugene P. (1930*a*). Berechnung der natürlichen Linienbreite auf Grund der Diracschen Lichttheorie. *Z. Physik*, **63**, 54–73. In German. `https://doi.org/10.1007/BF01336768`.

Weisskopf, Victor F. and Wigner, Eugene P. (1930*b*). Über die natürliche Linienbreite in der Strahlung des harmonischen Oszillators. *Z. Physik*, **65**, 18–29. In German. `https://doi.org/10.1007/BF01397406`.

Werner, Reinhard F. (1989). Quantum states with Einstein-Podolsky-Rosen correlations admitting a hidden-variable model. *Phys. Rev. A*, **40**, 4277–4281. `https://doi.org/10.1103/PhysRevA.40.4277`.

Weyl, Hermann (1910). Über gewöhnliche Differentialgleichungen mit Singularitäten und die zugehörigen Entwicklungen willkürlicher Funktionen. *Math. Ann.*, **68**, 220–269. In German. `https://doi.org/10.1007/BF01474161`.

Weyl, Hermann (1931). *The Theory of Groups and Quantum Mechanics*. Dover Publications, New York. translated by Howard Percy Robertson from the second (revised) German edition titled *"Gruppentheorie und Quantenmechanik"*, first published in 1928 by Hirzel (Leipzig).

Wheeler, John Archibald (1989). Information, physics, quantum: The search for links. In *Proc. III Internat. Symp. on Found. of Quantum Mechanics*, p. 354–368. Tokyo.

Widhalm, Laurenz (2005). CP violation in B decays: Experimental aspects. In *2005 European School of High-Energy Physics*, p. 199–214. `https://doi.org/10.5170/CERN-2006-014.199`.

Wieczorek, Witlef; Schmid, Christian; Kiesel, Nikolai; Pohlner, Reinhold; Gühne, Otfried; and Weinfurter, Harald (2008). Experimental Observation of an Entire Family of Four-Photon Entangled States. *Phys. Rev. Lett.*, **101**, 010503–010506. Preprint https://arxiv.org/abs/0806.1882. `https://doi.org/10.1103/PhysRevLett.101.`

010503.

Wiesner, Stephen J. (1983). Conjugate Coding. *SIGACT News*, **15**, 78–88. https://doi.org/10.1145/1008908.1008920.

Wigner, Eugene P. (1932). On the Quantum Correction For Thermodynamic Equilibrium. *Phys. Rev.*, **40**, 749–759. https://doi.org/10.1103/PhysRev.40.749.

Wigner, Eugene P. (1939). On Unitary Representations of the Inhomogeneous Lorentz Group. *Ann. Math.*, **40**, 149–204. https://doi.org/10.2307/1968551.

Wigner, Eugene P. (1961). Remarks on the mind-body question. In *The Scientist Speculates* (ed. I. J. Good), p. 284–302. London Heinemann.

Wigner, Eugene P. (1963). The Problem of Measurement. *Am. J. Phys.*, **31**, 6. https://doi.org/10.1119/1.1969254.

Wigner, Eugene P. (1970). On Hidden Variables and Quantum Mechanical Probabilities. *Am. J. Phys.*, **38**, 1005–1009. https://doi.org/10.1119/1.1976526.

Wilde, Mark M. (2017). *Quantum Information Theory* (2nd edn). Cambridge University Press, Cambridge, U.K.

Williamson, John (1936). On the algebraic problem concerning the normal forms of linear dynamical systems. *Am. J. Math.*, **58**(1), 141–163. https://doi.org/10.2307/2371062.

Wineland, David J.; Bollinger, John J.; Itano, Wayne M.; Moore, Fred Lee; and Heinzen, Daniel J. (1992). Spin squeezing and reduced quantum noise in spectroscopy. *Phys. Rev. A*, **46**, R6797. https://doi.org/10.1103/PhysRevA.46.R6797.

Witte, Frank M. C. and Trucks, Mathias (1999). A new entanglement measure induced by the Hilbert-Schmidt norm. *Phys. Lett. A*, **257**, 14–20. Preprint https://arxiv.org/abs/quant-ph/9811027. https://doi.org/10.1016/S0375-9601(99)00279-0.

Wolf, Michael M.; Eisert, Jens; and Plenio, Martin B. (2003). Entangling Power of Passive Optical Elements. *Phys. Rev. Lett.*, **90**, 047904. Preprint https://arxiv.org/abs/quant-ph/0206171. https://doi.org/10.1103/PhysRevLett.90.047904.

Wolf, Michael M.; Giedke, Geza; Küger, Ole; Werner, Reinhard F.; and Cirac, Juan Ignacio (2004). Gaussian entanglement of formation. *Phys. Rev. A*, **69**, 052320. Preprint https://arxiv.org/abs/quant-ph/0306177. https://doi.org/10.1103/PhysRevA.69.052320.

Wong, D. Pan (1979). Theoretical justification of Madelung's rule. *J. Chem. Ed.*, **56**, 714–718. https://doi.org/10.1021/ed056p714.

Wootters, William K. (1998). Entanglement of Formation of an Arbitrary State of Two Qubits. *Phys. Rev. Lett.*, **80**, 2245. Preprint https://arxiv.org/abs/quant-ph/9709029. https://doi.org/10.1103/PhysRevLett.80.2245.

Wootters, William K. (2001). Entanglement of formation and concurrence. *Quantum Inf. Comput.*, **1**(1), 27–44. https://doi.org/10.26421/QIC1.1.

Wootters, William K. (2004). Quantum measurements and finite geometry. Preprint https://arxiv.org/abs/quant-ph/0406032.

Wootters, William K. and Fields, Brian D. (1989). Optimal state-determination by mutually unbiased measurements. *Ann. Phys.*, **191**(2), 363–381. https://doi.

org/10.1016/0003-4916(89)90322-9.

Wootters, William K. and Zurek, Wojciech Hubert (1982). A single quantum cannot be cloned. *Nature*, **299**, 802–803. https://doi.org/10.1038/299802a0.

Woronowicz, Stanisław Lech (1976). Positive maps of low dimensional matrix algebras. *Rep. Math. Phys.*, **10**, 165–183. https://doi.org/10.1016/0034-4877(76)90038-0.

Wu, Jun-Yi; Kampermann, Hermann; Bruß, Dagmar; Klöckl, Claude; and Huber, Marcus (2012). Determining lower bounds on a measure of multipartite entanglement from few local observables. *Phys. Rev. A*, **86**, 022319. Preprint https://arxiv.org/abs/1205.3119. doi: 10.1103/PhysRevA.86.022319. https://doi.org/10.1103/PhysRevA.86.022319.

Wu, Shengjun; Yu, Sixia; and Mølmer, Klaus (2009). Entropic uncertainty relation for mutually unbiased bases. *Phys. Rev. A*, **79**, 022104. Preprint https://arxiv.org/abs/0811.2298. https://doi.org/10.1103/PhysRevA.79.022104.

Wu, Tai Tsun and Yang, Chen Ning (1975). Some remarks about unquantized non-Abelian gauge fields. *Phys. Rev. D*, **12**, 3843–3844. https://doi.org/10.1103/PhysRevD.12.3843.

Xu, Zhen-Peng; Steinberg, Jonathan; Nguyen, H. Chau; and Gühne, Otfried (2021). No-go theorem based on incomplete information of Wigner about his friend. Preprint https://arxiv.org/abs/2111.15010.

Yamagata, Koichi; Fujiwara, Akio; and Gill, Richard D. (2013). Quantum local asymptotic normality based on a new quantum likelihood ratio. *Ann. Statist.*, **41**, 2197–2217. Preprint https://arxiv.org/abs/1210.3749. https://doi.org/10.1214/13-AOS1147.

Yamasaki, Hayata; Morelli, Simon; Miethlinger, Markus; Bavaresco, Jessica; Friis, Nicolai; and Huber, Marcus (2022). Activation of genuine multipartite entanglement: beyond the single-copy paradigm of entanglement characterisation. *Quantum*, **6**, 695. Preprint https://arxiv.org/abs/2106.01372. https://doi.org/10.22331/q-2022-04-25-695.

Yang, Dong; Horodecki, Michał; Horodecki, Ryszard; and Synak-Radtke, Barbara (2005). Irreversibility for All Bound Entangled States. *Phys. Rev. Lett.*, **95**, 190501. Preprint https://arxiv.org/abs/quant-ph/0506138. https://doi.org/10.1103/PhysRevLett.95.190501.

Yang, K. and de Llano, M. (1989). Simple variational proof that any two-dimensional potential well supports at least one bound state. *Am. J. Phys.*, **57**, 85. https://doi.org/10.1119/1.15878.

Yang, Yuxiang; Chiribella, Giulio; and Hayashi, Masahito (2019). Attaining the ultimate precision limit in quantum state estimation. *Commun. Math. Phys.*, **368**, 223–293. Preprint https://arxiv.org/abs/1802.07587. https://doi.org/10.1007/s00220-019-03433-4.

Yin, Juan; Li, Yu-Huai; Liao, Sheng-Kai; Yang, Meng; Cao, Yuan; Zhang, Liang; Ren, Ji-Gang; Cai, Wen-Qi; Liu, Wei-Yue; Li, Shuang-Lin; Shu, Rong; Huang, Yong-Mei; Deng, Lei; Li, Li; Zhang, Qiang; Liu, Nai-Le; Chen, Yu-Ao; Lu, Chao-Yang; Wang, Xiang-Bin; Xu, Feihu; Wang, Jian-Yu; Peng, Cheng-Zhi; Ekert, Artur K.; and Pan,

Jian-Wei (2020). Entanglement-based secure quantum cryptography over 1.120 kilometres. *Nature*, **582**, 501–505. https://doi.org/10.1038/s41586-020-2401-y.

Young, Thomas (1804). I. The Bakerian Lecture. Experiments and calculations relative to physical optics. *Phil. Trans. R. Soc.*, **94**, 1–16. https://doi.org/10.1098/rstl.1804.0001.

Yu, Sixia (2013). Quantum Fisher Information as the Convex Roof of Variance. Preprint https://arxiv.org/abs/1302.5311.

Yu, Sixia and Oh, Choo-Hiap (2012). State-Independent Proof of Kochen-Specker Theorem with 13 Rays. *Phys. Rev. Lett.*, **108**, 030402. Preprint https://arxiv.org/abs/1109.4396. https://doi.org/10.1103/PhysRevLett.108.030402.

Zanardi, Paolo (2001). Virtual Quantum Subsystems. *Phys. Rev. Lett.*, **87**, 077901. Preprint https://arxiv.org/abs/quant-ph/0103030. https://doi.org/10.1103/PhysRevLett.87.077901.

Zanardi, Paolo; Lidar, Daniel A.; and Lloyd, Seth (2004). Quantum Tensor Product Structures are Observable Induced. *Phys. Rev. Lett.*, **92**, 060402. Preprint https://arxiv.org/abs/quant-ph/0308043. https://doi.org/10.1103/PhysRevLett.92.060402.

Zanardi, Paolo and Rasetti, Mario (1999). Holonomic quantum computation. *Phys. Lett. A*, **264**, 94–99. Preprint https://arxiv.org/abs/quant-ph/9904011. https://doi.org/10.1016/S0375-9601(99)00803-8.

Zauner, Gerhard (2011). Quantum designs: Foundations of a noncommutative design theory. *Int. J. Quantum Inf.*, **9**, 445–507. https://doi.org/10.1142/S0219749911006776.

Zeeman, Pieter (1897). The Effect of Magnetisation on the Nature of Light Emitted by a Substance. *Nature*, **55**, 347. https://doi.org/10.1038/055347a0.

Zeilinger, A (1999). A foundational principle for quantum mechanics. *Found. Phys.*, **29**, 631–643. https://doi.org/10.1023/A:1018820410908.

Zeilinger, Anton; Gähler, Roland; Shull, Clifford Glenwood; Treimer, Wolfgang; and Mampe, Walter (1988). Single- and double-slit diffraction of neutrons. *Rev. Mod. Phys.*, **60**, 1067–1073. https://doi.org/10.1103/RevModPhys.60.1067.

Zhang, Jing and Braunstein, Samuel L. (2006). Continuous-variable gaussian analog of cluster states. *Phys. Rev. A*, **73**, 032318. Preprint https://arxiv.org/abs/quant-ph/0501112. https://doi.org/10.1103/PhysRevA.73.032318.

Zhou, Xiao-Qi; Ralph, Timothy C.; Kalasuwan, Pruet; Zhang, Mian; Peruzzo, Alberto; Lanyon, Benjamin P.; and O'Brien, Jeremy L. (2011). Adding control to arbitrary unknown quantum operations. *Nat. Commun.*, **2**, 413. Preprint https://arxiv.org/abs/1006.2670. https://doi.org/10.1038/ncomms1392.

Żukowski, Marek; Zeilinger, Anton; Horne, Michael; and Ekert, Artur K. (1993). "Event-Ready-Detectors" Bell Experiment via Entanglement Swapping. *Phys. Rev. Lett.*, **71**, 4287–4290. https://doi.org/10.1103/PhysRevLett.71.4287.

Zurek, Wojciech Hubert (1991). Decoherence and the Transition from Quantum to Classical. *Physics Today*, **44**, 36. https://doi.org/10.1063/1.881293.

Zurek, Wojciech Hubert (2009). Quantum Darwinism. *Nat. Phys.*, **5**, 181. Preprint https://arxiv.org/abs/0903.5082. https://doi.org/10.1038/nphys1202.

Życzkowski, Karol and Bengtsson, Ingemar (2006). An Introduction to Quantum Entanglement: A Geometric Approach. Preprint https://arxiv.org/abs/quant-ph/0606228.

Copyright Notices

Index